P9-EED-160

**FOR REFERENCE**

Do Not Take From This Room

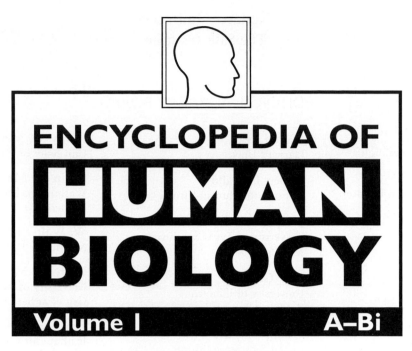

# ENCYCLOPEDIA OF HUMAN BIOLOGY

Volume I                                    A–Bi

## Second Edition

EDITOR-IN-CHIEF

## Renato Dulbecco
The Salk Institute

EDITORIAL ADVISORY BOARD

**John Abelson**
California Institute of Technology

**Peter Andrews**
Natural History Museum, London

**John A. Barranger**
University of Pittsburgh

**R. J. Berry**
University College, London

**Konrad Bloch**
Harvard University

**Floyd Bloom**
The Scripps Research Institute

**Norman E. Borlaug**
Texas A&M University

**Charles L. Bowden**
University of Texas Health Science
Center at San Antonio

**Ernesto Carafoli**
ETH-Zentrum

**Stephen K. Carter**
Bristol-Myers Squibb Corporation

**Edward C. H. Carterette**
University of California,
Los Angeles

**David Carver**
University of Medicine and
Dentistry of New Jersey

**Joel E. Cohen**
Rockefeller University and
Columbia University

**Michael E. DeBakey**
Baylor College of Medicine

**Eric Delson**
American Museum of Natural
History

**W. Richard Dukelow**
Michigan State University

**Myron Essex**
Harvard University

**Robert C. Gallo**
Institute for Human Virology

**Joseph L. Goldstein**
University of Texas Southwestern
Medical Center

**I. C. Gunsalus**
University of Illinois,
Urbana–Champaign

**Osamu Hayaishi**
Osaka Bioscience Institute

**Leonard A. Herzenberg**
Stanford University Medical Center

**Kazutomo Imahori**
Mitsubishi-Kasei Institute of Life
Science, Tokyo

**Richard T. Johnson**
Johns Hopkins University Medical
School

**Yuet Wai Kan**
University of California,
San Francisco

**Bernard Katz**
University College, London

**Seymour Kaufman**
National Institutes of Health

**Ernst Knobil**
University of Texas Health Science
Center at Houston

**Glenn Langer**
University of California Medical
Center, Los Angeles

**Robert S. Lawrence**
Johns Hopkins University

**James McGaugh**
University of California, Irvine

**Henry M. McHenry**
University of California, Davis

**Philip W. Majerus**
Washington University School
of Medicine

**W. Walter Menninger**
Menninger Foundation

**Terry M. Mikiten**
University of Texas Health Science
Center at San Antonio

**Beatrice Mintz**
Institute for Cancer Research, Fox
Chase Cancer Center

**Harold A. Mooney**
Stanford University

**Arno G. Motulsky**
University of Washington School
of Medicine

**Marshall W. Nirenberg**
National Institutes of Health

**G. J. V. Nossal**
Walter and Eliza Hall Institute of
Medical Research

**Mary Osborn**
Max Planck Institute for
Biophysical Chemistry

**George E. Palade**
University of California, San Diego

**Mary Lou Pardue**
Massachusetts Institute of
Technology

**Ira H. Pastan**
National Institutes of Health

**David Patterson**
University of Colorado Medical
Center

**Philip Reilly**
Shriver Center for Mental
Retardation

**Arthur W. Rowe**
New York University Medical
Center

**Ruth Sager**
Dana-Farber Cancer Institute

**Alan C. Sartorelli**
Yale University School of Medicine

**Neena B. Schwartz**
Northwestern University

**Bernard A. Schwetz**
National Center for Toxicological
Research, FDA

**Nevin S. Scrimshaw**
United Nations University

**Michael Sela**
Weizmann Institute of Science,
Israel

**Satimaru Seno**
Shigei Medical Research Institute,
Japan

**Phillip Sharp**
Massachusetts Institute of
Technology

**E. R. Stadtman**
National Institutes of Health

**P. K. Stumpf**
University of California, Davis
(emeritus)

**William Trager**
Rockefeller University (emeritus)

**Arthur C. Upton**
University of Medicine and
Dentistry of New Jersey, Robert
Wood Johnson Center

**Itaru Watanabe**
Kansas City VA Medical Center

**David John Weatherall**
Oxford University, John Radcliffe
Hospital

**Klaus Weber**
Max Planck Institute for
Biophysical Chemistry

**Thomas H. Weller**
Harvard School of Public Health

**Harry A. Whitaker**
Université du Québec á Montréal

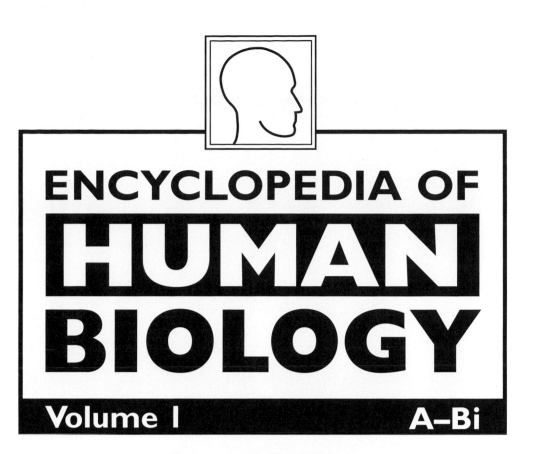

# ENCYCLOPEDIA OF HUMAN BIOLOGY

## Volume I     A–Bi

## Second Edition

Editor-in-Chief

**RENATO DULBECCO**

The Salk Institute

La Jolla, California

Riverside Community College
Library
MAR  '98 4800 Magnolia Avenue
Riverside, California 92506

**ACADEMIC PRESS**

San Diego   London   Boston   New York   Sydney   Tokyo   Toronto

REF   QP 11 .E53 1997 v.1

Encyclopedia of human
 biology

This book is printed on acid-free paper. ∞

Copyright © 1997, 1991 by ACADEMIC PRESS

All Rights Reserved.
No part of this publication may be reproduced or transmitted in any form or by any
means, electronic or mechanical, including photocopy, recording, or any information
storage and retrieval system, without permission in writing from the publisher.

Academic Press
*a division of Harcourt Brace & Company*
525 B Street, Suite 1900, San Diego, California 92101-4495, USA
http://www.apnet.com

Academic Press Limited
24-28 Oval Road, London NW1 7DX, UK
http://www.hbuk.co.uk/ap/


International Standard Book Number: 0-12-266971-3 (Volume 1)
International Standard Book Number: 0-12-266972-1 (Volume 2)
International Standard Book Number: 0-12-226973-X (Volume 3)
International Standard Book Number: 0-12-226974-8 (Volume 4)
International Standard Book Number: 0-12-226975-6 (Volume 5)
International Standard Book Number: 0-12-226976-4 (Volume 6)
International Standard Book Number: 0-12-226977-2 (Volume 7)
International Standard Book Number: 0-12-226978-0 (Volume 8)
International Standard Book Number: 0-12-226979-9 (Volume 9)
International Standard Book Number: 0-12-226970-5 (set)


PRINTED IN THE UNITED STATES OF AMERICA
97  98  99  00  01  02  EB  9  8  7  6  5  4  3  2  1

# CONTENTS OF VOLUME 1

*Contents for each volume of the Encyclopedia appears in Volume 9.*

# PREFACE TO THE FIRST EDITION

We are in the midst of a period of tremendous progress in the field of human biology. New information appears daily at such an astounding rate that it is clearly impossible for any one person to absorb all this material. The *Encyclopedia of Human Biology* was conceived as a solution: an informative yet easy-to-use reference. The Encyclopedia strives to present a complete overview of the current state of knowledge of contemporary human biology, organized to serve as a solid base on which subsequent information can be readily integrated. The Encyclopedia is intended for a wide audience, from the general reader with a background in science to undergraduates, graduate students, practicing researchers, and scientists.

Why human biology? The study of biology began as a correlate of medicine with the human, therefore, as the object. During the Renaissance, the usefulness of studying the properties of simpler organisms began to be recognized and, in time, developed into the biology we know today, which is fundamentally experimental and mainly involves nonhuman subjects. In recent years, however, the identification of the human as an autonomous biological entity has emerged again—stronger than ever. Even in areas where humans and other animals share a certain number of characteristics, a large component is recognized only in humans. Such components include, for example, the complexity of the brain and its role in behavior or its pathology. Of course, even in these studies, humans and other animals share a certain number of characteristics. The biological properties shared with other species are reflected in the Encyclopedia in sections of articles where results obtained in nonhuman species are evaluated. Such experimentation with non-human organisms affords evidence that is much more difficult or impossible to obtain in humans but is clearly applicable to us.

Guidance in fields with which the reader has limited familiarity is supplied by the detailed index volume. The articles are written so as to make the material accessible to the uninitiated; special terminology either is avoided or, when used, is clearly explained in a glossary at the beginning of each article. Only a general knowledge of biology is expected of the reader; if specific information is needed, it is reviewed in the same section in simple terms. The amount of detail is kept within limits sufficient to convey background information. In many cases, the more sophisticated reader will want additional information; this will be found in the bibliography at the end of each article. To enhance the long-term validity of the material, untested issues have been avoided or are indicated as controversial.

The material presented in the Encyclopedia was produced by well-recognized specialists of experience and competence and chosen by a roster of outstanding scientists including ten Nobel laureates. The material was then carefully reviewed by outside experts. I have reviewed all the articles and evaluated their contents in my areas of competence, but my major effort has been to ensure uniformity in matters of presentation, organization of material, amount of detail, and degree of documentation, with the goal of presenting in each subject the most advanced information available in easily accessible form.

*Renato Dulbecco*

# PREFACE TO THE SECOND EDITION

The first edition of the *Encyclopedia of Human Biology* has been very successful. It was well received and highly appreciated by those who used it. So one may ask: Why publish a second edition? In fact, the word "encyclopedia" conveys the meaning of an opus that contains immutable information, forever valid. But this depends on the subject. Information about historical subjects and about certain branches of science is essentially immutable. However, in a field such as human biology, great changes occur all the time. This is a field that progresses rapidly; what seemed to be true yesterday may not be true today. The new discoveries constantly being made open new horizons and have practical consequences that were not even considered previously. This change applies to all fields of human biology, from genetics to structural biology and from the intricate mechanisms that control the activation of genes to the biochemical and medical consequences of these processes.

These are the reasons for publishing a second edition. Although much of the first edition is still valid, it lacks the information gained in the six years since its preparation. This new edition updates the information to what we know today, so the reader can be confident of its full validity. All articles have been reread by their authors, who modified them when necessary to bring them up-to-date. Many new articles have also been added to include new information.

The principles followed in preparing the first edition also apply to the second edition. All new articles were contributed by specialists well known in their respective fields. Expositional clarity has been maintained without affecting the completeness of the information. I am convinced that anyone who needs the information presented in this encyclopedia will find it easily, will find it accessible, and, at the same time, will find it complete.

*Renato Dulbecco*

# A GUIDE TO USING
# THE ENCYCLOPEDIA

The *Encyclopedia of Human Biology, Second Edition* is a complete source of information on the human organism, contained within the covers of a single unified work. It consists of nine volumes and includes 670 separate articles ranging from genetics and cell biology to public health, pediatrics, and gerontology. Each article provides a comprehensive overview of the selected topic to inform a broad spectrum of readers from research professionals to students to the interested general public.

In order that you, the reader, derive maximum benefit from your use of the Encyclopedia, we have provided this Guide. It explains how the Encyclopedia is organized and how the information within it can be located.

## ORGANIZATION

The *Encyclopedia of Human Biology, Second Edition* is organized to provide the maximum ease of use for its readers. All of the articles are arranged in a single alphabetical sequence by title. Articles whose titles begin with the letters A to Bi are in volume 1, articles with titles from Bl to Com are in Volume 2, and so on through Volume 8, which contains the articles from Si to Z.

Volume 9 is a separate reference volume providing a Subject Index for the entire work. It also includes a complete Table of Contents for all nine volumes, an alphabetical list of contributors to the Encyclopedia, and an Index of Related Titles. Thus Volume 9 is the best starting point for a search for information on a given topic, via either the Subject Index or Table of Contents.

So that they can be easily located, article titles gener-ally begin with the key word or phrase indicating the topic, with any descriptive terms following. For example, "Calcium, Biochemistry" is the article title rather than "Biochemistry of Calcium" because the specific term *calcium* is the key word rather than the more general term *biochemistry*. Similarly "Protein Targeting, Basic Concepts" is the article title rather than "Basic Concepts of Protein Targeting."

## TABLE OF CONTENTS

A complete Table of Contents for the *Encyclopedia of Human Biology, Second Edition* appears in Volume 9. This list of article titles represents topics that have been carefully selected by the Editor-in-Chief, Dr. Renato Dulbecco, and the members of the Editorial Advisory Board (see p. ii for a list of the Board members). The Encyclopedia provides coverage of 35 specific subject areas within the overall field of human biology, ranging alphabetically from Behavior to Virology.

In addition to the complete Table of Contents found in Volume 9, the Encyclopedia also provides an individual table of contents at the front of each volume. This lists the articles included within that particular volume.

## INDEX

The Subject Index in Volume 9 contains more than 4200 entries. The subjects are listed alphabetically and indicate the volume and page number where information on this topic can be found.

## ARTICLE FORMAT

Articles in the *Encyclopedia of Human Biology, Second Edition* are arranged in a single alphabetical list by title. Each new article begins at the top of a right-hand page, so that it may be quickly located. The author's name and affiliation are displayed at the beginning of the article. The article is organized according to a standard format, as follows:

- Title and author
- Outline
- Glossary
- Defining statement
- Body of the article
- Bibliography

## OUTLINE

Each article in the Encyclopedia begins with an outline that indicates the general content of the article. This outline serves two functions. First, it provides a brief preview of the article, so that the reader can get a sense of what is contained there without having to leaf through the pages. Second, it serves to highlight important subtopics that will be discussed within the article. For example, the article "Gene Mapping" includes the subtopic "DNA Sequence and the Human Genome Project."

The outline is intended as an overview and thus it lists only the major headings of the article. In addition, extensive second-level and third-level headings will be found within the article.

## GLOSSARY

The Glossary contains terms that are important to an understanding of the article and that may be unfamiliar to the reader. Each term is defined in the context of the particular article in which it is used. Thus the same term may appear as a Glossary entry in two or more articles, with the details of the definition varying slightly from one article to another. The Encyclopedia includes approximately 5000 glossary entries.

## DEFINING STATEMENT

The text of each article in the Encyclopedia begins with a single introductory paragraph that defines the topic under discussion and summarizes the content of the article. For example, the article "Free Radicals and Disease" begins with the following statement:

A FREE RADICAL is any species that has one or more unpaired electrons. The most important free radicals in a biological system are oxygen- and nitrogen-derived radicals. Free radicals are generally produced in cells by electron transfer reactions. The major sources of free radical production are inflammation, ischemia/reperfusion, and mitochondrial injury. These three sources constitute the basic components of a wide variety of diseases. . . .

## CROSS-REFERENCES

Many of the articles in the Encyclopedia have cross-references to other articles. These cross-references appear within the text of the article, at the end of a paragraph containing relevant material. The cross-references indicate related articles that can be consulted for further information on the same topic, or for other information on a related topic. For example, the article "Brain Evolution" contains a cross reference to the article "Cerebral Specialization."

## BIBLIOGRAPHY

The Bibliography appears as the last element in an article. It lists recent secondary sources to aid the reader in locating more detailed or technical information. Review articles and research papers that are important to an understanding of the topic are also listed.

The bibliographies in this Encyclopedia are for the benefit of the reader, to provide references for further reading or research on the given topic. Thus they typically consist of no more than ten to twelve entries. They are not intended to represent a complete listing of all materials consulted by the author or authors in preparing the article.

## COMPANION WORKS

The *Encyclopedia of Human Biology, Second Edition* is one of an extensive series of multivolume reference works in the life sciences published by Academic Press. Other such works include the *Encyclopedia of Cancer, Encyclopedia of Virology, Encyclopedia of Immunology,* and *Encyclopedia of Microbiology,* as well as the forthcoming *Encyclopedia of Reproduction.*

# Abortion, Spontaneous

EWA RADWANSKA
*Rush Medical College, Chicago*

---

I. Physiology
II. Symptoms and Types of Spontaneous Abortion
III. Causes, Diagnosis, and Treatment of Spontaneous Abortion

## GLOSSARY

**Amniocentesis** Puncture of uterine cavity in pregnancy to obtain fluid for testing

**Blastocyst** Stage of embryo development during implantation

**Cerclage** Suture applied around cervix to prevent abortion

**Curretage** Minor surgical procedure for scraping the uterine cavity

**Gamete** Egg cell or sperm cell

**Gestation** Pregnancy

**Hysterosalpingogram** X-ray technique to visualize uterine cavity and tubes

**Hysteroscopy** Procedure to visualize uterine cavity with light source introduced through cervix

**Karyotyping** Testing for number and structure of chromosomes

**Myoma** Fibroid, benign tumor of the uterus

**Oocyte** Egg cell

**Septum** Tissue dividing two cavities

**Trimester** One-third of pregnancy

**Trophoblast** Early placenta consisting of chorionic villi

**Ultrasound** Imaging technique employing ultrasonc waves

**Zygote** Fertilized egg cell before cleavage

SPONTANEOUS ABORTION (MISCARRIAGE) IS DEfined as clinically recognized pregnancy loss prior to fetal viability. The minimum length of pregnancy beyond which a fetus may be viable [i.e., able to survive outside the uterus (womb) with currently available medical support] is 23 weeks from the last menstrual period. For this reason, spontaneous abortion has also been defined as pregnancy loss up to week 22. Most abortions occur during the first trimester of pregnancy (first 12 weeks from the last menstrual period). As a result of improved diagnostic techniques and because of the monitoring of conceptions assisted by new reproductive technology, the phenomenon of subclinical abortion (bleeding indistinguishable from menstrual period) has also been recognized as an entity.

## I. PHYSIOLOGY

The requirements for successful fertilization, implantation, and continuation of pregnancy are complex. Healthy and mature gametes (an egg cell and a sperm cell) are essential to combine into a normal zygote (fertilized egg), which then cleaves to form an embryo. The maturational process of an egg cell (also called oocyte) in the ovarian follicle as well as ovulation (release of the egg cell) are controlled predominantly by the follicle-stimulating hormone (FSH) and luteinizing hormone (LH) from the pituitary gland and by estradiol from the ovary. The processes of egg maturation and ovulation may be abnormal (dysfunctional) and may lead to abnormal fertilization and to pregnancy loss. Similarly, sperm cells, whose maturation is also controlled by pituitary hormones as well as by the male hormone testosterone, may not have normal structure, motility, or enzyme activity and may cause abnormal fertilization. Requirements for implantation include normal receptivity of the endometrium (an inner lining of the uterine cavity), which is depen-

ENCYCLOPEDIA OF HUMAN BIOLOGY, Second Edition, VOLUME I. Copyright © 1997 by Academic Press. All rights of reproduction in any form reserved.

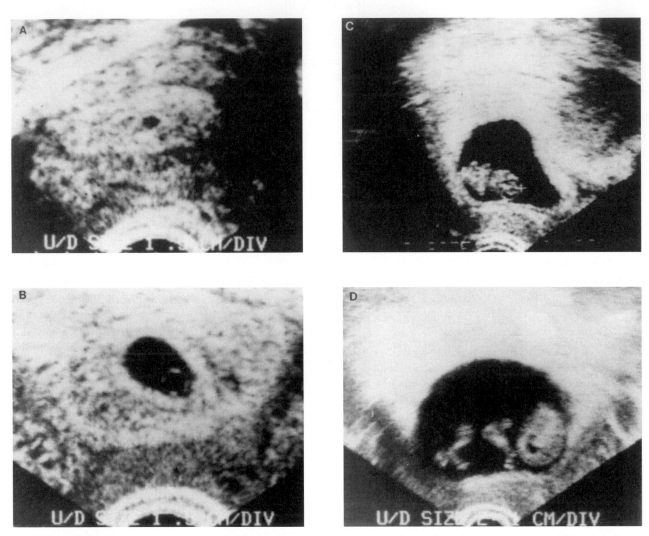

**FIGURE 1**   Ultrasound views of a normally developing pregnancy at (A) 6 weeks, (B) 7 weeks, (C) 8 weeks, and (D) 12 weeks.

dent on estradiol and progesterone produced by the corpus luteum, a gland that forms in the ovary each cycle after the ovulatory follicle has released an egg cell. Estradiol causes a necessary endometrial proliferation (estrogen priming); progesterone causes secretory and decidual changes in the endometrial tissue, necessary for implantation. These changes are very characteristic and have been used as a basis for histologic endometrial dating, often used to evaluate fertility. Progesterone also triggers an immune response in the uterus (suppressor cells), which protects the fetus. An embryo implants in the endometrium at the stage of blastocyst, 5–7 days after ovulation. The *gestational sac* can be located in the uterus by ultrasound examination from 6 weeks of pregnancy, 4 weeks after ovulation (Figs. 1A–1D). In addition to the requirements for normal estradiol and progesterone

levels, other hormones such as those derived from thyroid and adrenal glands play a role in reproduction. Endocrine disorders of these glands as well as diabetes, genetic abnormalities, environmental toxins, drugs, and infections may affect the reproductive process before or after implantation. Indeed, conceptus itself and its genetic makeup can affect implantation and the reproductive process. For these reasons, the causes of spontaneous abortion are multiple, and some may not be possible to diagnose. Types of abortion and their management are described in this article.

## II. SYMPTOMS AND TYPES OF SPONTANEOUS ABORTION

The first symptoms of abortion are bleeding and cramps; this stage is often referred to as threatened abortion. When the uterine cervix (neck of the womb) opens, abortion becomes inevitable. It is followed by an expulsion of the contents of the uterus. Bleeding may become very heavy, usually signifying incomplete abortion; placental tissue is not completely expelled and the uterus is not contracting efficiently. It is recommended that retained products of conception be gently removed using suction curettage (scraping) to prevent prolonged bleeding, which may lead to anemia and to decreased resistance to infection. When heavy bleeding stops and the uterus is empty and contracted, the abortion is complete and does not require curettage. If hormone measurements in blood for human chorionic gonadotropin (hCG) and ultrasound examinations are performed serially in early pregnancy, an abnormally developing pregnancy may be recognized before the symptoms of abortion appear. If hCG levels in the blood fail to increase normally, ultrasound examination often shows an empty gestational sac, which is termed a blighted ovum or anembryonic pregnancy (Fig. 2). If such an empty sac persists

for some time (typically past the 8th week) without triggering spontaneous abortion, it is called a missed abortion. Such nonviable pregnancies should be removed by dilatation and curettage (D&C) because of the increased risk of complications such as infection or abnormal blood clotting. The incidence of clinical spontaneous abortion in the general population is estimated to be approximately 15% of pregnancies. Abortions may be sporadic or recurrent. Sporadic abortions may occur as a result of chromosomal errors in the process of fertilization, delayed implantation, infection, or other incidental causes; they are nonrepetitive and generally inconsequential. When abortions are recurrent, the term *habitual* abortion is applied to couples who have experienced three or more consecutive abortions. The risk of another pregnancy loss is then two to three times higher than the risk of sporadic abortion. Statistical analysis indicates that habitual pregnancy failures are unlikely to be random events and that they tend to be caused by specific etiologic factors.

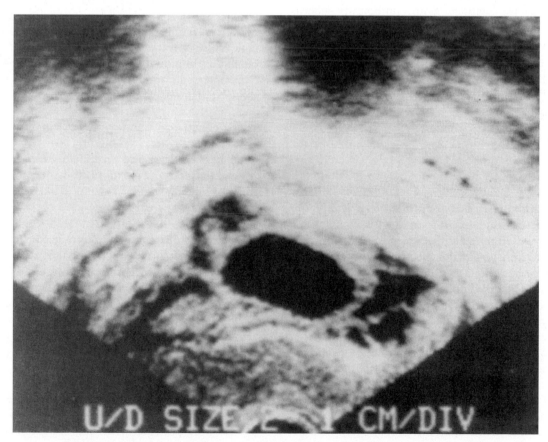

**FIGURE 2**   Blighted ovum (empty gestational sac) with partial separation before abortion.

## III. CAUSES, DIAGNOSIS, AND TREATMENT OF SPONTANEOUS ABORTION

### A. Luteal Deficiency

Luteal deficiency is caused by an abnormally low secretion of hormones, particularly progesterone, by the corpus luteum, leading to the inadequate preparation of endometrium for implantation and for the support of early pregnancy. Luteal deficiency appears to be the most common hormonal abnormality in women with recurrent abortions. It may be diagnosed by endometrial dating and by blood measurements of progesterone. Histologic endometrial retardation >2 days out of phase is a commonly accepted criterion for the diagnosis of luteal phase deficiency. Progesterone levels in blood <10 ng/ml have also been repeatedly described in association with luteal deficiency and with early pregnancy abnormalities. Luteal deficiency is often a part of the spectrum of ovulatory dysfunction (hormonal imbalance before as well as after ovulation). In these circumstances, the goal of the treatment is to improve the ovulatory process as well as the luteal phase during the conception cycle. This can be achieved by treatment with compounds such as clomiphene citrate, which stimulate the pituitary gland to produce increased levels of gonadotropins and, in turn, to stimulate the ovaries. Direct stimulation of the ovaries may be accomplished with exogenous gonadotropin injections such as human menopausal gonadotropins containing FSH and LH as well as with hCG. Progesterone may also be administered to support early pregnancy.

### B. Immune Factors

Abnormal immune response of the uterus to the implanting fetus may lead to spontaneous abortion because of the failure to activate suppressor cells and to protect the embryo against rejection. Such a situation may arise when the father and mother share antigen groupings. This condition may be investigated by human leukocyte antigen typing and by other immunologic testing. Treatment of the mother with leukocyte immunization has been recommended in some cases by authorities conducting such investigations. Sometimes the presence of abnormal antibodies in the mother causes severe fetal damage and death. An example of such a condition is systemic lupus erythematosus. The presence of abnormal antibodies (e.g., lupus anticoagulant, anticardiolipin) can be determined using available blood tests. Treatment involves immunosuppression with steroids such as prednisone and small doses of aspirin or heparin. The role of the immune factors in the success of pregnancy remains a subject of clinical debate.

### C. Genetic Factors

Chromosomal abnormalities are very common in aborted fetuses (approximately 50%); most occur sporadically and probably represent nondisjunction during gametogenesis. However, in some couples with recurrent pregnancy loss, balanced chromosomal translocations (exchanges between different chromosomes) may be present in the mother or father and may produce a chromosomally abnormal fetus due to the inheritance of an unbalanced form of a balanced parental rearrangement. Such chromosomal abnormalities include either so-called Robertsonian or reciprocal translocations, which may result in the transmission of either excessive or deficient chromosomal material to the embryo during the fertilization process. The incidence of chromosomal abnormalities among couples with repeated abortions is reportedly between 1 and 10%. Chromosomal abnormalities in parents are detected by the karyotyping of cultured peripheral lymphocytes, obtained by blood test. Genetic analysis of the fetus may be performed by chorionic villi sampling (CVS) or by amniocentesis and is recommended for high-risk couples. [See Chromosome Anomalies.]

### D. Uterine Factors

One important cause of pregnancy loss is uterine abnormalities. [See Uterus and Uterine Responses to Estrogen.]

#### I. Cervical Incompetence

Cervical incompetence is the term used for an abnormally wide or relaxed opening of the cervical canal, which may lead to premature expulsion of the pregnancy. Diagnosis of this condition is made by an X-ray—hysterosalpingogram (HSG)—or by calibration using dilators. The treatment involves placing a suture around the cervix (cerclage) during pregnancy, typically after 12 weeks of gestation.

#### 2. Congenital Uterine Anomalies

Congenital anomalies of the uterus are a well-known cause of repeated pregnancy wastage and occur in 10–15% of patients with habitual abortions. The most common uterine defects associated with spontaneous abortion are bicornuate and septate uterus (Figs. 3A and 3B). These anomalies arise before birth as a result of incomplete fusion of the Mullerian ducts

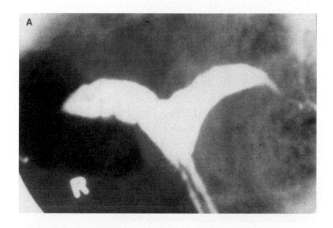

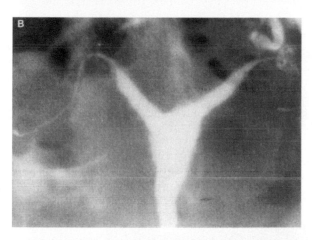

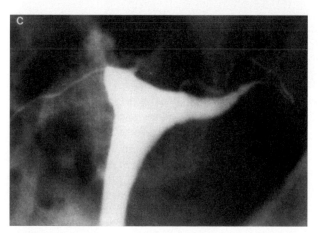

FIGURE 3   Hysterosalpingogram views of an abnormal uterus. (A) Bicornuate, before treatment; (B) septate, before treatment; and (C) septate, after treatment with hysteroscopic septum resection.

(embryonic structures from which the uterus is formed). Diagnosis is confirmed by HSG, and the treatment involves removal of the septum by hysteroscopy—a new and highly successful procedure (Fig. 3C). Unicornuate and didelphic (double) uterus represent less common uterine anomalies and are less likely to lead to abnormal pregnancies. Another problem with the uterus is diethylstilbesterol (DES) exposure of women whose mothers were treated during pregnancy with this compound during the 1950s and 1960s. A DES-exposed uterus may be slender, T-shaped, irregular, or hypoplastic, and the rate of pregnancy loss reaches 50%. Diagnosis is made by HSG. These women also appear to benefit from cerclage; greatly improved term pregnancy rates with this method have been reported.

### 3. Uterine Fibroids

Common benign uterine tumors, termed myomata (fibroids), may cause miscarriages and other pregnancy complications, particularly if they are compressing the uterine cavity. Diagnosis is made by examination, HSG, ultrasound, and, more recently, magnetic resonance imaging. The procedure to remove them is called myomectomy and requires major abdominal surgery (Fig. 4).

### 4. Intrauterine Adhesions

Intrauterine synechiae (adhesions) may be both a result and a cause of pregnancy loss. They represent scarring of the uterine cavity, which sometimes forms after curettage for incomplete or missed abortion (especially when complicated by infection) and, in turn, contributes to further pregnancy losses (Fig. 5A). Diagnosis is by HSG, and the treatment involves hysteroscopic removal of scar tissue (Fig. 5B).

### E. Endometriosis

Endometriosis is a progressive disease frequently associated with reproductive failure. Endometrial cells proliferate outside of their normal location in the uterine cavity, causing irritation and scarring, which involves ovaries, the fallopian tubes, and the uterus. Women with endometriosis are often infertile. Some may conceive but are at increased risk for pregnancy wastage. Endometriosis is diagnosed by laparoscopy with the inspection of the pelvic organs (uterus, fallopian tubes, and ovaries). The treatment may involve hormonal suppression, surgery, or both. Full restoration of fertility is often possible after treatment. [See Endometriosis].

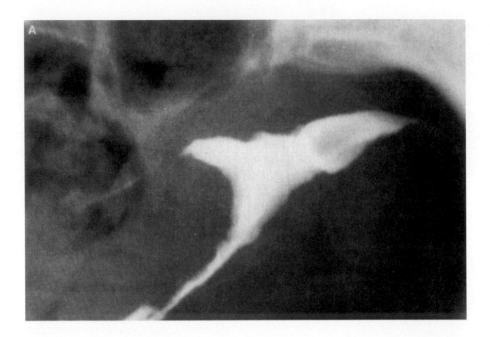

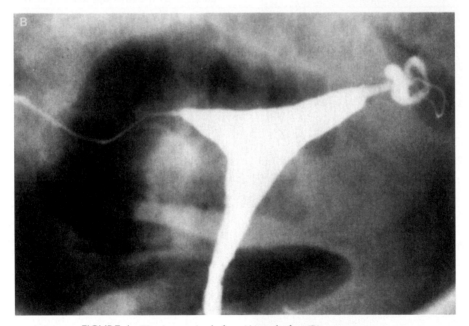

**FIGURE 4** Uterine cavity before (A) and after (B) myomectomy.

## F. Infections

Viral, bacterial, and parasitic infections have been associated with spontaneous abortions. Implicated viruses include rubella (German measles), cytomegalovirus (CMV), and herpes. These viruses tend to infect the fetus during an acute phase but, in the case of CMV, chronic infection may also become reactivated during pregnancy and cause recurrent abortions. Common bacterial infections include *Streptococcus* and *Escherichia coli*. Syphilis (a sexually transmitted disease) is also a well-known cause of fetal death and abortion. Parasitic diseases such as toxoplasmosis (transmitted by cats) and malaria (transmitted by mosquitoes) have been implicated in pregnancy losses. Diagnosis of these conditions is made by serologic

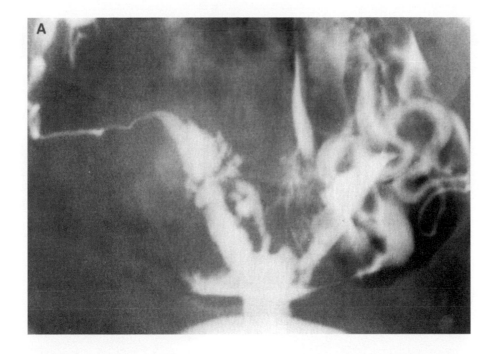

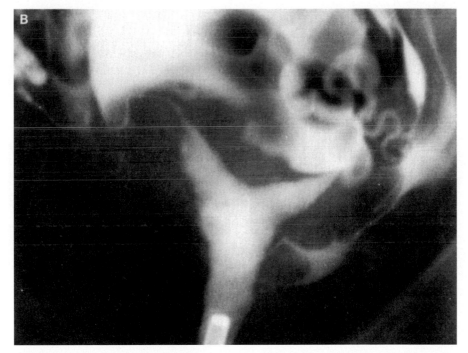

**FIGURE 5** Hysterosalpingogram view of a septate uterus with intrauterine adhesions before treatment (A) and after hysteroscopic treatment (B).

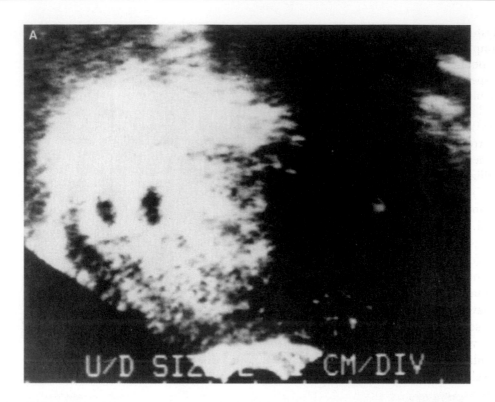

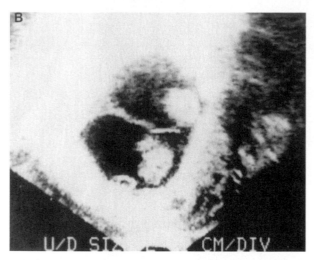
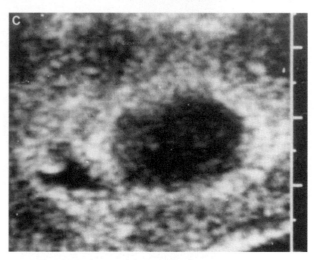

FIGURE 6    (A) Early twin pregnancy, (B) both twins developing normally, and (C) one normal and one "vanishing" twin sac.

blood testing and cultures. The treatment depends on an etiologic agent.

## G. Pregnancies after Infertility

A new concern relates to increased pregnancy wastage in women treated for infertility (inability to conceive). Infertility is widespread in the population and affects approximately 15% of married couples. In some women, both infertility and repeated pregnancy wastage may be part of the spectrum of reproductive difficulty. For example, myomata and endometriosis are progressive diseases and may result in abortions earlier in life and infertility during later years.

Ovulatory dysfunction may also lead to infertility or inadequate implantation and abortion. To improve reproductive potential, such couples may be treated by fertility specialists before conception; the resulting pregnancies are usually carefully monitored and supported. Unfortunately, some new problems are being uncovered as a result of the introduction of infertility treatments such as the induction of ovulation or new reproductive technologies such as *in vitro* fertilization (IVF), gamete intrafallopian transfer (GIFT), and related procedures. Abortion rates in pregnancies resulting from such therapies are often increased. Whether or not this phenomenon is due to the degree of reproductive difficulty in the women undergoing treatment or to factors related to treatment methods (e.g., increased number of multiple pregnancies) is not clear. It has been recently appreciated during early ultrasound examinations that in some multiple pregnancies one twin may "vanish" without causing miscarriage (Fig. 6). Ovarian stimulation with gonadotropins tends to cause the development of multiple follicles in the ovary as well as very high levels of estradiol and progesterone, which may fluctuate widely. This pattern may lead to a hormonal imbalance before or after implantation and trigger miscarriages in some cases. It is hoped that further progress in reproductive sciences and research in reproductive technology will lead to new improvements in the therapy of reproductive failures.

## BIBLIOGRAPHY

Coulam, C. B. (1992). Immunologic tests in the evaluation of reproductive disorders: A critical review. *Am. J. Obstet. Gynecol.* **167**(6), 1844–1851.

Daya, S. (1994). Issues in the etiology of recurrent spontaneous abortion. *Curr. Opin. Obstet. Gynecol.* 6(2), 153–159.

Fayez, J. (1988). Evaluation and management of recurrent early pregnancy losses. *Female Patient* **13**, 100.

Fraser, E. J., *et al.* (1993). Immunization as therapy for recurrent spontaneous abortion: A review and meta-analysis. *Obstet. Gynecol.* **82**(5), 854–859.

Giacomucci, E., *et al.* (1994). Immunologically mediated abortion (IMA). *J. Steroid Biochem. Mol. Biol.* **49**(2–3), 107–121.

Gleicher, N., and El-Roeiy, A. (1988). The reproductive autoimmune failure syndrome. *Am. J. Obstet. Gynecol.* **159**, 223.

Golan, A., *et al.* (1992). Obstetric outcome in women with congenital uterine malformations. *J. Reprod. Med.* **37**(3)L, 233–236.

Heap, R. B., and Flint, A. P. F. (1986). Pregnancy. *In* "Hormonal Control of Reproduction" (C. R. Austin and R. V. Short, eds.). Cambridge University Press, Cambridge.

Mishell, D. R., Jr. (1993). Recurrent abortion. *J. Reprod. Med.* **38**(4), 250–259.

Novy, M. J. (ed.) (1988). Recurrent spontaneous abortion. Seminars in Reproductive Endocrinology, Thieme Medical Publishers Inc., New York.

Radwanska, E., *et al.* (1988). Early endocrine events in induced pregnancies. *Int. J. Fertil.* **33**, 162.

# Acquired Immune Deficiency Syndrome, Epidemic

G. STEPHEN BOWEN
*Health Resources and Services Administration*

GARY R. WEST and LINDA GAUGER ELSNER
*Centers for Disease Control and Prevention*

## GLOSSARY

**Acquired immunodeficiency syndrome (AIDS)** The most severe clinical manifestations of infection with human immunodeficiency virus (HIV), AIDS is characterized by a CD4 lymphocyte count of less than 200, opportunistic infections (including three added in 1993—pulmonary tuberculosis, recurrent pneumonia, and, in women, invasive cervical cancer), unusual forms of cancer, central nervous system manifestations, and wasting syndrome. These manifestations are a result of the destruction of white blood cells and normal immunity by HIV and the direct neurotoxic effect of the virus

**Human immunodeficiency virus (HIV)** Causative agent of AIDS; it selectively attacks and destroys specific white blood cells that are essential for the body's immune system

**Protease inhibitors** AIDS drugs that block the activation of the protease enzyme that HIV needs to reproduce. Saquinavir is an example of this class of drug

**Reverse transcriptase inhibitors** Antiviral drugs that have shown usefulness in prolonging life for people with AIDS if not previously treated, and in slowing progression in AIDS in persons with asymptomatic HIV infection and milder HIV-related medical conditions. Examples are AZT (azidothymidine), also known as zidovudine (ZDV); ddI (didanosine); ddC (zalcitabine); d4T (stavudine); 3TC (lamivudine); and nevirapine

## I. IMPACT

The epidemic of human immunodeficiency virus (HIV) infection and the resulting acquired immunodeficiency syndrome (AIDS) is having a substantial impact on the public health of the United States. HIV/AIDS is now the leading cause of death for persons aged 25–44 years. By June 1996, more than half a million (548,102) persons had been diagnosed with AIDS and nearly 350,000 (343,000) had died. The Centers for Disease Control and Prevention (CDC) estimates that up to 900,000 people in the United States are currently infected with HIV. In the coming decade, hundreds of thousands of infected Americans will become ill, and most, if not all, will eventually die of AIDS or other HIV-related illnesses. [See Acquired Immune Deficiency Syndrome, Infectious Complications; Acquired Immune Deficiency Syndrome, Virology.]

A major impact of the epidemic in the United States has occured in medical and social support systems. In large urban areas, already hard hit by many of society's most difficult, persistent, and costly problems, the medical and social support systems have been especially strained by the large number of severely ill people. Ill children and AIDS orphans accompany the increasing numbers of women with HIV/ AIDS; families may have two or more infected or ill people requiring multiple types and unusual levels of social support, as well as medical services. Government has not yet been able to mobilize the resources in all areas of the country to provide adequate substance abuse treatment programs or to provide adequately for medical (including diagnostic tests and antiviral treatment), psychological, nutritional, housing, and other needs of persons with HIV infection and AIDS.

The HIV epidemic has had a major impact on the epidemiology, diagnosis, and treatment of many infectious diseases. Diagnosis and treatment of syphilis and tuberculosis (TB) may be more difficult in the presence of HIV infection, thus requiring a higher index of suspicion by clinicians. The incidence of tuberculosis was declining in the United States until 1984, when the decline leveled off. Increases in reported TB cases occurred in many areas of the country between 1985 and 1992, especially among young adults and certain minority populations. This reversal of the long-term trend in tuberculosis incidence is believed to be due in part to the HIV epidemic, as HIV-infected persons with latent tuberculosis infection have an extraordinary risk of developing active TB. Also, TB is an opportunistic infection for people with compromised immune systems. Progression to active clinical TB is much more frequent in people with HIV who have an infection with *Mycobacterium tuberculosis* than in people whose immune status is normal. Prevention of active disease and multi-drug-resistant disease for people dually infected with *M. tuberculosis* and HIV is best done with directly observed therapy similar to that done for people with active TB disease. [*See* Infectious Diseases.]

HIV has altered the epidemiology of many noninfectious diseases, including cancers such as Kaposi's sarcoma, lymphomas, and leukemia. Because of their association with HIV infection, such cancers are more common and occur in different age groups than previously. Morbidity and mortality of pneumonia and many fungal, bacterial, parasitic, and other viral diseases are also increasing with the growing prevalence of HIV infection. Furthermore, neurological and psychiatric illnesses and their treatment must be carefully distinguished from similar syndromes related to HIV infection. [*See* Leukemia; Lymphoma; Virology, Medical.]

In addition to the associated morbidity and mortality, the HIV epidemic has been, and will continue to

be, quite costly. The direct cost of health care for persons with AIDS, from the time of infection with HIV to death, has been estimated to be approximately $119,000 per patient. If this figure is accurate, the cumulative direct medical costs of medical care of people with AIDS already exceeds $30 billion. These estimates were made before viral load testing and combination antiviral therapy at all stages of HIV infection became standard of care in the United States. This lab test and drugs can add $7,000 to $10,000 to the annual costs of care. The development and use of improved but costly pharmacotherapeutic agents for prophylaxis and treatment of opportunistic infections, suppression of viral replication, and management of HIV and treatment-related anemia, leukocytopenia, and immune suppression [such as azidothymidine (AZT), zalcitabine (ddc), didanosine (ddI), stavudine (D4T), lamivudine (3TC), nevirapine, saquinavir, ritonavir, indinavir, foscarnet, gancyclovir, interleukin 2, epogen, neupogen, and various purified interferons] will dramatically increase the cost to provide care for persons with HIV infection or AIDS. Zidovudine (ZDV), ddI, and ddC, as well as other reverse transcriptase inhibitors and the protease inhibitors have been shown to prolong life for people with AIDS and may decrease the rate of progression to AIDS in people with mild HIV-related conditions. Prophylaxis for *Pneumocystis carinii* pneumonia, toxoplasmosis, cytomegalovirus, and *Mycobacterium avium intracellulare* is now recommended by the U.S. Public Health Service (PHS). Universal HIV counseling and testing of pregnant women and providing counseling about, and access to, ZDV perinatally to reduce mother-to-infant HIV transmission are also recommended by PHS. All of these interventions are costly. Direct individual care costs $2500 to $8000 per month during the last 6 months of life. Hundreds of thousands of persons are seeking services to prolong, and improve the quality of, their lives. There is currently no comprehensive way to finance the medical management of the 300,000 people with late-stage HIV disease and approximately 500,000–700,000 people with earlier-stage HIV disease who need it. Ten to 30% of people with HIV infection, depending on the stage of their illness, will not have health insurance, whether public or private. The impact of managed care, reduced funding for Medicaid and Medicare, and the moving of more people in these systems to managed care on quality, outcomes, and costs of HIV care remains to be evaluated.

In addition to straining the medical, economic, and social support systems of the United States, the HIV epidemic is exacerbating existing social and economic problems. Drug use is a major risk behavior for HIV infection. In 1994, more than half the AIDS cases reported by New York, Connecticut, New Jersey, and Puerto Rico were among persons who injected drugs. In some areas such as New York City; San Juan, Puerto Rico; and Newark, New Jersey, recent seroprevalence surveys indicate that 30–45% of current injecting drug users (IDUs) are infected with the virus. The parallel epidemic of "crack" cocaine use is paving the way for further spread of HIV through the exchange of sex for drugs.

The national response to the HIV epidemic has included a partial restructuring of the public health system and funding priorities. Many state and local public health authorities initially diverted personnel and funds from important existing programs, such as control of other sexually transmitted diseases, to respond to the needs of the new epidemic. As public funding of HIV prevention programs and HIV-related outpatient health care and support services for people with HIV increased, the need to divert resources from other programs diminished. Today, however, all health programs must compete for funds and qualified staff in an environment of level or shrinking funds.

At the grassroots level, community-based organizations—in their response to the HIV epidemic—are playing a much larger role in the public health arena than ever before. In many instances, community-based organizations have greater access to and credibility with populations of people engaging in high-risk behavior than do government agencies. In addition, community-based organizations may not be as constrained politically as governments may be in their prevention interventions and approaches to education.

In recognition of the important role that communities play in HIV prevention efforts, CDC issued guidance in December 1993 to the 65 state, territorial, and local health departments that receive HIV prevention funds, requiring them to initiate an HIV prevention community planning process in fiscal year (FY) 1994 to qualify for HIV prevention funding for FY 1995 and beyond. As defined by CDC, HIV prevention community planning is a process whereby the identification of high-priority prevention needs is shared between the health department administering HIV prevention funds and representatives of the communities for whom the services are intended. In addition, the community planning process embraces the conviction that the behavioral and social sciences must play a critical role in the development, implementation,

and evaluation of HIV prevention programs within a community. HIV prevention community planning is having a beneficial impact on local prevention efforts. CDC has initiated evaluation studies to validate the community planning process and to assess and measure the impact so that prevention successes can be identified and shared with other community planning groups across the country.

Another area affected by the epidemic is the education and training of health care professionals. Training of primary caregivers, including medical and non-medical personnel (such as police, fire fighters, correctional officers, barbers, beauticians, and morticians), is being expanded to reduce the already low risk of occupationally related HIV transmission. Medical and mental health personnel must now discuss sensitive subjects such as sex and drug use with their patients, and they must be capable of providing sensitive, confidential, and effective risk-reduction counseling, often coupled with HIV-antibody testing. Expanded training will be needed in these areas.

Although much of the impact of AIDS and the HIV epidemic has been disruptive, some of the changes may eventually result in improvements in the public health system. Among these are attempts to address drug abuse prevention and treatment more comprehensively and to integrate treatment of drug abuse with mainstream medical care. For several years the National Institute on Drug Abuse and the Health Resources and Services Administration (HRSA) have funded programs that successfully integrate drug treatment and other aspects of primary and HIV-related care by service providers in major cities.

To further demonstrate the benefits of service integration, HRSA and CDC first successfully pilot-tested a program in community health centers. Subsequently, this successful model of care became Title IIIb of the Ryan White CARE Act. This program uses expanded counseling and testing, partner notification, support services, and case management for each family, as well as extended individual, couples, and group counseling to reinforce safe behavior. Aggressive primary care and prevention prophylaxis for opportunistic infections for people with HIV, including periodic CD4+ lymphocyte monitoring, immunization, and PAP testing per PHS guidelines, are routinely carried out. These models of outpatient primary care, case management, and provision of support services are the basis for the entire Ryan White CARE Act. This landmark legislation is the only HIV-specific categorical care program; it pays for 15% of HIV care. It was authorized by the Congress in 1990 and funding has grown to $975 million for outpatient services in large metropolitan areas (Title I), states (Title II), and directly to service providers (Titles IIIb and IV). Legislatively mandated planning councils in more than 50 cities and more than 300 consortia in states plan, organize, and fund comprehensive care for over 300,000 people with HIV/AIDS who do not have private health insurance. Results of this program have included two- to fivefold increases in the number of HIV-infected persons in care; shorter waiting times for care; access to more diverse pharmaceuticals and outpatient services; decreased emergency room use; and shorter, less frequent hospitalizations, resulting in substantial cost savings. These are significant results for programs that fill in the gaps between individual, Medicare, and private insurance, and programs of the Department of Veterans Affairs and the Department of Defense.

Another improvement in the public health system resulting from AIDS and the HIV epidemic is increased support for the expansion of comprehensive school health education, including prevention of drug use and sexual behaviors that result in HIV infection, sexually transmitted diseases (STD), and pregnancy. Such support was not evident in the previous decade.

The epidemic of HIV infection will not be a short-lived phenomenon. Even the most optimistic agree that it will be a significant source of morbidity and mortality for decades. Its impact is being felt broadly, and public health and medical care systems have been disrupted. It is paramount that programs for the prevention of HIV transmission be established and sustained in every community and that their progress be closely monitored to ensure that they are being implemented successfully.

## II. ELEMENTS OF COMPREHENSIVE HIV PREVENTION PROGRAMS

The mission of HIV prevention programs is (1) to reduce or prevent further HIV transmission in all populations and communities and (2) to reduce the morbidity and mortality associated with HIV infection. Even if or when effective vaccines and curative therapies become available, it will be vitally important that public health programs—in addition to treating sexually transmitted diseases and drug addiction—be effective in preventing, reducing the frequency of, eliminating, or making safer those behaviors that result in HIV transmission. The behaviors associated with most HIV transmission are few and can be influenced by different interventions or combinations of interventions. These behaviors include:

1. needle sharing associated with drug injection;
2. failure to abstain from sexual activity or to use a condom every time when having sex with a person who is infected or whose infection status is unknown; and
3. donation of blood soon after engaging in risky behavior.

Behavior plays a role even in the health care setting, where compliance with recommendations for job site and laboratory procedures, proper use and decontamination of medical equipment, and safe disposal of needles and other sharp objects can be influenced to minimize the potential for occupationally related HIV transmission.

An important prevention breakthrough occurred in early 1994 when an interim review of the AIDS Clinical Trials Group (ACTG) Study 076 revealed that the risk of perinatal (mother-to-infant) HIV transmission could be substantially reduced when both the pregnant women and their newborns received zidovudine (ZDV or AZT) therapy. Research is continuing on the benefits and long-term risks of such therapy, but investigators have found that both mothers and infants tolerate ZDV treatment well, with no significant short-term side effects. These findings led the Public Health Service in 1994 to issue recommendations on the use of ZDV for the reduction of perinatal transmission, and further recommendations in 1995 calling for routine counseling and voluntary HIV testing for all pregnant women.

The findings from ACTG 076 and other studies that show that the use of antiretroviral drugs can suppress or reduce the amount of HIV in body fluids move the country in the direction of supporting voluntary HIV testing for all adults in primary care settings and hospitals, especially in higher seroprevalence settings. HIV prevention and care programs should work to ensure that all adults have access to HIV counseling and testing and that all people with HIV have follow-up care as a primary and secondary prevention strategy. It should be the goal of the public and private health care systems to attempt to have all people with HIV knowledgeable of their HIV serostatus and receiving regular primary care from a physician who knows how to care for people with HIV.

HIV prevention programs are evolving rapidly. More research is needed to reliably determine the effectiveness of specific program activities, either alone or in combination. Their effect on seroprevalence, seroconversion, specific risk behaviors, rates of sexually transmitted diseases and hepatitis B, and perinatal transmission must be measured. Comprehensive HIV/AIDS prevention programs will include the following essential components:

1. Participatory community planning that is evidence-based (i.e., based on HIV/AIDS epidemiologic surveillance and other data, ongoing program experience, program evaluation, and a comprehensive, objective needs assessment process) and incorporates the views and perspectives of the groups at risk for HIV infection/transmission for whom the programs are intended, as well as the providers of HIV prevention services.

2. HIV counseling, testing, referral, and partner notification (CTRPN) to provide, consistent with state laws, both anonymous and confidential client-centered opportunities for individuals to learn their serostatus and to receive prevention counseling and referral to other preventive, medical, and social services.

3. Epidemiologic and behavioral surveillance and research and collection of other health and demographic data to monitor the HIV/AIDS epidemic and behaviors/practices that facilitate HIV transmission and to project trends in the epidemic.

4. Individual-level interventions (e.g., street outreach and needle exchange prevention case management) that provide ongoing health education and risk-reduction counseling, assist clients in making plans for individual behavior change and ongoing appraisals of their own behavior, facilitate linkages to services in both clinic and community settings (e.g., substance abuse treatment settings) in support of behaviors and practices that prevent transmission of HIV, and help clients make plans to obtain these services.

5. Health education and risk-reduction interventions for groups to provide peer education and support, as well as to promote and reinforce safer behaviors and provide interpersonal skills training in negotiating and sustaining appropriate behavior change.

6. Community-level interventions for populations at risk for HIV infection that seek to reduce risk behaviors by changing attitudes, norms, and practices through health communications, social (prevention) marketing, community mobilization/organization, and communitywide events.

7. Public information programs for the general public that seek to dispel myths about HIV transmission, support volunteerism for HIV prevention programs, reduce discrimination toward individuals with HIV/AIDS, and promote support for strategies and interventions that contribute to HIV prevention in the community.

8. Evaluation and research activities necessary to conduct formative, process, and outcome evaluations of HIV prevention programs and to assess the cost-effectiveness and compare cost–benefits of strategies and interventions.

9. HIV prevention capacity-building activities, such as strengthening governmental and nongovernmental public health infrastructure in support of HIV prevention, implementing systems to ensure the quality of services delivered, and improving the ability to assess community needs and provide technical assistance in all aspects of program planning and operations.

10. HIV screening of blood products.

11. Prevention and treatment of sexually transmitted diseases.

12. Prevention of illicit drug use and treatment of drug addiction.

13. School-based drug, HIV, and sexually transmitted disease prevention education programs that are locally determined and consistent with community values. These interventions may be most effective when carried out within a more comprehensive school health education program that establishes a foundation for understanding the relationship between personal behavior and health and when students have opportunities to develop qualities such as decision making and communication skills, resistance to persuasion, and a sense of self-efficacy and self-esteem.

14. Strong legal protection against unauthorized disclosure of test results and clinical information.

15. For infected persons, prevention-related services.

16. Knowledgeable HIV-related health care, including the provision of information to infected pregnant women regarding the risks and benefits of ZDV therapy to reduce the risk of perinatal HIV transmission and arrangements for interested women to actually receive the ZDV therapy throughout the pre-, peri-, and postnatal period.

17. Prevention of discrimination against infected individuals, and the establishment of, and routine adherence to, procedures for responding if it does occur. People with HIV are covered under the federal Ameri-

cans with Disabilities Act. Many states have laws to protect against discrimination as well. If the legal protection aspects of prevention programs are not adequately addressed, persons at high risk for HIV infection may not participate in prevention programs such as counseling and testing.

Because of the need for broad community support, all program activities should be planned, implemented, and evaluated with the active participation and consultation of community leaders, especially leaders who are or represent persons at risk and persons with HIV infection. The HIV prevention community planning process and the planning councils, consortia, and statewide planning processes mandated under the Ryan White CARE Act are assisting in this process, and are addressing the need to improve mechanisms for providing funding and technical assistance to local government agencies and community-based organizations undertaking or interested in undertaking prevention activities.

## III. Public Health Strategies for Prevention of HIV Infection and Related Morbidity and Mortality

Seven basic strategies underlie current public health programs for the prevention of HIV infection and related morbidity and mortality. These different strategies can be implemented at an individual or clinical level, at an institutional (school, work site, prison) level, or at a community level. All of these strategies should be mutually reinforcing, implemented concurrently, and carried out over a long period of time.

The first strategy is *health promotion, including disseminating information about risky and safe behavior.* This encompasses active promotion of healthy behavior for individuals of all ages, but especially children and young persons. In the home, school, church, and workplace, people of all ages can learn a fundamental approach to living healthy lives based on adequate nutrition, exercise, mental health, and adoption of healthy behaviors (e.g., seat belt use, safe driving, maintaining monogamous sexual relationships, and having smoke alarms in houses).

The second strategy is to *prevent the initiation of specific risky behaviors* such as drug use (especially

injection drug use) and sex (especially anal sex) with an infected partner or one of unknown HIV status. To accomplish this, people must be educated about HIV transmission so they know which behaviors they should avoid. Primary care physicians, counselors, and care providers at family planning clinics, community health centers, and state and city primary care facilities should counsel patients about avoiding risky behavior. Opportunities to develop life-styles that eliminate or minimize risk of exposure to HIV must be provided, and social norms must be changed to support low-risk—and to discourage risky—behavior. Community organizations, business and labor groups, religious institutions, and schools play a central role in this prevention strategy. Research must be conducted to determine the most effective means for promoting health and influencing social norms. Recommendations for healthy behavior must be based on both this research and epidemiologic studies related to the prevalence and the known risk distribution of infection by persons engaging in specific behaviors. For example, encouraging homosexual men to use condoms always and to avoid anal sex with persons whose serostatus is unknown or who are infected is well founded and based on information compiled through epidemiologic research.

A third strategy is to *prevent the continuation of specific risky behaviors* among those practicing such behaviors. Interventions must be developed to reduce the frequency of the risky behavior, and/or prevent progression to even riskier behavior. For example, non-injection-drug users must be prevented from beginning to inject drugs and share needles. An important part of this strategy is to provide adequate treatment facilities for drug users and to implement street outreach to encourage users to obtain treatment. Voluntary counseling and testing should be provided at medical care facilities such as STD clinics, drug treatment facilities, prisons, and anonymous test sites where people who engage in risky behavior enter the health care system.

A fourth strategy is to *reduce risks associated with specific behaviors.* Some individuals will, for a variety of reasons, adopt or continue risky behaviors even when they are aware of the risks. For example, married couples in which one partner is infected (discordant couples) might continue to have sex without using condoms, or drug users who are not in treatment might continue to inject drugs. For such persons, condom use 100% of the time and the use of new, sterile injection paraphernalia ("works") are viable interim means to reduce the risks of specific behaviors and to provide time for other interventions to assist in eliminating the risks altogether.

A fifth strategy to reduce HIV transmission is *prevention and treatment of genital ulcer-producing STD, which facilitate HIV transmission.* Adequate resources and community support should be developed for STD programs. Public information and school-based HIV education programs should emphasize the role of STD in HIV transmission.

A sixth prevention strategy is to *reduce or eliminate environmental barriers to prevention* in communities where they exist. For example, treatment for substance abuse and STDs may not be readily accessible; contraceptive services for infected women may not be available; condom availability or sterile "works" possession, distribution, or sale may be restricted. Therefore, efforts are needed to identify such barriers and to remove or minimize their impact. Depending on the prevalence of HIV infection, reported AIDS cases, and risk behaviors, these efforts may include:

1. locating counseling and testing, drug treatment, family planning, and STD treatment clinics at easily accessible locations within communities;
2. increasing the capacity of existing clinics;
3. expanding distribution of condoms, sterile syringes ("works"), or, as a last resort, bleach for syringe disinfection directly to individuals at risk;
4. changing or repealing statutes that restrict the availability of sterile syringes for drug users,
5. expanding public information efforts in order to build community support for adequate funding of prevention activities; and
6. in the hospital or laboratory setting, providing adequate protective equipment and establishing safe procedures to prevent occupationally related transmission.

The seventh strategy is to *provide prevention-oriented and primary health care and support services,* including reinforcement of safe behavior, to infected individuals and their sex and needle-sharing partners to avoid further transmission. The range of prevention-oriented services provided to infected individuals includes:

1. Confidential notification of exposed partners to ensure they receive appropriate counseling. Studies have shown that extended counseling of discordant couples can be effective in achieving a high rate of safe behavior, including always using condoms, and reducing the rate of HIV transmission.
2. Establishment of systems to ensure that infected individuals are referred to and receive needed medical, psychological, and social support services, including treatment for substance abuse. For infected pregnant women, counseling should be offered on the risks and benefits of ZDV therapy for reducing the risk of transmission to their infants; providers should arrange for interested women to actually receive the perinatal treatment and follow-up regimen.
3. Contraceptive services for infected women who wish to use them and counseling regarding the need for condom use by husbands and sex partners.
4. Medical care and prophylactic therapy for HIV-related opportunistic infections.

In addition to preventing transmission, the other major goal of HIV/AIDS prevention programs is to reduce HIV-related morbidity and mortality. By ensuring that services are provided to infected individuals, prevention programs can make much progress toward attaining this goal. Confidential HIV-antibody counseling and testing programs—with the informed consent of persons receiving services—allow for early diagnosis and management of HIV-related illnesses. By monitoring the function of the immune systems of infected individuals, through means such as CD4 lymphocyte counts and provision of prophylactic treatment, some opportunistic infections can be prevented or have their severity reduced. Infection with *Mycobacterium tuberculosis* can be diagnosed and prophylactic treatment given before clinical tuberculosis develops. Bactrim can be given to prevent *Pneumocystis carinii* pneumonia. In addition, combinations of antiviral drugs, such as, ZDV, ddI, ddC, 3TC, D4T, protease inhibitors can be given to prolong life and delay immune system and neurologic deterioration. Viral load testing is now used to decide when to initiate and/or to change combination antiviral therapy. Aggressive diagnosis and management of HIV-related malignancies can prolong life and increase the quality of life of persons with HIV infection or AIDS. The benefit of early diagnosis has been clearly established; for pregnant women, it allows the further benefit of providing an opportunity, by use of ZDV perinatally, to reduce by approximately 67% the risk of HIV transmission to their infants. As these options become widely known, and as further bio-medical interventions including prophylaxis and treatment of other opportunistic infections are developed, acceptance of and demand for counseling and testing services will increase. In addition to aggressive medical management, psychosocial support services can also help people with HIV infection and AIDS to live happier and more productive lives.

## IV. IMPEDIMENTS TO IMPLEMENTATION OF EFFECTIVE HIV PREVENTION PROGRAMS

Numerous factors can impede the development and implementation of effective HIV prevention programs. In many communities, misunderstanding about how HIV is transmitted continues to be evident, especially among teenagers and people with low language or educational skills. This situation translates into a lack of support and adequate funding for public health programs targeted to individuals at high risk of acquiring, or those who already have, HIV infection, such as the establishment of neighborhood drug treatment facilities, expansion of services to all users willing to receive drug treatment, provision of sterile syringes to users who continue to inject, development of adequate programs for sexually transmitted disease prevention and treatment, getting and keeping all HIV-infected persons into primary care by willing and competent primary care providers, and implementation of controversial, but culturally relevant, programs targeted to men who have sex with men. Such misunderstanding also may lead to public support for unreasonable actions such as quarantine or isolation of infected individuals. Fear of infection from casual contact can lead to discrimination against persons who have HIV infection or AIDS and possibly against persons who are only perceived to be at risk. Actual acts of discrimination or perceptions that discrimination can occur may prevent disclosure of HIV infection status to sex partners or to health providers, and may curtail participation in public health programs by persons at risk, thus further impeding the public health response to the epidemic.

Another important impediment to the success of HIV prevention programs is that social norms in many communities support the continuation or even promote many of the behaviors that are associated with HIV transmission. Movies, television programs, commercial advertising, and popular music often glamor-ize sex or drug-using behaviors. Peer groups may encourage drug use or sexual behavior among adolescents and young adults. The existence of such norms not only makes it more difficult for individuals currently at risk to change their behavior, but also promotes the adoption of such behavior by persons not currently at risk, especially young people.

Implementation of effective prevention programs is hindered by the fact that many people at risk of HIV infection are not easily accessible by traditional means, such as through the media, at school, through health-care providers, or in the workplace. When they are reached, attempts to assist them in reducing risky behavior and adopting and maintaining safe behavior may be resisted. Prostitutes, drug users, their sex partners, and closeted gay men are often distrustful of government programs. In some communities, language itself may be an impediment, making it difficult to communicate prevention messages to migrant workers and others whose first language is not English.

The lack of information regarding determinants and patterns of sexual and drug-using behavior in the United States is a major impediment to the development and evaluation of public health programs. Without such information, it is difficult to target resources and to measure progress against the epidemic. The lack of understanding and support, from legislative bodies and the public, of the need to obtain data on the prevalence of risky behaviors and to conduct research regarding the causes and methods to prevent such behaviors has impeded prevention efforts.

Finally, a growing problem for HIV prevention programs is that without careful planning and support, caregivers and prevention and treatment program staff can become discouraged and suffer job-related burnout. For some categories of health-care workers, fear of occupation-related transmission contributes to this problem. Capable employees may leave, or job performance may suffer.

## V. EVOLVING HIV PREVENTION INTERVENTIONS

Although many aspects of HIV prevention programs are controversial, it has been particularly difficult to educate the public and obtain broad-based support for several prevention program activities. The following is a brief discussion of some of these activities and issues relevant to each of them.

## A. HIV Antibody Counseling and Testing

The goals of HIV counseling and testing programs include early diagnosis and medical management of HIV infection and related illnesses for the purpose of prolonging life and reducing mortality; changing or preventing risky behaviors that can lead to HIV transmission; providing access to an extended system of prevention-oriented health care; and, for pregnant women, to provide the opportunity for antiviral therapy that can reduce the risk of transmitting HIV to their fetus or newborn.

Counseling and testing have been controversial since the HIV antibody test was first licensed in 1985. During 1989–1993, more than half of the funds awarded by the Centers for Disease Control and Prevention were allocated to these activities. More recently, community planning bodies have determined the priorities for uses of prevention funds, and the support for counseling and testing has decreased. Opposition to counseling and testing has been based on challenges of the test's sensitivity and specificity; the limited evidence that knowledge of serostatus brings about a reduction in risky behavior; the assumption that because there is no cure for HIV infection, knowledge of antibody status has limited medical benefit; the belief that results cannot be kept confidential; and the fears of discrimination.

Many of the initial fears regarding counseling and testing have been proved to be unfounded. Most of the concerns regarding the accuracy, reliability, sensitivity, and specificity of test results have been resolved by ensuring that the correct sequence of screening tests—two reactive screening tests followed by a definitively positive confirmatory test performed on the same serologic specimen—is carried out before informing a person that he or she is "positive" for HIV antibody and therefore infected with the virus. Further concerns about the "truth" of test results, for people who do not admit to risky behavior or did not know about the risk behavior or infection status of their partner(s), can be resolved by procedures to evaluate the immune status of the person with positive test results (HIV antigen, T4 cell counts, tests for anergy, polymerase chain reaction, etc.).

Concerns persist about the benefits of counseling and testing as a behavior-change intervention. Limited data are available regarding heterosexuals and minority persons. A substantial amount of evidence has accumulated that homosexual and bisexual men, who have elected to receive counseling and testing and who have received their test results (especially those who are seropositive), are less likely to engage in risky behavior than those who are unaware of their antibody status. In one study, seropositive homosexual men were less likely to engage in risky behavior with men known to be seronegative. However, some studies have failed to show any behavioral impact from knowledge of serologic status, especially among those found to be seronegative. Others have shown decreased risk behavior among drug users after counseling and testing, especially when conducted by trusted people "on the street." Counseling and testing programs should work toward improving the quality of the counseling they provide, even as researchers continue to assess whether inducing sustained behavioral change is a realistic goal for these programs. More data are needed to document whether counseling can result in the maintenance of long-term behavior change among discordant, stable, or monogamous homosexual and heterosexual couples. Data from discordant partners studies have shown that transmission can be prevented by uniform practice of safe behavior (i.e., the correct and consistent use of condoms), although "partners studies" of men with hemophilia show that not all couples will adopt safer behaviors.

Ascertaining the effectiveness of counseling and testing as a tool for inducing behavior change is difficult for many of the same reasons that apply to the evaluation of behavioral interventions. These reasons are:

1. persons at risk are rarely exposed to a single intervention, and it is difficult to identify precisely the contribution of any specific intervention;
2. documentation of the additional effect of counseling and testing in the face of significant communitywide behavior change is difficult;
3. counseling procedures, the content of prevention messages, and the skills of the counselor vary significantly;
4. behavioral measures vary widely from study to study and are not always transmission-relevant (e.g., receptive oral intercourse for already infected people or insertive unprotected intercourse for known seronegative people); additionally, behaviors are not measured for defined fixed intervals after counseling of specified content and duration;
5. the data currently available are based primarily on studies of self-selected cohorts of self-identified gay men, who may not be representative of their communities; few data are available about

minority men and women, IDUs, and hetero-sexuals;

6. the immediate impact of learning test results and the long-term impact of knowledge of serostatus have not been carefully evaluated; and

7. the impact of follow-up and reinforcement of safe behavior by counselor, friends, spouse, sex partner, and family has not been thoroughly evaluated.

Evaluation of the behavior of both partners in discordant sexual relationships and their behavior outside their primary sexual relationship is especially important.

Concern about the medical benefits of knowing one's serostatus returns us to the earlier discussion of advances in prophylaxis and curative treatment for opportunistic infections that strongly support early diagnosis. Antiviral drugs (e.g., AZT, ddI, and ddC) have proved effective, especially in combination, in prolonging life for people with AIDS and may be effective in slowing the progression from asymptomatic and mild HIV infection to AIDS. Aggressive prophylaxis of opportunistic infections definitely reduces morbidity and prolongs life. Most recently, AZT has been proved effective in reducing the risk of mother-to-infant HIV transmission by approximately two-thirds. The need, therefore, for early diagnosis through counseling and testing will accelerate.

Unauthorized disclosure of test results will always be a concern. Experience indicates, however, that such disclosures of information are extremely rare in the public sector and infrequent in the private sector. Staff training and limited access to sensitive information, computer security procedures, and computer encryption of client identifying information are measures that can reduce unauthorized disclosure of HIV infection status. The most common situation in which information regarding the serostatus of an individual is disclosed is when that individual voluntarily informs another. Counselors should warn individuals with HIV infection about the risks of disclosing their serostatus to those who do not need to know. Sufficient state and/or federal legislation, such as the Americans with Disabilities Act, must be in place to ensure that infected persons cannot be the objects of discrimination, that test results cannot be disclosed without authorization, and that persons cannot be tested unless they provide informed consent. These protections greatly increase the acceptance and potential effectiveness of HIV counseling and testing as an intervention strategy.

There is continuing demand for publicly supported HIV counseling and testing services. In 1995 alone, nearly 2.5 million HIV tests were performed, of which more than 40,000 were positive for HIV antibody. A continuing concern is that large numbers of people fail to return for test results. Although health departments routinely conduct follow-up and provide results to most people found to be HIV positive, many uninfected persons at risk never receive needed counseling. A large number of additional people have received counseling about how to protect themselves from HIV. Reducing morbidity and mortality related to HIV infection is an important role for counseling and testing, as is the important additional role of helping reduce the risk of perinatal HIV transmission through early identification of HIV-infected pregnant women and the provision of ZDV therapy to those who request it. HIV counseling and testing for all adults, especially those who live in higher HIV seroprevalence areas, are becoming more prevalent and should be encouraged. These aspects of counseling and testing programs will lead to an increased demand in the future by providing access to medical care and follow-up for infected persons and their families so that morbidity and mortality can be reduced.

## B. Partner Notification

The sex partners of persons infected with HIV are at high risk of becoming infected themselves. Eleven to 40% of such persons may already be infected at the time the index "case" learns he or she is infected. People who are in continuing sexual relationships with infected persons require more education than the general message that "sex with an infected person is risky" or "Beware, many gay men and injecting drug users are infected in your community."

Partner notification is a prevention activity designed to provide focused educational and behavior-change messages to persons at highest risk of infection (spouses and other sex or needle-sharing partners of infected persons) so that they can protect themselves and elect to determine their own serostatus. Although partner notification continues to be controversial as a public health HIV prevention intervention, this public health activity has been carried out by STD programs for many years with careful protection of confidentiality and with the cooperation of infected persons. Partner notification has helped control outbreaks of syphilis and gonorrhea in many communities. Through this service, many sex partners have been diagnosed early in their infections and effectively treated.

There are two major partner notification strategies: (1) patient referral, in which infected persons are trained in ways to personally inform their partners of their exposure and let them know where they may receive counseling and testing services; and (2) provider referral, in which trained health care workers, working with the consent of infected individuals and without revealing the identities of those individuals, notify partners and ensure that they receive appropriate counseling and testing services.

Debate about partner notification has focused on confidentiality of records; the intrinsic nature of the process of partner notification; cost-effectiveness; lack of resources, especially in high seroprevalence areas; absence of data demonstrating the effectiveness of partner notification as an educational intervention in communities that are already educated about their risk; difficulty in locating some partners; and the large number of partners to be located because of the long incubation period of HIV infection.

Research efforts are now being directed to better determine the effectiveness of partner notification as a strategy and to compare the benefits of various approaches. Prevention program activities, such as partner notification, are rarely 100% successful. This does not negate the benefit of partial success in locating some partners to bring them into counseling, testing, and medical-care services, and should not deter the HIV program manager.

The priority for carrying out partner notification and the acceptance of the need to locate infected persons will increase as the medical benefits of knowing one's serostatus increase and as resources are allocated for antiviral drugs, pentamidine, tuberculosis therapy, and other medical benefits, as well as for extended individual and couple-oriented follow-up counseling and safe behavior reinforcement. Staff can be trained to carry out partner notification with discretion and sensitivity about confidentiality. We continue to need rigorous evaluations of the effectiveness of this intervention to determine whether behavior can be changed and seroconversion in partners at risk prevented.

In most states, resources for HIV prevention programs are limited, and the provider referral component of partner notification efforts can be expensive. All health care providers should encourage infected individuals to notify their partners and refer them for services. Some health departments undertake provider referral only in certain priority situations, such as for

female partners of infected men or sex partners in ongoing relationships. Women may not know that they or their unborn children are at significant risk. A second priority situation is to notify the spouse or long-term sex partner of an infected person where exposure will continue and transmission is likely unless safe behavior is understood and practiced by both partners. To receive federal funds, all state HIV prevention programs must emphasize index patient/client referral and implement provider referral if indicated.

## C. Street Outreach Programs

Street outreach programs carried out by "indigenous" street workers are common components of HIV prevention programs. Such programs often employ rehabilitated drug users, former prostitutes, and other "streetwise" people to communicate prevention messages and make referrals to treatment programs for persons at high risk for HIV infection, especially injecting drug users, their sex and needle-sharing partners, street youth who may be homeless, and prostitutes. These street workers may provide one-to-one counseling; refer persons at risk to drug treatment facilities or to counseling/test sites and STD treatment facilities; promote condom use by prostitutes; and provide information on the importance of using sterile syringes or the use of bleach to disinfect "works," or actually distribute sterile syringes, bleach, or condoms. More than 100 sites now have legal or underground needle exchange programs. Evaluations of several of these programs have concluded that they contribute significantly to reducing HIV transmission among injecting drug users. Acceptance of these programs is growing and, in many communities, street outreach workers are protected and supported by persons at risk and by the police. Data from research projects funded by the National Institute on Drug Abuse clearly show that many drug users not in treatment will seek treatment and many more will reduce their frequency of injection and needle sharing and increase the use of sterile needles/"works" after being contacted by street outreach workers. Such programs may be the only effective way to reach certain at-risk populations. [See Sexually Transmitted Diseases.]

## D. School-Based Education Programs

CDC has worked closely with national health and education organizations and state and local education agencies to establish HIV prevention policies and programs appropriate to schools and youth, to provide effective HIV prevention training to teachers and other school personnel, and to prepare and implement AIDS/HIV prevention curricula. Many schools integrate HIV prevention curricula into other health education efforts. The majority of schools have implemented HIV prevention policies. School-based policies and programs have been a positive step in reaching the majority of school-age youth in the nation. These programs can play a constructive role in promoting group/peer acceptance of abstinence and safer sexual behavior.

## E. Prevention-Oriented Follow-up and Treatment Programs for Infected Persons

Prevention-oriented follow-up and treatment interventions for infected persons are recognized as important parts of comprehensive HIV/AIDS prevention programs. Options for effective follow-up include:

1. referral for drug and STD treatment;
2. referral for clinical and immunologic evaluation;
3. long-term treatment for people with HIV—all persons with HIV should be receiving regular primary care, and suppression of viral load by combination antiviral therapy may reduce the risk of further transmission;
4. provision of contraceptives to infected women who wish to use them and counseling regarding condom use by their male partners;
5. counseling for infected pregnant women on the risks and benefits of ZDV therapy to reduce the risk of HIV transmission to their infants; arranging for the provision of the regimen for those who wish it is recommended by the Public Health Service (see "Morbidity and Mortality Weekly Report," August 5, 1994);
6. prevention of suicide and serious psychological deterioration;
7. tuberculosis prevention;
8. prevention of *Pneumocystis carinii* pneumonia (PCP) and other opportunistic infections; and
9. prevention of further transmission to uninfected partners by any or a combination of couples counseling, peer support, skills building, modeling, and extended professional counseling.

Case managers for infected clients or couples may be needed to ensure that all prevention services are obtained.

Couples counseling may be particularly beneficial for discordant couples in long-term relationships. The development of couples counseling, whereby (first separately, then together) discordant couples can be assisted to achieve completely safe or at least safer behavior, is an evolving public health concept. Data indicate a willingness to accept safe behavior on the part of both partners in most ongoing serologically discordant relationships.

## F. Funding and Technical Assistance for Communities

The community is an extremely important area for HIV prevention and care. Many successful community-based education and prevention programs were locally planned and financed, and many achieved success in reducing the prevalence of risky behaviors before the development of government programs and subsidies. In the past, local, state, and federal support for community-based HIV/AIDS prevention interventions focused on homosexual men and minority populations. CDC has provided assistance to more than 600 community-based organizations through both direct funding and cooperative agreements with states, cities, and the U.S. Conference of Mayors. HRSA, through the Ryan White CARE Act, funds 50 planning councils, 300 consortia, and more than 2000 community service providers to plan, organize, and deliver outpatient health care and support services for people with HIV.

Community organizations have neighborhood credibility and access to high-risk subpopulations that health departments or other governmental agencies and personnel may not have. They can plan and carry out a variety of interventions directed to infected people and people at high risk of infection, including youth, homosexual men, injecting drug users, prostitutes, women, and people in the criminal justice system. These community-based organizations continue to play an important role through the HIV prevention community planning process.

To assist in community planning efforts across the country, CDC is working with its prevention partners to provide technical assistance and training to health departments and community planning groups. In the first year, this technical assistance was delivered through a network of governmental, nongovernmental, and private providers and focused on the following areas:

1. parity, inclusion, and representation of affected populations;
2. the use of epidemiologic data in the planning process;
3. community planning processes and models;
4. access to behavioral and social science expertise, including information on effective and cost-effective HIV prevention efforts; and
5. conflict of interest and dispute resolution.

Based on experiences in the first year and input from project areas, technical assistance providers, and community consultants, CDC determined that for year two, a decentralized technical assistance network would help in meeting the needs of project areas and community planning groups. Although CDC maintains technical assistance providers at the national level, it also implemented a decentralized technical assistance program that encourages linkages with, and the development of, local resources. Enhancements of technical assistance in year two included providing tools and consultation to project areas to help assess technical assistance needs and priorities, accessing and building the capacity of local and regional organizations to provide technical assistance, implementing processes to establish common objectives and expectations for technical assistance, and improving the follow-up after technical assistance visits. With these enhancements, community planning groups and health departments are more directly involved in the assessment, selection, and provision of technical assistance in their areas.

HRSA, through a contract with a nongovernmental organization, funds site visits, telephone consultations, and group training by peer providers experienced in HIV care, program management, and technical assistance.

## VI. Evaluation

Evaluation of HIV/AIDS prevention programs for their effectiveness in preventing infection, morbidity, and mortality can focus on:

1. availability of resources;
2. effectiveness of administration and program direction, including community involvement;
3. process or quality assurance (e.g., surveillance and seroprevalence, counseling, testing, partner notification, and ensuring provision of or referral for clinical care being carried out with scientific rigor and careful monitoring); and
4. impact or outcome of (a) prevention programs in their entirety or (b) activities or interventions alone or in combination.

Activities designed to prevent infection should be evaluated separately from those designed to reduce morbidity and mortality among infected persons. Evaluation also must take into account geographic, racial/ethnic, gender, route of transmission, and community diversity.

For prevention programs as a whole, documentation of stabilization or reduction in seroincidence and seroprevalence in the diverse populations at risk alone and/or together is, perhaps, the most desirable evaluation measure over time. This outcome measure is expensive, requires ongoing or repeated seroprevalence surveys to document progress, and requires implementation of standardized protocols to collect baseline information from blinded surveys. Examples of serosurveys carried out according to standard protocols include heel-stick surveys of newborns to document seroprevalence in women delivering infants and surveys carried out at sentinel hospitals, STD clinics, drug treatment facilities, women's health clinics (including prenatal and obstetric facilities), tuberculosis clinics, colleges, universities, and prisons. Seroprevalence data are also available from routine screening programs for blood donors, military recruit applicants, and Job Corps entrants. Linked seroprevalence surveys are conducted among active duty military personnel by the Department of Defense and may also be conducted in populations, such as young gay and bisexual men, that cannot be accessed by anonymous, unlinked seroprevalence surveys.

Data on HIV incidence are limited, although new HIV incidence studies are under way in STD clinics and drug treatment centers. Some incidence data are available from vaccine study cohorts, although varying selection criteria and rapid declines in incidence after study enrollment limit the usefulness of these data for prevention evaluation. Incidence data are also available from active duty military personnel, but the few numbers of seroconverters in any given area and possible lack of representativeness of military personnel to the civilian population at large of the surrounding area severely limit the usefulness of these data at the local level.

Another important data set to evaluate and monitor trends in effectiveness of HIV/STD prevention programs includes surveillance data for sexually transmitted diseases such as syphilis, gonorrhea, and hepatitis B analyzed by sex, race, sexual orientation, and (for the first two illnesses) anatomic site. These analyses allow program managers to differentiate trends in behavior and infection among heterosexual and homosexual men and women in different cities, sections of cities, and separately for diverse racial and ethnic populations. Different trends among the three diseases mentioned may differentiate changes in behavior among homosexual and heterosexual populations and those who inject drugs or use "crack," and potentially between "fast track" homosexual men with large numbers of partners and other homosexual men.

The foregoing analyses allow resources to be used where they are needed, if sufficient resources are available. Analysis of admissions, retention, and recidivism of clients of drug treatment programs; street surveys of drug users to validate trends in admission to treatment, injection frequency, needle sharing, and the use of bleach and/or syringe exchange programs; and analysis of emergency room, coroner, and police records are all ways in which prevention programs targeting drug users can be evaluated. Urine testing among patients under treatment is an adjunctive evaluation method. As discussed in the following, pre- and postintervention behavior risk-assessment questionnaires also can be used to evaluate self-reported behavior change in this population. These pre/post designs are best carried out in a time series fashion (multiple measures of risk behavior over time) with control communities or neighborhoods to better assess program impact. These analyses need to be made by type of drug as well as by route of use. The pre- and postintervention analysis suffers, as do all self-reported behavior surveys, from the methodologic problems such as recall bias, desire to please the investigator, attrition, and volunteer bias. Questions regarding sexual behavior and HIV transmission must also be addressed.

An additional method to evaluate the outcome of HIV prevention interventions is pre- and postintervention risk behavior assessment, especially when a random or nonbiased experimental or quasi-experimental design or an intervention/enhanced intervention or immediate intervention versus delayed intervention design is used. Experimental designs with no "treatment" control groups and random assignment of participants to different interventions and to the control group are methodologically best for ascribing causality to the observed outcome. This approach, as well as others, is subject to the ethical considerations of offering no (control) intervention or perceived inferior versus superior interventions for a potentially fatal infection. Community organizations are especially concerned about "guinea pig" or experimental approaches to evaluation, but simultaneous or sequential implementation of equally plausible interventions may be acceptable.

## VII. CONCLUSION

HIV prevention does work. However, to be effective, HIV prevention programs must be multifaceted, ongoing, and flexible. Different communities, populations at risk, and individuals engaging in particular risky behaviors may need different interventions to prevent risky and reinforce safe behavior that ultimately will lead to reduction in the rate of HIV seroconversion. More intense or frequent interventions lead to the greatest reductions in risk behavior, and activities designed to build skills and modify community norms have been found to enhance behavior change. Also, the availability of prevention devices (such as condoms or sterile syringes) is an important factor. In addition, providing behavioral risk-reduction interventions and expert primary care for people already infected with HIV is an essential prevention strategy that reduces transmission and saves on costs of emergency room and hospital care.

The following characteristics of successful prevention programs have been identified through literature reviews:

1. basis in specific needs and community planning;
2. culturally competent;
3. clearly defined audiences, objectives, and interventions;
4. basis in behavioral and social science theory and research;
5. quality monitoring and adherence to plans;
6. use of evaluation findings and midcourse corrections; and
7. sufficient resources.

Changes in seroepidemiology, prevalence and patterns of risk behavior, and patterns of STD and drug use in different communities may demand rapid changes in interventions, resource allocation, and geographical distribution of prevention programs. Planned research; evaluation studies; regular review of AIDS case surveillance and seroprevalence data; behavior risk survey data of general and high-risk populations; and data from STD, hepatitis, tuberculosis, and drug surveillance programs will allow early and effective response by health departments and their community planning partners when changes are needed. Unfortunately, HIV prevention program interventions, as well as primary care, are expensive, and obtaining resources to implement new interventions or to change the focus of others may be difficult. To date, HIV prevention programs, especially drug treatment programs, continue to receive insufficient funds and treatment personnel. These deficiencies must be addressed if HIV transmission is to be reduced. Reductions in morbidity and mortality that may be brought about by new therapies also will be delayed if multiple means of financing health care for the uninsured are not developed.

## BIBLIOGRAPHY

Bandura, A. (1977). Self-efficacy: Toward a unifying theory of behavior change. *Psychol. Rev.* **84**, 191–215.

Bowen, G. S., Marconi, K. M., Kohn, S., Bailey, S. M., Goosby, E. P., Shorter, S., and Niemevy, K. (1992). First year of AIDS service delivery under Title I of the Ryan White Care Act. *Pub. Health Rep.* **107**(5), 491–499.

Cardo, D. M., Castro, K. G., Polder, J. A., and Bell, D. M. (1994). Management of occupational exposure to HIV. *In* "AIDS Testing: A Comprehensive Guide to Technical, Medical, Social, Legal, and Management Issues" (G. Schochetman and J. R. George, eds.), 2nd Ed., pp. 361–375. Springer-Verlag.

Centers for Disease Control and Prevention (1992). "National Center for Prevention Services AIDS Community Demonstration Projects; What We Have Learned, 1985–1990." CDC, Atlanta.

Centers for Disease Control and Prevention (1993). "Supplemental Guidance on HIV Prevention Community Planning for Noncompeting Continuation of Cooperative Agreements for HIV Prevention Projects." CDC, Atlanta.

Centers for Disease Control and Prevention (1994). "HIV Counseling, Testing, and Referral Standards and Guidelines." CDC, Atlanta.

Centers for Disease Control and Prevention (1994). Sexually transmitted diseases treatment guidelines (H. H. Handsfield, K. K. Holmes, A. O. Berg, W. E. Stamm, and T. MacKay, eds.). *Clin. Courier* **12**(17), 1–8.

Choi, K-H., and Coates, T. J. (1994). Prevention of HIV infection. *AIDS* **8**(10), 1371–1389.

Curran, J. W., Holtgrave, D. R., and Guinan, M. E. (1994). HIV prevention does work. *Issues Sci. Technol.* **10**, 16–17.

Diaz, T., Chu, S. Y., Conti, L., Sorvillo, F., Checko, P. H., Hermann, P., Fann, S. A., Frederick, M., Boyd, D., Mokotoff, E., Rietmeijer, C. A., Herr, M., and Samuel, M. C. (1994). Risk behaviors of persons with heterosexually acquired HIV infection in the United States: Results of a multistate surveillance project. *J. Acquir. Immune Defic. Syndromes* **7**(9), 958–963.

Doll, L. S., and Kennedy, M. B. (1994). HIV counseling and testing: What is it and how well does it work. *In* "AIDS Testing: A Comprehensive Guide to Technical, Medical, Social, Legal, and Management Issues" (G. Schochetman and J. R. George, eds.), 2nd Ed., pp. 302–319. Springer-Verlag, New York.

Edlin, B. R., Irwin, K. L., Faruque, S., *et al.* (1994). Intersecting epidemics: Crack cocaine use and HIV infection among inner-city young adults. *N. Engl. J. Med.* **331**, 1422–1427.

European Study Group on Heterosexual Transmission of HIV (1992). Comparison of female-to-male and male-to-female transmission of HIV in 563 stable couples. *Br. Med. J.* **304**, 809–813.

Fishbein, M., and Azjen, I. (1975). "Belief, Attitude, Intention, and Behavior: An Introduction to Theory and Research." Addison-Wesley, Reading, MA.

Francis, D. P., Anderson, R. E., Gorman, M. E., Fenstersheib, M., Padian, N. S., Kiser, K. W., and Conant, M. A. (1989). Targeting AIDS prevention and treatment toward HIV-1-infected persons: Concept of early intervention. *JAMA* **262**(18), 2572–2576.

George, J. R., and Schochetman, G. (1994). Detection of HIV infection using serologic techniques. *In* "AIDS Testing: A Comprehensive Guide to Technical, Medical, Social, Legal, and Management Issues" (G. Schochetman and J. R. George, eds.), 2nd Ed., pp. 62–102. Springer-Verlag, New York.

Groseclose, S. L., Weinstein, B., Jones, T. S., Valleroy, L. A., Fehrs, L. J., and Kassler, W. J. (1995). Impact of increased legal access to needles and syringes on practices of injecting drug users and police officers—Connecticut, 1992–1993. *J. Acquir. Immune Defic. Syndromes* **10**(1), 82–89.

Haverkos, H. W., and Jones, T. S. (1994). HIV, drug-use paraphernalia, and bleach. *J. Acquir. Immune Defic. Syndromes* **7**(7), 741–742.

Hellinger, F. J. (1993). The lifetime cost of treating a person with HIV. *JAMA* **270**, 474–478.

HIV/AIDS Surveillance Report (1996). *Centers for Disease Control and Prevention* **7**(2), 1–33.

Holtgrave, D., Qualls, N., Curran, J. W., Valdiserri, R. O., Guinan, M. E., and Parra, W. (1995). An overview of effectiveness and efficiency of HIV prevention programs. *Pub. Health Rep.* **110**(2), 134–146.

Jonz, N. K., and Berber, M. (1984). The health belief model: A decade later. *Health Edu. Quart.* **11**, 1–47.

Karon, J. M., Rosenberg, P. S., McQuillan, G., Khare, M., Gwinn, M., and Petersen, L. R. (1996). Prevalence of HIV infection in the United States, 1984 to 1992. *JAMA* **276**(2), 126–131.

Kirby, D., Short, L., Collins, J., Rugg, D., Kolbe, L., Howard, M., Miller, B., Sonenstein, F., and Zabin, L. (1994). School-based programs to reduce sexual risk behaviors: A review of effectiveness. *Publ. Health Rep.* **109**(3), 339–360.

Lehman, J. S., Allen, D. M., Green, T. A., and Onorato, I. M. (1994). HIV infection among non-injecting drug users entering treatment, United States, 1989–1992. *AIDS* **8**, 1465–1469.

Main, D., Iverson, D., McGloin, J., Banspach, S., Collins, J., Rugg, D., and Kolbe, L. (1994). Preventing HIV infection among adolescents: Evaluation of a school-based program. *Prev. Med.* **23**(4), 409–417.

McKinney, M. M., Weiland, M. K., Bowen, G. S., Goosby, E. P., and Marconi, K. M. (1993). States' responses to Title I of the Ryan White CARE Act. *Pub. Health Rep.* **108**(1), 4–11.

"Morbidity and Mortality Weekly Report" (1988). Partner notification for preventing human immunodeficiency virus infection—Colorado, Idaho, South Carolina, Virginia. *Centers for Disease Control and Prevention* **37**, 393–402.

"Morbidity and Mortality Weekly Report" (1992). Recommendations for prophylaxis against *Pneumocystis carinii* pneumonia for adults and adolescents infected with human immunodeficiency virus. *Centers for Disease Control and Prevention* **41**(RR-4), 1–11.

"Morbidity and Mortality Weekly Report" (1992). Projections of the number of persons diagnosed with AIDS and the number of immunosuppressed HIV-infected persons—United States, 1992–1994. *Centers for Disease Control and Prevention* **41**(RR-18), 1–29.

"Morbidity and Mortality Weekly Report" (1992). HIV counseling and testing services from public and private providers—United States, 1990. *Centers for Disease Control and Prevention* **41**(40), 743, 749–752.

"Morbidity and Mortality Weekly Report" (1993). Recommendations for HIV testing services for inpatients and outpatients in acute-care hospital settings and technical guidance on HIV counseling. *Centers for Disease Control and Prevention* **42**(RR-2), 1–17.

"Morbidity and Mortality Weekly Report" (1993). Update: Barrier protection against HIV infection and other sexually transmitted diseases. *Centers for Disease Control and Prevention* **42**(30), 589–591, 597.

"Morbidity and Mortality Weekly Report" (1993). Assessment of street outreach for HIV prevention—Selected sites, 1991–1993. *Centers for Disease Control and Prevention* **42**(45), 873, 879–880.

"Morbidity and Mortality Weekly Report" (1994). Recommendations of the U.S. Public Health Service Task Force on the use of zidovudine to reduce perinatal transmission of human immunodeficiency virus. *Centers for Disease Control and Prevention* **43**(RR-11), 1–20.

"Morbidity and Mortality Weekly Report" (1994). Health-risk behaviors among adolescents who do and do not attend school—United States, 1992. *Centers for Disease Control and Prevention* **43**(9), 129–132.

"Morbidity and Mortality Weekly Report" (1994). Update: Impact of the expanded AIDS surveillance case definition for adolescents and adults on case reporting—United States, 1993. *Centers for Disease Control and Prevention* **43**(9), 160–161, 167–170.

"Morbidity and Mortality Weekly Report" (1994). Zidovudine for the prevention of HIV transmission from mother to infant. *Centers for Disease Control and Prevention* **43**(16), 285–287.

"Morbidity and Mortality Weekly Report" (1994). Expanded tuberculosis surveillance and tuberculosis morbidity—United States, 1993. *Centers for Disease Control and Prevention* **43**(20), 361–366.

"Morbidity and Mortality Weekly Report" (1994). Birth outcomes following zidovudine therapy in pregnant women. *Centers for Disease Control and Prevention* **43**(22), 415–416.

"Morbidity and Mortality Weekly Report" (1995). U.S. Public Health Service recommendations for human immunodeficiency virus counseling and voluntary testing for pregnant women. *Centers for Disease Control and Prevention* **44**(RR-7), 1–15.

"Morbidity and Mortality Weekly Report" (1995). Notification of syringe-sharing and sex partners of HIV-infected persons—Pennsylvania, 1993–1994. *Centers for Disease Control and Prevention* **44**(11), 202–204.

"Morbidity and Mortality Weekly Report" (1995). Syringe exchange programs—United States, 1994–1995. *Centers for Disease Control and Prevention* **44**(37), 684–685, 691.

National HIV Serosurveillance Summary, Results through 1992 (1993). *Centers for Disease Control and Prevention* **3**, 1–51.

National Research Council and Institute of Medicine (1995). "Preventing HIV Transmission: The Role of Sterile Needles and Bleach" (G. Normand, D. Vlahov, and L. E. Moses, Eds.), Panel on Needle Exchange and Bleach Distribution Programs, Commission on Behavioral and Social Sciences and Education. National Academy Press, Washington, D.C.

Newman, R. G. (1987). Methadone treatment: Defining and evaluating success. *N. Engl. J. Med.* **317**, 447–450.

Nwanyanwu, O. C., Chu, S. Y., Green, T. A., *et al.* (1993). Acquired immunodeficiency syndrome in the United States associated with injecting drug use, 1981–1991. *Am. J. Drug Alcohol Abuse* **19**(4), 399–408.

Onorato, I. M., Markowitz, L. E., and Oxtoby, M. J. (1988). Childhood immunization, vaccine-preventable diseases, and infection with human immunodeficiency virus. *Pediatr. Infect. Dis. J.* **6**, 588–595.

Office of Technology Assessment (1995). "The Effectiveness of AIDS Prevention Efforts," OTA-BP-H-172. U.S. Govt. Printing Office, Washington, D.C.

Richards, S. B., and Horsburgh, C. R. (1994). Tuberculosis. *In* "Travelers' Health" (R. Dawood, ed.), pp. 86–91. Random House, New York.

Rolfs, R. T., Goldberg, M., and Sharrar, R. G. (1990). Risk factors for syphilis: Cocaine use and prostitution. *Am. J. Pub. Health* **80**(7), 853–857.

Rural Center for the Study and Promotion of HIV/STD Prevention (1995). Drugs for the treatment of HIV/AIDS (fact sheet). Indiana University, Bloomington.

Saracco, A., Musicco, M., Nicolosi, A., *et al.* (1993). Man-to-woman sexual transmission of HIV: Longitudinal study of 343 steady partners of infected men. *J. Acquir. Immune Defic. Syndromes* **6**, 497–502.

Selik, R. M., Chu, S. Y., and Buehler, J. W. (1993). HIV infection as leading cause of death among young adults in U.S. cities and states. *JAMA* **269**(23), 2991–2994.

Stamm, W. E., Handsfield, H. H., Rompals, A. M., *et al.* (1988). The association between genital ulcer disease and acquisition of HIV infection in homosexual men. *JAMA* **260**, 1427–1433.

Stone, K. M. (1994). HIV, other STDs, and barriers: An overview. *In* "Barrier Contraceptives: Current Status and Future Prospects" (C. K. Mauck, M. Cordero, J. M. Spieler, and R. Rivera, eds.), pp. 203–212. Wiley–Liss, New York.

Valdiserri, R. O., Aultman, T. V., and Curran, J. W. (1995). Community planning: A national strategy to improve HIV prevention programs. *J. Community Health* **20**(2), 87–100.

Valdiserri, R. O., and West, G. R. (1994). Barriers to the assessment of unmet need in planning HIV/AIDS prevention programs. *Pub. Admin. Rev.* **54**(1), 25–30.

Valleroy, L. A., Weinstein, B., Jones, T. S., Groseclose, S. L., Rolfs, R. T., and Kassler, W. J. (1995). Impact of increased legal access to needles and syringes on community pharmacies' needle and syringe sales—Connecticut, 1992–1993. *J. Acquir. Immune Defic. Syndromes* **10**(1), 73–81.

West, G. R., and Valdiserri, R. O. (1994). Understanding and overcoming obstacles to planning HIV-prevention programs. *AIDS & Pub. Policy Rev.* **9**(4), 207–213.

Williams, A. E., and Creedon, K. W. (American Red Cross Collaborative Study Group) (1988). Behavioral changes in former blood donors in the year following notification of HIV seropositivity. *In* "International Conference on AIDS Program and Abstracts" 8064: 464. Stockholm, Sweden.

# Acquired Immune Deficiency Syndrome, Infectious Complications

JONATHAN W. M. GOLD

*Bronx-Lebanon Hospital Center and Albert Einstein College of Medicine*

## GLOSSARY

**Acquired immune deficiency syndrome (AIDS)** A severe disease which first appeared in the late 1970s. It is characterized by extreme susceptibility to a variety of severe infections and malignancies. It is caused by infection with the human immunodeficiency virus (HIV), which results in progressive destruction of the T-helper lymphocyte

**Human immunodeficiency virus (HIV)** A human retrovirus that is the cause of AIDS. There are two types: HIV-1 which is the cause of most cases of AIDS and HIV-2 which is found in a relatively limited area of West Africa. This virus is present in the blood and other body fluids of infected patients and is transmitted sexually, by the transfusion of blood and blood products, between intravenous drug users who share their needles, and from infected mothers to their babies. It was not known prior to the early 1980s, but has now spread worldwide

**Immunocompromised patient** A person who has increased susceptibility to infection because of a breakdown in defenses against microorganisms. The defect may be the result of a disease such as cancer or AIDS or the result of treatment such as chemotherapy for cancer or corticosteroids

**Opportunistic infection** An infectious disease caused by an organism (sometimes called an opportunist) in a host whose susceptibility to infection has been increased by altered or weakened defenses. Some opportunists such as *Pneumocystis carinii* rarely, if ever, infect immunologically normal hosts, while others such as *Mycobacterium tuberculosis* do

**T-helper lymphocyte** A class of lymphocyte that plays a central role in regulating the response of cells of the immune system to a variety of microorganisms and neoplasms. They can be identified by their functions or by the presence of specific antigens (such as CD4) on their surfaces. It is these cells which are the primary targets of infection with HIV

INFECTIOUS DISEASES ARE BY FAR THE MOST IMportant causes of symptoms and death in patients with the Acquired Immune Deficiency Syndrome (AIDS). Progress in treating the underlying cause of AIDS, infection with the Human Immunodeficiency Virus (HIV), can ameliorate the disease and slow its progression. However, it is still not possible to stop the relentless development of immune deficiency, and patients with AIDS ultimately develop complications, most frequently a succession of severe infectious diseases. Much of the medical care of AIDS patients involves the diagnosis, treatment, and prevention of these infectious complications, which continue to undergo a remarkable evolution in response to advances in medical management (including antiretroviral therapies and the prevention and treatment of infectious diseases), the longer survival of AIDS patients with severe immune deficiency, and the appearance of AIDS in different populations. This article gives an overview of the infectious diseases that complicate AIDS. It discusses the susceptibility of AIDS patients,

ENCYCLOPEDIA OF HUMAN BIOLOGY, Second Edition, VOLUME 1.   Copyright © 1997 by Academic Press.   All rights of reproduction in any form reserved.

the clinical features and microbiology of AIDS-associated infections, and the evolution of the complications of AIDS.

## I. HUMAN IMMUNODEFICIENCY VIRUS AND IMMUNE DEFICIENCY IN AIDS

Since it was first recogized in the early 1980s, AIDS has become a major health problem of worldwide importance. AIDS is caused by infection with the Human Immunodeficiency Virus, a human retrovirus. This virus infects specific cells, most notably the T-helper lymphocyte, which plays a central role in the immune system. Progressive loss of T-helper lymphocytes leads to a profound weakening of the immune system, which in turn makes AIDS patients highly vulnerable to certain kinds of infections and malignancies. These *opportunistic infections,* so-called because they take advantage of immune deficiency, and malignancies (notably Kaposi's sarcoma and B-cell lymphomas) are the hallmarks of AIDS. Many of the opportunistic infections seen in AIDS patients were familiar to physicians prior to the AIDS epidemic as causes of diseases in other patients with immune deficiencies, such as those with cancer or those receiving immunosuppressive drugs. Indeed, it was the unexpected finding that homosexual men without any of the known causes of immune suppression were developing *Pneumocystis carinii* pneumonia (PCP) and Kaposi's sarcoma that led to the recognition of AIDS. PCP was a well-recognized complication of immunosuppression in cancer patients receiving steroids, but was unknown to occur in healthy individuals. Similarly, Kaposi's sarcoma, in addition to occurring in older men of Mediterranean or Eastern European background and in certain endemic areas in Africa, was well known as a complication in renal transplant recipients receiving immunosuppressive drugs. [*See* Acquired Immune Deficiency Syndrome, Epidemic; Acquired Immune Deficiency Syndrome, Virology.]

HIV is in the blood and body secretions of infected individuals. Infection most frequently occurs through sexual contact with an infected individual. It is also spread by transfusion of infected blood or blood products, between individuals who share hypodermic needles and syringes, such as intravenous drug users, and from infected mothers to their newborns before or during birth. Infection with the virus usually goes unrecognized, although it sometimes causes an acute syndrome similar to infectious mononucleosis with fever, swollen lymph glands, sore throat, rash, and other symptoms. For a period that may last many years, HIV-infected people have no symptoms of disease, although they can infect others. However, during this time there is progressive, irreversible destruction of the immune system. The mechanisms that cause this destruction are not yet fully understood. Proposed mechanisms include direct cellular lysis by the virus, cellular destruction mediated by the immune response, and programmed cell death (apoptosis). Most of this destruction occurs in the lymph nodes. T-helper lymphocytes (also called CD4+ lymphocytes, because of an antigen present on the cell surface) are gradually depleted, and when their number falls below a certain level, the immune system begins to fail. Patients become highly vulnerable to certain types of infections and the development of certain malignancies. It is at this point that patients with HIV infection begin to develop the signs and symptoms of illness. [*See* Acquired Immune Deficiency Syndrome, T-Cell Subsets.]

### A. Infections Associated with T-helper Lymphocyte–Mononuclear Phagocyte Deficiencies

Infections in AIDS patients may be caused by a number of organisms, including parasites, bacteria, viruses, and fungi (Table I). Although AIDS patients can be infected by organisms that infect healthy people, certain organisms are especially prone to take advantage of the immune deficiencies in AIDS. The principal immune deficiency in AIDS is in cell-mediated immunity. This refers to a complex system of defenses involving lymphocytes, cytokines, and mononuclear phagocytic cells. The defect is complex. It is primarily a consequence of T-helper lymphocyte depletion and is frequently referred to as a "T-cell defect." This T-cell defect is the hallmark of AIDS. The risk of developing opportunistic infections in AIDS patients correlates inversely with the CD4 count (Fig. 1). The normal CD4 lymphocyte count is about 800–1200 cells/mm$^3$. This risk of an "AIDS-defining" opportunistic infection or neoplasm starts to rise substantially when the CD4 count is less than 200/mm$^3$, although some of the complications (e.g., tuberculosis) may occur at higher levels, and some are associated with very low counts (e.g., cytomegalovirus and *Mycobacterium avium* complex infection). Some of the infectious complications of AIDS, such as tuberculosis,

TABLE I

Infectious Complications of AIDS

| Parasite | Bacteria | Fungi | Viruses |
|---|---|---|---|
| | Organisms taking advantage of T-cell defects | | |
| *Pneumocystis carinii* | *Mycobacterium* | *Cryptococcus* | *Herpes simplex* |
| *Toxoplasma gondii* | *tuberculosis* | *neoformans* | *Varicella zoster* |
| *Leishmania* | *M. avium* complex | *Candida albicans* | Cytomegalovirus |
| *Trypanosoma cruzii* | M. kansasii | *Candida*, other | Adenovirus |
| | Other myco- | species | JC virus |
| | bacteria | *Histoplasma* | |
| | *Salmonella* | *capsulatum* | |
| | *Bartonella henselae* | *Coccidioides immitis* | |
| | *B. quintana* | | |
| | Organisms taking advantage of B-cell defects | | |
| | *Streptococcus pneumoniae* | | |
| | *Hemophilus influenzae* | | |
| | Other bacteria causing infections in AIDS patients | | |
| | *Staphylococcus aureus* | | |
| | *Pseudomonas aeruginosa* | | |
| | *Enterococcus faecium* | | |
| | *Enterococcus faecalis* | | |
| | *Shigella flexneri* | | |
| | *Rhodococcus equi* | | |
| | Others | | |

histoplasmosis, cryptococcosis, salmonellosis, and cryptosporidiosis, regularly occur in patients with no identifiable immune deficiency. Others, such as *Pneumocystis carinii* pneumonia and disseminated *Myco-*

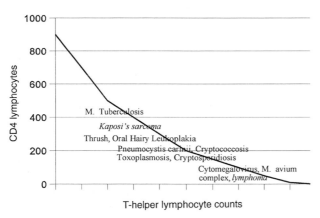

FIGURE I   The immune deficiency in AIDS becomes progressively more severe and the risk of developing an opportunistic infection increases as the CD4 lymphocyte count declines. Infections with more virulent organisms such as *M. tuberculosis* are likely to occur at higher CD4 counts. The risk of PCP starts to increase dramatically when the CD4 count is less than 200/mm³. The risk of *M. avium* complex and cytomegalovirus starts to increase with CD4 counts of less than 50–100/mm³.

*bacterium avium* complex infections are rare in immunologically normal hosts. Some organisms first achieved prominence in AIDS patients and were subsequently shown to be major public health problems in their own right. Cryptosporidia, for example, has been responsible for major waterborne outbreaks of diarrhea involving entire populations of both normal and immunosuppressed hosts served by contaminated municipal water supplies.

The organisms infecting AIDS patients share several characteristics. Many are "facultative" or "obligate" intracellular parasites; they form latent infections, and they have a tendency to cause relapsing infections and therefore to require prolonged treatment when they occur, for many months or even for life. Intracellular parasites survive inside cells, including those of the immune system, notably macrophages. Many, including the organisms that cause tuberculosis, toxoplasmosis, histoplasmosis, and all of the herpes viruses, have the ability to cause "latent" infections. The organisms remain alive in the host without causing symptoms for many years. When immunity wanes, the dormant organisms start multiplying and cause disease. Thus in many instances, infections occurring in AIDS patients were acquired many years before causing illness. In most instances, the infection occurred long before HIV infection, often in childhood. Often, these infections leave a footprint in the form of serum antibody (such as toxoplasmosis) or a delayed hypersensitivity reaction (the tuberculin skin test or Purified Protein Derivative used to detect infection with *Mycobacterium tuberculosis*). Identification of patients with evidence of past infection is important since these patients can receive preventive therapy (prophylaxis) or can be observed more closely for the recrudescence of disease. The propensity to relapse and the weakened host defenses in AIDS patients mean that the therapy of established infections must be prolonged in many cases. For some diseases, such as salmonellosis, toxoplasmosis, cryptococcosis, cytomegalovirus, and herpes virus infection, and candidiasis, treatment may be given for months, or for the lifetime of the patient.

## B. Infections in AIDS Patients Associated with Other Impaired Host Defenses

### 1. B-lymphocyte Deficiencies

In addition to the T-cell deficiency, AIDS patients have defective B-lymphocyte function. The principal function of B lymphocytes is to make antibodies in response to infection or challenge by antigens, including vaccines. The defect is secondary to the deficiency in T-helper cells, which normally play an important role in controlling antibody production by B cells. The ability of HIV-infected patients to make antibody normally in response to infections with certain bacteria and to certain vaccinations is markedly impaired. The consequence of this defect is a greatly increased susceptibility to certain bacterial infections, notably *Streptococcus pneumoniae* and *Hemophilus influenzae*, both of which cause pneumonia in AIDS patients. Antibodies that developed when the individual was immunologically intact tend to be maintained. This is useful in identifying patients with previously acquired latent infections, such as *Toxoplasma gondii*, who maintain detectable antibody levels for years.

### 2. Increased Susceptibility of Infection Caused by Disrupted Integument, Neutropenia, Medical Devices, and Altered Microbial Flora

In the terminal stages of AIDS there are additional factors that increase the susceptibility to infection. These include neutropenia (a low white blood cell count), skin breakdown from rashes, tumors, such as Kaposi's sarcoma, and decubitus ulcers (bedsores).

Neutropenia and prolonged antibiotic therapy are associated with an increased risk of infection with resistant bacteria and fungi in AIDS patients. Newly identified strains of *Enterococcus faecium* and *Enterococcus faecalis*, which are resistant to all antibiotics in clinical practice, are appearing with increasing frequency in AIDS patients in the late stages of their disease. *Aspergillus*, a mold that is ubiquitous in the environment and a known cause of invasive infections in cancer patients, causes pulmonary and disseminated disease late in the course of AIDS.

Disruption of the integument (skin breakdown) increases the susceptibility to a variety of bacterial infections, including staphylococci and gram-negative bacteria. The medications and medical devices used to treat AIDS patients can also significantly increase the risk of infection. Indwelling intravenous catheters used for the administration of drugs and for nutrition (parenteral nutrition) are major sources of infection. The use of antibiotics in AIDS patients is associated with a number of infectious complications, including severe enterocolitis caused by the toxin-producing bacteria *Clostridium difficile* and overgrowth with antibiotic-resistant bacteria, yeast, and molds.

### 3. Epidemiology of the Infectious Complications of AIDS

The incidence and types of infectious complications of AIDS have undergone profound changes in response to improved therapies, the development of prophylaxis, the longer survival of AIDS patients, and the emergence of AIDS in different populations. Awareness of the potential for such changes allows physicians to anticipate when and where new complications may emerge. *Pneumocystis carinii* pneumonia, originally the most common infectious complication of AIDS, is now nearly completely preventable with trimethoprim-sulfamethoxazole prophylaxis. Its frequency has greatly diminished in patients receiving adequate medical care, although it remains a significant problem in patients whose HIV infection is undiagnosed, notably children and inner city residents. Toxoplasmosis appears to be largely preventable by the same prophylaxis as PCP, and its incidence has substantially declined. The management of many other infectious complications, including herpes simplex, cytomegalovirus, tuberculosis, cryptococcosis, and candidiasis, has also improved. However, as was anticipated, with the introduction and widespread use of new antimicrobials, drug resistance has already emerged and is a major clinical problem. With antiretroviral therapy and better management of opportunistic infections, AIDS patients live longer with more profound immune deficiency. This has led to an increase in those infectious complications that occur at more profound levels of immune suppression, such as *Mycobacterium avium* complex and cytomegalovirus infections. It has also been associated with an increase in the incidence of lymphomas.

As HIV infection moves into different populations and different geographic areas, there are important interactions with certain infections common in those areas. Thus, infection with tuberculosis became a tremendously important problem when HIV infection entered the inner cities of the United States and underdeveloped countries of Africa and the Caribbean, whereas it was uncommon when AIDS was confined to middle-class gay males. The explanation for this phenomenon is that tuberculosis, including latent infection, is relatively uncommon among middle-class people in the United States, but is endemic in its inner cities and in developing countries. Leishmaniasis along the Mediterranean coasts of Spain, France, and Italy, and Chagas' disease, which is endemic in South America, are emerging in those areas as important complications of AIDS. A mold, *Penicillium marneffei*, has been reported as a common disseminated infection in AIDS patients in Thailand and other areas of Southeast Asia. The frequency and severity of these diseases are increased in AIDS patients and the response to standard therapy is often poor. [*See* Tuberculosis.]

## II. COMMON CLINICAL PRESENTATIONS OF THE INFECTIOUS COMPLICATIONS OF AIDS

Many of the infectious complications of AIDS present as typical syndromes (Table II). These include pneumonia or other pulmonary disease; gastrointestinal disease, including malabsorption, oropharyngitis, esophagitis, gastritis, cholecystitis, and enterocolitis and perianal disease; neurologic disease, including meningitis, brain abscess, and spinal cord disease (myelopathy); eye disease, including chorioretinitis and conjunctivitis; sinusitis; skin disorders; and systemic illnesses such as disseminated infection involving several organs, bacteremia, and the wasting syndrome (profound, intractable weight loss for which a specific etiology may not be recognizable).

For most infections, a specific etiologic diagnosis is made by detecting the responsible agent in the affected organ. For example, PCP or tuberculosis is diagnosed by detecting the organism in the sputum or bronchial secretions of a patient with pneumonia. In some diseases, it may be impractical or dangerous to obtain a specific diagnosis. Most cases of central nervous system toxoplasmosis are diagnosed presumptively from a typical appearance on computerized tomography (CT scan) or nuclear magnetic resonance imaging (MRI scan), the presence of specific serum antibody, and response to therapy. Only in instances where the patient does not respond or when there is significant doubt about the diagnosis is a brain biopsy performed.

Initially an infection that involves several organs may come to attention because its most striking effect is on one organ. For example, cytomegalovirus (CMV) infection, which is readily recognized by its effect on the retina, frequently turns out to be associated with CMV infection elsewhere. This may be recognized at the initial presentation or weeks or months later. The lungs, adrenals, esophagus, stomach, or intestines may also be significantly involved.

## III. MICROORGANISMS THAT CAUSE INFECTIONS IN AIDS PATIENTS

### A. Parasites (Protozoa)

#### 1. *Pneumocystis carinii*

Many of the first cases of AIDS were recognized in patients with *Pneumocystis carinii* pneumonia who had none of the recognized causes of immune suppression that were associated with that illness. *Pneumocystis carinii* was first identified in rats in the early 1900s and was first recognized as a cause of human disease after World War II, when it was responsible for outbreaks of pneumonia in malnourished, institutionalized infants. It was subsequently found to be an important cause of pneumonia in patients with neoplastic disease and those receiving corticosteroid therapy.

*Pneumocystis carinii* has been traditionally classified as a protozoan based on its microscopic appearance, growth characteristics, and antibiotic susceptibility patterns. However, more recent evidence, based on DNA homology studies, suggests that it is more closely related to fungi than to protozoa. Study of the organism has been complicated by the inability to

## TABLE II
### Clinical Presentations in AIDS Patients

| Clinical presentation | Parasites | Bacteria | Fungi | Virus |
|---|---|---|---|---|
| Pulmonary disease | *Pneumocystis carinii* <br> *Toxoplasma gondii* <br> Cryptosporidia | **Mycobacterium tuberculosis** <br> **Other mycobacteria** <br> **Streptococcus pneumoniae** <br> *Staphylococcus aureus* <br> *Pseudomonas* <br> *Bartonella henselae* | *Histoplasma capsulatum* <br> *Cryptococcus neoformans* <br> *Coccidioides immitis* | Cytomegalovirus (CMV) |
| Gastrointestinal disease <br> Oropharynegeal (thrush) | | | *Candida albicans* | Herpes simplex |
| Esophagus (esophagitis) | | | *Candida albicans* | Herpes simplex <br> CMV |
| Stomach (gastritis) | | | *Candida albicans* | CMV |
| Small and large intestine (enterocolitis) | **Cryptosporidia** <br> *Isospora belli* <br> **Microsporidia** | *M. avium complex* <br> *Salmonella* <br> *Campylobacter jejuni* <br> **Clostridium difficile** (toxin) | | CMV |
| Perianal ulcers | | | | Herpes simplex |
| Eye (chorioretinitis) | *Toxoplasma gondii* | *M. tuberculosis* | | CMV <br> *Varicella zoster* |
| Central nervous system disease (brain abscess and meningitis) | **Toxoplasma gondii** | *M. tuberculosis* | **Cryptococcus neoformans** | JC virus <br> Herpes simplex <br> CMV |
| Multiorgan disease | *Pneumocystis carinii* <br> *Toxoplasma gondii* | **M. avium complex** <br> **M. tuberculosis** <br> *Salmonella* <br> *Bartonella* | *Candida albicans* | CMV <br> *Varicella zoster* |

*Note:* Organisms in boldface type are especially frequent causes of the clinical presentations.

cultivate it *in vitro,* and hence most studies have to be done in animal models. Much of our understanding of the pathogenesis, treatment, and prevention of PCP is derived from these animal models.

Infection with *P. carinii* is probably very common during childhood, although it almost never causes recognizable symptoms. It is believed, although not proven, that most cases in immunosuppressed patients arise as a result of reactivation of such latent infections. In addition, there is some experimental evidence in animals that airborne organisms can infect other animals, thus the potential for person-to-person spread exists.

PCP usually presents with the progressive onset (over days to weeks) of fever, cough, and shortness of breath. Chest X rays typically show bilateral pneumonia ("interstitial-alveolar pattern"), and measurements of the partial pressure of oxygen ($pO_2$) in the blood ("arterial blood gas" determinations) usually show evidence of impaired diffusion of oxygen into the blood. Signs and symptoms can be very subtle, and the chest X ray may even appear normal. The diagnosis of PCP is proven by finding the cysts or trophozooites in expectorated sputum or pulmonary secretions obtained by bronchoscopy. Although it is principally a disease of the lungs, pneumocystis can disseminate to the bone marrow, lymph nodes, kidneys, liver, and retina.

A number of effective drugs exist for the treatment of PCP. These include trimethoprim-sulfamethoxazole (SXT), pentamidine, and atovaquone. Side effects are common with these drugs. One of the most startling, unexplained findings in AIDS is the extraordinarily high frequency of adverse drug reactions. About half of AIDS patients taking SXT for PCP develop rashes. Other common side effects include bone marrow suppression (manifested by anemia, low platelet counts, and low white blood cell counts). Pentamidine causes renal failure, marrow suppression, and both hypoglycemia and diabetes. Because it is not systemically absorbed, therapeutic failures with aerosol pentamidine may be associated with disseminated pneumocystis infection. Patients with severe PCP are generally given steroid therapy as this has been shown to improve survival, speed recovery, and decrease the likelihood and severity of chronic lung problems. However, steroid therapy also carries with it the risk of reactivating latent tuberculosis and must be used cautiously in patients suspected of prior exposure to this disease.

Prophylaxis of PCP has been one of the major advances in the management of patients with HIV infection. The risk of developing PCP starts to increase when the CD4 count falls below 200/mm$^3$, and prophylaxis is begun at that point. This is one reason why it is important to monitor CD4 lymphocyte counts regularly in HIV-infected patients. SXT is the most effective form of prophylaxis. Aerosol pentamidine and dapsone are also effective, although their use is associated with a higher failure rate. PCP has become rare among patients with AIDS who receive adequate medical care and who can tolerate and comply with the regimens of prophylaxis. Among medically underserved populations and patients with unrecognized HIV infection, PCP remains one of the most common opportunistic infections.

## 2. Toxoplasma gondii

*Toxoplasma gondii* is a protozoan that is found throughout the world. Humans are infected by ingestion of cysts in undercooked meat (lamb, pork, and beef) or cat feces (infected cats shed oocysts of *T. gondii* for a few weeks), or congenitally. Most infections in healthy adults are not associated with disease and are not recognized, although acute toxoplasmosis is a common cause of lymphadenopathy in young adults. The consequences of congenital toxoplasmosis range from undetectable to severe in neonates with hepatitis, pneumonia, retinitis, and encephalitis.

Toxoplasmosis results in the development of a latent infection and the appearance of specific antibody in the serum. Prevalence of latent *T. gondii* infection as determined by serologic studies varies in different populations. In the United States, the average prevalence is around 20%. In France it is up to 90%. In AIDS patients, toxoplasmosis is generally due to reactivation of a latent infection with *T. gondii*. It most commonly involves the central nervous system (CNS),

causing brain abscesses, although pneumonia, myocarditis, and other syndromes also occur. The presence of antibody in an AIDS patient indicates an individual at high risk for the development of CNS toxoplasmosis. Before the widespread use of SXT prophylaxis for PCP, 10–15% of toxoplasma antibody-positive AIDS patients would develop CNS toxoplasmosis each year.

CNS toxoplasmosis usually presents with fever, headache, alterations in mental function, and physical findings suggestive of a brain mass. Computerized tomography or nuclear magnetic resonance imaging of the brain usually shows multiple, round lesions with characteristics of an abscess. In AIDS patients, CNS toxoplasmosis and CNS lymphomas constitute the great majority of brain masses. Although definitive diagnosis would require a brain biopsy, this daunting procedure is usually deferred in favor of an empirical trial of treatment for CNS toxoplasmosis. A response to therapy usually occurs within 2 weeks. If there is no response, brain biopsy may be performed to determine the diagnosis.

Treatment with a combination of either sulfadiazine or clindamycin and pyrimethamine is effective, and the majority of patients respond. As with other infections complicating AIDS, relapses are frequent if therapy is stopped, and most patients are treated for a prolonged period, even for life.

The incidence of toxoplasmosis has declined in toxoplasma antibody-positive patients receiving adequate medical care. Retrospective studies have suggested that SXT prophylaxis for PCP also significantly reduces the risk of CNS toxoplasmosis. The combination of dapsone and pyrimethamine also reduces the risk of toxoplasmosis.

## 3. Cryptosporidia, *Isospora belli*, and Microsporidia

These three organisms have achieved medical prominence as a result of the AIDS epidemic. Cryptosporidia and *Isospora belli* have been found to have important public health implications beyond their roles in AIDS.

Cryptosporidia are widespread in nature, infecting many animals, including poultry and livestock. Prior to the AIDS epidemic, cryptosporidia in humans was confined to outbreaks of diarrhea among veterinary workers and a few cases in immunocompromised hosts. It emerged as a common pathogen in AIDS patients, in whom it causes diarrhea varying in severity and duration from acute, to chronic with a waxing and waning course, to being the cause of massive, uncontrolled, life-threatening watery diarrhea. Diagnosis is readily made by microscopic examination of the stool. Occasional cases of gallbladder disease (cholecystitis) and pneumonia have also been attributed to this organism. Despite trials of numerous agents, no specific treatment exists at this time. Medical management consists of efforts to control the diarrhea and replacement of fluid and electrolytes.

Cryptosporidia have been recognized in the past few years as major public health problems. They are not inactivated by chlorine and are difficult to filter. A recent outbreak in Milwaukee, Wisconsin, associated with contamination of the municipal water supply with runoff from cattle farms, infected some 400,000 people and caused deaths in immunosuppressed patients. In addition, cryptosporidia are now recognized in developing countries as a major cause of infant diarrhea and death from dehydration.

*Isospora belli* is also a cause of diarrhea in AIDS patients and in children in developing countries. It is readily diagnosed by stool examination and responds to treatment with SXT. Microsporidia have been identified as causes of malabsorption and diarrhea in AIDS patients. Diagnosis is by identification of the organism in intestinal biopsies, although techniques for detecting the organism in stool are being more widely accepted. There is no proven treatment.

## B. Bacteria

### 1. Mycobacteria

Mycobacteria are widely distributed in nature. Included among this family of organisms are the bacteria that cause tuberculosis, leprosy, and a number of other less familiar infectious diseases of humans and animals. A large number of mycobacteria have been associated with illness in patients with AIDS. The most important are *Mycobacterium avium* complex, *M. tuberculsis*, and *M. kansasii*. Infections due to *M. hemophilium*, *M. xenopi*, *M. gordonae*, and others have also been reported.

### 2. Tuberculosis and AIDS

It is estimated that 1.7 billion people worldwide are infected with *M. tuberculosis,* the bacterium that causes tuberculosis, and that there are 3 million deaths from this disease each year. Among AIDS patients in both the developed and developing countries, tuberculosis has emerged as one of the most significant complications of AIDS. Because it also can cause serious illness in immunologically normal hosts, and because

children are at special risk of developing severe forms of tuberculosis, the combined epidemics of tuberculosis and AIDS pose serious problems for the entire population.

From the beginning of the century through 1984, there had been a steady decline in the number of new cases of tuberculosis in the United States. This decline, one of the great triumphs of public health efforts, started to reverse in 1985, so that by 1991 an excess of 39,000 new cases, over what had been predicted based on the previous rate of decline, had occurred. Some of this increase was attributed to migration from areas where tuberculosis was highly prevalent, such as Southeast Asia, but the great majority of cases occurred among AIDS patients in the inner cities and prisons of the United States, where latent infection was prevalent and conditions conducive to the spread of tuberculosis, such as overcrowding and poor ventilation, were common. An important contributory factor in the resurgence of tuberculosis in the United States was the closing of Public Health Tuberculosis Control Programs, which supported clinics, case workers, provision of services and drugs, and laboratories. This occurred because of budget constraints and the perception that tuberculosis, like other infectious diseases, was no longer a major public health problem. This erosion of infrastructure for the management of tuberculosis in turn resulted in significant numbers of patients not completing therapy and developing recurrent tuberculosis, often due to drug-resistant strains. [*See* Tuberculosis, Public Health Aspects.]

When tuberculosis first emerged as an infectious complication in AIDS, it was characterized by late diagnosis, inadequate therapy because of failure to recognize the problem of multiple drug resistance, high infection rates in other hospitalized patients and prisoners, and spread of infection to health care workers and prison guards. The mortality rate among AIDS patients with tuberculosis was extraordinarily high, approaching 100% in some hospitals.

Infection with *M. tuberculosis* is usually acquired by inhalation of aerosolized organisms expectorated by another person with tuberculosis. The bacteria multiply in the lungs, frequently causing infiltrates and enlarged lymph nodes. These usually heal and calcify (forming a "Gohn complex" seen on the X ray). During the initial infection, the bacteria may spread to the meninges, kidneys, joints, bones, genitalia, and other organs. In most cases, the body's immune system controls the infection. The tuberculin reaction or PPD skin test is one manifestation of the immune response to *M. tuberculosis* and indicates prior infection, not necessarily active disease. Most cases of tuberculosis in adults occur when these latent, controlled infections reactivate. In some cases, especially in young children, the primary infection is not contained and pneumonia, meningitis, or other forms of tuberculosis occur. For this reason it is especially important to identify children who are contacts of patients with active tuberculosis for preventive therapy.

Tuberculosis in AIDS patients may present in a typical fashion with cough, fever, night sweats, and weight loss. A chest X ray shows infiltrates and cavities in the lungs and the organisms are identified in acid-fast stains of expectorated sputum. This is characteristic of reactivated tuberculosis, and most cases are in fact due to reactivation of previously acquired infection. However, X-ray findings are often atypical in AIDS patients. Patients may have features of acute infection, suggesting that, at least in some cases, infection is newly acquired. Severe, widely disseminated tuberculosis, with huge numbers of bacteria in many organs, occurs in some AIDS patients.

The optimal treatment of tuberculosis involves rapid diagnosis, institution of therapy directed at the likely susceptibilities of the organisms, isolation until the patient is no longer infectious, and completion of therapy. The first line drugs for the treatment of tuberculosis include isoniazid (INH), rifampin, ethambutol, pyrazinamide, and streptomycin. There are a number of second line drugs used principally in the treatment of drug-resistant tuberculosis. The response to therapy is best if INH and rifampin can be given. Regimens for the treatment of drug-resistant organisms need to be tailored to the specific isolate. A number of studies have documented that the administration of antituberculosis drugs under supervision (directly observed therapy, or DOT) is the best way to assure successful treatment of persons with tuberculosis. A major benefit in areas where DOT has been implemented has been an impressive decline in the numbers of new cases of drug-resistant tuberculosis.

### 3. *Mycobacterium avium* Complex

*Mycobacterium avium* complex (MAC) is one of the most common infectious complications of AIDS. Clinical and autopsy studies have suggested that as many as 50% of all AIDS patients develop this infection before they die. It is generally a very late complication, occurring when the CD4 count is very low (less than 100). For this infection to develop, the individual must survive for a long period with a relatively severe degree of immune suppression. This may explain the relative rarity of this infection in AIDS patients from developing countries, and it is also possible that diagnostic facilities are not adequate. The organism is ubiquitous in nature, found in soil and water, so that exposure is difficult to avoid. Skin tests suggest that exposure to this organism is common. It is not known whether there is such a thing as a latent infection in humans. At this time there is no evidence to suggest that MAC disease arises from a latent infection. The source of most infections is presumed to be the environment.

MAC has long been known to be a cause of chronic pulmonary disease, generally in elderly people with no obvious immune defects. It also is a cause of lymphadenitis in children. In contrast with *M. tuberculosis*, MAC is not transmissible from person to person.

The symptoms of MAC infection are nonspecific. Fever, night sweats, weight loss, abdominal pain from enlarged lymph nodes, and diarrhea may be present. Patients can sometimes have bacteremia with these organisms with relatively few symptoms. The organism may infect almost any organ, but the gastrointestinal tract, lymph nodes, liver, spleen, and bone marrow are especially common sites. In tissues, the organisms tend to be intracellular, within macrophages. Blood cultures are positive in nearly all cases. The organism can be present in large amounts in sputum, even with normal chest X rays. It is doubtful that MAC commonly causes significant lung disease in AIDS patients, but it is difficult to distinguish from *M. tuberculosis* microscopically, leading to significant confusion and unnecessary isolation of patients. MAC is found in up to 50% of AIDS patients at autopsy. In some instances, astonishingly large quantities of microorganisms fill the lymph nodes, spleen, liver, and other organs.

The diagnosis of MAC infection is most readily made through blood cultures. It can also be isolated from other involved tissues. Recent technological advances, including DNA hybridization, have led to more rapid identification of MAC and its differentiation from other mycobacteria.

The treatment of MAC is not standardized or based on controlled clinical trials. In fact, whether disseminated MAC infections should be treated remains controversial, although the preponderance of opinion is for treatment to attempt to decrease symptoms. There is no standard regimen, although combinations including clarithromycin, rifabutin, ethambutol, and amikacin, sometimes with additional drugs, have been used. *In vitro* susceptibility testing of patients' isolates is of little value in guiding therapy. In at least some cases, symptoms improve and blood cultures become sterile with treatment. Despite these improvements, the organisms are probably seldom eradicated. Treatment is associated with a considerable number of side effects, and the decision to treat is usually made individually with each patient.

Well-controlled clinical trials have shown that rifabutin prophylaxis can prevent MAC bacteremia and reduce symptoms. It is recommended for patients with fewer than 100 CD4 counts. However, there are side effects, including retinitis, hepatitis, gastrointestinal intolerance, and orange discoloration of tears and urine. There is concern that the widespread use of rifabutin may lead to the emergence of rifampin-resistant *M. tuberculosis*, especially in patients who are at high risk for reactivation of tuberculosis and who comply poorly with their treatment.

## 4. Other Mycobacteria

A number of other mycobacteria have been reported to cause disease in AIDS patients. Most of these organisms are widely distributed in nature, are difficult to distinguish from *M. tuberculosis* on microscopic exam alone, and are identified by their growth and metabolic characteristics. In contrast to *M. tuberculosis*, they usually do not spread from person to person. *Mycobacterium kansasii* is the cause of an illness that is clinically similar to tuberculosis, in both AIDS patients and normal individuals. It tends to be a later complication of AIDS. Skin is the most common site of infection with *M. haemophilium*, which may also involve joints, bone, lung, and lymph nodes.

## 5. Salmonella

There are over 2000 serotypes of the species *Salmonella*. Members of this group are most familiar as causes of typhoid fever and gastroenteritis. They also cause a variety of infections of other organs, including bacteremia, infections of blood vessels, endocarditis, arthritis, osteomyelitis, pneumonia, empyema, and urinary tract infections. Infection with *Salmonella* is usually from foods of animal origin or contaminated drinking water. Except for the organisms that cause typhoid fever, which are only carried by human beings, most cases of *Salmonella* are acquired from animals and animal products: eggs, poultry and meat, and pets. *Salmonella* has been isolated from up to 50% of chickens in the United States and the organism infects eggs developing in hens' ovaries. In immunosuppressed patients, especially those with AIDS, salmonella causes gastroenteritis, bacteremia, and inter-

nal abscess formation. The organism is adapted to survive inside macrophages, where it is protected against host defenses. In AIDS patients particularly, treatment of salmonella has to be prolonged, often lifelong, because of the high risk of recurrence. Treatment with trimethoprim-sulfamethoxazole, ampicillin, ceftriaxone, or ciprofloxacin is usually effective, depending on the susceptibility of the organism. Drug resistance is an important problem in salmonellosis, and it is important to determine the susceptibility of the patient's isolate.

## 6. Clostridium difficile

Pseudomembranous colitis is a very common cause of fever, abdominal pain, and diarrhea in AIDS patients. It is associated with the use of antibiotics and is caused by a toxin produced by *Clostridium difficile*, an organism in the same family as the bacteria that cause gas gangrene, botulism, and tetanus. Any antibiotic may lead to this complication. *Clostridium difficile* is found in the gastrointestinal tract of healthy people and may be spread by hospital staff from patient to patient. The disease is not confined to AIDS patients, although it is very common in them. Anyone who receives antibiotics and who carries the bacteria in the gastrointestinal tract may develop pseudomembranous colitis.

The "pseudomembrane" is in fact a sheet of polymorphonuclear leukocytes, which differs from the type of cells that compose "true" membranes. This pseudomembrane is easily recognized by endoscopic examination. The diagnosis of pseudomembranous colitis is generally made by detecting the toxin in the patient's stool. Treatment is with oral vancomycin or metronidazole. The disease may relapse despite adequate therapy. Widespread, indiscriminate use of oral vancomycin as treatment for pseudomembranous colitis may be a factor in the emergence of resistant enterococci.

## 7. Bartonella henselae and Bartonella quintana

Bacillary angiomatosis caused by infection with *Bartonella henselae* and *B. quintana* we first recognized in AIDS patients. These patients, usually in advanced stages of AIDS, develop fever and purplish lesions that bleed easily and can be confused with Kaposi's sarcoma. The bacteria are also present in blood and may infect a number of internal organisms, including lungs, liver, spleen, lymph nodes, heart, and bone marrow. The bacteria can be cultured from blood and tissues, but this is uncertain and slow. Diagnosis is usually made by biopsying the skin lesions. Treatment with erythromycin or doxycycline appears to be effective, although controlled studies have not been done. The source of infection is not known, but other members of the genus *Bartonella* are transmitted by insects or lice.

Bacillary angiomatosis is not a very common infectious complication of AIDS. It illustrates how previously unrecognized infectious disease have been discovered in AIDS patients. *Bartonella* (formerly called *Rochalimaea*) is a genus of bacteria that includes the organisms that cause trench fever (*B. quintana*) and Oroya fever (a severe, often fatal illness transmitted by the bites of sandflies in mountain river valleys in Peru, Colombia, and Ecuador). *Bartonella henselae* appears at this time to be the cause of cat scratch disease, a common cause of swollen lymph glands and fever in individuals, especially children, who have been exposed to young infected cats.

## C. Viruses

### 1. Herpes Viruses

Herpes viruses are familiar as a cause of a number of common human illnesses, such as chicken pox and shingles (*Varicella zoster* virus), infectious mononucleosis (Epstein–Barr virus; a similar syndrome is also caused by cytomegalovirus), cold sores (herpes simplex type 1), and genital herpes (herpes simplex type 2). Infections due to members of the herpes virus family are well recognized in immunosuppressed patients and are by far the most common viral infections in people with AIDS.

Herpes viruses are large, for viruses, have envelopes that are distinctive, and carry their genetic contents in DNA. Infection with these viruses tends to be highly species specific, although herpes B virus, a monkey virus, can cause severe encephalitis in people. Infection is highly prevalent, and latent infections can persist for life. [*See* Herpesviruses.]

### 2. Cytomegalovirus

Evidence of infection with cytomegalovirus in the form of serum antibody is found in over half of the adult population and in nearly all homosexual men with AIDS. Most infection is acquired during childhood and is asymptomatic. It is possible to isolate the virus from healthy people without symptoms. Transmission by intimate contact and by transfused blood or donated organs is well known. Intrauterine infection can produce severe fetal injury and neonatal illness and birth defects.

CMV in AIDS is a late manifestation that generally occurs in patients who have survived for a long time and have severe immunodeficiency. The CD4 count is generally less than 50. It is uncommon as the first infectious manifestation of AIDS. Its incidence has increased markedly as AIDS patients have lived longer. Effective treatment has become available, but it is not yet clear how to prevent this disease.

Patients can present with a number of different symptoms, depending on the organs that are infected. These include chorioretinitis, pneumonia, esophagitis, enteritis, adrenalitis (which is associated with adrenal insufficiency), encephalitis and transverse myelitis (inflammation of the spinal cord), and hepatitis, as well as other forms of disease. It frequently involves many organs, and the initial presentation in one organ is often a clue that other organ involvement may be present. CMV infection that was not clinically apparent is frequently recognized at autopsy, puzzling physicians as to its significance.

CMV chorioretinitis is one of the most feared complications of AIDS and is the most commonly recognized form of CMV infection. It may present with sight loss or may be asymptomatic and detected by routine ophthalmoscopic examination. It is generally progressive and leads to blindness if not recognized and treated. Effective treatment is available, but restoration of lost vision may not occur.

Pneumonia due to CMV appears as a patchy lung infiltrate that waxes and wanes. It may accompany other infections in the lung and may be associated with CMV disease in other organs. Esophagitis in AIDS patients is most commonly caused by *Candida albicans*, although CMV, herpes simplex, and esophageal ulcerations of unknown etiology may cause similar symptoms (see Section III,D,1). Gastrointestinal involvement with CMV also includes the stomach and small and large bowel. Gastritis presents with abdominal pain and occasionally with GI bleeding. CMV enterocolitis is one cause of the wasting syndrome seen in AIDS patients. These individuals have weight loss, fever, diarrhea, and abdominal pain and become extremely emaciated. CMV infection of the gastrointestinal tract is diagnosed by endoscopic biopsy. It appears to respond to anti-CMV therapy.

Adrenal involvement with CMV is very commonly found at autopsy, although clinical evidence of adrenal insufficiency (Addison's disease) is relatively uncommon. This most characteristically presents as fever, hypotension, and an elevated serum potassium. In practice the presentation can be quite nonspecific. Diagnosis of adrenal insufficiency is made by determining serum cortisol levels and measuring the ability of the adrenal glands to increase production of cortisol in response to stimulation with a synthetic form of adrenal cortical stimulating hormone. Treatment is to replace the missing steroids and to treat the underlying CMV infection.

Most central nervous system involvement in AIDS is thought to be a consequence of HIV itself, although the mechanisms are poorly understood. Nervous system involvement with CMV, other than retinitis, is not common and is difficult to diagnose. It includes encephalitis and spinal cord disease. It is rarely diagnosed early and the response to treatment has been difficult to evaluate, but has not appeared encouraging.

The diagnosis of CMV disease is either by its typical retinal appearance or by demonstrating the presence of CMV in affected tissues obtained by biopsy. At present, there is no useful blood test for diagnosing CMV disease. Even detection of the virus in blood does not effectively diagnose disease. The difficulty is that serologic tests and cultures are unable to distinguish infection from disease.

Several drugs are available for the treatment of CMV, including ganciclovir and foscarnet. Treatment is effective in suppressing the signs and symptoms of CMV infection, but must be continued indefinitely to prevent recurrence of the disease. These drugs must be given intravenously and are associated with substantial toxicities. In addition, they require the placement of permanent indwelling intravenous lines, which are prone to infection. Resistance to antiviral agents has emerged as a significant clinical problem. Oral forms of ganciclovir are now available for suppressive therapy. Ocular injections and implants of ganciclovir have been developed and are being studied, but they do not prevent disseminated disease. Studies are currently in progress to develop better oral drugs and to prevent CMV disease.

## 3. Herpes Simplex Virus

Severe perianal ulcerative lesions caused by herpes simplex virus (HSV) were among the first described manifestations of AIDS. There are two types of this virus, HSV 1 and HSV 2. Both viruses can cause similar conditions, but HSV 1 is typically the cause of fever blisters, herpetic whitlows, and, more rarely, herpes encephalitis. HSV 2 is the usual cause of genital herpes. In AIDS patients these viruses can reactivate to cause severe, deep, painful, often bloody ulcers, most commonly in the perianal area and sacrum and, less commonly, around the mouth and face. The diag-

nosis is suspected from the appearance and history of past episodes of recurrent herpes and confirmed by virus isolation from the lesions. Herpes simplex also causes esophagitis as discussed earlier. The development of acyclovir provided a highly effective treatment for these lesions, as well as effective suppressive therapy. Prior to this, these lesions posed extremely difficult clinical problems. Predictably, resistance to acyclovir in patients receiving long-term therapy has developed, and infection of other people with resistant virus has occurred. Foscarnet is an alternative treatment that is effective against acyclovir-resistant organisms at this time.

## 4. *Varicella zoster* Virus

*Varicella zoster* virus (VZV) is the cause of chicken pox. Reactivation of latent infection is the cause of shingles (Latin *cingulum*, a girdle), a painful eruption of vesicles in a band or belt-like pattern that is common in normal individuals and that occurs with greater frequency in patients with malignancy and HIV infection. It may appear as an early manifestation in HIV-infected patients who are not yet characterized as having AIDS (i.e., have more than 200 CD4 cells and have not yet had an AIDS-defining illness). The band-like pattern occurs because the virus lies latent in sensory nerve ganglia, which provide innervation to the skin in belt-like distributions called "dermatomes." Virus reactivates in the nerves supplying one of the dermatomes, and this leads to the characteristic distribution of the lesions. In severely immunosuppressed patients, VZV may disseminate to the entire skin and the internal organs. Secondary infection, notably with group A streptococcus, can occur. In addition, VZV in the skin lesions is infectious and can cause chicken pox in nonimmune contacts. This can lead to infection of health care workers who have not had chicken pox and other patients in the hospital if appropriate precautions are not taken.

The diagnosis of herpes zoster is made from its characteristic appearance and confirmed by virus isolation or characteristic cellular changes seen on histopathology. Treatment is with acyclovir. Again, viral resistance may emerge. Foscarnet is an alternative treatment.

## 5. JC (Polyoma) Virus

Progressive multifocal leukoencephalopathy (PML) is a progressive neurologic disease caused by infection of the brain with JC virus, a member of the polyoma family of viruses. These viruses have no envelopes, and their genetic material consists of DNA. It appears to be acquired in childhood and can be detected in the urine of pregnant women and immunosuppressed patients, such as renal transplant recipients. Most of these people never develop symptoms of disease attributed to JC virus. PML occurs almost exclusively in immunosuppressed patients, notably those with AIDS or malignancies. Patients develop a succession of severe lesions involving mostly the white matter of the brain. Paralysis, blindness, dementia, and aphasia may appear, and patients usually die within 3 to 6 months. The appearance of the brain on CT scan is very characteristic. A definitive diagnosis is made by demonstrating the virus particles in brain tissue obtained by biopsy or at autopsy. It is a rare complication, occurring in fewer than 1% of AIDS patients. There is no known therapy.

## D. Fungi

### 1. Candida albicans

*Candida albicans* is a yeast that is normally found in most humans as a harmless colonizer (commensal) of the skin, mouth, gastrointestinal tract, and genitourinary tract. There are, however, a large number of clinical manifestations of *Candida* that occur when host defenses are altered. Certain diseases, such as diabetes, cancer, intravenous substance abuse, malnutrition, and congenital immune deficiencies, are associated with different syndromes caused by *Candida*. In addition, candidiasis has become an increasingly common complication of medical progress: antibiotic therapy, corticosteroid drugs, immunosuppressive treatments leading to either neutropenia or defective T lymphocyte and mononuclear phagocyte function, intravenous catheters, urinary bladder catheters, and abdominal surgery all predispose to candidiasis. Syndromes associated with *Candida* vary with the types of deficiencies in host defenses. They include mucocutaneous candidiasis, oral candidiasis (thrush), esophageal candidiasis, candidemia, endocarditis (especially among intravenous drug users), endophthalmitis, disseminated candidiasis (in which several internal organs are infected), meningitis, and vaginitis.

Among patients with HIV infection, thrush and esophagitis caused by *Candida* are extremely common. These may be the earliest clinical indication of the presence of HIV infection. Thrush is manifested by the appearance of white, curd-like material on the inner surfaces of the mouth, yet it may be asymptomatic. *Candida* is by far the most common cause of esophagitis in AIDS patients. The clinical presentation

is very characteristic: people complain of pain behind the sternum when they swallow. This can be of sufficient severity to prevent eating and can lead rapidly to dehydration or severe weight loss. When due to candidiasis, there is frequently evidence of oral *Candida* infection (thrush). In general, the clinical approach to symptoms of esophagitis is to treat for *Candida*. Only if the symptoms do not improve as further steps taken to make a specific diagnosis, usually endoscopic examination of the esophagus during which biopsies are taken to establish the diagnosis. Less common causes of esophagitis in AIDS patients include CMV, HSV, and ideopathic ulcerations. *Candida* may also cause gastrointestinal ulcerations and disseminated candidiasis in AIDS patients. It may also cause candidemia, especially in patients who have permanent indwelling intravenous catheters inserted for the treatment of other AIDS-associated infections such as CMV or cryptococcal disease.

The diagnosis of candidiasis in AIDS patients is usually straightforward because of its characteristic clinical features of thrush or esophagitis. Confirmation of the diagnosis depends on microscopic identification of typical budding yeast forms with pseudohyphae. Catheter-associated infections are diagnosed by blood cultures obtained through the catheter and from a peripheral vein. Other forms of *Candida* infections are very uncommon in AIDS patients. The diagnosis of these forms, such as disseminated candidiasis, can be difficult and requires a high degree of clinical suspicion. Diagnosis depends on detection of the organisms in tissue biopsy specimens.

Several drugs exist for the treatment of *Candida* infections. Imidazole compounds (clotrimazole, ketoconazole, fluconazole, and itraconazole) are effective for thrush and esophagitis. Treatment generally has to be prolonged because of the high frequency of recurrence. The widespread use of imidazoles, especially fluconazole and ketoconazole, has led to the emergence of drug-resistant *Candida*, which takes two forms. Isolates of *C. albicans* have become resistant to drugs. In addition, in some patients, there is overgrowth with species of *Candida* (*C. glabrata* and *C. krusei*) that are inherently resistant to imidazoles. Amphotericin B, a drug associated with a high frequency of unpleasant side effects, is used for disseminated infections and may be useful when imidazoles are not effective.

### 2. Cryptococcus neoformans

*Cryptococcus neoformans* is a yeast with worldwide distribution. It is found in the nitrogen-rich feces of birds, most notably pigeons. Infection with cryptococcus is extremely common in AIDS patients, affecting perhaps 10% of them. The incidence may be lower in patients receiving antifungal therapy for other yeasts, such as *Candida*. Prior to the AIDS epidemic, cryptococcosis was a disease seen commonly in patients with defective T-cell mononuclear phagocyte function, those with neoplastic diseases such as Hodgkin's disease or lymphoma, and those receiving corticosteroids. Nearly half of all cases occurred in individuals with no recognizable immune defect. Cases in AIDS patients now account for the great majority of cases of cryptococcosis.

Exposure to cryptococcus is very common, although disease is relatively rare in normal hosts. Infection is acquired by inhalation of the yeast into the lungs, from where it disseminates by the bloodstream most typically to the central nervous system, where it causes meningitis or meningoencephalitis. Many other organs can be involved, including lungs, liver, bone marrow, lymph nodes, skin, and the prostate. Cryptococci can cause pulmonary nodules that can resemble lung cancer and require a lung biopsy for diagnosis.

*Cryptococcus neoformans* has a characteristic polysaccharide capsule that surrounds it. This capsule may impair the ability of host defenses to ingest and kill the organism, and also facilitates recognition of the organism as it is readily detected in blood and spinal fluid. The latex agglutination test for cryptococcal capsular polysaccharide ("cryptococcal antigen test") is a very sensitive and specific test for this organism. The diagnosis of cryptococcosis can also be made by isolation of the organism from blood, spinal fluid, and biopsy specimens.

In AIDS patients, the treatment of cryptococcosis needs to be prolonged, often for the life of the patient because of the high relapse rates when treatment is stopped too soon. Amphotericin B, which may be combined with 5-fluorocytosine, is the drug of choice for the treatment of cryptococcosis. These are associated with unpleasant side effects and significant toxicities. Fluconazole is an effective alternative for both treatment and maintenance therapy, and can be given orally as well as intravenously.

### 3. Histoplasma capsulatum

*Histoplasma capsulatum* is a fungus that is found in the soil, especially in areas where there is a high concentration of bird droppings. Starling roosts and chicken coops are typical sites that have been associated with outbreaks of histoplasmosis. The droppings

of bats in caves also harbor the organisms, which may infect spelunkers. The great river valleys of the United States have a high endemicity, and the organism also occurs in the Caribbean area.

Infection is caused by inhalation of the organisms, which most often is asymptomatic in normal people but may be associated with nonspecific upper respiratory symptoms. In a minority of infected people, histoplasmosis causes a variety of acute and chronic syndromes. Clinically in these cases it may closely mimic tuberculosis. In immunosuppressed patients, including those with AIDS, progressive disseminated histoplasmosis arises either as a result of an acute infection or, more typically, following reactivation of a dormant infection. Typical symptoms include fever, weight loss, malaise, cough, and shortness of breath. Severe, disseminated histoplasmosis may involve a number of organs, including lungs, liver, spleen, lymph nodes, bone marrow, skin, adrenals, and the central nervous system. The frequency of infection among AIDS patients from endemic areas approaches that of cryptococcosis. Infection may reactivate in individuals years after they have migrated from endemic areas.

Histoplasmosis should be suspected in febrile AIDS patients who are current or former residents of endemic areas. Antibody tests will identify some patients who have been previously exposed to histoplasma, but are generally not helpful in making the diagnosis. A polysaccharide antigen is detectable in the urine of up to 90% of patients with severe histoplasmosis. Definitive diagnosis depends on detecting the organisms in blood, bone marrow, or other tissues. Treatment of histoplasmosis is with amphotericin B. Ketoconazole and itraconazole are also effective and can be given orally. As with most other infections in AIDS patients, treatment must be prolonged, often for the life of the patient.

### 4. Coccidioides immitis

Coccidioidomycosis (San Joaquin Valley Fever) is endemic in the southwestern United States and many areas of Central and South America. The etiologic agent, *Coccidioides immitis*, inhabits the soil in hot arid areas with alkaline soil and little vegetation, characteristic of the Lower Sonoran Life Zone. Infection in apparently normal hosts is most often asymptomatic but may lead to a number of severe complications, including pneumonia, lung cavities, arthritis, myositis, dermatitis, osteomyelitis, and meningitis. It can be especially severe in pregnancy to both the mother and fetus. In immunosuppressed patients, especially those with AIDS and those receiving immunosuppressive drugs, it can reactivate long after the initial infection and rapidly disseminate.

Diagnosis depends on detecting the infecting organism in tissue biopsy specimens or expectorated sputum. It is usually difficult to isolate *C. immitis* from blood and spinal fluid cultures. Serum antibody tests and skin tests are available. In nonimmunosuppressed hosts, skin tests are helpful in identifying previously infected individuals. High titers of serum antibody indicate disseminated disease and changes in titers are useful in following the response to treatment. The role of these tests in the management of AIDS patients is uncertain.

Treatment of coccidioidomycosis is with amphotericin B. Ketoconazole, fluconazole, and itraconazole are effective against some forms of the disease. As with other infectious complications of AIDS, long-term therapy is generally necessary.

### 5. Other Fungi

*Aspergillus* species, *Blastomyces dermatitidis,* and *Penicillium marneffei* are fungi that have been associated with AIDS. *Aspergillus* is a mold that is ubiquitous in nature. It is associated with a number of syndromes in both normal and immunosuppressed patients, especially those with cancer, neutropenia, and prolonged antibiotic therapy. In AIDS patients with low CD4 counts and severe neutropenia or steroid treatment, *Aspergillus* can cause pulmonary disease or disseminated disease. Diagnosis requires demonstrating the organism in infected tissues. Cultures are often misleading. The organism is a common laboratory contaminant, leading to false-positive cultures, and false-negative cultures are also common. Although amphotericin B is effective, treatment often fails unless the predisposing factors can be eliminated.

*Blastomyces dermatitidis,* a yeast associated with a variety of syndromes, most commonly pulmonary disease, may cause late infectious complications of AIDS. The organism appears to inhabit decaying vegetation and is found in scattered locations in the eastern, central, and southern United States. It is not clear whether it is a truly opportunistic infection, but it appears more common in AIDS patients. Treatment is with amphotericin B.

*Penicillium marneffei* is a cause of a chronic, progressive, debilitating febrile illness in AIDS patients in Southeast Asia, notably Thailand and southern China. It is an interesting example of an infection that has emerged as AIDS moved into an area with different prevalent organisms. Diagnosis is made by detecting the yeast-like cells in pus draining from infected lymph nodes. Amphotericin B appears to be effective, but relapses are common.

## IV. SUMMARY

Infectious diseases that take advantage of impaired immunity remain the principal causes of morbidity and mortality in AIDS patients, and are likely to remain so until more effective treatments for HIV infection or for the immune defects it causes are developed. Much of the medical management of AIDS patients involves the diagnosis, treatment, and prevention of these infections. These infectious complications continue to change in response to advances in therapy. Different infectious complications will predictably emerge as AIDS affects new populations of people in different regions of the world.

## BIBLIOGRAPHY

Broder, S., Merigan, T. C., and Bolognesi, D. (1994). "Textbook of AIDS Medicine." Williams & Wilkins, Baltimore.
Cohen, P. T., Sande, M. A., and Volberding, P. A. (1994). "The AIDS Knowledge Base," 2nd Ed. Little Brown, Boston.
Mandel, G. L., Bennett, J. E., and Dolin, R. (eds.) (1995). "Principles and Practice of Infectious Diseases," 4th Ed. Churchill Livingstone, New York.
White, D. A., and Gold, J. W. M. (eds.) (1992). "Medical Management of AIDS Patients. Medical Clinics of North America." Saunders, Philadelphia.

# Acquired Immune Deficiency Syndrome, T-Cell Subsets

DONALD E. MOSIER

*The Scripps Research Institute*

## GLOSSARY

**CD4 T cells** A major subset of T cells expressing CD4 on their surface. They use CD4 to recognize class II major histocompatibility complex (MHC) molecules

**CD8 T cells** A major subset of T cells expressing CD8 on their surface. They use CD8 to recognize class I MHC molecules

**Cytokines** Proteins secreted by activated T cells that influence other cells in the immune system, e.g., interleukins, interferon, and growth factors

**HIV-1** Human immunodeficiency virus type I, the retrovirus that infects CD4 T cells and causes AIDS

**T cells** Thymus-derived lymphocytes that mediate cellular immunity and serve as regulators of the immune response

**Tc1/Tc2** CD8 T-cell subsets defined by their pattern of cytokine expression

**Th1/Th2** CD4 T-cell subsets defined by their pattern of cytokine expression

LYMPHOCYTES IN THE IMMUNE SYSTEM ARE DI-vided according to their origin, their function, the cell surface molecules that define their interactions with other cells, and the soluble products (cytokines or chemokines) they produce. T lymphocytes are the mediators of cellular immunity. They recognize cells infected with bacteria, viruses, and other pathogens and can also destroy tumor cells. One of the two major subsets of T cells, those cells bearing the CD4 molecule, are the primary target for infection by human immunodeficiency virus type 1 (HIV-1). Persistent infection with HIV-1 ultimately leads to a complete loss of CD4 T cells and the development of acquired immune deficiency syndrome (AIDS). At earlier asymptomatic stages of HIV-1 infection, partial depletion of CD4 T cells occurs and there are substantial changes in T-cell subsets. This article deals with those changes.

## I. DEFINITION OF T-CELL SUBSETS

### A. Developmentally Regulated T-Cell Subsets

T cells develop in the thymus, where they differentiate into two major subsets that express either the CD4 or the CD8 surface molecule. CD4 T cells recognize peptides bound to the major histocompatibility complex (MHC) class II molecule, whereas CD8 T cells recognize intracellularly processed peptides bound to the MHC class I molecule. CD4 T cells are often called "helper T cells" because they assist other lymphocytes in performing their immune function. HIV-1 uses the CD4 molecule as one target for entry into cells, so CD4 T cells are the primary cell type infected. CD8 T cells include "killer" or cytotoxic T cells that can kill infected or tumor target cells. These functional boundaries between CD4 and CD8 T cells are sometimes blurred, however, as when CD4 "killer" cells

ENCYCLOPEDIA OF HUMAN BIOLOGY, Second Edition, VOLUME I.   Copyright © 1997 by Academic Press.   All rights of reproduction in any form reserved.

or CD8 "helper" cells appear. CD4 T cells normally outnumber CD8 T cells by about 2 : 1. Most T cells utilize T-cell receptors composed of $\alpha/\beta$ subunits to recognize peptide–MHC complexes, but a subset of T cells uses the $\gamma/\delta$ T-cell receptor to recognize a more limited set of peptides. These $\gamma/\delta$ T cells are particularly important for manifestations of cellular immunity at epithelial or mucosal surfaces. The enormous diversity of $\alpha/\beta$ T-cell receptor sequences allows T-cell recognition of peptide fragments from the universe of potential disease-causing organisms, including the AIDS virus. [*See* Lymphocyte-Mediated Cytotoxicity; T-cell Receptors.]

## B. Activation-Induced T-Cell Subsets

Stimulation of nondividing T cells by recognition of foreign peptides bound to MHC molecules leads to proliferation and expression of additional surface molecules that distinguish activated from resting lymphocytes. Prior to the first stimulation, T cells are termed naive. Activated T cells express new cytokine receptors (e.g., the IL-2 receptor), new lymphocyte homing receptors (e.g., CD44) that control the recirculation patterns of T cells, new receptors that affect interaction with other cells in the immune system (e.g., CD28, CD40 ligand), and even receptors that influence susceptibility to cell death (e.g., Fas; Fas ligand). After cell proliferation has ceased, memory T cells (previously stimulated cells) continue to express unique markers (e.g., CD45RO, CD38) for a period of time.

Both CD4 and CD8 T cells are further divided by the selective production of certain cytokines, secreted molecules that regulate the immune response. Early after T-cell activation, a broad array of cytokines are produced. T cells at this stage are referred to as Th0 cells if they are CD4 helper cells and Tc0 if they are CD8 cytotoxic cells. Subsequently, T cells become more specialized and express unique subsets of cytokines, probably in response to different signals received from the macrophages or dendritic cells that generate and display peptide fragments to T cells. The unique cytokine expression patterns define Th1 and Th2 CD4 T cells and Tc1 and Tc2 CD8 T-cells (Table I). Tc1 and Tc2 express the same cytokines as Th1 and Th2 cells, except that IL-2 production by Tc1 cells is absent. The distinction between Th1 and Th2 T cells is clear if one measures cytokines secreted by activated T cells into tissue culture fluid, but less clear if cytokine messenger RNA levels are examined by more sensitive techniques. To the cautious observer,

**TABLE I**

Cytokines Associated with Th0, Th1, and Th2 T-Cell Subsets

| Cytokine | Th0 | Th1 | Th2 |
|---|---|---|---|
| Interleukin (IL)-2 | + | + | − |
| Interferon-$\gamma$ | + | + | − |
| IL-4 | + | − | + |
| IL-5 | + | − | + |
| IL-10 | + | − | + |
| IL-13 | + | − | + |

it is likely that the distinction between Th1 and Th2 T cells (or Tc1 and Tc2) is relative rather than absolute.

The importance of Th1 and Th2 CD4 T cells is that they regulate the balance between cellular and humoral (antibody-mediated) immunity. This balance can be critical in the response to some infections. Mice that are genetically biased to make a Th2 response die upon infection with the protozoan parasite *Leishmania major* whereas mice with a Th1 dominant response cure the infection. Allergic diseases in humans are often associated with a biased Th2 cytokine profile and elevated levels of immunoglobulin E, an antibody class that is dependent on IL-4 for its synthesis. Once a Th1 or Th2 bias is established, it is self-perpetuating because the cytokines both reinforce their own production and suppress the production of the opposing cytokines. This reciprocal regulation may be mediated either by direct effects on CD4 T cells or by indirect effects mediated by cytokines produced by macrophages or natural killer cells. Most immune responses involve a dynamic balance between Th1 and Th2 cytokines with both cellular and humoral immunity being generated. The "polarized" responses provide the clearest insight into the regulation of immunity, however, and these examples abound in the scientific literature. [*See* CD8 and CD4: Structure, Function, and Molecular Biology; Cytokines and the Immune Response.]

## II. RESTRICTIONS ON HUMAN IMMUNODEFICIENCY VIRUS TYPE I (HIV-1) CELL ENTRY AND REPLICATION

HIV-1 binds to target cells for infection via an interaction between the envelope glycoprotein of the virus (gp120) and CD4 expressed on the cell. It is now known that a second interaction between gp120 and chemokine receptors on CD4 T cells is essential for viral entry and infection (Fig. 1). Two alternative chemokine receptors are used by HIV-1. Viruses that infect both macrophages (which also express low levels of CD4) and CD4 T cells but not established T-cell lines are termed macrophage tropic. Such viruses enter cells by binding CD4 and CCR5, a chemokine receptor that normally binds macrophage inflammatory protein (MIP)-1α, MIP-1β, and RANTES (a chemokine that is regulated upon activation, normal T cell expressed and secreted). HIV-1 that infects T-cell lines and primary CD4 T cells, but not macrophages, is termed T-cell line tropic and enters cells by binding CD4 and fusin (a name likely to change to CXCR4), the chemokine receptor that normally binds stromal-derived factor (SDF)-1, a growth factor for immature B lymphoctyes. Macrophage-tropic HIV-1 is more frequently transmitted during primary infection and is found at all stages of HIV-1 infection. In contrast, T-cell line-tropic HIV-1 appears at later times after infection and frequently evolves into highly cytopathic virus variants that cause cell fusion and death. In the context of T-cell subsets, it is essential to understand the regulation of the two (or more) chemokine coreceptors for viral entry during T-cell activation. For example, downregulation of CCR5 following T-cell activation would lead to a population of T cells that was resistant to infection with macrophage-tropic HIV-1 but sensitive to infection with T-cell line-tropic virus.

The state of activation of T cells is known to influence their susceptibility to HIV-1 infection. Nondividing T cells are difficult to infect, and the few cells that are infected fail to progress beyond the proviral integration step of the virus replication cycle. Memory T cells identified by CD45RO expression are easier to infect than naive T cells not previously stimulated by antigen contact. Whether these differences relate to the expression of chemokine coreceptors for virus entry has yet to be determined. [See T-Cell Activation.]

## III. CHANGES IN T-CELL SUBSETS DURING HIV-1 INFECTION

### A. Changes in CD4 T Cells

Primary infection with HIV-1 leads to a variable decline in the numbers of CD4 T cells within the first 6 months of infection. Very high levels of virus (or viral RNA) are seen in the patient's blood before the onset of an effective immune response to HIV-1, and virus levels then decline to low or undetectable levels. CD4 T-cell numbers, which may have declined to less than half of their normal values during the first few months of infection, tend to stabilize and show a much slower rate of decline over the next several years of infection. Both naive (CD45RA) and memory (CD45RO) CD4 T cells decline during primary infection. This may reflect systemic infection with HIV-1 that reduces the thymic production of naive T cells as well as the death of T cells in lymph nodes, spleen, and other lymphatic tissues.

Long-term infection with HIV-1 eventually leads to a more rapid decline in CD4 T cells, and the opportunistic infections or tumors that constitute the clinical manifestations of AIDS appear when the numbers of CD4 T cells decline below 5% of normal values (below 50 CD4 T cells/mm³). At this point, the few surviving CD4 T cells all appear to be memory cells and many express the CD7 marker, which is normally expressed on few T cells in the blood. Although these T cells may produce IL-10, they are not typical Th2 T cells (see later).

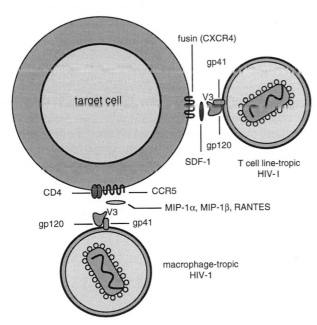

**FIGURE 1** Distinct mechanisms of HIV-1 entry for macrophage-tropic and T-cell line-tropic HIV-1. CD4 is a common receptor for both types of virus. The coreceptors differ. CCR5, a chemokine receptor for MIP-1α, MIP-1β, and RANTES, is the coreceptor for macrophage-tropic HIV-1. Excess chemokines block virus entry. Fusin (CXCR4) is the coreceptor for T-cell line-adapted HIV-1, and the SDF1 ligand blocks virus entry.

Patients treated with effective combinations of antivirals often show a remarkable and rapid increase in CD4 T cells. The rate of this increase has led to models of CD4 T-cell turnover during HIV infection that propose the replacement of about 2% of total T cells per day. The majority of these newly appearing CD4 T cells are memory cells, although small numbers of naive cells are also found. Chronic HIV infection thus is viewed as a balance between high rates of CD4 T-cell death due to infection and replacement due to the continued proliferation of uninfected cells. Intrathymic infection with HIV almost certainly reduces the generation of new T cells, so T-cell replacement probably comes from a dividing pool of cells in secondary lymphoid tissue. The slow decline in CD4 T cells over the years of HIV-1 infection would reflect a marginally higher rate of cell death over cell replacement, a view that provides a rationale for early and continuous antiviral therapy.

## B. Changes in CD8 T Cells

HIV-1 infection stimulates a strong CD8 T-cell immunity that is manifest as a large increase in cytotoxic T cells (CTL) capable of killing virus-infected target cells. The CTL response is responsible for limiting primary infection, as neutralizing antibodies to virus appear much later than the decline in virus levels. In addition, CD8 T cells that can suppress virus infection by nonlytic mechanisms also appear. One mechanism of CD8 T-cell antiviral activity is the secretion of the chemokines MIP-1$\alpha$, MIP-1$\beta$, and RANTES that block viral entry into cells (Fig. 1). Other nonlytic mechanisms of antiviral activity remain to be defined. The overall number of CD8 T cells increases during the course of HIV-1 infection, and the increase is almost totally due to previously activated, memory CD8 cells. The total number of T cells (CD4 + CD8) remains relatively constant until the onset of AIDS, when the absolute lymphocyte number declines. This has led to hypotheses of lymphocyte homeostasis that propose a mechanism to ensure a constant number of total CD4 + CD8 T cells, i.e., if CD4 T cells fall, CD8 T cells must rise (Fig. 2). However, an alternative explanation is simply that the ongoing mutation of HIV-1 genes leads to the stimulation of additional CD8 CTL precursors (as new stimulatory peptides are generated), and the pool of memory CD8 T cells continues to increase as the duration of virus infection increases. HIV-1 infection is unique in that the frequency of CTL is high enough to directly isolate such cells from blood without the intermediate step of re-

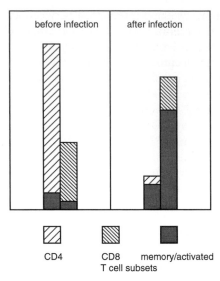

**FIGURE 2** Major changes in T-cell subsets following HIV-1 infection. CD4 T cells decline, CD8 T cells increase, and both populations contain a majority of activated and/or memory T cells that show evidence of previous stimulation.

stimulation with HIV-1-infected cells in tissue culture. A sustained and vigorous CTL response to HIV-1 peptides is also seen in long-term nonprogressors, individuals whose immune systems remain intact after many years of HIV-1 infection.

The rate of CD8 T-cell replacement during HIV-1 infection is unknown. It is likely that the intrathymic development of CD8 T cells is halted by infection and that the expansion of memory CD8 T cells reflects a continuous proliferation of cells in secondary lymphoid tissue. In contrast to CD4 T cells, however, HIV-1 does not cause the death of CD8 T cells, so the number of CD8 T cells continues to increase as long as a few CD4 T cells are present.

## IV. CHANGES IN CYTOKINES DURING HIV-1 INFECTION

Changes in cytokine profiles during primary HIV-1 infection seem to reflect mainly the activation of CD8 CTL. Increases in interferon-$\gamma$, IL-10, and tumor necrosis factor-$\alpha$ are observed. Primary infection stimulates both cellular immunity (CTL specific for HIV-1 proteins) and humoral immunity (antibodies to HIV-1), so there is no indication that a polarized Th1- or Th2-type response occurs during primary infection.

Isolation and characterization of individual CD4 T cells from patients chronically infected with HIV-1 have shown a relative increase in T cells with the Th0 pattern of cytokine production. These may reflect recently activated T cells that have yet to acquire a specialized cytokine secretion profile, as the frequency of activated T cells increases with HIV-1 infection and there is a rapid turnover of T cells. Individual T-cell clones with the Th1 cytokine profile are more difficult to infect with HIV-1 in tissue culture, suggesting that they may be more resistant to infection in patients. This is contrary to the proposal that the progression of HIV-1 disease is linked to a Th1 to Th2 shift. In fact, both cellular and humoral immunity decline during HIV-1 infection, so there is no convincing evidence for a general polarization of immune responses that would reflect a major imbalance between Th1 and Th2 cytokine production. The evidence for a Th2 bias comes primarily from observations of increased IgE levels in both pediatric and adult patients infected with HIV-1. However, elevated IgE is found only in a subset of patients (about 30%) and does not correlate with levels of CD4 T cells, although it is correlated with opportunistic infections. Increased IgE thus may reflect an appropriate response to other infectious agents rather than a bias imposed by HIV-1-induced cytokine changes. Tissue culture experiments have shown that the addition of IL-12, a cytokine that enhances Th1 responses, can improve the function of CD4 T cells isolated from HIV-1-infected patients. This finding suggests that Th1 cells are present but may be lacking the appropriate cytokines for optimum functional responses.

## V. FUNCTIONAL IMPAIRMENT OF T-CELL SUBSETS

Although the importance of Th1 and Th2 subsets in HIV-1 infection remains a subject of debate, it is clear that CD4 T-cell function is impaired when substantial numbers of CD4 T cells remain. T cells are more difficult to activate with vaccine antigens (e.g., tetanus toxoid) and are more easily triggered to undergo programmed cell death. The basis of this functional impairment remains unclear. HIV-1 infects macrophages and dendritic cells that present peptide antigens to T cells, and it is possible that contact of a T cell with such an infected cell partially activates the T cell, leading to a state of unresponsiveness. HIV-1-encoded proteins, particularly tat and nef, may also influence T-cell function. The rate of spontaneous cell death or apoptosis is high in both CD4 and CD8 T cells from HIV-1-infected individuals, suggesting that the high turnover rate in both populations is accomplished at the cost of normal regulation.

It is thus important to conclude that the known disturbances in T-cell subsets during HIV-1 infection do not fully account for the functional abnormalities observed during the course of the disease.

## BIBLIOGRAPHY

Clerici, M., and Shearer, G. (1993). A Th1 to Th2 switch is a critical step in the etiology of HIV infection. *Immunol. Today* **14**, 107.

Giorgi, J., Liu, Z., Hultin, L., Cumberland, W., Hennessey, K., and Detels, R. (1993). Elevated levels of CD38+ CD8+ T cells in HIV infection add to the prognostic value of low CD4+ T cells levels: Results of 6 years of follow-up. *J. Acquir. Immune Defic. Syndr.* **6**, 904.

Graziozi, C., Pantaleo, G., Gantt, K., Fortin, J.-P., Demarest, J., Cohen, O., Sékaly, R., and Fauci, A. (1994). Lack of evidence for the dichotomy of Th1 and Th2 predominance in HIV-infected individuals. *Science* **265**, 248.

Ho, D., Neumann, A., Perelson, A., Chen, W., Leonard, J., and Markowitz, M. (1995). Rapid turnover of plasma virions and CD4 lymphocytes in HIV-1 infection. *Nature* **373**, 123.

Levy, J. A. (1993). Pathogenesis of human immunodeficiency virus infection. *Microbiol. Rev.* **57**, 183.

Romagnani, S. (1994). Lymphokine production by human T cells in disease states. *Annu. Rev. Immunol.* **12**, 227.

# Acquired Immune Deficiency Syndrome, Virology

ERIC O. FREED

*National Institutes of Health*

## GLOSSARY

**Budding** Process by which a virus particle is released from the host cell

**Integration** Process by which the double-stranded DNA copy of the viral genome is inserted into the host cell chromosome

**Lentivirus** The genus of retroviruses which includes the primate immunodeficiency viruses: HIV-1, HIV-2, and SIV

**Reverse transcription** Synthesis of a double-stranded DNA copy of the viral RNA genome that occurs after infection and is mediated by the viral enzyme reverse transcriptase

**Virion** Virus particle

## I. INTRODUCTION

Since the early 1980s, the worldwide pandemic of acquired immune deficiency syndrome (AIDS) has driven an intensive international research effort aimed at understanding the causative agent of AIDS: the human immunodeficiency virus (HIV). This article describes the basic state of our current knowledge concerning HIV molecular virology. Since most AIDS cases worldwide have resulted from HIV type 1 (HIV-1) infection, this virus will be the focus of discussion. However, certain distinguishing properties of HIV-1 relative to the related HIV type 2 (HIV-2) and the simian immunodeficiency viruses (SIVs) will be mentioned.

## II. CLASSIFICATION AND TAXONOMY

HIV-1 and HIV-2 are the two human members of the lentivirus ("slow-acting") genus of the family Retroviridae. Retroviruses in general, and lentiviruses in particular, are characterized by a high degree of sequence heterogeneity from one isolate to another. The extent of variation is greatest in the *env* gene. Based on sequence relatedness, HIV-1 and HIV-2 have been organized into subtypes, or clades: nine (A–I) for HIV-1 and five (A–E) for HIV-2. HIV-1 shares less than 50% sequence identity with HIV-2 in the *env* gene. The prevalence of a particular clade often varies with geography, e.g., HIV-1 clade B is the most prevalent subtype in North America and Europe, whereas clade E is particularly common in parts of Asia. HIV-1 is found worldwide, whereas the majority of HIV-2 infection is confined to parts of West Africa.

A number of distinct SIVs have been isolated from nonhuman primates, including the sooty mangabey (SIV$_{SMM}$), macaque (SIV$_{MAC}$), African green monkey (SIV$_{AGM}$), mandrill (SIV$_{MND}$), sykes monkey (SIV$_{SYK}$), and chimpanzee (SIV$_{CPZ}$). With the exception of SIV$_{CPZ}$, which is closely related to HIV-1, the SIVs exhibit strong sequence relatedness with HIV-2 but are relatively distantly related to HIV-1. Some SIVs (e.g., SIV$_{MAC}$) cause an illness remarkably similar to

ENCYCLOPEDIA OF HUMAN BIOLOGY, Second Edition, VOLUME 1. Copyright © 1997 by Academic Press. All rights of reproduction in any form reserved.

human AIDS in their natural hosts, whereas other SIVs (i.e., $SIV_{AGM}$) are generally nonpathogenic.

## III. ORGANIZATION OF THE HUMAN IMMUNODEFICIENCY VIRUS TYPE I (HIV-I) GENOME

The genomes of all replication competent retroviruses, including HIV-1, are organized in a similar fashion. From 5' to 3' are located the *gag*, *pol*, and *env* open reading frames (Fig. 1). The initial translational product of the *gag* gene of HIV-1 is the Gag polyprotein precursor, Pr55[Gag]. This polyprotein contains all the information necessary to direct the assembly of noninfectious, immature virus particles. During or shortly after budding, the Gag precursor is proteolytically processed by the viral protease (PR) to produce the mature Gag proteins: p17 matrix (MA), p24 capsid (CA), p7 nucleocapsid (NC), and p6. In the virion, MA is located just inside the lipid bilayer of the viral envelope, CA forms the proteinaceous shell which defines the viral core, and NC is found within the core bound to the viral RNA. The location of p6 in the virion has not been definitively determined.

During translation of the *gag* gene, a ribosomal frameshift occasionally occurs which results in the synthesis of the Gag–pol polyprotein precursor (Pr160[Gag–pol]). The Pol portion of this polyprotein precursor contains the viral protease (PR), reverse transcriptase (RT), and integrase (IN). The PR catalyzes the proteolytic processing of the Gag and Gag–pol polyprotein precursors, RT directs the transcription of viral DNA from the RNA genome after virus entry, and IN mediates the integration of the DNA copy of the viral genome into the host cell chromosome. The frameshifting mechanism for Gag–pol synthesis en-

sures that the Pol proteins are synthesized at 5–10% the level of the Gag proteins.

The 3' portion of the HIV-1 genome encodes the envelope (Env) glycoproteins: the surface (SU) glycoprotein, gp120, and the transmembrane (TM) glycoprotein, gp41. The SU and TM glycoproteins are initially synthesized as a precursor protein, gp160, which is processed by a cellular protease during its transport to the cell surface. As their names imply, the SU glycoprotein is located entirely external to the cellular or viral membrane whereas the TM glycoprotein spans the lipid bilayer. SU and TM associate via hydrophobic, noncovalent interactions.

In addition to the *gag*, *pol*, and *env* open reading frames, lentiviral genomes encode a number of additional genes which serve a variety of functions in the virus life cycle. In the case of HIV-1, these additional open reading frames are *tat*, *rev*, *vif*, *vpr*, *vpu*, and *nef*. HIV-2 and SIV do not encode *vpu* but do encode another gene known as *vpx*. Space does not allow a detailed discussion of the functions of these regulatory and accessory proteins; however, a brief description of their major roles follows. The position of these genes in the viral genome is indicated in Fig. 1.

The Tat protein of HIV-1 functions to transactivate the viral promoter, located in the 5' long terminal repeat (LTR). In the absence of Tat, transcription from the HIV-1 LTR is very weak; mutations in Tat can therefore be lethal. The Rev protein functions posttranscriptionally to increase the amount of unspliced or singly spliced viral RNA transported to the cytoplasm. The mechanism of Rev function involves an interaction with a *cis*-acting RNA element known as the rev-responsive element (RRE). As a result of Rev function, the synthesis of viral structural proteins (e.g., Gag and Env) is significantly increased. Vif greatly enhances virus infectivity, apparently by acting at an early step in the virus life cycle. Interestingly,

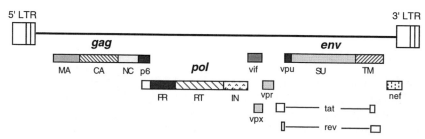

**FIGURE I** Genome organization of HIV. The positions of the open reading frames, and the gene products they encode, are indicated. The *vpu* gene is found only in HIV-1 and the *vpx* gene is found in HIV-2 and SIV but not in HIV-1.

the requirement for Vif is markedly cell-type dependent; in some cell types, mutational inactivation of Vif has no effect on virus infectivity, whereas in other cell types Vif mutations are lethal. The observation that the cell-type dependence is imposed by the virus-producing cell and not the target cell suggests that Vif may act to modify some component of the virion to increase its infectivity in the next round of infection. Studies indicate that Vif is found at low levels in the virion.

The Vpr protein is not essential for replication, although in some cell types (particularly primary monocytes) Vpr increases virus infectivity. Vpr has also been observed to arrest cells in the cell cycle; the biological implications of this property are unclear. Vpr is found abundantly in virions. The *vpx* gene, which is found in HIV-2 and SIV but not in HIV-1, encodes a virion protein which bears significant sequence homology with Vpr, suggesting that *vpr* and *vpx* arose by gene duplication. Although not essential for virus replication, Vpx expression enhances virus infectivity in nondividing cells and may play a role in the transport of the viral preintegration complex to the nucleus.

Vpu, which is encoded by HIV-1 but not HIV-2 or SIV, is not essential for virus replication but stimulates a severalfold increase in virus production from infected cells by an unknown mechanism. Vpu also degrades CD4, the major cellular receptor for HIV. The role Nef plays in virus replication is unclear and controversial. Although it was initially reported that Nef expression downmodulates transcription from the LTR, it now appears that Nef enhances virus infectivity. Like Vif, Nef appears to act early in the virus life cycle. Although Nef deletion typically has only minor effects on HIV-1 replication in tissue culture, studies with Nef-deleted mutants of SIV suggest that it may play a major role in replication and pathogenesis *in vivo*. Like Vpu, Nef possesses the ability to downmodulate CD4 expression. A summary of the major functions of the HIV/SIV proteins is provided in Table I.

## IV. HIV-1 LIFE CYCLE AND VIRAL PROTEIN FUNCTION

The HIV-1 life cycle (Fig. 2) begins with the binding of the SU envelope glycoprotein gp120 to the major cellular receptor, CD4. Following CD4 binding, conformational changes occur in both gp120 and in

### TABLE I
HIV Proteins

| Protein[a] | Primary function[b] |
|---|---|
| **Gag** | |
| Matrix (p17, MA) | Targeting of Gag to membranes |
| | Env incorporation into virions |
| Capsid (p24, CA) | Virus assembly |
| | Gag and Gag–pol interactions |
| Nucleocapsid (p7, NC) | RNA encapsidation |
| p6 | Promotes virus budding |
| | Vpr incorporation |
| **Pol** | |
| Protease (PR) | Cleavage of Gag and Gag–pol precursors |
| Reverse transcriptase (RT) | Reverse transcription of viral genome (RNA→DNA) |
| Integrase (IN) | Integration of viral DNA into host genome |
| **Env** | |
| Surface (gp120, SU) | Receptor binding |
| Transmembrane (gp41, TM) | Membrane fusion |
| **Accesory and regulatory** | |
| Tat | Transcription transactivation |
| Rev | RNA export from nucleus |
| Vif | Enhances virus infectivity |
| Vpr | Induces cell cycle arrest |
| | Nuclear transport (?)[c] |
| Vpu (HIV-1) | Stimulates particle release |
| | Degrades CD4 |
| Vpx (HIV-2/SIV) | Nuclear transport (?) |
| | Infection of nondividing cells |
| Nef | Enhances virus infectivity (?) |
| | Downmodulates CD4 expression |
| | T-cell activation |

[a] The two-letter standard abbreviations for the mature proteins are indicated.
[b] Partial list of the major functions of the indicated proteins.
[c] Indicates putative function.

the TM glycoprotein gp41. These conformational changes trigger a membrane fusion reaction between the lipid bilayers of the viral envelope and the host cell plasma membrane, thereby enabling the viral nucleocapsid to enter the cytoplasm of the host cell.

The binding of virus to the host cell, which is mediated by the SU envelope glycoprotein, determines the cell type which a particular virus isolate will infect. In culture, HIV-1 generally infects either T-cell lines or cells of the monocyte/macrophage lineage, leading to the designation of virus isolates as T-cell line tropic (T-tropic) or macrophage tropic (M-tropic), respec-

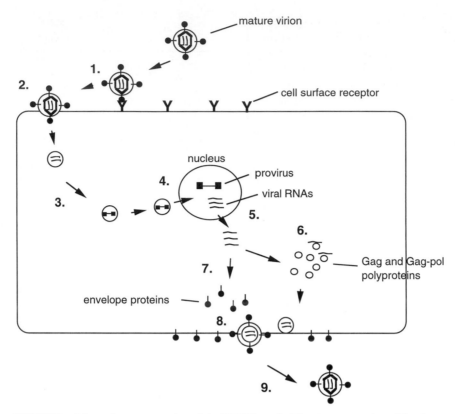

**FIGURE 2**  Schematic representation of the HIV life cycle. The virus enters the cell by direct fusion at the plasma membrane and utilizes the C-type assembly pathway (i.e., assembly at the plasma membrane). The steps indicated are (1) receptor binding, (2) membrane fusion and entry, (3) uncoating and reverse transcription, (4) nuclear transport and integration, (5) transcription and RNA transport, (6) translation, Gag, and Gag–pol transport, (7) Env transport, (8) virus assembly and budding, and (9) core condensation and virus maturation.

tively. HIV-1 strains capable of infecting both T-cell lines and macrophages (referred to as dual tropic) are relatively uncommon. *In vivo*, the M-tropic isolates are most prevalent early in infection (during the period of clinical latency); T-tropic strains tend to emerge late in infection, after the onset of disease. Some evidence suggests that M-tropic isolates are either transmitted more efficiently than T-tropic isolates or quickly predominate early after infection. A portion of the SU glycoprotein preferred to as the V3 loop (variable loop 3) largely determines the tissue tropism of a particular isolate. For years, investigators have speculated that the V3 loop might influence tropism by binding a secondary receptor, or coreceptor, distinct from CD4. Studies have revealed that in addition to CD4, several coreceptors are involved in the binding and fusion reactions. These cell surface proteins are members of a class of multiple membrane-spanning proteins whose physiological function is to serve as

receptors for a variety of chemokines. The expression of these chemokine receptors, at least in part, determines the type of cell which a particular virus isolate can infect. For example, the molecule known as fusin (also known as CXCR-4 or LESTR) facilitates infection by T-tropic viruses, whereas infection by M-tropic viruses may be controlled by the chemokine receptor CC-CKR-5.

Although a small amount of reverse transcription (the conversion of the single-stranded viral RNA genome to double-stranded DNA) may occur in virions, the vast majority of reverse transcription is carried out after entry of the nucleocapsid into the cytoplasm. This process is catalyzed by the viral enzyme RT, using a cellular tRNA (which is packaged into the virion during assembly) as a primer. The RT enzyme performs both the polymerization reaction (RNA → DNA) and the concomitant degradation of the genomic RNA. When discovered by Temin and Baltimore in the early 1970s, the conversion of RNA to DNA was viewed as a highly unusual type of polymerization reaction that violated the "central dogma" which dictated that information flows in biological systems from RNA→DNA→protein. This "backwards" flow of information gave retroviruses their name.

Following entry of the nucleocapsid into the cytoplasm, and during reverse transcription, a series of poorly understood uncoating steps occur. The double-stranded, linear viral DNA is transported into the nucleus as part of a high molecular weight structure known as the preintegration complex. Lentiviruses are unique among retroviruses in their ability to infect nondividing cells. The ability of the lentiviral preintegration complex to enter the nucleus in the absence of mitosis has led to the hypothesis that the protein(s) in this complex must contain signals which direct nuclear transport. The proteins which reportedly remain associated with the preintegration complex include MA, NC, Vpr, RT, and IN. At this time, the precise mechanism by which nuclear transport occurs is unclear.

After nuclear transport, the double-stranded viral DNA integrates into the host cell chromosome, a reaction that is catalyzed by the viral IN protein. Integration establishes the viral DNA, or provirus, as part of the cellular genome for the lifetime of the infected cell. Following integration, transcription is initiated from the 5' LTR. As mentioned earlier, LTR-driven transcription is markedly enhanced by the HIV-1 Tat protein via its interaction with a stem–loop RNA structure known as the TAR element. RNA splicing occurs to generate a number of singly and multiply spliced mRNAs. In general, the regulatory proteins (i.e., Tat and Rev) are translated from multiply spliced RNAs synthesized early in the infection. Unspliced and singly sliced RNAs, which are used to synthesize the viral structural proteins, are produced later. Their splicing and transport to the cytoplasm are regulated by the viral protein Rev via its interaction with a *cis*-acting element known as the rev-responsive element (RRE).

Lentiviruses, like C-type retroviruses, assemble at the plasma membrane. The Gag and Gag–pol polyprotein precursors (Pr55$^{Gag}$ and Pr160$^{Gag-pol}$, respectively) are transported to the plasma membrane where, in association with two copies of the single-stranded viral RNA genome, they interact to form electron-dense, budding structures. The MA domain of Pr55$^{Gag}$ plays a critical role in directing the transport to, and mediating the assocation with, the plasma membrane. The association between MA and plasma membrane is driven by a myristic acid moiety that is contranslationally attached to the amino-terminal residue of MA. Membrane binding may also be stabilized by electrostatic interactions between basic residues in MA and negatively charged phospholipids in the inner leaflet of the plasma membrane. The CA domain of Gag is largely responsible for mediating Gag–Gag interactions and for the association between Pr55$^{Gag}$ and Pr160$^{Gag-pol}$. The NC domain is required for the specific encapsidation of the viral RNA genome into the assembling particle. The p6 domain plays a role late in the budding process by enhancing the release of budding virions from the plasma membrane and is also responsible for the incorporation of Vpr into virus particles. The viral envelope glycoproteins, which are transported to the plasma membrane through the ER/Golgi secretory pathway, are incorporated into nascent budding virions as a heteromultimeric complex of SU (gp120) and TM (gp41). Their incorporation appears to be mediated by an interaction between the cytoplasmic domain of gp41 and the MA domain of Gag. During and shortly after budding, the viral PR cleaves the Gag and Gag–pol polyprotein precursors to generate the mature Gag and Pol proteins. The action of the viral PR can be readily visualized by electron microscopy as a rearrangement of the viral core from electron lucent to electron dense. This structural transition, which is often referred to as virion maturation, is required for virus infectivity. A highly schematic representation of a HIV virion is depicted in Fig. 3.

## V. DISEASE PREVENTION AND OUTLOOK

For many years, it appeared that little real progress was being made in preventing the onset of disease following HIV-1 infection. Although some gains were achieved in treating the opportunistic infections that plague HIV-infected patients after the onset of immune deficiency, effectively controlling the replication of HIV-1 itself remained an elusive goal. This difficulty was largely a result of the ease with which HIV escapes the human immune system and develops resistance to antiviral drugs. Increased understanding of HIV-1 at the molecular level, however, has sparked new-found optimism that preventing AIDS following HIV infection will soon be possible.

A variety of antiretroviral therapeutics are currently being tested. Nucleoside analogs, such as AZT, ddI, and ddC are designed to terminate polymerization

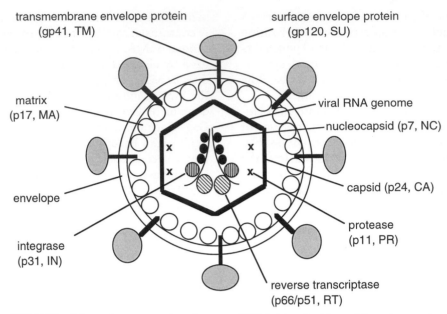

**FIGURE 3** Schematic representation of a mature HIV-1 virion. Positions of the major proteins are indicated. This diagram is not intended to provide information about the size, abundance, or exact location of the various proteins. Virion proteins whose positions in the mature virion are not known (e.g., p6 Gag, Vpr, and Vif) are not indicated.

during reverse transcription, thereby blocking infection at an early step. AZT has been used in AIDS patients for several years; unfortunately, when used alone, it is often poorly tolerated by the patient and AZT-resistant viral variants emerge rapidly. In an attempt to mitigate the latter problem, AZT is now being used in combination with other drugs. Particularly encouraging has been the dramatic response in patients treated simultaneously with two anti-RT inhibitors (e.g., AZT and ddI) and a new class of compounds which inhibit the viral PR. As discussed earlier, cleavage of the Gag and Gag–pol polyprotein precursors by PR is an essential step in the virus life cycle; the PR inhibitors block this step, thereby preventing virion maturation. When used alone, drug-resistant variants emerge, whereas when used in combination with nucleoside analogs, dramatic and long-term reductions in circulating virus loads are observed and patient health improves accordingly. Additional drugs which target other viral proteins (e.g., IN and NC) are being developed and tested. [*See* Chemotherapy, Antiviral Agents.]

A major drawback of the "combination" therapy just described is that the expense of these compounds makes widespread use in developing countries, some of which have very high rates of HIV-1 infection,

unlikely in the near future. An effective vaccine would be relatively inexpensive and might prevent infection rather than treating patients after infection. Thus far, attempts to develop an effective HIV-1 vaccine have not been successful. However, several advances suggest some grounds for optimism. A major stumbling block in HIV-1 vaccine development has been the lack of an animal model system. The only nonhuman primate that can be infected with HIV-1 is the chimpanzee; unfortunately, chimps are rare and expensive, and generally do not develop disease after HIV-1 infection. Several laboratories have developed animal model systems based on viruses, known as SHIVs, which are recombinants between SIV and HIV-1. These SHIVs contain HIV-1 Env coding sequences in a SIV background. They efficiently infect a variety of monkeys and in some cases are pathogenic. The SHIV model system will likely prove useful in developing and testing HIV-1 vaccines. Another promising but controversial approach to vaccine development involves the use of vaccines based on live but crippled (attenuated) virus strains. Much of this work derives from studies performed in monkeys which indicate that a SIV containing a deletion in the *nef* gene grew poorly *in vivo* but conferred resistance to superinfection with wild-type SIV strains. Despite these encouraging results, concern over possible long-term consequences of administering attenuated virus vaccines makes testing in humans unlikely in the near future. [*See* Retroviral Vaccines.]

Much work is also being focused on using gene therapy technology to develop a genetically based treatment for HIV-1 infection. The goal of this approach is to administer one or several vectors expressing nucleic acids or proteins which block virus replication. In theory, the infected individual could either be treated directly or cells could be removed, treated, and returned to the patient. Examples of this technology would employ RNA decoys, single-chain antibodies, ribozymes, anti-sense RNAs, or transdominant negative mutants of viral proteins (e.g., Gag, Rev, Env, IN, and PR) to interfere with HIV-1 replication. The major challenge currently faced by this approach is the difficulty in delivering the nucleic acid or protein of interest to the appropriate tissue with high efficiency.

A high degree of optimism has developed in the HIV research community since the early 1990s. It appears that a decade and a half of intensive research aimed at understanding HIV molecular biology has spawned a diversity of novel and creative approaches to treat HIV infection. Some of these therapies are currently extending the lives of AIDS patients; others will require further development and refinement but may prove effective at preventing or treating HIV infection in the future.

## BIBLIOGRAPHY

Coffin, J. M. (1996). Retroviridae: The viruses and their replication. *In* "Virology" (B. N. Fields, ed.), pp. 1767–1847. Lippincott-Raven, Philadelphia.

Freed, E. O., and Martin, M. A. (1995). The role of human immunodeficiency virus type 1 envelope glycoproteins in virus infection. *J. Biol. Chem.* **270**, 23883–23886.

Levy, J. A. (1993). Pathogenesis of human immunodeficiency virus infection. *Microbiol. Rev.* **57**, 183–289.

Luciw, P. A. (1996). Human immunodeficiency viruses and their replication. *In* "Virology" (B. N. Fields, ed.), pp. 1881–1952. Lippincott-Raven, Philadelphia.

Wills, J. W., and Craven, R. C. 1991. Form, function, and use of retroviral Gag proteins. *AIDS* **5**, 639–654.

# Acute Phase Response

JACK GAULDIE
*McMaster University*

STEVEN L. KUNKEL
*University of Michigan Medical School*

## GLOSSARY

**Acute phase** Early stages of the host response to injury. Typically, the response is seen within minutes and becomes maximum over the next few hours, and systems return to normal physiology within 24–48 hr

**Cytokine** Cell-derived polypeptide hormone-like mediators of small to intermediate molecular mass. They have multiple biological activities which are manifest through interaction with specific receptors on target cells. Included are interleukins, interferons, lymphokines, and growth factors

**Inflammation** Body's coordinated response to injury or infection aimed at halting damage and initiating repair

**Leukocytosis** Increase in the absolute number of leukocytes, primarily neutrophils, in the blood with the appearance of immature "band forms" in the white blood cell population

FOLLOWING TISSUE INJURY, TRAUMA, OR INFECTION, the body elicits a complex homeostatic response in an ordered and orchestrated manner to protect the organism and to return normal function to the tissues. Bacterial infections, burns, surgery, tissue necrosis, or immunological reactions are the most common initiators of this response, whose purpose is to halt the process of injury, protect the tissues from further damage, mediate the immune or memory component

of protection, and initiate the repair process aimed at returning the tissue to normal function. Associated with this response are the clinical signs of inflammation: calor (heat), tumor (swelling/edema), dolor (pain), rubor (redness), and loss of function. The overall process is called inflammation, and the rapid local and systemic changes are referred to as the acute phase response. While inflammation is associated with destructive and pathological events, we know that this acute phase response is part of the normal healing and repair mechanism of the body. Only when there is an overwhelming response with "innocent bystander" damage, or when chronicity becomes apparent, does this normal homeostatic response become pathological.

## I. INTRODUCTION

First described by the Greeks and the Romans, the humoral theory of inflammation held well into the 18th century. The work by Hunter, Virchow, Cohnheim, and Metchnikoff described in more detail the way the cells of the body contribute to this response. These descriptions formed the basis for our current understanding of the homeostatic response to injury. We are now aware that while the cellular components are integral to the response, a variety of humoral mediators orchestrate the response.

After the body has been damaged by infection or insult, a series of local and systemic events occur. At the local tissue level the blood vessels dilate and leak, particularly the postcapillary venules, giving rise to the redness due to the presence of intact red blood cells in the tissue or hemoglobin in the event of hemorrhage and hemolysis. There is aggregation of platelets

ENCYCLOPEDIA OF HUMAN BIOLOGY, Second Edition, VOLUME I.   Copyright © 1997 by Academic Press.   All rights of reproduction in any form reserved.

in the vessels and an early flux and accumulation of neutrophils and monocyte/macrophages, with associated activation of these cells and release of tissue-damaging proteinases and other lysosomal enzymes. In addition, most of the local tissue cellular components affected by the injury are activated with the release of significant amounts of various mediators, resulting in the initiation of a series of systemic responses. [See Macrophages; Neutrophils.]

It is these systemic changes, particularly the changes to the blood constituents, that are more appropriately termed the acute phase response. The changes include fever and pain and possible activation of the major biochemical pathways in the plasma, complement, and coagulation, as well as the kininogen/fibrinolysis sequence. Changes occur in the cellular content of the blood, with increases in leukocyte counts and activation of many of the phagocytic cells. This activation, particularly of the peripheral blood monocytes, results in the release of a number of protein hormone-like polypeptides or cytokines, along with highly potent metabolites of arachidonic acid: the prostaglandins (PG) and leukotrienes.

The most prominent change is seen in the plasma protein constituents as a result of the response of the liver. The plasma shows a marked decrease in iron and zinc levels and a dramatic increase in the levels of a series of proteins, synthesized by the hepatocyte, termed the acute phase proteins. At the same time, there is a marked uptake of amino acids by the liver, apparently transferred from the muscles.

Discoveries in the 1980s demonstrated that the local cells release cytokines that not only mediate the systemic humoral changes, but also result in the cellular influx and activation seen in the acute phase. Thus, resident cells "talk" to mobile cells via these cytokines and directly cause the accumulation of inflammatory cells in the tissues. The acute phase response is part of the overall homeostatic response involving inflammation, tissue repair, and immune regulation. It is a normal process which occurs frequently and has been preserved throughout phylogeny, implying its fundamental role in the survival of the organism. While documentation of the acute phase response in humans has depended on pathological examination, major recent advances have been made in our understanding of the response using animal models of inflammation. [See Cytokines in the Immune Response; Inflammation.]

One model involves the intraperitoneal or intravenous injection of killed bacteria, the bacterial cell wall, or purified lipopolysaccharides, otherwise known as endotoxins. While the purified material is still relatively complex, it represents a standardized way of determining an *in vivo* response to challenge, which can also be used *in vitro* for correlation. Moreover, there are species and strain differences in response to lipopolysaccharides which can be exploited to better define the cellular and molecular mechanisms in inflammation.

A second model involves local injection of a chemical irritant (e.g., turpentine, carrageenan, or silver nitrate) or the deposition in tissue of insoluble agents (e.g., talc or celite). These agents illicit both local and systemic acute major responses and, depending on the species, might establish complicated chronic and secondary adaptive responses. Turpentine appears to cause acute phase responses in all species and strains.

A third model uses direct tissue injury, such as that caused by thermal application or X-irradiation, resulting in tissue necrosis. It is important to recognize that some aspects of the acute phase response can only be studied *in vivo* (e.g., fever or leukocytosis), whereas other aspects, including cellular or molecular mechanisms, can be studied *in vitro*. Techniques such as isolated hepatocyte cultures and chemotaxis assays have shown that specific molecules in circulation during the acute phase response can mimic the *in vivo* response *in vitro*, and they are thereby implicated in the primary response of the body to challenge.

## II. CELLS AND MOLECULES INVOLVED

Historically, the hallmark of acute inflammation in tissues has been the demonstration of extravascular accumulation of fluid and plasma components, the intravascular activation of platelets, and the presence of polymorphonuclear leukocytes in the tissues. Should the response proceed to a more chronic phase, the cellular accumulation involves monocyte/macrophages, lymphocytes, eosinophils and mast cells, and plasma cells. Each of these cells has a primary role in the body's response to challenge, and the progression from an acute to a chronic response is seen as a continuum rather than the responses being distinct entities.

### A. Arachidonic Acid Metabolites

Recent advances in our understanding of the acute phase response have involved the discovery and characterization of the chemical and peptide mediators that the body uses to communicate with and regulate the cellular participants in the response. These molecules include various vasoactive compounds that immediately alter vascular permeability: preformed mediators such as histamine and serotonin; fragments from the activation of the plasma biochemical pathways of inflammation, including the complement components C3a and C5a, bradykinin and other kinin-like factors of the kallikrein pathway, and the products of fibrin degradation of fibrin-split products; elements of phospholipid metabolism, including platelet-activating factor; and products of the metabolism of arachidonic acid, including the prostaglandins and leukotrienes (Table I).

Of the vasoactive compounds, the factors generated by the action of phospholipase $A_2$ or C and released arachidonic acid have immediate and far-reaching consequences in the tissues. The actual mechanism depends on both the specific mediator and the tissue being affected. Most cells of the body, whether activated and mobile leukocytes or activated/stimulated stromal cells, release these mediators. Metabolism of arachidonic acid by cyclooxygenases results in the formation of thromboxane $A_2$ and other prostaglandins, whereas metabolism by lipoxygenases results in the formation of leukotrienes and a family of hydroperoxyeicosatetraenoic and hydroxyeicosatetraenoic acids. The metabolites have both stimulatory and inhibitory effects on the activation of inflammatory and stromal cells. Moreover, depending on the organ involved, major physiological changes result, such as bronchoconstriction and bronchodilation in the lung. Relevant to the acute phase response is the action of leukotriene $B_4$ as a potent chemotactic agent for leukocytes (Table II).

Following tissue injury there is immediate vasoconstriction followed by vasodilation, resulting in in-

### TABLE I

Vasoactive Mediators Involved in the Acute Phase Response

Histamine
Serotonin
Complement components (e.g., C3a and C5a)
Kinins, including bradykinin
Fibrin degradation products
Platelet-activating factor
Prostaglandins
Leukotrienes

### TABLE II

Arachidonic Acid Metabolites in the Acute Phase Response

| Compound | Action |
|---|---|
| Thromboxane $A_2$ | Vasoconstriction |
| Prostaglandins $I_2$, $E_2$, $D_2$, and $F_2\alpha$ | Vasodilation |
| Leukotrienes LTB$_4$ | Chemotaxis of phagocytes |
| LTC$_4$, LTD$_4$, and LTE$_4$ | Smooth muscle contraction |

creased blood flow to the tissues, which is the cause of the redness and warmth associated with the local acute phase response. The most prominent acute feature involves endothelial cell contraction and the opening of the gap junction between the cells, exposing the basement membrane. This increase in vessel permeability, particularly at the postcapillary venules, results in an increase of fluids and plasma components accumulating in the tissue, peaking about 30 min after tissue injury. Within the tissue the accumulation is referred to as edema, while subepithelial accumulation results in blister formation. Sometimes the fluid accumulation coincides with bacterial and leukocyte presence, resulting in the formation of pus. The increased permeability caused by some metabolites, coupled with the chemotactic action of others, represents a coordinated mechanism for the accumulation of leukocytes at tissues involved in the local acute phase response. Pain appears to be associated with the generation of molecules such as bradykinin upon activation of kallikrein action on kininogen during the clotting cascade.

## B. Cytokines

Within the last two decades and more particularly in the latter part of the 1980s, with the advent of molecular cloning techniques, there have been major advances in our understanding of the peptide mediators involved in the acute phase response. These peptides are synthesized and released from inflammatory and stromal cells and are variously referred to as lymphokines, monokines, interferons, growth factors, colony-stimulating factors (CSF), interleukins, or, collectively, cytokines to indicate their primary role in cellular communication. These peptides, many of which are glycosylated and have molecular masses in the 10- to 30-kDa range, are active at the pico- and nanomolar levels. They interact with specific high-affinity (i.e., $K_d = 10^{-9}$ to $10^{-12}$ $M$) receptors on the surface of various cells and elicit metabolic changes in the target cell. In this regard, and despite the fact that they are made and released by multiple dispersed cell types, they might be considered in the same way as other hormones (e.g., insulin or glucagon). Thus, while we now know the identity of many cytokines, we know little about tissue distribution or bioavailability, nor do we know how each signal is received and deciphered by the individual cell, whether simultaneous or sequential. In addition, similar to the endocrine system, there are cascades of networks and cytokines. These too must be deciphered before the full role of the cytokines can be understood in the context of the acute phase response.

Initial work in the mid-1970s identified leukocyte-derived material, called endogenous pyrogen or leukocyte endogenous mediator, which could induce most aspects of the acute phase response when injected into experimental animals. Further work proved that the effects could be induced by a single cloned polypeptide, termed interleukin-1 (IL-1). This cytokine, which exists in two forms, IL-1α and IL-1β, is produced by various cell types, but primarily by activated mononuclear phagocytes. The cytokine is pleiotropic in action, causing many cells to undergo inflammatory changes, most notable of which is the induction of prostaglandin synthesis and the secretion of further cytokines from the target cells. [*See* Phagocytes.]

A second cytokine that figures prominently in the acute phase response is tumor necrosis factor (TNF), which is identical to cachectin, the factor causing decreased lipoprotein lipase activity in endotoxin-treated animals. TNF has many of the same pleiotropic actions of IL-1 but, in addition, can cause the wasting or cachectic response seen in chronic debilitating inflammatory diseases. Like IL-1, TNF causes the release of prostaglandins from various stromal cells and induces the synthesis and release of other cytokines from these cells. These two cytokines appear to be involved with the initiation of the acute phase response, causing some changes directly and others through secondary cytokine induction.

A third cytokine that features prominently in the acute phase response is IL-6. This peptide, first described as a B lymphocyte-stimulating factor and as interferon-β₂, is the major factor eliciting the hepatic acute phase response. IL-6 is also released from activated mononuclear phagocytes, but more important is the fact that IL-6 is induced in fibroblasts and in

epithelial and endothelial cells by IL-1 and TNF. This secondary induction represents a potent recruitment and augmentation of the cytokine signal since there are so many more stromal cells in the body than inflammatory cells. Like IL-1, IL-6 is also pleiotropic in its actions, causing fever as well as stimulating the immune response, in addition to its action on the liver. Moreover, it is a potent thrombopoietin, causing the maturation of megakaryocytes and replenishing the platelet population, which can be used up in the early stages of the acute phase response. Leukemia inhibitory factor (LIF), first described as a T lymphocyte product, but known to be released by monocytes and stromal cells, was shown recently to be an additional cytokine that induces the hepatic acute phase response in a manner similar to that of IL-6.

Two other cytokines that have been identified recently have a major role in the accumulation and activation of neutrophils, monocytes, and lymphocytes. IL-8, or monocyte-derived neutrophil chemotactic factor, has potent *in vitro* and *in vivo* chemotactic activities for lymphocytes and neutrophils. IL-8 and its likely murine homolog macrophage inflammatory peptide 2 are members of a family of structurally related cytokines, which includes platelet factor 4 and β-thromboglobulin, both released from platelets upon aggregation, having chemotactic activity for neutrophils, monocytes, and fibroblasts.

Another chemoattractant cytokine, monocyte chemotactic and activating factor (MCAF), belongs to a second structurally related family of cytokines including the murine macrophage inflammatory peptide 1, which exhibits chemotactic and pyrogenic activities. These two cytokines, IL-8 and MCAF, are released by activated monocytes and, like IL-6, they can be released from fibroblasts and endothelial cells upon stimulation with IL-1 and TNF.

A factor that previously received attention as a transforming factor for nonneoplastic cells and was recognized subsequently as a cytokine with immunomodulatory roles is transforming growth factor β (TGF-β). Initially identified as coming from platelets, it is now recognized as a product of various hematopoietic cells, playing a role in the initiation of the acute phase response. Platelet aggregation leads to the degranulation and release of TGF-β, which has chemotactic activity for monocytes and enhances cytokine generation in these cells, including IL-1, TNF, and other growth factors. [*See* Transforming Growth Factor-β.]

Most of these cytokines and others with similar activity are released initially from activated mononu-

TABLE III

Cytokine Function in the Acute Phase Response

| Cytokine | Functions |
|---|---|
| IL-1α<br>IL-1β<br>IL-6<br>Tumor necrosis factor<br>Leukemia inhibitory factor | Induce fever; raise cortisol levels; induce acute phase; proteins; induce prostaglandins |
| IL-8 and monocyte chemotactic and activating factor<br>Macrophage inflammatory peptides 1 and 2 | Chemotaxis and activation of neutrophils, monocytes, and lymphocytes; induce fever |
| Transforming growth factor β | Chemotaxis and activation of monocytes |
| Granulocyte/macrophage and granulocyte colony-stimulating factors | Induce leukocytosis |

clear phagocytes (Table III). More recently, it has been recognized that the same cytokines are released from activated T lymphocytes, leukocytes, fibroblasts, and other stromal and mesenchymal cells. The biological activities of these cytokines and other related molecules are still being characterized, but we now recognize that there are families of these peptides which appear to be both functionally and structurally related. IL-1 and TNF appear to be distinct in that they can act directly to induce the acute phase response and indirectly by inducing stromal cell production of IL-6, IL-8, MCAF, and others.

## C. Cells

The initial response within the vessels is the activation of platelets, with aggregation and degranulation resulting in clot formation and the release of vasoactive components such as serotonin and thromboxane $A_2$, causing changes in vascular permeability and smooth muscle contraction. In addition, the release of ADP results in further platelet aggregation and activation of coagulation, while the release of platelet-derived growth factor results in the stimulation of cells (e.g., fibroblasts), giving rise to enhanced release of other cytokines, including IL-6 and IL-8. The release of TGF-β by platelets can induce monocyte accumulation and activation, resulting in enhanced release of inflammatory cytokines, including IL-1 and TNF. In turn, further activation and aggregation of platelets can be caused by the release of platelet-activating factor from mononuclear cells after the cells have been activated by trauma or infection, representing an amplification loop of cytokines and cells in the acute phase response. Since mononuclear cells also release

IL-6, a potent thrombopoietin, one begins to see the complexity of the acute phase response, with the same cell releasing factors that both activate and utilize the platelet as well as factors that cause the replacement of these important cells. [See Platelet-Activating Factor.]

Closely following the activation of platelets in the vessels in the accumulation of neutrophils within the involved tissue. We know that chemotactic molecules such as C5a and leukotriene $B_4$ or formyl peptides can directly recruit leukocytes to the tissue. We also know that mononuclear cell-derived cytokines (e.g., IL-1 and TNF) can induce the expression of surface leukocyte adhesion molecules on endothelial cells as well as inducing the synthesis of neutrophil-activating cytokines (e.g., IL-8 and other growth factors) by stromal cells.

The combination of increased expression of intracellular adhesion molecules 1 and 2 and endothelial leukocyte adhesion molecule 1, coupled with the local production of CSFs (granulocyte- and granulocyte/macrophage-CSF) and IL-8, can account for neutrophil margination (adherence to the vessel wall), extravasation, and chemotaxis, as well as leukocytosis from bone marrow activation and inflammatory cell maturation.

Thus, while it remains that a hallmark of acute inflammation is neutrophil accumulation, obviously platelet, mononuclear, and endothelial cell activation likely precede this accumulation, as does the local release of various cytokines with both microenvironmental and systemic effects. The multiple factors that can elicit neutrophil accumulation point to the importance of these cells in the survival of the host. Bacterial infection and monocyte activation lead to neutrophil chemotaxis via IL-8, while the same result would be seen through the release of platelet factor 4 by platelets in the acute phase associated with trauma.

The presence of mononuclear cell accumulation in the tissue has heretofore heralded the chronic stage of inflammation. However, work showing an early flux of monocytes into sites of inflammation, temporally equal to that of neutrophils, coupled with the knowledge that mononuclear phagocytes are the cells most likely releasing the early pulse of inflammatory cytokines, indicates that the tissue macrophages or the peripheral blood monocytes are the cells most likely to initiate the acute phase response. Not only are the macrophages aptly placed within the tissues to encounter pathogens (e.g., the alveolar macrophage), but the release of cytokines such as MCAF and granulocyte/macrophage-CSF at local sites ensures the entry of monocytes into the tissues and the regeneration of adequate numbers of peripheral blood monocytes from the bone marrow. Thus, while the polymorphonuclear neutrophil is a prominent participant in the acute phase response, the mononuclear phagocyte might well be the important initiator of the response.

A number of other cells, normally classed as inflammatory, play significant roles in the initiation or promulgation of the acute phase response. Mast cells and basophils are potent sources of vasoactive agents. Upon activation with anaphylatoxins such as C3a and C5a, these cells release their granular contents. These include histamine and serotonin, certain metabolites of arachidonic acid, including prostaglandin $D_2$ and leukotrienes $C_4$, $D_4$, and $E_4$, which are known as slow-reacting substances of anaphylaxis because of their action in causing the prolonged contraction of smooth muscle. Moreover, the mast cell is a major source of platelet-activating factor, resulting in platelet activation and degranulation, as well as releasing a highly active chemotactic factor for eosinophils. The mast cell and basophil contribute directly to the altered vascular permeability seen in the acute phase response.

## III. PHYSIOLOGICAL MECHANISMS IN ACUTE PHASE RESPONSE

### A. Fever and Leukocytosis

Fever is one of the main physiological changes seen in the acute phase response and is considered a major host defense mechanism. Activation of T lymphocytes by IL-1 is significantly greater at elevated tempera-

tures. T-cell killing of tumor cells and the activity of cytotoxic T cells are also enhanced. Moreover, fever reduces the replication of pathogenic organisms and leads to enhanced antibody formation by B cells. The temperature of the body is elevated by having the temperature set point in the hypothalamus raised above the normal range. The elevation is achieved through the local generation of arachidonate metabolites, particularly prostaglandin $E_2$, within the hypothalamus. [See Hypothalamus.]

A number of endogenous pyrogens, the most prominent of which is IL-1, can cause the induction of fever. Both forms of this cytokine induce a significant increase in body temperature, which is blocked by antipyretics such as aspirin. TNF is a pyrogenic cytokine with about the same potency as IL-1 and interferon-$\alpha$, which has been administered clinically to humans, and induces fever directly via the induction of brain prostaglandins. Other recently described factors, such as the family of mouse macrophage inflammatory peptides 1 and 2, induce fever by both prostaglandin-dependent and -independent mechanisms. Obviously, the activation of mononuclear phagocytes can result in the release of several of these endogenous pyrogens, and indirect activation of other cells by cytokine cascades can either augment these signals or deliver others in concert to the brain. The result is the febrile response caused by the same cytokines that initiate other aspects of the acute phase response.

Leukocytosis is the increase in total and relative numbers of circulating neutrophils, both mature and those undergoing maturation, or "band forms." The mechanism involves the action of CSFs, including granulocyte- and granulocyte/macrophage-CSF, on the proliferation and differentiation of bone marrow precursors for neutrophils, resulting in the accelerated release of neutrophils from the bone marrow. The precursors might circulate during maturation, giving rise to the appearance of band forms. Other forms of myeloid cell increases occur under certain circumstances (e.g., eosinophilia or mastocytosis, seen in some parasitic diseases). These too are caused by the release of different cytokines from inflammatory and stromal cells involved in the acute response to the invading organisms.

### B. Acute Phase Proteins

One of the more striking systemic changes that occur during the acute phase response is a rapid dramatic alteration in the plasma concentration of a series of proteins, commonly called acute phase proteins, which are primarily made by the liver. The changes appear to be independent of the cause of inflammation, whether it is trauma or burns; viral, bacterial, or parasitic infection; or tissue necrosis. The hepatic response is the most well studied and well understood aspect of the acute phase response. The fact that this reaction has been strongly conserved through evolution in vertebrate species indicate that this response plays a major and primary role in systemic homeostasis.

Cytokines involved in the other aspects of the acute phase response are also those involved with the alteration of liver metabolism. In considering the cytokines, which should be classed as hepatocyte-stimulating cytokines, we restrict ourselves to considering those molecules which interact specifically with hepatocytes through their own receptors and induce acute phase protein gene changes similar to those seen *in vivo* during inflammation.

The many acute phase proteins derived from the liver differ in their physicochemical properties and functional activities. Most of them are glycoproteins, and the kinetics and magnitude of plasma concentration changes are characteristic for each protein. An examination of their known physiological roles (Table IV) indicates that acute phase proteins can both mediate and be consumed by the inflammatory process. Two proteins in humans—C-reactive protein and serum amyloid A protein—rise up to several hundredfold from undetectable levels, whereas others, including fibrinogen, haptoglobin, and $\alpha_1$-acid glycoprotein, rise to four to five times the normal level. Other proteins, including albumin, transferrin, and $\alpha$- and $\beta$-lipoproteins, decrease in circulation during the acute phase response.

The physiological role of some acute phase proteins is well recognized. Complement C3 and C-reactive protein opsonize bacteria, immune complexes, and foreign particles, whereas fibrinogen is involved in blood coagulation and is consumed in the homeostatic response to trauma and therefore must be rapidly replaced. Other acute phase proteins play a major role as inhibitors of various serine and cysteine proteinases, most notably those released from activated neutrophils. $\alpha_1$-Proteinase inhibitor ($\alpha_1$-antitrypsin) specifically inhibits neutrophil elastase, whereas $\alpha_1$-antichymotrypsin inhibits leukocyte cathepsin G. The general nature of the acute phase protein response is that of controlling proteolytic activity and inhibiting the trauma and destruction associated with inflammation. In additon to mediating homeostatic responses, a number of the proteinase inhibitors also modulate immune reactions, at least *in vitro*. $\alpha_1$-Proteinase inhibitor and $\alpha_1$-chymotrypsin modify the activity of natural killer cells and antibody-dependent cell-mediated cytotoxicity. Complexes between inhibitors and specific proteinases can suppress macrophage activation and Ia antigen expression. [*See* Natural Killer and Other Effector Cells.]

One of the major acute phase proteins in humans, $\alpha_1$-acid glycoprotein, previously called orosomucoid, might also be an immunomodulatory molecule, although data supporting this contention are less solid than for the proteinase inhibitors. C-reactive protein, while appearing to have immunomodulatory activity,

TABLE IV

Physiological Role of Acute Phase Proteins

| Protein | Cytokine induction | Function |
|---|---|---|
| $\alpha_1$-Proteinase inhibitor | IL-6/LIF | Inhibition of proteinases |
| $\alpha_1$-Antichymotrypsin | IL-6/LIF | Inhibition of proteinases |
| Fibrinogen | IL-6/LIF | Blood clotting |
| Complement C3 | IL-1/TNF, IL-6/LIF | Opsonization |
| C-reaction protein | IL-1/TNF, IL-6/LIF | Opsonization |
| Serum amyloid A | IL-1/TNF, IL-6/LIF | ? |
| Serum amyloid P | IL-1/TNF, IL-6/LIF | ? |
| $\alpha_1$-Acid glycoprotein | IL-1/TNF, IL-6/LIF | Transport |
| Haptoglobin | IL-6/LIF | Binds hemoglobin |
| Hemopexin | IL-1/TNF, IL-6/LIF | Binds heme |
| Ceruloplasmin | IL-6 | Oxygen scavenger, transport |

as well as behaving as a scavenger and opsonizing factor for chromatin fragments released from damaged cells, is one of the acute phase proteins whose role and main physiological function remain undescribed.

By examining the overall acute phase protein response, it appears that the role of this response is to control the random destruction associated with inflammation and to mediate tissue repair and the return to normal function. In addition, there might be involvement in mediating the subsequent immune response, and, as such, the proteins play a broad protective role during the early stages of the host response to invasion prior to initiation of the specific immune response.

The great majority of the acute phase proteins are primarily synthesized by the hepatocyte. However, there is recent evidence that a number of these proteins, particularly the antiproteinases, might be synthesized by mononuclear cells. The control of synthesis within this extrahepatic population is under the same cytokine involvement as the liver, and while the level of synthesis is much lower than that of the hepatocyte, this mononuclear cell synthesis could be important in modifying the microenvironment around cells involved in activation or immune regulation.

Initially, proinflammatory cytokines such as IL-1 and TNF were thought to regulate the hepatic response since, when one injected the purified or recombinant material into an animal, plasma changes in the acute phase proteins were seen. However, with the availability of cloned and purified material and *in vitro* hepatocyte culture systems, other cytokines have been identified which appear to be the major regulators of acute phase protein gene expression.

## 1. IL-6

IL-6 was identified as causing increased fibrinogen synthesis in cultures of isolated rat hepatocytes and is the factor present in monocyte-conditioned media which causes major regulation of all of the acute phase proteins. The function of IL-6 *in vivo* or *in vitro* is augmented by corticosteroid, consistent with the known presence of raised levels of cortisol during the acute phase response. While IL-6 affects most of the acute phase proteins, when combined with other cytokines, its affect on some of the genes is modulated. IL-6 acts directly on the hepatocyte through the IL-6 receptor and modifies gene transcription and concentration of the appropriate mRNA species for the acute phase proteins. Here, we should note that hepatocytes undergoing an acute phase response do not express IL-6 themselves, but must have this cytokine delivered to them through the circulation. This action is reminiscent of endocrine hormone function.

## 2. LIF

LIF is a second molecule which regulates the expression of most of the acute phase proteins in the liver. Like IL-6, it causes transcriptional activation of the acute phase protein genes, yet works via a separate receptor on the hepatocyte membrane; moreover, the LIF RNA message is not present in hepatocytes undergoing acute phase responses, indicating that LIF must be delivered exogenously to the liver. While LIF and IL-6 act on the same sets of genes, but through different receptors, they can be additive in stimulating the acute phase proteins and each does not inhibit the action of the other.

## 3. IL-1/TNF

Both IL-1 and TNF, when injected *in vivo*, cause the induction of fibrinogen as an acute phase protein. However, when tested on isolated hepatocyte cultures, neither of these cytokines induces the expression of fibrinogen at either the mRNA or protein level. IL-1, either the $\alpha$ or $\beta$ form, and TNF cause the stimulation of complement components factor B and C3 as well as $\alpha_1$-acid glycoprotein in human hepatocytes and hepatoma cells. In addition, both cytokines suppress the synthesis of albumin and contribute to both the positive and the negative regulation of acute phase protein genes.

Contrary to the *in vivo* findings, particularly on the synthesis of fibrinogen, IL-1 and, to a lesser extent, TNF have a negative effect on the expression of fibrinogen *in vitro* and, indeed, when administered to hepatoma cells in the presence of IL-6, modulate the up-regulation caused by IL-6 and thereby effectively decrease the expected stimulation. Once again, the hepatocyte expresses receptors for IL-1 and TNF that are different from those for IL-6 and LIF.

## 4. Corticosteroid

One of the changes that have been recognized for some time as occurring in the plasma during the acute phase is an early and rapid rise in the level of cortisol after the onset of inflammation. While glucocorticoid has an expected negative and down-regulatory role to play in inflammatory cell activation, with respect to the liver, corticosteroid is a necessary component of the acute phase protein response. Corticosteroid does not exhibit hepatocyte-stimulating activity on its own; however, when administered in conjunction with IL-6 or the other cytokines, there is a synergistic effect on the induction of the acute phase protein genes. Similar to the findings with the other cytokines, most acute phase protein genes (e.g., $\alpha_1$-acid glycoprotein and haptoglobin) have glucocorticoid regulatory *cis*-acting elements in the 5′ region of the gene that are distinct from those in the regions coding for cytokine induction or for liver cell expression.

It is also noteworthy that two of the cytokines (i.e., IL-1 and IL-6) have been shown *in vivo* and *in vitro* to stimulate the adrenal–pituitary axis, resulting in an increase in the circulating level of corticosteroids. Thus, the factors that stimulate the liver, yet require raised corticosteroid levels for their full function, are also capable of inducing raised corticosteroid levels.

What started out as a relatively simplified view of leukocyte endogenous mediator or IL-1's being responsible for all of the induction of the acute phase is now considerably complex. Four major cytokines—IL-6, LIF, IL-1, and TNF—along with corticosteroid, appear to be able to account for all of the acute phase protein induction seen in mammalian systems. Other modulating cytokines occur, including interferon-$\gamma$ and epidermal growth factor, as well as TGF $\beta$. Each of these can modify the regulation caused by IL-6, but overriding stimulation of acute phase protein genes appears to be mediated by the primary cytokines. Many of these modulatory cytokines have not been demonstrated to be relevant *in vivo*. Situations involving the rapid release of IL-6 and IL-1 or TNF are likely to show a different pattern of plasma changes than situations involving the release of IL-1 and IL-6 along with raised levels of epidermal growth factor and TGF-$\beta$ or interferon-$\gamma$. This differential activity in modulation of the major stimulation could account for the variation that is seen in acute phase protein levels in different disease states.

Many of the cytokines that mediate acute phase protein genes also modify other aspects of the liver during the acute phase. Disturbances in carbohydrate metabolism, resulting in changes in the glycosylation pattern of many acute phase proteins, are due to alterations in the levels of enzyme involved in glycosyl transfer (e.g., sialyltransferase). Moreover, the normal hormonal control of hepatic glucose homeostasis is interrupted, resulting in hypo- and hyperglycemia, seen after severe trauma or septicemia. IL-6 appears to mediate the expression of phosphoenolpyruvate carboxykinase, a major enzyme in the gluconeogenic pathway. Other liver metabolic enzymes (e.g., catalase and fatty acid synthetase) are decreased in the acute phase; however, whether this decrease is mediated

by cytokines is not known. Concomitant with the increase in acute phase protein synthesis, the liver increases its uptake of precursor amino acids consistent with the increased protein being synthesized. This uptake can be directly stimulated by IL-6 as well as by other hormones (e.g., glucagon) and can be modified by other mediators (e.g., TGF-$\beta$ or epidermal growth factor).

In understanding how the various cytokines mediate the acute phase protein response, one should recognize that there are patterns of response. Each cytokine has a specific receptor on the hepatocyte, and mixtures of the cytokines result in different patterns of gene regulation from that caused by an independent cytokine. The sequence and kinetics of cytokine interaction at the hepatocyte, or "syntax," are important in determining the final outcome of acute phase protein gene regulation. The acute phase proteins can be separated into two main groups: one set includes the opsonins and transport proteins, which are stimulated by both IL-6 and IL-1, and a second set includes all of the antiproteinases and fibrinogen, which are stimulated only by IL-6 or LIF (Table IV). Combinations of cytokines result in stimulation that is both additive and synergistic, as in $\alpha_1$-acid glycoprotein, or inhibitory, as seen with fibrinogen. The secretion of acute phase proteins and their accumulation in plasma is thus the result of the liver receiving a complex mixture of signals derived from inflammatory tissue, and the effect and specific protein increase depend on the syntax of the message received by the liver.

## C. Cytokine and Cellular Interactions

In attempting to understand the coordinated aspects of the acute phase response, one must examine what goes on at the site of inflammation within the tissue or the vessel. The mononuclear phagocyte is the most likely cell involved in the initiation of the acute phase response. This cell, upon activation, can release all of the cytokines so far described as being involved in the acute phase response as well as many of the smaller mediators. Moreover, IL-1 and TNF are potent stimulators of the release of IL-6, IL-8, and MCAF, as well as other cytokines, from stromal cells such as fibroblasts and endothelial cells. These two cytokines stimulate stromal cells to release arachidonate metabolites, and it is likely that this activation and recruitment of the adjacent stromal cell population by locally released IL-1/TNF give rise to an augmented presence of IL-6, IL-8, MCAF, etc., causing a rapid rise in the circulating levels of these important mediators. Certainly, TNF and IL-6 as well as LIF have been detected in the circulation at appreciable levels after the induction of experimental inflammation, both in animals and in humans. IL-1 has also been found, but at somewhat lower levels. The IL-1 and IL-6 released at the tissue site are important in mediating the febrile response and in causing the raised cortisol levels associated with the acute phase response. The release of IL-8 and MCAF at the tissue site, along with other chemotaxins, explains the accumulation of neutrophils and monocytes associated with the acute phase response. Instead of acting simply as a phagocytic and clearing cell, the mononuclear phagocyte is a regulatory cell, recruiting stromal cells to participate in the overall induction of the acute phase response, resulting in changing levels of cytokines and vasoactive components and giving rise to altered vessel permeability, altered protein metabolism, altered carbohydrate metabolism, fever, and leukocytosis, all associated with the acute phase response.

Since many of the cytokines have short half-lives and the mRNAs for these mediators are short-lived, the acute phase response is a self-limiting response unless continual stimulation by a persistent pathogen leads to the chronic phase. There are, however, inhibitors of some cytokines that have been described, but they are not well documented at this time.

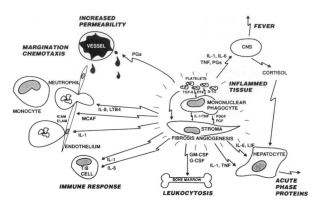

**FIGURE 1**    Cell and mediator interaction in the acute phase response. Mediators released from the local inflamed tissue interact with multiple tissues and organs to induce the acute phase response. Platelets, through mediators such as TGF-$\beta$, platelet factor 4 (PF$_4$), and $\beta$-thromboglobulin ($\beta$-TG), and mononuclear phagocytes, through cytokines IL-1, TNF, platelet-derived growth factor (PDGF), and fibroblast growth factor (FGF), involve and activate the local stromal cells in the inflamed tissue. The tissue-derived mediators effect other systemic responses. PGs mediate vascular permeability. IL-1, TNF, IL-6, and PGs affect the brain and induce fever as well as the synthesis of corticosteroids. IL-1, IL-6, TNF, and LIF along with corticol mediate the hepatic acute phase response. Granulocyte/macrophage- and granulocyte-colony stimulating factors (GM- and G-CSF, respectively) induce leukocytosis. IL-1 and IL-6 induce B- and T-cell activation and the immune response. IL-1 activates endothelial cells and causes the expression of intracellular adhesion molecules (ICAM) and endothelial leukocyte adhesion molecules (ELAM), and IL-8, leukotriene B$_4$ (LTB$_4$), and MCAF cause leukocyte margination and chemotaxis. CNS, central nervous system.

## IV. CONCLUDING REMARKS

Subsequent to invasion of the body by infectious organisms or activation by trauma, there is a series of cascading events involving cells, both inflammatory and stromal cells, and a family of acute phase cytokine hormones with multipotent and overlapping biological activities (Fig. 1). Within the tissue undergoing challenge, the cells and the cytokines interact along with vasoactive mediators to allow vessel permeability and the influx of inflammatory cells. The same signals act systemically at various target tissues, including the hypothalamus, to mediate fever; the bone marrow, to mediate leukocytosis; and the liver, to mediate acute phase protein synthesis. Interruption of the acute phase response pharmacologically can occur at various stages. Drugs such as aspirin and indomethacin can decrease the synthesis of prostaglandins and interrupt the febrile response. More potent drugs such as steroids have a broader spectrum of activity, inhibiting the activation of cells and the generation of the various cytokines.

It is clear that the acute phase response is an early protective response, controlled by multiple overlapping signals, which ensure that the response is swift and complete, since it is this early response that provides us with our main protection before the more sophisticated and specific immune response can play a role.

## BIBLIOGRAPHY

Beutler, B., and Cerami, A. (1988). Tumor necrosis, cachexia, shock and inflammation: A common mediator. *Annu. Rev. Biochem.* 57, 505.

Dinarello, C. A. (1986). Interleukin-1: Amino acid sequences, multiple biological activities and comparison with tumor necrosis factor (cachectin). *Year Immunol.* 2, 68.

Dinarello, C. A., Cannon, J. G., and Wolff, S. M. (1988). New concepts on the pathogenesis of fever. *Rev. Infect. Dis.* 10, 168, 34–64.

Fantone, J. C., and Ward, P. A. (1995). Inflammation in pathology. *In* "Essential Pathology" (E. Rubin and J. L. Farber, eds.). Lippincott, Philadelphia, PA.

Gauldie, J., and Baumann, H. (1990). Cytokines and acute phase protein expression. *In* "Cytokines in Inflammation" (E. H. Kimball, ed.). Telford, Caldwell, NJ.

Koj, A. (1985). Definition and classification of acute-phase proteins. *In* "The Acute-Phase Response to Injury and Infection" (A. H. Gordon and A. Koj, eds.), pp. 139–144. Elsevier, Amsterdam.

Mantovani, A., and Dejana, E. (1989). Cytokines as communication signals between leukocytes and endothelial cells. *Immunol. Today* **10**, 370.

Matsushima, K., and Oppenheim, J. J. (1989). Interleukin 8 and MCAF: Inflammatory cytokines inducible by IL-1 and TNF. *Cytokine* **1**, 2.

Wahl, S. M., McCartney-Francis, N., and Mergenhagen, S. E. (1989). Inflammatory and immunomodulatory roles of TGF-$\beta$. *Immunol. Today* **10**, 258.

Wolpe, S. D., and Cerami, A. (1989). Macrophage inflammatory proteins 1 and 2: Members of a novel superfamily of cytokines. *FASEB J.* **3**, 2565.

# Adaptational Physiology

C. LADD PROSSER

*University of Illinois at Urbana–Champaign*

## GLOSSARY

**Acclimation** Compensation for single environmental factors

**Acclimatization** Compensation to multiple factors such as seasonal

**Capacity adaptation** Normal functioning in slightly altered environments

**Conformity** Internal state same or similar to external state

**Ectothermy** Body temperature dependent on external heat production

**Endothermy** Body temperature dependent on internal heat production

**Environmental factors** Physical, biotic

**Heterotherms** Variable temperature of body regions and at different times

**Homeokinesis** Constancy of rate functions, especially energy production

**Homeostasis** Constancy of internal state

**Homeotherms** Constant body temperature in varying environment

**Human adaptation** Genetic and cultural

**Physiological adaptation** Functions permitting normal life processes in stressful environments

**Physiological species** Populations isolated by adaptations

**Poikilotherms** Body temperature varies with environment

**Regulation** Internal state maintained constant in varying environment

**Resistance adaptation** Functions at limits of environment

THE WORD ADAPTATION HAS SEVERAL MEANINGS according to context. In evolutionary biology, adaptation applies to morphological modifications; for example, from skeletal elements in fossils inferences are drawn concerning posture and modes of life. Sensory physiology uses adaptation to refer to accommodation of sensory responses to repeated stimulation. For general physiology, adaptation refers to the sum total of functions that favor normal life processes in a delineated environment.

The determinants of all animals are environmental, genetic, and developmental. Each of these three factors contributes to adaptation. This article is mainly concerned with environmental influences.

## I. CLASSES OF ADAPTATIONAL PATTERNS

Every organism is impinged upon by a variety of physical factors: temperature, oxygen, ions, water, nutrients, light, and mechanical and electrical forces. Every organism is also influenced by biotic factors: food, predators, prey, and conspecific individuals including mates and offspring. Biochemical, anatomical, and physiological adaptations to specific environmental factors have been described at cellular and molecular levels; interactions among environmental factors add complexity to the analyses. Adaptations at cellular levels are integrated by nervous and hormonal actions that provide for adaptations of whole organisms. More is known about cellular adaptations to single factors than to multiple factors, and more is known about cellular effects than about the integrated responses of whole organisms.

Two general classes of adaptations are (1) capacity adaptations, which provide for maintenance of normal function in response to stresses of an altered envi-

ENCYCLOPEDIA OF HUMAN BIOLOGY, Second Edition, VOLUME I.    Copyright © 1997 by Academic Press.    All rights of reproduction in any form reserved.

ronment, and (2) resistance adaptations, which provide for function, survival, and reproduction at environmental extremes. The two patterns merge according to environmental change, but the mechanisms of capacity and resistance adaptation are generally different.

Capacity adaptations include (a) internal states such as body temperature, osmotic and ionic balance, blood concentrations of sugar, and other metabolites, the so-called *milieu interior* of Claude Bernard, and (b) rate functions such as heart rate, respiration rate, and rate of energy production.

Two patterns of internal state as a function of external state are conformity and regulation (Fig. 1). In conformers, the internal state is similar to the external state; in regulators, the internal state is maintained at relative constancy. For conformers, the prefix *poikilo-* applies, as in poikilothermic and poikilo-osmotic. In conformers, the internal state rises or falls in parallel with the environment, but the internal state may be separated from the external state by a constant amount. As shown in Fig. 1 ($A_1$ and $A_2$), the range of normal function can be shifted by acclimation toward higher or lower levels. For temperature, body temperature rises or falls with ambient temperature; for oxygen, the measure of conformity is the level of metabolism that rises or falls with $P_{O_2}$.

In regulators, the internal state is maintained constant over a range of environmental variation. For regulators, the prefix *homeo-* applies, as in homeothermic and homeo-osmotic. Some animals may be homeo- in one environmental range and poikilo- in another range. At the ends of a midrange, mechanisms of regulation break down (dotted lines in Fig. 1). In general, conformers vary over a wider range of internal state, whereas in regulators the internal state varies little while the environment varies considerably.

The various patterns of conformity and regulation are well illustrated for osmotic adaptations of aquatic animals. Figure 2 shows the patterns of conformity (D), of regulation over a wide range (B), and of regulation in low osmotic concentrations (A) and in high concentrations (E).

## II. TEMPERATURE ADAPTATION

Special terms apply to temperature. Poikilotherms vary in body temperature ($T_B$) with environmental temperature ($T_A$) whereas homeotherms maintain constant $T_B$ in varying $T_A$. Heterotherms are intermediate with body temperature varying temporally (e.g., diurnally or seasonally) or spatially in different regions of the body as between skin and liver. Ectothermy and endothermy refer to the source of heat—external as by sunlight or internal as by metabolism. Therefore, ectothermy and endothermy are not synonymous with poikilothermy and homeothermy. [*See* Body Temperature and Its Regulation.]

Temperature regulation is maintained by a variety of mechanisms, which vary in relation to body size

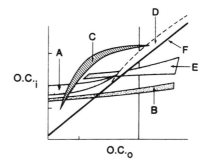

**FIGURE 2** Several patterns of internal osmoticity ($O.C._i$) as a function of osmoticity of the medium ($O.C._0$). (A) Strong hyperosmoregulators live in fresh water and are limited in capacity to live in brackish water; (B) animals that are hyperosmotic in fresh water and hypo-osmotic in seawater (e.g., euryhaline fish); (C) weak hypo-osmotic regulators live in estuaries and do poorly in seawater (e.g., estuarine crabs); (D) osmoconformers are marine invertebrates that maintain slight hypertonicity above and below seawater concentration; (E) strong hyperosmotic regulators at low external concentrations and hypo-osmotic regulators at high external concentrations (e.g., terrestrial crabs); and (F) line of osmotic equality. (From Prosser, Adaptational Biology, 1986.)

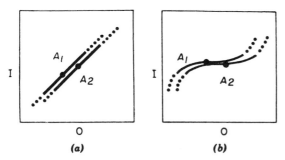

**FIGURE 1** Internal state (I) as a function of the same parameter outside (O) and inside the organism. Patterns for conformers (a) and regulators (b). $A_1$ and $A_2$ refer to two states of acclimation. Solid lines are ranges for normal function; dotted lines are ranges for unstable or limiting states. (Reproduced with permission from Comparative Animal Physiology, Vol. III, 1973.)

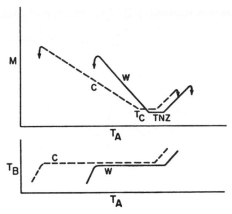

FIGURE 3   Changes in metabolism (M) and body temperature ($T_B$) of homeothermic animals at different ambient temperatures ($T_A$). TNZ is a thermoneutral zone of minimum metabolism. $T_c$ is critical temperature below which M increases; M also rises above TNZ. Curves for cold (C) and warm (W) acclimated animals at different $T_A$. Body temperature is maintained by vasomotor and metabolic reactions in midregion; regulation fails at both high and low extremes. (From Prosser, Adaptational Biology, 1986.)

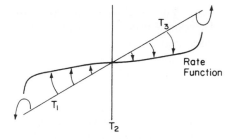

FIGURE 4   A reaction rate of a poikilothermic animal measured at three temperatures: $T_1$, $T_2$, and $T_3$. The thin line represents direct effect of temperature change. Arrows indicate compensation (rise in rate or fall in rate) during acclimation to low or high temperatures.

and habitat (Fig. 3). One mode of regulation is behavioral (i.e., seeking an optimal external temperature); another is circulatory-metabolic (i.e., metabolic responses, insulation, blood flow, and metabolic rate).

Constancy of internal state is homeostasis (e.g., constancy of body temperature, blood sugar, arterial $P_{O_2}$, blood sodium concentration). Homeostasis is characteristic not only of blood but also of cellular composition. For example, intracellular potassium concentration is extremely constant in variable blood potassium. [See pH Homeostasis.]

Constancy of energy production is homeokinesis. Many animals with variable body temperature (conformity) compensate energetically for changes in $T_B$. Homeokinesis or constant energy production can occur in the absence of homeostasis. Both homeostasis and homeokinesis are adaptive.

## III. TIME COURSE OF ADAPTATION

### A. Direct Responses

Adaptations can occur during each of three time courses. The first course is a direct response or immediate reactions to environmental change. Morphological responses can be changes in fur or feathers associated with warmth or cold. Poikilo-osmotic crabs and molluscs change serum concentrations of organic solutes more than inorganic ions.

## B. Acclimatory Changes

The second period is that of days or weeks. Acclimation occurs when single environmental parameters change, as in the laboratory. Acclimatization refers to changes with multiple factor alterations such as seasonal or geographical. A laboratory rat acclimates differently to cold than does a wild rat to winter conditions.

Acclimation in poikilotherms is compensation for temperature such that some constancy of functional capacity is ultimately attained. Energy-yielding reactions increase when cold and decrease when warm in compensation for changes in temperature (Fig. 4). Several patterns of acclimation have been identified (Fig. 5). The pattern of acclimation to temperature varies with measured rate function, with tissue, and

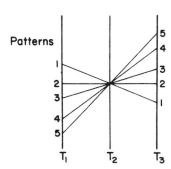

FIGURE 5   Patterns of rate functions of poikilotherms during temperature acclimation. $T_2$ is intermediate temperature; $T_1$ is lower, and $T_3$ is higher temperature. Rates shown after acclimation; pattern 4 shows no acclimation, only $Q_{10}$ effect; pattern 2 shows perfect acclimation with rate the same at each temperature after acclimation; pattern 3 shows partial acclimation; pattern 1 shows overcompensation; pattern 5 shows inverse or negative acclimation. (From Compar. An. Physiol. III, 1973.)

with species. Figure 5 shows that no compensation (pattern 4), complete compensation (pattern 2), or, most common, partial compensation (pattern 3) may occur. A few examples of overcompensation (pattern 1) and of inverse or paradoxical acclimation (pattern 5) are known, especially for hydrolytic reactions. Cellular mechanisms of acclimatory compensation are increases or decreases in enzyme activities, changes in amounts of specific proteins or of total cellular proteins, and changes in the proportion of saturated and unsaturated fatty acids in membrane lipids. Acclimatory changes occur only within genetically determined limits.

## C. Long-Term Selection of Genetic Changes

The third period for adaptation extends over long times—generations—by selection of genetic mutants and establishment of local races and varieties. In geographic clines and circles, the differences between terminal populations are highly adaptive and may result in differences at the species level.

Examples of molecular differences that are genetic include changes in the abundance of specific allozymes as distinguished by kinetic properties. Metabolic enzymes of deep-sea fish do not function at a pressure of one atmosphere. Selection of secondary and tertiary structures results in adaptation. The $Q_{10}$, or temperature dependence, of homeothermic enzymes is steeper than that of poikilotherms. Many examples of selection of morphological characters are well known, e.g., use of insulating fat in cetaceans and of hair in ungulates.

Interesting differences in adaptation to heat in mammals relate to body size. Figure 6 diagrams the heat exchanges and thermal responses of mammals of the size of dog or human. Large animals such as elephants and camels in a warm climate store heat subcutaneously with resulting constancy of temperature of deep organs; small mammals such as rodents with large surface–volume ratios rely more on vasomotor changes. Heat gain or loss varies with circulation in exposed structures, as in ears of rabbits. The fat of lower legs and feet of reindeer has lower melting points than internal body fat.

## IV. ADAPTATIONS AS KEY TO EVOLUTION

Adaptation is the key to understanding evolution. The word species has several meanings. The taxonomic

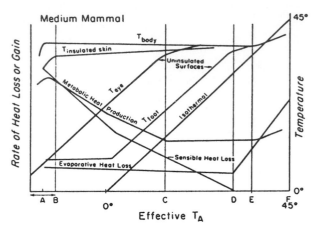

**FIGURE 6** Diagram of thermal responses and heat exchanges of a medium-size mammal, such as a dog or human, as a function of different effective $T_A$. Temperature of different regions: deep body, insulated skin (under fur), eye, foot. A–B, zone if cold at which regulation fails; C, lower critical temperature below which metabolic heat production increases; C–D, zone of evaporative control of $T_B$; E, Upper failure point at which $T_B$ begins to rise as does metabolism; F, upper survival limit approximately 45°C. (From Compar. An. Physiol. IV, 1990.)

species is that determined by the museum systematist; it is essential for physiologists to know the proper names of animals and plants on which they experiment. The cladistic species is determined by some arbitrary number of differences, usually computer-based. The biological species applies to kinds of animals (or plants) that are reproductively isolated populations between which *no* gene exchange can occur. The geographic species applies to organisms that reproduce parthenogenetically or by occurrence of hybrids. The physiological species is based on unique adaptations to ecological niche and geographic range. This is the species set off by adaptation and subsequent selection. In some animals, variation in adaptations leading to speciation is rapid (e.g., in field mice); in other animals, adaptive variation is very slow (e.g., in horse-shoe crabs).

## V. HUMAN ADAPTATION: GENETIC AND CULTURAL

The preceding sections summarize the principles of biological adaptation. Undoubtedly, human evolution has occurred by selection of many inherited characters, and humans in different environments can vary within genetically determined limits in critical characters much as can populations of other animals. Exam-

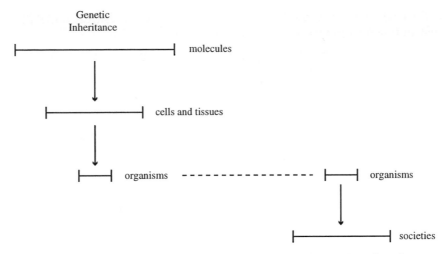

**FIGURE 7**  Ranges of genetically transmitted functional limits of molecules, cells and tissues, and organisms showing decreasing ranges with integration and of culturally transmitted functional ranges showing increasing ranges in complex social systems.

ples of structural adaptations include skeletal elements that allow for bipedialism with resulting freedom of use of forelimbs. Human metabolic adaptations allow for an omnivorous diet: enzymes for processing all classes of foods (except cellulose and related polysaccharides). In respiration and circulation, there are adaptations for airbreathing even at reduced $O_2$ levels and for vasomotor responses to physical stresses (exercise, altitude). The adaptations of the endocrines permit life under highly complex conditions. An important biological adaptation of humans is the long prepubertal period that allows extensive physiological development before reproduction is possible. The most distinctive biological features of humans is the speech area of the cerebral cortex. Development of Broca's and Wernicke's areas permits a variety of meaningful vocal communications not possible in subhumans. Genetic differences determine individuality (blood types, eye color, temperament); genetic differences are also ethnic (skin color, body size). However, despite genetic differences, all humans constitute a single species in that they can interbreed. Humans evolved in central Africa where a coat of hair would likely be a liability. However, use of animals skins as protective covering has allowed extension of range without extensive physiological acclimation of temperature tolerance. [*See* Evolution, Human.]

In parallel with the genetically based adaptations, humans have undergone social evolution, and adaptations based on cultural characters are fully as im-

portant as the biological adaptations. Many kinds of animals live in social groups, and behavior of these social groups has often been interpreted in anthropomorphic terms. Individuals in a colony of Hymenoptera perform different functions, and colonies of bees and termites have been called superorganisms. The social behavior of many birds and animals that live in colonies has a genetic basis with limited adaptive change during development and learning. Cultural evolution in humans is transmitted not genetically but by behavioral transmission from generation to generation. By selection of cultural traits, social adaptations have evolved.

A few examples of culturally transmitted adaptations are (1) societies for common functions—altruism, rules of conduct (moral behavior); (2) language that goes beyond the symbolic use of sound to abstract uses of words; (3) technology, including communal preparation of food, agriculture, and development of machines; (4) representation of nature and human moods in art, music, and dance; (5) belief in superhuman forces to explain events not explicable by human experience; and (6) population dispersal, migration, and exploration of the world, unlike migration for reproduction.

Cultural inheritance adds a new dimension to biological inheritance. In general, the range of genetically determined functions at environmental extremes is wide for molecules—proteins and lipids. The range is narrower when molecules are organized in cells and tissues and is still narrower for integrated whole organisms (Fig. 7). The characters of cultural inheritance widen the range of function and tolerance, i.e., reverse the trend of biological inheritance.

In its broadest context, adaptation to physical and biotic environment gives a basis for understanding biological evolution and also for the development of human civilization.

## BIBLIOGRAPHY

Prosser, C. L. (ed.) (1951, 1962, 1973). "Comparative Animal Physiology," 1st–3rd Eds. Saunders, Philadelphia. 4th Ed. 1991, Wiley-Liss, New York.

Prosser, C. L. (1986). "Adaptational Biology: Molecules to Organisms." Wiley, New York.

# Adaptation and Human Geographic Variation

ROBERTA HALL
*Oregon State University*

## GLOSSARY

**Adaptation** Change in an organism that allows it to function more effectively in a new or altered environment; it may be a short-term effect or may be developmental, affecting children's growth; genetic adaptation, involving natural selection, may apply to part of the population or the entire species

**Clinal distribution** Mapped gradient showing geographic variation of a physical or genetic trait; data points of a cline consist of values of metric traits (or allele frequencies) of local populations in specific geographic areas

**Cline** Gradient of morphological, physiological, or biochemical change exhibited by a group of related organisms usually along a line of environmental or geographic transition

**Ecogeographic rule** Principle that body size and shape of a species within its geographic range will follow thermoregulatory principles

**Local population** Group of people who live in the same area, interbreed, and share certain cultural features; an operationally defined category

**Major continental or geographic race** Collection of local populations that live within a continent or a defined part of a continent and are assumed to have ancestors in common

**Nasal index** Width of the broadest part of the nose at its base expressed as a proportion of the maximum height of the nose, multiplied by 100

**Regional continuity theory** Theory that anatomically modern humans evolved from *Homo erectus* populations in Africa, Asia, and Europe, with a great deal of intracontinent continuity and some genetic exchange between continental populations at borders

**Replacement theory** Theory that anatomically modern *Homo sapiens* evolved in one area within the last 200,000 years and dispersed to other areas, where it replaced previous human species (e.g., *Homo erectus*)

**Secular increase in size** General increase in body size that is secular (i.e., linear and continuing, as contrasted with random fluctuations around an average); it has occurred in industrialized countries in the past 150 years

**Thermoregulation** Principle that relationships between body surface and body mass exist to regulate the flow of heat, that is, to maximize heat retention in cold environments and heat dissipation in hot environments

## I. INTRODUCTION

Studies of human geographic variation address the biology of variation among humans worldwide within an evolutionary framework that proposes climatic, ecological hypotheses to explain observed distributions. The major emphasis in this article is on observable physical characteristics, although genetic variation is discussed briefly. The discussion emphasizes inherited features but also considers developmental changes that occur as a child grows up in a specific environment. Because of the extensive migrations of the past several centuries and the

ENCYCLOPEDIA OF HUMAN BIOLOGY, Second Edition, VOLUME I.   Copyright © 1997 by Academic Press.   All rights of reproduction in any form reserved.

intensely mobile quality of twentieth-century society, the geographic context in which ancestors of modern populations evolved must be considered. Older studies of human geographic variation that focus on continental "races" are contrasted with perspectives of contemporary physical anthropologists and human biologists that are based on variation among local populations.

## II. UNDERLYING THEORY

### A. Phenotypic and Genotypic Variation

Central to an understanding of human geographic variation is the principle of individual human uniqueness at the genetic level. At the phenotypic (observable) level, humans vary greatly in features that make it easy to identify individuals: hair color, eye color, skin color, height, body proportions, weight, and other features. At the same time, there is considerable genetic similarity within the modern human species, *Homo sapiens*—not just within families and small inbreeding populations, but across continents and worldwide. These observations express a paradox that is evident in comparing our species to its nearest living relative, the common chimpanzee, *Pan troglodytes*. In external features, the chimpanzee shows greater uniformity across its range than does the human species, but genetically the chimpanzee species encompasses greater differences.

This paradox has a relatively simple explanation with two primary components. The chimpanzee's native habitat is concentrated in tropical, forested Africa. Though some variation exists in the environment, it is much more uniform than that of the modern human species, which lives in hot as well as cold climates, in areas of little and of great rainfall, in humid and dry places, and from sea level to high altitudes. The great geographic and climatic range of human habitats has selected for variation in physique and for phenotypic plasticity, which facilitates adjustment to varied habitats. Second, genetic, paleontological, and archaeological studies suggest that the human species may be much younger than the common chimpanzee—on the order of 100,000 to 200,000 years of age as compared to an estimated 1,000,000 years for the common chimpanzee species. This could to some degree account for the low genetic variation found among modern humans.

### B. Origin of Anatomically Modern *Homo sapiens*

*Homo erectus,* the species ancestral to anatomically modern humans, is known earliest in Africa, where skeletal remains at 1.8 million years old have been identified. Some members of this population expanded out of Africa at approximately that time into tropical or near-tropical climates in Asia. *Homo erectus* individuals are thought to have reached today's Southeast Asian islands when low sea levels, which occurred during periods of glaciation, opened up a "land bridge" on the continental shelf. *Homo erectus* and its descendants eventually settled in colder parts of Asia and in Europe, possibly one million years ago. A theory of human evolution that was prominent for many years stated that these worldwide descendants of *Homo erectus* evolved on each of the three continents that comprise what has been called the Old World into the modern human species. This theory of regional continuity posits that evolutionary transition to *Homo sapiens* occurred essentially within each continent, but genetic contact between populations occurred at continental borders and kept the population from separating into diverse species.

An alternative theory of population replacement arose in part from studies of mitochondrial DNA and other DNA. Such studies found that human populations across the world—although quite variable phenotypically—are very similar genetically. This discovery indicated to many scholars that our human species is relatively young, probably less than 200,000 years, and that it likely developed in one part of the *Homo erectus* range and then spread worldwide. Because the most ancient skeletal fossils of anatomically modern humans have been found in Africa and because the largest amount of genetic diversity exists in Africa, most biologists and anthropologists who favor the replacement theory of modern human origins believe that the origin of modern humans—as of *Homo erectus*—lies in Africa. The replacement theory is sometimes called the "out of Africa theory." [*See* Human Evolution.]

Skeletal indicators of anatomically modern humans include an enlarged brain with a higher forehead and more rounded skull than in prior species, smaller teeth, the presence of a chin on the lower jaw, and alignment of the lower jaw under the face. Archaeological associations with anatomically modern human skeletons include diversified technologies based on bone as well as elaborated and diversified stone tool industries and the presence of art. Technologies and activities related to hunting, food-processing, and keeping warm (clothing and housing) are obvious cultural achievements that allowed a tropically based primate species to colonize cold areas. In the settlement of Australia, which is believed to have occurred at least 50,000 years ago, raft-building or boat-building rather than cold-weather technologies was required. Modern human languages are yet another possible innovation of modern *Homo sapiens,* although some anthropologists believe that capacity for language is much older than our species. [*See* Language, Evolution.]

## C. Adaptation to Nontropical Environments

Cultural traits, primarily effective hunting skills and technologies allowing manipulation of various natural resources, permitted the human species to enter nontropical environments (such as Europe, northern Asia, and, eventually, North America). If our species originated in Africa, its original physical attributes would have been adapted to African climates. Once it established itself in Asia and Europe, natural selection would work on its physical features, accommodating them to the climates in which the population lived. The replacement theory, together with archaeological evidence, suggests that adaptation to cold weather conditions by anatomically modern humans has an antiquity of about 50,000 years—not very long by evolutionary standards, but enough time to produce observable adaptations.

The theory presented here owes much to Charles Darwin's "Origin of Species" (1859) and refinements of Darwinian evolutionary theory that have engaged biologists over the last 150 years. This view of the basis of modern human diversity varies greatly from views of earlier European naturalists and other scholars who looked on living systems, including human populations, as essentially static. These earlier naturalists visited different continents several centuries before Darwin. Operating under Biblical creationist views, their perspective on the people as well as the flora and fauna that they saw was essentially typological; they tended to see the members of one continent as all of one type: created, and essentially unchanging. Though this view is no longer held in science, the categories that these naturalists created for human populations continue to be used.

## D. Analytical Categories: Local Populations, Major Continental Races, and Clines

The study of the adaptation of human populations to diverse climates in the past was based on the construct of human race categories referred to in the previous section. In the past, races were conceived as populations occupying continents, or large sections of continents, that shared a set of founding ancestors. Scholars differed in the number of groups recognized; traditional names for the three largest groups are "Mongoloid" for populations of North Asia and the Americas; "Caucasoid" for populations of Europe, North Africa, and the Middle East; and "Negroid" for most populations of sub-Saharan Africa. Yet these large populations occupy quite diverse environments, which evolutionary theory stipulates should produce diverse physical features over time. [*See* Population Differentiation and Racial Classification.]

Human biologists today prefer to examine climatic and geographic adaptations in which data are gathered for much smaller groups of individuals, characterized as local geographic populations. These are operationally defined population units that inhabit a given climatic or geographic region and participate in a cultural, economic, or religious system that is distinguished from other groups. This approach makes it possible to consider adaptations that may occur in small areas defined by a unique set of climatic, nutritional or behavioral traits.

For example, it is possible to consider the physical traits of the reindeer-herding Lapp people of northern Europe separately from the majority of fellow Scandinavians who have different languages and life-styles as well as different relationships with the environment. We can also differentiate the Ituri Forest Pygmies of Africa as a local population separate from their Bantu-speaking farming neighbors, who operate within a different cultural system and whose relationship to their environment is very different. The local population approach makes it possible to examine variation in skin color, for example, throughout Asia, by comparing the Southeast Asian peoples with northern Chinese, rather than having to consider them both as members of one "Mongoloid race" to be compared with other groups such as "Caucasoid" or "Negroid." In addition to genetic adaptations that have been selected over longer periods of time, and environmental effects that affect the growth and development of individuals, cultural features also contribute to physical differences among local populations. These cultural features include technologies, used in the performance of work, that thereby affect musculature as well as nutritional needs and nutritional resources, which in turn affect individual growth and development.

The local population approach thus assumes that climatic, ecological, and cultural factors in local environments produce different phenotypes within a continent. This perspective does not deny that common features can be identified among people of the same continent. After all, people of a continent share ancestors and relatives among themselves, although more barriers exist between some continents than between others, and populations at borders migrate and intermarry across continental divides. Major continental race concepts exist primarily as a legacy from the past that has been institutionalized into social systems. People learn to identify themselves and others by groups (e.g., Mongoloid, Caucasoid, and Negroid) even though these concepts provide information about few biological features or environmental adaptations.

Forensic anthropologists have the job of identifying human skeletal remains found in homicides and disasters. Social identification of individuals includes major continental race along with sex and age; all of these help narrow the search for the missing individual. In the United States, a society formed of descendants of populations primarily from Europe, Western Africa, and North America, these forensic specialists have sought physical features associated with the three continental race categories traditionally most numerous, that is, Native American, European-American, and African-American.

As skeletal biologists, forensic anthropologists do not have the benefit of soft tissues such as skin color, hair form, or body weight to make their estimations of "race," but instead they must rely on morphologies of the bone—both size and shape. Features that have proved useful in separating skeletal remains into one of three major continental populations focus primarily on a specific cranial attribute for each of the three groups. Individuals of European descent tend to have greater projection at the bridge of the nose than do Native Americans or African-Americans. African-American populations, by contrast, tend to have a nonprojecting, wider nose, and slight projection of the upper jaw. Native Americans tend to lack projection of both the nose and upper jaw, but have more prominent cheek bones than do people from the other continental groups. These three features of the skull constitute geographic variation in the sense that they are linked to three geographic areas (continents) and

the features have some connection to each ancestral population's past adaptations. But these cranial skeletal differences are subtle and often inconclusive, whether a forensic anthropologist uses statistical aids such as discriminant functions or a qualitative evaluation of traits, or a combination. Identification of the continental "race" of skeletal remains, therefore, is always an informed opinion, not an absolute truth; and in some cases a cranium simply defies categorization. This is due in part to the great variability within each of the socially identified major groups and to the overlapping of values between them. In addition, many Americans have ancestors from more than one of the continental populations, stemming from the five centuries of population contact and intermarriage that have occurred in the United States and elsewhere in the Americas. Possession of a feature that identifies a person as a member of one group or another may inaccurately reflect the individual's total genealogical (and genetic) ancestry, whether or not it correctly identifies the "major race" with which the individual is identified.

Traditionally, anthropologists have explained the prominent cheek bones of these northern Asians and Americans as deriving from selection for "padding" of the face as an adaptation to the extreme cold of northern Asia and possibly Beringia, the "land bridge" that existed between Siberia and North America during the last glacial period. Nasal projection may be an aid in accommodating to dry, cold environments, whereas a wider nose is an accommodation to hot, humid climates, where the free passage of moisture and air is advantageous. In later sections I discuss the adaptive value of greater nasal length in cold climates and the adaptive value of broad noses in moist, hot climates, both of which vary in local populations within continental groups and appear to be linked to local climates.

Overall, the major continental race categories contribute insights about ancient geographic adaptation but obscure and confound more recent variations. Faint traces of ancient adaptations allow tentative differentiation among three major groups, but differences are overlaid by many more recent, local adaptations to particular environments. The remainder of this article focuses on specific geographic adaptations in local populations. One of the weaknesses of the major continental race approach to human geographic variation is that it assumes that clusters of traits are inherited as a group and that such trait clusters define individuals and groups. Local population analysis, using clines, examines the geographic distribution of specific traits one at a time, relating a trait to one or more specific climatic conditions. Studies using local populations and clines have shown that specific traits can be selected in response to specific climatic or geographic factors.

## III. SKIN COLOR

### A. Background

All life is dependent on solar radiation, but the human species has some special features that affect its adaptation. One crucial feature is the lack of dense hair over most of the body and the reliance on sweating as the principal mechanism for dissipating heat. The naked skin, which is a radical departure from other members of the primate order, is believed to have its evolutionary origins in our ancestors' adoption of upright posture and bipedalism. According to this idea, the early hominid adaptation to the hot, dry climates of East Africa involved minimizing solar gain by vertical posture and reliance on sweating, an efficient cooling mechanism in dry heat provided it is not overused. Because human skin is exposed to solar radiation, its coloration has become a major focal point of geographic variation.

### B. Radiation

Geographic variation in skin color results primarily from two physiological processes that occur in opposition and depends on the amount of solar radiation that is provided in the environment. Though humans require some ultraviolet radiation, too much ultraviolet radiation can damage the skin and can lead to skin cancer. Ultraviolet radiation lies between the visible and X-ray regions of the electromagnetic spectrum. The narrow-wavelength band between 280 and 320 nm assists in the synthesis of vitamin D, which is needed for calcium absorption and bone mineralization. This band also is the part of the ultraviolet radiation that causes the most sunburn. Ozone screens out wavelengths below 280 nm, and wavelengths over 320 have weaker, though apparently not totally benign, effects on human health. [*See* Skin, Effects of Ultraviolet Radiation.]

### C. Selection for Dark Skin

Selection for darkly pigmented skin exists in tropical and semitropical environments and at high altitudes, where there is less atmospheric protection from ultraviolet radiation. Skin pigmentation provides protection against sunburn and, ultimately, skin cancer, particularly its most malignant form, melanoma. Sunburn starts when an untanned or light-skinned person is exposed to intense solar radiation, which results in reddening of the skin. Blood vessel dilation follows and leads to blistering, accompanied by infection or by reduction of the skin's ability to cool itself by sweating. If the skin peels, protective tanning does not take place. Repeated sunburns can lead to degenerative changes in the dermis and epidermis, resulting in premature aging of the skin, even if skin cancer does not follow. Although darkly pigmented skin is not immune to these processes, the risks are significantly lower.

### D. Selection for Lightly Pigmented Skin

The sun's rays that lead to sunburn and cancer also stimulate the synthesis of vitamin D. Deficiencies in vitamin D that occur in childhood result in rickets, whereas deficiencies in adults result in osteomalacia. In addition to bending limbs and increasing the incidence of fractures that result from these deficiency conditions, these two diseases can cause reproductive impairment in women owing to the deformation of pelvic bones. Though vitamin D is found in seafood, or today can be obtained in dietary supplements, most human populations in the past were dependent on sunlight in their environment to obtain vitamin D. Migration of human populations from equatorial or other tropical, open areas into cold or temperate environments, particularly those where cloud cover and vegetation restrict sunlight seasonally, is believed to have exerted a strong selection pressure for reduction in pigmentation.

Genetic models for skin pigmentation vary, but usually posit between two and five major genes and varying intensities of selective pressure. Thus, these models differ in the length of time required to produce human populations with light skin color from dark-skinned ancestors. Current models, using different assumptions, suggest that the existing range of human skin color differences could have evolved in as little as 24,000 years or as long as 45,000 years. Though differing greatly, both suggestions fit within the time framework of dispersal of anatomically modern *Homo sapiens,* referred to previously.

### E. Natural Experiments Testing These Models

Population migrations of the past several hundred years can be viewed as natural experiments, providing empirical support for the hypothesis that solar radiation exerts evolutionary effects on human skin color. Darkly pigmented populations who live in temperate zones and people from tropical areas who shield their bodies from the sun for religious or cultural reasons have been found to experience rickets if they do not take dietary supplements of vitamin D. The incidence of skin cancer in populations of European ancestry appears to double for every 10 degrees decrease in latitude, hitting a peak at the equator. A 5 to 10% increase in the effectiveness of ultraviolet rays in producing sunburn for each 500 m of altitude has been reported. At 2 km above sea level, the intensity of ultraviolet rays is approximately one-third greater than it is at sea level. Few environments with negative sea level exist, but at one of them, near the Dead Sea, sunburn is less of a problem than elsewhere. Ozone is a strong absorber of radiation between 220 and 300 nm. The recent reduction of ozone in the earth's atmosphere, which appears to be due to discharge of chlorofluorocarbons from sources such as aerosol propellants and refrigerants, may produce an increase in sunburns and skin cancers among light-skinned persons in regions that previously were subjected to lesser amounts of ultraviolet radiation.

## F.  Pigmentation of Hair and Eyes

In general, pigmentation in hair and in the iris of the eye follows the same pattern of geographic variation as in the skin. Genetically, hair and eye colors appear to be controlled independently. Blue eyes occur with either dark or light hair; the same is true for dark eyes. Dark eyes may be selected for in areas where solar radiation is intense, since ultraviolet light absorbed in the superficial layers of the eye can cause conjunctivitis and corneal inflammation (keratitis). Some Australian aborigines as well as some persons of European descent have lightly pigmented hair in childhood that becomes dark in adolescence or adulthood. Advanced age is associated with loss of hair pigmentation for people of various skin colors from many geographic areas.

## G.  Skin Color and Heat Tolerance

Intuitively it might seem that, so far as thermoregulation is concerned, dark skin color should be disadvantageous in a hot environment and advantageous in a cold one, since black absorbs heat. However, the distribution of color in various mammals and birds, as well as in humans, suggests that other factors (selection for protection against skin cancer and for the

ability to metabolize vitamin D) are of greater evolutionary importance. Although absorption of ultraviolet radiation depends largely on inherited skin color, research on thermoregulation shows that adaptation to heat depends on acclimatization rather than on skin color. Long-term acclimatization to hot climates is expressed in body shape and proportion as well as in physiological and behavioral changes.

Humans have very effective evaporative cooling systems, being equipped with two to four million eccrine sweat glands. Though sweating is an effective mechanism for cooling in dry climates, it is "expensive" because it requires a lot of water. Thus, long-term acclimatization to hot climates consists of behavioral and physiological changes that reduce the amount of sweating required. Adults who move to a hot climate from a cooler one may lose weight, whereas young children who move to the tropics may experience more pronounced effects upon growth and body composition. Physiological tests under hot and humid conditions, using subjects from African, Australian, and European populations, have shown that ethnic group is not as important in heat tolerance as the degree of heat acclimatization that the subject had achieved at the time of testing. Europeans are capable of achieving good heat tolerance, but this capability appears to rely more strongly upon sweating than does the capability of native Africans and Australians.

## IV.  ECOGEOGRAPHIC RULES

### A.  Thermoregulation Principles

Ecogeographic rules describe and attempt to explain body size and shape variation within mammalian species in line with basic physical principles related to the retention and dissipation of heat, that is, thermoregulation. Bergmann's Rule states this principle in respect to body size, whereas Allen's Rule states it in terms of appendages. Bergmann's Rule posits that populations of a species that live in the cold part of the species range are likely to be larger than those in hot parts of the range; Allen's Rule posits that populations in cold parts of a species' range will have shorter appendages and longer torsos compared to the relatively longer appendages found in hot environments.

The great range of climates that *Homo sapiens* occupies makes it an excellent test case for ecogeographic rules. A sphere is the most energy-conserving shape, whereas a linear form (particularly involving appendages) has more surface area whereby to lose heat into the atmosphere. Although no human can exist at either extreme (a sphere or a ribbon), ecological rules hypothesize that in very cold climates human forms would tend toward the sphere (longer torsos, shorter appendages, stockier build) and in very hot climates human forms would tend to be lean and have longer appendages (arms and legs, notably, and perhaps more protruding ears). These rules hold in general—all other things being equal. But particularly because human populations are very mobile, "all other things" are *not* equal. This article thus considers these "other things" that affect human geographic variation.

Climatic factors that directly affect human geographic variation include latitude, humidity, altitude, and the degree of seasonal variation in a given area. Indirect climatic effects on ecogeographic adaptations include factors affecting human growth patterns, such as nutritional resources, technologies relating to food-getting and food-processing, and cultural devices that stand between the individual and its environment. Some examples of the latter are housing, clothing, and labor-saving technologies affecting activity levels and mobility.

## B. Applications to Ancient Populations

In an innovative series of research papers, Christopher Ruff examined height, hip breadth, shoulder breadth, and body weight as thermoregulatory factors in prehistoric and modern populations. His work illustrates the interrelationships of these variables and shows how body shape adapts to geographic (primarily climatic) pressures. Ruff considered the famous Lucy skeleton (AL-288-1), discovered in Ethiopia, which dates to more than three million years ago. Because about 40% of Lucy's skeleton was found, including bones from all parts of her body, it was possible to reconstruct body proportions. Lucy was very short-statured, and would have stood about 107 cm, or 3-feet-6-inches. Ruff compared her body proportions with another famous early African hominid skeleton, that of the Kenyan Nariokotome boy (KNM-WT 15000), whose antiquity is estimated at about 1.53 million years. Although already about 160 cm (about 5-feet-3-inches), the Nariokotome boy had not completed his growth. He was estimated to be 11 or 12 years of age; had he lived to adulthood he was expected to reach a stature of about 185 cm, or slightly more than 6-feet tall. But his pelvic breadth was very similar to Lucy's.

Ruff's interest was in seeing how body size and body shape together affect thermoregulation. His calculations involved estimates of the surface area of

these two fossil hominids. Because they both lived in a hot, dry environment, Ruff posits that sweating would be an important means of maintaining thermal equilibrium. If these two specimens are put on the same scale, the small Lucy skeleton appears extremely stocky, whereas the Nariokotome skeleton appears lean, and in fact has hip and shoulder proportions common to peoples living in East Africa today. If the tall *Homo erectus* boy had simply been a larger version of Lucy, he would have had a relatively low ratio of surface to mass—and would have had a poor thermoregulatory system for a hot climate. Lucy's thermoregulation worked because she was short and had an appropriate surface to mass ratio. Modeling the body as a cylinder, Ruff argues that as long as the diameter of the cylinder is held constant, stature can vary—individuals can be as short as Lucy or as tall as the Nariokotome boy—and still maintain the same ratio of surface to body mass.

The next step Ruff took was to consider under what geographic conditions large body size (i.e., tall and lean) is adaptive, and under what conditions short body size works best. Being large offers the possibility of greater strength, but there are disadvantages too, such as the requirement for greater amounts of food. Being short offers the advantage of requiring fewer calories. Thermoregulatory concerns involving the humidity of the environment as well as the heat may be more important than energy itself. A tall, lean body has less heat gain from the sun when moving about in open, hot, dry environments, and is capable of dissipating heat, because sweating is an effective cooling mechanism in a hot and dry environment. Sweating is less effective as a cooling mechanism in humid environments such as heavily forested areas in the tropics. There, the best way to limit heat gain is to limit body size. Both temperature and humidity are crucial variables for explaining human geographic variation affecting body size and body shape. These results suggest that early African *Homo erectus* most likely inhabited dry open environments. This conclusion is supported by other evidence, including the shape of the nose and the paleoenvironment of the fossils.

# V. GROWTH PATTERNS AND BODY SIZE

## A. Cultural Influences

The discussion in the previous section provides connections between prehuman ancestors and contemporary humans living in nonindustrialized subsistence economies. Studies of children's growth in contemporary populations, both subsistence economies and industrial societies, offer insights into some of the "other things" that, along with strictly climatic variables, affect human phenotypes. These other things include food resources and technologies, along with other variables that collectively comprise life-style aspects of culture.

Body size increases in European and other industrialized societies over the past century generally have been explained by increase in food—both in absolute quantity and in the reduction of seasonal shortages that are common to subsistence economies that do not have modern food processing, storage and transportation systems. Secular increase in body size is evident in the faster tempos of child growth, lower ages at menarche for girls, and taller and heavier adults. Changes in body proportion, particularly involving lengthening of the legs, also may occur.

Preindustrial nutrition systems are characterized by greater seasonal variation in the type and amount of foods consumed than are nutritional systems in industrial societies. In addition, people in industrialized societies generally are much less physically active than are people in economies that are labor-intensive and nonmechanized. Thus, current geographic variation in body size, and variation in patterns of growth in children, are in large part related to the degree of industrialization in a country and to the abundance of nutrition year-round. Availability of public health measures in the community is an additional factor; for example, diarrheal disease in developing nations, which is largely responsible for mortality rates of up to 100 deaths per 1000 infants, and occasionally greater, also retards growth in children who survive infancy.

## B. Altitude

High-altitude populations, although they have greater chest dimensions, tend to be smaller-bodied than closely related populations who live at much lower altitudes. However, it is difficult to disentangle the various factors that contribute to these differences. Hypoxia (deficiency of oxygen reaching the tissues, due in high altitudes to low oxygen pressure) and cold stress associated particularly with night-time cooling in high-altitude areas are geographic and climatic features, but they are compounded by lower levels of nutrition and the greater seasonal scarcity of food, both being important indirect effects of altitude on the physical features of human populations. Evidence

exists that direct environmental effects occur in addition to effects of reduced nutrition; for example, high-altitude populations are reported to have later average menarche, even if nourishment levels are high.

## C. Migrations, Environment, and Body Size

All major continents have quite varied habitats and climatic features, providing their local populations with diverse possibilities for geographic variation. In addition, cultural and ethnic groups have unique histories of adaptation to different niches of a continent. Examination of each population's inherited physical features can provide some clues to the environments in which their ancestors lived. We are on more solid ground in studying recent migrations for which we know the population history than in reconstructing population migrations of the past, although those are of great interest; for the former, we can estimate the degree of intermarriage that has occurred. In 1912, the renowned anthropologist Franz Boas published studies comparing attributes of migrants to the Americas with those of their stay-at-home relatives (sedentees). This type of research design, featuring a kind of "natural experiment," has demonstrated that environmental factors can change physical measurements of migrant families within one generation. Following his lead, other studies have analyzed the physical dimensions of descendants of people who have moved from a variety of different environments and nutritional conditions into another. Variations on this theme have examined the effects of Westernization in populations such as native Samoans, both migrants and sedentees. [*See* Population Genetics.]

Population studies in Africa pose more difficult questions concerning variation in body shape and size than do studies of migrants from Europe or Africa to North America because of the great antiquity of human settlements in Africa. According to the theory of replacement outlined earlier, Africa is the birthplace of anatomically modern *Homo sapiens* and consequently is the continent where diverse adaptations and body sizes and shapes evolved in relation to different habitats, economies, and technologies. Since Africa has been populated by modern humans for many thousands of years, many migrations have occurred throughout the continent. Even though historians and anthropologists attempt to trace these movements and ascertain in which environments specific cultures and physical features developed, these reconstructions are incomplete. Populations are dynamic; furthermore, human populations in contact with each other tend to marry across borders, thus blurring evolutionary adaptations that may occur. Many contemporary African nations are composed of several ethnic groups whose ancestors adapted to different environments within the continent. Thus, local populations who live near each other have descended from groups who adapted to quite different environments in previous generations. Examination of human variation in Africa illustrates the importance of economic, technological, and cultural factors along with strictly geographic features such as latitude, precipitation, altitude, and climate on individual growth patterns and adult body size. Compilations of childhood growth data offer insights into these factors.

P. B. Eveleth and J. M. Tanner, in "Worldwide Variation in Human Growth," discuss data from five sub-Saharan African ethnic groups: "Nilotics and Nilo-Hamites," which include groups such as the Tutsi of Rwanda and the Ngisonyoka Turkana of Kenya; "Sudanese" populations, who live in Senegal, Nigeria, Somalia, and Gambia; "Bantu" populations, who live in many areas of equatorial, central, and southern Africa, and include groups such as the Hutu people of Rwanda and residents of Soweto, South Africa; "Khoisan" groups such as the San Bushmen and "Hottentot" peoples of southern Africa; and "Pygmy" peoples from equatorial countries such as Zaire. The authors qualify their five ethnic categories as "traditional, although by no means unequivocal" (1990: 63–67), indicating the difficulties of reconstructing population affiliations in a continent with a long history of population mobility. These populations vary greatly in body size and proportion. African habitats differ in temperature, precipitation, humidity, altitude, and vegetation, all of which are geographic features that potentially can affect human physical variation. The position of Africa with respect to the equator—which runs through the center of the continent—alone would lead to a hypothesis of lean body size in Africa compared to descendants of Africans who live in cool or temperate climates, such as in Northern Europe, Asia, or North America. Because almost all of Africa receives intense, direct solar radiation, we expect and find that most African populations have dark skin color, Khoisan groups being an exception, but this relative uniformity in one trait should not blind us to the great variation that exists in other physical traits. Disease and malnutrition due to droughts, civil wars, high population density, former colonial practices, and lack of industrialization feature very strongly along with climate in explanations

for the generally more slender bodies of Africans compared with descendant populations in the Americas. However, the health problems related to overnutrition in some of those African-American descendants should also be noted.

An example illustrates the diversity among African adaptations, since a discussion of the physical diversity in African populations is beyond the scope of this article. The African Efe Pygmy population is of special interest because of questions regarding factors that maintain their small size. Efe Pygmies of the Ituri Forest, the smallest pygmy peoples known, are reported to have adult heights averaging 143 cm for men and 136 cm for women (about 56 and 53 inches, respectively). Some researchers believe that Pygmy children have a growth pattern similar to that of neighboring, tall, Bantu-speaking populations until they reach adolescence, but lack an adolescent growth spurt because of a deficiency of a specific growth hormone. Eveleth and Tanner, in contrast, see a clear adolescent growth spurt in data on growth records of Efe Pygmy girls, and a rather smaller growth spurt in boys. Although physiological mechanisms mediating short stature are not yet resolved, selective pressures may explain the Pygmy adaptation to their hot, humid tropical forest home. Ruff's thermoregulatory explanation posits that in a hot, humid climate, sweating is an inefficient means of cooling; the best solution, which would be expected to evolve by natural selection if a population spent enough generations in a given environment and had the necessary genetic mutations, is to reduce the amount of heat produced. This can be done by selecting for short stature. Small body size and lower outputs of energy are also beneficial in reducing caloric requirements; and it has been postulated that small body size aids the Ituri Pygmies in negotiating their highly vegetated environment. The taller, Bantu-speaking neighbors of the Efe Pygmies are farming populations whose economy is based on clearing the forest, in contrast to the Ituri hunters who find their subsistence within the forest. Bantu occupancy of the area—and hence the period in which they could adapt to it—is considered to be more recent than that of the Pygmies. But their different nutritional resources, activity patterns, and ecological practices also help to explain their different body size.

Native populations of the Americas also vary tremendously in many physical features related to body size. Natives of the Amazon Basin are short-statured and small-bodied, as are many other residents of tropical forests. Several decades ago, Marshall Newman's analysis confirmed the application of ecogeographic rules to indigenous people of North and South America and demonstrated that food supplies also influence body size in these populations. Challenges for future studies of ecogeographic adaptation in an increasingly mobile world are to determine whether cultural factors promoting homogeneity of nutrition and technology will reduce the impact of natural selection and environmental factors on body shape and size.

## VI. NOSE SHAPE

The nose has a crucial role in channeling an adequate amount of air to the lungs, but it also adjusts the temperature and moisture of air going to the lungs. Cilia of the nose clean incoming air of dust, as well as add moisture to it in dry and dusty climates. The projection of the nose from the face, as well as length of the nose, contributes to this process. In cold, dry climates, a longer nasal chamber is advantageous to warm the air and conserve moisture, and it can reduce loss of body heat when air is expelled. Under hot and humid conditions, by contrast, a shorter, broader nose helps provide entry for large quantities of air. Nose morphology that allows maximum free passage of air can help to dissipate heat and water from the respiratory tract.

If the world's climates were of only two types—hot and humid versus dry and cold—we could expect nasal morphology of the world's peoples to come in two discrete forms, with broad and short in the former and long and narrow in the latter. However, climates are sometimes hot and dry, or cold and wet, and many people live in areas that are seasonally variable; other people move regularly between two or more climatic regimes. Therefore, we should expect compromise morphologies if we assume ecogeographic adaptation. Although latitude is a very important measure of climate and is easy to measure, other important determinants such as topography, proximity to ocean currents, altitude, and humidity are more complex. Thus human nose morphologies cannot be expected to fit simple geographic distribution models perfectly.

Data on nose shape have been gathered by anthropologists for many decades. Traditional measurements include the height of the nose, which is measured from the bridge of the nose to the junction of the nose with the face, and breadth, which is measured as the maximum distance across the fullest part of the nose. The nasal index is computed as the nasal

breadth divided by the nasal height, times 100. It thus can be compared among individuals and populations without consideration of absolute size. Nasal index is known to range from a broad 104 (in Africa's Ituri Forest Pygmy peoples) to a narrow 62 (in some Northwestern American Indians). Analysis of data gathered in the 1890s by Franz Boas in North American Indian populations shows a definite cline of the nasal index from north to south, with the narrowest noses in the north and the broadest in the south. Among these populations, average nasal indices go from a low of 62 for samples from the northwestern coast of Alaska to a high of 88 in southern California.

## VII. HEAD SHAPE

Studies of both fossil and living samples have found that head shape in humans corresponds to some degree with thermoregulatory expectations that relate to heat conservation and dissipation. In general, relatively broader (more spherical) heads are found in colder climates and more elongated heads in hotter climates. Head shape conventionally is described by the cranial index, which is computed as the maximum breadth of the head divided by the maximum length of the head, times 100. Studies show that migration to new environments, or change in nutritional resources, has a developmental effect in addition to a long-term population effect that presumably has a genetic basis. My research analyzing Franz Boas' anthropometric data on samples from western North American populations from the standpoint of geographic and climatic adaptation found that head length is positively (but weakly) correlated with shoulder width, shoulder height, and stature. All of these features in Boas' western North American samples appeared to increase in the nineteenth century as part of the secular trend toward increased body size, believed to be a result of increase in total annual calories.

Other measures of the head—head breadth, facial breadth, face height, nose height, and nose breadth—vary with latitude in Boas' samples. Local population average values are higher in the north and decline to lower values in the south for all of these measures except nose breadth. As noted in the previous section, the cline for nose breadth goes in the opposite direction, with broader noses in the south and narrower noses in the north. Nutrition clearly has an effect on both the head shape and on total body size that complicates analysis of the effect of ecogeographic rules, particularly when they are applied to populations undergoing cultural, technological, or economic changes. Nose shape, however, is less dependent on nutrition and is responsive to climatic factors, perhaps because its role in processing air is central to survival.

## VIII. GENETIC VARIATION

### A. Genetic Relationships among Populations

Human genetic variation, like morphological variation, has a geographic component. This section briefly illustrates some aspects to be considered in genetic geographic variation, but does not attempt to address the issue comprehensively. Although evolutionists several decades ago were disinclined to consider any genetic variation not linked to natural selection, evolutionary theorists today have accepted the concept that some genetic mutations are incorporated without the action of selection, that is, genetic changes due to mutations that are neutral so far as selection is concerned. Further, geneticists have developed models to reconstruct sequence changes in genes (DNA) and estimate the time required for them to occur. Genetic similarity and dissimilarity between populations can therefore suggest ancestral relationships.

Analyses of mitochondrial DNA, a portion of the genetic material that is inherited only through the mother and that is subject to relatively rapid genetic changes, have been used to estimate the antiquity and relationships of groups of people. For example, studies of the DNA of diverse indigenous populations from both South and North America have suggested to some scholars that the populations founding the American Indians must have existed (either in Asia or in the Americas) for many thousands of years longer than the customary 12,000 years often assumed as the beginning of the settlement of the Americas from Siberia. Similarly, as indicated in the introduction to this article, the greater genetic diversity among African populations than among human populations in other continents suggests to many scholars that Africa is the homeland of modern *Homo sapiens,* simply because genetic changes, on average, accumulate slowly. So far, morphological features in general have been better indicators of adaptation to the environment, whereas genes have been better indicators of genealogical relationships among groups. In their

different ways, both morphological and genetic features constitute human geographic variation.

### B. Example of Genetic Geographic Adaptation

However, genetic traits can reveal population adaptation. Other articles in this encyclopedia discuss genetic disease, but some genetic variants provide adaptations to diseases. They are geographically, climatically, and to some extent culturally based. One example is provided by genetic traits known collectively as abnormal hemoglobins, some of which have geographic distributions corresponding to the presence of malaria as a major disease of humans (while others appear unrelated to malaria). Though the origins of some abnormal hemoglobins are due to random mutational processes in DNA, their geographic distribution seems almost certainly to be due to the distribution of malaria, a scourge whose force can be mitigated by the possession of an abnormal hemoglobin that is resistant to the malarial parasite. [*See* Malaria.]

Malaria has probably killed more people than any other single disease. The malarial parasite is carried by several species of mosquitoes, but it has a complex life history requiring a mammalian host for part of its life. Malaria is endemic in tropical areas where carrier mosquitoes live year-round, but it has also produced epidemics in temperate areas of North America. It was responsible for the deaths of many native and nonnative people in the United States in the early 1800s. Control is easier in nontropical areas because mosquito vectors die off annually and because measures such as drainage and other elements of mosquito control are easier to achieve in countries with a strong industrial economy and funds for public health measures.

The historic movement of farming populations into forested environments in recent millenia was in part a product of the development of iron tools, which enabled the accelerated cutting of forests. These ecological upsets also created many pools of standing water, which permitted the proliferation of mosquitoes. As humans displaced other large mammals on which the mosquitoes fed, humans increasingly became subject to more mosquito bites, and human mobility helped to spread the parasite.

Best known of the abnormal hemoglobins is HbS, a genetic trait that causes a crippling disease of sickle-cell anemia when a person inherits the allele from both parents, thus carrying it in homozygous (i.e., same-zygote) condition. [*See* Sickle-Cell Hemoglobin.] By contrast, the heterozygous (i.e., different zygote) phenotype, called sickle-cell trait, has been found to be resistant to carrying the malarial parasite on the red blood cell. The heterozygous trait does not result in anemia at low-altitude conditions; hence it provides higher fitness than normal hemoglobin trait or the homozygous condition in environments where the most severe form of malaria is endemic. Many populations of tropical Africa carry a 10 to 20% frequency of this abnormal hemoglobin allele (HbS), even though 25% of the offspring produced by the marriage of two heterozygotes inherit a double dose of HbS, on average, causing sickle-cell anemia, a generally fatal condition. The HbS allele is found in malarial areas of Europe near the Mediterranean Sea, Madagascar, the southern part of the Arabian peninsula, and parts of India, as well as in malarial parts of Africa.

HbS is a mutation affecting the structure of the hemoglobin molecule, yet other mutations have produced altered molecules by determining the length and extent to which hemoglobin's polypeptide chains are formed. These Hb mutations produce several types of thalassemias, forms of anemia that are present in Africa, southern Europe, and Southeast Asia and that are known to offer some protection from malaria. When individuals carrying HbS or other abnormal hemoglobins leave regions where the trait is adaptive—for example, when Africans were brought to the United States for slavery—the trait ceases to be adaptive. Selection against HbS occurs slowly; in the meantime the trait exists as an ancestral legacy, indicating that at least one of an individual's ancestors came from a malarious area.

The various genetic adaptations to malaria are extreme examples of genetic adaptation that have been selected because of the extent and severity of the disease of malaria. Less costly genetic adaptations to other diseases may exist or may have existed in the past. Geneticists have pondered whether various genetically caused diseases that are geographically distributed might have been favored in the past because they conferred some adaptive value in a particular environment. If the environmental condition or disease selecting for the mutant allele has ceased to exist, it may be impossible to reconstruct its past adaptive value.

## IX. CONCLUDING COMMENTS

The fascinating geographic variation that occurs in humans has been difficult to decipher. Because it is multifaceted, the range of variation cannot be expressed by categorizing the species into continental groups as if all people and environments of a continent are alike. Though earlier human biologists assumed relative homogeneity of specific local populations, current thinking suggests much greater plasticity, both at the individual developmental level and involving long-term genetic change that occurs as populations live in specific environments. Despite the scholarly attention given to understanding our own species, we are only beginning to unravel our historical and adaptational mysteries, which new technologies (e.g., genetic sequencing) are facilitating. One of the challenges that human biologists of the future will investigate is the degree to which international travel and changing life-styles and technologies affect the adaptations that local populations make to their environments.

## BIBLIOGRAPHY

Bogin, B. (1988). "Patterns of Human Growth." Cambridge Univ. Press, Cambridge, England.

Eveleth, P. B., and Tanner, J. M. (1990). "Worldwide Variation in Human Growth," 2nd Ed., Cambridge Univ. Press, Cambridge, England.

Frisancho, A. R. (1993). "Human Adaptation and Accommodation." Univ. of Michigan Press, Ann Arbor.

Hall, R. L., and Hall, D. A. (1995). Geographic variation of native people along the Pacific coast. *Hum. Biol.* **67**, 407–426.

Harrison, G. A., Tanner, J. M., Pilbeam, D. R., and Baker, P. T. (1988). "Human Biology," 3rd Ed. Oxford Univ. Press, Oxford, England.

Robins, A. H. (1991). "Biological Perspectives on Human Pigmentation," Cambridge Studies in Biological Anthropology 7. Cambridge Univ. Press, Cambridge, England.

Ruff, C. B. (1993). Climatic adaptation and hominid evolution: The thermoregulatory imperative. *Evol. Anthropol.* **2**, 53–60.

Ruff, C. B. (1994). Morphological adaptation to climate in modern and fossil hominids. *Yearbook Physical Anthropol.* **37**, 65–107.

# Addiction

W. MILES COX

*University of Wales, Bangor, United Kingdom*

JOHN E. CALAMARI

*The Chicago Medical School*

I. Introduction
II. Overview of Addiction in America
III. Determinants of Addiction
IV. Treatment of Addiction

## GLOSSARY

**Alcoholics Anonymous** Organization of alcoholics helping other alcoholics not to drink; similiar organizations such as Narcotics Anonymous, are for people addicted to other drugs

**Endorphins** Category of opioid-like substances that are produced endogenously by the body and may play a role in the development and maintenance of addictions

**Motivational model** Model that integrates the various determinants of addiction showing that people's alcohol and other drug use is determined by the balance between the their expected emotional satisfaction from using alcohol and/or other drugs and their expected emotional satisfaction from other life areas

**Opioids** Drugs such as opium, morphine, and heroin that are derived from the opium poppy plant; sometimes referred to as narcotics, these drugs produce feelings of euphoria and are often used as pain killers

**Psychoactive drugs** Drugs that change the way people feel, either by enhancing positive feelings or counteracting negative feelings; many are highly addictive

**Relapse prevention** Techniques for helping addicted persons avoid becoming readdicted once they have changed their behavior; a primary goal is to teach previously addicted persons to cope effectively with the situations in which they are likely to resume their addictive behaviors

**Systematic motivational counseling** Counseling technique for helping alcohol and other drug addicts change their maladaptive motivational patterns in order to increase the emotional satisfaction that they obtain nonchemically, thereby reducing their motivation to seek emotional satisfaction by using alcohol or other drugs

**Tolerance** Decline in effectiveness of a psychoactive drug with its repeated use

**Withdrawal Syndrome** Set of adverse physical and psychological symptoms that accompany abrupt discontinuation of a psychoactive drug that a person has become accustomed to taking

ADDICTION IS A COMPLEX PHENOMENON, WITH biological, psychological, environmental, and sociocultural determinants. There is no single definition of addiction about which all scientists agree. Addiction is often said to occur when people continue to use excessive quantities of a psychoactive drug, even though doing so causes them physical and psychological harm. Addiction cannot be accounted for simply by people's desire to prevent the unpleasant withdrawal symptoms that are likely to occur should they abruptly stop taking the drug. Rather, drug use seems to be determined by the relative degree of satisfaction that they expect from drug use versus other avenues for gaining satisfaction. Treatment for drug addiction has been largely unsuccessful when success is defined in terms of permanent abstinence from the addicting drug. In consequence, efforts are now being focused on developing more effective treatments by helping addicts not to resume using alcohol or other drugs once they have stopped and to build a balanced lifestyle as an alternative means of achieving emotional satisfaction.

87

ENCYCLOPEDIA OF HUMAN BIOLOGY, Second Edition, VOLUME I.   Copyright © 1997 by Academic Press.   All rights of reproduction in any form reserved.

## I. INTRODUCTION

Addiction occurs when an individual is strongly motivated to use a drug (or some other substance) repetitively and excessively, even though there are clear negative consequences of doing so. How does this strong motivation come about? Historically, addiction to drugs has been explained in terms of the drug addict's desire to forestall withdrawal symptoms that might occur if the drug that he or she has grown accustomed to taking is abruptly discontinued. With certain drugs, withdrawal symptoms can be quite severe and unpleasant. For example, withdrawing from an opioid (one of several drugs derived from opium and sometimes referred to as narcotics) might cause a set of symptoms that resemble influenza, including a runny nose, chills and diarrhea. Withdrawal from alcohol can cause convulsions and hallucinations and can be life-threatening. Nevertheless, some drugs to which people are addicted do not seem to produce withdrawal symptoms at all. Moreover, even when people habitually take a drug that does lead to withdrawal, they might discontinue their drug use when their life circumstances change. For instance, while in Vietnam, many people in the military used drugs heavily, but stopped doing so when they returned to the United States. Withdrawal symptoms, therefore, cannot be the sole factor underlying addiction to drugs.

Biological, psychological, environmental, and sociocultural processes have each been investigated for the causal role that they play in drug addiction. The research on psychological factors that maintain people's drug use has demonstrated that by using drugs people attempt to achieve emotional benefits that they are unable to achieve through ordinary life experiences. Hence, when people lack effective alternative means of obtaining pleasure or ridding themselves of pain, they are more likely to turn to alcohol or drug use than would otherwise be the case. This point will be elaborated on later in the article.

What drugs do people abuse? They abuse psychoactive drugs—those drugs that have an immediate impact on the way people feel. In the "Diagnostic and Statistical Manual of Mental Disorders," The American Psychiatric Association has classified the psychoactive drugs to which people can become addicted into nine specific categories: (1) alcohol, (2) amphetamines and similarly acting drugs, (3) cannabis (i.e., marijuana), (4) cocaine, (5) hallucinogens (e.g., LSD), (6) inhalants (e.g., vapors from glue), (7) opioids, (8) phencyclidine (PCP) and similarly acting drugs, and (9) sedatives, hypnotics, and anxiolytics (i.e., drugs that induce sedation or tranquillity). Additionally people can become addicted to nicotine and caffeine, and some scientists maintain that people can become addicted to substances and activities other than psychoactive drugs. The different categories of drugs that people abuse have different effects on the way they feel. For example, alcohol and amphetamines cause people to feel less anxious and more optimistic. Marijuana makes pleasurable experiences seem more intense. Opioids dull the senses, making the user less aware of physical or psychological pain. Although debate continues on the causes and nature of addiction, most clinicians use the definitions of addiction that appear in the "American Psychiatric Association's Diagnostic and Statistical Manual." A committee of treatment providers and leading researchers has revised the definitions of substance abuse and substance dependence. Substance abuse is defined as psychoactive drug use to such an extent that the individual is often intoxicated throughout the day, fails to meet important life obligations, or fails in attempts to abstain from use of the substance. Further, the substance abuser may use alcohol or drugs in situations where to do so is dangerous (e.g., driving while intoxicated). Symptoms of physiological dependence are not among the criteria for defining the substance abuse syndrome.

In contrast, substance dependence is defined as a maladaptive pattern of substance use that leads to significant impairment in the individual's social or occupational functioning. To meet diagnostic criteria for this disorder, three or more of the following must have occurred at any time during a given 12-month period:

1. Tolerance, as defined by either of the following: (a) the need for markedly increased amounts of the substance to achieve intoxication or the desired effect, or (b) markedly diminished effects with continued use of the same amount of the substance.

2. Withdrawal, as manifested by either of the following: (a) the characteristic withdrawal syndrome for the substance, or (b) the same (or a closely related) substance is taken to relieve or avoid withdrawal symptoms.

3. The substance is often taken in larger amounts or over a longer period of time than the individual intended.

4. A persistent desire for the substance or unsuccessful efforts to cut down or control use of the substance.

5. A great deal of time is spent on activities necessary to obtain the substance (e.g., visiting multiple physicians to procure prescriptions or driving long distances to buy the drug), use the substance (e.g., chain smoking), or recover from the effect of having used the substance.

6. Important social, occupational, or recreational activities are given up or reduced because of the substance use.

7. Substance use is continued despite the individual's knowledge of having a persistent or recurrent physical or psychological problem that was likely to have been caused or exacerbated by the substance (e.g., current cocaine use despite recognition of cocaine-induced depression, or continued drinking despite recognition that an ulcer was made worse by alcohol consumption).

Despite the fact that clinicians agree on the symptoms that constitute drug addiction, there has been intense debate among scientists about the relative importance of the various factors involved in the etiology of drug addiction. Before examining the various kinds of explanations, it is useful to review the history of substance use and regulation in the United States, for doing so will help to elucidate current patterns of substance use. As will be shown, the regulation of drug use has changed significantly during the course of this country's history, with accompanying changes in the patterns of drug addiction. [See Nonnarcotic Drug Use and Abuse.]

## II. OVERVIEW OF ADDICTION IN AMERICA

Just 2 years after the United States was formed, its first law to regulate a psychoactive drug, an excise tax on alcohol, was enacted. However, passage of this law appears to have been motivated primarily by the desire of the federal government to establish its taxation authority over the territories west of the Appalachian Mountains. Protection of the public health and welfare, the stated rationale for governmental regulation of alcohol, probably was a secondary concern. Despite this early law, both the use and the sale of psychoactive substances (including such potent ones as narcotics) were unregulated throughout much of the country's history. Eventually, however, health and social problems produced by substance abuse led to shifts in public opinion and resulting political changes that precipitated increased governmental regulation.

The rise of substance abuse in the United States has been attributed to a variety of causes. One factor that has been implanted is the importation of large numbers of Chinese laborers into the United States during the 1850s, many of whom brought with them the practice of smoking opium. Smoking opium, even in small quantities, can be addicting, because doing so produces a pleasurable, dream-like state that can last for hours. The practice quickly spread, and "opium dens," public places where opium could be obtained and smoked, were established in cities on the west coast, such as San Francisco. These establishments functioned legally and were not subject to governmental regulation.

The use of opium and its more potent derivatives, morphine and heroin, was a major factor that eventually led to the governmental regulation of drug use in the United States. However, factors other than the smoking of opium by Chinese workers were more directly responsible for the abuse of opioids. For instance, physicians in the 1800s (who in comparison to present-day standards had a limited number of interventions to apply) often prescribed opium derivatives for their patients, particularly to treat ailments causing pain. Moreover, the introduction of the hypodermic syringe into the United States in 1856 made it possible for physicians to administer fast-acting dosages of morphine. The Civil War created a great need for medical intervention, and morphine was used extensively to treat both pain and dysentery (diarrhea). As a consequence, following the war, many veterans were addicted to the very drugs that had been medically administered to them. To make matters worse, physicians used heroin (a then newly discovered derivative of morphine) to treat morphine addictions. Originally, it was thought that heroin was not addicting, or at least much less addicting than morphine, but this belief was later proven to be quite erroneous.

The factor that is probably most directly responsible for the sharp increase in substance abuse in this country was the widespread use of patent medicines during the late 1800s and early 1900s. These substances sometimes contained large quantities of potent drugs such as alcohol, morphine, and cocaine, as well as various toxins. They were concocted by individuals or corporations who touted them as treatments for various physical and mental ailments, and they were available to anyone having the asking price. In fact, patent medicines became quite popular, with sales increasing from $3.5 million in 1859 to $74 million in 1904. As a result, a sizable proportion of the population of the United States is believed to have been addicted to drugs by the turn of the century.

The problems resulting from the uncontrolled use of addicting substances prompted increasing governmental regulation. In 1906, the Pure Food and Drug Act was passed which required that products containing addictive drugs such as alcohol, morphine, opium, cocaine, heroin, or marijuana be clearly labeled. One aim of this law was to deter the false therapeutic claims of manufacturers of patent medicines. In 1914, the Harrison Narcotics Act was passed which for the first time required dealers and dispensers of opiates and cocaine to register with the government and pay a small fee. Physicians, dentists, and veterinary surgeons were designated as the lawful distributors of these substances. The Harrison Act made it more difficult for drug addicts to legally obtain the drugs to maintain their addiction. Hence, many historians believe that its net effect was to drive the distribution of drugs "underground," with very serious consequences. Specifically, an illegal drug traffic arose that centered around organized crime.

Compared with other psychoactive drugs, alcohol has generally been more readily available and less regulated. This may be related to the fact that ordinary food substances, especially fruits and grains, can be used to produce alcohol through fermentation. The earliest attempts to regulate alcohol use in the United States were initiated by conservative religious groups. Combining their religious ideologies with the theme of self-reliance, hard work, and striving for personal accomplishment (the "pioneer ethic" of the early United States), these groups at first encouraged people to moderate their intake of alcohol. When, however, alcohol abuse became widespread and social problems associated with it mounted (especially among impoverished urban factory workers), both religious and secular elements of the society advocated total abstinence from alcohol. The "temperance movement" gained political momentum, leading eventually to legal prohibition of alcohol from American society. In 1851 the state of Maine was the first to pass a law to restrict the sale and use of alcohol. Other states followed suit so that by 1917, 64% of the population of the United States lived in "dry" areas. In 1920, the 18th Amendment to the United States Constitution, which prohibited the sale and transportation of alcoholic beverages throughout the United States, became effective. The Prohibition Amendment, however, did not eliminate alcohol problems from American society. There were unforeseen difficulties with enforcing the law, and an illegal industry, run largely by organized crime, was developed to supply that segment of the society that chose to continue to drink. In fact, because of the widespread disregard for the law and the profitable involvement of organized crime in the sale of alcohol, the 18th Amendment was eventually repealed by the 21st Amendment, which became effective in 1933.

Since World War II, social and political trends have significantly affected patterns of substance use and abuse in this country. The end of the war brought a general feeling of optimism among the American public that was accompanied by a period of relative affluence. During this time, the use of opioids was at a low level, except among poor people living in urban areas. However, during the late 1960s and early 1970s, there were reactions against political and social conservatism, and a "counter culture" developed where psychoactive drugs such as cannabis (marijuana and its derivatives) and hallucinogens were widely used. More recently, American society has again become relatively conservative, with a resulting reduction in the use of cannabis and hallucinogens. Concurrently, however, the use of cocaine and its derivatives has increased sharply, as has the number of people addicted to these drugs. Moreover, there has been a significant increase in the production and consumption of alcoholic beverages. Alcohol remains the most widely abused drug in the United States. In fact, studies published in the 1990s indicate that approximately one-fifth of the population of the United States consumes substantial amounts of alcohol and approximately 5% drinks heavily. Patterns of substance use and governmental regulation of it are partly determined by the philosophy of drug addiction that prevails in a particular society. Some societies consider substance abuse a moral problem that should be controlled through criminal penalities. Other societies view addiciton as a medical problem and the addict as someone in need of treatment or other rehabilitation. In the United States, both philosophies have been supported, although greater emphasis has been placed on criminalizing the use of psychoactive drugs than has been the case in other countries. For example, physicians in Great Britain can legally prescribe narcotics for heroin addicts, whose drug use might be maintained for many years, sometimes with minimal attempts to eliminate the addiction. It is interesting to observe that both the rate of narcotic addiction and the social problems associated with it (e.g., theft, organized crime, prostitution) are considerably lower in Britain than in the United States. This observation has prompted some people in the United States to propose a method to regulate the use of opioids that is similar to the one used in Great Britain. Others contend, however, that such an approach would result in rampant drug abuse in the United States. Hence, a consensus about an optimal solution to the current drug abuse problems in the United States has not been reached. [*See* Alcoholism.]

## III. DETERMINANTS OF ADDICTION

As indicated earlier, addiction has no simple explanation. It is a complex "biopsychosocial" phenomenon, with multiple determinants. This section reviews the major categories of explanations that have been offered to account for addiction and considers how these various explanations can be made congruous.

### A. Biological Determinants

There are wide variations in the ways that people react to psychoactive drugs, and some of this variability can be explained biologically. For example, different people metabolize alcohol and its metabolic by-products in different ways, due in part to genetically determined differences in their levels of liver enzymes for metabolizing alcohol and its metabolites. Acetaldehyde, one of the metabolites of alcohol, is a poisonous substance that can cause such symptoms as an upset stomach and flushing of the skin. People who have insufficient amounts of the liver enzyme aldehyde dehydrogenase, which is necessary for acetaldehyde to be metabolized, experience stronger negative effects from drinking than do other people. As a consequence of their biological "protection," these individuals are less likely to drink large quantities of alcohol—and hence less likely to become addicted to alcohol—than are people with an adequate level of aldehyde dehydrogenase.

Research on opioids further clarifies the role of biological factors in addiction. This research has identified specific receptors in the brain where opiates first begin to affect the central nervous system. Additionally, a group of opioid-like substances known as endorphins have been identified as chemicals that the body produces endogenously. These naturally produced substances are thought to play a critical role in the regulation of people's moods and psychological well-being and in their reactions to noxious, pain-inducing stimuli. What role do endorphins play in addictions? At this time the evidence is by no means conclusive, but one possibility is that people who are

addicted to opiates have imbalances in the levels of endorphins in their bodies, and they use opiates in order to bring about changes in their moods that the naturally occurring substances ordinarily would produce.

Similarly, the positive mood changes that people derive from drinking alcohol appear to be caused by the impact of alcohol on the neurotransmitters substances in the brain that give rise to pleasurable feelings. Thus, it seems likely that different people experience different positive effects from drinking alcohol because of differences in their brain chemistry. Finally, with regard to alcohol, it should be noted that although laboratory animals ordinarily refuse to drink alcohol, strains of animals that prefer drinking alcohol to water have been bred. The various lines of evidence that have been cited here leave little doubt that biological factors play a role in addiction.

### B. Psychological Determinants

The first efforts by psychologists to identify the psychological mechanisms involved in addiction were directed at identifying the "addictive personality." This approach involved administering personality tests to addicts in treatment in an effort to discover how their personality was different from that of other people. The ultimate aim was to identify the personality dynamics of addicts that would account for their addiction. Certain personality characteristics have, in fact, been frequently observed among people suffering from addictions. For example, they are often antisocial and impulsive, have a low tolerance for frustration, and seem to derive pleasure from activities and experiences that other people would regard as unusual or even dangerous. These personality characteristics, however, certainly do not describe all people who are addicted to psychoactive drugs. Accordingly, the idea that it is possible to identify an "addictive personality" that will serve as the sole explanation for the addict's addiction has not been supported.

At the same time, it has been clearly demonstrated that various other psychological mechanisms play an important role in drug addiction. One such mechanism is conditioning. When people habitually use a particular drug, they develop conditioned reactions to the stimulus cues associated with the drug-taking experience. As a result of such conditioning, when people encounter these same stimuli again in the future they are likely to "crave" the drug, thereby increasing their likelihood of using the drug and maintaining their addiction. Another example of a psychological mechanism involved in addiction is the emotional satisfaction—or lack of it—that people derive from their nondrug life experiences. When people encounter frustrations in life and do not have adequate psychological resources for coping with them, they are more likely than they would be otherwise to attempt to find emotional solace by using drugs. Similarly, if they do not have positive emotional experiences that preserve their sense of psychological well-being, they are more likely to try and find emotional satisfaction through chemical means.

## C. Sociocultural Determinants

Societies differ widely in their mores about the use of alcohol and other drugs. Certain societies, for example, tolerate and even encourage consumption of alcohol, whereas others completely prohibit drinking. Some Middle Eastern societies forbid alcohol consumption for religious reasons, yet they do not restrict the use of marijuana and hashish. Alcohol consumption rates in northern European countries and France are high, as is the prevalence of alcohol problems in these countries. Although alcohol consumption is also relatively high in Italy and Israel, these countries have low rates of alcohol problems. In short, societies differ widely both in the degree to which they encourage drug use and in the particular patterns of drug use that they foster. Moreover, the incidence of drug problems in a society is related to the particular drug use practices that the society instills.

How can we account for these wide variations in drug use among different societies? People's attitudes toward drugs—and the pattern of drug use that correspond to those attitudes—appear to be culturally transmitted through social learning mechanisms. People model their behavior after that of other people and are socially reinforced for doing so. When, on the other hand, people's behavior deviates from that of other people around them, they are socially ostracized for not following the established code of conduct. Accordingly, when individuals observe other people in their society using psychoactive drugs to induce pleasurable feelings and to counteract unpleasantness, they are likely to model these socially sanctioned uses of drugs.

The same kinds of differences in drug use that are observed among different societies can also be seen within particular societies. The rates of substance abuse vary significantly among the different subgroups in a given society. For example, as previously discussed, during the late 1960s and early 1970s, a sizable "subculture" in the United States used marijuana and hallucinogens extensively, although the majority of the society disapproved of this activity. Within diverse societies such as the United States, different ethnic groups use psychoactive drugs (even such generally socially sanctioned ones as caffeine) at levels significantly different from other segments of the society. These subgroup differences are related to cultural traditions, dietary habits, and religious beliefs. However, regardless of what other practices they are associated with, these differences can again be explained in terms of social learning principles. That is, people model their behavior after that of the groups to which they belong, such as their family and peers, and are influenced by the models whom they observe in the mass media and through advertising. People are reinforced for patterning their behavior after that of other people, but are socially disapproved when their behavior is different from that of other people around them.

## D. Integrating the Determinants of Addiction

Each of the determinants discussed in the preceding sections clearly plays a role in addiction, but none of the determinants taken alone can adequately account for addictive phenomena. How can we bring together the various determinants of addiction into a unifying model of addiction? One possibility for doing so is the motivational model—a model that originally was intended to apply to alcohol use, but which can easily be adapted to other forms of drug use and abuse. The model considers each of the kinds of variables that are known to affect drug use, considers ways in which these variables interact with one another, and suggests that the final pathway to drug use is motivational.

The motivational model interprets people's drug use in the context of their general motivational patterns. People are motivated to (a) acquire positive incentives that will bring them pleasure, and (b) get rid of—or avoid altogether—negative incentives that cause them discomfort. Thus, in an effort to achieve emotional satisfaction and to avoid discomfort, people organize their lives around striving for goals that will allow them to get, or get rid of, these two kinds of incentives. People are motivated to use drugs because drug use itself can serve as an incentive, e.g., people seek to find pleasure and to avoid discomfort by using drugs. This can occur in one of two ways. First, the drugs that people use can directly change the way they feel, either by enhancing positive feelings or counteracting negative feelings. Second, drug use can change the other incentives in peoples' lives. For exmaple, usings drugs can cause positive incentives to seem more attractive or it can cause negative incentives to seem more palatable.

How do the determinants of drug use discussed in the preceding sections figure into the motivational patterns discussed here? Each of the biological, psychological, environmental, and sociocultural determinants discussed earlier helps to define the incentive value of drug use and, in turn, the relative value of obtaining emotional satisfaction chemically and nonchemically. Thus, to the extent that each category of variable contributes to the positive feelings that a person expects to derive from drug use, the incentive value of using drugs will be enhanced. To the extent that each variable contributes to negative feelings that a person expects to derive from drug use, it will subtract from the incentive value of using drugs.

In short, a person might attempt to derive emotional satisfaction either chemically by using drugs or nonchemically through other life areas. People's decisions about using or not using drugs is (1) partly a rational process through which they weigh the positive and negative consequences of their behavior, and (2) partly an emotional process, based on the degree of emotional satisfaction that they expect to derive from drug use versus other life activities. In the final analysis, peoples' decision to use drugs is determined by the balance between the expected emotional satisfaction from using drugs and the expected emotional satisfaction from not doing so. A major advantage of the motivational model is that it offers promise for changing people's motivation for using drugs. Systematic motivational counseling, which is based on the motivational model, is designed to change drug abusers' motivational patterns, maximizing the emotional satisfaction that they derive from their nonchemical incentives, thereby reducing their motivation to seek emotional satisfaction by using drugs and, in turn, their actual drug use. The systematic motivational counseling technique is discussed in the following section.

## IV. TREATMENT OF ADDICTION

This section reviews some of the techniques that have been developed to treat drug-addicted people. Regardless of which treatment modality is used, the first step in assisting the addicted person must sometimes be detoxification—a medical procedure for helping ad-

dicts withstand the trauma and associated health risks that accompany abrupt withdrawal from the intoxicating drug. In fact, inpatient drug rehabilitation programs, which treat the more seriously addicted persons, often begin treatment with a period of detoxification. Nevertheless, detoxification must to some extent be tailored to the individual patient. That is, the attending physician must take into account the particular psychoactive drug or combination of drugs that the person has abused, the length of the abuse, and the medical complications that may have resulted from it. As indicated earlier, withdrawal from large doses of alcohol can be most severe. In order to prevent seizures and delirium tremens—visible tremors of the body along with hallucinations and delusions—the physician sometimes administers tranquilizing medications, such as Valium. Persons withdrawing from cocaine often are depressed, lethargic, and irritable. These symptoms are sometimes treated with antidepressant drugs or lithium, a medication that is ordinarily used to treat mood disorders. Individuals who have been addicted to drugs for many years or who have used them in high dosages sometimes have severe damage of the central nervous system and other internal organs. These conditions might require extensive rehabilitation and often are not completely reversible.

Following detoxification, the major goals of drug rehabilitation programs are (a) to teach addicts the skills that they need in order not to resume using drugs, and (b) to address the underlying problems that led to their using drugs in the first place. Different treatment programs deal with these two goals in very different ways. However, Alcoholics Anonymous (AA)—and adaptations of it, such as Narcotics Anonymous—is the intervention program that is most widely followed. In fact, many hospital programs, although using mulitple treatment strategies, depend heavily on the AA model and encourage patients to affiliate with AA when they leave the hospital.

Alcoholics Anonymous is a self-help program consisting of alcoholics helping other alcoholics. Established in 1934 by two recovering alcoholics, it has grown into a worldwide organization with millions of members. The members attend numerous meetings, sometimes as often as several times per day, with the frequency depending on individual members' needs and the stage of their recovery. The philosophy of AA consists of 12 guiding principles (the "12 steps"), which are discussed at the meetings, along with their application to individual members' current needs. The overriding goal of the AA program is for its members to remain completely abstinent from alcohol. Adherents admit that they have lost control both over alcohol and their lives, but they attempt to regain control by surrendering to a "higher power" and "working" the AA program. By doing so, they hope to develop a new, satisfying life-style that is free of alcohol.

Without doubt, many alcoholics have been helped by AA. At the same time, no scientific data exist on its effectiveness as a self-help organization because AA does not allow itself to be subjected to scientific scrutiny. What is clear, however, is that hospital programs that subscribe to the AA philosophy and whose effectiveness has been formally evaluated have been largely unsuccessful. Relapses (i.e., returning to problem drinking) are quite common among patients who complete these programs, with numerous alcoholics relapsing within a few months after having completed treatment. These discouraging results have prompted many professionals who work with alcoholics and other drug addicts to develop alternative treatment strategies.

Some of the alternative techniques are aimed at reducing the attractiveness of the abused drug. One example is aversive conditioning in which the abused substance is paired with an unpleasant, noxious stimulus, such as electric shock or nausea-inducing drugs. The goal is for the drug itself to acquire aversive qualities so that the addict will want to avoid it. Other techniques have been aimed at overcoming addicts' tensions and anxieties that may have propelled their drug use. For example, relaxation and stress management techniques have become an integral component of many treatment programs. Still other techniques have been employed to help addicts acquire skills (e.g., social or job-finding skills) to improve the quality of their lives in order to help them reduce their need to use drugs. Another technique—teaching alcoholics to drink moderately—has been surrounded by controversy because it challenges the most basic tenet of the AA philosophy, that alcoholics must maintain lifelong abstention from alcohol. Among professionals, the consensual opinion seems to be that controlled drinking is by no means advisable for the vast majority of chronic alcoholics, although it may be an appropriate strategy for a selected few. At the same time, controlled drinking has been used effectively with early-stage problem drinkers—those individuals who drink too much, but who are unwilling to enter a traditional treatment program where they would be diagnosed as "alcoholic" and told that they could never drink again.

The effectiveness of the newer interventions remains to be established, particularly when long-term abstinence is the criterion for successful treatment. Accordingly, professionals have begun to focus their efforts on "relapse prevention" during "aftercare." That is, once addicts have stopped using alcohol or other drugs with the aid of formal treatment and have returned to the environment of their everyday lives, they are taught skills for preventing relapses to abusive alcohol or drug use. The goal of relapse prevention is to help addicts overcome their maladaptive behaviors and cognitions that set the stage for eventual relapses to occur. Specifically, they are taught (1) how to deal with "high-risk" situations in which they are likely to resume their alcohol or drug use, and (2) how to develop a life-style that is balanced between obligatory and healthy, pleasurable activities. Helping patients to achieve a balanced life-style is also an aim of systematic motivational counseling, which was theoretically derived from the motivational model. According to the model, people's motivation to use alcohol or other drugs depends on the balance between the emotional satisfaction that they expect to achieve by doing so and by striving for nonchemical incentives in other life areas. However, whether people will succeed or fail in their goal strivings depends on how adaptive or maladaptive their motivational patterns are. For example, people are more likely to reach goals that are realistic (i.e., they have reasonable chances of achieving them), appropriate (i.e., reaching the goal is likely to produce the emotional satisfaction that was anticipated), and the person's various goals do not adversely affect one another (i.e., pursuit of one goal does not significantly interfere with attainment of another goal). Systematic motivational counseling helps alcohol and other drug addicts to formulate (1) appropriate realistic, and mutually compatible goals, and (2) plans for achieving them that will maximize their chances of success. In so doing, the aim of the technique is helping addicts find compelling sources of emotional satisfaction without the necessity to use alcohol or other drugs.

## BIBLIOGRAPHY

American Psychiatric Association (1994) "Diagnostic and Statistical Manual of Mental Disorders," 4th Ed. Washington, D.C.

Carroll, C. R. (1989) "Drugs in Modern Society." Brown, Dubuque, IA.

Cox, W. M. (ed.) (1987). "Treatment and Prevention of Alcohol Problems: A Resource Manual," Academic Press, Orlando, FL.

Cox, W. M. (ed.) (1990). "Why People Drink: Parameters of Alcohol as a Reinforcer." Gardner Press, New York.

Cox, W. M., and Klinger, E. (1988). A motivational model of alcohol use. *J. Abnorm. Psychol.* **97**, 168–180.

Hester, R. K., and Miller, W. R. (eds.) (1989). Handbook of Alcoholism Treatment Approches: Effective Alternatives. Pergamon Press, New York.

Institute of Medicine (1990). Broadening the Base of Treatment for Alcohol Problems." National Academy Press, Washington, D.C.

Maisto, S. A., Galizio, M., and Connors, G. J. (1995). "Drug Use and Misuse." Holt, Rinehart, & Winston, New York.

Marlatt, G. A., and Gordon, J. R. (eds.) (1985). "Relapse Prevention: Maintenance Strategies in the Treatment of Addictive Behaviors." Guilford Press, New York.

Marlatt, G. A. Larimer, M. E., Baer, J. S., and Quigley, L. A. (1993). Harm reduction for alcohol problems: Moving beyond the controlled drinking controversy. *Behav. Ther.* 24, 461–504.

Miller, W. R., and Heather, N. (eds.) (1986). "Treating Addictive Behaviors: Processing of Change." Plenum, New York.

Ray, O., and Ksir, C. (1992). Drugs, Society and Human Behavior, 6th Ed. MosbyYear Book, St. Louis, MO.

U.S. Department of Health and Human Services (1993). "Alcohol and Health." Washington, D.C.

# Adenosine Triphosphate (ATP)

LEOPOLDO de MEIS

*Universidade Federal do Rio de Janeiro*

## GLOSSARY

**Free energy of hydrolysis** Energy released during the hydrolysis of a compound that can be used to perform work

**Hydrolysis** Chemical reaction in which a compound reacts with water

ADENOSINE TRIPHOSPHATE (ATP) IS THE PRINCIPAL carrier of energy in the living cell. ATP consists of an aromatic base called adenine, a five-carbon sugar called ribose, and three phosphate groups, $\alpha$, $\beta$, and $\gamma$ with the $\alpha$-phosphate linked to the ribose by an ester bond and the $\beta$- and $\delta$-phosphates linked together by two phosphoanhydride bonds. Despite the complex structure of the ATP molecule, only the $\beta$- and $\gamma$-phosphates of ATP are used for the transfer of energy in the cell. This part of the molecule is similar to pyrophosphate ($PP_i$). It was previously thought that the standard free energy of hydrolysis ($\Delta G°$) of ATP and of $PP_i$ was the same. It is now known that the $\Delta G°$ of $PP_i$ hydrolysis is 3 to 4 kcal/mol less negative than that of ATP. It is not known why nature selected a complex molecule such as ATP as an energy carrier instead of a much simpler molecule as $PP_i$ or why the presence of adenine and ribose in the molecule may alter the energy of hydrolysis of the phosphoanhydride bond of ATP. In the cell, ATP is hydrolyzed to either adenosine diphosphate (ADP) and orthohphosphate ($P_i$) or to adenosine monophosphate ($AMP$) and $PP_i$. The hydrolysis of ATP is usually coupled with work. This can be machanical, as observed in muscle contraction; chemical, as for the synthesis of molecules; or osmotic, as when a gradient is formed across a membrane. Several processes of energy conversion are reversible. Thus, several membrane-bound ATPase hydrolyze ATP in order to build up an ionic gradient across a membrane. In the reverse process, the energy derived from the gradient can be used to synthesize ATP from ADP and $P_i$. [*See* Cell.]

## I. UTILIZATION BY ENZYMES OF THE ENERGY PROVIDED BY ATP AND OTHER HIGH ENERGY PHOSPHATE COMPOUNDS

It is still not known how energy flows from ATP to the enzyme during the process of energy interconversion. It had been thought that the energy of hydrolysis of ATP and other "high energy" phosphate compounds was the same regardless of whether they were in solution in the cytosol or bound to the enzyme surface and that energy would become available to the enzyme only after the phosphate compound had been hydrolyzed. This view was based in calorimetric

ENCYCLOPEDIA OF HUMAN BIOLOGY, Second Edition, VOLUME I. Copyright © 1997 by Academic Press. All rights of reproduction in any form reserved.

measurements performed at the beginning of this century which led to the conclusion that energy would only be released to the medium at the moment of cleavage of the phosphate bond. In this view the sequence of events in the process of energy transduction was thought to be: (1) the enzyme binds ATP; (2) ATP is hydrolyzed and energy is released at the catalytic site at the precise moment of cleavage of the phosphate bond; (3) the energy is immediately absorbed by the enzyme; and (4) the enzyme uses the energy absorbed to perform work.

For the synthesis of ATP from ADP and $P_i$, the sequence of events would be the same, but in the reverse order: (1) The enzyme would bind ADP and $P_i$; (2) energy would be released at the catalytic site of the enzyme and used to synthesize ATP; and (3) once formed, the ATP molecule would easily dissociate from the enzyme and diffuse into the cytosol without any further need of energy.

From these sequences, it was inferred that work would be performed in a part of the enzyme molecule where energy is released, i.e., in the immediate vicinity of the catalytic site.

The catalytic cycle of different enzymes has been elucidated since the mid-1970s. These studies revealed that the energy of hydrolysis of different phosphate compounds varies greatly depending on whether they are in solution, or bound to an enzyme (Table I). Reactions that were thought to be practically irreversible in aqueous solution, such as the phosphorylation of glucose by ATP, occur spontaneously when the reactants are bound to the enzyme (Table I). For en-

zymes involved in energy transduction, such as the $Ca^{2+}$-ATPase found in the sarcoplasmic reticulum of skeletal muscle, the energy becomes available for the enzyme to perform work before cleavage of the phosphate compound. During the catalytic cycle there is a large decrease of the equilibrium constant for the hydrolysis ($K_{eq}$) of the phosphate compound bound to the enzyme. In several enzymes studied, work is coupled with this transition of $K_{eq}$ and not with the cleavage of the compound (Table I). Hydrolysis seems necessary only to permit the dissociation of the nucleotide from the enzyme and not to provide energy to the system. In this view, the sequence of events for the hydrolysis of ATP is: (1) The enzyme binds ATP or other phosphate compounds; (2) the enzyme performs work without the phosphate compound being hydrolyzed, which is accompanied by a decrease in the energy level of the phosphate compound; (3) the phosphate compound is hydrolyzed in a process which involves relatively small energy change; and (4) the products of hydrolysis dissociate from the enzyme. In the reverse process, phosphate compounds such as ATP and acylphosphate residues are synthesized on the catalytic site of the enzyme without the need of energy. Energy is then needed for the conversion of the phosphate compound from "low energy" into "high energy." This transition usually occurs on the enzyme surface, before the phosphate compound being synthesized is released into the cytosol. A fair amount of information is now available on the structure of several energy-being transducing enzymes. These data indicate that work is being performed in a region of

TABLE I

Variability of the Energy of Hydrolysis of Phosphate Compounds during the Catalytic Cycle of Energy-Transducing Enzymes[a]

| Enzyme | Reaction | Solution or enzyme bound, before work | | Enzyme bound, after work | |
|---|---|---|---|---|---|
| | | $K_{kq}$ (M) | $\Delta G°$ kcal/mol | $K_{kq}$ (M) | $\Delta G°$ kcal/mol |
| $Ca^{2+}$-ATPase, $Na^+/K^+$-ATPase | Aspartyl phosphate hydrolysis | $10^6$ | $-8.4$ | 1 | 0 |
| $F_1$-ATPase, myosin | ATP hydrolysis | $10^6$ | $-8.4$ | 1 | 0 |
| Inorganic pyrophosphatase | $PP_i$ hydrolysis | $10^4$ | $-5.6$ | 4.5 | $-0.9$ |
| Hexokinase | ATP + gluc → gluc-P + ADP | $2 \times 10^3$ | $-4.6$ | 1.0 | 0 |

[a]For details, see de Meis (1993) and Romero and de Meis (1989). The relationship between the standard free energy of hydrolysis ($\Delta G°$) and the equilibrium constant of the reaction ($K_{eq}$) is $\Delta G° = -RT \ln K_{eq}$, where $R$ is the gas constant (1.981) and $T$ is the absolute temperature.

the tertiary structure of the protein that is distant from the protein region where the catalytic site is located and that conformational changes of the protein synchronize the sequence of events occurring in these two regions of the protein. The events responsible for the change in the energy level of phosphate compound at the catalytic site are not clearly understood at present. Experimental evidence suggests that one important factor is change in water activity in the environment of the catalytic site.

The difference between the two views is related to the contribution of enthalpy ($\Delta H°$) and entropy ($\Delta S°$) to the free energy of hydrolysis of ATP according to the equation:

$$\Delta G° = \Delta H° - T\Delta S°.$$

In the early view, the contribution of entropy was thought to be minimal, thus free energy and enthalpy would be practically the same. Possible interactions of the phosphate compound with solvent and physiological ions could not be evaluated in calorimetric measurements and thus were not taken into account. These interactions play an important role in determining the entropy of phosphate compound hydrolysis. It is now known that the $K_{eq}$ for the hydrolysis of $PP_i$ and ATP varies greatly depending on the water activity, pH, and divalent cation concentration of the medium. These changes are related mainly to changes of the entropy of the reaction. However, water activity, pH, and divalent cation have practically no effect on the $K_{eq}$ of phosphoesters, such as glucose-6-phosphate and phosphoserine. These measurements and the finding that enzymes may perform work before cleavage of ATP indicate that the contribution of entropic energy ($T\Delta S°$) may surpass the contribution of enthalpic energy.

In the subsequent sections, the events that led to this new concept will be described in a historical sequence.

## II. HEAT OF COMBUSTION AND ENERGY OF CHEMICAL BONDS

In order to understand the evolution of the concept of "high energy," one must go back to the work of Antoine Lavoisier at the end of the 18th century. Lavoisier was a meticulous accountant in charge of collecting the impost due to the king of France. It was his responsibility to see that the sum of the money collected matched the sum of tax money due to the

realm. Science was the major hobby of Lavoisier, and his activity as an accountant, which earned his living, played a major role in his approach to science. Before the studies of Lavoisier, the decrease of solid mass observed during the combustion of organic matter was attributed to the release of "phlogiston," an entity that the alchemists of the middle age believed to be related to the soul of the matter. It was thought that phlogiston was released both during the combustion of organic materials and during the respiration of animals. Lavoisier proved that both processes were chemical reactions. He weighed the gas and solid matter contained in a closed system before and after combustion and found that the total mass of the system did not vary during combustion. Thus, the hypothetical phlogiston should have a mass and be related to the gases found in the air before and after combustion. During the course of his experiments, Lavoisier discovered a new gas, oxygen, and demonstrated in both combustion and respiration that the oxygen of the atmosphere was consumed with the production of carbon dioxide ($CO_2$). The decrease of mass observed during the combustion of organic matter plus the mass of the oxygen consumed could be accounted by the sum of the masses of water and $CO_2$ produced. The central feature of the chemical revolution initiated by Lavoisier was the overthrow of the phlogiston theory and its replacement by a theory based on the role of oxygen and the new concept known as "conservation of matter" which explains that the total mass of all reacting substances in a chemical reaction must be identically equal to the total mass of the product substances.

Another important discovery was that the heat released by living animals was derived from the oxidative reactions involved in respiration. This conclusion was reached comparing respiration with combustion. In 1780, Lavoisier and Pierre Simon Laplace measured the amount of heat released when equal amounts of $CO_2$ were produced in the combustion of charcoal and in the respiration of guinea pig. For this experiment, Laplace devised an ice calorimeter. They found that the amounts of ice melted in the two processes were practically the same after burning charcoal to yield an amount of $CO_2$ equal to that exhaled by the animal during respiration. From these experiments, they concluded that "Respiration is therefore a combustion, very slow it is true. but otherwise perfectly similar to that of charcoal." The complete work on animal respiration was published in 1793. Unfortunately, Lavoisier was murdered in the guillotine on May 8, 1794.

The relationship between work and heat was not envisaged by Lavoisier. At that time it was not known that different forms of energy could be interconverted. In fact, at that time, energy was beginning to be regarded as a physical measurable entity. Before that, energy was referred to as a "natural force." The interconversion of energy was first observed in a mechanical system by Count B. Rumford, who studied the rise of temperature that accompanied the boring of a cannon. In 1798, Rumford concluded that the mechanical work involved in the boring was responsible for the heat produced. The subject did not attract any great interest until 1842 when Julius Robert Mayer enunciated the principle of the interconvertibility of energy. Mayer, who studied medicine at the Tubingen University, took his degree in 1838 after presenting a dissertation on the effect of santonin on worms in children, and in 1840 he signed on as a ship's doctor. According to Mayer's own story, his interest on the relation between heat and work began abruptly in Java, on the dock at Surabaya, when several of the sailors needed to be bled. The venous blood was such a bright red that at first he thought he had opened an artery. The venous blood of the sailors remained a bright red until they had acclimated themselves to the tropics. From the work done by Lavoisier, it was known that animal heat was generated by a combustion process. Mayer correlated the difference of color between the arterial and venous blood with oxidative reactions and heat production in the body. Because the animal heat was created by the oxidation of nutrients, Mayer put forth the question of what happened if, in addition to heat, the body also produced work. From an identical quantity of food, sometimes more and sometime less heat could be obtained. If a fixed total yield of energy from food is obtainable, then one must conclude that work and heat are interchangeable quantities of the same kind. By burning the same amount of food, the animal body can produce different proportions of heat and work, but the sum of the two must be constant. After this initial observation, Mayer started to analyze the values of heat absorbed by expanding gases and used them to compute the amount of heat equivalent to a given amount of work. From these studies, Mayer stated the principle of interconvertibility of energy and the conservation of energy. The relationship between the color of the venous blood of the sailors and the equivalence between heat and work is not readily apparent. Intuition probably played an important role in Mayer's discovery. The first paper he wrote was not accepted for publication because it seems that Mayer was confused about the distinctions between the concepts of force, work, and energy. His second paper was accepted by Liebing in

1842 for publication in the *Annalen der Chemie und Pharmazie*. In 1840, J. P. Joules commenced his classical experiments on the relationship between work expended and heat produced. However, the paper of Mayer appeared before any of Joule's results were published. The work of Mayer and of Joules aided in the deduction of the conversion factor between mechanical work ($w$) and heat ($q$). This is called the mechanical equivalent of heat $J$:

$$J = w/q.$$

In modern units, $J$ is usually given as joules per calorie. The combined work of Lavoisier and Mayer helped in deducing the first law of thermodynamics, which states that, "in nature nothing is created nor destroyed. All is transformed." Lavoisier demonstrated the conservation of the matter and that energy is released in biological reactions whereas Mayer demonstrated the conservation and transformation of energy. It is frequently stated that Lavoisier viewed a living system as a furnace to produce heat whereas Mayer saw it as a heat engine.

During the next 100 years it become apparent that the energy needed to sustain life should be derived from the cleavage of molecules in the tissues. In order to measure this energy, different substances were burned in a calorimeter and the heat released was measured as first reported by Lavoisier and Laplace. At the beginning of this century, long tables with the values of heat of combustion of many different substances were available. From these values, attempts were made to calculate the heat released after the cleavage of the different bonds which unite the atoms of a molecule. For instance, it was estimated that for every electron between C and C, and between C and H, 26,050 calories were generated during combustion and that 19,500 calories could be derived from a C:O cleavage.

At the time of the discovery of the ATP molecule in the 1920s, muscle contraction was the physiological model used to study the process of energy interconversion. In these experiments, calorimetric measurements were correlated with mechanical work. A paradigm emerged from these studies according to what energy would be released when a covalent bound between two atoms of a molecule was cleaved. When the cleavage was catalyzed by enzymes able to transduce energy, as in muscle contraction, then a part of the energy released would be absorbed by the protein and used to perform work whereas another part could be measured as heat. If the enzyme was not fit to perform work, then all the energy released would be dissipated as heat. These notions set the framework for the concept of an energy-rich bond that followed the identification of the different phosphate compounds of the cell. Although useful at an early stage of bioenergetics, the calorimetric measurement proved to be misleading later on, and the long lists with the values of heat of combustion of substances prepared at the beginning of the century were progressively abandoned and dropped. One of the bias introduced by calorimetry was to take the heat values indiscriminately as a true indication of the free energies.

## III. DISCOVERY OF THE ATP MOLECULE

The history of ATP began in 1847 with the studies of Justus Liebig on the content of meat extracts. Liebig was a precursor of biotechnology. At that time, meat storage and distribution on a large scale were practically impossible. Liebig prepared a meat extract that could be stored and commercialized. This extract made him famous. Liebig also become famous for his bitter criticism of Luis Pasteur elegant experiments on fermentation. Liebig crystallized a barium salt of inosinic acid from a meat extract. Analytical data for carbon, hydrogen, and nitrogen permitted the characterization of inosinic acid, but Liebig did not detect phosphorus in his measurements. The attachment of phorphorus to the adenosine molecule was first described in 1927 by Embden and Zimmermann who discovered adenosine-5'-phosphoric acid (AMP). ATP was finally isolated as a silver salt in the following year by Fiske and Subbarow. The biological role of ATP, however, was only characterized several years after its discovery.

## IV. USE OF ATP AS IMMEDIATE DONOR OF FREE ENERGY

The discovery of the physiologic role of ATP is intimately associated with that of creatine phosphate, a compound used to regenerate ATP in the cell:

ADP + creatine phosphate → ATP + creatine.

In fact, creatine phosphate, originally called phosphagen, was discovered by Fiske and Subbarow before ATP, and for several years most of the effects of ATP

were attributed to creatinine phosphate. Studying the energetics of muscle contraction, both the Eggletons and Fiske and Subbarow found that creatine phosphate was largely decomposed during a long series of contractions and was rapidly reconstituted during recovery of the muscle. At the same time, Meyerhof and Suranyi found that large amounts of heat were released by the enzymatic decomposition of creatine phosphate. In accordance with the paradigm prevailing at the time, these observations naturally led to the conclusion that energy would be released during cleavage of the phosphate bond of creatine phosphate. For the resynthesis measured during muscle recovery, energy should then be provided by the cleavage of other molecules. Accordingly, Nachmansohn observed that the creatine phosphate cleaved during muscle contraction could be resynthesized at the expense of glycolysis, and Lundsgaard found that through the breakdown of 0.5 mole of glucose to lactic acid, approximately 2 moles of creatine phosphate was reformed.

Evidence that creatine phosphate was not directly used for muscle contraction was first reported by Lohmann. He observed that muscle homogenates were able to catalyze the hydrolysis of creatine phosphate just as the intact tissue does. If the extract was, however, dialyzed, then the proteins were no longer able to cleave creatine phosphate, indicating that there was not a specific enzyme to catalyze its cleavage. The hydrolysis of creatine phosphate was restored if ADP was added together with creatine phosphate to the dialyzed extract. From this observation it was concluded that the true substrate of muscle contraction was ATP and that creatine phosphate was used to regenerate the ATP cleaved. This was confirmed in intact muscle cells with the use of 1-fluoro-2,4-dinitrobenzene (FDNB), a substance that stops glycolysis and also inhibits the transfer of phosphate from creatine phosphate to ADP. Muscles treated with FDNB contract normally, and their contraction is accompanied by the breakdown of ATP, with the content of creatine phosphate of the muscle remaining unchanged. Shortly after, different laboratories demonstrated that the main product of glycolysis and other catabolic routes is ATP and not creatine phosphate. This led to the conclusion that ATP is in fact the major immediate donor of free energy in biological systems. During catabolism ADP is phosphorylated to ATP and during anabolism ATP is hydrolyzed. Creatine phosphate is a storage form of free energy found in larger amounts in skeletal and cardiac muscles of vertebrates. In its place, most invertebrates contain phosphoarginine, which also serves to regenerate ATP from ADP.

## V. CONCEPT OF ENERGY-RICH PHOSPHATE BOND

The concept of "energy-rich" and "energy-poor" phosphate compounds was formalized by Lipmann in a review published in 1941 that become a classic in the bioenergetic bibliography. This review formally stated that entropy played a minor role in determining the thermodynamic parameter of a metabolic reaction. Thus, in the cell the value of free energy would be practically the same as that of enthalpy. The amount of energy that could be derived from the hydrolysis of a phosphate compound would be determined solely by the chemical nature of the bond that links the phosphate residue to the rest of the molecule. The possibility that some energy might be derived from the interaction of reactant and product with the environment (solvent, cations, etc.) was not taken into consideraton at that time because the energy of hydrolysis of most phosphate compounds was estimated using calorimetric measurements. The N~P bond of creatine phosphate and the phosphanhydride linkages P-O~P, carboxyl~P, and enol~P were identified as energy-rich phosphate compounds, thus having a $K_{eq}$ for the hydrolysis with a high value, ranging from $10^6$ to $10^9$ $M$ ($\Delta G°$ -8 to -12 kcal/mol). Phosphoesters such as glucose 6-phosphate and glycerol phosphate were referred to as "energy-poor" phosphate compounds. The $K_{eq}$ for the hydrolysis of phosphoester varies between 10 and 100 $M$ ($\Delta G°$ between $-1.5$ and $-2.5$ kcal/mol). The methodology used to determine the energies of hydrolysis of the two groups of phosphate compounds was different. For the energy-poor phosphoester, the molar concentrations of reactants and products available after the reaction reached equilibrium were measured, and from the value of the $K_{eq}$ for hydrolysis, the $\Delta G°$ of the reaction was calculated. Thus, for the hydrolysis of glycerolphosphate measured by Cori and associates:

glycerolphosphate + $H_2O$ → glycerol + $P_i$

$K_{eq}$ = [glycerol] × [$P_i$]/[glycerolphosphate]

and $\Delta G° = -RT \ln K_{eq}$,

where $R$ is the gas constant (1.981) and $T$ is the absolute temperature in which the reaction was per-

formed. This method could not be used for ATP and creatine phosphate because the amount of the energy-rich phosphate compound remaining in solution after the reaction reaches equilibrium was very small and impossible to measure with the methods available at that time. Therefore, free energies were calculated from the heat of combustion and heat release during hydrolysis.

In his review, Lipmann introduced the term "group potential" which strongly influenced the course of the theoretical studies in subsequent years. Linkages designed to transfer groups with loss of energy were called "weak" linkages, including the "energy-rich" phosphate bounds. Lipmann reasoned that if large amounts of energy could be made free with cleavage, then the tendency to burst the linkage would be relatively great and the linkage would thus be weak, i.e., the phosphate group would have a small affinity for the rest of the molecule. The energy-rich, weak phosphate bond was designated by a squiggle (~ph or N~P). If little energy would be freed with cleavage or if energy has to be furnished, then the linkage was called strong (large affinity). According to the Lipmann proposal, the amount of energy that could be derived from the hydrolysis of a phosphate compound would be determined solely by the chemical nature of the bond cleaved. The transfer of phosphate from one molecule to another would be determined by the energy of hydrolysis of the bond involved (Fig. 1). The γ-phosphate of ATP could be transferred to a molecule of glucose forming glucose 6-phosphate, a compound with a lower energy of hydrolysis than ATP, but the reverse reaction could not occur without an extra input of energy. Thus, the ATP hydrolyzed in the cell could only be regenerated from phosphate

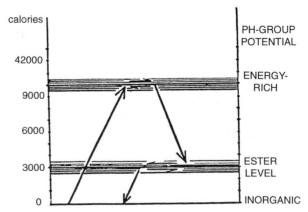

**FIGURE 1** Lipmann's schematic representation for the energy level of different chemical bonds.

compounds having the same or a higher energy of hydrolysis than ATP itself, such as creatine phosphate.

## VI. THEORETICAL APPROACH

From 1941 until 1969 the theoretical studies of high-energy compounds naturally followed the proposal of Lipmann, focusing in particular on the affinity of the phosphate group for the rest of the molecule. Thus, it was thought that intramolecular effects such as opposing resonance, electrostatic repulsions, and electron distribution along the P–O–P backbone were the dominant factors contributing to the large negative-free energies of hydrolysis of high energy phosphate compounds such as pyrophosphate and ATP:

$$R-\overset{\overset{\displaystyle O^-}{|}}{\underset{\underset{\displaystyle O}{\|}}{P}}-\overset{+}{O}=\overset{\overset{\displaystyle O^-}{|}}{\underset{\underset{\displaystyle O}{|}}{P}}-OH \rightarrow R-\overset{\overset{\displaystyle O^-}{|}}{\underset{\underset{\displaystyle O}{\|}}{P}}-O-\overset{\overset{\displaystyle O^-}{|}}{\underset{\underset{\displaystyle O}{\|}}{P}}-OH$$

$$\rightarrow R-\overset{\overset{\displaystyle O^-}{|}}{\underset{\underset{\displaystyle O}{|}}{P}}=\overset{+}{O}-\overset{\overset{\displaystyle O^-}{|}}{\underset{\underset{\displaystyle O}{\|}}{P}}-OH$$

The negative charges on either side of the linkage would repel each other, creating tension within the molecule, and the opposing resonance would generate points of weakness along the P–O–P backbone which tends to stabilize the products relative to reactants. Thus, it would be easy to cleave the molecule and difficult to bring together the products of the hydrolytic reaction. In these formulations, water was ignored or regarded as a continuous dielectric for the purpose of calculating repulsion energies.

In 1970, George and colleagues used a totally different approach. They were the first to propose that the interaction of reactants and products with the solvent might play an important role in determining the $K_{eq}$ of a reaction. George and associates reasoned that phosphate compounds interact strongly with water (Table II). In aqueous solutions, water molecules will organize around the phosphate compound and will both shield the charges of the molecules, thus neutralizing the electrostatic repulsion, and form bridges between different atoms of the molecule, thus reinforcing the weak points generated along the molecule backbone by apposing resonances. Therefore, George and co-workers proposed that the energy of hydrolysis of a phosphate compound would be determined by

the differences in solvation energies of reactants and products. Solvation energy is the amount of energy needed to remove the solvent molecules that organize around a substance in solution. Thus, a more solvated molecule would be more stable, i.e., less reactive, than a less solvated molecule and the $K_{eq}$ for hydrolysis would have a high value because the products of the reaction are more solvated than the reactant. The solvation energies of orthophosphate and pyrophosphate are shown in Table II. In totally aqueous medium and depending on the experimental conditions used, the observed standard energy of hydrolysis ($\Delta G°$) measured for the hydrolysis of pyrophosphate varies between $-3$ and $-6$ kcal/mol (Table III). This represents a very small fraction of the total solvation energy of either orthophosphate or pyrophosphate (Table II). Thus, a small change in the organization of solvent around the molecules of reactants and products might easily lead to a significant change in the thermodynamic parameters of a reaction.

In 1978, Hayes and colleagues calculated the energy of hydrolysis of several phosphate compounds in the gas phase and compared these values with those measured in water (Table III). In an aqueous solution, acetylphosphate and the N~P bonds in both phosphocreatine and phosphoarginine are of a high-energy nature. However, in the gas phase this is no longer true. On the contrary, the large positive $\Delta H$ of hydrolysis indicates that when reactants and products are not solvated, acetylphosphate and phosphocreatine are more stable than the products of their hydrolysis and, according to the definition of Lipmann, they behave as "energy-poor" phosphate compounds. From these data, Hayes and associates concluded that the solvation energies of reactants and products are

### TABLE II

Solvation Energies of Different Ionic Forms of Inorganic Phosphate and Pyrophosphate[a]

| Molecule | Solvation energy (kcal/mol) |
|---|---|
| $H_2PO_4^-$ | 76 |
| $HO_4^{2-}$ | 299 |
| $PO_4^{3-}$ | 637 |
| $H_3P_2O_7^-$ | 87 |
| $H_2P_2O_7^{2-}$ | 134 |
| $HP_2O_7^{3-}$ | 358 |
| $P_2O_7^{4-}$ | 584 |

[a]Values reported by George *et al.* (1970).

**TABLE III**

Energy of Hydrolysis of Different Phosphate Compounds in Totally Aqueous Medium, Gas Phase, and Mixtures of Water with Organic Solvents

| Compound | $\Delta G°$ kcal/mol | | |
|---|---|---|---|
| | Water | Gas phase | Organic solvent mixtures |
| ATP | −7.0 to −9.0 | — | +0.3 |
| $PP_i$ | −3.0 to −6.0 | −0.4 to −0.9 | −1.0 to +2.0 |
| Aspartyl phosphate | −9.0 to −11.0 | +5.0 to +32.0 | +0.3 to +2.3 |
| Creatine phosphate | −9.0 to −11.0 | +9.0 to +212.0 | — |
| Glucose 6-phosphate | −1.5 to −3.0 | −1.5 to −2.5 | −1.5 to −2.5 |

by far the most important factors in determining the energies of hydrolysis of creatine phosphate, ATP, $PP_i$, and acetylphosphate. The same conclusion has been reached by Ewing and Van Wazer, who calculated the energy of hydrolysis of $PP_i$ in the gas phase. Solvation energy, however, does not seem to determine the energy of hydrolysis of all phosphate compounds. According to the calculations of Hayes and co-workers (Table II), the energy of hydrolysis of phosphoesters such as glucose 6-phosphate is the same in the water phase and in the gas phase, indicating that the interaction of reactant and products with the solvent does not play a significant role in determining the $K_{eq}$ of hydrolysis of a phosphoester. A peculiar situation then becomes apparent. In aqueous solutions, an acylphosphate residue such as that of aspartyl phosphate or creatine phosphate has a much higher energy of hydrolysis than glucose 6-phosphate, but in the gas phase the situation is reversed and glucose 6-phosphate becomes a compound with a higher energy of hydrolysis than either creatine phosphate or aspartyl phosphate.

## VII. PHOSPHOANHYDRIDE BONDS OF HIGH AND LOW ENERGY

The simplest known "high-energy" phosphate compound is pyrophosphate. The only possible product of $PP_i$ hydrolysis is $P_i$, which greatly facilitates the measurement of $K_{eq}$ and the calculation of $\Delta G°$ as compared with more complex molecules such as ATP where the ADP produced after hydrolysis can be further cleaved to AMP. Therefore, because of the similarities between the polyphosphate chain of ATP and $PP_i$, the thermodynamic parameters measured for $PP_i$

are often extrapolated to ATP. Measurements of the $K_{eq}$ revealed that the $\Delta G°$ of $PP_i$ hydrolysis varies greatly depending on the pH and divalent cation concentration in the medium (Fig. 2). A similar variability was found for ATP hydrolysis. By adding the free energy changes of hydrolysis of several different reactions, Alberty calculated that in the pH range of 6.0 to 8.0 and depending on the $Mg^{2+}$ concentrations, the $\Delta G°$ of ATP hydrolysis may vary between −5.0 and −9.9 kcal/mol. However, pH and $Mg^{2+}$ have practically no effect on the $\Delta G°$ of phosphoester hydrolysis (Fig. 2). Thus, depending on the experimental conditions used, the $\Delta G°$ of $PP_i$ hydrolysis can be either several kilocalories more negative than that of glucose 6-phosphate or have practically the same value as that of glucose 6-phosphate. A similar pattern was observed when the water activity ($w_a$) of the medium

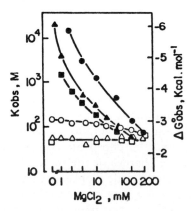

**FIGURE 2** Effect of $Mg^{2+}$ and pH on the energies of hydrolysis of $PP_i$ and phosphoesters. The $K_{eq}$ for hydrolysis of $PP_i$ (filled symbols) and phosphoserine (open symbols) were determined at pH 6.1 (○, ●), 7.0 (△, ▲), and 7.8 (□, ■).

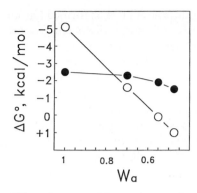

**FIGURE 3** Effect of water activity on the energies of hydrolysis of $PP_i$ (○) and glucose 6-phosphate (●).

was decreased with the use of organic solvents (Fig. 3). The solvation of molecules in aqueous solution varies depending on the bulk water activity. A decrease of $w_a$ promotes a drastic increase of the $\Delta G°$ of both $PP_i$ (Table III and Fig. 3) and ATP, but has practically no effect on the energy of hydrolysis of phosphoesters such as glucose 6-phosphate. Thus, in a totally aqueous medium ($w_a = 1$), the compounds ATP and $PP_i$ have a larger energy of hydrolysis than glucose 6-phosphate, but after a small decrease of $w_a$, the situation is inversed and glucose 6-phosphate become a phosphate compound having a larger energy of hydrolysis than either ATP or $PP_i$.

The values of $\Delta G°$ in Table III and Fig. 3 are related to the total concentrations of all ionic species of $PP_i$ and $P_i$, including those that are free and those that are in the form of a complex with divalent cations. The change of $\Delta G°$ found at the different pH and $Mg^{2+}$ concentrations reflects the balance of the different ionic reactions found in each experimental condi-

tion. By measuring the $K_{eq}$ of a reaction at different temperatures, it is possible to calculate the $\Delta H°$ and $\Delta S°$ values of each ionic reaction. Note in Table IV that the values of $\Delta H°$ for the different ionic reactions of $PP_i$ hydrolysis vary little. What does vary significantly among the different reactions is the $\Delta S°$ value. According to the equation $\Delta G° = \Delta H° - T\Delta S°$, this finding indicates that the large variability of the $\Delta G°$ of $PP_i$ hydrolysis observed when the conditions of the medium are changes is related mostly to entropic changes of the ionic reactions prevailing in the system and not to a significant change of enthalpy. This is not observed for phosphoesters. The $K_{eq}$ varies little with temperature, and the contribution of entropy to the value of $\Delta G°$ is minimal.

The effects of $w_a$, pH, and magnesium on the $\Delta G°$ of $PP_i$ and glucose 6-phosphate can be easily interpreted according to the solvation energy theory, but difficult to accommodate with the Lipmann proposal. If the energy of the reaction could be determined solely by the nature of the phosphate bond cleaved, then the energy of hydrolysis of $PP_i$ should always be higher than that of glucose 6-phosphate, regardless of the conditions used. In addition, the effect of pH and magnesium on the energy of hydrolysis of $PP_i$ should be related to the $\Delta H°$ of the different ionic reactions and not to different $\Delta S°$ values as observed (Table IV).

## VIII. TRANSITION BETWEEN BULK SOLUTION AND ENZYME SURFACE

From the data of Fig. 2 and Table I and III it can be deduced that the thermodynamic parameters of a reaction may vary greatly depending on whether the

**TABLE IV**
Ionic Reactions of $PP_i$ and Glucose 6-phosphate Hydrolysis

| Reaction | $\Delta G°$ (kcal/mol) | $\Delta H°$ (kcal/mol) | $\Delta S°$ (e.u.) |
|---|---|---|---|
| $HP_2O_7^{3-} + HOH \rightleftharpoons H_2PO_4^- + HPO_4^{2-}$ | −5.2 | −8.2 | −10.1 |
| $MgP_2O_7^{2-} + HOH \rightleftharpoons MgHPO_4 + HPO_4^{2-}$ | −3.3 | −8.4 | −16.8 |
| $MgP_2O_7^{2-} + HOH \rightleftharpoons 2HPO_4 + Mg^{2+}$ | −0.6 | −11.4 | −35.8 |
| $MgHP_2O_7^{2-} + HOH \rightleftharpoons MgHPO_4 + H_2PO_4^-$ | −3.6 | −8.8 | −17.5 |
| $MgHP_2O_7^{2-} + HOH \rightleftharpoons H_2PO_4^- + HPO_4^{2-} + MG^{2+}$ | −0.8 | −11.8 | −36.1 |
| $Glucose\text{-}P + HOH \rightleftharpoons glucose + H_2PO_4^-$ | −2.4 | −1.5 | −3.1 |
| $Glucose\text{-}P + HOH \rightleftharpoons glucose + HPO_4^{2-}$ | −2.4 | −1.5 | −3.1 |
| $Glucose\text{-}P + Mg^{2+} + HOH \rightleftharpoons glucose + MgHPO_4$ | −2.4 | −1.7 | −2.5 |

reactants and products are free in solution or bound to the surface of a protein. The enzyme selectively binds specific forms of the substrate.

Physiological solutions contain different ionic forms of ATP, both free and forming complexes with magnesium. However, most energy-transducing enzymes selectively bind the complex $Mg \cdot ATP$. Thus, the $\Delta G°$ of ATP hydrolysis at the catalytic site will be different from that of ATP in solution and similar to that measured in nonphysiological solutions containing a large excess of magnesium. One of the major differences between bulk solution and the surface of proteins is the activity of the solvent. The water molecules that organize around a protein solution have properties that are different from those of medium bulk water, e.g., a lower vapor pressure, a lower mobility, or a greatly reduced freezing point. Similar changes in the properties of water are observed in mixtures of organic solvents and water. In addition, the properties of the solvent in a given region of a protein may vary greatly as a consequence of a conformational change. Thus, although the water activity of the bulk solution remains constant, the $w_a$ in the microenvironment of the catalytic site may vary during the catalytic cycle of the enzyme. This has been measured in different transport ATPases.

## IX. ENTROPIC ENERGY AND TRANSPORT ATPases

In equilibrium thermodynamics, as proposed by Ludwig Boltzmann in 1877, entropy is a measure of disorder. In a closed system, entropy increases as the system flows from the initial state toward equilibrium. For an osmotic gradient, the entropy increases as the concentration of solutes is equalized in the two compartments. To restore the gradient, and thus to decrease once again the entropy, energy must be provided to the system. Transport ATPases operate based on this principle. During active transport, these enzymes use the energy derived from ATP to form a gradient across the membrane. In the reverse process, the increase of entropy associated with the decrease of the gradient is used to synthesize ATP. This was first shown by Peter Mitchell who demonstrated that the ATP $F_1–F_0$ complex of mitochondria, chloroplasts, and bacteria catalyzes the synthesis of ATP from ADP and $P_i$ using the energy of the electrochemical proton gradient derived from electron transport. In these organeles, ATP is first spontaneously formed from ADP and $P_i$ at the catalytic site of the enzyme without the need of an energy imput. The ATP thus formed has a low energy of hydrolysis and cannot dissociate from the enzyme. The energy derived from the electrochemical proton gradient is needed both to increase the $K_{eq}$ for the hydrolysis of the tightly bound ATP from 1 to a value higher than $10^{-6}$ $M$ and to permit the dissociation of the ATP synthesized from the enzyme.

The catalytic cycle of transport ATPases is simpler than that of the ATP synthase and is better understood at present. This includes the $Ca^{2+}$ transport ATPase found in the sarcoplasmic reticulum of skeletal muscle:

This $Ca^{2+}$ ATPase can catalyze both the hydrolysis and the synthesis of ATP. The chemical energy derived from the hydrolysis of ATP is used to form a $Ca^{2+}$ gradient across the membrane and, in the reverse process, the energy derived from the osmotic gradient is used to synthesize ATP from ADP and $P_i$. For the hydrolysis of ATP, the catalytic cycle is initiated after the ATP and $Ca^{2+}$ ions bind to two different domains of the enzyme (steps 1 and 2 in the sequence). The catalytic site of the enzyme, which selectively binds the complex $Mg \cdot ATP$, is located in a hydrophylic region of the protein protruding from the membrane into the cytosol. Two calcium ions bind in a different region of the protein immersed in the hydrophobic moiety of the membrane. After the binding of $Mg \cdot ATP$ and $Ca^{2+}$, an aspartyl residue located at the catalytic site is phosphorylated by ATP forming an acylphosphate residue (step 3). The $K_{eq}$ of this reaction is close to 1, and the $K_{eq}$ for the hydrolysis of the acylphosphate residue formed is higher than $10^6$ $M$, which is therefore similar to that of ATP in solution. After phosphorylation, the protein undergoes a spontaneous conformational change that is felt both at the catalytic site and at the $Ca^{2+}$-binding domains of the ATPase (step 4). During this event, the calcium ions bound to the protein are translocated across the membrane. This is followed by a large decrease of the affinity of the protein for calcium which permits the two calciums to be translocated to dissociate from the protein on the other side of the membrane (step

5). Simultaneously with the translocation, there is a large decrease of the $K_{eq}$ for the hydrolysis of the acylphosphate residue at the catalytic site. This is accompanied by a decrease of water activity in the microenvironment of the catalytic site. The calculated value for the energy of hydrolysis of an acylphosphate residue in the gas phase is similar to that measured for the acylphosphate residue after the conformational change of the protein (Tables I and III), indicating that the decrease of the energy was promoted by a change of solvation of the reactant and products at the catalytic site. The final events are the hydrolysis of the acylphosphate residue, dissociation of $P_i$ from the enzyme (steps 6 and 7), and the spontaneous return of the enzyme to the original conformation (step 8). There is little or no energy exchange in these final steps and the enzyme is ready to initiate a new cycle.

The sequence of events for the synthesis of ATP is the same as that for hydrolysis but in the reverse order. This is observed when there is a large difference of $Ca^{2+}$ concentration on the two sides of the membrane, less than $10^{-6}$ $M$ in the cytosolic side, and higher than $10^{-3}$ $M$ on the other side of the membrane. The enzyme is initially phosphorylated by $P_i$ forming a low energy acylphosphate residue (steps 7 and 6). There is no need of energy for this step; the $K_{eq}$ for this reaction is close to unity. $Ca^{2+}$ must then bind on the noncytosolic side of the membrane (step 5). The energy derived from this binding permits the conformational change (step 4) which leads to an increase of the water activity in the catalytic site and conversion of the phosphoenzyme from low into high energy. This acylphosphate residue transfers its phosphate to ADP-forming ATP which dissociates from the enzyme (steps 3 and 2). The calcium is simultaneously translocated across the membrane in a process which involves a large increase in the affinity of the protein for the calcium bond (step 4). A very low $Ca^{2+}$ concentration in the cytosolic side of the membrane is then needed to permit the dissociation from the ATPase of the tightly bound calcium (step 1) and initiation of a new catalytic cycle (step 8). In this system, the transduction of energy occurs during the conformational change which changes the energy level of the phosphoenzyme (step 4). During the hydrolysis of ATP, the conformational change that permits the conversion of the phosphoenzyme from high into low energy occurs spontaneously, but in the reverse direction, the energy derived from the binding of $Ca^{2+}$ to the low affinity sites of the ATPase (step 5) is needed to permit the conversion of the phosphoenzyme from low energy

into high energy. Notice, however, that in both cases the process of energy transduction is not associated with the cleavage or formation of the acylphosphate bond. During the active transport, calcium is translocated across the membrane *before* the cleavage of the bond. During synthesis, energy is provided by the binding of calcium *after* the formation of the acylphosphate bond. A curious feature of the $Ca^{2+}$ transport ATPase is that the enzyme can synthesize ATP not only when a $Ca^{2+}$ gradient is formed across the membrane but also when a $H^+$, tempeature, or water gradient is formed. As discussed earlier, the energy of hydrolysis of phosphoanhidrides varies depending on the divalent cation concentration, pH, $w_a$, and temperature. These parameters change the contribution of entropic energy (T $\Delta S°$) of the reaction. Therefore, it seems that the $Ca^{2+}$-ATPase can detect different entropic changes on the two sides of the membrane and use it to synthesize ATP from ADP and $P_i$. The conformational change which permits the conversion of the phosphoenzyme from low into high energy is mediated by the binding of $Ca^{2+}$ to the enzyme. The affinity of the enzyme for $Ca^{2+}$ varies depending on the pH and temperature in contact with each side of the membrane. A 2 pH unit increase of the medium leads to a 100-fold increase of the enzyme affinity for $Ca^{2+}$. A similar increase is observed when the temperature is decreased from 30° to 0°C. Thus the ATPase can synthesize ATP when there are equal $Ca^{2+}$ concentrations on the two sides of the membrane but a different pH or temperature which alters the $Ca^{2+}$ affinity and thus the $Ca^{2+}$ concentration needed to promote the conversion of the phosphoenzyme from low to high energy. Changes of water activity operate in a different manner. The conformation of the enzyme which permits the formation of the low energy phosphoenzyme is only available when there is no binding of $Ca^{2+}$ on the cytosolic side of the membrane. A decrease of $w_a$ on the cytosolic side of the membrane greatly facilitates the partition of $P_i$ from the medium into the catalytic site of the ATPase, thus increasing the affinity of the enzyme for $P_i$ more than five orders of magnitude. In this condition the ATPase can be phosphorylated by $P_i$ even when there are equal $Ca^{2+}$ concentrations on the two sides of the membrane. An hydrophobic–hydrophylic transition is then needed to permit the change of water activity inside the catalytic site and conversion of the phosphoenzyme from low to high energy. Thus the ATPase can catalyze the synthesis of ATP when the $Ca^{2+}$ concentration is maintained high on the two sides of the membrane and the $w_a$ changes in cycles around the enzyme.

The reaction sequence described for the sarcoplasmic reticulum of skeletal muscle is characteristic of different transport ATPases. These include the $(Na^+ + K^+)$ ATPase and calmodulin-regulated $Ca^{2+}$ transport ATPase of plasma membrane, the $H^+/K^+$ ATPase of plants. These enzymes are referred to as $E_1$–$E_2$ enzymes because all of them undergo a conformational change during the catalytic cycle with formation of an acylphosphate residue having different energies of hydrolysis: $E_1$ high energy and $E_2$ low energy.

## X. ATP-REGENERATING SYSTEMS OF HIGH AND LOW ENERGY

Physiological conditions exist in which the cell may need to suddenly increase the consumption of ATP. An example is a rapid and intense muscle contraction. In these conditions the ATP hydrolyzed can be readily regenerated from other phosphate compounds, such as creatine phosphate or ADP:

$$ADP + \text{creatine phosphate} \rightarrow ATP + \text{creatine}$$

$$2ADP \rightarrow ATP + AMP$$

These reactions are catalyzed by creatine kinase and adenylate kinase, respectively, and permit the rapid recovery of the cytosolic ATP level before the resynthesis through more complex metabolic routes, such as glycolysis in the cytosol or the oxidative phosphorylation in mitochondria are activated. These are the classic ATP-regenerating systems and for several years it was thought that ATP could only be regenerated from phosphate compounds having a higher, or at least the same, energy of hydrolysis than ATP itself. This notion was derived from the original proposal of Lipmann in 1941 which assumed that the energy of hydrolysis of a phsophate compound would not vary in the cell. Within this view, hexokinase could not catalyze the reaction

$$\text{glucose 6-phosphate} + ADP \rightarrow ATP$$
$$+ \text{glucose}(\Delta G° + 4.5 \text{ kcal/mol})$$

because the enzyme can only bind one molecule of ADP for each glucose 6-phosphate molecule bound, and the unfavorable $K_{eq}$ of the reaction ($6 \times 10^{-4}$) would require that more than a thousand molecules of glucose 6-phosphate for each ADP molecule should be crowded inside the catalytic site of hexokinase to

permit the formation of one molecule of ATP. Thus the enzyme would only be able to promote the phosphorylation of glucose from ATP but never the reversal, i.e., synthesis of ATP from glucose 6-phosphate and ADP. Contrasting with this view, Wilkinson and Rose discovered that the $K_{eq}$ of the reaction on the surface of hexokinase is one (Table I), indicating that glucose 6-phosphate can easily transfer its phosphate to ADP because the energy of ATP hydrolysis has a similar value to that of glucose 6-phosphate, as shown by the sum of the following equations:

$$\text{glucose 6-phosphate} + HOH \rightarrow P_i$$
$$\underline{+ \text{glucose} \quad ADP + P_i \rightarrow ATP + HOH}$$
$$\text{glucose 6-phosphate} + ADP \rightarrow ATP + \text{glucose}$$

The finding of Wilkinson and Rose is readily explained by the solvation theory. As shown in Table II and Fig. 3, a decrease of water activity promotes a decrease of the energy of ATP hydrolysis to a level similar or even lower than that of glucose 6-phosphate. This permits the enzyme to promote the transfer reaction without the need of energy, but after dissociation from the enzyme, the concentration of ATP that can be accumulated in the solution after equilibrium is very small compared to that of ADP because in water the difference between the energies of hydrolysis of ATP and glucose 6-phosphate once again become large. Several enzymes involved in energy tranduction, however, possess a very high affinity for ATP. Examples are the transport ATPases, such as the $Ca^{2+}$-ATPase and the ATP synthase complex of mitochondria and chloroplasts. The $K_a$ for ATP binding at the catalytic site of these enzyme is $10^{-7}$ and $10^{-12}$ $M$, respectively. The affinity of these two enzymes is sufficiently high to permit the formation of the enzyme–substrate complex even in the presence of the very small concentrations of ATP formed from ADP and glucose 6-phosphate. After each catalytic cycle, the ATP hydrolyzed by the transport ATPases is rephosphorylated by glucose 6-phosphate to maintain the equilibrium concentration of ATP. In steady-state conditions, the concentrations of ATP remain constant, and the work performed by the two transport ATPases is coupled to a decrease in the glucose 6-phosphate concentration, a compound that has a smaller energy of hydrolysis than does either ATP or creatine phosphate. Once formed, the energy derived from the gradients can in turn be used to promote either (1) the synthesis of ATP from ADP and $P_i$ to a concentration much higher than that possible after the equilibrium of the hexokinase reaction, which was measured with the Ca$^{2+}$ transport ATPase; or (2) the uphill electron transfer from succinate to NAD$^+$ in mitochondria. These reactions have an energy requirement higher than that derived from the simple cleavage of glucose 6-phosphate.

## XI. CONCLUSIONS

The high and low energy character of a phosphate molecule is determined by the K$_{eq}$ of its hydrolysis reaction. The energy of hydrolysis of ATP and other phosphate compounds possessing a phosphoanhydride bond such as PP$_i$ and acylphosphate residue varies greatly depending on the water activity and ionic composition of the medium in which they are found. This variability is related to changes of the entropy of the hydrolysis reaction. In aqueous solutions, these compounds have a high K$_{eq}$ for the hydrolysis, but in conditions similar to those found on the surface of proteins they have a low K$_{eq}$ for the hydrolysis, thus behaving as "low energy" phosphate compounds. The work performed by proteins is associated with this transition of the K$_{eq}$ in the enzyme surface, i.e., with the conversion of the phosphate compound from high into low energy, and not with the cleavage of the phosphoanhydride bond at the catalytic site of the enzyme. After dissociation from the enzyme, the products of ATP hydrolysis ADP and P$_i$ can no longer react spontaneously to reform ATP because after returning to the aqueous medium, the K$_{eq}$ of ATP hydrolysis once again has a very high value. Thus, instead of the classical definition of ATP as being a high energy phosphate compound, a more accurate definition would be that ATP and other phosphate molecules possessing phosphoanhydride bonds are molecules that permit the use of entropic energy. Contrasting with phosphoanhydride bonds, the entropy for the hydrolysis of phosphoesters such as glucose 6-phosphate and phosphoserine does not vary after large variations of water activity, salt concentrations, or pH values of the medium. Therefore, the energy of hydrolysis of these compounds will be the same during enzyme catalysis, i.e., in the transition from the aqueous medium to the surface of the enzyme and then back again to the solution. Instead of referring to phosphoesters as "low energy," they may be defined

as phosphate compounds that do not permit the use of entropic energy. While having a K$_{eq}$ for the hydrolysis much smaller than that of ATP in aqueous solution, in the surface of enzyme phosphoester may have an energy of hydrolysis higher than that of ATP. This variability permits the transfer of the phosphate from glucose 6-phosphate to ADP, forming ATP in the surface of hexokinase. Therefore, phosphate compounds having quite different energies of hydrolysis in water, such as glucose 6-phosphate and creatine phosphate, can be used by enzymes to regenerate ATP from ADP. The difference between the two systems is, however, the ATP:ADP ratio found in the solution after equilibrium is reached. With creatine phosphate, practically all of the nucleotide is in the form of ATP, whereas with glucose 6-phosphate, only a very small fraction is ATP and most of the nucleotide found in solution is ADP. In the particular case of enzymes having a very high affinity for ATP, this does not represent an impediment for the use of glucose 6-phosphate and hexokinase as an ATP-regenerating system because they can recognize and use the very low ATP concentration available in the medium. Thus, in steady-state conditions the work performed by an enzyme involved in energy transduction such as the Ca$^{2+}$ transport ATPase of skeletal muscle can be ultimately coupled with the cleavage of glucose 6-phosphate, a phosphate compound that in aqueous solution has energy of hydrolysis smaller than that of creatine phosphate and ATP.

## BIBLIOGRAPHY

de Meis, L. (1989). *Biochim. Biophys. Acta* **973**, 333.
de Meis, L. (1993). *Arch. Biochem. Biophys.* **306**, 287.
de Meis, L., Behrens, M. I., Petretski J. H., and Politi, M. J. (1985). *Biochemistry* **24**, 7783.
de Meis, L., and Inesi, G. (1985). *Biochemistry* **24**, 922.
George, P., Witonsky, R. J., Trachtman, M., Wu, C., Dorwatr, W., Richman, L., Richman, W., Shuray, F., and Lentz, B. (1970). *Biochim. Biophys. Acta* **223**, 1.
Hayes, M. D., Kenyon, L. G., and Kollman, A. P. (1978). *J. Am. Chem. Soc.* **100**, 4331.
Pedersen, P. L., and Carafoli, E. (1987). *TIBS* **12**, 46.
Penefsky, H. S., and Cross, R. L. (1991). *Adv. Enzymol.* **64**, 174.
Romero, P., and de Meis, L. (1989). *J. Biol. Chem.* **264**, 7869.
Schlenk, F. (1987). *TIBS* **12**, 367.
Wolfenden, R., and Williams, R. (1985). *J. Am. Chem. Soc.* **107**, 4345.

# Adenoviruses

*Howard Hughes Medical Institute, Princeton University*

---

## GLOSSARY

**Interferon** Cytokine that induces an antiviral response when it binds to its receptor on the surface of a virus-infected cell. There are three members of the interferon family: $\alpha$, $\beta$ and $\gamma$

**Serotype** Adenoviruses are categorized into groupings termed serotypes based on their resistance to neutralization by antisera to other known adenoviruses. The most extensively studied adenovirus serotypes are types 2, 5, and 12

**Transcription unit** Unit of genetic information expressed from a single promoter. The unit may encode one or more proteins

**Transformation** Conversion of cultured cells to an oncogenic or cancer-like phenotype

**Tumor suppressor protein** Cellular protein that generally imposes a constraint on cellular growth. Loss of tumor suppression activity by mutation is tumorigenic

**Virion** Virus particle that in the case of adenovirus is composed of the viral double-stranded DNA chromosome surrounded by a protein shell

ADENOVIRUSES WERE DISCOVERED IN THE EARLY 1950s by investigators searching for the etiologic agents of acute respiratory infections. The first isolates were called acute respiratory disease, adenoid degeneration, adenoidal–pharyngeal–conjunctival, or respiratory illness agents. These agents were eventually grouped under the term "adenovirus" in recognition of the original isolation of this class of agents from adenoid tissue. After their initial discovery in humans, adenoviruses were isolated from a wide range of animal hosts. The human isolates have been divided into about 50 distinct groupings, termed serotypes, based on immunological criteria. All adenoviruses are composed of a linear, double-stranded DNA contained within an icosahedral protein shell measuring 70 to 100 nm in diameter. Although infection of humans with these viruses generally results in respiratory or gastrointestinal symptoms, some human adenovirus serotypes induce a variety of benign and malignant tumors if injected into a rat or hamster. Since these viruses are tumorigenic under certain conditions and contain DNA, they are classified as DNA tumor viruses even though they are not known to be oncogenic in their normal human host.

## I. DISEASE ASSOCIATIONS, DIAGNOSIS, AND TREATMENT

Adenoviruses enter their human host through the mouth and sometimes through the eye. The most common sites of infection are the respiratory tract, gastrointestinal tract, and the eye. Less frequently, the viruses can infect other organ systems, including the liver, pancreas, and central nervous system. Most adenovirus infections in otherwise healthy individuals are either subclinical or result in mild illness followed by complete recovery without sequelae.

Adenoviruses are responsible for about 5% of the acute respiratory disease in young children, and the

ENCYCLOPEDIA OF HUMAN BIOLOGY, Second Edition, VOLUME I.   Copyright © 1997 by Academic Press.   All rights of reproduction in any form reserved.

symptoms of adenovirus infections resemble those of other respiratory viruses such as influenza, parainfluenza, and respiratory syncytial virus. These infections are most commonly caused by adenovirus types 1, 2, and 5; antibodies to these serotypes are found in about 50% of 2-year-old children in the United States and Europe. Adenovirus respiratory infections generally are not life-threatening, although some epidemics of adenovirus type 7 have resulted in significant mortality. Acute respiratory disease that has been described in military recruits is also due to adenovirus, most commonly adenovirus type 4 or 7. Adenovirus types 40 and 41 are associated with gastrointestinal disease. Adenovirus-related gastroenteritis occurs most often in children under 4 years of age, and some studies have described populations in which as many as 50% of children in this age group have antibodies to the common enteric adenoviruses. The adenovirus disease is difficult to distinguish, based on clinical criteria, from rotavirus diarrhea, which is more prevalent. Eye infections, limited to the conjunctiva, are caused by adenovirus types 3 and 7. The virus can be spread in swimming pools, where large numbers of children have contracted "swimming pool conjunctivitis." A more serious eye disease, epidemic keratoconjunctivitis, is caused by adenovirus types 8 and 37. [*See* Respiratory Viruses.]

Adenovirus infections are generally difficult to discriminate clinically from infections by other agents. As a result, laboratory diagnosis is needed for their positive identification as etiologic agents of respiratory, gastrointestinal, and other diseases. Traditionally, specimen samples have been propagated in cultured cells and shown to contain adenovirus by observation of characteristic cytopathogenicity followed by serological confirmation and classification in terms of serotype. More recently, polymerase chain reaction (PCR) assays that do not require propagation of the patient sample in tissue culture have been employed to identify and type adenoviruses. [*See* Polymerase Chain Reactions.]

Attempts to control adenovirus infection have focused primarily on vaccination programs in the military. Attenuated live vaccines have been developed that prevent disease by adenovirus types 4 and 7. Vaccine strains are not available for the adenovirus serotypes that are primarily responsible for respiratory disease in children, that is, types 1, 2, and 5. The relatively mild nature of the illness and continuing concerns about the administration of a virus with oncogenic potential to children have argued against their development for routine use.

## II. VIRION STRUCTURE AND CHROMOSOMAL ORGANIZATION

The adenovirus virion is an icosahedral particle (Fig. 1) that ranges from 70 to 100 nm in diameter in different serotypes. It is composed of DNA and protein. The DNA is complexed with four virus-coded basic proteins that facilitate its compaction to form a DNA–protein complex, known as the core. The core is surrounded by a shell, termed the capsid, that is composed of seven virus-coded proteins. The capsid is distinguished by projections extending from each of its 12 vertices. These projections are composed of the virus-coded fiber protein and play a key role in the interaction of the virus with its receptor. The three-dimensional structure of the virion has been determined at a resolution of 35 Å by image reconstruction from cryoelectron micrographs.

The viral chromosome consists of a single molecule of linear, double-stranded DNA with a viral protein, termed with terminal protein, covalently attached to each 5′ end. Viral DNA replication is initiated within identical repeated sequences at the two ends of the chromosome, and there is a packaging sequence encoded near one end of the chromosome that is required for encapsidation of the viral DNA into a capsid to form a virion. The viral chromosome contains eight transcription units that are transcribed by the cellular RNA polymerase II (Fig. 2). Each of these units gives rise to multiple mRNAs due to differential splicing and poly(A) site utilization. As a result, the eight transcription units encode more than 40 proteins. The viral DNA also contains one or two (varies with serotype) small units transcribed by RNA polymerase III. These RNAs are known as virus-associated (VA) RNAs, and they do not code for proteins.

## III. REPLICATION CYCLE

The replication cycles of most DNA viruses, including adenoviruses, are divided by convention into early and late phases that are separated by the onset of viral DNA replication. During the early phase of infection, the virus binds to the cell surface and enters the cell, its chromosome travels to the nucleus, and its early set of transcription units is expressed. The early proteins encoded by these units modulate cell growth, protect the infected cell from antiviral responses of the host, and activate viral DNA replication. The late phase of infection begins concurrently

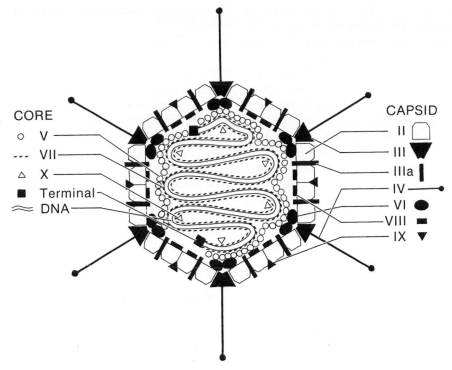

**FIGURE 1** Diagrammatic cross section of the adenovirus virion that is based on our current understanding of the arrangement of its polypeptides and DNA. Virion proteins are designated by their polypeptide numbers with the exception of the terminal protein (terminal). [Reproduced, with permission, from P. L. Stewart and R. M. Barnett (1993). *Japan. J. Appl. Phys.* **32,** 1342–1347.]

with the onset of viral DNA replication and includes the activation of the late transcription unit. Most of the late proteins encoded by this unit serve as building blocks for the assembly of progeny virions. [*See* DNA Replication.]

## A. Adsorption and Entry

The attachment of the virus particle to the surface of a cell is termed adsorption. The initial interaction of the virus with the cell is mediated by the fiber protein,

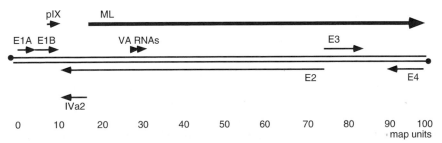

**FIGURE 2** Map of the adenovirus chromosome showing the locations of its transcription units. The viral DNA molecule is represented by two parallel lines in the center of the diagram; the two molecules of covalently attached terminal protein are marked by solid circles. Transcription units are designated by arrows, which indicate the direction of transcription relative to the conventional adenovirus map. Early (E1A, E1B, E2, E3, and E4), delayed early (pIX and IVa2), and major late (ML) units are transcribed by RNA polymerase II, whereas the two VA RNA units are transcribed by RNA polymerase III.

which, as noted earlier, projects from the vertices of the icosahedral virion. The cell-surface receptor to which the fiber protein binds has not yet been identified. Attachment of the adenovirus fiber to its cell-surface receptor does not lead to efficient internalization of the virion unless a second virus–cell interaction occurs. This interaction involves the viral penton protein, which is located in the virion at the base of each fiber projection, and members of a family of cell-surface receptors termed integrins. When the fiber and penton proteins both interact with their cognate receptors, the virus is internalized by receptor-mediated endocytosis. The internalization process is highly efficient and quite rapid: up to 85% of adsorbed virus eventually penetrates the cell and half of newly adsorbed virus appears in endosomes within 10 minutes.

As the pH in the newly internalized endosome drops, the virus escapes into the cytosol; and about 40 minutes after infection, partially disassembled virus particles can be seen at nuclear pores by electron microscopy, presumably releasing the viral DNA molecule into the nucleus. When the viral chromosome reaches the nucleus, it associates with the nuclear matrix through its terminal protein. Viral mutants with alterations in the terminal protein do not interact with the nuclear matrix and fail to efficiently activate the expression of viral transcription units.

## B. Early Gene Expression

E1A is the first transcription unit to become active after the adenovirus genome reaches the nucleus of the infected cell. Transcription of this unit is controlled by an enhancer that responds to endogenous cellular transcription factors, and it produces two related proteins early after infection. The E1A proteins have two principal functions. First, they control the expression of the adenovirus chromosome. This regulatory activity of the E1A proteins was first discovered when viral mutants were studied with alterations in the E1A coding region. These mutants failed to activate the transcription of all of the remaining viral genes. The E1A proteins activate transcription by binding to a variety of cellular transcription factors and modifying their activity. For example, one of the E1A proteins has been shown to bind to the DNA-binding subunit of the TFIID auxiliary transcription factor and activate transcription through the DNA-binding element recognized by this factor, the TATA motif. This motif is found about 25 base pairs upstream of many adenovirus and cellular transcriptional initiation sites.

The second function of the E1A proteins is to induce the host cell to enter the S phase of the cell cycle. Most cells that the virus enters in natural infections are not actively growing or synthesizing DNA. By forcing quiescent cells that are residing in the G0 compartment of the cell cycle to enter the S phase, the virus induces the expression of a broad array of cellular proteins important for the replication of DNA, creating an environment optimal for viral DNA replication. The E1A proteins modulate the cell cycle in part by targeting the retinoblastoma protein. This protein is a cellular tumor suppressor protein that is named for the cancer in which mutant derivatives of this protein were first described. The retinoblastoma protein inhibits cell-cycle progression by binding to the cellular E2F transcription factor and inhibiting its ability to activate transcription. E2F plays a key role in cell-cycle progression, activating a series of S phase-specific cellular genes. The ability of the retinoblastoma protein to bind and regulate E2F is normally controlled by its phosphorylation state. The adenovirus E1A protein short-circuits this normal control by binding to the same position on the retinoblastoma protein as E2F and releasing the transcription factor, which then promotes progression of the infected cell into the S phase. [See Cell Cycle.]

E1A proteins expressed from a transfected plasmid DNA can force the cell to enter the S phase and begin replicating DNA, but outside of the context of a viral infection the expression of E1A proteins induces apoptosis, which is a process of programmed cell death or suicide. Cells infected with viruses sometimes undergo apoptosis, and this helps to inhibit the spread of the infecting agent. Even though the adenovirus E1A proteins can induce programmed cell death, adenovirus-infected cells do not die of apoptosis. This is because two additional viral proteins, encoded by the E1B unit, block apoptosis. One of the E1B proteins binds to the cellular tumor suppressor protein, p53. The intracellular level of p53 rises in response to the activity of the E1A proteins, and an elevated level of p53 can induce apoptosis. The E1B protein that binds to p53 blocks its ability to signal cell death. The second E1B protein is a functional mimic of the cellular Bcl2 protein, and like Bcl2, it inhibits apoptosis induced by p53 as well as many other inducers of the process. [See Tumor Suppressor Genes: p53]

In addition to inhibiting the process of apoptosis, early gene products antagonize a variety of additional host antiviral responses. For example, an adenovirus protein encoded by the E3 transcription unit interferes with the destruction of infected cells by cytotoxic T lymphocytes (CTLs). One of the functions of CTLs is the recognition and killing of virus-infected cells. For recognition, viral peptide antigens must be displayed on the surface of the infected cell in a complex with a major histocompatibility complex (MHC) class I antigen. The adenovirus E3 protein is a transmembrane protein that resides in the endoplasmic reticulum. It binds to the antigen-combining pocket of newly synthesized MHC class I antigen and retains it in the endoplasmic reticulum. As a result, the level of class I antigen on the cell surface drops, reduced quantities of adenovirus antigens are exhibited on the cell surface, and recognition and lysis of the infected cell are inhibited. [See Major Histocompatibility Complex (MHC).]

The small VA RNAs antagonize another host response to infection: the $\alpha$ and $\beta$ interferons. Interferons block viral replication through a variety of mechanisms, including the induction of a protein kinase, PKR, that is activated by double-stranded RNA that accumulates in cells infected by many different viruses. When activated, PKR phosphorylates the $\alpha$ subunit of eIF2, a translational initiation factor. Phosphorylation of eIF2 blocks its activity and inhibits protein synthesis in the infected cells, a process that is very bad for the individual cell but good for the host since it inhibits virus spread. VA RNAs are designed to bind to PKR and block its activation. As a result, adenovirus is relatively resistant to the action of interferon. [See Interferons.]

Early adenovirus gene products also mediate the replication of viral DNA. The E2 unit encodes all of the viral proteins that function directly in the replication process. The E2-coded terminal protein serves as a primer for DNA replication. The terminal protein forms a covalent complex with deoxycytidine 5'-triphosphate (dCTP) through an ester bond that links the $\beta$-OH of a serine residue in the protein to the $\alpha$-phosphoryl group of the deoxynucleoside triphosphate. The terminal protein–dCTP complex, in conjunction with the E2-coded DNA polymerase and two cellular proteins, interacts with the replication origin residing in the terminal sequences of the viral DNA to initiate DNA synthesis. An E2-coded single-stranded DNA-binding protein plus a third cellular protein function together with the viral DNA polymerase to mediate chain elongation. The early phase of the infection ends with the onset of viral DNA replication.

## C. Late Gene Expression and Virus Assembly

The adenovirus late proteins are encoded by the major late transcription unit. The major late promoter resides at about 16 map units on the conventional adenovirus map, and it is activated as DNA replication begins. It directs the synthesis of a 29,000-nucleotide primary transcript that is processed by differential polyadenylation and splicing to generate at least 18 late mRNAs. As the expression of late gene products begins, the virus blocks further expression of host cell proteins. This allows the virus to dominate macromolecular synthesis in the doomed cell. One component of this block is the inhibition of the cytoplasmic accumulation of cellular mRNAs. Cellular genes continue to be transcribed and cellular transcripts are processed, but they no longer exit the nucleus and accumulate in the cytoplasm. The study of mutant viruses has shown that a complex of two early viral proteins, one from the E1B unit and the other from the E4 unit, is responsible not only for the block to the accumulation of host cell mRNAs but also for efficient cytoplasmic accumulation of late viral mRNAs.

The late viral mRNAs code primarily for constituents of the virion, and assembly of progeny virus begins as viral DNA and late proteins accumulate in the nucleus of infected cells. The outer shell of the virion, which is termed the capsid, is assembled from seven late polypeptides; and the viral DNA in conjunction with four late core proteins is thought to enter preformed, empty capsids. A sequence on the viral DNA called the packaging element mediates DNA–capsid recognition and prevents accidental packaging of nonviral DNA molecules. Presumably one or more proteins bind at the packaging element to mediate the DNA–capsid interaction, but they have not been identified. Virus multiplication eventually kills the host cell. When adenovirus is propagated in human HeLa cells in the laboratory, about 10,000 progeny virions are generated in each infected cell over the course of about 24 hours. [See HeLa Cells.]

## IV. ONCOGENESIS

Extensive screens have failed to correlate the presence of adenoviruses with malignant disease in the human. However, all human adenoviruses that have been tested can transform cultured rodent cells to an oncogenic phenotype, and some adenovirus serotypes can induce tumors directly in newborn rats and hamsters. Adenovirus-transformed cells generally form dense, multilayered foci; they exhibit the hallmarks of cancerous cells, including growth in medium containing reduced amounts of serum and anchorage-independent growth. Most tumorigenic adenoviruses, such as Ad12 or Ad18, induce sarcomatous tumors at the site of injection. Ad9 is an exception to this rule; it induces only estrogen-dependent mammary tumors, independent of the site of injection. It is not clear why some adenoviruses can induce tumors in rats and hamsters but are not oncogenic in humans.

The adenovirus E1A and E1B units encode oncoproteins that cooperate to transform rodent cells to an oncogenic phenotype. The ability of E1A and E1B proteins to deregulate the cell cycle and block the onset of apoptosis by antagonizing the function of the retinoblastoma protein and p53 is consistent with their ability to disrupt the control of cell growth. It is interesting to note that the simian virus 40 transforming protein, T antigen, binds to both the retinoblastoma protein and p53, blocking their function; and the human papillomavirus oncoproteins, E6 and E7, cause the degradation of p53 and block function of the retinoblastoma protein. Thus, inactivation of tumor suppressor proteins is a common practice of DNA tumor viruses. [See Tumor Suppressor Genes.]

Recently, several additional adenovirus proteins have been shown to exhibit oncogenic potential. A protein encoded by open reading frame 6 of the E4 transcription unit antagonizes p53 function. This protein cooperates with the E1A protein to transform cells and increases the frequency of transformation by E1A plus E1B proteins. One additional oncoprotein, which appears to be unique to Ad9, has been described. This oncoprotein is encoded by open reading frame 1 of the Ad9 E4 unit, and it can transform cells independently of the E1A and E1B oncoproteins. This protein is also essential for mammary tumorigenesis by Ad9. So far, its mode of action is not understood.

## V. ADENOVIRUS AS A VECTOR FOR GENE THERAPY

Recently, there has been intense interest in the use of adenovirus as a vector to deliver nonviral genes to target cells. Several features of the adenoviruses predict their utility in gene therapy applications. First, the genome of adenovirus is easily manipulated and propagated in the laboratory, facilitating the substitu-

tion of therapeutic genes for viral sequences and the production of large quantities of recombinant viruses. Second, the viral genome can accommodate quite large segments of foreign DNA. Theoretically, only the viral replication origins and packaging elements need to be retained, allowing for the insertion of at least 35,000 base pairs of nonviral DNA. Third, the virus can express its genes in cells that are not actively growing, opening the door to direct *in vivo* therapies.

First-generation vectors that contain marker or potential therapeutic genes in place of the E1A and E1B transcription units have been produced and extensively tested. Although it has been possible to efficiently infect many different types of cells with these substituted vectors, their performance has been disappointing. In general, *in vivo* expression of the reporter genes has been limited to several days, during which an inflammatory reaction leads to the destruction of the transduced cells. The inflammation is due in part to residual viral gene expression, and second-generation vectors appear to induce considerably less inflammation. These vectors lack a functional E2-coded DNA-binding protein or the entire E4 transcription unit as well as the E1A and E1B units. Adenovirus vectors containing only the terminal viral replication origins and packaging element surrounding nonviral sequences are currently under development and, unless the incoming virion proteins are sufficient to induce inflammation, these third-generation vectors might substantially overcome the deficiencies of the earlier adenovirus vectors.

Adenovirus vectors have been extensively evaluated for their potential to deliver the cystic fibrosis transmembrane conductance regulator (CFTR) protein to lung epithelial cells. Since many adenoviruses exhibit a respiratory tropism, it seemed possible that the virus might perform especially well in the delivery of wild-type CFTR protein to compensate for mutations in the CFTR gene that lead to cystic fibrosis. Unfortunately, these studies revealed an unexpected problem. Although adenoviruses infect many different cell types with high efficiency, they proved to enter epithelial cells in the lung very inefficiently, apparently because these cells lack a subunit of the cell-surface integrin that serves as one of the two receptors for adsorption and internalization of the virus. In retrospect, it is not surprising that we have evolved a lining of cells in our lungs that are difficult for a viral pathogen to infect.

The utility of adenovirus vectors will be limited unless it proves possible to achieve long-term expression of therapeutic genes in transduced cells. Nevertheless, adenovirus vectors have proven efficient and effective vehicles for the *in vivo* delivery of genes to tumor cells in animal models. The transgene either inhibits the growth of tumor cells (e.g., p53) or allows them to be killed by relatively nontoxic drugs (e.g., introduction of the herpes simplex virus thymidine kinase gene followed by treatment with gangcyclovir, which kills cells expressing the viral enzyme). Thus, a likely niche for adenovirus as a gene therapy vector is already evident.

## BIBLIOGRAPHY

Horwitz, M. S. (1996). Adenoviruses. *In* "Fields Virology" (B. N. Fields, D. M. Knipe, and P. M. Howley, eds.), 3rd Ed. Lippincott–Raven, Philadelphia.

Nevins, J. R., and Vogt, P. K. (1996). Cell transformation by viruses. *In* "Fields Virology" (B. N. Fields, D. M. Knipe, and P. M. Howley, eds.), 3rd Ed. Lippincott–Raven, Philadelphia.

Shenk, T. (1996). Adenoviridae: The viruses and their replication. *In* "Fields Virology" (B. N. Fields, D. M. Knipe, and P. M. Howley, eds.), 3rd Ed. Lippincott–Raven, Philadelphia.

# Adipose Cell

E. R. BUSKIRK
*The Pennsylvania State University*

## GLOSSARY

**Biopsy** Process of removing a tissue sample from the body for subsequent analysis

**Brown fat** Distinctive brown-colored fat that is metabolically quite active; its color develops because of its extensive vascularity, mitochondrial density, and cytochrome content

**Fine structure** Micro- or ultrastructure of the cell as revealed by electron microscopy

**Hyperplasia** Process of development of more cells

**Hypertrophy** Process by which cells become larger

**Receptors** Specific sites on the cell surface that bind stimulating substances such as hormones

**Thermogenesis** Process of heat generation from metabolic processes arising within the cell

**White fat** Usual form of adipose tissue with the cells containing stored lipid that comprises a substantial energy store for the body

THE FAT CELL, ADIPOSE CELL OR ADIPOCYTE WHEN mature, is approximately 95% triglyceride by weight. The cell consists of a large lipid droplet surrounded by a thin layer of cytoplasm and contained within a cell envelope. The cell is active metabolically and is responsive to nerve and hormonal control as well as to nutritional influences. The cell converts glucose and triglyceride taken up from the circulating blood and extracellular space to fatty acids and subsequently to stored triglyceride. Stored triglyceride is released by the cell as fatty acids and glycerol—the former used as fuel by working muscle and other tissues. The fine structure of the cell, as revealed by electron microscopy, is complex but reasonably understood in terms of function. Both cell size and number show considerable intraindividual variation. The stored lipid in the cell contains not only triglyceride but also free cholesterol, some cholesterol esters, and a small pool of free fatty acids. Although white adipose tissue is by far the most dominant form in mature individuals, brown fat exists in the newborn with perhaps some vestiges remaining in the adult. Brown fat has distinctly different features than white fat and is thought to play a role in body temperature regulation, particularly in the newborn.

## I. HISTORY

In 1857, benign and malignant tumors in adipose tissue were first described with the concept that adipose tissue was made up of connective tissue cells burdened with fat. Although some researchers supported this view, others thought that adipose tissue was unique; this uniqueness was traced to the embryonic development and further maturation of fat cells in people. Around 1926, studies on the metabolism of adipose tissue were initiated. The upshot of this early work established the individuality of adipose tissue with respect to its development and metabolism. Perhaps the first suggestion that fat might be synthe-

ENCYCLOPEDIA OF HUMAN BIOLOGY, Second Edition, VOLUME I.
Copyright © 1997 by Academic Press. All rights of reproduction in any form reserved.

sized in the adipose tissue itself was provided in 1902 and 1903 based on observations of force-fed geese.

Several investigators have demonstrated the importance of neuroinnervation to the accumulation of stored lipid in adipose tissue. Denervation of adipose tissue, for whatever reason, led to the accumulation of stored lipid. Wertheimer and Shapiro published a classic review in which they emphasized the incorporation of metabolic substrates into stored lipid in adipose tissue and inferred fatty acid synthesis as an effect of insulin. Subsequently, the important role played by insulin was clarified. Furthermore, glucose incorporation into triglyceride stored in adipose tissue has been demonstrated.

Equally important to the storage of lipid in adipose tissue is the release of substrates from it. Various researchers have independently observed free fatty acids released from adipose tissue with fasting and epinephrine administration. Others found that adipose tissue incubated *in vitro* also released free fatty acids, a process that was inhibited by glucose and insulin but enhanced by several hormones. In 1962, glycerol release from adipose tissue was demonstrated to track lipolysis, and glycerol and free fatty acids were found to be released together. In 1964, adipose tissue was shown to be disrupted by collagenase so that single adipocytes were released. These fat cells (adipocytes) could be incubated in an appropriate medium and studied *in vitro*. Much of what is known about the function of adipocytes is dependent on this technique.

Another important methodological advance was the sizing and counting of adipocytes. Thus, the number of lipid-containing cells could be calculated, which led to the concepts of hypertrophy (large cells) and hyperplasia (large number of cells) of adipocytes. Still, a further technological contribution came from the revival of the connective tissue issue with the discovery that a special cell line (3T3 cells) resembling connective tissue cells could take on the characteristics of adipocytes; these cells have since served as a model system for the study of enzymes in the control of adipocyte metabolism.

## II. ADIPOCYTE DEVELOPMENT

The human fetus in the second trimester of gestation already demonstrates morphological differentiation of adipose tissue. The signals required for adipocyte formation are present and neovascularization accompanies the development of fat lobules. The newly formed adipocytes can respond to various lipolytic hormones. Presumably, the ability to develop new cells proceeds throughout life; the turnover rate of mature adipocytes is slow. Once formed, adipocytes tend to remain viable even though lipid stored within them may be depleted.

The development of a region of adipose tissue starts with penetration of a capillary bud onto which an adipocyte lobule differentiates. Primitive fat cells may arise from pericapillary primordial cells, which are difficult to separate from endothelial cells by light microscopy. Such "adipogenic reticular cells" seem to sequester small sudanophilic (incorporating Sudan dyes) lipid droplets, which appear near the center of the cell and also near the endoplasmic reticulum and mitochondria. Presumably, the next step is the accumulation of more lipid droplets, which ultimately coalesce. Staging of the cells into adipoblasts and preadipocytes may well take place with the former containing less lipid than the latter. The mature adipocyte contains the major single lipid droplet (unilocular) and assumes a spherical configuration surrounded by the thin layer of cytoplasm and a displaced nucleus that identifies the cell as a "signet ring" in appearance. [*See* Lipids; Reticuloendothelial System.]

A more recent view of adipocyte origin involves the stem cell as being indistinguishable from a fibroblast. While the origin of adipocytes is under investigation, researchers apparently agree that capillaries and some form of precursor adipocyte appear together in developing adipose tissue.

Several techniques have been utilized to clarify adipocyte development. These include studies of DNA synthesis by presumed adipocyte precursor cells, adipocyte sizing and counting, lipid accumulation by cultured precursor cells including the mouse fibroblast cell line 3T3, and incorporation *in vitro* of metabolic substrates by cultured precursor cells. [*See* DNA Synthesis.]

The presence of adipogenic activity in human plasma has been established both in cell lines and in primary cultures of stromal-vascular cells from rat adipose tissue. Growth hormone and glucocorticoid hormones as well as insulin have been identified as promoting the differentiation of adipocyte precursor cells.

## III. MEASUREMENT OF THE SIZE AND NUMBER OF ADIPOCYTES

A variety of methods have been utilized to determine the number and size of adipocytes, and a universally accepted measurement regimen has not been established. Measurement of the quantity of DNA in adipose tissue overestimates the number of adipocytes simply because other cells contribute to the measurement. Conventional histologic sections are not acceptable because of cellular retraction during specimen preparation. With thin sections, the plane of the section passes through several cells but only a few at their maximum dimension. Measurement of cell size on the periphery of a fat lobule underestimates cell size because the peripheral cells are smaller than those near the center of the lobule.

Samples of adipose tissue can be taken at surgery from almost any site, but only from subcutaneous sites with minor surgery in healthy people. Because the latter involves some risk, the needle transcutaneous biopsy technique has been employed under local anesthesia, as has the transcutaneous punch biopsy technique. A common site for taking these subcutaneous samples of adipose tissue is the superior quadrant of the gluteal region, although sites on the upper arm, thigh, abdomen, and upper back have been employed.

Fat cells are isolated from the small quantities of adipose tissue obtained by transcutaneous techniques using digestion with the enzyme collagenase, which breaks down the matrix connecting the cell. Cell separation can be done either before or after tissue fixation (i.e., treatment with substances that modify cellular molecules, holding them together and preventing their decay). Once the fat cells are isolated and appropriately fixed, the average diameter of the cells can be calculated by the direct measurement of a sufficiently large number under the microscope. The assumption is made that the cells are spherical to calculate a mean cell volume. Mean cell weight can be calculated from mean cell volume assuming the density of the contained lipid to be that of triolein, or 0.915. Alternatively, a cell counter such as a Coulter counter can be employed using a known volume of a cell suspension. Some investigators have used a thick section (100–200 $\mu$m) of adipose tissue and then have repeatedly focused the microscope to measure individual cell sizes for those cells that remain intact.

The total number of adipocytes in the body can be calculated from the value of the total body fat mass and either the average cell weight or the number of adipocytes per milligram of adipose tissue. Average cell weight varies by subject, site, and measurement technique and ranges from about 0.020 to 0.300 $\mu$g of triglyceride per cell. Cell diameters range from about 40 to 160 $\mu$m. It should be emphasized that none of these techniques take preadipocytes into account. Lipid must be present within the cell in identi-

fiable amounts for the cell to be recognized as an adipocyte.

## IV. ADIPOSE CELL ULTRASTRUCTURE

Each adipocyte consists of a large lipid droplet surrounded by a thin layer of cytoplasm, thus forming a ring-like unilocular structure. Electron microscopy reveals subcellular organelles along with a variety of lipid droplets plus an extensive system of membranous organelles within the cytoplasm of the cell. The central lipid droplet is surrounded by a fenestrated envelope. Cytoplasmic lipid droplets, occurring separately or as aggregates, are also frequently surrounded by fenestrated envelopes but may occur without them. The system of membranes stems from invaginations of the cell membrane. Features such as vesicles, simple and vesiculated vacuoles, a smooth-surfaced endoplasmic reticulum, and Golgi complexes are found.

The mature white adipocyte is roughly 95% stored lipid (e.g., triglyceride) by weight. The cell is active metabolically, although the metabolic rate (e.g., energy consumption) of adipose tissue relative to many other tissues is low. Unfortunately, little is known about the cells function in relation to the fine structure revealed by electron microscopy.

Usually, the perinuclear cytoplasm is thin and the nucleus flattened. Fenestrations that divide the flattened envelope vary considerably in dimension with cytoplasm filling each fenestration and separating the envelope from the stored lipid. The contained lipid droplets vary in size, structure, and distribution. Three classes of inclusions have been identified within the adipocytes cytoplasm: lipid droplets of moderate and relatively uniform size, lipid droplets not associated with a fenestrated envelope (generally small and in groups), and packed aggregates of very small droplets.

An additional ultrastructural feature is the system of membranous organelles. A system of cytoplasmic vesicles and vacuoles appears to invaginate from the membranous structure that surrounds the cytoplasm of the adipose cell. A basement membrane also surrounds each cell. Distortions of the cell membrane (invaginations) occur at irregular intervals and range from slight indentations to developed vesicles connected to the membrane by a narrow neck. Rosette-like invaginations (i.e., complexes of fused vesicles) are observed. The structural similarities among invaginations, rosettes, small vacuoles, and other vesicles suggest a common origin from the cell membrane. Nevertheless, technical artifacts may produce some of the observed diversity. The ultrastructure of the adipocyte may be related to the demand of a bidirectional transport and metabolism of fatty acids and triglyceride synthesis and mobilization. Presumably, the adipocyte is prepared to respond rapidly to lipogenic (movement in) or lipolytic (movement out) hormones that control movement of the lipid transport molecules.

## V. STORAGE AND RELEASE OF LIPID

Circulating lipoprotein triglyceride is the main precursor of lipid within the adipocyte, and assimilation of triglyceride requires hydrolysis by the enzyme lipoprotein lipase (LPL). Although LPL is considered the key enzyme regulating lipid deposition, the mobilization of lipid is importantly regulated by hormone-sensitive triglyceride lipase. The activity of LPL in adipose tissue is relatively high in the fed state and is less in the fasted or insulin-deficient state. Thus, assimilation of stored lipid occurs when circulating triglycerides in the blood are abundant. The hormone gastric inhibitory peptide, secreted after glucose or fat ingestion, increases LPL activity. Insulin may act on the adipocyte by enhancing LPL synthesis. Released LPL migrates to the adjacent capillaries and binds to endothelial cells where it exerts lipolytic action. At present, it is unknown whether or not the enzyme turnover associated with capillary endothelium is a function of lipoprotein triglyceride flux through the capillary bed. The distribution of lipoprotein triglyceride in tissues such as the myocardium, skeletal muscle, and adipose tissue may well be a function of the mutational state and timing of meals rather than diurnal variations of adipocyte LPL, but quantitation of such distribution has not been accomplished.

Investigations of the control lipolysis suggest that free fatty acids in plasma released by hydrolysis of triacylglycerols by lipoprotein lipase may act in the feedback control of lipoprotein lipase activity. This has been demonstrated at the vascular epithelium and may also occur at the adipocyte epithelium.

The messages that cells exchange involve multiple proteins, principally hormones and growth factors. Specialized receptors on the cell surface receive the messages. For example, in insulin the signal can bypass the cell nucleus to impact the machinery of translation, the processes whereby the information in mRNA is used to build protein and store lipid.

The message from insulin instructs adipocytes to make more of the enzymes that catalyze the synthesis of fat. The process is not simple. Once the receptor sees insulin, a reaction path involving several intermediate enzymes and proteins is initiated. When activated by MAP kinase, a phosphate group is attached to a newly identified intermediate called PHAS-1 which constitutes the crucial link between insulin and the translation process. Although PHAS-1 presumably suppresses initiation of translation, the phosphorylation of PHAS-1 by MAP kinase interrupts such action and translation proceeds. Other second messenger systems may yet be identified via this growing area of research.

Adipose tissue adapts to positive caloric balance by the hydrolysis of lipoprotein triglyceride to fatty acids through the capillary endothelium into the adipocytes, where the new triglyceride is stored. The entry of fatty acids into adipocytes may well involve a saturable transport system with diffusion coming into play at high concentrations of unbound fatty acids. Some of the crucial enzymes involved in triglyceride synthesis are associated with the processes of phosphorylation–dephosphorylation and insulin induction. The adipocyte contains a lipid-binding and acyl-CoA synthetase which are proteins involved, respectively, in fatty acid binding and activation. Apparently, long chain fatty acids not only activate but reversibly modulate gene expression encoding these proteins at a transcriptional level. Whether these *in vitro* findings apply *in vivo* remains to be demonstrated. Coordination of these several processes is quite important but poorly understood (e.g., the localization of gene coding of the enzymatic proteins remains to be accomplished).

At least three hormones exhibit antilipolytic properties: insulin, E-prostaglandins, and the $\alpha$2-receptor-specific catecholamines. Insulin is probably the most important hormone in the mobilization of stored lipid. Insulin has several known effects including inhibition of adenylate cyclase and activation of soluble cyclic adenosine monophosphate phosphodiesterase and may directly block free fatty acid efflux. Mediators of lipid flux include AMP-dependent kinase and phosphoprotein phosphatase. The alteration of cyclic AMP concentrations or phosphatase activation may result. The inhibition of lipolysis by $\alpha$2-agonists occurs through inhibition of adenylate cyclase. The respective importance of these regulatory processes remains unknown *in vivo*. Suffice it to say that adipocytes are usually exposed to a variety of signals that carry contradictory messages. The intake of food, for example, induces both insulin and catecholamine release, and exposure to lipolytic agents produces production of prostaglandins along with lipolysis. The integrated responses may well change as adipocytes enlarge or shrink or display different characteristics in diverse storage areas. [*See* Insulin and Glucagon.]

Although *de novo* fat synthesis in man is a slow and relatively minor process in relation to total daily energy turnover, studies of weight cycling brought about by fasting and refeeding in animals suggests that a preferential accumulation of saturated and monosaturated fatty acids takes place. This presumably occurs with little long-term effect on total fat mass. Thus, lipid storage in adipose tissue is dependent on environmental variables such as dietary composition, amount, and weight cycling.

Some insight into the storage of lipid within the adipocyte can be gleaned from twin studies designed to ascertain genetic input. Thus, differential genetic contributions to various obesity indices have been found. For example, the correlation coefficient among monozygotic twin pairs for accumulation of abdominal visceral fat storage with overfeeding was 0.72. Nevertheless, interpretation of such findings is not simple and has been summarized to indicate that genetic influences establish whether a person can become obese, but environment determines whether obesity occurs and its extent.

## VI. ADIPOCYTES AS SECRETORY CELLS

Adipocytes have been demonstrated to secrete not only lipoprotein lipase, but cholesterol ester transfer protein, angiotensinogen, adipsin or factor D, factor $C_3$ and factor B as well, and possibly apolipoprotein E. In addition, sex steroids are secreted. Thus, adipocytes in secreting peptides and peptide factors become involved in not only signaling other organs, but with metabolic events as well as physiopathological events such as hypertension and dyslipoproteinemia. Exciting research is underway in these areas.

## VII. LIPID COMPOSITION OF ADIPOSE TISSUE

Although triglyceride is the major storage form, adipose tissue is also a major site for cholesterol storage. Low-density lipoproteins add to cholesterol storage, whereas high-density lipoproteins both add and remove cholesterol. Apparently, human adipocytes syn-

### TABLE I

Adipose Tissue Fatty Acid Composition in Nonobese Subjects and Obese Patients[a]

| Fatty acid[b] | Obese (N = 14) | Nonobese (N = 21) |
|---|---|---|
| Myristic, 14:0 | 3.7 ± 0.6[c],*** | 4.7 ± 0.9 |
| Palmitic, 16:0 | 22.9 ± 1.2* | 23.3 ± 1.9 |
| Palmitoleic, 16:1 | 9.0 ± 2.1** | 6.9 ± 1.6 |
| Stearic, 18:0 | 2.9 ± 0.8*** | 4.8 ± 0.7 |
| Oleic, 18:1 | 48.1 ± 2.1 | 46.8 ± 2.9 |
| Linoleic, 18:2 | 11.7 ± 1.9 | 11.0 ± 3.3 |
| Linolenic, 18:3 | 2.0 ± 0.6 | 1.9 ± 0.7 |
| Arachidonic, 20:4 | 0.6 ± 0.1 | 0.5 ± 0.3 |

[a]Adapted from Rossner *et al.* (1989).
[b]For each fatty acid, the number of carbon atoms and double bonds are indicated.
[c]Mean ± standard deviation percentage.
*$P < 0.05$, **$P < 0.01$, ***$P < 0.001$.

thesize little cholesterol but derive it from interstitial lipoproteins. [*See* Cholesterol.]

The fatty acid composition of lipid stored in human adipose tissue differs somewhat between nonobese and obese individuals. In samples taken from subcutaneous adipose tissue by needle biopsy, the dominant fatty acid, oleic, was similar in both the obese and nonobese subjects, but the obese had a significantly lower percentage of the saturated myristic, palmitic, and stearic fatty acids. In contrast, the obese had a higher percentage of palmitoleic, a partially unsaturated fatty acid. No significant differences were found between the two groups for any of the polyunsaturated fatty acids. Because of the significant group differences in the saturated fatty acids, the polyunsaturated to saturated (P/S) ratio was 0.49 in the obese and 0.41 in the nonobese ($P < 0.05$) (Table I). Lipid composition of adipose tissue will vary somewhat with a person's habitual diet. [*See* Fatty Acid Uptake by Cells.]

Such variation has been found in the adipocytes of rabbits fed different mixes of fatty acids in their diets. Adipose stores of essential fatty acids reflect the composition of the diet as shown in Table II.

## VIII. ADIPOCYTE DISTRIBUTION

Gender effects are quite obvious in adipose tissue distribution in humans, with women having a larger fat

TABLE II

Fatty Acids (FA) in Different Diets Fed to Rabbits
and Disposition in Adipose Tissue (AT)[a]

| Fatty acid | Range in diet[b] (% by wt) | Range in AT[c] (% of total FA) |
|---|---|---|
| Palmitic | 0.2–9.5 | 13.4–30.4 |
| Stearic | 0.2–6.7 | 4.5–12.4 |
| Oleic | 2.9–7.3 | 16.7–49.7 |
| Linoleic | 0.7–13.3 | 11.9–57.9 |
| Linolenic | 0.02–10.1 | 1.0–31.1 |

[a]Adapted from Lin *et al.* (1993).
[b]Five different diets were utilized and the mean FA calculated.
[c]Mean FA content from AT of six rabbits in each dietary group.

depot generally, particularly in the femoral-gluteal region. Such gender differences are associated with differences in sex hormone concentrations. In contrast, men, particularly as they grow older, tend to have relatively enlarged abdominal depots. However, such gender differences are not always clear-cut. Excess abdominal obesity, the so-called android type, appears to present an increased risk of cardiovascular disease, whereas the femoral-gluteal accumulation, or gynoid type, appears to pose no known risk. With gross obesity, both the internal and the subcutaneous depots become quite large, with the latter relatively larger than the former. Although some differences in the regional mobilization of stored fat are possible, it is generally accepted that all depots participate with dietary restriction associated with considerable body weight loss. Nevertheless, there is evidence, at least among premenopausal women, that weight loss induced by underfeeding was associated with a somewhat larger loss in subcutaneous as compared to visceral adipose tissue. Thus, moderate regional fat redistribution was observed.

A stable isotope labeling ([$^{14}$C]oleic acid) technique has been devised to follow *in vivo* triglyceride fatty acid uptake into abdominal and femoral adiocytes in relation to LPL activity after different meals and periods of time. Adipose tissue biopsies revealed a greater early uptake of the label around the waist than around the hips. LPL activity was higher in the fed state in the abdominal area as compared to the femoral area, but not when expressed per unit of adipocyte surface area. The latter observation reflects a background of a greater uptake of dietary lipid in the abdominal region. The fasting plasma insulin concentration was related to lipid uptake in both sites. It was concluded that LPL plays a role in regulating triglyceride uptake into adipocytes, but that other factors such as hormonal status (e.g., insulin) and regional blood flow are important as well.

## IX. BROWN ADIPOSE TISSUE

What has been discussed thus far is white adipose tissue, which is by far the dominant form in mature humans. Brown fat exists as well in the newborn and particularly in rodents. Brown fat has distinctly different features than white fat and is regarded as the "flash heater," important as part of the central heating system that functions in an animal's arousal from hibernation and provides nonshivering thermogenesis in cold environments, the latter undoubtedly a protective mechanism in the newborn. Brown fat has considerable innervation and is quite responsive to adrenergic stimuli. It is noteworthy that denervated brown adipose tissue (BAT) assumes the morphology of white adipose tissue.

Brown adipose tissue has been identified by the color brought about by its extensive vascularity, mitochondrial density, and associated cytochrome content. It is characterized by round nuclei and multivesicular fat droplets. The most extensive sites of BAT are the perirenal and axillary–deep (i.e., at relatively central and internal sites) cervical areas.

Studies of mitochondrial energetics have revealed that BAT is specialized for sympathetically regulated

TABLE III

Some Known and Unknown Alterations in
Brown Adipose Tissue (BAT) Function[a]

| Condition | Animals | Human adults |
|---|---|---|
| Congenital obesity | Decreased | Unknown |
| Noninsulin-dependent diabetes mellitus | thermogenesis and increased | Unknown |
| Excess corticosteroids | lipid content in BAT | Unknown |
| Advanced age | | Unknown |
| Pheochromocytoma (adrenal medullary tumor) | Increased thermogenesis and decreased | Increased thermogenesis and activation |
| Catecholamines | lipid content in BAT | of BAT |

[a]Adapted from Leon and Trayhurn (1987).

**TABLE IV**
Features Distinguishing Brown from White Adipose Tissue[a]

| | White | Brown |
|---|---|---|
| Primary function | Energy storage | Thermogenesis |
| Anatomical distribution | Extensive, internal, and subcutaneous | Limited, internal sites |
| Fat droplet | Unilocular | Generally multilocular |
| Ultrastructure | Few mitochondria | Numerous mitochondria |
| Vasculature | Limited | Extensive |
| Sympathetic innervation | Limited (mainly to blood vessels) | Extensive (to both blood vessels and adipocytes) |
| Fatty acids | Exported | Largely oxidized *in situ* |
| Mitochondria | Coupled | Regulated uncoupling |
| Uncoupling protein | Absent in all but former BAT sites | Present (up to 15% of mitochondrial protein) |
| Responses to cold | Slight | Very extensive changes |

[a]Adapted from Trayhurn and Ashwell (1987), Trayhurn and Milner (1987), and Leibel and Hirsch (1987).

heat production associated with uncoupled oxidative phosphorylation in which energy provided by metabolic substrates, such as free fatty acids, is used to produce heat instead of being used to generate ATP, the energy currency of the cell. In the infant, 30 g of BAT (1% of body weight) may well account for the thermogenesis response to cold exposure or norepinephrine challenge. By measuring blood flow along with temperature gradients, it has been calculated that the perirenal adipose tissue of young men, which contains brown fat, produced 25% of the thermogenesis brought about by oral ephedrine administration. Thus, this small amount of BAT-containing tissue had a measurable thermic effect.

A summary of the modifications in BAT function produced by different conditions in animals and man appears in Table III. Sites that contained BAT in the child presumably retain a biochemical distinction from white adipose tissue by virtue of the fact that the protein responsible for uncoupling oxidative phosphorylation can be identified at these sites in the adult. Also, evidence indicates that intra-abdominal fat is more biochemically active in terms of lipolysis and lipogenesis (i.e., in response to insulin and catecholamines). A comparison of features that distinguish brown from white adipose tissue appears in Table IV.

# BIBLIOGRAPHY

Ailhaud, G., Grimaldi, P., and Negrel, A. (1992). A molecular view of adipose tissue. *Int. J. Obesity* **16**(Suppl. 2), S17–S21.

Bonnet, F. P. (1981). Measurement of the size and number of adipose cells. *In* "Adipose Tissue in Childhood," pp. 29–47. CRC Press, Boca Raton, FL.

Bouchard, C., Tremblay, A., and Despres, J. (1990). The response to long term overfeeding in identical twins. *N. Engl. J. Med.* **322**, 1477–1482.

Cahill, G. F., Jr., and Renold, A. E. (1983). Adipose tissue: A brief history. *In* "The Adipocyte and Obesity: Cellular and Molecular Mechanisms" (A. Angel, C. H. Hollenberg, and D. A. K. Roncari, eds.), pp. 1–7. Raven Press, New York.

Cushman, S. W. (1970). Structure-function relationships in the adipose cell. I. Ultrastructure of the isolated adipose cell. *J. Cell Biol.* **46**, 326–341.

Hollenberg, C. H., Roncari, D. A. K., and Djian, P. (1983). Obesity and the fat cell: Future prospects. *In* "The Adipocyte and Obesity: Cellular and Molecular Mechanisms" (A. Angel, C. H. Hollenberg, and D. A. K. Roncari, eds.), pp. 291–300. Raven Press, New York.

Jimenez, J. G., Fong, B., Julien, P., Despres, J. P., Rotstein, L., and Angel, A. (1989). Effect of massive obesity on low and high density lipoprotein binding to human adipocyte plasma membranes. *Intl. J. Obesity* **13**, 699–709.

Leibel, R. L., and Hirsch, J. (1987). Site and sex-related differences in adrenoreceptor status of human adipose tissue. *J. Clin. Endocrinol. Metab.* **64**, 1205–1210.

Leon, M. J., and Trayburn, P. (1987). Brown adipose tissue in humans. *J. Obesity Wt. Reg.* **6**, 234–253.

Lin, D. S., Connor, W. E., and Spenler, C. W. (1993). Are dietary saturated, monounsaturated, and polyunsaturated fatty acids deposited to the same extent in adipose tissue of rabbits. *Am. J. Clin. Nutr.* **58**, 174–179.

Renold, A. E., and Cahill, G. F., Jr. (eds.) (1965). "Handbook of Physiology, Section 5: Adipose Tissue." American Physiological Society, Washington, D.C.

Rössner, S., Walldius, G., and Bjorvell, H. (1989). Fatty acid composition in serum lipids and adipose tissue in severe obesity before and after six weeks of weight loss. *Int. J. Obesity* **13**, 603–612.

Trayburn, P., and Ashwell, M. (1987). Control of white and brown adipose tissues of the autonomic nervous system. *Proc. Nutr. Soc.* **46**, 135–142.

Trayburn, P., and Milner, R. E. (1987). Mechanisms of thermogenesis: Brown adipose tissue. *J. Obesity Wt. Reg.* **6**, 147–161.

# Adolescence

ROBERT ATKINSON
*University of Southern Maine*

---

## GLOSSARY

**Formal operations** Piaget's last stage of cognitive development, during which logical, abstract, and hypothetical thought processes are possible

**Identity** Gradual process of developing a clear sense of self-direction, self-understanding, and a commitment to moral, political, and vocational values

**Puberty** Stage of biological growth during which the individual becomes physiologically capable of sexually reproducing

**Rite of passage** Initiation or experience through which one gains recognition of having attained a new status or role in life

**Self-image** Way in which individuals feel about themselves

ADOLESCENCE IS A PSYCHO–SOCIAL–BIOLOGICAL stage of development occurring between childhood and adulthood. It usually starts with puberty and ends when the person gains a reasonable degree of parental independence. This can be a complex, challenging, and sometimes confusing transformation, largely because there are no longer clearly defined lines between these developmental stages as there once was when socially prescribed rites of passage separated these periods of life. Significant growth in biological, cognitive, and psychosocial areas is universal among teenagers, but individual timing and understanding of what is occurring vary greatly. The length of adolescence itself also varies greatly; it can include the ages from 12 to 20, or even beyond. While puberty and sexual maturation, abstract thinking, and forming an identity were once thought to be the source of considerable conflict and stress for the adolescent, recent research shows that adolescence is no more a period of turmoil for the vast majority than is any other time of life. While a great range of personal experience and adjustment patterns is evident during the teen years, adolescents do share a commonality in confronting the same developmental tasks of adjusting to their new bodies, their new ways of thinking, and their maturing identity.

## I. ADOLESCENCE AS A DEVELOPMENTAL STAGE

### A. Cultural Background

In traditional cultures of the past, initiation ceremonies, or rites of passage, were used to guide the individual through the necessary transition from one social status or life stage to another. Marriages and funerals are two common examples of this. At around the onset of menarche for girls and puberty for boys, a special puberty rite was held to initiate the youth into adulthood. Upon completion of this dramatic and often perilous ordeal, which included tests of bravery and endurance as well as separation from one's family and community, the youth would return a new person, an adult with a new status and new responsibilities. In this cultural context adolescence usually did not exist at all, and if it did it was clearly a liminal, or limbo, period that lasted anywhere from a few days to a few months. The important point about these community-wide ceremonies is that they made it very clear how the youth was to become an adult and exactly when this transition would take place, as well as when it was completed.

ENCYCLOPEDIA OF HUMAN BIOLOGY, Second Edition, VOLUME I.   Copyright © 1997 by Academic Press.   All rights of reproduction in any form reserved.

## B. Historical Background

Unfortunately, there has not been as clear a beginning and ending to adolescence since the loss of the socially prescribed rite of passage. The biological events that signal the beginnings of the transformation of the child into adult have always been in place, but the social and economic factors that interact with these have been constantly changing.

The popular concept of adolescence began to take shape in the 18th century. Prior to this, the average life span was considerably shorter than it is today, and the help of people of all ages was needed in the work force, both of which made adolescence a very short period of life. Compulsory public education, however, began to extend the adolescent years while obscuring the distinctions among childhood, adolescence, and young adulthood. This also caused confusion among the terms "puer," "juvenile," "adolescent," "teenager," and "youth." Only "adolescent" implies change and process, while the others refer to status or product.

In 1904, G. Stanley Hall, president of Clark University, published the first comprehensive study of adolescence. In his two-volume work, *Adolescence,* Hall introduced the phrase "sturm und drang" (i.e., "storm and stress") to characterize the development of adolescents. He took this concept from the German romanticists, primarily Goethe and Schiller, who focused on idealism, rebellion against established ways, and the expression of deep passion. This became referred to as "adolescent turmoil" and was accepted as typical for all adolescents, even though Hall's work was based on his own personal unsystematic observations.

It was thought that both disturbed and normal adolescents experienced this turmoil or significant emotional oscillation between the extremes of psychological functioning. This disruption in equilibrium led to mood fluctuations, thought confusion, and changeable and unpredictable behavior, such as feeling happy and altruistic one day and hopeless and depressed the next. Hall saw adolescence as the last of four stages of development—the final transitional stage between childhood and adulthood, and the one requiring open rebellion against the established values in order to separate from the parents and become independent. Anna Freud also expanded on this theme, drawing from her own psychoanalytic analysis of disturbed children, and generalized the phenomenon of turmoil to all adolescents.

## C. A New Perspective on Adolescence

The idea of normal adolescent turmoil has made the distinction between serious psychopathology and mild crisis among adolescents difficult to determine. However, new empirical studies have found that adolescents are no more intrinsically disturbed than are adults or children. Surveys of over 25,000 normal adolescent students taking the Offer Self-Image Questionnaire have resulted in findings that confirm that the percentages of disturbance among adolescents is similar to that found among adults. Consistently, 20% of the adolescents—clearly a minority—reported disturbing feelings of loneliness, emptiness, or confusion. These adolescents, however, are far outnumbered by the 80% who do cope well with the teenage years and make a relatively smooth transition to adulthood. There was no evidence of extreme mood swings, unpredictability of behavior, or deep-rooted social pessimism. The 80% adjusting well to the adolescent transformation represent the norm who are generally relaxed under everyday circumstances, can control their day-to-day trials, and have confidence in their ability to deal with stress.

It is now widely accepted that adolescence does present a special burden to the individual experiencing it, but it is seen equally as a challenge and an opportunity. The youth, even with an unclear beginning and ending point in the journey to adulthood, has to individuate, establish self-confidence, make important decisions concerning the future, and become independent from attachments to parents. The majority of teenagers do this well. By no means, however, is adolescence being made out to be an easy time, any more than life as a whole is easy. Adolescence is a period of rapid and profound change in the body and the mind. It is a time to find out who one is and where one is going in the future. Most conflict, particularly the family bickering that increases during this time, is useful to the adolescent in that it contributes to developing a sense of individuality. The turmoil that Hall and Freud emphasize, though, is only relevant today to a small subgroup of the adolescent population that includes psychiatric patients, juvenile delinquents, and other social deviants.

## D. Contemporary Approaches to the Transition of Adolescence

One thing missing for today's teenagers that was central in the life of all those in a traditional community is a socially prescribed, culturally meaningful rite of passage. This loss of tradition, this breakdown of significant symbols to live by, is one of the primary contributors to not only the floundering of youth, but the breakdown of society in general. Without clearly defined guidelines to assist one in a difficult passage of life, that transition will be met with confusion, uncertainty, and inconclusiveness.

In an effort to provide what is missing for today's adolescents, some in the helping professions are simulating traditional rites of passage experiences in therapeutic and educational settings to help teenagers better understand themselves and their world as a result of the transition they are undergoing. Such contemporary versions of the rite of passage, or vision quest, allow the teenager to experience a modern equivalent of what was a difficult passage but which had a clear purpose and end point. These rites of passage were designed to facilitate a necessary transformation that had to take place prior to the youth being able to assume adult responsibilities in the community. The contemporary versions are seen as creative approaches to the necessary crises of adolescence, which guide teenagers through the same three elements of the ageless pattern, defined variously as separation–initiation–return or, simply, beginning–middle–resolution. The result of experiencing this pattern in one's life is an important feeling of accomplishment. Traditional youth and contemporary adolescents both know from living this pattern that their transition has been successful and complete.

Adolescents today who participate in such a structured, experiential exercise live out a symbolic adventure which brings alive timeless mythological motifs and archetypes for them. This can provide them with an even more meaningful, purposeful, and well-defined journey into adulthood than most other secular or sacred transitions they will experience in today's world because their overall place and role in their life and development are made very clear. Contemporary rituals are designed primarily to serve the same function as traditional rites of passage, to guide the individual to a deeper understanding of himself or herself in relation to others and the world so they are better prepared to carry out their adult roles and responsibilities.

## II. ADOLESCENT SELVES

### A. Background

Even though the self is regarded as that which a person really and intrinsically is, the self is also seen as having successive and varying states of consciousness. In other words, there are many selves developing at varying levels simultaneously. William James, one of the first to explain this, held that each of the selves that constitutes the person has its own vulnerability, its own time of ascendancy, and its own reason for being. For James, the self consists of three parts: the material self (one's body), the social self (one's roles and relationships), and the spiritual self (one's inner or subjective being). The growth and development of these selves throughout the adolescent years are particularly important. Individually, they provide an effective way to view adolescent development since the adolescent is more likely than the adult to have many selves competing for recognition and calling for ultimate integration. Adolescence is a time of trying out new roles, discarding or retaining old roles, and establishing a sense of coherence. A major goal of adolescent development is to achieve a balanced stable integration of selves that a teenager becoming an adult can own as "myself." Collectively, these selves are integrated to form a total self from which a person draws a many-layered answer to the question "Who Am I?". [See Development of the Self.]

### B. The Biological Self

According to Sigmund Freud, the young adolescent is coming out of a latency period during which psychosexual activities were secondary to new social interests. With adolescence, however, a gradual resurgence of sexuality occurs following the onset of puberty. This new capacity for sexual reproduction requires that the adolescent master sexual and aggressive drives in socially acceptable ways. The biological and physical changes that accompany adolescence also bring about an awakening of new feelings toward one's own body. [See Psychoanalytic Theory.]

Puberty is a complex process, typically beginning between the ages of 9 and 14, which is characterized by a physical growth spurt and the maturation of primary and secondary sex characteristics. It not only has a biological impact on the adolescent, but psychological and social ones as well. A growth spurt, first in weight, then in height and strength, can occur at the rate of 3.5 inches for a girl to 4 inches for a boy during the year of fastest growth. The sex organs grow larger, and menarche in girls and ejaculation in boys usually signal reproductive potential, although peak fertility is reached several years later. Secondary sex characteristics (i.e., breasts; pubic, facial, and body hair; even changes in voice) appear for both boys and girls at varying times. [*See* Puberty.]

During these pubertal changes, more calories and vitamins are needed than at any other time in life. This is a highly critical time nutritionally as unbalanced diets can prevent normal growth. The serious, sometimes life-threatening, problem of anorexia nervosa, or self-starving, has become a well-publicized issue for some teens, especially girls.

The timing of sexual maturation, or the age at which an individual reaches and passes through pubertal changes, varies considerably and is determined by the individual's gender, genes, body type, and nutrition. Boys, thin children, and malnourished children typically reach puberty later than their counterparts. Usually, boys are about 6 months behind girls, with the average girl reaching menarche at about age $12\frac{1}{2}$ and the average boy ejaculating at age 13. Nevertheless, hormone signals from the brain (i.e., the hypothalamus) to the pituitary gland to the gonads always occur in the same sequence, creating a similar pattern of pubertal events for most young people.

Among the effects of early and late maturation can be significant differences in psychological adjustment. These relate most to body image, moods, relationships with parents and members of the opposite sex, and even school achievement. Recent studies have found that being early or late to mature can affect adolescents' satisfaction with their appearance and their body image. For seventh- and eighth-graders especially, girls who were physically more mature were generally less satisfied with their weight and appearance than their less mature classmates. While girls tend not to like being early to mature, and even become embarrassed and ashamed, boys, on the other hand, feel better if they are early maturers.

More physically mature boys tend to be more satisfied with their weight and overall appearance than their less mature peers. Developing earlier can give some boys a feeling of superiority. Boys who reach puberty usually report positive moods more often than their prepubertal male classmates. For girls, puberty often affects how they get along with their parents. Girls whose physical development is advanced tend to talk less to their parents and have less positive feelings about family relationships than do less developed girls. Early maturers also tend to get higher grades than later maturers in the same class.

While boys and girls have opposite feelings about their pubertal changes (i.e., generally it is a positive experience for boys), these feelings are usually temporary and balance out over the years. The biological events of puberty and adolescence cannot be changed, but the social and cultural attitudes toward variations in these events can become less rigid, which could make the adolescent's passage to adulthood even smoother. Normal biological development during adolescence includes primarily adjusting to pubertal changes, maintaining healthy relationships with parents and peers, establishing a healthy body image, and adopting age-appropriate sexual attitudes.

## C. The Cognitive Self

The intellectual maturation that takes place on an inner level for the adolescent is more subtle than biological changes, but just as important. Taken together, changes in thinking, along with earlier physiological changes of puberty, constitute what can be seen as a "psychic revolution." Cognitive development in adolescence signals the beginning of a new level of thought in which a greater reasoning and problem-solving capacity prepares the maturing teenager to become a philosopher of sorts, able to speculate, hypothesize, fantasize, and build elaborate systems of thought.

The most influential cognitive theorist is Piaget, whose stage-based view is that when adolescents between the ages of 12 and 15 reach the level of formal operational thought, they gain the ability to think logically. Scientific principles can be articulated, logical arguments can be engaged in, and social problems can be reasoned about, while drawing implications from many related propositions. Adolescents are thus capable of combining thought processes into self-reflection about vocational goals, personal satisfaction, and social responsibility. Maturing adolescents can utilize whatever innate or acquired knowledge they have, as well as newly developed capacities for logic, orderly analysis, and reflection. They thereby give greater cohesiveness and meaning to experience. For Piaget, mental life evolves toward a final form of equilibrium. The self becomes a true personality as self-reflective thoughts and feelings are integrated into a total life perspective.

With formal operational thought comes the tendency toward a particular form of egocentrism. For adolescents this takes the form of overestimating their significance to others. Because they can conceptualize the thoughts of others as well as their own, they might falsely assume that other people are preoccupied with their thoughts or behavior, resulting in a self-consciousness about physical appearance and interpersonal behavior. Adolescents also tend to create an imaginary audience in social settings that gives them the illusion of being under constant scrutiny. Egocentrism can also give young adolescents a sense of the heroic or mythical with the creation of a personal fable, or a sense of being immune to the laws of mortality and probability through creating an invincibility fable. Thus, adolescent thought processes are usually a mixture of the abilities to imagine many logical possibilities and to try to reshape reality when it interferes with hopes and fantasies. This heightened sense of self-consciousness usually peaks at about age 13 and diminishes during late adolescence.

There is also a relationship between formal thinking and moral development, or moral reasoning. Being able to imagine alternative solutions to various problems in science, logic, or social issues means being able to apply the same types of mental processes to thinking about right and wrong. Moral judgments, and their development through stages, are an interdependent component of cognitive developmental stages. Formal operational thought, or cognitive maturity, is usually a necessary but insufficient condition for principled morality. As a result, adolescents gradually come to see moral questions more broadly, loosening their hold on narrow personal interests while gradually looking at the values of their society and beyond. The tendency is for young people to reason at a higher level about moral issues in their own experience, or about those issues they have discussed with others. Giving adolescents the chance to discuss moral issues and to make their own moral choices can help them develop more complex ethical and moral thinking. Thus, the highest level of moral understanding (i.e., postconventional) is sometimes available to certain adolescents if principled morality is part of their experience.

A difference between the way males and females make moral judgments is also usually evident. Most girls and women tend to base their moral choices on the human relationships involved. Care and responsibility are the primary considerations for females, whereas it is usually rights and rules for males. These differences are not absolute, nor is one necessarily better than the other. The best moral thinking would synthesize both approaches.

Cognitive maturity would enable adolescents to arrive at more rational and healthy decisions concerning the major issues they face, such as sexuality, nutrition, substance use, and delinquency. The intellectual growth of the teenager includes the ability not only to memorize and recite ideas, but to think reflectively about those ideas and about one's self as well. Because of its expansiveness, adolescent thinking is well beyond childhood thinking. For the adolescent, a world of possibilities opens, both concrete and abstract, real and hypothetical. The distinction between false and true is more evident; thoughts about thoughts, meditative reflection, and introspection occur; and thoughts about the future, planning and exploring personal career options, begin in earnest, while other horizons broaden, including religion, justice, and identity.

## D. The Psychosocial Self

As the adolescent moves closer to maturity, certain dimensions of adequacy become more important. First, there is a sense of individual maturity, which includes self-control, self-esteem, and self-initiative. Next is interpersonal maturity, covering the ability to communicate, trust, and understand and manage relationships with others. Finally, there is social maturity requiring a general openness to the idea that things change and an acceptance or tolerance of differences among people. Thus, the psychosocial self means the total configuration of the individual and the personality mechanisms that integrate him or her.

A coherent sense of personal identity is the broadest expression of successful mastery of these areas. Successful identity formation can be characterized as the process of gradually bringing into accord the variety of changing self-images that have been experienced during childhood. Identity is therefore the bridge between individual and social reality that gives the individual a sense of meaningfulness and self-continuity. Identity eventually includes establishing a sexual, political, moral, religious, and vocational identity that gives one a sense of direction, commitment, trust in a personal ideal, and individual uniqueness. Identity achievement, or the resolution of the identity crisis accompanied by a healthy secure sense of self, might not occur until adulthood or until the values and goals set by parents and society have been fully explored and accepted or abandoned on one's own.

The dangers for the adolescent are role confusion (i.e., a failure to arrive at a consistent, coherent, and integrated identity) and identity diffusion (i.e., an inability to commit oneself, even in late adolescence, to an occupation or ideological position and assume a responsible stance in life). While some identity confusion is considered to be a normative and necessary experience, protracted confusion can lead to disturbance and possible pathology. Another danger for the adolescent is negative identity, or adopting the opposite of what parents and society expect, because this can lead to a debased self-image and social role.

Finally, another less obvious problem that can sometimes have a delayed reaction is that of foreclosure, or committing to an identity or vocational role too early without sufficiently exploring alternatives. A moratorium, or "time-out" specifically for self-exploration and experimentation with alternative identities, can sometimes be extremely valuable.

Religious beliefs, values, and a sense of one's spiritual nature can become more clearly focused during late adolescence. Traditionally, the adolescent years were when societies would clarify religious beliefs for the individual through specially designed initiation ceremonies, such as the vision quest. Today, while the Christian confirmation or the Jewish bar mitzvah would be the equivalent, more adolescents seem to be less drawn to strict observance of religious customs. The tendency, however, especially during late adolescence, is toward an independent search for truth, a reexamination and reevaluation of many of the beliefs and values they have grown up with. After a period of exploration, a more personalized spiritual orientation, usually toward their original affiliation, is often the result.

The self, self-knowledge, and self-image therefore become primary issues during adolescence. Important shifts occur in the way teenagers think about and characterize themselves. Their self-conceptions become more sophisticated and differentiated, often consisting of abstract, psychological, and interpersonal descriptors. Teenagers become more interested in understanding themselves and why they behave the way they do or what influences shaped their personality. They become concerned with matters of confidence, the ability to perform well, a sense of worth, a sense of personal control, low levels of anxiety, and feeling good about one's self. These, in fact, are the components of a good self-concept and are also important contributors to psychological well-being.

Self-image, or how one sees one's self, is therefore an extremely crucial aspect of the psychosocial self, as well as the biological and cognitive selves. Based on recent studies, while adolescents' feelings about themselves fluctuate, and actual self-image might be lower during the early adolescent years, generally self-image gradually becomes stable and more positive by late adolescence. Older adolescents are more self-confident, more open to the feelings and opinions of others, and seem to have a more balanced view of their families than do younger adolescents. This reflects the common notion that adolescence encompasses a process of increasing maturity, knowledge, and self-confidence.

## E. Gender Issues in Adolescence

It is true that the adolescent experience is not the same for females as it is for males in regard to biological development or emotional development. Only in the mid-1970s did researchers begin to question such differences and to look at how girls grow and develop. Prior to this, research subjects consisted mostly of males, and it was assumed that what was found to be true for boys was also true for girls.

Current research shows, however, that there can be many important differences. In early adolescence, girls begin to have their own unique struggles, with depression, body image, eating disorders, and lower self-esteem and self-confidence. They also begin to lose their own voice, think less of their own needs, and make their priority the relationship itself. The passage to womanhood for girls is often a journey into silence and disconnection. As they move from early adolescence into middle and late adolescence, girls are more likely to let themselves live with the pain of a relationship than trying to end it. Girls seem to be influenced by conventional thought that says women should be concerned about others at all costs, even to themselves. Boys, on the other hand, tend to take care of themselves first.

Girls, therefore, may face the additional developmental task, more so than boys, of learning how to maintain their own voice while still having healthy relationships with others. In order to be successful at this, researchers have found that girls may need to be in engaging relationships with others, particularly their mothers, where they are allowed and encouraged to speak their mind and to fight for what they believe, if necessary. They may need to be actively supported in finding and speaking in their own voice.

Identity development may be different for girls for the same reason. If girls are more concerned with their relationship with others, and maintaining that at whatever cost, what can get lost in the process is a clear sense of self, which is the essence of a positive identity. Instead of getting to know their own values, beliefs, and goals in life, they could overlook these in favor of keeping a relationship with others going. Again, girls may have to make more of an effort than boys to be honest with themselves and to explore their own interests and needs in order to arrive at an understanding of their own values and goals by the time they come out of adolescence so that they will have developed a sense of personal identity.

## III. THE CORE ADOLESCENT

Because adolescents experience the same biological, cognitive, and psychosocial changes and face the same developmental tasks, it is reasonable to expect that certain aspects of their experience will be common to all adolescents. Because of the growing number of studies using self-report questionnaires, even internationally, areas in which the adolescent experience is similar are now identifiable.

### A. Universal Aspects of the Adolescent Experience

When attempts to understand adolescents are made using standard psychological measurements, many common developmental patterns, as well as feelings, concerns, and interests, are found. In fact, adolescents seem to have little difficulty understanding each other. Teenagers today have a body of knowledge that is shared across many cultures, due largely to the emergence of what can be seen as a world culture. With television and other media often having global audiences, one event or idea can influence an entire global cohort of adolescents in the same way at the same time. Recent empirical studies have verified that, worldwide, today's teenagers do share a collective personality as well as a collective consciousness. They have assimilated common elements of human nature, culture, and civilization, as well as a common pattern of meanings that have been dispersed and spread throughout the world. The media transmit ideas and events to all corners of the globe, defining what is new or desirable, and they are adopted by young developing minds.

The result of this process is that it is now possible to provide a self-portrait of the universal adolescent. Teenagers who have the most in common with their peers describe themselves as being happy most of the time. They enjoy life, perceive themselves as able to exercise self-control, are caring, and are oriented toward others. They care about how others might be affected by their actions, prefer not to be alone, derive a good feeling from being with others, and like to help a friend whenever they can. They feel there is plenty they can learn from others. They value work and school. They enjoy doing a job well, think about the kind of work that they will do in the future, and would rather work than be supported.

Sexually, they feel confident about their body image and hold age-appropriate sexual attitudes. They do not feel far behind their peers, are not afraid to think or talk about sex, and do not feel they are boring to the opposite sex.

In the family they have positive feelings toward their parents. They feel that both parents are basically good and will not be disappointed or ashamed of them in the future. They do not carry a grudge against their parents and feel that their parents are usually patient and satisfied with them most of the time.

They cope well with life's vicissitudes, are able to make decisions, feel talented, like to put things in order and make sense of them, do not give up after their first failure, try to prepare in advance for new situations, and feel that they will be able to assume responsibilities for themselves in the future.

This profile of the core adolescent contrasts with popular conceptions of adolescence as a time of alienation from one's parents and as a time of self-centeredness and directionlessness. Instead, adolescents do generally accept their parents' attitudes and values, respecting them as well as their own responsibilities. Importantly, aspects of their common experience that teenagers most agree on are values, goals, and relationships.

### B. The Well-Adjusted Adolescent

Teenagers are fundamentally family oriented. To one degree or another, they change their relationships with their families of origin, both physically and psychologically, as they increasingly become more invested in peer relationships. This does not necessarily mean rebelling from or becoming antagonistic toward their immediate family. Contrary to previous thought, an adolescent is able to become more independent from his or her family of origin without bitterness or disavowal. Cognitively and psychologically, adolescents are able to express both love and respect for their parents and affirm good feelings toward their peers at the same time. In fact, new friends do serve to facilitate the needed separation from their parents and also aid in the subsequent identity formation within the larger context of the social network they are moving into.

Sources for the core adolescent experience are the biological, cognitive, and psychosocial changes that occur during these years. Biological development is universal, offering clear characteristics that distinguish childhood from adolescence, but variation in timing requiring different types and degrees of personal adjustment to these changes. Similarly, cognitive development offers the universal of formal operational thinking, but there is considerable individual difference as to when this is achieved and to what end.

Psychosocial development is also similar for all adolescents. The universal tasks are to form a clear coherent view of self, to separate from one's family of origin, to relate well to others of a like age, to prepare to form a conjugal family of one's own, and to develop a viable social, as well as personal, identity that will synthesize personal characteristics with an acceptable social role. There is also a great deal of individual variation in accomplishing these tasks, but the added difficulty for the adolescent lies in the lack of clearly defined social recognition of the status of adulthood and when this is finally achieved. The well-adjusted adolescent will be the one who accomplishes and understands these developmental changes and who also has the help of parents who offer support when it is needed, but allow the teenager enough independence to be challenged by these tasks.

## BIBLIOGRAPHY

Adams, G. R., and Gullotta, T. (1994). "Adolescent Life Experiences." Brooks/Cole, Pacific Grove, CA.
Brown, L. M., and Gilligan, C. (1992). "Meeting at the Crossroads: Women's Psychology and Girls' Development." Harvard, Cambridge, MA.
Cobb, N. J. (1992). "Adolescence: Continuity, Change, and Diversity." Mayfield, Mountain View, CA.
Elkind, D. (1994). "A Sympathetic Understanding of the Child: Birth to 16." Allyn, New York.
Esman, A. (1990). "Adolescence and Culture." Columbia, New York.
Lerner, R. M., *et al.* (eds.) (1991). "Encyclopedia of Adolescence." Garland, New York.
Offer, D., Ostrov, E., Howard, K., and Atkinson, R. (1988). "The Teenage World: Adolescents' Self-Image in Ten Countries." Plenum, New York.
Orenstein, P. (1994). "School Girls: Young Women, Self-Esteem, and the Confidence Gap." Doubleday, New York.
Steinberg, L. (1992). "Adolescence." McGraw, New York.

# Adrenal Gland

LAWRENCE N. PARKER

*University of California, Irvine and VA Medical Center, Long Beach*

## GLOSSARY

**Androgens** Steroid hormones that cause development of the male phenotype during embryonic life and male secondary sex characteristics during and after puberty

**Catecholamines** Hormones such as epinephrine (adrenaline), secreted by the adrenal medulla, and norepinephrine (noradrenaline), secreted by sympathetic nerve fibers, as well as the adrenal medulla; they promote cardiovascular and metabolic actions

**Glucocorticoids** Steroid hormones that have multiple actions, including the stimulation of glucose production (gluconeogenesis) and glycogen synthesis by the liver, and metabolism of triglycerides to free fatty acids and glycerol (lipolysis) in adipose tissue

**Mineralocorticoids** Steroid hormones that regulate mineral metabolism, such as that of sodium and potassium

**Sympathetic nervous system** Part of the autonomic nervous system, which mediates reactions to stressful stimuli and includes the adrenal medulla.

HUMAN ADRENAL GLANDS ARE PAIRED STRUC-
tures located at the upper pole of each kidney. They are compound glands composed of an outer cortex and an inner medulla. The cortex secretes three types of steroid hormones: androgens, glucocorticoids, and mineralocorticoids. In humans, the main examples of these hormones are cortisol (glucocorticoid), dehydroepiandrosterone (DHA or DHEA) (androgen), and aldosterone (mineralocorticoid). The medulla secretes nonsteroidal hormones with a catechol nucleus, such as epinephrine and norepinephrine, which are primarily involved in the biological alarm mechanism of the sympathoadrenal system. The embryological development of the cortex and medulla is different, along with their control mechanisms and functions. In some nonhuman vertebrates, including fish, the cortex and medulla are not located together in the same structure. Diseases of the adrenal gland, therefore, cause different syndromes depending on the part of the gland affected.

## I. DEVELOPMENT AND STRUCTURE OF THE ADRENAL GLANDS

### A. Adrenal Cortex

The adrenal cortex is derived from the mesoderm and can be identified in the human fetus by the age of 6 weeks near the cephalic end of the mesonephros. By 2 months, a capsule of connective tissue surrounds each gland. In the fetus, the gland is composed of an inner fetal zone and an outer definitive zone, the latter of which is similar to the adult cortex.

The fetal adrenal increases in size rapidly, largely due to growth of the fetal zone, and by midgestation is larger than the fetal kidney. After the 20th week of gestation, growth of the adrenal is centrally directed, as fetuses with anencephaly are born with an atrophic fetal zone. In comparison to total body mass, the adrenal gland is disproportionately larger in the fetus than in the adult. At birth, the fetal zone, which produces mostly dehydroepiandrosterone sulfate (DHAS), involutes, and within days, circulating concentrations of DHAS fall. Within several months, the fetal zone is not detectable. It is not clear why involution occurs. Although the fetal adrenal is responsive

ENCYCLOPEDIA OF HUMAN BIOLOGY, Second Edition, VOLUME 1.   Copyright © 1997 by Academic Press.   All rights of reproduction in any form reserved.

to adrenocorticotropin (ACTH), involution occurs in the presence of ACTH.

The location and surrounding structures of the human adrenal glands are shown in Fig. 1. The weight range of each adrenal gland is approximately 3.5–5 g, and the cortex comprises 90% of the gland volume. Chronically increased ACTH secretion causes the adrenal weight to increase. The blood supply to the adrenal glands is mostly from small arteries arising from the aorta and the renal arteries. These branch over the surface of the adrenal cortex, penetrate the outer capsule, and reach the adrenal medulla after forming a network of capillary sinusoids. This anatomical arrangement causes blood reaching the adrenal medulla to contain very high concentrations of adrenocortical steroids. As mentioned later, one of these steroids, cortisol, is a stimulus in extremely high concentrations for epinephrine secretion by the adrenal medulla. Blood leaves the adrenal gland via the adrenal veins, which empty into the inferior vena cava on the right and the renal vein on the left. Occasionally, additional adrenal tissue (accessory adrenocortical rests) may be found in the connective tissue near the main glands or further away, for example, in the gonads. The existence of this extra adrenal tissue is thought to represent the remnant of migrating primitive adrenal cortical cells during early embryological development.

By light microscopy, three zones can be identified in the cortex, as shown in Fig. 2. The outer zone, the zona glomerulosa, is relatively thin and contains cells

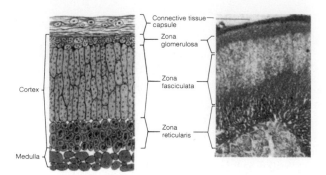

**FIGURE 2**   Histology of the human adrenal gland.

that secrete aldosterone. The middle zone, the zona fasiculata, is usually the thickest layer of the adrenal cortex and has a columnar structure. Its cells are relatively clear because they are large and have a high lipid content. The inner zone, the zona reticularis, surrounds the medulla. Its cells are relatively dark staining and compact in appearance, and often contain lipofuscin pigment granules. Although there is not a distinct boundary between the zona fasiculata and the zona reticularis, under electron microscopy, from the outside of the zona fasiculata to the inside of the zona reticularis, there is a gradual increase of smooth endoplasmic reticulum and number of lysosomes. Both the zona fasiculata and the zona reticularis produce cortisol and androgens, but in the human, the zona reticularis has sulfotransferase activity and is the probable source of DHAS. Chronically increased ACTH concentrations result in lipid depletion from the zona fasiculata and an increase in the width of the zona reticularis.

## B. Adrenal Medulla

In the fetus, the sympathetic nervous system arises from cells (sympathogonia) of the neural crest. By the fifth week of gestation, these cells migrate from the primitive spinal ganglia in the thoracic region to form the sympathetic chain. By the sixth week, large groups of neural cells migrate along the central vein into the adrenal gland, and thereby form the primitive adrenal medulla. Storage granules, which stain brown with chromic acid due to oxidation of catecholamines to melanin, are found by the twelfth week, which give the cells which contain them the name chromaffin or pheochrome cells. These cells also are found on both sides of the aorta and comprise the paraganglia. The largest collection of these cells is found near the infe-

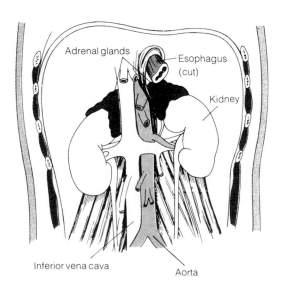

**FIGURE 1**   Location and surrounding structures of the human adrenal glands.

rior mesenteric artery, where they fuse to form a fetal structure termed the Organ of Zuckerkandl, which undergoes involution within the first year of life.

The remainder of the chromaffin cells in the paraganglia and adrenal medulla persist. In the adrenal medulla, the cells are arranged in an irregular network with a rich blood supply and are in contact with sympathetic ganglia. The cells of the adrenal medulla are innervated by preganglionic fibers of the sympathetic nervous system. As mentioned earlier, the blood supply of the adrenal glands enters through the cortex and drains into the medulla, except for a small number of vessels which supply the medulla directly.

## II. ACTIONS OF ADRENAL GLAND HORMONES

### A. Adrenal Cortex

Cortisol exerts major effects on glucose metabolism in several different ways. In many tissues, cortisol combines with cytosolic glucocorticoid receptor proteins and interacts with glucocorticoid responsive elements (GRE) of DNA to activate appropriate hormone-responsive genes. In the liver, cortisol increases hepatic gluconeogenesis markedly, partly by the stimulation of transcription and translation to synthesize enzymes which convert amino acids to glucose and also by increasing amino acid mobilization from muscle and other tissues. Cortisol also causes an increase in glycogen synthesis by the liver and decreases the rate of peripheral glucose utilization. Taken together, these effects of cortisol are the basis for the observation that insulin is less effective in a milieu of increased cortisol concentrations, an effect termed insulin antagonism.

With respect to protein and fat metabolism, cortisol stimulates the catabolism of nonhepatic proteins and increases the rate of amino acid uptake into liver. In adipose tissue, cortisol stimulates lipolysis and the subsequent release of glycerol and free fatty acids, while enhancing the rate of fatty acid oxidation. It is known from observations of the effects of cortisol deficiency that normal levels of cortisol are necessary to maintain a normal degree of cardiac output and blood flow to the kidney.

In addition, cortisol is secreted in so many forms of stress that an increase in cortisol secretion is often considered the result of what may be defined as a stressful stimulus. It has been speculated that cortisol functions to help an organism survive a stressful stim-ulus by means of the energy-producing and biosynthetic pathways noted earlier. The inflammatory process is common in illness and injury, and cortisol may also help minimize damage to the body due to excessive inflammation by stimulating mechanisms such as lysosomal membrane stabilization, to avoid release of proteolytic enzymes, and by reducing capillary permeability, to avoid leakage of plasma and blood cells into an inflamed area. Some of this latter effect is mediated by the blockage of effects of substances such as histamine and bradykinin.

At high concentrations, cortisol inhibits many immune mechanisms. Numbers of lymphocytes, mainly T lymphocytes, decrease, and there is usually a decreased antibody production from B lymphocytes. Meanwhile, increased numbers of polymorphonuclear leukocytes are released from the bone marrow, but their mobility is decreased, and fewer are able to migrate to inflammatory sites. Many inhibitory effects of high concentrations of cortisol and other glucocorticoids on immune and inflammatory reactions are mediated by their effects on cytokines, including a decreased T lymphocyte production of interleukins and interferons.

DHAS and other adrenal androgens circulate in young adults at concentrations much higher than those of cortisol. However, their functions have not been as clearly elucidated as have those of cortisol. Evidence has been presented that these steroids have a hepatic receptor and that they may prevent osteoporosis (loss of bone), facilitate the birth process by causing cervical softening, mediate female libido, and serve as precursors for more potent sex steroids. In animal studies, these steroids have been shown to protect against obesity, diabetes, and certain types of infections and tumors. DHAS is present in the human brain and has been shown to be synthesized in rat brain, but its function in the nervous system of either species is not known.

Aldosterone and other mineralocorticoids maintain normal sodium and potassium concentrations and extracellular volume. Aldosterone combines with an intracellular cytosolic mineralocorticoid receptor, which has significant similarity with the ligand-binding domain of the glucocorticoid receptor, especially its DNA-binding domain. Aldosterone exerts actions at the nuclear level of the target cell by the induction of protein synthesis and subsequent activation of a sodium pump, whose effect is to transport sodium across cell membranes. In the kidney, aldosterone acts primarily on the principal cells of the cortical collecting duct. At this site, sodium is reabsorbed and potassium is secreted. Other cells of the cortical collecting duct, the intercalated cells, are the site of hydrogen ion and bicarbonate absorption or secretion. Under the influence of aldosterone, over 99% of sodium filtered by the glomerulus is conserved, whereas potassium and hydrogen ion are excreted into the urine. Aldosterone has similar effects in sweat glands, salivary glands, and in the intestinal lumen. Cortisol also has some effectiveness as a mineralocorticoid, at least partly due to the similarity between glucocorticoid and mineralocorticoid receptors.

## B. Adrenal Medulla

Catecholamines exert their actions through two types of receptors, alpha and beta. Norepinephrine stimulates alpha receptors, which results in a variety of actions, such as vasoconstriction and blood pressure elevation, iris dilatation, and bladder sphincter contraction. Epinephrine stimulates both alpha and beta receptors. The latter interaction results in responses such as vasodilatation, acceleration of the heart rate, bronchodilatation, glycogenolysis, and lipolysis. These actions are part of the physiological response to stressful stimuli and complement those of cortisol, secreted by the cortex. Dopamine, the precursor to norepinephrine, weakly stimulates alpha and beta receptors, and also stimulates specific dopamine receptors. In the heart, kidney, brain, and gastrointestinal tract, these receptors mediate vasodilatation. Dopamine, norepinephrine, and, to a smaller extent, epinephrine are neurotransmitters in the central nervous system. The three groups of adrenergic receptors have been further subdivided by their anatomic location, binding of natural and synthetic ligands, and their response to antagonists.

## III. CONTROL OF SECRETION OF ADRENAL GLAND HORMONES

### A. Adrenal Cortex

Cortisol is produced by the adrenal cortex in response to ACTH secreted by the human pituitary gland. ACTH secretion occurs in response to decreased circulating concentrations of cortisol, as part of a negative feedback system, and in response to stressors of many types, including surgery, hemorrhage, thermal injury, and hypoglycemia. In addition, there is a circadian rhythm of pulsatile ACTH and cortisol secretion that results in increased secretion toward the end of the sleep period and therefore higher levels of circulating cortisol in the morning than at night.

ACTH is a 39 amino acid peptide derived from the larger molecule pro-opiomelanocortin (POMC), whose secretion is under the control of neurotransmitters and corticotropin-releasing hormone (CRH), a 41 amino acid peptide of hypothalamic origin. CRH and CRH receptors are also found outside of the hypothalamus, in the limbic system, brain stem, and spinal cord. A number of neuroamines, neuropeptides, and cytokines have been found to increase CRH secretion. These include acetylcholine, norepinephrine, serotonin, neuropeptide Y, and interleukins 1 and 6. Inhibition of CRH secretion has been caused by the GABA/benzodiazepine system, endogenous opioids, and nitric oxide. CRH exerts its effect by binding to specific pituitary corticotroph membrane receptors, which are coupled to guanine nucleotide-binding proteins. These stimulate the release of ACTH by an adenyl cyclase-dependent mechanism. CRH also stimulates the synthesis of POMC in pituitary corticotrophs. Arginine vasopressin (AVP) is also synthesized by the hypothalamus and is secreted into the hypophyseal portal system. AVP stimulates ACTH secretion either by itself or in synergy with CRH. Both CRH and AVP are secreted in stress and can stimulate secretion of each other.

The relationship of ACTH to POMC is shown in Fig. 3. Although there is considerable interspecies variability of POMC structure in many parts of the molecule, the first (N-terminal) 24 amino acids of ACTH are highly conserved among species and are biologically active. Evidence from animal experiments shows that non-ACTH POMC peptides may synergize with ACTH in controlling glucocorticoid secretion.

ACTH binds to specific adrenal cell membrane receptors, which are coupled to G-proteins, and activates adenyl cyclase, thereby causing an increase in the intracellular concentration of cyclic $3',5'$-monophosphate (cAMP), which in turn causes an increase in cellular protein kinase activity and an increase in the activity of cholesterol ester hydrolase. This produces free cholesterol for the rate-limiting conversion of cholesterol to pregnenolone. This reaction is mediated by the mitochondrial side chain cleavage enzyme. Plasma lipoproteins also provide cholesterol for steroidogenesis. [See Cholesterol; Plasma Lipoproteins.]

As shown in Fig. 4, pregnenolone can be converted to mineralocorticoids, glucocorticoids, or androgens.

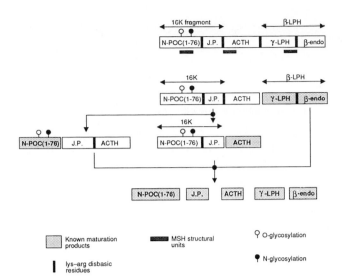

**FIGURE 3** Structure and processing of pro-opiomelanocortin (POMC; POC). LPH, lipotropin; endo, endorphin; JP, joining peptide. [Reproduced with permission from N. Seidah, J. Rochemont, J. Hamelin, S. Benjannet, and M. Chretien (1981). The missing fragment of the prosequence of human POMC: Sequence and evidence for C-terminal amidation. *Biochem. Biophys. Res. Commun.* **102,** 710.]

Many of the microsomal enzymatic steps are controlled by ACTH by means of regulation of the rate of steroidogenic enzyme synthesis. In contrast to the situation in humans, in rodents there is little 17-hydroxylase activity (Fig. 4), and therefore the main glucocorticoid is corticosterone. In humans, a single cytochrome $P$-450 ($P$-$450_{17\alpha}$) has two distinct enzymatic activities (Fig. 4).

Adrenal androgens are also secreted in response to acute ACTH stimulation, but their control is more complex because in some situations they are not secreted in conjunction with cortisol. These situations include adrenarche, puberty, aging, polycystic ovarian syndrome, stress, and starvation. Adrenarche is the process of adrenal gland maturation, which occurs before puberty at approximately age 7, that involves increased secretion of DHA and DHAS with constant secretion of cortisol. During aging, while basal levels of cortisol are constant, those of DHAS decrease markedly after a peak in the third decade of life (Fig. 5). Evidence indicates that at least part of the explanation for the dissociation in cortisol and adrenal androgen concentrations may be due to the influence of non-ACTH POMC-related peptide secretion.

Aldosterone is produced only by the zona glomerulosa because it is the only zone with 18-hydroxylase

and 18-dehydrogenase activity. The zona glomerulosa does not contain 17-hydroxylase activity and does not produce cortisol. ACTH causes acute stimulation of aldosterone secretion. However, other important control mechanisms are found in addition to ACTH.

Angiotensin II is the major regulator of aldosterone secretion. Secretion of angiotensin II is controlled by renin, as shown in Fig. 6. Renin is a proteolytic enzyme secreted by the juxtaglomerular (JG) apparatus of the kidney, which cleaves angiotensinogen, synthesized by the liver. The release of renin is controlled primarily by the sodium concentration of fluid in contact with the renal juxtaglomerular cells and the renal blood pressure, as sensed by renal baroceptors. Increased renin release is caused by decreased sodium concentration or blood pressure.

Renin mediates the conversion of hepatic renin substrate (angiotensinogen) to the 10 amino acid peptide angiotensin I, which in turn is converted to the 8 amino acid peptide angiotensin II by angiotensin-converting enzyme (ACE) in lung and other tissues. More than 80% of angiotensin I is converted to angiotensin II during a single pass through the lung. Angiotensin II is also a very potent vasoconstrictor. The 7 amino acid peptide angiotensin III is also bioactive in stimulating aldosterone secretion.

Angiotensins II and III bind to specific angiotensin receptors in zona glomerulosa cells and stimulate aldosterone secretion by a calcium-dependent mechanism. The signal transduction mechanism is G-protein dependent and includes the activation of protein kinase C, with at least a transient increase in concentrations of inositol triphosphate. The action of angiotensins may be mediated by prostaglandins, especially of the E series, and inhibitors of prostaglandin synthesis inhibit the effects of angiotensin II on aldosterone secretion.

Potassium ions also influence aldosterone secretion. An increase in serum potassium ion concentration of 1 meq/liter may triple the rate of aldosterone secretion. This is a direct effect on zona glomerulosa cells, does not include a cell surface receptor, and forms the basis for a feedback mechanism to regulate the concentration of extracellular potassium ions. Concentrations of potassium ion have the opposite effect on renin concentrations, but the direct effect on aldosterone secretion is predominant. In addition, there are other control mechanisms. Atrial natriuretic peptides (ANP) and dopamine have been shown to inhibit aldosterone secretion. Also, as in the case of cortisol and adrenal androgen secretion, evidence shows that

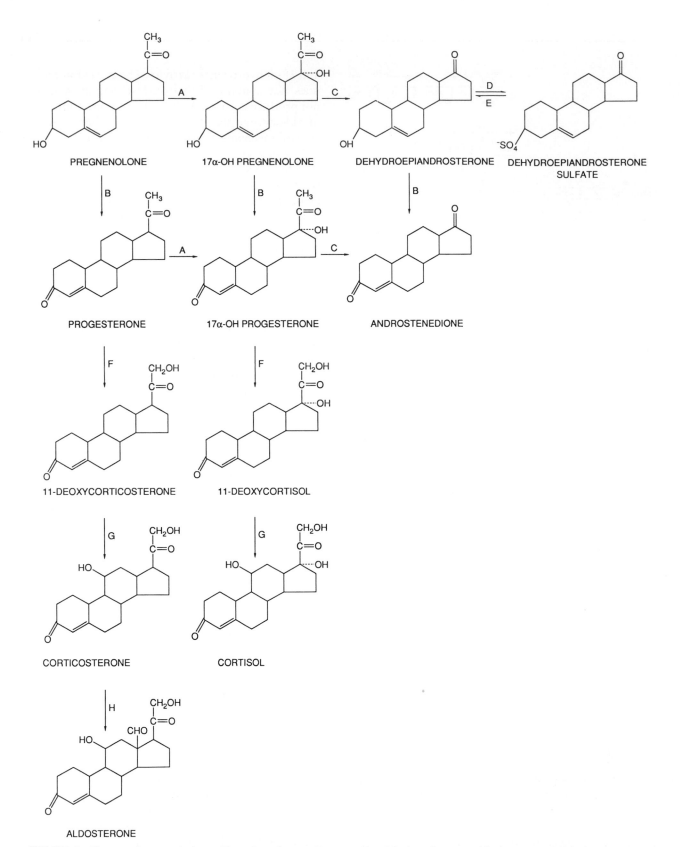

**FIGURE 4**  Human adrenocortical steroidogenic pathways. Enzymes: A, 17-hydroxylase; B, 3β-hydroxysteroid dehydrogenase-isomerase; C, C17-20-desmolase; D, steroid sulfotransferase; E, steroid sulfatase; F, 21-hydroxylase; G, 11-hydroxylase; H, 18-hydroxylase, 18-dehydrogenase; A and C, cytochrome $P\text{-}450_{17\alpha}$ ($17\alpha$-hydroxylase/C17-20-desmolase).

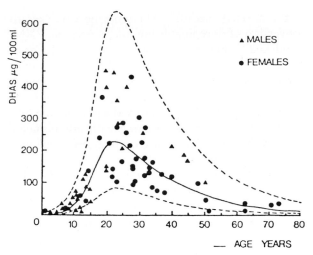

**FIGURE 5** Serum concentrations of DHAS in normal subjects 1–73 years of age. [Reproduced with permission from M. Smith (1975). A radioimmunoassay for the estimation of serum DHAS in normal and pathological sera. *Clin. Chim. Acta* **65**, 5.]

the stimulation of aldosterone secretion may be exerted by non-ACTH POMC-related peptides. With respect to aldosterone secretion, the most potent stimuli have been shown to be from the 31-kDa POMC fragment.

## B. Adrenal Medulla

Control of secretion of the adrenal medulla is best understood in context of the mechanism of function of the sympathetic nervous system. Whereas preganglionic fibers of the parasympathetic branch of the

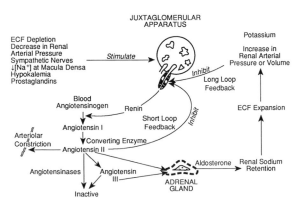

**FIGURE 6** Control of aldosterone secretion by the renin–angiotensin system. ECF, extracellular fluid. [Reproduced with permission from P. Bondy and L. Rosenberg (1980). "Metabolic Control and Disease." Saunders, Philadelphia.]

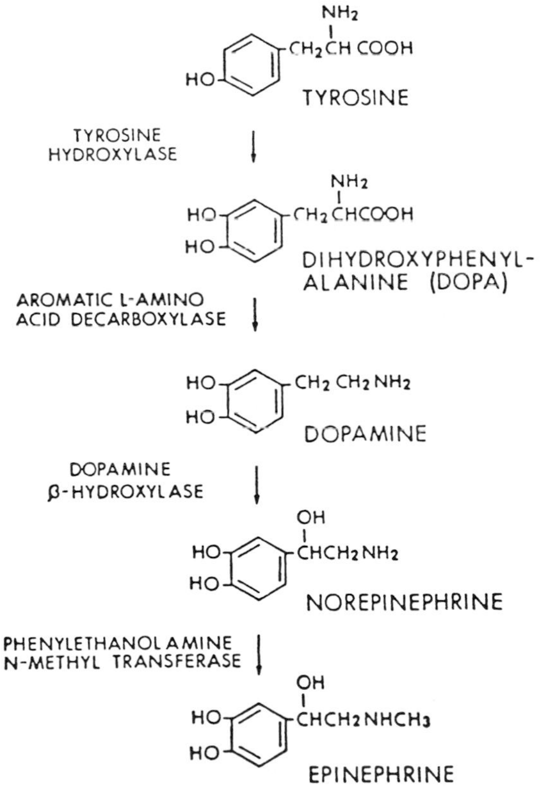

autonomic nervous system emerge from cranial and sacral spinal nerves, those of the sympathetic nervous system emerge from thoracic and lumbar spinal nerves and innervate many organs, including the adrenal medulla, as shown in Color Plate 1. These fibers then terminate in ganglia of the paraspinal sympathetic trunk, in nearby plexuses, or in the adrenal medulla.

Preganglionic nerve impulses are transmitted to postganglionic fibers by the liberation of acetylcholine at nerve terminals. This results in the secretion of catecholamines by the peripheral sympathetic nervous system and by the adrenal medulla. Norepinephrine is the major secretory product of the peripheral nervous system. In the human adrenal medulla, the ratio of epinephrine to norepinephrine secretion is approximately 4:1.

Catecholamine biosynthetic pathways are shown in Fig. 7. The rate-limiting step in catecholamine biosynthesis is the initial conversion of tyrosine to dihydroxyphenylalanine (DOPA) by tyrosine hydroxylase, an enzyme found only in neural tissue. Tyrosine itself

**FIGURE 7** Catecholamine biosynthetic pathway of the sympathetic nervous system. [Reproduced with permission from P. Cryer (1987). Diseases of the sympathochromaffin system. *In* "Endocrinology and Metabolism" (P. Felig, ed.). McGraw-Hill, New York.]

is derived from the diet or is converted in the liver from phenylalanine by phenylalanine hydroxylase. DOPA is decarboxylated by DOPA decarboxylase (aromatic L-amino acid decarboxylase). This enzyme, which forms dopamine, is a widely distributed cytosolic enzyme, which also decarboxylates other amino acids. Dopamine is taken up from the cytoplasm into chromaffin granules by a stereospecific process. In the chromaffin granules, dopamine is converted to norepinephrine by dopamine-$\beta$-hydroxylase (DBH). The enzyme DBH is found in the vesicles of catecholamine-containing cells and is secreted along with catecholamines. Dopamine and norepinephrine are found in sympathetic neurons and in the adrenal medulla. The major difference between the pathways of the adrenal medulla and the peripheral sympathetic nervous system is the presence of the cytosolic enzyme phenylethanolamine-$N$-methyl transferase (PNMT)

in the adrenal medulla. This enzyme catalyzes the conversion of norepinephrine to epinephrine. PNMT is induced by high concentrations of cortisol that are present in the portal circulation which flows from the adrenal cortex to the medulla. Once formed, epinephrine joins dopamine in secretory granules, and both amines are released together by the adrenal medulla. Other substances released along with adrenomedullary catecholamines include chromogranins, enkephalins, neuropeptide Y, and ATP.

A large percentage of synaptically released catecholamines are inactivated by reuptake into storage granules. Metabolism of circulating catecholamines occurs via two main pathways, mediated by the enzymes catechol-$O$-methyl transferase (COMT) and monoamine oxidase (MAO), as shown in Fig. 8. The end product of norepinephrine and epinephrine metabolism, after conversion by both enzymes, is 3-methoxy-

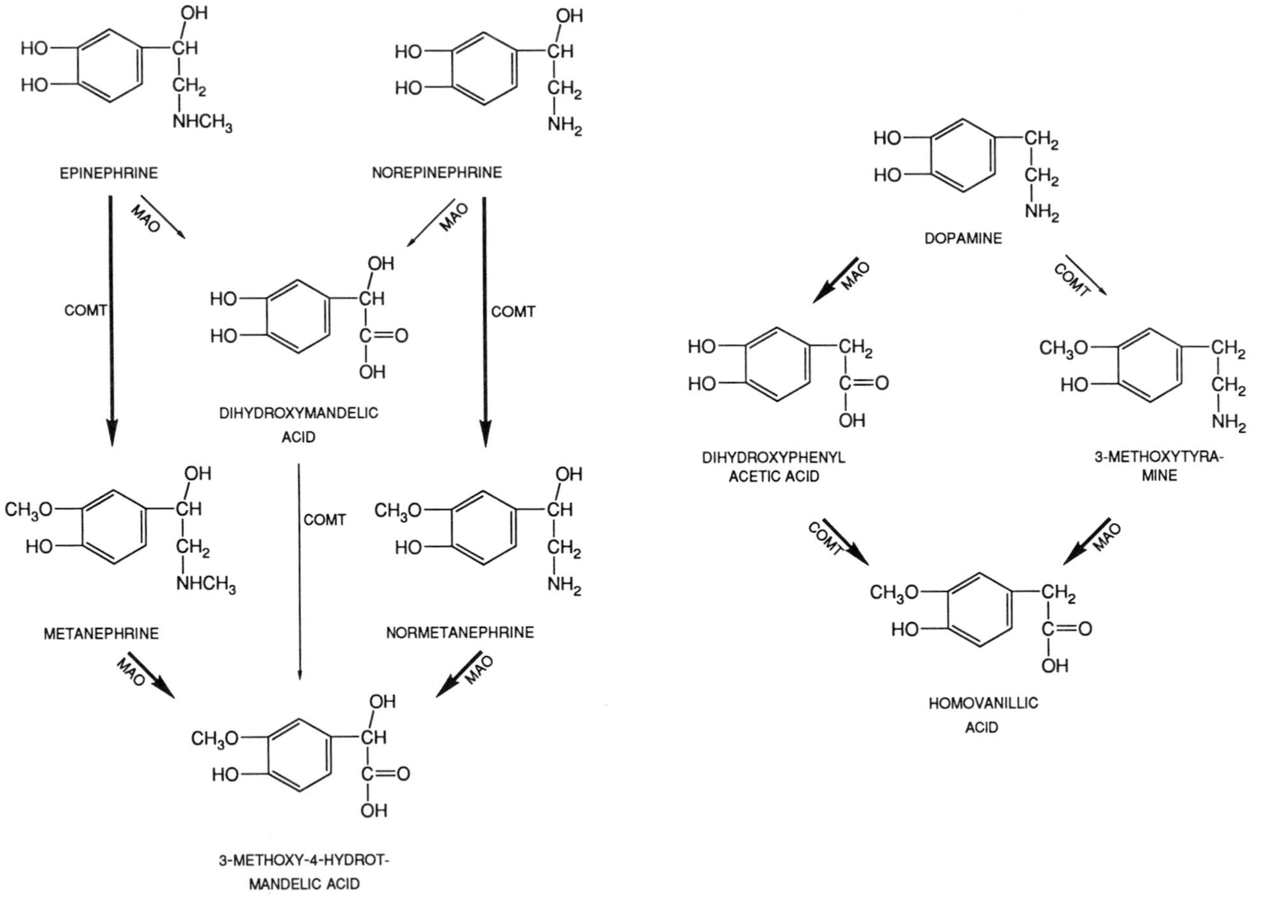

FIGURE 8    Metabolism of catecholamines by catechol-$O$-methyltransferase (COMT) and monoamine oxidase (MAO). [Reproduced with permission from A. Goldfien (1986). The adrenal medulla, *In* "Basic and Clinical Endocrinology" (F. Greenspan and P. Forsham, eds.), p. 326. Lange Medical Publications, Los Altos, CA.]

4-hydroxymandelic acid (vanillylmandelic acid; VMA).

The hypothalamus is the main regulator of sympathetic nervous system function. Impulses from the posterior and lateral hypothalamus result in generalized discharge of the sympathetic nervous system, including the adrenal medulla. As discussed earlier, this occurs in response to a variety of noxious, threatening, or stressful stimuli. In addition, the sympathetic nervous system is instrumental in maintaining an appropriate circulating volume and cardiac output during changes of posture from supine to upright. These feedback systems are mediated by sensors in the carotid sinuses, aorta and medulla, which detect changes in circulatory volume and blood pressure.

## IV. ADRENAL GLAND DISORDERS

### A. Adrenal Cortex

#### 1. Steroid Deficiency Syndromes

These disease states are due either to a primary disorder of the adrenal glands or to a secondary or tertiary disorder in the pituitary or hypothalamus, respectively. The most common causes of primary adrenal disorders are a congenital enzymatic block or destruction of the adrenal glands because of autoimmunity, infection, hemorrhage, tumor, drugs, surgery, or radiotherapy. In the case of autoimmunity, antibodies may also be directed against other glands, including the thyroid.

Signs and symptoms of these disease states vary with the pattern of steroid pathways affected and the degree to which they are altered. The most common form of congenital adrenal enzymatic block is 21-hydroxylase deficiency (Fig. 4), in which the lack of synthesis of cortisol and mineralocorticoids causes hypoglycemia and electrolyte imbalance. Overproduction of adrenal androgens causes masculinization of a female fetus with resultant ambiguous genitalia at birth. The lack of cortisol feedback on ACTH secretion causes hypersecretion of ACTH and enlargement of both adrenal glands, which gives the syndrome its name, congenital adrenal hyperplasia (CAH). Other forms of CAH result from the decreased activity of other adrenal enzymes. Primary steroid deficiency syndromes are verified by the measurement of appropriate steroid hormones in the blood before and after ACTH stimulation. Many synthetic glucocorticoid and mineralocorticoid hormones are now available, which make lifesaving replacement therapy possible.

Secondary and tertiary forms of adrenal insufficiency occur due to the therapeutic usage of glucocorticoid therapy for conditions such as obstructive and asthmatic lung disease, or arthritis, and to pituitary or hypothalamic tumors, surgery, or radiotherapy. The use of high dosages of glucocorticoids for long periods of time results in atrophy of the pituitary–adrenal axis. In ACTH deficiency, there is usually minimal electrolyte abnormality because the renin–angiotensin system is still operative. The opposite situation occurs in some patients with kidney disease and renin deficiency who have decreased aldosterone but normal cortisol secretion.

Apparent glucocorticoid or mineralocorticoid deficiency syndromes with *increased* circulating concentrations of glucocorticoids or mineralocorticoids have been found. In endocrinology, similar types of disorders have been described in many other hormone systems. In this case, these disorders are due to congenital abnormalities in glucocorticoid or mineralocorticoid receptors or in postreceptor mechanisms. In the case of glucocorticoids and aldosterone, this is called primary cortisol resistance and pseudohypoaldosteronism, respectively.

#### 2. Steroid Excess Syndromes

States of circulating adrenocortical steroid excess are caused by taking glucocorticoids or by either primary disease of the adrenal glands or excessive pituitary or hypothalamic secretion. If there is an excess of glucocorticoids in the circulation, the disorder is known as Cushing's syndrome.

Primary Cushing's syndrome results from adrenocortical adenomas and carcinomas. These can cause hyperglycemia, muscle-wasting, and osteoporosis, due to overproduction of cortisol, and masculinization in women, due to adrenal androgen overproduction. If there is excessive pituitary secretion of ACTH, this is called Cushing's disease, which in some cases may be due to excessive CRH. Some tumors regain their previously suppressed genetic ability to secrete peptide hormones, and if ACTH is secreted, often by lung tumors, this is known as ectopic ACTH syndrome. Cushing's syndrome is verified by measurement of the appropriate hormones before and after attempted suppression with a synthetic glucocorticoid, dexamethasone. Cortisol and ACTH are measured in this type of testing. An adrenocortical carcinoma often has enzyme blocks and hypersecretes

DHAS. Treatment depends on the source of steroid excess. For example, an adrenal tumor or a pituitary tumor is usually treated by surgical removal. Alternatively, medications that inhibit adrenal enzymes or block glucocorticoid receptors can be used in some cases.

Increased secretion of aldosterone can be caused by hyperplasia or tumor formation of the zona glomerulosa. Because cortisol secretion is normal, signs and symptoms are due to an electrolyte imbalance. The most common problems associated with increased mineralocorticoid secretion are sodium retention, volume overload, and increased blood pressure (hypertension). Renal potassium hypersecretion, with resultant serum potassium deficiency, also occurs. A potassium deficiency is readily diagnosed by routine blood tests. It causes muscular weakness and, if severe or prolonged, decreased kidney function and life-threatening cardiac rhythm disorders. If a tumor of the adrenal cortex is the cause of excessive aldosterone secretion, measurements of plasma renin activity are decreased because the tumor is autonomous. An excess of aldosterone secretion is sometimes treatable with the aldosterone antagonist spironolactone, and a tumor of the zona glomerulosa (aldosteronoma) may be surgically removed. Removal of an aldosteronoma often cures hypertension.

## B. Adrenal Medulla

Adrenalectomized patients are deficient in epinephrine, but if the rest of the sympathetic nervous system is normal, symptoms of autonomic insufficiency do not develop. However, symptoms of catecholamine excess occur in the presence of hypersecretion by chromaffin cells. Tumors of chromaffin cells may occur in association with sympathetic ganglia, anywhere in the body from the neck to the pelvis, but most often occur as tumors of the adrenal medulla, which are called pheochromocytomas. Tumor cells contain catecholamine storage granules. Although most are benign, they cause disease due to their oversecretion of catecholamines.

Pheochromocytomas cause signs and symptoms due primarily to hypertension. They increase blood pressure through excessive catecholamine secretion, mediated by alpha and beta stimulation, which results in increased peripheral vasoconstriction and cardiac output. They also cause episodic nonspecific symptoms such as headache, palpitations, and sweating,

which can be mistaken for severe anxiety. The hypertension caused by pheochromocytomas is usually resistant to treatment with the usual antihypertensive medications, which may serve as a diagnostic clue. Another clue is that patients are usually younger than most others with severe hypertension. If blood pressure elevation persists, it may result in the usual hypertension-related complications of heart attack, (myocardial infarction) heart failure, or stroke. Stimulation of glycogenolysis may result in hyperglycemia and in a mistaken diagnosis of primary diabetes. [*See* Hypertension.]

In some cases, pheochromocytomas occur as part of the autosomal dominant syndrome of multiple endocrine neoplasia, Type II, which also includes calcitonin-secreting tumors of the C cells of the thyroid gland, and parathyroid hormone-secreting tumors of the parathyroid gland. The diagnosis of pheochromocytoma is verified by catecholamine measurements. If drug treatment is desired, alpha and beta receptor-blocking medications are used. If this is unsuccessful, or if surgery is more appropriate for the patient, surgical removal of the tumor after blood volume expansion, and during alpha and beta blockade, is the treatment of choice.

## BIBLIOGRAPHY

Baxter, J., and Tyrrell, J. (1987). The adrenal cortex. *In* "Endocrinology and Metabolism" (P. Felig, ed.). McGraw-Hill, New York.

Biglieri, E., and Kater, C. (1987). Disorders of the adrenal cortex. *In* "Internal Medicine" (J. Stein, ed.). Little, Brown and Company, Boston.

Keiser, H. (1995). Pheochromocytoma and related tumors. *In* "Endocrinology" (L. DeGroot, ed.). Saunders, Philadelphia.

Guyton, A. (1986). The adrenocortical hormones. *In* "Textbook of Medical Physiology." Saunders, Philadelphia.

Grossman, A. (1995). Corticotropin-releasing hormone: Basic physiology and clinical applications. *In* "Endocrinology" (L. DeGroot, ed.). Saunders, Philadelphia.

Haynes, R. (1990). ACTH; adrenocortical steroids and their synthetic analogs; inhibitors of the synthesis and actions of adrenocortical hormones. *In* "The Pharmacological Basis of Therapeutics" (A. Gilman, ed.). Pergamon Press, New York.

Mortensen, R., and Williams, G. (1995). Aldosterone action. *In* "Endocrinology" (L. DeGroot, ed.). Saunders, Philadelphia.

Parker, L. (1989). What is the biological role of adrenal androgens? *In* "Adrenal Androgens in Clinical Medicine." Academic Press, San Diego.

Pescovitz, O., Cutler, G., and Loriaux, D. (1990). Synthesis and secretion of corticosteroids. *In* "Principles and Practice of Endocrinology and Metabolism" (K. Becker, ed.). Lippincott, Philadelphia.

# Adrenergic and Related G Protein-Coupled Receptors

ROMAN L. ZASTAWNY
*University of Toronto*

SUSAN R. GEORGE
*University of Toronto and Addiction Research Foundation, Toronto*

MICHEL BOUVIER
*University of Montreal*

BRIAN F. O'DOWD
*University of Toronto and Addiction Research Foundation, Toronto*

## GLOSSARY

**Adenylyl cyclase** Membrane-bound enzyme responsible for the formation of cyclic adenosine monophosphate (cAMP) from adenosine triphosphate (ATP); the function of the enzyme is regulated by its interaction with stimulatory ($G_s$) and inhibitory ($G_i$) G proteins

**Adrenergic receptors** Membrane-bound receptors that bind and are activated by epinephrine and norepinephrine

**Desensitization** Process by which the magnitude of a biological response wanes even in the face of a stimulus of constant intensity

**Down-regulation** agonist-induced decrease in receptor number that occurs upon prolonged exposure to agonist

**G protein-coupled receptors** A superfamily of cell surface receptors that upon activation by extracellular stimuli modulate the activity of a variety of effectors such as enzymes and ion channels via intermediary G proteins

**G protein receptor kinases** A family of serine/threonine kinases that specifically phosphorylate activated G protein-coupled receptors at serine/threonine residues

**Guanine nucleotide-binding regulatory proteins** Referred to as G proteins because they bind and hydrolyze the guanine nucleotide, guanosine triphosphate; they functionally couple receptors to various biochemical effectors such as enzymes and ion channels

**Protein kinase A** cAMP-dependent protein kinase

**Protein kinase C** Calcium/phospholipid-dependent protein kinase

**Sequestration** Rapid agonist-induced translocation of receptors away from the plasma membrane into distinct intracellular vesicular compartments

ADRENERGIC RECEPTORS (AR) ARE MEMBERS OF a superfamily of cell surface receptors, referred to as G protein-coupled receptors (GPCR), that mediate the signals of a diverse array of extracellular stimuli via guanine nucleotide-binding proteins (G proteins) to intracellular second messenger systems. Once transduced to second messengers (e.g., adenylate cyclase, phospholipase C, ion channels) these signals lead to profound effects on many aspects of normal cell physiology. The adrenergic family of receptors binds and responds to the catecholamine neurotransmitters epinephrine and norepinephrine; however, endogenous activators for GPCRs include peptide hormones, lipids, cytokines, and chemoattractants as well as sensory stimuli such as light and odorants. Despite the

ENCYCLOPEDIA OF HUMAN BIOLOGY, Second Edition, VOLUME I.   Copyright © 1997 by Academic Press.   All rights of reproduction in any form reserved.

great diversity in the types of ligands which interact with these receptors, the cloning of genes encoding for these receptors has revealed that a remarkable similarity exists in the structure of these GPCRs.

The adrenergic receptor family, which consists of nine members based on molecular cloning, can be divided into three main types according to sequence similarities, pharmacological profiles, and coupling to second messenger systems. The $\alpha_1$ class of adrenergic receptors, which includes $\alpha_1B$, $\alpha_1C$, and $\alpha_1D$, activate phospholipase C to generate the second messengers, inositol triphosphate and diacylglycerol. The $\beta$ class of adrenergic receptors, which includes $\beta_1$, $\beta_2$, and $\beta_3$, activates adenylate cyclase (AC) to form the second messenger cyclic AMP (cAMP) whereas the $\alpha_2$ class, which includes $\alpha_2A$, $\alpha_2B$, and $\alpha_2C$, inhibits adenylate cyclase. Substantial evidence in the literature indicates that other $\alpha_1$- and $\alpha_2$- subtypes exist; however, their existence has yet to be confirmed by molecular cloning.

Of all GPCRs cloned to date, the $\beta_2$-adrenergic receptor ($\beta_2AR$) has been one of the most extensively investigated. Hence, $\beta_2AR$ has served as the prototypical receptor for understanding GPCR structure–function relationships and has provided much insight into the biochemical mechanisms underlying signal transduction across cellular membranes to second messenger systems. New members belonging to the GPCR superfamily are constantly being identified and characterized. It is becoming increasingly apparent that although many of the proposed mechanisms describing adrenergic function and modes of cellular signal transduction apply to many other members of the GPCR superfamily, important differences have also begun to emerge. Thus, with the genes encoding over 200 GPCRs cloned to date, our understanding of $\beta_2AR$ structure and function has and will undoubtedly continue to add to our knowledge about GPCRs in general.

This article reviews the current knowledge of adrenergic and related G protein-coupled receptors with emphasis on some of the most recent developments in the field. In addition, this review discusses some of the recent evidence linking a host of human disease states to abnormally functioning GPCRs.

## I. STRUCTURE OF ADRENERGIC AND OTHER G PROTEIN-COUPLED RECEPTORS

Comparison of the primary structure for cloned GPCRs including the adrenergic, muscarinic, dopa-

minergic, and serotonergic family of neurotransmitter receptors, rhodopsin, thyroid-stimulating hormone (TSH) receptor, follicle-stimulating hormone receptor, angiotensin receptor, and many others (see Table I for complete list) has revealed that a common structural organization exists. The hallmark of this superfamily of receptors is the characteristic membrane topology consisting of an extracellular amino terminus, seven membrane-spanning hydrophobic domains interconnected by three extracellular and intracellular loops, and an intracellular carboxy terminal (see Fig. 1). The highest sequence similarity among the receptors is found in the transmembrane (TM) domains which have been shown for the monoamine receptors to associate to form a ligand-binding pocket. The identification of amino acid residues within the TM domains that are involved in ligand binding has been determined experimentally for several G protein-coupled receptors (see Section II). Mutagenesis studies of the $\beta_2$-adrenergic receptor also indicate that amino acids in TM 1 and 7 interact, suggesting a specific spatial arrangement of TM domains within the plasma membrane (see Fig. 1). The ability of GPCRs to bind a broad range of structurally different ligands seems to be paradoxical, considering the structural similarity seen among members of the GPCR superfamily. However, it does suggest that the tertiary structure formed by the seven TM domains within the plasma membrane bilayer must be well suited for transmitting a signal across the cell membrane upon ligand binding, which has resulted in evolutionary conservation. Hence, in order to transmit the signal across the cell membrane to G proteins, the conformational changes which occur within GPCRs upon ligand binding are probably quite similar for all GPCRs. Although notable progress is being made in this area using molecular strategies largely based on deletion, chimeric, and site-directed mutagenesis studies of adrenergic and other GPCRs, our knowledge in this area of receptor research is still limited. [*See* G Proteins.]

Aside from the similarity observed in the seven putative TM domains among different receptors, other regions of G protein-coupled receptors show much less similarity. The amino terminus varies greatly in size, ranging from as few as 12 amino acids for the human $\alpha_2$-adrenergic receptor to nearly 400 amino acids for the thyrotropin (TSH) receptor. In the case of peptide hormone receptors, the extracellular domains also play a role in ligand binding. The most unusual of the peptide receptors is the thrombin receptor, to which thrombin binds and results in the proteolysis of the receptor's long extracellular amino terminus

**TABLE I**

Cloned G Protein-Coupled Receptors

| Receptor | Ligand | Subtypes | Effectors[a] |
|---|---|---|---|
| Acetylcholine (muscarinic) | Acetylcholine | M1; M2; M3; M4; M5 | AC, PI; $Ca^{2+}$; $K^+$ |
| Adenosine | Adenosine | $A_1$, $A_{2A}$, $A_{2B}$, $A_3$ | AC |
| Adrenergic | Adrenaline; noradrenaline | $\alpha_{1B}$; $\alpha_{1C}$; $\alpha_{1D}$; $\alpha_{2A}$; $\alpha_{2B}$; $\alpha_{2C}$; $\beta_1$; $\beta_2$; $\beta_3$ | AC; $Ca^{2+}$; $K^+$ |
| Adrenomedullin | Adrenomedullin | AM-R | AC |
| Angiotensin II | Angiotensin II | $AT_{1A}$; $AT_{1B}$; $AT_2$ | PI |
| Bombesin | Bombesin; neuromedin B; GRP | $BB_1$; $BB_2$; $BB_3$ | PI |
| Bradykinin | Bradykinin; kallidin; ornitho-kinin | $B_1$; $B_2$ | PI; AC; $PLA_2$ |
| C3a anaphylatoxin | C3a | C3a-R | |
| C5a anaphylatoxin | C5a | C5a-R | PI |
| Calcitonin-gene-related peptide | CGPR1 | CGPR1-R | |
| Cannabinoid | Anandamide | CB1; CB2 | AC |
| C-C Chemokine | MCP; MIP-1$\alpha$; MIP-1$\beta$; RANTES; Eotaxin | CKR1; CKR2; CKR3; CKR4; CKR5 | $Ca^{2+}$ |
| Cholecystokinin/gastrin | CCK-8; CCK-33; CCK-4 | $CCK_A$; $CCK_B$ | PI |
| Dopamine | Dopamine | D1; D2; D3; D4; D5 | AC; $Ca^{2+}$; $K^+$ |
| Endothelin | ET-1; ET-2; ET-3 | $ET_A$; $ET_B$ | PI |
| Formyl-methionyl peptide | fMet-Leu-Phe (fMLP) | FPR1; FPR2 | PI; AC; PLC |
| Galanin | Galanin | GALR1 | AC; $Ca^{2+}$; $K^+$ |
| Glycoprotein hormone | TSH; FSH; LH/hCG | FSH-R; LHHCG-R; TSH-R | AC; PI |
| Gonadotropin-releasing hormone | GnRH | GnRH-R | PI |
| Histamine | Histamine | $H_1$; $H_2$ | AC; $PLA_2$ |
| 5-Hydroxytryptamine | 5-HT (serotonin) | $5\text{-HT}_{1A}$; $5\text{-HT}_{1B}$; $5\text{-HT}_{1D}$; $5\text{-HT}_{1E}$; $5\text{-HT}_{1F}$; $5\text{-HT}_{2A}$; $5\text{HT}_{2B}$; $5\text{-HT}_{2C}$; $5\text{-HT}_4$; $5\text{HT}_{5A}$; $5\text{-HT}_{5B}$; $5\text{-HT}_6$; $5\text{-HT}_7$ | AC; PI; $K^+$ |
| Interleukin-8 | IL-8 | IL8AR; IL8BR | PI |
| Interferon-gamma inducible protein 10 | IP-10; Mig | GPR9 | $Ca^{2+}$ |
| Melanocortin | ACTH; $\alpha$-MSH; $\beta$-MSH; $\gamma$-MSH | MC1; MC2; MC3; MC4; MC5 | AC |
| Melatonin | Melatonin | $ML_{1A}$; $ML_{1B}$ | AC |
| Neuropeptide Y | NPY; PP; PYY | Y1; Y2; Y4; Y5 | AC; $Ca^{2+}$ |
| Neurotensin | Neurotensin | NTR1; NTR2 | AC; PI |
| Odorant | Odors | >50 | AC; PI |
| Opioid | $\beta$-endorphin; dynorphin A; leu-enkephalin; Met-enkephalin | $\mu$; $\kappa$; $\delta$ | AC; $Ca^{2+}$; $K^+$ |
| Opsin | 11-*cis*-retinal | blue; red; green; rhodopsin | |
| Orphanin | Orphanin-FQ | ORL-1 | |
| Oxytocin | Oxytocin | OT-R | PI |
| $P_2$ Purinoceptors | ATP; UTP | $P_{2Y1}$; $P_{2Y2}$; $P_{2Y3}$; $P_{2Y4}$; $P_{2Y5}$; $P_{2Y6}$; $P_{2Y7}$ | |
| Platelet-activating factor | PAF | PAF-R | PI |
| Prostanoid | Prostaglandins; thromboxanes | $EP_1$; $EP_2$; $EP_3$; $EP_4$; FP; IP; TP; DP | AC; PI |
| Protease-activated receptor | Tethered peptide ligands | PAR1; PAR2 | AC; PI |
| Stromal cell-derived factor 1 | SDF-1 | LESTR | $Ca^{2+}$ |
| Somatostatin | SS-14; SS-28 | $sst_1$; $sst_2$; $sst_3$; $sst_4$; $sst_5$ | |
| Tachykinin | Neurokinin A (substance K) Neurokinin B (neuromedin K) Substance P | NK1; NK2; NK3 | PI |
| Thyrotrophin-releasing hormone | TRH (thyoliberin) | TRH-R | PI |
| Vasopressin | Vasopressin | $V_{1A}$; $V_{1B}$; $V_2$ | AC; PI |

[a] AC, adenylate cyclase; $Ca^{2+}$, calcium channel; cGMP-PDE, cyclic GMP-phosphodiesterase; $K^+$, potassium channel; PI, phosphoinositol phosphate; $PLA_2$, phospholipase $A_2$; PLC, phospholipase C.

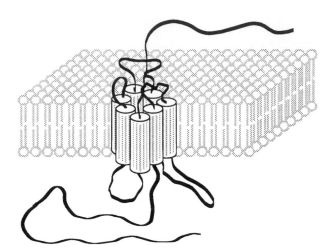

**FIGURE 1** Model demonstrating the transmembrane topology of a G protein-coupled receptor.

domain; the new amino terminus subsequently acts as a tethered peptide ligand which then autoactivates the receptor. Sequence analysis of extracellular domains reveals that most receptors contain consensus sequences for N-linked glycosylation. The $\beta_2$-adrenergic, rhodopsin, NK1 neurokinin, and $D_1$ and $D_2$-dopamine receptors are glycosylated; however, site-specific mutagenesis studies have shown that nonglycosylated forms of these receptors do not exhibit any alterations in ligand binding or G protein coupling. Although these results suggest that glycosylation is not required for receptor function, it may be required for optimal membrane expression and trafficking since receptors mutated at their glycosylation sites usually are expressed at lower levels on the cell surface membranes when compared to their wild-type counterparts.

The extracellular loops which join the TM loops are typically short and quite divergent in sequence except for several conserved cysteine residues in the second and third loops which have been implicated in forming disulfide bonds. Studies which dissociated putative disulfide bonds using reducing agents or mutated specific cysteine residues (e.g., Cys-106 and Cys-184 in $\beta_2$AR) resulted in either the loss or the decrease of ligand-binding activity. These results suggested that disulfide bonds are essential for maintaining the $\beta_2$-adrenergic receptor in a proper functional conformation.

On the intracellular side of the plasma membrane, the three loops joining the TM domains and the carboxy-terminal tail show very little sequence similarity among receptors when compared to sequences within the highly conserved TM domains. These intracellular domains are essential for G protein coupling and are the site of modification by several posttranslational events, most notably phosphorylation and palmitoylation. As discussed in a subsequent section, phosphorylation of these domains is involved in regulating receptor responsiveness. More recently, palmitoylation has been suggested to play an important regulatory role for G protein-coupled receptor functions. $\beta_2$AR, $\alpha_2$AR, rhodopsin, $D_1$, and $D_2$ dopamine, and the 5HT1B receptors have been shown to be palmitoylated. It has been proposed that palmitoylation anchors part of the carboxy terminal domain of these receptors to the plasma membrane to form a fourth intracellular loop that is important for G protein coupling (Fig. 2). However, not all GPCRs contain cysteine residues at this location. Substituting Cys-341 with glycine in $\beta_2$AR markedly reduces its ability to couple to G proteins, suggesting that this modification is critical for efficient G protein coupling for this receptor. More recently, $\beta_2$AR lacking Cys-342 has been shown to be constitutively phosphorylated and desensitized, further suggesting that palmitoylation in some manner modulates the accessibility of kinases to specific receptor residues. The observation that many other G protein-linked receptors have cysteine residues at equivalent positions suggests that palmitoylation may be involved in regulating the functions of other GPCRs. However, the effects on coupling of G protein-coupled receptors are not universal, following mutation of a potentially palmitoylated cysteine residue. Indeed, abolition of palmitoylation by mutation of Cys-621 and Cys-622 for the luteinizing hormone/choriogonadotropic hormone (LH/hCG) receptor and of Cys-442 for the $\alpha_{2A}$-adrenergic receptor are without effect on their respective coupling to phosphoinositol phosphate turnover and cAMP production. However, these mutations greatly affect the agonist-promoted internalization of these two receptors. It is noteworthy that neither of these receptors harbor phosphorylation sites near the palmitoylated cysteine. Therefore, it may be proposed that, for these receptors, palmitoylation does not modulate the accessibility to phosphorylation sites, but rather to domains involved in receptor internalization and trafficking.

Studies involving receptor chimeras, deletions, point mutations, and competition experiments using peptides that inhibit or mimic receptor–G protein interactions have identified four regions of receptors that are thought to be important in determining G protein specificity. These include the N- and C-terminal region of intracellular loop three, the C-terminal regions of intracellular loops two and three, and part

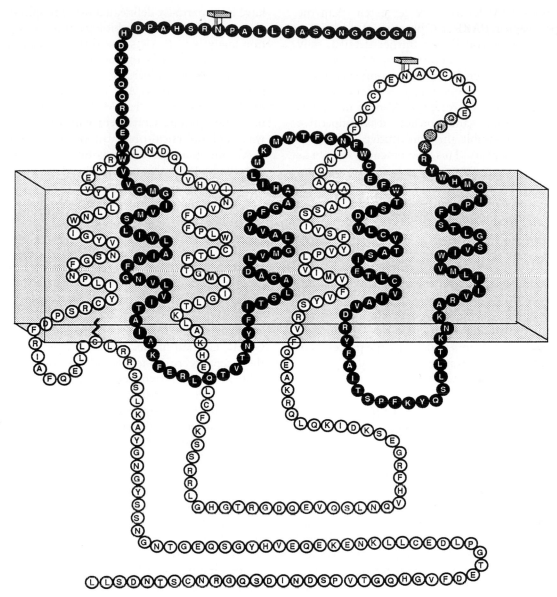

**FIGURE 2** Schematic diagram of the predicted seven transmembrane secondary structure of the human $\beta_2$-adrenergic receptor. The boxed shaded area represents the plasma membrane. Putative sites of N-linked glycosylation are indicated.

of the carboxy terminus. Although these regions of the receptors have been defined, little is known about how these regions confer G protein receptor subtype specificity to receptors or how a signal is relayed from the receptor to the G protein. However, it was discovered that if a small stretch of amino acids in the C-terminal of the third intracellular loop of the $\beta_2$AR were mutated, the resultant receptor was constitutively active. That is, the basal signaling activity of an agonist unoccupied receptor was comparable to that of a fully

agonist-occupied wild-type $\beta_2$-adrenergic receptor. Interestingly, the constitutively active mutant receptor had much higher affinity for agonists but not for antagonists. It is speculated that the mutation spontaneously isomerizes the mutant receptor into an active form, hence mimicking the conformational effect of an agonist binding to a wild-type receptor. This premise was further strengthened by the observation that the constitutively active $\beta_2$AR is phosphorylated, in an agonist-independent fashion, by $\beta$-adrenergic receptor kinase ($\beta$ARK) to a greater extent than the wild-type receptor. $\beta$ARK, a G protein-coupled receptor kinase (see Section III for further details), only phosphorylates active agonist-occupied receptors, hence this further suggests that the conformation of the constitutively active mutant $\beta_2$AR mimics an agonist-occupied wild-type receptor. Also of interest is the observation that ligands considered as competitive antagonists can inhibit the constitutive activity of these mutant receptors. This phenomenon, known as inverse agonism, suggests that compounds with negative intrinsic activity can promote destabilization of the active receptor conformation. These ligands can also inhibit the spontaneous (agonist-independent) activity of overexpressed wild-type receptors. This suggests that receptors can spontaneously isomerize between inactive and active conformations and that ligands can favor or inhibit the interconversion. Future studies of the constitutively active mutant $\beta_2$AR will undoubtedly further our understanding of the factors which maintain receptors in a particular conformation and the factors which link the processes of activation and desensitization.

## II. RECEPTOR–LIGAND INTERACTIONS

Initial efforts to define regions of GPCRs involved in ligand binding focused on rhodopsin. The ligand for rhodopsin is a photon of light, and the receptor is activated by a retinal chromophore covalently attached to rhodopsin. Under nonstimulating conditions (in the dark), retinal acts as an antagonist, bound to the receptor but not able to activate it; however, when exposed to light, retinal changes from an antagonist-like molecule to an agonist-like molecule which has the ability to activate rhodopsin. On exposure to light, the retinal molecule undergoes conversion from a 9-*cis* to an all-*trans* conformation which subsequently causes rhodopsin to take on an active conformation. Detailed biochemical and biophysical studies of rhodopsin, including structural modeling based on the known structure of the functionally related bacteriorhodopsin molecule, have defined the binding site of retinal to be situated deep inside the membrane bilayer in a cleft formed by the TM domains. These studies of rhodopsin have provided the groundwork for defining the ligand-binding domains for other GPCRs. [*See* Receptors, Biochemistry.]

The ligand-binding domains of $\beta_2$AR, which bind the biogenic amines epinephrine and norepinephrine,

have been partially delineated using biophysical, biochemical, and molecular biological techniques. Deletion analysis of $\beta_2$AR showed that extracellular and intracellular domains and loops could be deleted, individually, without adversely affecting the ligand-binding properties of the $\beta_2$AR. This suggested, by process of elimination, that the TM domains were involved in ligand binding. Consistent with these conclusions was the biophysical observation that the fluorescent $\beta_2$AR antagonist carazolol is buried at least 10.9 Å into the hydrophobic core of the receptor.

Although deletion studies were important in implicating the TM domains in ligand binding, they had limited use in more precisely defining which TM domains were directly involved in ligand binding since deletion of any one TM domain from $\beta_2$AR invariably resulted in a loss of expression of receptor at the plasma membrane. This suggested that the TM domains were absolutely critical for receptor integrity. The use of chimeric receptors, i.e., receptors genetically engineered to contain regions from two related GPCR, has had limited utility in defining regions of the receptors critical for ligand binding as incompatibility between the two receptors often leads to reduced expression and binding affinity, most likely due to significant disruption in the tertiary structure of the chimeric receptors. Consequently, the most significant information regarding the structure of ligand-binding domains of $\beta_2$AR and other GPCRs has come from introducing single amino acid substitutions. Residues that have been examined and mutated were chosen on the basis of several criteria. First, analysis of the structure–function relationship of adrenergic agonist and antagonists revealed that biogenic amines have protonated nitrogens that are critical for receptor–ligand binding. Hence, the binding site for these compounds must contain a residue that can function as a counter ion for binding the protonated amine. Second, molecular modeling of the TM domains using a helical wheel model showed that the transmembrane $\alpha$-helices are amphipathic, i.e., one side of the helix consists mostly of hydrophobic residues while the opposite side consists of hydrophilic residues. It is postulated that the hydrophobic sides of each of the $\alpha$-helices face the membrane bilayer while the hydrophilic sides face each other and form the ligand-binding pocket. Hence, mutations within the TM regions have been limited to residues comprising the hydrophilic sides of the $\alpha$-helices. Third, sequence comparison of $\beta_2$AR with other GPCR including other biogenic amine receptors, which included the muscarinic, dopamine, serotonin, and histamine receptors, not only showed that the TM regions were highly conserved, but that certain residues within the TM regions were invariant, suggesting that these residues were universally important functionally. Moreover, a subset of these invariant residues was found only in GPCRs which bind biogenic amines, suggesting that these residues may be important for interactions with biogenic amines. For example, it was shown that if Asp-113, a highly conserved residue among biogenic amine receptors, in the third TM domain of the $\beta_2$AR was substituted with a neutral residue such as serine, ligand affinity decreased 10,000-fold, whereas introducing a glutamate residue into this same position, which retains the counter ion charge but changes its relative position, resulted in only a 100-fold decrease in ligand affinity. This suggested that the negatively charged Asp-113 provides the counter ion for an electrostatic interaction with the positively charged amino group of cationic amine ligands, thus stabilizing their binding to the receptor. Substitution of the analogous aspartic acid in the muscarinic, $\alpha_2$-adrenergic, and histamine receptors has further confirmed the important role of this residue in the ligand binding of biogenic receptors.

Interestingly, even though agonist and antagonists bound to the mutated $\beta_2$AR mentioned earlier with much lower affinity, agonists such as isoproterenol were still able to fully activate the receptor. Although Asp-113 is involved in ligand binding, it does not affect receptor activation. This pointed to other residues being involved in receptor activation. A second aspartate group, Asp-79, found in the second TM domain of $\beta_2$AR was also observed to be highly conserved in all GPCRs. Substitution of this Asp-79 with alanine in $\beta_2$AR resulted in a mutant receptor which displayed a decrease in affinity for agonists together with a decreased agonist efficacy. However, there was no appreciable change in antagonist binding. These results suggested that Asp-79 plays a role in the conformational change which occurs on agonist-mediated receptor activation. To date, a similar role for the conserved aspartic acid in the second TM domain has been described for $\alpha_2$-adrenergic, angiotensin AT1, serotonin 5HT2, and several other receptors. This suggests that this aspartate residue may be part of a general mechanism among many GPCRs for relaying the agonist binding-induced conformational change in the receptor which leads to G protein activation.

Similar strategies, as described earlier, have been used to define other residues within the TM domains critical for normal ligand binding and $\beta_2$AR activa-

tion, including Ser-204 and Ser-207 in the fifth TM domain and Phe-290 in the sixth TM domains. In all cases, mutations of these residues affected agonist binding and agonist-dependent receptor activation, but antagonist binding was not affected. The Conservation of these residues in other biogenic amine receptors suggests that these residues have been evolutionarily conserved to provide ligand specificity to biogenic amine receptors.

Interestingly, the fifth and sixth TM domains are linked together by the third intracellular loop which has been implicated in G protein coupling. Since the previously mentioned residues are only important for agonist binding, it has been speculated that the interaction of agonists with these residues might cause a conformational change in the fifth and sixth TM domains which is subsequently transmitted to the third intracellular loop to trigger G protein activation. Antagonists do not activate receptors because they do not interact with these residues in the fifth and sixth TMs. If this hypothesis is proven to be true, then it will be interesting to see whether the fifth and sixth TMs of other GPCRs are also involved in G protein activation.

The application of molecular biological techniques has greatly increased our understanding of receptor–ligand interactions. As discussed earlier, the critical residues involved in ligand binding to biogenic amine receptors has been partly delineated; however, our knowledge is still too limited to make any firm conclusions about mechanisms involved in agonist-mediated receptor activation. The limiting factor has been the lack of high resolution structural data for any GPCR; however, the electron diffraction structure of rhodopsin has been obtained at a resolution of 9 Å. This suggests that higher resolution analysis of rhodopsin and other GPCR structures will be possible in the future. The ultimate goal in these studies is to understand the detailed interactions which occur between receptor and ligand. This knowledge will have tremendous impact on the future design of new classes of selective therapeutic agents specifically targeted to individual GPCR subtypes.

## III. REGULATION OF RECEPTOR SIGNALING

Upon binding of endogenous, extracellular ligands, the primary role of GPCRs is to activate heterotrimeric ($\alpha\beta\gamma$) G proteins. Although it is well established that $G_\alpha$ subunits transmit the receptor-generated signal to cellular second messenger systems, evidence has shown that G$_{\beta\gamma}$ dimers also play important roles in signaling cellular effector systems. Much is known about the multiplicity of proteins involved in G protein-coupled signal transduction. Each of the G protein subunits are members of families of highly homologous proteins which include $\alpha$, $\beta$, and $\gamma$ subunits. A tremendous diversity in cellular signaling can be achieved through various combinations of the GPCRs, G protein $\alpha\beta\gamma$ subunits, and effector proteins. Prolonged exposure of GPCRs to agonist leads to a loss of responsiveness or desensitization of a receptor. Despite the continuous presence of a stimulus, the signal normally transmitted to the cell interior by the receptor begins to taper off to prevent stimulatory overload. The molecular mechanisms underlying receptor desensitization include phosphorylation and uncoupling of the receptor from its G protein. Evidence from numerous studies has shown that uncoupling of $\beta_2$AR from G proteins and desensitization are strongly associated with the rapid phosphorylation of active receptors (ligand-occupied receptor complex). Two different types of kinases—second messenger kinases and G protein-coupled receptor kinases (GRK)—have been implicated in playing a significant role in this process. Second messenger kinases, which include cAMP-dependent protein kinase (PKA) and protein kinase C (PKC), have been shown to phosphorylate specific serine and threonine residues within the third internal loop and carboxy-terminal of $\beta_2$AR. It is thought that phosphorylation of these residues changes their net charge and thus interferes with G protein coupling. Second messenger kinases are considered to represent mechanisms for classical feedback regulatory loops to terminate signals from receptors which directly activated the kinase (homologous desensitization) or to modulate the activity of other GPCRs (heterologous desensitization). G protein-coupled receptor kinases represent a more recently discovered family of serine/threonine kinases with the specialized function of phosphorylating active GPCRs. To date, five such kinases have been cloned, rhodopsin kinase (GRK1), $\beta$-adrenergic receptor kinase 1 ($\beta$ARK1 or GRK2), $\beta$-adrenergic receptor kinase 2 ($\beta$ARK2 or GRK3), and, most recently, IT11 (GRK4) and GRK5. $\beta$ARK1 and rhodopsin kinase are the most extensively studied members of the GRK family. These kinases were initially named after their preferred receptor substrate; however, they have since been shown to be able to phosphorylate a number of GPCRs. For instance, it has been shown that $\beta$ARK1 is capable of phosphorylating not only the $\beta_2$-adrenergic receptor, but also rhodopsin, $\alpha_2$-adrenergic, and muscarinic receptors. This ability of $\beta$ARK1 to phosphorylate multiple GPCRs is consistent with its wide distribution throughout the entire mammalian central nervous system. In contrast, rhodopsin kinase is only found in the retina and pineal, suggesting a more specific receptor selectivity. [*See* Cell Signaling.]

Our current understanding of $\beta$ARK involvement in $\beta_2$AR function suggests that following dissociation of the GTP-bound G$\alpha\beta\gamma$ heterotrimer from the $\beta_2$AR, cytosolic $\beta$ARK is recruited to the vicinity of the membrane-bound $\beta_2$AR, whereupon it phosphorylates multiple serines and threonine residues located in the carboxy terminus of the $\beta_2$-adrenergic receptor. Subsequently, a second protein, $\beta$-arrestin, recognizes and binds to the phosphorylated receptor. It is thought that $\beta$-arrestin binding interferes with G protein binding, thereby disrupting further signal transmission between the $\beta_2$AR and its cognate G protein. Experiments that removed serine and threonine residues from the carboxy tail of the $\beta_2$AR substantially decreased the homologous (agonist-specific) component of desensitization.

As mentioned earlier, besides playing important roles in signaling cellular effector systems, G$\beta\gamma$ dimers also play a critical role in agonist-induced phosphorylation and desensitization of the $\beta_2$-adrenergic receptor. G$\beta\gamma$ dimers have been implicated in mediating the activation and translocation of $\beta$ARK to the plasma membrane where it phosphorylates activated $\beta_2$AR. Hence, the ability of $\beta$ARK, a cytosolic protein, to phosphorylate and desensitize activated $\beta_2$-AR is highly dependent on its binding to membrane-anchored G$\beta\gamma$ dimers. It has been shown that the role of G$\beta\gamma$ dimers can be mimicked by mutating $\beta$ARK such that it can be isoprenylated like G$\gamma$ and therefore presumably becomes membrane anchored. *In vitro*, isoprenylated $\beta$ARK showed an marked increase in enzymatic activity and membrane association as compared to wild-type $\beta$ARK. The G$\beta\gamma$ binding domain on $\beta$ARK was initially mapped to a region comprising 125 amino acids near the C terminus of $\beta$ARK. This region was first suspected to be the binding site for G$\beta\gamma$ since it represented a unique domain that is not present in rhodopsin kinase, a GRK that does not interact with G$\beta\gamma$. Further experiments defined the $\beta$ARK–G$\beta\gamma$ binding region to a smaller domain that represents the last 28 amino acids of the original 125 amino acid domain. Although this binding domain has been defined, details describing the mechanisms by which G$\beta\gamma$ activates $\beta$ARK and specifically targets $\beta$ARK to activated receptors are limited.

Interestingly, an artificially isoprenylated $\beta$ARK mimics a naturally occurring situation that occurs with another GRK rhodopsin kinase. Rhodopsin kinase is the only GRK that is itself isoprenylated and translocated to the membrane in a stimulus-dependent fashion without associating with G$\beta\gamma$. A point mutation of Cys-558 to Ser in rhodopsin kinase, which prevents isoprenylation, substantially reduces its enzymatic activity and prevents light-stimulated translocation to outer segments of the retinal rod cells. Hence, these data demonstrate that activity and membrane association of GRKs is highly dependent on isoprenoid moieties, whether it involves direct isoprenylation or involves the binding of GRKs with isoprenylated G$\beta\gamma$ dimers.

Acylation (palmitoylation) of the receptors may also be part of the agonist-promoted desensitization. Studies with $\beta_2$AR demonstrated that agonist stimulation leads to an increased rate of receptor depalmitoylation. This depalmitoylation is believed to increase the accessibility of the carboxyl tail to regulatory kinases like PKA and $\beta$ARK. A change in the palmitoylation status of the receptor might be part of the structural changes which allow selective recognition of the active form of the receptor by $\beta$ARK.

Sequestration, the rapid-induced internalization of functional receptors from the cell surface, though not part of the desensitization process, occurs concurrently with receptor desensitization. It has been suggested that sequestration is involved in the mechanism that removes the agonist and phosphate groups from the receptor and recycles the receptors back to the cell surface as part of a process termed resensitization. For $\beta_2$AR, sequestration occurs after phosphorylation by PKA and $\beta$ARK; however, phosphorylation is not required for sequestration as mutant forms of the $\beta_2$AR lacking phosphorylation sites are sequestered normally. It has been shown that Tyr-326 in $\beta_2$AR is essential for sequestration. Replacing this tyrosine with an alanine residue completely abolished agonist-mediated sequestration without affecting the ability of $\beta_2$AR to activate adenylate cyclase or to undergo rapid desensitization and to down-regulate. The only defect observed with this mutated receptor is the complete loss of the ability of the receptor to resensitize. This experiment provided direct evidence that sequestration is associated with receptor resensitization. Interestingly, Tyr-326 situated just outside TM seven of $\beta_2$AR resides within a highly conserved sequence

motif, NPXXY (X represents an aliphatic residue), which is present in many other GPCRs. This suggests that the mechanisms by which these GPCRs are sequestered may be similar to the mechanism by which $\beta_2$AR is sequestered. However, the same motif is also found in receptors such as $\beta_1$ and $\beta_3$AR which do not undergo agonist-promoted sequestration, thus suggesting that the NPXXY sequence is necessary but not sufficient to support this internalization process.

While $\beta_2$AR desensitization and sequestration are rapid processes, occurring within minutes, longer time intervals of agonist exposure lead to a distinct process termed down-regulation. This is a prevalent mechanism underlying long-term receptor signal attenuation. This process results in a net decrease in the total receptor number in the cell, presumably due to modulation of the rate of receptor degradation and synthesis. Several lines of evidence have shown that PKA-dependent phosphorylation of $\beta_2$AR is important for down-regulation. $\beta_2$AR mutants lacking PKA phosphorylation sites are down-regulated at a slower rate than wild-type receptors. However, mutant receptors lacking $\beta$ARK phosphorylation sites down-regulate similarly to wild-type receptors, suggesting that $\beta$ARK-dependent receptor phosphorylation is not involved in down-regulation. In addition, some data have shown that if Tyr-350 and Tyr-354 residues located in the C terminus of $\beta_2$AR are replaced with alanine residues, the ability of the mutant receptor to down-regulate is dramatically decreased; however, sequestration is not affected. These data suggest that Tyr-350 and Tyr-354 in the cytoplasmic tail of $\beta_2$AR are important for proper receptor down-regulation but do not play a role in receptor sequestration. Hence, even though down-regulation and sequestration both involve receptor internalization, it appears that these processes also involve other mutually exclusive mechanisms. Interestingly, both sequestration and down-regulation involve tyrosine residues about which little is known at the present time. Tyrosine residues have been shown to play crucial roles in the trafficking from and to the plama membrane for numerous proteins. For receptors such as the mannose-6-phosphate receptor, tyrosine residues have been shown to promote interaction with adaptor proteins located in the clathrin coat of the endosomes mediating their internalization. Sequestered $\beta_2$AR have been colocalized with transferrin receptors in clathrin-coated endosomes. Whether or not the tyrosine residues identified in $\beta_2$AR are involved in interactions with similar adaptor proteins remains to be deter-

TABLE II

G Protein-Coupled Receptor-Linked Diseases in Humans

| Receptor | Chromosome | Disease state | Type and function of mutation [loss (−) or gain (+) of function] |
|---|---|---|---|
| Rhodopsin | 3q21 | Retinitis pigmentosa | (−) apoptosis of rod cells |
| Rhodopsin | 3q21 | Stationary night blindness | (+) missense mutation |
| Blue opsin | 7q22 | Tritan color blindness (tritanopia) | (−) |
| Red opsin | X | Protan color blindness (protanopia) | (−) chromosome rearrangement |
| Green opsin | X | Deutan color blindness (deuteranopia) | (−) chromosome rearrangement |
| $V_2$ vasopressin | X | Nephrogenic diabetes insipidus type II | (−) |
| ACTH (MC2) | 18p11.2 | Isolated glucocorticoid deficiency | (−) |
| TSH | 14q31 | Hyperthyroidism | (+) missense; somatic mutation |
| LH | 2p21 | Familial precocious puberty Leydig cell agenesis Leydig cell hypoplasia | (+) missense; males only |

mined. In other studies, internalized $\beta_2$AR were found in noncoated vesicles believed to be caveoli. The precise role played by these distinct populations of vesicles in the sequestration and down-regulation pathways is not clear at this time.

## IV. G PROTEIN-COUPLED RECEPTORS IN DISEASE

Our present understanding of the structure–function relationships of the GPCR has advanced mainly because of the use of molecular biology techniques. Many investigators have shown by site-directed mutagenesis that substituting particular amino acid residues affects normal receptor function and sometimes even completely abolishes receptor function. These

TABLE III

G Protein-Coupled Receptors Associated with Human Diseases

| Receptor | Disease state |
|---|---|
| Adrenergic | Malignant hypertension |
| Angiotensin AT2 | Hypertension |
| Dopamine | Schizophrenia, alcoholism |
| Endothelin ET$_B$ | Hirchsprung disease type II |
| 5-HT (serotonin) | Anxiety, depression, migraine |
| Tachykinin | Bronchial asthma |
| Thromboxane | Impaired platelet aggregation (bleeding disorder) |

type of experiments raise the question of what happens if such mutations occur in nature? As mentioned earlier, GPCRs process a diversity of extracellular signals that are central to the regulation and maintenance of normal cell physiology, therefore it would be easy to envision how aberrant functioning GPCRs may underlie a variety of disease states. Indeed, a number of acquired or hereditary human diseases such as certain types of blindness, diabetes insipidus, hyperthyroidism, and neoplasia have been linked to defective GPCRs (see Tables II and III).

The mechanisms by which abnormalities in G protein-mediated signal transduction cause disease are diverse and are not limited to abnormal GPCRs alone. Defects in other molecules involved in the signal transduction pathway of GPCRs, such as the endogenous ligands, G proteins, and effector molecules, have also been linked to human diseases. However, this review focuses only on disease states associated directly with GPCRs. In theory, diseases could result from genetic defects in GPCRs that lead to constitutively active receptors resulting in increased signal transduction. On the other hand, mutations may lead to a variety of defective receptor functions, including synthesis and membrane targeting, ligand binding, G protein coupling, or phosphorylation resulting in decreased signal transduction. The clinical manifestation of these defects would depend on the severity of the receptor dysfunction and on the cellular distribution and normal function of the receptor. If the receptor is widely expressed and its function is ubiquitous to many tissue types, then mutations may be lethal. However, if the receptor is expressed in highly specific tissues, the clinically associated abnormalities may be more tissue-specific. Moreover, the severity of the clinical abnormality directly correlates with the position and the number of mutations on the receptor. In some cases individuals may be totally asymptomatic even though they carry a particular mutation.

One of the first examples of mutated GPCRs was observed in patients with autosomal dominant retinitis pigmentosa, a hereditary form of retinal degeneration that leads to blindness. Initial studies found that 18% of 150 patients with autosomal dominant retinitis pigmentosa had one of three different mutations present in the rhodopsin gene. Since that study, more than 20 other mutations have been found in the rhodopsin gene in patients with autosomal dominant retinitis pigmentosa. Similar mutations have been found in the blue, green, and red opsin genes in patients with different types of color blindness. Although mutations found in rhodopsin are associated with autosomal dominant retinitis pigmentosa, the mechanism by which every mutation affects rhodopsin function is not yet known. However, one study found that rhodopin receptors incorporating any one of the three mutations in the cytoplasmic domain that were similar to ones found in patients with autosomal dominant retinitis pigmentosa, when functionally expressed, were impaired in their ability to couple to the G protein transducin. These results suggest that autosomal dominant retinitis pigmentosa is due to decreased signal transduction via the rhodopsin receptor pathway because of defective receptor–G protein coupling.

Nephrogenic diabetes insipidus (NDI), an inherited disease that is transmitted as a X-linked recessive trait, has been linked to defects in the $V_2$ vasopressin receptor. NDI is characterized by failure of the kidneys to concentrate urine despite the presence of adequate concentrations of the neurohypophyseal hormone vasopressin. In normal kidneys, vasopressin stimulates water reabsorption in the renal collecting ducts. However, individuals with NDI do not respond to vasopressin and thus water is not reabsorbed, resulting in large volumes of dilute urine. Extensive evidence has linked NDI to defects in the $V_2$ vasopressin receptor. With the recent cloning of the $V_2$ vasopressin receptor cDNA, numerous investigators have tested individuals with NDI for mutated $V_2$ vasopressin receptor genes. A definitive link was made between X-linked NDI and mutations in the $V_2$ vasopressin receptor gene and, to date, 60 distinct mutations in the $V_2$ vasopressin receptor gene have been found to occur in NDI patients, including missense, nonsense, deletions,

and insertion mutations. Interestingly, the mutations are not clustered in a particular location but occur along the whole sequence of the $V_2$ vasopressin receptor. Four of the missense mutations have been functionally analyzed and, in each case, the mutant receptors are impaired in their ability to couple to $G_s$ and stimulate adenylate cyclase. The nonsense and deletion mutants in the $V_2$ vasopressin receptor gene appear to cause premature termination of the receptor protein during translation. Hence, data suggest that NDI is due to decreased signal transduction via the $V_2$ vasopressin receptor pathway. Moreover, in contrast to rhodopsin, the defects that decrease signal transduction via the $V_2$ vasopressin receptor pathway include defective receptor–G protein coupling and expression of truncated receptors. As other mutations in the $V_2$ vasopressin receptor are investigated, it will not be surprising that other types of defects may be revealed which contribute to NDI.

In contrast to the $V_2$ vasopressin receptors, a single mutation in the LH receptor, which causes constitutive activation of the LH receptor found in individuals with familial male precocious puberty, has been described. A similar situation has been described for the thyrotropin receptor in which a somatic mutation of the receptor gene produces a constitutively active receptor which leads to functioning thyroid adenomas with hyperthyroidism. These are the first cases of constitutively active receptors being linked to human disease states. It is inevitable that the number of examples of defective GPCRs will increase greatly as more GPCRs are examined for mutations, particularly in families with inherited diseases. This exciting new era of discovery will not only teach us about the causes of acquired or inherited human diseases and possible therapies for them, but gives us new insight into the fundamental processes of receptor activation and signal transduction.

## BIBLIOGRAPHY

Clapman, D. E., and Neer, E. J. (1993). New roles for G-protein $\beta\gamma$-dimers in transmembrane signaling. *Nature* **365**, 403–406.

Lefkowitz, R. J. (1993). G protein-coupled receptor kinases. *Cell* **74**, 409–412.

Raymond, J. R. (1994). Hereditary and acquired defects in signaling through the hormone-receptor-G protein complex. *Am J. Physiol.* **266**, F163–F174.

Strader, C. D., Fong, T. M., Tota, M. R., Underwood, D., and Dixon, R. A. F. (1994). Structure and function of G protein-coupled receptors. *Annu. Rev. Biochem.* **63**, 101–132.

# Affective Disorders, Genetic Markers

DANIELA S. GERHARD

*Washington University School of Medicine*

I. Introduction
II. Description of the Illness
III. Evidence of Genetic Basis of Affective Disorders
IV. Family and Association Studies
V. Future Prospects

## GLOSSARY

**Affective disorder** An illness of mood; depending on the disease subtype, the patient may have symptoms of mania or depression that can cycle with either a well state or another ill episode

**Genetic linkage** An event when the alleles of two distinct genes, located on the same chromosome, are inherited together the majority of the time; the closer the two genes are, the more frequently they will be linked

**Marker** Genetic polymorphism with a simple mode of inheritance and therefore useful in family studies and linkage analysis

**Penetrance** Frequency of expression of a genotype; if less than 100%, the trait is said to exhibit reduced penetrance

**Phenotype** The physical, biochemical, and physiological makeup of an individual that is determined by genetics and environment

AFFECTIVE DISORDERS ARE ILLNESSES OF MOOD that include mania, hypomania, schizoaffective disorder, major depression, and cycles of manic and depressive episodes.

## I. INTRODUCTION

The ultimate aim of genetic research is to understand, at the level of DNA composition (the building block of genes), the cause of inherited human diseases. In inherited diseases, a mutation (a small change in the standard DNA sequence) is present in a gene which alters its function. It is passed on from either (or both) of the parents. In autosomal disorders, only one copy of a mutated gene is sufficient to cause the illness whereas in recessive disorders both parents have to transmit a mutated gene to their child. The first step toward the isolation of a gene harboring a disease-causing mutation is the identification of its location on 1 (or possibly more) of the 23 chromosome pairs that contain the genetic blueprint of every cell. This process is called gene mapping. The rapid pace of progress in the development of the human genome map means that we can expect to know the location of most monogenic disease mutations by the year 2000. The knowledge of a disease gene's location allows the molecular geneticist to apply technology, termed positional cloning, for the identification of its sequence (the genetic blueprint). For example, the genes for Huntington's chorea, myotonic dystrophy, fragile X mental retardation, breast cancer, retinoblastoma, and cystic fibrosis have recently been cloned. Scientists are now busy performing experiments that will explain how the mutations cause these genes to malfunction, thereby resulting in disease symptoms. It is hoped that better patient treatment will be developed once the function of the genes is known. [*See* Genetic Diseases.]

Progress in understanding psychiatric illnesses lags far behind other genetic diseases because the diagnostic nuances and the mode of inheritance are not well resolved. For example, in the 1950s, schizophrenia was diagnosed for almost all psychotic patients, even though in many cases they had affective disorder. In addition, not all patients with affective disorder have inherited the illness—they are ill due to environmental effects. Finally, the inheritance pattern is quite confus-

157

ENCYCLOPEDIA OF HUMAN BIOLOGY, Second Edition, VOLUME I. Copyright © 1997 by Academic Press. All rights of reproduction in any form reserved.

ing. It is not clear if a single mutation is sufficient for the illness to manifest itself or if both parents must pass on a "disease-predisposing" gene. Alternatively, it may be necessary to have mutations in more than one gene (this last case is an example of polygenic inheritance) for disease symptoms to occur. It is very possible that genetic forms of affective disorders are due to all three mechanisms. The difficulty in determining the inheritance patterns is partially due to the incomplete penetrance of the genes (i.e., the gene is present in an individual, but she/he does not have any of the symptoms) as well the presence of nongenetic cases (phenocopies).

This article reviews the current status of major affective disorder (MAD) research using modern molecular genetic and statistic approaches. Affective disorder is a name given to a heterogeneous group of illnesses that include mania, hypomania, schizoaffective disorder, major depression, and cycles of manic and depressive episodes. The difficulties encountered in genetic studies of affective disorders include the accurate identification of individuals ill due to genetic predisposition, the hypothesized (and probably present) genetic heterogeneity (i.e., different genes will cause the disease in specific families), and the uncertain mode of inheritance. During the last few years a number of studies have reported a linkage between a chromosomal marker and MAD, but a strict replication of any one finding has not yet been achieved. The article concludes with a short presentation of new alternatives that can be used to resolve the problems presented by these diseases. [*See* Genetics and Mental Disorders.]

## II. DESCRIPTION OF THE ILLNESS

Affective disorders are illnesses of mood that are usually cyclical with either well and ill or manic and depressive periods alternating. They are broadly divided into bipolar and unipolar types. The diagnoses rely on a series of symptoms reported by the patients themselves. Since the early 1970s, the American Psychiatric Association, together with trained diagnosticians, has been developing criteria for the most accurate determination of psychiatric diagnosis. The "Diagnostic and Statistical Manual of Mental Disorders" (4th ed.) (DSM-IV) provides the most up to date guidelines, yet efforts are already underway to more accurately parse the disease phenotypes. The problem of diagnostic undertainty is exemplified by the instructions in the chapter of DSM-IV devoted to MAD: "other biochemical and physical causes of the symptoms must be ruled out before the diagnosis of primary affective disorder is warranted." Therefore, unlike such diseases as sickle-cell anemia or heart disease, the diagnosis is made on the basis of exclusion rather than by seeing distinct clear-cut physiological changes.

### A. Depression

Unipolar affective disorder is another term for major depressive disorder (MDD). The lifetime prevalence is estimated at 10–20% and seems to be increasing for individuals born after World War II. Table IA lists the symptoms that represent the change from a patient's previous functioning and result in the diagnosis of MDD. These include feelings of depression, a change in eating pattern, worthlessness, and the inability to cope with everyday life decisions. Five of the nine symptoms, present persistently for at least 2 weeks or more, are necessary for diagnosis. Some depressions are secondary effects of either alcohol and drug abuse or some other illness (e.g., Huntington's chorea or Alzheimer's disease). Suicide is the major cause of death of depressed individuals. It is estimated that up to 50% of people who successfully commit suicide had depressive symptoms prior to death. Ironically, a patient commits suicide when the severity of the symptoms is easing. [*See* Depression.]

### B. Mania

Bipolar affective disorder is a name given to either mania alone or a cycle that includes depression with a manic state. Table IB lists the symptoms that are necessary for the diagnosis of mania. They include feelings of euphoria, increased energy or irritability, a decreased need for sleep, rapid and staccato speech, discontinuous thoughts or flight of ideas, distractibility, hypersexuality, and grandiosity of ideas. The mood disturbance is sufficiently severe that social and occupational function markedly changes. Mania is diagnosed if three or four of the seven symptoms last at least 2 weeks or if the patient requires hospitalization to prevent harm to oneself and/or others. Mania, either alone or cycling with MDD, is termed bipolar I disorder. The estimates of lifetime prevalence vary from 0.5 to 2.0%. Patients with bipolar II disease usually have MDD and a less severe form of mania called hypomania, either because the symptoms are present for a shorter period or they are not as severe. Other causes for the manic state (e.g., drug abuse) must be ruled out.

TABLE I

A. Criteria for major depressive episode

  A. Five (or more) of the following symptoms have been present during the same 2-week period and represent a change from previous functioning; at least one of the symptoms is either a (1) depressed mood or (2) loss of interest or pleasure.

  *Note:* Do not include symptoms that are clearly due to a general medical condition or mood-incongruent delusions or hallucinations.

  1. Depressed mood most of the day, nearly every day, as indicated by either a subjective report (e.g., feels sad or empty) or an observation made by others (e.g., appears tearful). *Note:* In children and adolescents this can be an irritable mood.
  2. Markedly diminished interest or pleasure in all, or almost all, activities most of the day, nearly every day (as indicated by either a subjective account or an observation made by others).
  3. Significant weight loss when not dieting or weight gain (e.g., a change of more than 5% of body weight in a month) or a decrease or an increase in appetite nearly every day. *Note:* In children, consider failure to make expected weight gains.
  4. Insomnia or hypersomnia nearly every day.
  5. Psychomotor agitation or retardation nearly every day (observable by others, not merely subjective feelings of restlessness or being slowed down).
  6. Fatigue or loss of energy nearly every day.
  7. Feelings of worthlessness or excessive or inappropriate guilt (which may be delusional) nearly every day (not merely self-reproach or guilt about being sick).
  8. Diminished ability to think or concentrate, or indecisiveness, nearly every day (either by a subjective account or as observed by others).
  9. Recurrent thoughts of death (not just fear of dying), recurrent suicidal ideation without a specific plan, or a suicide attempt or a specific plan for committing suicide.

  B. The symptoms do not meet criteria for a mixed episode.
  C. The symptoms cause clinically significant distress or impairment in social, occupational, or other important areas of functioning.
  D. The symptoms are not due to the direct physiological effects of a substance (e.g., a drug abuse, a medication) or a general medical condition (e.g., hypothyroidism).
  E. The symptoms are not better accounted for by bereavement, i.e., after the loss of a loved one, the symptoms persist for longer than 2 months, or are characterized by marked functional impairment, morbid preoccupation with worthlessness, suicidal ideation, psychotic symptoms, or psychomotor retardation.

B. Criteria for manic episode

  A. A distinct period of abnormally and persistently elevated, expansive, or irritable mood, lasting at least 1 week (or any duration if hospitalization is necessary).
  B. During the period of mood disturbance, three (or more) of the following symptoms have persisted (four if the mood is only irritable) and have been present to a significant degree:
  1. inflated self-esteem or grandiosity
  2. decreased need for sleep (e.g., feels rested after only 3 hr of sleep)
  3. more talkative than usual or pressure to keep talking
  4. flight of ideas or subjective experience that thoughts are racing
  5. distractibility (i.e., attention too easily drawn to unimportant or irrelevant external stimuli)
  6. increase in goal-directed activity (either socially, at work or school, or sexually) or psychomotor agitation
  7. excessive involvement in pleasurable activities that have a high potential for painful consequences (e.g., engaging in unrestrained buying sprees, sexual indiscretions, or foolish business investments)

  C. The symptoms do not meet criteria for a mixed episode.
  D. The mood disturbance is sufficiently severe to cause marked impairment in occupational functioning or in usual social activities or relationships with others, or to necessitate hospitalization to prevent harm to self or others, or there are psychotic features.
  E. The symptoms are not due to the direct physiological effects of a substance (e.g., a drug of abuse, a medication, or other treatment) or a general medical condition (e.g., hyperthyroidism).

  *Note:* Manic-like episodes that are clearly caused by somatic antidepressant treatment (e.g., medication electroconvulsive therapy, light therapy) should not count toward a diagnosis of bipolar I disorder.

## C. Other Subtypes

In affective disorders, psychotic symptoms, auditory hallucinations, and delusions can be present and they are appropriate for the mood of the patient (negative auditory hallucinations would occur during depression). If the psychoses occur together with the affected symptoms, they are a part of MAD and the illness is called schizoaffective. Schizoaffective patients are as severely ill as the bipolar I patients. Less severe forms of mood disorders include minor depression, dysthymia, and cyclothymia.

## D. Treatment

Bipolar and unipolar disorder patients are treated with a combination of drugs that includes lithium, antidepressants such as the tricyclics and selective seretonin reupdate inhibitors, antipsychotics, and monoamine oxidase inhibitors; the latter are used the least because of systemic side effects. Most of the efficacious drugs modulate monoamine molecules, such as dopamine and seretonin, or influence the level of their receptors. However, the biochemical basis of drug treatment in the reduction of a patient's symptoms is unknown. Electrotherapy has been found to be helpful for the alleviation of symptoms of chronic major depression. Psychotherapy is administered in conjunction with most drug treatments. About 80% of patients respond well to a combination of these regimens; however, for the other 20%, the treatment produces unpleasant side effects without ameliorating the symptoms.

## E. Research Diagnostic Criteria

Illnesses that have so much flexibility in interpretation as does MAD require a common standard of assessment if results between different studies can be compared. Clinically, the diversity in diagnoses remains, therefore the psychiatric community has expended a great effort to establish a common ground for scientific investigations. The American Psychiatric Association updates its "Diagnostic and Statistical Manual of Mental Disorders" every 5–10 years based on the latest clinical findings. In conjunction with these efforts, the Schedule for Schizophrenia and Affective Disorder, Lifetime version (SADS-L) and the accompanying Research Diagnostic Criteria (RDC) were developed. The SADS-L is a structured questionnaire that takes up to 2 hr to administer by trained personnel. It has been instrumental in the establishment of uniformity in the diagnostic interviews. The training of interview personnel is critical in ensuring reliability and reproducibility of diagnosis.

## III. EVIDENCE OF GENETIC BASIS OF AFFECTIVE DISORDERS

"Mendelian Inheritance in Man" lists about 4000 disease states, the large majority of which are inherited as a dominant (one gene is defective), recessive (both genes are defective), or X-linked (the mutated gene is found on the sex chromosome) trait. Some diseases, like hypertension and diabetes, are clearly multigenetic, i.e., they are causes by a combination of a number of genes. For affective disorders, as well as other psychiatric diseases, the inheritance pattern is still unknown. A number of factors are responsible for this uncertainty: (1) MAD belongs to the group of "common" diseases, i.e., it has a population frequency of 20% or more; (2) MAD is phenotypically heterogeneous, i.e., any two random patients with a diagnosis of bipolar I probably have a different combination of symptoms; and (3) the environment, such a death in the family, can trigger depression and each person has a specific induction threshold. Nonetheless, changes in sleeping cycle and appetite and the response of the illness to neuropharmacological drugs have long suggested biological cause. In addition, family, twin, and adoption studies all suggest genetic involvement in the disorders.

In twin studies, up to 72% of the monozygotic twin pairs were concordant for the disease, whereas only 14% of the dizygotic twin pairs were concordant. Complete concordance in the identical twins, expected for a simple, fully penetrant trait, was never observed. The Collaborative Study funded by the National Institutes of Mental Health found that if a bipolar I patient has a first degree relative with MAD, she/he has a greater risk of having an affected child than a member of the general population. Finally, in Scandinavian adoption studies, the disease was more prevalent in adoptees born to ill parents than to parents without affective disorders.

To determine the mode of inheritance in the population or a family, the complete ascertainment of ill and well individuals over their lifetime, age of onset, and the disease subtype(s) is required. However, even with the DSM-IV guidelines, as well as the SADS-L and the RDC, the best diagnosis may not always be the correct one. In addition, the identification of the ill individuals can vary due to the study design. For example, different patients will be identified in a study in which the patients are interviewed at only one point in time versus another that also includes follow-up a few years later. The other confounding effect, incomplete penetrance, has already been mentioned.

Given these uncertainties, it is not surprising that research on the mode of transmission of MADs has not reached a consensus. All possible results, including dominant sex-linked, autosomal dominant with incomplete penetrance, additive (or multigenic), and non-Mendelian transmissions, have been found. As already stated, different modes of inheritance will probably be correct in independent families. The same spectrum of phenotypes may be due to different genes or to a combination of genes that exert their effects in a different manner. The two independent studies, both involving single, very large extended pedigrees of more than 300 members, showed that simple autosomal Mendelian inheritance with reduced penetrance best explains the genetic contribution in these families. This result may be unique for large kindreds, but these pedigrees provide a good resource for genetic study.

## IV. FAMILY AND ASSOCIATION STUDIES

### A. Genetic Markers and Methodologies

The genetic blueprint of a cell, its DNA, is organized on 23 pairs of chromosomes. Almost every cell of the body has the same DNA content, 50% of which is contributed by the mother and 50% by the father. One of the aims of the Human Genome Project is the generation of reagents with which the identification of which chromosomal region is inherited from each parent is possible. In 1995 this goal was almost 100% realized. About 99.9% of the DNA between nonre-

lated individuals is identical. In other words, we differ in about 0.1% of the DNA sequence; approximately three million nucleotides. A marker is an assay that can detect the difference at a distinct position on a chromosome. A genetic map is composed of a few thousand of these markers whose relative order, distance, and chromosome location are known. [*See* Gene Mapping.]

For a simple autosomal dominant disorder, the aim is to identify a marker that is transmitted together with the disease phenotype and since the location of each marker is known, therefore the location of the disease is also known. It is important to remember that a marker is not the disease gene itself, rather it is only a tag for a chromosome region. Figure 1 shows an example of a very large family with a hypothetical autosomal dominant disease and a marker pattern in each individual. Note that the allele called "1" cosegregates with the disease.

The other important issue in family studies is the use of correct statistical methods to analyze the genetic data. Some cosegregation can occur at random, since each parental chromosome has a 50% chance in being transmitted. In the example shown in Fig. 1, the probability is close to one in a thousand that this inheritance pattern has occurred by chance. However, this

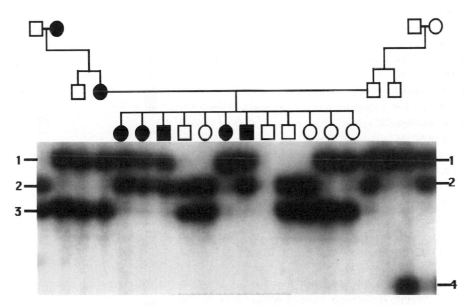

**FIGURE I**   Genotypes of a large three-generation pedigree that segregates a hypothetical autosomal dominant illness. The circles denote females and squares denote males. A filled pattern indicates that the individual is affected. There are four alleles segregating in this pedigree, the largest is labeled 1 and the smallest 4. One male is missing a genotyping. Allele 1 is being transmitted by both of the parents and is linked to the disease only on the mother's side of this family.

family is fairly unique; it is not often that three generations with 12 grandchildren are available for study. Therefore, this method of gene identification requires that the genetic contribution of many individuals and/or families be examined before the correct pattern is discerned.

## B. Results of Analyses

The majority of the research performed to date relies on the method of linkage analysis to identify the location of the MAD genes. Linkage analysis is a statistical method that requires that the inheritance model be specified and then asks if a given marker is transmitted together with the disease under that model. The results are presented in terms of a lod score, a statistical measure of the strength of the results. A lod score of +3 has a 1 in 20 chance of occurring at random. A negative lod score of −2 indicates that a given location cannot harbor a gene predisposing to MAD in that family. A number of possible genomic locations have been published, but a confirmation of any of them has not yet been achieved.

### 1. Chromosome 11

The first study using modern genetic methods and showing positive lod scores was published in 1987. A large Old Order Amish kindred was studied first for the mode of genetic transmission and then for linkage. The Amish population has a number of advantages for psychiatric studies. Drug and alcohol abuse, which confound diagnoses, are prohibited by religious tenets. The life-style resembles that of 19th century Europe and behavioral extremes are easily identified by the community. Paternity has never been a question and the families are large. The prevalence of lifetime severe forms of affective illness is about 0.8%, similar to the rest of the American population. The families have been very cooperative and participate in annual diagnostic follow-ups. The mode of transmission was predicted to be autosomal dominant, with reduced penetrance. A lod score of close to +4 was obtained, suggesting that a gene predisposing individuals for MAD in this kindred was on the 11th human chromosome, close to one of its ends (the short arm, 11p). The result was doubly exciting when it was discovered that the gene for tyrosine hydroxylase (TH), the first enzyme in dopamine biosynthesis, was located exactly in the region of highest significance. However, two subsequent expansions of the pedigree (an increase in the number of individuals provides a better statistic) could not replicate this finding, and

neither did studies using other families. A further search for the predisposition gene throughout the genome in this pedigree has not yet revealed the true location of this gene (unpublished work) (Fig. 2).

However, a gene for MAD on chromosome 11p cannot be completely dismissed. Two studies, using the association method, found a significant linkage between a specific nucleotide change in the tyrosine hydroxylase gene and MAD. In an association study, two groups are examined. One group consists of only ill individuals and the second consists of matched well controls. It is desirable to have the same number of individuals that are of the same sex, age, and ethnic group in each. The hypothesis tested in an association study is whether the most common genotype(s) of the ill group is different from the well controls and if the result is statistically significant. The association between the TH and the MAD is also controversial since this result could not be replicated in certain

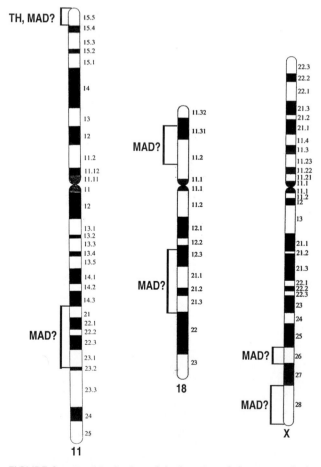

**FIGURE 2** Graphic display of the location of the proposed affective disorder genes on chromosomes 11, 18, and X.

populations. However, considering that the original portion of the Amish pedigree still retains small evidence of a gene on 11p, the resolution must await further research.

It has been suggested that another chromosome 11 gene could predispose individuals to affective disorder. A patient with a bipolar I disorder and a balanced translocation chromosome between the long arms of chromosome 9 and 11 has been reported. The patient was a member of a small family in which there were at least six cases of MAD. Most of the ill individuals in this kindred have this distinct chromosome. Chromosome translocations can result in the mutation of genes that are in the vicinity of the rearrangement. Therefore, it is possible that a gene located on the long arm of the chromosome 9 or 11 is mutated and is the cause of illness in this family. Balanced translocation chromosomes were helpful in the identification of genes that cause different forms of cancers because they immediately identify the important region of the genome to be examined. Investigation of the sequence of this translocation is under way.

## 2. The X Chromosome

A report of an X-linked gene predisposing to manic-depressive illness was originally published in 1930. Since then, several studies, using different pedigrees and at times different diagnostic methods (it must be remembered that it is only in the past couple of decades that standardized diagnostic criteria became widely utilized), claimed a linkage between affective disorder and markers on the short arm (one) or on the long arm (two). Determination of a linkage to a marker located on the X chromosome is unique since a father cannot transmit his X chromosome to his son and therefore pedigrees with a male–male transmission can be excluded at the outset. This means that only a subset of the pedigrees is analyzed. [See X Chromosome.]

The proposed gene on the short arm did not hold up to reanalysis of the original data; however, the situation of linkage to the long arm loci is more complicated. The large study reporting linkage to markers on the distal long arm in five Israeli pedigrees could not be confirmed when better genetic markers were used to characterize the family chromosomes. However, the report of a gene that is some distance from this location (still on the long arm) persists. This finding has been buttressed by a study of a large Finnish family showing statistical evidence for a gene in this region. The researchers involved in this study have returned to this region of Finland so that they can

screen all the extended members of this large kindred to determine if this finding will be replicated. If it is replicated, it will be possible to apply the molecular genetic techniques that were successful in identifying other disease genes to clone the predisposition gene on the X chromosome. It will be important to identify the biochemical change in these patients with the ultimate goal (probably many years in the future) of providing better treatment.

## 3. Chromosome 18

Another curious situation has arisen with linkage and association studies looking for MAD genes on human chromosome 18. A recent study using both standard linkage analysis and affecteds only in a series of 32 small- and medium-sized pedigrees found that in a few families there is evidence for a potential MAD gene on the short arm of chromosome 18. In an effort to check this result, another group, using 28 different families, tested markers on the short and long arm of this chromosome. The sib-pair method of analysis used in this study does not require that the mode of transmission be specified. The results were somewhat surprising. When the entire data set was analyzed, there was weak evidence for linkage to the short arm markers; however, in the subset of families in which the illness was transmitted through the father, the best statistical evidence for linkage came with the long arm markers. The author's own laboratory has been studying the Old Order Amish pedigree as well as the one large American pedigree, and linkage to either chromosome 18 short or long arm markers is not found.

## 4. Other Chromosomes

A report showing the association of major histocompatibility antigen (HLA) haplotypes, coded for by genes on the short arm of chromosome 6, to affective disorders was published in the mid-1980s; however, subsequent studies, using different families, have been unable to corroborate this finding. Three different research groups have tried to resolve these inconsistencies; however, each group reached a different conclusion. Preliminary results from a study using schizoaffective patients suggest that a gene for MAD may be present on the short arm of chromosome 6, but not as close to the HLA genes as initially estimated. Therefore, this region of the genome requires further investigation.

Another report of a possible linkage between a polymorphic marker and a major affective disorder came from Wales, where the investigators found a family in which five out of six patients with a genetic skin disease (Darier's) also had MAD. Since Darier's gene is located on the long arm of chromosome 12, these data suggest that there may be a locus that predisposes individuals in this family to MAD. Linkage analysis using 45 other European pedigrees did not provide sufficient evidence for the presence of a predisposition gene. In neither the Old Order Amish pedigree nor in the American pedigree is the predisposition to MAD due to a gene found in this region of chromosome 12 (unpublished results).

Finally, two reports suggesting linkage to the long arm of chromosome 21 have been published. In the first report, one medium-sized family contributed the majority of the statistically significant results (out of 47), with five other families showing modest positive lod scores with the same markers. Analysis of the same data with a nonparametric method showed the same trend. Surprisingly, statistical evidence for genetic heterogeneity could not be found and performing a post hoc separation of the pedigrees based on linkage results is not an accepted procedure. Further research will be necessary to clarify this interpretation of this result. In the second report, the authors could not exclude this region of the genome, but their statistical support value did not reach significance.

## V. FUTURE PROSPECTS

Psychiatric genetics has recently suffered from unfulfilled expectations. A number of wildly exaggerated claims have been made on the basis of a few positive linkage results, first in affective disorders and then in schizophrenia from both within and outside the field. Since a strict replication of any of the linkage results remains to be obtained, some have stated that this approach has failed. However, the expectations raised were out of proportion to the scientific conclusions. There is no reason to expect that the complexities of diagnosis, genetic transmission, and disease heterogeneity will disappear. These issues continue to confound this research and probably account for the failure to replicate the existing results.

There is a compelling need to pursue this avenue of research as it has been the most productive approach to the identification of disease genes. At the time of this writing, two of the genes that are mutated in Alzheimer's disease have been identified on the basis of their location. Alzheimer's disease had similar difficulties of diagnosis, mode of inheritance, and genetic heterogeneity as other psychiatric disorders, yet the mutated genes are being identified. Significant progress has been made in the three vital areas of diagnosis, analysis, and polymorphism typing. For example, improvement has been made in the diagnostic methodologies so that true cases of illness are identified on the basis of proportional weights given to the type and severity of symptoms. Analytical methods based on the number of symptoms in combination with the genotype are also being developed. Extension of the sib-pair approach allows the determination of linkage between markers and disease without prior knowledge of the form of genetic transmission. Even though this method may not be as robust as classical linkage analysis, it compensates for it with the ability to analyze complex, not fully penetrant traits. The combination of linkage and affected pedigree methods can provide the needed analytical depth to a study.

A new approach has been applied to the search for genes that cause recessive diseases. The fundamental requirement is an availability of a large single family whose ancestry and genealogy are precisely determined. All the ill people will have the same chromosomal region which they inherited from the common ill ancestor. This approach is possible due to the excellent comprehensive genetic maps now available. It was designed to minimize laboratory efforts and to maximize the results obtained. The Old Order Amish and other population isolates such as those being collected in Costa Rica and Finland will provide the ideal resources for this new method. Experiments are under way in our and other laboratories to see how quickly a gene(s) can be identified.

In some genetic diseases the identification of linked genetic markers can lead to a presymptomatic test. Clearly, this option is not available to psychiatric disorders because the prenetrance is not complete. It is unethical to tag someone as developing an illness when there is a finite chance that it will not manifest itself. It must be remembered that treatment for these illnesses exists and should be sought upon diagnosis.

## ACKNOWLEDGMENTS

The author's research is being supported by a grant from the National Institute of Mental Health. I want to thank Carissa M. Smith for a careful reading of the manuscript.

## BIBLIOGRAPHY

American Psychiatric Association (1994). "Diagnostic and Statistical Manual of Mental Disorders," 4th Ed. American Psychiatric Association, Washington, D.C.

Clerger-Darpoux, F., Falk, C. T., and MacCluer, J. W. (eds.) (1989). Genetic analysis of complex traits: Insulin-dependent diabetes mellitus and affective disorders. *Genet. Epidemiol.* **6**(1), 161–310.

Dib, C., Fauré, S., Fizames, C., Marc, S., Vignal, A., Heilig, R., Lathrop, M. Morrissette, U. J., Gyapay, G., and Weissenbach, J. (1995). The final version of the genethon human linkage map." Submitted for publication.

Egeland, J. A., Gerhard, D. S., Pauls, D. L., Sussex, J. N., Kidd, K. K., Allen, C. R., Hostetter, A. M., and Housman, D. E. (1987). Bipolar affective disorders linked to DNA markers on chromosome 11. *Nature* **325**, 783–787.

Gerhard, D. S., LaBuda, M. C., Bland, S. D., Allen, C., Egeland, J. A., and Pauls, D. L. (1994). Initial report of a genome search for the affective disorder predisposition gene in the old order Amish pedigrees: Chromosomes 1 and 11. *Am. J. Med. Genet.* **54**, 398–404.

Goodwin, D. W., and Guze, S. B. (1989). "Psychiatric Diagnosis," 4th Ed. Oxford University Press, New York.

LaBuda, M. C., Maldonado, M., Marshall, D., Otten, K., Gerhard, D. S. (1996). A follow-up report of a genome search for affective disorder predisposition loci in the Old Order Amish. *Am. J. Hum. Gen.*

McKusick, V. A. (1994). "Mendelian Inheritance in Man," 11th Ed. Johns Hopkins University Press, Baltimore, MD.

Murray, J. C., *et al.* (1994) A comprehensive human linkage map with centimorgan density. *Science* **265**, 2049–2054.

Rice, J., Reich, T., Andreasen, N. C., Endicott, J., Van Eerdewegh, M., Fishman, R., Hirschfeld, R. M. A., and Klerman, G. L. (1987). The familial transmission of bipolar illness. *Arch. Gen. Psychia.* **44**, 441–447.

Rice, J. P., Endicott, J., Knesevich, M. A., and Rochberg, N. (1987). Estimation of diagnostic sensitivity using stability data: Application to major depressive disorder. *J. Psychiatric Res.* **21**, 337–345.

Spitzer, R., Endicott, J., and Robins, E. (1975). "Research Diagnostic Criteria Instrument No. 58." New York Psychiatric Institute, New York.

# Affective Responses

SUSAN T. FISKE
*University of Massachusetts–Amherst*

JANET B. RUSCHER
*Tulane University*

## GLOSSARY

**Emotions** Sensitive terms to differentiate among qualitatively different affects, even those with similar pleasantness or unpleasantness; could have physical or physiological manifestations

**Evaluations** Simple positive or negative reactions to specific stimuli

**Mood** Low-intensity and enduring affect having a less specified target than do evaluations

**Preferences** Mild subjective reactions that are essentially pleasant or unpleasant

**Schema** Cognitive structure that represents one's general knowledge about a particular concept or stimulus domain

AFFECTIVE RESPONSES, WHICH INCLUDE THE ENtire range of preferences and emotions, are studied by virtually every domain of psychology. The theories described here primarily address the relationship between cognitive processes (e.g., thought, memory, and reasoning) and affect. Early theories of emotion, physiological theories, and the more recent social cognitive theories are described. The latter theories address how cognitions can underlie emotions and how affect can influence cognition, as well as the premise that affect and cognition are independent.

## I. GENERAL BACKGROUND

### A. Differentiating among Affects

Characterizing the rich variety of affective responses has been a longstanding scientific problem. Affects can be characterized in one of two ways that emerge consistently across analyses of the structure of emotion: dimensions and categories. An illustrative dimensional approach is shown in Fig. 1. The lines in the figure indicate that two common dimensions are pleasantness/unpleasantness and high/low arousal (i.e., engagement). When people are asked to describe their current affective responses or are asked to sort emotion words according to similarity, these two dimensions emerge reliably. Thus, at a given moment people might express contentment and pleasure, but be unlikely to experience loneliness as well because these terms are at the opposite end of the same dimension. This structure is especially applicable to simple, intense emotions.

In contrast, when people are asked about their experience of emotion over time, reports of positive and negative affect are curiously independent. That is, over time, whether one has felt distressed, fearful, or hostile is unrelated to whether one has also at other times felt elated, enthusiastic, and excited. In the short run, people rarely feel simultaneously good and bad, but longer-term summaries of emotional experience show no correlation between positive and negative emotions. This structure especially applies to less intense and more complex emotions. In both cases, however, a two-dimensional structure captures the current scientific understanding of the structure of emotion.

In general, negative emotions, analyzed separately, have a more complex dimensional structure than do

ENCYCLOPEDIA OF HUMAN BIOLOGY, Second Edition, VOLUME 1.   Copyright © 1997 by Academic Press.   All rights of reproduction in any form reserved.

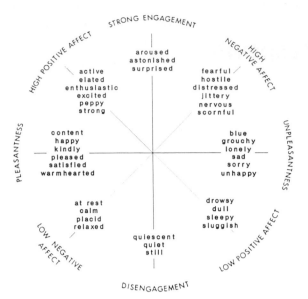

**FIGURE 1** Two basic dimensions of emotion are pleasantness and engagement (arousal). [After Watson, D., and Tellegen, A. (1985). Toward a consensual structure of mood. *Psychol. Bull.* 98, 219–235. Modified figure reprinted by permission of the authors and the American Psychological Association. Copyright, 1985.]

positive emotions analyzed separately; there are greater differences among anger, sadness, and fear than among love, pride, and joy.

The dimensional analyses of emotion are exceedingly useful, but even they do not completely capture the characteristics of particular emotions; as shown in Fig. 1, fear and scorn, although subjectively experienced as distinct, are part of the same cluster. Alternative analyses (e.g., taxonomic category systems) communicate both the concept of emotion and the meaning of particular emotions.

In the category view, deciding what constitutes an emotion is simpler if emotion is not required to be a classical concept with both necessary and sufficient defining features. Instead, there might be prototypical emotions, to which experienced emotions conform in varying degrees. In this view there is an appraisal of particular scripted events that elicit emotions, which in turn consist of expressions, action tendencies, subjective feelings, and physiological states. For example, prototypic joy begins with a desired outcome; the joyous person then seeks to communicate, has a positive outlook, feels energetic, and smiles. However, although people conceive of a prototypic experience of joy, with a good example containing all or most of these elements, poorer examples containing fewer elements might still be classified as joy. From all this

work one cannot be certain whether common individual experiences cause shared prototypes or whether the culture defines certain experiences as certain emotions, which individuals then enact as emotions. That is, the direction of causality between individual emotional prototypes and cultural expectations is not clear, but this does not undermine the prototype approach to defining emotion.

Some theorists maintain that emotion, including its valence, meaning, and physiological manifestations, cannot be fully understood without reference to culture. This social constructivist view of emotions (e.g., that of James Averill) regards emotions as transitory social roles represented cognitively in members of a shared culture. Thus, emotions are identifiable roles that people assume, each with variations on a central (or prototypic) theme. In this view, emotions are defined by social rules prescribed in cultural scripts and are important interpersonal phenomena that are interwoven with other actors as part of the larger cultural story that gives them meaning.

Emotions involve choice regarding one's participation as an actor, require training for skilled performance, and require identification with the role in order to experience intensity. As such, emotions entail a coherent organized syndrome of social, biological, and psychological responses. This entire set of characteristic responses identifies the emotion, within a certain set of related reactions. Other social constructivist views emphasize the individual's experience of emotion in relationship to society, although not all of these views focus on the cognitive structures for storing social rules and roles.

## B. Early Theories

One century ago William James proposed that the experience of autonomic feedback (e.g., heart rate and stomach tension) and muscular feedback (e.g., posture and facial expressions) itself constitutes emotion; Conrad Lange invented a similar theory at the same time. In this James–Lange view the physiological patterns unique to each emotion reveal to people what they are feeling. James stated that when people see a bear in the woods, they experience fear because they tremble and run away: the physical responses cause the emotion. The James–Lange theory of emotion downplayed the role of cognition or mental activity as a sole basis for emotion. In a similar manner, Charles Darwin proposed that the relevant muscular activity can strengthen or inhibit emotion.

Decades later the James–Lange theory was devastated by Walter Cannon's dual argument that visceral sensations are too diffuse to account for all of the different emotions and that the autonomic system responds too slowly to account for the speed of emotional response. Following this critique, many psychologists assumed that physiological contributions to emotion were limited to diffuse arousal and did not include specific patterns of bodily sensation. Assuming this undifferentiated view of arousal, then, a basic problem still remained: If arousal is diffuse and simply ranges from high to low, how can the rich texture of emotional experience be accounted for? One set of answers is physiological; the other is cognitive.

## II. PHYSIOLOGICAL THEORIES

Physiology can explain the richness of emotional experience, most importantly through that multifaceted and highly sensitive organ, the face. Alternatively, arousal can intensify existing positive or negative feelings, in effect providing the second dimension of emotion. A third possibility is that arousal is itself differentiated, contrary to the Cannon critique. These three possibilities are not independent, for the facial and arousal possibilities are related.

### A. Facial Feedback Theory

The original facial feedback hypothesis held that emotional events directly trigger certain innate configurations of muscles and that people become aware of feelings only upon feedback from the face. Although this view is compatible with the James–Lange theory, it does not assume that arousal is fast and differentiated, only that facial responses are. According to facial feedback theorists (e.g., Paul Ekman, Carroll Izard, and Silvan Tomkins), development and upbringing constrain the range of social expressions peoples adopt and so also the range of emotions they can feel. Thus, over time, people assemble a repertoire of emotions on the basis of the facial muscles society allows them to use in expressing their emotions. While variants exist, the core hypothesis is that feedback from facial expressions influences emotional experience and behavior.

Facial expressions reflect both pleasantness and intensity, the two basic dimensions of emotion. Observers agree on the core emotions communicated by certain facial expressions across cultures. Moreover, facial expressions are related to other physiological responses in emotion. The pleasantness of a person's facial expression (as rated by observers) can be directly related to heart rate (i.e., extreme pleasantness with acceleration, extreme unpleasantness with deceleration); observers' ratings of expression intensity is related to skin conductance. Thus, visceral responses in facial expression, heart rate, and skin conductance reflect basic dimensions of emotions and form an integrated configuration of physiological responses in emotion. Moreover, the face reliably reflects the pleasantness and intensity dimensions of emotion, even when overt expressions are not noticeable to observers. That is, electrodes attached to the face can detect minute muscular (i.e., electromyographic) activity too subtle or fleeting to be seen, and this activity parallels the muscles used in overt facial expressions, under similar circumstances.

There is also evidence for the fundamental idea that facial expressions exert a direct effect on mood, emotion, and evaluations. In the typical research paradigm, subjects are induced to adopt specific facial expressions (e.g., a smile or a frown) without labeling them as such. For example, the experimenter might instruct subjects to contract and hold the relevant muscles one by one, for a supposed electromyographic recording, until they have assumed an emotional expression. The assumed facial expression can then influence subsequent perceptions (e.g., an assumed smile can increase the amount of humor perceived in cartoons).

The facial feedback hypothesis is controversial, for not all researchers replicate the finding that posed facial expressions change emotion. One likely explanation of the contradiction is that instructing people to exaggerate their spontaneous facial expressions produces changes in emotion, while rigidly posed expressions only sometimes produce changes in emotion; the conditions under which posed expressions do produce emotion are unclear. Another controversy revolves around whether facial feedback effects are cognitively mediated; that is, people's affect might change because at some level they are aware of the expression they make. Finally, subjects might realize that their expressions are being manipulated and so respond on the basis of the experimenter's apparent expectations of them. Nevertheless, facial feedback theory and related evidence regarding feedback from other nonverbal (e.g., postural) channels have provided one physiological solution to the problem of how emotions are differentiated into their complex variety.

## B. Vascular Theory

Another theory considers the face, but differs in virtually all other respects from facial feedback theory. In the current context it serves as a preliminary example of a fundamentally noncognitive theory of emotional differentiation. Robert Zajonc has recently resurrected a theory posited by Israel Waynbaum at the beginning of this century; briefly, this vascular theory of emotion holds that the facial muscles serve to regulate blood flow to the brain. This view begins with the observation that emotions potentially disrupt cerebral circulation with changes in the heart rate. Facial expressions restore balance by changing air flow to the nasal cavity and by changing particular aspects of blood flow. Jointly, these alter the temperature of cerebral blood flow and thereby brain temperature, producing the subjective experiences of emotion. Thus, flushing with anger or turning pale with fear exemplify changes in blood flow to the face, which regulate the temperature of blood to the brain. This revived theory has sparked considerable debate, and empirical support is just beginning to emerge. Nevertheless, it is at least a useful counterpoint to the cognitive approaches and the facial feedback theory.

## C. Hard Interface: Emotion in the Muscles

This muscular theory of emotion uses the entire body, not just the face, and it posits that the body stores motoric emotional responses (e.g., clenching a fist in anger) *and* motoric cognitive responses (e.g., subvocalizing while thinking). Thus, most relevant here, motoric memories act as a noncognitive direct representation of the emotion. In this view, postural and facial configurations are "hard representations" of internal events, analogous to the hardware (i.e., electronics) of computers. In the emotion realm the heard interface idea accounts for people's imitating each other's facial expressions and postures when empathizing, and it might account for convergence in the physical appearance of spouses over years of muscular mimicry. In less emotional realms it also describes one's muscular memory for specific procedural behaviors (e.g., typing and skating).

## D. Excitation Transfer

Thus, affect can be differentiated into discrete emotions without requiring that arousal be differentiated.

That is, the complexity of the human face and the entire human motoric system can provide specific patterns of emotion, while arousal need only be a diffuse intensifier. Regardless of whether arousal plays a differentiated or diffuse role in emotion, it is important to know how arousal originates and how it influences emotion.

Dolf Zillmann starts his explanation from the assumption that arousal has both automatic and learned origins; for example, a startle response is automatic, whereas fear of airplane travel is learned. To explain both the learned and unlearned features of arousal, Zillmann maintains that emotion depends on three initially independent factors. The first two components create immediate emotional reactions: (1) a dispositional component consists of an immediate motor reaction (e.g., uncontrolled facial reactions) and then (2) an excitatory component energizes the organism through arousal. Finally, (3) the experiential aspect involves people's assessment of their initial reactions and interpretations of the situation, which can modify subsequent reactions. The core of the theory is that arousal is nonspecific and slow to decay and that people are inept at partitioning the sources of their arousal. Hence, residual arousal from a previous setting can combine with arousal in a new situation and intensify emotional reaction.

There is much evidence that arousal from otherwise innocent sources can intensify affect toward seemingly irrelevant people and objects that one encounters. For example, arousal resulting from fear-provoking situations can transfer to sexual arousal, and arousal from exercise can transfer to anger and aggression. Similarly, arousal can intensify evaluation.

Excitation transfer theory argues that arousal influences emotion even when people do not consciously feel aroused or are not fully aware of the stimulus, but physiological measures indicate arousal. For example, subliminally (i.e., unconsciously and rapidly) perceived affect-laden stimuli can produce physiological changes, although subjects report no change in subjective feelings.

## III. SOCIAL AND COGNITIVE FOUNDATIONS OF AFFECT

Both facial muscles and a differentiated arousal system can explain the variety of emotions. Nevertheless, physiology can hardly account for the whole of emotion, and in social psychology the view of arousal as undifferentiated and inadequate to account for the variety of emotions led to a parallel effort to see how cognition could explain emotional complexity; these ideas worked essentially without reference to developing ideas about physiology. Here they serve to describe the critical roles of social and cognitive factors in emotions.

## A. Emotion as Arousal plus Cognition

Stanley Schachter's theory of emotional lability proposes that when people are aroused, they explain their arousal in different ways, depending on previous experience, socialization, and context. This theory posits that diffuse physiological arousal catalyzes cognitive interpretation within the current social context, so emotions are mediated by this cognitive activity. This differs from the previous views that focus more directly on unmediated physiological responses (e.g., facial feedback theory and excitation transfer theory). The superficially most similar theory, Zillmann's excitation transfer theory, differs importantly from Schachter's theory of emotion, as well as its derivatives, which posit that emotion results when people try to explain their consciously perceived arousal; Zillmann's theory does not require consciously perceived arousal. Preliminary evidence from people with spinal cord injuries indicates that the perception of arousal might not be necessary for emotional experience, as Schachter's theory assumes, although it might intensify experience, as Zillmann's theory holds. Another important difference between Zillmann's excitation transfer theory and Schachter's theory is that the latter originally applied only when the initial source of arousal is ambiguous, which is not the case for the Zillmann theory. Despite considerable controversy, Schachter's theory has had considerable impact on psychological thinking about emotion. Recent theories take Schachter's arousal-plus-cognition viewpoint several steps further.

George Mandler's schema congruity theory, for example, explicitly builds on Schachter's theory, in that autonomic arousal combines with evaluative cognition to produce emotion. Visceral activation provides the intensity and particular emotional "feel" of the experience, while evaluative cognitions provide the quality of differentiated emotional experience. However, unlike Schachter's theory, Mandler's theory locates an origin for arousal—namely, in cognitive and behavioral discrepancy and interruption. Most arousal, in this view, follows from a perceptual or cognitive discrepancy or from the interruption or blocking of ongoing action. Greater discrepancy or interruption of complex action sequences generates higher arousal.

In Mandler's view, arousal also initiates cognitive interpretation. For example, with complex action sequences an interruption might be interpreted as hindering a goal or unexpectedly advancing it; this determines whether emotions are positive or negative. For instance, spilling a cup of coffee on oneself might be interpreted as indicative of dispositional clumsiness or as a clever way to excuse oneself from an uncomfortable social situation. Whatever the cognitive interpretation, it shapes not only the quality of one's immediate affect, but also one's lasting mental representation of the event.

## B. Emotion in Close Relationships

Ellen Berscheid presents a compelling case for the application of Mandler's theory to emotion in close interpersonal relationships. Berscheid's analysis proceeds as follows: The more intimate the relationship, the more two people's goals are intermeshed with each other. Greater goal enmeshing creates more opportunity for and severity of goal interruptions, which in turn generates more opportunities for intense emotion. The greater the interdependence, then, the greater the potential for intense negative emotion if one partner leaves, withdraws, or dies. Similarly, there is greater potential for positive emotions if one partner becomes suddenly more attuned, considerate, and helpful. However, if the intermeshed sequence continues as usual, as occurs when a pair is in synchrony, there are no interruptions and little emotion.

Hence, Berscheid notes, there is the pardox that the most intimate, involved, and interdependent relationships might show as little emotion as distant, parallel, uninvolved relationships, simply because the intimate one is running fairly smoothly. Finally, new relationships show more emotions than do older more established relationships because the newer ones provide more possibility of interruption, both facilitating and interfering, while there is less interruption and surprise in well-established relationships.

## C. Cognitive Structures and Affect

Other lines of research posit that cognitive structures cue affect, but unlike the preceding theories, they emphasize the successful application of a knowledge structure rather than its interruption and consequent arousal. This work examines cognitive structures such as social schemata, interpretations of outcomes obtained, the imagined outcomes that might have been, and the broad interplay between emotions and goals.

One approach stems in part from the observation that emotions can result from the successful application of affect-laden schemata. That is, some people (e.g., outgroup members) and situations (e.g., giving a talk to a large audience) inspire emotion without necessarily interrupting any goals. Originally termed "schema-triggered affect" by Susan Fiske, this idea was later incorporated into the broader distinction between category- and attribute-based responses. Schemata can carry immediate affective tags; when a new instance fits the schema, not only does prior cognitive knowledge apply, but so might prior affect.

Other approaches focus on the features of the schema itself and their implications for affective responses. Patricia Linville's work on the affective consequences of informational complexity is an example of this approach. Generally, the greater the complexity of a schema, the more moderate the affect it elicits. An alternative view of the link between knowledge structures and affect is Abraham Tesser's theory that thought polarizes feelings because it leads to perceiving a tighter organization of the attributes of a given stimulus in those people who possess a schema for thinking about it. Over time, people tend to make an instance fit the schema, so evaluation becomes more extreme as the attributes become more organized.

Just as schema theories emphasize prior expectations about events, other theories focus on post hoc explanations for events. Cognitions about one's already achieved outcomes underlie many common emotional experiences. Bernard Weiner's attributional theory of achievement motivation posits basic causal dimensions that people use to understand and react to their successes and failures: locating a cause as internal or external, with a certain amount of stability over time, and having various degrees of controllability. The locus and controllability factors determine the quality of emotions whereas the stability factor tends to exaggerate them. These dimensions in turn provoke basic emotions, as well as expectations for future outcomes. Together, the emotions and expectations guide behavior. For example, pride follows from a positive outcome that is attributed to oneself (i.e., internally) and that is seen as controllable, as when one works hard and succeeds; pride then leads to continued task performance.

According to Weiner, people implicitly understand these rules of causal attribution and emotion and use them to control the emotions of others. When providing excuses for interpersonal failures, people often attribute their behavior to external, uncontrollable, and unintentional factors. This is especially true when the real reasons are internal, controllable, and intentional because people know which excuses provoke anger and which defuse it, which elicit help, and so forth. The general principles of the attribution–emotion–action relationships develop in the naive psychology of children as young as 5 years old.

Just as Weiner's work focuses on cognitions (i.e., attributions) about outcomes already obtained, other psychologists studying emotion have focused on cognitions about outcomes that might have been. When one can easily imagine that outcomes might have been otherwise, the actual state of affairs seems like a fluke, a result of abnormal causes. In contrast, when it is difficult to imagine alternatives, the current situation seems inevitable. According to norm theory (work by Daniel Kahneman and Dale Miller), events with abnormal rather than ordinary causes elicit stronger emotions; hence, the former situations produce more affect. For example, one feels more sympathy for people victimized by freak accidents than for those harmed in the course of their generally hazardous occupations.

Cognitive structural theory (work by Gerald Clore, Andrew Ortony, and Allan Collins) views emotions as valenced reactions to events, agents, and objects that are relevant to one's concerns. There are three main types of emotion in this view. The desirability of an event is evaluated with regard to its facilitation or hindrance of one's goals; this class of emotions pertains to pleasure (e.g., satisfaction or resentment). The praiseworthiness of a person's behavior is evaluated in terms of standards (e.g., norms); this class of emotions relates to the degree of approval (e.g., pride or shame). Finally, the appeal of an object is based on one's attitudes (essentially evaluative reactions).

Emotions elicited in these ways all vary in intensity according to the inducing situation's perceived reality, proximity, and unexpectedness, as well as one's prior physiological arousal. The three specific classes—goal, standard, and attitude emotions—also have respectively specific factors that increase intensity (e.g., degree of likelihood, effort, and realization for goal-based emotions). This description of emotions is designed to give a detailed account of the cognitive antecedents of emotion, potentially suitable for computer simulation.

Another complementary approach examines the emotional effects of alternative future worlds—what might be or what ought to be. Certain emotions can result from the mere possibility of disruption or abnormality in a sequence of goal-directed actions; such disruption can occur when two mutually exclusive alternatives are entertained simultaneously. For example, feelings of conlict result from incompatible alternative goals.

Another set of approaches posits that emotions essentially manage people's priorities. According to Herbert Simon, emotions cause interruptions and thereby act as controls on cognition, alerting people to changes in the status of important goals. In effect, emotions are alarm signals that divert people from pursuing one goal and point them toward pursuing another goal that has meanwhile increased in importance. This view follows from the premise that people are capacity-limited information processors. That is, although people can pursue basically only one goal at a time, survival depends on the organism's being able to interrupt ongoing goals before completion if other environmental contingencies demand it. In this view the physiological arousal that accompanies emotion comes from the interruption itself. In a related view, emotions provide transitions between plans when there is a change in the evaluation of a plan's likely success; given that people have multiple goals, emotions provide a way to coordinate among them. Emotions are a form of internal communication about changes in the relative priority of goals and a way to maintain those priorities.

Finally, another concept of emotion as interruption emerges from Charles Carver and Michael Scheier's cybernetic theory of self-regulation, which describes self-attending people as noticing discrepancies between their current state and the achievement of some goal or standard. When people notice the discrepancy, they attempt to adjust their behavior in order to reduce it. When people succeed, they can proceed with another goal; if they fail, however, the theory states that people persist in the original goal. But in the case of repeated failure, emotion eventually interrupts ongoing behavior, causing a reassessment of the probability of success; people then might redouble effort or withdraw accordingly.

## D. Appraisal Theories

People appraise situations to ascertain how these situations might have an impact on them. An older set of cognitively oriented emotion theories revolves around such a concept of environmental appraisal. According to Magda Arnold, people immediately and automatically appraise everything encountered, as a fundamental act of perception, producing tendencies to act. An important basis for appraisal is memory of similar past experiences, along with associated affect, and an important element of plans for action is expectation about the consequences of actions.

In Richard Lazarus' more recent approach, appraisal consists of evaluating any given stimulus according to its personal significance for one's own well-being. The assignmnet of personal meaning is viewed as a type of cognition, but not implying that cognition is necessarily conscious, verbal, deliberate, or rational. In this view, appraisal is a process of relating one's goals and beliefs to environmental realities. Primary appraisals are the first step in which people assess motivational relevance (i.e., with regard to their own goals) as well as motivational congruence (i.e., regarding whether the stimulus facilitates or thwarts goals). The emotional consequences of primary appraisal are relatively primitive, being simple reactions to potential harm or benefit.

The consequences of secondary appraisal, in which people consider how to cope, are more specific emotions. Two primary types include problem-focused coping, which is an attempt to change the relationship between the person and the environment, and, if that fails, emotion-focused coping, which attempts to adjust reactions through avoidant attentional strategies or changing the meaning of the threat. Two particularly relevant secondary appraisal processes include attributions of past accountability (i.e., credit or blame) and change in one's future expectancies, which alters the psychological situation in terms of its perceived motivational congruence. The most effective secondary appraisal coping strategies depend on the degree of realistic control the person has and on the stage of the threat involved.

The cognitive appraisal approach of Phoebe Ellsworth and Craig Smith focuses more explicitly on people's knowledge of their circumstances and how these cognitive appraisals lead to emotion; hence, it is more explicitly cognitive than that of Lazarus. A core idea of this approach is that people appraise various dimensions of the situation, and these dimensions determine their specific emotional reaction. After people appraise the pleasantness of the situation, more specific emotions then result from evaluating agency (e.g., control by self, other, or circumstances), knowledge uncertainty, and degree of attention required. For example, agency due to circumstances typifies sadness, whereas other agency often characterizes anger. As shown in other areas of affect research, people differentiate negative emotions more distinctly than positive emotions; a situation can be unpleasant in more ways than it can be pleasant. Moreover, some preliminary evidence suggests that particular appraisals influence physiological concomitants of emotion; specifically, anticipated effort influences heart rate and perceived obstacles influence eyebrow frown. In any case, certain appraisal dimensions appear to be central to certain emotions.

Many compatible formulations of emotion, essentially as based on cognitive appraisal, have been offered. Some theories provide taxonomies of appraisals for specific emotions, revealing considerable overlap across theories. Most dimensional theories of appraisal name the following as important in distinguishing the emotions: pleasantness, agency, certainty, and attention.

Other cognitive theories of emotion examine appraisals made in the course of coping with particular stressors. The essentials of Howard Leventhal's perceptual—motor theory of emotions are contained in a hierarchical system consisting of expressive motor processing, perceptual memory of emotional events, and conceptual memory that aids volitional processing. The theory assumes that emotions, which are cognitions that develop from and are fundamentally based in the perceptual system (although they can involve abstract reasoning), convey information about both one's organismic state and the state of the environment. Subjective experience is the proper starting point for the study of emotion, as emotional meaning systems change with individual development. As these are best studied in specific contexts, Leventhal's own empirical work has focused on illness cognitions and coping.

In a related vein, Seymour Epstein's cognitive–experiential self-theory describes the preconscious construals that give rise to specific emotions, with particular emphasis on potential response options rather than on stimulus situations as affecting appraisal. His work touches on fear, anger, sadness, joy, and affection in everyday life. Still other theories focus on more specific kinds of appraisal (e.g., perceived self-efficacy as a cause of emotion) or an accessible discrepancy between one's actual self and one's "ideal," or "ought," self. [See Development of the Self.]

This collection of theories addresses the process of appraisal. When individuals appraise the meaning of motivationally relevant events, emotions ensue. The intensity of both positive and negative emotion depends on the perceived facilitation or interference with one's well-being. Emotions are then organized and regulated to protect the person's well-being.

# IV. AFFECTIVE INFLUENCES ON COGNITION

Moods have various influences on behavior, memory, judgment, decision-making, and persuasion. Before discussing these, it is useful to bear in mind that (1) people are, in general, moderately optimistic, (2) laboratory-induced moods are mild compared to some moods produced by real-life events, and (3) the effects of positive moods are more predictable, consistent, and interpretable than those of negative moods.

## A. Mood and Social Behavior

The work by Alice Isen and others indicates that good moods lead people to help others. Pleasant experiences (e.g., success on a task) generate postive moods, which in turn increase such prosocial altruistic behavior. Several hypotheses have been proposed to account for these effects, but generally people in good moods will help if the situation makes salient their needs for rewards and emphasizes the rewards of helping. Changing people's focus of attention to their own good fortune, emphasizing the benefits of helping rather inducing guilt, and improving social outlook all enhance helping. In addition, people are concerned with mood maintenance and are less likely to help if doing so would ruin a good mood.

In a related vein, cheerful people are more sociable: They initiate interactions, express liking, aggress less, and cooperate more; this does not seem to be due to their being generally more compliant when cheerful. When in a good mood, people are also nicer to themselves, rewarding themselves and seeking positive feedback; this is not merely due to loss of self-control.

The effects of negative moods are less clear. For example, people who are in a bad mood can be more helpful than people in a neutral mood, but only under particular conditions. In effect, the negative mood conditions that do encourage helping are those in which guilt, rather than anger or self-pity, is operating. According to the responsibility/objective self-awareness view, unhappy people who perceive themselves to be the cause of a negative event are helpful, assuming prosocial norms are salient. In contrast, according to the focus-of-attention explanation, people in a bad mood who perceive themselves as the target of a negative event are less helpful to others. The attentional explanation thus applies to helping that is instigated by both positive and negative moods; when helping focuses attention away from one's mood, the mood is weakened, but when helping focues attention on the conditions producing one's mood, the mood is enhanced.

A final explanation for the effects of bad moods on helping is more controversial. According to a negative state–relief hypothesis, unhappy people help when it could dispel their negative mood; even children are aware of the personally salutary effects of helping. It seems clear that people attempt to regulate their moods, so some version of this hypothesis is likely to hold true. However, there is disagreement about the interpretation of such results.

## B. Mood and Memory

Present mood has important effects on memory of past experiences. Two essential phenomena have formed the core of research on mood and memory: mood-congruent memory and mood state-dependent memory. [*See* Learning and Memory.]

Under many circumstances, people more easily remember material whose valence fits their current mood state; this is termed mood-congruent memory. Across a variety of settings and procedures, people recall positive material in positive moods and sometimes recall negative material in negative moods. Some researchers argue that the effect is located primarily at information retrieval, whereas others argue that the evidence is stronger for effects at encoding information.

Most mood congruence research finds uneven effects for negative moods, and efforts at mood repair might account for these failures to obtain strong effects for induced negative moods. Alternatively, the weakness of negative mood effects might result from the store of negative material in memory being less extensive and less integrated, so negative moods might not as effectively cue congruent material. Possibly, the weakness of negative mood effects is also due to there being more differentiation among negative moods than among positive moods so that the negative mood states do not match as easily and there are fewer associations to any one of them than to an overall positive mood.

One important class of negative mood congruence does occur reliably—namely, in people who are clinically depressed, for whom negative events are presumably mood congruent. For example, when depressed and nondepressed people experience a series of experimenter-controlled successes and failures, depressed people underestimate successes. The effect thus appears more likely to be a retrieval bias than an encoding deficit. As another example, depressed people "underremember" positive words and phrases and overrecall negative ones. These negative mood congruence effects might occur only when people focus on the applicability of the material to themselves, but otherwise the conditions under which they have been obtained are numerous and varied. Depressed persons might deliberately focus on negative material to rebut it, to improve themselves, or to confirm their self-image so the effects might or might not be automatic.

Altogether, a considerable amount of evidence supports the facilitating effects of congruence between mood and the material to be remembered, although there are a few isolated examples of advantages to mood-*in*congruent memory. Still, the real-world effects are especially robust: Clinically depressed patients reliably show the mood congruency effects, and real-life events show stronger mood congruence effects than do experimenter-provided items.

Mood state-dependent memory research adresses the separate mood-and-memory phenomenon that concerns the congruence between the mood context in which material is learned and the mood context in which material is retrieved. Mood-dependent memory ignores the valence of the material itself, focusing only on the fit between the two contexts. It is fairly well established that state-dependent memory exists for drug-induced states; for example, something learned while intoxicated is easier to remember while intoxicated than while sober. The reliability of drug-induced state-dependent memory led researchers to look for a similar phenomenon in mood states. Evidence for such an effect is actually quite weak. Studies that associate half the material with one mood and the other half with another mood (i.e., interference designs) show fairly strong effects compared to studies that associate all the material with one mood state only, but the possibility of demand (i.e., subjects ascertaining how the experimenter expects them to behave) is then raised. One possibility if that the clear state dependency effects of drugs operate more similarly to arousal-dependent memory rather than mood-dependent memory, an effect that does appear reliable.

Gordon Bower originally proposed a network model of mood and memory to account for the various effects of mood on memory. The theory posits that emotion is simply a retrieval cue like any other. This means that memories or events that come to mind at the same time as a given emotion are linked to that emotion and, hence, indirectly, to other emotion-congruent memories or events. Mood-congruent memory thus has an advantage because the emotion provides an additional route to the item in memory.

In this view both mood-congruent memory and mood-dependent memory would be based on the retrieval advantages, respectively, of similar affect attached to the mood and the inherent valence of the item to be recalled, or of similar affect associated with the item at learning and at retrieval. However, subsequent research was disappointing in its attempts to support the facilitating effects of mood on perception of similarly toned material and on mood-dependent retrieval, as noted earlier in this section. Moreover, the combined effects of conceptual and emotional relatedness in memory networks are not well supported. Overall, the network model of mood and memory has fared poorly, suggesting that new frameworks are needed.

## C. Effects of Mood on Judgment and Decision-Making Style

One of the clearest effects in the mood literature is that cheerful people evaluated many things more positively than when they are in a neutral mood. However, there are some limits on this phenomenon; cheerful people do not overvalue the criminal and the unattractive, for example. The converse—whether unhappy people dislike everything—is less clear for many of the reasons noted in the previous section regarding studies of negative mood and memory. Some effects have emerged from well-controlled studies; for example, people judge other people according to negative applicable traits more when temporarily depressed than when in a neutral mood. Similarly, both temporary depression and a chronic negative outlook lead people to perceive themselves as having less social support.

Other intriguing puzzles remain in this line of work. For example, not only do positive (and sometimes negative) moods show congruence in judgments, but aroused moods show congruence effects as well. According to work by Margaret Clark, when people are physiologically aroused (e.g., by exercise), they make arousal-congruent judgments, viewing another person's ambiguously positive facial expression as more joyous than serene and interpreting ambiguous statements in the same way.

Mood not only influences memory and evaluation, but it also influences the manner in which judgments are made. According to Isen, elated people are expansive, inclusive, and somewhat impulsive; they make decisions quickly, they group more varied things into the same category because they see more unusual connections among things, and they are willing to take more risks if the possible losses are small, although losses loom larger to them. Again, the effects of negative moods are less conclusive.

## D. Mood and Persuasion

Along with being expansive and inclusive and generally pleasant to others, cheerful people are more compliant with persuasive communications, whereas angry uncomfortable people are generally less compliant. The positive mood results might explain the effectiveness of marketing efforts that include free samples, soothing music, and friendly banter. Much of the mood and persuasion research was originally conducted under a classical Pavlovian conditioning paradigm, but subsequent studies have suggested some roles for cognitive processes. For example, not all persuasion is automatically enhanced by positive mood—perhaps only persuasion under conditions of low involvement and low cognitive activity. Positive moods themselves can be distracting and can reduce cognitive capacity, leading subjects to a cognitively superficial processing of messages. Under conditions of moderate involvement, however, affect might enhance thought because of affect's arousing and attention-getting impact. Under conditions of high involvement, mood might serve an informative function relevant to one's possible reactions or it might bias retrieval of relevant supporting information.

## E. Separate Systems View

Despite the scientific and common sense idea that people think about things in order to ascertain affect, there is a case for affect preceding cognition rather than vice versa; people can make major life decisions on the basis of emotional preferences guided by no apparently relevant cognitive data. Robert Zajonc's separate systems view suggests that affective and cognitive processes proceed in parallel without influencing each other much. Affective processes are argued to occur at a more basic level than cognitive processes, in several respects.

One line of research that addresses this view is the mere exposure effect, in which people grow to like an initially unobjectionable but novel stimulus (e.g., Chinese character) the more frequently it is encountered, even if they can only recognize it at levels approximating chance guessing. Thus, it would appear that affective processes more than cognitive ones underlie the mere exposure effect. Another line of research shows that evaluative impressions (one kind of affect) can be independent of memory for the details on which they were based (one kind of relevant cognition). In general, this occurs when impressions are formed on-line, at the time of the initial encounter. Thus, as people form an impression of someone, their affective responses are likely to occur independently of their later ability to remember details about the person. If affective judgments are not necessarily based on recallable cognitions, then affective reactions are better characterized as immediate direct responses.

Several theorists have responded to Zajonc's provocative view of emotion as a system separate from cognition, but space precludes covering the debate here. The resolution to the cognition—emotion debate depends not only on how cognition is defined, but also on how emotion is defined. For example, Zajonc's original statement focused on preferences (e.g., evaluations, affective judgments, and liking) rather than on moods or true emotions. The latter, many would argue, intrinsically depend on cognitively driven appraisal processes. Others suggest that the entire distinction is largely definitional and therefore not constructively pursued per se. Inherent problems exist in comparing cognition and affect; affect and cognition might not be comparable. Judgments representing affect have included evaluation, preference, and differentiated emotions, whereas reactions representing cognition have included attention, inference, and memory. Which are the relevant cognitions and which are the comparable affective responses remain unclear.

Trying to establish the independence of affect and cognition is essentially the same as trying to establish the null hypothesis. To the extent that one argues that they are independent, one is trying to establish the absence of a relationship. The more sensible task is to show on what each is based, if not entirely on each other. To the extent that the separation is not complete, another task is to show the ways in which they do relate, as the work reviewed here endeavors to do.

## BIBLIOGRAPHY

Fiske, S. T., and Taylor, S. E. (1991). Affect and cognition. *In* "Social Cognition," 2nd Ed. McGraw-Hill, New York.

Frijda, N. H. (1986). "The Emotions." Cambridge University Press, Cambridge.

Hamilton, V., Bower, G. H., and Frijda, N. H. (eds.) (1988). "Cognitive Perspectives on Emotion and Motivation." Kluwer Academic, Norwood, MA.

Isen, A. M. (1987). Positive affect, cognitive processes, and social behavior. *Adv. Exp. Social Psychol.* **20,** 203–254.

Lazarus, R. S. (1991). "Emotion and Adaptation." Oxford Press, New York.

Mandler, G. (1984). "Mind and Body: Psychology of Emotion and Stress." Norton, New York.

Oatley, K. (1992). "Best Laid Schemes." Cambridge University Press, New York.

Ortony, A., Clore, G. L., and Collins, A. (1988). "The Cognitive Structure of Emotions." Cambridge University Press, Cambridge, England.

Shaver, P. (ed.) (1984). "Review of Personality and Social Psychology: Emotions, Relationships, and Health." Sage, Beverly Hills, CA.

Strongman, K. T. (1978). "The Psychology of Emotion," 2nd Ed. Wiley, New York.

# Aggression

IRWIN S. BERNSTEIN
*University of Georgia*

---

## GLOSSARY

**Agonistic behavior** Aggression and responses to aggressive behavior

**Intervening variable** Hypothetical process or state that accounts for variation in the effect of an independent variable on a dependent variable

**Proximal cause** Those events immediately preceding the response of interest that are thought to trigger the response

**Punishment** Process by which the behavior of an actor reduces the future probabiliy of the preceding behavior of the recipient

**Socialization** Effect of social interactions on future social behavior

## I. THE CONCEPT

### A. Usage

Aggression, like many familiar words we are all certain we understand, is extremely difficult to define precisely. The criteria used to identify aggressive acts are generally subjective, based on assumptions about the internal motivational state of another. We also use aggression metaphorically to imply initiator (socially aggressive), persistent (aggressive salespeople), or active participant in an interaction (aggressive conversa-

tionalists who speak more than their share of the time). The term "aggression" is also value laden with connotations of evil motivations and undesirable behavior.

A certain amount of ambiguity is tolerable in our everyday use of language. We can appeal to common experiences for clarification and simply define terms by "common usage." The scientific use of a term, however, requires much greater precision, and all measurement must be preceded by rigorous definition of that which is to be measured and compared. Niko Tinbergen, in "A Study of Instinct" published in 1951, stressed the necessity of limiting definitions to descriptions of observable structure. For behavior, this means description of the motor acts and sequences. Questions about the proximal cause (what triggered the response), function (what are the consequences of the actions), evolution (what are the genetic contributions to the behavior), and ontogeny (what life experiences make this behavior more likely in an individual) should all follow from descriptions of behavior rather than being incorporated into the definition of the behavior.

### B. Sequential Inferences

The problem with aggression, however, is that there is no unequivocable act that is always seen in aggressive behavior and only seen in aggressive behavior. Aggression is a concept invoked as an intervening variable to explain why an individual emits a particular response to a specific stimulus; other individuals, or the same individual at other times, may emit a different response to the same stimulus. The judgment of aggression is thus context dependent and we must examine entire sequences of responses to decide if aggression was involved in a particular response.

Sequences that result in injury to one or more parti-

ENCYCLOPEDIA OF HUMAN BIOLOGY, Second Edition, VOLUME I.   Copyright © 1997 by Academic Press.   All rights of reproduction in any form reserved.

cipants are especially likely to be examined for aggressive components. The existence of an injury itself, however, is neither necessary nor sufficient to define the acts leading to the injury as aggressive. When someone steps on another's foot we may think that this was an aggressive stamp, an unintentional act due to negligence, or an unavoidable accident inasmuch as the perpetrator was pushed or falling at the time. In fact, even if the act were acknowledged to be intentional, we would not classify it as aggressive if the actor smiled and we felt this was "playful" behavior. On the other hand, we often classify as "aggression" acts that result in no physical harm. If we believe the actor intended harm but the action was inept or otherwise unsuccessful, the act is nonetheless perceived as aggressive. Threats, vocal or otherwise, are considered aggressive enough though the threat itself produces no injury. If we believe that a threat indicates that future actions in the sequence may cause injury, then the threat is aggressive because it occurs in sequences where injuries are a more likely consequence than in sequences lacking such behavior. Likewise, injuries following from sequences that we do not believe ordinarily result in injury are not considered to be embedded in aggressive contexts.

In order to include injury as a component of all aggressive sequences, including those not involving physical damage, some would define "injury" to including being deprived of something, suffering a limitation of freedom, or receiving a psychological "injury," such as an insult. All of these may occur where physical force is an implied threat, but some limitations of freedom and deprivations result from situations that are not intuitively aggressive. If someone purchases the last theater tickets before we arrive at the box office in complete ignorance of our intentions, our goals are thwarted, and we have fewer degrees of freedom, but we would hardly regard the buyer as "aggressive."

## C. Intentions

Once again, we have relied on the presumed intentions of the actor to decide if an action was aggressive. It is the nature of intentions that makes the study of human behavior both exciting and frustrating. We often guess at motivation by examining functional consequences and assuming that the consequence was intentional, i.e., it was the goal of the behavior. The danger in such assumptions is exemplified by a simple example. Driving this morning may surely have resulted in my being stranded on the road with an empty gas tank, but I assure you that that was not why I started out driving my car this morning. I really had intended to come directly to work.

It is, nonetheless, our presumptions of intention that would make us argue that when one individual strikes another, and we approve, that the act was in defense or punishment, if we think hitting the other was provoked or "deserved." The action and consequence are the same as in aggressive behavior, but we interpret the eliciting stimuli in a special way. We say that spanking a child for doing something that could be self-injurious, like running into the roadway, is for the child's own good and, that since the net effect is beneficial to the child, the spanking was not an example of aggression. On the other hand, it is often anger that is the prevalent immediate emotion when you spank your child for having run into the road after you told him or her to stay on the sidewalk. You feel anger because you were disobeyed, relief because the child was not injured, fear because the child might do it again and be injured, and hope that you will benefit the child. All of this is present when the act consists simply of a quick swat across the child's bottom, but the child certainly experiences it as an injury.

## D. Predation

Predation is also generally recognized as distinct from aggression. A wolf killing and eating a rabbit certainly injures the rabbit. A mosquito, or other parasite, may damage its prey, but the behavior producing the injury is thought to be embedded in a feeding context where ingestion is the goal and not the production of the injury. The behavior producing the injury is not an instrumental act solely to modify the behavior of another. It certainly does modify the behavior of the rabbit in a more or less permanent fashion, but I am certain that the wolf would gladly eat the rabbit without chasing and killing it if the rabbit would only hold still. Somehow, killing the rabbit to make it hold still is not the same as punishing a child to make it hold still and stop putting its fingers into electrical outlets. The injuries in predation are regarded as incidental to eating, much as the injury a falling person inflicts on you is incidental to the falling and not an act designed to injure you.

## E. Agonistic Behavior

Defensive behavior may indeed result in injuries to another that we would regard as integral to the behavior. Aggression and defense are often indistinguishable except with regard to sequences and inferences regarding proximal causes and eliciting stimuli. An individual may inflict grievous injury on another in self-defense, and wars have been initiated and waged "in defense" of one thing or another. Aggression and the responses to aggression are referred to as "agonistic" behavior (as opposed to antagonistic behavior). Agonistic behavior includes aggression, but also includes distinctive elements that we may call submissive, i.e., they signal that the emitter is unlikely to initiate an injurious act spontaneously. They do *not* mean that the signaler will not inflict injury on the other if further provoked. Submission does not necessarily imply total passivity—only that responses which might inflict an injury on the other will not be forthcoming unless there is further provocation. Submissive individuals are quite capable of retaliatory behavior.

## II. CAUSES

Focusing on intentions and injurious consequences can lead to serious errors in studies of aggression. The human ability to project into the future and produce behavior "in order to" achieve an intentional goal is not the same as the functional consequence, or the future, being the cause of behavior in the present. Intentions are in the present and, after all, despite our intentions at the time, the future does not always turn out as we predicted. The intentions of others are, however, not readily observed and so we tend to infer them from the consequences of behavior, thereby inviting a confusion of function (the consequences that follow) with the cause of behavior (the situation that provoked the activity). It is all too easy to believe that one animal attacks another in order to drive it away (the end point of the interaction) rather than observing that one animal attacks the other because it responded to the location and behavior of the first just before the attack.

In searching for the cause of aggression, or any form of behavior, we can begin by considering internal individual components contributing to the behavior or to the external environmental conditions eliciting the behavior. Internal individual components contributing to aggressive behavior might include genetic predispositions to behave in certain ways. If aggression were a type of reflex motor act, we could look for genes that were responsible for this reflex. Because aggression is a complex intervening variable, we cannot expect to find genes for it, but we may ask what

genetic influences exist on the expression of aggression. Naturally, physical attributes influencing motor abilities influence the form of behavioral expression. We might also hypothesize inherited neural structures that produce personality predispositions that influence aggressive behavior.

Searches for physiological substrates of aggressive behavior have revealed brain loci in the amygdala and other areas in the limbic system that do appear to be involved in aggressive expression. Classical studies of rage and attack induced by brain stimulation have indicated some localization of function, but repeated stimulation of the same centers does not always produce the same responses. In fact, the social and physical environment of the subject at the time of stimulation can radically alter the form of behavioral expression. We do not yet have a sufficient understanding of neuroanatomical mechanisms to allow more than a theoretical appreciation of the integrative function of the nervous system in filtering external stimuli, interpreting these signals, and selecting from available responses.

In a similar vein, some authors have suggested that there are strong hormonal influences on aggressive expression, but the idea that high circulating levels of testosterone are a cause of aggressive behavior is controversial, at best. Males may have higher levels of circulating levels of testosterone than females, and male aggression may have more serious consequences in sexually dimorphic species or may take certain forms of expression more commonly, but there is no indication that testosterone is the cause of differences in aggressive behavior in general, nor the cause of many of the differences noted in aggressive expression. One can readily challenge the assumption that individuals with higher testosterone levels (males) are necessarily more frequently involved in all forms of aggressive behavior. In several published studies of nonhuman primates, females have been noted to be more frequently involved in aggressive interactions involving other group members than are males. Whereas male aggression often produces more damaging consequences (as a function of size, strength, and sexually dimorphic physical features such as larger canine teeth) and male aggression is more noticeable when directed towards extra-group individuals (such as human observers), female aggressive participation may far exceed that of males within normal social groups. [*See* Hormonal Influences on Behavior.]

Other hypothesized internal causes of aggression are neither physiological nor hormonal, but psychological. Some have argued that perceived pain is an invariant cause of aggression. Whereas it is undoubtedly true that inflicting an injury on another may provoke an attack on oneself and whereas it is undoubtedly true that an individual experiencing pain may attack anything that it perceives to be the source of (responsible for) the pain, individuals suffering prolonged and profound pain are not in a continuously aggressive state. Aggression can be used as an instrumental act to terminate pain or avoid pain, but that does not mean that pain is either a necessary or sufficient cause for aggression.

Pain may be seen as one element in a larger psychological set, and a popular theory several years ago was that any situation producing frustration produced aggression. Frustration was the thwarting of a goal. Identifying goals thus became necessary and the psychological state of frustration was sometimes inferred by the expression of aggressive behavior. This tautology was not intended by the original proponents of the model, but the difficulties in measuring psychological states like frustration have never been overcome. Moreover, we can all point to experiences in ourselves, or witnessed in others, where failures to achieve goals and losing in competitions produced no obvious aggressive response. Redefining depression as aggression may save a theory, but muddles our terminology irretrievably.

## III. ONTOGENY

Some forms of psychological variables are somewhat more amenable to study and measurement. Learning is an intervening variable accounting for a change in behavior as a function of past experience so that responses to the same stimuli are changed as a function of that experience. Typical forms of behavioral expression in recognized situations (personalities) may be influenced by inherited neural networks but they are also influenced by ontogenetic experiences. [See Learning and Memory.]

Patterns of past experience may certainly be expected to influence the frequency and form of expression of aggressive behavior. In their review of the development of aggression in children, published in the "Handbook of Child Psychology" in 1983, Parke and Slaby suggested that inconsistent punishment experienced in childhood can increase the frequency of later aggressive expressions. It was the inconsistency, rather than the frequency of experiencing punishment, that correlated with increased aggressive behavior. Aggression can be an instrumental act that individuals learn to use to achieve specific ends. It can also be

used as a punishment to decrease the frequency of selected responses, including aggressive responses. Individuals can, therefore, learn that aggressive behavior is, or is not, an effective instrumental act in competitive situations. As a consequence, aggression can be triggered by competitive interactions where it will persist until the rival ceases to compete or until the consequences to the aggressor exceed what the aggressor is willing to bear in that situation.

## IV. FUNCTIONS

### A. Social Behavior

Aggression as an instrumental act to modify the behavior of another brings us to a consideration of some of the social uses of aggression and social consequences of aggressive interactions. K. R. L. Hall wrote that the single most common cause of aggression in primate societies was a perceived infraction of the social code. Aggression was seen most commonly as a response to a violation of expected social behavior. Aggression, as behavior directed toward (or against) another, is a social interaction and an example of social behavior. Since it does not obviously promote further social interaction, some people call it "antisocial." There is, however, no a priori reason to believe that aggression drives individuals apart or weakens social bonds.

Harry Harlow long ago indicated that maternal punishment of an infant does not necessarily drive the infant away from its mother. Mother is the only source of comfort and security for a young infant, and infants are attached to their mothers because of the comfort and solace mothers provide. Mother remains the infant's only source of comfort even when mother is the source of pain that the infant experiences. Mother, in holding and comforting the crying infant after punishing it, seems to counteract any negative affect generated by the punishment. Even as adults, monkeys in social groups display well-developed mechanisms serving to reconcile group members after a moment of conflict.

### B. Socialization

In his analysis of aggression in primate societies, Hall stressed that one important function of aggression is to "correct" or punish unacceptable behavior in another and to shape behavior to expected standards. Infringing on the space of another, eating out of turn, failing to gain approval before taking liberties with another, and similar acts are likely to provoke aggressive responses. Significantly, the violation has to be perceived, but need not be necessarily real. If undetected, socially disapproved behavior provokes no aggression; if the recipients believe an act has occurred that is in violation of accepted norms, they may respond with aggression, even if no such act has actually taken place.

The social modification of the social behavior of another is the socialization process. Aggression can modify the behavior of others, and such modification may actually benefit the recipient in the long run. Although we may prefer to positively reinforce behavior based on ethical principles, avoidance training is indeed a powerful means of modifying behavior. Being able to avoid punishment by responding to the threat of punishment may be a far more powerful learning experience than either punishment itself or positive reinforcement. Ehardt and Bernstein have suggested that in monkey societies, where verbal instruction and verbal behavior are not available options, mother monkeys modify the behavior of their infants by punishing unacceptable behavior. As a consequence, socialized monkeys do not casually bite another monkey, try to reach into another monkey's mouth, or use another monkey as a climbing apparatus. Infant monkeys do all these things and mother monkeys tolerate this for only a few months before responding with controlled bites and rough handling. Bites and rough handling seem to sensitize the infant to threat signals warning that physical punishment is imminent. As the developing monkey responds more readily to threat, the incidence of physical punishment decreases. Physical aggression seems to function to get the young monkey to respond to animate social partners, to realize that there are consequences to their actions on others, and to learn to attend to the signals of others that indicate likely forthcoming behavior. The net effect may be a powerful socialization process, but in the course of this learning, baby monkeys are punished more by their mothers and close kin than by any other monkeys. This experience may save them considerable grief as adults, and it may be precisely the lack of such experience that produces the bizarre and inappropriate behavior of monkeys reared in social isolation. These isolates seem oblivious to the social consequences of their behavior and ignore the warning signals of other monkeys.

## C. Societies

In addition to its socializing function, aggression can serve to protect individuals and groups from external threats of disturbance. Joint defense against predators, other rival groups, or other sources of disruption can elicit strong united action by group members. Strong cohesion in aggression against a common enemy has long been recognized in human societies. All other differences are laid aside as a group protects its members, and even its very integrity, against disruption. Human societies mobilize the strongest aggressive potential against such threats—armies and police forces are the aggressive arms of a society that are used to repel or suppress external and internal threats to established order. Such aggression is seldom seen as "aggression" in the negative sense.

Charles Southwick has suggested that xenophobia, a fear of strangers, is the most potent elicitor of primate aggression. The more powerful the threat the stranger is seen to pose to a group, the more vigorous is the aggressive response. Societies are generally well prepared to marshal the most extreme aggressive "defense" against a perceived threat to the established social organization.

## V. EVOLUTIONARY PERSPECTIVES

Aggression thus may serve functions benefiting the individual, the group, and any subgroup, such as a family, within a group. One may argue that evolution would favor the selection of aggressive behavior that resulted in increased genetic fitness of the individuals displaying the aggression. This may well be so, and individual and inclusive fitness may account for any genetic propensity to engage in specific types of aggressive response to specific stimuli. Benefits to the group can also be understood as benefits to the individual when groups contain mating partners, offspring, and the allies necessary for success in competitions with other groups or in solving environmental problems. Mutualism is beneficial only when mutualists survive, and the benefits of mutualism may make some limited instances of altruism toward partners in mutualism beneficial in the long run.

## VI. CONCLUSION

The most desirable goal may not be to eliminate or even reduce human aggression, but rather to understand and control it. Aggression may be a powerful positive force molding society. A total lack of aggression may make social living impossible. Uncontrolled aggression, on the other hand, may cause more destruction than can be compensated for by any other positive force.

# BIBLIOGRAPHY

Bernstein, I. S. (1981). Dominance: The baby and the bathwater. *Behav. Brain Sci.* **4,** 419.

Bernstein, I. S., and Ehardt, C. L. (1985). Age–sex differences in the expression of agonistic behavior in rhesus monkey (*Macaca mulatta*) groups. *J Comp. Psychol.* **99,** 115.

Bernstein, I. S., and Ehardt, C. L. (1986). Modification of aggression through socialization and the special case of adult and adolescent male rhesus monkeys (*Macaca mulatta*). *Am. J. Primatol.* **10,** 213.

Mason, W. A., and Mendoza, S. P. (eds.) (1993). "Primate Social Conflict," pp. VI, 419. State University of New York Press, Albany.

McKenna, J. J. (1983). Primate aggression and evolution: An overview of sociobiological and anthropological perspectives. *Bull. Am. Acad. Psychiatry Law* **11,** 105.

Parke, R. D., and Slaby, R. G. (1983). The development of aggression. *In* "Handbook of Child Psychology" (P. H. Mussen, ed.), Vol. 4. Wiley, New York.

Waal, F. B. M. de, and Yoshihara, D. (1983). Reconciliation and redirected affection in rhesus monkeys. *Behav.* **85,** 224.

# Aging and Language

ELLEN BOUCHARD RYAN
*McMaster University*

SHEREE KWONG SEE
*University of Alberta*

---

I. Language Differences
II. Communication Disorders
III. Understanding Variability in Aged Language

## GLOSSARY

**Aphasia** Language-specific disturbance resulting from (focal) brain damage

**Apraxis** Speech disorder resulting from an impaired capacity to program the positioning of muscles used in speech execution following neurological damage

**Cross-sectional studies** Studies that investigate maturational change by comparing the relative performance of two or more age groups on some specified measure

**Dementia** Age-associated pathological condition involving progressive deterioration of memory, intellect, personality, and communicative functioning due to central nervous system degeneration

**Dysarthria** Speech disorder resulting from weakness or incoordination of the muscles responsible for speech execution following neurological damage

**Information processing** Reception, storage, retrieval, and utilization of sensory information relevant to the communication process

**Language competence** Current level of the ability to interpret and express messages in the verbal and nonverbal symbolic code of language, where the level is determined by genetic potential, cumulative life experience, and possibly age-associated pathological conditions

**Language performance** Performance on a measure of language, where performance is determined by language competence as well as environmental variations, including the specific testing situation

**Longitudinal studies** Studies that investigate maturational change by comparing some measure of performance on one age group relative to one or more previous levels of performance on the same measure

**Presbycusis** Age-associated, progressive, bilaterally symmetrical, sensorineural deterioration of auditory functioning

THE STUDY OF AGING AND LANGUAGE FOCUSES primarily on isolating the influences of normal aging processes and/or age-associated pathological conditions (e.g., dementia) on the language performance of older adults. Researchers in this field are also interested in understanding the variability of language performance seen among older adults. A model is presented here which defines variability in terms of individual differences and environmental variations.

## I. LANGUAGE DIFFERENCES

The study of language in healthy elderly people residing in the community has focused primarily on identifying specific areas of change in late adulthood. A number of methodological limitations require that generalizations based on studies of language among older adults remain tentative. First, all of the major studies of language are cross-sectional, except when vocabulary and other verbal measurements have been administered as part of intelligence batteries.

In addition to the lack of longitudinal research in this area, studies tend to involve small samples, a variable representation of males and females, and variable definitions of "old." Moreover, despite the large number of separate reports, only a few investigators systematically address questions of aging across a series of cumulative studies. Controls for educational

185

ENCYCLOPEDIA OF HUMAN BIOLOGY, Second Edition, VOLUME 1.   Copyright © 1997 by Academic Press.   All rights of reproduction in any form reserved.

level, socioeconomic status, English as a native versus second language, activity level, and health are sporadic and have not been simultaneous.

## A. Receptive Language

### 1. Hearing the Message

Many older adults experience at least mild hearing difficulty, which has an impact on the sensory reception of the message. Hearing loss experienced by older individuals, typically referred to as presbycusis (see Section II,A), is commonly manifested in a reduced sensitivity to higher frequency sounds and a decreased ability to discriminate among adjacent frequencies and among consonants which are higher in frequency than vowels. The higher frequency consonants (e.g., "f," "g," "s," "t," and "z") are particularly difficult to distinguish. In conversation this results in difficulty in hearing a spoken word as "sat" or "fat," more so than distinguishing "sat" from "sit." Since most women, young children, and teenagers tend to speak at a high frequency, their speech might be particularly difficult to understand.

Older people with hearing loss often describe speech as sounding "fuzzy," with words running together. This distortion in speech perception can occur even if the volume of speech is loud enough to be heard and is compounded when the speech rate is rapid and when there is competition from background noise.

### 2. Vocabulary

Vocabulary knowledge has been extensively studied, primarily as part of psychometric batteries of intellectual behavior, with the conclusion that recognition of vocabulary words does not decline with age, especially when educational background is controlled. Tasks not requiring the retrieval of specific words (e.g., multiple-choice vocabulary tests) are most likely to elicit good performance by older adults. Age group differences in the rate of word access or activation of associated meanings are revealed through speeded tasks.

### 3. Sentence Comprehension

Knowledge of grammatical structures remains constant across the adult age range from the 20s to the 70s. However, utilization of complex grammatical structures is reduced among older adults in various situations. Studies showing age associations in grammatical performance frequently place high demands on participants for sensory processing and/or memory.

Middle-aged adults perform better on a variety of sentence comprehension tasks than adults over age 60, especially for complex grammatical constructions and semantically improbable sentences. It is generally agreed that elders find loss of redundancy (i.e., an overlap in meaning among components of a sentence) a bigger problem in comprehension than do younger adults, largely related to subtle hearing problems and to slower processing. Redundant information improves online guessing of meaning to support ease of understanding. The sentence-processing difficulties of older adults are exacerbated by noisy conditions and by increased speech rate.

### 4. Reading Comprehension

The literature on prose comprehension and recall across the life span is extensive. However, the findings regarding age differences are highly variable and complex. Generally, older adults are more likely to show lower scores than are younger counterparts in the following circumstances: when participants are poorly educated and have low intelligence, text materials require organizational effort, materials are youth oriented, memory demands are high, inferences or logical reasoning is required, delayed testing is involved, or free recall is assessed. An illustration of the interaction among these factors is the finding that age differences are likely for both main ideas and details for individuals with low verbal ability, but only for details among the highly verbal.

## B. Expressive Language

### 1. Speech and Voice

Anatomical, physiological, and neurological changes associated with normal aging result in gradual changes in specific acoustic characteristics of the human voice. Acoustic changes in aging identified in preliminary studies include reduced speech rate, lowered vocal intensity, increased fundamental frequency (for men only), slight hoarseness, and vocal jitter. Fluency differences have also been addressed in terms of increased hesitations and revisions.

Vocal changes can be used by trained and naive listeners to estimate speaker age, with relatively reliable accuracy within a decade. The role of physical health status is highlighted by research showing the clearest age-associated acoustic changes for elders in ill health.

### 2. Vocabulary

In contrast to the passive understanding of words, the active use of vocabulary does show a decline across age groups (from middle age to the seventh decade to the ninth decade of life). The aging person's major difficulty seems to be with accessing word names, as in giving the word for a definition or in retrieving names of people or places. The quality of word definitions is better for younger adults, with usage of the most appropriate synonyms being less common among older people. Age differences appear on the frequently used clinical tests of verbal fluency, which require that the timed naming of words meet a structural (e.g., words beginning with the letter "a") or semantic (e.g., names of animals) criterion.

## 3. Sentence Production

As with sentence comprehension, sentence repetition tasks have shown age sensitivity when the sentences are long and complex. Studies of the spontaneous speech and writings of older adults have exhibited a narrower repertoire of grammatical constructions, avoidance of constructions imposing high memory demands, and more errors in the use of simple sentence structures. The degree of grammatical complexity of oral and written statements has been linked to individual differences in memory among older adults.

## 4. Conversation

Although little research has focused specifically on conversational skills, this is clearly a vital domain for the everyday life of older adults. Laboratory studies based on analogous situations illustrate the ways in which the memory problems of some older adults can influence conversational skills: difficulty keeping track of which speaker made which statement in a multi-speaker interaction, repetition due to poorer judgment about whether something has been said already, and poorer recollection of diverse topics addressed in a complex discussion. Ambiguity of reference (i.e., the use of pronouns with unclear antecedents) and reduced efficiency in conveying information are the two aspects of storytelling and retelling that seem to differentiate older from younger people. In contrast, storytelling is seen to be part of the role of elders, and the identical story is often better appreciated when told by an older person.

Loquaciousness (i.e., excessive talkativeness) has been identified as a problem for some older persons. A major hypothesis offered by clinicians—that verbose individuals were especially lonely and were more demanding in social interactions to make up for fewer interactions—has not been supported. Verbose elders who stray away from the conversational topic were

more extroverted, older, and less physically mobile; experienced more stress; sought more social contacts; and expressed less concern for making favorable impressions on others.

## C. Nonverbal Communication

The reception and production of nonverbal cues (e.g., facial expressions, gestures, and interpersonal distance) are an important part of overall communication. Subtle nonverbal cues are necessary for understanding and negotiating roles within conversation as well as for distinguishing apparent from true communicative intention. Nonverbal cues can become particularly important for elders with hearing difficulties. Preliminary work indicates that elders decode nonverbal cues less effectively than do those in younger age groups, but they perform significantly better with age peers and after practice.

## D. Summary

Age-associated language differences have been identified in a number of domains of speech and language performance. The key areas of age group differences to be targeted for continuing research are vocal characteristics, word retrieval, sentence processing, memory for components of conversation and for text, reference and efficiency in talking, and nonverbal comprehension. Nevertheless, disparities from youthful standards are often subtle, usually unlikely to handicap an individual in everyday communicative exchanges, and are highly linked to variables not systematically studied simultaneously.

## II. COMMUNICATION DISORDERS

In contrast to the majority, some older adults incur age-associated pathological conditions which interfere with the adequate reception and expression of verbal and nonverbal language.

## A. Hearing Disorder

Hearing loss is the most widespread age-associated disorder affecting the communication process. In the United States it has been estimated that 24% of noninstitutionalized people between 65 and 74 years of age and almost 40% of those over 75 are affected by hearing loss. Institutionalized older adults have even higher prevalence rates, with estimates in the 80–90% range.

Presbycusis is the term most often used to refer to hearing loss associated with aging. It involves a progressive bilaterally symmetrical sensorineural hearing loss resulting from the degeneration of hair cells or nerve fibers in the cochlea as well as fibrous changes in the small blood vessels that supply the cochlea. Although presbycusis is widely held to be due to normal aging, the fact that it does not affect everyone has led some researchers to view it as a disease process to which the aging biological system is more susceptible. Research exploring this possibility is currently underway.

The effect of severe hearing loss on communication is far-reaching. A diminishing capacity to hear, compounded by diminishing confidence in the ability to hear, often results in inappropriate conversational behaviors, such as speaking too loudly, standing too close, and excessively requesting repetitions. These inappropriate compensations irritate unimpaired communicators and serve to limit further communicative interaction. In relation to this, it is not uncommon for impaired elders to withdraw socially to avoid the embarrassment of not understanding what is said.

Interventions for hearing loss need to take into account not only the level of severity of the loss, but also the particular auditory manifestation of the loss. Depending on these factors, electronic hearing aids, cochlea implants, special training in lip reading, and alternate communication devices might be prescribed.

## B. Disorders of Language and Speech

Cerebrovascular accidents (i.e., strokes), Parkinson's disease, and cancer are clinical disorders that occur more frequently with advancing age and are the underlying causes for a number of communication disorders. Strokes are the most common cause for a number of expressive and receptive communication disorders. [*See* Speech and Language Pathology.]

### I. Aphasia

Aphasia is a disorder seen most frequently in old age and refers to a language-specific disturbance resulting from damage primarily in the left cerebral hemisphere. Aphasia is considered primarily a language disorder, not a speech disorder, since it is the efficacy with which the symbolic code of language is interpreted (i.e., decoded) and formulated (i.e., encoded) that is affected rather than the motor programming or execution of muscle movements involved in speech. As a result, aphasia is characterized by some reduction in language comprehension (heard and read) and/or expression (written and spoken). The degree of language impairment varies considerably from mild word-finding difficulties in speech to more severe involvement in all language modalities.

Lesions involving the third frontal convolution result in a disorder called Broca's aphasia. Broca's aphasics generally comprehend spoken or written language and know what they want to communicate, but have difficulty finding the words and organizing them into meaningful communication. Speech is typically characterized as nonfluent or "telegraphic," in which content words are present but function words such as articles, pronouns, prepositions, and auxiliary verbs are omitted. With the exception of automatic or rote phrases, oral struggle in an effort to speak is usually noted. This language impairment often mistakenly projects the image of an individual unable to understand speech, which adds to the frustration of the Broca's aphasic.

Lesions involving the first temporal gyrus result in what is typically called Wernicke's aphasia. In this type of aphasia, contrasted with the Broca's type, auditory comprehension is impaired, while the ability to speak fluently is not. Grammatical structure and sentence intonation patterns are maintained, but what is said is often meaningless. Related to the predominant comprehension difficulties, Wernicke's aphasics seem oblivious to the nonsensicality of their utterances and do not engage in self-correction. The use of pictures, photographs, and gestures might help in communicating with the Wernicke's aphasic.

Age of onset plays only a minor role in predicting spontaneous or treated recovery from aphasia, but does seem to affect the type of aphasia observed. With increasing age there is a trend toward a greater degree of severity of aphasic symptoms and aphasic syndromes, with more pronounced auditory comprehension impairments (Wernicke's aphasia).

### 2. Dysarthria and Apraxia

Dysarthria and apraxia are two classes of motor speech disorders. They are so called because, unlike aphasia, command of the symbolic code of language remains intact, while the motor expression of speech is impaired.

Dysarthria results from a lesion to the central or peripheral nervous system which causes weakness or incoordination of the muscles used in speech execution. The result is impairment in any or all of the basic motor speech processes of respiration, phonation, resonance, articulation, and prosody.

Apraxia results from damage to the base of the third frontal convolution in the left hemisphere (Broca's area). The result is an impaired capacity to program the positioning of muscles used in speech execution, despite the fact that muscle strength is undiminished. Speech behavior is characterized by variable articulatory patterns, disturbed prosody, oral struggle, and inappropriate phonemic sequencing. Typically, apraxic patients show no disturbance in producing rote sequences or automatic phrases.

### 3. Right Hemisphere Communication Disorders

At present, communicative disturbances associated with left hemisphere lesions are fairly well defined and most can be attributed to specific lesion sites. In contrast, communication disturbances associated with right hemisphere damage are more subtle and less clearly identifiable, with specific sites of lesion due to the more complex neurological organization of the right hemisphere.

Impairments following right brain damage are manifested in inappropriate communicative behavior and deficiencies in pragmatic skills (i.e., the use of language in social situations). Pragmatics require the speaker to be sensitive to the amount of shared information between communicators, the social situation, and appropriateness in selection, introduction, maintenance, and change of topics.

### 4. Parkinson's Disease

Parkinson's disease is a neurological syndrome resulting from degeneration of the corpus striatum or substantia nigra. Symptoms affect primarily the non-verbal aspects of communication. Most notable is the characteristic "mask-like" freezing of the facial features, which deprives the communicator of the modulation of messages obtained by facial mimicry. Body posture also suffers from this freezing effect, depriving the communication situation of the added contextual information communicated by gestures. As a result the affective aspects of the message are poorly communicated, since the message is deficient in the emotional nuances on which communicators rely to help understand and empathize during communication. [*See* Parkinson's Disease.]

### 5. Cancer of the Larynx

Cancer of the larynx is primarily a problem of old age, and, as such, removal of the larynx and the incurred loss of speech ability are problems experienced by some older individuals.

Management methods following laryngectomy include artificial larynges, which serve as a sound source for speech production, esophageal speech, or a combination of the two. For many cancer patients who also require surgery for cancers in other articulatory structures, the prognosis for speech rehabilitation is poor. For these individuals, alternate means of non-oral communication (e.g., writing or electronic communication devices) are options.

### C. Disorders of Cognition

Dementia is a pathological condition resulting from central nervous system degeneration and is most often associated with old age. Many causes for the characteristic deterioration of memory, intellect, personality, and communicative functioning have been identified. Some causes (e.g., drug toxicity, metabolic and endocrine disorders, and infections) are reversible with appropriate intervention undertaken within a reasonable time following the onset of symptoms. A number of other clinical disorders (e.g., multiinfarct dementia) and diseases (e.g., Alzheimer's and Parkinson's) can result in irreversible dementias. Of these, Alzheimer's disease is the most common dementing illness [*See* Alzheimer's Disease; Dementia in the Elderly.]

Dementia of the Alzheimer's type typically involves atrophy (especially in the temporoparietal and anterior frontal regions) and characteristic neurofibrillary tangles, senile plaques, and granulovascular degeneration. Although the progression of communication breakdown varies, a predominant pattern fitting most patients emerges.

In the early stages of dementia, changes in communicative efficiency are not overly debilitating. Some naming and word-finding difficulties are seen, but discourse is still informative, although at times overly long. Comprehension for single sentences is usually intact, but deterioration is seen for longer paragraphs. Patients can usually write a paragraph and read aloud, although there might be stuttering on occasional words, and comprehension for what is read tends to decrease as material increases in length and complexity.

Middle stages of dementia are characterized by more apparent communication deficiencies. Many naming errors are made, and speech discourse is often rapid, incoherent, hard to stop, and empty of meaning. In conversation, patients do not appear to fully comprehend what is being said, picking up only the occasional content word. They can no longer tell coherent stories nor respond appropriately to questions requiring specific answers. Often, no more than one sentence can be written, and grammar might be impaired. In reading aloud, words might be substituted or nonsense words might be produced, and there is apparent difficulty in comprehending paragraph-length material.

At late stages of illness, there is a severe breakdown in the ability and intention to purposefully communicate. Frequently, patients are mute or discourse is severely limited. In the discourse emitted there are usually excessive repetitions and clang associations (i.e., word strings of phonetically similar words that are often unrelated in meaning). At this stage patients do not appear to comprehend what is said to them.

Although at present there is a limitation on what can be done clinically in the case of irreversible dementia, much can be done to help both the patient and his caregivers make effective use of remaining communication competencies. The promotion and training of active coping strategies for caregivers (e.g., learning to extract intended messages from contextual cues) should be encouraged. Often, despite obvious decrements in verbal comprehension, comprehension of nonverbal cues (e.g., tone of voice and facial expression) might still be available to the patient, a form of communication that caregivers can use to express nonverbal messages even in the most advanced stages of illness.

## D. Summary

Communication disorders of old age include disturbances in hearing, language functioning, speech, and cognition. Communication disorders reviewed here have their impairing effects on reception and/or expression in communicative exchange. Moreover, it is not uncommon for more than one of these communication impairments to coexist. In these instances the total handicapping condition might be greater than the sum of each individual condition.

## III. UNDERSTANDING VARIABILITY IN AGED LANGUAGE

### A. Three Basic Principles

A balanced understanding of language in later life must incorporate three principles. The first is that of heterogeneity among older adults in their cognitive and language performance. Individual differences based on genetic inheritance and life histories as well as differential vulnerability to age-associated diseases and environmental changes increase with age. Studies comparing older and younger groups focus on group differences, but the great variability within age groups cannot be ignored. Despite averages in favor of younger adults, a considerable number of healthy elders invariably exceed the younger group's average performance.

The second principle is that adult age differences in language competence (i.e., knowledge of the language presumably acquired early in life) are minimal among healthy elders, whereas age-associated sensory and cognitive changes lead to noticeable age group differences in language performance in some domains. Linguistic and pragmatic knowledge acquired by adolescence is maintained, and there are, no doubt, important increases based on the wealth of experiences throughout life in dealing with communication challenges. However, performance in language tasks can be limited by other age-associated changes (e.g., hearing difficulties, decreased speed, and poor memory).

The third important principle is that communicative success depends not only on a person's abilities, but also on the interpersonal and environmental situation in which communication occurs. Until recently, the social context of language use has been ignored in the study of the psycholinguistic and clinical aspects of language change in aging. Whereas older adults perform best under conditions of moderate environmental challenge with appropriate support, they are frequently in environments which understimulate. In addition, they are often asked to perform tasks for test purposes with little relevance to their usual environmental challenges.

### B. Framework

Integrating the three basic principles just mentioned, a framework for understanding variability among elders in language performance is presented here. The schematic model given in Fig. 1 depicts the major contributors to language performance as a function of individual differences in language competence, information-processing factors, and social strategies, as well as variations in the immediate situation. Correspondingly, the language-relevant individual differences are based on individual life histories and current abilities and predispositions. The immediate situation is dependent on the sociocultural environment within which individuals have lived, especially environmental variations in their current lives. In addition, the

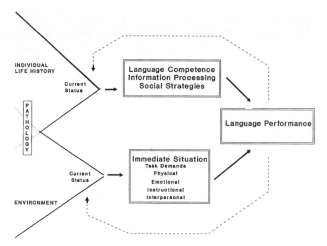

**FIGURE 1**   Factors influencing variability in language performance among elders.

specific demands of the assessment context impose constraints on participants' opportunities to exhibit their language competence. The specific influence of age-associated pathological conditions (e.g., severe hearing impairment, dementia, and aphasia) is highlighted in Fig. 1 as primarily affecting individual life histories, but also having an impact on the communication environment experienced by an individual.

## 1. Individual Life History

Within a given group of older people, some are more likely to exhibit communication difficulties than others. Critical background variables associated with individual life history include education, socioeconomic status, multilingualism, health, gender, and personality. These factors can influence the level of language competence, information processing, and social skills which individuals bring to their old age. Moreover, several researchers have interpreted studies of reading comprehension, for example, as evidence of a threshold hypothesis according to which those with strong verbal skills appear to be less susceptible to decline with age.

Current variations among individuals in old age can have major influences on language competence, information-processing strategies, and social strategies. Chronological age is obviously one key factor, in that older adults in their 90s are likely to show more language changes than those in their 60s or 70s. Neurological conditions (e.g., dementia and aphasia) directly affect language competence and information-processing skills and indirectly affect social strategies. Performance difficulties can be caused or exacerbated

by general medical illnesses as well as by the medications used to treat them. Problems associated with nutritional status, alcohol abuse, and poor physical fitness can also reduce a person's information-processing skills. Depression, the most prevalent mental health problem among older people, limits communication performance since it is associated with poor social relationships, social withdrawal, apathy, poor concentration, and memory complaints. [*See* Depression.]

The key information-processing constraints for language performance relate to sensory abilities, speed of processing, memory, and selective attention. Sensory abilities are critical to the processing of linguistic messages as well as the supporting contextual information. Hearing difficulties pose the most prevalent limitations associated with aging. In addition, vision changes associated with aging (e.g., lower acuity, the need for greater illumination, sensitivity to glare, and diminished color discrimination) can inhibit the perception of nonverbal gestures or the use of lip reading.

The gradual slowing of perceptual, cognitive, and psychomotor processes is one of the best established findings in the psychology of aging. Intensive research on memory has identified particular difficulties for elders. Of most importance here are age group differences in memory for new information, especially when effortful processing is required, active retrieval (versus recognition) is tested, and the task is particularly difficult. With respect to attention, older adults have more difficulty selectively attending to target information during language processing. Elders' greater distractibility is thought to reflect an age-related decline in the ability to inhibit irrelevant information.

Social strategies are also an important contribution to variation in language performance. Communication is a resource for coping and adapting to the stresses of life through the satisfaction of instrumental needs and the development and maintenance of social support networks. Some communication styles might represent strategies to compensate for declines in hearing and memory (e.g., avoidance of talking and conversational domination), whereas others might reflect changing objectives for social interchange (e.g., different goals for new friendships and contrasting social purposes in intergenerational conversations).

## 2. Environment

Individual life histories are obviously greatly affected by the sociocultural environment within which one matures. Of greatest relevance here is the influence of the current environment on the ability of aging people to use language skillfully. According to discussions of person–environment fit, performance and growth in aging are maximized in an environment that challenges people to perform slightly above their current levels.

Stereotypes of old age and ageist bias in interpreting the behaviors of older people have been well documented. In particular, older adults are expected to show memory decline, to be slower to learn, to be less successful in achievement-oriented situations, and to show more helpfulness and friendliness than do younger adults.

Negative expectations and overprotection can pose barriers to successful communication. The self-appraisal of older adults is based to some extent on social expectations, and meeting challenges with the appropriate effort and strategies depends on a strong sense of one's capabilities. Expectations that older individuals with some health limitations cannot think or speak adequately can lead to avoidance of communication, overly constrained communication opportunities, diminished corrective feedback about negative behaviors, overprotection, or patronizing behaviors (e.g., oversimplified speech).

The opportunity for continued use and growth of communication skills within one's current environment can clearly influence language performance in an assessment situation. For example, the maintenance of communication skills is hindered in environments in which an older person is relegated to a position outside regular social contacts. Older adults are at risk for loss of contact with members of their social network through retirement, illness, death, geographic relocation, and institutionalization. Communication in community settings, in lifelong friendships, and especially within a marriage occurs naturally and is spontaneously nurtured. If these everyday opportunities are lost, the older person is challenged to establish new friendships and intimate relationships. Reductions in the traditional social network of an older person as well as increasing dependence can also diminish the opportunities for reciprocal exchanges of support.

The level and the type of activities engaged in by older adults are reciprocally negotiated within their environment. Extroverted and experienced elders can alter the environment for themselves and for others in terms of initiating activities or encouraging others to take part. Elders participating in activities which challenge them socially, mentally, and physically are more likely to practice and extend their social and communication skills.

Language performance is a function of various characteristics of the immediate situation. With re-gard to the specific task, time constraints and memory demands are probably the best identified factors increasing age group differences. In addition, older adults seem to exhibit relatively greater motivation when the task is familiar and clearly relevant to everyday life and when an opportunity for practice is available. Emotionally, testing situations that are highly unusual or critical for life decisions are likely to arouse anxiety among older adults.

Testing situations often fail to consider possible information-processing issues particularly salient to elders (e.g., small print, noisy or busy settings, glare, poor lighting, and insufficient intensity of language stimuli). The increased likelihood of fatigue must be taken into account in terms of the length of the testing sessions. From an instructional perspective the appropriateness and manner of conveying information about what is expected in an assessment situation can greatly influence a participant's understanding of the task at hand as well as their comfort in participating.

## 3. Conclusion

An individual's performance in a given language situation can be seen from Fig. 1 to involve contributions of the individual and environmental factors just discussed. Within this framework, chronological age is merely one of many influences on assessed behavior. To the extent that the situation calls for language behaviors that are a typical part of the older individual's activities in a manner that accommodates possible limitations in informing-processing functions, performance is likely to be a good measure of everyday language skills. The interactive nature of the components can be illustrated by considering the apparently egotistical tendency of some older individuals to dominate a conversation. Such behavior might result from self-centeredness which is lifelong or from a decline in the perspective-taking ability. However, this behavior could also arise as a social strategy for coping with hearing or memory problems or as an acceptance of expectations about declining social involvement with aging.

Language performance has reciprocal influences on the individual and the environment. For example, positive experiences can enhance a person's self-efficacy, challenge negative social expectations, increase one's social network, and set the stage, through practice, for subsequent improved performance. Correspondingly, an unsuccessful performance can confirm negative social expectations, undermine one's sense of communication efficacy skills, lead to avoidance of similar situations, and set the stage, through failure, for even poorer subsequent performance.

## BIBLIOGRAPHY

Bayles, K. A., and Kaszniak, A. (1987). "Communication and Cognition in Normal Aging and Dementia." Little, Brown, Boston, MA.

Hummert, M. L., Wiemann, J. M., and Nussbaum, J. F. (eds.) (1994). "Interpersonal Communication in Older Adulthood." Sage Publications, Thousand Oaks, CA.

Kemper, S. (1992). Language and Aging. *In* "The Handbook of Aging and Cognition" (F. I. M. Craik and T. A. Salthouse, eds.), pp. 213–270. Lawrence Erlbaum, Hillsdale, NJ.

Obler, L. K., and Albert, M. L. (1985). Language skills across adulthood. *In* "Handbook of the Psychology of Aging" (J. E. Birren and K. W. Schaie, eds.), 2nd Ed., pp. 463–473. Van Nostrand Reinhold, New York.

Ryan, E. B., Giles, H., Bartolucci, G., and Henwood, K. (1986). Psycholinguistic and social psychological components of communication by and with the elderly. *Lang. Commun.* 6, 1–24.

Shadden, B. B. (ed.) (1988). "Communication Behavior and Aging: A Sourcebook for Clinicians." Williams & Wilkins, Baltimore, MD.

# Aging, Molecular Aspects

PETER J. HORNSBY
*Baylor College of Medicine*

---

I. Primary and Secondary Changes in Aging
II. Genetic Instability in Aging
III. Extrinsic Hazards
IV. Germ Line and Soma in Aging
V. Genes that Affect Aging
VI. Cellular Senescence
VII. Summary

## GLOSSARY

**Aging** The gradual and progressive decline in fitness of the individual, causing increased susceptibility to death from internal causes (e.g., chronic disease) and external causes (e.g., pathogens). Although molecules, cells, and extracellular materials may correctly be said to "age," i.e., to show gradual and progressive deterioration, these processes do not necessarily cause, and are not necessarily related to, aging of the body as a whole (e.g., because the aged components are efficiently turned over and do not accumulate over extended periods of time), therefore, the term "aging" is best confined to the definition used in this article

**Cellular senescence** The process of permanent exit from the cell cycle that eventually occurs in all normal human cells in culture; also termed clonal senescence or clonal attenuation. Senescent cells are in a permanently nondividing, but viable, state. Although defined as a cell culture phenomenon, it is not an artifact of inadequate growth conditions and reproducibly occurs in all human cells in culture. It occurs in some kinds of cells during human aging *in vivo*

**Genetic instability** Primary changes in the structure of DNA and chromatin that result from the action of metabolism-related and cell division-related hazards to which cells are exposed

**Telomerase** A ribonucleoprotein DNA polymerase complex that maintains telomere length

**Telomeres** Specialized tandemly repeated DNA sequences at the ends of the linear chromosomes, and associated proteins, that serve to maintain the integrity of the chromosomes

**Segmental progeroid syndrome** Human genetic diseases that show features of accelerated aging. "Segmental" indicates that even those that most closely resemble normal aging fail to mimic it in all respects; others are much more limited in their resemblance

THE MOLECULAR BASIS OF HUMAN AGING IS A complex subject for scientific study. It is unclear whether aging is due to a small number of universal molecular processes or whether the aging of each organism, and each organ and tissue, occurs via unique mechanisms. It is not known to what extent age-related changes in the fitness of the organism as a whole—the well-known declines in organ function—result from permanent changes in the properties of cells in the older individual. This question has gone unanswered because age-related molecular changes in cells and tissues are the net result of numerous concurrent events that may obscure the primary causes of aging. These problems and others are listed in Table I.

## I. PRIMARY AND SECONDARY CHANGES IN AGING

Many molecular changes in aging, involving differences in the levels of individual proteins, mRNAs, and lipids, are secondary to changes occurring elsewhere in the body. Notably, changes in circulating hormones have profound effects on the levels of molecules in their target cells. In other cases, changes in molecules are secondary to changes in cell popula-

ENCYCLOPEDIA OF HUMAN BIOLOGY, Second Edition, VOLUME I.    Copyright © 1997 by Academic Press.    All rights of reproduction in any form reserved.

### TABLE I
Problems and Pitfalls in Elucidating Molecular
Aging Processes

Primary versus secondary changes

Role of intrinsic cellular changes versus
   Changes in homeostasis
   Noncellular changes (extracellular matrix, etc.)
   Tissue composition (cell populations)

Species differences

Individual heterogeneity

Tissue heterogeneity

Disease versus aging

### TABLE II
Characteristics of Primary Age-Related Changes in Cells[a]

*Cumulative* (the effect of the process is added to that which has previously occurred in the tissue)

*Universal* (all members of the species show the same changes, to distinguish from specific disease processes)

*Progressive* (change goes in one direction, does not reverse)

*Intrinsic* (a true change in cellular phenotype, not a temporary/reversible effect of environmental factors such as hormones, local factors, and extracellular matrix to which the cell is exposed)

*Deleterious* (leads to a lowering of the performance of the cell/tissue/organ/organism with respect to the ability of the organism to survive challenges in its environment)

[a]Adapted from Arking (1991).

tions, such as increases in connective tissue or fat within an organ; subpopulations of cells of a different phenotype that were less abundant in the young individual may expand in aging (e.g., in the immune system).

In contrast, primary changes in aging (defined in Table II) include changes in the structure of DNA and changes in those proteins or lipids that have a sufficiently long half-life that they can cause changes in the properties of cells and tissues. Such changes include mutations in DNA, crosslinks in proteins, and formation of peroxidized lipid molecules. At present it is not known which organs, in which primary age-related changes take place, are the most important for human aging overall.

In general, primary molecular changes during aging result from the accumulated and ongoing hazards to

which the body is exposed. These hazards are of three principal varieties (Fig. 1.).

The first class is time dependent. Purely time-dependent changes are unusual and are probably not important in molecular changes in aging. However, one example is the slow racemization of L-amino acids to D-amino acids that has been recorded in very long-lived proteins, e.g., in the tooth.

The second category is metabolism related, i.e., changes that are proportional to metabolic rate. It has been known for some time that, within groups of mammals, maximal life span is inversely proportional to the specific metabolic rate. It is generally thought that this is because side effects of metabolism cause

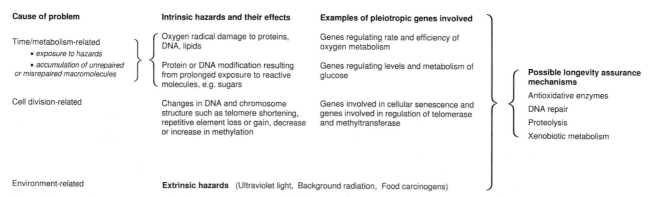

FIGURE I    A "cell's-eye view" of events in aging. The hazards to which the cell is exposed are classified as intrinsic if they originate from the operation of processes within the organism and extrinsic if they originate from the environment of the organism. Cause of problem means why the cell is exposed to the hazard. These causes include the deleterious effects of genes that have pleiotropic effects. The defense processes of the body, the potential longevity assurance mechanisms, are grouped because all of them, to varying extents, act to protect against both intrinsic and extrinsic hazards.

changes in molecules that are not efficiently repaired and have cumulative effects. A major source of damage to nucleic acids, lipids, and proteins is the release of oxygen radicals, such as superoxide anion, by oxygen-metabolizing systems, principally the respiratory chain but also other oxygen-using proteins such as cytochrome P450 enzymes. Ongoing oxygen radical damage to DNA results in the production and excretion of damaged bases, such as 8-hydroxyguanine, and may be a cause of genetic instability. Other DNA damage products, resulting from oxygen radical attack and carcinogen damage, accumulate with aging and are detectable as abnormal *I-spots* by two-dimensional chromatography.

Proteins may also undergo metabolism-related damage by the nonenzymatic reaction of glucose and other sugars with proteins; the products of this reaction, complexes called advanced glycation and oxidation end products (AGE), have been shown to accumulate with aging and to damage organs such as the kidney. Pentosidine is a major AGE and accumulates as cross links in proteins of long half-life such as collagen. Oxygen radical damage to proteins includes other forms of crosslinks and the addition of carbonyl groups. Crosslinking may substantially change the properties of proteins.

Subcellular bodies containing fluorescent complexes called lipofuscin, or age pigment, accumulate in aging. The formation of these complex products depends on the peroxidation of lipids and the incomplete digestion of proteins (protcolysis). However, it is not known whether the presence of lipofuscin impairs cell function. [*See* Metabolic Regulation; Lipids.]

A third class of changes is cell division related. Such changes result from the fact that somatic cells are unable to maintain normal function when they divide repeatedly over long periods in aging. They comprise several types of genetic instability.

## II. GENETIC INSTABILITY IN AGING

### A. Point Mutations

Mutations increase in somatic cell DNA in aging, but do not appear to contribute in a major way to changes in cellular properties that contribute to aging; in persons with high levels of point mutations in their DNA, resulting from exposure to radiation, aging is not substantially affected.

### B. Mitochondrial DNA

Mitochondrial DNA becomes progressively damaged with aging, probably because of continuous damage by oxygen radicals produced by the respiratory chain coupled with the inefficient repair of mitochondrial DNA. Although no specific deletion or other change in the circular mitochondrial DNA molecules is very common, the damage is of such a wide variety that few totally normal molecules persist in old age. However, the presumed effect on energy metabolism resulting from this damage has not yet been proven to cause age-related defects in cell function.

### C. Telomeres

Telomeres are subject to a replication-dependent shortening in aging. This probably results from the absence of active telomerase in most somatic cells; however, the germ line and some subset of stem cells do have active telomerase and maintain constant telomere lengths. The progressive shortening of telomeres might eventually result in an inability for further division (see Section VI).

### D. DNA Methylation

DNA methylation changes during development and differentiation. In aging, mechanisms that normally maintain methylation may not be adequate to maintain normal methylation patterns, leading to shifts (decreases or increases) in methylation, in turn possibly leading to permanently inherited cellular changes in gene expression. [*See* DNA Methylation in Mammalian Genomes.]

### E. Repetitive Elements

Tandem DNA repeats may change in number during long-term cell division, probably by slippage occurring during replication. Interspersed repeats have also been found to appear as extrachromosomal circular DNA elements. Any functional consequences of these changes are unclear.

### F. Transcription Factors

An interesting concept, with a theoretical basis but no direct data, is that aging may involve changes in transcription factors that act in autoregulatory positive feedback loops, i.e., they act as transcription factors for their own genes. These factors are sometimes called "master" gene regulatory proteins. Such loops are metastable; over repeated divisions or long time periods, a loop could become disrupted and might then assume a different metastable state, permanently altering the cell's pattern of gene expression.

## G. Cell Death

The role of cell death (apoptosis) in aging is not clear. The death of the individual is not caused by or accompanied by much cell death. Indeed, cell turnover (both proliferation and cell death) mostly slows down in aging, in the absence of neoplastic changes. However, the gradual elimination by apoptosis of populations of critical cells, particularly neurons, might occur over long periods, but the current evidence is too limited to determine its importance. [*See* Apoptosis (Programmed Cell Death).]

## III. EXTRINSIC HAZARDS

Intrinsic hazards to which cells are exposed in aging are accompanied by extrinsic hazards to which the body is exposed, and which may contribute to age-related changes. Although progressive and deleterious, such changes should perhaps not be described as aging because, in theory, they are avoidable (i.e., not universal; see Table II). By tradition, processes such as the action of repeated exposure to ultraviolet light in the skin are called aging (photoaging). Another example is the continuous exposure to carcinogens and other xenobiotics in food and the environment to which more or less all members of the species are uniformly exposed and which might contribute to genetic instability in aging, as well as cancer.

## IV. GERM LINE AND SOMA IN AGING

The concept of the "disposable soma" is important in understanding the role of genes in aging. The germ line in humans and other species does not age; over long periods of time (millions of years) there is no degradation in the ability of the germ line to perform its function of passing the genetic information, in the form of DNA, from one generation to the next. Note that this applies to the germ line per se; germ cells in

individuals do show changes in aging. Rather than suffer a progressive loss of its function, the germ line "improves" over evolutionary time by the operation of the processes of recombination, via sexual reproduction, and natural selection. In contrast, the soma (the body except for the germ cells) is needed only temporarily, from the point of view of the genome, to efficiently pass on the genetic information to the next generation. Thus from the point of view of the interests of the genome, now safely housed in the next generation, the soma can be allowed to deteriorate past reproductive age.

Why does the genome not preserve the soma longer against aging, allowing it to survive longer, thus enabling it to pass on the genetic information to more progeny? This question raises the concept of "trade-offs;" how much of its resources should the genome devote to the survival of the soma versus reproduction? In human aging, female reproduction is limited to the time from puberty to menopause. At a maximum rate of about one offspring per year, the reproductive rate is relatively slow (in comparison with most other species), thus limiting the number of potential offspring. However, the reproductive period is also very long in comparison to other species, thus ensuring that, on balance, there will be enough children who survive to form the next generation (recall that the human species had, for most of its history, very high infant mortality and an average life span of about 30 years). The human genome has therefore set the trade-off balance well in favor of somatic maintenance versus rapid reproduction. This complex subject is covered in more detail in the works cited in the bibliography.

## V. GENES THAT AFFECT AGING

If the disposable soma concept is valid, then it is highly unlikely that there are genes whose sole function is to *cause* aging because there are no conceivable mechanisms whereby such genes could evolve and be preserved in the genome. The soma is a temporary entity, not required to survive indefinitely, yet there is no need (from the point of view of the genome) to eliminate the soma by a deliberate aging process that leads to death. The aging and death of an individual obviously do not contribute to the survival of its own genome and could only contribute to the survival of copies of its genome (in its offspring) in the case of nonmotile organisms where death of the parent frees up an ideal site for their survival, a situation obviously not applicable to human aging.

Nevertheless, it is clear that there is a strong genetic influence on aging. Humans and other mammals have species-specific maximal life spans, like other species-specific characteristics such as body size, metabolic rate, or gestation period. Maximal life span is a biological constant, whereas average life span is determined not only by maximum life span but also by a variety of environmental influences.

Several classes of genes affect the molecular aging processes described earlier and thereby contribute to the determination of maximal life span. One class comprises those genes whose products contribute to those aging processes. A second comprises those genes whose products in various ways protect the organism against those events. A third (possibly overlapping with these two) comprises genes that increase the rate of aging when defective. However, no human genes are known to *decrease* the rate of aging when defective, probably because the human genome has weighted the trade-off of reproduction versus somatic maintenance heavily toward the latter, as discussed earlier.

## A. Antagonistic Pleiotropy

Genes in the first class exhibit trade-offs in life history, i.e., their good effects in early life outweigh their bad effects, which occur mostly after reproductive age, or do not significantly affect reproductive potential or potential for survival prior to reproductive age. Because the force of selection varies inversely with the time of expression in the life span, such negative effects are not selected against over evolutionary time. These genes are said to exhibit antagonistic pleiotropy; their pleiotropic effects are antagonistic (good at one part of the life span and bad at a later time). For example, the genes that encode the proteins involved in oxygen metabolism in mitochondria may be a good example of such genes because they have evolved in order to increase the efficiency of ATP generation, which is important for most aspects of metabolism and survival. But their tendency for harmful effects, via the generation of oxygen radicals, although limited by evolution, has not been completely suppressed.

## B. Longevity Assurance Genes

Longevity assurance genes are genes whose effects tend to protect against or negate the deleterious molecular changes described earlier. They either have a clear-cut protective effect (e.g., antioxidative enzymes such as superoxide dismutase or DNA repair enzymes) or protect indirectly (e.g., proteolysis, which eliminates damaged proteins as well as being responsible for normal protein turnover).

Note that longevity assurance genes cannot be defined as those genes that are associated with a shorter life span when inactivated or deleted (e.g., by homologous recombination) because such a definition would include a large fraction of the known genome. True longevity assurance genes specifically protect the soma against age-dependent deterioration (the hazards that cause aging), thus contributing directly to the preservation of the soma in the trade-off between somatic maintenance and reproduction.

## C. Genes That Affect the Rate of Aging When Defective

Another class of genes is defined by the fact that their inactivation causes segmental progeroid syndromes ("accelerated aging"). The syndrome that has drawn the most attention as most resembling a speeding up of aging is Werner syndrome, but Hutchinson–Gilford progeria (resulting from a defect in an unknown gene) gives a more dramatic phenotype at an early age. Other human genetic diseases have been ranked as resembling accelerated aging; one with a lower level of resemblance is Cockayne syndrome. The Cockayne syndrome gene is a helicase involved in transcription and DNA repair, and although the biochemical function of the Werner syndrome gene product has not at the time of writing been elucidated, it too resembles a helicase. The nature of these genes implicates DNA damage as a cause of accelerated aging in these syndromes, but this has not been proven.

It is not clear that these genes are major determinants of the rate of aging when normal. A parallel example is of genes that cause or predispose to age-related diseases (note that such diseases are not aging, but increase in incidence with age). For example, mutations in the $\beta$-amyloid precursor protein (APP) can cause Alzheimer's disease, but this does not mean that APP in its normal unmutated state plays a major role in determining the life and death of central nervous system neurons. Similarly, until more is known, it should not be assumed that the Werner gene has a large role in determining the rate of aging in its normal unmutated state.

## VI. CELLULAR SENESCENCE

The limited clonal proliferation of normal human cells in culture was first described in detail by Leonard Hayflick and is often referred to as the Hayflick limit. The senescent state is an alternate exit from the cell cycle, distinguished from terminal differentiation, reversible $G_0$ arrest, and cell death (apoptosis), which are other ways in which cells can cease cycling. Cessation of proliferation occurs after cells have undergone 30 to 100 divisions in culture; the number is quite reproducible, but varies according to the starting cell population. Clearly, therefore, there is a counting process which eventually triggers a genetic program of commitment to the nondividing state, i.e., cellular senescence. In normal human cells, the cessation of proliferation involves the increased expression of an inhibitor of cyclin-dependent kinases, p21 or senescent cell-derived inhibitor (SDI-1). The best current candidate for the counting process is telomere shortening, which appears to occur both *in vivo* and in culture because of the absence of telomerase activity in somatic cells. However, the mechanisms by which telomere shortening could trigger senescence are unknown. [*See* Cell.]

Apart from the loss of telomere sequences, changes occur in DNA methylation and repetitive sequences, both in cells in culture and *in vivo*. However, cellular senescence is not caused by damage to or loss of the genetic material needed for proliferation, as evidenced by the effects of immortalizing oncogenes. A diverse group of cancer-causing genes enable cell populations to undergo indefinite proliferation. Most of these genes act by preventing senescence rather than by preventing cell death (as unfortunately implied by the term "immortalization"). Most immortalized lines have active telomerase, but some that do not appear to have developed an alternate mechanism for avoiding telomere shortening. Immortalized cell lines fall into four complementation groups; cell hybrids whose parental cells are from two different groups senesce, whereas those whose parental cells are from within the same group do not. This indicates that are at least four different genes which normally contribute to the genetic program of senescence, the inactivation of any of which abrogates senescence and leads to immortalization.

Several observations suggest that cellular senescence occurs *in vivo* during aging. Clones of fibroblasts have a lower total proliferative potential in culture as a function of the age of the donor, and proliferative potential is greatly reduced in cells from Werner syndrome and other segmental progeroid syndromes. An unusual form of $\beta$-galactosidase unique to senescent fibroblasts *in vitro* also appears in subsets of cells in the skin during aging. Many other cell types show donor age effects on proliferative potential in culture; notably, clones of T lymphocytes that are senescent and have short telomeres are observed both in normal old age and in immune disorders where the abnormal proliferation of T cells occurs. Cellular senescence may be deleterious in old age because the ability to proliferate is an essential function of many cells, but the extent to which it contributes to the overall changes in organ function in aging is unknown. However, cellular senescence (and the suppression of telomerase in somatic cells) is probably also beneficial in early life because it may provide an antitumor mechanism, thereby contributing to survival prior to reproductive age. If so, the genes responsible for cellular senescence are examples of antagonistic pleiotropy.

## VII. SUMMARY

In summary, the knowledge of molecular events in aging is still lacking a major structural framework. It must be emphasized that none of the molecular changes described in this article has been shown to have an unequivocal effect on age-related changes in organ function. Thus, although many molecular events during aging have been described, and the causes of some are known, the extent to which they actually contribute to aging is unknown.

## BIBLIOGRAPHY

Arking, R. (1991). "Biology of aging: Observations and Principles." Prentice-Hall, Englewood Cliffs, NJ.

Dawkins, R. (1989). "The Selfish Gene," 2nd Ed. Oxford University Press, Oxford.

Finch, C. E. (1991). "Longevity, Senescence, and the Genome." University of Chicago Press, Chicago.

Hodes, R. J., McCormick, A. M., and Pruzan, M. (1996). Longevity assurance genes: How do they influence aging and life span? *J. Am. Geriatr. Soc.* **44**, 988–991.

Holliday, R. (1987). "Understanding Ageing." Cambridge University Press, Cambridge.

Hornsby, P. J. (1996). Genes, hormones, and aging. *In* "Advances in Cell Aging and Gerontology," (P. S. Timiras and E. E. Bittar, eds.), Vol. 1, pp. 31–61. JAI Press, Greenwich, CT.

Kirkwood, T. B., and Rose, M. R. (1991). Evolution of senescence: Late survival sacrificed for reproduction. *Phil. Trans. R. Soc. London B* **332**, 15–24.

Ricklefs, R. E., and Finch, C. E. (1995). "Aging: A Natural History." Freeman, San Francisco.

Smith, J. R., and Pereira-Smith, O. M. (1996). Replicative senescence: Implications for in vivo aging and tumor suppression. *Science* **273**, 63–70.

# Aging, Psychiatric Aspects

BENNETT S. GURIAN
*Harvard University*

I. Demography
II. Epidemiology
III. Geropsychiatry

## GLOSSARY

**Anxiety** Subjective state of internal discomfort, dread, and foreboding accompanied by nervous system arousal

**Gerontology** Study of aging from its broadest perspectives

**Geropsychiatry** Study of mental illnesses of late life and their treatments

**Mood disorder** Disturbance of mood (i.e., a prolonged emotion that colors the whole psychic life) involving either depression or elation

**Organic mental disorder** Psychological or behavioral abnormality associated with transient or permanent dysfunction of the brain

**Paranoia** Presence of a persistent delusion

A RAPIDLY EXPANDING BODY OF KNOWLEDGE IS related to the psychology of aging which encompasses the "normal," or usual, aspects of thinking and feeling behaviors. In contrast, psychiatry is concerned with pathological or unusual behaviors and with the interventions used to try to return persons with such disorders to a state of health (i.e., decreased symptoms and improved function).

Geropsychiatry (also called geriatric psychiatry or psychogeriatrics) is a well-defined specialty within psychiatry, based on knowledge in demography, epidemiology, psychopathology, and therapeutics. Just as child psychiatry is not adult psychiatry practiced on children, geropsychiatry is not just adult psychiatry practiced on older people. The application of research data to clinical practice is in its early stages, and the number of formally trained geropsychiatrists is far below that necessary to address the sizable mental health needs of a rapidly aging population.

## I. DEMOGRAPHY

Most gerontologists currently believe that a human being has an average life span of about 115 years. In the United States in 1900, the average life expectancy was 47 years, and 4% of the population at that time were over the age of 65. Now the average life expectancy has increased to 75 years (72 for men, 78 for women), and 12% of the population are elderly. It is expected that 50 years from now, the average life expectancy might be 78 years, with close to 20% of the population above the age of 65. The greatest percentage of increase is expected among those 75 years of age or older.

There are five widows for every one widower, i.e., most older women in the United States are widowed, whereas most older men are married. Twenty-one percent of the elderly population in this country also number among the poor. With dramatically reduced perinatal death and with acute illness being more susceptible to correction through appropriate treatment, chronic illness now accounts for most death in the United States. As life expectancy has increased, so has the prevalence of chronic illness. There are 24,000 nursing homes which provide 1.4 million beds, most of which are filled by the elderly. There are more long-term care beds than there are hospital beds for the acutely ill in the United States.

## II. EPIDEMIOLOGY

Fifty percent of Americans over the age of 65 have at least one chronic illness: 48% have arthritis, 39%

201

ENCYCLOPEDIA OF HUMAN BIOLOGY, Second Edition, VOLUME 1.   Copyright © 1997 by Academic Press.   All rights of reproduction in any form reserved.

have hypertension, 29% have hearing impairment, and 14% have cataracts. Of those living in nursing homes, close to 50% have a dementing illness, almost 30% have spent some time in a psychiatric hospital, and 9% have come directly to nursing homes from psychiatric hospitals. About half of all patients in mental hospitals, both public and private, are over the age of 65. With regard to those elderly living in the community, 15% are estimated to have a mental illness and 15% have some degree of organic mental disorder. Five percent have a severe dementia, which amounts to almost 3 million people.

About 5% of the elderly are totally home bound. At any one time 4% of the elderly are living in an institutional setting, yet for older people there is a 20% chance of institutionalization during the latter part of their lives. In recent years one-third of the total federal health budget is spent on the elderly, although they account for only 12% of the population. They also use 40% of all hospital bed days and purchase 25% of all prescription drugs.

There is a well-recognized underreporting of illness among the aged in the United States, perhaps due to the lingering myth held by both the elderly and their caregivers that to be old is to be sick. Old age free from disease is a time of feeling well and a time during which elderly people can perform usual daily functions. Older people become sick because of disease, not because of age. Underreporting might also be a result of fear that something bad will be found with one's health and, as a result, the older person will be moved from home to an unfamiliar institutional setting. Identification of a feared illness is perceived as comparable to a death sentence.

Traditionally, we have thought of "growth" and "development" as terms applicable to the early phase of life until a peak in young adulthood, followed by the decline of senescence. There is a major thrust in contemporary gerontology that attempts to put these terms back into the period following the peak of physical prowess. As more facts are accumulated about the continued creativity and acquisition of new skills during this latter phase of life, it will do a great deal to reverse the historical negativism associated with aging. [See Gerontology.]

## III. GEROPSYCHIATRY

As the body of knowledge of gerontology evolves, we can look at mental illnesses in late life and develop an intelligent rationale for therapeutics based on our understanding of normal, or usual, aging as well as on our own practical clinical experience. The geropsychiatrist most often is confronted by patients with organic mental disorders, mood disorders, paranoia, sleep disturbances, and anxiety.

## A. Organic Mental Disorders

### 1. Definitions

Although the central nervous system shows a progressive loss of nerve cells in certain cortical and subcortical regions of the brain, this condition is not necessarily correlated with the clinical manifestation of an organic mental disorder. Dementia is an age-related selective degeneration of brain cells (i.e., neurons) with a progressive loss of intellectual abilities (especially those higher-order functions measured by memory, judgment, abstract thinking, reasoning, and visual–spatial relationships), with little sensory or motor loss until late in the disease, all in the context of preserved alertness. [See Dementia in the Elderly.]

Primary dementia must be differentiated from delirium, which is a clouding of consciousness, with a decreased awareness of both external and internal environments and a decreased ability to sustain attention, manifested by disordered thinking and agitation. Dementia and delirium can also be differentiated from the most common aberration of cognitive function in the elderly: acute confusional states. These syndromes are characterized by the acute onset of an inability to maintain a coherent stream of thought, speech, or action. If delirium and confusional states are diagnosed early, and treatment is undertaken quickly, usually there is a return to some degree of normal function.

It is estimated that of the progressive irreversible dementias, those of the Alzheimer's type account for 70–75%. The next major cause of dementia is multiple small cerebral infarcts (i.e., strokes), accounting for 15–20%. The remaining fraction of organic mental disorders can be accounted for by such neurological problems as Parkinson's disease, Huntington's chorea, Creutzfeldt-Jakob encephalopathy, tumors, trauma, alcoholism, and acquired immunodeficiency syndrome (AIDS). Two additional forms of dementia are (1) frontal dementia in which memory is preserved but awareness, insight, and judgment are impaired, and (2) diffuse Lewy body disease in which the Lewy bodies, which are traditionally seen in the substantia nigra in Parkinson's disease, are seen throughout the cortex in this disorder.

## 2. Etiology

Research into the possible etiology of Alzheimer's disease has focused on autoimmune dysfunction, aluminum toxicity, slow viral infection, and genetic predisposition. Alzheimer's disease probably reflects neuronal degeneration at both the cortical and subcortical levels of the brain, involving nerve cells in the frontal cortex, the hippocampus, the basal nucleus of Meynert, and the locus coeruleus. Decreased levels of certain neurotransmitters (e.g., acetylcholine, nonadrenaline, serotonin, and somatostatin) in the cerebrospinal fluid have been reported. The abnormal paired helical filaments found in the neurons, called neurofibrillary tangles, have been characterized as insoluble protein polymers not found in normal brains. [See Alzheimer's Disease.]

The abnormal proteins of the paired helical filaments have caused the appearance of monoclonal antibodies in rabbit cerebrospinal fluid, which do not react with normal brain protein. This observation suggests either that a previously repressed gene has given rise to the abnormal protein or that new genetic material has been introduced into the cell by a viral or other infection.

Genetics also play an important role, as it is known that in 30–40% of Alzheimer's disease patients there is a positive family history. This role is emphasized by the similarity of the disease with Down's syndrome, which is caused by a genetic defect (i.e., three, rather than two, copies of chromosome 21). In fact, by the time people with Down's syndrome reach age 25, 80% have histopathological changes characteristic of Alzheimer's disease, and by age 50 virtually all Down's syndrome patients show these abnormalities, but not all such patients become clinically demented. It has been suggested that there are common structural and functional deficits in the neurons of people with Down's syndrome and those with senile dementia of the Alzheimer's type (SDAT). [See Down's Syndrome, Molecular Genetics.]

Some years ago there was a report of a high concentration of aluminum in the brains of Alzheimer's disease patients, suggesting heavy metal toxicity as a possible etiology. Also, there is aluminum in the antacid used in kidney dialysis fluid, and there is a well-documented "dialysis dementia" related to aluminum toxicity. However, in this disorder there are no observed neurofibrillary tangles. Regarding SDAT, it seems more likely that the aluminum concentrates in an already damaged neuron as a secondary event rather than as the cause of damage.

The immune system changes with aging, leading to an increase in autoimmune disorders. Titers of antibodies capable of reacting with the brain cells are higher in the sera of patients with Alzheimer's disease than in those of age-matched controls. It is not clear whether more antibody is produced or whether more antibody appears because of an altered blood–brain barrier. [See Autoimmune Disease.]

## 3. Diagnosis

A well-done history and physical exam yield a diagnostic accuracy of 85%. Persuasive arguments exist for discernible subgroups of dementia currently classified under the term "primary neuronal degenerative dementias." Work using brain electrical activity mapping has shown clear differences between demented patients with presenile or senile onset and normal people. A number of dementia-rating scales help differentiate Alzheimer's disease from other forms of cognitive impairment as well as provide a way of quantifying the impairment over time. The combination of analysis of brain tissue density by computerized tomography (CT) scan, using X-rays, with a quantitative measure of ventricular volume yields a diagnostic accuracy of 94% for SDAT.

Also of interest is the use of inhaled xenon gas with CT measurement of blood flow through the cerebral arteries, which often allows differentiation among normal age-related brain changes, multi-infarct dementia, and SDAT. The resolution of positron emission tomography is still not specific enough to diagnose SDAT. This procedure is expensive and its availability is limited to a few major medical centers. However, this treatment, used on SDAT patients, shows consistently decreased metabolism in the frontal and temperoparietal cortices. Single-photon emission computerized tomography (SPECT) uses a rotating gamma camera and radioactively labeled iodine to produce a three-dimensional map of cerebral blood flow, which shows that patients with SDAT have a decreased activity at the parietal lobe and at the temporoparietal junction. Magnetic resonance imaging (MRI) can detect tumors, abscesses, strokes, and multiple sclerosis. There are some reports of its use in demonstrating abnormalities in the brains of patients with SDAT, and it will likely be of increasing diagnostic value as technology is refined. [See Magnetic Resonance Imaging.]

The dementing illnesses, especially Alzheimer's disease, are devastating to millions of elderly victims and their families and to all levels of society because knowledge of their causes is primitive and the possibilities for therapeutic intervention are extremely limited.

### 4. Therapeutics

Careful analyses of treatment with hyperbaric oxygen, vitamins, antipsychotic drugs, central nervous system stimulants, choline precursors or cholinesterase inhibitors, and primary vasodilators show that there is no significant improvement in behavioral measures in progressive SDAT. Since the neurotransmitter acetylcholine is reduced in the brains of patients with SDAT and since it is closely associated with memory functions, efforts to increase its availability by adding choline or by slowing down its metabolic degradation seemed rational. Although increased levels of acetylcholine can be produced by these techniques, there has been no report of sustained clinical benefit.

## B. Mood Disorders

### 1. Epidemiology

The conventional wisdom that states that "with increasing age comes an increase in depression" is now being challenged. Although depressive neurosis shows little change over the life cycle, major depressive episodes show a decline with age. Research using a large sample and rigorous diagnostic criteria indicates that although the elderly population manifests more psychological and somatic symptoms of depression than do younger people, the incidence of major clinical depression is lower. [See Depression; Mood Disorders.]

For several reasons depression is likely to be underdiagnosed and undertreated. Some elderly, their families, and their caregivers still believe the myth that depression is an expected consequence of aging and therefore do not deal with it as a disease. Older people might not volunteer their feelings of depression unless carefully interviewed by their family physician. A "masked" depression is one in which denial, rationalization, and the presence of physical complaints might cover up the symptoms of depression. The concurrent existence of one or more medical disorders, as well as their pharmacological treatments, can complicate the picture. Many of those who are now elderly were forced by social pressure not to admit mental illness or seek psychiatric help. Clinicians have long been

aware that although the full-blown presentation of major depression declines with age, the presentation of certain depressive symptoms is increasingly present among the old–old and among those elderly with concomittant physical disease and disability. The predominant presenting symptoms are a loss of the zest for life and withdrawal from people and activities. Various labels have been applied to this condition, including "depletion syndrome of the elderly" and "amotivational syndrome." Data suggest that such patients seem to have a rather rapid and sustained response to low doses of methylphenidate (Ritalin).

### 2. Etiology

Attempts at understanding the biological basis of depression have focused on genetic determinants, neurotransmitter abnormalities, and endocrinopathy.

Genetic factors seem to be prominent among a subgroup of patients with depression that occurs early in life. Genetics is clearly involved with bipolar disorders (e.g., alternating manic–depressive) and not quite so clearly involved with unipolar disorders (e.g., simple depression).

The transmission of signals among neurons through chemical neurotransmitters is altered: brain norepinephrine and serotonin are depleted, $\alpha$-adrenergic receptor activity is decreased and $\beta$-receptor responsiveness is increased. The opposite changes in activity of the two kinds of adrenergic receptors seem to occur normally with aging. These differences could explain the finding that major depression shows a decrease with aging. [See Depression: Neurotransmitter and Neuropeptide Receptors.]

The term "pseudodementia" has been used for over 25 years and describes a group of patients with functional disorders, usually depression, who present with cognitive impairment. This condition occurs in about 10% of the elderly with a major depression. This group can also include malingerers, those who refuse to relate to the examiner and who are intensely preoccupied with other thoughts. The intellectual decline seen in certain depressed elderly people might represent a defensive response to what they perceive to be a threat. Patients with pseudodementia perform at a much higher level on neuropsychological examination than would be expected based on their subjective experience and complaints. Their cognitive impairment tends to plateau, is nonprogressive, and has a rapid onset. There is often a past history or family history of depressive illness. There are clinicians who believe all elderly with an early dementia deserve a trial of psychotherapy and antidepressants to treat a potentially reversible process.

It has long been observed that some patients with endocrinopathies have an associated depression. Over one-half of elderly depressed patients fail to suppress the production of their own cortisol (a steroid hormone) when they are given an artificial steroid (i.e., dexamethasone); this is called the dexamethasone suppression test. Patients who do not suppress cortisol during this test and who also show decreased time from onset of sleep to rapid eye movement sleep (i.e., REM latency) seem more likely to suffer from biogenic depression (i.e., a depression with a clear biological etiology) than those who exhibit only one of these signs.

Data gathered by the National Center for Health Statistics compared the suicide base rates for various age groups in 1970, 1975, and 1980. The rates plateau in midlife for women and for nonwhite males. The prevalence of suicide is highest for white males; there has been a progressive decline over the last 10 years for white males between the ages of 65 and 74, while there has been a comparable increase for those older than 85. All suicides are not necessarily attributable to depression, but this condition could increase the probability of suicide for those that are old, poor, sick, and who have suffered a major loss. When an older person has a major depressive episode, personality changes might occur. However, the personality usually returns to premorbid structure as the mood disorder improves. Dementing illness might also present as depression, with little evidence of cognitive impairment. A diagnostic approach is needed to differentiate accurately and consistently between dementia and depression. [*See* Suicide.]

## C. Paranoia

Many of today's elderly are of foreign birth, English being their second language, who may have experienced severe hardship and deprivation earlier in their lives. There is significant poverty and multiple chronic illness among this population. All of these factors can predispose for suspicion. There are other substantial grounds for suspicion in older people, including multiple compounded losses, decreased sensory functions, shrinking life space, exclusion, neglect, and rejection from many parts of society. In addition, with an extension of life expectancy, more dementing illness appears, with paranoid ideation in over 50% of cases.

The continuum of suspicion/paranoia significantly limits the ability of these elderly people to maintain their independence, to have access to needed services, and to function effectively in the community. Most caregiving agencies, as well as families, would agree that persons with paranoid thinking and behavior are among the most difficult to work with, precisely because of their distrust and noncompliance. Without accurate identification and early intervention, the behavior can worsen and lead to major health and mental health crises. The frail elderly are often evicted, involuntarily hospitalized, and placed on antipsychotic medications, which, despite their benefits, can cause serious irreversible neuromuscular dysfunction. There is great cost to everyone in both human and economic terms.

Paranoia is a troublesome and not uncommon symptom of older people associated with a variety of disorders, including delirium, dementia, depression, mania, personality disorders, hearing loss, and schizophrenia as well as delusional disorders. An acute paranoid reaction can be either drug induced or part of the presentation of an underlying metabolic, infectious, or traumatic disorder. It can also be a symptom of major depression. All of these disorders deserve active therapeutic intervention and offer considerable hope of reversibility. Paranoia can persist as a symptom from disorders occurring earlier in life or can appear for the first time late in life.

Paranoid fears can lead to social isolation and help-rejecting behaviors, which can result in poor nutrition, untreated medical problems, and health and safety hazards. This, in turn, can hasten an otherwise avoidable functional decline in those elderly who live in the community. When referred for medical and/or psychiatric care, assessment and intervention can be extremely difficult. The paranoid person may refuse to visit a doctor, to allow entry into their home, or to accept needed treatment and support services.

Paranoia among the elderly is an insidious and difficult problem to address and treat. Early intervention can help prevent the functional decline and incidence of other problems that occur when paranoia is left untreated.

## D. Sleep Disorders

Elderly patients often complain about sleep problems and might seek help for them. Sleep disturbance is frequently one symptom of depression and other psychopathology. [*See* Sleep Disorders.]

There are two basic forms of sleep: sleep during which there are rapid eye movements (REM), and sleep without REM. There are also stages of sleep from light (stages 1 and 2) to deep (stages 3 and 4).

Stages 3 and 4 are characterized by the appearance of large slow $\delta$ waves in an electroencephalogram. Dreams occur during REM sleep. As we age, $\delta$ waves diminish and there is less deep sleep and less dream time. The elderly also have more waking periods and a shorter REM latency (i.e., the length of time from onset of sleep to REM sleep). They take longer to adjust to sleep–wake schedule changes. Frequent waking can be due to urinary urgency, nocturnal myoclonus (i.e., involuntary muscle jerks), or sleep-related breathing disturbances.

The elderly report considerable changes in sleep patterns: more time to fall asleep, fewer hours of sleep at night, more frequent awakening during the night, and more frequent daytime napping. They also report feeling more tired during the day and spending more total time in bed, and they often have a general dissatisfaction with the quality of their sleep.

Many elderly, especially in nursing homes, go to bed after dinner and can be asleep by 7:00 or 8:00 PM. Waking at 4:00 AM is therefore not necessarily a symptom of depression. Once an accurate record has been obtained of a patient's sleep pattern over several weeks, a rational care plan can be evolved. This might include the use of sedative–hypnotic medication, in which case the short-acting benzodiazepines have been widely prescribed. There is an appropriate concern regarding the use of benzodiazepines in geropsychiatry. They may accumulate rapidly to toxic levels, they can cause postural hypotension (a fall in blood pressure when one stands up), they may impair cognitive function, and they are difficult to withdraw from and may even be addicting.

### E. Anxiety

Anxiety is a common symptom in older people, but an uncommon syndrome. It seems reasonable to question whether elderly people manifest anxiety in the same way and with the same frequency as do younger patients. There is little consensus among experts as to whether there is a "classic" general clinical presentation of anxiety in older patients. As seen in clinical practice, many patients manifest a wide range of symptoms, simultaneously or sequentially.

Much has been learned about neurochemical and neurophysiological mechanisms in anxiety disorders. Many new antianxiety (i.e., anxiolytic) agents have been identified and marketed, and important epidemiological information has been obtained. However, there are difficulties with differential diagnosis, with the persistence of symptoms despite intervention, and with the difficulty in withdrawing medication.

Anxiety can be defined as a subjective state of internal discomfort, dread, and foreboding, accompanied by autonomic nervous system arousal. Different from fear, anxiety tends to occur without a conscious or apparent stimulus. The physical symptoms include excessive breathing, palpitation, sweating, diarrhea, trembling, dizziness, headache, restlessness, and muscle aches. Certain cognitive changes are also associated with anxious states: impaired attention, poor concentration, and memory problems. For elderly patients, accurate diagnosis is complicated by a similar presentation of common geriatric medical illnesses. Many patients with anxiety disorders are not seen in psychiatric practice because they seek treatment from their internist or family physician, especially when there is somatization of their anxiety (e.g., chest pain, headache, or fatigue).

One difficulty in the identification of anxiety in older people is the tendency for some elderly to attribute agitation, fears, or aches and pains to aging per se, therefore denying and underreporting such symptoms. Another difficulty relates to the altered presentation of certain diseases in older people due to anatomic and physiological changes that occur in usual aging. A reduction in the physical signs and symptoms of anxiety therefore could reflect the decrease in autonomic nervous system (i.e., sympathetic) activity associated with biological aging.

Little is known about the effects of age per se on anxiety disorders. A small number of older persons might be functionally impaired by anxiety for the first time late in life or, more likely, certain experiences late in life might activate or exaggerate a preexisting anxiety disorder. Panic can begin in the 20s and continue into old age as somatic complaints or hypochondriasis or mixed with depression. Being 85 is clearly different from being 35, not only because of 50 years of biological change, but also because of 50 years of experience with life.

Traditionally, theory has held that anxiety in older people can appear as a response to a compounding of losses, increasing dependency, loneliness and fear of isolation, increasing health problems, declining vigor, diminished sensory and functional capacities, changes in economic or social status, feelings of uselessness, one's awareness of cognitive impairment early in dementia, and the approach of dying and death. No studies have been done to substantiate that any of these variables are causally related to the more serious and persistent conditions described earlier as anxiety disorders.

The subcategories of anxiety disorders can overlap with each other and with depression. Low morale, life dissatisfaction, and feelings of helpless dependency are critical components of the psychiatric picture of older people presenting with mixed anxiety and depression. This dual presentation has been frequently observed, particularly in the demented elderly.

When anxiety is related to some clear environmental stress, a strong interpersonal supportive relationship, with identification and working through of the issues, might lead to symptom reduction. When this is inadequate, the use of antianxiety agents is called for.

Alcohol is the most widely self-prescribed drug for anxiety reduction, but it is contraindicated because it produces sleep disturbance and behavioral disorders and might cause unwanted side effects in combination with other medications. In larger amounts it produces habituation, malnutrition, and damage to the central and peripheral nervous system.

Barbiturates and bromides are no longer used because their action is primarily that of sedation based on central nervous system suppression, which carries a high risk of toxicity and addiction. The benzodiazepines constitute the most often prescribed group of anxiolytic agents (10% of the U.S. population has had them prescribed). They can be subdivided based on how long they remain active in the body. Those with a long half-life tend to accumulate and can lead to toxicity in elderly patients. In geropsychiatry, therefore, we usually tend to prescribe drugs with a shorter half-life.

As is true in many other areas within psychiatry, we are beginning to learn more about genetic/biological determinants as well as the importance of life experience in the etiology of anxiety. Although much of what has been studied in younger adults can be stretched to an imperfect fit with elderly patients, little research has been directed at the impact of aging on the course of these disorders. Not only are there problems inherent in following a large sample of anxious adults into their later years in a longitudinal fashion, but the diagnostic criteria themselves undergo continual modification as the result of our evolving body of knowledge.

## F. Service Delivery

The elderly present multiple problems, requiring a comprehensive approach combining mental health, medical, and social services as part of a well-managed human service system. The full range of services is often referred to as the continuum of care. Unfortunately, while some of the services are available in some communities, the full range is rarely available in any one community. The services needed might be provided in the public sector at the national, state, or local level of government, each with its own set of regulations and eligibility requirements. There are over 100 federal departments involved with some aspect of life necessities for the older American (e.g., medical, housing, social, and financial). There is also a multitude of private agencies and programs that address the specific needs of the elderly. This complex maze befuddles the well-educated young person and becomes an enigmatic morass for many elderly who have not grown up in the system and might not have the verbal skills and education necessary to find their way through this complicated matrix. Therefore, even though a service might exist on paper, access to that service is limited in part by a lack of education of the elderly consumer, as well as absence in many parts of the country of good case management.

Gerontological research in the United States continues to refute the notion that families abandon their elders and indicates that when there are families they generally maintain high levels of involvement and caring. A number of factors in American society, however, affect a family's ability to care for its aged members. Increasing numbers of the very old mean that there are more people in need of greater assistance. The declining birth rate leaves fewer young people to support this increasing portion of the population. Traditionally, middle-aged women have assumed responsibility for the care of the aged, but their participation in the work force is increasing, leaving less time and energy for the caretaking role. The very old might have elderly children themselves who are coping with their own problems of late life. The tasks required in caring for an old person can be emotionally taxing and technically difficult. Difficulties can arise from the nature of the task itself (e.g., helping to use the toilet, which evokes discomfort). Negative feelings can be triggered by the change in roles between parent and child, caused by the elder's increased dependency.

## G. Abuse, Sexuality, and Ethics

In addition to the current emphasis on dementing illnesses, there are many other areas of concern to geropsychiatrists and gerontologists, including elder abuse, sexuality, and ethics. Recognition and acknowledgment of the problem of elder abuse are relatively recent, so information about this behavior is scarce. Abuse can be divided into four types: physical, psychological, financial, and neglect. Dependent elders, particularly those with cognitive and mobility impairments, are at higher risk for abuse, as are those under the care of people who themselves abuse alcohol or other drugs. Research indicates that adults who abuse their children were more likely to have been abused by their parents; however, no such correlation seems to exist with regard to abuse of one's elders.

American society has tended to view the aged as being sexually incapacitated or disinterested. For some unclear reason we have been uncomfortable with any overt expression of an active interest in sexual functions by an older person. It is another example of how the elderly have been victimized by what has been termed "ageism." Few well-done studies have been performed to determine sexual attitudes among the elderly and caregivers alike with regard to sexual behavior. Data from these studies indicate that the elderly consider themselves and were considered by others to be below average in desire, capacity, frequency, and social opportunity for sexual activity. There is also some degree of compliance with social expectations on the part of the elderly. It is hoped that as our acceptance of sexuality in late life increases, we can expect special aspects of sexuality in old age to become generally understood and the diagnosis and treatment of sexual problems to be refined to a much greater degree.

Geriatrics and geropsychiatry present the physician with a variety of value conflicts and ethical dilemmas. Conflict exists between traditional medical training, and the patients' rights to self-determination. The right of the patient to participate in the decisions that initiate or withhold treatment is complicated by the patient's competency and by the physician's understanding of the life course of that illness and the potential reversibility of it with proper intervention.

Other ethical dilemmas exist with regard to the distribution of scarce resources (e.g., the use of renal dialysis in young versus older people). There are also times when the wishes of the patient or the best judgments of other health workers might conflict directly with the judgment of the physician.

Competency seems to be at the heart of many ethical issues, yet neither medicine nor the courts have arrived at a standard definition for determining competency. The right to accept or refuse treatment, the right to privacy, and the right to die are all subjects of intense discussion, which is important for protecting the quality of life of our elder citizens.

Although all patients are at risk of having their independence and autonomy compromised, the elderly are particularly vulnerable. This generation of old people has been socialized to passivity in the health care system. They are often unaware of their rights, are afraid of losing their health care if they "make waves," and carry a sense of general disenfranchisement. In addition, the aged frequently do not have adequate physical, financial, or psychological access to or choice among the needed services.

The informed consent doctrine requires that a patient's consent to treatment be voluntary, informed, and competent. Voluntariness implies that consent be given without coercion, even implicit. Informedness indicates that a full explanation of all benefits and potential risks of treatment and the consequence of no treatment be given to the patient. There is general agreement among ethicists and clinicians that competency implies that the patient is capable of making a reasoned decision integrating and weighting all of the presented information.

There is an energetic thrust in American medical education which underscores the correctness and importance of active therapeutic intervention in geriatrics and geropsychiatry. It has been a disservice to elderly patients and their families to have approached this field in the past with therapeutic nihilism or, at best, with an attitude that supported humane but custodial care. Elderly patients have the right to expect that their primary-care physicians, be they internists or psychiatrists, have current knowledge of the illnesses of late life and their interventions so that these patients are assured the highest quality of care available.

## BIBLIOGRAPHY

Fitten, L. J., Morley, J. E., Gross, P. L., Petry, S. D., and Cole, K. D. (1989). Depression (UCLA geriatric grand rounds). *J. Am. Geriatr. Soc.* 37, 459–472.

Gurian, B., and Auerbach, S. (1984). Psychiatric syndromes of old age. *In* "Clinical Neurology of Aging" (M. Albert, ed.). Oxford Univ. Press, New York.

Gurian, B., and Miner, J. (1990). Clinical presentation of anxiety in the elderly. *In* "Anxiety in the Elderly," (C. F. Salzman, and B. Lebowitz, eds.). Springer-Verlag, New York.

Gurian, B., and Rosowsky, E. (1993). Methylphenidate treatment of minor depression in very old patients. *Am. J. Geriat. Psychiat.* 1, 171–174.

Martin, R. L. (1989). Update on dementia of the Alzheimer type. *Hosp. Community Psychiat.* 40, 593–604.

Walker, J. I., and Brodie, H. K. (1985). Paranoid disorders. *In* "Comprehensive Textbook of Psychiatry" (H. I. Kaplan and B. J. Sadock, eds.), Vol. IV. Williams & Wilkins, Baltimore, MD.

# Agronomy

THOMAS A. LUMPKIN
LARRY E. SCHRADER
*Washington State University*

## GLOSSARY

**Agronomy** Field of agriculture dealing with the development and practical application of crop and soil science to produce abundant, high quality food, feed, and fiber crops

**Biotechnology** Application of biological organisms (i.e., microorganisms, plants, animals), systems, or processes to provide desirable goods and services. Agricultural biotechnology is based on the application of advanced concepts and techniques of biological science such as recombinant DNA, genetic engineering, some enzyme processes, plant cell and tissue culture, cell fusion, clonal propagation, monoclonal or polyclonal antibodies, embryo and other germ cell manipulations, and process or system engineering (i.e., fermentations that are relevant to specific agricultural programs)

**Chromosome** Small rod-shaped strands of DNA in the nucleus of a cell that contain genetic information for that cell

**Cultivar** Cultivated variety. The term cultivar can refer to any one of several entities, often depending on the method of propagation employed for the particular crop. With crops that are generally asexually propagated (e.g., potato and various fruits) the term refers to a particular clone. In self-pollinated plants (e.g., wheat or tomato) the term cultivar usually means a particular inbred or pure line that breeds true naturally. In cross-pollinated crops (e.g., alfalfa), the term usually refers to a population of plants distinguished on some morphological or physiological basis and maintained by selection and isolation. The term hybrid cultivar may refer to a particular combination of inbred lines (e.g., corn hybrids)

**DNA** Deoxyribonucleic acid is a double helix composed of strands of bases: adenine, guanine, cytosine, and thymine; referred to as the language of life

**Eutrophication** Process by which a body of water becomes rich in nutrients; may occur naturally or may result from agricultural run off and industrial or municipal wastewater flowing into streams, rivers, and lakes

**Forages** Plant materials consumed by livestock; the most common forage crops are grasses and legumes

**Gene** Unit of information on chromosomes that specifies the composition of a protein

**Herbicide** Chemicals used to kill plants, usually weeds; commonly called weed killers

**Hybrid** Product of cross-pollination between genetically different plants

**Hybridization** Process of cross-pollinating genetically different plants to obtain a hybrid

**Legumes** Family of plants that in association with bacteria in the genus *Rhizobium* participate in biological nitrogen fixation. Common legumes include alfalfa, peas, beans, and soybeans

**Macronutrients** Essential elements required in relatively large amounts by all higher plants

**Micronutrients** Essential elements required in relatively small amounts by all plants

**Pedology** Branch of soil science concerned with the morphology, genesis, classification, and geography of soils and their relation to landscapes in the field

**Recombinant DNA** Process by which specific segments of DNA—genes—are excised from a complex DNA molecule of one organism and inserted or recombined with the DNA of another organism; sometimes referred to as gene splicing

ENCYCLOPEDIA OF HUMAN BIOLOGY, Second Edition, VOLUME I.   Copyright © 1997 by Academic Press.   All rights of reproduction in any form reserved.

AGRONOMY IS THE DEVELOPMENT AND PRACTICAL application of crop and soil science to produce abundant, high quality food, feed, and fiber crops. It is a broad field of study that in a sense embraces the whole of plant and soil biology. Agronomy encompasses numerous basic and applied biological and physical sciences that are related to more efficient production of food and other consumer products and to protection of the environment.

## I. BROADER DEFINITION OF AGRONOMY

Agronomy, from the Greek words *agros* (field) and *nomos* (to manage), is a broad field of study that includes both crop and soil sciences. It is based on knowledge from many fundamental sciences and supplies the principles for soil and crop management, protection of the environment, and development of new crops and more efficient crop varieties that will provide abundant, high quality food, feed, and fiber crops. The means by which minerals, carbon dioxide, water, and sunlight are converted into useful products are studied by agronomists. Land resources must be carefully managed to maintain and enhance the productivity of areas suitable for crops and for pasture and grazing, as well as to ensure effective use of rainfall so water needs can be met.

Crop science relates primarily to genetics, breeding, physiology, biochemistry, production, management, and molecular biology of field and turf crops. It also deals with the production of quality seeds, improvements in the nutritional value of crops, and the effects of environmental changes on crop and food production.

Soil science includes soil physics, soil chemistry, pedology/genesis, soil microbiology and biochemistry, soil mineralogy, soil fertility, and soil and water management as they apply to the growth of plants. In addition, protection of the environment (e.g., air and water quality, soil conservation), foundations for buildings and highways, reclamation of mined soils and those disturbed by military use, land use and urban planning, recreation, and waste disposal are now integral to the discipline.

## II. BRIEF HISTORY OF AGRONOMY

The scientific study of crop improvement and management is a recent phenomena, although many of the important field crops grown today, such as wheat, barley, rice, peas, and lentils, were domesticated over 9000 years ago (Figs. 1 and 2). The written record of agriculture begins with Egyptian hieroglyphs dating before 3400 BC. The Greeks and Phoenicians were the first to describe agriculture in detail. Historical works of this period which describe agriculture include Hesoid's "Works and Days" (800 BC), Herodotus (ca. 420–484 BC), Xenophon's "Oeconomicus" (431–352 BC), and writings by Aristotle's student Theophrastus (300 BC). These histories were followed by Roman writings such as Cato the Censor's "De Re Rustica" (234–149 BC), Varro's "De Re Rustica" (37 BC), Virgil's "Georgics" (30 BC), Pliny (50 AD), and Columella's "De Re Rustica and De Arboribus" (first century AD).

Chinese agricultural books have a separate origin which dates from the Zhou (1030–221 BC) and Han (202 BC–220 AD) dynasties when treatises such as "Shen Nong Shu" and "Fan Sheng Chi Shu" were

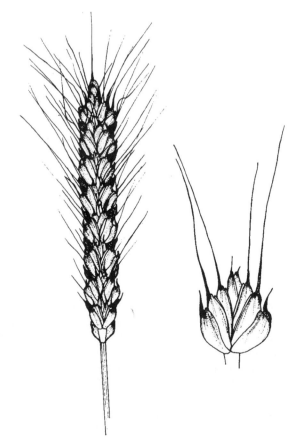

**FIGURE I** Wheat (*Triticum aestivum*). (Reprinted with permission of Cambridge University Press.)

FIGURE 2    Pea (*Pisum sativum*). (Reprinted with permission of Cambridge University Press.)

written, respectively. These books did not survive but are quoted in important later works such as "Chi Min Yao Shu" by Jia Sixie (535 AD). In contrast to European writings, the Chinese have maintained a nearly continuous record of agricultural publication ever since its commencement during the Zhou dynasty and have discussed many more crops.

These European and Chinese descriptions of plants and management practices had a great impact on food production. Successful plants or practices developed and described in one location were read about and implemented elsewhere. For example, Romans recognized the value and promoted the use of legumes in crop rotations to improve soil fertility, although they did not realize that symbiotic nitrogen fixation was the primary source of this benefit.

Following the Greco-Roman period, European agricultural literature remained nearly static for over 1000 years. European writings on agronomy consisted of little more than copying the works of the early Latin historians. The first evidence of new and critical thinking by a European agronomist came with the writing of "Opus Ruralium Commodorum" by Pietro de Crescenzi in 1304, which reviewed the Latin authors and added observations on farming practices in Italy, especially the causes of soil erosion and its

control. This book was considered so important that it became the first printed agricultural book in 1471, and the printing process became one of many emerging stimuli to agricultural evolution and thought.

The Arab invasion of southern Europe and the infusion of new ideas, crops, and techniques that ensued brought significant evolution. One of the great books emerging from this period was the 'Kitab al Filāhah" (12th century) by Ibn al-'Awwān of Seville, which includes extensive discussion of irrigation and exotic crops such as citrus, sugarcane, and rice. The voyage of Columbus and the era of exploration provided another major impetus to European agriculture, namely exchange of new crops and management practices between the New and Old World, between Europe and Asia—a classical approach to agricultural innovation.

One of the earliest fruits of the resulting cross-cultural pollination of thought came from Jethro Tull (1674–1741), an Oxford University graduate. He published the book "Horse Hoeing Husbandry" in 1733 which proposed an integrated system of row-cultivation for dryland cereals, the basis for the highly mechanized agricultural systems used in the West today. Many of the ideas in his book, including the seed drill and ideas of other Europeans for the curved iron moldboard plow and hoe cultivators, probably came from Asia, especially the north China plain where these implements and practices had been in use for over 1000 years. A major source for the new ideas about agriculture which circulated in Europe during this period was the 46-volume "Annals of Agriculture" published by Arthur Young (1741–1820).

Modern agriculture has made dramatic improvements in food production through a new approach, i.e., scientific research. Scientific advances in botany, chemistry, and physics, with their eventual application to agriculture, provided the greatest impetus for the emergence of agronomy as a science. Botanical discoveries and inventions by pioneering researchers such as Robert Hooke, Antoni Van Leeuwenhoek, Rudolph Jacob Camerarius, Abbot Gregor Mendel, Carl von Linne, Stephen Hales, Joseph Priestley, and Louis Pasteur created the technical foundation for the initiation of plant breeding, crop physiology, and plant pathology.

The first agricultural experiment station is thought to have been started in the Alsace region of France by J. B. Boussingault in 1834. This development was followed by the establishment of the famous agricultural research station at Rothamsted in England by John Bennet Lawes and Joseph Henry Gilbert in 1843.

After 1870, agricultural research in the United States began to flourish as a result of the establishment of the land grant agricultural colleges.

The American Society of Agronomy was established in 1908. Agronomy was first recognized as a distinct science in the United States with the formation of the Department of Agronomy at the University of Illinois in Urbana–Champaign in 1899. The faculty were dedicated to establishing a system of permanency in which superior crops were necessary to use fertile soils to their fullest advantage and to better the social and economic position of the owner and tiller of the soil. Farmers were viewed as stewards of the soil, and it was their duty to pass on to posterity land that was richer, not poorer, than when they took over its management. Early emphasis was on solving applied problems of crop production, preparing a comprehensive survey of Illinois soils and chemical analyses of different kinds of soils, and establishing experimental fields at numerous locations throughout the state where new research findings could be demonstrated to farmers. Similar patterns of development have occurred at the other land grant agricultural colleges.

The next major organizational advance in agronomic research came with the establishment of international agricultural centers by the Ford and Rockefeller Foundations. The International Rice Research Institute in the Philippines was established first in 1960, and was followed in 1966 by the Centro Internacional de Mejoramiento de Maiz y Trigo in Mexico. As the number and scope of international institutes began to increase, the World Bank helped establish the Consultative Group on International Agricultural Research (CGIAR) in 1971.

CGIAR is an informal association of 42 donors from public and private sectors that currently supports and advises 18 international agricultural research centers. The research programs at the centers fit into six major categories: productivity research, management of natural resources, improving the policy environment, institution building, germplasm conservation, and building linkages. CGIAR also supports centers designated to meet training and research needs in forestry, living aquatic resources, and livestock management. The following descriptions are of centers that are active in crop improvement.

Centro Internacional de Agricultura Tropical (CIAT), Columbia: Supports research in germplasm development in beans, cassava, tropical forages, and rice for Latin America as well as in resource management in humid agroecosystems in tropical America.

Centro Internacional de Mejoramiento de Maiz y Trigo (CIMMYT), Mexico: Supports research to increase the production of maize and wheat in developing countries.

Centro Internacional de la Papa (CIP), Peru: Supports research programs on potato and sweet potato as well as providing worldwide collaborative training aimed at helping scientists meet the changing demands in agriculture.

International Center for Agricultural Research in the Dry Areas (ICARDA), Syria: Committed to improving the productivity of winter rainfed agricultural systems in harsh environments with attention to soil degradation and water use efficiency.

International Crops Research Institute for the Semi-Arid Tropics (ICRISAT), India: Supports research for crops in the semi-arid tropics including sorghum, finger millet, pearl millet, chickpea, pigeonpea, and groundnut.

International Food Policy Research Institute (IFPRI), Washington, D.C.: Dedicated to improvement of policies governing production and land use, consumption and income levels of the poor, links between agriculture and other economic sectors, and trade conditions.

International Irrigation Management Institute (IIMI), Sri Lanka: Geared toward development, dissemination, and adoption of lasting improvements in irrigated agriculture in developing countries.

International Institute of Tropical Agriculture (IITA), Nigeria: Supports research on food production in the humid and subhumid tropics, specifically on maize, cassava, cowpea, plantain, soybean, and yam.

International Network for the Improvement of Banana and Plantain (INIBAP), France: Conducts research to increase production and stability of banana and plantain grown on smallholdings. Concentrates specifically on the exchange of information and disease-free genetic material and training for scientists and technicians from developing countries.

International Plant Genetic Resources Institute (IPGRI), Italy: Focuses on the conservation and use of plant genetic resources while providing training and information, especially for developing countries.

International Rice Research Institute (IRRI), Philippines: Founded to improve the well-being of rice farmers and consumers through the generation and dissemination of knowledge, technology, and enhanced research.

International Service for National Agricultural Research (ISNAR), The Netherlands: Dedicated to the improvement of the national agricultural research systems in developing countries by supporting and promoting institutional development, funding policies, and improved research techniques.

West Africa Rice Development Association (WARDA), Côte d'Ivoire: Conducts research to improve rice varieties, production, and processing and to increase the options available to smallholder farm families in the upland/inland swamp continuum, the Sahel, and the mangrove swamp environments of Africa.

## A. Focus of Agronomic Research

Primary research focus during the early years in crop science was placed on the domestication of crop plants and the development of new crop cultivars with high yield potential, disease resistance, and other desired agronomic traits. Virtually all agronomic crops have been derived from wild species. Cultivated plants as they are known today have undergone extensive modification from their wild progenitors due to continual improvement efforts. Selection for improved growth habit, adaptation to environment, and fruiting characteristics have changed some species (e.g., corn or maize) so much that wild ancestors have become obscure. However, some plants such as hickory, persimmon, Brazil nut, black walnut, and various medicinal plants are the same as they were originally found in nature.

Although humans have attempted to improve plants by selection from earliest times, the application of genetics led to dramatic changes in the gene pool of certain crop plants. Plant breeding flourished during the early 1900s after the rediscovery in 1900 of Gregor Mendel's classic genetic discoveries. Plant breeding has had a great impact on the increased productivity of present-day agriculture. The most vivid example of genetic improvement through hybridization is corn (Fig. 3). In about 1930, commercial corn hybrids became available; by 1945, about 90% of the acreage was planted to hybrids. Today, hybrids are used on virtually all the corn acreage in the United States. Corn yields have progressively increased during the past 60 years because of genetic improvement and use of better cultural and agronomic practices. On average, U.S. corn yields have increased about 100 kg/ha annually (about 2 bushels/acre). For years, soybeans were largely used as a forage crop, but research and extension efforts resulting in new, im-

**FIGURE 3** Maize (*Zea mays*). (Reprinted with permission of Cambridge University Press.)

proved cultivars with higher grain yields allowed farmers to rapidly increase the acreages planted to soybeans (Fig. 4). Soybeans were recognized as a rich source of both protein and oil. The protein is used for animal feeds and flours, and the oil is used for

**FIGURE 4** Soybean (*Glycine max*). (Reprinted with permission of Cambridge University Press.)

edible products—mainly shortenings, margarine, and salad oils—and for industrial applications such as in paints, plastics, inks, varnishes, and linoleum. Large quantities of U.S.-grown soybeans are exported to other countries each year.

Although plant breeding is still important for crop improvement, the latest techniques in molecular biology and molecular genetics are being employed in crop improvement today. Recombinant DNA technology has permitted the transfer of a human gene into a plant, after which the plant produced a human hormone. A genetically engineered soil bacterium may help prevent frost damage on fruits and vegetables. Herbicide-resistant crop plants are being developed, and crop plants are being made resistant to certain pests by introducing agents that lead to the formation of naturally occurring substances that provide biological control (i.e., chemical pesticides are replaced by natural compounds synthesized by the plant itself).

Not until the 1950s did chemical weed control become a major management factor in crop production. Annual losses to weeds are estimated to reduce crop productivity in the United States by 10%, amounting to a loss of $12 billion. Farmers spend about $3.6 billion for chemical weed control and about $2.5 billion for cultural, ecological, and biological methods of weed control. Thus, total losses caused by weeds and the cost of their control are estimated at more than $18 billion in the United States annually.

In the early years, soil scientists were concerned with the classification of soils through soil surveys. One of the most useful end products of the soil survey is the soil map. Traditionally, the soil was classified according to origin to obtain meaningful mapping units. Soil testing became important for recommending appropriate amounts of fertilizer and other soil amendments such as lime to correct soil acidity. Soil management is an area in which fundamental discoveries of soil science are applied with the aim of maximizing agricultural crop yields. More recently, soil scientists have played a major role in the development of techniques to protect the environment and/or to reclaim soils that have been damaged by humans. Soil erosion has been studied and techniques have been devised to minimize it and to conserve water.

## III. CROP SCIENCES

A major objective of crop science is crop and food production. The challenge of feeding the world's populace is indeed a major responsibility.

## A. Food, Feed, Fiber, and Natural Product Production

Crop plants include those used directly for food, feed, or fiber, such as cereal grains, soybeans, and citrus; those converted biologically to products of use, such as forage plants, hops, and mulberry; and those used for beverages, medicinals, or special products, such as digitalis, opium poppy, coffee, and cinnamon (Fig. 5). Plant products such as crambe and rubber are used in industries in which synthetic products are unsatisfactory.

### 1. Leading World Food Crops

Plant products constitute about 90% of the human diet, with about 30 crop species providing most of the world's calories and protein. Although the number of crops in use worldwide is in the thousands, only eight food crops (maize or corn, wheat, rice, barley, potato, sweet potato, cassava, and soybeans) are produced in amounts exceeding 100 million metric tons/year (Table I). Eight species of cereals collectively account for 52% of the world food supply with wheat, rice, and corn constituting about 45% of that. Total world production of cereals increased 15% between 1982 and 1992. The proportion of the world's cereal production provided by North America declined from 23% in 1982 to 22% in 1992.

### 2. Food From Animal Products

Animal products, constituting 7% of the world's diet, come indirectly from plants. Animals consume forages

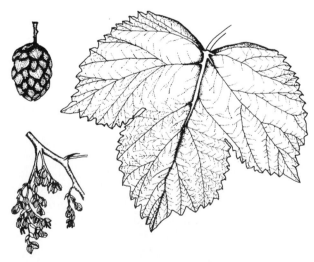

**FIGURE 5** Hop (*Humulus lupulus*). (Reprinted with permission of Cambridge University Press.)

## TABLE I

World Food Production for 1982, 1987, and 1992 (in Millions of Metric Tons)[a]

| Commodity | Gross production | | | % increase 1982–1992 | % of total in North America | | |
|---|---|---|---|---|---|---|---|
| | 1982 | 1987 | 1992 | | 1982 | 1987 | 1992 |
| Total cereals | 1710 | 1787 | 1961 | 15 | 23 | 19 | 22 |
| Wheat | 483 | 517 | 565 | 17 | 21 | 16 | 18 |
| Rice, paddy | 424 | 454 | 528 | 25 | 2 | 1 | 2 |
| Maize (corn) | 450 | 457 | 528 | 17 | 48 | 44 | 50 |
| Barley | 164 | 179 | 165 | 1 | 15 | 14 | 13 |
| Sorghum | 68 | 59 | 70 | 3 | 37 | 41 | 41 |
| Oats | 45 | 47 | 34 | (24) | 27 | 18 | 21 |
| Roots and tubers | 557 | 594 | 589 | 6 | 4 | 4 | 5 |
| Potatoes | 266 | 285 | 277 | 4 | 7 | 7 | 9 |
| Sweet potatoes | 106 | 135 | 124 | 17 | 1 | 1 | 1 |
| Cassava | 126 | 137 | 153 | 21 | <1 | <1 | 1 |
| Total pulses (legumes) | 45 | 53 | 54 | 20 | 4 | 4 | 7 |
| Fruits[b] | 316 | 324 | 377 | 19 | 8 | 8 | 13 |
| Oilseeds and nuts[c] | 181 | 204 | 238 | 31 | 38 | 31 | 31 |
| Soybeans | 92 | 98 | 114 | 24 | 66 | 54 | 54 |
| Sugarcane/sugar beets | 103 | 103 | 118 | 15 | 5 | 7 | 18 |
| Vegetables and melons | 371 | 421 | 462 | 25 | 8 | 7 | 9 |
| Animal products Meat, milk, and eggs | 647 | 706 | 748 | 16 | 16 | 15 | 18 |

[a]Source: "FAO Production Yearbook," Vol. 47. (1993). Food and Agricultural Organization of the United Nations, Rome, Italy.
[b]Major fruits include grape, citrus fruit, banana, and apple.
[c]Includes soybean, palm kernels, sunflower, rapeseed, linseed, cottonseed, olive oil, and coconut.

and grain crops to produce meat, milk, and other animal products. Total meat, milk, and egg production increased 16% between 1982 and 1992 (Table I), and the proportion provided by North America also increased.

### 3. Fiber and Forest Crop Production

Many crop plants are harvested for their cellulose and lignin, compounds known for their intrinsic structural properties. Up to 50 billion metric tons of cellulose are estimated to be produced by land plants annually. Much of this goes to yield paper, wood, fiber, and fuel as the world's biomass is recycled. As petroleum reserves are exhausted in the future, more energy and chemical feed stocks will probably be derived from cellulose and starch. [See Nutrition, Dietary Fiber.]

Examples of important fiber crops include cotton, flax, hemp, jute, kenaf, and ramie (China grass) (Fig. 6). These fibers are used in the textile industry and as filling fibers in stuffing mattresses and upholstering. Broomcorn is used to make brooms.

Although forests serve many useful purposes (i.e., recreational sites, habitat for wildlife), they are mainly noted for the wood products that are produced from them. Wood was for centuries the most important fuel; in recent times, fossil fuel (coal and oil) and now atomic energy are surpassing wood in importance. Worldwide, the second greatest use (after fuel) of wood is for production of lumber. Wood is also the chief raw material for conversion to pulp, which is principally used for making paper.

### 4. Beverages

Plants also provide important beverages. The beverage crops include coffee, tea, cacao (the source of cocoa and chocolate), and others of lesser importance. Beer, next to water, is the world's most popular beverage. Beer is not the product of a single plant species, but rather the end result of fermentation by yeast of any number of carbohydrate sources. Barley, rice, sorghum, wheat, corn, potatoes, or even cassava can be added as an adjunct to the malt during brewing

FIGURE 6    Cotton (*Gossypium hirsutum*). (Reprinted with permission of Cambridge University Press.)

(Fig. 7). Modern beers are usually flavored by hops. Wine making consumes about 40 million metric tons of grapes annually. The production of distilled liquors consumes millions of tons of grain annually.

## 5. Vegetable Oils, Fats, and Waxes

Vegetable oils, fats, and waxes are extracted from numerous crops and are used for both food and industrial purposes (about two-thirds and one-third, respectively). For some oil crops, the residue remaining after extraction is of value. For example, the cake remaining after extraction of oil from soybeans is used as a high-protein supplement for livestock. Major sources of oil include soybean, corn, sunflower, palm, peanut, cottonseed, coconut, olive, canola, sesame, linseed, castor, and safflower (Fig. 8). Some newer specialty oils, such as from crambe and jojoba, are of interest because of unique chemical properties. Crambe oil is useful as a lubricant in the continuous casting of steel, as a rubber additive, and as a raw material in the production of various chemicals. Jojoba bears a seed that contains about 50% liquid wax that is prized as a lubricant because it substitutes for sperm whale oil. By hydrogenation, liquid jojoba wax

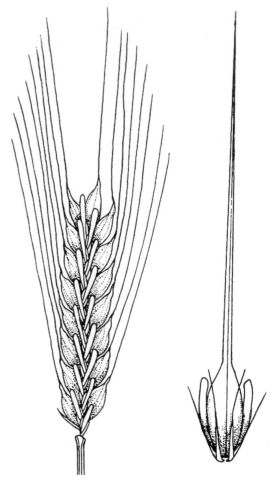

FIGURE 7    Two-rowed barley (*Hordeum vulgare*). (Reprinted with permission of Cambridge University Press.)

can be transformed into a substitute for carnauba wax and other hard waxes of commerce. [*See* Nutrition, Fats and Oils.]

Vegetable fats differ only slightly from oils, having acidic constituents that are more or less solid at ordinary temperatures. Whereas oils are fatty acid esters of trihydroxy glycerol, waxes are the fatty acid esters of monohydroxy alcohols. Waxes are found mostly as protective coatings on the leaves and stems of plants, where they serve to retard water loss. They are a less voluminous commodity than the vegetable fats and oils and are generally more expensive. Waxes are used chiefly for polishes (e.g., carnauba wax), carbon paper, and, to a lesser extent, in products such as candles.

## 6. Spices, Flavorings, Perfumes, and Other Essential Oils

The essential oils are highly aromatic substances that are benzene or terpene derivatives or straight-chain

FIGURE 8 Sunflower (*Helianthus annuus*). (Reprinted with permission of Cambridge University Press.)

hydrocarbon compounds of intermediate length. The aromatic nature of these substances provides the taste of flavorings, the "zing" of spices, the fragrance of perfumes, and the "clean" smell of antiseptics and medicinals.

Plants are the source of many of the spices and flavorings used today. Examples include cloves, black pepper, cinnamon, mints such as spearmint and peppermint, vanilla, oregano, and sage. Essences used in perfumes and soaps are usually derived from flowers. A number of aromatic oils also serve industrial purposes. Camphor and turpentine are good examples. Camphor is used in medicinals, liniments, and insecticides. Turpentine has many industrial uses, including solvents, wetting agents, and inhibitors of bacterial growth.

### 7. Medicinal and Related Derivatives

Plants are a source of many other natural products that have medicinal properties. Some are used as laxatives, ointments, lotions, antiseptics, stimulants, depressants, and psychedelics. Natural insecticides such as rotenone, pyrethrum, and nicotine are extracted

from plants. Natural plant growth regulators may also be obtained by routine extraction procedures.

### 8. Other Natural Products

Latex products, resins, gums, and related exudates are important products of plants. Many plants yield latex, a milky colloidal secretion that occurs in specialized cells that drain when matured. Several types of rubber latex and gums are obtained from plants in large quantities, as are tannins and dyes.

## B. Effects of Environment on Crop Production

A number of environmental factors influence crop production. Inadequate soil moisture is the environmental factor that most frequently limits crop production worldwide. Temperature is another environmental variable over which humans have little control. Crops are subjected to either low or high temperature stresses annually in many parts of the world. Frequently, crops are destroyed by extreme temperatures. In other instances, crops are stressed sufficiently to reduce yields. Scientists are working to incorporate tolerance to these stresses into crop plants, but only limited success has been achieved to date.

The effects of pollution and increased ultraviolet radiation (caused by depletion of the ozone layer) on crop plants are not well understood. The progressive increase in carbon dioxide levels in the atmosphere is beneficial to many crop plants because the fixation of carbon dioxide into sugars through the process of photosynthesis is enhanced by higher concentrations of atmospheric carbon dioxide. However, increasing concentrations of carbon dioxide and other gases (e.g., methane and chlorofluorocarbons) in the atmosphere are now thought to cause the greenhouse effect leading to an increase in temperature. If the warming trend continues, major changes in weather patterns may occur and crop production worldwide may be affected.

Acid rain has resulted from emissions of sulfur dioxide and nitrous oxides from coal-burning stacks. Much of the U.S. crop-growing area of the Midwest and Northeast is subjected to acid rain having a pH between 4 and 5. Research at the University of Illinois in Urbana–Champaign and elsewhere has shown that corn and soybeans, two of the major agronomic crops grown in the Midwest, are not deleteriously affected by acid rain. However, forests are more sensitive to it.

An increasing world population will produce more sewage sludge, wastes for landfills, and industrial by-products. Agronomists are concerned about these mounting problems and are conducting research on the application of sewage sludges and industrial by-products to farmland. Sludge is highly variable in plant nutrient content; it is generally sufficient in nitrogen and phosphorus levels, but limiting in the amount of potassium. Frequently, sludge contains high levels of barium, cadmium, chromium, copper, nickel, lead, and zinc. Some of these elements can accumulate to high concentrations in plants and become toxic when consumed by animals. Power plant fly ash from bituminous coal also contains many elements needed for plant growth. Boron, molybdenum, zinc, phosphorus, and potassium are available to plants, but high application rates of fly ash to cropland can lead to toxicity from boron and soluble salts.

## C. Disciplines within Crop Science

### 1. Plant Breeding and Genetics

Plant breeding is the systematic improvement of plants through selections resulting from the crossing of two parents. Genetics, the science of heredity, has placed plant breeding on a firm theoretical basis. For several decades, plant breeders and geneticists have made major advances in improving crops. Yields have increased, resistance to several diseases has been widely incorporated into the germplasm, nutritional quality has been improved, and crop plants have responded better to fertilizer and are now able to withstand environmental stresses better than in the past.

### 2. Plant Physiology and Biochemistry

Plant physiology is a major subfield of botany, the study of plants, which in turn is a subfield of biology, the study of organisms. Physiology, which is the study of how living organisms function, examines the processes that make life possible. Although plants have much in common with animals and microorganisms, numerous distinct functions of plants make plant physiology a unique science. Thus, photosynthesis, water relations, mineral nutrition, growth and development, and environmental physiology are some of the major subfields of plant physiology. Plant biochemistry deals with living processes and involves the study of how molecules comprising living organisms interact with each other during metabolism and other processes. Energy transformations in living cells, chemical reactions catalyzed by enzymes, regulation of cell reactions, and the processes by which living organisms replicate themselves with nearly perfect fidelity are examples of areas of plant biochemistry.

### 3. Molecular Biology and Genetic Engineering

This relatively new field involves the use of the techniques of recombinant DNA, tissue culture, gene transfer, embryo manipulation, and other biotechnologies to understand and manipulate life processes at the molecular level in ways that were unknown and unattainable only a few years ago. These new tools hold great promise for increasing the efficiency and sustainability of production agriculture and for assuring the safety, quality, variety, and quantity of the products desired.

The first genetically engineered crop approved for marketing by the Food and Drug Administration was Calgene's Flavr Savr tomato. This tomato, approved in early 1994, has been modified to ripen on the vine and to resist softening during transport or storage on supermarket shelves.

In late 1994, seven additional genetically engineered plants were approved for sale:

- A squash from Asgrow Seed Company that resists infection by two deadly plant viruses.
- Three delayed-ripening tomatoes from Monsanto Company, DNA Plant Technology Corporation, and Zeneca Plant Sciences.
- A Calgene-developed cotton plant that tolerates the herbicide bromoxynil; these plants will be used to produce cottonseed oil.
- A Monsanto-developed soybean plant that tolerates the herbicide glyphosate (marketed as Roundup).
- A genetically engineered canola plant rich in laurate, a fatty acid used to make soaps, detergents, and shampoos.

Scientists employing genetic engineering techniques have inserted the gene for Bt-toxin (from the bacteria *Bacillus thuringiensis*) into plants, making them resistant to some insects.

These genetically engineered crops represent the beginning of a large number of products to come. In the future, we can expect to see crops that are more tolerant of environmental stresses such as heat, cold, drought, and salt, and products whose composition has been modified to meet a specific need. The incorporation of resistance to additional insects, diseases, and other pests will permit farmers to rely more on biological forms of control, and therefore depend less on chemical pesticides for control. Plants will also be used to produce pharmaceuticals for animals and humans.

### 4. Crop Production and Management

Specialists in this area focus on increasing food production at home and abroad while using limited food production resources more efficiently. In a competitive global economy, researchers in crop production and management are challenged to learn how to reduce inputs of fertilizer and chemicals so that the cost per unit produced can be reduced. Although beyond the scope of this article, crop protection is crucial for crop production. It is the prevention or reduction of the damage caused by pests to useful plant species. A pest is any organism that is economically harmful and, in the context of crop protection, involves a large array of bacteria, viruses, and fungi as well as nematodes, mollusks, insects, mites, and vertebrates.

### 5. Crop Modeling

Sophisticated computer programs are used to predict the performance or growth of plants when the environment and/or inputs (chemicals, etc.) are varied. These computer-generated models can predict responses to the environment and can help identify what factors limit crop plant growth and development. Crop modeling provides a framework for interdisciplinary interactions in research. For example, the study of integrated pest management systems requires models of system components. Specifically, models of the focal crop, major insect pests, beneficial organisms, and management tactics must be designed with special attention to the interactions of these components so that interfacing the models will provide a meaningful tool for studying pest management systems. Some of the crop models are so sophisticated that they require the use of artificial intelligence and expert systems.

### 6. Weed Science

This is a relatively young field. In the 1950s, independent discoveries in the United States and United Kingdom showed that certain synthetic analogs [e.g., (2,4-dichlorophenoxy)acetic acid] of a natural plant growth hormone (i.e., indoleacetic acid) could kill broadleaf weeds. Application of this discovery to the selective control of dicotyledonous weeds revolutionized chemical weed control and crop production technology. The synthesis, formulation, application, uptake and translocation, modes of action, environmental responses, and degradation of these herbicides have attracted the interest of scientists from many disciplines. Because herbicides are now applied in greater quantities than all other agricultural pesticides, great concern exists about their effect on the environment.

### 7. Crop Ecology

The study of the relation between crop plants and their environment is known as crop ecology. It includes studies of climatic factors such as moisture, light, and temperature; edaphic factors including parent material and soil; and biotic factors such as the effects of other organisms, as all these relate to plant adaptation, distribution, and production.

## IV. SOIL SCIENCES

Soil is a complex, naturally occurring substance commonly defined as the weathered upper surface of the earth's crust, capable of supporting plant growth. Soil is a living entity. The viable nature of the biologically active outer crust of the earth (pedosphere) distinguishes soil science from geology, the study of the inactive or nonliving parts of the earth (lithosphere). Soil is at the interface between the atmosphere and lithosphere (i.e., the mantle of rocks making up the earth's crust). It also interferes with bodies of fresh and salt water (referred to as hydrosphere). The soil sustains the growth of many plants and animals and therefore forms a part of our biosphere.

The demands for soil in the future will be for more than just food and fiber production. Soils will be used even more for engineering construction (i.e., roads, buildings, reservoirs), recreation, watersheds, and disposal of many wastes that people accumulate as garbage and sewage.

### A. Characteristics of Soils

There are many different kinds of soil that differ in suitability for crop production. All soil is composed of three basic parts: minerals, organic matter, and the biotic component.

The mineral component comes from weathering of the parent material from which soil is formed and determines important chemical and physical properties of the soil. There are three basic soil particles of which the mineral part of most soils is composed: sand is largest, silt is medium sized, and clay is the smallest particle. Because clay particles are chemically the most active, they have the greatest impact on soil fertility.

Organic matter comes from dead plant and animal matter. Organic matter content varies greatly among different soils and influences the inherent fertility of a soil, its structure, and how the soil adsorbs and holds water. Organic matter also helps protect the soil from erosion.

The biotic component consists of living plants and animals including microorganisms such as bacteria and fungi. Microorganisms play a key role in the decay of organic matter. Some fix nitrogen biologically from the atmosphere.

In addition to these three parts of the soil, the soil pore space (that part not occupied by solids) is of major importance. Water is held on the surface of the soil particles, and gases such as oxygen also occupy these pores.

## B. Function of Soils

Soils serve three major functions for plants. First, they anchor the roots and thus support the plants. Second, soils serve as a reservoir to store and supply essential moisture to plants. Third, with the exception of carbon, oxygen, and hydrogen, all the chemical elements essential for plant growth are provided by the soil. These include the following macronutrients: nitrogen, phosphorus, potassium, calcium, magnesium, and sulfur. The following elements are considered as essential micronutrients: manganese, iron, boron, zinc, copper, molybdenum, and chlorine. Nickel appears to be essential for the metalloenzyme urease and is considered by some to be essential for crops such as soybeans. Others consider cobalt, vanadium, and silicon to be essential, although their roles are not well understood.

## C. Preservation of Natural Resources

A major thrust of soil scientists has been the preservation of our natural resources. Considerable emphasis is now given to ground and surface water quality. It has been estimated that over half of the world's population does not have reasonable access to safe and adequate water supplies and that contaminated water and poor sanitation are responsible for 80% of the illness in the 100 or more countries in which there are food shortages.

Contamination of water by pesticides and fertilizers has been detected in several states in the United States. The extent of contamination substantially varies and is influenced by many factors, such as the nature of the chemical and the types of soil through which the chemical moves. For example, chemicals move quickly through sandy soils, whereas certain chemicals are bound to particles of soils that contain high amounts of clay and organic matter. Nitrate–nitrogen percolates through soil easily, whereas ammonium–nitrogen can be bound to negatively charged particles of clay. Phosphate fertilizers do not readily reach through soils but are found in runoff waters and cause eutrophication of lakes and streams. Algae and other organisms thrive in bodies of water polluted with phosphate from runoff waters. These organisms are able to obtain their nitrogen through biological nitrogen fixation.

Soil and water conservation have been emphasized by soil scientists for several years. Farmers have been encouraged to build waterways, terraces, and dams to slow the run off of water and to reduce soil erosion. More and more farmers are adopting conservation tillage practices in which plant residues are left on the surface of the soil to reduce erosion caused by wind and rain. Many types of conservation tillage practices exist, ranging from no-till, in which virtually all crop residue is retained on the soil surface, to chisel plowing, in which substantially less crop residue remains. Formerly, most farmers used conventional tillage in which moldboard plows were used to turn under crop residues. In addition to conserving soil, conservation tillage requires less energy inputs for tillage and may lead to more efficient use of water by the crop.

## D. Restoration of "Damaged" Natural Resources

Surface mining occurs in many states and affects a wide range of soils, vegetation, ecosystems, and climatic zones. Because mining is confined to a narrow area, it can totally devastate those environments. Soil scientists have been involved in the reclamation of land disturbed by strip mining for coal and other minerals, and techniques are being developed through research to restore the productivity of the mined lands to their original levels.

Instances occur in which water supplies have been contaminated in rural and urban areas by pesticide spills or accidents near distribution centers and on farms. Methods are currently being developed to clean up these spills before they lead to the contamination of water supplies.

## E. Disciplines within Soil Science

The diverse characteristics and functions of soils have given rise to several disciplines of study within soil science. Historically, the study of soils dealt primarily with agriculture and related problems, but now a wide range of environmental problems are studied. The preservation of our natural resources and the restoration (reclamation) of those resources damaged by humankind are also challenging areas of investigation. Waste disposal (e.g., human wastes, sludge, industrial wastes), urban planning, recreation, and other nonagricultural activities are studied. Because the scope of soil science is broad and could be subdivided in many ways, the seven commissions listed by the International Society of Soil Science will be used as a means of discussing the subdisciplines of soil science.

## 1. Soil Chemistry

Soil chemistry is the study of chemical reactions associated with soil systems. It is studied to improve the availability of nutrients to plants, to use soil microbial populations, to avoid toxicities of elements to plants, to improve the physical condition of the soil, to eliminate the pollution of soil and water, to improve soil stability, and to reduce the corrosion of pipes and cement. The chemical properties of soils include kinds and amounts of clay minerals, mineral solubility, nutrient availability, soil reaction (pH), cation exchange, buffering action, and sorption of anions, micronutrient cations, and pesticides.

## 2. Soil Physics

Soil physics is the study of the physical characteristics of the soil system. Soil physics involves the derivation and application of physical principles and laws that govern the behavior of soils. Examples of physical properties are bulk density, water-holding capacity, hydraulic conductivity, porosity, pore-size distribution, and aggregation. The retention and movement of water, gases, solutes, and heat in soil are studied as they relate to the soil environment of plants and the relation between the soil and its surroundings. Emphasis is generally given to soil water, as this is the aspect of soil physical behavior that most often limits, controls, and unifies the other aspects.

## 3. Soil Mineralogy

Soil mineralogy involves the nature, properties, and reactions of the inorganic fractions of the soil system. About 92 chemical elements are known to exist in the earth's crust, and some 2000 minerals have been recognized. Approximately 98% of the crust of the earth is composed of eight chemical elements: oxygen, silicon, aluminum, iron, calcium, sodium, potassium, and magnesium. The first two comprise nearly 75% of the earth's crust. Most of the elements in the earth's

crust have combined with one or more other elements to form compounds called minerals, and these minerals generally exist in mixtures to form the rocks of the earth. The physical, chemical, and biological properties of the soil are strongly influenced by the kinds and amounts of clay minerals that are present, and the clay mineral fraction has been extensively investigated.

## 4. Soil Genesis, Classification, and Cartography

This branch of soil science, commonly equated in the United States to pedology, is concerned with the morphology, genesis, classification, and geography of soils and their relation to landscapes in the field. The proper use and management of soils are determined by their properties, distribution, and combinations in the landscape. Pedologists frequently are involved in soil surveys, classification of soils, distribution and movement of chemical elements in the soil profile in relation to soil fertility and genesis, land use, reclamation of mined land, rooting patterns of select crops in different soil types, mineralogy studies, and productivity of soils under different management systems.

## 5. Soil Biology

Soil biology is the study of macro- and microanimals and plants in soils and their associated reactions and processes influencing soil systems. Much of soil biology deals with microbiology. Soil microbiology encompasses studies of the isolation and identification of microorganisms, their ecological relations, and their associations with higher plants (i.e., rhizosphere, mycorrhizal fungi). Living organisms in the soil range from macrofauna (i.e., vertebrate animals of the burrowing type) to mesofauna (i.e., invertebrates such as mites, springtails, insects, earthworms, and nematodes) to microorganisms (i.e., bacteria, fungi, actinomycetes, algae, and protozoa). Many of these organisms are involved in the decomposition of organic residues in soil whereas others, such as the *Rhizobium*, are involved in symbiotic dinitrogen fixation. Other studies are concerned with the microbial degradation of xenobiotics (i.e., human-made compounds that are recalcitrant to biodegradation and/or decomposition).

Soil biochemistry is related to soil microbiology and is concerned with biochemical transformations of nitrogen, phosphorus, and sulfur; kinetics of nutrient cycling processes; mechanisms of biological dinitrogen fixation; biochemical pathways of pesticide degradation; and other processes mediated by microorganisms.

## 6. Soil Technology

Soil technology refers to the practical management of the soil in the field and includes studies of soil and water management. Soil and water management involves the wise application of principles of soil science to develop an integrated system for the production of abundant, high quality food and fiber crops while protecting the environment. Tillage operations, cropping practices, fertilization, water management, crop residue management, pest control, and other management practices applied to a soil for production of crops are studied. Maintaining water quality through the protection of ground and surface water from pesticide contamination is a high priority.

## 7. Soil Fertility

Soil fertility research is concerned with soil–plant relations, particularly with the mineral nutrition of plants grown in soils and the efficient use of fertilizers. The inherent fertility of the soil and diagnostic techniques for available forms of plant nutrients are important parameters.

## V. TRAINING AND JOB OPPORTUNITIES IN AGRONOMY

The U.S. Department of Agriculture conducted a national assessment of graduates and employment opportunities in 1990. The report issued is entitled "Employment Opportunities for College Graduates in the Food and Agricultural Sciences" and substantiates the numerous job opportunities found in the food and agricultural system. Agronomy's pursuit of scientific and technological developments offers impressive challenges to future graduates. The success of America's food and agricultural system is known throughout the world, with many benefits stemming from our science-based food and agricultural industry. The need for most cost-efficient production and the demand for agricultural products will continue to grow in the years ahead as global requirements for food and fiber increase and as production resources decrease. Highly qualified scientists and professionals working to advance the frontiers of knowledge and technology will be essential.

The 1990 national assessment of employment opportunities for college graduates in the food and ag-

ricultural sciences indicates that U.S. colleges and universities will not produce a sufficient number of graduates with food and agricultural expertise to fill important scientific and professional positions. Significant shortages of college-educated individuals are projected in the scientific and business specialties associated with the U.S. food and agricultural system. Through 1995, scientists, engineers, managers, sales representatives, and marketing specialists will account for about 75% of the total annual U.S. employment openings for new college graduates with expertise in agriculture, natural resources, and veterinary medicine. More than 14,000 new jobs are projected annually for food and agricultural scientists, engineers, and related technicians. Data on degrees granted by U.S. colleges and universities indicate that about 11,900 qualified graduates will be produced each year, leaving a shortfall of some 2100 individuals. Nearly two-thirds of this projected shortage of scientists, engineers, and related specialists is specific to occupations that will likely require graduates with a master's or doctorate degree.

As indicated earlier, agronomy is a broad field of study that includes both crop and soil sciences and embraces a wide range of applied and basic sciences. At many major universities in the United States, the soil science discipline is combined with plant sciences to form an agronomy department, although at some institutions, soil science is a separate department. Agronomy today is much more than farming, and diverse job opportunities are readily available.

Most research and teaching agronomists have had graduate training. Faculty members at universities and many agronomists employed by industry have doctoral degrees. Many with the master of science degree are employed by the Cooperative Extension Service and by industry. Numerous agronomists with the bachelor of science degree are employed as consultants, soil conservationists, field representatives, and sales persons for chemical and fertilizer companies. Many other roles exist in which technical knowledge of field crop production is needed; these include jobs such as advisers on large commercial farms, technical supervisors of food or seed production firms, technicians for researchers, and farm managers. Many U.S. farmers now have formal training as agronomists.

As agriculture becomes more dependent on sophisticated technologies to produce and process food, fiber, and forest products, more highly trained agronomists will be needed to sustain the effectiveness of those technologies. Improvements in modern computer systems and the development of new tools in molecular biology will provide agronomists with exciting opportunities for addressing these issues. Agronomists trained in the more traditional areas (i.e., plant breeding, crop production, weed control, soil and water management, and pedology) will also be needed to sustain the production of adequate quantities of high quality food for the world's population.

## VI. WORLD POPULATION AND FOOD PRODUCTION

As we approach the 21st century, the challenges of feeding the world are becoming increasingly difficult. Currently, world population is increasing at the rate of 1.6% per year. If this rate of increase continues, by the year 2040 world population will exceed 11 billion, nearly double the current 5.7 billion.

Total food production increased 39% during the 1980s, but per capita production only increased by 13%. More alarming, however, are the statistics that per capita production has decreased in 75 countries, including three-fourths of Africa, two-thirds of Latin America, and half of Asia. As the 1980s came to a close there was less food per person produced in these countries than at the beginning of the decade. Indeed world population is now growing at about 3% per year while food production is increasing at only 2% or less. Rapidly developing large countries like China will be buying an increasing proportion of available food stocks, in competition with food aid organizations trying to supply people in food-deficit poor nations. Because of rising affluence, the demand for feed grains to produce animal products is increasing at twice the rate of population growth. A major drought in the United States, which is the source for about half of world grain exports, or in major consuming nations like China or India could have a catastrophic effect on world grain markets.

Although statistically there is a sufficient amount of food produced to provide about 2700 calories per person per day, over 700 million do not have access to adequate nourishment. Despite the amount of food that we are *able* to grow, people are starving because access to this food is impossible. For example, per capita grain consumption in developing countries like Kenya and India is less than 200 kg while consumption in developed countries like France averages more than 450 kg and in the United States more than 850 kg. Thus, it is essential that those who are unable to purchase food be assisted to increase their income or food self-sufficiency. Increases in food production must result from more efficient use of the land, while maintaining natural resources, rather than as a result of area expansion which is now economically and ecologically impossible in most areas of the world. Hunger associated with the failure to "integrate the poor into the economic development process" "will accelerate unless policies to alleviate poverty, generate employment, and raise incomes are vigorously pursued."

## ACKNOWLEDGMENT

We acknowledge Suzanne M. Farrow for her diligent efforts in copy editing and manuscript preparation.

## BIBLIOGRAPHY

Borlaug, N. E., and Dowswell, C. R. (1988). World revolution in agriculture. *In* "1988 Britannica Book of the Year." Encyclopedia Britannica, Chicago, IL.

Bray, F. (1984). Part II: Agriculture. *In* "Science and Civilisation in China," Vol. 6. Cambridge University Press, Cambridge.

CGIAR Annual Report 1993–1994. (1994). "Consultative Group on International Agricultural Research," Washington, D.C.

Coulter, K. J., Goecker, A. D., and Stanton, M. (1990). "Employment Opportunities for College Graduates in the Food and Agricultural Sciences: Agriculture, Natural Resources, and Veterinary Medicine." Higher Education Programs, Cooperative State Research Service, U.S. Department of Agriculture, Washington, D.C.

Fairbridge, R. W., and Rinkl, C. W., Jr. (1979). "The Encyclopedia of Soil Science, Part 1: Physics, Chemistry, Biology, Fertility, and Technology." Dowden, Hutchinson, and Ross, Stroudsberg, PA.

Foth, H. D. (1978). "Fundamentals of Soil Science," 6th Ed. Wiley, New York.

Harlan, J. R. (1975). "Crops and Man." American Society of Agronomy and Crop Science Society of America, Madison, WI.

Hayes, J. (1983). Using our natural resources. *In* "1983 Yearbook of Agriculture." U.S. Department of Agriculture, Washington, D.C.

Janice, J., Schery, R. W., Woods, F. W., and Ruttan, V. W. (1981). "Plant Science." An Introduction to World Crops," 3rd Ed. Freeman, San Francisco.

Langer, R. H. M., and Hill, G. D. (1991). "Agricultural Plants," 2nd Ed. Cambridge University Press, Cambridge.

Martin, J. H., Leonard, W. H., and Stamp, D. L. (1976). "Principles of Field Crop Production," 3rd Ed. Macmillan, New York.

McDonald, K. A. (1996). Standing room only? *Chronicle of Higher Education* February 9, Section A.

Metcalfe, D. S., and Elkins, D. M. (1980). "Crop Production: Principles and Practices," 4th Ed. Macmillan, New York.

Pinstrup-Andersen, P., and Pandya-Lorch, R. (1994). Enough food for future generations? *CHOICES* Third Quarter 9, 13–16.

# Alcohol, Impact on Health

RICHARD M. MILLIS
*Howard University College of Medicine*

## GLOSSARY

**Agnosia** Loss of the ability of the cerebral cortex to recognize persons or objects and their meaning

**Aphasia** Loss of the ability of the cerebral cortex to process language appropriately

**Cerebral cortex** The superficial layer of the brain's cerebral hemispheres where the highest level (most complex) integration of neural information occurs that make a person aware of thoughts and perceptions, including understanding and reasoning (cognition). Separate regions of the cerebral cortex are devoted to mapping sites of specific sensations (e.g., touch, temperature, pain), for initiating skilled muscle movements, for perceiving visual images, for perceiving sound (audition), for creating sounds (speech), for perceiving and writing language, and for expressing emotions as personality traits

**Cirrhosis** A disease of the liver in which the cells (hepatocytes) responsible for carrying out the functions of the liver are progressively destroyed. There is usually evidence of regeneration of liver cells and an increase in the amount of nonfunctioning connective tissue (fibrosis). Cirrhosis is most common in alcoholics with inadequate diet (malnutrition), but is also seen in infectious diseases, parasitic diseases (e.g., malaria), and blood vessel or bile duct obstructions of the liver

**Excitatory neurotransmitter** Chemicals synthesized and released by presynaptic neurons at synapses produce depolarizations and increase the excitability of the postsynaptic membrane. When sufficient excitatory neuro-transmitter is present, an action potential signal is produced in the postsynaptic membrane. Neuronal centers (nuclei) and circuits in the brain containing the following excitatory neurotransmitters appear to be affected by alcohol: glutamate, dopamine, serotonin, acetylcholine, and norepinephrine

**Inhibitory neurotransmitter** Chemicals synthesized and released by neurons at synapses that produce hyperpolarizations decrease excitability and inhibit postsynaptic neuronal signaling. Neuronal centers and circuits in the brain containing the following inhibitory neurotransmitters appear to be affected by alcohol: $\gamma$-aminobutyric acid (GABA) and the endogenous opiates (endorphins, enkephalins). The absence of substrates and enzymes for synthesis and release of the chemical neurotransmitter, postsynaptic receptor, or enzymes that metabolize the neurotransmitter can also produce the inhibition of neuronal signaling

**Neuropathy** A term indicative of the abnormal functioning of nerves outside of the central nervous system (peripheral nerves). Abnormalities include deficits in sensory and/or motor functions such as tingling, pricking, painful sensations, and muscle spasms or weakness

**Nucleus** There are two biological meanings of the term "nucleus" that are often confused: (1) Most cells have nuclei containing chromosomes that regulate the growth and metabolism of the cell and (2) nuclei of neurons (cell body, soma) are also seen as a cluster at a single site in the central nervous system to serve as centers for integrating neural information and regulating a specific function or control reflex, e.g., the nucleus accumbens is a brain center that appears to reinforce and reward behaviors related to the frequent ingestion of drugs (e.g., cocaine, nicotine) and drinking of alcoholic beverages (ethanol)

**Pellagra** A nutritional deficiency of niacin (nicotinic acid), one of the vitamin B complexes, in alcoholics. Pellagra can be exhibited by a wide spectrum of signs and symptoms such as skin disorders, digestive disturbances, spinal pain, emotional depression, idiocy, and seizures

**Receptor** All cell membranes have specialized proteins that act as attachment sites for regulatory molecules such as neurotransmitters and hormones. There are two biologi-

ENCYCLOPEDIA OF HUMAN BIOLOGY, Second Edition, VOLUME 1.   Copyright © 1997 by Academic Press.   All rights of reproduction in any form reserved.

cal meanings of the term "receptor" that are often confused: (1) At a synapse, the postsynaptic neuronal membrane contains specialized proteins that act as specific receptors for the neurotransmitter. After attachment of the neurotransmitter to a receptor, sodium, potassium, chloride, or calcium ion channels are either opened (activated) or closed (inactivated) to produce changes in the resting electric potential of the postsynaptic membrane that moves it to one that is less negative (depolarization) or more negative (hyperpolarization). The $N$-methyl-D-aspartate receptor for the excitatory neurotransmitter glutamate appears to have an exquisite sensitivity to ethanol. There are also receptors on the presynaptic membrane that control synthesis and release of the neurotransmitter. (2) Specialized neural receptors, different from the aforementioned membrane receptors, synapse with or are directly connected to sensory neurons for receiving information about changes in the external or internal environments of the body to produce reflex responses to environmental stimuli

**Risk factor** Exogenous or endogenous conditions, including those like drinking alcoholic beverages and cigarette smoking, that can be related to the genetic makeup or life-style of a person that increases the likelihood of contracting a particular disease, e.g., the frequent drinking of alcohol is a risk factor for emotional depression

**Synapse** A space (synaptic cleft) that serves as a region of virtual contact between adjacent neurons where the nerve impulse (action potential, signal) in the first (presynaptic) neuron is transmitted to the second (postsynaptic) neuron by a chemical neurotransmitter; the chemical neurotransmitter is synthesized in and released by the presynaptic neuron

**Tolerance** A progressively increasing resistance to the effects of a frequently administered drug, thereby requiring progressively larger amounts (dosages) of the drug to produce a desired effect. Tolerance is commonly associated with drug dependence and drug addiction. The larger drug dosages required for an effect might approach a level at which the drug produces dangerous side effects (untoward effects) or poisoning (toxicity) that place the person at risk for serious illness and death

**Wernicke–Korsakoff syndrome** A syndrome combining the features of Wernicke's disease of the brain (encephalopathy) and Korsakoff's psychosis. Wernicke's encephalopathy is seen in chronic alcoholics with inadequate nutrition having thiamine (vitamin $B_1$, aneurine) deficiency, including signs and symptoms of vomiting, muscle weakness and coordination problems, and visual disturbances. The failure to be oriented to the environment, increased susceptibility to suggestion, falsification of memory, and hallucinations are exhibited by alcoholics with Korsakoff's psychosis

**Withdrawal** A constellation of unpleasant symptoms associated with sudden stopping of the frequent ingestion or administration of a drug. The most commonly recognized signs and symptoms of drug withdrawal include dilated pupils, excessive tearing of the eyes (lacrimation), watery stools (diarrhea), hyperactive bowel sounds, gooseflesh, restlessness, muscle cramps, rapid heart rate (tachycardia), high arterial blood pressure (hypertension), yawning, and an inability to sleep (insomnia)

ETHYL ALCOHOL (ETHANOL) IS THE ACTIVE INGREdient in alcoholic foods and beverages that is produced by the fermentation of sugars and grains. Ethanol is a chemical compound quite distinct from other alcohols that are used in laboratories, industries, and homes, e.g., methyl alcohol (methanol) or wood alcohol and isopropyl or rubbing alcohol are both substantially more toxic than ethanol and should therefore never be ingested.

## I. ALCOHOL-RELATED PROBLEMS

Excessive consumption of ethanol is a serious public health problem. Alcohol usage is a factor in innumerable fatal motor vehicle and aviation accidents, drownings, suicides, and homicides. Much research indicates that alcohol-related problems often coexist with various psychiatric illnesses. There appears to be a special linkage among depression, the severity of alcohol-related problems, and the frequency of hospitalizations. People with psychiatric symptoms seem to drink alcohol as a form of self-medication to relieve their symptoms. In addition, families and physicians are often hindered from identifying those persons who are experiencing alcohol-related problems and who should be referred for medical treatment. This is because the physiologic effects of alcohol are often masked or mistakenly attributed to the normal aging process, the side effects of prescription medication being taken, or to a chronic disease that is affecting the individual. The chronic effects of alcohol are associated with malnutrition because the excessive drinking of alcoholic beverages often provides sufficient calories, thereby affecting the balance between the brain's perceptions of hunger and satiety. This occurs at the expense of an adequate diet of nutrients.

## II. RISK FACTORS FOR ALCOHOL-RELATED HEALTH PROBLEMS

Genetics plays a significant role in alcohol-related problems. Identical twins who are alcoholics share similar drinking behaviors and personality traits that seem to make a person susceptible to alcoholism. Some evidence also shows that genetics plays a role in the development of a taste preference for alcohol.

Alcoholism is associated with tolerance to the effects of alcohol. Tolerance is brought about by changes in the way the brain and other organs respond to alcohol. An increase in the ability of the liver to metabolize alcohol produces higher levels of toxic metabolites and lowers the blood alcohol level more quickly in alcoholics demonstrating tolerance than in others. Usage of alcohol by parents and peers has a profound influence on an adolescent's desire for alcohol. Alcohol-related problems seriously impair physiologic, psychologic, and social maturation. Young aspiring athletes should be aware of the findings that chronic ethanol intake can reduce the amount of (type II) muscle fibers used for distance running and other athletic events requiring endurance. Alcohol was by far the drug used most often by high school seniors surveyed in 1990. However, legislation making alcohol consumption in persons below 21 years of age illegal seems to have had the positive effect of reducing drinking and alcohol-related traffic accidents in adolescents and young adults. In people 65 years of age and older, alcohol abuse and the adverse health effects of alcohol are not as severe as those found in younger persons. Females seem to metabolize alcohol less efficiently than males. Thus, for the same alcohol intake, women develop greater blood alcohol levels over a shorter time period than men. Although this could make women more susceptible to the adverse acute and chronic health effects of alcohol, numerous studies have shown that men drink more heavily and suffer more alcohol-related adverse health effects than women. Alcohol abuse has also been shown to be a causative cofactor in cancers of the mouth, lung, stomach, liver, pancreas, colon, rectum, and breast.

## III. EFFECTS OF ALCOHOL ON BODY SYSTEMS AND BEHAVIOR

Alcohol injures virtually every body organ and system. The liver is the primary site of alcohol metabolism. Fat accumulation, hepatitis, fibrosis, and cirrhosis are the most common alcohol-induced liver injuries. Cirrhosis of the liver is the primary chronic health hazard associated with alcohol use and is responsible for substantial illness and death among alcoholics. [See Alcohol Toxicology.]

The cardiovascular system is also affected adversely by alcohol intake. Hypertension, heart muscle atrophy, cardiac arrhythmias, and stroke contribute to an increased risk for sudden death in alcoholics. However, substantial evidence shows that the risk for coronary artery disease might actually be decreased by moderate alcohol intake. It is thought that this positive effect is brought about by blocking the formation of clots in the blood vessels, raising the blood levels of high-density lipoproteins that clear cholesterol from the blood, and increasing the blood estrogen levels in postmenopausal women (whose susceptibility to coronary artery disease increases substantially after menopause). Alcohol also decreases the ability of the body to develop an appropriate immune response to infectious microorganisms, an effect that is a life threatening problem in individuals with a compromised immune system such as those infected with the human immunodeficiency virus. Blood hormone levels and the regulatory actions at their specific target tissues are also adversely affected by alcohol intake. The chief hormonal effects of alcohol can produce infertility by decreasing blood testosterone levels in men and producing menstrual cycle disturbances in women. This could contribute to the increased incidence of spontaneous abortions found in alcoholic women who become pregnant.

The most striking neurologic effects of acute alcohol consumption are the impairment of muscle coordination, mental judgements, and ability to process information by the cognitive mechanisms of the brain. At blood alcohol concentrations (BAC) ranging from 0.02 to 0.05 g/dl (1 drink containing $\frac{1}{2}$ oz of alcohol, a 12-oz beer, a 4-oz glass of wine, a shot of 80-proof spirits), novice drinkers experience euphoria and a reduction in anxiety. Judgement and motor coordination are impaired at a BAC level of 0.06 to 0.1 g/dl (3–5 drinks); this BAC level is also associated with aggressive and violent behavior. At BAC levels of 0.2 to 0.25 g/dl (10–13 drinks) staggering gait is observed, speech is slurred, vision is impaired, and the blood glucose level is lowered (hypoglycemia). At a BAC of 0.3 g/dl, loss of memory and consciousness can occur. At a BAC of 0.4 to 0.5 g/dl, breathing is depressed and coma can result.

Chronic alcohol usage is associated with alcohol withdrawal syndrome after cessation of drinking. Alcohol-induced liver disease and nutritional vitamin deficiencies are also associated with specific neurologic disorders such as Wernicke–Korsakoff syndrome, pellagra, and damage to peripheral nerves (neuropathy). Admonitions against alcohol usage in pregnant women must be taken seriously because of the fetal alcohol syndrome. Alcohol-induced malformations of the head, growth retardation, and cognitive and motor disorders associated with alcohol withdrawal continue to have adverse effects on infants and children born to alcoholic mothers for years after their intrauterine exposure to alcohol. Newborns exposed to alcohol prenatally have a weak sucking response that impairs their ability to feed effectively. Children born with fetal alcohol syndrome are hyperactive, distractable, impulsive, and have short attention spans. Some of these behaviors are similar to those of infants and children with perinatal hypothyroidism or the attention-deficit hyperactivity disorder.

## IV. ALCOHOL AND THE BRAIN

Susceptibility to alcoholism is thought to be determined by complex interactions among genetic, environmental, and neurophysiologic factors. Nerve transmission between adjacent neurons (at synapses) requires the neuronal secretion of chemical neurotransmitters. Experimental studies in rats suggest that individuals with low levels of the neurotransmitter serotonin may be susceptible to alcohol addiction. The fact that acute administration of alcohol increases brain levels of serotonin indicates that drinking alcohol could be a behavior that attempts to normalize brain serotonin, a neurotransmitter important in regulating mood and sleep. In humans, dysfunctions of serotonin systems in the brain have been linked with depression, anxiety, aggressiveness, and antisocial behavior. Alcohol-induced antisocial behavior is commonly recognized and strengthens the argument for the involvement of serotonin in alcoholism. [*See* Serotonin in the Central Nervous System.]

In addition to the capacity of alcohol to increase brain serotonin levels, alcohol penetrates the cell membranes of all neurons and changes their permeability characteristics (fluidity). Alcohol-induced changes in the permeability of the membrane to sodium, potassium, chloride, and calcium ions at specific channels in neuronal membranes probably interferes with neural signaling. Abnormalities of electrical signaling mechanisms in the heart (arrhythmias) undoubtedly contribute to the high incidence of sudden death in alcoholics.

Acute exposure to alcohol increases the activity of a specific type of neuronal membrane receptor for the neurotransmitter $\gamma$-aminobutyric acid (GABA). The $GABA_A$ receptor regulates chloride ion transport into neurons and is largely responsible for the capacity of

the brain to produce the inhibition of neural control circuits. The effects of alcohol on the $GABA_A$ receptor appear to contribute to the anxiety-reducing and sedative actions of alcohol by mechanisms mediated by the medial septal nucleus, a brain center regulating the amount of emotional input necessary to activate the higher cognitive centers in the cerebral cortex. Effects of alcohol on the cerebellum seem to be important for muscle coordination and movement disorders associated with alcohol intoxication.

Alcohol also inhibits the $N$-methyl-D-aspartic acid (NMDA) receptor for the brain's excitatory neurotransmitter glutamate. Activation of NMDA receptors produces the long-term potentiation (LTP) of electrical signals necessary for learning and memory. The capacity of alcohol to inhibit NMDA receptors, therefore, appears to be linked to the acute memory losses (amnesia) and difficulties in learning new information associated with alochol intoxication. However, the amnesia associated with chronic alcohol usage in humans is thought to involve several different brain centers making acetylcholine (basalis of Meynert) and norepinephrine (locus ceruleus). Addiction to alcohol is thought to be produced by a reinforcing or rewarding effect on the brain. Alcohol ingestion increases the level of the excitatory neurotransmitter dopamine in a brain region known as the nucleus accumbens. The nucleus accumbens is an important reinforcement–reward center where dopamine levels are increased after the administration of highly addicting drugs such as cocaine and alcohol. As previously mentioned, brain serotonin increases after alcohol ingestion, and serotonin also appears to modulate dopamine levels in the nucleus accumbens.

A class of inhibitory neurotransmitters known as the endogenous opiates (endorphins and enkephalins) are also key components of the reinforcement–reward system of the brain. They are so named because their (analgesic) actions on reducing the brain's perception of pain and increasing a person's feeling of well-being (euphoria) mimic those of other opiates such as heroin and morphine. Increased levels of the brain's endogenous opiates after alcohol ingestion are probably responsible for the euphoria that serves as an important reinforcement and "reward" for repetitive alcohol-drinking behaviors.

## V. DIAGNOSIS AND TREATMENT OF ALCOHOLISM

The treatment of alcoholism requires specialized strategies for stopping the drinking behaviors and ameliorating the multitude of effects associated with the chronic craving of alcohol. Common indicators for recognizing harmful effects of alcohol usage have been recognized as the alcohol dependence syndrome. A strong desire or compulsion to drink and difficulty in controlling the onset or termination of drinking are usually found in combination with a physiologic withdrawal state consisting of anxiety, tremors, hyperthermia, sleep disturbances, hallucinations, or seizures. There is usually evidence of tolerance such that increased doses of alcohol are required to achieve effects originally produced by lower doses. A progressive neglect of alternative pleasures or interests, including a progressively increased amount of time spent drinking or recovering from the effects of drinking, is also common. Persistence in drinking alcohol despite evidence that the individual is experiencing the harmful effects of alcohol (e.g., liver injury, mood disorders) is a clear indication of alcohol dependence.

Self-reports about alcohol consumption are usually unreliable. Consequently, substantial research is being undertaken to identify valid biological markers, indicative of the level of alcohol consumption, that do not depend on the patient's cooperation. Several blood proteins ($\gamma$-glutamyl transpeptidase, carbohydrate-deficient transferrin, aspartate aminotransferase) as well as the mean red blood cell (corpuscular) volume and uric acid level have been identified as potential markers for alcohol consumption. The most common treatments for alcoholism include psychological, behavioral, and pharmacological interventions. Inpatient and outpatient rehabilitation strategies include marital–family therapy, coping skills training, stress management classes, employee assistance programs, and the Alcoholics Anonymous 12-step mutual help program.

## VI. DRUGS FOR TREATMENT OF ALCOHOLISM

Drugs that have been found to be useful in alcoholism are (1) agents that aid in the management of the alcohol withdrawal syndrome, (2) those that cause unpleasant reactions to alcohol consumption, (3) those that decrease the desire or craving for alcohol, (4) antagonists of the acute intoxicating and depressant effects of alcohol, and (5) medications for the treatment of coexisting psychiatric disorders such as depression and anxiety.

Primary medications for the management of alcohol withdrawal are the benzodiazepine tranquilizers

such as diazepam (Valium) lorazepam, or oxazepam and the tricyclic anticonvulsant carbamazepine. Drugs that antagonize the effects of stimulating neural receptors for the neurotransmitter norepinephrine (adrenergic blockers) include propranolol, atenolol, and clonidine, which appear to be useful for managing tachycardia, sweating, and tremors associated with alcohol withdrawal and seem to increase the effectiveness of treating alcohol withdrawal when they are used in combination with a benzodiazepine.

Disulfiram (Antabuse) is an alcohol-sensitizing agent that produces unpleasant effects such as flushing, nausea, vomiting, and heart palpitations to deter alcoholics from drinking. Findings that have implicated several neurotransmitters, neuropeptides, and hormones in the craving of alcohol and other drinking-related behaviors have stimulated research on drugs that inhibit the neuronal uptake of serotonin, mimic the effects of dopamine, or antagonize effects of the endogenous opiates (endorphins and enkephalins). The serotonin uptake inhibitors fluoxetine, citalopram, zimelidine, and viqualine seem to decrease alcohol consumption by a mechanism thought to be mediated by brain centers involved in the regulation of food intake. Bromocriptine is a drug that mimics the effects of dopamine in the brain and also produces an alcohol anticraving effect. Naltrexone reduces alcohol consumption by inhibiting the opiate receptors of the brain. A similar opiate antagonist, naloxone, has been effective in reversing alcohol-induced comas. However, the search for an agent that has a specific antialcohol effect is fraught with dangers and ethnical concerns because of the potential for abuse as a "sobering-up" pill. Acupuncture and transcranial electrical treatments are examples of some current speculative investigations that have shown some limited promise for reducing the consumption of alcohol.

## VII. SUMMARY

A number of genetic factors and personality traits can make a person susceptible to alcoholism and alcohol-related health problems. Risk-taking behaviors characteristic of an adolescent's life-style, a genetic predisposition to have a taste preference for alcohol, and differences in enzyme levels for the metabolism of alcohol seem to contribute to interindividual differences in the susceptibility of a person to the effects of alcohol. It is thought that drinking alcohol might represent a form of self-medication for various behavioral disturbances and psychiatric illnesses. Depression and anxiety are the most common psychologic disorders associated with alcohol-related problems.

There is currently no known site of alcohol's actions on the nervous system that can explain all of its behavioral effects. However, there is considerable evidence for important effects of alcohol on neuronal receptors for the neurotransmitters known as endogenous opiates (endorphins and enkephalins), GABA, glutamate (NMDA receptors), serotonin, dopamine, brain regions known to be rich in norepinephrine, and acetylcholine. The nucleus accumbens is a brain center for substantial interactions between serotonin and dopamine that appears to produce a reinforcement–reward effect for drinking alcohol and for the craving of a number of other drugs of abuse. The most common acute behavioral effects of alcohol are the impairment of muscle coordination and the processing of new information. Chronic effects of alcohol include tolerance and the loss of learning, memory, and information processing functions associated with the Wernicke–Korsakoff syndrome. The liver is a primary site for the chronic effects of alcohol, and alcohol-induced cirrhosis often leads to serious illness and death. The chronic effects of alcohol are associated with malnutrition because the excessive drinking of alcoholic beverages often provides sufficient calories at the expense of an adequate diet of nutrients. Infants born to mothers of alcoholics often exhibit growth and behavioral abnormalities of the fetal alcohol syndrome, which have adverse consequences on their health well into childhood. The treatment of alcoholism is limited by the ability of family, friends, and physicians to recognize the alcohol dependence syndrome as an indicator of disease separate and distinct from the effects of other concurrent diseases and prescription medications. An armamentarium of tools for treating alcoholics includes aftercare following acute withdrawal, psychological counseling, and drugs to manage withdrawal, mood, and behavioral disorders that are commonly associated with alcoholism.

## BIBLIOGRAPHY

Appelbaum, M. G. (1995). Fetal alcohol syndrome: Diagnosis, management, and prevention. *Nurse Pract.* **20,** 24–33.

Guerri, C., Montoliu, C., and Renau-Piqueras, J. (1994). Involvement of free radical mechanism in the toxic effects of alcohol: Implications for fetal alcohol syndrome. *Adv. Exp. Med. Biol.* **366,** 291–305.

Hunt, W. A. (1996). Role of acetaldehyde in the actions of ethanol on the brain: A review. *Alcohol* **13,** 147–151.

Kannel, W. B., and Ellison, R. C. (1996). Alcohol and coronary heart disease: The evidence for a protective effect. *Clin. Chim. Acta* **246,** 59–76.

Komura, S., Fujimiya, T., and Yoshimoto, K. (1996). Fundamental studies on alcohol dependence and disposition. *Forensic Sci. Int.* **80,** 99–107.

Kostowski, W. (1995). Recent advances in the GABA-A-benzodiazepine receptor pharmacology. *Pol. J. Pharmacol.* **47,** 237–246.

Lohr, R. H. (1995). Treatment of alcohol withdrawal in hospitalized patients. *Mayo Clin. Proc.* **70,** 777–782.

Luijckx, G. J., Nieuwhof, C., Troost, J., and Weber, W. E. (1995). Parkinsonism in alcohol withdrawal: Case report and review of the literature. *Clin. Neurol. Neurosurg.* **97,** 336–339.

Miller, N. S. (1995). Pharmacotherapy in alcoholism. *J. Addict. Dis.* **14,** 23–46.

Peters, D. H., and Faulds, D. (1994). Tiapride: A review of its pharmacology and therapeutic potential in the management of alcohol dependence syndrome. *Drugs* **47,** 1010–1032.

Reynolds, J. D., and Brien, J. F. (1995). Ethanol neurobehavioural teratogenesis and the role of L-glutamate in the fetal hippocampus. *Can. J. Physiol. Pharmacol.* **73,** 1209–1223.

Taylor, A. N., Ben-Eliyahu, S., Yirmiya, R., Chang, M. P., *et al.* (1993). Actions of alcohol on immunity and neoplasia in fetal alcohol exposed and adult rats. *Alcohol Alcohol. Suppl.* **2,** 69–74.

U. S. Department of Health and Human Services. (1993). "Eighth Special Report to the U. S. Congress on Alcohol and Health." NIH Publication No. 94-3699, Government Printing Office, Washington, D.C.

Weinberg, J. (1993). Neuroendorine effects of prenatal alcohol exposure. *Ann. N.Y. Acad. Sci.* **697,** 86–96.

White, N. M. (1996). Addictive drugs as reinforcers: Multiple partial actions on memory systems. *Addiction* **91,** 921–965.

# Alcohol Toxicology

W. MILES COX
*University of Wales, Bangor, United Kingdom*

WEI-JEN W. HUANG
*University of Illinois at Chicago*

## GLOSSARY

**Acetaldehyde** Poisonous substance that is produced when alcohol is metabolized by the body

**Acute effects** Immediate, short-term consequences of ingesting alcohol, e.g., a person's ability to drive an automobile is impaired during intoxication

**Alcohol blackout** Period of time during which a person is intoxicated that he or she cannot later recall while sober

**Blood alcohol level** Concentration of alcohol in the blood; expressed in terms of the number of milligrams of alcohol in each 100 ml of blood and reported as mg%

**Congeners** Substances (some of which are quite toxic) that alcoholic beverages contain other than alcohol. Congeners give each alcoholic beverage its distinctive color, taste, and aroma

**Chronic effects** Long-term, cumulative consequences of heavy drinking, e.g., brain atrophy and cirrhosis of the liver are two consequences that might result from years of heavy drinking

**Cirrhosis** A serious liver disease, usually resulting from years of excessive drinking, that is largely irreversible and frequently leads to death

**Dependence** Condition whereby a person has withdrawal symptoms when he or she is unable to get the alcohol that he or she is accustomed to drinking

**Distillation** Process by which an alcoholic solution is heated and the resulting vapors are condensed in order to provide a stronger concentration of alcohol

**Ethyl alcohol** Type of alcohol that humans usually consume; it is less toxic than other alcohols

**Fetal alcohol syndrome** Pattern of serious birth defects, including growth deficiencies and facial abnormalities, that have been observed among children of alcoholic mothers who drank heavily during their pregnancy

**Fermentation** Process by which yeast converts sugar and water into ethyl alcohol

**Tolerance** Decline in effectiveness of a drug (alcohol, in this case) after repeated administration

**Wernicke–Korsakoff's syndrome** Symptoms associated with brain damage from severe alcoholism that include memory disorders, disorientation with regard to time, place, and people, and a profound inability to learn new information

**Withdrawal** A set of symptoms (including headache, sweating, and nausea) that occur when a person who is accustomed to drinking large amounts of alcohol suddenly stops

ALCOHOL TOXICOLOGY DEALS WITH THE POISonous effects of alcohol on the body, mind, and behavior. These effects can be divided into two broad categories: (a) those occurring during the time that a person is intoxicated, and (b) the long-range, cumulative effects of alcohol use. Further, there are toxic effects both for the individual drinker and for the society in which he or she lives. The primary purpose of the present article is to delineate the toxic effects of

ENCYCLOPEDIA OF HUMAN BIOLOGY, Second Edition, VOLUME 1. Copyright © 1997 by Academic Press. All rights of reproduction in any form reserved.

prolonged, excessive use of alcohol on the individual person's bodily and psychological functioning.

## I. INTRODUCTION

The term toxicology is defined by Webster's New Collegiate Dictionary as the "science that deals with poisons and their effects and with the problems involved." Thus, alcohol[1] toxicology deals with the poisonous effects of alcohol on the body and the consequences of those effects. There are various ways in which a discussion of alcohol toxicology could be approached. For example, we could focus on the negative consequences of drinking during the time that a person is intoxicated or we could discuss the long-range, cumulative detrimental effects of alcohol use. We could discuss both kinds of negative consequences for the individual drinker or we could broaden our discussion to include the consequences of excessive drinking for a society at large. In the present article, however, we will focus mainly on the long-range toxic effects of alcohol consumption on the human body and the psychological consequences of those long-range effects.

Alcohol, which has been consumed by humans since before recorded history, is currently one of the most widely used mood-altering drugs in the world.

Today in the United States, approximately 20% of the population drinks substantial amounts of alcohol and approximately 5% of the population drinks heavily. People use alcohol for many different reasons. Some use it to cope with negative emotional states such as tension and anxiety. Others use alcohol because it suppresses their inhibitions and temporarily allows them to feel more attractive, optimistic, and powerful. Still others use alcohol because it gives them access to other things that they value, such as approval by their peers.

Despite the popularity of alcohol as a mood-altering drug, its prolonged, excessive use has been shown to have significant harmful (i.e., toxic) effects on the human body and on people's mental functioning. Moreover, alcohol alone (and not simply the malnutrition that often accompanies excessive drinking) can have direct toxic effects on the body. However, the exact amount of alcohol that must be consumed before it causes noticeable damage to the body is

not entirely clear. In fact, there is some evidence that consuming alcohol in small quantities can actually have beneficial effects. For instance, drinking one alcoholic beverage per day seems to have desirable effects on cholesterol levels in the body, increasing levels of beneficial cholesterol [high-density lipoproteins (HDL)], and decreasing levels of harmful cholesterol [low-density lipoproteins (LDL)]. Accordingly, moderate drinking may reduce the risk of cardiovascular diseases. Some people maintain, moreover, that drinking a glass of red wine with meals aids the digestion. [*See* Cholesterol.]

To understand alcohol toxicology it is useful first of all to know some basic facts about alcoholic beverages and how they are metabolized by the body. This information is reviewed in the following section.

## II. ALCOHOL BEVERAGES

The alcohol that people drink is called ethyl alcohol (or ethanol), which is distinguished from other alcohols (e.g., isopropyl alcohol, methanol) that are unfit for consumption because they are highly toxic to the body. Ethyl alcohol is made by a process called "fermentation," a chemical process that requires three ingredients: yeast, sugar, and water. Yeast are living organisms, and as they grow in a solution containing sugar and water, they chemically transform the sugar into ethyl alcohol and carbon dioxide.

Different kinds of alcoholic beverages are made from different sources of sugar. For instance, wine is usually made from grapes and rum is made from sugar cane. Beer is made from grain (often barley) and hops (a pungent plant of the mulberry family); the latter is added to provide a distinctive flavor. Whiskey is also made from grains (often corn), whereas vodka is often made from potatoes. If a grain product is used as the basis for fermentation, however, an extra step, called "malting," is required to transform the starch in the grain product into sugar. Malting involves soaking the grain in water and heating it until it sprouts. In so doing, enzymes are formed that change the starch into sugar.

Through the simple process of fermentation, no beverage can have an alcohol concentration higher than about 15%. When the concentration reaches this level, the yeast begin to die so that additional alcohol cannot be produced. In other words, the yeast are eventually killed by the alcohol that they produce. In order for a beverage to contain a higher concentration of alcohol, an additional process, called distillation,

---

[1]Unless otherwise stated, the use of the world "alcohol" refers to ethanol.

is used. Distillation involves heating the alcoholic solution in a vat until it boils. The evaporating vapors from the heated alcohol are siphoned into another container, where they are cooled and condensed. The resulting solution is both higher in alcoholic content and purer than the original fermented solution.

There are three basic classes of alcoholic beverages: beer, wine, and distilled spirits. The ethyl alcohol that each of these beverages contains is exactly the same, but it is found in different concentrations in the three types. Beer has the lowest alcoholic content of the three. Most beer manufactured in the United States has approximately 3 to 5% alcohol, although beer made in certain other countries (e.g., Belgium) typically has a considerably higher concentration. Wines are intermediate in alcoholic content. Table wines (e.g., white wines, red wines, and rosés) typically have from 10 to 15%. However, a special category of wines, called fortified wines, have a higher concentration of alcohol. Fortified wines are made by adding enough of a distilled substance to fermented wine to increase its alcoholic content to about 20%. Sherry, port, and Madeira are examples of fortified wines. Distilled spirits are highest in alcoholic concentration. The alcoholic content of distilled spirits typically ranges from 40 to 50% and is indicated by the "proof" on the bottle. The proof equals twice the percentage of alcohol, so that, for instance, a 100 proof beverage contains 50% alcohol. Vodka, rum, and bourbon are examples of distilled spirits. Liqueurs or cordials are also distilled spirits containing from 40 to 50% alcohol. They are distinguished, however, by the fact that sweet and aromatic substances have been added to them, making them especially suitable as "after-dinner" drinks. Creme de menthe, Benedictine, and Grand Marnier are examples of liqueurs. A "standard-size" alcoholic beverage in the United States—whether beer, wine, or distilled spirits—contains the same amount of absolute alcohol. Thus, 12 ounces of beer, 4 ounces of wine, and 1 ounce of 100-proof distilled spirits each has about one-half ounce of absolute alcohol.

In addition to ethyl alcohol, alcoholic beverages contain water and small amounts of other chemicals called "congeners," which are formed at the same time that alcohol is made through fermentation. The congeners in alcoholic beverages are largely responsible for giving each its distinctive color, aroma, and taste, and some of them are quite toxic. Alcoholic beverages vary widely in the level of congeners that they contain. For instance, vodka has a very low level of congeners, consisting largely of pure alcohol and water. Bourbon, at the other extreme, has a very high congener level. Whether or not the long-range harmful effects of consuming alcohol are also related to the beverage congener level has not been systematically investigated.

## III. METABOLISM OF ALCOHOL

As a food substance, alcohol is unique. It requires no digestion and is absorbed unchanged from the stomach and small intestine into the bloodstream. Although alcohol is rich in calories, they are "empty" calories, for although they might be a source of energy for the body, they do not provide essential nutrients.

After alcohol has entered the bloodstream, it is distributed throughout the body. Small amounts of it are then excreted from the body through the perspiration glands, kidneys, and lungs. As a result of the last method of excretion, we can detect whether or not another person has been drinking by smelling his or her breath. As a result of the latter two methods of excretion, a urine test or a breath test can be taken to estimate not only whether or not a person has been drinking, but also the concentration of alcohol that is in his or her bloodstream. In fact, breath tests, are commonly used by police officers to judge whether or not a driver is at the legal level of intoxication.

The majority of alcohol that enters the bloodstream must be metabolized (i.e., chemically transformed) in order to be removed from the body. Several steps are involved in this process. First, alcohol is transformed into another chemical called acetaldehyde (a substance that itself is toxic to the body). Next, acetaldehyde is transformed into acetic acid (i.e., vinegar). Finally, acetic acid is metabolized into carbon dioxide and water. Two enzymes in the liver play a crucial role in metabolizing alcohol. One enzyme, alcohol dehydrogenase, transforms alcohol into acetaldehyde, and the second enzyme, aldehyde dehydrogenase, converts acetaldehyde into acetic acid.

A fixed amount of time is required for a given amount of alcohol to be metabolized. Typically, 1 hr is required for each one-half ounce of pure alcohol that has been consumed. People commonly believe that drinking black coffee, exercising, or taking a cold shower can speed up the sobering process. These beliefs, however, have proven to be myths, for nothing can be done to facilitate the rate at which alcohol is metabolized in the body.

## IV. FACTORS AFFECTING THE LEVEL OF INTOXICATION

A person's level of intoxication at any given time is proportional to the concentration of alcohol in the blood, which, in turn, is determined by the rate at which alcohol has been absorbed into the bloodstream. The faster alcohol is absorbed, the higher level of intoxication that is attained. The rate of absorption is affected by a number of characteristics of the alcoholic beverage that is consumed, how it is consumed, and characteristics of the person who consumes it.

### A. Characteristics of Beverages and How They Are Consumed

For a given quantity of alcoholic beverage that is consumed in a given period of time, the higher the concentration of alcohol in the beverage, the higher the level of intoxication that will be reached. Thus, a person would become more intoxicated by drinking a given amount of undiluted distilled spirits in a certain period of time than he or she would by drinking the same amount of beer in the same period of time. This does not mean, however, that a person cannot become intoxicated by drinking only beer nor that people do not develop problems with alcohol if they drink only beer. Usually, people consume a higher quantity of beer than of distilled spirits, and thus they can become equally intoxicated from beer as from distilled spirits.

If alcohol is mixed with something else (e.g., water, fruit juice, a "mixer"), the concentration of alcohol will, of course, be reduced. Thus, for a given quantity of alcohol, a lower level of intoxication will be achieved if a diluted rather than an undiluted beverage is consumed. It is also important to know what the mixer is and especially whether or not it contains carbon dioxide. This is because carbon dioxide rapidly moves the contents of the stomach into the small intestine, and alcohol in the small intestine is absorbed into the bloodstream more rapidly than is alcohol in the stomach. For this reason, carbonated beverages cause a person to become intoxicated faster than uncarbonated ones. Consequently, if all other factors are equal, alcohol mixed with a carbonated beverage (such as a commercial soft drink) will result in a higher level of intoxication than alcohol mixed with a noncarbonated beverage (such as tap water). For the same reason, a person can become more intoxicated by drinking champagne than table wine. Although champagne might contain the same concentration of alcohol as table wine, the carbon dioxide in champagne facilitates the absorption of alcohol into the bloodstream.

It is also important to consider the manner in which a person consumes alcoholic beverages. The amount consumed per unit time is one critical variable. Thus, if a person gulps a drink instead of sipping it slowly, a higher level of intoxication will be attained. Another critical variable is whether or not food is eaten when alcoholic beverages are consumed. When alcohol is drunk with or just after a meal, the alcohol is absorbed more slowly than if it is drunk on an empty stomach. The particular food that is eaten is also a contributing factor. For instance, the protein in food retains alcohol and prevents it from being rapidly absorbed from the stomach into the bloodstream. Foods rich in protein are more effective in retarding absorption than foods containing little protein.

### B. Characteristics of the Drinker

#### 1. Bodily Characteristics

There are two major characteristics of a person's body that are related to the level of intoxication that he or she will achieve from drinking alcohol: (1) the weight of the body and (2) the proportion of the body that is fat relative to the proportion of muscle tissue. In general, the greater one's body weight, the greater the quantity of blood and other body fluids one has in his or her body. Because of the larger volume of fluids that can dilute the concentration of alcohol, the heavier person can consume more alcohol than the lighter person before he or she reaches the same level of intoxication or, for a given quantity of an alcoholic beverage, a heavier person will become less intoxicated from consuming it than will a lighter person.

If two people have the same body weight, but different proportions of adipose (or fatty) tissue relative to muscle tissue, a given quantity of alcohol will have a stronger effect on the person with the higher proportion of body fat. This is because fatty tissue contains less bodily fluids than muscle tissue. Hence, muscle tissue, to a greater extent than fatty tissue, allows alcohol to become diluted, leaving relatively less alcohol to be distributed in the bloodstream.

For both of these reasons, the average female will become more intoxicated from consuming the same amount of alcohol than will be the average male. Females, on the average, weigh less than males, and feamles' bodies contain a higher proportion of fatty tissue relative to muscle tissue than males' bodies. In addition, females, unlike males, fluctuate during the course of the month in the concentration of alcohol in the bloodstream that they achieve from consuming a given amount of alcohol (and hence the level of intoxication they will experience). These fluctuations coincide with different phases of the female's menstrual cycle, water retention, and the level of estrogens (or female sex hormones) in the bloodstream. Females become most intoxicated from consuming a given quantity of alcohol immediately before menstration and become least intoxicated during menstration. Finally, some evidence suggests that even after adjustments are made for differences in bodily characteristics, women achieve higher blood alcohol levels (BAL) and are more susceptible to liver disease than are men. This difference appears to result from men's stomachs having a greater capacity to oxidize alcohol than women's, thereby allowing less alcohol to pass directly into men's bloodstream.

## 2. Degree of Tolerance to Alcohol

Some people can tolerate alcohol better than others. Tolerance to alcohol is partly inherited and is partly acquired through one's drinking experiences. That is, a habitual drinker needs to have a higher dose of alcohol in order to produce the same effect that was previously produced with a lower dose. Stated otherwise, if the same quantity of alcohol is consumed by a habitual drinker and an infrequent drinker or a nondrinker, the habitual drinker will likely become less intoxicated than will the person who is not so accustomed to drinking. In fact, research has confirmed that at a given blood alcohol level, the performance of chronic heavy drinkers is less impaired than that of moderate drinkers.

## 3. Ethnicity

Different ethnic groups metabolize alcohol at different rates, largely as a result of having inherited different levels of liver enzymes that are necessary for alcohol to be metabolized. In particular, some Asian people (and native Americans who are thought to be descended from Asians) lack a critical form of the enzyme aldehyde dehydrogenase to metabolize alcohol efficiently. As a result, acetaldehyde, which has considerable toxic effects on the body, accumulates instead of being converted into acetate. Thus, Asians and native Americans who metabolize alcohol in this manner will experience stronger negative physical effects (such as flushing of the skin and feeling sick to the stomach) from a given amount of alcohol than will people of other ethnic backgrounds.

It is interesting to note, however, that factors that encourage people to drink alcohol can sometimes be strong enough to override protection that they have inherited due to ethnicity. For instance, rates of alcohol consumption and problems associated with drinking run rampant among native Americans. This phenomenon can apparently be explained by the fact that the native American culture strongly encourages heavy drinking and that native Americans are motivated to drink in an attempt to escape from their unpleasant life circumstances. Moreover, although rates of alcohol consumption and problem drinking are relatively low among Asians, a certain percentage of people in Asian societies do, in fact, become alcoholic. Alcoholism among Asians probably occurs more frequently among those who are not deficient in aldehyde dehydrogenase. Nevertheless, it appears that factors that promote heavy drinking are sometimes strong enough to outweigh the biological protection against becoming alcoholic. [*See* Alcoholism.]

# V. ACUTE EFFECTS OF ALCOHOL INTOXICATION

As we have just seen, there are many variables that influence the level of intoxication that a person will achieve from consuming alcohol. Moreover, predictable behavioral and psychological consequences are associated with given levels of intoxication. This section discusses these acute effects of alcohol intoxication. Acute effects are those associated with a particular drinking bout—effects that are apparent while a person is intoxicated from alcohol or while he or she is withdrawing from it.

## A. Effects On the Central Nervous System

Once alcohol has been absorbed into the bloodstream, it is distributed throughout the body and has noticeable effects on each of the major organs and systems of the body. Alcohol crosses the blood–brain barrier and enters the brain, especially the cerebral cortex, the most highly developed part of the brain. In fact, the most obvious effects of alcohol intoxication are on the central nervous system. [*See* Central Nervous System Toxicology.]

Because alcohol is a depressant drug, it slows the activity of the central nervous system. Nevertheless, as alcohol begins to be absorbed into the bloodstream on a given drinking occasion, the drinker may appear initially to be stimulated rather than depressed. This paradoxical effect appears to result from the depressant action of alcohol on inhibitory centers in the brain that ordinarily serve to restrain a person's behavior. When the inhibitory centers are depressed, the net behavioral effect is for the person to be stimulated.

The behavioral effects of intoxication are clearly related to the concentration of alcohol in the bloodstream. The term blood alcohol level refers to the number of milligrams of alcohol in each 100 ml of blood, and the degree to which a person is impaired from having drunk alcohol can be predicted from his or her BAL. Nevertheless, relationships among BAL and the degree of impairment depend on each person's initial sensitivity to alcohol and the degree of tolerance that he or she has developed. Hence, the relationships that are discussed here should be regarded only as average approximations. When the BAL is approximately 20mg% (roughly equivalent, for the average person, to having consumed 1 drink), the person feels relaxed and mellow, but it will probably not be apparent to other people that he or she has been drinking. As the BAL rises above 20mg%, however, the depressant effects of the alcohol become apparent. At approximately 50mg% (roughly equivalent to having drunk 3 drinks), the average person begins to feel euphoric and uninhibited, is less alert, and experiences impaired judgment. At a BAL of 100mg% (i.e., equivalent to having consumed approximately 5 drinks and the legal criterion for intoxication in most states in the United States), the person's reaction time is slowed and motor coordination is impaired. At 200mg% (equivalent to 10 drinks), a person's bodily reactions are markedly depressed. At 300mg% (or 15 drinks), a person is stuporous and has no comprehension of the world around him or her. At 400mg% (or 20 drinks), alcohol may depress the medulla (a center in the lower part of the brain that controls breathing) and cause respiratory failure. In fact, this level of intoxication is lethal for 50% of the population.

It is interesting to note that people experience different effects from alcohol depending on whether their BAL is rising or falling. During a particular drinking occasion, we could think of a person's BAL as gradually rising from zero, reaching a peak level, and then gradually returning to zero. For any particular point on the ascending limb of the blood alcohol curve, there is a corresponding point on the descending limb where the concentration of alcohol in the bloodstream is exactly the same. Nevertheless, for a given point

on the ascending limb, a person is more likely to feel euphoric and be behaviorally impaired than at the corresponding point on the descending limb. In fact, as the BAL falls, a person might well feel depressed and experience other discomfort.

Finally, it is important to note that impairment from alcohol intoxication has been documented through neuropsychological testing. These tests have confirmed that a person is more impaired when the BAL is rising than when it is falling, and they have identified the following specific kinds of impairment: losses in attention; inability to concentrate; decreased performance on auditory, visual–spatial, and verbal tasks; impairment in fine-motor coordination; and, if a large amount of alcohol is consumed, retardation of motor speed.

## B. Alcohol and Hangovers

When people overindulge in alcohol, they frequently experience a "hangover" as the alcohol is being removed from the bloodstream. The hangover is characterized by such symptoms as headache, fatigue, nausea, stomachache, thirst, depression, and anxiety. In fact, it resembles a mild withdrawal syndrome. Two factors have been found to be related to the number of different hangover symptoms that a person experiences and the severity of each symptom: (1) the quantity of alcohol that is consumed, and (2) the level of congeners in the alcoholic beverage. The greater the quantity of alcohol consumed and the higher the level of congeners, the greater the severity of the hangover. Thus, a person is more likely to experience a hangover after drinking bourbon than after drinking vodka because bourbon has a higher level of congeners than vodka.

Scientists have identified two factors that contribute to the increase in thirst that is experienced during hangovers. First, alcohol decreases the output of the antidiuretic hormone that is responsible for retaining fluids in the body. Second, alcohol causes cellular dehydration because it forces fluid inside cells to move to the outside. Nausea, another symptom of hangover, appears to be caused by an accumulation of acetaldehyde in the bloodstream. As mentioned previously, acetaldehyde is quite toxic and, under normal conditions, is rapidly transformed by the enzyme aldehyde dehydrogenase into acetic acid. However, when the rate of drinking exceeds the rate at which alcohol is metabolized, acetaldehyde can accumulate in the bloodstream. Even in small quantities, it can cause nausea.

## C. Physical Dependence and Withdrawal from Alcohol

It is well known that physical dependence on alcohol can develop when a person drinks regularly and heavily over an extended period of time. People who are physically dependent on alcohol will experience withdrawal symptoms if they suddenly stop drinking. Withdrawal symptoms are similar to but much more intense than those of the hangover. In severe cases of physical dependence, the person withdrawing from alcohol will experience delirium tremens (DTs). During DTs, the person's body reacts as if it were overstimulated, and the person may experience frightening hallucinations and delusions, disorientation, agitation, and seizures of the body. Withdrawal from alcohol can be quite dangerous, and medical attention is usually required. Because drinking more alcohol will alleviate withdrawal symptoms, many alcoholics report that they continue to drink in order to forestall the discomfort of withdrawal.

## D. Alcohol Blackouts

Alcohol-induced blackouts are regarded by many authorities as a dangerous sign that one's drinking has gotten out of hand. Blackouts refer to sober alcoholics' loss of memory for events that occurred during periods of time when they were under the influence of alcohol. Although intoxicated, the drinker appears to function normally. He or she remains conscious, interacts with other people, and carries out routine activities. Later, while sober, however, the person cannot recall the earlier events.

## VI. CHRONIC TOXIC EFFECTS OF ALCOHOL ON MAJOR ORGANS AND SYSTEMS OF THE BODY

As already discussed, ample evidence exists to indicate that alcohol is toxic to the human body. Much of this evidence is related to chronic (as opposed to acute) negative consequences, i.e., the cumulative, harmful effects that come from regular, heavy use of alcohol. The following sections review this evidence, exploring the chronic effects of alcohol on each of the major organs and systems of the body.

## A. Central Nervous System

As with the acute effects of alcohol intoxication, the chronic toxic effects of alcohol are quite apparent on the central nervous system, particularly the brain. Research has clearly demonstrated that prolonged, repetitive drinking can alter the structure of the brain and how it functions. This conclusion was reached through several different methods of research.

### 1. Brain Atrophy Research

Brain atrophy (loss of brain cells) has long been considered a major consequence of alcoholism. In order to assess the extent of brain atrophy, the brains of chronic alcoholics have been compared with those of nonalcoholics. These studies have been conducted using techniques such as computer-assisted tomography (CAT scans), as well as through autopsies performed on deceased alcoholics. The CAT scan is a computerized technique for visualizing and measuring the structure of the brain. Studies utilizing CAT scans and those utilizing autopsies have both found a number of differences between the brains of chronic alcoholics and those of "normal" people to support the hypothesis that excessive drinking can cause structural damage to the brain. For instance, these studies have reported enlargement of the brain ventricles (i.e., the empty spaces in the brain) and widening of sulci (the space between folds in brain tissue) in alcoholics. Moreover, there is evidence that the extent of brain atrophy in alcoholics is "dose dependent." In other words, the more alcohol that the alcoholic has consumed, the greater is the observed damage to the brain.

In contrast to the studies that have identified damage to the brain generally, other studies have found abnormalities in specific regions of the brains of alcoholics, including the frontal cortex, temporal cortex, hippocampus, and the cerebellum. The cerebral cortex (of which the frontal cortex and the temporal cortex are a part) is the most highly developed part of the brain in humans and is responsible for the higher mental processes that are peculiar to humans. The hippocampus, a small area of the brain beneath the cerebral cortex, plays an important role in memory. In contrast, the cerebellum controls bodily movements.

When given tests to measure their intellectual functioning, alcoholics have shown specific patterns of deficits when compared to nonalcoholics. Namely, it is their ability to solve problems and think abstractly that is most impaired, whereas their ability to use language remains relatively intact. Although authorities have long believed that brain damage from chronic alcohol abuse is irreversible, recent CAT scan studies show that some alcoholics who stop drinking have a decrease in cerebral atrophy and a reduction in average ventricular and sulcal width.

## 2. Neurochemical Research

Neurotransmitters are chemicals in the brain that are responsible for transmitting messages between nerve cells. Recent research to investigate how drinking alcohol affects the neurotransmitters has advanced our understanding of the toxic effects of alcohol on the brain.

Norepinephrine is a neurotransmitter that is associated with a person's moods and level of arousal. Research has identified a relationship between alcohol consumption and the level of norepinephrine in the brain. Specifically, at low doses, alcohol facilitates the release of norepinephrine from individual nerve cells, whereas at higher doses, alcohol inhibits the release of norepinephrine. This finding helps to explain the paradoxical dose-related effect of alcohol on the central nervous system that was discussed earlier. That is, when alcohol is consumed initially and in small amounts it is stimulating, but when larger quantities are ingested, it has a depressant effect.

γ-Aminobutyric acid (GABA) is a neurotransmitter that inhibits the transmission of nerve impulses, and interruption of its normal functioning seems to be responsible for certain types of convulsions. Research on GABA has advanced our understanding of the relationship between seizures and heavy drinking. Laboratory studies have found that animals chronically given alcohol have a reduced number of receptor sites to which GABA can bind. As a consequence, these animals have less GABA available to control seizures than animals not given alcohol. This finding may help explain why seizures occur more frequently among alcoholics than other people.

Finally, other research has investigated the effects of alcohol on cellular membranes. Animal studies have suggested that acute alcohol exposure reduces the viscosity of the membranes of nerve cells, causing irregularities in the transmission of nerve impulses. Other studies have found that prolonged and regular ingestion of alcohol decreases these effects of alcohol on nerve cell membranes. This latter finding helps explain the development of tolerance and physical dependence on alcohol.

## 3. Research with Blood Flow Measures

Blood flow measurements have been used to determine rates of metabolism in different parts of the brain. Using this technique, scientists have discovered that alcoholics, when compared to normal people,

have a significantly lower level of cerebral blood flow. Moreover, older alcoholics (those aged 45 and over) have a significantly reduced blood flow compared to younger alcoholics, particularly at the lower frontal and anterior temporal areas of the cerebral cortex. Alcoholics' decreased blood flow reduces the ability of the brain to obtain sufficient nutrients and oxygen and to remove wastes. This impairment may be one factor that accounts for the brain damage among alcoholcis that was discussed earlier.

## 4. Research Using Event-Related Potentials

Another important method used to study the brain functioning of alcoholics uses event-related potentials (ERPs). Like the CAT scan, ERPs involve an advanced computerized method of research. This method involves measuring responses of the brain (i.e., brain waves) to repetitive visual (e.g., light) and auditory (e.g., sound) stimulation. ERPs are very sensitive measures that can detect disorders of the central nervous system.

Studies using ERPs to compare alcoholics and normal people have identified a number of differences between the two groups. For example, because the right hemisphere of the cerebral cortex of the brain seems to be more susceptible to the influence of persistent, heavy drinking, the amplitude of alcoholics' ERPs at the right frontal lobe of the cerebral cortex has also been found to be significantly reduced compared to that of normal people. This deficit seems to be related to alcoholics' impaired ability to understand visual-spatial relationships, such as that measured by block-design tasks on intelligence tests. Other research has found that the ERPs of young alcoholics are similar to those of elderly normal people. This finding supports the notion that excessive alcohol consumption causes the brains of alcoholics to age prematurely.

## 5. Alcohol and Wernicke–Korsakoff's Syndrome

The actual physical damage to the brain discussed in the previous sections is correlated with impairments in alcoholics' psychological functioning. One severe form of cognitive impairment resulting from years of heavy drinking is Wernicke–Korsakoff's syndrome. Patients with Wernicke–Korsakoff's syndrome are disoriented with regard to time and place, are unable to recognize familiar people, and have a loss of short-term memory (e.g., they are unable to memorize telephone numbers). These patients also confabulate (i.e., devise imaginary stories to fill the gaps in their memory). Perhaps most remarkably, Wernicke–Korsakoff's patients have a profound inability to learn new information. The memory problems associated with the syndrome is probably caused by damage to the hippocampus. Wernicke–Korsakoff's syndrome can be treated with large doses of vitamin B (thiamine) in order to improve significantly such symptoms as confusion and ataxia (the inability to coordinate voluntary muscular movements). However, rarely can a Wernicke–Korsakoff's patient achieve complete recovery.

## B. Toxic Effects of Alcohol on the Liver

As discussed earlier, alcohol is metabolized almost entirely in the liver. Therefore, ingestion of large amounts of alcohol affects the liver's ability to perform its bodily functions efficiently. Consequently, one of the toxic effects of excessive alcohol use is liver damage. Three types of liver damage can occur: fatty liver, hepatitis, and cirrhosis.

### I. Fatty Liver

One consequence of heavy alcohol consumption is a "fatty liver." Normally, fatty acids (lipids) are fuel for the liver, but when alcohol enters the liver, it has a higher priority for being metabolized than fatty acids from food substances. As a result, the unmetabolized fatty acids are accumulated and stored in liver cells, resulting in a fatty liver.

A fatty liver does not necessarily cause a major threat to health because the condition is reversible, i.e., if an individual stops drinking, the liver can utilize the stored fatty acids for energy. However, if the liver accumulates too much fatty acids, cell membranes may rupture, causing liver cells themselves to die. Furthermore, fatty acids from the ruptured cells entering the bloodstream may cause serious danger to the cardiovascular system (e.g., blockage of blood vessels or stroke). In extreme cases, a fatty liver can cause liver failure and even death, especially among younger people.

### 2. Hepatitis and Cirrhosis

Although it may have other causes, hepatitis is often related to prolonged, heavy consumption of alcohol, but it is not clear whether a fatty liver causes hepatitis. In any event, alcohol hepatitis is a serious liver disease which results in inflammation and impairment of liver function.

Cirrhosis, another liver disease that is related to prolonged heavy alcohol intake, is the seventh leading cause of death in the United States. With this disease, liver cells are replaced by fibrous tissue, which, in turn, changes both the structure and the function of the liver. Specifically, the loss of liver cells and the resulting structural changes in the liver impair liver function by decreasing blood flow. When the liver is unable to function properly, toxins start to accumulate in the blood. Bile pigments from the liver are distributed throughout the body, causing jaundice. Cirrhosis of the liver is irreversible, although cessation of alcohol intake can retard its development. It is clear that alcohol plays an important role, along with diet and heredity, in the development of cirrhosis. However, the extent to which the toxic effects of alcohol on the liver are direct or indirect has not been entirely resolved. [*See* Liver.]

## C. Other Toxic Effects of Alcohol on the Endocrine System

### I. Male Sex Hormones

Many male alcoholics have a variety of primary and secondary sexual problems, including sexual impotence, loss of libido, enlargement of breasts, loss of facial hair, and testicular atrophy. In fact, research has shown that chronic alcohol intake may lead to "feminization" of the male. Alcohol has this effect by breaking down testosterone (a male hormone that is responsible for sexual drive and secondary sex characteristics) and simultaneously increasing the rate at which androgens (testosterone and other male sex hormones) are converted into estrogens (the female sex hormones). Several studies have also indicated that alcohol administered either acutely or chronically can lower serum testosterone.

### 2. Other Endocrine and Metabolic Problems

Frequently, alcoholics are poorly nourished because they derive a major portion of their calories from alcohol rather than well-balanced meals. As a result of poor nutrition, alcoholics often have a low store of glucose, which is fuel for the body. Alcohol also inhibits the conversion of amino acids into glucose, and these two factors working together often contribute to the development of hypoglycemia. This is a condition involving reduced blood sugar that is often found in persons with diabetes who have taken too much insulin. In addition to inhibiting the conversion of amino acids into glucose, alcohol also interferes with the conversion of amino acids into other essential proteins that the liver produces.

Disorders of the pancreas also seem to be associated with heavy drinking. In fact, it has been found that about 75% of patients with chronic pancreatitis or pancreatic cancer are moderate to heavy drinkers. Chronic pancreatitis is a disease of the pancreas that causes abdominal pain and vomiting.

## D. Toxic Effects of Alcohol on the Cardiovascular System

Both alcohol and its highly toxic metabolite acetaldehyde have detrimental effects on the heart muscle. Heavy consumption of alcohol can lead to diseases such as alcoholic cardiomyopathy, coronary heart disease, and cardiac arrhythmias.

### 1. Cardiomyopathy

Cardiomyopathy (damage to the heart muscle) often develops after 10 years or more of heavy drinking. Patients with cardiomyopathy sometimes exhibit chronic shortness of breath, swelling of the ankles, and other signs of congestive heart failure, such as enlargement of the heart, noisy breathing, and disturbances of cardiac rhythm and conduction.

### 2. Cardiac Arrhythmias

Cardiac arrhythmias are irregularities of the heartbeat that occur both during and after alcohol intoxication and can lead to heart failure. For people who already have heart disease, ingestion of even small quantities of alcohol can suppress the heart's pacemaking abilities and endanger their lives. People whose hearts are ordinarily normal can sometimes develop cardiac arrhythmias after a weekend of heavy drinking and frequently are taken to hospital emergency rooms. For this reason, cardiac arrhythmias are sometimes referred to as the "holiday heart syndrome."

### 3. Heart Rate, Hypertension, and Strokes

Consuming alcohol can increase both pulse rate and blood pressure, and it can adversely affect the contractility of the heart muscle. Indeed, some studies have shown that blood alcohol levels associated with mild to severe intoxication decrease the strength of the pumping action of the heart. It is not surprising, therefore, that a large proportion of patients with unexplained heart disease, including heart muscle failure, are chronic heavy drinkers.

Drinking alcohol can be life-threatening for patients already suffering from heart disease. In fact, research has indicated that cardiac patients' vulnerability to both heart attacks and strokes is significantly increased after consuming even one alcoholic drink. Moreover, strong epidemiological evidence indicates that chronic heavy consumption of alcohol is significantly correlated with a higher prevalence of hypertension. Finally, several studies have found a positive link between chronic heavy drinking and the occurrence of strokes.

### 4. Peripheral Blood Effects

One interesting effect of alcohol on the cardiovascular system is on the peripheral blood vessels. Alcohol causes the peripheral blood vessels to dilate, making the drinker feel warm. In point of fact, the warm sensation actually results from loss of heat from the interior of the body. Therefore, contrary to popular belief that alcohol keeps people warm, it actually is quite dangerous to give alcohol to a person to drink who has been exposed to extreme cold. Such a person would require warmth in the internal organs of the body in order to survive.

## E. Toxic Effects of Alcohol on the Gastrointestinal Tract

Increasing evidence indicates that alcohol can damage or cause abnormalities to the esophagus, stomach, and small intestine, particularly when high concentrations of alcohol are consumed for a long period of time.

### 1. Esophagus

Alcohol can irritate the interior lining (mucosa) of the esophagus. It can also damage the esophagus by (1) inducing severe vomiting that tears the mucosa or (2) causing an upward movement of stomach acid into the esophagus, which can erode the esophageal tissue. These difficulties can lead to local pain, difficulty in swallowing, and hemorrhage. It has also been found that cancers of the mouth, tongue, and esophagus are more common among alcoholics than normal people.

### 2. Stomach and Small Intestine

The toxic effects of alcohol on stomach tissue are well established. Specifically, alcohol and the stomach acid that it induces can erode stomach tissue, causing inflammatory and bleeding lesions. With regard to the small intestine, it has been found that alcohol changes intestinal motility, speeding up the rate of propulsion of material through the small intestine and causing digestive disturbances such as diarrhea. Finally, alcohol contributes to the malabsorption of calcium and thiamine from the small intestine.

## F. Toxic Effects of Alcohol on the Fetus

Fetal alcohol syndrome, first identified by physicians at the University of Washington in 1973, describes a pattern of serious birth defects that have been observed among children of alcoholic mothers who drank heavily during their pregnancy. Collectively, the term refers to the following four categories of abnormalities: (1) central nervous system dysfunction, including mental retardation, poor motor coordination, irritability during infancy, and hyperactivity during childhood; (2) growth deficiencies, such as reduced body weight and length; (3) a characteristic cluster of facial abnormalities, including a short, upturned nose, an underdeveloped ridge between the base of the nose and the upper lip, a sunken nasal bridge, and retarded growth of the jaw; and (4) other major and minor malformations, such as septal defects, heart murmurs, limited joint movements, and renal (kidney) anomalies.

A growing body of evidence supports the link between maternal alcohol abuse and fetal alcohol syndrome. Several studies have found that maternal alcohol abuse during pregnancy is associated with low birth weight of the offspring. Low birth weight is frequently associated with mental retardation and neurological defects in newborn babies. In addition, studies have found that mothers who abuse alcohol during pregnancy have children with low IQs, retarded motor development, short attention spans, and slow reaction times in comparison to children born to nondrinking mothers. For ethical reasons, experimental research on fetal alcohol syndrome has been conducted only with laboratory animals. This research, however, leaves little doubt that heavy consumption of alcohol during pregnancy causes birth defects very similar to those found in humans.

Although there is strong evidence to link alcohol intake during pregnancy to fetal alcohol syndrome, it should be noted that alcohol may not be the only factor involved. Alcoholic mothers often are malnourished, smoke heavily, use drugs other than alcohol, and have life-styles that are less conducive to the normal development of the fetus than those of nonalcoholic mothers. These factors, together with excessive use of alcohol, may work together in the development of fetal alcohol syndrome.

Moreover, it should be recalled that fetal alcohol syndrome has been observed only in the offspring of mothers who clearly drank excessively during their pregnancy. The exact consequences of moderate alcohol use during pregnancy have not been clearly established and are currently a subject of controversy. Nevertheless, because safe levels of alcohol use during pregnancy have not been identified, it would seem prudent to caution pregnant mothers not to drink at all during pregnancy or, for that matter, not to use any other unnecessary medication. The use of any medication during pregnancy should be under the supervision of a physician.

## VII. CONCLUSIONS

We have seen that excessive consumption of alcohol has chronic deleterious effects on the major organs and systems of the body. These effects, along with acute intoxication from alcohol, are responsible for much physical and psychological suffering. Alcohol abuse is often associated with family violence, sexual abuse, child abuse, and broken homes, in addition to homicide, suicide, and other deaths. Human consumption of alcohol is also costly to society when measured monetarily. Tremendous sums of money are spent each year on alcoholic beverages, much of it by people who drink excessively. In addition, manufacturers of alcoholic beverages spend large sums to advertise their products. Besides the money spent directly on alcoholic beverages, however, alcohol costs society in terms of alcohol abusers' lost time from work and decreased productivity. Additional costs are incurred from alcohol-related accidents, including traffic and industrial accidents. Medical costs are also associated with alcohol abuse—those resulting both from alcohol-related accidents and from the treatment of alcoholism. In short, alcohol when drunk excessively is indeed a toxin, both to the human body and to society at large.

## BIBLIOGRAPHY

Carroll, C. R. (ed.) (1989). Alcohol: Drinking, alcohol abuse, and alcoholism. *In* "Drugs in Modern Society," 2nd Ed., pp. 97–152. William C. Brown, Dubuque, IA.

Cox, W. M. (ed.) (1990). "Why People Drink: Parameters of Alcohol as a Reinforcer." Gardner, New York.

Frezzo, M., di Padova, C., Pozzato, G., Terpine, M., Baraona, E., and Lieber, C. S. (1990). High blood alcohol levels in women: The role of decreased gastric alcohol dehydrogenase activity and first-pass metabolism. *N. Engl. J. Med.* **322**, 95–99.

Goodwin, D. W. (1985). Alcoholism and alcoholic psychosis. *In* "Comprehensive Textbook of Psychiatry" (H. I. Kaplan and B. J. Sadock, eds.), Vol. IV, pp. 1016–1026. Williams and Wilkins, Baltimore.

Institute of Medicine (1990). Broadening the Base of Treatment for Alcohol Problems." National Academy Press, Washington, D.C.

Pattison, E. M., and Kaufman, E. (1982). "Encyclopedic Handbook of Alcoholism." Gardner Press, New York.

Ray, O., and Ksir, C. (1992). Alcohol. *In* "Drugs, Society and Human Behavior," 6th Ed. Mosby Year Book, St. Louis, MO.

U.S. Department of Health and Human Services (1993). "Alcohol and Health." Washington, D.C.

# Alkaloids in Medicine

ARNOLD BROSSI

*National Institutes of Health*

## GLOSSARY

**Antagonist** Drug acting in opposition to the action of another drug

**Antipodes** Two nonsuperimposable mirror-image structures of the same molecule with opposite optical rotation; also called optical isomers or enantiomers

**Inhibitor** Agent that represses physiological activity

**Isomers** Substances with identical chemical composition, but different physical properties, such as geometrical isomers, optical isomers, and conformational isomers

**Opioids** Natural or synthetic compounds that have morphine-like pharmacological activity

**Organophosphates** Organic compounds containing phosphorus, including insecticides and nerve gases

**Psychotomimetics** Drugs altering the mind and the psyche

**Racemate** Equal mixture of $(-)-$ and $(+)-$ antipodes

**Receptor** Cell constituent that combines with a specific drug, effecting a change in cellular function

**Substrate** Any substance on which an enzyme acts

THE TERM "ALKALOID," ORIGINALLY RESERVED FOR nitrogen-containing substances with basic properties found in plants, has a much broader meaning today and includes such substances occurring in mammals, fish, and mushrooms as well. A clear separation of alkaloids from amino acids, amino sugars, purines, and peptides remains highly arbitrary. Alkaloids are compounds of complex structures and great structural variety; most of them are optically active.

Alkaloids are classified on the basis of chemical and biogenetic relationships. Two-thirds of the more than 10,000 alkaloids known are found in flowering plants (i.e., angiosperms).

Many alkaloids were used in traditional medicine in the form of extracts and powders. Serious research on alkaloids started with the isolation of morphine by Sertürner in 1817. The continued study of alkaloids has stimulated the development of synthetic drugs (i.e., analgesics, antimalarials, and anticholinergics). Many alkaloids are toxic, and intoxication of humans and animals by alkaloids is well known (e.g., by *Atropa belladonna, Amanita muscaria,* and strychnine). The formulas of alkaloids presented in this article are shown with their correct absolute configurations. Biological activity in most cases depends on a defined three-dimensional molecular expression (left- and right-handed). Alkaloids are isolated from plant materials and plant extracts by treatment with acids and are purified in various ways. Alkaloids are recognized on thin-layer chromatography plates by their fluorescence or upon spraying with alkaloid reagents (e.g., iodoplatinate or Dragendorff reagent), which gives colored spots. For quantitative analysis, gas chromatography and high-performance liquid chromatography are often used, and the substances are characterized by their physical properties (e.g., by specific rotations $[\alpha]_D$, optical rotatory dispersion, or circular dichroism). No general procedure is available for the technical production of alkaloids.

ENCYCLOPEDIA OF HUMAN BIOLOGY, Second Edition, VOLUME I. Copyright © 1997 by Academic Press. All rights of reproduction in any form reserved.

## I. PHENYLALKYLAMINES

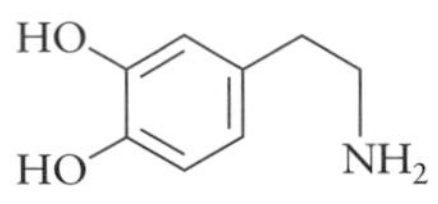

Dopamine

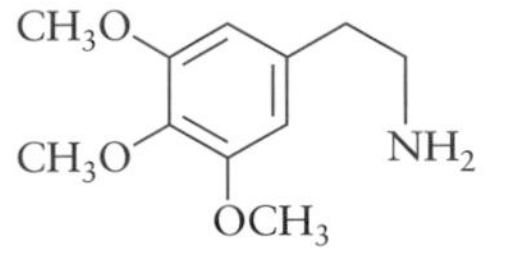

Mescaline

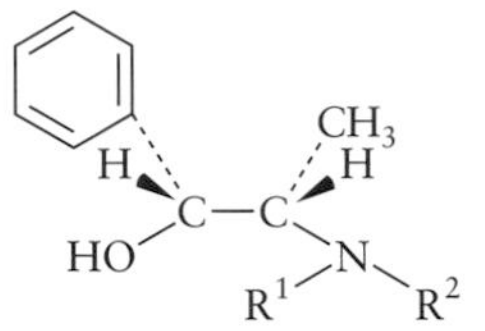

(–)-Ephedrine
($R^1$ = H; $R^2$ = $CH_3$)

The phenylalkylamines include the neurotransmitter dopamine, biochemically related to other biogenic amines such as norepinephrine and epinephrine, the hallucinogen mescaline from certain cactus species (e.g., peyote), and *Ephedra* alkaloids present in plants of the genus *Ephedra*. Ephedrines were used in China for centuries. Natural ephedrine, now prepared by synthesis, releases norepinephrine from adrenergic nerve endings and is widely used as a bronchodilator and as an orally active sympathomimetic agent.

## II. PYRIDINE ALKALOIDS

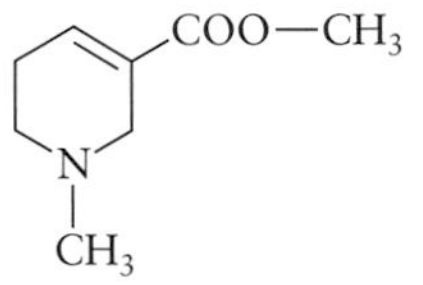

Arecoline

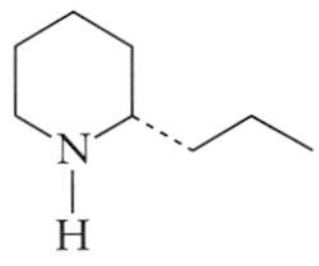

(+)-Coniine

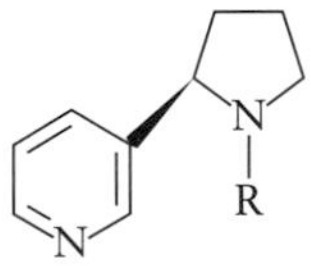

(–)-Nicotine
(R) = $CH_3$

The pyridine alkaloids include compounds containing a pyridine ring that can be partially or fully reduced. The betal nut (i.e., seeds of *Areca catechu* L.) is widely used in the Middle East as a stimulant, and most of its effects are caused by the alkaloid arecoline. Toxic effects of poison hemlock (*Conium maculatum* L.) originate from coniine, the toxin that poisoned Socra-

tes. Most important, however, is the alkaloid nicotine, present in the tobacco plant. It is produced in large quantity by the tobacco industry and is used in the form of a crude extract as an insecticide. Pure nicotine is as toxic as hydrogen cyanide and affects the peripheral and central nervous systems. It has a stimulating effect, but intake of larger doses can lead to paralysis, vasoconstriction, and tachycardia.

## III. TROPANE ALKALOIDS

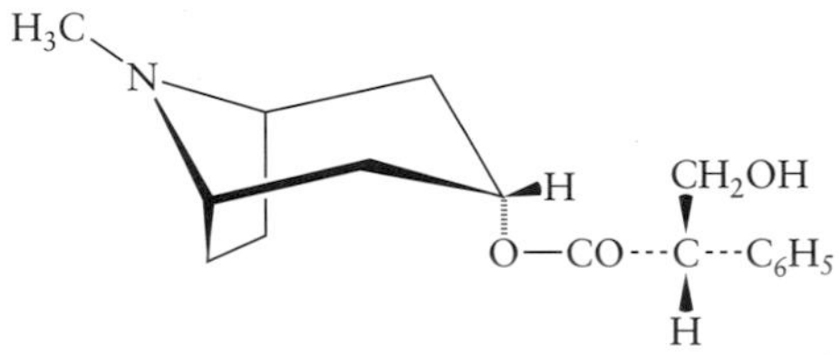

(–)-Atropine

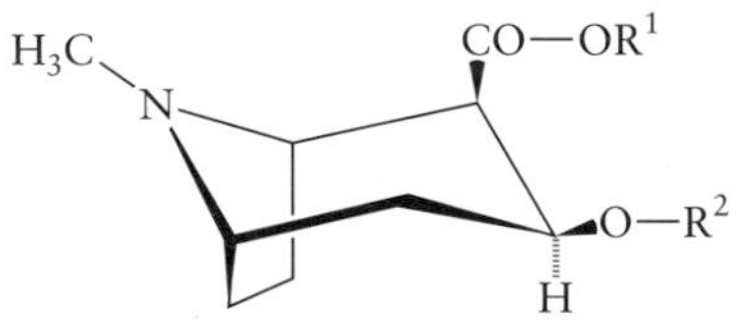

(–)-Cocaine
($R^1$ = $CH_3$; $R^2$ = —CO—$C_6H_5$)

Toxic effects produced by extracts of *Atropa belladonna* and dilating effects of extracts of *Hyoscyamus niger* L. have been known since ancient times, but the local anesthetic effect of cocaine from *Coca* leaves was only discovered a century ago. Atropine now prepared by synthesis is widely used to dilate the pupils (i.e., mydriasis) and as an antidote in organophosphate poisoning (e.g., insecticides and nerve gases) and is the active principle of *A. belladonna*. Natural hyoscyamine, the active principle of *H. niger* L., has medical uses similar to those of synthetic atropine. Cocaine today is a drug of abuse of enormous proportions. It inhibits the uptake of norepinephrine in nerve endings. Its use in society as a mood stimulator is associated with addiction (i.e., cocainism).

## IV. ISOQUINOLINE ALKALOIDS

The isoquinoline alkaloids contain an isoquinoline moiety and are among the most abundant plant alkaloids. They are divided into several subgroups; two of these are presented below.

## A. Morphine Alkaloids

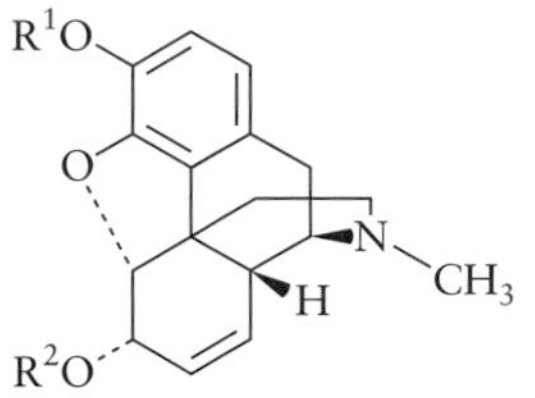

Morphine
(R¹ = R² = H)
Codeine
(R¹ CH₃; R² = H)
Heroine
(R¹ = R² = —CO—CH₃)

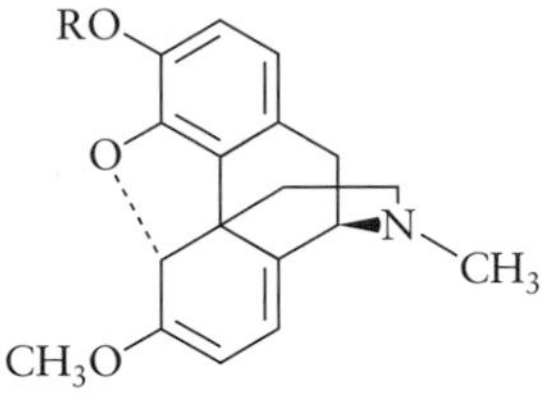

Thebaine
(R = CH₃)

The most important morphine alkaloids are thebaine, morphine, and codeine. Both thebaine and morphine are major alkaloids of the poppy plant (*Papaver somniferum* L.). Morphine is extracted from opium, the dried juice of the poppy. This plant is cultivated in Turkey, Pakistan, and other Eastern countries. Codeine, widely used as an analgesic and antitussive agent, is chemically prepared from morphine by O-methylation. Heroine, an illict drug in many countries, is prepared from morphine by acetylation. Thebaine is a crucial intermediate in the biosynthesis of morphine in the poppy and in mammals (i.e., mammalian morphine). Changing the N-methyl group in morphine into an N-allyl group converts the agonist into an antagonist (nalorphine). Most natural morphine alkaloids show high affinity for opioid receptors. Several total syntheses of morphine were achieved, and that by Rice at the National Institute of Health seems technically feasible.

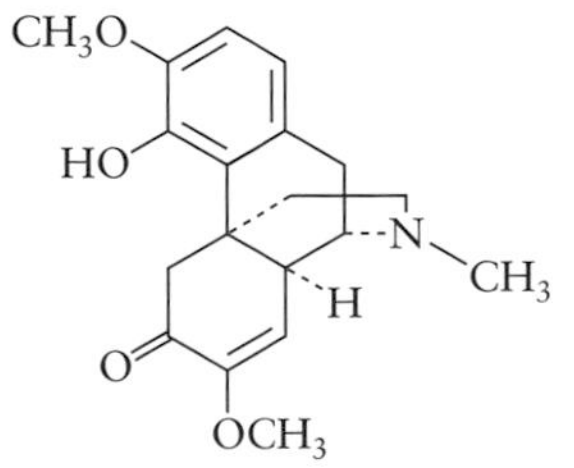

Sinomenine

Sinomenine, from the plant *Sinomemium acutum*, belongs to the unnatural series of morphine alkaloids (i.e., right-handed), and simpler analogs prepared by synthesis are widely used as antitussive agents (e.g.,

dextromethorphan). The unnatural isomers do not bind to opiate receptors and have little addiction and abuse potential.

## B. Biosynthesis of Opium Alkaloids

Knowledge on which alkaloids are being synthesized in plants affords valuable information for their synthesis with biotechnological methods (e.g., fermentation and tissue culturing). The early stages in the biosynthesis of reticuline in poppy plants, required as the (R)-enantiomer for its conversion into morphine, have now been fully elucidated by Zenk at the University of Munich, Germany. The synthesis proceeds via several intermediates prepared in enzymically controlled reactions, as illustrated below:

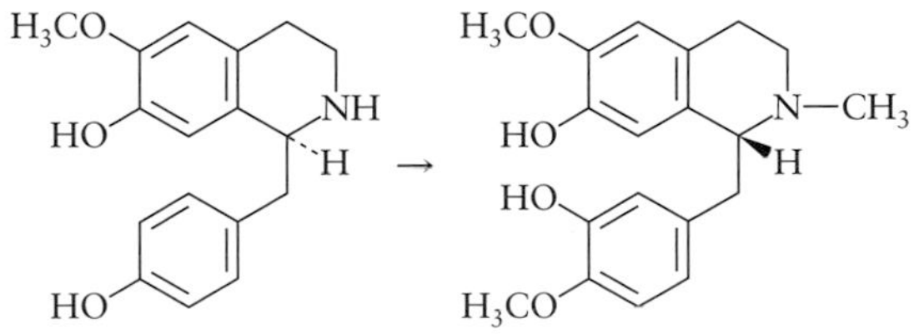

R-(−)-Reticuline

## C. Mammalian Alkaloids

Evidence has accumulated that isoquinolines and β-carbolines, closely related in structure to plant alkaloids, are trace chemicals in mammalian fluids and tissues. Representative mammalian alkaloids are salsolinol and 3′,4′-dideoxynorlaudanosoline-1-carboxylic acid (DNLCA).

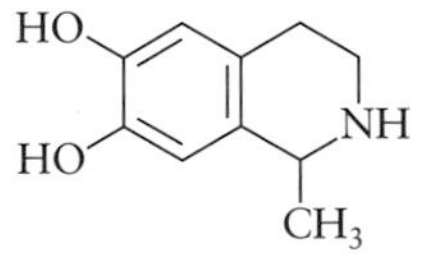

Salsolinol

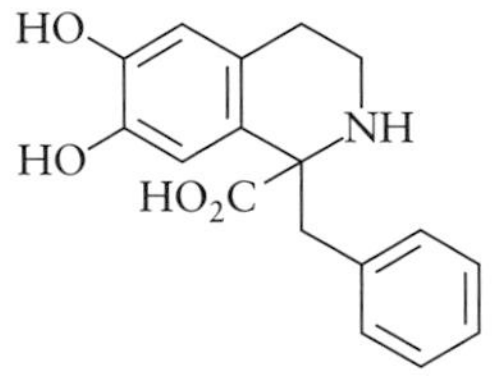

DNLCA

Salsolinol was detected in the urine of parkinsonian patients undergoing treatment with L-dopa, and its levels are greatly enhanced in alcoholics. DNLCA is present in significant amounts in the urine of phenylketonurics. Only alkaloids derived from dopamine, tryptamine, and possibly serotonine have so far been identified to occur in mammals. The carbonyl substrates of these amines are aldehydes and $\alpha$-keto acids. Although mammalian alkaloids have a wide range of physiological and behavioral effects in experimental animals, their possible role in biology remains open to question. Another mammalian alkaloid recently found to be present in trace amounts in human brain tissue is natural morphine, and it is believed that it originates by a pathway similar to that used by the poppy plant.

## V. INDOLE ALKALOIDS

Indole alkaloids are presented by more than 1000 alkaloids of great structural variety and are derived from the amino acid L-tryptophan.

### A. Simple Indole Alkaloids

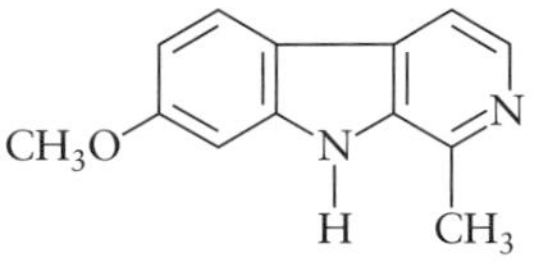

Harmine

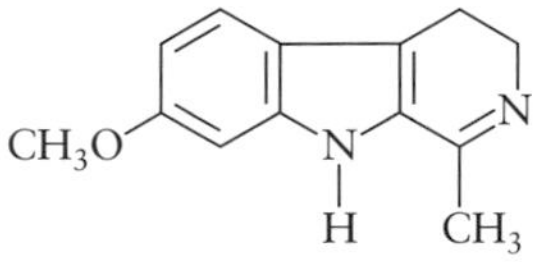

Harmaline

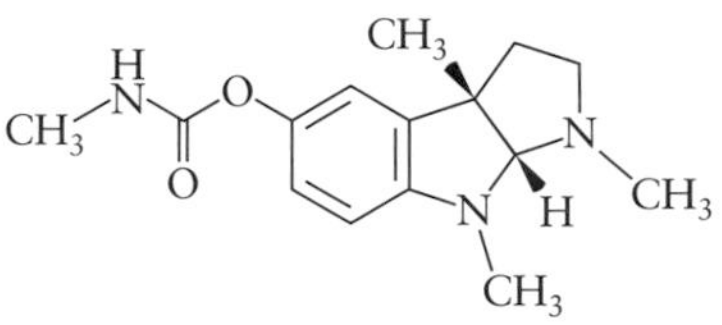

Physostigmine

Harmine and harmaline are present in *Peganum harmala* and can be obtained by synthesis. They are potent irreversible inhibitors of the enzyme monoamine oxidase, which biodegrades amines. Both have been used clinically as antidepressant drugs. Physostigmine, isolated from seeds of the West African vine *Physostigma venenosum* (Calabar bean), inhibits acetylcholinesterase, the enzyme that controls the conversion of acetylcholine into choline. It is clinically used to suppress intraocular pressure (i.e., glaucoma), is used against organophosphate poisoning (e.g., insecticides and nerve gases), and seems, when given over an extended period, to have a beneficial effect in Alzheimer's disease patients. Total syntheses of natural physostigmine and its unnatural isomer, which has no effect on acetylcholinesterase, have been achieved.

### B. Ergot Alkaloids

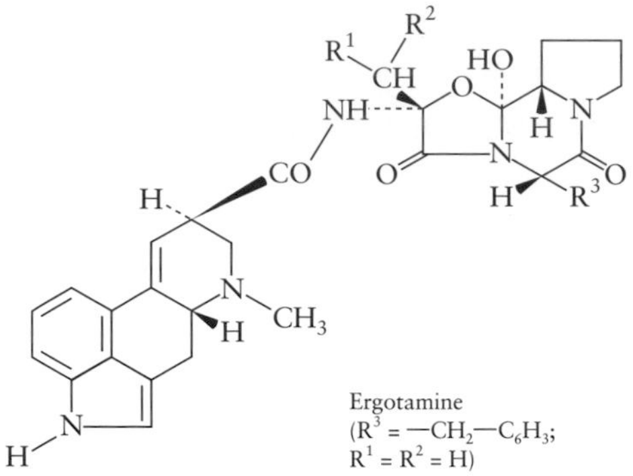

Ergotamine
($R^3 = $ —$CH_2$—$C_6H_3$;
$R^1 = R^2 = $ H)

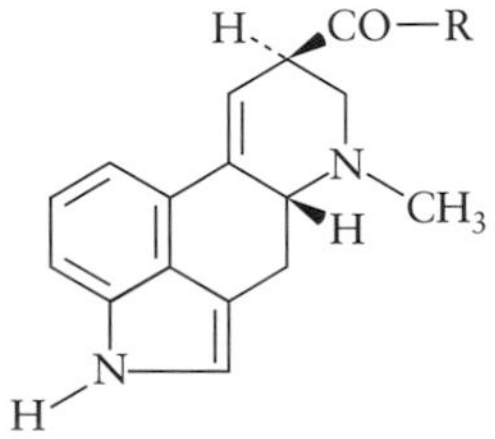

N, N-Diäthyllysergamide
(R = N $(C_2H_5)_2$)

The fungus *Claviceps purpurea*, which infects rye and corn, was for many years a source for the production of ergot alkaloids (Sandoz, Basel, Switzerland). Ergotamine stimulates smooth muscle, especially of blood vessels and the uterus, and stimulates contraction. Hydrogenation of the 9,10-double bond produces compounds which show marked changes in the pharmacological properties. The hydrogenated alkaloids are useful for the treatment of essential hypertension and migraine. Lysergic acid diethylamide is a potent psychotomimetic agent and a dangerous drug of abuse.

### C. Bisindole Alkaloids from *Catharanthus roseus* L.

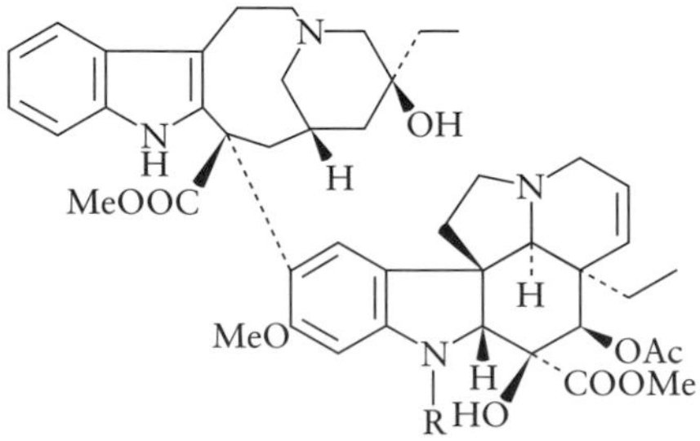

Vinblastine
(R = $CH_3$)
Vincristine
(R = CHO)

The most important alkaloids used in medicine today are bisindole alkaloids from *C. roseus* (vinca alkaloids). They are present in relatively small amounts and have complex chemical structures. The alkaloids vinblastine and vincristine are useful for the treatment of malignant and nonmalignant diseases (e.g., leukemia, breast cancer, and Hodgkin's disease). They are given by intravenous injection and often are used in combination with other antitumor agents. The mechanism of action involves entry into the cell, binding to tubulin, and interference with cellular metabolic functions.

## D. Physostigmine (Eserine)

Physostigmine is present in Calabar beans and is used in medicine to treat glaucoma and urinary retention, is used in the management of myasthenia gravis, and seems to be beneficial in Alzheimer's disease. Recent research conducted by Brossi at the National Institutes of Health has led to the finding that phenserine, the phenylcarbamate analog of physostigmine, is more selective in its inhibition of cholinesterases, less toxic, and much longer acting. Total synthesis of natural physostigmine and optically active phenserine has been achieved.

R = CH₃   Physostigmine
R = Ph    Phenserine

## VI. CINCHONA ALKALOIDS

Quinine                    Quinidine

Quinine obtained from the bark of the Cinchona tree, cultivated in India and Java, has found useful application as an antimalarial agent. Quinine has been obtained by total synthesis. Quinidine, prepared from quinine by isomerization, and dihydroquinidine, obtained from quinidine by reduction of the double bond, are both widely used to control cardiac arrhythmias.

## VII. COLCHICINE

Colchicine
(R¹ = —CO—CH₃; R² = H)
Demecolcine
(R¹ = CH₃; R² = H)

Alkaloids from *Colchicum autumnale* (meadow saffron) and *Gloriosa superba* are represented by colchicine and demecolcine. Crude extracts of *Colchicum* plants were used for centuries against gout, and the less toxic demecolcine was found to be clinically useful in chronic leukemia. It is believed that the antimitotic effect of colchicine and its congener alkaloids originates by interfering with cell division (mitosis) through binding to tubulin and inhibition of the formation and function of the mitotic spindle. The therapeutic effect of the alkaloids as antitumor agents is close to their toxic dose. Colchicine is the drug of choice in acute gout and in familial mediterranean fever. The structure of colchicine was determined by X-ray, and several syntheses exist. Synthetic analogues of colchicine have recently been developed.

## BIBLIOGRAPHY

Bernauer, K. (1973). Alkaloide. *In* "Ullmann's Enzyclopädie der technischen Chemie," Vol. 7. Chemie GmbH, Weinheim, Federal Republic of Germany.
Brossi, A. (ed.) (1983–1991). "The Alkaloids," Vols. 20–40, and with Cordell, G. A. (coed.) (1992 and 1994), Vols. 41 and 45. Academic Press, San Diego.
The Chemical Society (1971–1983). "The Alkaloids," Specialist Periodical Reports, Vols. 1–13. Chem. Soc., London.
Cordell, G. A. (1981). "Introduction to Alkaloids." Wiley, New York.
Cordell, G. A. (ed.). "The Alkaloids," Vols. 42–44 and 46 and 47. Academic Press, San Diego.
Hesse, M. (1981). "The Alkaloids." Wiley, New York.
Manske, R. H. F. (ed.) (1950–1973). "The Alkaloids," Vols. 1–20. Academic Press, New York.
Pelletier, S. W. (1970). "Chemistry of the Alkaloids." Van Nostrand-Reinhold, New York.
Pelletier, S. W. (ed.) (1983–1992). "Akaloids: Chemical and Biological Perspectives," Vols. 1–8. Wiley (Interscience)/Springer-Verlag, New York.
Southon, I. W., and Buckingham, J. (1989). "Dictionary of Alkaloids." Chapman & Hall, London.

# Allergy

SANDRA C. CHRISTIANSEN
BRUCE L. ZURAW
*The Scripps Research Institute*

## GLOSSARY

**Antibody** Immunoglobulin molecule produced in response to antigen exposure with the ability to specifically recognize and combine with the inciting agent

**Antigen** Molecule capable of inducing antibody formation

**Atopy** Clinical expression of allergic, or Type I, hypersensitivity reactions

**Epitope** Single antigenic determinant or recognition unit for the antibody molecule

**Idiotype** Unique determinant expressed on the variable region of the heavy or light chains of the antibody, where antigen is recognized

**Isotype** Constant region determinant on immunoglobulins defined by separate constant region genes; heavy-chain isotypes define the immunoglobulin class or subclass

THE TERM "ALLERGY" WAS ORIGINATED TO DESIGnate a changed reactivity of the host when reencountering a specific agent. It can be used broadly to encompass any adverse hypersensitivity reaction. A general mechanistic classification of hypersensitivity responses has been provided by Gel and Coombs (Table I). The definition of allergy has evolved to become virtually synonymous with Type I or immunoglobulin E (IgE)-mediated immediate, hypersensitivity. The expression of allergy begins with the recognition of antigen (known as the allergen) by specific IgE molecules affixed to high-affinity receptors on mast cells and basophils. The ensuing local or systemic reaction involves the interplay of released preformed and newly generated mediators from these cells acting in concert with recruited cellular effectors. These early events can occur within seconds of allergen exposure. The importance of more protracted late phase inflammatory reactions, which may take place hours following allergen recognition, has also been investigated. [*See* Inflammation.]

## I. PATHOPHYSIOLOGY OF THE ALLERGIC RESPONSE

### A. Structure

IgE shares the same basic four-chain structure of other immunoglobulins, being composed of two heavy and two light chains joined by disulfide bonds (Fig. 1). The molecular weight is estimated at 190,000, with 10.7–11.7% contributed by carbohydrates. The $\varepsilon$ isotypic determinants are located on the heavy chain, which is composed of five domains, one variable and four constant. Each domain represents an individual exon at the gene level. The light chains are common to the five classes of immunoglobulins, either of the $\kappa$ or $\lambda$ type. Again, these can be divided into domains, one variable and one constant. The antigen recognition sites are located at the amino-terminal ends of the molecule. This region has also been designated the $F(ab')_2$ fragment and can be isolated following pepsin digest. Within this area are the unique, or idiotypic, determinants contributed by heavy and light chains.

The term "reagin" was used prior to identification of the IgE molecule to describe a heat-labile factor in serum able to confer immediate hypersensitivity. The heat-labile property of IgE is accounted for by the Fc

ENCYCLOPEDIA OF HUMAN BIOLOGY, Second Edition, VOLUME I.   Copyright © 1997 by Academic Press.   All rights of reproduction in any form reserved.

TABLE I

Classification of Immune Responses

| Type I | Immediate hypersensitivity reactions; dependent on specific triggering of IgE sensitized mast cells and basophils resulting in release of inflammatory mediators |
|---|---|
| Type II | Antibody dependent cytotoxicity; antibody directed against tissues or cells may engage cellular effectors or activate complement resulting in damage |
| Type III | Immune complex disease; immune complex formation leading to triggering of inflammatory processes (often complement amplified), either local or systemic |
| Type IV | Cell mediated hypersensitivity; transferred by T-cells sensitized to a particular antigen |

region of the molecule. This area can reversibly bind to IgE receptors. The regional site for binding to the high-affinity receptor is located on the third constant domain (C$\varepsilon$3), formed by a ridge composed of three loops containing charged amino acids. Only a single $\varepsilon$ chain is thought to interact directly with the receptor.

## B. Receptors

Because of its location on mast cells and basophils, the high-affinity IgE receptor (Fc$\varepsilon$RI) plays a central role in the allergic response. Fc$\varepsilon$RI has additionally been identified on eosinophils where it participates in the IgE-dependent killing of schistosomes *in vitro*. Langerhans cells and activated monocytes also display Fc$\varepsilon$RI where its function is still subject to some speculation. The receptor is composed of three subunits ($\alpha$, $\beta$, and $\gamma$), with the IgE-binding site located on the $\alpha$ chain. Modeling studies based on amino acid distribution of the $\alpha$ chain predict a transmembrane configuration with most of the molecule exposed on the extracellular surface. The $\beta$ subunit is strongly hydrophobic, with four apparent transmembrane segments, and is nonhomologous with other known proteins. In contrast, the $\alpha$ chain shares sequence similarity with the binding chain for the IgG receptor and has two immunoglobulin-like domains. There are two $\gamma$ subunits, linked by disulfide bonds. The $\gamma$ as well as the $\beta$ chains are required for signal transduction to occur.

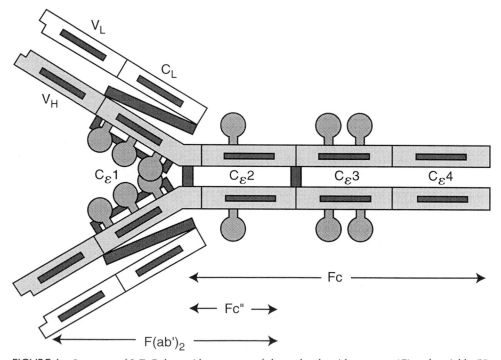

**FIGURE I** Structure of IgE. Polypeptide structure of the molecule with constant (C) and variable (V) domains. Two heavy (H) and two light (L) chains are connected by inter- and intrachain disulfide bonds. Oligosaccharide units are depicted as spheres. The F(ab′)$_2$ region contains the antigen binding sites and is generated by enzymatic digestion of the molecule. The Fc portion refers to the binding site, and Fc″ refers to the region of overlap between the two segments. [Reprinted from I. M. Roitt, J. Brostoff, and D. K. Male, "Immunology." Gower, London, 1985. Courtesy of Roitt, Brostoff, and Male and Gower Medical Publishing.]

The low-affinity receptor for IgE (FcεRII or CD23) is found primarily on B lymphocytes. The binding site of FcεRII is mapped to the Cε3 of the IgE molecule but appears distinct from the FcεRI-binding site. Low-affinity receptors have also been identified on monocyte/macrophages, Langerhans cells, eosinophils, platelets, and a small percentage of T cells. Two distinct forms of FcεRII have been described, a and b. The a form is exclusively found expressed in antigen-activated B cells prior to differentiation into plasma cells. FcεRIIb is inducible by interleukin 4 (IL-4) in all cell types listed to carry the receptor. The FcεRII is a single-chain glycoprotein divided into a short cytoplasmic domain, a transmembrane segment, and a large extracellular domain. It has no homology with FcεRI. Functionally, FcεRII is involved in the regulation of IgE synthesis (discussed in the following sections), as well as in other aspects of both the humoral and cellular immune response. FcεRII participates in the phagocytosis of immune complexes *in vitro* by monocytes, B-cell IgE-dependent antigen presentation to T cells, and possibly through its location on follicular dendritic cells, homing of B cells to germinal centers of secondary follicles.

## C. Synthesis

As with all immunoglobulins, IgE is produced by plasma cells derived from B lymphocyte precursors. The remarkable diversity of these molecules is governed by a family of immunoglobulin genes. The composition of each immunoglobulin chain is determined by the internal rearrangement of germline DNA during the differentiation of precursor B cells. A complete immunoglobulin heavy-chain molecule requires the juxtaposition of selected variable (V), joining (J), diversity (D), and constant (C) regions. During this process, interspersed segments or introns of the initially linear DNA are spliced out. These areas may participate in the process of gene rearrangement by annealing to one another, allowing for juxtaposition of the selected segments. In the light chain, V and J genes are then linked to the appropriate C region. A D region gene appears between V and J regions in the production of the heavy chain, and again is spliced to the appropriate C gene. Further potential for variability in the immunoglobulin molecule is introduced by alteration in the exact site of recombination of the gene segments. Somatic mutation has also been observed in the V gene and intervening sequences.

## D. Regulation of Synthesis

Regulation of the immune response involves a complex array of specific cells, tissues, and soluble components. IgE is subject to the same influences as other immunoglobulins, as well as to certain unique controls. The general framework of antigen-presenting cells, balance of T-cell helper and suppressor functions, cytokines, and their interactions with B cells therefore affect IgE production. The immunoglobulin itself may also participate in its overall control by serving as an antigen. Additional antibodies raised against the idiotypic portion of the immunoglobulin are considered to play a regulatory role in antibody specificity. [*See* B-Cell Activation; T-Cell Activation.]

IgE synthesis is regulated at several different levels. Isotype switching to synthesis of IgE is now recognized to be differentially regulated by cytokines released by T helper (Th) 1 or Th2 cells. When activated, Th2 cells release IL-4, IL-5, IL-10, and IL-13 while Th1 cells release IFN-γ and IL-2. IL-4 is required for IgE synthesis, providing the essential first signal for isotype switching.

Mice that are genetically engineered by homologous recombination to be IL-4 deficient cannot synthesize IgE. In contrast, IFN-γ inhibits IL-4-dependent IgE synthesis. Furthermore, allergen-specific T cells obtained from sites of allergic inflammation in atopic subjects have been shown to be predominantly of the Th2 subtype whereas Th2 cells are rarely found in nonatopic subjects. The mechanisms regulating Th2 development are not clearly known; however, the presence of cytokines in the microenvironment at the time of allergen presentation appears to be an important factor in the development of Th cells. IL-4 also promotes the differentiation of Th2 cells, while IL-12 (an inducer of IFN-γ) promotes the differentiation of Th1 cells. Interestingly, mast cells can both act as antigen-presenting cells and synthesize IL-4, suggesting that mast cells may enhance IgE synthesis.

Activation of allergen-specific B cells occurs following the uptake of allergen via surface Ig molecules. The allergen is then processed and presented on the B-cell surface in association with MHC class II molecules. A cognate T–B-cell interaction occurs between the allergen peptide/MHC complex and the T-cell receptor, resulting in the activation of both cells and the secretion of cytokines from the activated T cell. Noncognate T–B-cell interactions mediated through the engagement of B-cell CD40 by T-cell CD40-L provide a necessary second signal for optimal IgE synthesis. [*See* Major Histocompatibility Complex.]

FcεRII has pleiotropic effects on IgE synthesis. FcεRII on the surface of B cells may participate in IgE-dependent allergen presentation. Proteolysis of FcεRII generates multiple soluble fragments (sFcεRII), all retaining specificity for binding IgE. These fragments may have important regulatory functions in the control of IgE synthesis. Those of higher molecular mass ($\geq$29 kDa) promote synthesis of IgE, while a smaller fragment (16 kDa) appears to result in inhibition. Interestingly, IgE itself inhibits the release of soluble FcεRII fragments. Larger sFcεRII fragments are thought to up-regulate IgE synthesis by cross-linking membrane

IgE and complement receptor 2 (CR2) on the surface of B cells. Through its ability to act as a ligand for CR2, FcεRII also participates in the rescue of germinal center B cells from apoptosis. A complex network for the feedback control of IgE synthesis has been proposed by Sutton and Gould and is presented in Fig. 2.

The isotypic regulation of IgE synthesis in mice has been described by Ishizaka. Two IgE-binding factors have been isolated from T cells: one enhances IgE synthesis whereas the other suppresses it. Both are glycoproteins in the molecular weight range of 13,000–15,000 and differ only in their carbohydrate

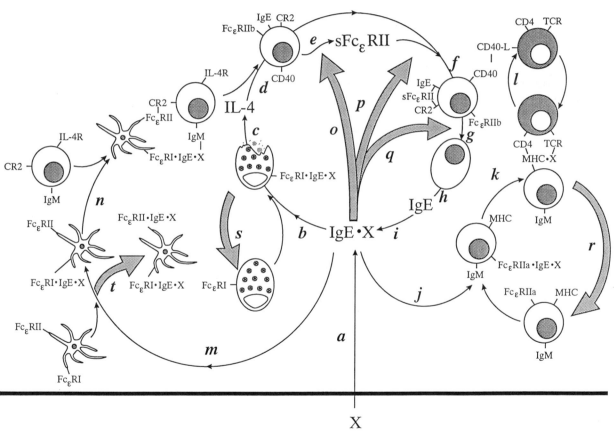

**FIGURE 2** Feedback control of IgE synthesis. Positive signaling (thin arrows): (*a*) Antigen X enters the system and forms a complex IgE·X; (*b, j,* and *m*) alternative fates for IgE·X at low concentrations of IgE. (*b*) Binding of IgE·X to FcεRI on a mast cell. (*c*) Secretion of IL-4 (and other inflammatory mediators that are not shown) activated by binding of IgE·X to FcεRI. (*d*) Induction of FcεRIIb and CD40 expression, and class switching to IgE, in an antigen-activated B cell by IL-4. (*e*) Release of sFcεRII from the B cell itself or associated Langerhans cell. (*f*) Binding of sFcεRII to CR2 and mIgE, promoting survival of the IgE-committed B cell. (*g*) Differentiation of the cell into a plasmacyte, stimulated by the ligand of CD40 (CD40-L). (*h*) Secretion of IgE. (*i*) Binding of IgE to the antigen X. (*j*) Binding of IgE·X to FcεRIIa on the surface of a B cell. (*k*) Antigen presentation to a T cell specific for the processed antigen X. (*l*) Induced expression of CD40-L by the activated T cell and binding to CD40 on an IgE-committed B cell. Expression of CD40-L by mast cells (not shown) as well as T cells may obviate a strict requirement for the latter cells in the periphery. (*m*) Binding of IgE·X to FcεRI on a Langerhans cell. (*n*) Antigen presentation to a B cell involving the suggested coligation of FcεRI-IgM and FcεRII-CR2. Negative signaling (thick arrows): (*o*) Inhibition of the cleavage of FcεRIIb on the B cell (or Langerhans cell) by IgE·X. (*p*) Inhibition of the binding of sFcεRII to CR2 on the B cell by IgE·X. (*q*) Inhibition of the secretion of IgE, caused by IgE·X cross-linking of FcεRIIb on the B cell. (*r*) Degradation of IgE in the course of antigen presentation and (*s*) after internalization by mast cells. (*t*) Switching off the Langerhans cell by saturation of FcεRII at a high IgE concentration. (Reprinted with permission from *Nature*. Copyright 1993 Macmillan Magazines Limited.)

moiety. A series of studies has led to the conclusion that these factors share a common polypeptide precursor, with functional distinction subsequently conferred by their glycosylation. A common antigenic determinant is shared by the two binding factors as well as Fcε receptors on T and B cells. The binding factors are formed by Fcε-bearing T cells under the combined influence of IFN-γ released from activated antigen-primed helper T cells and two additional factors released from discrete activated T-cell subsets. The glycosylation-enhancing factor (GEF) is secreted from Lyt-1$^+$ T cells and promotes the assembly of N-linked oligosaccharides. This factor is also characterized by kallikrein-like activity. Angiten-primed Lyt-2$^+$ T cells release glycosylation-inhibiting factor (GIF), which inhibits this assembly. GIF is homologous with a fragment of phosphorylated lipomodulin. GEF acting with IFN-γ on unprimed FcεR$^+$ T cells results in an IgE-potentiating factor, whereas GIF costimulation results in an IgE-suppressive factor. The balance between GEF and GIF therefore determines the nature of the IgE-binding factor produced and, hence, IgE synthesis. [See T-Cell Receptors.]

IL-4 induces the formation of GEF by a subset of T cells, thereby paving the way for increased production of the IgE-potentiating factor. Reciprocal control is exerted by IFN-γ in this system. In mice, strains that are high responders with respect to IgE synthesis have also been shown to preferentially form GEF, while poor responders form GIF. Similar isotypic regulation of IgE synthesis appears to be operative in humans. Potentiating IgE-binding factors have been described from T cells of patients with hyper-IgE syndrome and atopic dermatitis. Suppressive factors have also been found in the sera of individuals with low IgE. T-cell lines bearing FcεR have been established from patients with hyper-IgE syndrome. Factors derived from the supernatants of these cells bind IgE and can enhance IgE synthesis by B cells. The molecular weight of the factors lies in the range of 15,000–60,000.

## I. Mast Cells and Basophils

Basophils originate from bone marrow pluripotential hematopoietic CD34$^+$ stem cell precursors. In keeping with other leukocytes, the basophil completes differentiation in the bone marrow and then enters the circulation, comprising approximately 0.5% of the total pool. They can then be recruited into tissue during inflammatory events. Mast cells also originate from bone marrow precursors. They leave the bone marrow in precursor form and take up residence in tissue. Unlike the basophil, which has a life span

on the order of 2–3 days, the mast cell survives for weeks to months after differentiation and can continue to undergo mitosis under certain circumstances. Typically, they are found in connective tissue, often clustered in areas interfacing with environmental antigens such as in the gastrointestinal or respiratory tract or in the skin. They are also concentrated along nerves, glandular ducts, and blood and lymph vessels.

*In vitro* investigation of murine mast cell lines has provided insight as to certain mast cell growth requirements and regulatory factors. IL-3, derived from T-helper cells, promotes proliferation of the murine mast cells. IL-4 can promote the maturation of the mast cells in the presence of IL-3. Granulocyte–macrophage colony-stimulating factor (GM-CSF) inhibits the differentiation and transforming growth factor β (TGFβ) inhibits the proliferation of IL-3-dependent mast cells. The stem cell factor (SCF) and the c kit ligand induce murine mast cell proliferation *in vitro* and *in vivo*. IL-9 in conjunction with L-3 (but not alone) enhances the proliferation of murine mast cells in culture. Similarly, IL-10 will promote mast cell proliferation if used in combination with IL-3 or 4 but not in isolation. A mastocytoma line of human mast cells has now been reported. Human mast cells can be derived from CD34$^+$ progenitor cells. Similar to the murine model, c kit and mast cell growth factor (MCGF; SCF) act in concert with IL-3 to promote mast cell proliferation.

Morphologically, the mast cell is approximately 10–30 μm in diameter and contains an oval nucleus, multiple cytoplasmic granules, and prominent membrane folds. The basophil is smaller (10–14 μm), has fewer granules (approximately 80 versus 1000 for the mast cell), and contains a segmented nucleus and cytoplasmic glycogen. Both cells stain with metachromatic dyes, although the basophil does so less intensely. The two classes of cells can be further separated by granule content and biological products. The predominant proteoglycan in basophils is oversulfated chondroitin sulfate versus heparin in the mast cell. Basophils also lack the neutral proteases present in mast cell granules. Functionally, basophils do not produce prostaglandin D$_2$ (PGD$_2$) upon stimulation, although leukotriene C$_4$ (LTC$_4$) is present. Both cells contain histamine as well as a variety of other potent inflammatory mediators. Mast cells contain growth factor receptors IL-3R, IL-4R and c kit. Basophils lack the c kit receptor.

A significant heterogeneity exists within the differentiated mast cell populations. At least two types can be distinguished in humans. Connective tissue type mast cells contain heparin and the neutral proteases chymase and tryptase. They stain intensely with metachromatic dyes and are the predominant type in skin and small intestinal submucosa. The mucosal mast cell populates the gut mucosa and the respiratory tract. It stains less intensely and contains tryptase, but not chymase, in its granules. In addition to heparin, chondroitin sulfate E is present. Functionally, the connective tissue mast cell in skin produces $PGD_2$ with minimal $LTC_4$ following activation, whereas the mucosal type of the gut and respiratory tract secretes $LTC_4$ and near-equivalent amounts of $PGD_2$. Skin mast cells have also been differentiated by histamine release following exposure to the neuropeptide substance P. A variety of factors may account for the differences between mast cells. Some of the most interesting speculations arise from work examining the microenvironment of the cell, including cell–cell interaction and the effects of growth factors and cytokines.

The activation sequence for mast cells and basophils begins with the cross-linking of IgE molecules affixed to the high-affinity receptor by a multivalent allergen (i.e., an allergen bearing more than one antigenic determinant or epitope), triggering a cascade of membrane and cytoplasmic events. The binding of ligand to the FcεRI results in phosphorylation of the $\gamma$ and $\beta$ chains in addition to other proteins (including phospholipase C isoforms) by a tyrosine-specific kinase. Inositol phospholipids are metabolized, and cytosolic free $Ca^{2+}$ is increased. Ultimately, preformed mediators are released from the cell along with the generation and release of lipid-derived compounds.

The mast cell products possess an array of potent biological properties. Table II outlines the biochemical and functional characteristics of some of these mediators. Additional subdivision is useful between those preformed in the mast cell, thereby available for immediate release, and those generated from precursors upon cell activation. Mast cells synthesize a variety of cytokines, including IL-4, IL-5, IL-6, and IL-8. [*See* Cytokines and the Immune Response.]

## 2. Immediate and Late Phase Responses

The immediate response is characterized by its prompt onset, often within seconds, in a sensitized individual following allergen exposure. The recognition of a multivalent allergen by specific IgE molecules displayed on mast cells and basophils initiates cell activation, ultimately with the release of preformed mediators and the elaboration of newly generated materials. The consequent local or systemic effects can include smooth muscle spasm, edema formation, stimulation of afferent nerves, cellular chemotaxis, and mucus secretion. The symptomatic expression and clinical findings are dependent on the target organ(s) affected. In turn this variable is influenced by endogenous host factors, the type of antigenic stimulus, and the route of exposure.

Late phase reactions are delayed in onset following resolution of the immediate response. Isolated late reactions to allergen provocation occur rarely. Generally, onset of the late phase response begins 3–4 hr after exposure, resolving between 12 and 24 hr. They are typified by the appearance of an inflammatory infiltrate composed of eosinophils and polymorphonuclear and mononuclear cells, as well as reportedly increased numbers of metachromatic cells. There are additional distinctions with regard to the array of mediators present when compared to the immediate response.

The development of an experimental nasal lavage model has allowed the study of the immediate and late phase upper respiratory reactions in some detail. Intranasal installation of a relevant allergen results in increased concentrations of histamine, kinins, $N$-$\alpha$-tosyl-L-arginine methyl ester esterase activity, leukotrienes, and $PGD_2$ in nasal lavage fluid, correlating with the onset of symptoms and increased nasal airway resistance. Measurements return to baseline values; however, 3–11 hr postchallenge a second wave of release is detected. The profile is similar to the immediate reaction, but is lacking in $PGD_2$. This difference has been interpreted as evidence for involvement of the basophil in the late phase response. Cytological examination of the nasal passage 11 hr after antigen challenge confirms the presence of increased numbers of metachromatic cells. Using morphological criteria, 80% of these cells were basophils. The most striking increase in cell type was the eosinophil, with an additional increase in neutrophils.

An endogenous elevation in mediators has also been shown in the lower airway of atopic asthmatics following allergen provocation. Clinically, within minutes following allergen exposure, there is development of tightness, dyspnea, and wheezing consistent with asthma and correlating with a measurable decrease in the flow rates on pulmonary function tests. These parameters usually return to baseline within the hour. During the late phase response there is a resurgence of symptoms and objective findings approximately 3–4 hr after the initial exposure. Typically, these reactions peak at 4–8 hr and take 12–24 hr to resolve. Bronchoalveolar lavage at late phase time points reveals increased numbers of activated

## TABLE II
Human Mast Cell Products

| Preformed mediators | Biologic functional characteristics |
|---|---|
| Histamine | Biogenic amine, acts through $H_1$ and $H_2$ receptors: $H_1$ receptor: Smooth muscle contraction; edema formation; increase cGMP; increase nasal mucous; stimulation of afferent nonmylinated nerves ($H_2$ contribution). $H_2$ receptor: Down regulation of cellular responses (cytokine production; T-cell cytotoxicity; lymphocyte proliferation; lysosome release; gamma interferon production); enhancement of T-cell suppressor activity; expression of C3b receptors on human eosinophils; secretion of gastric acid; vascular permeability; increase cAMP; increase airway mucous; idioventricular responses. $H_1$ and $H_2$: Pruritis; vasodilation |
| Acid hydrolases | General role; participation in degradation of glycoprotein/carbohydrate |
| Exoglycosidases | |
| $\beta$-hexosaminidase | Cleavage of $\beta$-linked hexosamines |
| $\beta$-D-galactosidase | Hydrolyses $\beta$-linked galactose from carbohydrate |
| $\beta$-glucuronidase | Cleaves $\beta$-linked glucuronic acid from carbohydrate sidechains |
| Arylsulphatase | Isoenzyme $\beta$, hydrolyses aromatic sulphate esters |
| Chemotactic factors | |
| Eosinophil | Chemotaxis and activation; oligopeptide; heterogeneous MW range 300–5000 ECF-A MW@500 |
| Neutrophil | Chemotaxis and activation; high molecular weight NCA; glycoprotein 600–750 kD |
| | Heat-labile NCA (low MW species; oligopeptides) |
| Oxidative enzymes | |
| Superoxide dismutase | Conversion of $O_2$ to $H_2O_2$; high isoelectric point; binds to heparin |
| Peroxidase | Binds to heparin with increased activity; inactivates dihydroxy and sulfidopeptide leukotrienes; converts $H_2O_2$ to $H_2O$; contributes to generation of lipid mediators (functional data mainly from rat) |
| Proteoglycan | |
| Heparin | Highly acidic, 60 kD molecule; binds/stores preformed mediators; enzymes modulating activity; anticoagulant (enhanced antithrombin 3 activity); anticomplement activity; inhibits eosinophil cytotoxicity; stimulates endothelial cell migration; fibronectin binding to collagen |
| Neutral proteases | |
| Chymase | Hydrolyses angiotensin I to angiotensin II; cleavage of type IV collagen; glucagon; fibrinogen; neurotensin; fibronectin; 30 kD molecule |
| Tryptase | Cleavage of C3 to C3a,b in the presence of heparin; degrades C3a; cleavage of fibronectin; kininogen; Tetramer, $\alpha2\beta2$; 144 kD |
| Carboxypeptidase B | Cleavage of aromatic amino acids |
| Kininogenase | May be identical to tryptase; releases kinin from kininogen at pH 5.5 |
| Newly generated | |
| Cyclooxygenase products | |
| $PGD_2$ | Major cyclooxygenase product; broncho-constriction; dilatation of peripheral vessels; coronary and pulmonary vasoconstrictor; neutrophil chemokinesis; inhibits platelet aggregation; increase vascular permeability; increase cAMP level |
| $9\alpha,11B$-$PGF_2$ | Biologically active metabolic product of $PGD_2$ |
| $TXA_2$ | Bronchoconstriction; constriction of microvasculature and pulmonary vasculature; stimulation of platelet adherence and aggregation; 0.5 min. half life |
| Lipoxygenase products | |
| $LTB_4$ | Leukocyte chemotaxis; chemokinesis neutrophils and eosinophils; leukocyte activation |
| Sulfidopeptide series | |
| $LTC_4$, $LTD_4$, $LTE_4$ | Collectively $LTC_4$, $LTD_4$, $LTE_4$; bronchoconstriction; increased permeability; mucous and electrolyte secretion; gastric acid secretion; $LTC_4$ main lipoxygenase product from human mast cells |
| Monohydroxyeicosatetraenoic acids (HETE) | Mucous secretion; leukocyte chemoattractant |
| Platelet activating factor, alkylglyceryl-etherphosphorylcholine (AGEPC) | Vasopermeability; bronchospasm; confers bronchial hyperreactivity on normals; chemoattractant eosinophils; neutrophils; aggregation of platelets; neutrophils; eosinophils; pulmonary vasoconstrictor; mucous secretagogue |
| Adenosine | Nucleoside; intereacts with $A_2$ receptors; bronchoconstrictor; enhances mast cell mediator release; inhibits platelet aggregation |

eosinophils as well as increased neutrophils, lymphocytes, and activated macrophages. Augmented bronchial hyperresponsiveness is also demonstrable following the induction of late phase reactions. This can be quantified by a reduction in the threshold dose, eliciting provoked responses to methacholine or histamine inhalation. The late phase reaction is felt to most closely mimic the clinical picture of chronic asthma, which is currently viewed as an inflammatory disease.

The skin of sensitized individuals has been studied for mediator release using blister chambers. Results confirm the release of mast cell products histamine and $PGD_2$ during immediate reactions, with absence of $PGD_2$ during the late phase. The classic wheal and flare response with symptoms of pruritus develop within seconds of introduction of allergen in the skin. These macroscopic findings resolve within 30–60 min. The late phase cutaneous response appears indurated, warm, and erythematous, with similar timing to the respiratory reactions. Histologically, following allergen cutaneous challenge the earliest findings are edema and slight engorgement of the blood vessels. As the late phase reaction ensues, a mixed cellular infiltrate appears, with eosinophils, neutrophils, and mononuclear cells as the predominant cell types. Degranulated basophils are often seen, and fibrin deposition has been reported. Mediators have also been measured in tear fluid following allergen provocation. During systemic anaphylactic reactions, increases in plasma histamine and tryptase have been documented, confirming participation of the mast cell. Clinically, late recurrences of systemic manifestations are reported hours after the initial event.

## 3. Cellular Effectors

Among the array of biological effects initiated by mast cell activation, chemotactic factors are released and elaborated, thereby recruiting additional effector cells. Eosinophils are classically present at sites of IgE-mediated inflammatory responses. They are attracted by the combined effects of eosinophil chemotatic factors preformed in the mast cell, epithelial cell-derived RANTES, and the newly generated lipid-derived products $LTB_4$ and platelet-activating factor (PAF). The eosinophil can participate in the ensuing inflammatory response by the release of granule contents, oxidation, and generation of mediators. Major basic protein (MBP) is characterized by its high isoelectric point (10) and is the quantitatively predominant granule constituent. It has been localized to the core of the large granules resident in the cell. The molecule is without enzymatic activity; its release, however,

results in the desquamation of respiratory epithelium, ciliary dysfunction, and noncytolytic histamine release from human basophils. MBP has also been shown to damage a wide variety of cell types *in vitro*, including intestinal, splenic, cutaneous, and peripheral blood mononuclear cells. In allergic diseases MBP has been found in the tissue of patients with eczema, urticaria, and asthma by indirect immunofluorescence. An additional cationic protein found in the large granules and relevant to allergic disorders is eosinophil peroxidase, which catalyzes oxidation in the presence of halide and hydrogen peroxide. This system also results in the release of mediators from mast cells. The eosinophil can also participate by the production of PAF and lipoxygenase products of arachidonic acid, predominantly $LTC_4$. The Charcot-Leyden crystals found in a variety of allergic disorders are composed of lysophospholipase derived from membrane and cytoplasmic structures of the eosinophil.

Neutrophils have not typically been considered to be involved in allergic reactions, although a variety of neutrophil chemotactic factors are released during allergic reactions, including chemokines and $LTB_4$. At present, data from allergen-provoked reactions in humans generally support the presence of increased neutrophil numbers during respiratory or cutaneous late phase reactions. Potential participation of the neutrophil in the inflammatory response might include the generation of reactive oxygen species, which, as in the case of the eosinophil, can result in the direct release of mast cell products. Neutrophils are also capable of generating PAF and $LTB_4$.

Lymphocytes are also found in increased numbers at the site of local allergic reactions. T cells modulate the local immunological response through the generation of potent lymphokines. Involvement of T lymphocytes in asthmatic disease is supported by both airway biopsy and bronchoalveolar lavage studies. Increased numbers of irregularly shaped lymphocytes are found in airway biopsies of symptomatic asthmatics with increased lymphocyte activation as evinced by IL-2 receptor (IL-R2) expression. Atopic asthma has also been associated with activation of the IL-3,4,5, GM-CSF gene cluster, similar to findings in late phase skin responses to allergen.

## 4. Amplification of the Allergic Inflammatory Response

Potential avenues for amplification of the allergic inflammatory response include neuropeptides, kinin generation, and involvement of cytokines. Neurogenic inflammation due to the release of neuropeptides from sensory nerves may result in bronchoconstriction, vasodilatation, plasma exudation, mucus secretion, and hyperemia. When injected into the skin, substance P elicits a wheal and flare response; additionally, it has been shown to contract smooth muscle and to release histamine from skin mast cells. In contrast, vasoactive intestinal peptide (VIP) is normally present in lung nerve fibers, where it acts to relax bronchial smooth muscle. Postmortem studies of human lung, however, revealed an absence of VIP in tissue from pateints with asthma. This finding has been interpreted to support a dysregulation of this system, possibly allowing substance P to act unchecked in the lung, thereby contributing to the bronchial hyperreactivity in asthma.

Kinins are pluripotent mediators liberated from the kininogen substrate by the action of kallikreins. Their direct effects include smooth muscle contraction and edema formation. In animal models, the release of neuropeptides from nonmyelinated C fibers and histamine release from mast cells by kinin have been described. Kinins also augment the production of lipid-derived mediators via the stimulation of phospholipase, which liberates the arachidonate substrate from phospholipid. Investigations of animal organ explants, tissue cell lines, and human platelets demonstrate increased arachidonate release and production of cyclooxygenase and 5-lipoxygenase products. An increase in PAF has been shown following kinin stimulation of human endothelial cells. Kinins have been found in elevated amounts during local IgE-mediated reactions involving the eye, upper and lower respiratory tract, and skin.

Histamine-releasing factors (HRFs) have been demonstrated in a variety of biological fluids, including bronchoalveolar lavage, nasal lavage, and skin blister fluids of allergic individuals. HRF production from human alveolar macrophages and U937 cell lines has been established, as has release from neutrophils, platelets, vascular endothelial cells, T and B lymphocytes, and peripheral blood mononuclear cells. HRFs are capable of inducing mediator release from basophils, which define their functional activity. Mast cells are reported to be less responsive, although exceptions exist. This heterogeneous group of factors remains to be precisely defined with regard to its biochemical characteristics. Partial neutralization of HRF activity has been demonstrated by antibodies to IL-1, IL-3, IL-8, MCSF, and TNFα. Several chemokines of the CC family exhibit patent HRF activity. Another HRF has been identified which has dependence on cell-bound IgE to effect the release of histamine, however,

this is not a general feature of this group of mediators. This HRF was found to be homologous to P23.

Cytokines play an influential role in allergic inflammatory responses. IL-3, IL-5, and GM-CSF are involved in eosinophil activation, viability, degranulation, and enhanced $LTC_4$ synthesis. IL-4 promotes IgE synthesis (isotype switching), induction of FcεRII on B cells, and display of adhesion molecules. IL-1β, TNFα, IL-4, and possibly IL-13 result in endothelial activation and leukocyte adhesion. Cell adhesion molecules control the removal of leukocytes from circulation into sites of inflammation. An increased expression of adhesion molecules has been identified in allergic inflammatory responses. The chemokine RANTES produced by epithelial cells is an additional important augmentor of the allergic response, serving as a potent chemoattractant for eosinophils.

The L-arginine–nitric oxide pathway appears to have a ubiquitous involvement in homeostasis and host defense. It may also participate in inflammatory responses. Histamine, leukotrienes and bradykinin promote the synthesis of arginine-derived nitric oxide. In animal models, nitric oxide is released during anaphylactic challenge and may play a role in reducing the vascular tone during the course of the reaction.

## II. CLINICAL ALLERGY

Allergic or IgE-mediated diseases are common health problems that are estimated to affect 40 million people in the United States alone. The individual risk for the development of allergy is strongly influenced by hereditary factors. The mode of inheritance for allergy is evidently complex and remains to be precisely delineated. Immunogenetic experiments have been able to demonstrate strong associations between genetic loci mapping with the major histocompatibility complex and response to a specific allergen. It has been shown in these investigations that the production of IgE and IgG following exposure to a ragweed allergen (Ra5) is highly correlated with the presence of human leukocyte antigen *Dw2*. Other genetic control not linked to the major histocompatibility locus is also postulated to act on the regulation of serum IgE levels. These and possibly other genetic factors may be responsible for the allergic phenotype. Environmental factors (e.g., infection with microorganisms, air pollution, pharmacological agents, and inhaled or ingested exposure to allergens) may exert modulatory influences.

The clinical expression of allergy depends on the specific sensitivities of the endogenously predisposed individual, the route and the dose of allergen exposure, and coexisting influences such as concurrent disease, medication, or prior treatments affecting the immune response (i.e., immunotherapy). More than 80% of affected individuals manifest symptoms prior to the age of 30. Diseases in which allergy can play a primary or participatory role affect the eye, skin, and respiratory or gastrointestinal tract or present systemically in the case of anaphylaxis. Table III summarizes the local and systemic symptoms characteristic of allergic disorders. Physical findings also depend on the target organ(s) affected; these are summarized in Table IV. The major allergic diseases are discussed

in the following sections in the context of evaluation and treatment.

## III. EVALUATION

### A. History

The diagnostic approach to allergic disorders begins with a detailed history. This often provides clues as to the inciting allergen by linking the timing, locale, and predictable provoking factors to the elicitation of symptoms. The effectiveness of prior treatment (e.g., avoidance of allergens, pharmacological agents, and immunotherapy) also provides valuable insights. Medications should always be carefully reviewed, as side effects can mimic certain allergic disorders without invoking an immunological mechanism. The historical review must also cover alternative explanations for patients' complaints which might masquerade as allergy.

### 1. Diseases Related to Aeroallergens

Conjunctivitis, rhinitis, and asthma represent local inflammatory events that can be initiated by contact with a specific allergen. In ocular or respiratory disease the allergen is usually airborne, typically 2–60 μm in diameter. Year-round, or perennial, symptoms frequently relate to indoor aeroallergens, including animal protein, house dust mite (*Dermatophagoides farinae* or *pteronyssimus*), cockroach, or mold spores. Unique occupational exposures should also be considered such as the investigational use of biological materials or sensitizing chemicals (e.g., isocyanates). Seasonal symptom patterns—classically, rhinoconjunctivitis, or hay fever—usually follow defined growing seasons, such as spring tree, summer grass, or fall weed pollens. In the midwestern and eastern states a dramatic example of this is the explosive onset of symptoms during late summer and fall months, correlating with ragweed pollenosis. Mold spore or insect debris are also possible seasonal provocateurs. Airborne allergens such as dust mites have also been implicated in sustaining cutaneous inflammation in atopic eczema. [*See* Asthma.]

Allergic reactions might be the only etiological basis for a patient's complaints or can coexist with other participatory factors. An excessive reactivity to irritants or vasoactive stimuli, for example, could be responsible for nasal symptoms. In the allergic patient this reflects an enhanced sensitivity of the inflamed

#### TABLE III
Symptoms of Allergic Disease[a]

| | |
|---|---|
| Ocular | Pruritus; conjunctival edema and injection; lid edema, tearing, and occasional gelatinous or mucopurulent discharge |
| Nasal | Sneezing; congestion; rhinorrhea; postnasal drainage; snoring; pruritus of nose, palate, and pharynx; diminished olfactory sense |
| Eustachian tube | Perception of pressure; popping; decreased auditory acuity |
| Sinus | Headache; pain; postnasal drainage |
| Pulmonary (asthma) | Cough; tightness; wheezing; dyspnea; pruritus; pain; sputum or congestion |
| Skin | Atopic dermatitis: Intense pruritus with characteristic skin lesions<br>Urticaria: Evanescent pruritic or painful (burning) raised skin eruption<br>Angioedema: Well demarcated asymmetric swelling rarely pruritic |
| Gastrointestinal | Pain; bloating; nausea; vomiting; diarrhea |
| Anaphylaxis | Any of the above; often laryngeal symptoms; inability to talk and swallow; "lump in throat"; cardiovascular symptoms of orthostasis; weakness; palpitations; metallic taste; "sense of doom" |
| Other | Fatigue from nasal obstruction; sleep disruption; chronic allergies may affect mood and/or school performance; diminished sense of well being |

[a]Reprinted with permission from Christiansen, S. C. (1988). "Evaluation and Treatment of the Allergic Patient." Little Brown and Co., Boston, Massachusetts.

## TABLE IV
Physical Findings in the Allergic Patient[a]

| | |
|---|---|
| General | "Allergic facies" with mouth breathing; allergic shiners and frequent rubbing of the nose (allergic salute); degree of tachypnea, tachycardia, pulsus paradoxis and use of accessory muscles as a gauge to asthmatic severity; orthostasis in systemic anaphylaxis |
| Ocular | Dennie's line (prominent fold on lower eyelid extending from inner canthus 1/2 to 2/3rds of lower lid); glassy appearance; injection or chemosis of conjunctivae; papillomatous changes; ropey mucous; corneal and conjunctival scaring; anterior cataracts and keratoconus in atopic keratoconjunctivitis |
| Nasal | External nasal crease; unilateral or bilateral impairment to airflow; speculum examination of mucosa for edema; coloration (typically pale or "bluish," may be erythematous); engorgement of turbinates; presence of polyps; discharge |
| Sinus | Tenderness; diminished transillumination |
| Ear | Effusion; diminished drum mobility (pneumatic otoscope) |
| Oral cavity, pharynx | Tonsillar; adenoid hypertrophy; postnasal drainage; ET obstruction |
| Neck | Anterior cervical adenopathy; auscultation over larynx for stridor |
| Chest | Increased A-P diameter; decreased expansion; diaphragmatic excursion with inspiration; hyperresonance; prolonged expiratory phase; inspiratory/expiratory wheezing (may be silent with poor air movement in severe asthma attack) |
| Cardiac | Arrhythmia as a complication of asthma/anaphylaxis; evidence of increased pulmonary pressure (P2); differential exam for dyspnea |
| Abdomen | Localized angioedema may mimic bowel obstruction; hypermotility; distention; discomfort |
| Extremities | Nails: Clubbing does not occur in asthma; buffer appearance from constant scratching; cyanosis |
| Skin | Atopic dermatitis: White dermatographism; chronic eruption in early life erythema; papulovesicular; exudative; later dry; lichenified; superimposed excoriation; exudative superinfected areas<br>Urticaria: Circumscribed evanescent raised wheal with surrounding erythema; examine for dermatographism; presence of urticaria pigmentosa (Darier's sign); telangiectasia macularis eruptiva perstans suggest underlying mastocytosis<br>Angioedema: Well demarcated swelling; nonpitting; asymmetric; may be evanescent or last as long as three days |

[a]Reprinted with permission from Christiansen, S. C. (1988). "Evaluation and Treatment of the Allergic Patient." Little Brown and Co., Boston, Massachusetts.

tissue. Allergic reactions also heighten the airway hyperreactivity in an asthmatic patient, as experimentally demonstrated following the late phase response to inhaled allergens. Asthma can additionally be triggered, however, by a variety of non-IgE-mediated stimuli (e.g., exercise, irritants, neurological factors, viral respiratory infections, underlying sinusitis, emotion, hormonal influences, or gastroesophageal reflux). Severe asthmatic reactions can also occur in subsets of patients following the ingestion or inhalation of sulfite preservatives. Allergic mechanisms have only rarely been implicated in these reactions. Ocular symptoms can also occur via non-IgE-mediated pathways. Illustrative of this is the development of giant papillary conjunctivitis related to contact lens wear. The eosinophilic infiltrate in these cases is much the same as in severe allergic reactions.

The importance of considering medication side effects is well exemplified in ocular and respiratory tract diseases. Topical products in the eye can cause local symptoms via an immunological or irritant mechanism. The repeated use of vasoconstrictive nasal sprays often presents a picture similar to that of allergic rhinitis by inducing rebound congestion. Exogenous hormonal agents (e.g., contraceptives) and medications used in the treatment of cardiac disease or hypertension (e.g., hydralazine, $\alpha$-adrenergic blockers, and $\beta$-adrenergic blockers) can also accentuate nasal congestion. $\beta$-adrenergic blockers have additionally been shown to trigger asthmatic attacks, and angiotensin-converting enzyme inhibitors can result in cough imitating asthma. In the latter case, symptoms may relate to a reduction in the breakdown of kinin(s) and neuropeptides ordinarily cleaved by an angiotensin-converting enzyme. Individuals with the triad of eosinophilic sinusitis, nasal polyps, and asthma also appear to be uniquely susceptible to severe reactions to ingested aspirin or nonsteroidal anti-inflammatory agents. These reactions do not involve specific IgE; in fact, the patients only rarely have a significant allergic component to their disease.

## 2. Diseases Related to Ingested or Injected Allergens

In atopic eczema, acute urticaria, gastrointestinal reactions, or anaphylaxis, food allergens should be considered a possible provoking factor. Common culprits are peanuts, nuts, eggs, milk, soy, wheat, and seafood. In cases of IgE-mediated reactions to foods, a cause and effect relationship needs to be established by history and confirmed by diagnostic testing. Typically, symptoms should occur within the first 2 hr of exposure. Chronic symptoms, either allergic or from alternative cause, are frequently misattributed to food sensitivity. Venom or drugs, particularly $\beta$-lactam or penicillin drugs and the venom of *Hymenoptera* species, are often the inciting allergens for the elicitation of systemic anaphylaxis or occasionally limited cutaneous reactions (e.g., urticaria or angioedema). When an IgE mechanism is a consideration, symptoms commonly begin immediately after the exposure, and an alternative mechanism should be considered if more than 2 hr have elapsed.

Anaphylactoid reactions are identified as anaphylactic-type symptoms not resulting from an immediate-type hypersensitivity reaction. Examples include symptoms following exercise, administration of ionic contrast media for radiological procedures, high molecular weight dextran, or opiate medications. Urticaria or angioedema can also result from nonallergic mechanisms. Urticaria, particularly when lasting more than 6 weeks, is often without an identifiable cause and is only infrequently allergic. Physical stimuli, including heat, cold, mechanical trauma, solar exposure, vibration, or pressure, can be the primary factor in certain cases. Underlying vasculitis, connective tissue disease, hereditary conditions, thyroiditis, intestinal parasites, and prodromes of hepatitis are also diagnostic considerations. Isolated angioedema without urticaria could represent either a hereditary or an acquired deficiency of the inhibitor of the first component of complement. Such cases are important to identify due to unique prognostic and treatment considerations. Atopic eczema is also aggravated by a variety of nonallergic factors, including infectious, irritant, hormonal, or emotional stimuli.

Respiratory or ocular reactions are rarely the sole symptoms in allergic reactions to ingested substances. Individuals with pollen sensitivity, however, could complain of oral pruritus when eating certain foods.

This reflects the antigenic cross-reactivity between selected inhaled pollens and ingested substances. Documented examples include ragweed with melon, mugwort with celery, and birch pollen with apples.

### a. Laboratory Evaluation

i. *Testing for Specific IgE* The profile of a patient's specific sensitivities can be established by cutaneous or *in vitro* serum testing for IgE antibody. In skin testing, small amounts of allergenic extracts are introduced below the epidermis, and a wheal and flare response is observed in sensitive subjects within 15–20 min. The amount of specific IgE is graded by the wheal size elicited by relevant concentrations of allergen. This approach is useful for aeroallergens, venoms, and certain drugs by either the scratch or the intradermal technique. For food allergens, testing by the intradermal route is not used due to elicitation of irritant or nonspecific responses. Negative responses to the scratch test for foods virtually preclude the possibility of significant food allergy. Positive food allergen tests, however, still need to be confirmed as to clinical relevance by a double-blind placebo-controlled food challenge. Cutaneous testing procedures must include appropriate controls in all cases with histamine, saline, and diluent.

Precautions must also be taken as to the starting dilution due to the risk of inducing anaphylactic reactions in exquisitely sensitive patients. The radioallergosorbent test is the most common *in vitro* test. It provides semiquantitative information similar to that from the skin tests; however, it is generally less sensitive, more expensive, and takes longer for results. There is rarely indication for the measurement of total IgE in the evaluation of routine allergic problems. Values can be within the normal range in patients with significant amounts of specific IgE or elevated in conditions unrelated to allergy. At times the total IgE level is a helpful parameter of disease activity in cases of atopic dermatitis or allergic bronchopulmonary aspergillosis.

ii. *Cytological Evaluation* Materials can be obtained from the conjunctiva or the nasal mucosa by gentle scraping. These are routinely stained with a Wright–Giemsa stain, allowing for the detection of metachromatic or eosinophilic cells. Although similar findings of predominant eosinophilic cytology can be found in patients with vernal keratoconjunctivitis, papillary disease from contact lens wear, or aspirin sensitivity, the presence of these cell types lends additional support to the diagnosis of allergic disease. The absence of typical cytological findings should also invoke alternative diagnostic considerations, such as predominantly irritant-related disease or confounding health conditions or medications. The presence of a neutrophilic infiltrate is also suggestive of a primary or complicating bacterial infectious process. Analysis of expectorated sputum from asthmatic patients or of bronchoalveolar lavage fluid, when available, is amenable to a similar approach. In addition to eosinophils, Charcot-Leyden crystals of lysophospholipase, Curschmann's spirals of mucus, and Creola bodies of desquamated epithelium can also be present in the asthmatic sputum.

iii. *Ancillary Investigations* Central to the evaluation of asthma is the demonstration of reversible obstruction of the airways by either response to treatment or provocative challenge. In adult patients an improvement of 15% in forced expiratory volume during the first second ($FEV_1$) as a result of bronchodilator treatment is generally accepted as diagnostic for the disease. In individuals with normal baseline spirometry, the diagnosis can be confirmed by provocation, resulting in a decrease in FEV1 of more than 20%. A variety of techniques can be used, including methacholine inhalation, histamine, cold air, hyperosmolar solutions, or exercise to demonstrate airway hyperreactivity.

Roentgenographic evaluation is indicated primarily when a complicating infection is suspected in the respiratory tract, either the sinus cavity or the chest. In a subset of asthmatic patients, pulmonary infiltrates can also occur, which are related to a condition known as allergic bronchopulmonary aspergillosis. This disorder is confirmed by diagnostic criteria, including a history of asthma, pulmonary infiltrates, elevated total serum IgE, peripheral eosinophilia, proximal bronchiectasis, and immediate cutaneous reactivity and precipitating antibodies to *Aspergillus* species in the serum.

A variety of other laboratory tests may be applicable to aid in supporting a primary allergic diagnosis, investigating for complications, directing medical care, or exploring other possibilities for the patient's complaint.

## IV. TREATMENT

The mainstays of allergic treatment fall into the categories of avoidance, pharmacological agents, and immunotherapy.

### A. Avoidance of Allergens

In cases of immediate hypersensitivity to food allergens, avoidance is the only currently accepted approach to treatment. Periodic reevaluation is warranted due to the large percentage of patients who outgrow their clinical reactions (approximately 44% over a 1- to 7-year time frame) to many foods. Drugs to which the patient is known to be hypersensitive should also be avoided. In cases in which a $\beta$-lactam agent would be lifesaving, temporary desensitization can be accomplished by a protocol of careful drug administration either orally or intravenously.

Avoidance of aeroallergens, when possible, is also critical in the overall treatment program of specifically sensitized individuals. As emphasized in preceding sections, allergen exposure results in inflammatory late phase reactions in asthmatic subjects with accentuated airway hyperreactivity. This increased lability to stimuli and enhancement of symptoms can be extrapolated to other local events in the eye, skin, or nasal passages. Avoidance of animal protein, insects, or occupational exposures can be orchestrated by removal of the offending agent. Other household allergens (e.g., dust mites) are often the primary factor in ocular, respiratory, and cutaneous disorders. Targeted high-yield measures can significantly reduce exposure (e.g., removal of carpets or encasement of mattresses and pillows in plasticized material). Exposures to mold spores and pollens can be reduced by dehumidifying and air-conditioning units, respectively.

### B. Pharmacological Agents

#### 1. Antihistamines

Agents formulated to block the H1-histamine receptor have long been in use to control the symptoms of itching, sneezing, ocular watering, rhinnorhea, and urticarial reactions. Available agents can have distinguishing features, such as a lack of a sedative side effect or suppression of mast cell mediator release. H2-histamine receptor-blocking agents have been useful as an adjunct treatment for urticaria and during anaphylaxis. H2 blockers are of little efficacy in the control of atopic dermatitis or respiratory tract symptoms.

#### 2. Cromolyn Sodium and Nedocromil

Cromolyn is used as a primary treatment for asthma and rhinoconjunctivitis. Pretreatment in challenge models of allergic disease prevents the occurrence of both immediate and late phase responses. Topical preparations are available and are virtually devoid of side effects. Frequent application is necessitated due to the short half-life of the drug (i.e., 4 hr). The mechanism of cromolyn action remains to be precisely defined, but may involve a specific cromolyn-binding protein which, when occupied, prevents calcium influx and thus mast cell activation.

Nedocromil inhibits eosinophil, neutrophil, and macrophage activation. Immediate and late phase responses to airway antigen challenge are inhibited. Mast cell mediator release is prevented. Topical preparations are available.

### 3. Corticosteroids

Corticosteroids have a diverse spectrum of action, including potent anti-inflammatory effects. Contributory to their anti-inflammatory impact in the treatment of allergic disorders is the ability to attenuate the production of newly generated lipid-derived mediators. Corticosteroids also restore $\beta$-adrenergic responsiveness and prevent mediator release from basophils. The accumulation of inflammatory cells is reduced, as is mucus secretion and vasopermeability. Steroids inhibit the release and/or production of both eosinophil (IL-3, IL-5, GM-CSF, and RANTES) and endothelial (IL-1, TNF, IL-4, and IL-13) activators, thereby reducing the recruitment, adhesion, activation, and survival of eosinophils during allergic reactions. Experimentally, pretreatment with oral corticosteroids abrogates late phase responses following allergen challenge. For reasons that remain obscure, topical nasal treatment with corticosteroids prevents both immediate and late phase reactions in the nasal challenge model. These drugs are available in systemic or topical preparations for ocular, respiratory, or skin application. [See Steroids.]

### 4. Sympathomimetics and Bronchodilators

$\beta$-adrenergic agonists are usually used as a first line treatment in asthma and anaphylaxis. Epinephrine is the drug of choice in the treatment of anaphylaxis; its advantages include prompt onset of action and combined $\beta 1$ (i.e., cardiovascular) and $\beta 2$ effects. Selective $\beta 2$ agonists have been developed and are usually preferred in the treatment of asthma. Stimulation of the $\beta 2$ receptor causes smooth muscle relaxation and reduced vascular permeability. The $\beta 2$ agonists

prevent immediate, but not late, phase responses in experimental challenges. Anticholinergic agents provide an additive bronchodilatory effect with the sympathomimetics in the treatment of respiratory tract disease. Methylxanthine drugs are also used in asthma treatment, with properties including smooth muscle relaxation, acceleration of mucociliary transport, antagonism of adenosine action, and improved diaphragmatic contractility.

## C. Immunotherapy

Immunotherapy is beneficial in the treatment of IgE-mediated rhinoconjunctivitis, asthma, and anaphylactic reactions to venom. The success of treatment is specific for the allergen administered and is dose dependent. The precise immunological changes occurring in the recipient and responsible for treatment success remain to be clearly elucidated. Closely correlated with the remission of symptoms or protection against anaphylaxis, however, is the appearance of increasing allergen-specific IgG (predominantly IgG4 subclass). Institution of immunotherapy is appropriate in cases of life-threatening anaphylactic reactions to venom. It should be considered an adjunct to care in rhinoconjunctivitis and asthma when the history is consistent with IgE-mediated disease and when significant levels of specific antibody have been documented. Immunotherapy should be reserved, however, until appropriate environmental controls and pharmacological therapy have been instituted and were either insufficient to control diseases or limited by drug side effects.

## BIBLIOGRAPHY

Busse, W. W., and Holgate, S. T. (eds.) (1995). "Asthma and Rhinitis." Blackwell, Cambridge, MA.

Ishizaka, K. (1989). Regulation of IgE biosynthesis. *Hosp. Pract.* **24,** 51.

Middleton, E., Reed, C. E., Ellis, E. F., Adkinson, N. F., Yunginger, J. W., and Busse W. W. (eds.) (1993). "Allergy Principles and Practice," 4th Ed. Mosby, St. Louis, MO.

Roitt, L. M., Brostoff, J., and Male, D. K. (1985). "Immunology." Gower, London.

Sutton, B. J., and Gould, H. J. (1993). The human IgE network. *Nature* **336,** 421.

# Altruism

C. DANIEL BATSON
*University of Kansas*

KATHRYN C. OLESON
*Reed College*

## GLOSSARY

**Altruism** Motivational state with the ultimate goal of increasing another organism's welfare

**Egoism** Motivational state with the ultimate goal of increasing one's own welfare

**Empathy** Other-oriented emotional response congruent with the perceived welfare of another

**Sociobiology** Systematic study of the biological basis of social behavior.

ALTRUISM REFERS TO THE MOTIVATION OF ONE organism, usually human, for benefiting another. Although some biologists and psychologists speak of altruistic *behavior*, meaning behavior that benefits another, this use of the term *altruism* is not recommended as it fails to consider the motivation for the behavior. Motivation is the central issue in discussions of altruism. If one organism's ultimate goal in benefiting another is to increase the other's welfare, then the motivation is altruistic. If the ultimate goal is to increase the organism's own welfare, then the motivation is egoistic. Since antiquity, there has been debate over the existence of altruism. Advocates of universal egoism claim that we act to benefit others only as an instrumental means to reach the ultimate goal of benefiting ourselves. Advocates of altruism do not deny that much human activity is directed toward self-benefit, but they claim that at least some people, under some circumstances, to some degree, act with the ultimate goal of benfiting another.

## I. THE BASIC QUESTION: IS ALTRUISM PART OF HUMAN NATURE?

Interest in altruism centers around the question of whether or not it exists in humans. Clearly, we humans devote much time and energy to helping others. We send money to rescue famine victims halfway around the world. We work to save whales. We stay up all night to comfort a friend who has just suffered a broken relationship. We stop on a busy highway to help a stranded motorist change a flat.

But why do we help? Often, of course, the answer is easy. We help because we have no choice, because it is expected, or because it is in our own best interest. We may do a friend a favor because we do not want to lose the friendship or because we expect to see the favor reciprocated. But it is not for such easy answers that we ask ourselves why we help; it is to press the limits of these answers. We want to know whether our helping is *always and exclusively* motivated by

ENCYCLOPEDIA OF HUMAN BIOLOGY, Second Edition, VOLUME I. Copyright © 1997 by Academic Press. All rights of reproduction in any form reserved.

the prospect of some benefit for ourselves, however subtle. We want to know whether anyone ever, in any degree, transcends the bounds of self-benefit and helps out of genuine concern for the welfare of another. We want to know whether altruism is part of human nature, whether it is within the repertoire of normal humans living in at least some societies. This is the altruism question.

As Charles Darwin made clear in "The Descent of Man," how we answer this question has wide-ranging implications. If altruistic motivation is within the human repertoire, then both who we are as a species and what we are capable of doing are quite different than if it is not. How we answer tells us something fundamental about the role of others in our lives and of us in theirs. It tells us about our capacity for involvement with and caring for one another.

The question of the existence of altruism is not new. It has been central in Western thought for centuries, from Aristotle (384–322 BC) and St. Thomas Aquinas (1225–1274), through Thomas Hobbes (1588–1679), the Duke de la Rochefoucauld (1613–1680), David Hume (1711–1776), Adam Smith (1723–1790), and Jeremy Bentham (1748–1832), to Friedrich Nietzsche (1844–1900) and Sigmund Freud (1856–1939). The majority view among Renaissance and post-Renaissance philosophers, and more recently among biologists and psychologists, is that we are, at heart, purely egoistic, that we care for others only to the extent that their welfare affects ours. A persistent minority has, however, claimed that altruism exists, that in some circumstances and to some degree, we help others for their benefit and not simply for our own.

Many forms of self-benefit can be derived from helping. Some are obvious, as when we get material rewards and public praise or when we escape public censure. But even when we help in the absence of obvious external rewards, we may still benefit. Seeing a person or animal in need may cause us to feel distress, and we may act to relieve the other's distress as an instrumental means to reach the ultimate goal of relieving our own distress. Or we may benefit by feeling good about ourselves for being kind and caring or by escaping guilt and shame.

Even heroes and martyrs can benefit from their acts of apparent selflessness. Consider the soldier who saves his comrades by diving on a grenade or the man who dies after relinquishing his place in a rescue craft. These persons may have acted to escape anticipated guilt and shame for letting others die. Or they may have acted to gain rewards, either the admiration and praise of those left behind or the benefits expected in a life to come. Or they may simply have misjudged the situation, never dreaming that their actions would cost them their lives. The suggestion that heroes' noble acts could be motivated by self-benefit may seem cynical, but it must be faced if we are to answer the altruism question.

## II. THE ALTRUISM QUESTION CLARIFIED

Whether or not altruism exists is an empirical question; it concerns what *is*. Yet attempts to answer this question have often failed because of conceptual confusion. Therefore, before considering the empirical evidence, it is essential to understand what is and is not at issue in the egoism–altruism debate.

### A. Distinguishing Altruism from Egoism

*Altruism* is a motivational state with the ultimate goal of increasing another organism's welfare; *egoism* is a motivational state with the ultimate goal of increasing one's own welfare. Three key phrases in each of these definitions deserve comment.

1. *"motivational state."* Motivation here refers to a goal-directed force within an organism. Goal-directed motivation has the following four features: (1) The organism desires some change in his or her experienced world; this is what is meant by a goal. (2) A force of some magnitude exists, drawing the organism toward the goal. (3) If a barrier prevents direct access to the goal, alternative routes will be sought. (4) The force disappears when the goal is reached. Note that the goal-directed motivation involved in altruism and egoism is not within the repertoire of many species; to set and seek goals requires high-level perceptual and cognitive processes generally associated with a developed neocortex of the sort found in higher mammals, especially humans.

2. *"with the ultimate goal."* An ultimate goal is an end in itself and not just an intermediate means for reaching some other goal. If a goal is an intermediate means for reaching some other goal and a barrier arises, then alternative routes to the ultimate goal will be sought that bypass the intermediate goal. Moreover, if the ultimate goal is reached without the intermediate goal being reached, the motivational force will disappear. If, however, a goal is an ultimate goal, it cannot be bypassed in this way.

3. *"of increasing another organism's welfare"* or *"of increasing one's own welfare."* These phrases identify the specific ultimate goal of altruistic and egoistic motivation, respectively. Increasing another's welfare is an ultimate goal if an organism (a) perceives some desired change in another organism's world and (b) experiences a force to bring about that change as (c) an end in itself. Increasing one's own welfare is an ultimate goal if an organism (a) perceives some desired change in his or her own world and (b) experiences a force to bring about that change as (c) an end in itself.

Altruism and egoism, as defined here, have much in common. Each refers to goal-directed motivation; each is concerned with the ultimate goal of this motivation; and, for each, the ultimate goal is increasing someone's welfare. These common features provide the context for highlighting the crucial difference: Whose welfare is the ultimate goal, another's or one's own?

## B. Implications of These Definitions of Altruism and Egoism

Like most definitions, these definitions of altruism and egoism have some implications that may not be apparent at first glance. We shall mention eight:

1. The distinction between altruism and egoism is qualitative, not quantitative; it is the ultimate goal, not the strength of the motive, that distinguishes altruistic from egoistic motivation.

2. A single motive cannot be both altruistic and egoistic. This is because to seek to benefit both self and other implies two ultimate goals (as long as self and other are perceived to be distinct), and each new ultimate goal defines a new motive.

3. Both altruistic and egoistic motives can exist simultaneously within a single organism. This is because an organism may have more than one ultimate goal at a time, and so more than one motive. If the altruistic and egoistic goals are of roughly equal attractiveness and lie in different directions so that behaviors leading toward one lead away from the other, then the organism will experience motivational conflict.

4. As defined, altruism and egoism apply only to the domain of goal-directed activity. If an organism acts reflexively or automatically without any goal, then no matter how beneficial to another or to the self the act may be, it is neither altruistic nor egoistic.

5. Focusing on the human level, a person may be altruistically motivated and not know it, may be egoistically motivated and not know it, may believe his or her motivation is altruistic when it is actually egoistic, and vice versa. This is because we do not always know—or report—our true motives. We may have a goal and not be aware of it or we may mistakenly believe that our goal is *A* when it is actually *B*.

6. Both altruistic and egoistic motives may evoke a variety of behaviors or no behavior at all. A motive is a force. Whether this force leads to action will depend on the behavioral options available in the situation, as well as on other motivational forces present at the time.

7. As defined here, altruistic motivation need not involve self-sacrifice. Pursuing the ultimate goal of increasing another's welfare may involve cost to the self, but it also may not. Indeed, it could even involve self-benefit and the motivation would still be altruistic, as long as obtaining this self-benefit is an unintended consequence of benefiting the other and not the ultimate goal.

Some scholars assume that altruism requires self-sacrifice, citing as examples cases in which the absolute cost of helping is very high, often involving loss of life. These scholars apparently believe that in such cases the costs of helping must outweigh the rewards so the helper's goal could not be self-benefit.

There are at least two problems with including self-sacrifice in the definition of altruism. First, it shifts the focus of attention from the crucial question of motivation to a focus on consequences. What if the helper had no intention of risking death, but things got out of hand? Is the motivation altruistic? Or what about a cost-free comforting hug for a friend? It may involve no self-sacrifice, but the ultimate goal may still have been to increase the friend's welfare.

Second, a definition based on self-sacrifice overlooks the possibility that some self-benefits for helping increase as the costs increase. The costs of being a hero or martyr may be very great, but so may the rewards. To avoid these two problems, it seems best to define altruism in terms of benefit to other, not cost to self.

8. At least among humans, there may be motives for helping that are neither altruistic nor egoistic. For example, a person might have an ultimate goal of upholding a principle of justice. This motive could lead the person to help anyone he or she perceived to be unjustly in need. The help might benefit both the needy and the self, but these benefits would be unintended consequences, not the ultimate goal. If the

ultimate goal is niether benefit to another nor benefit to self, the motive is neither altruistic nor egoistic.

## C. Relating Altruism and Egoism to Helping

From the foregoing discussion, it is clear that helping another person may be altruistically motivated, egoistically motivated, both, or neither. To ascertain that some act was beneficial to another and was intended, which is what is meant by helping, does not in itself say anything about the nature of the underlying motivation. As Fig. 1 indicates, if we are to know whether the motivation is altruistic or egoistic, we must determine whether benefit to the other is (a) an ultimate goal and any self-benefits are unintended consequences or (b) an instrumental means to reach the ultimate goal of benefiting oneself.

But how are we to determine a helper's ultimate goal? We cannot know the ultimate goal with absolute certainty because we do not observe goals or intentions directly; they must be inferred from behavior. Moreover, if we observe only a single behavior that has different potential ultimate goals, the true ultimate goal cannot be discerned. It is like having one equation with two unknowns; a clear answer is impossible. We can, however, draw reasonable inferences about the true ultimate goal if we can observe potential helpers' behavior in different situations that involve a change in the relationship between potential ultimate goals. The behavior should always be directed toward the true ultimate goal.

Employing this general strategy, two steps are required to arrive at an answer to the altruism question. First, we need to have a clear idea of potential egoistic motives for helping and of potential sources of altruistic motivation. Unless we know that a given self-benefit may have been a helper's goal, there is little likelihood of concluding that it was. Similarly, unless we

know where to look for altruism, there is little likelihood of finding it. Second, we need to observe potential helpers' behavior in circumstances systematically varied in a way that disentangles the relationship between relevant egoistic and altruistic ultimate goals, making it possible to obtain the self-benefits without benefiting the other or vice versa. Potential helpers' behavior across these situations should always be directed toward the ultimate goal, revealing the nature of the underlying motivation.

## III. POTENTIAL EGOISTIC AND ALTRUISTIC MOTIVES FOR HELPING

### A. Egoistic Motives

Research with humans suggests two broad classes of egoistic motives for helping: (1) gaining rewards and avoiding punishments, and (2) reducing aversive arousal. These two egoistic motives are summarized in Paths 1 and 2 of the flow chart in Fig. 2. Path 1 is further subdivided to differentiate (a) reward-seeking and (b) punishment-avoiding motives.

### 1. Instigating Situation

Each path in Fig. 2 begins with perception of another organism in need. Perception of the other's need is all that is required to instigate motivation along Path 2. Before motivation can be instigated along Path 1, the potential helper must also expect to receive rewards for helping, punishments for not helping, or both in the particular situation. These expectations are the result of the potential helper's prior learning history, including rewards and punishments received in similar situations, as well as rewards and punishments others have been observed to receive.

### 2. Internal Response

On Path 1, expectation of reward and punishment, combined with perceiving the other's need, leads to anticipating rewards or punishments in the current situation. The anticipated rewards and punishments may be obvious and explicit, such as being paid, gaining social approval, or avoiding censure, or they may be more subtle, such as receiving esteem in exchange for helping, complying with social or personal norms, seeing oneself as a good person, or avoiding guilt. On Path 2, perceiving the other's need evokes aversive arousal, including feelings of distress, anxiety, and uneasiness.

Reward and punishment anticipation (Path 1) and

| When we help ... | A | B |
| --- | --- | --- |
| | We benefit the other ... | and receive self-benefits. |
| | | |
| **Why do we help?** | | |
| | | |
| Altruistic account: | Ultimate goal | Unintended consequence |
| Egoistic account: | Instrumental goal | Ultimate goal |

**FIGURE 1**   Formal structure of the altruism question.

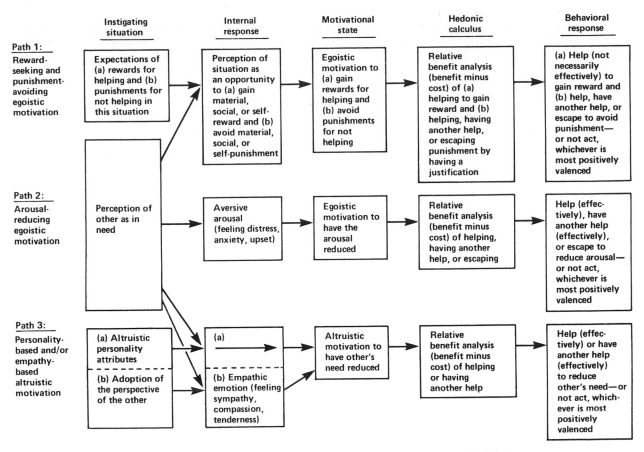

FIGURE 2 Three-path model of potential egoistic and altruistic motives for helping.

feeling aversive arousal (Path 2) are distinct but not mutually exclusive internal responses to perceiving another in need. In many helping situations, such as emergencies, both responses are likely. In other situations, such as making a routine annual contribution to a charity, one may be very aware of the rewards for helping and punishments for not, yet feel little aversive arousal. In still other situations, such as witnessing a gory automobile accident, one may experience much aversive arousal but pay little or no attention to possible rewards and punishments.

## 3. Form of Motivation

Anticipated reward, anticipated punishment, and aversive arousal each evoke their own form of egoistic motivation: motivation to gain the reward (Path 1a), avoid the punishment (Path 1b), or have the arousal reduced (Path 2). These motives are distinct but not mutually exclusive. When experienced simultaneously, their goals may be compatible or incompatible. Sometimes actions that enable one to gain rewards

or avoid punishments also reduce aversive arousal, as when one returns a lost child to his mother. At other times, acting to gain rewards or avoid punishments increases aversive arousal, as when one comforts a badly injured accident victim.

## 4. Hedonic Calculus

Before acting on any of these motives, a hedonic calculus, or relative–benefit analysis, is performed: Benefit is weighed against cost for each potential behavioral response. The magnitude of the benefit in this analysis is a function of the strength of the motive because the benefit is to reach the goal. The magnitude of the cost is the sum of the various costs perceived to be associated with the behavior. Perhaps the simplest way to think about these costs is in terms of conflict with other egoistic motives, such as motives to avoid pain, save time, keep one's money, and so on.

The behavioral responses for which one computes the hedonic calculus are not the same on each egoistic path. On Path 1a the desired rewards are contingent on being helpful so the hedonic calculus focuses on a single behavioral response: helping. On Path 1b, possible punishments may be avoided by three different means: helping, having someone else help, or having good justification for not helping. Similarly, on Path 2, aversive arousal may be reduced by three different means: helping, having someone else help, or escaping exposure to the need situation. Note that on Path 1 simply trying to help is often sufficient to gain rewards or avoid punishments, even if the effort is unsuccessful. As people say, "It's the thought that counts." On Path 2, however, the helping must be effective; only if the other's suffering ends will the stimulus causing one's aversive arousal be terminated.

## 5. Behavioral Response

As a result of the hedonic calculus, the egoistically motivated person will help, let someone else help, justify not helping, or escape, whichever available response will most efficiently reach the person's ultimate goal. If, however, the anticipated cost of each available response exceeds the benefit, then the person will pursue some unrelated goal or will do nothing.

Together, the two egoistic paths described provide a plausible general account of the motivation to help. Each path makes considerable intuitive sense; each is internally consistent; each is complex, yet permits relatively precise behavioral predictions; each is based on a classic approach to motivation—reinforcement for Path 1, arousal reduction for Path 2; and each is supported by much empirical research. Yet, despite these virtues, advocates of altruism claim that these two egoistic paths do not provide a full account of why we help. They claim that an altruistic path exists as well.

## B. Altruistic Motives

The two most commonly proposed sources of altruistic motivation in humans are summarized in Path 3 of Fig. 2. One source is what has been called the *altruistic personality*. Advocates of this source claim that certain personality attributes evoke altruistic motivation whenever individuals having these attributes perceive another to be in need. Among the attributes claimed to make up the altruistic personality are high self-esteem, a sense of personal or social responsibility, well-developed moral principles, and an inclination to feel high levels of empathy or sympathy for those in need.

The second frequently mentioned source of altruistic motivation is *empathy,* an other-oriented emotional response congruent with the perceived welfare of another. Advocates of the *empathy–altruism hypothesis* claim that feeling sympathy, compassion, softheartedness, tenderness, and the like for another in need evokes altruistic motivation to relieve that need. As indicated in the lower part of Path 3, these empathic feelings are usually considered to be a product of (a) perceiving the other as in need and (b) adopting the perspective of the other, which means imagining how the other is affected by his or her situation.

The empathic feelings referred to by the empathy–altruism hypothesis should not be confused with the general inclination to have such feelings that is claimed to contribute to the altruistic personality. The feelings themselves are a situation-specific response to perceiving a particular other in need, whereas the inclination to have such feelings is a general personality disposition. The former need not be related to the latter.

As outlined in Path 3, if the altruistic motivation evoked by one of these sources is above some minimal threshold, then the individual will proceed to consider behavioral means for reaching the goal of reducing the other's need. Paralleling the egoistic paths, the altruistically motivated individual will perform a hedonic calculus before acting, seeking the least costly means to this goal.

To suggest a hedonic calculus for altruistic motivation may seem contradictory because the goal of this calculus is clearly egoistic: to reach the desired altruistic goal while incurring minimal costs to self. Yet, existence of this egoistic goal does not mean that the motivation to have the other's need reduced has now become egoistic; it only means that the impulse to act on this altruistic motivation is likely to evoke an egoistic motive as well. Existence of the latter motive need not negate or contaminate the former, although it complicates the relationship between the altruistic motive and behavior.

The magnitude of the benefit in the altruistic hedonic calculus is, as on the egoistic paths, a function of the strength of the motive because the benefit is to reach the goal. The magnitude of the cost is the sum of the various costs associated with the behavior, including physical harm or risk, discomfort, exertion, mental strain, time, and monetary expense.

The behavioral responses considered in Path 3 are restricted to helping or having someone else help; no consideration should be given to justifying not helping or escaping because these are not viable behavioral means of reaching the altruistic goal of reducing the other's need. As in Path 2 but not Path 1, helping in Path 3 must be effective if the goal is to be reached. As in Paths 1b and 2 but not Path 1a, having someone else help should be as viable, but no more viable, a means of reaching the altruistic goal as being the helper oneself.

Reaching the altruistic goal is likely to enable the helper to gain rewards (Path 1a), avoid punishments (Path 1b), and reduce aversive arousal (Path 2). But to the extent that the motivation is altruistic, these benefits to self are not the ultimate goal of helping, they are unintended consequences.

Considering all three paths at once, the instigating conditions arousing Path 3 altruistic motivation are also likely to arouse Path 1 and Path 2 egoistic motives. These altruistic and egoistic motives are distinct because they have different ultimate goals, but to the extent that the different goals are compatible, the force of the motives should sum.

Overall, the three-path model in Fig. 2 specifies potential egoistic motives for helping and potential sources of altruistic motivation. Moreover, it specifies behavioral responses that enable a potential helper to reach one or more of the egoistic goals without having the other's need reduced. These behavioral responses permit systematic variation of the helping situation so that one or more of the egoistic goals can be reached without reducing the other's need or vice versa. Observing the effect of this variation provides the basis for inferring helpers' ultimate goals, the second step in answering the altruism question. This step takes us from conceptual analysis to methods of research.

## IV. METHODS FOR ADDRESSING THE ALTRUISM QUESTION

Three general methods have been used to address the altruism question: deductive argument, field observation, and laboratory experiment. The first of these methods involves argument for the logical necessity of either universal egoism or altruism, given some general premise about what human nature is or should be. If empirical evidence is used in the argument at all, it tends to be used either illustratively or to establish the general premise.

In the hands of Renaissance and post-Renaissance philosophers like Hobbes, Hume, Smith, Bentham, and Kant, deductive argument has added much to our understanding of the implications of the altruism question and has provided a rich array of possible answers. It has not, however, provided a persuasive answer. The problem is that deduction from a general premise cannot persuade someone who does not already agree with the premise. Deductive arguments for egoism or altruism seem inevitably to hinge on assertions that begin "It cannot be doubted that. . . ." But what is beyond doubt for someone on one side of the issue is likely to be far from clear to someone on the other side. Indeed, it often seems that the only assertion about the egoism–altruism debate not to be doubted is that persons of good will and sincerity can and do disagree.

## V. DEDUCTION FROM NATURAL SELECTION: SOCIOBIOLOGY

In biology, the theory of natural selection has provided a powerful premise for understanding human nature. Predictably, this theory has been used as the basis for deductive arguments about altruism. Even though Darwin believed that altruistic impulses based on sympathy were part of human nature, the general picture of natural selection as "survival of the fittest" and "red in tooth and claw" has often been invoked as proof of universal egoism. After all, if we are genetically programmed to optimize our own survival, then where would we get the capacity for investment in another's welfare that altruism requires?

Over the past three decades the new subdiscipline sociobiology has made a major contribution to the egoism–altruism debate by showing that altruism need not be inconsistent with the theory of natural selection. Sociobiologists have pointed out that natural selection occurs at the level of the gene, not the organism, and that under certain circumstances, genes can enhance their own survival by leading the organism carrying them to risk its own survival to benefit another.

Think, for example, of a circumstance in which the benefactor and the benefited share the same genes and the benefited has a greater chance of placing these genes in the next generation. In this case, it is to the shared genes' advantage to have the benefactor pay more attention to survival of the benefited than to survival of self. William Hamilton used this logic of *inclusive fitness* to explain the self-sacrificial behavior of the sterile worker castes among the social insects, including social bees, wasps, ants, and termites. A second circumstance in which helping another survive could be in the best interest of the benefactor's genes is when the benefited is likely to return the favor, either to the benefactor or to another organism carrying the same genes. Robert Trivers used this logic of *reciprocal altruism* to explain symbiotic relationships across the phylogenetic spectrum.

Sociobiology has made an important contribution by showing that altruism need not contradict the theory of natural selection, but it has not provided a persuasive answer to the altruism question. Two problems exist with most sociobiological arguments. First, theoretical possibility is substituted for empirical reality. The question asked by sociobiologists is whether an altruistic allele, or genetic alternative, would tend to increase in a population relative to a nonaltruistic allele. But logical, even mathematical, demonstration that an altruistic allele, if it existed, would increase is not the same as demonstrating that this allele exists. Not every potentially successful allele actually exists. In this regard, sociobiological arguments show the same weakness as earlier deductive arguments for and against altruism.

Second, sociobiologists tend to apply the term altruism to self-sacrificial behavior across the phylogenetic spectrum, regardless of the nature of the underlying motivation or, indeed, the presence of any motivation at all. The self-sacrificial behavior of social insects or the symbiotic relationship between cleaner fish and their hosts, although interesting in its own right, is almost certainly not a product of the kind of goal-directed motivation at issue in the egoism–altruism debate. Many sociobiologists use the language of the egoism–altruism debate, but they have no real intention of addressing the altruism question. They carefully avoid motivational issues.

## VI. EVIDENCE FOR THE EXISTENCE OF ALTRUISTIC MOTIVATION

Two other research methods have proved more useful than deductive argument in the search for an answer to the altruism question: field observation and laboratory experiment. Field observations have been conducted on a wide range of species, whereas laboratory experiments have been conducted primarily on humans. These two more productive research methods can and should serve complementary functions. Field observation can be useful in indicating the range and pattern of helping that occurs in different species in their natural habitat. But field observation does not permit the systematic variation of the helping situation needed to infer the nature of the underlying motivation. Laboratory experiments lack the ecological

validity of field observation, but they permit the needed systematic variation and, as a result, inference about the nature of the underlying motivation.

### A. Field Observation

Observation of the behavior of a range of species, including humans, has provided numerous examples of beneficial acts that researchers have interpreted as directed toward the goal of increasing another organism's welfare. Here is a brief selection:

1. Robins, thrushes, and titmice have been observed emitting cries to warn other birds of the approach of a hawk. Mother grouse will risk capture by feigning a broken wing and attracting attention to lead a predator away from chicks in the nest.

2. Elephants injured by a falling tree, weapon, or in a fight may be aided by other elephants, who cluster around and use their foreheads, trunks, and tusks to help the injured elephant to rise. Once on its feet, the injured elephant may be supported by others walking or running alongside.

3. African wild dogs return from hunting and regurgitate pieces of meat to feed both the young pups and the adults who stayed behind to care for the litter. Pet dogs can be highly protective of family children, including children who do not feed or pet them. Pet dogs sometimes even make a nuisance of themselves by the persistent "rescue" of children who are swimming, despite repeated punishment.

4. Porpoises have been seen risking their lives to support a harpooned porpoise on the surface so that it can breathe. There are also reports of porpoises rescuing drowning humans in the same way.

5. Chimpanzees share food with lower status chimps who beg. Orphaned infant chimps may be adopted and reared by their adult brothers or sisters. Chimps in captivity have been observed pulling other chimps' hands back as they reach toward potential danger.

6. Examples of human helpfulness are well known. There are the celebrated acts of Mother Teresa, Albert Schweitzer, and Mahatma Gandhi; of rescuers of Jews from the Nazis; of organ donors; and of heroes who risk their lives in emergencies. There is also the less celebrated kindness and thoughtfulness that is part of each of our daily lives.

Even though these and other observations of helpful behavior have been interpreted as evidence for altruism, this evidence is weak. For, in each case, we are left with the question of the nature of the underlying motivation. The observed behavior certainly sharpens this question, and it often leaves the observer convinced that the motivation is altruistic. But testimonials, even from trained and trusted observers, are not enough. We need more objective evidence concerning the nature of the underlying motivation. For that, we are best advised to turn to the laboratory.

## B. Laboratory Experiments

Laboratory experiments with humans during the past decade have begun to test the nature of the motivation associated with assumed sources of altruism. Employing the logic outlined in Section II,C, these experiments involve systematically varying a helping situation, permitting some potential helpers to reach one or more of the possible egoistic goals of helping without having to help. The experiments have tested the validity of the two major proposed sources of altruistic motivation: the altruistic personality and empathy (see Section III,B).

### 1. Evidence for an Altruistic Personality

Several experiments have been conducted in which, under varied circumstances, individuals scoring low or high on the personality attributes claimed to reflect an altruistic personality have been given a chance to help a person in need. The most likely egoistic goal of helping associated with these personality attributes is avoidance of anticipated punishment in the form of shame and guilt for not living up to one's positive moral self-image. Therefore, for some participants in these experiments, avoiding the negative consequences to self of failure to help was made relatively difficult; if they chose not to help, they anticipated having to stay and be reminded of their decision. For others, avoidance was relatively easy; if they chose not to help, they anticipated being able to leave.

Across these two experimental conditions, the rate of helping associated with three "altruistic" personality attributes—self-esteem, personal responsibility, and dispositional empathy—suggests that the motivation to help associated with these attributes is not altruistic but is directed toward the egoistic goal of avoiding shame and guilt. Higher scores on these attributes were associated with increased helping when potential helpers anticipated being reminded of their failure to help but not when they did not anticipate being reminded.

These experiments contradict the claim that "altruistic" personality attributes actually evoke altruistic motivation. But not all of the relevant personality attributes have been tested. Moreover, across experiments, the pattern of results has not always been consistent. Attempts to find experimental evidence for the altruistic personality continue.

### 2. Evidence for the Empathy–Altruism Hypothesis

Clear experimental evidence exists of an empathy–helping relationship; feeling empathy or sympathy for a person in need increases the likelihood of helping that person. But is the ultimate goal of this helping to reduce the sufferer's need, as the empathy–altruism hypothesis claims, or is it some form of self-benefit? During the past decade and a half, more than two dozen experiments have addressed this question. Three different egoistic explanations, one based on each egoistic path in Fig. 2, have been tested against the empathy–altruism hypothesis.

The most commonly proposed egoistic explanation of the empathy–helping relationship has been aversive arousal reduction (Path 2). According to this explanation, the goal of the empathically aroused helper is to reduce the empathic emotion itself, which, as a congruent response to the perceived suffering of another, is experienced as aversive. Benefiting the sufferer is simply an instrumental means to this self-serving end.

Experimental evidence does not support this aversive arousal reduction explanation. As noted in Fig. 2, when escape from exposure to a suffering victim is easy and therefore less personally costly than helping, escape should be the preferred means of reducing aversive arousal caused by witnessing the suffering. So, if empathy evokes aversive arousal reduction, then high-empathy individuals should help less when escape is easy than when it is difficult. Contrary to this prediction, experiments reveal that individuals experiencing a high level of empathic emotion are as likely to help when escape is easy as when it is difficult. Only individuals experiencing low empathy help less when escape is easy.

The other two egoistic explanations for the empathy–helping relationship are based on punishment avoiding (Path 1b) and reward seeking (Path 1a), respectively. The punishment-avoiding explanation claims that we have learned to expect special social or self-punishment in the form of censure or guilt for a failure to help those for whom we feel empathy. The reward-seeking explanation claims that we have learned to expect special social or self-rewards in the form of commendation or enhanced self-image after helping those for whom we feel empathy.

Once again, recent experimental evidence does not support either of these explanations. Contrary to the predictions of the punishment-avoiding explanation, empathically aroused individuals who are provided with a ready justification for not helping are as likely to help as those not provided a justification. Contrary to the predictions of the reward-seeking explanation, empathically aroused individuals are as pleased when the sufferer's need is relieved by chance as when it is relieved by their own action.

In experiment after experiment testing these egoistic alternatives, results have patterned as predicted by the empathy–altruism hypothesis. Overall, these experiments have provided impressive evidence that feeling empathy for another in need does evoke altruistic motivation to help that person. It appears that this form of altruism is indeed within the human repertoire.

## VII. FURTHER RESEARCH QUESTIONS

Even if, as it seems, we now have an experimentally based affirmative answer to the question of the existence of altruism, research questions remain. Indeed, because an affirmative answer was not what our biological and psychological models of motivation led us to expect, we now have more questions than before. Consider four of the more obvious:

1. If empathy instigates altruism, then what instigates empathy? Are we more likely to feel empathy when we have experienced or could experience the same need ourselves, when we feel secure ourselves, when our attention is directed toward a single individual in need as opposed to a great number or vice versa? Are we more likely to feel empathy for kin, friends, similar others, innocent victims, the helpless, someone who makes a dramatic need display, or someone who does not?

2. What are the neurophysiological processes involved in the experiences of empathy and altruism? The possibilities for neurophysiological research on empathy remain severely limited. Many neurophysiological research procedures cannot be applied to humans. At present, we only have speculation that empathic feelings, like other emotions, are linked to the limbic system and, perhaps, to the prefrontal cortex and to the operation of brain opoids. One possible line of future research may be to pursue the neurophysiology of sociopathy since sociopaths seem to feel relatively little empathy.

Some understanding of the hormonal and biochemical factors related to aggression has come from research on infrahuman animals using brain lesions, brain stimulation, and chemical intervention. Whether similar techniques can be used to study altruism is not yet clear. To the extent that higher brain functions are involved in altruism than in aggression, altruism is likely to prove more intractable than aggression to study by brain intervention techniques.

3. Is empathic feeling for another in need the only source of altruistic motivation in humans or are there others?

4. Are animals besides humans capable of experiencing altruistic motivation and, if so, what is the source of their altruism?

We now have fairly strong empirical evidence that empathy-induced altruism is part of human nature. Yet these formidable questions, and similar ones, still need answers before we can claim to have a real understanding of our apparent capacity to, at times, seek another's welfare.

## VIII. IMPLICATIONS OF THE EXISTENCE OF ALTRUISM

The growing experimental evidence for altruistic motivation, if substantiated in future research, could have broad implications, both theoretical and practical. Universal egoism has long been the dominant view of human nature in the behavioral and social sciences. But if feeling empathy can lead humans to act with an ultimate goal of increasing the welfare of others, then universal egoism must be replaced by a more complex view of human motivation that allows for altruism too. Such a shift implies that we humans may be more social than we have thought: Others may be more to us than sources of information, stimulation, and reward—of facilitation and inhibition—as we each seek our own welfare. We may have the potential to care about their welfare as well.

This shift in theoretical perspective suggests, in turn, wide-ranging practical implications. Imagine, for example, the implications for child-rearing practices and moral education if, instead of assuming that sensitivity to the needs of others must always be made worth a child's while, we assume that the child is capable of altruistic concern for the welfare of at least some others. Or imagine the implications for economic theory and models of social change if we drop the assumption that self-benefit alone defines utility and value.

One may thus build on the recent evidence that altruism exists, trying to develop practices that encourage it. But this is not the only possible response. Instead of embracing altruism as a positive dimension of human nature to be encouraged, one might, following Friedrich Nietzsche and Ayn Rand, oppose it as a defect to be eradicated. Alternatively, one might, following Garrett Hardin, accept altruism as positive but as so weak and limited in scope as to be of little practical significance in attempts to improve the human condition.

Whether our altruistic potential can, or should, be harnessed and put to significant practical use to create a more compassionate, humane society is, as yet, unclear. But in a world so full of fear, insensitivity, suffering, and loneliness, the possibility certainly seems worth considering.

## BIBLIOGRAPHY

Batson, C. D. (1991). "The Altruism Question: Toward a Social-Psychological Answer." Erlbaum Associates, Hillsdale, NJ.

Campbell, D. T. (1978). On the genetics of altruism and the counterhedonic components in human culture. *In* "Altruism, Sympathy, and Helping: Psychological and Sociological Principles" (L. Wispé, ed.). Academic Press, New York.

Dawkins, R. (1976). "The Selfish Gene." Oxford University Press, New York.

Hardin, G. (1977). "The Limits of Altruism: An Ecologist's View of Survival." Indiana University Press, Bloomington.

Hoffman, M. L. (1981). Is altruism part of human nature? *J. Personality Soc. Psychol.* **40**, 121–137.

Hunt, M. (1990). "The Compassionate Beast: What Science Is Discovering about the Humane Side of Humankind." William Morrow, New York.

MacIntyre, A. (1967). Egoism and altruism. *In* "The Encyclopedia of Philosophy" (P. Edwards, ed.), Vol. 2. Macmillan, New York.

Oliner, S. P., and Oliner, P. M. (1988). "The Altruistic Personality: Rescuers of Jews in Nazi Europe." The Free Press, New York.

Wilson, E. O. (1978). "On Human Nature." Harvard University Press, Cambridge.

# Alzheimer's Disease

TONI PALADINO

*University of California, San Diego*

## GLOSSARY

**Dystrophic neurites** Fusiform neurites that mainly contain paired helical filaments, some dense bodies, and a variety of proteins including amyloid precursor protein, tau, ubiquitin, neurofilaments, synaptophysin, epidermal growth factor receptors, protein kinase C, GAP43, brain spectrin, and neurotransmitters

**Neurodegeneration** The atrophy of neurons. In particular, the cholinergic neurons of the basal forebrain undergo degeneration in Alzheimer's disease and their loss correlates with plaque accumulation and the severity of dementia

**NMDA receptor** An ionotropic subtype of excitatory amino acid receptors which, when present in excess or when the cell is metabolically compromised, leads to toxicity

**Positron emission tomography** A functional brain imaging technique providing information by the noninvasive monitoring of brain metabolism

**Pyramidal cells** Large efferent neurons in the entorhinal cortex and in the CA 1 field of the hippocampus severely affected by neurofibrillary tangles and vulnerable to degeneration in Alzheimer's disease

ALZHEIMER'S DISEASE (AD) IS A PROGRESSIVE DEgenerative brain disorder involving memory and cognitive deficits, severe behavioral abnormalities, and, finally, death. This type of dementia is characterized by major neuronal losses in specific regions of the brain, especially the hippocampus and the association cortices, a reduction in the levels of many neurotransmitters, the deposition of Aβ peptide as extracellular plaques, and the formation of neurofibrillary tangles within degenerating neurons. It can be broadly classified into early onset, occurring before the age of 50–55, and late onset, occurring after 60 years of age. On average the duration of the disease is 7–10 years but may be quite variable. Currently, over 4 million people in the United States are affected with the disease and upwards of 25 million worldwide. AD affects all ethnic groups but there appears to be a slight predilection for the occurrence of the disease in females. The single most important and proven risk factor for the disease is age. Other putative risk factors include a family history of AD, head injury, and, possibly, a low level of education. There is currently no definitive treatment for this devastating disease.

## I. GROSS MORPHOLOGICAL CHANGES

The cerebral atrophy occurring in AD involves a reduction in the volume of AD gray matter, causing a shrinkage of the brain. At the gross morphological level, Alzheimer's disease is displayed by a thickening and opacification of the leptomeninges corresponding to the addition of fibroblasts and collagen to this site, by the shrinkage of the gyri, a widening of the sulci, and an enlargement of the ventricles. [*See* Brain, Central Gray Area.]

## II. CELLULAR PHYSIOLOGY

The neurodegeneration that occurs in AD is neither random nor ubiquitous. Neuronal loss is mostly confined to the midfrontal cortex, rostral superior tempo-

ENCYCLOPEDIA OF HUMAN BIOLOGY, Second Edition, VOLUME I. Copyright © 1997 by Academic Press. All rights of reproduction in any form reserved.

ral cortex, inferior parietal regions, pyramidal cells in the hippocampus, layer 2 of the entorhinal cortex, cholinergic basal nucleus of Myenert, and pigmented neurons of the locus ceruleus. Along with a decrease in neuronal number, there is an increase in the number of fibrous astrocytes, especially in layers 2 and 6 of the neocortex. Thus, in general, cortical neurons involved in corticocortical and hippocampal projections are particularly susceptible to degeneration while primary sensory areas and the motor cortex are relatively spared.

The neurotransmitter systems involved are likewise neither random nor ubiquitous. The cholinergic inputs are substantially depleted in the brain of AD patients while dopamine, catecholamines, and noradrenaline inputs are relatively spared. Specifically, the muscarinic receptors located on presynaptic cholinergic terminals (M2 and nicotinic receptors) are decreased whereas M1 receptors are not; $\alpha_2$-adrenergic receptors on terminals of nonadrenergic projections are decreased whereas there is little change in postsynaptic adrenergic receptors; $\gamma$-aminobutyric acid receptors are decreased; excitatory amino acid receptors are decreased; and $N$-methyl-D-aspartate (NMDA) and adenyl cyclase-linked metabotropic receptors are more affected in the cortex and the hippocampus. Most profoundly, acetylcholine levels are reduced by 90% in AD patients, although the levels of other neurotransmitter such as serotonin, somatostatin, and noradrenaline are decreased. In the hippocampus, glutamatergic pyramidal neurons in the CA fields are decreased in number and the remaining spared neurons show an irregular mophology. Loss of glutamatergic function contributes to the memory dysfunction prevalent in AD. Thus there is selective neuronal vulnerability not only with respect to the affected brain region, subregion, or layer but also at the level of specific neuronal subpopulations. [*See* Neurotransmitters and Neuropeptide Receptors in the Brain.]

## III. MICROSCOPIC LESIONS

At the microscopic level, the hallmark of AD pathology is the presence of amyloid plaques and neurofibrillary tangles (NFTs). Plaques, most often found in the cerebral cortex, are composed of A$\beta$ fibrils sometimes with a central amyloid core surrounded by dystrophic neurites. NFTs are composed of fibers, called paired helical filaments (PHFs), within the cytoplasm typically of large neurons.

The criterion for the diagnosis of AD set by the National Institute of Aging requires the presence of a minimal number of neocortical plaques per square millimeter while accounting for specific age groups. Most pathologists include the additional consideration of NFT accumulation for the diagnosis of AD. However, although there is an increase in the density of plaques and tangles in the AD brain compared to age-matched controls, a poor correlation exists between the accumulation of either of these entities and the clinical severity of AD. Plaques or NFTs are not restricted to AD pathology: plaques are frequently found in Down's syndrome, inclusion body myositis, Guamanian parkinsonism–dementia, and severe head trauma and NFTs are common in Down's syndrome, progressive supranuclear palsy, dementia pugilistica, leprosy, Niemann–Pick type C disease, Lewy body disease, Guamanian parkinsonism–dementia, motor neuron disease, and subacute sclerosing panencephalitis. Whether amyloid plaques are part of normal aging or represent presymptomatic or unrecognized early symptomatic AD stages is controversial.

## A. Plaques

One of the major neuropathological features of AD is the formation of extracellular plaques composed of amyloid surrounded by abnormal neurites, particularly in the cerebral cortex. The principal component of amyloid in plaques is the insoluble A$\beta$ peptide of 40–42 amino acids, approximately 4.2 kDa. Molecular cloning identified amyloid precursor protein (APP) on chromosome 21 as the precursor from which A$\beta$ is derived by proteolytic attacks at its N and C termini by $\beta$ and $\gamma$ secretases, respectively. In addition, a nonamyloidogenic pathway of APP processing exists in which A$\beta$ is cleaved by an $\alpha$ secretase within the middle of its sequence, liberating a large extracellular domain and leaving a 100 amino acid C-terminal fragment attached to the membrane. A$\beta$ forms a fibril with a $\beta$-pleated sheet structure. Plaques are classified as diffuse or neuritic depending on the absence or presence of dystrophic neurites, respectively. A$\beta$ is thought to be first deposited in diffuse plaques which are reactive with anti-A$\beta$ antibodies but are not associated with dystrophic neurites nor do they contain amyloid cores. The development of plaques progresses from diffuse plaques to spherical neuritic plaques with dense amyloid cores associated with dystrophic neuritic processes, NFTs, astrocytes, and microglia. [*See* Atherosclerosis.]

## B. Neurofibrillary Tangles

The other major neuropathological feature in AD brains is the neurofibrillary tangle, principally composed of highly insoluble paired helical filaments. Tangles are most abundant in the hippocampus and entorhinal cortex and involve the cytoplasm of large pyramidal neurons. NFTs remain as an insoluble residue upon the death of the neuron. PHFs consist of two strands of subunits which twist around one another in a helical fashion. By electron microscopic examination, the width of the two strands alternates between 8 and 20 nm, and the period is 80 nm. The consensus finding from the isolation of PHFs via a variety of methods has identified tau, a microtubule-associated protein in a hyperphosphorylated state, as the major, if not the only, component of PHFs. The human brain contains six tau isoforms ranging from 352 to 441 amino acids arising from a single gene by way of alternative mRNA splicing. The six isoforms differ from each other by the presence or absence of three inserts. Each tau contains three or four tandem repeats of 31 or 32 amino acids located near the C terminus of the molecule thought to be the microtubule-binding domains.

The phosphorylation of tau is developmentally regulated. Developmentally immature tau is highly phosphorylated whereas the adult form is relatively less phosphorylated. The fetal phosphorylation pattern of tau is reminiscent of that associated with AD. In AD, there is a shift in the pool of tau from that soluble in association with microtubules to that insoluble and hyperphosphorylated in PHFs. *In vitro* studies have identified likely candidates for the responsible kinases: mitogen-activated protein kinase, glycogen synthase kinase 3, and cyclin-dependent kinase 5. It is likely that other kinases in the AD brain will be shown to be involved in the conversion of normal tau to hyperphosphorylated tau. The hyperphosphorylated state of tau could also result from decreased phosphatase activity, the majority of which appears to be accounted for by protein phosphatase 2A.

## C. Other Lesions

In addition to the two characteristic lesions of plaques and tangles, other evidence of pathology in the AD brain also exists. Neuropil threads scattered throughout the hippocampus and entorhinal cortex are mostly dendritic in origin. These elements arise from degenerating NFT-bearing neurons and are composed of PHFs. Amyloid angiopathy is a common feature of AD pathology and typically occurs in small-to-medium sized vessels, especially those in the cerebral cortex and cerebellum. These vascular amyloid deposits occur in the media adventitia–tunica media junction in the outer basement membrane of vessels. Hirano bodies are intracytoplasmic, paracrystalline, rodlike structures frequently observed in the CA1 area of the hippocampus. Hippocampal neurons are also the site of granulovacular degeneration. [*See* Hippocampal Formation.]

## IV. THEORIES AS TO CAUSES

The etiology of AD is heterogeneous. The majority of AD cases are classified as sporadic and are linked to the apolipoprotein E4 locus on chromosome 19 as a major risk factor. About 20% of all cases are of genetic origin, termed familial AD (FAD), with the onset of symptoms occurring relatively early in individuals prior to the age of 65. To date, three genetic loci have been identified in which autosomal dominant mutations give rise to AD. Mutations in the APP locus on chromosome 21 account for less than 1% of AD cases. Loci bearing numerous mutations on chromosomes 14q24.3 and 1q31-42, termed presenilin 1 and 2, respectively, account for the greater majority of FAD cases. Despite the genetic linkage to AD noted earlier, the underlying cause of the pathology is still unknown. The following molecular mechanisms offer possible theories as to the cause of the disease, taking into account current knowledge of the pathophysiology, biochemistry, and genetics implicated in AD.

## A. Abnormal Processing of APP

Accumulation of A$\beta$ is one of the earliest pathological features of AD. Processing of the APP molecule via the amyloidogenic pathway yields A$\beta$ species of 1–40 and 1–42 amino acids in length. It has been proposed that the balance between these two forms may be key to the deposition of amyloid. Both A$\beta$1–40 and A$\beta$1–42 are formed by the normal metabolism of APP and are found in cerebrospinal fluid of healthy individuals. However, a shift in this balance yielding increased amounts of A$\beta$1–42 may be sufficient to cause fibril formation and the pathological consequences. The evidence in support of altered APP processing causally related to AD is as follows: (a) In Down's syndrome, trisomy 21, 1.5-fold expression of APP invariably causes plaque formation at much earlier stages than other signs of pathology such as NFT formation; (2) mutations in the APP gene (e.g., at codons 670, 671, 692, 713, 717) cosegregate with early onset FAD; (c) mutations at codons 670 and 671, identified in Swedish FAD kindreds, cause an increase in A$\beta$1–40 and A$\beta$1–42 levels; (d) mutations at codon 717 cause a selective increase in A$\beta$1–42 levels, which is the form more prone to amyloid fibril deposition; (e) fibrillar A$\beta$ is toxic to neurons (*in vivo* and *in vitro*); (f) transgenic mice expressing mutations of APP develop plaques in appropriate brain regions and show signs of cognitive deficits; and (g) AD subjects with mutations in the presenilin 1 and 2 genes produce elevated levels of A$\beta$1–42. In summary, under pathological conditions, an increased production of A$\beta$1–42 occurring by abnormal processing of APP may be sufficient, if not the primary cause, for amyloid fibril deposition and its ensuing neuropathological defects. In addition, A$\beta$ exacerbates the toxicity of a variety of other modes of damage, including glucose deprivation, excitotoxicity, and oxidative damage, as outlined in the following sections.

## B. Abnormal Phosphorylation of Tau

Abnormal phosphorylation of tau represents one of the earliest changes leading to AD neurofibrillary pathology and may have deleterious effects on neurons over time. In PHFs, all six tau isoforms are present in abnormally phosphorylated forms, containing about 12 moles of phosphate per mole of protein, approximately four times the amount of normal brain tau. The normal phosphorylation state of tau reflects a balance between kinases and phosphatases. Most likely, both of these enzymatic pathways are altered in AD, resulting in the excessive phosphorylation of tau.

*In vitro,* excessive phosphorylation inhibits the ability of tau to bind to and promote the assembly of microtubules. Conversely, the *in vitro* dephosphorylation of hyperphosphorylated tau renders it competent in microtubule assembly. Thus, it seems that the abnormal phosphorylation of tau disrupts the assembly of microtubules in the AD brain and, indeed, quantitation of the number of microtubules reflects their reduced numbers in AD compared to controls. A vital function of microtubules is the intracellular transport of molecules and organelles between the cell body and its neurites. Thus, disruption of axonal transport along with dissociation of tau from microtubules may be principal consequences of tau hyperphosphorylation, with the final outcome being neuronal dysfunction and/or death.

Once dissociated from microtubules, tau proceeds toward the formation of AD lesions. Formation of the PHF structures present in AD requires the self-assembly of microtubule-dissociated, soluble tau. Filaments with characteristics resembling PHFs have been shown to form readily *in vitro* from the self-assembly of full-length tau and the microtubule-binding domain of tau, supporting the notion that a similar process occurs *in vivo.* Subsequent processing and ubiquitination of tau fragments result in the formation of highly insoluble PHFs that accumulate in neuronal cell bodies as NFTs, in neurites as neuropil threads, and in dystrophic neurites surrounding amyloid plaques.

## C. Apolipoprotein E

Evidence shows that apolipoprotein E (ApoE), a cholesterol transport protein, plays an important role in the development of AD pathology. The ApoE gene is located on chromosome 19q13.2 and is characterized by three common alleles, epsilon 2, epsilon 3, and epsilon 4, giving rise to six phenotypes. Of these, the ApoE 4 allele is strongly associated with late-onset FAD and sporadic AD. Frequency of the ApoE 4 allele in this group of patients is significantly higher compared with that in age-matched controls. In contrast, the ApoE 2 allele appears to protect against the risk of AD.

ApoE genotyping combined with positron emission tomography has shown promise as a diagnostic tool in preclinical subjects. Asymptomatic ApoE 4 homozygotes display reduced glucose metabolism in the same brain regions as clinically diagnosed AD subjects. However, the presence of the ApoE 4 allele is not diagnostic in itself; it is neither sufficient nor necessary for AD to develop.

Several mechanisms have been put forth as to how ApoE 4 leads to the development of AD pathology. *In vitro,* ApoE 4, but not ApoE 3, binds tightly to A$\beta$ and can be found, *in vivo,* complexed with A$\beta$ in plaques. This association suggests that ApoE 4 may be causing A$\beta$ to precipitate out of solution and consequently form plaques. Furthermore, ApoE 4 homozygotes bear a greater amyloid burden in their blood vessels and plaques than ApoE 3 homozygotes. ApoE is also present in NFTs. *In vitro,* ApoE 3, but not ApoE 4, binds tightly to tau. It has been proposed that the binding of ApoE 3 and ApoE 2 to the microtubule-binding domain of tau interferes with the processes of tau hyperphosphorylation or self-assembly into PHFs. However, ApoE 4, which does not bind tau, cannot prevent the formation of PHFs and thus enhances the formation of neurofibrillary lesions. ApoE 3 and ApoE 4 exert opposite effects on neurite outgrowth in neuronal cultures. ApoE 3 enhances neurite outgrowth whereas ApoE 4 inhibits this process, suggesting that it may be inhibiting the remodeling of neurons after injury in the AD brain and thus contributing to the progression of the neuropathology seen in AD.

## D. Abnormal Energy Metabolism

A defect in energy metabolism may play a role in the pathogenesis of AD. The brains of symptomatic AD subjects show reduced cerebral blood flow, glucose metabolism, and oxygen utilization. Correspondingly, AD pathology presents defects at the level of brain capillaries, glucose metabolism, and mitochondrial respiration. First, there are extensive changes in brain capillaries in AD: ultrastructural changes as well as molecular changes in the endothelium compromising glucose and oxygen delivery to neural cells. Most notably, the densities of glucose transporters and mitochondria are reduced in cerebral microvessels.

Second, severely disturbed brain glucose metabolism is a consistent feature of AD. In the hippocampus and dentate gyrus, regions greatly affected by the neurodegeneration occurring in AD, there is a significant reduction in the number of neuron-specific glucose transporter 3 molecules. Positron emission tomography studies in AD patients have been useful in the detection of this glucose hypometabolism. Other glycolytic functions are also affected: increased levels of pyruvate are closely associated with the severity of the dementia and a reduction in the rates of oxidation of glucose and glutamine, processes requiring mitochondrial function, is noted whereas lactate metabolism does not seem to be affected in AD.

Third, defects in the electron transport chain, specifically a deficiency in cytochrome oxidase activity, could lead to a reduction in energy stores. Message levels for the mitochondrial-encoded cytochrome oxidase subunits I and III are significantly reduced in the AD brain, which could greatly contribute to the reduced energy metabolism in AD. Additionally, in the brains of AD, the production of ATP steadily declines as the disease state progresses. Both decreased glucose transport and utilization and increased glucocorticoid levels are known to occur with aging and may lead to decreased energy supplies and ATP depletion. Thus it can be argued that a decrease of the oxidative energy metabolism and the resulting ATP

deficit may critically affect neuronal functions such as protein metabolism, neurotransmission, and ion homeostasis, with dire consequences for the survival of the cell.

## E. Oxidative Damage

Oxidative stress arises from an imbalance between the generation of reactive oxygen species (ROS) and the antioxidant defense systems. Evidence at the protein, lipid, and nucleic acid level shows that free radical damage contributes to the pathogenesis of AD. The generation of ROS, the formation of peroxidation products, the activities of oxidative defense systems, and the action of iron attest to the relevance of oxidative stress in AD. The majority of ROS in a cell is generated by mitochondrial oxidative phosphorylation. Leakage of high energy electrons along the mitochondrial electron transport chain causes the formation of superoxide radicals and hydrogen peroxide. AD brain mitochondria generally exhibit defects in the electron transport chain complexes. Most markedly, there is a reduction in the activity of the terminal complex of the mitochondrial respiratory chain, cytochrome oxidase, a marker for oxidative metabolism, in the brain regions affected by AD neurodegeneration. Monoamine oxidase, located in the outer membrane of mitochondria, is responsible for the oxidation of amines to aldehydes, forming hydrogen peroxide in the process. Deprenyl, a monoamine oxidase-B inhibitor with antioxidant properties, has been shown to slow down the rate of cognitive impairment in AD, suggesting that an overproduction of hydrogen peroxide production can have detrimental consequences in AD. Furthermore, measurements of oxidative damage to mitochondrial DNA assessed by monitoring the levels of the oxidized nucleoside, 8-hydroxy-2'-deoxyguanosine, is increased in AD compared to age-matched controls.

Iron levels in the AD brain are increased. The production of hydroxyl radicals is facilitated by iron via the Fenton reaction. In culture, A$\beta$-mediated toxicity is exacerbated in the presence of iron, suggesting that oxidative mechanisms are involved in the cell death *in vitro* and, perhaps, *in vivo*. That antioxidants, such as 17$\beta$-estradiol, catalase, and superoxide dismutase, offer protection against A$\beta$-mediated toxicity further strengthen this notion. In turn, *in vitro* oxidation promotes the aggregation of A$\beta$ as it does that of tau and increases the binding of ApoE 4 to A$\beta$, implying that oxidative damage of these proteins *in vivo* promotes the formation of plaques and NFTs. Also, the direct generation of ROS by both A$\beta$ and tau remains a possibility.

Lipid peroxidation is increased in the AD brain. Damage to membrane lipids could potentially alter the processing of APP at the membrane, leading to increased production of A$\beta$. *In vitro,* A$\beta$ is a potent lipoperoxidation initiator. Thus a positive feedback system is set up which could escalate damage over time. NFT formation may also be enhanced by oxidative damage to membrane lipids. A product of lipid peroxidation, E-4-hydroxy-2-nonenal, has been shown to crosslink proteins, especially tau.

Calcium imbalances and glutamate receptor abnormalities are both features of AD. Calcium-dependent metabolisms of arachidonic acid and xanthine result in the production of superoxide anions. The reaction of superoxide anions with nitric oxide produces peroxynitrite species which attack tyrosine residues in proteins. The formation of nitric oxide is catalyzed by the calcium-dependent enzyme nitric oxide synthetase, which is activated by the NMDA receptor. Identification of nitrotyrosine residues in NFTs links oxidative damage resulting from calcium imbalances and glutamate receptor abnormalities to pathological features of AD. Another indication that the AD brain is under increased oxidative stress is its elevated activities of antioxidant enzymes, namely superoxide dismutase, catalase, glutathione peroxidase, and glutathione reductase, presumably to compensate for the imbalance.

## F. Abnormal Calcium Homeostasis

A disturbance in the homeostasis of cytosolic calcium concentration is common to a number of pathological conditions of various origins. The inability of neurons to maintain appropriately low levels of calcium may contribute to neurodegeneration in AD. An excitatory amino acid mechanism of cell death has been proposed for AD. Glutamate is the principal neurotransmitter of the corticocortical association fibers and the major hippocampal pathways, which are the earliest to degenerate in AD. This observation has led to the proposal that calcium fluxes in AD could be driven by glutamate receptor activation. In particular, it has been demonstrated that stimulation of the NMDA receptor leads to an influx of calcium, raising intracellular levels to toxic concentrations. Further support for abnormal calcium homeostasis contributing to the lesions in AD comes from the finding that neuronal cultures treated with glutamate or calcium ionophores develop cytoskeletal alterations similar to those observed in AD.

Furthermore, A$\beta$ causes an elevation of intracellular calcium levels leading to neurotoxicity. Structurally, A$\beta$ is able to form calcium channels in lipid bilayers with defined characteristics potentially capable of disrupting calcium gradients across the membrane and allowing for toxic concentrations of calcium to be reached intracellularly. Once inside the cell, calcium could trigger neuronal death via the activation of calcium-dependent enzymes, putatively the calcium-dependent kinases involved in the hyperphosphorylation of tau. A$\beta$-induced toxicity can be prevented in cells overexpressing the calcium-binding protein calbindin D28k. Furthermore, cultured hippocampal neurons, which display resistance to A$\beta$ toxicity, express the calcium-binding protein calretinin, suggesting that the survival of these cells is due to their enhanced capacity to buffer calcium fluxes. *In vivo,* calretinin-positive neurons are not associated with plaques or NFT, implying that their enhanced capacity to buffer calcium may be sparing them from death. A$\beta$ also increases the vulnerability of neurons to glutamate, which induces toxicity via the activation of EAA receptors and the ensuing influx of toxic levels of calcium. It should be apparent that defects in energy metabolism, excitotoxicity, oxidative stress, and calcium homeostasis are interrelated.

## G. Heavy Metal Accumulation

An accumulation of zinc, aluminum, iron, selenium, lead, and silicon occurs in discrete brain regions in AD. It is possible that the accumulation of heavy metals exacerbates the development of AD pathology by enhancing the formation of A$\beta$ or tau aggregates. Zinc, aluminum, lead, iron, and silica enhance A$\beta$ fibril polymerization and deposition. Similarly, incubation of tau with aluminum salts induces the aggregation of tau into macromolecular complexes. The ability of these heavy metals to alter the solubility of A$\beta$ or tau may also facilitate the proteins' resistance to enzymatic breakdown. By interfering with the catabolism of A$\beta$ and tau, heavy metals may contribute to the stability of plaques and NFTs. Additionally, elevated iron levels could exacerbate neuronal degeneration by enhancing reactive-free radical formation. [*See* Minerals in Human Life; Zinc Metabolism.]

## BIBLIOGRAPHY

Coyle, J. T., and Puttfarcken, P. (1993). Oxidative stress, glutamate, and neurodegenerative disorders. *Science* **262,** 689.

Evin, G., Beyreuther, K., and Masters, C. L. (1994). Alzheimer's disease amyloid precursor protein (AβPP): Proteolytic processing, secretases and βA4 amyloid production. *Amyloid Int J. Exp. Clin. Invest.* **1**, 263.

Goedert, M., Spillantini, M. G., Jakes, R., Crowther, R. A., Vanmechelen, E., Probst, A., Gotz, J., Burki, K., and Cohen, P. (1995). Molecular dissection of the paired helical filament. *Neurobiol. Aging* **16**, 325.

Strittmatter, W., and Roses, A. D. (1995). Apolipoprotein E and Alzheimer disease. *Proc. Natl. Acad. Sci. USA* **92**, 4725.

Terry, R. D., Katzman, R., and Bick, K. L. (eds.) (1994). "Alzheimer Disease." Raven Press, New York.

# Amoebiasis, Infection with *Entamoeba histolytica*

JONATHAN I. RAVDIN
*Case Western Reserve University School of Medicine*

## GLOSSARY

**Axenic** In the absence of microbes

**Capping** Aggregation of cell surface membrane proteins, often due to the attachment of divalent molecules such as antibodies or plant lectins

**Chitinous** Consisting of an impermeable polymer of *N*-acetyl-D-glucosamine, a sugar

**Colitis** Inflammation of the mucosal lining of the large bowel, associated with ulcerations, blood in feces, and inflammation in the submucosal area

**Cyst** Round excreted form which contains a rigid chitinous outer cell wall, allowing survival in the environment outside the host and resistance to stomach acid

**Cytolytic activity** Lethal effect on target cells which results in tissue lysis; can be contact dependent or due to secreted cytotoxins

**Cytotoxins** Molecules which are harmful to cells, usually interfering with normal cell functions, such as synthesis of new proteins, or maintaining normal nontoxic concentrations of intracellular ions, such as calcium

**Encystation** Change from trophozoite to cyst form

**Excystation** Release of a trophozoite form from a cyst

**Filopodia** Filamentous extension from the ameba, often used to anchor the cell to a substratum

**Glycocalyx** A carbohydrate found on the exterior surface of a cell membrane

**Lectin** A protein that, by virtue of its carbohydrate binding activity, agglutinates cells or mediates cell–cell attachment

**Phagocytosis** Ingestion of large particles by a cell due to invagination of the surface membrane; also referred to as endocytosis

**Sialic acid** A carbohydrate moiety; also referred to as neuraminic acid

**Trophozoite** Amoeboid form that survives only within the host, residing in the large bowel

**Xenic** Cocultivation of amoebae with more than one bacterial species

**Zymodeme** Unique patterns of isoenzyme (enzyme variants) on starch gel electrophoresis of fecal isolates, differentiates *E. dispar* from pathogenic *E. histolytica*

*ENTAMOEBA HISTOLYTICA* IS A PROTOZOAN (single-celled organism) parasite that infects the human gastrointestinal tract. Infection results from ingestion of the *cyst* form, whose chitinous outer wall resists desiccation outside the human body and the harmful effects of stomach acid. Excystation occurs in the small bowel; the *trophozoite* form infects the colon and may either invade tissue or encyst, the latter case resulting in disease transmission. Infection occurs in approximately 1.0% of the world's population, resulting in a substantial disease burden in underdeveloped and tropical areas. *Entamoeba histolytica* trophozoites are 10–60 μm in size, are highly phagocytic, lack mitochondria, and have a complex relationship with colonic bacterial flora. *Entamoeba histolytica* must be differentiated from the morphologically iden-

ENCYCLOPEDIA OF HUMAN BIOLOGY, Second Edition, VOLUME 1. Copyright © 1997 by Academic Press. All rights of reproduction in any form reserved.

tical *Entamoeba dispar,* which does not cause disease. The main clinical syndromes resulting from *E. histolytica* infection include asymptomatic cyst passage, invasive inflammatory colitis with bloody diarrhea, and amoebic liver abscess, which is usually manifest as localized abdominal pain and fever. Pathogenesis of invasive amoebiasis is apparently dependent on amoebic adherence proteins, cytolytic activities, and proteinases. Human immunity to invasive amoebiasis may develop following pharmacological cure; serum and mucosal antiamoebic antibodies and amoebicidal cell-mediated responses appear to be contributory. Diagnosis currently relies on the detection of amoebae in stool or antiamoebic antibodies in serum and non-invasive imaging studies of the liver. Prevention of amoebic infection is best accomplished by proper sanitation, availability of uninfected water and food, and avoidance of direct fecal–oral contamination. A vaccine is currently not available, but research is promising.

## I. LIFE CYCLE AND EPIDEMIOLOGY OF *E. Histolytica*

The life cycle of *E. histolytica* is straightforward (Fig. 1); only two forms of the parasite exist: the cyst and the trophozoite. Infection results from ingestion of the cyst form, which can exist for months outside the human host. In contrast, trophozoites rapidly disintegrate within feces after excretion and would also be destroyed by stomach acid upon ingestion. The infective dose (i.e., the minimum number required for initiating infection in a human) may be as low as a single cyst; however, the greater the number of cysts ingested, the higher the likelihood of infection and the shorter the incubation period prior to the onset of symptomatic disease. Excystation occurs in the small

bowel; its regulation by parasite and environmental factors is undefined. *In vitro,* excystation can be induced by low oxygen tension, inorganic salts, and an osmolality that supports the growth of trophozoites. Excystation is followed by a metacystic development; mature quadranucleate cysts most frequently undergo four cycles of cell division, producing eight uninucleate trophozoites.

Trophozoites reside in the large bowel, most likely attached to the colonic mucus blanket, feeding on bacteria. Approximately 1 in 10 infected with *E. histolytica* develops symptomatic invasion of the colonic mucosa; however, the relative prevalence of *E. dispar* and *E. histolytica* infection varies tremendously among different geographic areas. For example, the probability of *E. histolytica* infection compared to *E. dispar* is low in Canada, but is much higher in Mexico or India. Encystation may be stimulated by an inhospitable local gut environment, but little information is available. Feces may contain up to 45 million cysts per day per individual, but cysts are infrequently excreted during active amoebic colitis. *In vitro,* encystation of xenic *E. histolytica* cultures can be induced by magnesium sulfate and serum at a particular osmolality. This process is associated with the development of a chitinous (trimers and tetramers of N-acetyl-D-glucosamine) cyst wall and the appearance of stage-specific proteins containing sialic acid. *Entamoeba histolytica* possesses a chitin-binding lectin having a size of 220,000 D which has been hypothesized to have a role in encystment. Chitin synthetase inhibitors, such as Polyoxin D or Nikomycin, inhibit *in vitro* encystation by *Entamoeba invadens* (a parasite of reptiles); their use provides a potential strategy for the prevention of encystation in human amoebiasis.

Transmission of amoebic infection results from poor sanitation, with fecal contamination of water and food, or from direct fecal–oral contact due to oral–anal sexual practices or poor personal hygiene. Insects such as cockroaches and flies have been postulated to act as vectors of transmission. Although *E. histolytica* cysts can attach to the external surface of flies or survive in the cockroach gut, no epidemiological data support this means of spreading. In the least developed countries and in the tropics, where *E. histolytica* is most prevalent, disease is most intense among the poorest members of society who live in a crowded unsanitary environment or have cultural practices that favor transmission; a prevalence of amoebic infection of up to 50% has been observed in some countries. There are more than 50 million cases of invasive amoebiasis in the world and 50,000 fatalities per year.

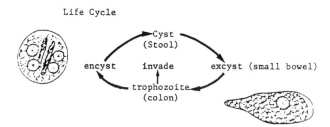

**FIGURE 1**   Life cycle of *E. histolytica.* Ingestion of cysts results in excystation in the small bowel and colonization of the large bowel with trophozoites. Encystment and excretion of the chitinous cyst form result in the transmission of disease.

The disease is highly endemic in Mexico, India, western and southern Africa, and Southeast Asia. In Western developed countries, *E. histolytica* infection is mainly confined to certain high-risk groups, such as institutionalized people (especially the mentally retarded), sexually promiscuous male homosexuals, foreign travelers, and emigrants or migrant workers previously residing in highly endemic areas. Unusual health practices, such as colonic irrigation with improperly sterilized equipment, or unsanitary communal living conditions have also resulted in outbreaks of amoebiasis.

Among those infected with *E. histolytica*, certain subsets are at greater risk for occurrence of severe invasive amoebiasis. High-risk groups include pregnant women, children below age 2, individuals receiving high doses of corticosteroids, and the malnourished. All of these groups may have impaired cell-mediated immune mechanisms; however, their susceptibility to invasive amoebiasis may have a multifactorial origin. *Entamoeba histolytica* is a frequent cause of symptomatic diarrhea in North Americans with acquired immunodeficiency syndrome; however, fulminant invasive amoebiasis has not been noted in these individuals, despite their profound defects in cell-mediated immunity.

## II. PHYLOGENY AND STRAIN VIRULENCE

*Entamoeba histolytica* and *E. dispar* belong to the family Entamoebidae of the order Amoebida in the subphylum Sarcodina of the pseudopod-forming protozoan superclass Rhizopoda. In the past, taxonomy of *Entamoeba* spp. was based on morphology, *in vitro* growth requirements, *in vivo* virulence, antigen analysis, amino acid content, *in vitro* drug susceptibility, and *in vivo* host specificity. Recent studies using restriction enzyme digestion of genomic DNA and DNA hybridization are replacing older, less rigorous, criteria. Other species in this family include *E. hartmanni*, *E. polecki*, *E. coli*, and *E. gingivalis*. The noninvasive *E. hartmanni* was previously known as "small race" *E. histolytica* because of its identical morphology but smaller size. It has been distinguished as a separate species by isoenzyme and antigen analyses and by DNA hybridization criteria. The *Entamoeba*-like Laredo amoeba is a commensal organism which grows at lower temperatures *in vitro* and has genomic DNA distinct from that of *E. histolytica*.

As first suggested by isoenzyme analysis, *E. histolyt-ica* has been demonstrated to consist of a species complex. Starch gel electrophoresis of clinical isolates has revealed more than 20 different mobility patterns for the *Entamoeba* hexokinase, phosphoglucomutase, L-malate, NADP$^+$ oxidoreductase, and glucose-6-phosphate isomerase enzymes. Unique patterns of isoenzyme mobility are referred to as zymodemes; certain zymodemes were found to be reproducibly, but not exclusively, associated with the occurrence of either asymptomatic noninvasive intestinal infection or clinically symptomatic invasive amoebiasis. As mentioned, 10% of individuals with asymptomatic amoebic infection excrete an isolate that has a zymodeme characteristic of *E. histolytica*. Some experts have concluded that infection with *E. dispar* is without risk to the host or others (by transmission).

This is supported by epidemiological studies of *E. dispar* infection and recent work with DNA and ribosomal RNA probes which conclude that distinct species do exist. However, *in vitro* a cloned *E. histolytica* isolate has been demonstrated by some researchers to change from a nonpathogenic to a pathogenic zymodeme (isoenzyme pattern) when cocultured with bacterial flora from a patient with invasive amoebic colitis. This new "pathogenic" amoeba was apparently due to incomplete cloning of the fecal isolate. Isoenzymes do not directly participate in parasite virulence, but appear to be phenotypic markers of this process. The distinction of *E. histolytica* from *E. dispar* provides useful applications to clinical practice. Further studies of DNA homology and a more complete understanding of the molecular regulation of parasite pathogenicity are needed. [*See* Isoenzymes.]

## III. STRUCTURE AND METABOLISM

*Entamoeba histolytica* trophozoites are 10–60 $\mu$m in diameter, with a spherical nucleus usually occupying less than one-fifth of the cell. Stained trophozoite nuclei demonstrate a central karyosome and dark chromatin granules in a symmetrical nonclumped distribution on the inner aspect of the nuclear membrane. Trophozoites are highly vesiculated and demonstrate classic ameboid motility by the extension of pseudopodia (Fig. 2). Amoebae are actively phagocytic and contain ingested erythrocytes or bacteria. Cysts of *E. histolytica* average 12 $\mu$m in size, with a range of 10–20 $\mu$m, and may contain up to four nuclei having a morphology identical to that of trophozoite nuclei. Immature cysts containing a single nucleus may contain masses of glycogen, which will stain with iodine.

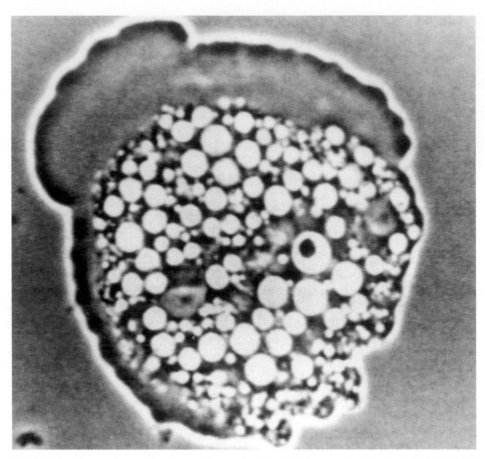

**FIGURE 2**   Phase micrograph of a motile axenic *E. histolytica* trophozoite (strain HM1–IMSS) photographed with a Ziess Axiomat (×1000). [Original magnification (×3).] Note pseudopod extension and highly vesiculated cytoplasm.

Dark-staining rod-shaped chromatoid bodies present in cysts apparently consist of ribosomal particles packed into crystalline arrays.

Studies of trophozoite ultrastructure by electron microscopy reveal a polarized structure with a uroid (tail) region, characterized by a large number of filopodia. Binding by the plant lectin concanavalin A or antiamoebic antibodies indicates that microfilament-dependent capping of the surface membrane occurs in the uroid area, with shedding of amoebic membrane material by motile trophozoites. Short filopodia are present on the outer edge of the amoebic basal surface, which is adherent to a substrate. The amoebic surface trilaminar (three-layered) membrane has an externally attached fuzzy glycocalyx that is 20–30 nm thick in xenic trophozoites and 5 nm thick on an axenically cultured amoeba.

Cytoplasmic structures include acid pH vesicles which have characteristics of classic lysosomes (e.g.,

they contain acid phosphatase). There appear to be at least two vesicle populations: large endocytic vesicles, which can be studied by endocytosis of fluorescein-conjugated dextrose, and small vesicles, which label with acridine orange (Fig. 3). Actin-containing mitochondria are absent; ribosomes are present in helical arrays rather than in a classic rough endoplasmic reticulum. Tubular structures are present that can be associated with endocytic vesicles. Actin-containing microfilament-like cytoskeletal structures are demonstrable; amoebic actin is one of the many parasite proteins that have now been cloned in bacteria. A tubular smooth endoplasmic reticulum is evident and may be functioning as a Golgi apparatus; typical Golgi apparatuses have not been found in *E. histolytica*. Viral particles have been demonstrated in trophozoite cytoplasm; however, there is no evidence of virus-mediated transfer of parasite virulence properties.

The nucleus is surrounded by a membrane 120 nm

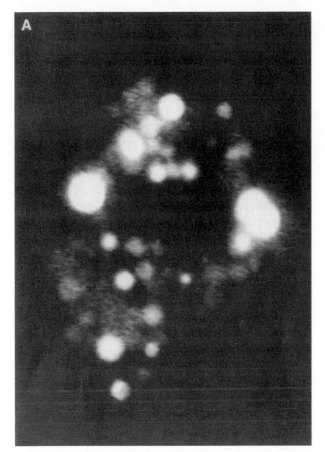

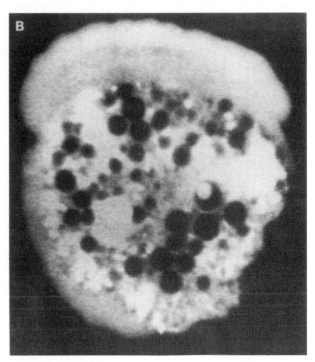

**FIGURE 3** Acid pH vesicle population of *E. histolytica.* (A) Fluorescence micrograph of fluorescein-conjugated dextran endocytosed by *E. histolytica* trophozoites. (Vesicle pH was 5.1, as estimated by spectrofluorimetry.) (B) Concentration of the acid-vesicle label acridine orange in a distinct small vesicle population. Both ×1000.

thick, containing numerous 50-nm pores; the chromatin present on the inner surface consists of RNA–protein complexes and DNA. Mitosis in *E. histolytica* has only very recently been demonstrated with true spindle formation. Detailed analysis of *E. histolytica* chromosomes has not been performed.

Cyst ultrastructure is less well defined; cyst walls are 40–86 nm in thickness initially, but during maturation may increase to over 700 nm. Vacuoles are located peripherally and are reduced in size and number compared to trophozoites. X-ray diffraction studies of the cyst wall reveal patterns identical to that of standard chitin.

Axenic culture of *E. histolytica* has provided the opportunity for studying parasite metabolism. *Entamoeba histolytica* degrades glucose to pyruvate, in the absence of oxygen, by the anaerobic glycolytic pathway; the end products are ethanol and $CO_2$. Amoebae contain high concentrations of inorganic pyrophosphate $(P_3OPO_3)$, including membrane-bound phosphate, which apparently substitutes for ATP in several glycolytic reactions. Trophozoites and cysts contain abundant glycogen, despite the absence of glycogen synthetase. *Entamoeba histolytica* is most likely a facultative anaerobic organism, i.e., it can live with or without oxygen; trophozoites do consume oxygen, despite a lack of mitochondria, and oxidize pyruvate by several pathways, using molecular oxygen; acetate is the end product. Only a small number of carbohydrates can be metabolized by *E. histolytica* and thus support its growth *in vitro*.

In *E. histolytica,* the electron transport through the respiratory chain, which is required for oxidation, is markedly different from that of the mitochondria of mammalian eucaryotic cells. There are no heme proteins; nonheme iron and sulfur proteins appear to substitute. Therefore, amoebic respiration is not inhibited by actinomycin, cyanide, or azide; however, metal chelators are inhibitory. In addition, unlike mammalian cells, amoebic aerobic electron transport is apparently not coupled to ATP synthesis. The absence of mitochondria may account for amoebic survival without glutathione, which is thought to be essential to protect against oxygen toxicity resulting from mitochondrial oxidations. *Entamoeba histolytica* does contain superoxide dismutase, which protects against oxygen toxicity; its presence possibly allows survival of the amoeba in tissues such as lung and brain, which have high oxygen tension.

## IV. CLINICAL DISEASE SYNDROMES

The clinical syndromes that may result from human infection by *E. histolytica* are outlined in Table I. Overwhelmingly, the most common outcome is asymptomatic intestinal infection, apparently noninvasive. Depending on the geographic location and host risk factors, over 90% of infected individuals have asymptomatic amoebiasis. Lack of tissue invasion is recognized by the absence of blood in the stool, fecal excretion of amoebic cysts but not trophozoites, and a normal colonic mucosa observed during lower gastrointestinal fiberoptic endoscopy. However, the latter is rarely performed unless differentiation from

**TABLE I**

Clinical Syndromes Associated with *E. histolytica*[a]

---

Intestinal disease
  Asymptomatic infection
  Symptomatic noninvasive infection
  Acute rectocolitis (dysentery)
  Fulminant colitis with perforation
  Toxic megacolon
  Amoeboma
  Chronic nondysenteric colitis
Extraintestinal disease
  Liver abscess
  Liver abscess complicated by
    Peritonitis
    Empyema
    Pericarditis
  Lung abscess
  Brain abscess
  Genitourinary disease

---

[a]Adapted from J. I. Ravdin and W. A. Petri, (1989). *Entamoeba histolytica. In* "Principles and Practices of Infectious Diseases" (G. L. Mandell, R. G. Douglas, and J. E. Bennett, eds.), 3rd Ed., pp. 2036–2040. Churchill-Livingstone, New York.

an alternative disease entity, such as idiopathic inflammatory bowel disease, is desired. Usually this is due to *E. dispar* infection, but 9 of 10 *E. histolytica* infections are also asymptomatic. Patients with noninvasive *E. histolytica* infection may have nonspecific gastrointestinal complaints, such as diarrhea, bloating, or mild abdominal pain, which may respond to antiamoebic therapy; whether this reflects subclinical invasive colitis or an alternative microbial etiology is unknown.

Acute amoebic colitis or dysentery presents over 1–3 weeks with bloody stool, diffuse abdominal pain, and tenderness in the area of the intestine involved. White blood cells may be absent in the stool, despite colonic mucosal inflammation, because they are lysed on contact by *E. histolytica* trophozoites. Fulminant colitis is infrequent and generally occurs in the high-risk groups defined in Section I. Such individuals are severely ill, with explosive bloody mucoid diarrhea, severe abdominal pain, and signs of diffuse peritonitis. Colonic perforation may be present as a consequence of colonic dilatation, with thinning and necrosis of colonic tissues. Amoeboma is a localized chronic lesion that presents as a tender palpable single- or multiple-mass lesion(s), often in the right or ascending colon, which may be confused with carcinoma. Invasive amoebic colitis can manifest in an indolent intermittent form which is clinically indistinguishable from idiopathic inflammatory bowel disease (i.e., ulcerative colitis). Amoebic infection must always be ruled out prior to the use of corticosteroids to treat presumed inflammatory bowel disease, as such agents markedly exacerbate amoebic infection.

Extraintestinal amoebiasis (Table I) usually results from trophozoites metastasizing to the liver via the portal venous drainage of the large intestine. An amoebic liver abscess generally presents with the clinical triad of fever, right upper quadrant abdominal pain, and exquisite tenderness over the liver. However, a more chronic presentation, with weight loss rather than fever, is common. Liver abnormalities are reflected by an elevation in serum transaminase enzymes and alkaline phosphatase and a reduction in serum albumin. Extension of the amebic liver abscess may result in diffuse peritonitis following abscess rupture into the peritoneal cavity, pleuropulmonary disease after abscess penetration through the diaphragm, or acute pericarditis by spread of a left lobe liver abscess into the pericardial space. Such events are often difficult to diagnose and add considerably to morbidity and mortality rates. Localized lung or brain abscess due to hematogenous spreading is quite rare and presents in a manner similar to other etiologies of abscesses in those tissues. Genitourinary disease is extremely unusual, but is important due to confusion of this easily treatable disease with carcinoma of the penis or uterine cervix. Suspicion for extraintestinal amoebiasis must be maintained in individuals having epidemiological risk factors.

## V. PATHOGENESIS OF AMOEBIASIS

Pathogenesis can be defined as the cellular events, reactions, and other mechanisms that occur during development of a disease. The pathogenesis of invasive amoebiasis can be divided into three stages: (1) intestinal colonization by a virulent *E. histolytica* strain, with adherence of trophozoites to the intestinal mucus layer; (2) disruption of intestinal barriers by amoebic enzymes or toxic products; and (3) amoebic lysis of intestinal cells and host inflammatory cells, leading to mucosal interruption, colonic ulcers, and deep-tissue invasion.

### A. Intestinal Colonization and Adherence

*Entamoeba histolytica* infection is species specific; humans and Old World primates are the only reservoir for natural infection. Host species specificity may be related to intrinsic parasite properties or to intestinal environmental factors such as microflora and colonic redox potential. Younger animals have increased susceptibility to infection, consistent with reports of rapidly fatal invasive amoebiasis in human infants. *Entamoeba histolytica* is closely associated with intestinal bacterial flora; in experimental models of intestinal amoebiasis, *E. histolytica* virulence is dependent on the presence of viable bacteria. Bacteria that have mannose-binding lectins at their surface adhere to and are ingested by trophozoites; amoebae adhere to other bacteria which contain galactose or *N*-acetyl-D-galactosamine (GalNAc) residues on their surface. Virulence of axenic *E. histolytica* trophozoites increases following a brief (30-min) *in vitro* incubation with live adherent bacteria. Bacteria may function as oxygen scavengers, stimulating the electron transport system of *E. histolytica*; whether bacteria induce gene switching or alter amoebic mRNA transcription or translation is unknown.

*Entamoeba histolytica* trophozoites adhere to the colonic mucosal blanket and mucosal lining cells prior to invasion. *In vitro* studies of adherence of *E. histolytica* trophozoites to target cells have identified a relevant parasite adherence molecule, a surface lectin that requires galactose or GalNAc. Galactose or GalNAc, but not numerous other monosaccharides, completely inhibits adherence of axenic trophozoites to Chinese hamster ovary cells (Fig. 4). This lectin also mediates parasite adherence to erythrocytes, neutrophils, mononuclear cells, numerous other mammalian cells in tissue culture, and bacteria having galactose-containing lipopolysaccharide or attached antibacterial antibodies. The *E. histolytica* lectin has been isolated and characterized. It consists of a single heavy subunit and a light subunit of 170,000 and 35,000 D, respectively (Fig. 5). The heavy subunit apparently mediates the binding of the lectin to carbohydrate. The multigene family encoding the heavy subunit it has recently been characterized, a cysteine-rich protein comprises the galactose-binding region. The lectin

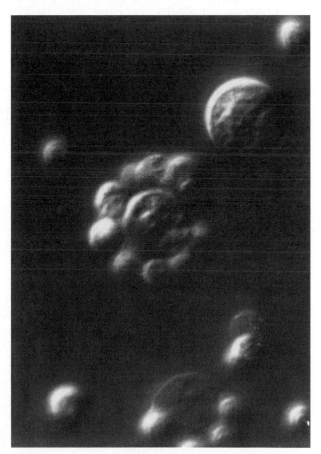

FIGURE 4 *Entamoeba histolytica* trophozoite (center) with multiple adherent Chinese hamster ovary cells following sedimentation at 4°C ×1000.

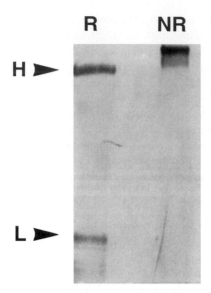

**FIGURE 5** Coomassie brilliant blue-stained sodium dodecyl sulfate–polyacrylamide gel electrophoresis of a monoclonal antibody affinity-purified galactose or GalNAc lectin, reduced (R) with 10% β-mercaptoethanol or nonreduced (NR). Relative migrations of the reduced heavy and light subunits (170,000 and 35,000 molecular weight, respectively) are indicated. [Reprinted from *J. Biol. Chem.* **264**, 3007–3012 (1989) by permission.]

binds optimally to β(1–6)-branched N-linked cell surface carbohydrates which lack terminal sialic acid residues. It mediates parasite attachment to colonic mucins, which cover epithelial cells in the large bowel. In this way, colonic mucins facilitate *E. histolytica* intestinal colonization; inhibition of mucin binding by the antilectin secretory immunoglobulin A (IgA) antibody is a potential strategy for the prevention of amoebic intestinal infection. Intra-Peyer's patch immunization of rats with native lectin and oral immunization of mice with a recombinant portion of the heavy subunit elicits an adherence-inhibitory intestinal IgA antibody response. Another lectin of 220,000 molecular weight, which recognizes the sugar chitotriose, apparently participates in the attachment of *E. histolytica* to erythrocytes. Additional surface molecules may be involved in amoebic phagocytosis. Monoclonal antibodies that inhibit amoebic ingestion of erythrocytes recognize a 112,000 molecular weight amoebic protein.

## B. *E. histolytica* Secretory Enzymes and Toxins

Numerous proteolytic enzymes have been isolated from *E. histolytica*, including trypsin, pepsin, hyal-

uronidase, gelatinase, and hydrolytic enzymes for casein, fibrin, collagen, and hemoglobulin. Many of these enzymes are inhibited by serum and require direct parasite contact for release onto host tissue. A 56,000 molecular weight amoebic thiol proteinase and a 26,000 molecular weight cysteine proteinase have been isolated and characterized. The genes encoding the cysteine proteinases have been characterized one gene ACPG, is associated with pathogenicity. Both enzymes degrade specific connective tissue matrices, such as fibronectin, laminin, and type I collagen, and destroy layers of cells in tissue culture, probably by degrading cell-anchoring proteins. *Entamoeba histolytica* contains a serum-inhibitable collagenase that is demonstrable only when amoebae make direct contact with collagen substrates; collagenase activity has been shown to correlate with parasite virulence in three strains of axenic amoebae. Parasite proteinases apparently contribute to the pathogenesis of amoebiasis by the disruption of tissue barriers, releasing epithelial cells from basement membrane anchors. These enzymes also make it difficult to isolate nondegraded *E. histolytica* proteins.

Secreted cytotoxins have been hypothesized to contribute to the diffuse mucosal damage and/or watery diarrhea that may accompany invasive colonic amoebiasis. However, studies of human pathology and experimental animals reveal colonic mucosal damage at sites of adherent amoebae. *Entamoeba histolytica* protein fractions have enterotoxic activity under certain conditions in rats or rabbits; amoebic enterotoxigenic activity is inhibited by fetuin or high concentrations of indomethacin, which inhibits the formation of inflammatory prostaglandins. A recently partially purified toxin was found to be heat labile, to be inhibited by sialoglycoproteins (fetuin and mucin) and *p*-chloromercuribenzoate, and to have an apparent molecular weight of 30,000. Lysates of *E. histolytica* applied to the peritoneal side of rabbit serosa apparently induce an ion- and fluid-secreting effect. It remains to be determined whether viable *E. histolytica* trophozoites elaborate an enterotoxin responsible for enhanced intestinal secretion during amoebic adherence to or invasion of the colonic submucosa.

## C. *E. histolytica* Cytolytic Activity

The characteristic lytic pathology of invasive amoebiasis, for which the organism was named, consists of amoebae surrounded by necrotic debris (Fig. 6). *In vitro* studies of the cytolytic mechanisms of *E. histolytica* have provided information regarding

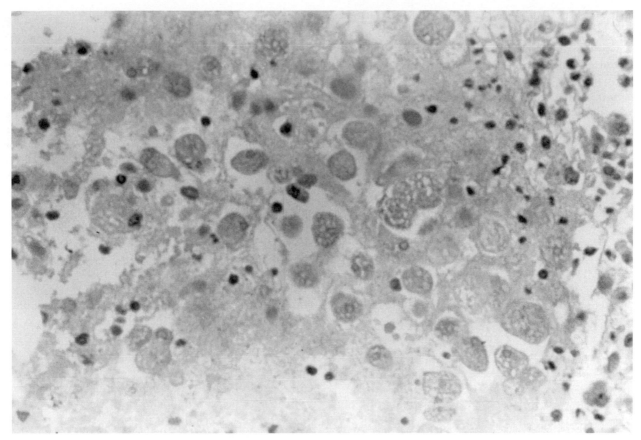

**FIGURE 6** Light micrograph of colonic biopsy tissue from a patient with invasive amoebic colitis, showing the dissolution of recognizable tissue architecture immediately surrounding numerous *E. histolytica* trophozoites and the acute inflammatory response at the margin of this "microabscess." Hematoxylin–eosin staining ×400. [Courtesy of Dr. Charles Schleupner, Veterans Administration Hospital, Salem, Virginia. Reprinted from *J. Infect. Dis.* **143**, 83–93 (1981) by permission.]

the molecular basis for host tissue necrosis. The cytolethal effect of *E. histolytica* occurs only on direct contact with target cells; within seconds following amoebic adherence, mediated by the galactose-inhibitable adherence lectin, a "lethal hit" is delivered to the target cell. The primary role of this lectin in initiating amoebic cytolysis is indicated by the fact that galactose, GalNAc, or purified colonic mucins completely inhibit amoebic *in vitro* lysis of target cells. In addition, glycosylation-deficient target cell mutants which lack galactose and GalNAc residues on cell surface carbohydrates are entirely immune to amoebic cytolytic activity.

Cytolysis of adherent target cells by *E. histolytica* requires parasite microfilament function, phospholipase A enzyme activity, and maintenance of an acid pH in amoebic endocytic vesicles. Amoebic cytolytic activity is stimulated by phorbol esters, direct activators of protein kinase C enzymes.

*Entamoeba histolytica* trophozoites contain two phospholipase A enzymes; one is calcium dependent and associated with the surface plasma membrane, and the other is independent of calcium. These enzymes, which hydrolyze certain phospholipids, have a central role in *in vitro* E. histolytica cytolytic activity. In fact, phospholipase A enzyme inhibitors and antibodies block amoebic cytolytic activity and inhibit amoebic adherence and lysis of target cells. [*See* Enzyme Inhibitors.]

*Entamoeba histolytica* vesicle pH has been measured using fluorescein isothiocyanate linked to dextran; the endocytic vesicle pH is $5.1 \pm 0.2$ by spectrofluorometry. Weak bases such as $NH_4Cl$, at concentrations sufficient to increase the vesicle pH to $\geq 5.7$, inhibit amoebic killing of target cells *in vitro*. Amoebic vesicle exocytosis, which is temperature, microfilament, and calcium dependent, is also stimulated by phorbol esters. Phorbol esters such as phorbol 12-

myristate, 13-acetate (PMA) induce a greater than twofold enhancement in amoebic killing of target cells, enhancing amoebic cytolytic activity without promoting adherence. Sphingosine, a specific inhibitor of protein kinase C, totally blocks PMA-stimulated or basal *E. histolytica* cytolytic activity. Apparently, an *E. histolytica* protein kinase C regulates parasite cytolytic activities; mechanisms responsible for *in vivo* activation of the amoebic kinase are unknown, but may relate to the galactose-inhibitable adherence lectin.

A role for calcium ions in *E. histolytica* cytolytic activity was first suggested by the finding that calcium chelators, pharmacological calcium antagonists, and calcium channel blockers inhibit amoebic cytolytic activity. Free intracellular $Ca^{2+}$ concentrations ($[Ca^{2+}]_i$) have been measured in amoebae and in target cells using the $Ca^{2+}$ probe Fura-2 and computer-enhanced digitized microscopy. Although motile *E. his-*

*tolytica* trophozoites demonstrate random cyclic increases in $[Ca^{2+}]_i$ at the leading edge or tail regions of cell cytoplasm, there is no increase in regional or total amoebic $[Ca^{2+}]_i$ upon contact with a target cell. In contrast, target cells exhibit a marked irreversible increase in $[Ca^{2+}]_i$ within seconds following contact by an amoeba (Fig. 7). In the presence of galactose, the contact of amoebae does not alter the target Chinese hamster ovary cell $[Ca^{2+}]_i$. Therefore, $[Ca^{2+}]_i$ does not function as a "secondary messenger" for the initiation of parasite adherence or cytolytic activities. Binding by the galactose-inhibitable lectin-mediated initiates a rapid lethal rise in target cell $[Ca^{2+}]_i$. Target cell death occurs within 20 min by an apoptosis-like mechanism, removal of extracellular $Ca^{2+}$ ions prevents target cell death. A putative mediator of such ion flux and target cell death is the *E. histolytica* pore-forming protein, a vesicle protein which induces rapid ion flux ($Na^+$, $K^+$, and, to a lesser extent,

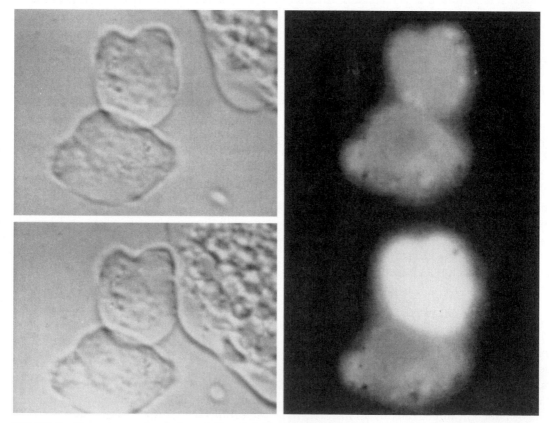

**FIGURE 7**    Photomicrographs of phase (left) and digitized (right) fluorescent images of a Fura-2-loaded Chinese hamster ovary (CHO) cell precontact (top) and at 30 sec postcontact (bottom) with an *E. histolytica* trophozoite. Fura-2 fluorescence is directly proportional to the free intracellular $Ca^{2+}$ concentration ($[Ca^{2+}]_i$). The two trypsinized CHO cells were in suspension and not directly contiguous; there was a marked rise in $[Ca^{2+}]_i$ only in the CHO cell in direct contact with the amoeba, despite the proximity of the other target cell. [Reprinted from *Infect. Immun.* **56**, 1505–1512 (1988) by permission.]

$Ca^{2+}$) across lipid bilayers or vesicles and can depolarize and kill eukaryotic cells. However, the most important function of the protein is as a parasite defense mechanism against ingested bacteria. [*See* Calcium Antagonists.]

## VI. HOST IMMUNE RESPONSE TO *E. histolytica*

### A. Nonimmune Defense Mechanisms

To cause disease, invasive enteric organisms must overcome the low gastric pH, digestive enzymes, competition by normal intestinal bacterial flora, and the protective mucosal blanket covering the gut epithelium. As mentioned earlier, stomach acid is ineffective against the chitinous *E. histolytica* cyst. Bile salts in combination with human breast milk can lyse trophozoites *in vitro,* but this is unlikely to be relevant to *in vivo* disease. Intestinal bacteria compete for iron and may elaborate toxic metabolites, such as short-chain fatty acids or enzymes, harmful to amoebae; these potential defense mechanisms have not been well studied.

Colonic mucins appear to be the major nonimmune host defense mechanism against invasion by *E. histolytica* trophozoites by preventing the trophozoites from attaching to and lysing colonic epithelial cells. *Entamoeba histolytica* trophozoites exhibit a potent mucosal secretagogue effect; depletion of the protective mucosal blanket is apparently required before the parasite invades the colonic mucosa. The ability of the mucin gel to physically impede amoebic motility may vary in different areas of the large bowel, causing regional increases in susceptibility. Trapping of trophozoites in colonic mucus with expulsion may be increased in immune individuals, as observed in experimental nematode infection.

### B. Acquired Resistance to Invasive Amoebiasis

Noninvasive *E. histolytica* infection is spontaneously cleared over a period of 6–12 months; whether this is due to development of a local gut immune response or is followed by host resistance to rechallenge is unknown. Long-term immunity to asymptomatic intestinal infection does not appear to occur as the prevalence of *E. histolytica* infection increases with age in endemic areas. Individuals cured of invasive amoebiasis have substantial resistance to the recurrence of invasive disease. In experimental animal models, cure of invasive infection or immunization with a total parasite antigen preparation results in protection against intestinal or intrahepatic challenge with *E. histolytica* trophozoites, indicating the feasibility of eliciting effective host immune defense mechanisms. Apparently, both antibody- and cell-mediated mechanisms contribute to the immune resistance to recurrent invasive amoebiasis. [*See* Immunology of Parasitism.]

### 1. Serum and Intestinal Secretory Antibody Response

Patients with invasive amoebiasis develop a high-titer serum antiamoebic antibody response to a well-defined set of conserved *E. histolytica* antigens, including the galactose-inhibitable adherence lectin. Antilectin antibody inhibits amoebic *in vitro* adherence to colonic mucins or target epithelial cells. Uncharacterized coproantibodies of the IgA class have been identified in patients with amoebic colitis: A secretory antilectin IgA antiamoebic antibody response occurs during amoebic liver abscess and intestinal infection. Although the serum antiamoebic antibody response does not result in the spontaneous resolution of established amoebic colitis or liver abscess, serum antiamoebic antibodies may contribute to the resistance to subseqeunt invasive disease. Asymptomatic infection with *E. histolytica,* in contrast to *E. dispar* infection, is associated with a serum antibody response, identical to that observed in amoebic liver abscess patients. Prospective field studies are necessary to determine the significance of a serum or intestinal antiamoebic antibody during asymptomatic infection.

### 2. Complement-Mediated Resistance

*Entamoeba histolytica* trophozoites are lysed *in vitro* by alternative or classical antibody-dependent complement-mediated pathways. However, trophozoites isolated from human liver abscesses are complement resistant; such amoebae can also be selected by *in vitro* long-term culture with increasing concentrations of serum. The complement membrane attack complex attaches to the plasma membrane of the parasite; parasite resistance to complement-mediated lysis may relate to the ability of trophozoites to rapidly turn over their surface membrane (capping).

### 3. Polymorphonuclear Leukocytes

Polymorphonuclear white blood cells are ineffective against *E. histolytica* trophozoites; in fact, their response to amoebic invasion may be deleterious to the host by increasing tissue destruction. *Entamoeba histolytica* trophozoites are chemoattractant for human polymorphonuclear neutrophils. Early in the formation of an experimental amoebic liver abscess, parasite contact-dependent lysis of neutrophils has a deleterious effect on distant hepatocytes. *In vitro,* amoebae lyse neutrophils on contact (Fig. 8), with release of the cytotoxic nonoxidative neutrophil constituents. Soluble *E. histolytica* proteins inhibit human neutrophil oxidative activities, which may account for the immunity of the parasite to neutrophil oxidative effector activity. Neutrophils activated with lymphokines, such as tumor necrosis factor are able to kill amoebae *in vitro.*

### 4. Cell-Mediated Immune Responses

Patients cured of amoebic liver abscess demonstrate a competent cell-mediated immune response to the parasite. *In vitro,* an antigen-specific T lymphocyte proliferative response with production of lymphokines can be demonstrated. These secreted lymphocyte products elicit *in vitro* amoebicidal activity in human monocyte-derived macrophages. *Entamoeba histolytica* trophozoites kill human neutrophils, mononuclear cells, and nonactivated monocyte-derived macrophages *in vitro* without loss of parasite viability. However, immune T lymphocytes stimulated *in vitro* with total soluble *E. histolytica* antigen or purified adherence lectin are capable of killing trophozoites; nonstimulated immune cells or antigen-exposed nonimmune lymphocytes are ineffective.

How activated macrophages and antigen-stimulated T lymphocytes can overcome amoebic cytolytic mechanisms and recognize trophozoites as target cells remains undefined. [*See* Lymphocytes.]

Cell-mediated immunity plays an important role in preventing or limiting invasive amoebiasis, but is not relevant to the initial penetration of the parasite through colonic mucosal defenses.

## VII. DIAGNOSIS AND THERAPY OF AMOEBIASIS

### A. Diagnosis of Amoebic Colitis and Liver Abscess

Intestinal amoebiasis can be diagnosed by finding *E. histolytica* cysts or trophozoites in the stool. Hematophagous trophozoites are characteristically present during invasive amoebic colitis. Examination of at least three separate stool samples is required to reach 95% sensitivity; use of antibiotics or barium should be avoided prior to the collection of stool samples. Virtually all patients with active invasive amoebic colitis will test positive for occult blood in feces; this is a helpful inexpensive screening test. Despite the inflammatory nature of amoebic colitis, leukocytes may not be seen in feces due to the cytolytic ability of *E. histolytica* trophozoites. After 7 days of symptoms,

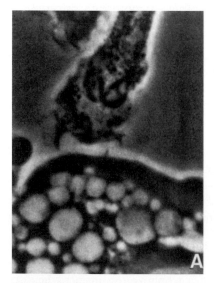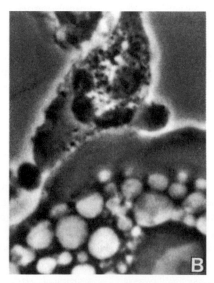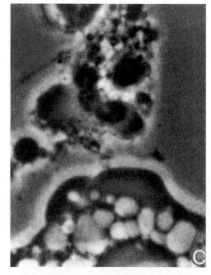

**FIGURE 8**    Time-lapse phase-contrast cinemicrographs of a polymorphonuclear neutrophil approaching and being killed by an axenic trophozoite of *E. histolytica* strain HM1–IMSS. A polymorphonuclear neutrophil approaches the amoeba (A), establishes contact (B), and undergoes membrane blebbing and granule disappearance (C). A Zeiss photomicroscope, Sage cinemicrographic apparatus, and Bolex 16-mm camera were used. ×2000. [Reprinted from *J. Infect. Dis.* **143,** 83–93 (1981) by permission.]

95% of patients with invasive colitis or amoeboma will have serum antiamoebic antibodies detectable by standard methods. Serology is useful, especially in the setting of chronic amoebiasis, which may be confused with idiopathic inflammatory bowel disease.

To avoid confusion due to past disease, serological methods in highly endemic areas should be used which only detect antiamoebic antibodies for brief periods following invasive infection. Lower gastrointestinal endoscopy may be helpful in establishing a diagnosis by rapidly examining the intestinal mucosa and obtaining biopsies of involved tissue. Localized disease (amoeboma) or right-sided colonic involvement may require endoscopy with biopsy for diagnosis. The differential diagnosis of amoebic colitis generally includes inflammatory bacterial etiologies such as *Shigella, Campylobacter, Salmonella,* or *Yersinia* species. Nonspecific gastrointestinal complaints due to giardiasis, cryptosporidiosis enterotoxigenic infection, or viral gastroenteritis must be differentiated from amoebiasis in an individual infected with *E. histolytica.*

The diagnosis of amoebic liver abscess is based on clinical presentation and the identification of epidemiological risk factors. The key diagnostic studies are a noninvasive ultrasound evaluation of the biliary tract and liver and a serum antiamoebic antibody assay. The finding of a nonhomogeneous cystic lesion in the liver immediately focuses attention on amoebic or bacterial abscess and away from infection of the biliary tract (which is more common and has a similar clinical presentation). Serum antiamoebic antibodies are almost always present if the patient has been symptomatic for at least 7 days; if unanticipated, their absence during a more acute presentation may be misleading. Computerized tomography scans add little by way of sensitivity or specificity to the ultrasound, are much more costly, and result in substantial radiation exposure. If the diagnosis is in doubt and antibody studies are pending or unavailable, a percutaneous fine-needle biopsy can be used to rule out bacterial abscess by Gram's stain and culture. Amoebic abscess fluid is usually yellow, with few neutrophils, and amoebae are not commonly seen on microscopy. However, Echinococcal cysts are also in the differential diagnosis and are potentially hazardous to biopsy; a serological diagnostic test is available.

## B. Therapy of Amoebiasis

A detailed discussion of antiparasitic agents and their use in treating amoebiasis is outside the scope of this article. The interested reader should consult one of the authoritative references provided. Therapy is complicated by the use of multiple agents and by the need to select drugs based on the site of infection. Certain oral agents, such as diloxanide furoate, paromomycin (a nonabsorbable aminoglycoside), and diiodohydroxyquin, are only effective against amoebic forms within the human bowel. The tetracyclines and erythromycin are active against colonic disease but not liver abscess. Nitroimidazoles, such as metronidazole, are preferred for the treatment of colitis or liver abscess, but often are ineffective against cyst forms in the gut. Emetines are also active at all tissue sites, but their use is limited by drug toxicity.

## VIII. PREVENTION OF INFECTION

Presently, prevention relies mainly on adequate sanitation, with availability of clean water and uncontaminated food. Avoiding sexual practices or poor hygiene which result in fecal–oral contamination can prevent infection. The disease burden is concentrated in the poorest areas of the world, in which complex socioeconomic and political problems are unlikely to be solved in the near future. Residents or visitors to such an environment can take precautions such as boiling drinking water, avoiding all uncooked vegetables or fruits that cannot be peeled, and drinking carbonated beverages. Vegetables must be soaked in acetic acid or vinegar for prolonged periods (at least 30 min) to ensure the eradication of cysts.

There is great need for a vaccine which results in immunity to *E. histolytica*; biomedical research leading to the development of a vaccine which prevents invasive disease or, better yet, provides resistance to intestinal colonization by *E. histolytica* is the best hope. Experimental animal models demonstrate that immunization with *E. histolytica* proteins can provide protection against invasive amoebiasis. Recent characterization of the genes encoding parasite adherence proteins, proteolytic enzymes, and antigenic proteins provides many potential approaches for vaccine development. Identification of immunogenic epitopes and novel vaccine delivery systems will provide the maximum likelihood of creating a safe effective amoebiasis vaccine.

## BIBLIOGRAPHY

Abd-Alla, M., Jackson, T. F. H. G., Gathiram, V., El-Hawey, A. M., and Ravdin, J. I. (1993). Differentiation of pathogenic from nonpathogenic *Entamoeba histolytica* infection by detection of galactose-inhibitable adherence protein antigen in sera and feces. *J. Clin. Microbiol.* **31,** 2845–2850.

Chadee, K., Petri, W. A., Innes, D. J., and Ravdin, J. I. (1987). Rat and human colonic mucins bind to and inhibit the adherence of lectin of *Entamoeba histolytica. J. Clin. Invest.* **80,** 1245–1254.

Kain, K. C., and Ravdin, J. I. (1995). Galactose-specific adherence mechanisms of *Entamoeba histolytica,* a model for the study of enteric pathogens. *Methods Enzymol.* **253,** 424–439.

Leippe, M., Tannich E., Nickel R., van der Goot, G., Pattus, F., Horstmann R. D., and Muller-Eberhard H. J. (1992). Primary and secondary structure of the pore-forming peptide of pathogenic *Entamoeba histolytica. EMBO J.* **10,** 3501–3506.

Mann, B. J., Torian, B. E., Vedvick, T. S., and Petri, W. A., Jr. (1991). Sequence of a cysteine-rich galactose-specific lectin of *Entamoeba histolytica. Proc. Natl. Acad. Sci. USA* **88,** 3248–3252.

Petri, W. A., Jr., Chapman, M. D., Snodgrass, T., Mann, B. J., Broman, J., and Ravdin, J. I. (1989). Subunit structure of the galactose and N-acetyl-D-galactosamine-inhibitable adherence lectin of *Entamoeba histolytica. J. Biol. Chem.* **264,** 3007–3012.

Ravdin, J. I., and Petri, W. A. (1995). Introduction to Protozoal Diseases. *In* "Principles and Practices of Infectious Diseases" (G. L. Mandell, R. G. Douglas, and J. E. Bennett, eds.), 4th Ed., pp. 2395–2407. Churchill-Livingstone, New York.

Soong, C. J. G., Abd-Alla, M., Kain, K. C., Jackson, T. F. H. G., and Ravdin, J. I. (1995). A recombinant cysteine-rich section of the *Entamoeba histolytica* galactose-inhibitable adherence lectin is efficacious as a subunit vaccine in the gerbil model of amebic liver abscess. *J. Infect. Dis.* **171,** 645–651.

Tannich, E., Horstmann, R. D., Knobloch, J., and Arnold, H. H. (1989). Genomic DNA differences between pathogenic and nonpathogenic *Entamoeba histolytica. Proc. Natl. Acad. Sci. USA* **86,** 5118–5122.

Tannich, E., Ebert, F., and Horstmann, R. D. (1991). Primary structure of the 170-kDa surface lectin of pathogenic *Entamoeba histolytica. Proc. Natl. Acad. Sci. USA* **88,** 1849–1853.

# Amyotrophic Lateral Sclerosis

EDITH G. McGEER

*University of British Columbia*

---

I. General Description
II. Pathology
III. Theories as to the Cause
IV. Animal Models
V. Treatment

## GLOSSARY

**Anterior horn cells** Motor neurons located in the anterior horn of the spinal cord and, like all motor neurons, using acetylcholine as a neurotransmitter

**Axonal transport** Processes by which proteins synthesized in the cell body are transported to axon terminals (antero-grade transport) or by which trophic factors and debris picked up or accumulated by the axons and dendrites are transported to the cell body (retrograde transport)

**Cholinergic** Adjective for neurons that use acetylcholine as a neurotransmitter

**Dementia** Loss of mental faculties, particularly memory

**Excitotoxin** A material that excites neurons and is toxic to them

**Glutamate** Major excitatory neurotransmitter in brain; excessive glutamate or some analogs can be excitotoxins

**Inclusion bodies** A variety of abnormal swellings visible by histological techniques in degenerating neurons

**Intrathecal injection** Into the subarachnoid space of the spinal cord

**Myoclonus** Shock-like contractions of muscles

**Neurotransmitter** One of a group of specific chemicals used by the nervous system to carry messages from one neuron to another

**Neurotrophic factors** Materials contributing to the development and/or maintenance of neurons; the best known example is nerve growth factor

**Peptide** Compound made up of a chain (polymer) of amino acids

**Prion** A naturally occurring protein in brain which, however, through mutation or other change appears to become a causative factor in scrapie disease of sheep and Creutzfeld-Jakob disease in humans

**Substantia nigra** A major nucleus in the extrapyramidal system involved in motor control

**Supraspinal** Above the spinal cord, i.e., in the brain

**Transgenic mice** Mice carrying an artificially introduced gene, usually of human origin, in order to study its effects

AMYOTROPHIC LATERAL SCLEROSIS (ALS) IS THE most common subtype of the motor neuron diseases, a heterogeneous group of disorders which produce muscle weakness and atrophy by their effects on the anterior horn cells of the spinal cord. ALS also affects supraspinal motor neurons. The worldwide annual incidence of ALS is about 1–2/100,000, with a somewhat greater risk for men than for women. No effective treatment is available, and the course of the disease is generally rapid, with death occurring within 3 years of diagnosis in about 50% of the patients. About 5–10% of the cases seem to be familial, with the rest being sporadic, although clusters of cases have occasionally appeared in nonfamilial individuals who shared a common environment many years before onset of the disease. A mutation in a specific enzyme called superoxide dismutase has been identified as the cause in some families, but the cause in most familial and all sporadic cases is unknown. Increasing evidence suggests that there may be multiple causes leading to similar symptomatology. Some cases have been attributed to exposure to specific neurotoxins or identified enzymic deficiencies, and there is considerable evidence that some disturbance in the immune system may play a role in many. Oxidative stress and deficiency of a necessary neurotrophic factor are other possibilities. A very high incidence of ALS, often asso-

ENCYCLOPEDIA OF HUMAN BIOLOGY, Second Edition, VOLUME 1.   Copyright © 1997 by Academic Press.   All rights of reproduction in any form reserved.

ciated with Parkinsonism and/or dementia, is seen in some native populations in the Western Pacific, notably among the Chamorros on Guam. In these populations, however, the incidence appears to be decreasing fairly rapidly (from 87/100,000 in 1958–1962 to 28.5/100,000 in 1978–1982), while the incidence of sporadic ALS in many countries in the world seems to be increasing, especially among the elderly.

## I. GENERAL DESCRIPTION

Degeneration of the motor neurons leads to weakness and wasting of the muscle which they innervate. Muscular weakness often begins in the hands or the upper arm and spreads to adjacent body regions. Stiffness or clumsiness of movement or cramp-like pains in the limbs may be an early symptom. In whatever part of the body muscular wasting begins, in most cases it soon becomes generalized. In the final stages, the symptoms are devastating, with weakness of the appropriate muscles making it impossible for the patient to sit up in bed, to speak, to swallow, or, finally, to breathe.

## II. PATHOLOGY

### A. Histological Pathology

The large motor neurons of the spinal cord, brain stem, and motor cortex are characteristically attacked but there are often also degenerative changes in nonmotor systems of the spinal cord and in noncholinergic systems in the brain. Frequently, some degeneration is seen in the substantia nigra, the site of the pathology characteristic of Parkinsonism. [*See* Parkinson's Disease.] Affected neurons of the spinal cord and motor nuclei often show Lewy bodies and similar abnormal inclusions akin to those seen in degenerating neurons in Parkinson's disease. Although there are frequently some disorders of cognition, overt dementia and the pathological signs of Alzheimer's disease rarely occur except in the ALS–Parkinson–dementia complex of the Western Pacific.

### B. Chemical Pathology

Identification of chemical neuropathology allows a more precise definition of the neuronal networks affected in a disease and may suggest possible etiological factors. The marked losses of cholinergic indices, such

as in the activities of the enzymes that synthesize (choline acetyltransferase) and destroy (acetylcholinesterase) acetylcholine, in the spinal cord and brain stem are not surprising in view of the degeneration of motor neurons which use acetylcholine as a neurotransmitter. Similar larger losses in the spinal cord in binding sites (receptors) for a number of transmitters such as glycine, substance P, and thyroid-releasing hormone (TRH) suggest that these are normally located on the motor neurons that are lost. The marked increase in spinal cord-binding sites believed to correspond to one type of serotonin receptor is more difficult to explain, particularly since measurements of the serotonin metabolite, 5-hydroxyindoleacetic acid, in cerebrospinal fluid (CSF) indicate no striking abnormality.

Of more interest are reports, often controversial, that have been used as the basis for hypothesizing a mechanism of neuronal loss or a possible therapeutic approach. Several examples have been used to support the general hypothesis that the excessive activity of excitatory amino acids may play a role in the neuronal destruction (see Section III). Perhaps the most interesting finding is of a factor in the CSF in ALS patients which significantly reduces the survival of rat neurons in culture, apparently by acting at one type of glutamate receptor. The widespread distribution of glutamate receptors, however, makes it difficult to explain the selectivity of the neuronal loss in ALS on the excitotoxic hypothesis.

Also of present interest are the reports of abnormal patterns of gangliosides in ALS tissue. Because gangliosides can act as neurotrophic factors, these reports, together with data on antibodies to gangliosides in ALS blood (Section III), suggest a possible etiology. [*See* Gangliosides.]

## III. THEORIES AS TO THE CAUSE

The fundamental problem in this and other neurodegenerative disorders is why do these particular neurons die? The cause in most cases of ALS is completely unknown. Various theories have been advanced, including viruses, the action of specific neurotoxins or heavy metals, failure of DNA repair mechanisms, altered axonal transport, some immune system abnormality, or lack of a necessary neurotrophic factor. It seems highly probable that there are a number of different causes which can lead to similar symptoms and pathology. Such heterogeneity may explain many of the inconsistencies in the literature.

A minor proportion (5–10%) of ALS cases appear to be familial, with the pattern of inheritance in families suggesting an autosomal dominant gene with age-dependent penetrance. Linkage studes in about 25% of such families have identified the gene as that coding for Cu/Zn superoxide dismutase, an enzyme active in the metabolism of toxic oxygen radicals. Although no abnormality in this enzyme appears to exist in most cases of ALS, this finding has focused considerable attention on the possibility that oxidative stress may play a major role in the neuronal death in ALS.

Other rare, familial forms seem to be recessive. One of these, seen in adults, has been linked to the long arm of the X chromosome but the gene product is still unknown. Some cases in young adults are apparently due to a recessively inherited deficiency in the enzyme hexosaminidase, but it is not yet understood why an equivalent lack of this enzyme in two members of a family can cause the devastating brain disease Tay-Sachs in an infant and a later-onset, relatively slowly progressing disorder of spinal motor neurons with normal mentation in a first-degree relative.

Various neurotrophic factors of a peptide nature have been found to support the survival of motor neurons in culture. Hence, as with other neurodegenerative diseases, there is considerable research based on the possibility that a deficiency in some such factor may be a contributing cause. Ciliary neurotrophic factor (CNTF) is receiving particular attention since it has been found to support survival of motor neurons in wobbler mice and other rodent models of ALS.

Clues as to the possible mechanism of neuronal loss in sporadic ALS have been sought by careful studies on particular populations where related conditions are endemic. One such population is the Chamorros on Guam where an extremely high incidence of an ALS–Parkinson–dementia complex has been studied since 1945. Since that time, there has been a striking decline in incidence with no change in the clinical or pathological manifestations. An environmental factor is therefore suspected and there are two main candidates: a metallic imbalance or a toxin in the cycad seeds of the false sago palm. Because the soil of Guam is low in calcium and magnesium and high in aluminum, it has been suggested that deposition of aluminum derivatives in neurons might disrupt axonal transport and lead to the disorder. The decline in incidence would be explained on the basis of increased calcium in the diet and the consumption of many imported vegetables and other foods. The other hypothesis depends on the facts that flour made from cycad seeds was a mainstay of the diet when the disease was prevalent and that neuronal degeneration in monkeys was said to occur on feeding such flour or an unusual amino acid found in the seeds, $\beta$-N-methylamino-L-alanine (BMAA; Fig. 1). There has been some difficulty in confirming these reports, however, and the basic cause of the Guamanian disease is still unknown.

The situation is clearer with regard to the lathyrism that is endemic in parts of India and other countries where flour made from the chickling pea is often consumed, particularly during droughts to which this plant has unusual resistance. The victims show a spastic paraplegia and pathology of the upper motor neurons. It seems clear that this condition is caused by another unusual amino acid found in the chickling pea. Chronic feeding of this amino acid, $\beta$-N-oxalylamino-L-alanine (BOAA), to monkeys induces an upper motor neuron disorder. BOAA is structurally related to glutamate (Fig. 1) and acts as an excitotoxin in animal experiments. Like sodium glutamate, it enhances the flavor of foods, and this is one reason why it has been difficult to discourage consumption of the chickling pea. The example of lathyrism is a major basis for the hypothesis that neuronal loss in ALS may be due to an excitotoxin, but it is notable that the lower motor neurons are not involved in the pathology of lathyrism although they are in ALS.

The appearance of isolated cases of ALS after exposure to some insecticides or solvents has led to suggestions of other types of organic compounds which might cause the disease, but none are proven.

For more than a decade, starting in the mid-1960s, most research on the possible cause of ALS was concentrated on the possibility of a persistent virus infection. One reason was the demonstration by Gajdusek and his associates that Creutzfeldt-Jacob disease (CJD) is due to a transmissible agent, then called a slow virus but now believed to be a prion. [*See* Prions.]

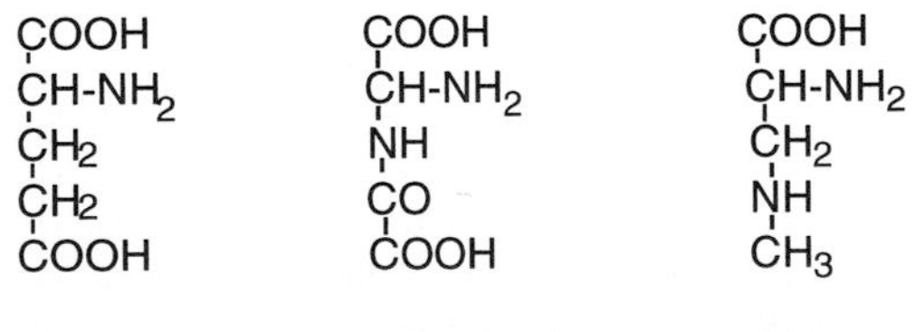

FIGURE 1  Structures of the excitatory neurotransmitter glutamate, the excitotoxic BOAA from chickling peas, and BMAA which is found in cycad seeds. The lack of two acidic groups in BMAA makes it highly unlikely that it is an excitotoxin, although it may be metabolized to one.

Although dementia and myoclonus are dominant features of CJD, some cases show motor neuron pathology similar to that seen in ALS. Another reason to suspect a persistent virus was the knowledge that poliomyelitis attacks motor neurons and there were reports, later denied, that people who had survived polio were at increased risk of ALS. The viral hypothesis was largely abandoned by 1984 because repeated attempts to transmit ALS to primates had failed, as had most attempts to detect viral nucleic acids in ALS tissue by *in situ* hybridization. Recently, however, some interest in possible viral factors has been revived by data indicating that some retroviruses, such as the one involved in AIDS, may affect motor neurons.

There is considerable circumstantial and often controversial evidence for a role of autoimmune factors in ALS. Controversial reports included some on an increased incidence of immune-related diseases in subjects with ALS, the occurrence of abnormalities in histocompatibility antigens, immune complexes, complement deposition, macrophage migration and lymphocyte transformation, and the presence in serum of factors toxic to myelin or motor neurons. Many small clinical trials using immunosuppressant drugs with or without plasma exchange, or plasma exchange by itself, have been carried out with a minimal reported benefit. A number of findings have focused new interest on autoimmunity in ALS. These include reports of high concentrations of MHC class II antigens in degenerating tissue and, particularly, a high incidence of paraprotinemia (i.e., the existence of high levels of abnormal monoclonal serum immunoglobulin bands), with antibodies to gangliosides being important components. [*See* Major Histocompatibility Complex (MHC).] Abnormally high levels of such ganglioside antibodies have been found in more than 75% of patients with ALS in some studies. Since similar antibodies are found in patients with a variety of autoimmune disorders, and even in some healthy individuals, it is not immediately clear how their presence could be directly related to the occurrence of ALS. Although these antibodies react with the carbohydrate moieties of gangliosides, it has been suggested that the primary antigens may be other carbohydrates, and the true, as yet unknown, specificity of the antibodies may determine their pathogenicity. The relatively small benefit seen on immunosuppressant treatment might depend on the fact that immune responses to carbohydrates are, as a rule, T-cell independent responses of B cells; such processes are generally less susceptible to immunosuppressant treatment than are T-cell-dependent responses. Intensive research into

this possible mechanism of ALS will undoubtedly renew attempts at immunosuppression with more powerful or more specific agents.

The possible importance of an autoimmune factor in ALS does not rule out a virus or environmental toxin as the initial agent which would precipitate the process in susceptible individuals. The occurrence of statistically improbable clusters of sporadic cases in individuals contracting ALS many years following a period in which they had been closely associated argues for some such precipitating factor. Examples include the group of 22 men who served on Canadian warships based in Halifax in the 1950s and four members of the San Francisco '49er football team of about the same era, all of whom developed ALS a quarter of a century later. Three unrelated persons who taught in the same classroom for 2–5 years in a small town in Ohio came down with ALS up to 18 years after they left that locale. Such clusters are difficult to explain without postulating some infectious or environmental factor.

In summary, the cause of ALS is unknown and may be multifactorial but much attention is presently being directed to the possibilities that an unusual type of autoimmune reaction, oxidative stress, and/or loss of a neurotrophic factor may be important.

## IV. ANIMAL MODELS

The existence of animal models of a disease can give insight into the etiology, pathogenesis, and biochemical mechanisms underlying the human disorder, and many attempts have been made to produce good animal models of ALS. These have included exposure to toxins, including inhibitors of axonal transport, excitatory amino acids, and pesticides; administration of metallic ions such as aluminum, lead, or mercury; infection with polio or other viruses; and immunization with motor neuron extracts. Some animal mutant strains such as the wobbler mouse have been used. Even though none of the models so far provides an exact copy of ALS, some reproduce certain aspects of the pathology. An exception may be transgenic mice expressing a human Cu/Zn superoxide dismutase mutation; such mice seem to be good models for this familial form of ALS (see Section III).

## V. TREATMENT

No effective treatment for ALS is yet known. Several authors reported an immediate benefit in uncontrolled trials of TRH, or analogs of this peptide, given intravenously or intrathecally. However, no substantial effect was found in controlled trials and the reports of high incidences of hypothroidism in ALS, on which the initial trials were based, have not been confirmed.

As mentioned previously, many attempts at immunosuppressant therapy have been tried with very limited success. However, reports of some clinical improvement in some subgroups, such as those with very high paraprotinemia or men with recent onset of the disease, together with increasing evidence that autoimmunity may play a role (Section III), will undoubtedly spur further clinical trials with more intense immunosuppression.

Hypotheses that glutamate, acting as an excitotoxin, or that oxidative stress may play important roles in neuronal death in ALS have led to some clinical trials of agents which inhibit either glutamate release or the formation of free radicals. Most reports have indicated little or no clinical improvement in such trials, but an initial trial of riluzole, a glutamate release inhibitor, was sufficiently promising that a multicenter clinical trial is now underway in the United States. Major research on possible therapies is now being directed at neurotrophic factors and a clinical trial of CNTF is underway.

## BIBLIOGRAPHY

Baringam M. (1994). Neurotrophic factors enter the clinic. *Science* **264**, 772–774.

Davies, A. M. (1993). Promoting motor neuron survival. *Curr. Biol.* **3**, 879–884.

Eisen, A., and Krieger, C. (1993). Pathogenic mechanisms in sporadic amyotrophic lateral sclerosis. *Can. J. Neurol. Sci.* **20**, 286–296.

McNamara, J. O., and Fridovich, I. (1993). Did radicals strike Lou Gehrig? *Nature* **362**, 20–21.

# Anaerobic Infections in Humans

*Veterans Administration Medical Center, West Los Angeles and University of California, Los Angeles*

---

## GLOSSARY

**Bronchiectasis** Chronic dilatation of the bronchi with superimposed infection

**Cholesteatoma** Cyst-like mass, most commonly in the middle ear or mastoid, filled with desquamating debris, often including cholesterol

**Debridement** Removal of devitalized or contaminated tissue and foreign bodies

**Decubitus ulcer** Bed sore; pressure sore, ulcer caused by prolonged pressure due to lying too long in one position

**Empyema** Accumulation of pus in a body cavity

**Hemolysis** Liberation of hemoglobin from red blood cells; the substance responsible for such liberation is known as hemolysin

**Metastatic** Refers to new foci of disease (i.e., infection) in a distant part of the body, usually by spreading via the bloodstream

**Mucosa** Mucous membrane

**Necrotic** Dead, devitalized

**Oxidation–reduction potential** Tendency of a system to accept or give up electrons. When a substance is oxidized, it gives up electrons; when it is reduced, it accepts electrons. Measurement is done by using a measuring electrode and a reference electrode

**Pathogenic** Disease-producing

**Pilonidal sinus** Sinus containing a tuft of hair, usually near the coccyx

ANAEROBIC BACTERIA ARE ORGANISMS THAT LIVE at low oxygen tension or in situations with a low oxidation–reduction potential. These organisms are prevalent on all mucosal surfaces of the body and can be found on the skin as well. When there is a break in the mucosal surface (or translocation of bacteria across this surface), as occurs following surgery or trauma or in relation to certain disease states, and when the oxidation–reduction potential is low enough, anaerobic bacteria invade and multiply in adjacent tissues or body spaces and produce infection.

The five most common types of infection produced by anaerobic bacteria are oral and dental, pleuropulmonary, intraabdominal, female genital tract, and skin and soft tissue infections. Special procedures are necessary to collect and transport specimens for examination for these organisms, and special techniques are required for their culture. In terms of therapy, the two basic approaches are (1) surgery to remove the dead tissue and drain collections of pus, both of which are characteristic of most anaerobic infections, and (2) antimicrobial therapy. There are a number of antimicrobial drugs with good activity against various anaerobes.

## I. ANAEROBIC BACTERIA

### A. Definition of Anaerobiosis

Definition of the word "anaerobiosis" is difficult. A practical operational definition for laboratory purposes is that anaerobes require reduced oxygen tension for growth and fail to grow on the surface of solid media in 10% carbon dioxide in air (18% oxygen). Truly obligate anaerobes generally do not possess catalase or superoxide dismutase. Some less fastidious anaerobes that can survive some exposure to air have been demonstrated to produce superoxide

ENCYCLOPEDIA OF HUMAN BIOLOGY, Second Edition, VOLUME 1.    Copyright © 1997 by Academic Press.    All rights of reproduction in any form reserved.

dismutase. This enzyme protects organisms that metabolize oxygen from the detrimental effects of the superoxide-free radical (an intermediate that results from the univalent reduction of molecular oxygen). There is a general correlation between the amount of superoxide dismutase and the degree of aerotolerance of an anaerobic organism.

The importance of the oxidation–reduction potential, independent of the presence of oxygen, is demonstrated by studies in which anaerobic bacteria were found to be capable of growing in broth through which streams of air (i.e., oxygen) were passed; this was accomplished simply by holding the oxidation–reduction potential at low levels by electrical means. The absence of oxygen (or very low levels of oxygen), independent of the oxidation–reduction potential, can be important to certain anaerobes; for example, three pathogenic anaerobes were inhibited by oxygen even at a relatively low oxidation–reduction potential of $-50$ mV. On the other hand, in the absence of oxygen, these organisms were not inhibited, even when this value was as high as 325 mV.

Anaerobic bacteria vary tremendously in their sensitivity to oxygen or air. An organism occasionally found in septicemia, *Butyrivibrio*, is largely killed if a thin layer of broth containing it is exposed to air for just 6 min. On the other hand, a virulent anaerobe, *Clostridium perfringens*, can withstand similar exposure for many hours. Surface cultures of this organism on solid media can be exposed to air for 4 days or more and still be subcultured successfully. Under one classification, strict anaerobes are defined as those being incapable of growing on the surface of agar at oxygen levels over 0.5%; moderate anaerobes, which include most of the anaerobes pathogenic for humans, tolerate oxygen levels as high as 2–8%.

## B. Classification

All morphological types among bacteria are represented among the anaerobes. Included are bacilli, cocci, spirochetes, spirilla, and vibrios. Organisms are further subdivided according to whether they take up and retain the crystal violet color of the Gram stain (gram positive) or whether they can be destained and then accept the counterstain in the Gram reaction (gram negative). The gram-negative cocci are generally not clinically significant. Cocci can be arranged in chains (streptococci) or in irregular groups or masses, but these characteristics are not generally used for classification. The size of the cocci varies, and this is a factor in species determination. Gram-positive

bacilli can produce spores (genus *Clostridium*) or can be nonspore forming. Anaerobic bacteria can also be motile or nonmotile. Table I lists common anaerobic pathogens.

## C. Characteristics

As noted in the previous section, various morphotypes exist among the anaerobes. It has also been noted that anaerobes differ in their tolerance to oxygen and to various oxidation–reduction potentials. In general, anaerobic bacteria are rather fragile and must be protected from exposure to oxygen during the transport of clinical specimens that might contain them to the laboratory for culture on artificial media. Most anaerobic bacteria grow more slowly than do the nonanaerobes, although certain of them (e.g., the *Bacteroides fragilis* group) can show good growth in 18–24 hr. Some anaerobes, however, can take as long as 1 week to grow from clinical specimens.

Some anaerobic bacteria are quite fastidious and have special growth requirements, not all of which are well known. Many grow better or grow only in the presence of blood, serum, or ascitic fluid in the medium. Some are stimulated by vitamin $K_1$ whereas others are stimulated by bile, carbon dioxide or hydrogen, or various short-chain fatty acids or organic acids. Certain anaerobes might depend on other organisms (anaerobic or otherwise) present in a mixture to supply them with some necessary growth factor; thus, one can see a satelliting of colonies around colonies of another organism that is providing the growth factor.

Microscopically, the morphology can be distinctive. Some of the gram-positive nonspore-forming bacilli show branching. These organisms and clostridia may destain relatively easily and appear to be gram negative. Spores in clostridia can vary in size, shape, and location or can be difficult to demonstrate at all. Certain organisms in the genus *Fusobacterium*, *F. nucleatum* in particular, may show tapering of the ends of the bacilli. Moderate to marked shape differences (pleomorphism) can be seen among gram-negative bacilli, especially *F. mortiferum* and *F. necrophorum*. Pale, irregular staining is common among anaerobic gram-negative bacilli. Flagella may be demonstrable with special stains of motile anaerobes; their arrangement and location can be useful in classification.

The morphology of colonies of various anaerobic bacteria can also be distinctive and useful in characterizing or classifying organisms. Some organisms (e.g., *Bacteroides ureolyticus*) produce an agarase, which

**TABLE I**

Major Anaerobes Encountered Clinically

Gram-negative bacilli
  *Bacteroides fragilis* group
    Especially *B. fragilis, B. thetaiotaomicron, B. distasonis,*
      *B. ovatus, B. vulgatus*
  Other *Bacteroides*
    *B. gracilis*
    *B. ureolyticus*
    *B. splanchnicus*
  *Campylobacter* (*C. concisus, C. recta, C. curva*)
  *Porphyromonas* species (*P. asaccharolytica, P. gingivalis,*
    *P. endodontalis*)
  Pigmented *Prevotella* species (*P. corporis, P. denticola, P. in-*
    *termedia, P. loescheii, P. melaninogenica, P. nigrescens*)
  Other *Prevotella* species
    *P. oris*
    *P. buccae*
    *P. oralis* group
    *P. bivia*
    *P. disiens*
  *Fusobacterium* species
    *F. nucleatum*
    *F. necrophorum*
    *F. mortiferum*
    *F. varium*
  *Bilophila wadsworthia*
  *Sutterella wadsworthensis*
Gram-positive cocci
  *Peptostreptococcus*
    Especially *P. anaerobius, P. micros, P. magnus, P. asacchar-*
      *olyticus, P. prevotti*
  Microaerophilic streptococci (especially *S. anginosus, S. con-*
    *stellatus, S. intermedius*)[a]
Gram-positive spore-forming bacilli
  *Clostridium perfringens*
  *C. ramosum*
  *C. septicum*
  *C. novyi*
  *C. histolyticum*
  *C. sporogenes*
  *C. sordellii*
  *C. bifermentans*
  *C. fallax*
  *C. difficile*
  *C. innocuum*
  *C. botulinum*
  *C. tetani*
Gram-positive nonspore-forming bacilli
  *Actinomyces* (*israelii, meyerii, naeslundii, odontolyticus, vis-*
    *cosus*)
  *Propionibacterium propionicum* (*Arachnia propionica*)
  *Propionibacterium acnes*
  *Bifidobacterium dentium* (*B. eriksonii*)
  *Eubacterium* (*lentum, nodatum*)

[a]Not true anaerobes.

breaks down agar so that colonies are actually located in pits in the agar medium. *Fusobacterium nucleatum* may produce a colony with a speckled internal structure; greening around these colonies also exists. At other times, *F. nucleatum* produces colonies resembling bread crumbs. The colonies of *Actinomyces israelii* resemble molar teeth. Colonies of motile clostridia can show swarming at the edges. Some organisms (e.g., *C. perfringens*) produce characteristic types of hemolytic reactions on blood-containing agar. Certain organisms produce a tan to black pigment on blood-containing media (the pigmented *Prevotella* and *Porphyromonas* group). Table II shows the key characteristics for differentiation of various genera of anaerobic bacteria.

## D. Usual Habitats

On the skin, the dominant anaerobic organisms belong to the genus *Propionibacterium*. *Propionibacterium acnes* dominates, but *P. granulosum* and *P. avidum* are also seen with some frequency. *Propionibacterium acnes* and *P. granulosum* are found in the hair follicles and sebaceous glands. They are found in adults in greatest numbers in skin with a high sebum content (e.g., the scalp, forehead, and sides of the nose). Unlike the other two species, *P. avidum* is rarely found in the lipid-rich areas of the skin, but, rather, in moist areas (e.g., the axilla and the anterior nose). Strains of *Peptostreptococcus* species are also found with some frequency on the skin. Other anaerobes do not appear to be resident skin flora, but can be seen as a function of contamination of the skin from orifices such as the anus. This can be an important factor in certain types of infections. Gas gangrene (clostridial myonecrosis), for example, is a rare but dreaded complication of hip surgery. The organisms involved in this infection have their origin in the colonic flora. Various nonspore-forming bacteria that can be important in infected ulcers of the feet in diabetics or others with impaired circulation can, similarly, have their origin in the bowel flora.

The flora of the nose is similar to that of the skin. Anaerobes commonly isolated from the oropharynx include anaerobic cocci, *Prevotella, Porphyromonas, Bacteroides* species, and *Fusobacterium*.

In the oral cavity, anaerobes are found in the tonsillar crypts, the crypts of the tongue, plaque forming on the surfaces of the teeth, and the gingival crevices (i.e., the space between the teeth and gums). Areas with low oxidation–reduction potentials have the greatest colonization of anaerobes. In periodontal

**TABLE II**
Differentiation of Genera of Anaerobes[a]

Gram-negative bacilli
I. Nonmotile or peritrichous flagella
  A. Produce butyric acid without isoacids — *Fusobacterium*
  B. Produce major lactic acid — *Leptotrichia*
  C. Produce acetic acid, reduce sulfate — *Desulfomonas*
  D. Not as above — *Anaerorhabdus*
  *Bacteroides*
  *Bilophila*
  *Megamonas*
  *Mitsuokella*
  *Porphyromonas*
  *Prevotella*
  *Tissierella*
  *Sutterella*
II. Motile, not peritrichous flagella
  A. Fermentative
    1. Produce butyric acid — *Butyrivibrio*
    2. Produce succinic acid
      a. Spiral-shaped cells, single polar flagellum — *Succinivibrio*
      b. Spiral-shaped cells, bipolar tufts of flagella — *Anaerobiospirillum*
      c. Ovoid cells — *Succinimonas*[b]
    3. Produce propionic and acetic acids
      a. Single polar flagellum — *Anaerovibrio*[b]
      b. Tufts of flagella on concave side — *Selenomonas*
      c. Flagella in a spiral path along cell body — *Centipeda*
    4. Produce acetic acid, twitching motility — *Mobiluncus*
  B. Nonfermentative
    1. Produce succinic acid from fumarate — *Campylobacter*
    2. Produce acetic acid, reduce sulfate — *Desulfovibrio*

Gram-negative cocci
I. Produce propionic and acetic acids — *Veillonella*
II. Produce butyric and acetic acids — *Acidaminococcus*
III. Produce isobutyric, butyric, isovaleric, valeric, and caproic acids — *Megasphaera*
Gram-positive cocci
I. Require a fermentable carbohydrate
  A. Produce butyric (plus other acids) — *Coprococcus*
  B. No butyric produced — *Ruminococcus*
II. Do not require a fermentable carbohydrate
  A. Lactic acid sole major product — *Streptococcus* *Gemella*
  B. Not as above — *Peptostreptococcus* *Peptococcus*
Gram-positive spore-forming bacilli — *Clostridium*
Gram-positive nonspore-forming bacilli
I. Produce propionic and acetic acids as major end product — *Propionibacterium*
II. No propionic acid produced
  A. Produce acetic and lactic acids (A ≥ L) — *Bifidobacterium*
  B. Produce lactic acid as sole major end product — *Lactobacillus*
  C. Produce moderate acetic acid plus one of the following: — *Actinomyces*
    1. Major succinic and lactic acids
    2. Major succinic acid
  D. Other: butyric ± others, acetic or no major acids — *Eubacterium*

[a]Modified from Summanen *et al.* (1993).
[b]No known human isolates.

pockets this value can vary from −48 to −300 mV. The clean enamel surface of the teeth has an oxidation–reduction potential of 200 mV, but this drops to −141 mV after 7 days of plaque development. Dental plaque is a deposit of bacteria imbedded in an adhesive matrix made up of salivary glycoproteins and extracellular bacterial polymers. Counts of anaerobes in saliva and elsewhere in the oral cavity reach $10^7$ to $10^8$ per milliliter. Table III lists various anaerobic bacteria isolated from the oral cavity of humans.

Normally, counts of anaerobes and other bacteria in the stomach and the upper small bowel are quite low. The organisms derive primarily from swallowed oral flora. In the terminal ileum, counts of bacteria vary from $10^4$ to $10^6$ per milliliter, and the flora is rather diverse, resembling that of colonic flora. In the colon the bacterial population is the greatest of any inhabited region of the human body. Bacterial counts can exceed $10^{11}$ organisms per gram of dry weight of colonic contents. Anaerobes outnumber nonanaerobes by a ratio of about 1000:1. Members of the *Bacteroides fragilis* group dominate; the two most frequently encountered species are *B. vulgatus* and *B. thetaiotaomicron*, but *B. distasonis*, *B. fragilis*, and

## TABLE III
Anaerobic Bacteria Isolated from the Human Oral Cavity[a]

| Category | Genus |
|---|---|
| Gram-positive rods | *Actinomyces, Bifidobacterium, Eubacterium, Lactobacillus, Propionibacterium* |
| Gram-negative rods | *Bacteroides, Campylobacter, Fusobacterium, Leptotrichia, Selenomonas, Prevotella, Porphyromonas* |
| Gram-positive cocci | *Peptostreptococcus, Streptococcus* |
| Gram-negative cocci | *Veillonella* |
| Spirochetes | *Treponema* |

[a]Modified with permission from D. J. Hentges (1989). *In* "Anaerobic Infections in Humans" (S. M. Finegold, and W. L. George, eds.), p. 41. Academic Press, San Diego.

*B. ovatus* are also quite common. Various nonspore-forming gram-positive rods, *Peptostreptococcus* spp., and *Clostridium* spp. are also found in high counts. Table IV lists various anaerobic organisms isolated from human feces.

Various anaerobes, including *Bacteroides, Fusobacterium, Peptostreptococcus, Eubacterium,* and *Clostridium*, can be found in the urethral flora in counts of $10^2$ to $10^4$ per milliliter. In normal vaginal flora, lactobacilli predominate, but various anaerobic cocci, *Prevotella, Porphyromonas, Bacteroides,* and clostridia are also commonly found. Table V lists various anaerobes isolated from normal vaginal flora.

*Propionibacterium acnes,* found as normal con-

## TABLE IV
Anaerobic Bacteria Isolated from Human Feces[a]

| Category | Genus |
|---|---|
| Gram-positive rods | *Actinomyces, Bifidobacterium, Clostridium, Eubacterium, Lachnospira, Lactobacillus, Propionibacterium* |
| Gram-negative rods | *Bacteroides, Butyrivibrio, Desulfomonas, Fusobacterium, Leptotrichia, Succinimonas, Succinivibrio, Prevotella, Porphyromonas, Bilophila* |
| Gram-positive cocci | *Coprococcus, Gaffkya, Gemmiger, Peptococcus, Peptostreptococcus, Ruminococcus, Sarcina, Streptococcus* |
| Gram-negative cocci | *Acidaminococcus, Megasphaera, Veillonella* |

[a]Modified with permission from D. J. Hentges (1989). *In* "Anaerobic Infections in Humans" (S. M. Finegold, and W. L. George, eds.), p. 46. Academic Press, San Diego.

## TABLE V
Anaerobic Bacteria Isolated from the Human Vagina[a]

| Category | Genus |
|---|---|
| Gram-positive rods | *Actinomyces, Bifidobacterium, Clostridium, Eubacterium, Lactobacillus, Propionibacterium* |
| Gram-negative rods | *Bacteroides, Fusobacterium, Prevotella, Porphyromonas* |
| Gram-positive cocci | *Gaffkya, Peptococcus, Peptostreptococcus* |
| Gram-negative cocci | *Acidaminococcus, Veillonella* |

[a]Modified with permission from D. J. Hentges (1989). *In* "Anaerobic Infections in Humans" (S. M. Finegold, and W. L. George, eds.), p. 50. Academic Press, San Diego.

junctival flora, plays a role in the infection of intraocular lens implants.

## II. PATHOGENESIS OF ANAEROBIC INFECTION

### A. Virulence Factors in Anaerobes

The three major virulence factors in anaerobes are the ability to adhere to or invade epithelial surfaces, the production of toxins or enzymes that play a pathogenic role, and surface constituents of organisms such as capsular polysaccharide or lipopolysaccharide.

The ability to adhere to epithelial cells is vital to the establishment of colonization or infection. Both *Prevotella melaninogenica* and *F. nucleatum* adhere to the crevicular epithelium in the oral cavity, the former showing an ability to attach to certain gram-positive organisms *in vitro. Porphyromonas gingivalis,* thought to be an important organism in human periodontal disease, possesses fimbriae that facilitate attachment. The three different types of structures shown to be responsible for the adherence of *B. fragilis* to various epithelial structures are the capsule, negative-staining structures consistent with pili, and lectin-like adhesins. Binding and degrading of human fibrinogen by *P. gingivalis* could mediate colonization with this organism in the gingival crevice. Enzymes felt to be important with regard to invasion include phospholipase A, collagenase, and hyaluronidase.

Numerous toxins and enzymes play a role in bacterial virulence. The importance of superoxide dismutase in permitting anaerobic bacteria to survive exposure to oxygen was discussed in Section I,A.

*Clostridium perfringens* serves as a model for toxin production among anaerobes. Its major toxin is $\alpha$ toxin, a phospholipase C. This enzyme hydrolyzes lecithin and sphingomyelin in the cell membranes of a number of cell types, including red blood cells, platelets, endothelial cells, and muscle cells. This toxin and others produced by *C. perfringens* affect capillary permeability and destroy polymorphonuclear leukocytes and/or prevent their diapedesis into infected tissues. This organism also produces a collagenase. Other toxins and enzymes produced by anaerobes include neuraminidase, DNase, phosphatase, heparinase, leukocidin, hemolysins, hemagglutinins, lysophospholipase, proteinases, protease, sulfatase, sialidase, various enterotoxins, tetanus neurotoxin, tetanolysin, and botulinal toxin.

Surface constituents include capsules and lipopolysaccharide or endotoxin. The capsular polysaccharide of *B. fragilis,* free of other components of the bacterial cell, is capable of inducing abscess formation. Table VI summarizes the potential virulence factors of gram-negative anaerobic bacilli.

## B. Host Defense Mechanisms

Certain gram-negative anaerobic bacilli are killed directly by serum complement. Random migration of polymorphonuclear leukocytes does not differ significantly under aerobic and anaerobic conditions; however, anaerobes attract polymorphonuclear leukocytes into their immediate area by activation of complement and by direct mechanisms. It is also likely that anaerobes are susceptible to killing by macrophages. In the polymorphonuclear leukocyte, both oxidative and nonoxidative mechanisms contribute to killing of the anaerobes.

Acquired immunity involves both humoral and cell-mediated immune mechanisms. Circulating antibody and complement protect against bacteremia associated with experimental intraabdominal infection and T lymphocytes contribute to resistance against abscess formation. [*See* Lymphocytes.]

Anaerobes can exert adverse effects on humoral and cellular host defense mechanisms. Some anaerobes bind or deplete opsonins (this prevents the binding of opsonins to nonanaerobes and thereby prevents the phagocytosis of nonanaerobes) and, under certain conditions *in vitro,* might directly depress the function of polymorphonuclear leukocytes, macrophages, and lymphocytes.

**TABLE VI**
Potential Virulence Factors of Gram-Negative Anaerobic Bacilli

| Putative virulence factor or property | Possible significance |
|---|---|
| Adherence to peritoneal mesothelium | Factor in the development of peritonitis |
| Adherence to gingival crevicular epithelium | Factor in the development of periodontal disease |
| Capsule | Inhibits macrophage migration; antiphagocytic for aerobes and anaerobes; interference with T-cell function; promotes abscess formation |
| Superoxide dismutase and catalase | Confer oxygen tolerance |
| Immunoglobulin proteases | Resist host defenses |
| Hyaluronidase, collagenase, chondroitin sulfatase, neuraminidase, and fibrinolysin | Tissue digestion or dissolution (i.e., "spreading factors") |
| Heparinase and other coagulation-promoting factors | Impairment of blood supply to the infected area |
| Lipopolysaccharide | Inflammation and bone resorption in periodontal disease |
| Leukotoxin | Cytopathic for a variety of mammalian cell types |
| Butyrate | Cytotoxic substance |
| Soluble inhibitors of chemotaxis | Blunting of the inflammatory response |

## C. Factors Predisposing to Infection

As noted in Section I,D, anaerobes are prevalent as indigenous flora on all mucosal surfaces. The mucosal barrier can be disrupted by surgery, trauma, or various disease states, thus affording an opportunity for these organisms to penetrate deeper tissues and to set up infection. In other cases (e.g., aspiration pneumonia), anaerobic bacteria from a site of normal carriage (oropharynx in this case) can move into another area normally free of organisms to produce infection at that site. Tissue necrosis or a poor blood supply lower the oxidation–reduction potential, favoring the growth of anaerobic organisms. Vascular disease, cold, shock, trauma, surgery, foreign bodies, malignancy, edema, and gas production by bacteria therefore can predispose to anaerobic infection significantly, as can previous infection with nonanaerobic bacteria.

Antimicrobial agents to which anaerobes are notably resistant (e.g., aminoglycosides) can facilitate anaerobic infection. Anaerobic organisms that are more aerotolerant (e.g., those producing superoxide dismutase) are more likely to survive after the normally protective mucosal barrier is broken and until conditions are satisfactory for multiplication and invasion by these organisms. As anaerobes multiply, they maintain their own reduced environment by means of their end products of fermentative metabolism. Table VII summarizes conditions predisposing to anaerobic infection.

## D. Nosocomial Infection

The only nosocomial infection involving anaerobic bacteria is antibiotic-associated pseudomembranous colitis due to *Clostridum difficile* or, much less commonly, other clostridia. *Clostridium difficile* is a prolific spore producer. The spores are highly resistant to environmental influences and can survive for months in a hospital environment unless eliminated during the cleaning process. The small percentage of the patient population that are normal carriers of *C. difficile* in their gastrointestinal tracts do not represent

a risk to others. It is the patient with overgrowth of *C. difficile* and the disease related to it (with diarrhea) who can produce significant contamination of his or her environment. Subsequent ingestion of these spores by a susceptible patient (i.e., one receiving an antimicrobial agent that suppresses significant elements of the indigenous bowel flora and that is not active against *C. difficile*) can lead to disease in this other patient.

## III. ANAEROBIC INFECTIONS AND INTOXICATIONS

### A. Incidence

Anaerobic bacteria can be recovered from one-third to one-half of the specimens processed in a clinical bacteriology laboratory. The specific incidence, of course, varies with the nature of the infection being studied. Table VIII summarizes a number of such studies.

The two most common groups of organisms recovered from clinical specimens are gram-negative anaerobic bacilli and gram-positive anaerobic cocci. Together, they account for about two-thirds to three-quarters of all anaerobic isolates. The group that is next most prevalent is the gram-positive nonspore-forming bacilli. Clostridia usually account for 5–10% of the isolates from clinical specimens, and gram-negative anaerobic cocci are found in only about 2% of the specimens.

### B. Types

Table VIII lists many of the most commonly encountered anaerobic or mixed infections. Chronic mastoiditis, endometritis, actinomycosis, pseudomembranous colitis related to antimicrobial therapy, *C. perfringens* food poisoning, and the two major anaerobic intoxications, tatanus and botulism, should be added to this list.

### C. Unique Features

Distinctive features of anaerobic infections are given in Table IX. Of the clinical clues, the only one that is specific for anaerobic infection is foul or putrid odor to tissues or discharges. No other organism causing infection in humans produces this type of odor; on the other hand, absence of such an odor does not rule

---

#### TABLE VII
Conditions Predisposing to Anaerobic Infection

General
  Diabetes mellitus
  Corticosteroid therapy
  Leukopenia
  Hypogammaglobulinemia
  Immunosuppression
  Cytotoxic drugs
  Splenectomy
  Collagen disease
Decreased redox potential
  Tissue anoxia
  Tissue destruction
  Aerobic infection
  Foreign body
  Calcium salts
  Burns
  Peripheral vascular insufficiency
Specific clinical situations
  Malignancy
    Colon, uterus, and lung
    Leukemia
  Gastrointestinal and female pelvic surgery
  Gastrointestinal trauma, disease
  Human and animal bites
Therapy with aminoglycosides, trimethoprim/
  sulfamethoxazoles, most quinolones

## TABLE VIII
Infections Commonly Involving Anaerobes

| Infection | Incidence (%) | Proportion of cultures positive for anaerobes yielding only anaerobes |
|---|---|---|
| Bacteremia | 20[a] | 4/5 |
| Bacteremia secondary to tooth extraction | 84 | 21/45 |
| Ocular infections | 38 | 10/43 |
| Corneal ulcers | 7 | 9/11 |
| Central nervous system | | |
|   Brain abscess | 89 | 1/2 to 2/3 |
|   Extradural or subdural empyema | 10 | — |
| Head and neck | | |
|   Chronic sinusitis | 52 | 4/5[b] |
|   Acute sinusitis | 7 | — |
|   Chronic otitis media | 56 | 1/10 |
| | 59 | 11/115 |
| | 33 | 0 |
| Cholesteatoma | 92 | 1/11 |
| Neck space infections | 100 | 3/4 |
| Wound infection following head and neck surgery | 95 | 0 |
| Peritonsillar abscess | 76 | 6/28 |
| Bite wounds | 47 | 1/34 |
| Dental and oral | — | — |
| Orofacial, of dental origin | 94 | 4/10 |
| Root canal infection | 95 | 13/18 |
| | 100 | 18/55 |
| Periodontal abscess | 100 | 0/9 |
| Dental abscess, endodontic origin | 100 | 8/12 |
| | 90 | 6/9 |
| Thoracic | | |
|   Aspiration pneumonia | 93 | 1/2[c] |
| | 62 | 1/3 |
| | 100 | 1/2 |
|   Lung abscess | 93 | 1/2 to 2/3 |
| | 85 | 3/4 |
|   Bronchiectasis | — | — |
|   Empyema (nonsurgical) | 76 | 1/3 |
| | 62 | 1/2 |
| Abdominal | | |
|   Intraabdominal infection (general) | 86 | 1/10 |
| | 90 | 1/3 |
| | 81 | 1/3 |
| | 94 | 1/7 |
|   Appendicitis with peritonitis | 96 | 1/100 |
|   Liver abscess | 52 | 1/3 |
|   Other intraabdominal infection (postsurgery) | 93 | 1/6 |
|   Wound infection following bowel surgery | — | — |
|   Biliary tract | 45 | 0 |
| | 41 | 2/117 |
| Obstetric–gynecological | | |
|   Miscellaneous types | 100 | 1/3 |
| | 74 | 1/3 |
| | 72 | — |
|   Pelvic abscess | 88 | 1/2 |
|   Vulvovaginal abscess | 75 | 1/4 |

TABLE VIII    (*Continued*)

| Infection | Incidence (%) | Proportion of cultures positive for anaerobes yielding only anaerobes |
|---|---|---|
| Vaginal cuff abscess | 98 | 1/30 |
| Septic abortion, sepsis | 67 | — |
|  | 63 | — |
| Pelvic inflammatory disease | 25 | 1/14 |
|  | 48 | 1/7 |
| Soft tissue and miscellaneous |  |  |
|    Nonclostridial crepitant cellulitis | 75 | 1/12 |
|    Pilonidal sinus | 73+ | — |
|    Diabetic foot ulcers | 95 | 1/20 |
|    Infected diabetic gangrene (deep tissue culture) | 85 | 1/11 |
|    Soft tissue abscesses | 60 | 1/4 |
| Cutaneous abscesses | 62 | 1/5 |
| Decubitus ulcers with bacteremia | 63 | — |
| Osteomyelitis | 40 | 1/10 |
| Gas gangrene (clostridial myonecrosis) | — | — |
| Breast abscess | — | — |
| Perirectal abscess | — | — |

[a]Each set of figures represents data from a separate specific study of the type of infection noted.

[b]Twenty-three of 28 cultures (82%) yielding heavy growth of one or more organisms had only anaerobes present.

[c]Aspiration pneumonia occurring in the community, rather than in the hospital, involves anaerobes to the exclusion of aerobic or facultative forms two-thirds of the time.

out the possibility of an anaerobic process. The other clues, although not specific, are helpful in permitting clinicians to suspect the possibility of an anaerobic infection. There are three types of malignant disease most likely to be complicated by anaerobic infection: bronchogenic carcinoma, carcinoma of the uterus, and colon cancer.

## D. Other Pathology or Pathophysiology Induced by Anaerobes

Anaerobic bacteria are important in bowel bacterial overgrowth syndromes. Background factors for this type of problem include stagnation of the bowel contents (e.g., via surgically created blind loops), disease leading to impaired motility or stricture, large diverticula of the small bowel, decreased gastric acidity, and "contamination" of the small bowel by virtue of a gastrocolic or enterocolic fistula or an infected biliary tract draining into the duodenum. Manifestations of small intestinal bacterial overgrowth are diarrhea, often with malabsorption, including steatorrhea (i.e.,

TABLE IX

Clinical Clues to Anaerobic Infections

Foul odor of lesion or discharge

Location of infection in proximity to mucosal surface

Tissue necrosis, abscess formation

Infection secondary to human or animal bite

Gas in tissues or discharges

Classical clinical picture (e.g., gas gangrene)

Previous therapy with aminoglycoside antibiotics (e.g., neomycin, gentamicin, and amikacin), trimethoprim/sulfamethoxazole, and most quinolones

Black discoloration or red fluorescence of blood-containing exudates under ultraviolet light (pigmented *Prevotella* or *Porphyromonas* infection)

Septic thrombophlebitis

Presence of "sulfur granules" in discharges (actinomycosis)

Unique morphology on Gram stain of exudate (pleomorphic or otherwise distinctive)

Failure of the culture to grow, aerobically, organisms seen on Gram stain of original exudate increased fat in the stool) and vitamin $B_{12}$ deficiency. There may also be protein malnutrition and an impaired absorption of sugars. Impaired motility and stagnation of bowel contents, in particular, tend to favor overgrowth of anaerobic bacteria. The steatorrhea is related to bacterial deconjugation of the bile acids, which renders them ineffective in the formation of micelles required for fat absorption. Vitamin $B_{12}$ deficiency is probably related to bacterial binding of the vitamin, thus rendering it unavailable for absorption. *Bacteroides fragilis* is one organism that can bind this vitamin.

Certain mechanical abnormalities of the bowel typically related to the resection of large amounts of bowel or bypass procedures for obesity can lead to metabolic acidosis caused by D-lactic acid. Patients with this problem manifest repeated episodes of stupor or coma, and some have been committed to psychiatric institutions. These patients have a grossly abnormal fecal flora characterized by high counts of gram-positive nonspore-forming anaerobic bacteria. These organisms produce significant amounts of D-lactic acid, which is absorbed from the gut but is not metabolized by humans. *Bacteroides* spp., normally the dominant flora element, are absent from the stool samples of these patients.

Anaerobes and other organisms in the bowel flora can exert important effects by metabolizing drugs that have been ingested by patients. For example, *Eubacterium lentum* reduces digoxin to products with markedly decreased cardiac activity. Fecal anaerobes have been shown to metabolize deoxycorticosterone.

## E. Importance of Specific Anaerobes

The *B. fragilis* group is the most commonly encountered and among the most resistant of all the anaerobes to antimicrobial agents. This group accounts for about one-fourth of all anaerobic bacteria recovered from clinical specimens. The pigmented *Prevotella* and *Porphyromonas* spp. are rarely found in pure culture because of their specialized nutritional needs; however, they appear to be important in infection. *Sutterella* is much more virulent than the phenotypically similar organisms *Bacteroides gracilis* and *Bacteroides ureolyticus* and it is resistant to a number of antimicrobial agents. *Fusobacterium necrophorum* is clearly the most virulent of the nonspore-forming anaerobes. It is seen much less frequently now than it had been prior to the availability of antimicrobial agents. Despite its susceptibility to many antimicrobial agents, however, it often produces overwhelming

sepsis and metastatic infections. *Fusobacterium nucleatum* is the most commonly encountered of the fusobacteria. A newly described anaerobe, *Bilophila wadsworthia,* is found in approximately one-half of patients with gangrenous or perforated appendicitis. It has been found in other intraabdominal and other types of infections as well. This anaerobe usually produces $\beta$-lactamase. *Peptostreptococcus magnus* seems to be particularly pathogenic among the anaerobic cocci.

Among the clostridia, *C. perfringens* is the most commonly isolated and is extremely virulent. *Clostridium ramosum*, although much less virulent, is seen with about the same frequency as *C. perfringens* and is more resistant to antimicrobial agents. *Clostridium septicum* is of particular importance because of its association with malignancy of the bowel, particularly the cecal area. *Actinomyces, Propionibacterium propionicum, P. acnes, Eubacterium nodatum,* and *Bifidobacterium dentium* (*eriksonii*) are the best-documented pathogens among the gram-positive nonspore-forming bacilli.

## IV. DIAGNOSIS

### A. Laboratory Procedures

#### 1. Collection and Transport of Specimens

Because anaerobes are so prevalent among the normal flora of the body, it is imperative that specimens for determination of the causative agents of an infection be collected in such a way as to bypass the normal flora. Specimens must be transported under anaerobic conditions to avoid die-off of oxygen-sensitive delicate forms. The best type of specimen for anaerobic culture is a piece of tissue that has been debrided or biopsied, but purulent secretions are also suitable. Specimens obtained on cotton swabs are the least desirable, but, when obtained and transported in proper fashion, they can be used.

#### 2. Direct Examination

The gram stain was discussed in Section I,B. It is a valuable tool for the direct examination of clinical specimens because it gives immediate information about the types and the relative numbers of different organisms present in the specimen. This is particularly true of the anaerobes since so many of them have a distinctive morphology.

Other direct examinations include the noting of any foul or putrid odors, which would indicate that anaerobes must be present in the specimen, or the presence of pigment produced by certain *Prevotella* or *Porphyromonas* species. Examination of the specimen under dark-field microscopy may be useful for detecting motile organisms or spores. [*See* Bacterial Infections, Detection.]

### 3. Culture

A variety of solid- and liquid-enriched selective and differential media is available for culture of anaerobic organisms. Selective media often facilitate the recovery of specific anaerobes from complex mixtures and lead to early identification of these special groups. Certain anaerobes have specific growth requirements that must be met in the media provided for their growth. Inoculation of culture plates and broths in the conventional way on the lab bench, with subsequent incubation in an anaerobic atmosphere (e.g., an anaerobic jar that uses hydrogen to combine with oxygen to eliminate it from the environment), is satisfactory for many types of clinical specimens. However, there may be an advantage, at least in the initial culture, to process the specimen entirely in an anaerobic chamber to avoid exposure to oxygen.

For the most fastidious anaerobes it is desirable to use so-called prereduced anaerobically sterilized media (i.e., media that have not been exposed to oxygen while they are put into solution and subsequently autoclaved). Anaerobic chambers can be equipped with incubators so that incubation can be carried out in the chamber atmosphere. As indicated in Section I,A, increased carbon dioxide in the environment, along with anaerobiosis, is desirable.

### 4. Identification Techniques

Identification techniques include the notation of colonial and microscopic morphology (as described in Sections I,B and I,C), determination of metabolic end products by gas–liquid chromatography, utilization of various carbohydrates and other substrates, stimulation or inhibition by various compounds (e.g., bile), and analysis of cellular medium-chain fatty acids.

### 5. Antimicrobial Susceptibility Determination

A number of anaerobic bacteria have been increasingly resistant to antimicrobial agents in recent years. Important strain differences exist, however, so one cannot presume that an organism is susceptible or resistant by knowing its speciation. Various testing techniques are in use at present, but the best and most popular are those using a serial twofold dilution of antimicrobial agents in either agar or broth or gradi-

ent end point techniques and subsequent inoculation with the test organism to see at what drug concentration the organism is inhibited. This, of course, is compared with the usual levels achievable in blood and other body fluids.

## B. Clinical Clues

Clinical clues, discussed in Section III,C, are useful in suspecting, and in some cases confirming, the diagnosis of anaerobic infection and, at times, specific anaerobic infections.

# V. THERAPY

## A. Surgery and Related Approaches

Because anaerobic bacteria produce significant tissue destruction and are prone to produce abscesses, surgical management is commonly needed for debridement of the necrotic tissue and for drainage of the collections of pus. On occasion, particularly with intraabdominal abscesses, it may be feasible to perform drainage percutaneously under the guidance of ultrasound or computerized tomography.

## B. Antimicrobial Agents

### 1. Usual Susceptibility Patterns

Although there are strain differences, as noted in Section IV,A,5, since initial therapy of anaerobic infections must be empiric (it takes considerable time for information to become available from the anaerobic cultures), it is important to know not only the usual bacteriology of the infection being treated but also the usual susceptibility patterns. In a very sick patient, then, one can use one of the agents with the most consistent activity against particular types of anaerobes. Table X gives patterns of susceptibility of various anaerobes to the more promising antimicrobial agents.

### 2. Pharmacological Considerations

The blood and tissue levels achievable by various agents must be taken into account in deciding among several agents that appear to be active against the infecting organism(s). In the case of brain abscess or subdural empyema, for example, one should not use clindamycin because it does not cross the blood–brain barrier well. Metronidazole would be one good choice

**TABLE X**

Susceptibility of Anaerobes to Antimicrobial Agents[a]

| % susceptible[b] | B. fragilis | Other B. fragilis groups[d] | B. gracilis | Other Bacteroides spp. | Prevotella sp. | Porphyromonas | Sutterella wadsworthensis | F. nucleatum | F. mort/varium |
|---|---|---|---|---|---|---|---|---|---|
| >95 | Ampicillin (Amp) + sulbactam (sulb)<br>Piperacillin (Pip) + tazobactam (tazo)<br>Ticarcillin (Ticar) + clavulanate (clav)<br>Cefoperazone/sulb<br>Imipenem<br>Chloramphenicol<br>Clinafloxacin<br>Metronidazole | Ampicillin/sulb<br>Cefoperazone/sulb<br>Pip/tazo<br>Ticar/clav<br>Imipenem<br>Chloramphenicol<br>Clinafloxacin<br>Metronidazole<br>Minocycline | Piperacillin<br>Amoxicillin (Amox)/clav<br>Pip/tazo<br>Ticar/clav<br>Cefoxitin<br>Ceftizoxime<br>Ceftriaxone<br>Imipenem<br>Meropenem<br>Ciprofloxacin<br>Fleroxacin<br>Clindamycin<br>Metronidazole<br>Minocycline<br>Tetracycline | Amp/sulb<br>Piperacillin<br>Ticar/clav<br>Cefoperazone<br>Cefoper/sulb<br>Cefotaxime<br>Cefoxitin<br>Imipenem<br>Chloramphenicol<br>Clinafloxacin<br>Clindamycin | Amox/clav<br>Ceftizoxime<br>Imipenem<br>Clindamycin<br>Chloramphenicol<br>Clinafloxacin<br>Metronidazole | Amox/clav<br>Ceftizoxime<br>Imipenem<br>Chloramphenicol<br>Clinafloxacin<br>Metronidazole<br>Minocycline | Amox/clav<br>Ticar/clav<br>Cefoxitin<br>Ceftizoxime<br>Ceftriaxone<br>Clindamycin | Amox/clav<br>Ceftizoxime<br>Imipenem<br>Chloramphenicol<br>Clinafloxacin<br>Clindamycin<br>Metronidazole<br>Minocycline<br>Tetracycline | Imipenem<br>Chloramphenicol<br>Clinafloxacin<br>Metronidazole<br>Minocycline |
| 85–95 | Cefotetan<br>Cefoxitin<br>Ceftizoxime<br>Piperacillin<br>Clindamycin<br>Minocycline | Piperacillin<br>Ceftizoxime | | Cefotetan<br>Ceftazidime<br>Ceftizoxime<br>Ceftriaxone<br>Minocycline | | Ciprofloxacin<br>Clindamycin | Piperacillin<br>Pip/tazo | | Amox/clav |
| 70–84 | Moxalactam | Cefoxitin<br>Clindamycin | | Penicillin G<br>Moxalactam | Minocycline | | | Ciprofloxacin | Clindamycin<br>Tetracycline |
| 50–69 | Cefoperazone<br>Cefotaxime<br>Ceftazidime<br>Ceftriaxone | Cefoperazone<br>Cefotetan<br>Moxalactam | | Ciprofloxacin<br>Tetracycline | Ciprofloxacin<br>Tetracycline | Tetracycline | Metronidazole | | Ciprofloxacin |
| <50 | Penicillin G[c]<br>Ciprofloxacin<br>Tetracycline | Penicillin G<br>Cefotaxime<br>Ceftazidime<br>Ceftriaxone | | | | | | | Ceftizoxime |

| % susceptible[b] | Other Fusobacterium | B. wadsworthia | Peptostreptococcus | C. difficile[e] | C. ramosum | C. perfringens | Other Clostridium | NSF-GPR[f] |
|---|---|---|---|---|---|---|---|---|
| >95 | Amp/sulb<br>Ceftizoxime<br>Penicillin G<br>Piperacillin<br>Pip/tazo<br>Imipenem<br>Chloramphenicol<br>Clinafloxacin<br>Clindamycin<br>Minocycline | Amox/clav<br>Amp/sulb<br>Penicillin G<br>Piperacillin<br>Ticarcillin<br>Cefoxitin<br>Ceftizoxime<br>Imipenem<br>Chloramphenicol<br>Ciprofloxacin<br>Clindamycin<br>Metronidazole<br>Minocycline<br>Tetracycline | Amp/sulb<br>Penicillin G<br>Piperacillin<br>Ticar/clav<br>Cefoperazone<br>Cefoper/sul<br>Cefotetan<br>Ceftazidime<br>Ceftriaxone<br>Imipenem<br>Chloramphenicol<br>Ciprofloxacin<br>Clinafloxacin<br>Metronidazole<br>Moxalactam | Amox/clav<br>Amp/sulb<br>Ticar/clav<br>Ampicillin<br>Ticarcillin<br>Cefotetan<br>Imipenem<br>Clinafloxacin<br>Metronidazole | Amox/clav<br>Pip/tazo<br>Ticar/clav<br>Cefotaxime<br>Ceftizoxime<br>Imipenem<br>Chloramphenicol<br>Clinafloxacin<br>Metronidazole | Amp/sul<br>Amox/clav<br>Pip/tazo<br>Ticar/clav<br>Ampicillin<br>Piperacillin<br>Ticarcillin<br>Cefotetan<br>Cefoxitin<br>Ceftizoxime<br>Imipenem<br>Chloramphenicol<br>Ciprofloxacin<br>Clinafloxacin<br>Clindamycin<br>Metronidazole | Ampicillin<br>Amp/sulb<br>Amoxicillin<br>Carbenicillin<br>Penicillin G<br>Piperacillin<br>Ticarcillin<br>Imipenem<br>Chloramphenicol<br>Clinafloxacin<br>Minocycline | Amp/sul<br>Amox/clav<br>Penicillin G<br>Piperacillin<br>Ticar/clav<br>Cefotaxime<br>Ceftizoxime<br>Imipenem<br>Chloramphenicol<br>Clindamycin<br>Clinafloxacin<br>Minocycline |
| 85–95 | Ticar/clav<br>Cefoperazone<br>Cefoper/sulb<br>Cefotaxime<br>Cefotetan<br>Cefoxitin<br>Ceftriaxone | | Clindamycin<br>Minocycline | Chloramphenicol | Amp/sulb<br>Ampicillin<br>Piperacillin | | Moxalactam | Cefoper/sul<br>Cefotetan<br>Cefoxitin |
| 70–84 | Ceftazidime<br>Moxalactam | | | | Cefoxitin<br>Clindamycin | Minocycline | Clindamycin<br>Tetracycline | Cefoperazone<br>Moxalactam<br>Tetracycline |
| 50–69 | | | Tetracycline | Clindamycin<br>Minocycline<br>Tetracycline | Minocycline<br>Tetracycline | Tetracycline | Cefoperazone<br>Cefotaxime<br>Cefoxitin<br>Ceftizoxime<br>Ceftriaxone | Ciprofloxacin<br>Metronidazole |
| <50 | | Amoxicillin<br>Ampicillin | | Cefoxitin<br>Ceftizoxime<br>Ciprofloxacin | Ciprofloxacin | | Ceftazidime<br>Ciprofloxacin | |

[a] The order of listing of drugs within percent susceptible categories is not significant.
[b] According to the NCCLS-approved breakpoints (M11-A3), using the intermediate category as susceptible.
[c] NCCLS approved breakpoint 4 μg/ml. However, the breakpoint should probably be lowered to 1 μg/ml, which will considerably lower the values for percent susceptible. For example, at 1 μg/ml, no strains of the B. fragilis group were susceptible.
[d] Excluding B. fragilis.
[e] Breakpoint is used only as a reference point. C. difficile is primarily of interest in relation to antimicrobial-induced pseudomembranous colitis. These data must be interpreted in the context of level of drug achieved in the colon and the impact of agent on indigenous colonic flora.
[f] Nonspore-forming gram-positive rod.

in these situations because it does cross well and has good killing activity against anaerobes. [*See* Blood–Brain Barrier.]

### 3. Resistance to Antimicrobial Agents

As noted from Table X, resistance to the top-line drugs (e.g., chloramphenicol, metronidazole, imipenem, and combinations of β-lactam drugs with β-lactamase inhibitors) is extremely low, except that *Actinomyces* and other gram-positive nonspore-forming bacilli are resistant to metronidazole about 75% of the time and *Sutterella* less frequently. There has been increased resistance to penicillin G so that now only about 2% of the *B. fragilis* group strains are susceptible to this drug at achievable levels and a number of other anaerobic gram-negative rods and a few clostridia are β-lactamase producers. Other drugs that previously were active against 95–100% of the *B. fragilis* group are now less effective because of the development of resistance. Twenty to 25% of the *B. fragilis* strains are now resistant to cefoxitin and clindamycin and 15% to broad-spectrum penicillins (e.g., piperacillin). [*See* Antimicrobial Drugs.]

The principal mechanisms of resistance in anaerobic bacteria are listed in Table XI. A large number of anaerobic bacteria now produce β-lactamases, which inactivate certain β-lactam drugs (penicillins and cephalosporins), and some produce chloramphenicol acetyltransferase or nitroreductase, which inactivate chloramphenicol. Changes in the porin channels in the outer membranes prevent penetration of certain antimicrobial agents into the bacterial cells. In the case of tetracycline resistance, the drug can get into bacterial cells but can be pumped out. Changes in the targets for drug activity, such as the penicillin-binding proteins in the case of β-lactam agents, are also seen. With metronidazole the organism may fail to reduce the drug to an active intermediate form. In some cases

the resistance mechanism is inducible upon exposure of the organism to the drug.

## C. Other Procedures

Oxygen above usual concentrations (hyperbaric oxygen) may be useful in gas gangrene. It is difficult to be certain as to how effective this therapy is since patients with this infection invariably receive all other possible modes of therapy and adequate controlled trials have not been possible. There appears to be at least some improvement with regard to the rapid spread of the gangrene and a demarcation of the process so that surgeons can better judge where to amputate in the case of an extremity that cannot be saved. On occasion, other oxygen-releasing compounds (e.g., zinc peroxide) may be useful topically in managing certain resistant infections.

In the case of septic thrombophlebitis, in addition to surgery that might be required, anticoagulation could be useful; it does, however, pose some potential risk to the patient. Exchange transfusion has been recommended in postabortion or other septicemia due to *C. perfringens* when there is significant intravascular hemolysis. Obviously, general supportive measures are required in patients who are quite ill with anaerobic infections. Included are such things as blood transfusion, maintenance of fluid and electrolyte balance, adequate immobilization of the infected injured part, treatment of shock, relief of pain, and management of renal failure. The use of antibodies to a toxin is important if it is a key factor in pathogenesis, as in tetanus and botulism.

## VI. PROPHYLAXIS

The principles of prophylaxis (and therapy) are (1) controlling the environment so that anaerobic bacteria find it difficult or impossible to proliferate, (2) checking the spread of anaerobic bacteria into healthy tissues, and (3) neutralizing toxins produced by anaerobes. Antimicrobial agents are an important factor in limiting the spread of anaerobes into healthy tissues. Furthermore, clindamycin and metronidazole may inhibit alpha-toxin production by *Clostridium perfringens*. Toxin neutralization is achieved primarily with specific antitoxin, applicable to tetanus and botulism. The patient who has potential exposure to tetanus should receive passive immunization with antitoxin if he or she has not been actively immunized previously and the threat of disease is high, or the

**TABLE XI**
Mechanisms of Resistance in Anaerobes

Inactivation of drugs
Failure of drugs to get into bacterial cells
Pumping drug out of bacterial cells
Changes in targets for drugs
Failure to reduce drugs to active form

patient could receive a booster dose of tetanus toxoid (inactive but immunogenic toxin) for active immunization. For those not previously immunized, active immunization with toxoid should be started when antitoxin is administered.

## VII. PROGNOSIS

In general, the prognosis for most anaerobic infections is much improved in recent years as a result of better information about the causes and pathogenesis, the availability of superior diagnostic tests, rapid techniques for the presumptive identification of anaerobic bacteria, marked improvements in surgical management, and the availability of potent antimicrobial agents. The prognosis depends not only on the nature and the severity of the infection but on the speed with which diagnosis is made and treatment initiated. In certain cases the process is so overwhelming that the prognosis remains poor. Included in this category would be serious soft tissue infections (e.g., gas gangrene), infection in patients with impaired host defense mechanisms, and disease involving the production of toxins, such as tetanus (in less well-developed countries the mortality rate remains over 50% in such patients).

## VIII. *Helicobacter pylori*

It has been appreciated that a microaerophilic organism (an organism that requires reduced oxygen tension in the atmosphere, but not strict anaerobiosis), *Helicobacter pylori*, is an important pathogen in human disease. The disease process induced by this organism is unique in that the bacterium colonizes the gastric mucosa for years or decades and causes continuous low-grade inflammation.

### A. Microbiology

*H. pylori* is a gram-negative curved or spiral, microaerophilic bacterium that is actively motile by means of multiple flagella at one pole of the organism. It is a strong urease producer, producing ammonia and carbon dioxide from urea. The organism resides in the semipermeable mucus layer of the stomach in humans; the ammonia probably serves as a survival mechanism for this organism in the acidic environment of the stomach.

### B. Epidemiology of Infection with *H. pylori*

This infection is worldwide. It clusters in families and is associated with low socioeconomic status. In the United States, about half of 60-year-old adults are infected, whereas in developing countries, most people are infected by the age of 10. Although the specific modes of transmission are unknown, it is likely that most transmission is from person to person; no nonhuman reservoir has been identified.

### C. Pathogenesis of Infection with *H. pylori*

The organism produces chronic superficial gastritis in both the antrum and the fundus of the stomach, often resulting in hypergastrinemia. The bacterium remains in the gastric mucus layer and does not invade the epithelium. Although specific antibodies are produced by the host, they do not lead to eradication of the organism.

The mechanisms by which *H. pylori* produces inflammation are not well understood. Both the ammonia produced and a cytotoxin of the organism are injurious to eukaryotic cells. Certain components of the bacterium are antigenically cross-reactive with host tissues and could lead to autoimmune damage to the host. The inflammatory response to the organism itself may lead to the injury of epithelial cells. Approximately 60% of the strains of *H. pylori* contain the *cagA* gene; it has been noted that there is an association between the presence of this gene in an infecting strain and the presence of duodenal ulceration.

### D. Disease States Induced by *H. pylori*

*H. pylori* is the principal cause of chronic superficial gastritis, but this entity usually produces no clinical symptomatology. The possible role of this organism in nonulcer dyspepsia, a poorly defined clinical syndrome with upper gastrointestinal complaints, is uncertain. Similar symptoms are induced by a variety of entities such as gastroesophageal reflux and motility disorders.

A striking association exists between chronic superficial gastritis and peptic ulcer disease and between *H. pylori* infection and peptic ulcer disease. Infection with this organism is significantly more frequent in individuals with peptic ulcer disease than in age-matched controls. It is interesting to note that both gastric and duodenal ulcers are strongly associated with infection with *H. pylori*, at least partly related to islands of gastric tissue present in the duodenum (gastric metaplasia). Treatment with a wide variety of agents to which *H. pylori* is susceptible leads to healing of duodenal ulcers and significant reduction in the recurrence rates of ulcers.

Chronic superficial gastritis may progress, over decades, to chronic atrophic gastritis, a well-recognized risk factor for gastric adenocarcinoma. Several studies have shown a significant association between infection with *H. pylori* and gastric cancer, particularly cancer of the distal portion of the stomach. [*See* Gastrointestinal Cancer.]

## E. Diagnosis of Infection with *H. pylori*

Biopsy or other specimens obtained during endoscopy may be used to demonstrate the organism with special stains, to culture the organism, or to detect urease activity. Urea breath tests and determination of antibodies to the organism in the serum of patients afford means of diagnosis without endoscopy. When properly performed, all of the major diagnostic tests (culture, histology, breath testing, and serology) have >95% accuracy.

## F. Therapy

Various therapeutic approaches have been utilized. At present, triple therapy is favored; this regimen includes a bismuth salt, a nitroimidazole such as metronidazole, and either amoxicillin or tetracycline for a total of 2 to 4 weeks. The nitroimidazole seems to be the most crucial part of the regimen; if a patient's strain is resistant to this agent, failure or relapse is likely. Evidence suggests that shorter courses of therapy, such as for 1 week, may be effective. Other regimens, such as omeprazole plus amoxicillin, are useful but not quite as effective as the triple regimen just described.

## BIBLIOGRAPHY

Beerens, H., and Tahon-Castel, M. (1965). "Infections Humaines à Bactéries Anaérobies Non-toxigènes." Presses Acad. Eur., Brussels, Belgium.

Blaser, M. J. (1992). *Helicobacter pylori*: Its role in disease. *Clin. Infect. Dis.* 15, 386–393.

Cover, T. L., Glupczynski, Y., Lage, A. P., *et al.* (1995). Serologic detection of infection with cagA+ *Helicobacter pylori* strains. *J. Clin. Microbiol.* 33, 1496–1500.

Dubois, A. (1995). Spiral bacteria in the human stomach: The gastric helicobacters. *Emerg. Infect. Dis.* 1, 79–85.

Duerden, B. I., and Drasar, B. S. (eds.) (1991). "Anaerobes in Human Disease." Wiley-Liss, New York.

Finegold, S. M. (1977). "Anaerobic Bacteria in Human Disease." Academic Press, New York.

Finegold, S. M., and George, W. L. (eds). (1989). "Anaerobic Infections in Humans." Academic Press, San Diego, CA.

Finegold, S. M., George, W. L., and Rolfe, R. D. (eds.) (1984). International symposium on anaerobic bacteria and their role in disease. *Rev. Infect. Dis.* 6 (Suppl. 1).

Finegold, S. M., and Goldstein, E. J. C. (eds.) (1993). Proceedings of the First North American Congress on Anaerobic Bacteria and Anaerobic Infections. *Clin. Infect. Dis.* 16 (Suppl. 4).

Holdeman, L. V., Cato, E. P., and Moore, W. E. C. (1977). "Anaerobe Laboratory Manual," 4th Ed. Virginia Polytechnic Inst. and State Univ., Blacksburg, Virginia.

Kasper, D. L., and Finegold, S. M. (eds.) (1979).Virulence factors of anaerobic bacteria. *Rev. Infect. Dis.* 1, 245–400.

Prévot, A.-R. (1972). "Les Bactéries Anaérobies." Crouan & Roques, Lille, France.

Rosebury, T. (1962). "Microorganisms Indigenous to Man." McGraw-Hill, New York.

Smith, L. DS., and Williams, B. L. (1984). "The Pathogenic Anaerobic Bacteria," 3rd Ed. Thomas, Springfield, IL.

Summanen, P., Baron, E. J., Citron, D. M., Strong, C. A., Wexler, H. M., and Finegold, S. M. (1993). "Wadsworth Anaerobic Bacteriology Manual," 5th Ed. Star, Belmont, CA.

Willis, A. T. (1969). "Clostridia of Wound Infection." Butterworths, London.

# Anesthesia

EDWARD CHEN
*Massachusetts General Hospital and Harvard Medical School*

## GLOSSARY

**Agonists** Drugs that activate receptors

**Antagonists** Drugs that bind to receptors without activating them. They also prevent agonists from activating the receptor

**Minimum alveolar concentration** Partial pressure of an inhaled anesthetic at 1 atmosphere that prevents skeletal muscle movement in response to a noxious stimulus in 50% of patients

**Pharmacodynamics** Analysis of the responsiveness of the receptor to the drug and eventually the drug's effect on the body

**Pharmacokinetics** Analysis of the relationship between the dose of a drug and its concentration. It describes the absorption, distribution, metabolism, and excretion of drugs

ANESTHESIA IS DEFINED AS THE LOSS OF SENSA-tion with or without loss of consciousness. Although the modern history of anesthesia dates back to the middle of the nineteenth century, the practice of anesthesiology has changed considerably in the last 50 years. This transformation is the result of the development of new drugs and new delivery systems as well as the expansion of surgical subspecialties. More recently, the anesthesiologist as a medical specialist has expanded his or her role from intraoperative care to a variety of tasks both inside and outside the operating room.

According to the American Board of Anesthesiology, anesthesia is defined as the practice of medicine dealing with but not limited to (1) the assessment of, consultation for, and preparation of patients for anesthesia; (2) the provision of insensibility to pain during surgical, obstetric, therapeutic, and diagnostic procedures, and the management of patients so affected; (3) the monitoring and restoration of homeostasis during the perioperative period, as well as homeostasis in the critically ill, injured, or otherwise seriously ill patient; (4) the diagnosis and treatment of painful syndromes; (5) the clinical management and teaching of cardiac and pulmonary resuscitation; (6) the evaluation of respiratory function and applications of respiratory therapy in all its forms; (7) the supervision, teaching, and evaluation of performance of both medical and paramedical personnel involved in anesthesia, respiratory, and critical care; (8) the conduct of research at the clinical and basic science levels to explain and improve the care of patients; and (9) the administrative involvement in hospitals, medical schools, and outpatient facilities necessary to implement these responsibilities.

## I. HISTORY

Before the introduction of anesthesia, surgery was approached with trepidation by the patient, who compared the experience with awaiting execution. In addition to the pain and mental anguish, other risks from the procedure included hemorrhage, shock, and infection. Surgeons attempted to relieve pain with various adjuncts, including hypnosis, local application of pressure or ice, and the ingestion of wine, whiskey, herbs, opium, and coca leaves. To this day, credit to

**319**

ENCYCLOPEDIA OF HUMAN BIOLOGY, Second Edition, VOLUME I.   Copyright © 1997 by Academic Press.   All rights of reproduction in any form reserved.

the person who made the actual discovery of the first modern anesthetic, diethyl ether, remains controversial. Ether was described as far back as the eighth century by an Arabian philosopher who prepared it through the distillation of sulfuric acid with wine. It was noted that animals fell asleep upon exposure and awakened without pain. Ether was used initially for the treatment of minor diseases and then was also used for recreational purposes.

In 1842, Dr. Crawford W. Long, a physician from rural Georgia, was the first physician to give ether to a patient for surgery, but he did not report this event. Nitrous oxide, another anesthetic, was first discovered by Joseph Priestley, a chemist who also purified oxygen in the late eighteenth century. This gas was initially used also for recreational purposes, but in 1844 an American dentist, Dr. Horace Wells, noticed that a soldier did not complain of any pain after injuring his leg while under the influence of nitrous oxide. His attempted public demonstration of this gas ended in failure when an insufficient amount of nitrous oxide was administered and the patient subsequently complained of pain during the procedure.

Credit for the first public demonstration of surgical anesthesia with ether is given to Dr. William T. Morton, a dentist from Connecticut. The procedure, which took place at the Massachusetts General Hospital in Boston, was the removal of a tumor below the jaw by the surgeon, Dr. John C. Warren, on October 16, 1846. The audience included surgeons, medical students, and a reporter whose account appeared the following day in the *Boston Daily Journal*.

## II. PHARMACOLOGY OF ANESTHETIC AGENTS

An understanding of clinical pharmacology provides the foundation for rational selection and use of anesthetic agents. Pharmacokinetics describes what the body does to the drug, and pharmacodynamics reflects what the drug does to the body.

### A. Pharmacokinetics of Intravenous Drugs

With pharmacokinetics, the concentration of the drug is affected by the drug's absorption, distribution, and clearance. Factors affecting absorption include route of administration and blood flow. In general, intravenous administration results in relatively predictable plasma concentrations, but with oral or intramuscular injections the absorption is not as easily predictable and is more dependent on blood flow.

Distribution is more readily expressed as the volume of distribution (or $V_d$), which equals the dose of drug administered intravenously divided by the plasma concentration. This can be applied to the two-compartment (central and peripheral) pharmacokinetic model in which a high plasma concentration of drug results in a relatively low calculated $V_d$. Drugs that remain in the central compartment tend to be ionized, highly bound to plasma protein, and with limited lipid solubility. These all limit passage of the drug to tissues (or peripheral compartment). Conversely, drugs that easily pass between the two compartments tend to be nonionized and lipid soluble.

Clearance is defined as the volume of plasma (in the central compartment) cleared of drug over time (ml/min) by liver metabolism and renal excretion. Although the liver is the primary site of metabolism of the administered drug, the lungs, kidneys, and gastrointestinal tract also play a role. The goal is to convert a pharmacologically active lipid-soluble drug to an inactive water-soluble drug for renal excretion. Microsomal enzymes located in the hepatic smooth endoplasmic reticulum are responsible for this metabolic process. The level of this enzyme activity is genetically determined, which explains the variability in rate of metabolism among patients. Rate constants describe both the transfer of drug between the central and peripheral compartments and the elimination of the drug from the central compartment. The latter is often labeled the elimination half-life ($t_{1/2}$), which is defined as the time for 50% decrease in the plasma concentration of the drug. It should be noted that this constant is independent of the concentration and dose of the drug.

### B. Pharmacodynamics of Intravenous Drugs

There are many types of receptors in the body, all of which are large protein complexes that act as a signaling system across the cell membrane. Substances that bind to the receptor include agonist drugs, hormones, and neurotransmitters. Examples of receptors include a simple protein channel (e.g., gamma aminobutyric acid receptor and its associated flow of chloride ions) and a multicomponent system composed of a receptor protein, a guanine nucleotide binding protein (G protein), and an effector. This system may or may not utilize a secondary messenger. An example of a secondary messenger is cyclic adenosine monophosphate, or cyclic AMP. G proteins can be either stimulatory or inhibitory, which can then control the level of the secondary messenger, cyclic AMP. Following the drug-receptor interaction, a series of cellular events eventually results in a physiologic response.

Stereospecificity plays a role in the response of the body to the drug. When a drug is synthesized, two isomers are formed, dextro ($d$) and levo ($l$). A racemic mixture describes an equal amount of each. Despite the same chemical composition of both, one isomer is often more biologically active than the other; thus, the inactive isomer has no benefit and can often cause unwanted side effects. The presence of a racemic mixture versus a mixture containing only the active isomer can have to dramatically altered effects.

Another major factor affecting the pharmacodynamics of a drug is the number of receptors. The actual number of receptors in the cell membrane is not static: it either increases or decreases depending on the presence of certain stimuli. For example, the prolonged exposure of an agonist can result in a decrease in the number of receptors. Thus, tachyphylaxis occurs in which a larger dose of drug is required to produce the same desired effect.

## C. Pharmacokinetics of Inhaled Anesthetics

Similar to what is seen with intravenous anesthetics, the pharmacokinetics of inhaled anesthetics describes their uptake (absorption), distribution, and elimination. To establish anesthesia, a partial pressure gradient is needed. Such a gradient is formed when the inhaled anesthetic is delivered from the anesthetic machine to the brain by way of passage in arterial blood. Thus, the gradient consists of the inspired partial pressure of the anesthetic with equilibration of the alveolar (lung) partial pressure ($P_A$), the partial pressure in arterial blood, and the partial pressure in the brain ($P_{br}$). Stated another way, the path of the gradient travels from the anesthetic machine to the lung, the arterial blood, then finally to the brain. If the alveolar partial pressure reflects the partial pressure of the anesthetic gas in the brain, then $P_A$ can be used as an index of anesthetic depth. The goal is to establish a desired partial pressure of the anesthetic in the brain over time.

Factors that affect these gradients can be divided into what can affect each transfer point as the anesthetic gas is delivered to the brain. These factors include the inspired partial pressure of the anesthetic, the patient's pulmonary status, certain aspects of the anesthetic breathing system, the solubility of the gas, and the patient's cardiac performance.

## D. Pharmacodynamics of Inhaled Anesthetics

There are no precise mechanisms to describe how general anesthesia actually works. One proposed mechanism, the Meyer–Overton theory (or critical-volume hypothesis), relates to the association between the lipid solubility of the anesthetic and its potency. Potency can be defined as the amount of anesthetic dissolved in a membrane needed to achieve a desired effect. Another theory, the protein receptor hypothesis, describes the presence of receptors in the central nervous system that can form complexes with the anesthetic agent. Ongoing research reveals that inhaled anesthetics have a variety of effects in the central nervous system, including changes in neurotransmitter release, alterations in enzyme and ion channel activity, and variable sensitivities based on genetic factors. At this time, no single theory predominates, and the establishment of general anesthesia can be explained at best with a combination of more than one mechanism.

Potency of an inhaled anesthetic is described clinically in terms of the minimum alveolar concentration, or MAC. Strictly defined, MAC is the partial pressure at 1 atmosphere of an inhaled anesthetic that prevents movement of skeletal muscle in response to skin incision in 50% of subjects. This allows a comparison of these anesthetics relative to their inspired concentrations.

With the successful use of ether, nitrous oxide, and chloroform in the mid-1800s, no subsequent anesthetics were introduced until almost 80 years later when advances in technology resulted in improvement of the profile of the anesthetic gas. Each anesthetic that was developed revealed enhancement of characteristics of an ideal inhaled anesthetic: absence of flammability, potency, low blood solubility for rapid induction and recovery from anesthesia, minimal metabolism, skeletal muscle relaxation, and minimal toxic effects to major organs.

The inhaled anesthetics used today include one gas (nitrous oxide) and the vapors of five volatile liquids (halothane, enflurane, isoflurane, desflurane, and sevoflurane). Earlier anesthetics created problems such as instability in oxygen, flammability, toxicity related to the formation of extensive metabolites, and the high incidence of unpleasant side effects such as nausea and vomiting. The current volatile liquids in use are halogenated compounds. It is believed that such a structure allows for stability, nonflammability, and relatively low blood solubility. The halogens chlorine and bromine contribute to the anesthetic potency. Each anesthetic differs in structure, molecular weight, MAC, vapor pressure, and blood:gas solubility. In addition to the induction of anesthesia, these anesthetics can profoundly influence the function of various organs, particularly the heart, lung, brain, liver, and kidney.

## III. GENERAL ANESTHESIA

The principal phases of care provided by the anesthesiologist include preoperative preparation, induction, maintenance, emergence, and postoperative care. In addition to attaining an optimal depth of anesthesia for acceptable surgical conditions, the anesthesiologist is responsible for the safety of the patient throughout the perioperative period.

Preoperative preparation consists of a thorough evaluation of the patient's medical history, physical exam, and laboratory data. Informed consent is obtained following discussion of not only what the patient can expect but also the risks and complications of anesthesia. A premedication can be given to help allay the anxiety of the patient, and the patient is placed on various monitors to follow the patient's cardiac and pulmonary status throughout the procedure.

The induction period is a critical phase for the anesthesiologist because it involves the transition from the awake phase to the unconscious state. Drastic changes in airway patency and circulatory hemodynamics can occur during this period. Usually an intravenous agent is given to produce loss of consciousness, and once patency of the airway is established, ventilation can be accomplished with either a mask or an endotracheal tube. This tube is inserted with a specialized instrument that allows the tube to be passed through the mouth or nose into the trachea.

There are four major stages of general anesthesia, defined following the observation of the responses of patients during induction with diethyl ether. Stage I (amnesia) covers the period from the awake state to loss of consciousness. There is no change in pain perception during this stage. Stage II (delirium) describes the uninhibited excitement in response to noxious stimuli. This is a relatively unstable stage in which the transition can be accelerated with various induction agents. Stage III (surgical anesthesia) is the phase in which optimal surgical conditions are met in that there is no deleterious response to painful stimulation. Stage IV (overdosage) describes a depth of anesthesia that is too deep with adverse hemodynamic conditions. If this occurs, the anesthetic plane is lightened as soon as possible.

Although an inhaled anesthetic can establish anesthesia by producing loss of consciousness, amnesia, analgesia, and skeletal muscle relaxation, it is possible to substitute it with a mixture of intravenous anesthetics, particularly if the patient is unable to tolerate the various hemodynamic changes produced by the inhalational agent. These intravenous agents include barbiturates, benzodiazepines, opioids, muscle relaxants, and nonbarbiturate induction drugs. They all produce their desired effects by interacting with various membrane receptor proteins, but each one differs on the basis of its mechanism of action, pharmacokinetics, pharmacodynamics, and profile of adverse reactions.

Barbiturates have a depressant effect on the central nervous system by inhibiting the release of excitatory neurotransmitters and enhancing the effect of inhibitory ones. Benzodiazepines also depress the brain by enhancing the inhibitory effect of gamma aminobutyric acid. Opioids bind with specific receptors located throughout the body. The primary desired effect is the suppression of pain. Despite providing analgesia, undesirable side effects do occur: nausea, respiratory depression, excessive sedation, pruritis, and urinary retention. Muscle relaxants provide optimal conditions for airway management and surgical procedures. The mechanism of these neuromuscular blocking drugs occurs at the neuromuscular junction. They cause either sustained depolarization of the acetylcholine receptor (resulting in decreased muscle excitability) or blockage of the acetylcholine receptor, thus preventing the effect of acetylcholine itself. This results in the prevention of end-plate depolarization and no contraction of muscle.

Emergence at the end of the surgical procedure consists of the discontinuation of the anesthetics used during the maintenance of anesthesia. Inhalational agents are removed by ventilation of the lung and from metabolism by the liver. Intravenous agents are either metabolized by the liver or redistributed before they are excreted by the kidney. Most neuromuscular blocking drugs require a pharmacologic reversal agent. The endotracheal tube is removed when the patient is fully awake and demonstrates adequate muscle strength and ventilation. Postoperative care requires the continuation of oxygen supplementation and monitoring of the patient's cardiac and pulmonary status.

## IV. REGIONAL ANESTHESIA

Regional anesthesia can be divided into major conduction blockage (spinal and epidural anesthesia) and peripheral nerve blockade (for superficial operations on the extremities). The mechanism of action consists of the binding of the local anesthetic to the sodium channel of the nerve membrane. As a result, there is no conductance of sodium down the membrane, and no action potential is formed. There are two types of local anesthetics: esters (procaine, chloroprocaine, and tetracaine) and amides (lidocaine, mepivacaine, bupivacaine, and etidocaine). They differ in their metabolism and potential to produce an undesirable allergic reaction. Selection of the local anesthetic can also be based on differences found in their onset as well as duration of action.

A regional anesthetic can be considered when it is desirable or advantageous for the patient to remain awake during the procedure (an example of this is the obstetrical patient in labor). In contrast with a general anesthetic, a regional anesthetic is selective for certain areas of the body, including the lower abdomen and lower extremities. Often a regional anesthetic can be combined with a general anesthetic to minimize hemodynamic fluctuations produced by the general anesthetic and to augment postoperative pain control when the regional anesthetic is continued after the surgical procedure. Contraindications include lack of patient consent, infection, and bleeding disorders. Despite the fear of nerve injury and damage to the spinal cord, the incidence of this is quite rare.

## V. FUTURE DIRECTIONS

Changes in the financing of health care and delivery of this care have drastically altered the environment in which medicine is practiced. Managed care has affected the anesthesiologist as new clinical and administrative roles are being defined. The anesthesiologist's role has expanded into the title of "perioperative physician." For example, formal preoperative evaluation programs have been established to reduce the number of surgical delays and cancellations. In addition to the anesthesiologist's role in the operating room and recovery room, he or she is also directly involved in both the care of patients in the intensive care unit setting and the management of acute and chronic pain. Anesthesiologists also function as managers who oversee policy decisions in scheduling, budgeting, staffing, and administration.

Future advances in anesthesia will include further understanding of the molecular basis of anesthetic agents; the development of new drugs with higher potency, faster recovery, and fewer side effects; the use of computerized data bases for risk stratification and outcomes research; continued development of new medical devices that will enhance faster recovery with minimal injury and lower cost; and a transformation of various practices that will allow greater flexibility in a managed care environment.

## BIBLIOGRAPHY

Alpert, C. C., Conroy, J. M., and Roy, R. C. (1996). Anesthesia and perioperative medicine: A department of anesthesiology changes its name. *Anesthesiology* 84, 723–715.

Barash, P. G., Cullen, B. F., and Stoelting, R. K. (eds.) (1992). "Clinical Anesthesia," 2nd Ed. Lippincott, Philadelphia.

Davison, J. K., Eckhardt, W. F., III, and Perese, D. A. (eds.) (1993). "Clinical Anesthesia Procedures of the Massachusetts General Hospital," 4th Ed. Little, Brown, Boston.

Miller, R. D. (ed.) (1990). "Anesthesia," 3rd Ed. Churchill Livingstone, New York.

Stoelting, R. K., and Miller, R. D. (1994). "Basics of Anesthesia," 3rd Ed. Churchill Livingstone, New York.

# Animal Models of Disease

*Southwest Foundation for Biomedical Research*

Revised by
BARBARA A. JOHNSTON
*Fred Hutchinson Cancer Research Center*

---

I. Introduction
II. Legislation and Policies
III. Alternatives to the Use of Animals
IV. Animal Rights Movement
V. Laboratory Animal Care
VI. Animals as Models for Research
VII. Summary

## GLOSSARY

**Atherosclerosis** Deposition of fatty substances in, and fibrosis of, the inner layer of the arteries

**Hepatitis B** Viral disease often transmitted by injection of infected blood products or by use of contaminated needles

**Immunodeficient** Condition of compromised immunity due to genetic, infectious, or environmental causes

*In vitro* Outside of the living body and in an artificial environment

**Macaques** Old World monkeys belonging to the genus *Macaca*

**Neoplasm** An abnormal growth, such as a cancer

**Oncogenic** Tending to cause cancer

**Pathogenesis** Origination and development of a disease

**Teratogenic** Tending to produce fetal malformation or congenital body defects

**Xenotransplant** Transplant from one species to another

MOST OF TODAY'S MEDICAL AND SURGICAL advances in human and animal medicine have been made possible through research using experimental animals. These research efforts have increased the human life span and have improved the quality of life for both humans and animals. Increasingly alternative methods to the use of animals as disease models are being used, but for the foreseeable future animal experimentation will still be necessary. The majority of animals used in experimentation are bred specifically for research. Institutions are required by law to provide humane care and veterinary treatment to their laboratory animals. In the United States an extensive network of regulations governs the use of laboratory animals. Several well-organized groups oppose the use of animals in experimentation and actively attempt to impede biomedical research by preventing the use of animal models. Some of the major medical advances due to laboratory animals are presented here.

## I. INTRODUCTION

The goal of biomedical research is to improve the welfare of both humans and animals by the application of knowledge gained through experimentation. The majority of the medical advances enjoyed by society have depended on the use of animal models in research, and it is estimated that more than 90% of all new medical knowledge is derived from studies involving experimental animals. In the United States, animal experimentation has contributed to an increase in the average life expectancy of about 25 years since 1900. A partial list of these contributions is shown in Table I.

Alternative methods which replace animals in research are used whenever possible; however, for the

ENCYCLOPEDIA OF HUMAN BIOLOGY, Second Edition, VOLUME 1. Copyright © 1997 by Academic Press. All rights of reproduction in any form reserved.

TABLE I
Medical Achievements Made Using Experimental Animals

Benefits to humans
  Immunization against polio, distemper, diphtheria, mumps, measles, rubella, and smallpox
  Broad-spectrum antibiotics and other anti-infective drugs
  Anesthetics and analgesics
  Blood transfusion
  Bone marrow transplantation
  Intravenous feeding and medication
  Radiation therapy and chemotherapy for cancer
  Diagnostic techniques, such as the electrocardiogram, electro-encephalogram, and angiogram; cardiac catheterization; endoscopy
  Open-heart surgery: coronary bypass, valve replacement, correction of congenital defects
  Surgical treatment for atherosclerosis: to open blocked arteries, repair aneurysms, replace damaged arteries with artificial ones
  Insulin for diabetes management
  Medications to treat asthma and other respiratory diseases
  Medications to control epileptic seizures
  Medications for arthritis management
  Medications to treat colitis, ileitis, ulcers, and other gastrointestinal diseases
  Organ transplantation and drugs to prevent organ rejection
  Kidney dialysis
  Artificial joints
  Diagnosis and treatment of rhesus factor disease and phenylketonuria in newborns
  Medications to treat mental illness
  Cataract removal and artificial lens implants
  Medications to treat hypertension and other cardiovascular diseases
  Microsurgery to reattach severed limbs
  Rehabilitation of stroke and brain-damaged accident victims
Benefits to animals
  Vaccination against distemper, rabies, parvovirus, infectious hepatitis, anthrax, and tetanus
  Treatment for parasites
  Corrective surgery for hip dysplasia in dogs
  Orthopedic surgery and rehabilitation for horses
  Treatment for leukemia and other cancers in pets
  Detection and control of tuberculosis and brucellosis in cattle
  Detection and control of hog cholera
  Control of heartworm infection in dogs
  Improved nutrition for pets
  Treatment of osteoarthritis in dogs

immediate future, animal experimentation will be critical for the advancement of both human and animal medicine.

Animals are used in biomedical research as surrogates for humans in order to better understand the functioning of the healthy body, to determine the effects of disease on the body, and to discover treatments for various maladies. In the United States, ethical considerations usually prevent the use of humans as disease models without first conducting preliminary studies in animals.

According to the Congressional Office of Technology Assessment, 20 million animals are used annually in biomedical research, testing, and education in the United States. Included are mammals, birds, reptiles, amphibians, and fish. This is a small portion of the 5 billion animals used annually for food and clothing, and other purposes. More than 95% of all animals used for experimental purposes are mice and rats bred specifically for research by licensed suppliers. Approximately 120,000 dogs and cats are used each year in biomedical research, just a fraction of the number (15 million) which are euthanized in pounds and shelters. There is considerable controversy regarding the use of pound animals, and certain states and local communities have passed laws preventing their use in biomedical research. Other animals commonly used in biomedical research are rabbits, amphibians, and birds. Farm animals, such as pigs, cattle, sheep, and goats, are often supplied from agricultural sources. Nonhuman primates are largely derived from scientific breeding centers.

A fundamental requirement in using animals in research is selection of the appropriate animal model as part of the experimental design. Scientists must be knowledgeable regarding the physiologic characteristics of the various animal species in order to select the species best suited to serve as a model of the biologic process or disease under study. Laws require that research protocols be reviewed for appropriate model selection before experimental investigation begins.

## II. LEGISLATION AND POLICIES

Federal legislation protecting laboratory animals, commonly known as the Animal Welfare Act (PL89-544), was first enacted in 1966 and was amended in 1970, 1976, and 1985. The purpose of this act is to ensure that animals used in research, exhibition, and the wholesale pet trade receive humane care and adequate veterinary treatment. The act, administered by the U.S. Department of Agriculture (USDA), sets minimum standards for handling, housing, feeding, and watering laboratory animals. It also establishes regular review requirements for animal facilities as a whole and for individual protocols. Animals currently covered under the act include dogs, cats, hamsters, guinea pigs, rabbits, nonhuman primates, and marine mammals. Institutional animal care programs are subject to unannounced on-site inspection by USDA veterinary officers, who are empowered to take legal action against programs not in compliance with federal regulations.

Two laws were enacted in 1985 that apply to the regulation of animals used in research. The Health Research Extension Act of 1985 (PL99-158) applies to all research funded by the United States Public Health Service (USPHS). This legislation transformed into law many of the guidelines contained in the PHS Policy of Humane Care and Use of Laboratory Animals. This legislation requires that:

- Research facilities establish institutional animal care and use committees including at least one veterinarian and one individual not associated with the institution.
- Animal care committees review the care and treatment of animals at least semiannually.
- Scientists and animal care personnel be trained in the proper methods of animal care and use.
- Applicants for National Institutes of Health (NIH) funds file assurances with NIH certifying that investigations adhere to the NIH guidelines.

The second law passed was the Improved Standards for Laboratory Animals Act. This was an amendment to the Animal Welfare Act and was part of the Food Security Act of 1985 (PL99-198). This law has many of the same requirements as PL99-158, with additional requirements as follows.

- The institutional animal care and use committees are required to visit the animal facilities twice per year and report any deficiency to the institution.
- Investigators are required to consider alternatives to animal use.
- The standards to be issued include provisions for the exercise of dogs, an environment adequate for the psychological well-being of nonhuman primates, pre- and postsurgical care, the use of pain relieving drugs, and prohibition of the use of an animal for more than one major surgical procedure.

In addition, the Food and Drug Administration and the Environmental Protection Agency have established good laboratory practices that affect the use and care of laboratory animals.

## III. ALTERNATIVES TO THE USE OF ANIMALS

Alternative methods that do not involve the use of animals can be used in certain research protocols. Concentrated efforts are underway to develop methods to replace the use of animals when feasible, to reduce the number of animals used, and to refine procedures so that pain and suffering are minimized. Some of these methods include *in vitro* techniques using tissue culture, mathematical computer models for the analysis of certain biological systems, and physical and chemical test procedures. Alternative methods also include the replacement of mammalian species with other species, such as fish, amphibians, reptiles, birds, insects, and microorganisms. Scientists use these techniques whenever possible because they are less invasive, less expensive, faster, and more precise. Centers focusing on the development of alternative methods have been established at Johns Hopkins and other universities. Information on alternatives to animal use is available from the Animal Welfare Information Center, Beltsville, Maryland, and the Biological Models and Materials Research Program, Bethesda, Maryland.

Advances in medicine, however, will continue to depend, for the foreseeable future, on experiments using living animals. It is impossible to imitate in any other way the immensely complex, and often poorly understood, system of interactions among different organs and systems that exist in living humans and animals. To understand how such a system functions in a particular set of circumstances, experiments on animals become necessary.

## IV. ANIMAL RIGHTS MOVEMENT

The use of animals as models for the study of human diseases has been contested on various grounds for many years. Contemporary animal activists can be divided into two major groups: (1) individuals concerned with *animal welfare* but not opposed to the use of animals in research, provided that animals are treated humanely and are used only when absolutely necessary, and (2) individuals concerned with *animal rights,* who believe that animals have inherent rights equal to those of humans and who are therefore completely opposed to biomedical research using animals. The term "rights" in our society refers to moral and legal relationships among humans and, strictly speaking, has not been applied to either nonhuman animals or other living organisms. Society does, however, acknowledge that living things have inherent values and that animals in particular require our stewardship and protection. It is generally held that we all have ethical obligations regarding the humane treatment of animals. Many members of the animal rights movement believe that animals should not be used for food, clothing, or the advancement of medical knowledge through experimentation. Extreme activists of the animal rights movement have used terrorist techniques, including the destruction of property, the theft of laboratory animals used in legitimate research projects, and the destruction of data. Opponents of the use of animals in research often do not realize that most of the important medical and surgical advancements in both human and animal health have been made possible through animal experimentation.

Members of the animal rights movement accuse scientists of inflicting undue pain and suffering on laboratory animals. This is certainly contrary to the motives of scientists and veterinarians, who devote many years of schooling and their professional careers to improving animal health. Today there are over 150 well-funded U.S. organizations dedicated to ending the use of animals in biomedical research. In 1990, a conservative estimate of the combined operating budgets of the 10 largest animal rights organizations was $61 million.

The general public should recognize that the aims of the animal rights extremists would severely limit future advances in medicine and that an interest in animal welfare need not preclude supporting the judicious use of animals in biomedical research.

## V. LABORATORY ANIMAL CARE

The scientific community recognizes that it has an ethical responsibility to assure that laboratory animals are treated humanely and that all research subjects receive proper care. In addition to the humanitarian concerns, there are scientific reasons for the provision of proper care and treatment. Animals stressed by inadequate housing, nourishment, or social interaction cannot provide reliable scientific data.

The care and treatment of laboratory animals used in experimentation is a highly developed scientific discipline in the United States. Animal use programs are usually under the supervision of veterinarians with postgraduate specialty training in laboratory animal medicine, and animal experimentation is carried out under their direction. In addition, the technical staff

responsible for the daily care of animals receives extensive training in the husbandry of the various animal species (Fig. 1). Training of scientists and animal care personnel is required by the PHS Policy on the Humane Care and Use of Laboratory Animals.

The general requirements for the proper use of laboratory animals can be found in the National Research Council publication entitled "Guide for the Care and Use of Laboratory Animals." This publication is widely used by research institutions as the basis for establishing standards for their animal care programs. NIH requires that investigators receiving NIH funds adhere to the guide standards. Numerous publications are available on the environmental, nutritional, and clinical care requirements of all commonly used laboratory animals. In addition to the guidelines and standards for proper animal care, veterinarians and scientists must exercise professional judgment in the care and treatment of various species.

A private nonprofit organization known as the Association for Assessment and Accreditation of Laboratory Animal Care (AAALAC) has been established to review and accredit the animal care programs of research institutions worldwide. NIH and other Federal agencies strongly urge that research organizations be accredited by AAALAC. In order to evaluate research facilities, comprehensive on-site reviews by teams of laboratory animal veterinarians and scientists are conducted at regular intervals. High standards have been established, and accreditation by

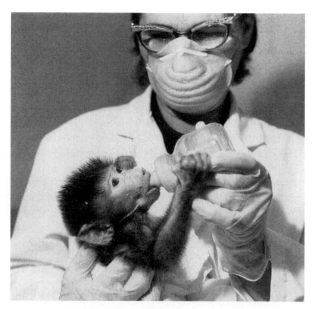

**FIGURE 1**    An animal technician feeding an infant baboon.

AAALAC indicates the availability of excellent animal facilities and an outstanding level of animal husbandry and veterinary care.

## VI. ANIMALS AS MODELS FOR RESEARCH

### A. Infectious Diseases

Experimental animals have been used as models for studying infectious diseases for more than 100 years and have played a major role in the development of drugs, vaccines, and antibiotics for the control of these diseases. Historically, infectious diseases have been responsible for major epidemics, resulting in high mortality among human populations.

The development of vaccines and therapeutic agents for many infectious diseases, such as yellow fever, poliomyelitis, and measles, has depended on the use of experimental animals, primarily nonhuman primates. However, a broad spectrum of animal species have contributed to our greater understanding of infectious disease mechanisms. Spontaneously occurring animal models of viral disease, such as those for hepatitis B virus or AIDS, have been pivotal in revealing information about the disease pathogenesis in humans. Similarly, research on such historically devastating conditions as leprosy or malaria has been dramatically furthered by contributions from unusual research animals.

Hepatitis B virus is a major cause of acute and chronic hepatitis, cirrhosis of the liver, and liver cancer in humans. This hepadnavirus is transmitted primarily by blood products or sexual contact and is a serious health problem worldwide. Woodchucks and Beechy ground squirrels have been useful animal models in hepadnavirus research. Naturally occurring woodchuck hepatitis virus and ground squirrel hepatitis virus are closely related to hepatitis B and induce a similar clinical spectrum in their appropriate hosts. Characterization of these viruses has contributed to understanding the steps in viral replication and expression of viral gene products and has been used to test the efficacy of antiviral compounds. Chimpanzees and other nonhuman primates also have historically played an important role in furthering human antihepatitis B vaccine development.

For centuries humans have attempted to eradicate the protozoan blood parasite, *Plasmodium* spp., which causes malaria. The mosquitoes that transmit this disease have developed insecticide resistance and the parasite has acquired resistance to most anitmalarial therapeutic agents. Vaccine development and testing are currently under way for the prevention of this parasitic disease. Research on Golden hamsters infected with *Plasmodium berghei* has demonstrated physiologic and pathologic alternations similar to humans. [*See* Malaria.]

Leprosy, a chronic infectious disease caused by *Mycobacterium leprae*, is still prevalent in many parts of the world. Researchers since the late 1800s have attempted to identify a suitable animal model for this disease of an estimated 10 million people. Successful experimental transmission of leprosy to hamster ear tissue, mouse footpad tissue, and nine-banded armadillos has allowed for the study of immunity, chemotherapy, and the transmission of leprosy. Because *M. leprae* does not grow *in vitro,* armadillos currently provide the most important source of the leprosy bacillus. Naturally acquired leprosy was first reported in the sooty mangabey monkey in 1981. The sooty mangabey monkey is a very effective model of leprosy as it is the first nonhuman primate to demonstrate high susceptibility to leprosy; it develops both naturally acquired and experimental leprosy; and the disease closely simulates leprosy in humans. Researchers studying the sooty mangabey can monitor clinical, pathologic, and immunologic parameters of the host response to inoculation from the time of infection through advanced disease and posttherapy. Valuable data on the efficacy of antileprosy vaccines under controlled conditions can be obtained through the study of this important animal model.

Lyme borreliosis or "lyme disease" is an infection transmitted by ticks. This disease has been diagnosed in Europe and North America with increasing frequency since the early 1990s. The causative organism is a spirochete, *Borrelia burgdorferi*. Clinically, infection can manifest in multisystem disease, involving the nervous and cardiovascular systems, skin, and joints. Infection of the nervous system by the spirochetes is a serious disease manifestation and is poorly understood in regard to pathogenesis and therapy. Neuroborreliosis research utilizing mouse and hamster animal models is hindered and inconsistent central nervous system infection and inflammation. Rhesus monkeys infected with *B. burgdorferi* are useful models of the human disease as they demonstrate similar clinical and pathological features, including a skin rash, mycoarditis, and central nervous system involvement. Ongoing research of the rhesus monkey model has advanced understanding of the pathogenesis of neuroborroliosis, improved diagnostic techniques, and demonstrated the protective effect of vaccines in preliminary studies [See Lyme Disease.]

Acquired immune deficiency syndrome (AIDS) has reached epidemic proportions since the mid-1980s. Several animal models for this disease have been developed, including naturally occurring simian immunodeficiency virus (SIV), a feline immunodeficiency virus (FIV), and human immunodeficiency virus (HIV) infection of chimpanzees and highly specialized strains of mice. Xenotransplantation of human cells or tissues into genetically tolerant mice allows detailed study of the human response to HIV. [See Acquired Immune Deficiency Syndrome, Virology.]

Animal models of human infectious diseases have been and will continue to be essential for the elucidation of disease pathobiology, treatment, and control. In addition to the examples cited, experimental animals have been useful models for the study of many infectious diseases, including rabies, viral encephalitis, measles, Chagas's disease, dengue fever, herpes virus infections, and Creutzfeldt–Jakob Disease. Our increasingly refined ability to simulate the human immune system through genetic manipulations of immunodeficient rodents will assist in developing ever more faithful models of the human disease condition.

## B. Cancer

Cancer is a complex biological process which includes the development of malignant tumor cells, invasion of various tissues, and eventual death of the host. Cancer occurs in both humans and various animals. Therefore, experimental animals provide excellent models for studying the etiology, development, and treatment of cancer.

A significant proportion of human cancer results from chronic exposure to certain chemicals in our environment, usually after long periods. Animals exposed to various chemical carcinogens develop cancer as humans do, thereby permitting the study of the natural history and biological behavior of the disease. In addition, various *in vitro* techniques using animal tissues are employed to study the complex phenomenon of carcinogenesis.

Mice and rats are frequently used in cancer research, due to the availability of well-established genetic strains that are naturally susceptible to various neoplasms. Immunodeficient mice have allowed the transplantation of cancerous tissues from humans and have proven to be excellent models for studying the biological characteristics of noeplastic tissues and their response to chemotherapeutic agents. The most

fundamental understanding of the role of genes in cancer comes from studies of oncogenic viruses and DNA subsegments in chickens, mice, rats, and primates.

Cancer chemotherapy involves the development of drugs that destroy tumor cells. Animals, are being used to test the effectiveness of each drug against a number of cancers. The body's defense mechanisms against the formation of tumors, which hold much promise for cancer therapy, can be studied only in animals. [See Chemotherapy, Cancer.]

## C. Cardiovascular Disease

Cardiovascular disease is the leading cause of death and a major disabling factor among persons in the United States: Some 1.5 million individuals have heart attacks each year. The cost of cardiovascular disease is estimated at $80 billion annually in the United States alone. Experiments with large animals have been essential for the development of therapeutic procedures for many cardiovascular diseases. These include transplantation of the heart, the treatment of congenital cardiac disorders, the development of artificial blood vessels, the treatment of acute myocardial infarction, and the development of cardiac valve prostheses. Many of these advances can be attributed to experiments using the dog as the animal model.

The major underlying cause of cardiovascular disease is atherosclerosis, a disease that begins in early life. Nearly half of all adult Americans have severe atherosclerotic lesions at the time of death. The development of atheroscelerosclerosis is slow and silent, involving progressive hardening of the arteries. Plaques composed of fat, cholesterol, and calcium accumulate on the inner walls of the arteries, impairing the flow of blood and depriving the heart and other organs of oxygen and nutrients. The deadly effects of atheroscelerosis include heart trouble, stroke, aneurysms, and gangrene of the legs. [See Atherosclerosis.]

Prior to 1960, rabbits, chickens, dogs, and swine were the laboratory animals being used in cardiovascular research. Subsequently, it was found that atherosclerotic plaques are essentially identical in both monkeys and humans, making several species of nonhuman primates the animal models of choice in cardiovascular research. Primates are used to (1) determine the course and progress of atherosclerotic disease; (2) identify the mechanisms of atherosclerotic development at the cellular and molecular levels; and (3) define the influence of risk factors such as hypertension, diabetes, tobacco, alcohol, gender, fats, obesity, and heredity on disease progression, regression, and prevention. Diet-induced atherosclerosis in primate models has shown the importance of dietary lipids and lipoproteins as risk factors and, likewise, studies have shown that diets low in cholesterol can reverse atherosclerosis in monkeys. In other primate studies, by varying the amount of cholesterol in the diet, it is possible to induce atherosclerotic lesions that mimic the various levels of disease progression in humans. The rhesus monkey and the baboon have been used extensively because they have naturally occurring atherosclerotic lesions and a cholesterol metabolism quite similar to that of humans.

## D. Metabolic and Nutritional Disease

Animals are ideal for studying nutritional deficiency diseases in humans. Studies are being conducted on the effects of malnutrition in early life on brain development and the effects of vitamin, mineral, and amino acid deficiencies on growth and metabolism. Rats, guinea pigs, dogs, swine, and nonhuman primates are commonly used for these studies. Diabetes mellitus, resulting from inadequate insulin production, occurs naturally in baboons, squirrel monkeys, and several species of macaque monkeys. It has been extensively studied in the Celebes macaque, *Macaca nigra,* which is an excellent model for studying many clinical aspects of the disease.

## E. Reproductive Disorders

Many species of animals are used as models to study human reproduction. Selection of an animal model depends on whether the research is for testing drug therapy, safety assessment of drugs, or basic research on reproduction. Nonhuman primates have played a major role in reproductive studies, based on similarities to humans to endocrine control of the reproductive cycle, and perinatal development. Although many would conclude that the most similar animal species would always be the best model, this is not always true. Researchers utilize species' similarities and differences in physiology and disease to better understand human reproduction. The complex and dynamic interaction of systemic hormones with multiple organ involvement frequently justifies the use of animal models for reproductive studies over less interdependent *in vitro* systems.

Models used in male reproduction vary depending on the objectives of the research. Three areas that

exemplify the use of animal models would be spermatogenesis, male contraception, and prostatic problems. When studying spermatogenesis, monkeys and humans have similar complex patchwork of spermatozoa maturation in the seminiferous epithelium. Although primates are good for assessing the testicular effects of many compounds, researchers will often use rats as models. Rats not only are far less expensive but also have more discrete zones of spermatozoal maturation, making it easier to study toxicity and/or cellular physiology. Animal models dealing with vasectomies, the most common form of male contraception, are widespread. Following vasectomy, different species (including man) may develop secondary problems. Rats frequently form sperm granulomas at the surgery site and rhesus monkeys often have marked dilations of the efferent ductules postsurgery. The choice of animal models would again depend on the objectives of the study. The third example of male reproductive models is prostatic hyperplasia, common in aged men and intact male dogs. Although prostatic enlargement in dogs is predominantly a glandular proliferation, humans have predominantly a muscular–stromal proliferation. Dogs are still widely used as models because they are readily available, develop the lesion spontaneously, and share several features of the human condition.

Models of female reproduction are also varied. Three examples of conditions using animal models include contraception, *in vitro* fertilization, and endometriosis. Much of the work on contraceptive drugs uses rhesus monkeys because they have similar endocrine cycles. Although gorillas and chimpanzees have a more similar reproductive tract to humans, use of endangered species as animal models is rare. *In vitro* fertilization studies frequently use nonhuman primates; however, much of the basic research in this area uses rats, mice, and hamsters. Endometriosis is another example of a reproductive condition in humans with animal models. Rhesus monkeys have spontaneously occurring endometriosis and are used as a model to study predisposing causes, therapy, and detection methods. Endometriosis does not naturally occur in rodents, and induced models in rats have had limited application in research. [*See* In Vitro Fertilization.]

For studies involving fetal development, the more common animal models include primates and sheep. Although nonhuman primates share many characteristics of fetal development and endocrine control of pregnancy with humans, surgical manipulation of the uterus in these animals can result in abortion. Models using chronically catheterized fetal sheep have been used to study maternal–fetal interactions, fetal lung development, initiation of parturition, and abortions. Nonhuman primates are used for studies of fetal stress and fetal respiratory disorders. Many species of animals are used in research on developmental biology and teratology. By understanding basic mechanisms of normal development, we gain insight into how mistakes in embryology occur (birth defects). Similarly, animal models allow us the opportunity to develop therapies and corrective surgical procedures on the developing fetus. The screening of compounds for teratogenic effects is part of a battery of safety tests. These studies are extremely important for drugs intended for use in females of childbearing age. Thalidomide is an example of a human teratogen that had devastating effects in the 1950s. The compound was initially tested on animal species that were not susceptible to its teratogenic effects. Nonhuman primates and several strains of rabbit are the only models with similar teratogenic effects to humans. This example highlights the dangers inherent in the extrapolation of data from animal models to humans.

Additional examples of the use of a wide variety of reproductive (and nonreproductive) animal models can be further studied in a series of bulletins produced by the Armed Forces Institute of Pathology.

## F. Behavioral Modification

Human behavior is extremely complex and is influenced by many social, environmental, and genetic factors. Various aspects of human behavior can be studied only with animal models that demonstrate similar behavioral characteristics. The rat is commonly used because of its size and adaptability and because it displays certain neurological conditions similar to those of humans. The dog is a highly social animal with a rich behavioral repertoire and is used in genetic, developmental, neurological, and psychopharmacological studies. Nonhuman primates are an ideal animal model for behavioral studies because they have a large brain, a relatively long period of development, and a prolonged dependence on others. In addition, they display highly complex behavior, which is modified by learning and social factors. Behavioral studies with primates include social organization, mother–infant interactions, aggression, growth and development, puberty, communication, and learning.

Nonhuman primates are used as models to study the physiological and chemical mechanisms in the brain's response to drugs. New pharmacological agents designed to treat human behavioral disorders can be screened in nonhuman primates. Thus chimpanzees and monkeys, which readily become addicted to alcohol, are used to study certain aspects of alcohol addiction.

Nonhuman primates are also useful in the area of communication because monkeys can transmit information by vocalization and great apes have been able to learn and communicate with American sign language and a computer-operated keyboard of word symbols that represent objects. These studies have enabled scientists to develop and evaluate language systems used by handicapped humans.

## G. Aging

Aging can be defined as the sum total of progressive changes in clinical, biological, and behavioral processes with the passage of time. The aging process is influenced by heredity, sex, medical history, nutrition, and environmental conditions. Aging research involves long-term studies that can best be carried out in experimental animals. Rats and mice are commonly used, as they have a short life span and are easy to maintain in the laboratory. In germ-free rats, the aging process can be studied in the absence of potentially distorting disease episodes. Studies in this species have helped define nutritional components of longevity.

Monkeys allow the study of characteristics of aging unique to long-lived human and nonhuman primates. With a life span about 10 times that of mice, but still only one-third that of humans, nonhuman primates experience many of the same age-related changes in anatomy, physiology, and mental functions as humans. Macaques, such as rhesus monkeys and pig-tailed macaques, are used to study the influence of dietary restriction, antioxidants, and other factors on the rate of aging in long-lived species.

Because some of the most marked differences between the primates and shorter-lived species have to do with the brain, reproduction, and social behavior, many studies of aging in nonhuman primates parallel those in humans, making these animals excellent models for studying the intellectual and social aspects of aging.

The use of nonhuman primates in aging research is limited by the availability of monkeys and apes that can be classified as "old." Several primate research centers involved in longitudinal aging research projects have colonies of these animals; it is hoped that they will shed light on the aging process in humans.

## H. Surgery

The majority of surgical procedures used in human medicine today have been developed and refined using animal models. Some examples in the field of orthopedic surgery include the use of dogs in the development of bone plating and pinning techniques to treat fractures. Much of the research leading to the implantation of artificial hips and knees into people was also done in dogs. This research has also resulted in hip replacement becoming a clinical option for treating the many pet dogs affected with hip dysplasia. Dogs were also used to develop techniques for repairing torn anterior ligaments, frequently needed by athletes.

Heart diseases are a significant cause of death in people, and animal experimentation has produced a variety of treatments that are resulting in increased vitality and prolonged life. Thousands of people undergo bypass surgery every year to alleviate the problems caused by blocked vessels in their limbs or coronary arteries in their heart. These procedures were developed in animals. Artifical heart implantation and heart transplantation are currently being developed by research performed on cattle. When defective or diseased heart valves must be replaced, pig valves or metal valves perfected in animal research are used.

Unsightly scars and many of the complications following surgery, such as infection, pain, and bleeding, are associated with the incision that is made to expose the tissue to be repaired or removed. Research using pigs and other animals has opened a new type of surgery, referred to as "minimally invasive surgery." This is a term used to describe new methods of performing surgery without making a large incision. Instead, small instruments are introduced through the skin which allow the surgeon to visualize the tissue to perform the operation. Surgery using such techniques in joints, such as the knee, is called arthroscopy and is called laparoscopy in the abdomen. Research is underway to extend this technology to other operations, including heart and spinal cord surgery.

Research in a variety of animal species, including rats, pigs, dogs, sheep, and cattle, has resulted in the development of materials that are well tolerated as implants into the body. This has allowed the development of dental implants, artificial joints, metal bone plates, vascular grafts, and many artificial organs. Studies in sheep led to the development of catheters that can be implanted through the skin and into blood vessels to allow for the infusion of chemotherapeutic agents into cancer and AIDS patients, into the peritoneal cavity to facilitate dialysis of patients with kidney disease, and into the spinal cord for infusion of pain-relieving drugs.

With an increasing demand for organs to transplant, research is in progress to develop xenotransplantation, the transplantation of organs from other species, as an option. Such tissues are almost always recognized as foreign by the recipient and rejected. However, through genetic engineering, it may be possible to grow organs and tissues in animals that would be compatible when implanted into people.

## VII. SUMMARY

The use of animals for research on the cause, prevention, and cure of both animal and human diseases is essential to assure continuing improvement in health care. Federal regulations governing the use of experimental animals ensure that laboratory animals are receiving humane care and treatment. The curtailment of animal experimentation would seriously hinder future advancements in both human and animal medicine.

## BIBLIOGRAPHY

Animal and Plant Health and Inspection Service, U.S. Department of Agriculture. (1996). "Animal Welfare Enforcement, Fiscal Year 1995: Report of the Secretary of Agriculture to the President of the Senate and the Speaker of the House of Representatives." USDA Office of Communications, Washington, D.C.
Armed Forces Institute of Pathology. (1970 to present). "Comparative Pathology Bulletin." Washington, D.C.
National Institutes of Health. (1996). "Guide for the Care and Use of Laboratory Animals." National Academy Press, Washington, D.C.
National Institutes of Health. (reprinted 1996). "Public Health Service Policy on Humane Care and Use of Laboratory Animals." National Institutes of Health, Rockville, MD.
National Research Council. (1995). "Nutrient Requirements of Laboratory Animals," 4th revised Edition. National Academy Press, Washington, D.C.
Olfert, E. D., Cross, B. M., McWilliam, A. A. (eds.) (1993). "Guide to the Care and Use of Experimental Animals," 2nd Ed. Canadian Council on Animal Care, Ottawa, Ontario, Canada.

# Animal Parasites

IRWIN W. SHERMAN
*University of California, Riverside*

## GLOSSARY

**Commensalism** Association of two species of organisms, in which one member benefits, while the other neither benefits nor is harmed

**Host** Organism in or on which the parasite lives

**Parasite** An organism that benefits by living in or on another species of organism, the host, and does so at the expense of the host; the host is usually not killed

**Predator** An animal larger than its prey, which it kills outright

**Symbiosis** Close association of two organisms of different species. If the association is beneficial, it is termed mutualism, whereas if one organism benefits at the other's expense, it is termed parasitism

PARASITISM CAN BE DEFINED AS THE INTIMATE relationship between a parasite and its living host; the parasite lives at the expense of the host and may cause harm, damage, or kill its host. When a parasite lives on the surface of its host, it is called ectoparasitism, whereas, if internal, the condition is called endoparasitism.

## I. ECOLOGY OF PARASITISM

Each of us is a biological island, serving as a habitat for millions of other creatures. If, like Alice in Wonderland, you could be reduced in size to about the height of the letter "i" on this page, and then were able to wander on the inside and the outside of a normal-sized human, a grand panorama of life would appear. Suppose an "i"-sized you were placed on the skin surface of a normal-sized person, what kinds of things would be seen? The epidermal cells paving the skin surface would appear like flat translucent stones; among these would stand hairs like tree trunks and the openings of the oil and sweat glands. Grouped around the hairs and the gland openings would be clusters of iridescent beads closely resembling caviar. There would also be bundles of soft translucent sticks lying about—growing, dividing, and forming chains and clusters. The "caviar" and "sticks" on the skin surface—cocci and bacilli, respectively—are bacteria that are normal inhabitants of our skin. Indeed, they are so numerous that if it were possible to make an aerial survey of the skin, it would resemble an air view of the crowds seen at Coney Island Beach on a sunny Sunday in August. Our skin is not unique. If we wandered into the mouth and took a peek at the fleshy crevices between the teeth we would see spirochetes and vibrios in furious movement, and corkscrew-like spirilla would glide back and forth. No matter where we travel in the body—skin, gut, genital system, lungs, liver, blood, and so on—a rich array of foreign life would be encountered. In some cases, we would find the "guests" to be more than just tiny microbes. Lice in the hair, worms in the gut and liver, and protozoa in the blood are a few of the resident aliens we might encounter. These foreign residents in and on our body make up small communities, and each member competes with the others for food and habitat.

"Symbiosis" is the general term given to an association between two different kinds of organisms; the word literally means "living together." There is a

ENCYCLOPEDIA OF HUMAN BIOLOGY, Second Edition, VOLUME I.   Copyright © 1997 by Academic Press.   All rights of reproduction in any form reserved.

whole range of associations that can be termed symbiotic, but generally we recognize three: mutualism, commensalism, and parasitism.

In mutualism, both species involved in the association derive some benefit from it. For example, the bacteria that live in our large intestine provide us with large quantities of B vitamins, so much so that we rarely need them in our diet. The intestine provides the microbes with darkness, moisture, warmth, and a rich supply of nutrients.

Commensalism involves an association in which one member derives benefit, while the other receives neither benefit nor harm. This association has been described by one biologist (Pierre van Beneden in "Animal Parasites and Messmates," 1876) in more nautical terms: "A commensal is a mess-mate who lives aboard the vessel of its host, partaking of the excess provisions. . . ." In a sense, a commensal lives on the crumbs from a rich man's table. *Entamoeba coli,* a common intestinal protozoan (ameba), is an example of commensalism between a human and a protozoan. The *E. coli* feed on the bacteria in the gut, but do us no significant harm.

There are no sharp boundaries between commensalism and mutualism; likewise, these associations grade into parasitism. What exactly is a parasite? A parasite is an organism that lives on or in another organism, known as the host, and at the host's expense. The biologist Pierre Van Beneden described the parasite as one whose profession it is to live at the expense of his host, partaking of his superfluities and practicing the precept not to kill the goose to get the eggs.

The word "parasite" in everyday usage has a negative connotation. Nobody likes to be called a parasite; it means "a lazy good-for-nothing." Yet, in an ecological sense, parasitism is just another way by which organisms make a living, and it is inappropriate to apply value judgments, designating some parasites as "good" and others as "bad." It is true that the environment of the parasite (the host) is alive and that the parasite may prosper as the host is harmed, but very often a balance between the species is struck such that the host survives without excessive damage. Indeed, it is a distinct disadvantage for a parasite to kill its host with any regularity, for destruction of the host species very often leads to extinction of the parasite.

The parasite and the host are in a state of dynamic equilibrium, achieved and balanced by a matching of host resistance and parasite pathogenicity. Most commonly, balance is encountered in long-established host–parasite relationships. Thus, the longer the association of the parasite with the host, the less severe the harm inflicted. However, there are occasions when the equilibrium between the host and its parasite can be disturbed, and one may be favored at the expense of the other; if the parasite is favored, disease develops.[1]

Parasites probably evolved from free-living organisms, many of which began as commensals, living on the body surface; only later did they invade the body proper. Selective pressures have promoted a strong host preference for some parasites, and there is a considerable degree of host specificity. Thus, the tapeworm living in the intestine of a rat cannot invade or develop in a dog. Parasites are not degenerate organisms; rather, during the course of evolution, they have developed special adaptations which have enabled them to survive and propagate their kind. Great reproductive capacity, reduced sensory organs, elaborate attachment devices such as suckers and hooks, and special glands for penetration are all well developed in multicellular animal parasites.

Parasites do not comprise a single (taxonomic) group, rather they are quite varied: Some are bacteria, fungi, or one-celled animals (protozoa), while others are roundworms or flatworms. All viruses are parasitic.

## II. THIS WORMY WORLD

Some years ago, the retiring presidential address for the Society of Parasitologists was delivered by Norman R. Stoll, entitled "This Wormy World." Stoll estimated the number of worm diseases in humans and concluded that for every man, woman, and child living on the face of the Earth there was at least one worm-caused case of disease. If, in some part of the world, an individual was not so afflicted, then it meant that another person harbored at least two or more different parasitic worm species. What was true at the time Stoll spoke still applies today, although we rarely give it notice. The insidious and chronic nature of worm diseases does not have the journalistic value of viral or bacterial epidemics, and so these diseases often go unnoticed by many of the developed coun-

---

[1] "Disease" literally means "without comfort" and signifies some deviation from normal function. Disease may be due to nutritional deficiency, cancer, organic failure due to congenital malfunction, aging, poison, or invasion by foreign agents, such as viruses, bacteria, or animal parasites.

tries of the world. Yet they persist and produce untold human agony.

## A. Snail Fever and Blood Flukes

In 1972, Chairman Mao Tse-tung wrote a poem called "Farewell to the God of Plague":

> Green streams, blue hills—but all to what avail?
> This tiny germ left even Hua To powerless;
> Weeds choked hundreds of villages, men wasted away;
> Thousands of households dwindled, phantoms sang with glee.
> On earth I travel eighty thousand li a day
> Ranging the sky I see myriad rivers.
> Should the cowherd ask tidings of the God of Plague,
> Say: Past joys and woe have vanished with the waves.
> The spring wind blows amid ten thousand willow branches,
> Six hundred million in this sacred land all equal Yao and Shun.
> Flowers falling like crimson rain swirl in waves at will,
> Green moutains turn to bridges at our wish;
> Gleaming mattocks fall on heaven-high peaks;
> Mighty arms move rivers, rock the earth.
> We ask the God of Plague: Where are you bound?
> Paper barges aflame and candlelight illuminate the sky.

Mao's poem refers to snail fever, a parasitic disease related to the use of human excrement as fertilizer, a most serious health problem in both China and Egypt. Snail fever or schistosomiasis (also called bilharziasis) is caused by the blood fluke *Schistosoma,* a flatworm (Platyhelminthes) that inhabits the blood vessels near the intestines, liver, and bladder. Estimates of the number of infected individuals run to more than 200 million humans. Indeed, from all indications, blood fluke disease is on the rise: It rages the entire length of the Yangtze and Nile rivers, and its distribution is worldwide. In humans, it produces obstruction and rupturing of blood vessels, damaging the liver, small intestine, rectum, and urinary bladder.

The schistosomes live in the blood vessels close to the bladder and the small intestine. The female worm deposits hundreds to thousands of eggs each day in the blood vessels, and, by an inflammatory response, these are discharged into the surrounding tissues and are ultimately evacuated with the feces or urine. In the process, tissue damage is extensive, the liver becomes cirrhotic, blood is found in the stools and urine, abdominal pain is severe, the abdomen becomes bloated, there is dilation of superficial veins, and dysentery is quite common. Loss of appetite, emaciation, and death from exhaustion or other infections may result.

The eggs laid by the females contain embryos. When these are evacuated with the feces or urine and the egg comes into contact with water, it hatches, releasing a ciliated larva, the miracidium. This swims about for a time and then penetrates the soft tissues of a suitable snail host. As the miracidium penetrates the body of the snail, it sheds its outer garment of cilia, becoming a sac-like sporocyst. The sporocyst is literally a reproductive machine, whose sole function is the daily production of thousands of fork-tailed cercaria. From a single miracidium that penetrates a snail, up to 10,000 cercaria can be produced. The cercaria leave the snail, swim free in the water by lashing their tails to and fro, and wait in readiness for a human. Upon contact with the exposed human skin, the cercaria penetrate the skin and the tail drops off; the immature schisotosomes enter the bloodstream, migrate in 1 day to the lungs, and after 2 weeks are found in the blood vessels, where they mature. Schistosomes are of separate sexes, and the male "hugs" the female in the special female-carrying groove. As far as we know, they remain wedded for life, living 5 years and infrequently up to 30 years, producing large numbers of embryo-containing eggs. Schistosomiasis is not a new disease of humans. Eggs have been recovered from Egyptian mummies over 5000 years old. The problem persists in Egypt.

Antimony-containing drugs can be used as a treatment only in the earliest stages of infection, but there are severe side effects. Praziquantel is an effective drug, relatively free of side effects, but treatment is expensive. Eradication of snail fever could be effected most easily by controlled sterilization of all human excrement before it is deposited in the waters or used as fertilizer. Yet, in countries where hygienic practices are virtually unknown, this is a major obstacle. Attempts to destroy snails have only been partially successful. No immunization has as yet been developed. Sanitation and avoidance of skin contact with cercaria-infested waters are simple measures for breaking the transmission cycle of schistosomiasis, but these are still to be accomplished.

## B. Hookworm Anemia

Earlier than 2000 BC the disease caused by hookworm was recognized by the Egyptians. The ancient Chinese were also familiar with the condition and called it "the able to eat, but lazy to work, yellow disease." The Greek physician Hippocrates described the hookworm condition: "yellowing of the skin, intestinal disturbance and a desire to eat dirt."

Today, hookworm disease remains, afflicting more than 630 million people worldwide. The total blood lost from victims of the disease amounts to the blood volume of 1.5 million people. As a consequence, the disease signs and symptoms are those of a severe iron-deficiency anemia: pallor, edema, dullness, listlessness, and sapped vitality, and in children there may be profound retardation of mental and physical development. The "shiftless, lazy, good-for-nothing" white poor of the southern United States, who received much scorn in years gone by, were not genetically inferior, but were the unfortunate victims of hookworm disease. Hookworm was a major disease in the United States in the early 1900s, but due to an eradication program begun in 1910 under the Rockefeller Foundation, it is virtually unknown in this country today.

How can an eradication program be so successful in so short a time? The answer lies in understanding how humans become infected. Indeed, the life cycle of this parasitic worm is simple and direct. The adult roundworms (Nematoda) live in the small intestine. By means of their cutting teeth and a pharyngeal sucking bulb, they attach to the wall of the intestine, where they are nourished by ingested mucosal tissue as well as blood. Blood pours through their slender bodies at a fantastic rate, and a single worm can "drink" about 0.25 ml/day. Because blood is also lost by hemorrhages at the site of attachment, the amount of blood loss from the intestine is considerable. Heavy infection in an infant may be fatal. Each female may produce 10,000 embryonated eggs per day. These embryos, enclosed in the eggshell, pass out of the intestine with feces and the eggs hatch in the moist soil, liberating a young infective larva. The larva, when it comes into contact with the human skin (e.g., by being stepped on by bare feet), penetrates the skin, burrows until it reaches a blood vessel, is carried to the lungs, and then is coughed up, swallowed, and ultimately reaches the intestines. The travel period is 3–5 days, and the adult worm may live for up to 14 years, although most survive only 10 years.

Hookworm disease is principally found in warm moist climates, where the soils are sand–loam and where agricultural practices involve the use of "night soil" (i.e., feces) for fertilizer, or where sanitation is low. The reason for the distribution of hookworm in moist climates is related to the fact that the eggs are very sensitive to low temperature and to desiccation.

Hookworm disease can be treated with tetrachloroethylene or levamisole, albendazole, and pyrantel. However, control measures involve the elimination of skin contact with contaminated soil, and this involves the wearing of shoes, sanitation, and the treatment of feces. Since protein deficiency contributes to the severity of the disease, as well as in aiding the victim to ward off a severe infection, it is necessary that control measures are implemented with improved nutrition. The irony of hookworm disease is that it afflicts those who can least bear its burden—the poor, the malnourished, and the inhabitants of underdeveloped nations, where agricultural practices require the use of night soil and where sanitation is virtually unknown.

## C. Tapeworms

Tapeworms, or cestodes, are flattened, ribbon-like, creamy-white flatworms (Platyhelminthes) that most commonly occupy the intestinal tract. They are reported in ancient descriptions from Egypt and Greece. The adult tapeworm consists of a scolex, or head, which contains hooks and/or suckers for attachment, a neck region just behind this, and a long series of proglottids (i.e., segments). The proglottids contain the reproductive organs and serve as a reservoir for eggs. A tapeworm is an absorptive reproductive machine, taking in nutrients across its body surface (they lack a digestive tract) and fabricating eggs.

Beef and pork tapeworms are common parasites of humans. In all cases, humans become infected by eating meat containing a larval stage called the "bladder worm." The bladder worm consists of an invaginated scolex, a short neck, and a fluid-filled capsule. When infected meat is eaten, the head evaginates from the bladder and attaches to the intestinal wall. In 3 months, an adult worm is formed and eggs are discharged in the feces. In order for the life cycle to be completed, the eggs must be ingested by cattle or hogs, in which they hatch in the duodenum and embryos migrate from the intestine through the blood, reaching the muscles, where, in 2–3 months, they are transformed into bladder worms.

Although tapeworms may range in size from 15 feet to a record length of 75 feet, they often cause no digestive impairment, except for the inconvenience of having proglottids shed continuously from the anus. However, there can be diarrhea, pain, weight loss, anemia, and intestinal obstruction.

The control of tapeworms involves the sanitary disposal of feces and the proper treatment of meat before eating it, either by thorough cooking or freezing to kill off the encysted bladder worms. Persons infected with the tapeworm can be freed of the worm by drug treatment, using Yomesan (niclosamide).

## D. Trichinosis

The worm diseases thus far considered are spread by fecal contamination—eggs are distributed to the soil, and these are directly or indirectly transferred to humans. Trichinosis, in contrast, is spread not by feces, but by flesh. The adult trichina roundworm lives in the intestine, and after the mature worms copulate, the female gives birth to live young. The young worms penetrate the intestinal wall, enter the bloodstream, and are carried to the striated muscles, where they coil up and become encapsulated. The entry of the young worms into the blood and their subsequent migration cause inflammatory reactions, edema, and possible toxic reactions.

In nature, trichinosis is ordinarily propagated by cannibalism in brown and black rats. Rats containing encapsulated larvae are eaten by hogs, and these then become a source of infection for humans. By eating infected pork that has not been adequately cooked, humans become victims of the trichina worm. In humans, the cycle ends, unless human flesh is fed upon, but by this time the damage by migrating larvae will have been done. Probably 4% of the adult population in the United States is infected with trichinosis. Control is easily effected by freezing and cooking meat, preventing swine from eating infected sources, and examining pork for encapsulated trichina. There is no treatment for trichinosis.

Trichina is the exception to the rule that parasites are specific for a particular host. Trichinosis occurs in humans, mice, cats, dogs, pigs, bears, whales, walruses, and almost all other carnivorous mammals. [See Trichinosis.]

## E. River Blindness

A familiar scene in parts of western Africa is a young child leading a blind adult man to the field where the crops are to be tended. This is literally a case of the blind being led by those who are to become blind. The condition is river blindness, or onchocerciasis which also occurs in Guatemala and other parts of Central and South America.

River blindness, which affects up to half a million people a year, is transmitted from human to human by the biting blackfly, *Simulium damnosum,* which requires high humidity and streamside vegetation and whose larvae require clear fast-running streams for maturation. When the female blackfly takes a blood meal, she also ingests infective nematode larvae (called

microfilariae) from the skin of the infected human. These microfilariae migrate from the gut of the fly to the thoracic muscles, where during the next month, they undergo several molts to become infective third-stage larvae. At her next blood feeding, the infective larvae emerge from the mouthparts of the blackfly, crawl on the skin, and, through their own activity, enter the wound. The larvae then molt into adults, which remain localized in the skin. The groups of knotted thread-like adults become encapsulated by a host reaction, and while they are able to live and copulate, they are unable to leave the skin. The adult female worms are viviparous, i.e., they give birth to millions of live, unsheathed, microscopic microfilariae which spread out in the skin. When the microfilariae reach the infected person's head, they frequently invade the eye, and it is the accumulation of dead parasites that causes the lesions that ultimately result in blindness. Blindness in humans is not the result of a single exposure, but is due to the cumulative effects of unremitting reexposure as well as a severe inflammatory response on the part of the host.

The control of river blindness involves treatment of diseased humans with the drug ivermectin. This drug, which acts on $\gamma$-aminobutyric acid (GABA) and cholinergic receptors at the neuromuscular junctions of the microfilaria, leads to microfilarial paralysis and death. Since ivermectin does not readily pass the mammalian blood–brain barrier and GABA synapses in mammals are confined to the central nervous system, the drug has little adverse effect on the host. In addition, control of the blackfly vector is being undertaken using the biocontrol agent *Bacillus thuringiensis* and/ or larvicides, such as Abate or chlorphoxim.

## III. VAN LEEUWENHOEK'S ANIMALCULES AS PARASITES

### A. Giardiasis

It was November 4, 1681, and with considerable excitement a letter was written: ". . . animalcules a-running very prettily; some of 'em a bit bigger, others a bit less than a blood globule; but all of one and the same make. Their bodies were somewhat longer than broad, and their belly, which was flatlike furnisht with sundry little paws, wherewith they made a stir in the clear medium . . . and albeit they made a quick motion with their paws, yet for all that they made but slow progress." It was signed Anton van Leeuwenhoek. By training his homemade microscope on his own excrement, van Leeuwenhoek discovered a new ecosystem and the first "animalcule" parasite of humans. van Leeuwenhoek had suffered from a mild case of diarrhea and, by looking at a sample of his own feces, found the agent responsible for his discomfort—a one-celled protozoan called *Giardia lamblia;* the "paws" of *Giardia* were its flagella used in swimming.

*Giardia* is still with us. In 1973, the girl's drill team from a U.S. high school visited Russia for 5 days; half the time was spent in Leningrad and half in Moscow. Of the 215 people in the group, 199 had gastrointestinal illness, and over half of these were diagnosed as having *Giardia.* Tap water contaminated with cysts of *Giardia* is probably the main source of infection, but it is not limited to the former Soviet Union. Epidemics have been recorded in Colorado, Utah, Oregon, Washington, and New York. *Giardia* resembles a horseshoe crab in general contour and has eight flagella. Each *Giardia* has an adhesive disk with which it attaches to the microvilli of the small intestine, and it is the attachment and detachment of the disk that probably cause increased peristalsis, inflammation, and diarrhea. When about to leave the small intestine, *Giardia* rounds up, pulls in its flagella, secretes a resistant cyst wall, and passes out with the feces. The cyst wall is highly resistant to chlorination and drying, and therefore the disease is spread when cysts are ingested with drinking water. Another source is cysts on the hands, which are spread by hand-to-mouth contact. The treatment of water by boiling or with iodine is an effective preventive measure. Metronidazole is the current drug of choice.

In the more than 300 years since van Leeuwenhoek saw a parasitic organism too small to be seen with the naked eye, we have become aware of hundreds of other such disease-producing organisms. These unicellular animals can occupy virtually every watery recess of our body—blood, urogenital spaces, the alimentary tract, and so on. Unlike parasitic worms, protozoa reproduce themselves within the body, producing large numbers of progeny and often debilitating the host.

## B. Amebic Dysentery

Most students of biology are familiar with the free-living freshwater *Amoeba.* Fortunately, only a few of us have encountered its parasitic cousin, *Entamoeba histolytica,* the causative agent of amebic dysentery. Worldwide, however, up to 400 million people harbor this ameba but more than 80% are carriers, showing no outward symptoms of the parasite. Like the worm diseases, hookworm and schistosomiasis, amebic dysentery is found mainly in areas where the climate is tropical and the sanitary conditions are poor.

The life cycle of *Entamoeba* is simple and direct. The ameba lives in the large intestine, and, by means of its pseudopods (i.e., false feet), it engulfs bacteria or dead cells. For some unknown reason, these feeding amebae become round in shape, stop moving and feeding, and secrete a cyst wall. The encysted amebae pass out of the intestine with the feces. Such feces are deposited on the soil or in the water and contaminate food; the amebae enter a new host through contaminated food and drink. Once in the intestine, the cyst wall breaks down and the active feeding amebae emerge.

The feeding amebae eat not only bacteria and dead cells, but they penetrate deep within the crypts of the large intestine, and by the secretion of digestive enzymes they begin to devour the lining of the intestine—thus, the name "*histolytica,*" or tissue destroyer. Local damage produces small ulcers, and bleeding often accompanies these carnivorous amebae. The ulcerated area also provides a place where bacteria can enter the body to set up a bacterial infection. Some of the ulcers heal, but the repair involves the laying down of scar tissue, which interferes with normal absorptive processes. The individual suffers from abdominal discomfort, diarrhea, bloody stools, and emaciation.

From the surface of the intestines, the amebae may erode their way into the blood vessels and lymph channels and in this way can be transported to the liver and other organs, such as the lung and the brain. Here, the invasive amebae continue to digest and destroy the tissues, producing large amebic abscesses. Symptoms usually include fever, abdominal tenderness, and enlargement of the liver; if the lung is affected, the symptoms may resemble those of bronchitis or pneumonia, and if abscesses are severe or localized in the brain, death may result.

Treatment of the intestinal phase of amebic dysentery and liver abscess involves treatment with metronidazole. Control of the spread of amebae is of considerable public health importance since, in any large community, many carriers exist without apparent disease symptoms. Control measures include the careful inspection of food handlers and the restricted spread of contaminated feces in the water or soil and by flies.

## C. Malaria

Amebic dysentery involves the intestinal ecosystem and is spread by fecal contamination. In contrast, malaria is a disease of another ecosystem and has another mode of transmission. Almost 2500 years ago, Hippocrates associated the fever disease malaria with swamps and stagnant water, where the air was bad, and thus the name given to the disease was "bad air," or malaria. The causative agent of the disease turned out not to be the air itself, but rather a small one-celled animal called *Plasmodium*.

In 1880, Alphonse Laveran, a physician in Algeria, while studying the blood of a patient with a fever, found a small patch of pigment in a globule that was within a red blood cell. He had discovered the malarial parasite. Within the red blood cell, the parasite grows and multiplies, ultimately destroying the red blood cells that provide its nourishment. Every cycle of parasite multiplication (48–72 hr) sees the destruction of red blood cells and the liberation of a new brood of parasites to begin the red blood cell invasion anew. At this point, there is marked fever and chill in the patient. As the parasites continue to destroy blood cells faster than the host can manufacture them, there is severe anemia, emaciation, debilitation, and death. Today, there are more than 250 million cases of malaria, more than 100 million in India alone. Each year, 2 million people die from malaria, and it has been estimated that over $2 billion annually is lost from productive labor due to malaria. Today, malaria ranks as one of the major killers of mankind.

But, if malaria is a disease of the blood, what does it have to do with "bad air?" In 1898, Sir Ronald Ross, an army surgeon in India, found that the blood forms of malaria were derived from the bite of an infected mosquito. The mosquito, Ross showed, contained infective stages of the parasite, sporozoites, in its salivary glands, and when these were injected into the bloodstream during feeding, the human became infected. Now the connection between the bad air of swamps, marshes, or stagnant waters and malaria became clear, for these damp regions were the breeding grounds for the mosquitos that transmitted the disease. Thus, mosquitos picked up infected blood while feeding, became infected, and could then infect a human by reinoculation of the sporozoites during a subsequent blood meal.

A strange aspect in the life cycle of the malarial parasite was that, when sporozoites were inoculated into a human, they did not directly invade the red blood cells. Indeed, it took days before parasites were evident in the blood. Where did the sporozoites go? It took until 1948 (68 years after the human parasite was first described) to find a third cycle of the malarial parasite. In this cycle, sporozoites first invaded the liver cells in which they multiplied and then they proceeded to enter the blood cells. Thus, in malaria, there is a mosquito cycle; an erythrocytic, or red blood cell, cycle; and a preerythrocytic, or liver, cycle. Once the life cycle was understood, work could begin on breaking the vicious cycle of disease transmission. Mosquito breeding sites were destroyed, mosquitoes were killed by insecticides, notably DDT (i.e., dichlorodiphenyltrichloroethane, now banned in the United States) and dieldrin, and humans were treated with antimalarial drugs, such as quinine, atebrine, chloroquine, and primaquine phosphate.

A major program to eradicate malaria began in 1945 and, by the extensive use of DDT and antimalarial drugs, the incidence dropped. Despite such a marked reduction of malaria in many regions, the disease is not yet eradicated. Indeed, today control is talked of more than eradication. Why? Mosquitoes have shown DDT resistance, caused by genetic changes, which makes them capable of surviving the spraying of DDT and continuing to transmit the disease. Chloroquine, once an effective drug, no longer kills malarial parasites since these too have undergone genetic changes that make them resistant. No effective vaccine exists against malaria, but its preparation is the subject of intensive research. To completely defeat the mosquito and the malarial parasite, the ecology of both must be changed such that they cannot survive. This has not yet been achieved. [*See* Malaria.]

## D. Toxoplasmosis

Toxoplasmosis is caused by the protozoan parasite *Toxoplasma gondii* and probably evolved from the cat coccidian *Isospora bigemina*. In the cat, *Toxoplasma* undergoes a coccidian developmental cycle: asexual reproduction in the cells of the intestinal epithelium, followed by a phase of sexual reproduction that results in the formation of egg-like oocysts, which pass out with the feces of the cat. In order for the oocysts to become infective, they must be exposed to oxygen and a temperature lower than the body temperature of a mammal; when these conditions are satisfied, as they are in the soil, the oocyst will contain infective sporozoites. When ingested, the sporozoites are released from the oocyst only if it has been exposed to trypsin, bile, carbon dioxide, anaerobiosis, and

37°C—conditions found in the mammalian intestine. The released sporozoites invade the epithelial cells, and a typical coccidian cycle of development takes place. However, if an oocyst of *Toxoplasma* is ingested by a mammalian host other than a cat or by a bird, then the parasite may develop extraintestinally—that is, in the liver, lymph nodes, lungs, heart, and brain—but no oocysts are formed. In these hosts, *Toxoplasma* rapidly multiplies intracellularly by a budding process (i.e. endodyogeny) and forms 8 to 32 tachyzoites. The host cell disintegrates, permitting the released tachyzoites to reinvade other cells. In 5–10 days, the parasites develop into slowly multiplying forms called bradyzoites. These stages, which are enclosed in a cyst wall, can persist for long periods, and a single cyst may contain thousands of parasites.

Ingestion of bradyzoite- or tachyzoite-infected meat most commonly serves as the means of transmission from carnivore to carnivore; but only if such infected meat is ingested by a cat will intestinal development and oocyst production occur. The cycle between cats and the rodents and birds on which they prey depends on the prolonged survival of the parasite in the tissues of these (intermediate) hosts. Humans are accidental intermediate hosts.

In adults and older children, *Toxoplasma* gives rise to an asymptomatic infection because of antibodies acting on extracellular parasites and T cell factors, such as interferon-γ, acting on intracellular parasites. However, in immunocompromised individuals or during pregnancy, disease can result. Congenital human toxoplasmosis can occur by the transplacental transmission of tachyzoites from the pregnant mother to the fetus, in which it might cause fetal blindness, encephalitis, microcephaly, hydrocephaly, or death.

*Toxoplasma* is usually transmitted by the ingestion of oocysts from feces-contaminated soil, from cysts in raw or undercooked meat, or by organ transplantation. Dissemination in the body is via the lymphatics, lymph nodes, and blood. In humans, congenital toxoplasmosis can be prevented by cooking meat, by washing hands thoroughly after preparing raw meat, and by not feeding the cat raw meat; also, cat litter trays should be handled with gloves and treated with boiling water for 5–10 min.

Diagnosis of toxoplasmosis is usually made by the demonstration of antibodies to the toxoplasmas, seldom by the demonstration of the parasite itself. Humans usually continue to have antibodies after exposure to the parasite, despite the absence of an active infection. Depending on the particular location, 20–80% of the human population may have antibodies to *Toxoplasma.*

Pyrimethamine and sulfonamides are drugs commonly used against *Toxoplasma,* except during the first trimester of pregnancy.

## E. Trichomoniasis

The causative agent of trichomoniasis, *Trichomonas vaginalis,* is a flagellated protozoan first described by Donné in 1836. The body of *T. vaginalis* is easily recognizable because of its shield shape, an anterior tuft of four flagella, a stout supportive rod (axostyle), and its undulating membrane. There are no sexual stages, as reproduction occurs by binary fission. The parasite forms no cysts. This parasite is found in the vagina and urethra of women, where it causes vaginitis characterized by an intense itching and a copious white discharge containing both trichomonads and epithelial cells. *Trichomonas vaginalis,* despite its species name, can occur in the male, where it may infect the urethra, prostate gland, and seminal vesicles. However, in men, the infection is usually asymptomatic and only rarely is there urethritis and prostatitis.

Trichomoniasis is a sexually transmitted disease that affects 5–20% of women in an unselected population. Infections are spread by direct contact with the trichomonads in the vaginal discharge, and 200 million people worldwide may be infected.

*Trichomonas vaginalis* does not grow well at pH values lower than 5.0, and because of this, the vaginal pH (4.0–4.5) ordinarily discourages its growth; however, during menstruation, the vaginal pH increases and this may favor development of the parasite. In addition, once established, *T. vaginalis* itself tends to alkalinize the vagina, encouraging further proliferation.

Diagnosis of trichomoniasis is made on the basis of finding *T. vaginalis* in the secretions of the vagina. Treatment involves administrations of metronidazole, as well as vaginal douches and suppositories, to restore the acid pH of the vagina and to discourage parasite growth.

*Trichomonas vaginalis* lives in an oxygen-poor environment, and the parasite has no mitochondria. However, they do have a special organelle, the hydrogenosome, that contains the enzymes involved in hydrogen formation from pyruvate; the parasite also degrades carbohydrate to acetate, $CO_2$, and lactate.

Metronidazole is an effective oral treatment for trichomoniasis since, by reduction of its nitro group, it becomes cytotoxic; *Trichomonas,* which has a ferredoxin-like electron transport compound in its hydrogenosome, is especially efficient in reducing metronidazole. Because mammalian cells have no such reducing system, the drug itself is not toxic to humans.

## IV. INTESTINAL ROUNDWORMS

### A. Pinworms

The pinworm, *Enterobius vermicularis,* is perhaps the most common nematode parasite of humans, with estimates of prevalence ranging near 1 billion people infected. Pinworms are so named because the posterior end of the worm is extended into a long, slender, pointed end. The sexes are separate, and both females and males live in the lower part (ileocecal region) of the intestine, where they feed on bacteria and tissue debris. The male transfers the sperm to the female and soon dies. The gravid females migrate down the colon, passing out of the anus and crawling onto the perianal region, where the embryonated eggs are deposited. A female worm may deposit 5000–10,000 eggs and, soon after egg deposition, dies.

The eggs of the pinworm, which are deposited outside the body, contain a larval worm and can be infective within 6 hr; infestation of others occurs by swallowing these eggs, which then hatch in the small intestine. A third-stage larva is released, which molts twice and migrates to the ileocecal region to become fully mature in 2 weeks to 1 month.

Discomfort due to infestation with pinworms is due to the nightly migration of the gravid females to deposit eggs, which causes perianal inflammation and itching. Heavy infestations, especially in children, result in insomnia, nightmares, weight loss, perianal pain, and inflammation of the bowel.

Infection usually occurs when egg-soiled fingers or other objects are placed in the mouth or by contact of the mouth with egg-contaminated bedding, clothing, towels, etc.

Pinworms are more common in temperate regions of the world. The frequency of infestation is highest in orphanages, mental hospitals, and schools, where the conditions for transmission and reinfection are most favorable.

Diagnosis is made by finding eggs or adult females (which are cream colored and 1 cm in length) in the perianal area. Effective treatment includes the drugs mebendazole (Vermox) pyrvinium pamoate (Povan); washing of egg-contaminated linens, clothing, towels, and utensils is a useful preventive measure against pinworm infestations.

### B. Trichuriasis

Trichuriasis, or "whipworm disease," is a worldwide disease, predominantly found in children, which afflicts up to 800 million people throughout tropical and temperate regions. Having fewer than 100 worms rarely causes clinical symptoms; however, with numbers greater than this, a chronic infection may develop, resulting in dysentery, anemia growth retardation, and rectal prolapse. The most consistent feature of trichuriasis is inflammation of the colon.

Trichuriasis is caused by nematodes and is so named because the body of the adult worm is threadlike for most of its length, but at the posterior end becomes thickened, resembling a whip with a handle. The anterior end of the worm burrows into the mucosa of the intestine, causing trauma and small hemorrhages. Secondary infections may lead to further inflammation. The sexes are separate, and the worms are about 30–50 mm in length, the male being somewhat smaller than the female. The infection is initiated by the ingestion of eggs. An infective larval worm hatches from the egg in the small intestine and, after a short period of development in the crypts of Lieberkühn, migrates to the ileocecal regions. Here, it becomes fully mature in 3 months. The worms do not multiply in the intestine, but the female may produce 1000–7000 eggs per day. The larvae within the eggs are passed out of the body with the feces, and development to an infective stage occurs in about 21 days in the soil. Ingestion of eggs in fecal-contaminated food, water, or fomites leads to infection.

Diagnosis is based on the identification of worms or eggs in the feces. The drugs most frequently used are mebendazole or albendazole, both of which affect depolymerization of the microtubules; as a result, the adult worms are killed and expelled.

### C. Ascariasis

About three and a quarter centuries ago, the English physician Edward Tyson demonstrated that the anatomical structure of the nematode *Ascaris lumbricoides* was entirely different from that of the earthworm. Indeed, Tyson described *Ascaris* as "that common roundworm which children usually are troubled with." It is estimated that there are a billion cases of ascariasis, or a global prevalence of 22%, making it one of the most common helminthic infestations of humans. Typically, the prevalence is higher in children than in adults, and often it is more common in females than in males.

The adult cream-colored roundworms live in the jejunum, and because of their large size—female and male worms can be up to 30 and 20 cm in length, respectively—they can cause intestinal obstruction. They may also cause bile duct obstruction or hepatic abscesses. Occasionally, the worms migrate into the appendix or the esophagus, where they may be expelled in vomitus or through the nose or mouth. A migratory worm that becomes lodged in the trachea can cause suffocation, and one in the bile duct may cause jaundice and elevated levels of bilirubin in the blood. In children, heavy infestations may result in malnutrition and stunted growth.

Adult females may live up to 1.5 years and can produce 240,000 eggs per day, which are passed in the feces. The diagnosis of ascariasis is usually made by examining the feces for eggs. When an individual is infected with only female worms, the eggs are unfertilized and they can be distinguished microscopically from fertilized eggs.

Fertilized eggs that pass out of the body require about 70 days for complete maturation. Embryonic development and the first molt occur within the resistant chitinous eggshell; this takes 10–14 days. Hatching of the eggs requires a temperature of 37°C, a low oxidation–reduction potential, high concentrations of carbon dioxide, and neutral pH—conditions found in the gut. Under these conditions, hatching may also be carried out *in vitro* with considerable success. The hatching stimuli activate the larval worm to secrete a chitinase that digests a hole in the eggshell, allowing release of the worm. Four days after ingestion the larvae are found in the liver, and by 7 days they are in the lungs; by being coughed up and swallowed, they return to the intestine by day 17. During the migration of the larvae from the intestine to the liver and lungs, hypersensitivity and respiratory complications (eosinophilic pneumonitis) may develop.

Transmission of ascariasis is by fecal–oral contact. The prevalence of the disease is higher in countries with a humid climate and is more common in rural, rather than urban, environments.

Diagnosis is, as mentioned, by detection of eggs in the feces, but there are also immunological tests based on the detection of immunoglobulin M antibodies to the worms. Treatment is by the oral administration of anthelmintic drugs, such as mebendazole, albendazole, or pyrantel pamoate.

## BIBLIOGRAPY

Grove, D. I. (1990). "A History of Human Helminthology." CAB International, Wallingford, Oxon, United Kingdom.

Jordan, P., Webbe, G., and Sturrock, R. F. (1993). "Human Schistosomiasis." CAB International, Wallingford, Oxon, United Kingdom.

Miyazaki, I. (1991). "An Illustrated Book of Helminthic Zoonoses." International Medical Foundation of Japan, Tokyo.

Neva, F., and Brown, W. (1994). "Basic Clinical Parasitology," 6th Ed. Appleton and Lange, Norwalk, CT.

Schmidt, G. D., and Roberts, L. S. (eds.) (1996). "Foundations of Parasitology," 5th Ed. W. C. Brown Publishers, Dubuque, IA.

Simpson, A. J. G., and Rollinson, D. (eds.) (1987). "The Biology of Schistosomes: From Genes to Latrines." Academic Press, Orlando, FL.

Smyth, J. (1994). "Introduction to Animal Parasitology," 3rd Ed. Cambridge Univ. Press, Cambridge, United Kingdom.

Stephenson, L. S. (1987). "Impact of Helminth Infections on Human Nutrition." Taylor & Francis, London.

Trager, W. (1986). "Living Together: The Biology of Animal Parasitism." Plenum, New York.

# Antibiotic Inhibitors of Bacterial Cell Wall Biosynthesis

DONALD J. TIPPER

*University of Massachusetts Medical School*

## GLOSSARY

**Autolysins** Murein hydrolases produced by bacteria, uncontrolled activation of which can lead to lethal self-lysis

**$\beta$-Lactam antibiotics** Antibiotics containing a four-membered lactam ring that inhibit murein transpeptidases

**Cell wall** Insoluble structures exterior to the bacterial cytoplasmic membrane

**Lysozymes** Endo-$N$-acetylmuramidases that hydrolyze murein by fragmenting the glycan, causing lysis of bacteria

**Murein** Cross-linked peptidoglycan, the unique and essential structural component of bacterial cell walls

**$N$-Acetyl muramic acid** The 3-O-D-lactyl derivative of $N$-acetyl-D-glucosamine, the point of peptide attachment to glycan in murein

**Peptidoglycan** Polysaccharide (glycan) with attached peptides

**Transpeptidation** A two-step transacylation reaction that cross-links murein peptides, essential for murein integrity and strength

ANTIBIOTICS USEFUL FOR THE TREATMENT OF human bacterial infectious diseases must be toxic to bacteria but have minimal toxicity for their host. For a major group of antibiotics, the basis of this selective toxicity lies in the inhibition of bacterial cell wall biosynthesis. The targets of these antibiotics are the enzymes involved in the synthesis of murein, a unique peptidoglycan found only in bacteria. Murein is the essential structural component in the cell walls of all known human bacterial pathogens and inhibition of its synthesis in growing bacteria is usually lethal: the best understood but not necessarily exclusive mechanism of lethality involves the triggering of an imbalance between murein synthesis and hydrolysis (autolysis), leading to lysis and death. This class of antibiotics is, therefore, bactericidal for almost all sensitive organisms. The most important members of this group of antibiotics are the $\beta$-lactams, of which penicillin G is the best known and earliest representative. This group of antibiotics inhibits the transpeptidation reactions which cross-link the peptide component of peptidoglycan creating a functional, three-dimensional murein.

## I. ANTIBACTERIAL CHEMOTHERAPY AND THE DEVELOPMENT OF ANTIBIOTICS

Paul Ehrlich coined the term chemotherapy near the start of the 20th century. Based on his studies of

ENCYCLOPEDIA OF HUMAN BIOLOGY, Second Edition, VOLUME I. Copyright © 1997 by Academic Press. All rights of reproduction in any form reserved.

tissue-specific dyes (some still used in histochemistry), he hoped to find synthetic chemotherapeutic agents that would selectively bind and kill animal parasites. The search for less toxic derivatives of arsenates, already in use in veterinary medicine, led to the development of Salvarsan, an agent with moderate effectiveness against syphilis but retaining considerable toxicity. However, not until the introduction of Prontosil as an antistreptococcal agent in 1935, 20 years after Ehrlich's death, did his dream of a "magic bullet," an agent essentially harmless to the patient but selectively toxic to life-threatening microorganisms, approach fruition. Neither Salvarsan nor Prontosil were active *in vitro*, i.e., against pure cultures of bacteria. Thus when sulfanilamide was isolated from the urine of treated patients and identified as the active metabolite of Prontosil, it was the first simple, readily available chemotherapeutic agent active *in vitro*. By providing a simple assay of function, this at once opened the door both to the rapid development of the sulfonamides (sulfa drugs), improved versions of sulfanilamide, and to study of their mode of action. This itself was a tortuous process, starting with the identification of *p*-aminobenzoic acid (PABA; Fig. 1A) as the component of yeast extract which antagonized sulfonamide action *in vitro*. Because of the obvious chemical similarity of sulfanilamide and PABA (Fig. 1A), the concept of antimetabolic activity through mimicry was codified well before the essential role of PABA as a biosynthetic precursor of folic acid was established. The role of PABA in folic acid synthesis finally provided an explanation for the selective toxicity of the sulfa drugs: folic acid is a vitamin for mammals, an essential dietary component not made in our cells. Bacteria cannot use extracellular sources and must make their own.

The history of sulfa drugs illustrates several concepts that remain central to the relentless search for novel antibacterial antibiotics and improved versions of existing antibiotics. First, most antibiotics in current use have been identified by empiric screens and their mechanisms have been elucidated only after years of subsequent work. Second, most useful antibiotics are semisynthetic derivatives of natural compounds with improved activity and pharmacological properties. Third, many of the most useful antibiotic targets are enzymes involved in biosynthetic reactions unique to bacteria. Cell wall synthesis is in this category. Other targets such as DNA, RNA and protein synthesis are complex essential cell processes where differences between bacteria and mammals can be exploited by selec-

tive antibiotic inhibitors. Fourth, the inhibitors are usually mimics of the normal enzyme substrates. This concept of molecular mimicry has been vital to the elucidation of the mode of action of many antibiotics, including the cell wall synthesis inhibitors. It also provides the basis of "rational drug design," a process in which inhibitors are sought for a known target, as in the development of trimethoprim, a selective inhibitor of bacterial dihydrofolate reductase which acts synergistically with sulfa drugs. In a common amalgam of these approaches, an *in vitro* assay for inhibitors of a specific set of targets, such as late stages in murein synthesis, is used in empirical screening.

Antibiotics, as originally defined by Waksman, are substances produced by microorganisms that are capable, in low concentrations, of killing or inhibiting the growth of other microorganisms. The phenomenon of bacterial antagonism was first recorded by Pasteur in contaminated cultures of *Bacillus anthracis* and was later described by many workers. Penicillin (benzyl penicillin, penicillin G), the progenitor of the $\beta$-lactams (Fig. 1D), was discovered serendipitously by Fleming in 1929 as the unknown component present in *Penicillium* culture filtrates that caused lysis of *Staphylococcus aureus*. This potent but labile compound defied purification and characterization, and interest in antibiotics waned in the 1930s; however, the development of the sulfonamides rekindled interest in penicillin. In 1939, the purification and use in veterinary medicine of gramicidin, a mixture of related peptide inhibitors of bacterial protein synthesis, was an important contemporary example of successful clinical usage of an antibiotic. Although gramicidin proved too toxic for use in human medicine, its efficacy certainly stimulated the work on penicillin in Oxford as Europe moved inexorably toward the second world war. The purification of penicillin and demonstration of its unequaled potency and clinical efficacy soon followed; resources were pooled with the scientific community of the United States so that the many novel problems of penicillin chemistry and production could be solved as rapidly as possible. [*See* Antibiotics.]

This success stimulated the search for additional antibiotics among soil microorganisms, resulting in the discovery of streptomycin by Waksman and his colleagues in 1944, the first agent active against tuberculosis and many gram-negative bacteria, and the first of many thousands of antibiotics investigated in the past 50 years by the pharmaceutical industry. Among these are large numbers of diverse $\beta$-lactam antibiot-

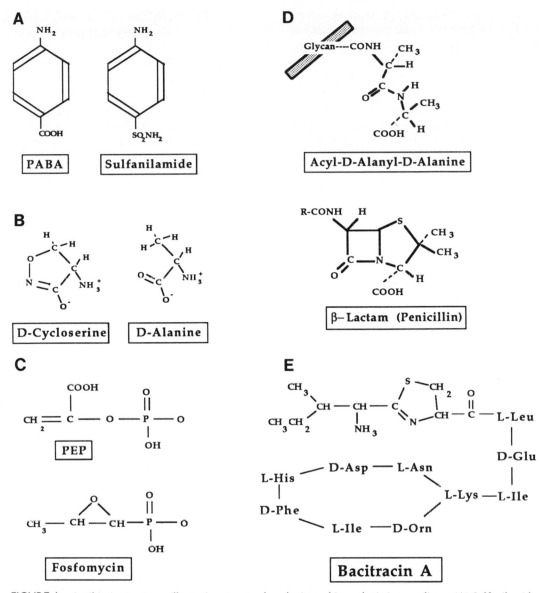

**FIGURE I** Antibiotic structures illustrating structural analogies to biosynthetic intermediates. (A) Sulfanilamide and *p*-aminobenzoic acid (PABA). (B) D-Cycloserine and D-alanine. (C) Fosfomycin and phosphoenolpyruvate (PEP). (D) Penicillin (a *β*-lactam antibiotic) and the acyl-D-alanyl-D-alanine substrate for peptidoglycan transpeptidase (see Figs. 4 and 5). (E) Bacitracin A.

ics, from which are derived the two or three novel semisynthetic (chemically modified) *β*-lactams that are introduced into human medicine in an average year. Each of these "modern" *β*-lactam antibiotics has some potential advantage in potency, antibacterial spectrum, pharmacological characteristics, or resistance to inactivation by bacterial *β*-lactamases (see Section IX), but all have the same mode of action.

## II. PENICILLIN AND BACTERIAL CELL WALL PEPTIDOGLYCAN (MUREIN)

Early observations on the mode of action of penicillin implicated the cell wall of sensitive cells as the primary target. The grossly visible effects of cell swelling and lysis suggested a weakening of the wall, whose chemical composition and properties were, at that time, completely unknown. Bacterial cell wall structures are now known to be varied and complex, but one component is common to all penicillin-sensitive bacteria and provides the strength and shape of the walls of these bacteria. This component, called murein by Weidel, is a peptidoglycan that resists proteolytic and mild chemical treatments which strip off other cell wall polymers. The murein is left as an intact "bag-shaped macromolecule," or sacculus, retaining the shape of the cell. The biosynthesis of this polymer is the target of the currently available antibiotics active against bacterial cell walls. Its ubiquity accounts for the broad antibacterial spectrum of these antibiotics. Because murein (and therefore its biosynthetic pathway) is unique to bacteria, selective toxicity of these antibiotics is explained. Because cell wall integrity is essential to bacterial survival, a simple explanation for the bactericidal activity of these agents is apparent.

The principal role of bacteriostatic agents is to contain bacterial multiplication until adaptive immune defenses can eliminate the infection. This is perfectly adequate for the treatment of most instances of bacterial infectious disease in normal individuals. However, a bactericidal response is of special importance in the successful treatment of infection where normal immune defenses are locally or systemically impaired. Modern $\beta$-lactam antibiotics are the primary agents used in prophylaxis against infection acquired during surgery and in the treatment of bacterial infections in patients with cancer or AIDS. Given the low cost, low toxicity, and continued efficacy of penicillin G for many instances of several life-threatening diseases, the continued popularity of this drug is also easily explained. For example, its unrivaled efficacy against pneumococci, the most common cause of bacterial pneumonia, has been instrumental in saving many lives and in changing the pattern of hospital care by eliminating a major class of inpatients. It is inevitable but unfortunate, however, that common pathogens such as the pneumococcus have developed a high level resistance in the face of decades of heavy use of penicillin and related $\beta$-lactam antibiotics (see Section IX). Total annual production of $\beta$-lactam antibiotics now approximates 12,000 tons. Streptomycin and newer aminoglycoside antibiotic inhibitors of protein synthesis and the quinolone inhibitors of bacterial gyrase are important alternative bactericidal agents. However, toxicity or a high frequency of acquired resistance frequently limits their utility.

## III. MUREIN AND BACTERIAL TAXONOMY

Comparison of ribosomal RNA sequences demonstrates that three main cell types diverged several billion years ago. They were the progenitors of the modern eukaryotes and the two groups of prokaryotes, Eubacteria and Archebacteria. The Eubacteria comprise all known human bacterial pathogens and all contain murein in their walls. Mycoplasma are Eubacteria that have dispensed with cell walls during evolution. They are, of course, intrinsically resistant to antibiotic inhibitors of cell wall synthesis. The Archebacteria do not contain murein, although some contain a related peptidoglycan called pseudomurein. Thus the synthesis of an extracellular murein evolved very early in prokaryotes and has proven to be an extremely successful solution to the problem of survival in a hostile environment. Murein structure has remained remarkably constant through the ensuing eons. Eubacteria diverged somewhat later into organisms with rigid or flexible walls, differing in mechanisms of motility. The former, if motile, use external rotating flagella to propel themselves whereas the flexible bacteria use other means. The only important known representatives of this group among human pathogens are the spirochetes (syphilis) and borrelia (relapsing fever, Lyme disease). The vast majority of human bacterial pathogens have rigid cell walls, and these again diverged in the distant past into two major groups, distinguished by their cell wall architecture as seen in cross section (schematically illustrated in Fig. 2) and by response to the gram stain. Because the response to this simple staining procedure reflects this fundamental difference in cell wall structure, it plays a primary role in bacterial classification and identification. The wall structures determining the response to the Gram stain also correlate with major differences in pathogenic mechanisms, antibiotic susceptibility, and propensity for acquisition of antibiotic resistance, enhancing the importance of this simple diagnostic procedure. [*See* Bacterial Infections, Detection.]

## IV. CELL WALL STRUCTURE, RESPONSE TO THE GRAM STAIN, AND ANTIBIOTIC SENSITIVITY

In the Gram stain, heat-fixed bacteria are stained with crystal violet which is then cross-linked by brief oxida-

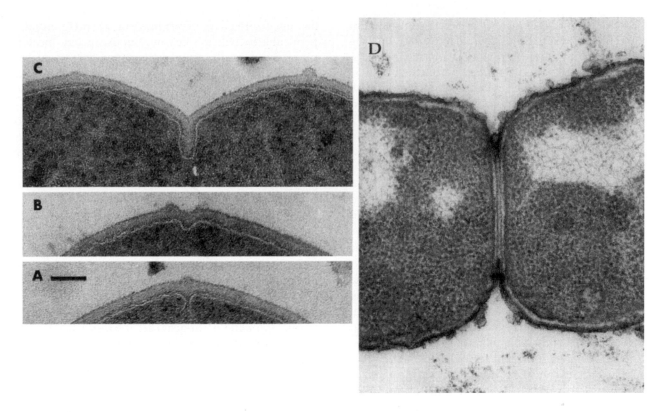

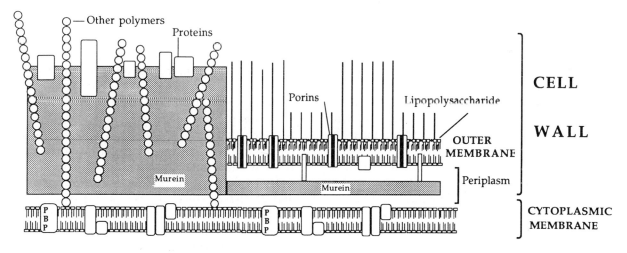

**E**  **GRAM POSITIVE**     **GRAM NEGATIVE**

FIGURE 2   The cross-sectional profile of gram-positive and gram-negative bacterial cell walls. (A–C) Stages in synthesis of the cross septum wall segment in *Enterococcus faecalis* (gram positive; courtesy of Dr. Michael Higgins). The bar represents 100 nm. (D) Cross septum wall synthesis in *Escherichia coli* (gram negative; courtesy of Dr. R. G. E. Murray) at approximately the same magnification. (E) Schematic model of wall structures. The thicker, more porous gram-positive wall has a high murein content; other carbohydrate polymers and proteins comprise 20–60%. The gram-negative wall has a thin murein layer within the periplasm. The periplasm is surrounded by the outer membrane whose outer surface is composed of lipopolysaccharide. The outer carbohydrate chains of this polymer extend into the medium. The active sites of the penicillin-binding proteins (PBPs), the transglycosylase and transpeptidase targets of vancomycin and the β-lactam antibiotics, are on the outer surface of the cytoplasmic membrane.

tion with iodine. Following destaining with ethanol or a similar solvent, gram-positive bacteria retain the dye and remain blue. Gram-negative bacteria are decolorized and are counterstained with a second dye, usually Safranin (red). Cross sections of the cell wall of a gram-positive bacterium (*Enterococcus faecalis*) are shown in Figs. 2A–2C. The cross section of a gram-negative bacterium (*Escherichia coli*) is shown in Fig. 2D. Schematic interpretations of these cross sections are shown in Fig. 2E.

The gram-negative bacterial cell wall contains an outer membrane of unique composition (Figs. 2D and 2E). The murein layer in these walls, which is thin and presumably relatively porous, resides within the periplasm, the space between this outer membrane and the cytoplasmic membrane (Fig. 2E). The periplasm contains a significant fraction of total cellular protein, including species involved in nutrient uptake, assembly of flagella and pili and antibiotic resistance. Access to and exit from the periplasm is controlled by the porosity of the outer membrane. The combination of outer membrane and periplasm provides gram negative bacteria with effective control of the immediate environment of the cytoplasmic membrane. The proteins and murein in the periplasm may be sufficiently concentrated to produce a gel-like state. All nutrients must be able to cross the outer membrane and periplasm to reach the cytoplasmic membrane. This is also true for access of all antibiotics to their targets, with the exception of the polymyxins, which directly disrupt the outer membrane itself. Typical cell membranes are bilayers of phospholipids with the polar phosphate groups exposed to water at the two surfaces and containing embedded proteins. However, only the inner leaflet of the gram-negative outer membrane is composed of normal phospholipids. The lipid component of the outer leaflet is the lipid A component of a lipopolysaccharide unique to gram-negative bacteria (Fig. 2E). This lipopolysaccharide is the gram-negative bacterial endotoxin, a major stimulator of the immune response and the primary cause of life-threatening endotoxic shock in gram-negative sepsis. Because of the low fluidity of the lipid A leaflet, the gram-negative outer membrane is essentially impermeable to hydrophobic compounds that penetrate normal cell membranes with ease. Antibiotics and nutrients can only permeate this membrane through aqueous channels within the hollow, barrel-like structure of porins, transmembrane proteins specifically located in the outer membrane (Fig. 2E). The majority of porins in most gram-negative pathogens resemble those in *E. coli* and have relatively nonspecific chan-

nels lined with charged residues that are preferentially permeable to hydrophilic compounds. Specific mechanisms exist for the uptake of certain larger solutes such as disaccharides, cobalamins (vitamin $B_{12}$), and iron chelates, and a few antibiotics can exploit these more specific pores. However, in general, the antibiotic sensitivity of gram-negative bacteria is limited to low molecular weight, relatively hydrophilic compounds. Outer membrane permeability is actually highly variable among gram-negative genera, and major differences occur in the variety and characteristics of their porins. Thus, the outer membranes of *Neisseria* are the most permeable known, and these bacteria are sensitive to erythromycin (an inhibitor of bacterial ribosome function) and the more hydrophobic β-lactams such as penicillin G, whose activity is otherwise restricted principally to gram-positive bacteria. In contrast, *E. coli* and related members of the "enterics" (gastrointestinal gram-negative bacilli, including both normal flora and pathogens) are intrinsically resistant to both of these antibiotics but are sensitive to ampicillin, a zwitterionic derivative of penicillin G. Certain other enterics (e.g., Enterobacter) have more limited porin permeability and an even narrower sensitivity spectrum. *Pseudomonas aeruginosa* is an aerobic gram-negative rod whose very low outer membrane permeability renders it intrinsically resistant to most antibiotics, contributing to its role as a major opportunistic pathogen, especially in hospitalized patients and burn units. It lacks the common type of porin found in enterics and relies on a series of more specific channels. One of these, probably used for the uptake of basic amino acids, is exploited by the carbapenem β-lactam derivative imipenem, a major drug for the treatment of pseudomonas infections. Unfortunately, loss of this single channel leads to high level resistance and is an increasingly common cause of antibiotic-resistant infection in hospitalized patients.

The walls of gram-positive bacteria (Figs. 2A–2C and 2E) have a much higher peptidoglycan content in a multilayered structure to which other polymers are covalently attached. This produces a wall that is much thicker than in gram-negative bacteria, but also much more permeable, with a sieving limit sufficient to allow easy passage not only of all β-lactam antibiotics, but also of larger antibiotics such as bacitracin, vancomycin, daptomycin, and moenomycin, which are inactive against gram-negative bacteria. The spectrum of activity of these and related antibiotics is, therefore, limited to gram-positive bacteria. Gram-positive bacteria are far more heterogeneous than gram-negatives and include the Streptomycetes, one group of which form the acid-fast bacteria. Acid-fast bacteria fail to stain with crystal violet, but can be stained in hot 5% phenol and resist destaining with acidified ethanol (the Ziehl–Neelsen procedure). This staining response is also a reflection of cell wall composition: the acid-fast cell wall contains large quantities of waxy glycolipid polymers linked to their murein. Like the gram-negative outer membrane, this waxy layer is impermeable to hydrophilic and hydrophobic compounds alike, and acid-fast bacteria rely on wall-spanning porin proteins for the delivery of nutrients. However, in these slow-growing organisms the rate of permeation through these channels is slow and antibiotic penetration is selective. In addition, these pathogens multiply within phagocytic cells whose membranes provide an additional barrier to antibiotic access. The major human acid-fast pathogens, including *Mycobacterium tuberculosis* (TB) and *Mycobacterium leprae* (leprosy), are resistant or tolerant (see Section VIII) to all of the cell wall synthesis inhibitors, with the exception of D-cycloserine, and are sensitive only to a very limited group of antibiotics, most of which are specific for mycobacteria.

## V. STRUCTURE OF MUREIN PEPTIDOGLYCAN

The structure of *E. coli* cell wall peptidoglycan is illustrated in Fig. 3. The glycan structure, shown in detail in Fig. 3A, is universal in murein: it consists of alternating, β-1,4-linked residues of *N*-acetyl-D-glucosamine (GlcNAc) and *N*-acetylmuramic acid (MurNAc), the 3-*O*-D-lactyl ether derivative of GlcNAc. The glycan is thus a modified form of chitin, although spectral analyses suggest that it has a helical conformation with peptide substituents extending in all directions rather than the flat ribbon conformation of chitin. MurNAc provides the carboxylate group to which the peptide chains are attached to the glycan. The strict alternation of GlcNAc and MurNAc (represented as G and M in Fig. 3C) in the glycan is ensured by synthesis of the disaccharide-repeating subunit prior to polymerization.

Mureins initially carry a peptide subunit on each MurNAc residue. Figure 3A illustrates the L-Ala-D-Glu-mDAP-D-Ala-D-Ala pentapeptide (mDAP is *m*-diaminopimelic acid, the immediate biosynthetic precursor of L-lysine in bacteria) that constitutes the nascent peptide subunit of *E. coli* peptidoglycan. This murein structure is probably common to all gram-negative bacteria and also occurs in many gram-posi-

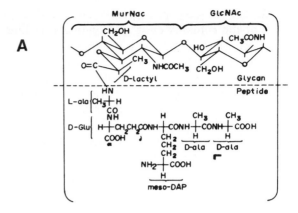

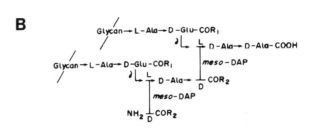

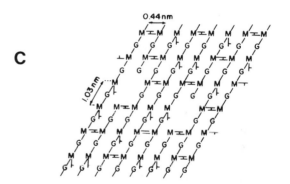

**FIGURE 3** The structure of *E. coli* murein. (A) The repeating unit. The glycan consists of alternating residues of GlcNAc and MurNAc (see text). In a newly polymerized murein peptidoglycan, the D-lactyl group of each MurNAc residue is linked to the pentapeptide shown (L-Ala-D-Glu-mDAP-D-Ala-D-Ala). mDAP, *m*-diaminopimelic acid. (B) The cross-linked peptide dimer produced by transpeptidation between D-Ala and the free amino group of mDAP in a second peptide subunit. (C) A two-dimensional *E. coli* murein. About 40% of the peptide subunits exist as dimers, producing the cross-links represented as horizontal bars.

tive bacteria, with variation in amidation at D-Glu and mDAP. In the mature peptidoglycan of *E. coli*, transpeptidation converts about 40% of the nascent peptide subunits shown in Fig. 3A into the dimers shown in Fig. 3B. Each dimer is cross-linked between the penultimate D-alanine and the free amino group of mDAP in a second peptidoglycan subunit. Most of the rest of the subunits exist as monomers, lacking one or both of their D-alanine residues. The first is removed by the action of D,D-carboxypeptidases, enzymes related in mechanism to the transpeptidases (see Section VII), and the second by L,D-carboxypeptidase action. Several minor variants in the subunit structure exist, but are not shown. The mature peptidoglycan, shown schematically as a two-dimensional net in Fig. 3C, is cross-linked into a continuous, cell-sized polymer. The gram-negative murein has the simplest known peptidoglycan structure, and a thin layer is clearly all that is needed for cell viability when it is accompanied and protected against enzymatic attack by the gram-negative outer membrane.

The gram-positive peptidoglycan contains many variants in its peptide structure, but all derive from the pentapeptide (A)-D-Glu-(B)-D-Ala-D-Ala where (A) is usually L-alanine and D-Glu is always $\gamma$ linked to (B), which is usually a dibasic amino acid (mDAP, L-lysine, L-ornithine, etc.). This tripeptide is linked to the ubiquitous D-Ala-D-Ala C-terminal dipeptide. In most gram-positive mureins, the cross-link between D-alanine and the dibasic amino acid, instead of being direct as in *E. coli* (Fig. 3B), involves an intervening cross-bridge peptide. In *S. aureus,* for example, the pentapeptide is L-Ala-D-Glu-L-Lys-D-Ala-D-Ala and most of the cross-bridges are pentaglycine. In the lipid cycle stage of synthesis (see Section VII), the five glycine residues are added to the $N^{\varepsilon}$-amino group of L-Lys and act as the acceptor for subsequent transpeptidation.

## VI. MUREIN HYDROLASES: AUTOLYSINS AND LYSOZYMES

Bacterial cells (with the exception of the mycoplasmas) generally have considerable turgor pressure. This is contained by the murein sacculus which is moderately flexible but nonextensible. In a growing bacterium, therefore, the murein must be in a dynamic state of controlled expansion. This involves both insertion and cross-linkage of new peptidoglycan strands and the controlled hydrolysis of preexisting murein to allow expansion and morphogenesis. Septation and cell separation clearly require these processes. The murein hydrolases, whose unbridled activity can lead to suicidal lysis, are called autolysins. They include a variety of endoglycanases and endopeptidases. Lytic bacteriophage may produce similar enzymes in order to escape from their host cell.

Lysozyme, also discovered by Fleming, is a major component of human antibacterial defenses found in phagocyte granules, serum, and secretions such as tears and saliva. It hydrolyzes the bond between Mur-NAc and GlcNAc in murein, causing bacterial lysis. Gram-negative murein is sensitive, but is protected by the outer membrane. Disruption of this membrane by complement or phagocyte granular proteins allows rapid killing of gram-negative bacteria by lysozyme. Acid-fast bacteria are also protected against lysozyme by their waxy cell wall. To be successful as pathogens, however, gram-positive bacteria must have relatively lysozyme-resistant murein. This is achieved in *Staphylococci,* for example, by O-acetylation of MurNAc residues. The lytic activity of lysozyme illustrates the essential role of murein in cell wall integrity.

## VII. BIOSYNTHESIS OF PEPTIDOGLYCAN AND SITES OF INHIBITION BY ANTIBIOTICS

The biosynthesis of peptidoglycan takes place sequentially in three locations: the cytoplasm, the inner surface of the cytoplasmic membrane, and the outer surface of this membrane, where the murein is assembled (Fig. 4). The entire process is integrated into the cell growth and division cycle and is controlled in three-dimensional space so that cell shape, which is defined by the cross-linked shape of the murein sacculus, is maintained. Knowledge of the related control mechanisms is currently primitive, but the polymerizing enzymes, recognized as penicillin-binding proteins (PBPs), are almost certainly of prime importance in this control.

### A. Cytoplasmic Phase: Fosfomycin, Cycloserine, and Related Antibiotics

The cytoplasmic phase of murein synthesis (Fig. 4A) initiates with the first step in muramic acid synthesis and culminates in the production of UDP-MurNAc-pentapeptide, the "Park" nucleotide. The isolation and characterization of this compound from *S. aureus,* by James T. Park and colleagues, were crucial clues to the structure of murein, the mechanism of its synthesis, and the mode of action of penicillin. The terminal D-alanyl-D-alanine of the pentapeptide is added as a preformed dimer, ensuring the presence of the donor substrate for transpeptidation on each subunit of nascent murein. Both steps in its synthesis

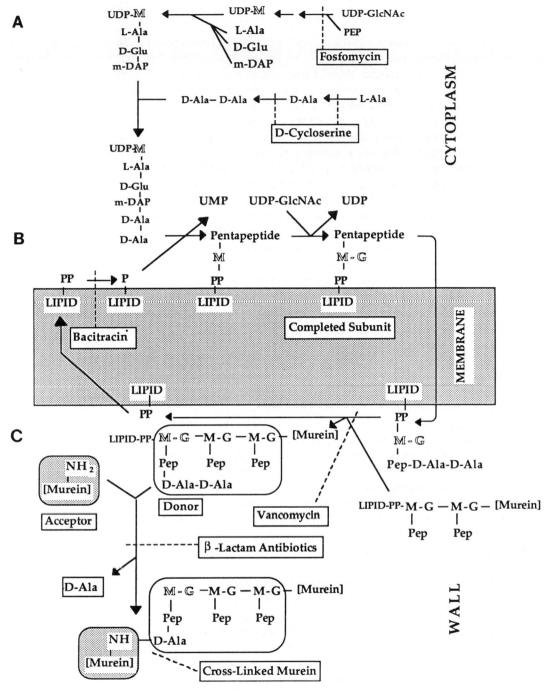

**FIGURE 4** The biosynthesis of murein peptidoglycan and the sites of enzyme inhibition by antibiotics. (A) In the cytoplasmic phase, *N*-acetyl muramic acid (MurNAc, M) is synthesized from *N*-Acetyl-D-glucosamine (GlcNAc, G) in two steps. The first step involves the phosphoenolpyruvate (PEP) addition to UDP-GlcNAc and is inhibited by fosfomycin, an analogue of PEP (Fig. 1C). The terminal D-Ala–D-Ala dipeptide of the pentapeptide is presynthesized from L-alanine in two steps, both inhibited by D-cycloserine, an alanine analogue (Fig. 1B). The final product of this phase is the UDP-MurNAc-pentapeptide precursor. (B) In the lipid cycle phase, P-M-pentapeptide is transferred from the UDP nucleotide precursor to undecaprenol-phosphate (Lipid-P). The subunit of murein is completed by the addition of GlcNAc (G) and any cross-bridge amino acids (not shown). The completed subunit of murein is then transferred to the exterior surface for polymerization. "Pep" represents a generic peptide subunit, whose terminal D-Ala residues may have been removed by carboxypeptidase action. (C) The final phase is initiated by polymerization of the peptidoglycan by vancomycin-sensitive transglycosylases. The released Lipid-PP must be hydrolyzed to Lipid-P by the bacitracin-inhibited pyrophosphatase before it can function again as an acceptor. The nascent polymerized peptidoglycan is then cross-linked by transpeptidases, the targets of β-lactam antibiotics.

from L-alanine (alanine racemase and D-Ala-D-Ala ligase) are inhibited by antibiotic alanine analogues. D-Cycloserine (Fig. 1B), the only example of this class of antibiotics in current clinical use, preferentially inhibits the ligase at its minimal growth inhibitory concentration (MIC) for a given bacterium. At higher concentrations, D-cycloserine inhibits alanine racemase and also D-glutamate synthesis from D-alanine by transmidation (not shown). The selective toxicity of this antibiotic is readily understood since D-alanine and D-glutamate play no known role in mammalian metabolism (indeed, animal D-amino acid oxidases are probably designed to destroy these bacterial products). D-Cycloserine has a broad spectrum of activity, being most active against gram-positive bacteria. It is also unique among murein synthesis inhibitors in being active against mycobacteria. Unfortunately, however, neurological toxic effects unrelated to its activity against murein synthesis limit the usefulness of this antibiotic: it is, at best, only a third-line drug for treatment of TB.

The first unique step in murein synthesis, the addition of enolpyruvate to UDP-GlcNAc (Fig. 4A), is inhibited by phosphonomycin (fosfomycin, Fig. 1C), a structural analogue of phosphoenolpyruvate (PEP). Although PEP is a common metabolic intermediate, the action of fosfomycin is apparently limited to inhibition of its utilization by this unique bacterial enzyme, accounting for the selective toxicity of this antibiotic for bacteria. Being small and hydrophilic, fosfomycin has a broad activity spectrum that includes most gram-negative bacteria. Unfortunately, however, its target is cytoplasmic, and organophosphates can only traverse the cytoplasmic membrane by exploiting a few specific permeation mechanisms. Those used by fosfomycin are the glycerophosphate and glucose-6-phosphate permeases. Because the loss of permease activity by mutation is nonlethal, *in vitro* resistance readily occurs. Principally for this reason, this antibiotic has not been released for clinical use in the United States. It appears, however, that the permease mutants are crippled as *in vivo* pathogens, and this antibiotic is successfully used in other countries such as Japan. Resistant mutants in the target enzyme also occur. More recently reported high-level resistance in gram-negative bacteria and Staphylococcus epidermidis is due to plasmid-mediated enzymes that inactivate fosfomycin. The gram-negative enzymes catalyze the production of glutathione adducts.

Daptomycin, a cyclic lipopeptide active only against gram-positive pathogens, showed excellent bactericidal activity *in vitro* against major pathogens such as methicillin-resistant *S. aureus* (MRSA) and multiply resistant enterococci, but has given disappointing results in clinical trials. Daptomycin, at its MIC, inhibits the synthesis of lipoteichoic acid, a polymer found only in gram-positive bacterial walls that has been implicated in the control of autolysin activity in several genera. At slightly higher concentrations, daptomycin inhibits synthesis of the nucleotide–pentapeptide precursor of peptidoglycan. Both effects may result from the dissipation of the membrane electrical potential, inhibiting the uptake of amino acids required for synthesis of cell wall polymers. Eradication of life-threatening enterococcal infections may require the use of a mixture of bactericidal antibiotics, and daptomycin has been reported to act synergistically with both fosfomycin and aminoglycosides against enterococci. While vancomycin tends to enhance the nephrotoxicity of the aminoglycosides, daptomycin appears to ameliorate this problem.

## B. Second Phase of Murein Synthesis: The "Lipid Cycle" (Bacitracin, Vancomycin, and Related Antibiotics)

The second phase of peptidoglycan biosynthesis takes place on the inner surface of the cytoplasmic membrane and is initiated by transfer of phospho-MurNAc-pentapeptide from the Park nucleotide to undecaprenylphosphate (Lipid-P; Fig. 4B), the membrane-bound anchor for construction of the peptidoglycan subunit. The disaccharide repeating unit of the glycan is produced by the addition of GlcNAc (Fig. 4B), followed by other species-specific modifications such as the addition of cross-bridge amino acids and the amidation of free amino groups (not shown). Ramoplanin, a new lipoglycopeptide antibiotic active against gram-positive bacteria, inhibits GlcNAc addition to the lipid intermediate. Although toxicity may limit use of this antibiotic to topical applications, its activity against MRSA and vancomycin-resistant enterococci is potentially promising. Following transfer of the lipid-linked completed subunit to the exterior of the cytoplasmic membrane, the third phase is initiated by polymerization of the glycan by transglycosylation, the target of vancomycin action (Fig. 4C). The chain grows from its reducing end (Fig. 3C), and the addition of each disaccharide–peptide subunit releases a molecule of Lipid-PP. Lipid-PP must be hydrolyzed by a membrane-bound pyrophosphatase before it can be reused as a carrier for glycan polymerization (Fig. 4B). This is prevented by bacitracin, which binds to polyisoprenyl-pyrophosphate derivatives, particularly the Lipid-PP substrate for the pyrophosphatase.

Bacitracin A (Fig. 1E), the active and principal component in commercial bacitracin, a secreted product of *Bacillus licheniformis,* is a hydrophobic cyclic dodecapeptide containing several D-amino acids. Its activity is restricted to gram-positive bacteria, treponemes and *Neisseria,* and its use is confined to topical use since it is too toxic to kidney function for systemic use. It is bactericidal for sensitive bacteria, and both the inhibition of murein synthesis and the nonspecific membrane disruption caused by formation of the lipid–bacitracin complex probably contribute to the lethal activity of this antibiotic. Since permeation of the cytoplasmic membrane is not required for bacitracin to reach its target and no direct interaction with an enzyme is involved, the only obvious mechanism for resistance to bacitracin is cell wall impermeability, as in gram-negative bacteria. There are no reports of significant clinically acquired resistance in sensitive gram-positive bacteria.

Vancomycin and teicoplanin, a newer member of this class of glycopeptide antibiotics, were regarded as minor weapons in the antibiotic armamentarium until quite recently. They are large molecules and, therefore, limited in spectrum to gram-positive bacteria. The β-lactams provide a less expensive, less toxic, and broader spectrum alternative for treating most gram-positive bacterial infections. If allergy to β-lactam antibiotics prevents their use, then a bacteriostatic agent frequently suffices as an alternate. Several circumstances have changed this picture. First, the rapidly increasing population of patients with severely compromised immunity has increased the demand for use of bactericidal antibiotics, particularly in tertiary care hospitals. Second, the appearance of high-level resistance to β-lactam antibiotics in diverse gram-positive species (see Section IX) requires the use of alternative bactericidal agents. Vancomycin is currently the best choice. In 1989, vancomycin was already second among antibiotics in total cost of use in hospitals in the environs of Paris.

Like bacitracin, vancomycin does not bind directly to an enzyme, but rather to a substrate, in this case the D-Ala-D-Ala C termini of murein biosynthetic precursors. This binding, presumably to nascent lipid intermediates and nascent peptidoglycan, inhibits transglycosylation (Fig. 4C) and consumes the available pool of undecaprenyl phosphate. Transpeptidation would presumably be inhibited if glycan polymerization persisted. Unexpectedly, increasing numbers of vancomycin-resistant enterococci have been re-

ported. These carry a conjugal plasmid encoding structural genes called *vanA, vanH,* and *vanY. vanA* encodes a D-Ala-D-Ala ligase with broad substrate specificity, *vanH* encodes a dehydrogenase that produces D-lactate, and *vanY* encodes a carboxypeptidase that destroys the normal D-Ala-D-Ala dipeptide. This results in the production of nucleotide peptidoglycan precursors terminating in D-Ala-D-lactate. These depsipeptides are functional in transpeptidation but fail to bind vancomycin. A resistance mechanism dependent on such extensive remodeling of crucial elements of peptidoglycan synthesis was completely unanticipated; this pathway presumably evolved in some environment where it provides selective advantage. Teicoplanin action is less affected by this resistance mechanism; however, it does not provide a solution to the serious clinical problem presented by vancomycin resistance in enterococci.

Moenomycin and related glycolipid antibiotics are structural analogues of the lipid intermediate in peptidoglycan synthesis which inhibit peptidoglycan transglycosylase. Their large size and hydrophobicity restrict their activity to gram-positive pathogens. Because of toxicity, their use is restricted to veterinary medicine.

## C. Third Phase of Murein Synthesis: Cross-Linkage of Peptidoglycan and Its Inhibition by β-Lactam Antibiotics

Polymerized nascent peptidoglycan is soluble and only becomes functional when it is cross-linked to preexisting insoluble murein by transpeptidation. The reaction, as shown in Fig. 4C, results in cleavage of the terminal D-alanyl-D-alanine bond of the stem pentapeptide in a donor peptide unit and linkage to an amino acceptor on an acceptor peptide unit, presumably attached to a different glycan strand. As shown in Fig. 5, this reaction, which takes place outside of the cytoplasmic membrane in the absence of an energy source, takes place in two stages. Initial formation of an acyl-D-alanyl-enzyme intermediate (Fig. $5_1$) conserves the peptide bond energy. Release of D-alanine is followed by binding of the amino acceptor and a second transacylation reaction, creating the peptide cross-link (Fig. $5_2$). In an analogous reaction, use of water as an acceptor results in a D,D-carboxypeptidase action (Fig. $5_3$). Many bacteria have both transpeptidases and D,D-carboxypeptidases; the balance of their activities may control the extent of murein cross-linkage. Both activities are inhibitable by β-lactam antibiotics. The β-lactam antibiotics were hypothesized to

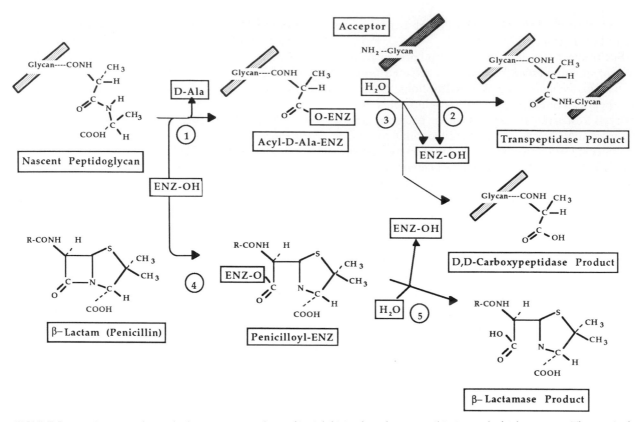

**FIGURE 5** Mechanisms of peptidoglycan transpeptidase, of its inhibition by β-lactam antibiotics, and of β-lactamases. The terminal acyl-D-alanyl-D-alanine component of a peptidoglycan subunit, newly incorporated into a growing glycan chain, is shown at the upper left (nascent peptidoglycan). (1) This donor substrate for transpeptidation reacts with a transpeptidase (a PBP: ENZ-OH) releasing D-alanine and producing an acyl-D-Ala-ENZ intermediate. (2) Binding of the amino acceptor of a second peptide subunit on a separate strand results in formation of the transpeptidase product (cross-linked murein). The transpeptidase is regenerated. (3) D,D-Carboxypeptidases (ENZ-OH) use the same mechanism, but the acyl-D-Ala-ENZ intermediate is hydrolyzed to release the product (lacking its terminal D-alanine) and free enzyme. A β-lactam antibiotic, such as the penicillin (R = benzyl, etc.) shown in the lower left, mimics the acyl-D-alanyl-D-alanine donor substrate for transpeptidases (Fig. 1D) and D,D-carboxypeptidases. (4) Penicillin reacts with ENZ-OH to produce penicilloyl-ENZ, which is inactive and relatively stable. (5) A class A or C β-lactamase uses the same reaction mechanism, but the pencilloyl-ENZ intermediate is very rapidly hydrolyzed to produce inactive antibiotic, regenerating active β-lactamase.

be transition state analogs of the acyl-D-alanyl-D-alanine substrates for these enzymes (Fig. 1D), having the appropriate conformation for facile acylation of the active sites of both transpeptidases and D,D-carboxypeptidases (Fig. 5₄). They act as substrates for the initial step that normally produces the acyl-D-alanyl-enzyme adduct (Fig. 5₁). Sequence analysis has demonstrated that an identically located serine residue in both types of enzyme is acylated by both the normal substrate and β-lactams. X-ray crystallographic structural analyses demonstrate that acyl-D-alanyl-D-alanine peptide substrates and β-lactam inhibitors bind in identical fashion to the active site of a streptomycete D,D-carboxy-

peptidase, exactly as predicted by the structural analogy hypothesis. The β-lactams act as suicidal substrates, producing a stable inactive enzyme adduct (Fig. 5₄). The use of labeled penicillin (or other β-lactam antibiotics), therefore, allows the transpeptidases and D,D-carboxypeptidases to be identified as penicillin-binding proteins. The analysis of PBPs, most advanced in *E. coli* (see below), has demonstrated the highly complex organization of murein synthesis catalyzed by these enzymes. Certain smaller PBPs are capable of acting as potentially autolytic D,D-endopeptidases by reversal of the transpeptidation mechanism. Other high molecular weight, autolytic endo-*N*-acetylmuramidases have also been postulated to be PBPs, acting by a reversal of their transglycosylase role (see below).

## D. Penicillin-Binding Proteins of *E. coli*

The PBPs of *E. coli* are all localized within the cytoplasmic membrane as transmembrane proteins with their larger, catalytically active N-terminal domains exposed on the periplasmic face. They are numbered in order of decreasing molecular weight, and the smaller PBPs (4, 5, and 6) are principally D,D-carboxypeptidases. PBP 4, however, also functions to modulate the cross-linkage of preformed murein as either a transpeptidase or a D,D-endopeptidase. The activities of these smaller PBPs are not essential to cell survival. In contrast, the high molecular weight PBPs, 1a, 1b, 2, and 3, are all essential for cell survival, although 1a and 1b have a considerable redundancy of function. Inhibition of PBP 1a and 1b, 2, or 3 is lethal for growing cells. PBP 1a and 1b function in cylindrical wall synthesis in the elongation of the rod-shaped *E. coli* cell, whereas PBPs 2 and 3 function in septum formation and morphogenesis. It was predicted that efficient cross-link formation would require direct coupling of transpeptidation to peptidoglycan polymerization by transglycosylation to ensure efficient juxtaposition of polymeric donor and acceptor substrates. It was subsequently shown that PBPs 1a, 1b, 2, and 3 are all bifunctional transglycosylase–transpeptidases. This helps to explain why vancomycin, binding to the carboxy-terminal end of the peptidoglycan precursor, is able to inhibit transglycosylation at the distal end of this precursor.

The morphological consequences of exposure of growing *E. coli* cells to β-lactam antibiotics reflect the relative affinity of each antibiotic for individual PBPs. Thus, preferential inhibition of PBPs 1a and 1b (e.g., by cephaloridine) causes immediate lysis, whereas preferential inhibition of PBP 3 (e.g., by ampicillin and many other β-lactam antibiotics) causes inhibition of septation, chain formation, and swelling and lysis at normal sites of septation. Inhibition of PBP 2 (mecillinam and imipenem) prevents initiation of cell elongation, leading to the production of rounded cells which eventually lyse. *Escherichia coli* thus possesses multiple independently essential PBPs and therefore has multiple targets for β-lactam antibiotic action. While the general structure and functions of PBPs are conserved in all bacterial genera investigated, cell cycle-specific functions have been described for very few of them. Thus, while most Eubacteria are sensitive to β-lactam antibiotics and critical PBP targets have been identified in many, their more precise role in cell wall growth is usually ill-defined. As described in Section IX, the resistance of MRSA to all currently available β-lactam antibiotics is the result of acquisition of the gene for a novel PBP called PBP 2a. This single enzyme is, presumably, able to perform all of the complex functions required for growth of this organism.

β-Lactamases are bacterial enzymes that hydrolyze and inactivate β-lactam antibiotics, frequently providing resistance to these antibiotics. A natural extension of the predicted role of β-lactams in inhibition of transpeptidases was the proposal that β-lactamases might be evolutionarily related to the transpeptidases, employing the same catalytic mechanism, except that the acyl enzyme intermediate would be subject to rapid hydrolysis (Fig. $5_5$). This mechanism has been demonstrated for Class A and Class C β-lactamases and has led to the design of efficient inhibitors. Such inhibitors (clavulanic acid, etc.), by sparing β-lactamase-susceptible β-lactam antibiotics, act synergistically in combination with these antibiotics against β-lactamase producing bacteria. This class of β-lactamases has little significant sequence homology to the PBPs. Nevertheless, X-ray crystallographic analysis demonstrates that they share the same overall structure, consistent with evolution from a common ancestral gene. A minor class of β-lactamases, found in a few gram-positive bacteria, are zinc-containing enzymes that use a quite different mechanism of hydrolysis.

## VIII. MECHANISMS OF LETHALITY: AUTOLYSINS, PERSISTENCE, AND TOLERANCE

Inhibitors of murein synthesis cause lysis of growing cells of most sensitive bacteria. However, certain species, such as group A streptococci, are killed without gross lysis. Lysis is a consequence of autolysin activity and is clearly irreversible; autolysin activation is frequently invoked as the direct cause of lethality of cell wall synthesis inhibitors. Like many generalizations in bacterial physiology, this is probably an oversimplification. Pneumococci are normally exquisitely sensitive to penicillin and lyse rapidly at their MIC. Mutants lacking their major autolysin remain just as sensitive to penicillin, their MIC is unchanged, but cell death is retarded. Growth inhibition occurs, but is reversible for some time. This is called tolerance;

additional mutations can prolong survival, enhancing tolerance. Tolerance could affect the outcome of infectious disease only if immune mechanisms at the site of infection are grossly defective. Both naturally tolerant bacteria, such as commensal oral streptococci (e.g., *S. sanguis*), and tolerant mutants of other streptococci have been implicated as occasional causes of treatment failure in bacterial endocarditis; phagocytic defenses function poorly in the avascular tissue of infected heart valves so that an effective bactericidal antibiotic response is needed for successful treatment.

A well-documented phenomenon related to tolerance is the antagonism between bactericidal antibiotics, such as the β-lactams, and bacteriostatic antibiotics, such as the majority of protein synthesis inhibitors (tetracyclines, chloramphenicol, etc.). Autolysin activation is a normal and necessary event in the bacterial division cycle and is required both for morphogenesis to accommodate growth and for cell separation. This activation is prevented if growth stops because of nutrient deprivation or the action of a bacteriostatic antibiotic. Growth inhibition causes a repression of autolysin activity that is dominant over the unbalanced autolysin activation induced by the β-lactams in growing bacteria. Growth inhibition, likewise, must abrogate alternative lethal mechanisms.

Persistence, a phenomenon noted early in penicillin-treated staphylococci, results in survival of about 1 in $10^6$ of a growing culture of bacteria during prolonged exposure to normally lethal antibiotic concentrations. These "persisters" can repopulate the culture on antibiotic withdrawal. Because the new population behaves like the former, they are not mutants. They may represent the prokaryotic equivalent of a resting ($G_0$) cell state. Reentry of these nongrowing cells into the normal cell cycle is apparently prevented in the presence of murein synthesis inhibitors. Mutations in two distinct genetic loci in *E. coli* can increase the frequency of persisters to about $10^{-3}$. The mechanism is not understood, and clinical problems related to such a phenotype have yet to be reported. In general, whereas tolerant or persistent mutants may cause problems in individual patients, the clinical impact is insignificant in comparison to acquired resistance to β-lactam antibiotics.

## IX. ACQUIRED RESISTANCE TO β-LACTAM ANTIBIOTICS: A RUNNING BATTLE

Classical mechanisms for acquired resistance to antibiotics include reduced access to the target, reduced target affinity, and antibiotic inactivation. While examples of all of these mechanisms are known for the β-lactam antibiotics, inactivation by β-lactamases was, for many years, the only mechanism of recognized importance. The modern antibiotic era was initiated by the widespread use of penicillin following the second world war. Efficacy against many life-threatening infections, such as pneumococcal pneumonia and *S. aureus* bacteremia, was dramatic. The ability of common bacterial pathogens to adapt to such an apparently novel antibiotic challenge was starkly illustrated when, within a few years, penicillin resistance in *S. aureus* became so prevalent that this antibiotic became almost useless for the treatment of staphylococcal infections. Resistance was due to the acquisition of plasmids carrying the genes for inducible β-lactamase production. Clearly, the genetic mechanisms already existed, having presumably evolved during the eons of trench warfare in the soil between antibiotic producers, such as the streptomycetes and fungi, and their potential bacterial competitors. The rapidity with which these genes appeared in staphylococci attests to the efficiency of the mechanisms designed for the transfer of genes between bacteria: mobilization of resistance genes on composite transposons, transfer to plasmids, and transfer between cells by conjugation, phage transduction, and transformation. Fortunately, the staphylococcal β-lactamase proved to have a rather narrow substrate specificity, and the search for resistant semisynthetic penicillin derivatives soon produced methicillin and related antibiotics such as cloxacillin and nafcillin. When the cephalosporins were developed, they also proved to be relatively resistant to the staphylococcal β-lactamase, providing a wide choice for the treatment of staphylococcal infections. Because the staphylococcal β-lactamase also seems incapable of mutating to hydrolyze these modified antibiotics efficiently, they remain effective. Moreover, β-lactamase-mediated resistance is confined to the staphylococci and *Enterococcus faecalis* among gram-positive pathogens. It is fortunate and also surprising, for example, that β-lactamase production has not been reported in streptococcus pyogenes ("group A strep") or in pneumococci.

This simple pattern of β-lactamase production in gram-positive pathogens is in stark contrast to the gram-negative pathogens, which produce a bewildering array of plasmid-borne β-lactamases of widely variant specificity, capable of rapid transmittal within related bacterial genera. These plasmids are examples of the "R" (resistance) factors prevalent in gram-negative bacterial pathogens. They mediate "infectious" antibiotic resistance, the rapid spread of resistance through a bacterial population by plasmid-encoded conjugation mechanisms which promote direct cell to cell plasmid transfer. Some of these plasmids are also capable of transfer between widely separated genera at low efficiency. For example, enyzmes initially recognized in enteric gram-negative bacilli have appeared in *Neisseria* and *Haemophilus,* greatly complicating the treatment of meningitis and gonorrhea.

If β-lactamases are to provide protection to bacteria, they must destroy β-lactam antibiotics before they reach their PBP targets. β-Lactamases must, therefore, be secreted enzymes. In staphylococci, because of the relatively porous cell wall structure, the enzyme diffuses freely into the medium. Large amounts must be produced to be effective, and because of the large energy demand, production is, necessarily, inducible. Gram-negative bacteria have a much more efficient mechanism: secretion into the periplasm (Fig. 2E) provides an extremely high local concentration of β-lactamase. Its efficacy against any β-lactam antibiotic depends both on the rate of antibiotic hydrolysis and on the rate at which the antibiotic permeates the outer membrane of the cell. Where permeation is slow, as in *Pseudomonas,* even an inefficient β-lactamase will provide protection. Several genera (*Enterobacter, Pseudomonas,* etc.) produce chromosomally encoded β-lactamases, but only at low levels unless induced by exposure to a β-lactam. In *Enterobacter,* mutations causing high-level constitutive production of these enzymes, coupled with the relative impermeability of their outer membranes, render these bacteria resistant to all but a few of the available β-lactams. Such mutant strains are now prevalent in some modern tertiary care hospitals.

The variety of β-lactamases in gram-negative enterics, together with the high level of intrinsic resistance in *Pseudomonas aeruginosa* has, for many years, provided the major battle ground driving the search for β-lactam antibiotics resistant to all of these enzymes and better able to permeate the gram-negative cell wall. This has led to the development of literally hundreds of penicillin and cephalosporin derivatives, carbapenems such as imipenem and monobactams such as aztreonam. Porin mutations resulting in reduced uptake and PBP mutants reducing sensitivity are also reported as contributors to clinical resistance. However, unlike the β-lactamase mutations, such mutations are likely to reduce bacterial survival in the absence of antibiotic selection, preventing them from becoming fixed in the bacterial population. Thus the major battle front among antibiotic-resistant gram-negative bacteria has remained the β-lactamases. Al-

though the race has always been close and constant modifications of these antibiotics have never provided more than temporary respite, some complacency has resulted from the apparent success of the pharmaceutical industry in keeping pace with bacterial adaptation. Of late, however, this complacency has been shaken by the appearance of high level resistance in previously sensitive gram-positive pathogens (see below), by the *Enterobacter* mutants described earlier and by the appearance of point mutations in the β-lactamases prevalent in all enterics allowing them to hydrolyze the very latest "β-lactamase-resistant" antibiotics. The process by which resistant organisms are selected has been accelerated by the very success of antibiotic treatment. This has allowed the survival of an ever-increasing population of immune-compromised patients, susceptible to disease caused by such opportunistic pathogens as the enterics, *Pseudomonas* and enterococci, and who represent the ideal milieu for selection of resistance in these organisms. Complacency has been replaced by watchful trepidation in the face of the appearance of new classes of resistance, especially among the familiar gram-positive pathogens which had seemed to be well under control. This problem has been compounded by the apparent universal failure in the search for new classes of antibiotics of comparable potential utility.

The pneumococcus remains the major cause of community-acquired bacterial pneumonia, just as *Neisseria gonorrhea* remains a major cause of sexually transmitted disease. A slow increase in the dose of penicillin necessary for treatment of both types of infection was observed over many years of use, probably resulting from the accumulation of multiple point mutations in PBPs. Relatively high cell wall permeability makes β-lactamase production a less effective strategy for resistance in these genera than in enterics. More recently, however, profound β-lactam resistance has been found in both genera and results from major alterations in critical PBPs. The appearance of similar mutants in *N. meningitidis* is another cause of grave concern. The high transformation efficiency of both pneumococci and neisseria allows the efficient spread of chromosomal mutations within each genus, and the modified genes are mosaics formed by multiple independent recombination events with PBP genes from distantly related organisms. High-level resistance in pneumococci, first recognized in South African isolates in 1977, reached a peak of 50–60% of isolates in Hungary and Spain in the late 1980s. Clinicians in these countries had for years relied on the heavy use of penicillin in therapy. Following a switch to alternate drugs, the frequency of this type of pneumococcal mutant has declined significantly, perhaps indicating a slight growth disadvantage for the mutants. Although this is a cause for hope, it may represent another temporary respite.

MRSA arise from the acquisition of a novel β-lactam-resistant PBP on a large DNA fragment of unknown source. This event appears to have occurred only once, but has been transferred to *S. epidermidis* (MRSE) and other coagulase-negative staphylococci of increasing importance as opportunistic pathogens. MRSA are a major problem worldwide; such strains are frequently resistant to almost all common antibiotics due to the presence of multiple resistance genes. When first introduced, the fluoroquinolones such as ciprofloxacin were highly effective against MRSA. Unfortunately, resistant mutants appeared rapidly, and vancomycin is the only remaining effective bactericidal antibiotic currently available. For this reason, the probably inevitable transfer of vancomycin resistance from enterococci, already seen in the laboratory, is awaited with deep concern. Experimental drugs such as daptomycin and ramoplanin seem unlikely to be useful alternatives. *Streptomyces pyogenes*, the cause of strep throat, has received recent prominence as the "flesh-eating" bug causing rapidly fatal infections in a small number of individuals. Although it is not clear that a real increase in the frequency of this type of pathology has occurred, indicating a significant increase in pathogenicity, *S. pyogenes* remains a dangerous pathogen if not treated effectively. Fortunately, it is an anomaly in having failed to develop significant penicillin resistance. If both vancomycin and β-lactam resistance become widespread in staphylococci and streptococci, we may be forcibly reminded of some of the worst aspects of the preantibiotic era.

In conclusion, the continued efficacy of the β-lactam antibiotics will depend on better control of their use, on the continued development of new modifications, and perhaps on increased sophistication in clinical laboratory methods for assessing sensitivity. New classes of murein synthesis inhibitors would obviously be welcome, although the PBP remains the ideal target because of its location. Potentially useful β-lactam-like antibiotics with five-membered (γ) lactam rings have been isolated. Drugs capable of directly triggering autolysis might be ideal antibiotics, but a better understanding of the mechanisms involved is probably needed before effective criteria for search or development can be developed. Since lipopolysaccharide production is essential for gram-negative cell viability and ketodeoxyoctanate is a universal and unique component, inhibitors of its synthesis have been developed as agents specific for gram-negative bacteria. They work well *in vitro,* but their utility is compromised by the lack of reliable delivery to their cytoplasmic target and hydrolysis in serum. Recently described inhibitors of lipid A biosynthesis hold more promise.

## BIBLIOGRAPHY

Actor, P., Daneo-Moore, L., Higgins, M. L., Salton, M. R. J., and Shockman, G. D. (eds.) (1988). "Antibiotic Inhibition of Bacterial Cell Surface Assembly and Function." American Society for Microbiology, Washington, D.C.
Queener, S. F., Webber, J. A., and Queener, S. W. (eds.) (1986). "Beta-Lactam Antibiotics for Clinical Use." Dekker, New York Basel.
Tipper, D. J. (ed.) (1987). "Antibiotic Inhibitors of Bacterial Cell Wall Biosynthesis." Pergamon Press, Oxford, England.
Cohen, M. L., Bloom, B. R., Murray, C. J. L., Neu, H. C., Krause, R. M., and Kuntz, I. D. (1992). The Crisis in Antibiotic Resistance (and related articles). *Science* **257,** 1050–1082.
Travis, J., Culotta, E., Stone, R., Gabay, J. E., Davies, J., Nikaido, H., and Spratt, B. G. (1994). Resistance to Antibiotics. *Science* **264,** 359–393.

# Antibiotics

JOHN H. HASH
*Vanderbilt University*

## GLOSSARY

**Gram-positive, gram-negative** Bacterial reaction to a staining procedure developed by Christian Gram for the microscopic observation of bacteria. In his method, gram-positive bacteria appear blue and gram-negative bacteria pink; Gram's stain identifies two groups of bacteria with different surface components

**Microbial antagonism** Inhibition of one species of microbe by chemicals produced by another species

**Mutation** Change in the sequence of DNA

**Plasmid** Autonomously self-replicating extrachromosomal circular DNA

**Ribosomes** Subcellular ribonucleoprotein organelles that are the sites of protein synthesis; in bacteria the 70S ribosome consists of a 30S and a 50S subunit

THE ORIGINAL DEFINITION OF AN ANTIBIOTIC (GR., *anti*, against; Gr., *bios*, life) was any chemical substance produced by a microorganism that is capable of killing or inhibiting the growth of another microorganism. This definition now requires some revision because of synthetic modifications or even total chemical synthesis of some antibiotics. As used in this article, the term antibiotic includes all antimicrobial substances of microbial origin, even though chemical synthesis may have replaced natural production for some, while others may have been chemically modified to form semisynthetic derivatives. The definition may require additional revision should useful antibiotics be found in higher plants and animals.

## I. HISTORY

The discovery of penicillin by Sir Alexander Fleming in 1929 is widely regarded as the beginning of the "Era of Antibiotics" or the "Age of Antibiotics." Whichever term is used, there is no question that antibiotics have revolutionized the practice of human and veterinary medicine, have contributed to improved animal nutrition and to crop and plant protection, and have aided in the understanding of basic cellular metabolism by providing powerful tools in elucidating metabolic pathways.

Prior to the discovery of penicillin, early microbiologists, especially Pasteur and Koch, demonstrated that infectious diseases were caused by specific microbes that could be isolated in pure culture, thus paving the way for the experimental investigation of infectious diseases. The principle of microbial antagonism between different species of bacteria was recognized by Pasteur and ways were explored to use this antagonism, later called antibiosis, in infectious disease therapy. Crude extracts of the antagonizing organism were applied topically to treat infected wounds. One such attempt produced mixed results when a preparation called pyocyanase, produced by a bacterium now known as *Pseudomonas aeruginosa*, was used topically and parenterally to treat infections. The preparation was considered to contain an enzyme, hence its name, but it is clear that whatever efficacy pyocyanase had was due to antibacterial substances produced by this organism. Today, pseudomonic acid (mupirocin), a minor antibiotic from *Pseudomonas fluorescens*, is

ENCYCLOPEDIA OF HUMAN BIOLOGY, Second Edition, VOLUME 1. Copyright © 1997 by Academic Press. All rights of reproduction in any form reserved.

used in the topical treatment of skin infections. The early use of pyocyanase contributed a very significant idea; namely, microbes can produce substances that have an effect on bacterial cells that is different from effects on cells of the human host. This idea of selective toxicity is a cardinal requisite for all antibiotics used in present-day chemotherapy. Also of historical interest is the possibility that some folk medicine practices (such as the application of soy-flour poultices to infected wounds) may have owed their efficacy, if any, to antibiotics produced by molds growing in the poultices.

In 1929, after observing that a mold contaminant, *Penicillium notatum,* lysed a staphylococcal colony on an agar plate, Fleming grew the *Penicillium* and found that it secreted a soluble substance that was highly inhibitory to gram-positive bacteria. He also noted that this substance, which he named penicillin (now penicillin G), did not harm human white blood cells, but he failed to recognize its chemotherapeutic potential. It was not until 1939 that penicillin was isolated and its full therapeutic potential was recognized. Methods were developed for its large-scale production, which allowed it to play an important role in World War II. During the war years, additional agents (actinomycin and streptomycin) were discovered as the result of a deliberate search, and the word antibiotic was coined to describe these naturally occurring antimicrobial compounds. Following the war, efforts were intensified to find new antibiotics, and much of the effort shifted from university research laboratories to pharmaceutical companies. In rapid succession, new antibiotics were found, including chloramphenicol, the tetracyclines, and erythromycin. The first two were notable because they were broad-spectrum antibiotics that inhibited the growth of many species of gram-positive and gram-negative bacteria, thereby extending therapy beyond the gram-positive spectrum observed for penicillin G.

The search for antibiotics turned up thousands of compounds that had antimicrobial properties. But most of these lacked the requisite selective toxicity (i.e., toxicity to bacterial but not human cells) to be used as antimicrobial agents in human and veterinary medicine. A few exceedingly cytotoxic antibiotics have found limited use as antitumor agents. However, the dozen or so antibiotics that do have selective antimicrobial properties have truly revolutionized the practice of medicine. The search continues for new antibiotics, although on a diminished scale, and

sources such as higher plants and marine animals are being investigated as potential sources of new antibiotics. [*See* Antimicrobial Drugs.]

The importance of antibiotics in global terms can be appreciated by the amounts used on a worldwide basis. Figures for 1980 estimate that 16 million kilograms of penicillins alone were produced for medical and veterinary use. Annually, over 100 million kilograms of antibiotics are used for all purposes.

## II. CLASSIFICATION

Antibiotics represent an extremely broad group of natural compounds, mostly of relatively low molecular weight. Many types of organic structures are found in antibiotics and often their presence in these molecules represents their only occurrence in nature. Although a rigid chemical classification of the therapeutically useful antibiotics is difficult, there are families of antibiotics where a useful classification based on their common chemistry can be made. Table I gives a list of families and examples of major antibiotics. The suffixes -*mycin* and -*in* are part of the systematic nomenclature used to identify the majority of antibiotics.

In addition to sharing common chemistry, the members of a family of antibiotics also have a common mechanism of action that can be used to classify them. Some antibiotics, such as chloramphenicol, have sufficiently simple chemical structures that they can be produced by chemical synthesis. But for most antibiotics, commercial production uses living microorganisms because chemical synthesis is either not possible or is much more expensive. However, some antibiotics lend themselves to chemical modification, and semisynthetic antibiotics are the result. In this way thousands of semisynthetic penicillins and cephalosporins have been made, many of which have improved therapeutic properties. Some semisynthetic $\beta$-lactam antibiotics are listed in Table I and more can be expected to be developed in the future.

The ease of synthetic modification is not the same for all antibiotics and usually only minor chemical modifications are possible in the ansamacrolides, nonpolyene macrolides, polyene macrolides, cyclic peptides, and tetracyclines without loss of antibacterial activity (see Table I). As a result, the number of modified or semisynthetic antibiotics in these groups is far fewer than for the $\beta$-lactam group.

## TABLE I
### Families of Antibiotics

| Family | Examples |
| --- | --- |
| Aminoglycosides | Streptomycin, kanamycin, tobramycin, paromomycin, gentamicin, neomycin, amikacin |
| Ansamacrolides | Rifamycin SV, rifampicin |
| β-Lactams | |
|   Natural penams | Penicillin G, penicillin V |
|   Semisynthetic penams | Ampicillin, carbenicillin, amoxicillin, oxacillin, ticarcillin, methicillin |
|   Natural cephems | Cephalosporin C |
|   Semisynthetic cephems | Cephalothin, cephaloridine, cefazolin, cefotaxime, cefamandole, ceftazidime |
|   Carbapenems | Thienamycin, imipenem |
|   Clavams | Clavulanic acid[a] |
|   Oxacephems | Moxalactam |
|   Monobactams | Aztreonam |
| Chloramphenicol | Chloramphenicol, thiamphenicol |
| Cyclic peptides | Gramicidin, tyrocidine, polymyxins |
| Lincosamides | Lincomycin, clindamycin |
| Nonpolyene macrolides | Erythromycin, oleandomycin, tylosin, spiramycin, virginiamycin |
| Polyene macrolides | Amphotericin B, nystatin |
| Steroids | Fusidic acid |
| Tetracyclines | Tetracycline, chlortetracycline, minocycline, oxytetracycline, demeclocycline, doxycycline |
| Others | Bacitracin, cycloserine, fosfomycin, vancomycin, griseofulvin, novobiocin, spectinomycin |

[a] Clavulanic acid is weakly antibacterial, but is a powerful β-lactamase inhibitor. It is used in conjunction with β-lactam antibiotics to extend their lives in the presence of β-lactamase producing bacteria.

## III. BIOSYNTHESIS

The question of why microorganisms produce antibiotics has been posed repeatedly since the discovery of penicillin, but no completely satisfactory answer has emerged. As more and more antibiotics have been discovered, several unifying themes have been noted. First, the vast majority of antibiotics is produced by certain groups of organisms that produce spores as a survival mechanism (actinomycetes, fungi, and bacilli). Second, antibiotics are produced late in the culture cycle when limiting growth conditions have been reached due to the depletion of essential nutrients and the culture has entered the stationary phase prior to decline and death. It is in this period that sporulation occurs to provide a survival mechanism for better times. Last, it is also in this period that the organism produces secondary metabolites, a class of compounds that includes antibiotics. Secondary metabolites, in contrast to primary metabolites (amino acids, sugars, purines, pyrimidines, fatty acids, and all the other building blocks of macromolecular structures), are not essential for the growth of the organism. Seemingly, they have no vital role in metabolism, and mutants lacking the ability to produce antibiotics grow well in culture although they seem to have an inferior ability to survive under adverse conditions. A characteristic feature of secondary metabolites is that a microorganism usually produces a series of related compounds, including several chemically related antibiotics.

Various ideas have been put forth to explain why a cell makes these secondary metabolites. One idea is that they may play a role in protecting the producing microorganism from adverse conditions in the stationary phase of growth, thereby allowing its survival until environmental conditions again support growth. An adverse factor may be a toxic effect of normal primary metabolites, which are kept at low, nontoxic concentrations by continual use during growth. Because they continue to be produced but not utilized when growth is slowed or stopped, their concentrations might reach toxic levels. Secondary metabolism might then be a way to convert some of the primary metabolites as a process of detoxification. The antimicrobial properties of some secondary metabolites would then be accidental, and the particular antibiotic produced would simply be a function of the enzymatic complement of the organism. As previously noted, the majority of useful antibiotics is produced by spore-forming organisms; many attempts were therefore made to find a role for these metabolites in spore formation, but without success. In nature, the ability to excrete a secondary metabolite with antimicrobial properties would also provide survival value for the organism by giving it a selective growth advantage over the surrounding organisms that were inhibited. Such inhibition would be an example of the microbial antagonism noted by Pasteur.

The entire issue of the significance of secondary metabolites is extremely complex. The biosynthetic pathway required to produce an antibiotic is unique for each antibiotic family and, in many cases, very complicated. For example, the biosynthesis of streptomycin requires 15–20 enzymes to convert the primary metabolites glucose, ribose, and glucosamine to streptomycin. The enzymes involved, which are unique to streptomycin biosynthesis, catalyze standard biochemical reactions. Another example is the biosynthesis of tetracycline, which requires more than a dozen enzymes to convert the primary metabolites acetate and malonate to the antibiotic. In both cases the enzymes are unique to the biosynthesis of the specific antibiotic.

Antibiotic-producing organisms usually produce antibiotics in low concentrations, presumably because their synthesis is tightly regulated. To produce antibiotics on a commercial scale, the producing organisms are grown in large aerated tanks under optimal growth conditions of temperature, aeration, pH, and media composition. Also, for large-scale production, efforts are directed to selecting mutant strains that produce higher yields. The selection is based on the observation that structural genes for the enzymes involved in antibiotic biosynthesis, encoded in the DNA of the producing organism, are normally repressed by the activity of regulatory genes until unfavorable conditions arise in the stationary phase of growth prior to sporulation. Derepression of the genes then allows the synthesis of the enzymes. The selection is generally aimed at finding natural or artificially induced mutants altered in their regulatory genes in such a way that the structural genes are continually derepressed and the enzymes are made constitutively. When selection is successful, antibiotic production can be vastly increased, in some instances from micrograms per liter to grams per liter.

## IV. MECHANISM OF ACTION

Early in its use, penicillin was observed to cause the lysis of susceptible bacteria. However, the knowledge of cellular structure and physiology was too rudimentary for investigators to understand the molecular events involved. It became apparent that antibiotics could be used as tools in elucidating metabolic pathways and in understanding structure and function. Not only the therapeutically useful antibiotics but also some that are too toxic for medical use have played important roles in illuminating bacterial physiology.

Our current understanding of bacterial cell wall structure and function, membrane structure and function, ribosomal structure and function, and the synthesis of nucleic acids and proteins has depended greatly on the use of antibiotics. In turn, as bacterial metabolism has become better understood, prospects have been increased for the rational design of antimicrobials for specific target sites.

The selective toxicity that is required for therapeutic use of antibiotics depends on the presence of macromolecular structures in bacteria that are different from their mammalian counterparts. Targets of antibiotic inhibition are given in Table II. The major antibiotics either inhibit the synthesis of nucleic acid, protein, and cell walls or interfere with membrane function.

**TABLE II**
Modes of Antibiotic Action

| Activity inhibited | Antibiotic | Molecular action |
|---|---|---|
| DNA synthesis | Novobiocin | Inhibits DNA topoisomerase II |
| RNA synthesis | Rifampicin | Inhibits RNA polymerase |
| Protein synthesis | | |
| 30S ribosomal subunit | Streptomycin | Complexes with 30S protein |
| | Spectinomycin | Complexes with 30S protein(s) |
| | Tetracyclines | Inhibits binding of tRNA |
| 50S ribosomal subunit | Chloramphenicol | Complexes with 50S protein |
| | Erythromycin | Complexes with 50S protein(s) |
| | Lincomycin | Complexes with 50S protein(s) |
| Nonribosomal | Fusidic acid | Inhibits elongation factor G |
| Cell wall synthesis | Fosfomycin | Inhibits synthesis of muramic acid |
| | Cycloserine | Inhibits D-Ala-D-Ala synthetase |
| | Bacitracin | Inhibits undecaprenol pyrophosphatase |
| | Vancomycin | Inhibits murein polymerase |
| | Penicillin | Inhibits murein transpeptidase(s) |
| Membrane function | Polymyxin E | Dissolves lipid membranes |
| | Amphotericin B | Complexes sterols in membranes |

Perhaps the best understood example of selective toxicity is provided by the β-lactam antibiotics. All members of this class inhibit the synthesis of the rigid macromolecular structure of the bacterial cell wall, a structure that has no counterpart in mammalian cells. The cell wall structure, called peptidoglycan or murein, consists of polysaccharide chains cross-linked by short peptide bridges to yield a strong protective external layer. Any agent that interferes with the synthesis or integrity of the cell wall will cause lysis and death due to unbalanced growth, as was initially observed for penicillin. The bacterial cell wall is therefore a natural target for chemotherapy. Other antibiotics that owe their efficacy to inhibition of reactions involved in cell wall synthesis include fosfomycin, cycloserine, bacitracin, and vancomycin. Each of these agents inhibits a different reaction in cell wall synthesis. [See Antibiotic Inhibitors of Bacterial Cell Wall Biosynthesis.]

Antibiotics that inhibit protein synthesis owe their selective toxicity to differences between bacterial and mammalian ribosomes. Bacterial 70S ribosomes, the sites of protein synthesis, consist of two ribonucleoprotein subunits, 30S and 50S. Mammalian ribosomes are 80S and consist of 40S and 60S subunits. All of the therapeutically useful antibiotics in this group owe their efficacy to action on 70S ribosomes but not 80S ribosomes. It is of some interest that mammalian mitochondrial ribosomes are of the 70S variety and that protein synthesis in mitochondria is subject to inhibition by the antibiotics that inhibit bacterial protein synthesis. As shown in Table II, protein synthesis may be inhibited by action on either the 30S or the 50S subunits or on proteins that are related to protein synthesis but are independent of ribosomal structure. In most cases it has not been possible to describe in molecular terms the exact nature of the inhibition other than to identify a ribosomal subunit protein to which a particular antibiotic binds. In the case of tetracyclines, the transfer RNA bearing an amino acid is prevented from binding to the ribosome, thereby interrupting protein synthesis. [See Ribosomes.]

It has been difficult to find antibiotics that inhibit nucleic acid synthesis and yet have sufficient selective toxicity to be therapeutically useful as antimicrobial agents. Novobiocin, an antibiotic whose use is severely limited because of adverse side effects, interferes with DNA synthesis by inhibiting topoisomerase II, an enzyme concerned with supercoiling of DNA. The rifamycins inhibit RNA synthesis by forming complexes with RNA polymerase, thus inactivating the enzyme. [See DNA Synthesis.]

The cyclic peptides are surface active agents and act by dissolving the lipid components of membranes, thus killing the cell. The polyene macrolides, which are antifungal rather than antibacterial antibiotics, also act on the cytoplasmic membrane. They act by forming complexes with ergosterol in fungal membranes, creating channels that cause death of the cells.

## V. RESISTANCE

Within a short time after the introduction of penicillin into therapeutic use, it was noted that several resistant bacterial species produced the enzyme β-lactamase, which inactivates penicillins by hydrolysis of the β-lactam ring. This enzyme was found to be widely distributed in nature, raising the possibility that the usefulness of penicillin in medicine might be limited. However, semisynthetic penicillins, which are resistant to β-lactamase, have helped alleviate the problems of resistance and have extended the usefulness of the β-lactam antibiotics. As other antibiotics were discovered and used in therapy, resistance to them followed as a seemingly inevitable outcome.

In regard to resistance, it is useful to consider natural and acquired resistance. Natural resistance refers to inherent properties of an organism that confer insensitivity to an antibiotic. Examples include the insensitivity of bacteria to amphotericin B because they lack sterols in their membranes and the insensitivity of Mycoplasma to penicillin because they lack a cell wall. Acquired resistance represents the emergence of resistant organisms from sensitive populations following the use of antibiotics. Resistance, once acquired, remains a stable inheritable characteristic, indicating it has a genetic basis. Mutations normally occur in bacteria with a frequency of 1 in every million to 100 million cellular divisions ($10^{-6}$–$10^{-8}$). A major question was whether mutations to resistance are caused by the antibiotic or whether the antibiotic, by inhibiting sensitive cells, is simply selecting for a preexisting, naturally resistant mutant. The issue was settled decisively in favor of the latter mechanism: in the presence of the antibiotic the resistant organisms have a selective growth advantage. Widespread use of antibiotics in medicine fosters conditions that favor the survival of such resistant organisms. Bacterial resistance has proven to be a limiting factor in the usefulness of antibiotics, raising the specter that it might outstrip the ability to find new or modified antibiotics and return the treatment of infectious diseases to the preantibiotic era.

Acquired resistance to antibiotics may be either mediated by genes on chromosomes or plasmids or both. Examples of antibiotics where chromosomal mutations provide the basis for resistance include novobiocin, rifampicin, streptomycin, chloramphenicol, and erythromycin. For each of these antibiotics, mutations in the structural gene for the target protein (enzyme or ribosomal, Table II) produce an altered protein that no longer binds the respective antibiotic. The organism is then resistant to that antibiotic.

A new dimension in bacterial resistance was introduced in the late 1950s with the discovery of antibiotic resistance transfer factors. These resistance transfer factors are plasmids that carry antibiotic resistance genes in addition to genes necessary for their own propagation. On rare occasions a single plasmid may carry resistance genes for as many as eight antibiotics. These resistance genes are, for the most part, structural genes for enzymes that inactivate the antibiotic. Examples are enzymes that hydrolyze penicillins and cephalosporins ($\beta$-lactamase, which may also be chromosomal), enzymes that acetylate chloramphenicol and the aminoglycosides, and enzymes that phosphorylate or adenylylate the aminoglycosides. Resistance to the tetracyclines is also plasmid mediated but by a different mechanism: an alteration of the permeability of the organism prevents the antibiotic from reaching the ribosomal target site.

The presence of resistance plasmids that confer resistance to multiple classes of antibiotics compounds the problem. But the problem does not end there. The resistance plasmids can be transferred to other bacteria not only of the same species but also to other species, both related and unrelated. For example, pathogenic *Salmonella* or *Shigella* species infected with a plasmid carrying multiple antibiotic resistance genes might transfer it by conjugation to a normal inhabitant of the intestines, *Escherichia coli*, which in turn might transfer it to other enteric organisms. The spread of resistance by this mechanism is extremely rapid because it does not depend on cellular division, as is the case with chromosomal resistance. Although the $\beta$-lactamase gene may be carried on a plasmid and transferred among gram-negative organisms, its transfer in gram-positive organisms such as staphylococci is mediated either by bacteriophages (transduction) or by conjugal cell to cell transfer.

In practice, while both are important, resistance transfer plasmids pose a greater threat to the continued use of antibiotics than do chromosomal mutations.

## VI. MEDICAL USES

Following the initial success of penicillin in treating infections caused by gram-positive organisms such as staphylococci and streptococci, and the successes of broad-spectrum antibiotics such as chloramphenicol and the tetracyclines in treating a variety of infections due to gram-positive and gram-negative bacteria, antibiotics were labeled "miracle drugs." Indeed, it appeared that humankind might be substantially freed from much of the morbidity and mortality that has accompanied bacterial infections throughout human history. Chemotherapy, together with two other important methods of controlling infectious diseases—namely, environmental measures (safe water and milk supply, sanitary disposal of sewage, vector control, etc.) and augmentation of the immune system through vaccination—was widely expected to control, if not eliminate, most infectious diseases caused by bacteria. This hope, however, has not been realized despite the enormous progress made in the treatment of bacterial diseases. The discovery of the antifungal antibiotics amphotericin B and griseofulvin has added powerful agents against fungal diseases. There are still no useful antiviral antibiotics, but there are a few synthetic antiviral agents. [*See* Antimicrobial Drugs.]

An ideal antibiotic for human use would be one that selectively destroys an invading pathogen without affecting the normal bacterial flora or imposing any adverse reactions on the host. Such an ideal antibiotic has not yet been found; all antibiotics discovered to date have some undesirable side effects. Adverse reactions include allergy (penicillins, cephalosporins, demeclocycline, and novobiocin); nephrotoxicity (cyclic peptides, aminoglycosides, vancomycin, and amphotericin B); ototoxicity (aminoglycosides); blood disorders, including fatal aplastic anemia (chloramphenicol); intestinal disorders, including diarrhea and colitis (tetracyclines, lincosamides, and some of the oral penicillins and cephalosporins); and photosensitivity (some of the tetracyclines). Oral broad spectrum antibiotics often lead to disruption of the normal intestinal flora and subsequent suprainfection by naturally resistant opportunistic intestinal organisms such as yeasts and *Clostridium difficile,* whose growth is normally held in check by microbial antagonism of the normal flora. Such suprainfection may require additional antibiotic therapy with suitable antibiotics.

Practically every bacterial pathogen for which useful vaccines cannot be made is susceptible to one or more antibiotics, which provide the first line of defense against them. While there are many antibiotics for most gram-positive and gram-negative pathogens, relatively few are effective against mycobacteria, the causative agents of tuberculosis and leprosy. Streptomycin and rifampicin are effective against tuberculosis, but only rifampicin has any activity against leprosy. Because the tubercle bacillus has developed widespread chromosomal resistance against streptomycin, antibiotics are usually used in combination with synthetic antimycobacterial agents for both tuberculosis and leprosy. Another problem is seen with intracellular bacterial pathogens such as rickettsiae (spotted fever, typhus) and chlamydia (trachoma, psittacosis); these are affected by a narrow range of antibiotics, such as the tetracyclines and chloramphenicol, which can penetrate the host cell.

Bacterial resistance, more than adverse reaction, is the largest problem facing the future of antibiotic therapy. Resistance caused by chromosomal mutations can be dealt with more easily than plasmid-mediated resistance. As noted earlier, chromosomal mutation rates in bacteria are of the order of $10^{-6}$. For an organism to become resistant to two antibiotics simultaneously the rate would be $10^{-12}$, requiring an improbably large bacterial population to find a double mutant. Therefore, compatible antibiotics are generally given in combination to reduce the selection of resistant organisms. An example is the combination of an aminoglycoside with a $\beta$-lactam; because there is synergy between these two classes of antibiotics, their combined effect is even greater than the sum of their individual effects. Also, combination therapy is often employed to broaden empiric coverage in the case of polymicrobial infections. Such combinations allow lower doses of the antibiotics, thus minimizing adverse affects while making improbable the selection of doubly resistant organisms. Combinations of antibiotics may in some cases be antagonistic, one reducing the effect of the other. An example is the combination of penicillin and tetracycline, and such combinations are avoided in clinical medicine. Resistance acquired through resistance transfer plasmids is more threatening than chromosomal resistance because of the speed through which resistance to multiple classes of antibiotics may be transferred.

Sensitivity testing of the isolated pathogen is always important in order to identify an antibiotic to which the resistant organism is sensitive. Semisynthetic derivatives have extended the lives of many antibiotics because many derivatives are unaffected by the reactions that inactivate the parent antibiotic. An example is the $\beta$-lactam family. Many of the semisynthetic penicillins and cephalosporins (in contrast to penicillin G) have improved acid stability and thus can be given orally, have a broadened spectrum that includes gram-negative as well as gram-positive organisms, and also have improved resistance to $\beta$-lactamases. These improved properties have vastly extended the life of penicillin and cephalosporin antibiotics. Semisynthetic derivatives for other antibiotics may also be more resistant to the enzymes specified by genes carried on resistance plasmids, as is the situation with amikacin, a kanamycin derivative that is resistant to some of the enzymes capable of inactivating kanamycin.

General principles have gradually emerged to maximize the advantages of antibiotics and to minimize the emergence of resistant organisms. Whenever possible the pathogen is isolated and identified. Until this happens, empiric therapy with broad-spectrum antibiotics is instituted. Sensitivity testing of the pathogen to different antibiotics then permits the selection of the most effective antibiotic. In general, it is better to use specific therapy with a single antibiotic based on knowledge of the sensitivity of the organism and the pharmacological properties of the antibiotic than to continue therapy blindly with combinations of antibiotics—the latter can lead to the selection of resistant organisms or to the development of suprainfection. Many hospitals frequently hold an antibiotic in reserve to handle cases of bacterial resistance.

In global terms, bacterial resistance poses far greater problems in the developing nations than in industrialized countries. The developing nations lack adequate resources to monitor antibiotic usage. Frequently there is self-prescribing with freely available, over-the-counter antibiotics, conditions that contribute to the emergence of resistant organisms. There are no national boundaries to infectious diseases and the spread of antibiotic-resistant organisms. Consequently, international efforts are being directed toward uniform and effective means of controlling bacterial resistance, thus bringing the benefits of antibiotic therapy to all the people of the world.

## VII. AGRICULTURAL USES

Just as antibiotics have revolutionized the therapy of human infectious diseases, they have also revolutionized the therapy of animal infectious diseases. Moreover, antibiotics have been incorporated into animal feed at low or subtherapeutic concentrations in order to improve growth and production performance. Antibiotics also have found other important uses in agriculture in the protection of plant crops.

Diseases caused by infectious agents are widespread in domesticated animals. Antibiotics used in veterinary medicine to treat such infections include the penicillins, cephalosporins, aminoglycosides, macrolides, tetracyclines, and chloramphenicol. Diseases treated include localized infections as well as generalized infections such as septicemia, peritonitis, pneumonia, and mastitis. Of animals treated, the bovine species accounts for three-fourths of the amounts of all antibiotics used in animal chemotherapy, principally for mastitis. The remainder is used in equine and porcine species and in various other domestic animals, including pets.

Veterinary medicine employs a technique in antibiotic therapy that is not used in human medicine; namely, herd medication. Farm animals such as cattle are bred for milk or meat production and are maintained in large herds under conditions in which infectious agents may be introduced and easily spread. Economically, it is sound practice to treat the entire herd for an infectious disease when it is introduced rather than treating isolated cases as they appear. Veterinarians the world over use this technique, which is actually a mixture of therapeutic and prophylactic use of antibiotics. Herd medication is used also in poultry flocks, in beekeeping, and in fish farming.

A second important use of antibiotics in agriculture is the practice of incorporating low or subtherapeutic concentrations of antibiotics in animal feed to improve animal growth and performance. On such regimens, animals are healthier, have a higher weight gain, and produce more eggs, milk, etc., per unit of food consumed than nontreated animals. From a volume standpoint, antibiotics used as feed additives far outweigh those used in animal disease therapy. Antibiotics registered with the U.S. Food and Drug Administration (FDA) as antibacterial feed additives include bacitracin, bambermycins, chlortetracycline, erythromycin, lincomycin, neomycin, novobiocin, oleandomycin, oxytetracycline, penicillin, tylosin, and virginiamycin. Chloramphenicol is not approved by the FDA as a feed additive for use in meat-producing animals because of concern for toxic residues. The use of antibiotics as feed additives in meat-producing animals alone has been calculated to save world consumers billions of dollars per year.

Since the introduction of antibiotics as feed additives in the early 1950s, two important concerns over their use have emerged: the fear that the development of antimicrobial resistance to antibiotics would reduce the efficacy of antibiotics in human disease therapy, and the fear that antibiotic residues in food used for human consumption could lead to human health problems.

The concern over bacterial resistance has two components, as shown by the following example. It has been shown that isolates of the common intestinal organism *E. coli* increased from near zero resistance to as much as 90% resistance to oxytetracycline when this antibiotic was used as a feed additive in swine. The first concern is that flora that have become resistant to an antibiotic through feed additives might be transferred directly to humans. Animal pathogens that might be directly transferred are staphylococci and salmonella. The second concern is that resistance genes might be transferred from a nonpathogenic to a pathogenic organism. These possibilities and their effects on human disease have been subjects of much controversy, but the issues remain unsettled. The antibiotics receiving the most attention are the penicillins and tetracyclines, which are very important in human disease therapy. Proposals have been made to eliminate these antibiotics from use in animal feed, but the evidence to justify their total removal is still generally regarded as insufficient. Even so, most antibiotic manufacturers, who have a large financial stake in the outcome of this controversy, have moved steadily to find antibiotics that are not used or used only slightly in human medicine for use as feed additives. Examples of antibiotics not used in human medicine are the bambermycins and virginiamycin, which are used in poultry and swine. Antibiotics used slightly or not at all in human medicine include bacitracin, lincomycin, and tylosin. A new antibiotic, monensin, with little antibacterial activity, was discovered to have activity against the protozoal agent of coccidiosis in swine and poultry and has become the dominant anticoccidial agent in the world. Many such agents are under active investigation and testing, and more can be expected to be used in the future.

Most countries of the world have regulatory agencies charged with the responsibility of ensuring that residues of drugs from feed additives do not render foods derived from animals unsafe for human consumption. In the United States, this responsibility resides with the FDA, which has set strict limits for maximum levels of antibiotics that can be used in animal feeds and the minimum time interval between the last use of food containing antibiotics and processing, thus allowing time for the natural elimination of antibiotics. In general, because many antibiotics are inactivated by cooking, antibiotic residues in food do not currently appear to pose a significant problem.

Another use of antibiotics in agriculture, which has not been as fully developed as therapeutic and feed additives uses, is in crop protection. Many plant diseases are of bacterial, fungal, and viral origin and some of these are susceptible to antibiotics, including many that are toxic for humans and animals. The greatest success in plant protection with antibiotics has been with fungal diseases; the method of application is generally by spraying. Aureofungin, a broad-spectrum fungicide, has been used successfully with a host of plant fungal diseases including, for example, citrus gummosis, mango rot, rice blast disease, and apply mildew. Many antibiotics, including some that are used in human and animal medicine such as chloramphenicol, tetracyclines, and erythromycin, have been tested against plant pathogens. Inasmuch as plant pathogens are not directly transmissible to humans, the main concern about resistance relates to reduced efficacy against the pathogens rather than transfer to humans; however, just as with feed additives for animals, antibiotic residues have been a concern in edible plant products. There has been little evidence to date of deleterious effects from the use of antibiotics in plants. The use of antibiotics for plant and crop protection is still largely experimental but it seems likely that their use will increase.

## BIBLIOGRAPHY

Bryan, L. E. (ed.) (1984). "Antimicrobial Drug Resistance." Academic Press, Orlando.

Chin, G. J., and Marx, J. (eds.) (1994). Resistance to antibiotics. *Science* **264**, 359–393.

Franklin, T. J., and Snow, G. A. (1981). "Biochemistry of Antimicrobial Action," 3d Ed. Routledge, Chapman, and Hall, New York.

Glasby, J. S. (1979). "Encyclopedia of Antibiotics," 2nd Ed. Wiley, New York.

Joklik, W. K., Willett, H. P., Amos, D. B., and Wilfert, C. M. (eds.) (1988). "Zinsser Microbiology," 19th Ed. Appleton and Lange, Norwalk, CT.

Kuchers, A., and Bennett, N. M. (1987). "The Use of Antibiotics," 4th Ed. J. B. Lippincott Co., Philadelphia.

Levy, S. B., Burke, J. P., and Wallace, C. K. (1987). Antibiotic use and antibiotic resistance worldwide, *Rev. Infect. Dis.* **9** (Suppl. 3).

Moats, W. A. (ed.) (1986). "Agricultural Uses of Antibiotics," American Chemical Society, Washington, D.C.

Zähner, H., and Maas, W. K. (1972). "Biology of Antibiotics," Springer-Verlag, Berlin.

# Antibody–Antigen Complexes: Biological Consequences

NICHOLAS R. StC. SINCLAIR
*University of Western Ontario*

## GLOSSARY

**Antibody** A protein product of lymphocytes and plasma cells which binds antigen through its variable regions and triggers various inflammatory and regulatory reactions

**Antigen** Any chemical to which antibody can bind; as an immunogen, it induces immune responses

**Antiidiotypes** Specific immune products that recognize autoantigens associated with the variable regions of other, complementary, immune products

**Autoimmunity** Formation of immune products, such as antibody or B- and T-cell receptors, that recognize autoantigens; this may be either a normal process or one associated with autoimmune disease

**Fc portion** A part of the antibody molecule which triggers various inflammatory reactions, controls the metabolic breakdown and distribution of antibody, and regulates immune responses

**Immune response** Activity that the lymphoid system engages in when responding to an immunogenic stimulus; both specific antibodies and various cells are generated, which react with the antigen to initiate inflammatory and regulatory functions

IMMUNE COMPLEXES MADE UP OF ANTIBODY AND antigen play an important role in the genesis of in-

flammatory reactions which provide immunity to microbiological attack as well as cause various immunological diseases. When antibody combines by its variable region with antigen, this complex activates membrane-damaging enzymes (complement), induces the release of pharmacologically and enzymatically active inflammatory agents, and/or causes the ingestion of this immune complex by phagocytic cells of the reticuloendothelial system. These inflammatory reactions take place because of activities associated with the Fc portion of the antibody molecule and with Fc receptors and complement receptors on inflammatory cells. The way that the host orchestrates this inflammatory reaction and the nature of the invading microorganism, be it bacteria, virus, or various other parasites, determine whether the host will be protected. The Fc portion is also involved in regulatory mechanisms which control the level of antibody formed against foreign antigens and self components. Defects in these regulatory mechanisms lead to autoimmune disease.

## I. COMPOSITION AND INFLAMMATORY CONNECTIONS

### A. Nature of Antibody–Antigen Complexes

The lymphoid system generates immune responses to a multitude of chemicals which it recognizes. Antibodies are one of the products of an immune response, and the chemicals to which antibodies bind are called antigens. Antibodies are proteins (immunoglobulins) produced by B lymphocytes and by plasma cells. Antigens may have almost any chemical structure, but

ENCYCLOPEDIA OF HUMAN BIOLOGY, Second Edition, VOLUME 1.   Copyright © 1997 by Academic Press.   All rights of reproduction in any form reserved.

include proteins, carbohydrates, nucleic acids, lipids, and many other compounds.

When antibodies and antigens combine, they form a lattice arrangement of varying sizes and shapes. In the test tube, this reaction is used to detect the presence of either antibody or antigen and has proven the methodological usefulness of immunology. When this reaction occurs in the body, it signals the onset of inflammation. Understanding the reactions in the body has shown—and will continue to illuminate— important interactions between the host's immune system and the world of antigens around us, both the external world of foreign substances and parasites as well as the internal world of autoantigens which the immune system readily and obligatorily recognizes.

## B. Activation of Inflammatory Reactions

The binding of antibody to antigen elicits powerful inflammatory processes which exert widespread effects on both the antigen and the surrounding host tissue. If the antigen is an invading microorganism, it may be destroyed by these antibody-directed inflammatory reactions. [*See* Inflammation.]

The binding of antibody to antigen marks the antigen for destruction by phagocytic cells which have receptors for the Fc portion of antibody (Fig. 1). These Fc receptors are classified according to their molecular structure, cells on which they are found, and biological function. Another major Fc-dependent mechanism in inflammatory reactions is complement activation.

Complement is a series of blood proteins activated in the presence of antibody–antigen complexes. Aside from destroying biomembranes, complement coats antigens recognized by antibody so that they can be more readily destroyed by phagocytic cells possessing receptors for complement (Fig. 1). Complement activation also generates a series of small peptides (chemotatic factors) to attract inflammatory cells (the polymorphonuclear neutrophils) which then ingest antibody–antigen complexes and release hydrolytic enzymes (Fig. 2). [*See* Complement System.]

If either the phagocytic or polymorphonuclear cell systems or the initial complement components are missing, the individual is deficient, not because specific antibodies and effector cells are not produced in response to immunogen, but because the inflammatory mechanisms, described here, are required for these specific immune products to deal effectively with infections. While neutralization of toxins and viruses undoubtedly occurs in the test tube, effective removal of these noxious agents from the body may require more than the simple binding of antibody to antigen in order to protect the host from disease. For example, acquired immunodeficiency syndrome (AIDS) is diagnosed by finding antibodies capable of neutralizing the human immunodeficiency virus (HIV) *in vitro*; nevertheless, the host is not protected. The virus may evade a potentially neutralizing antibody by cell-to-cell transmission (without entering the intercellular space where antibody is) and/or by changing its anti-

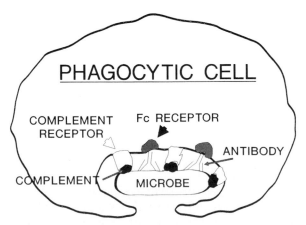

FIGURE 1   Phagocytosis of a foreign particle (microbe). Binding of antibody and/or complement to the surface of the particle increases the rate of uptake of the particle by the phagocytic cell possessing both Fc receptors and complement receptors.

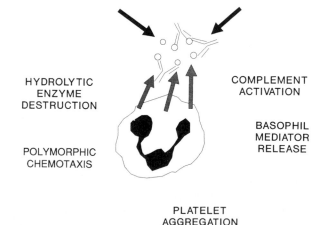

FIGURE 2   Activation of inflammatory mechanisms by antibody–antigen complexes. These mechanisms show many interrelationships, both at the level of induction and at the level of their effects on both the antibody–antigen complex and the surrounding tissues, which often become damaged in the inflammatory process.

gens. Also, antibody-coated HIV enters cells by a phagocytic process, but, instead of being destroyed, it resists the intracellular digestive processes and continues to replicate. [*See* Acquired Immunodeficiency Syndrome, Virology.]

Antibody, through its Fc portion, binds to Fc receptors on many cells in the body, either alone or in complexes with antigen. The immunoglobulin E (IgE) antibody, involved in allergic reactions, binds to tissue cells (mast cells) and to circulating cells (basophils) through Fc receptors specialized for IgE (FcεR). When this cell-fixed IgE antibody binds antigen (now called an allergen), an explosive release of chemicals, including histamine, causes the outward signs of allergy. Although the effects can range from annoying to fatal, allergic reactions can protect the host, particularly in parasitic infections, in which a parasite-destroying cell, the eosinophil, is activated and attacks IgE antibody-coated parasites after binding to the Fc portion of attached IgE via its FcεR. [*See* Allergy.]

Antibody–antigen complexes are taken up by cells throughout the body, specialized to ingest particulate and aggregated material. These cells form the reticuloendothelial system and include circulating monocytes and tissue macrophages (Fig. 1). These cells have large numbers of Fc receptors for IgG (FcγR) and complement receptors. This accounts for the phagocytic destruction of many antigenic particles, to which antibody and complement have become attached (Fig. 1). [*See* Reticuloendothelial System.]

The granulocyte, especially the polymorphonuclear neutrophil, is another cell with Fc receptors and complement receptors which busies itself with antibody–antigen–complement complexes. It has both phagocytic properties and the ability to release lytic enzymes from its granules (Fig. 2), so that, since it need not ingest the foreign material, it can take on targets larger than itself.

Another type of effector cell that interacts with antibody-coated target cells is the K cell (Fig. 3). This K cell reacts with antibody on target cells through an Fc receptor, then delivers lytic signals which destroy the target cell through a nonphagocytic process.

## II. ROLE IN IMMUNITY TO INFECTIONS

### A. Phagocytosis and Immune Clearance

The importance of antibody is obvious when one considers its ability to bind to an invading microorganism

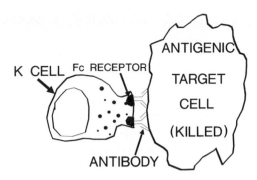

**FIGURE 3** Cytotoxic destruction of an antigenic target cell by the nonphagocytic K cell, a large lymphocyte (neither B nor T) with cytoplasmic granules and Fc receptors.

to cause its rapid destruction in the phagocytic cells of the body. This phagocytic process is the one we classically attribute to a widespread and well-functioning reticuloendothelial system. Not only does phagocytosis rid the body of unwanted guests, it also helps clear the body of cells that have outlived their usefulness. Here again, antibodies to these effete cells, even if self cells, are involved. If the invader is large so that phagocytosis is not possible, antibody attached to the invader will activate complement and cause the release of small complement fragments, which attracts neutrophils that release destructive enzymes to remove the invader, be it a multicellular parasite or a sliver of wood. [*See* Phagocytes.]

### B. Prevention of Disease

By eliminating microorganisms from the body, the lymphoid and phagocytic systems provide the host with a greater degree of immunity. Many forms of immunization have been devised to increase the immunity of the human population so that the ravages of certain infections can be kept to a minimum. The strategy has been to observe the natural resistance conferred on those infected but surviving the first infection and to mimic this resistance artificially through the use of immunizing agents (i.e., antigens or immunogens), commonly called "vaccines" in honor of Jenner's cowpox vaccination. In most cases, active immunization programs are employed to increase the production of antigen-specific antibody and cells; these products then utilize the complement and phagocytic/polymorphonuclear cell systems to rid the host of the infection. One may also give antibodies passively to attain a rapid, but short-term, immunity.

## III. PRESENCE IN CHRONIC DISEASE

### A. Mechanisms of Induction

The immune system reacts against and eliminates an invader, or ceases to react to it; thus, the host succumbs. The latter event was common in the preantibiotic era and is often forgotten as a natural outcome of infection in our modern world. If, however, an infection is too efficient—so as to destroy all of its hosts in the vicinity—it would be at an evolutionary disadvantage. Many infectious agents have opted for a longer relationship with a particular host, which is often achieved by stimulating the host to continue its immune responses to the parasite so that fulminating infections do not occur.

This is achieved by influencing the regulatory elements in the immune system so that a prolonged immune response occurs even in the continued presence of parasitic antigen. The parasite uses two general strategies to achieve this end: (1) Parasites undergo antigenic variation so that the initial immune responses eliminates the bulk of the parasitic load, but not those that have changed to a new antigenic form. (2) Agents causing chronic infections also interfere with the immunoregulatory elements so that a prolonged immune response is possible. Various adjuvants of microbial origin exert this immunomodulatory function.

In many natural states, initial exposure does not lead to a form of immunity resulting in the ejection of the parasite, even though the immune system is demonstrably activated and antibodies are produced. The antibodies generated bind with antigens from the microorganism, but cannot bring about elimination of the invader. The production of antibody, which cannot eliminate the offending microorganism, may, however, prevent the invader from ravaging the body in a rapid and fatal infection. This leads to the combined presence of both antibody and antigen, a situation in which the generation of antibody–antigen immune complexes is virtually assured.

### B. Contributor to Chronicity

The presence of antibody–antigen complexes leads to the many manifestations of immune complex pathology in chronic infections. When the host cannot eject the parasite, be it a bacteria, virus, worm, tumor, or organ transplant, immune complexes circulate, settle in tissues, activate complement, attract circulating neutrophils, and cause damaging inflammation. While we have been able to reduce the mortality and morbid-ity due to acute infections, we have had much less success with chronic infections (e.g., malaria, leprosy, hepatitis, and many others), which abound in the Third World. These diseases are characterized by tissue inflammation due to complement, platelet, basophil, and neutrophil activity, following the deposition of antibody–antigen complexes (Fig. 2). [*See* Neutrophils.]

A more common representative of this reaction often occurs when we receive a second, or booster, injection in immunization. While reaction to the first injection is mild, a painful swelling at the site of injection may occur with succeeding injections. The reason is that the injected antigen combines with preformed antibody to form antibody–antigen complexes which activate complement and chemical mediator-containing basophils, aggregate platelets, cause the formation of microthrombi (small clots), and attract neutrophils which attack antibody–antigen complexes deposited near blood vessel walls, thus damaging the blood vessel (Fig. 2). Another less common example is seen as an occupational lung disease in, for example, farmers and bird fanciers. As a result of chronic inhalation of otherwise harmless antigens, inflammation occurs, triggered by antibody–antigen complexes which form where internally produced antibody meets externally derived antigen, and may lead to a life-threatening destruction of the lungs.

When the potential target is self, immune responses are normally regulated to achieve a state of immunological peace. If this does not occur, many forms of autoimmune disease arise. In systemic lupus erythematosus, widespread destruction of blood vessels, known as vasculitis, is caused by nuclear antigens complexed with antibody specific for these antigens. Vasculitis in various tissues results in diseases, such as glomerulonephritis, a leading cause of kidney destruction and disease. In rheumatoid arthritis, antibodies to the Fc portion of IgG bind to IgG, form immune complexes, and induce, at least in part, the joint problems and vasculitis seen in this disease. In all of these examples of autoimmune tissue damage, inflammatory reactions (Figs. 1–3) are found to varying degrees. [*See* Autoimmune Diseases.]

In chronic infections and autoimmune diseases, there are two main questions, one of which is: What makes certain antibodies disease causing, while others are not? It seems that IgG is more damaging than IgM. Certain properties, such as avidity for antigen, ability to avoid uptake in the reticuloendothelial system [except for autoimmune destruction of red blood cells and platelets, in which inappropriate phagocytosis (Fig. 1) of circulating blood elements is part of the problem], electrostatic charge, and cross-reactivity to various tissue antigens, make anti-self antibody–autoantigen complexes more damaging. An increased ability to harness, in inappropriate ways, the inflammatory mechanisms (Fig. 2) is associated with clinical abnormalities. The nature of blood flow and tissue architecture in the target organs affects the likelihood that the anti-self antibody, once formed, will damage certain organs. The other main question is why these anti-self responses occur in the first place. Are there normal anti-self responses? What makes normal anti-self responses abnormal? These questions bring us to the subject of how immune responses are regulated, particularly by antibody–antigen complexes.

## IV. ROLE IN THE REGULATION OF IMMUNE RESPONSES

### A. Fc Signaling Mechanisms

The immune system is often, but not invariably, inhibited by the presence of antibody–antigen complexes. Antibody–antigen complexes inactivate lymphocytes by Fc-dependent mechanisms. The Fc-dependent inactivation of the two major classes of lymphocytes (B and T cells) involves two distinct mechanisms. [*See* Lymphocytes.]

In B cells, which are responsible for antibody responses, linking of the antigen receptor with the Fc receptor by antibody–antigen complexes (Fig. 4) induces changes in intracellular "second messenger" systems so that the B cell cannot respond to an immunogenic stimulus. Inositol triphosphate, which is normally generated when B cells are activated, is rapidly

broken down to the inactive monophosphate. Overproduction of diacylglycerol takes place, which, along with dissociation of G proteins from the antigen receptor, prevents the receipt of normal activation signals through the B cell antigen receptor. Many early response genes that are normally activated on B-cell stimulation do not turn on; however, protein tyrosine phosphorylation, an early step in the activation pathway, occurs even in the presence of Fc signaling. Therefore, regulatory signals from antibody–antigen complexes through the antigen receptors and Fc receptors alter the ability of B cells to respond to antigen.

T cells, which are responsible for cell-mediated immune responses and regulatory activities, are also inhibited by antibody–antigen complexes through a Fc-dependent process which may involve mechanisms equivalent to those found in B cells or are more akin to an inflammatory reaction. When T cells are caught with antibody–antigen complexes on their surface, they are phagocytized (Fig. 5) by the host's reticuloendothelial system as if they were foreign invaders. Complement receptors are also found on lymphocytes; some of these receptors are associated with the antigen receptor complex on B cells and may contribute to lymphocyte activation.

### B. Importance in Induction of Immune Responses

Regulatory effects of antibody were considered to occur late in immune responses or as the result of immunosuppression, such as during the administration of anti-Rhesus factor-positive (Rh⁺) antibodies to a woman who is Rh⁻ carrying a Rh⁺ fetus. These passive anti-Rh⁺ antibodies prevent the mother from re-

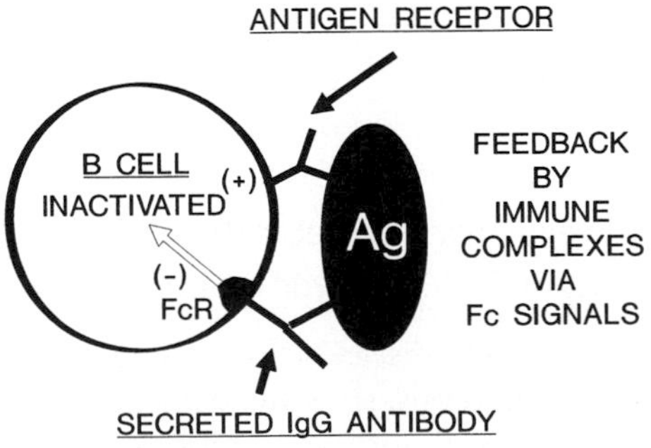

FIGURE 4  Regulation of B lymphocytes. The antigen receptor transduces a positive (+) signal. Cross-linking of the antigen receptor and Fc receptor (FcR) for immunoglobulin G (IgG) by an IgG antibody–antigen complex results in a negative (−) outcome, or inactivation.

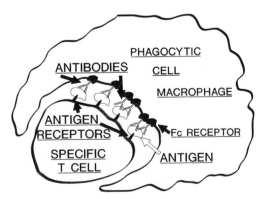

FIGURE 5  Regulation of T lymphocytes. Antibody–antigen complexes on the surface of T lymphocytes induce the uptake of these T cells by the reticuloendothelial system.

sponding to the Rh$^+$ fetal red blood cells with the active production of anti-Rh$^+$ antibodies that cross the placenta to attack the fetus. Hemolytic disease of the newborn used to kill, or severely damage, 1 in 1000 fetuses which neared term.

Fc signaling by IgG antibody is an early event in immune responses that begins with exposure of the immune system to antigen. The IgG antibody to antigen is formed constitutively in low amounts sufficient to regulate. When immunogen reaches specific lymphocytes which respond to it, immunogen exists as antibody–antigen complexes which link antigen receptors with Fc receptors. Helper T lymphocytes interfere with the Fc-dependent suppression of immune responses by antibody to antigen. Interleukin-4 is the only T-cell lymphokine so far shown to interfere with this negative signaling.

Antibodies with a narrow specificity for the Fc portion of *secreted* IgG, the most suppressive of the immunoglobulin classes, overcome the B lymphocyte's requirement for most T cells in the induction of a normally T-cell-dependent response. Rheumatoid factor (an antibody against the Fc portion of IgG, but of the nonsuppressive IgM class) also reduces the requirement for a large majority of T lymphocytes. Since rheumatoid factor is produced early in immune responses, the blockade of early Fc signaling during the induction of an immune response may be a normal biological function for this frequently occurring auto-antibody.

## C. Importance in the Control of Anti-Self Responses

While it is clear that there are many cases in which autoreactive lymphocytes are eliminated or otherwise made nonresponsive, there are numerous examples in which anti-self B lymphocytes are present, and can even be in a state of partial activation, without any evidence of overt autoimmune disease. Furthermore, low levels of anti-self IgG antibody are normally present in the absence of autoimmune disease. Variable region genes encoding antibodies with reactivity to self are frequent, both in the inherited germ line and generated by somatic mutations in mature lymphocytes. By being able to recognize autoantigens early and throughout life, the immune system regulates itself, thus preventing high levels of damaging immune responses to self-antigens.

Immune complexes have been identified in most people. These are made up of IgG antibodies to self-antigens and IgM antibodies to the antigen-binding sites of the IgG antibodies, i.e., the IgM antibodies are anti-idiotypes. The IgG antibodies in these immune complexes are polyreactive, generally not of high avidity, and do not accumulate somatic mutations characteristic of IgG antibodies to foreign antigens. Since these immune complexes are not found in autoimmunity, it is likely that these complexes limit B-cell activity in nonautoimmune individuals. To regulate, the IgM anti-idiotypic antibody would bind secreted IgG anti-self antibodies and antigen receptors on self-reactive B cells, while the Fc portion of the secreted IgG antibody in the immune complex engages the Fc receptor. This mechanism could prevent the activation of anti-self B cells (negative feedback) or somatic mutation of activated and isotype-switch B cells before they produce high avidity anti-self IgG antibodies (negative feedforward).

## D. Characteristics of Autoimmune Disease

There are two general types of autoimmune disease: those in which immune responses are directed against widely distributed antigens (e.g., nucleic acids, histones, IgG, and cytoskeletal proteins) and those in which the immune response is directed to specific tissues (e.g., antigens in the thyroid, adrenal glands, stomach, and $\beta$-islet cells in the pancreas). In either type of disease, general (i.e., antigen-nonspecific) dysregulation is associated with the autoimmune disease. IgG autoantibodies are overproduced, including many IgG autoantibodies that have not yet resulted in damage to their particular targets. While specific variable region genes are used in autoimmunity, the selection of these variable region genes appears normal. This is in keeping with a general regulatory defect rather than a random process or one in which autoimmune disease-prone individuals possess a higher representation of variable region genes specific for autoantigen recognition.

## E. Regulatory Defects Are a Primary Cause of Autoimmune Disease

Since autoimmune disease is grounded on a general autoimmune diathesis, the control of general immune responsiveness is abnormal. A series of observations has emerged in recent years which strongly implicates defects in Fc signaling [i.e., defects in immunoregulation by IgG antibody–antigen complexes (Fig. 4)] in the genesis of autoimmune disease. These include

(1) abnormalities of carbohydrate residues in the Fc portion of IgG molecules in certain types of autoimmune disease—notably rheumatoid arthritis—these residues being necessary for the immunoregulatory functions of the Fc portion of IgG antibody, (2) defects in Fc receptors for IgG in autoimmune strains of mice that prevent efficient Fc signaling to B cells, (3) uncontrolled production of endogenous Fc signal-blocking agents in autoimmune disease, especially rheumatoid factor and various immunoglobulin binding factors, (4) increased T-cell activity of the type which reduces control by Fc signaling, and (5) the presence of Fc-binding proteins on the various microbial triggers of autoimmunity. The "moth-eaten" mouse strain, which has an intense form of B-cell autoimmunity, lacks a phosphatase enzyme normally associated with the Fc receptor that is required for Fc receptor regulatory functions. These defects result in reduced Fc signaling by IgG antibody, leading to an increased tendency for the production of large amounts of IgG autoantibody. Therefore, because of the abnormalities in Fc signaling by IgG antibody demonstrated in autoimmune disease and the presence of small amounts of natural anti-self IgG antibodies in nonautoimmune prone subjects, these IgG antibodies appear to prevent high-level production of damaging autoimmune responses so that autoimmune disease is avoided.

Defective phagocytic capacity, due to abnormal Fc receptor function on macrophages (Fig. 1), occurs in autoimmunity as either a cause or an effect of circulating antibody–antigen complexes. This defect may prevent the normal clearance (Fig. 5) of autoreactive T cells that escape censoring mechanisms in the thymus.

## V. NEW FUNCTIONS FOR CELLS OF THE IMMUNE SYSTEM

### A. New Functions for Old Cells

Until recently, little importance has been paid to the immunoregulatory role of IgG antibody–antigen complexes through the generation of Fc signals (Fig. 4). We must now reevaluate the elements that control immune responses against foreign and autoantigens. The three main cells involved in antibody responses are B lymphocytes (the direct progenitors of antibody-producing cells), T lymphocytes (which often "help" B lymphocytes in their responses to "T-cell dependent" antigens), and antigen-presenting cells. The following sections look at the function of each of these cells in light of immunoregulation by antibody–antigen complexes.

### B. Autoregulatory B Cells

Formerly, B lymphocytes were considered entities that required the firm regulatory hand of helper and suppressor cells, but did not control themselves in any important way, either before or after activation by immunogen. Most immunologists, who worry about B cells losing control and attacking host antigens, think that antigen receptors receive only negative signals and that helper T cells change this negative result into a positive signal for activation. Helper T cells do favor B cell activation by cell–cell contact and by releasing activating molecules known as lymphokines. With the elucidation of the immunoregulatory consequences of cocross-linking antigen receptors with Fc receptors (FcγR) via antibody–antigen complexes (Fig. 4), it is now possible that antigen receptors can transmit positive signals as well as negative ones. Antigen receptors of mature B lymphocytes appear capable of inducing different outcomes in B cells ranging from deletion to activation. The outcome depends on the initial antigen binding (avidity, amount of antigen receptor cross-linking, length of time) and recognition of other ligands by various B-cell receptors. Some nonantigenic ligands are found on regulatory T cells and on antigen-presenting cells (favoring various cell–cell interactions), while other ligands define antigenic signals as microbes, as complement bound to antigen, or as damaged host cells expressing various stress proteins (all favoring a B-cell response). In order to allow B cells numerous options, in terms of the response which occurs, B cells not only have surface immunoglobulin, but many associated chains (Ig-$\alpha$ and Ig-$\beta$/$\gamma$, the CD19–CR2–Leu13–TAPA1 complex, and various other adhesion/signaling molecules) that allow the B cell a wide range of autoregulatory functions. Some of these functions depend on the localization and state of antibody-antigen complexes. [*See* B-Cell Activation.]

### C. T Cells Prevent Fc Signaling

Helper T lymphocytes promote B-cell development in the bone marrow and activation by antigen. In the activation of B cells, T cells help in positive signaling through the antigen receptor complex as well as limit inactivation of B cells, either through the antigen receptor or when antigen receptors and Fc receptors are cocross-linked. These helpful T-cell influences result from cell contact between T cells and B cells and involve defined surface molecules on these cells or because T cells liberate chemicals called lymphokines which promote B-cell differentiation. T cells act at various stages of B-cell activation in response to antigen; however, the major effect of T–B cell interaction appears to be the switch from IgM antibodies with low avidity for antigen to high-avidity IgG (IgA or IgE) antibodies. Part of the involvement of T cells in the switch from IgM to IgG is that T cells, either via cell–cell contact or via lymphokines such as IL-4, reduce the negative impact of IgG antibody–antigen complexes on the B-cell response.

Further regulatory loops increase the ability of T cells to act as helper cells. As an example, B cells take up antigen and present fragments of antigen to T cells. B cells accomplish this by endocytosing antigen via their antigen receptor. Can antibody–antigen complexes be taken up by the Fc receptor on B cells? The answer is no. The Fc receptor on B cells is odd, it does not associate with vacuoles going into the B cell, but remains on the cell surface near the antigen receptor to be involved in negative signaling when the antigen receptor and Fc receptor are cocross-linked. Antibody–antigen complexes only gain access into B cells with antigen receptors that recognize either antigen or the Fc portion of antibody. This later recognition is what B cells producing rheumatoid factors do; these B cells are active in immune responses to many foreign antigens, especially when the stimulus is prolonged. The rheumatoid factor B cells can obtain help from many T cells which recognize the myriad of antigens in antibody–antigen complexes, thus they are easily activated to produce IgM rheumatoid factor which binds Fc portions of IgG and blocks negative Fc signals from antibody–antigen complexes.

## D. Antigen-Presenting Cells and Fc Signaling

The main site for antigen presentation to B lymphocytes is in secondary follicles, a spherical aggregation of various cells in lymph nodes, where antibody–antigen complexes accumulate and are retained. Here, immune complexes are held on large cells with many cytoplasmic extensions, termed "follicular–dendritic cells." These cells are loaded with Fc receptors and thus can take up antibody–antigen complexes avidly. Provided that follicular–dendritic cells can bind all Fc portions of antibody in the antibody–antigen complex, they serve as an important site for the stimula-

tion of antibody responses. Because antibody–antigen complexes bind better than antigen (administered without antibody) to these follicular–dendritic cells, which block negative Fc signals, antibody–antigen complexes are sometimes better stimulators of immune responses than antigen alone. If Fc signals do escape, follicular–dendritic cells become sites for the inactivation of immune responses by "supertolerogens," made up of certain antigens with attached antibody. Other than continued antigen signaling in secondary follicles, cellular interactions and elaboration of lymphokines and other molecules favor the maturation of B cells to high avidity antibody-forming cells and memory cells.

## VI. HOW THEY ARE PERTURBED IN AIDS

AIDS is not the first infectious disease which has thrown the immune system into disarray. Chronic syphilis provided many examples of immune dysfunction, including the production of autoantibodies detected in the serological diagnosis of syphilis. Many parasitic infections demonstrate numerous immune responses, including destructive reactions against surrounding tissue without the elimination of the parasite. In many cancer patients, immune responses occur, do not eject the tumor, but lead to the deposition of immune complexes. Cancer in cells of the immune system is often associated with autoimmunity and immunodeficiency.

In AIDS, the HIV virus induces a series of abnormalities in immunological control mechanisms. It is then difficult, perhaps impossible, to naturally achieve an immune response that defends the host against HIV, but does not damage the host. Abnormalities in the control mechanisms, discussed here, contribute to the damage seen.

Some HIV-infected individuals demonstrate high amounts of immunoglobulin production, the occurrence of autoantibodies, and deficiencies in specific antibody responses to microbial antigens. Therefore, in these individuals, B lymphocytes are dysregulated and hyperactive, even though helper T lymphocytes are severely depleted. Proteins of HIV show abnormal binding to immunoglobulins generally (suggestive of Fc binding); thus, like other microbial triggers of autoimmunity, they may interfere with negative Fc signaling by immunoregulatory antibody–antigen complexes (Fig. 4). IgG antibodies to the Fab region have also been noted in AIDS. These antibodies augment negative Fc signaling and are associated with poor B-cell responses and loss of T cells in AIDS, but also with the control of transplant rejection and clinical autoimmunity.

Complexes, made up of antibody to the protein coat of HIV and the protein coat itself, will bind *all* T lymphocytes and mark them for phagocytosis, a process which normally accounts for the regulation of antigen-specific T lymphocytes by antibody–antigen complexes (Fig. 5). Antibody against only the site recognizing T cells on HIV could be protective; however, this antibody may prevent the proper functioning of T lymphocytes by binding to class II major histocompatibility complex antigen (which helper T cells must recognize in conjunction with foreign antigen).

Phagocytosis of HIV following the binding of anti-HIV antibody to the HIV surface (Fig. 1) does not guarantee destruction of the virus. In fact, there is evidence that the phagocytic process enhances the spread of HIV within the body. HIV infection of antigen-presenting cells prevents these cells from functioning properly.

This state of affairs has led some immunologists to suggest that all immune responses and inflammatory reactions (including those depicted in Figs. 1–3) to HIV should be suppressed. This is unlikely since the experience with HIV infection in patients, with lowered immune responses due to other causes, is dismal. HIV can damage the host by ways other than the generation of antibody–antigen complexes and autoimmunity.

Most immunization strategies have been based on models of naturally induced resistance to the disease, and none has been defined in HIV infection. Immunologists must develop forms of immune responses, which will confer resistance to HIV infection, without having a natural counterpart to follow. It was a mystery what this protective immune response would be, but, recently, T-cell stimulation to induce macrophage activation (delayed-type hypersensitivity), but not antibody formation, seems to have some association with resistance to the development of full-blown AIDS.

## VII. CONCLUSIONS

In the majority of cases, immune responses to damaging invading antigens are generated and those to innocent auto- and foreign antigens are avoided. The destruction of these invaders involves coordination

### TABLE I

Effector and Regulatory Mechanisms of Antibody–Antigen Complexes and Their Possible Outcomes[a]

| Mechanism | Outcome | |
|---|---|---|
| | Protective | Harmful |
| Neutralization | Toxin–virus interference | Hormone–receptor blockade |
| Phagocytosis | Foreign/effete cells destroyed | Self (e.g., red blood) cells killed |
| Neutrophil attack | Removal of foreign material | Destruction of self structures |
| Mediator release | Antiparasitic activity | Responsibility for many allergies |
| B-cell inactivation | Control of anti-self antibody | Low response to pathogens |
| T-cell phagocytosis | End of T cell activation to self | Prevention of tumor destruction |

[a]This list is not meant to be complete, but to be illustrative of the nature of effector and regulatory mechanisms engaged in by antibody–antigen complexes and some of their outcomes.

between specific products of the immune response and the reticuloendothelial system. Antibody–antigen complexes also induce varying degrees of inflammation, which may range from useful to disastrous. We do not know how to control the effects of immune complexes with any degree of certainty. Specific products of the immune response are major regulators, by virtue of their ability to bind to antigen and react, as a complex, with antigen-specific B and T lymphocytes. Immunological effector or regulatory mechanisms are never exclusively useful or harmful; each mechanism may have either outcome depending on the circumstances (Table I). Most especially, we do not yet know how antibody–antigen complexes can be harnessed to prevent disease.

## BIBLIOGRAPHY

Abbas, A. K., Lichtman, A. H., and Pofer, J. S. (1994). "Cellular and Molecular Immunology," 2nd Ed. Saunders, Philadelphia.

Anderson, C. C., Rahimpour, R., and Sinclair, N. R. StC. (1993). Mutual antagonism between antigen- and lipopolysaccharide-induced antibody production. *Immunol. Invest.* **22**, 531.

Avrameas, S. (1991). Natural autoantibodies: From 'horror autotoxicus' to 'gnothi seauton'. *Immunol. Today* **12**, 154.

Boros, P., Odin, J. A., Chen, J., and Unkeless, J. C. (1994). Specificity and class distribution of FcγR-specific autoantibodies in patients with autoimmune disease. *J. Immunol.* **152**, 302.

Clerici, M., and Shearer, G. M. (1993). A Th1 → TH2 switch is a critical step in the etiology of HIV infection. *Immunol. Today* **14**, 107.

D'Ambrosio, D., Hippen, K. L., Minskoff, S. A., Mellman, I., Pani, G., Siminovitch, K. A., and Cambier, J. C. (1995). Recruitment and activation of PTP1C in negative regulation of antigen receptor signaling by FcγRIIb1. *Science* **268**, 293.

Fridman, W. H. (1993). Regulation of B-cell activation and antigen presentation by Fc receptors. *Curr. Opin. Immunol.* **5**, 355.

Hutchison, I. V. (1980). Antigen-reactive cell opsonization (ARCO) and its role in antibody-mediated immune suppression. *Immunol. Rev.* **49**, 167.

Janeway, C. A., and Travers, P. (1994). "Immunobiology: The Immune System in Health and Disease." Current Biology Ltd./Garland Publishing Ltd., London/New York.

Kepler, T. B., and Perelson, A. S. (1993). Cyclic re-entry of germinal center B cells and the efficiency of affinity maturation. *Immunol. Today* **14**, 412.

Kim, K.-M., Alber, G., Weiser, P., and Reth, M. (1993). Signalling function of the B-cell antigen receptors. *Immunol. Rev.* **132**, 125.

Muta, T., Kurosaki, T., Misulovin, Z., Sanchez, M., Nussenzweig, M. C., and Ravetch, J. V. (1994). A 13-amino-acid motif in the cytoplasmic domain of FcγRIIb modulates B-cell receptor signalling. *Nature* **368**, 70.

Noelle, R. J., and Snow, E. C. (1993). Helper T cell signaling of B cell growth and differentiation. *Adv. Mol. Cell. Immunol.* **1B**, 133.

Nossal, G. J. V. (1994). Negative selection of lymphocytes. *Cell* **78**, 229.

Nossal, G. J. V. (1992). Cellular and molecular mechanisms of B lymphocyte tolerance. *Adv. Immunol.* **52**, 283.

Parker, D. C. (1993). T cell-dependent B cell activation. *Annu. Rev. Immunol.* **11**, 331.

Parry, S. L., Hasbold, J., Holman, M., and Klaus, G. G. B. (1994). Hypercross-linking surface IgM or IgD receptors on mature B cells induces apoptosis that is reversed by costimulation with IL-4 and anti-CD40. *J. Immunol.* **152**, 2821.

Paul, W. E., and Seder, R. A. (1994). Lymphocyte responses and cytokines. *Cell* **78**, 241.

Roitt, I., Brostoff, J., and Male, D. (1996). "Immunology," 4th Ed. Mosby, London.

Schwartz, R. S., and Stollar, B. D. (1994). Heavy-chain directed B-cell maturation: Continuous clonal selection beginning at the pre-B cell stage. *Immunol. Today* **15**, 27.

Shoenfeld, Y. (1993). Pathogenic natural autoantibodies. *Israel J. Med. Sci.* **29**, 142.

Sinclair, N. R. StC. (1993). Natural history of signaling events in B cells. *Adv. Mol. Cell. Immunol.* **1B**, 145.

Sinclair, N. R. StC., and Anderson, C. C. (1994). Do lymphocytes require calibration? *Immunol. Cell Biol.* **72**, 508.

Sinclair, N. R. StC., and Challis, J. R. G. (1993). Tentativeness and fervor in cell biology require negative and positive feedforward control. *Life Sci.* **52**, 1985.

Sinclair N. R. StC., and Panoskaltsis, A. (1988). The immunoregulatory apparatus and autoimmunity. *Immunol. Today* **9**, 260.

Sinclair, N. R. StC., and Panoskaltsis, A. (1989). Rheumatoid factor and Fc-signaling: A tale of two Cinderellas. *Clin. Immunol. Immunopathol.* **52**, 133.

Snapper, C. M., and Mond, J. J. (1993). Towards a comprehensive view of immunoglobulin class switching. *Immunol. Today* **14**, 15.

Splawski, J. B., Fu, S. M., and Lipsky, P. E. (1993). Immunoregulatory role of CD40 in human B cell differentiation. *J. Immunol.* **150**, 1276.

Süsal, C., Lewin, I. V., Stanworth, D. R., Terness, P., Daniel, V., Oberg, H.-H., Huth-Kühne, A., Zimmerman, R., and Opelz, G. (1992). Anti-IgG autoantibodies in HIV-infected hemophilia patients. *Vox Sang.* **62**, 224.

Terness, P., Berteli, A., Süsal, C., and Opelz, G. (1992). Regulation of antibody response by an IgG-anti-Ig autoantibody occurring during alloimmunization. II. Selective inactivation of antigen receptor-occupied B cells. *Transplantation* **54**, 92.

Terness, P., Kirschfink, M., Navolan, D., Dufter, C., Kohl, I., Opelz, G., and Roelcke, D. (1995). Striking inverse correlation between IgG anti-F(ab')$_2$ and autoantibody production in patients with cold agglutination. *Blood* **85**, 548.

Tsubata, T., Murakami, M., and Honjo, T. (1994). Antigen-receptor cross-linking induces peritoneal B-cell apoptosis in normal but not autoimmunity-prone mice. *Curr. Biol.* **4**, 8.

Van Den Herik-Oudijk, I. E., Westerdaal, N. A. C., Henriquez, N. V., Capel, P. J. A., and Van De Winkel, J. G. J. (1994). Functional analysis of human FcγRII (CD32) isoforms expressed in B lymphocytes. *J. Immunol.* **152**, 574.

Via, C. S., and Shearer, G. M. (1989). Autoimmunity and the acquired immune deficiency syndrome. *Curr. Opin. Immunol.* **1**, 753.

Wade, W. F., Davoust, J., Salamero, J., André, P., Watts, T. H., and Cambier, J. C. (1993). Structural compartmentalization of MHC class II signaling function. *Immunol. Today* **14**, 539.

Weiss, A., and Littman, D. R. (1994). Signal transduction by lymphocyte antigen receptors. *Cell* **78**, 263.

# Antibody Diversity (Clonal Selection)

DAVID TARLINTON

*The Walter and Eliza Hall Institute of Medical Research*

## GLOSSARY

**Allele** One of the two chromosomal copies of a gene in a diploid cell

**Anergy** A state of B-cell nonresponsiveness when exposed to antigen

**Antibody** A globular protein secreted into the serum in response to an antigen (also called immunoglobulin)

**Antigen** Any substance which elicits antibody upon introduction into an animal

**Cognate** An antibody–antigen interaction of sufficient affinity that binding is stable

**Clone** Population of identical cells resulting from the binary division of a parent cell

**Locus** Chromosomal location of a gene, at which different alleles may occur

**Tolerance** Lack of reactivity of self-antigens

THE HALLMARK OF THE IMMUNE SYSTEM IS THE ABILity to recognize and respond to things that are foreign while not responding to things that are self. A substance that can be the target of an immune response is called an antigen. Introducing an antigen into an animal results in the production of antibodies specific for that antigen, meaning that the antibodies will bind that antigen and not other unrelated substances. There is a potentially unlimited number of foreign antigens in the universe and our immune systems must have the potential to deal with them all. The question of whether all antibodies necessary to recognize all antigens exist without exposure to the antigen or whether exposure to the antigen generates the appropriate antibody was the subject of considerable research during the first half of the 20th century. The current concept is that all antibody specificities preexist antigen exposure and that antigen selects the antibody-producing cell whose antibody best binds the antigen. This concept of antigen selecting a preexisting cell is the basis of clonal selection. The immune system is thus able to generate a sufficiently diverse set antibodies such that an enormous number of foreign antigens can be recognized while simultaneously preventing the production of antibodies that would bind to self. [*See* Immune System.]

## I. GENERATING ANTIBODY DIVERSITY

### A. Antibody Structure

Antibody molecules are globular proteins composed of two types of protein chains: heavy and light. Antibodies exist either on the surface of the cells which produce them, specialized cells of the immune system called B lymphocytes, or as serum proteins. On the B-cell surface, antibodies are composed of two identical heavy chains and two identical light chains. Each chain is composed of a variable region unique to that antibody and a constant region which defines the isotype of the antibody. There are seven different heavy chain constant regions in humans, each one defining an immunoglobulin isotype or subclass. The isotypes are IgM, IgD, IgG (comprising the IgG1, IgG2, IgG3, and IgG4 subclasses), IgE, and IgA (comprising IgA1

ENCYCLOPEDIA OF HUMAN BIOLOGY, Second Edition, VOLUME 1.   Copyright © 1997 by Academic Press.   All rights of reproduction in any form reserved.

and IgA2 subclasses). Similarly, there are two isotypes for the light chains: $\kappa$ and $\lambda$. The light chains are not restricted in the heavy chains with which they can pair. Thus IgM molecules may contain either $\kappa$ or $\lambda$ light chains, but never both in the one antibody molecule. The variable region of the antibody, which is the N-terminal part, is the part of the antibody which binds to antigen and is unique to each antibody. The variable region of the heavy chain and that of the light chain come together in the assembled antibody protein to form the antibody-combining site, the part of the antibody which binds antigen.

## B. Immunoglobulin Gene Rearrangement

The enormous degree of variability in antibody-combining sites is a consequence of the mechanism by which complete antibody genes are generated. Unlike almost every other gene, antibody genes are not complete in the gamete. Rearrangement of the DNA is required to generate the complete immunoglobulin variable (or V) region genes, for both the heavy and the light chains. The joining together of the separate elements to make a functional V gene occurs only in B lymphocyte precursor cells. Muscle cells, for example, do not rearrange their immunoglobulin variable region gene elements. The V region gene of the heavy chain ($V_H$ gene) is composed of three mini-genes: a V element, a D (for diversity) element, and a J (for joining) element. Multiple V, D, and J elements exist in the unrearranged chromosome, but only one of each type is present in the final $V_H$ gene. This allows for enormous combinatorial diversity to be generated. For example, 51 V genes, approximately 30 D elements, and 6 J elements ($51 \times 30 \times 6$) give a total of 9180 possible combinations from only 86 germline elements. When the additional variability due to imprecision in joining and nucleotide addition and deletion at the site of joining is taken into account, there are considerably more variants possible than the strict combinatorial calculation. Furthermore, the same process of rearranging germline elements occurs at the light chain loci, although the $V_L$ genes are formed from two rather than three germline segments, one V element joining with one J element. Thus, at the immunoglobulin $\kappa$ and $\lambda$ loci, combinatorial possibilities are limited to ($40 \ V\kappa \times 51 \ J\kappa$) plus ($35 \ V\lambda \times 4 \ J\lambda$), a total of 340. Each of these 340 different light chains, however, can pair with any of the 9180 possible heavy chains, giving a total of over 3 million possible combinations from less than 170 different elements. When the additional variability that can occur

at the sites of joining is factored into such a calculation, the total possible number of combinations goes up an additional order of magnitude. In this way the immune system of humans has evolved a mechanism that is able to generate an enormous diversity of antibody variable region combining sites at the cost of relatively little genetic space being used.

Three features of the system used to generate immunoglobulin diversity, however, are not necessarily beneficial to the organism. First, because there is no control over the specificity of the antibody that is being generated, self-reactive antibodies will result at some frequency. These autoreactive antibodies (called autoantibodies) could be harmful and the B cells expressing them need to be eliminated. Second, the B cells in which Ig gene rearrangement is occurring need to be able to stop the process when they have successfully rearranged a single heavy chain and a single light chain variable region. This ensures that each B-cell clone expresses a single specificity which, in turn, is fundamental to the proper functioning of the immune system. This phenomenon of B cells expressing only one functional heavy chain and one functional light chain is called allelic exclusion. This means that expression of the second allele at each immunoglobulin locus is excluded by the productive rearrangement at the first. Third, again due to the random nature of the joining, the majority of junctions will not maintain the correct reading frame of the different gene segments. B-cell precursors that fail to make a single productive heavy chain or light chain gene rearrangement need to be removed from the developmental pathway as they are essentially useless cells. [*See* B-Cell Activation.]

## II. B-CELL DEVELOPMENT IN THE BONE MARROW

B cells, which are responsible for the production of antibodies, are generated throughout the life of vertebrates. B-cell development occurs in the bone marrow in adult animals, although in the fetus it occurs predominantly in the liver. B cells are generated from an uncommitted precursor, a cell which has no attributes of the B-cell lineage and which, in principle, has the potential to become hematopoietic cell types other than a B cell. The process of lineage commitment of precursor cells, while being a common theme in developmental biology, is at present poorly understood. Commitment to the B-cell lineage is evidenced by the expression of cell surface markers, which are characteristic of B cells. This includes cell surface proteins such as CD19 and CD10. Development then proceeds and can be monitored by the regulated alteration in the expression of a number of cell surface proteins, many of which are associated with B-cell function. Use of such markers has allowed a scheme of development to be formulated where various stages are defined by a unique distribution of marker proteins. Precursor B cells move from one stage to the next in an ordered manner, i.e., cells isolated from one stage will proceed under the appropriate stimulus through the successive stages, eventually giving rise to a mature B cell. In general terms, the earliest identifiable precursor committed to the B-cell lineage is referred to as a pro-B cell. Pro-B cells give rise to pre-B cells, which in turn become virgin or immature B cells. Virgin B cells, which are the first cells of the B-cell lineage to express a fully assembled immunoglobulin molecule on the cell surface, then become mature B cells. Mature B cells coexpress IgD and IgM, whereas virgin B cells express only IgM. Both IgM and IgD on a given mature B cell utilize the identical variable region, generated by alternative splicing of the one RNA transcript. Mature B cells are the predominant type of B lymphocyte in the peripheral lymphoid system.

The developmental stages defined by cell surface markers have been found to correlate with gene rearrangements at the heavy and light chain loci. As noted earlier, in order to generate a functional immunoglobulin variable region gene, the germline encoded segments need to be joined by DNA recombination. This rearrangement only occurs in B lineage cells and is a highly regulated process. First the heavy chain variable region is assembled, joining a D and J element on both chromosomes, followed by a V element being joined to the DJ junction on one chromosome. If this junction is productive, in that a functional antibody heavy chain protein results, further rearrangement at the heavy chain locus stops and the B-cell precursor differentiates to the next stage. If this rearrangement is unsuccessful, then V to DJ joining occurs on the second chromosome. If this joining is successful, it again results in differentiation of the precursor. If, however, this rearrangement is also unsuccessful, such that the precursor now has two failed heavy chain rearrangements and therefore no possibility of making a functional immunoglobulin molecule, then the further development of that particular cell is blocked and it undergoes an apoptotic cell death, thus removing it from the pathway. If rearrangement was successful at one heavy chain locus, then light chain gene re-arrangement is initiated. Again this occurs in an ordered pattern, first at one κ locus, then the second if the first is unsuccessful. If both κ light chain rearrangements are nonproductive, then rearrangement starts at the λ locus. As soon as a productive rearrangement is made, meaning an antibody chain can be synthesized, further rearrangement at that locus stops and the precursor differentiates into the next compartment. A productive heavy chain gene rearrangement results in a pro-B cell becoming a pre-B cell. A successful light chain gene rearrangement results in a pre-B cell becoming a virgin B cell. Limiting gene rearrangement to a single productive V gene for each type of antibody chain, heavy and light, means that the virgin B cell that results at the end of the process expresses antibodies with only a single specificity. The antibody specificity is said to be clonally restricted, meaning that only that B cell and its progenitors will contain that particular combination of immunoglobulin gene rearrangements and consequently that particular antibody molecule. Thus the end product of the early stages of B-cell development is a population of cells expressing antibody molecules with an incredibly diverse array of combining sites but in which the antibodies expressed by an individual cell (and its progeny) are all identical.

## III. TOLERANCE TO SELF

### A. Clonal Deletion

One outcome of generating the antigen-combining sites of antibody molecules by a random combinatorial process is that anti-self reactivity is possible. Therefore, there must be mechanisms to deal with such B cells if and when they arise because autoimmunity is relatively rare. Such mechanisms for eliminating or silencing self-reactive B-cell clones are collectively referred to as tolerance. Two specific mechanisms are commonly invoked to account for immunological tolerance to self: clonal deletion and clonal anergy. As the name implies, clonal deletion is the removal of a clone of B cells from the repertoire. In the theory of clonal selection, Macfarlane Burnet proposed that B cells went through a particular stage of development within which they were susceptible to being killed by binding their cognate antigen, a process called "clonal deletion." After this tolerance sensitive stage, the same interaction of B cell and antigen would result in activation and differentiation of the B cell. It was originally suggested that this tolerance-susceptible stage was restricted to the neonatal period. When it subsequently became clear that B cells were generated throughout life, the proposal then became that the immature or virgin B-cell stage was the tolerance-susceptible phase of B-cell development. Animal experiments using transgenic technology have confirmed this proposal. Thus, B cells expressing an antibody which binds strongly to a self-antigen present in the bone marrow will initiate what is now called activation-induced cell death, a process that results in the death of the B lymphocyte. In this way dangerous self-reactive clones are deleted as they appear. Susceptibility to antigen-mediated deletion must be restricted to a narrow period of development, otherwise B cells would be unable to respond to foreign antigens. It is currently thought that such a tolerance-susceptible state is that of the virgin B cell.

## B. Clonal Anergy

Not all self-antigens, however, are expressed in the bone marrow. Proteins specific to the pancreas, for example, represent potential antigens to which virgin B cells would never be exposed in the bone marrow and to which they could therefore never be tolerant. That is, there must also be mechanisms ensuring that self-antigens in the periphery are not the target of autoimmune attack. At present how this is achieved is not completely clear. Various model systems have been established to analyze this situation, and a number of potentially important mechanisms have been revealed. For example, it appears that in normal situations, B cells do not encounter most of the non-lymphoid organs of the body. B-cell migration normally consists of the lymphoid organs, the lymphatic system, and the blood. B cells do not go to other locations in the body such as the solid organs, unless these are sites of inflammation. This process of keeping the B cells segregated from potential self-antigens is referred to as "clonal ignorance." There also appears to be a particular state of lymphocyte nonresponsiveness called "clonal anergy" which can be induced in lymphocytes if they encounter an antigen in a particular context. In an anergic state, the self-reactive B cell persists but is unable to respond to the antigen. Clonal anergy may be another way in which self-reactive lymphocytes are silenced. If, however, a B cell does initiate a response to a self-antigen, then that response will be quite limited in the absence of T-cell help. T cells, which are another kind of lymphocyte that develop in the thymus, are the controlling cells of the immune response, as outlined in more

detail later. For this reason, T-cell tolerance, which operates along similar principles to that described for B cells, is critical for maintaining an overall state of nonresponsiveness to self-antigens. Through the imposition of tolerance on B cells and T cells, animals are able to generate and maintain a large pool of lymphocytes with a huge degree of variability in antigen specificity but with a limited potential for self-reactivity. These mature lymphocytes circulate through the blood and lymphoid system, ready to respond when they encounter the appropriate foreign antigen. [*See* T-Cell Activation.]

The end product of B-cell development is a pool of mature B lymphocytes expressing a huge array of different specificities, but purged of self-reactivity. These specificities are clonally distributed, meaning that all of the antibody molecules expressed on the surface of a particular B cell are identical. Mature B cells will remain in a quiescent state unless they encounter an antigen to which their antibody binds with an affinity above some critical threshold. The antigen has to be "seen" by the B cell in a particular context, so that the B cell can distinguish foreign from self-antigens, as discussed earlier. The antigen is thus said to select the responding B cell from among the total population of B cells by virtue of binding to the surface immunoglobulin. This event is the clonal selection originally proposed by Macfarlane Burnet and Talmage. One of the consequences of the interaction of the B cell with a foreign antigen can be the further diversification of the antibody molecules expressed by the responding B cells.

## IV. ANTIBODY DIVERSIFICATION IN RESPONSE TO ANTIGEN

### A. The B-Cell Response to Antigen

Foreign antigens that contain a protein component elicit a response that involves both T and B lymphocytes. Indeed the production of antibody molecules to such antigens is entirely dependent on the presence of T cells and the "helper" factors (e.g., cytokines) they provide to antigen-specific B cells. Interaction of the antigen-specific B and T cells results in a number of differentiative events occurring. First, there is a period of clonal expansion in which the number of antigen-specific B and T cells increases enormously. A fraction of these expanded B cells differentiate into antibody-secreting cells called plasma cells. The antigen-specific antibody secreted by these cells is important in combating the foreign antigen, be that by neutralizing a virus or by targeting the complement cascade to the surface of a bacterium. Other members of the expanded clones form histologically recognizable structures called germinal centers, which are unique to T-cell-dependent responses.

## B. Germinal Centers and Somatic Hypermutation

Germinal centers are formed in the B-cell areas of the lymphoid tissues and comprise mainly B cells, a small number of T cells, macrophages, and a highly specialized cell called a follicular dendritic cell. Within the germinal center there is extensive B-cell proliferation and a certain amount of cell death. A number of experiments have determined that the function of the germinal center is twofold: (i) the improvement in antibody affinity for antigen, a process known as affinity maturation, and (ii) the generation of memory B lymphocytes. Immunological memory describes the ability of animals to respond to an antigen upon a second exposure to that antigen both more rapidly and with a better quality antibody than in the first response. The quality of the antibody is improved both by a change in the isotype (e.g., from IgM to IgG1) and by an increase in binding affinity. The increase in antibody affinity has been determined to be the result of a process called V gene somatic hypermutation and represents a postantigenic form of antibody diversification. Somatic hypermutation occurs only during T-cell-dependent responses and is almost completely restricted to the V region genes of B cells in germinal centers. Although there have been some reports of somatic hypermutation of T-cell receptor V region genes, the significance of this is not yet clear.

Somatic hypermutation of V genes describes the phenomenon in which nucleotide exchanges are introduced in a random fashion into the V gene segments of the immunoglobulin heavy and light chain genes. These nucleotide exchanges may result in amino acid exchanges within the V region of the antibody protein, which in turn may influence the ability of the antibody to bind antigen, i.e., antibody variable regions are composed of subregions called the hypervariable regions and the framework regions. As the name suggests, the framework regions provide a scaffold upon which the part of the antibody which binds antigen rests. Even though all variable regions are different, the frameworks regions of V segments are the most similar between different V regions. The hypervariable regions, however, are always different. The hy-

pervariable regions form loops that stick out from the framework of the V region, and these loops, or the structure they form, are the major contact site for antigen binding. Thus the hypervariable regions are both critical in determining antigen specificity and also the affinity with which the antibody binds the antigen. Amino acid changes within the hypervariable regions would therefore be expected to have the most significant effects on the binding characteristics of the antibody. Some exchanges may make binding worse, others may have no effect, and yet others may improve the affinity of the antibody for the antigen. Exchanges outside the hypervariable regions may also influence antigen binding, not so much by improving the "fit" of the antibody for antigen, but by changing the scaffold on which the antigen-combining site sits. Amino acid exchanges in the framework regions would not generally be favored as they would most likely alter the basic structure of the variable region and thus reduce antigen binding rather than improve it. When the variable region genes of B cells that have been through rounds of somatic mutation and selection are examined and compared to the starting or unmutated form, it has been found that DNA mutations which result in amino acid exchanges occur more often than by chance alone in the hypervariable regions and less often than by chance alone in the framework regions. This nonrandom distribution of mutations is evidence for the selection of mutated antibodies on the basis of antigen binding.

## C. B-Cell Selection in the Germinal Center

The introduction of mutations into immunoglobulin V genes is an essentially random process, meaning that there must be strong selection for certain outcomes if it is to be useful to the immune response, i.e., those V gene mutations which improve the binding of the antibody to antigen need to be selectively retained in the system at the expense of those which either have no or a deleterious effect. This selection also occurs in the germinal center and is thought to involve the contact of B cells with follicular dendritic cells. A unique feature of follicular dendritic cells is that they retain antigen in the form of immune complexes (aggregates of antibody and antigen) on their cell surface for extensive periods. It is thought that these immune complexes act as depots of antigen on which the cell surface immunoglobulin derived from the newly mutated V genes in germinal center B cells can be tested. Those mutated B cells whose cell surface antibody binds better than that of the other B cells in the germinal center will compete better for access to the antigen on the follicular dendritic cell surface and thereby be given a survival signal. This survival signal will allow the B cell to divide and for its clonal progeny to undergo additional rounds of mutation and selection. In this way, those mutations which improve the affinity of the B cell for antigen will be retained in the germinal center at the expense of those which either have no effect or diminish binding. The overall effect is for an improvement in affinity of the whole antigen-specific B-cell population. The nonrandom distribution of somatic mutations within the V region, outlined earlier, is evidence for the efficiency of the selection process.

## V. B-CELL TOLERANCE IN THE GERMINAL CENTER

Although certain biases exist in the nature of the mutations introduced into V genes during somatic hypermutation (such as one DNA strand being preferred and certain base combinations being more frequently targeted than others), there is no prior selection on the basis of antigen binding, i.e., it is possible for any amino acid to be substituted for any other during the mutational process. As noted earlier, a selection process exists in the germinal center to favor those B cells whose V genes have been mutated to give improved binding to antigen over other B cells. Another possible outcome of V gene mutation, however, needs to be considered. This being the situation in which antibody diversification through V gene mutation in the germinal center leads to self-reactivity. The first section outlined how self-reactive B cells could be generated during the rearrangement of immunoglobulin gene segments. It was also apparent that the immune system had developed methods for removing self-reactive B cells either as they were generated or at the time they encountered their cognate self-antigen. Does some mechanism exist, then, by which B cells which may acquire self-reactivity in the germinal center can be rendered harmless?

The question of whether tolerance can be maintained and/or imposed in the germinal center has been difficult to address experimentally. It is only recently that model systems have been developed which go some way to answering this question. Immunization with a T-cell-dependent antigen results in the formation of both germinal centers and clusters of antibody-secreting cells. The antibody-secreting cells are terminally differentiated, nondividing, and have a limited life span. In order to address the question of tolerance in the germinal center, a system was developed that asked whether something that was normally a foreign antigen could be administered in such a way that it would behave like a self-antigen. A number of early experiments in immunology had shown that protein antigens administered in a deaggregated or soluble form were toleragenic rather than immunogenic, i.e., in this form the antigens behaved like self-antigens. The approach was therefore taken to initiate an immune response to a protein antigen by administering it in an immunogenic, aggregated form and then later giving a large dose of the same antigen in a soluble or toleragenic form. Thus the whole response could be considered to have converted to self-reactivity at the time at which the antigen was given in a toleragenic form. When this was done it was found that the soluble antigen induced a wave of death among the antigen-specific germinal center B cells. No increase in cell death was observed in the antibody-secreting cells, indicating that the cell death being induced by soluble antigen was specific for germinal center B cells. Although experiments such as this are highly contrived, the results do suggest that if a germinal center B cell mutated to an anti-self specificity, and if that self-antigen was present in a nonimmunogenic form in the germinal center, then that B cell would be killed, thereby maintaining B-cell tolerance. This tolerance susceptibility in the germinal center is referred to as a "second window" of tolerance, the first being early in development in the bone marrow.

## VI. ANTIBODY ISOTYPE SWITCHING

The final aspect of antibody diversity to consider does not involve the antigen-combining site, but rather the other end of the protein, the constant region. There are eight heavy chain constant regions in humans. Early in development B cells express only IgM, then IgM together with IgD. After exposure to antigen, a fraction of the responding B cells will differentiate into antibody-secreting cells. Initially, such cells will secrete antigen-specific IgM which can be detected in the blood. As the response progresses, the nature of the secreted antibody changes in two ways. First, the affinity for antigen improves as described in a previous section. Second, the isotype of the secreted antibody changes from IgM to one of the other classes. This change, called isotype switching, involves DNA recombination at the immunoglobulin heavy chain locus. Unlike gene rearrangements occurring in the bone marrow during B-cell development, the variable region is not affected. The genes encoding the various heavy chain constant regions are arranged in a linear fashion along the chromosome. The genes encoding the IgM constant region (called $C\mu$) are closest to the variable region elements. Downstream of $C\mu$ is $C\delta$ (encoding IgD), then $C\gamma3$ (IgG3), $C\gamma1$ (IgG1), $C\alpha1$ (IgA1), $C\gamma2$ (IgG2), $C\gamma4$ (IgG4), $C\varepsilon$ (IgE), and finally $C\alpha2$ (IgA2). The constant region genes have no promoter and therefore cannot be expressed as RNA. Rearrangement of the variable region gene segments in the bone marrow brings the V region promoter close to the IgM constant region and allows its expression. IgD is very close to IgM and is expressed off the same mRNA molecule by alternate splicing. In this way a single B cell is able to express both IgM and IgD with the same variable region and with the same light chain. Upon antigenic stimulation and the B cell undergoing isotype switching, the already rearranged DNA encoding the heavy chain variable region recombines with DNA adjacent to one of the more distant constant regions. The DNA in between the variable region and the target constant region is deleted from the chromosome, meaning that isotype switching is irreversible. After recombination, the B cell expresses the same heavy chain variable region but fused to a different constant region. The light chain is unchanged and thus the specificity of the B cell is unchanged. The choice of downstream isotype is not random but is rather directed by the circumstances in which the B cell is activated. T cells, for example, can influence the outcome via the helper factors (cytokines) they secrete. In this way the immune response produces the antibody isotype which is most appropriate for a given antigen.

What advantage is there in diversifying the constant region of antibody molecules? The reason for the immune system developing this feature lies in the function of the constant region. The antibody constant region determines what the so-called effector functions of the antibody will be. The different antibody classes specialize to some extent in the biological properties they mediate. Some, such as IgM and IgG3, are very efficient in complement fixation by the classical pathway. Others like IgA1 efficiently activate complement by the alternative pathway. IgG2 and IgG4 cross the placenta and thereby provide antibody protection to newborn infants. Not all of the effector properties of the antibodies are necessarily good. IgE antibodies, for example, bind via their constant regions to mast cells and basophils. Crosslinking of this IgE can lead to triggering of these cells which results in the symptoms of allergy.

## BIBLIOGRAPHY

Clevers, H. C., and Grosschedl, R. (1996). Transcriptional control of lymphoid development: Lessons from gene targeting. *Immunol. Today* **17**, 336–343.

Cornall, R. J., Goodnow, C. C., and Cyster, J. G. (1995). The regulation of self-reactive B cells. *Curr. Opin. Immunol.* **7**, 804–811.

Kelsoe, G. (1995). The germinal center reaction. *Immunol. Today* **16**, 324–326.

Maclennan, I. C. M. (1995). Autoimmunity: Deletion of autoreactive B cells. *Curr. Biol.* **5**, 103–106.

Melchers, F., Rolink, A., Grawunder, U., Winkler, T. H., Karasuyama, H., and Ghia, P. (1995). Positive and negative selection events during B lymphopoiesis. *Curr. Opin. Immunol.* **7**, 214–227.

Rajewsky, K. (1996). Clonl selection and learning in the antibody system. *Nature* **381**, 751–758.

Tarlinton, D. (1994). B-cell differentiation in the bone marrow and the periphery. *Immunol. Rev.* **137**, 203–229.

# Antidepressants

CHARLES L. BOWDEN
*University of Texas Health Science Center*

## GLOSSARY

**Anorexia nervosa** An ingrained misperception of body image, wherein the person subjectively feels fatter than in fact and deliberately eats inadequate amounts of food in an effort to further reduce an already low body weight

**Bulimia** A condition characterized by alternative excessive eating and self-induced vomiting in a maladaptive effort to keep weight at a subjectively perceived ideal level

**Monoamine oxidase inhibitors** Monoamine oxidase is a key enzyme involved in the degradation of biogenic amines; drugs effective in blocking this enzyme are generically referred to as monoamine oxidase inhibitors

**Tricyclic antidepressants** Sometimes called TCAs or standard antidepressants, their name comes from the three linked benzyl rings that are common to their structure

ANTIDEPRESSANTS ARE DRUGS USED TO RELIEVE or prevent psychic depression.

## I. INTRODUCTION

Antidepressants are a relatively new group of drugs, the first specifically effective drugs having been introduced in the early 1950s as the agents that heralded the age of modern, effective drug treatments for mental disorders. Antidepressant drug research is an exceptionally strong area of scientific investigation. Advances in antidepressant treatment have been aided by the ability to classify depressive disorders into more homogeneous subgroups by relatively strong evidence that CNS amine metabolism is disturbed in some depressive disorders and by the fact that drugs effective in treatment of depression alter some aspects of amine metabolism. Based on this vigorous activity, several new antidepressant drugs have been approved. These and other drugs under development will be marketed based on evidence of a more salutary side effect profile, or different point of action within the brain, with the consequent possibility of improvement in patients unresponsive to other treatments.

## II. CHARACTERISTICS OF RESPONSE TO ANTIDEPRESSANTS

Effective treatment of major depressive disorder is particularly valuable because functional capability is significantly impaired by the condition, risk of suicide is increased, and untreated depressive episodes often exceed a year in duration. Several characteristics are common to most effective antidepressants. Response is not immediate, usually requiring around a week or more at adequate doses before improvement begins. Data indicate that response in many patients does occur earlier than had previously been thought, with the greatest percentage reduction in symptoms during the first week of treatment. Improvement with antidepressants is often evident earlier to outside persons than to the patient. Improvement is not a generic process, with all symptoms improving in tandem. Rather, certain features of depression appear to be specifically likely to show improvement. These early indicators of response include improvement in anxiety, depressed mood, and concentration. For many antidepressants the response appears to be associated with attaining threshold levels of drug concentrations.

ENCYCLOPEDIA OF HUMAN BIOLOGY, Second Edition, VOLUME I. Copyright © 1997 by Academic Press.
 All rights of reproduction in any form reserved.

Furthermore, the dosages needed for prophylaxis are essentially the same as needed for resolution of an episode. These facts have resulted in clinical utility for quantitative measurement of drug concentrations in blood. Once recovery is attained, the likelihood of relapse is reduced if the patient continues treatment for at least a year and, for many patients, indefinitely. Determination of which patients may then do well with medication discontinuance and which will need indefinite maintenance is largely empirical at present. In general, chronicity of prior depressive condition is associated with a need for long-term treatment.

Most clinical studies of antidepressants have focused on 1- to 2-month responses. It appears that a longer-term response is about the same as acute response, i.e., with some few exceptions the drugs do not lose their effectiveness with time. The same dosages necessary for acute treatment are required for longer-term maintenance treatment.

Around 75% of patients are definitely improved by antidepressant treatment. The improved group is actually composed of two subgroups. The first, comprising about 50% of all patients, is essentially well. The second, comprising about 25%, is improved but retains significant symptomatology and some functional impairment. It is this smaller group of partially improved patients, as well as those who are treatment refractory, that impels efforts to develop more effective treatments.

## III. INDICATIONS FOR USE

The primary indication for antidepressant drugs is major depressive disorder. This condition entails depression of at least moderate severity. It does not include transient depressive states, understandable reactions to loss, or depressions that alternate with manic episodes (bipolar depression). Bipolar depression is also treated with antidepressants, although many authorities recommend lithium alone for initial trial instead of tricyclic-type antidepressants because the latter, when used alone, may cause rapid cycling from depression to mania. Less severe depressions, as seen in dysthymia or adjustment disorders, also respond in some instances. These have been less systematically studied than major depressive disorder. Given the evident circumstantial and psychological factors in these conditions, many authorities prefer to address these through advice and psychotherapy rather than to employ antidepressants initially. [*See* Depression; Mood Disorders.]

Antidepressants are also beneficial in a variety of secondary depressions. These are conditions that follow other major medical disorders, such as strokes, heart disease, or any condition with chronic, debilitating, or discomforting effects. The major therapeutic task in such conditions is of course attention to the primary condition. Nevertheless, treatment of the depressive state is often vital. For example, over half of stroke patients develop significant depressions that, untreated, interfere with ability to participate in rehabilitation efforts and ultimate recovery potential.

It has become apparent that several disorders other than depression respond to antidepressant treatment. In some instances, these may co-occur with depression (referred to as comorbidity). In others, there is no evidence of concurrent depression, although preliminary evidence of some of the same underlying biochemical disturbances has been reported. These conditions include bulimia, anorexia nervosa, anxiety disorders, obsessive-compulsive disorder, panic disorder, and disorders characterized by chronic pain, such as low back pain.

## IV. CLASSES OF ANTIDEPRESSANTS

### A. Tricyclic Antidepressants

These antidepressants (the largest and most widely used group) are also called "iminodibenzyl derivatives" and "standard antidepressants." All share some mix of effects on blocking the mechanism by which amine neurotransmitters, particularly norepinephrine and/or serotonin, enter nerve cells. The functional effect of this, as well as that of other types of antidepressants, is at least short-term enhancement of actions mediated by the particular neurotransmitter systems. [*See* Depression; Neurotransmitter and Neuropeptide Receptors.]

The dosage ranges of most of these drugs are relatively similar: 50–300 mg/day (see Table I). Since the effects of the drug are relatively sustained, they are generally prescribed only once daily. A lag time of several days to a few weeks occurs before response, even when initial dosage is adequate. Blood level determinations may be useful as indicators of safe and effective dosage. The side effects of these drugs are relatively similar, although there are quantitative differences. On average, for example, amitriptyline or doxepin is more sedating than nortriptyline or desipramine. Common side effects include dry mouth, blurred vision, dizziness, sedation, weight gain, and

**TABLE I**
Drugs Used in the Treatment of Depression

| | Daily dosage range (mg) |
|---|---|
| Drugs that block uptake of amine neurotransmitters | |
| Amitriptyline | 50–300 |
| Amoxapine | 50–300 |
| Desipramine | 50–300 |
| Doxepin | 50–300 |
| Fluoxetine | 10–80 |
| Imipramine | 50–300 |
| Maprotiline | 75–225 |
| Nortriptyline | 50–150 |
| Paroxetine | 20–50 |
| Protriptyline | 15–60 |
| Sertraline | 50–200 |
| Trimipramine | 50–300 |
| Venlafaxine | 75–375 |
| Drugs that inhibit monoamine oxidase | |
| Phenelzine | 15–90 |
| Tranylcypromine | 10–60 |
| Drugs with unestablished mechanisms | |
| Bupropion | 100–450 |
| Lithium carbonate[a] | 150–2700 |
| Trazodone | 100–400 |

[a]Use limited to bipolar depression. Also benefits mania. May help as adjunct in major depressive disorder.

constipation. Delayed conduction of electrical signals in the heart can cause problems with heart rhythm, especially in elderly persons with preexisting heart disease. Elderly patients often require lower doses to achieve adequate clinical effects because of their slower metabolic breakdown of these drugs. In most instances, side effects with these and other antidepressants are controllable by dosage adjustment or by a change to another drug with a different side effect profile.

## B. Specific Serotonin Reuptake Inhibitors

These drugs (fluoxetine, paroxetine, sertraline) are effective strictly through their blockade of the neuronal reuptake of serotonin. Because they lack effects on several other neurotransmitter systems, they do not cause weight gain, constipation, dry mouth, or several other side effects commonly seen with the tricyclic antidepressants. Their side effects include nervousness, nausea, sedation, diminished sexual response, and difficulty sleeping. In part because of a favorable benefit to the side effect ratio and because of beneficial effects on impulsive behavior, these drugs

are now often employed as first-line agents for depression.

## C. Other Drugs Affecting Amine Transport

Bupropion does not directly affect either norepinephrine or serotonin. It has some dopamine-enhancing effects, but it is not clear that its clinical effectiveness is through this mechanism. Its side effect profile is similar to fluoxetine and also may include a tendency to cause psychotic symptoms in susceptible patients. Venlafaxine effectively acts on both norepinephrine and serotonin, but is free of several other adverse neurotransmitter effects characteristic of tricyclic antidepressants. Because of its effects on two transmitter systems, it has been postulated to alleviate depression in otherwise treatment refractory cases, but has not been adequately assessed regarding this action. Trazodone also affects serotonin metabolism, but by a different mechanism than fluoxetine. Except for more sedating properties and a greater tendency to lower blood pressure, it has a side effect profile somewhat similar to fluoxetine.

## D. Lithium

Lithium is generally used for control of manic symptoms, but it has two modes of use in depression. In bipolar depression it may be effective in more than one-half of cases, thus allowing effective treatment of bipolar disorder with one drug rather than a combination of drugs. Lithium also enhances the effects of other antidepressants in patients only partially responsive to the antidepressant alone. Lithium is associated with side effects of shakiness, nausea, increased frequency of urination, memory impairment, and weight gain.

## E. Monoamine Oxidase Inhibitors

Monoamine oxidase inhibitors (phenelzine, tranylcypromine) also enhance biogenic amine activity, but by the mechanism of blocking the enzymatic breakdown of the active substances. These compounds are often referred to as MAOIs. Their side effects include weight gain, nervousness, and dizziness. Because of its enzyme blockade effect, tyramine (a normally innocuous protein component of certain foods) can build up to levels that cause a dangerous increase in blood pressure. Therefore, adherence to a diet which eliminates foods high in tyramine is necessary while taking these drugs.

Other compounds also appear to have antidepressant properties in certain instances, but have not been conclusively established as effective and safe. Stimulant compounds such as dextroamphetamine and methylphenidate may improve certain depressive symptoms, especially low energy. Alone, they lose their effectiveness over time and also pose risks of psychological dependence.

## BIBLIOGRAPHY

Bowden, C. L., Koslow, S., Maas, J. W., Davis, J., Garver, D. L., and Hanin, I. (1987). Changes in urinary catecholamines and their metabolites in depressed patients treated with amitriptyline or imipramine. *J. Psychiat. Res.* **21,** 111–128.

Bowden, C. L., Schatzberg, A. F., Rosenbaum, A., Contreras, S. A., Samson, J. A., Dessain, E., and Sayler, M. (1993). Fluoxetine and desipramine in major depressive disorder. *J. Clin. Psychopharmacol.* **13,** 305–311.

Fontaine, R., Ontiveros, A., Elie, R., Kemsler, T. T., Roberts, D. L., Kaplita, S., Ecker, J. A., and Faluok, G. (1994). A double-blind comparison of nefazodone, imipramine and placebo in major depression. *J. Clin. Psychiat.* **55,** 234–241.

Katz, M. M., Koslow, S. H., Maas, J. W., Frazer, A., Bowden, C. L., Casper, R., Croughan, J., Kocsis, J., and Redmond, E., Jr. (1987). The timing, specificity, and clinical prediction of trycyclic drug effects in depression, *Psychological Medicine* **17,** 297–309.

Keller, M. B., Lavori, P. W., Coryell, W., Andreasen, N. C., Endicott, J., Clayton, P. J., Klerman, G. L., and Hirschfeld, R. M. A. (1986). Differential outcome of pure manic, mixed/ cycling, and pure depressive episodes in patients with bipolar illness, *Journal of the American Medical Association* **255,** 3138–3142.

Khan A., Fabre L. F., and Rudolph, R. (1991). Vanlafaxine in depressed outpatients. *Psychopharmacol. Bull.* **27,** 141–144.

Kupfer, D. J., Frank, E., Perel, J. M., Cornes, C., Mallinger, A. G., Thase, M. E., McEachran A. B., and Grochocinski, V. J. (1992). Five-year outcome for maintenance therapies in recurrent depression. *Arch. Gen. Psychiat.* **49,** 679–773.

# Anti-inflammatory Steroid Action

HENRY N. CLAMAN
*University of Colorado School of Medicine*

STEPHEN TILLES
*Oregon Health Sciences University*

HELEN G. MORRIS
*Sandoz Research Institute*

## GLOSSARY

**Cytokines** Polypeptides produced by lymphocytes (lymphokines) or mononuclear phagocytes (monokines) that regulate the function of other cells

**Fibroblast** Stellate connective tissue cell found in fibrous tissue, responsible for producing collagen

**Glucocorticoid** Corticoid that affects glucose metabolism; secreted principally by the adrenal cortex

**Inflammation** Local tissue response to injury characterized by redness, swelling, pain, and heat

**Leukocyte** Colorless, ameboid blood cell having a nucleus and granular or nongranular cytoplasm

**Lymphocyte** Agranular leukocyte formed primarily in lymphoid tissue; occurs as the principal cell type of lymph and composes 20–30% of the blood leukocytes

**Monocyte** Large, agranular leukocyte with a relatively small eccentric, oval or kidney-shaped nucleus, produced in the bone marrow

RECOGNITION OF THE REMARKABLE ANTI-INFLAM-matory action of adrenal cortical steroids began in the late 1940s when a small quantity of Compound E (cortisone) was isolated from adrenal cortical ex-tracts. When patients with rheumatoid arthritis were given this material, there was a remarkable improve-ment in their disease. These events were followed quickly by the development of methods for large-scale synthesis of cortisol analogs and these were intro-duced as pharmacologic agents. Knowledge of the mechanisms of anti-inflammatory steroid action and the physiologic role of endogenous steroids evolved as information on immunologic regulatory mecha-nisms increased.

## I. ROLE OF STEROIDS IN HOST DEFENSE

Inflammation is a host response to tissue injury caused by infectious, chemical, thermal, physical, or immu-nologic factors. The inflammatory response serves to confine the damage to a localized area, to eradicate the noxious agent, to repair the damage to tissue, and, at the same time, to protect the host from the deleterious effects of the injury. The mobilization of defenses includes recruitment of leukocytes and the release of a large number of powerful, proinflamma-tory mediators that interact with each other and with the recruited cells to produce a cascade effect that amplifies the response to injury. [*See* Inflammation.]

In many instances, the inflammatory response is highly beneficial. However, the mobilization of defense mechanisms and the diverse effects of the pro-inflammatory mediators sometimes result in a re-sponse that extends the initial injury and itself threat-ens homeostasis. An important aspect of host defense,

ENCYCLOPEDIA OF HUMAN BIOLOGY, Second Edition, VOLUME 1.    Copyright © 1997 by Academic Press.    All rights of reproduction in any form reserved.

therefore, involves modulation of pro-inflammatory mechanisms, as well as those that limit or counteract inflammation.

Endogenous glucocorticoids are part of an integrated system of neuroendocrine—immunologic regulation of host-defense mechanisms. Their role is to modulate the inflammatory cascade by preventing the unbridled release of pro-inflammatory mediators and their effects on cellular function. Pharmacologic steroids exert their effects by the same mechanisms as those involved in physiologic actions. The enhanced anti-inflammatory action of the synthetic analogs results from the quantitative effects of giving supraphysiologic amounts of steroids with greater potency and a more sustained duration. [*See* Neural–Immune Interactions; Neuroendocrinology.]

## II. STEROID EFFECTS ON INFLAMMATION

### A. Suppression of the Response

Glucocorticoids suppress inflammation irrespective of cause by interfering with numerous steps in the inflammatory cascade. They suppress redness, heat, swelling, and pain in the acute inflammatory response by constricting the microvasculature and by preventing the local increase in vascular permeability, leakage of fluid and protein into the extravascular space, and the influx of leukocytes in the area of injury. They also suppress the recruitment of cells for chronic inflammatory reactions, the mobilization of fibroblasts for repair processes, and the manifestations of both delayed hypersensitivity and the allergic response. In addition, they suppress systemic responses including fever, cardiovascular effects, and synthesis of acute-phase proteins by the liver.

### B. Influence of Corticosteroids on Leukocyte Traffic

Within hours after administration of steroids, lymphocytes (particularly T helper cells), monocytes, basophils, and eosinophils move from the circulation and are sequestered in bone marrow and lymphoid organs. In contrast, the number of circulating neutrophils increases as a result of mobilization from bone marrow and tissue storage sites. Nevertheless, neutrophils are less able to enter the tissues when steroids are given. These changes in the availability of circulating leukocytes lead to their reduced entry into inflamma-

tory foci. Since the development of inflammation requires recruitment of circulating leukocytes, steroids have a profound inhibitory effect on the development of inflammatory responses. [*See* Steroids.]

The mechanisms by which steroids affect leukocyte traffic are not completely understood, although recent studies indicate that inhibition of pro-inflammatory cytokine expression is a primary feature. The absence of these cytokine signals results in poor growth and differentiation of leukocyte marrow precursors and decreased expression of endothelial cell adhesion molecules which are essential for recruitment of leukocytes to sites of inflammation. Steroids also have a more direct influence on neutrophil demargination by decreasing neutrophil expression of an adhesion molecule from the integrin family called Mac-1. Decreased expression of Mac-1 presumably results in less adhesion of neutrophils to endothelial surfaces. Thus, steroid-influenced neutrophils are less able to marginate to the vessel walls and thus are less likely to leave the circulation and enter nascent inflammatory foci. [*See* Cytokines and the Immune Response.]

### C. Suppression of Inflammatory Mediators

The primary anti-inflammatory action of steroids is to block production of the chemical mediators of inflammation. Many of the other effects of steroids are secondary to the reduction in mediator activity. Steroids also modulate the responses to some of the mediators, but these effects are of lesser importance in reducing inflammation and can often be abrogated experimentally by supplying the missing mediator.

Steroids enlist a variety of mechanisms in blocking the production of inflammatory mediators. For protein mediators (e.g., the cytokines), the primary mechanism involves a decrease in DNA transcription. This may involve direct binding of the steroid to the DNA in the promoter region of cytokine genes or it may involve interaction with transcription activating factors required for cytokine gene synthesis. In addition, steroids may increase the expression of other gene products which suppress cytokine synthesis. Steroids can accelerate the breakdown of cytokine messenger RNA and can also inhibit the expression of certain cytokine receptors. Although the details of the exact mechanisms have not been worked out for each of the cytokine mediators, steroids have been shown to inhibit the expression of interleukins-1, 2, 3, 4, 6, and 10, as well as tumor necrosis factor $\alpha$ and interferon-$\gamma$. The nonprotein mediators of inflammation which are inhibited by steroids include histamine, bradykinin, and the arachidonic acid metabolites (leukotrienes, prostaglandins, thromboxanes, and platelet-activating factor). The mechanisms for these effects are not well understood.

## D. Influence on Cellular Function

Glucocorticoids modify the functions of many cells that participate in inflammatory responses by direct actions on cells and by their effects on the chemical mediators that influence cellular functions. Steroids also modify cellular function by decreasing glucose utilization, reducing the uptake and incorporation of amino acids into proteins, and regulating the intracellular transport of calcium ions.

### 1. Monocytes/Macrophages

Monocytes–macrophages have a central role in host defense as they are among the first cells to respond to an injury. They aid in the recruitment and activation of other leukocytes, engulf and destroy bacteria and other pathogens, debride injured tissue, remove the products of inflammation, and promote fibrosis and tissue repair. Steroids have profound effects on monocytes and macrophages. Acute dosing with steroids causes a monocytopenia, thus decreasing the circulating pool of macrophage precursors which can enter inflammatory foci. Steroids cause an inhibition of the release of interleukin-1, interferon-$\gamma$, tumor necrosis factor, and a variety of other products. In addition, steroids suppress some of the phagocytic and antimicrobial properties of monocytes and macrophages. Steroids also inhibit their ability to stimulate lymphocytes by decreasing the macrophage expression of MHC class II. [*See* Macrophages.]

### 2. Lymphocytes

Steroids also have profound effects on lymphocytes. As mentioned earlier, steroids decrease the circulating numbers of these cells, including T cells (both CD4 and CD8) and B cells. Steroids are also potent inhibitors of the proliferative response of lymphocytes to antigen and mitogen stimulation. This is primarily due to the inhibition of production of interleukin-2 (initially called T-cell growth factor), interleukin-4 (initially called B-cell growth factor), and interferon-$\gamma$. Interleukin-2 and interferon-$\gamma$ are important for the differentiation of helper T lymphocytes into "$T_H$-1" effector cells. $T_H$-1 cells are pro-inflammatory in conditions such as delayed-type hypersensitivity. Interleukin-4 is important for the differentia-

tion of lymphocytes into the "$T_H$-2" phenotype. $T_H$-2 cells are the classical T cells which "help" B cells make antibody. Antibody responses are somewhat dependent on both $T_H$-1 and $T_H$-2 cells. In the presence of steroids, B lymphocytes alone are able to differentiate relatively normally and can secrete specific immunoglobulin in response to antigen challenge. Nevertheless, substantial doses of systemic steroids over weeks can lower serum immunoglobulin levels in humans. This result probably reflects ongoing inhibition of helper T-cell functions. [*See* Lymphocytes.]

### 3. Other Leukocytes

Steroid effects on mast cells, basophils, and eosinophils are mainly indirect as these cells are essentially paralyzed by the steroid-induced decrease in necessary growth, differentiation, and chemotactic factors. One exception to this is steroid enhancement of the expression of endonucleases which mediate eosinophil programmed cell death (apoptosis).

## III. IMMUNOSUPPRESSIVE PROPERTIES OF CORTICOSTEROIDS

Corticosteroids are used both to suppress inflammation and to down-regulate immune responses. It is often difficult to distinguish these two activities because many of the same cells, pathways, mediators, and cytokines are used in both processes.

It is important to note definite differences between species in terms of their ability to be immunosuppressed by corticosteroids. For example, the lymphoid cells and tissues of rats, mice, and rabbits are readily damaged by corticosteroids at doses that have much less effect on the lymphocytes of man, monkeys, guinea pigs, and sheep. These inhibitory and immunosuppressive effects include apoptosis, redistribution out of the circulation, and suppression of intracellular metabolic processes. The molecular basis of this species difference in sensitivity to steroids is not yet known. The net result is that antibody formation, allograft rejection, and delayed hypersensitivity can all be far more easily inhibited in mouse, rat, and rabbit than in the more resistant species. However, the differences are relative. Sufficient glucocorticoids given to humans for long enough can down-regulate all of the just-mentioned immunologic processes.

## IV. STEROID TREATMENT

One of the great virtues of glucocorticosteroids, considered as therapeutic agents, is that they can be given by almost any route and in a bewildering array of physical formulations. Thus, corticosteroids are among the most versatile of all classes of medications.

## A. Systemic Steroids

Corticosteroids were first used clinically in the treatment of severe rheumatoid arthritis. There were given orally and were exceedingly effective. Since then, all inflammatory connective tissue illnesses have been successfully treated with corticosteroids. These illnesses include systemic lupus erythematosus, scleroderma, dermatomyositis/polymyositis, polymyalgia rheumatica, cranial arteritis, and polyarteritis nodosa. Systemic corticosteroids are also often indicated in the treatment of systemic immunologic and allergic disorders, including vasculitides, serum sickness, and systemic drug reactions.

Systemic corticosteroids are also used for organ-specific inflammatory diseases where local steroid treatment would not be possible. Such conditions include the treatment of renal, gastroenterological, pulmonary, cardiac, hepatic, thyroid, and generalized inflammation of the skin.

## B. General Principles of Systemic Steroid Therapy

Clinicians have begun to consider various phases of a course of treatment. These include, in order:

1. An induction phase using high doses in an attempt to halt inflammation.
2. A consolidation phase at a lower dose when inflammation has been controlled.
3. A tapering phase, hoping to stop steroids.
4. A maintenance phase (at the lowest possible dose) if tapering to zero is not possible.

## C. Methods for Giving Systemic Corticosteroids

### 1. Intravenous

This route is often used to treat systemic inflammation. As absorption is bypassed, it is, in theory, the most rapid and efficacious route. However, it is also the most expensive and cumbersome.

### 2. Intramuscular Depot Steroids

Preparations of slowly released steroids can be given once a month. While this strategy works, it can be harmful if repeated because the plasma levels are steady and there are no daily troughs to help turn on the hypothalamic–pituitary–adrenal (HPA) axis. Thus, HPA axis suppression can be deleterious when these preparations are repeatedly used.

### 3. Oral Corticosteroids

These are the cheapest formulations available. Prednisone, although it is a pro-drug (which must be converted to prednisolone in the liver), is 100% absorbed. It also has a short plasma half-life, thus minimizing HPA axis suppression. Many physicians consider prednisone to be the systemic corticosteroid of choice.

## D. Local Corticosteroids

Obviously, if the target tissues of corticosteroids are localized, then local therapy would be best, minimizing systemic side effects. It is to the credit of the pharmaceutical industry that it has provided a multitude of local preparations. These have been applied to virtually every tissue in the body, such as the lungs (aerosol), skin, nose, eyes, ears, and gastrointestinal tract. These agents tend to be more lipophilic than intravenous or oral preparations. They are successful in inhibiting cytokine production, are vasoconstrictive, and can result in a diminution of inflammatory cells, such as neutrophils, eosinophils, basophils, and mast cells, away from the sites of inflammation.

## V. STEROID SIDE EFFECTS AND COMPLICATIONS

Corticosteroid therapy has the potential of resulting in a wide variety of complicated side effects. Paradoxically, this is due to the same mechanisms which enable corticosteroids to have such potent anti-inflammatory effects, i.e., the steroid receptor is present in all nucleated cells and therefore virtually all cells in the body are potentially susceptible to steroid action. Accordingly, steroid side effects may involve any organ system.

Many of the specific mechanisms for steroid-induced side effects remain to be elucidated. However, years of clinical experience have taught us which side effects are likely to occur in various steroid treatment scenarios. Both the dosage and the duration of treatment are the main determinants of the expected side effect profiles. Although many side effects and complications are expected, others are rare and even unpre-

## TABLE I
### Corticosteroid Side Effects

Expected with high-dose sustained therapy
    Cushing's syndrome
    Hypothalamic–pituitary–adrenal axis suppression
    Weight gain (fluid and fat)
    Mood disturbance
    Impaired wound healing
    Increased risk of infection
    Hypercalciuria
    Increased protein catabolism

Expected with large cumulative doses
    Osteoporosis
    Posterior capsular cataracts
    Skin atrophy, ecchymoses
    Growth retardation (in children)
    Atherosclerosis

Exacerbated by therapy, dose dependent
    Hypertension
    Glucose intolerance
    Peptic ulcer disease (disputed)
    Acne vulgaris

Occasionally seen, generally dose dependent
    Avascular necrosis of bone
    Proximal myopathy
    Fatty liver
    Hirsutism

Rarely seen, unpredictable
    Psychosis
    Lipomatosis
    Steroid allergy
    Pseudotumor cerebri
    Glaucoma
    Pancreatitis

dictable. Table I lists many of the side effects of systemic corticosteroids. Locally administered steroids have fewer systemic effects, although local absorption into the circulation has the potential to result in the same side effects as systemically administered drugs. More commonly, locally administered steroids result in local side effects that tend to be dose and potency dependent. Because of the tremendous therapeutic potential of corticosteroid therapy, the pharmaceutical industry continues to develop newer steroid formulations which can provide greater benefits with fewer undesirable side effects.

## BIBLIOGRAPHY

Axelrod, L. (1989). Side effects of glucodorticoid therapy. *In* "Anti-inflammatory Steroid Action: Basic and Clinical Aspects" (R. P. Schleimer, H. N. Claman, and A. L. Oronsky, eds.), pp. 377–408. Academic Press, San Diego.

Boumpas, D. T., Chrousos, G. P., Wilder, R. L., Cupps, T. R., and Balow, J.E. (1993). Glucocorticoid therapy for immune-mediated diseases: Basic and clinical correlates. *Ann. Intern. Med.* **119**, 1198–1208.

Kunicka, J. E., Talle, M. A., Denhardt, G. H., Brown, M., Prince, L. A., and Goldstein, G. (1993). Immunosuppression by glucocorticoids: Inhibition of production of multiple lymphokines by *in vivo* administration of dexamethasone. *Cell. Immunol.* **149**, 39–49.

Robertson, D. B., and Maibach, H. I. (1989). Topical glucocorticoids. *In* "Anti-inflammatory Steroid Action: Basic and Clinical Aspects" (R. P. Schleimer, H. N. Claman, and A. L. Oronsky, eds.), pp. 494–524. Academic Press, San Diego.

Schleimer, R. P. (1993). Glucocorticosteroids: Their mechanisms of action and use in allergic diseases. *In* "Allergy, Principles and Practice" (E. Middleton, Jr., C. E. Reed, E. F. Ellis, N. F. Adkinson, Jr., J. W. Yunginger, and W. W. Busse, eds.), pp. 893–925. C.V. Mosby Co., St. Louis, MO.

Schleimer, R. P., Claman, H. N., and Oronsky, A. (eds.) (1989). "Anti-inflammatory Steroid Action: Basic and Clinical Aspects." Academic Press, San Diego.

Truhan, A. P., and Ahmed, A. R. (1989). Corticosteroids: A review with emphasis on complications of prolonged systemic therapy. *Ann. Allergy* **62**, 375–390.

Williams, T. J., and Yarwood, H. (1990). Effect of glucocorticosteroids on microvascular permeability. *Am. Rev. Respir. Dis.* **141**, S39–S43.

# Antimicrobial Agents, Impact on Newborn Infants

RUTGER BENNET
*Karolinska Institute*

## GLOSSARY

**Aerobic** Living in the presence of oxygen. Most "aerobic" intestinal bacteria are, in fact, facultative (i.e., living under both aerobic and anaerobic conditions)

**Anaerobic** Living in the absence of oxygen

**Gram negative** Refers to a method for staining bacteria, introduced by Danish physician H. C. Gram; common gram-negative bacteria include *Escherichia coli, Klebsiella,* and *Bacteroides*

**Lumen** Space inside a tubular organ (e.g., the intestine)

**Microflora** Various microorganisms (mostly bacteria and fungi) populating a certain space. A more correct, but less used, term would be "microbiota"

**Mucus** Substance, composed mainly of glycoproteins, which lines most tubular organs, with a great number of biological functions

**Neonatal** The first 28 postnatal days of life

**Preterm** Born before completion of the 37th week of gestation

**Very low birth weight** Birth weight under 1500 g, usually corresponding to approximately 31 weeks of gestation

ANTIMICROBIAL AGENTS MAY INTERACT WITH HUMAN biology in a number of ways, including allergic reactions, liver toxicity, and interactions with leukocyte functions. This article, however, deals only with the impact of these agents on the microflora normally present in the intestinal tract. In newborn infants this microbial ecosystem is different from that of older children and adults.

## I. INTESTINAL MICROFLORA OF NEWBORN INFANTS

### A. General Principles of Intestinal Microecology

#### 1. Methods of Study

##### a. Sampling

The most commonly used method for studying intestinal microflora is the culture of feces. It is important to realize that this method gives a simplified picture of conditions in the parts of the intestine closer to the stomach. Populations important to these areas (e.g., the small intestine) are washed further out into the colon, where they might "drown" in the much larger bacterial populations that prevail there. Also, important bacterial species can be strictly adherent to the intestinal wall and not appear in high enough numbers in the luminal contents to appear in cultures. It is impossible to say whether a given fecal bacterial species has its natural habitat in the lumen of the colon or elsewhere. Ideally, cultures should be collected from as many parts of the intestinal tract as possible, and preferably both luminal content and mucus should be cultured. In studies of humans, this poses large practical problems.

##### b. Culture Techniques

Many bacterial species, especially of the colon, are strictly anaerobic and perish in the presence of oxy-

ENCYCLOPEDIA OF HUMAN BIOLOGY, Second Edition, VOLUME I.    Copyright © 1997 by Academic Press.    All rights of reproduction in any form reserved.

gen. These bacteria often have other, very specific growth requirements because they are generally highly adapted to their normal habitat. Thus, a large number of intestinal bacteria have not yet been cultured *in vitro*. In fact, only a handful of laboratories in the world have the facilities to perform all of the procedures for identification and enumeration of bacteria in fecal cultures. Such procedures might involve several hundred bacteriological platings!

## 2. Species Composition
### a. Aerobic Bacteria

Aerobic bacteria are the most well known because they frequently cause disease and are easy to culture. However, they constitute only a small fraction of the intestinal microflora. In the colon, aerobic bacteria make up only 0.1% of the bacteria present. *α-Streptococci* are frequently found in the upper intestinal tract, whereas bacteria belonging to the Enterobacteriaceae family prevail in the distal part of the small intestine and in the colon. Most of these species can cause disease if they enter the bloodstream.

### b. Anaerobic Bacteria

Anaerobic bacteria constitute the bulk of the microflora in the lower intestinal tract. Most species belong to the *Bacteroides, Bifidobacterium,* and *Lactobacillus* genera. *Bacteroides fragilis* is frequently involved in mixed abdominal infections, but anaerobic bacteria rarely cause disease compared to the number of infections associated with aerobic bacteria. See Table I for data on bacterial species and bacterial concentrations in various paarts of the intestine. [*See* Anaerobic Infections in Humans.]

### c. Ecological Considerations

Attachment or adherence to structures lining the intestinal tract is required for permanent colonization. In some cases the molecular mechanisms of attachment have been defined. The receptor molecules on the intestinal wall, to which bacteria attach, might be part of the epithelial cell membrane or of the mucus layer covering the entire epithelial surface of the intestinal tract. In most cases, however, details of attachment are unknown. Certain bacteria might also influence the growth of other species. The aerobic bacteria of the lower intestinal tract are considered necessary for creating a suitable environment for growth of the dominating anaerobic microflora. Some bacteria (e.g., *Lactobacillus*) produce substances, so-called bacteriocins, which inhibit the growth of other bacterial species. Many bacteria produce organic acids as a meta-

bolic end product, thereby increasing acidity, which has an inhibiting effect on most bacteria. Details of this undoubtedly immensely complex ecological interaction system, however, remain to be elucidated.

## 3. Functions
### a. Colonization Resistance

It has been shown in several animal species that a normal anaerobic microflora acts as a barrier to colonization by potentially pathogenic bacteria from the environment. Antimicrobial agents that reduce or completely destroy this anaerobic microflora frequently lead to death of the animals from infection by gram-negative aerobic microorganisms. The chain of events is thought to be initial colonization, overgrowth of the organism in high numbers, "translocation" of the organism across the intestinal wall, and invasion of the bloodstream. Agents that spare the anaerobic microflora have a much smaller tendency to cause such effects. In humans, however, the results are not so clear-cut, and although there are some studies which implicate similar functions of the human anaerobic microflora, there are also studies with conflicting results. It seems likely that colonization resistance in humans is a complex phenomenon that can vary in different parts of the intestine and also for different invading species. Other compromising factors such as major trauma, radiation, and cancer also seem to be important.

### b. Metabolic, Immunological, and Nutritional Functions

The cells that compose the intestinal microflora are more numerous than the entire number of cells of the human body, and it is clear that the metabolism in these cells must influence the host in a number of ways. The microflora participates in the enterohepatic circulation between the intestine and the liver of a large number of steroid compounds, produces several vitamins, and stimulates the immune system. It is, however, important to realize that none of these functions has been conclusively shown to be of vital importance to the host and that germ-free animals, which never come into contact with bacteria, are at least as healthy as their conventional counterparts, as long as they are kept in a sterile environment.

## B. Development in Normal Infants

Sterile until the moment of rupture of the fetal membranes, during birth the infant encounters an enor-

TABLE I

Bacterial Species and Their Approximate Numbers[a] at Various Levels of the Gastrointestinal Tract in Adults

| Species | Intestinal region | | | | |
| --- | --- | --- | --- | --- | --- |
| | Mouth | Stomach | Proximal small | Distal small | Large |
| Aerobic | | | | | |
| Gram positive | | | | | |
| α-Streptococcus | 5 | 1 | 3 | | |
| Enterococcus | | | 2 | 5 | 8 |
| Staphylococcus | 2 | | 1 | | 7 |
| Gram negative | | | | | |
| Escherichia coli | | | 2 | 4 | 9 |
| Anaerobic | | | | | |
| Gram positive | | | | | |
| Bifidobacterium | | | | 7 | 10 |
| Eubacterium | 5 | | | 3 | 10 |
| Lactobacillus | 7 | 4 | 4 | 7 | 7 |
| Peptococcus | | | | | 10 |
| Peptostreptococcus | | | | | 10 |
| Ruminococcus | | | | | 9 |
| Clostridium | | | | | 9 |
| Gram negative | | | | | |
| Bacteroides | | | | 2 | 11 |
| Fusobacterium | | | | | 9 |
| Veillonella | 6 | 2 | 3 | 4 | 5 |

[a]Logarithm to base 10 of the number of colony-forming units per gram of wet weight.

mous number of bacteria in the vaginal canal. These enter the mouth and the stomach of the infant, and immediately after birth a large number of species can be cultured from the stomach contents. The aerobic bacteria *Escherichia coli* and *Enterococcus* and the anaerobic bacteria *Bacteroides, Bifidobacterium,* and sometimes *Lactobacillus* make up the dominating part of the intestinal microflora as long as breast-feeding continues. Rectal cultures are sterile immediately after birth, but often *E. coli* and sometimes also *Bifidobacterium* are already found in the first stool. During the first days of life, *E. coli* reaches high numbers, which decrease somewhat when subsequently more anaerobic bacteria and *Enterococci* are established. It was earlier held that *Bifidobacterium* in the breast-fed infant always dominated after that time. It has been established, however, that nowadays *Bacteroides* sometimes dominate in breast-fed infants. At any rate, in normal, breast-fed infants there is a rather well-defined balance between anerobic and anaerobic bacteria, appearing in approximately equal numbers.

Compared to the intestinal microflora in adults, the infant microflora is much simpler, with perhaps 5 to 10 species isolated in each fecal culture. The adult 100- to 1000-fold dominance of anaerobic bacteria is not obtained until after weaning. Another typical trait of neonatal microflora is the tendency to fluctuate; only rarely is the same *Bifidobacterium* or *Bacteroides* strain isolated more than once or twice in serial cultures, perhaps due to continuous new seeding and replacement of preexisting strains.

## C. Influence of Feeding, Mode of Delivery, and Preterm Birth

### I. Feeding

If cow milk is given instead of breast milk to newborn infants, the quality of feces changes from the typical light yellow, semiliquid, acid-smelling breast milk feces to one that is more similar to adult feces. Constipation, which never occurs during breast milk feeding, frequently appears. The dominance of *Bifidobacterium* frequently seen in breast milk feces is replaced by a microflora more similar to the adult one, with a dominance of *B. fragilis,* and species diversity also increases. With the refinement of modern infant feeding formulas, which now are similar to breast milk in a number of qualities, differences between breast-fed and formula-fed infants are diminishing.

## 2. Mode of Delivery

Cesarean section profoundly influences the colonization of the newborn gut. Anaerobic bacteria appear later, sometimes not until 10 days of age. The gram-negative aerobic bacteria colonizing the gut come from the hospital environment, not from the maternal intestinal microflora, and often belong to potentially pathogenic species such as *Klebsiella* or *Pseudomonas*. *Bacteroides* tend to be absent for several weeks from the intestine of the cesarean section-born infants, whereas *Bifidobacteria* are more rapidly established.

## 3. Preterm Birth

The preterm infant is frequently subjected to antimicrobial treatment, and cesarean section is common among these infants. Often, more than 50% of the infants in a neonatal intensive care unit have been delivered by cesarean section. However, if a preterm infant is vaginally delivered and not subjected to antibiotic treatment, the intestinal microflora develop in a way similar to that seen in term infants. A delay might be seen in infants below 1000 g, due to the fact that these infants often cannot tolerate more than minute amounts of oral feeding, thus diminishing the amount of available substrate for bacteria in the intestine.

## II. ANTIMICROBIAL AGENTS

### A. Spectrum

Antimicrobial agents used in newborn infants frequently have a broad antibacterial spectrum. Combinations are often used, providing even broader coverage. The reason for this is that newborn infants, especially those with low birth weight, are prone to develop serious infections with virtually any of the microbes present in the intestine and in the environment. The clinical appearance is similar regardless of which organism causes the disease. Treatment has to be instituted early because if unequivocal symptoms have developed, it might be too late to save the infant. Typically, there is an insidious onset of reduced spontaneous activity, poor feeding, and some pallor, which, if unnoticed, might proceed to profound shock and death within a couple of hours. A list of antimicrobial agents and their efficacy against the most com-

mon pathogenic bacteria in newborn infants is given in Table II.

### B. Pharmacology

Orally administered antibiotics are absorbed to various degrees. After intravenous administration, excretion takes place mainly through the kidneys or through the bile system. Usually, excreted compounds retain their antimicrobial effects. The aminoglycosides are almost exclusively excreted through the kidneys, whereas some modern cephalosporins are almost solely excreted through the bile, reaching high concentrations in the intestines. In the case of penicillins, both routes are used. Little is known about other routes of excretion (e.g., with intestinal secretions of various kinds or transepithelially). The gut epithelium of the newborn is more permeable than in adults, making transepithelial excretion theoretically possible. [*See* Antimicrobial Drugs.]

## III. IMPACT OF ANTIMICROBIAL AGENTS ON NORMAL MICROFLORA

### A. Principles of Action

A broad antimicrobial spectrum and high concentration levels in the intestine increase the ecological effects of a given antimicrobial agent. Low absorption of an oral compound or high bile concentrations of intravenous agents as well as secretion through saliva, through pancreatic juice, or transepithelially increase concentrations.

There are several ways in which antimicrobial agents can be inactivated in the intestine. It is well known that some bacteria produce enzymes which inactivate antibiotics (e.g., $\beta$-lactamases of various kinds). Another way of inactivating antibiotics is adsorption to waste compounds or to bacteria themselves. Thus, sometimes extremely high concentrations of an antibiotic (e.g., erythromycin) might be found in feces with almost no effect on the growth of the bacteria sensitive to this antibiotic.

### B. Results of Studies in Infants

#### 1. Aerobic Microflora

In adults the impact of antibiotics on the normal microflora depends closely on the pharmacological properties and antimicrobial spectrum of the drug. Narrow spectrum antibiotics (e.g., phenoxymethylpenicillin)

**TABLE II**

Activity of Commonly Used Antimicrobial Agents against Important Pathogens in Neonatal Care[a]

| Antimicrobial agent | Gram-negative enteric rods | Group B *Streptococcus* | *Enterococcus* | *Staphylococcus* | *Listeria* |
|---|---|---|---|---|---|
| Phenoxymethylpenicillin | − | ++ | (+) | (+) | − |
| Cloxacillin | − | + | (+) | + | − |
| Ampicillin | + | ++ | ++ | (+) | ++ |
| Cephalosporins | ++ | ++ | − | + | − |
| Aminoglycosides | ++ | + | − | + | − |

[a]Combinations must be used when the causative agent is not known; −, no effect; (+), sometimes effective; +, usually effective; ++, effective.

cause little or no detectable changes, whereas broad spectrum agents (e.g., erythromycin, ampicillin) cause diminishing levels of anaerobic bacteria and overgrowth of certain aerobic species.

In infants, changes are more dramatic, including when narrow spectrum antibiotics are used (Fig. 1). During treatment aerobic bacteria are suppressed according to the spectrum of the drug. Thus, *E. coli* disappears when ampicillin is used. *Enterococci,* which form part of the normal microflora of most infants, disappear during ampicillin but, conversely, increase during cephalosporin treatment. Cephalo-

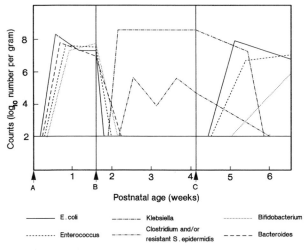

**FIGURE I** Typical changes of the intestinal microflora in a newborn infant during antibiotic treatment. The lower detection limit is $10^2$ bacteria per gram of feces. (A) Vaginal delivery. (B) Start of antibiotic treatment. (C) End of antibiotic treatment. Rapid initial colonization is followed by prompt eradication of normal microflora during treatment. Colonization with potential pathogens and carriers of antibiotic resistance ensues. The normalization after treatment is slow.

sporins are known for their lack of effect against *Enterococci.*

*Staphylococcus epidermidis* has emerged as an important pathogen in intensive-care units for both adults and newborn infants. This species causes low-grade, usually rather benign, infections, but it has a marked tendency to rapidly develop resistance to recently introduced broad spectrum antimicrobial agents. It tends, therefore, to colonize the intestine of a newborn infant in large numbers during antibiotic treatment.

There is frequently a shift from antibiotic-sensitive gram-negative bacteria (e.g., *E. coli*), to resistant potentially pathogenic bacteria (e.g., *Klebsiella, Proteus, Pseudomonas, Citrobacter,* or *Enterobacter* species). Usually one of these species dominates in each intensive-care unit.

When antimicrobial treatment is withdrawn, the microflora reverts to normal in 1–2 weeks. It was formerly thought that the antibiotic-sensitive strains, typical of normal microflora, have distinct ecological advances over the multiply antibiotic-resistant ones seen during treatment. Recently, however, such strains have been shown to sometimes persist in fecal flora for many months.

## 2. Anaerobic Microflora

Regarding anaerobic bacteria, changes are much more pronounced in newborn infants than what is usually seen in adults. Even when narrow spectrum antibiotic regimens are used, the anaerobic microflora is reduced to undetectable levels in 80–90% of treated infants. If any anaerobic bacteria remain, they are usually either *Clostridium difficile* or *Clostridium perfringens,* which never occur alone in normal infants.

After treatment the anaerobic microflora is slowly reestablished. This normalization process is similar to the slow anareobic colonization seen after cesarean section delivery. In most infants a completely normal microflora is found 1–2 weeks after stopping antibiotic treatment. However, in some cases anaerobic colonization does not occur until more than 1 month after treatment, and in these infants an abnormal aerobic microflora is also frequently seen.

## IV. CONSEQUENCES OF IATROGENIC CHANGES OF NORMAL MICROFLORA

### A. Superinfection and Selection of Resistant Strains

A sequence of events, consisting of colonization and overgrowth of potentially pathogenic bacteria from the hospital environment during antibiotic treatment, subsequent invasion of the bloodstream, and septic infection caused by the same strain, has been well described in both newborn infants and adults. Usually the invader is a gram-negative, aerobic, rod-like bacterium (e.g., *Klebsiella pneumoniae* or *Pseudomonas aeruginosa*). Infection caused by such gram-negative rods is frequently serious. If the infection is complicated by meningitis, death or permanent neurological sequelae are extremely common. Another frequent ecological consequence of antibiotic treatment is the proliferation to enormous numbers of stains resistant to many antibiotics. The feces of such infants constitute a reservoir of resistant bacteria which might easily be transmitted to and contaminate equipment used for invasive treatment and monitoring, characteristic of modern intensive care. Antibiotic resistance may also be transmitted from harmless to potentially pathogenic bacteria.

Antibiotic-induced pseudomembraneous colitis (i.e., inflammation of the large bowel), caused by a toxin produced by *C. difficile,* is a feared complication, especially among elderly patients. Newborn research animals and probably also infants are fortunately not susceptible to this toxin. A severe bowel inflammation in preterm infants, necrotizing enterocolitis, has been associated with the growth of *C. difficile* and other *Clostridium* species. However, it is not linked to antibiotic treatment, and other bacteria (e.g., *Klebsiella* and *S. epidermidis*) have been implicated.

## V. PROPHYLACTIC AND THERAPEUTIC MEASURES

There are many anecdotal reports about the health-promoting effect of ingestion of *Lactobacillus, Bifidobacterium, Enterococcus,* other species, and mixtures of several species for the correction of antibiotic-induced changes and other disturbances of gut function and as general health promoters. Reliable data about this so-called "probiotic" therapy are scarce and they do not have the same scientific founding as required for most other medical interventions. Methods are now emerging which make it possible to study this interesting field in a more stringent way, and in the near future we hope to know more about the potential benefits of this approach.

So far, the only way to minimize ecological disturbances is a rational use of antimicrobial therapy, which should be instituted only when clear indications exist. However, such indications might exist at one time or another in nearly all very low birth weight infants, and there is no doubt that modern broad spectrum antibiotic therapy has contributed enormously to the present-day high survival rates of these infants, which were unthinkable only a few decades ago.

## BIBLIOGRAPHY

Bennet, R., Eriksson, M., Nord, C. E., and Zetterström, R. (1986). Fecal bacterial microflora of newborn infants during intensive care management and treatment with five antibiotic regiments. *Pediatr. Infect. Dis.* **5,** 533–539.

Bennet, R., and Nord, C. E. (1989). The intestinal microflora during the first weeks of life: Normal development and changes induced by caesarean section, preterm birth, and antimicrobial treatment. *In* "The Regulatory and Protective Role of the Normal Microflora" (R. Grubb, T. Midtvedt, and E. Norin, eds.), pp. 19–34. Macmillan, London.

Bennet, R., Nord, C. E., and Zetterström, R. (1992). Transient colonization of the gut of newborn infants by orally administered bifidobacteria and lactobacilli. *Acta Pediatr.* **81,** 784–787.

Gaya, H., and Verhoef, J. (eds.) (1988). Current topic: Colonization resistance. *Eur. J. Microbiol. Infect. Dis.* **7,** 91–113.

Mackowiak, P. A. (1982). The normal microbial flora. *N. Engl. J. Med.* **307,** 83–93.

Nord, C. E., Kager, L., and Heimdahl, A. (1984). Impact of antimicrobial agents on the gastrointestinal microflora and the risk of infections. *Am. J. Med.* **76,** 99–106.

Savage, D. C. (1977). Microbial ecology of the gastrointestinal tract. *Annu. Rev. Microbiol.* **31,** 107–133.

# Antimicrobial Drugs

JOHN H. HASH
*Vanderbilt University*

I. History
II. Classification
III. Mechanism of Action
IV. Resistance
V. Medical Uses
VI. Agricultural Uses

## GLOSSARY

**Gram-positive, gram-negative** Two groups of bacteria distinguished by a staining procedure developed by Christian Gram for the microscopic observation of bacteria. The Gram stain depends on differing surface components of the two groups of bacteria and is an important aid in diagnosis.

**Mutation** Structural change in the sequence of DNA

**Mycolic acids** Long-chain, $\alpha$-substituted, $\beta$-hydroxy fatty acids that occur either free or ester bound in mycobacteria

**Plasmid** Autonomously self-replicating extrachromosomal circular DNA present in bacteria

**Structural analog** Chemical compound similar in structure to another, but differing from it in respect to a certain component

ANTIMICROBIAL DRUGS, AS OPPOSED TO ANTI-biotics, are defined here as compounds derived wholly by chemical synthesis. Like antibiotics, these compounds either kill or inhibit the growth of microorganisms. They are distinguished from antibiotics only in that they are synthesized by humans instead of microorganisms.

## I. HISTORY

The recorded history of *Homo sapiens* includes the never-ending struggle with infectious diseases. The victims most often are individuals or small groups, but occasional epidemics of infectious diseases (e.g., the bubonic plague or influenza) have decimated large populations. Understandably, people responded with folk remedies and various nostrums against infection. Survival from most infections, however, depended primarily on individual immunity. This situation began to change in the late 19th and early 20th centuries as the isolation and identification of bacteria causing specific diseases were accomplished. As a result the effects of individual chemicals could then be studied on pure bacterial cultures and, following this, on infections in experimental animals and finally in humans. The discovery that organic dyes could be used to stain bacteria further aided in their microscopic examination.

The newly acquired knowledge of inhibitory effects of chemicals on bacteria was used by Lister when he selected phenol to sterilize surgical instruments and dressings and therefore greatly reduced the morbidity and mortality associated at that time with surgery. Similarly, this knowledge was applied to the treatment of infected animals and humans. Certain dyes (e.g., gentian violet and methylene blue) had been noted to have inhibitory effects on living bacteria. Ehrlich began to evaluate poisonous dyes which he thought might attach to and subsequently kill an invading microorganism but not harm the human host, in much the same manner as a dye might attach to wool but not cotton. He screened known dyes for such "magic bullets" and found that trypan red and trypan blue could kill the protozoal agent of cattle trypanosomiasis. Although some success was obtained in treating cattle with this disease, these agents were too toxic for general use. Encouraged by these initial findings, Ehrlich turned to arsenic compounds, hundreds of which were synthesized and tested against the agents of trypanosomiasis and syphilis.

In 1910 the 606th compound synthesized was

ENCYCLOPEDIA OF HUMAN BIOLOGY, Second Edition, VOLUME 1. Copyright © 1997 by Academic Press. All rights of reproduction in any form reserved.

found to cure syphilis. This compound, arsphenamine, and its derivative, neoarsphenamine, became the mainstays for the treatment of syphilis until being replaced by penicillin in the 1940s. Ehrlich was the first to use the term "chemotherapy," and posterity has accorded him the title "Father of Chemotherapy." While Ehrlich was not successful in finding general antibacterial agents, his modest success with arsphenamine initiated the era of modern chemotherapy with synthetic drugs.

Little progress in finding agents active against bacteria was made until 1935, when Domagk reported that the azo dye Prontosil rubrum p-[2,4-diaminophenyl)azo]benzenesulfonamide (sulfamidochrysoidine), which had been synthesized in 1931, could cure streptococcal infections in the bloodstreams of mice. Tried in humans, it also cured bloodstream streptococcal infections. It remained for others to show that this azo dye was effective, because in the body it was metabolized to sulfanilamide, the first of over 5500 compounds to be synthesized and known generically as sulfonamides or sulfa drugs. Of this large number of compounds, less than 30 had sufficient antibacterial activity to be useful clinically. Although sulfa drugs have been largely replaced by antibiotics, they still form an important class of antimicrobials and continue to be effective in human and veterinary medicine.

Continuing research on the development of new synthetic antimicrobial agents has resulted in the discovery of many compounds with therapeutic value. However, synthetic antimicrobials make up a relatively small portion of the chemotherapeutic armamentarium as compared to antibiotics.

## II. CLASSIFICATION

Synthetic antimicrobial drugs represent broad groups of organic compounds. After a new compound is found to have antimicrobial activity, numerous derivatives are synthesized in the hopes of creating new agents with improved properties, including increased potency, stability, absorption, serum levels, and half-lives; decreased toxicity; and a broader antibacterial spectrum. As a result, families of synthetic antimicrobial agents are generated. Inasmuch as members of these chemical families generally have the same mechanism of action, it is conventional to group the families according to their common mode of action of

### TABLE I
Families of Synthetic Antimicrobial Agents

| Family | Examples |
| --- | --- |
| Antifolates | |
| Sulfonamides | Sulfadiazine, sulfamerazine, sulfamethoxazole, sulfamethazine, sulfisoxazole, sulfaphenazole |
| Trimethoprim | Trimethoprim |
| Antifungals | |
| Imidazoles | Clotrimazole, fluconazole, ketoconazole, itraconazole |
| Nucleosides | 5-fluorocytosine |
| Antimycobacterials | |
| For tuberculosis | Isoniazid, ethambutol, thiacetazone, p-aminosalicylic acid, pyrazinamide, ethionamide |
| For leprosy | Dapsone, acedapsone, clofazimine |
| Urinary tract agents | |
| Quinolones | Nalidixic acid, oxolonic acid, cinoxacin, norfloxacin, ciprofloxacin, ofloxacin |
| Nitrofurans | Nitrofurantoin, nitrofurazone, furazolidone, nifuratel |
| Antivirals | Acyclovir, ribavirin, adenine arabinoside, AZT (zidovudine) |

their general use. Table I lists families of synthetic antimicrobial compounds.

## III. MECHANISM OF ACTION

The mechanisms of action of synthetic antimicrobial agents are less well understood than those for antibiotics. This field is still under active investigation.

The sulfonamides, as well as trimethoprim, inhibit the synthesis of folic acid, a cofactor required by both humans and bacteria for the synthesis of nucleic acids and proteins. Humans cannot synthesize folic acid and therefore must obtain it from dietary sources. Bacteria, on the other hand, synthesize their own folic acid from p-aminobenzoic acid, yet are unable to use external sources of folic acid. This fact accounts for the selective inhibitory action of the sulfonamides and trimethoprim. Sulfonamides are structural analogs of p-aminobenzoic acid and thus inhibit the enzyme dihydropteroate synthetase, which catalyzes the synthesis of dihydropteroic acid from dihydropteridine and p-aminobenzoic acid.

Inhibition of this enzyme deprives bacteria of the metabolite needed to synthesize folic acid. The following step in the synthesis of folic acid involves the addition of glutamic acid to form dihydrofolic acid, which is then reduced to tetrahydrofolic acid. Trimethoprim, a structural analog of the pteridine portion of dihydrofolic acid, inhibits the enzyme dihydrofolic acid reductase, which catalyzes the reduction. Trimethoprim is able to act more rapidly on bacteria than do the sulfonamides because the inhibitory effect of the sulfonamides cannot occur until a large pool of dihydrofolate is depleted. Conversely, the rapid depletion of a much smaller tetrahydrofolate pool is felt immediately.

Other mechanisms of action of synthetic antimicrobial agents are less well understood. For example, the mechanism of action of the imidazole antifungal agents (e.g., ketoconazole) appears to be related to their effects on cytoplasmic membranes of fungi which contain the plant sterol ergosterol. *De novo* synthesis of ergosterol is inhibited by the imidazole agents, and lanosterol, a precursor methylsterol, accumulates. This precursor is probably incorporated into fungal cytoplasmic membranes, adversely affecting their structure and permeability. In addition, imidazoles disturb fatty acid synthesis, a property that also adversely affects membrane structure and function.

The nucleoside analog 5-fluorocytosine has an unique antifungal activity restricted to yeast-like fungi. The yeast cells are able to deaminate the compound to 5-flurouracil, which is then phosphorylated and incorporated into RNA. The resulting RNA does not function correctly.

A number of synthetic agents are specific for mycobacteria, but for the most part their mechanism of action is poorly understood. Mycobacteria are characterized by a high content of mycolic acids, compounds which confer on them a waxy nature. The specificities and activities of isoniazid (i.e., isonicotinylhydrazine), ethambutol, and ethionamide for these organisms are probably related to interference with the synthesis of mycolic acids. However, it cannot be assumed that this is the only mechanism of action involved. *p*-Aminosalicylic acid (PAS) is a structural analog of *p*-aminobenzoic acid, as are the sulfonamides. However, while the sulfonamides are effective against many bacteria, they are ineffective against mycobacteria. PAS, which is generally effective against mycobacteria, is ineffective against most other bacteria. The reasons for these specificities are probably related to the permeability of the respective organisms to PAS and sulfonamides and to different affinities of the enzyme dihydropteroate synthetase for PAS and sulfonamides in mycobacteria as compared to other bacteria. The mechanisms of action for the antitubercular agents

pyrazinamide and thiacetazone have not been determined.

Dapsone (i.e., diaminodiphenylsulfone), like the sulfonamides and PAS, is yet another structural analog of *p*-aminobenzoic acid. However, dapsone or its diacetyl derivative acedapsone is more effective against the leprosy bacillus than the tubercle bacillus. This specificity is related to permeability of the leprosy bacillus to dapsone and to affinity of the dihydropteroate synthetase of the leprosy bacillus for dapsone. Clofazimine, a synthetic dye, is bactericidal to the leprosy bacillus, but is mechanism of action is not known.

All of the quinolone antibacterial agents have a mechanism of action similar to that of the antibiotic novobiocin. The quinolones inhibit topoisomerase II, the bacterial enzyme responsible for the supercoiling of DNA, as does novobiocin. Topoisomerase II consists of two nonidentical subunits and, although their inhibitory action on bacteria is the same, the quinolones and novobiocin act on different subunits. There is no cross-resistance between the quinolones and novobiocin. [See Quinolones.]

Nitrofurans have an inhibitory activity that is poorly understood. It is known that several enzymes are inhibited, but whether these are primary or secondary events is not known. It has been found that nitrofurans cause strand breakage of the bacterial DNA. A normal cellular reductase reduces the nitrofuran and in the process generates free radicals, including the hydroxyl radical, which is the likely agent for DNA strand breakage.

Viruses use cellular enzymes for their replication. This use of host enzymes makes it difficult to find agents with antiviral activity that do not interfere with host cellular metabolism. A few useful synthetic antiviral drugs do exist; most of those that have any utility are nucleoside analogs. Some of these analogs are activated by cellular enzymes and are incorporated into either viral DNA (e.g., acyclovir, ribavirin, and adenine arabinoside) or viral RNA (e.g., ribavirin), thereby either terminating synthesis or creating dysfunctional nucleic acids. AZT acts as a chain terminator of DNA synthesis by RNA-directed DNA polymerase (reverse transcriptase). [See Chemotherapy, Antiviral Agents.]

## IV. RESISTANCE

Bacterial resistance to synthetic antimicrobial compounds develops as a consequence of their use, just as it does for antibiotics. Similarly, resistance can be either natural or acquired, and acquired resistance can be either chromosomally or plasmid mediated.

Chromosomally mediated resistance to the sulfonamides and trimethoprim can be of more than one kind. The resistant organism might have mutations, causing the synthesis of an altered dihydropteroate synthetase with a reduced affinity for sulfa drugs, or it might produce increased amounts of $p$-aminobenzoic acid, thus overcoming the inhibitory effects of the sulfa drugs. Chromosomally mediated resistance to trimethoprim is similar, with either the synthesis of an altered dihydrofolate reductase or the increased production of enzymes involved in dihydrofolate biosynthesis.

Plasmid-mediated resistance to both the sulfonamides and trimethoprim is related primarily to reduced permeability of the bacterial cytoplasmic membrane to these agents rather than to the synthesis of enzymes which chemically modify the drugs. In rare cases, plasmids might transfer altered genes for dihydropteroate synthetase or dihydrofolate reductase, which cause the synthesis of enzymes resistant to sulfonamides and trimethoprim, respectively.

Fungi resistant to the imidazole class of synthetic antifungals have been slow to develop. An exception appears to be the appearance of yeasts resistant to fluconazole, particularly in patients with acquired immunodeficiency disease syndrome. The development of resistance to the imidazole antifungals is similar to that of the antifungal antibiotic amphotericin B, both of which act on the cytoplasmic membrane. Resistance develops rapidly to the nucleoside antifungal agent 5-fluorocytosine and is chromosomally mediated. Resistance to 5-fluorocytosine is related to altered enzyme systems, which lead to increased synthesis of competing pyrimidines.

Resistance to the synthetic antimycobacterial agents is variable. Because therapy with a single antimycobacterial drug favors the development of resistance, these agents are normally used in combination with other antimycobacterial drugs; under these conditions the development of resistance has been low. Resistance to isoniazid is chromosomally mediated and is related to an altered catalase-peroxidase gene product and to reduced cellular permeability, resulting in less uptake of isoniazid by the resistant organism. Resistance to PAS and dapsone is also chromosomally mediated and is similar to the resistance to the sulfonamides—namely, altered dihydropteroate synthetase or increased production of $p$-aminobenzoic acid. When it occurs, multiple drug resistance in mycobacteria is mediated by multiple chromosomal mutations rather than by plasmids.

Resistance of bacteria to the quinolone family of antimicrobial agents is similar to their resistance to novobiocin. It is chromosomally mediated. The resistant organism has an altered subunit of topoisomerase II which does not bind the quinolone.

Resistance to nitrofurans is unusual and, in practice, has not been a problem. When resistance occurs, it is related to the cellular reductase which, through chromosomal mutation, is either altered or eliminated completely. There has been no cross-resistance with other antimicrobials. Plasmid-mediated resistance occurs rarely, if at all.

Viruses become resistant to the nucleoside antiviral agents through mutations in the genes of the viral chromosome which code for the viral DNA polymerase (e.g., acyclovir and adenine arabinoside), RNA-dependent DNA polymerase (zidovudine), and the viral thymidine kinase (e.g., acyclovir). Acquired viral resistance to ribavirin has not been encountered.

## V. MEDICAL USES

The sulfonamides have had a greater impact on medicine than any of the other synthetic antimicrobial agents. Because of their broad antibacterial spectrum, which includes gram-positive and -negative bacteria, chlamydia, nocardia, and even a few protozoa, these agents are commonly used in treating human disease. After their introduction in the 1930s, the sulfonamides were used for the treatment of many diseases caused by gram-positive and -negative organisms. The subsequent decline in their use as individual agents resulted from the development of increased bacterial resistance and from the discovery and use of antibiotics. A limited number of sulfonamides are currently used in human medicine in the United States, primarily for treating urinary tract infections. Their toxicity has been reduced and their effectiveness has been increased by combining different sulfonamides or by combining sulfonamides with trimethoprim. An example of the former is a triple sulfa consisting of sulfadiazine, sulfamerazine, and sulfamethazine (1:1:1) and an example of the latter is a mixture of sulfamethoxazole and trimethoprim (5:1), known generically as cotrimoxazole. The use of sulfonamides, either alone or in combination with other antimicrobial agents, will continue for the treatment of susceptible bacterial infections.

Initially, trimethoprim was considered simply as a potentiator of sulfonamide action and was used only with a sulfonamide because the two agents blocked sequential reactions in folate synthesis and acted synergistically. Their combined use reduced the emergence of organisms resistant to either one. In recent years trimethoprim has been used alone for urinary tract infections.

Other synthetic antimicrobials have a more limited antimicrobial spectrum. These agents are used against fungi, mycobacteria, a few viruses, and infections of the urinary tract. In some cases synthetic antimicrobials provide effective alternatives to antibiotics, especially when the development of resistance to an antibiotic is a serious problem.

Mycobacterial diseases present special problems in chemotherapy because of the nature of tubercle and leprosy bacilli. Only a few antibiotics (e.g., streptomycin and rifampicin) and a few synthetic antimicrobials (e.g., isoniazid and ethambutol) are highly effective against mycobacteria. Accordingly, these compounds are the major agents used against tuberculosis. In addition, secondary antitubercular drugs (e.g., PAS, pyrazinamide, ethionamide, and thiacetazone) are used. An alarming trend is that some of the new wave of resistant tuberculosis organisms are insensitive to as many as 11 antituberculous agents. For leprosy the major agents are rifampicin, clofazamine, and dapsone. To avoid resistance, which develops readily to these agents when used alone, therapy virtually always includes two such drugs.

Just as for antibiotics, an ideal synthetic antimicrobial agent is one that will selectively destroy an invading microorganism with minimal damage, if any, to the host. As is also true for antibiotics, the ideal synthetic antimicrobial has not been developed. For many synthetic agents the therapeutic dose is close to the toxic level, and practically all of them have untoward or adverse reactions in the host. These reactions include gastrointestinal disturbances (sulfonamides, trimethoprim, quinolones, nitrofurans, imidazoles, and most of the antimycobacterial agents), allergies, and hypersensitivities (sulfonamides, quinolones, nitrofurans, and many antimycobacterial agents), blood disorders (trimethoprim, thiacetazone, dapsone, isoniazid, and zidovudine), and central nervous system toxicities (nitrofurans, isoniazid, ethambutol, and dapsone). Additional side effects that are peculiar to each agent exist and stringent medical supervision is required. Most of the adverse effects are reversible when therapy is discontinued.

## VI. AGRICULTURAL USES

Synthetic antimicrobials are used in veterinary medicine to treat infectious diseases, much as they are in human medicine. The principal synthetic agents used are the sulfonamides, trimethoprim, nitrofurans, and imidazole antifungals. The use of antimicrobial drugs in veterinary medicine causes the selection of resistant organisms, just as it does in human medicine. Moreover, in veterinary medicine, therapy is often initiated without a full knowledge of the etiological agents. Choices of therapeutic agents are made on the basis of professional experience. Under these circumstances there is a greater opportunity for the selection of resistant microbes than there would be if the organism and its sensitivity to antimicrobials were determined prior to the initiation of specific therapy. As a consequence, the development of resistance threatens the continued use of synthetic antimicrobials in veterinary medicine as much as, if not more than, it does for antibiotics. To reduce the development of resistance in shared pathogens to the same antimicrobial agent, proposals have been made for the use of one group of antimicrobial agents in human medicine and another group in veterinary medicine. None of these proposals have been adopted.

The synthetic antimicrobials principally used as feed additives to enhance growth and health are the sulfonamides and the nitrofurans. The total amounts of synthetic antimicrobial agents and antibiotics used in subtherapeutic levels in livestock feed are large. In the United States it is estimated that essentially all chickens and turkeys, most swine, and more than one-half of the beef cattle receive either antibiotics or synthetic antimicrobial agents in their feeding during some part of their growth period.

The concerns over the use of synthetic antimicrobials as feeds additives are the same as for antibiotics—namely, the development of resistant organisms that might be transferred directly or indirectly to humans. In addition, the possibility exists that residues of antimicrobial drugs in consumable animal products might lead to human health problems. To date, evidence for the transfer of resistant animal pathogens to humans is generally considered insufficient to warrant discontinuation of the routine use of synthetic antimicrobials in animal feeds. Residues of synthetic drugs in animal products do not appear to constitute a problem, especially when adequate withdrawal periods are allowed before the animals are processed for consumption.

## BIBLIOGRAPHY

Jukes, T. H., DuPont, H. L., and Crawford, L. M. (eds.) (1984). "Antibiotics, Sulfonamides, and Public Health," Vol. 1, Sect. D. CRC Press, Boca Raton, FL.

Kuchers, A., and Bennett, N. M. (1987). "The Use of Antibiotics," 4th Ed. Lippincott, Philadelphia.

Mandell, G. L., Douglas, R. G., Jr., and Bennett, J. E. (eds.) (1992). "Handbook of Antimicrobial Therapy." Churchill Livingstone, New York.

Moats, W. A. (ed.) (1986). "Agricultural Use of Antibiotics." Am. Chem. Soc., Washington, D.C.

Pratt, W. B., and Fekety, R. (1986). "The Antimicrobial Drugs." Oxford Univ. Press, Oxford.

# Antinuclear Antibodies and Autoimmunity

ENG M. TAN

*Scripps Clinic and Research Foundation*

I. Nature of Systemic Autoimmune Diseases
II. Disease-Related Specificity of the Autoimmune Response
III. Cloning of cDNAs and Genes That Encode Autoantigens
IV. Nature of Immunogens
V. Factors Inducing Autoimmunity

## GLOSSARY

**Antinuclear antibodies** Generic term used to describe antibodies reactive with intranuclear components, including DNA, RNA, and proteins

**Autoimmunity** Perturbation of the immune system, usually recognized by the presence of autoantibodies reactive with self-components

**Epitope** Distinct region of an antigen that is recognized by the combining site of an antibody. In the case of a protein antigen, this epitope might be a linear sequence of amino acids or might be conformational, formed by the apposition of discontinuous sequences

**Immunological specificity** Usually refers to antibodies that are peculiar or special to a particular antigen or characteristic of a particular disease

**Systemic autoimmune diseases** Autoimmune disorders characterized by the presence of autoantibodies that are reactive with antigens which are ubiquitous and present in all cell types of the host

"ANTINUCLEAR ANTIBODIES" IS A GENERIC TERM for antibodies that are reactive with antigens such as DNA, RNA, and proteins usually resident in the nucleus. These antigens are ubiquitous and common to all cell types. Systemic autoimmune diseases should be distinguished from organ-specific autoimmune diseases in which antibodies are reactive with antigens present only in certain organs. The interest in antinuclear antibodies has been two-fold: (1) Some individual autoantibodies and distinct groups of related autoantibodies are disease specific and are useful as diagnostic markers, and (2) they have proved to be powerful reagents for the characterization of subcellular particles, which are engaged in important cellular functions. The interactions of clinical science with molecular and cell biology have contributed to our understanding of factors inducing autoimmunity and to elucidation of cellular processes such as DNA replication, precursor mRNA splicing, and transcription.

## I. NATURE OF SYSTEMIC AUTOIMMUNE DISEASES

The systemic autoimmune diseases include systemic lupus erythematosus, or lupus; mixed connective tissue disease; scleroderma; Sjögren's syndrome; and dermatomyositis/polymyositis. These diseases are characterized by illnesses of a more systemic or generalized nature in contrast to organ-specific autoimmune diseases such as thyroiditis and pancreatitis, in which the illnesses are restricted to involvement of the thyroid and the pancreas, respectively. Lupus is considered to be the prototype of a systemic autoimmune disease, and physiological abnormality can involve such widely separated organs as the skin, joints, kidneys, lungs, brain, and heart. The pathogenesis (i.e., the mechanisms by which damage to these organs occurs) is not completely known, except that involvement of the kidneys in lupus glomerulonephritis is in part related to the formation of immune complexes in the glomeruli. The immune complexes trigger a number of inflammatory cascades, the most prominent of which is the initiation of complement activa-

ENCYCLOPEDIA OF HUMAN BIOLOGY, Second Edition, VOLUME 1.   Copyright © 1997 by Academic Press.   All rights of reproduction in any form reserved.

tion, resulting in the liberation of complement anaphylatoxins. In the other systemic autoimmune diseases, even the role of immune complex-mediated inflammation has not been clearly established. [*See* Autoimmune Disease.]

Autoantibodies reactive with nuclear antigens are a distinctive hallmark of systemic autoimmune diseases. They include autoantibodies reactive with DNA, RNA, and scores of nuclear proteins, some of which are complexed with intranuclear RNAs to form RNA–protein subcellular particles. Historically, the first description of an antinuclear antibody can be traced to the discovery in 1948 of the lupus cell phenomenon, caused by antihistone antibodies which react with the nuclei of damaged cells. This antigen–antibody reaction is followed by binding of complement, a process called opsonization. The opsonized nuclei are then ingested by living polymorphonuclear leukocytes, resulting in a distinctive lupus cell.

Over the next two decades there was successive immunological identification and characterization of other types of antinuclear antibodies, showing that they reacted with DNA, histones, and many nonhistone proteins of the nucleus. The characterization of these intranuclear antigens was based on their distinctive interactions with antibodies present in the sera of patients with lupus and other diseases. Many of these antigens were also differentiated from other proteins on the basis of their physicochemical properties. Lately, the tools of molecular and cell biology have been used to elucidate the molecular structure of these antigens and their functional properties.

Immunofluorescence microscopy has played a fundamental role in the identification and partial characterization of autoantibodies and the nuclear antigens they recognize. A representative example is shown in Fig. 1, in which sera from three patients were reacted with cryostat sections of mouse kidney (Figs. 1a, 1c, and 1e) and with monolayer human HEp-2 tissue culture cells (Figs. 1b, 1d, and 1f). Three patterns of reactivity are demonstrated: a homogeneous or patchy pattern of nuclear staining (Figs. 1a and 1b), a speckled pattern (Figs. 1c and 1d), and nucleolar staining (Figs. 1e and 1f), each reflecting the reactivity of a different type of antibody combining with its intranuclear antigen.

Certain sera, especially those containing high titers of antibodies, can frequently be further characterized by immunoprecipitation with their respective soluble antigens. This type of study results in the production of an immunoprecipitation line that is highly specific for a particular antigen–antibody reaction. The coupling of immunofluorescence microscopy with immunoprecipitation resulted in the identification of several antigen–antibody systems.

Figure 1 also depicts the conserved nature of the intracellular antigens with which autoantibodies react. In this case the antigens are present in mice (mouse kidney sections) and humans (HEp-2, an epidermal cell line). Some antigens reactive with the same autoantibodies are present in plants, yeast, protozoa, amphibia, birds, and other mammals. [*See* Antibody–Antigen Complexes: Biological Consequences.]

## II. DISEASE-RELATED SPECIFICITY OF THE AUTOIMMUNE RESPONSE

### A. Lupus, Mixed Connective Tissue Disease, and Sjögren's Syndrome

Table I reports the types of autoantibodies (primarily antinuclear antibodies) that are detected in patients with lupus, the molecular identity of the antigens, and the frequency of their reactive antibodies. Lupus is a disease characterized by polyclonality of the autoimmune response. There are antibodies to native (i.e., double-stranded) DNA as well as denatured (i.e., single-stranded) DNA, histones, and several nonhistone nuclear proteins. Some of the antibody specificities are restricted to lupus; these have been useful to clinicians as diagnostic markers of the disease. They include antibodies to native DNA and to intranuclear RNA–protein complexes called Sm antigen. However, these antibodies are not detected in all patients with lupus, and therefore their absence from a patient does not rule out the diagnosis.

A second distinctive feature of this disease is that most patients have an average of three to four types of antibodies at the same time. It should also be observed from Table I that antibodies to histone and to denatured DNA are commonly detected in the sera of patients with a form of lupus induced by drugs such as procainamide and hydralazine. Antibodies to nuclear ribonucleoprotein (RNP) occur in the sera of patients with mixed connective tissue disease, without association with other autoantibodies. Antibodies to SS-A/Ro and SS-B/La (both intracellular RNP complexes) also occur in the sera of patients with Sjögren's syndrome.

The fine structure of many of the nonhistone protein antigens has been elucidated. An important group of nuclear antigens consists of subcellular particles called small nuclear RNPs (sn RNPs). The Sm and

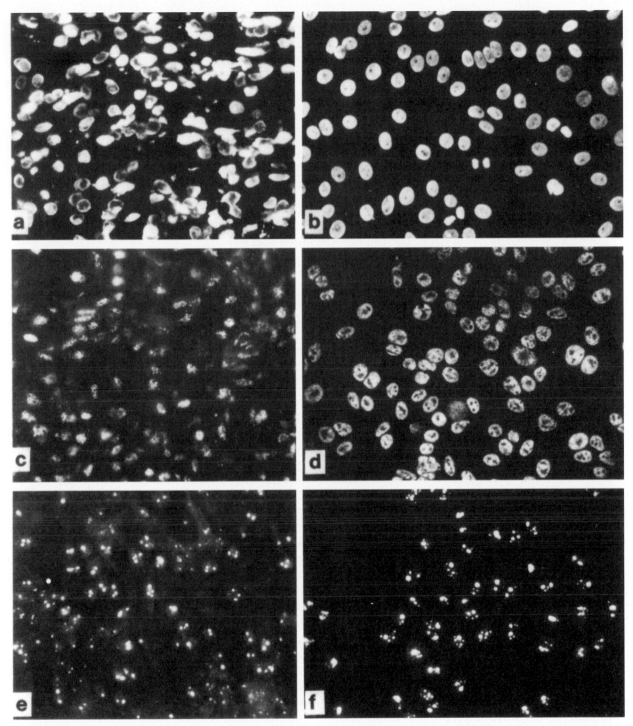

FIGURE I    Fluorescence micrograph comparing patterns of antinuclear antibody staining on cryopreserved acetone-fixed sections of mouse kidney (a, c, and e) and the human epithelial cell line HEp-2 (b, d, and f). A homogeneous pattern (a and b) is characteristically produced by sera containing antibodies to DNA and/or histones. Antibodies to nuclear ribonucleoprotein and Sm antigens give a speckled pattern of immunofluorescence (c and d). An antinucleolar antibody gives the pattern shown in e and f. The large nucleoli of the HEp-2 cells allow easy identification of the nucleolar pattern (f), although they are also recognized in the mouse kidney section (e). Original magnification ×450.

TABLE I

Autoantibodies and Antigens in Lupus and Other Diseases

| Antigen | | Autoantibody frequency (%) | |
| --- | --- | --- | --- |
| Clinical designation | Molecular identity | Lupus | Other diseases[a] |
| Native DNA | Double-stranded DNA | 40 | — |
| Denatured DNA | Single-stranded DNA | 70 | 80 (DLE) |
| Histones | H1, H2A, H2B, H3, H4 | 70 | >95 (DLE) |
| Sm | 29 (B')-, 28 (B)-, 16 (D)-, 13 (E)-kDa proteins complexed with U1, U2, U4–U6 snRNAs | 30 | — |
| Nuclear RNP | 70-, 33 (A)-, 22 (C)-kDa proteins complexed with U1 snRNA | 32 | >95 (MCTD) |
| SS-A/Ro | 60-, 52-kDa proteins complexed with Y1, Y3–Y5 RNAs | 35 | 60 (SS) |
| SS-B/La | 48-kDa protein complexed with nascent RNA polymerase III transcripts | 15 | 40 (SS) |
| Ku | 86, 66-kDa DNA-binding proteins | 10 | |
| Ki | 29.5-kDa protein | 14 | |
| PCNA/Cyclin | 36-kDa auxiliary protein of DNA polymerase $\delta$ | 3 | |
| Ribosomal RNP | 38-, 16-, 15-kDa proteins associated with ribosomes | 10 | |
| Hsp 90 | 90-kDa heat-shock protein | 50 | |

[a]DLE, drug-induced lupus erythematosus; MCTD, mixed connective tissue disease; SS, Sjögren's syndrome.

nuclear RNP antigens fall into this class of snRNPs, which consist of particles comprising small nuclear RNAs (called U RNAs because they are relatively uridine rich) and small proteins.

One of the techniques that has been useful in elucidating the molecular structure of the autoantigens is Western blotting, a procedure in which proteins are separated by electrophoresis on a slab of polyacrylamide gel and the separated proteins are transferred to a solid adsorbent such as nitrocellulose paper. The nitrocellulose is then flooded with serum, and the antibodies that bind specifically to the separated protein antigens are demonstrated with a detecting reagent such as radiolabeled *Staphylococcus* protein A.

In Fig. 2, lane 1, serum from a patient with Sjögren's syndrome contained antibodies that bind to 60-, 52-, 48-, and 43-kDa protein bands. The 60- and 52-kDa proteins are different protein species and are components of the SS-A/Ro antigen (see Table I). The 43-kDa protein is a degradation product of the 48-kDa protein, which is the antigen of SS-B/La (Table I). Lane 2 was the reaction of a different serum containing antibodies reactive with 60- and 52-kDa SS-A/Ro antigen. Lanes 3–5 represent the reactivity of antibodies affinity purified by elution from the 60-, 52-, and 48-kDa bands, respectively, of lane 1. These affinity-purified antibodies reacted only with their re-

spective antigens and did not cross-react with other proteins, showing that each antibody possessed separate immunological specificities. In this manner the types of autoantibodies that occur in patient sera can be classified.

An example of the fact that certain autoantigens are engaged in important or essential cellular functions is proliferating cell nuclear antigen (PCNA) (Table I). Autoantibody to this intranuclear protein is detected in only 3% of patients with lupus and was first detected by immunofluorescence microscopy and immunodiffusion. In immunofluorescence, the antibody reacted with the nuclei of tissue culture cells in a variegated pattern, illustrated in Fig. 3. Some nuclei reacted strongly, others weakly or not at all, and the positively reacting cells showed different patterns of staining. This is related to the fact that the antigen is a protein of 36 kDa associated with DNA polymerase $\delta$ and is engaged in DNA replication. Cells that stained positively with antibody to PCNA were actively engaged in DNA replication.

## B. Dermatomyositis and Polymyositis

Another type of autoimmune disease, dermatomyositis/polymyositis, is characterized by a totally different set of autoimmune response (Table II). An inter-

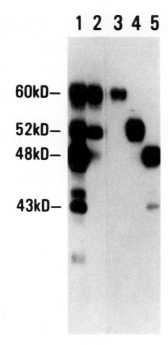

**FIGURE 2**  Western blot analysis of the reactivities of whole sera and affinity-purified antibodies to 60-kDa, 52-kDa, and SS-B proteins using HeLa cell extract as the antigen source. Lane 1 represents serum from a patient with Sjögren's syndrome containing antibodies to 60- and 52-kDa proteins as well as to the 48-kDa SS-B protein and its 43-kDa degradation product. Lane 2 shows reactivity of serum positive mainly for anti-60- and anti-52-kDa antibodies. Lanes 3–5 demonstrate affinity-purified antibodies to each component: 60 kDa, 52 kDa, and SS-B, respectively. The serum in lane 1 served as a source for the isolation of affinity-purified antibodies. [From E. Ben-Chetrit, E. K. L. Chan, K. F. Sullivan, and E. M. Tan (1988). *J. Exp. Med.* 167, 1560–1571.]

esting feature is that tRNA synthetases, which are enzymes engaged in the charging reaction of tRNAs with amino acid, are the targets of autoantibodies in dermatomyositis/polymyosistis. Autoantibodies to histidyl, threonyl, and alanyl tRNA synthetases have been identified. The PM–Scl antigen occurs in patients with a polymyositis–scleroderma overlap syndrome, hence its designation. This antigen is a complex of 11–13 proteins immunoprecipitated by specific antibodies. The SRP antigen is a protein of 54 kDa present in the signal recognition particle engaged in polypeptide synthesis and its translocation across the endoplasmic reticulum. In contrast to lupus, these target antigens in dermatomyositis/polymyositis are resident in the cytoplasm, with the PM–Scl antigen localized in nucleolus as well as cytoplasm.

Certain insights into the nature of the epitopes on antigens have come from studies on the tRNA synthe-

tase antigens. For example, autoantibodies to threonyl tRNA synthetase inhibit the function of the synthetase, whereas experimentally induced antibodies, such as those obtained by immunization of laboratory animals with purified antigen, do not. This result implies that the epitopes recognized by autoantibodies to tRNA synthetases are related to the catalytic sites of the synthetases, a theme that recurs with other autoantibodies.

## C. Scleroderma

The autoantibodies and antigens encountered in scleroderma are given in Table III. Scleroderma is generally divided into two large subgroups: one is characterized by diffuse skin involvement, often associated with diseases of the kidney and the heart, and the other, of a milder nature, is described with the acronym CREST (for calcinosis, Raynaud's phenomenon, esophageal dysmotility, sclerodactyly, and telangiectasia). The majority of patients with diffuse scleroderma have autoantibody to DNA topoisomerase 1, a nuclear enzyme of 100 kDa. The native protein is rapidly degraded by intracellular enzymes in preparative procedures, and a degradation product of 70,000 kDa retaining antigenicity for antibody is often isolated. Hence, the early designation of this antigen was Scl-70 (scleroderma-70). Patients with CREST make autoantibodies to proteins associated with the centromeric regions of chromosomes. Three antigenic proteins are the targets of autoantibodies in CREST patients: 17, 80, and 140 kDa. Other antigens are RNA polymerase 1, fibrillarin, PM–Scl, To, and NOR-9, the characteristics of which are described in Table III.

Elucidation of the structure and function of antigens in scleroderma has provided some important insights into the nature of immunogens in systemic autoimmunity. Immunofluorescence microscopy shows that the antigens RNA polymerase 1, fibrillarin, To, and NOR-90 are all nucleolar in location, whereas PM–Scl is both nucleolar and nucleoplasmic. After the characterization of the Scl-70 antigen as DNA topoisomerase 1, it became clear that Scl-70 is also nucleolar as well as nucleoplasmic in location. DNA topoisomerase 1 is engaged in relaxation of supercoiled DNA during phases of DNA replication and transcription. In certain cell types in which such a function appears to be particularly active in the nucleolus, topoisomerase 1 can be demonstrated by immunofluorescence to be prominent in the nucleolus. Centromere antigens (Fig. 4), as identified by anticentromere antibodies in CREST patients, have been

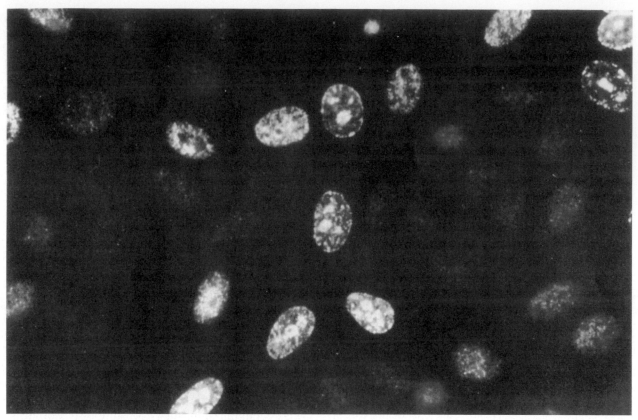

**FIGURE 3** A variegated staining pattern is obtained on nonsynchronized human amnion tissue culture cells by a monoclonal antibody to proliferating cell nuclear antigen. Staining is restricted to the nucleus and corresponds to sites of DNA synthesis, as determined by autoradiography of thymidine uptake. Original magnification ×1000.

shown to be present in the nucleolar organizer regions (NORs) of dividing cells. Thus, many of the antigens associated with scleroderma appear to be nucleolus associated, either exclusively or at certain phases of cell cycling or cell differentiation. In other words, it appears that nucleolus-associated antigens in sclero-

derma might be driving the immune response. [*See* DNA in the Nucleosome.]

Immunofluorescence microscopy can also serve to distinguish antinucleolar antibodies of the RNA polymerase 1, PM–Scl, and fibrillarin varieties, as illustrated in Fig. 5. Careful observation demonstrated

**TABLE II**

Autoantibodies and Antigens in Dermatomyositis/Polymyositis

| Clinical designation | Antigen Molecular identity | Autoantibody frequency (%) |
|---|---|---|
| Jo-1 | 50-kDa protein, histidyl tRNA synthetase | 25 |
| PL-7 | 80-kDa protein, threonyl tRNA synthetase | 4 |
| PL-12 | 110-kDa protein, alanyl tRNA synthetase and alanyl tRNA | 3 |
| PM–Scl | Complex of 11 proteins 110 to 20 kDa | 8 |
| Mi-2 | 61-, 53-kDa proteins | 5 |
| SRP | 54-kDa signal recognition particle protein complexed with 7 SL RNA | Rare |

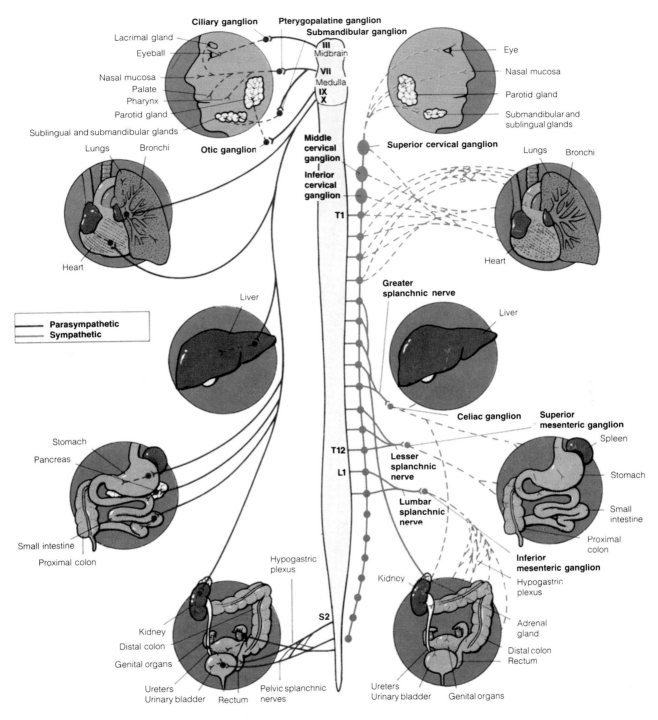

**COLOR PLATE I** The sympathetic nervous system. [Source: Gaudin, A. J., and Jones, K. C. (1989). "Human Anatomy and Physiology." Harcourt Brace Jovanovich, San Diego, p. 345. Reproduced with permission.] [*See* Adrenal Gland.]

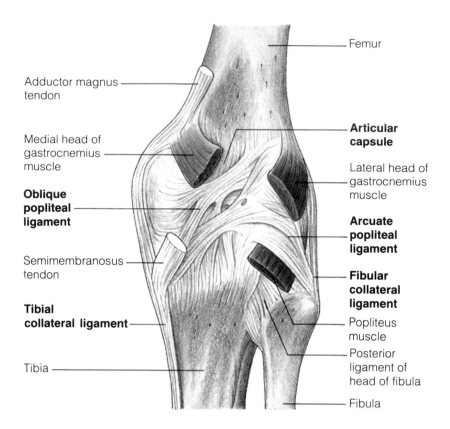

COLOR PLATE 2 Frontal view of the knee joint with the patella removed. [Source: Gaudin, A. J., and Jones, K. C. (1989). "Human Anatomy and Physiology." Harcourt Brace Jovanovich, San Diego, p. 175. Reproduced with permission.] [*See* Articulations, Joints.]

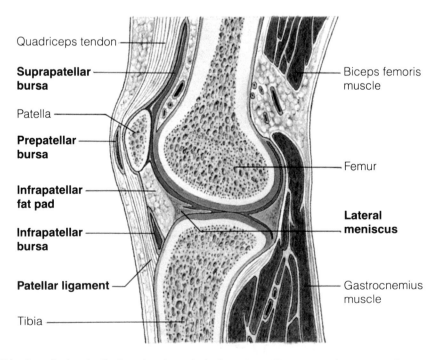

COLOR PLATE 3 Side view of a longitudinal section through the knee joint. [Source: Gaudin, A. J., and Jones, K. C. (1989). "Human Anatomy and Physiology." Harcourt Brace Jovanovich, San Diego, p. 175. Reproduced with permission.] [*See* Articulations, Joints.]

**COLOR PLATE 4** Calves with artificial hearts should be "happy" calves. When they have a normal growth curve, it indicates that all is well. [*See* Artificial Heart.]

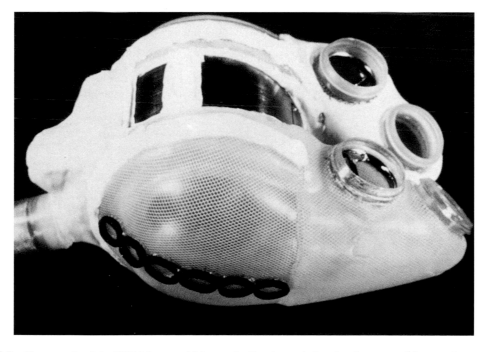

**COLOR PLATE 5** Photograph of the ERDA heart, which was the first heart designed to be powered by atomic energy. Driven by an electromotor, it sustained a calf for 37 days. [*See* Artificial Heart.]

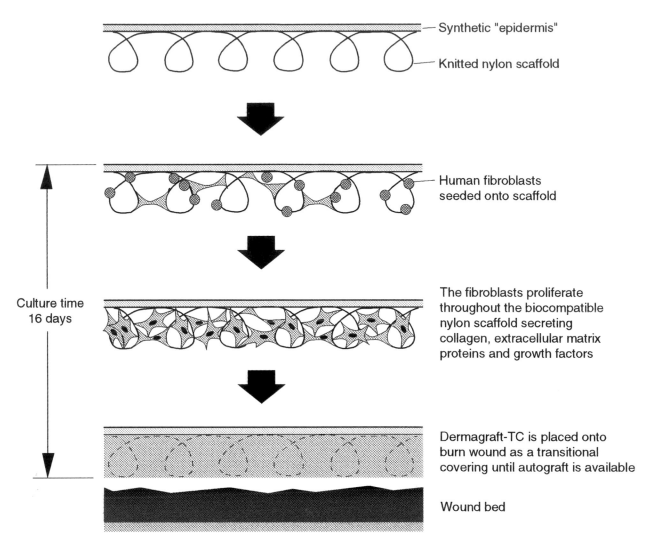

**COLOR PLATE 6** Dermagraft-TC is grown by seeding human dermal fibroblasts onto a knitted nylon scaffold where they proliferate and lay an extracellular matrix composed largely of collagen. The final product is used as a covering for severe wounds. Related systems use a scaffold composed of a biodegradable polymer, such as vicryl or collagen. In some systems the "synthetic epidermis" used in the Dermagraft-TC system is replaced with living keratinocytes to generate a tissue-engineered analog of full thickness skin. [*See* Artificial Skin.]

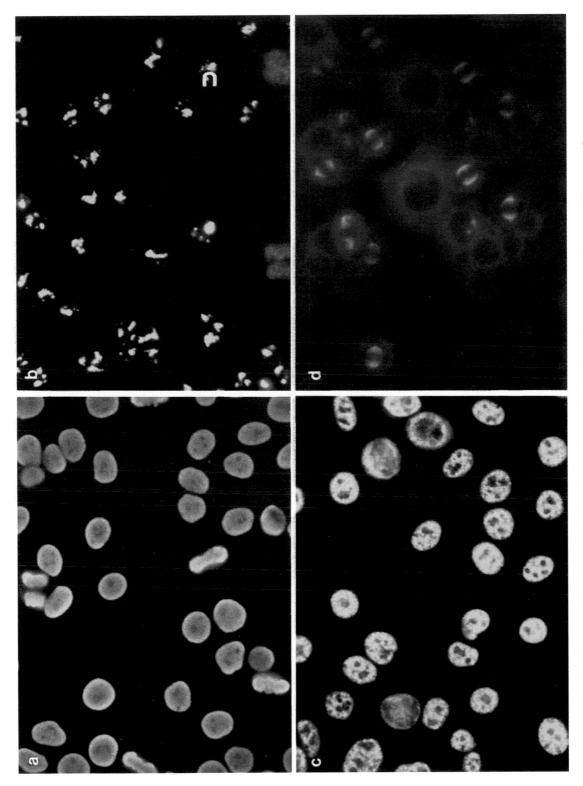

COLOR PLATE 7  Different specificities of antinuclear antibodies give different patterns when tested by immunofluorescence. These patterns are broadly classified into (a) homogenous, (b) nuclear, and (c) speckled. (d) Autoantibody reacting with the spindle apparatus of dividing cells. [See Autoantibodies.]

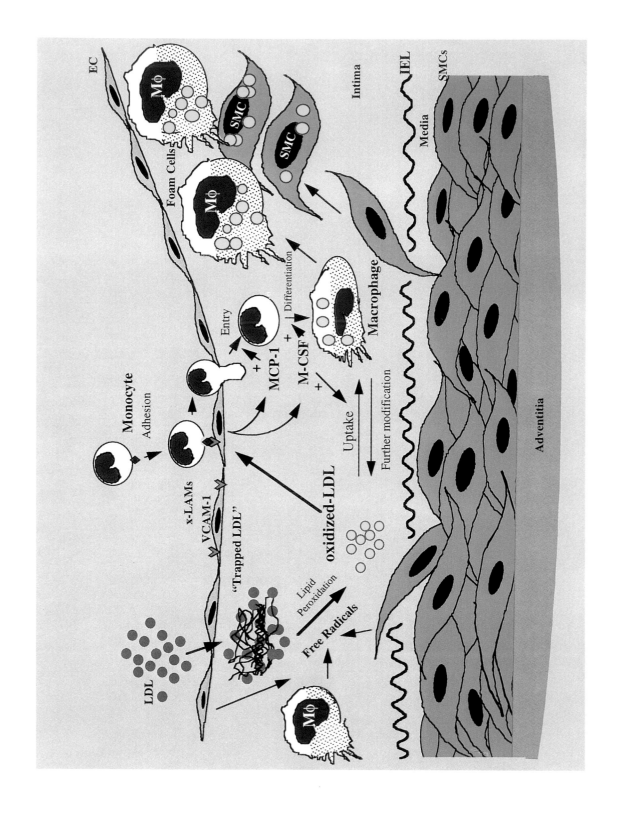

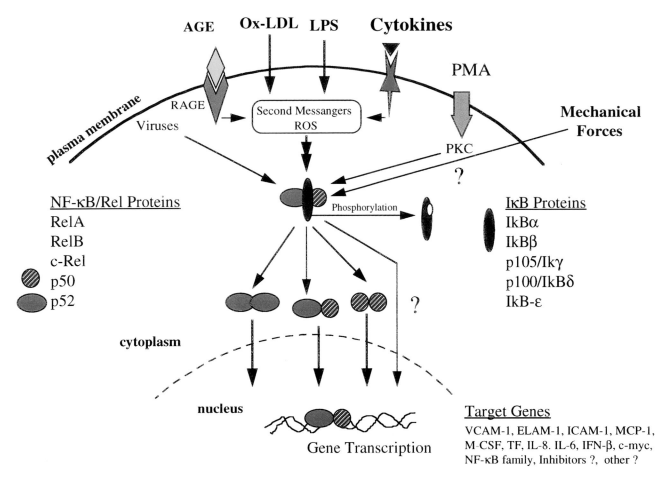

**COLOR PLATE 9** The activation of genes by NF-κB. The NF-κB is a family of DNA-binding proteins consisting of the members RelA, RelB, c-Rel, p50, and p52. NF-κB is normally bound in the cytoplasm by its inhibitor, IκB. Phosphorylation of IκB allows NF-κB to dissociate from the cytosolic complex and move to the nucleus, where it binds to its recognition sequence in the regulatory regions of specific genes and thereby activates gene transcription. Various stimuli act in a receptor dependent or independent pathway and lead to the production of reactive oxygen species (ROS) in the cytoplasm. ROS serve as secondary messengers and activate NF-κB. Some stimuli may activate NF-κB directly or by mechanisms other than through ROS. NF-κB is involved in the activation of many genes that may play a critical role in inflammation. [See Atherosclerosis: From Risk Factors to Regulatory Molecules (Figure 5).]

**COLOR PLATE 8** A model for the role of lipids in atherogenesis. Circulating LDL enters the subendothelial space and becomes trapped in a meshwork of structural proteins. In this microenvironment, the LDL is exposed to free radical products of ECs, SMCs, and macrophages (Mφ's) and becomes oxidized to varying degrees. Mildly oxidized forms of LDL stimulate ECs to produce adhesion molecules (X-LAMs and VCAM-1), cytokines (M-CSF), and chemokines (MCP-1). These molecules mediate adherence, transmigration, differentiation, and activation of monocytes. M-CSF promotes scavenger receptor-mediated endocytosis of ox-LDL by Mφ's, which become foam cells as they accumulate lipid. SMCs migrate from the vascular media to the intima and synthesize extracellular matrix structural proteins. SMCs also phagocytose ox-LDL and become foam cells. These events lead to the formation of the fatty streak, the earliest pathologic lesion in atherogenesis. [See Atherosclerosis: From Risk Factors to Regulatory Molecules (Figure 4).]

that antibodies against RNA polymerase 1 demonstrate a distinctive speckled staining of the nucleolus (Fig. 5a), with strong staining of NORs in mitotic cells (arrows). Antibodies to fibrillarin (U3 RNP) demonstrate a clumpy pattern of nucleolar staining which, in the mitotic cell, is manifested as circumferential staining of the condensed chromosomes (Fig. 5c, arrows). This type of staining of the mitotic cell can be readily differentiated from the staining demonstrated by antibody to RNA polymerase 1. Antibody to PM–Scl also gives nucleolar staining, but is of a homogeneous pattern, and in the mitotic cell there are no distinctive features.

## D. Immunodiagnostic Markers

Careful analysis of the data presented in Table I–III, together with studies of a more clinical nature, shows that there are some significant associations between autoantibodies of certain specificities and distinct systemic autoimmune diseases. As mentioned previously, antibodies to native DNA and Sm antigen are restricted to lupus and are therefore reliable diagnostic markers. In contrast, antibodies to denatured DNA and histones, although present in systemic (idiopathic) lupus, are also and more frequently present in the drug-induced form of lupus (Table I) and are therefore not reliable as diagnostic markers for idiopathic lupus. Importance, however, should be placed on clusters of antibodies; in fact, the association of antibodies to denatured DNA and to histones without other antibodies strongly points to drug-induced lupus.

To carry this type of analysis further, antibody to nuclear RNP (Table I) is present in idiopathic lupus, but also in mixed connective tissue disease, where it

occurs in more than 95% of patients. Presence of this antibody without antibodies of other specificities strongly points to mixed connective tissue disease. Finally, the cluster of antibodies to SS-A/Ro and SS-B/La is frequently seen in patients with Sjögren's syndrome, in the absence of other antibodies. Thus, immunodiagnostic markers are of two major categories. One category consists of individual antibodies that are restricted to one disease and are, themselves, diagnostic markers. Among these are the tRNA synthetase antibodies in dermatomyositis/polymyositis, anti-DNA topoisomerase I in diffuse scleroderma, and anticentromeres in CREST. The second category consists of clusters of antibodies and the negative association of these clusters with other autoantibodies.

The utility of antinuclear antibodies in clinical diagnosis has encouraged international organizations to pool resources in order to set up standardized reference reagents, which are available to investigators and research and clinical laboratories worldwide and they can be obtained by writing to Arthritis Foundation of America/Centers for Disease Control–Antinuclear Antibody, Reference Laboratory, Immunology Branch, 1-1202 A25, Centers for Disease Control, Atlanta, GA 30333.

## III. CLONING OF cDNAS AND GENES THAT ENCODE AUTOANTIGENS

Spontaneously occurring antinuclear antibodies and antibodies to other intracellular antigens have been used as reagents to clone the genes encoding the antigens. For example, human autoantibodies can be used to purify the desired antigen. From the amino acid

### TABLE III
Autoantibodies and Antigens in Scleroderma

| Clinical designation | Antigen<br>Molecular identity | Autoantibody frequency (%) |
|---|---|---|
| Scl-70 | 100-kDa native protein and 70-kDa degradation product, DNA topoisomerase I | 70% in diffuse scleroderma |
| Centromere | 17-, 80-, 140-kDa proteins localized at inner and outer kinetochore plates | 70–80% in CREST |
| RNA polymerase I | RNA polymerase I complex of subunit proteins 210 to 11 kDa | 4 |
| Fibrillarin | 34-kDa protein, component of U3 RNP particle | 8 |
| PM–Scl | Complex of 11 proteins 110 to 20 kDa | 3 |
| To | 40-kDa protein complexed with 7-2 and 8-2 RNAs | Rare |
| NOR-90 | 90-kDa protein localized in nucleolus organizer region | Rare |

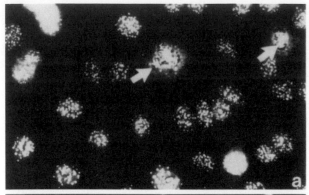

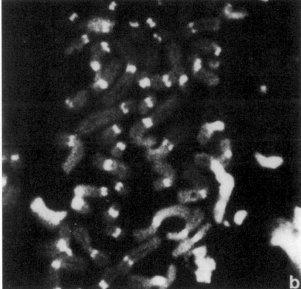

**FIGURE 4** Human autoantibody to centromere antigens is detected by immunofluorescence as dots in the nucleoplasm of interphase cells. (a) In metaphase cells (arrows), antigen is associated with the condensed chromosomes of the dividing cells. Original magnification ×550. (b) In a chromosome spread, autoantibody is localized at the centromeric regions or the primary constrictions of individual chromosomes. Original magnification ×1650. Autoantibodies of this type are present in the sera of patients with the CREST subset of scleroderma.

sequence of the antigen, oligonucleotide probes are synthesized according to the genetic code and are then used to clone the gene. Antibody-containing sera have also been used to screen and isolate cDNA clones expressing the desired protein. By these and other methods, DNAs encoding the following nuclear and other cellular antigens have been obtained:

1. Small RNA–protein antigens. These include the 68/70-kDa proteins A, B, B′, B″, C, D, and E of the Sm–RNP antibody systems, SS-B/La, and 52- and 60-kDa SS-A/Ro. RNAs which have been cloned include hY1 and hY3 RNAs.
2. Other nuclear antigens. These include the proteins PCNA and CENP-B (centromere protein B), the 66/70- and 80/86-kDa Ku proteins, the 29.5-kDa Ki protein, the 75- and 95-kDa PM–Scl proteins, and lamin B protein.

The amino acid sequences of the proteins derived from the DNA sequences have given important information on the properties of the autoantigens. Many members of the RNA–protein category contain common sequences involved in RNA recognition. Other proteins contain regions that, upon binding zinc, produce a finger-like loop, through which they bind DNA or other proteins (so-called zinc finger motif). The lamin B protein contains regularly spaced repeats of the amino acid leucine, which are involved in protein–protein interactions (so-called leucine zipper motif).

The 68/70-kDa protein contains a region of homology with a protein present in retroviruses whose gene is present in many mouse strains. On the basis of such homologies, it has been proposed that viral agents are the inducing stimulus in the generation of autoantibodies, but this concept remains rather tenuous because the homologies might be incidental.

Attempts have been made to use the clones to define the epitopes that react with the autoantibodies. Preliminary results suggest that there is usually more than one epitope on a single protein antigen and that many of the epitopes are conformational because they are made up of discontinuous stretches of amino acid sequences.

The cloning of DNAs encoding autoantigens also has the goal of providing recombinant antigens in purified form for the detection of autoantibodies as diagnostic markers in clinical work. The practicability of this approach has been repeatedly demonstrated for many of the cloned antigens described earlier. [*See* DNA Markers as Diagnostic Tools.]

## IV. NATURE OF IMMUNOGENS

### A. Immunogens Are Subcellular Particles Engaged in Essential Cellular Functions

The information on the identity and function of many of the intranuclear antigens demonstrates that the majority are component parts of subcellular particles which comprise aggregates of proteins often associated with RNA or DNA. This is an important reason for considering that the immune response in an auto-

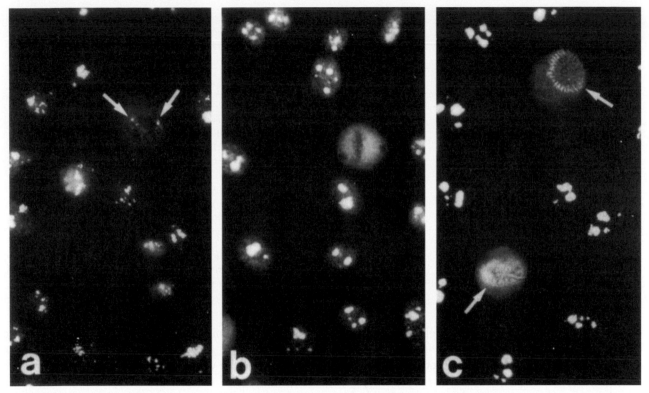

**FIGURE 5** Nucleolar staining produced by sera with antinucleolar antibodies using HEp-2 cells as substrate in immunofluorescence. (a) Speckled (punctuate nucleolar staining for RNA polymerase I). Note that staining in metaphase cells is confined to several dots in the area of the condensed chromosomes (arrows), delineating the putative nucleolar organizing regions. (b) Homogeneous nucleolar staining produced by anti-PM–Scl serum. Immunofluorescence in metaphase cells is diffusely distributed throughout the nucleoplasm and the cytoplasm, with little staining of the condensed chromosomes. (c) Clumpy nucleolar staining produced by a representative antifibrillarin serum. In metaphase cells (arrows), significant staining of the condensed chromosomes is present, but is of a different nature from that produced by anti-RNA polymerase I. Original magnification ×550. [From *Arthritis Rheum.* **31**, 525–532 (1988).]

immune patient is driven by certain subcellular particles. Another reason is that the immune response is usually polyclonal, not only because there are several epitopes in a single protein, but also because autoantibodies to different proteins can be detected in each given disease. Host factors would also be expected to have important roles in determining the total outcome of the immune response to such particles. For example, genetic factors related to the major histocompatibility complex genes would be expected to regulate T-cell involvement and/or macrophage processing. Therefore, the observed diversity of the immune response would be determined not only by the complexity of the antigens, but also by host factors.

## B. Epitopes Are Functioning Sites

The functions of many of the autoantigens are known and are shown in Table IV. The functions of some of

them have been clearly demonstrated, whereas others have been deduced.

An important observation is that many "auto"-epitopes appear to be the active sites, catalytic centers, or functioning regions of these subcellular particles. In fact, autoantibodies are capable of inhibiting splicing (autoantibody to snRNPs), DNA replication (autoantibody to PCNA), unwinding of supercoiled DNA (autoantibody to DNA topoisomerase 1), transcription of ribosomal DNA (autoantibody to RNA polymerase 1), and amino acylation of tRNAs (autoantibody to tRNA synthetases).

Even more revealing is the comparison of the effect of autoantibodies with antibodies generated in experimental animals by immunization with purified antigen. For example, in the analysis of DNA polymerase $\delta$-regulated nucleotide synthesis, autoantibodies to PCNA inhibit the processivity of nucleotide synthesis, whereas experimentally induced antibodies to purified

**TABLE IV**

Functions of Some Autoantigens

| Antigen | Function |
|---|---|
| Demonstrated functions | |
| Sm (U1, U2, U4–U6 snRNPs), nuclear RNP (U1 snRNP) | Splicing of pre-mRNA |
| PCNA (DNA polymerase $\delta$ auxiliary protein) | DNA replication |
| Poly(ADP-ribose) polymerase | DNA repair and cell differentiation |
| Scl-70 (DNA topoisomerase I) | Transcription and DNA replication |
| RNA polymerase I | Transcription of rDNA |
| tRNA synthetases | Aminoacylation of tRNA$^{His,Thr,Ala}$ |
| Probable functions | |
| SS-B/La | Processing of RNA polymerase III transcripts |
| Fibrillarin (U3 RNP) | Processing of RNA polymerase I transcripts |
| Ku | DNA-binding proteins |
| Nuclear lamins | Nuclear architecture and chromosome-binding proteins |
| Centromere proteins | Cell division and spindle attachment |
| Ribosomal RNP | Ribosomal protein synthesis |
| Signal recognition particle | Protein translocation across the endoplasmic reticulum |

PCNA, polyclonal as well as monoclonal, although highly reactive with the antigen, do not. Human autoantibodies to PCNA recognize different epitopes on the protein than those recognized by experimentally induced polyclonal or monoclonal antibodies. Similarly, human autoantibodies to threonyl tRNA synthetase inhibit amino acylation of tRNA, whereas highly reactive rabbit antibodies to purified synthetase do not.

Other studies have shown that human autoantibodies recognize more highly conserved epitopes than antibodies generated by immunization with purified antigens, again showing that there is a special property of the autoepitope compared to other immunogenic sites on the same protein. The sequences corresponding to epitopes needed for important or essential cellular functions are generally also evolutionarily conserved.

In summary, several observations indicate that epitopes recognized by human autoantibodies are the active sites of proteins or catalytic centers of subcellular particles. These epitopes are not the only immunogenic determinants on the antigens since, experimentally, one can induce antibodies recognizing other determinants on the antigens. However, the autoepitopes appear to be more highly conserved and are engaged in important cellular functions.

## V. FACTORS INDUCING AUTOIMMUNITY

An important concept concerning autoantigens is that many of them are enzymes or other catalytic proteins engaged in essential biological functions. At different phases of cell differentiation, during progression through the cell cycle, or in response to pathological or physiological stimuli, these enzymes are assembled with other cellular components into functioning particles to perform certain required biological responses. During such periods some enzymes might be preferentially concentrated in certain cell compartments, such as nuclear, nucleolar, or cytoplasmic.

An important question is whether such organelles might, under certain circumstances, become immunogenic and give rise to the production of autoantibodies. This could happen when the immunogenic subcellular particles are assembled *de novo* in response to abnormal stimuli or augment preexisting intracellular pools to meet increased demands for the biological functions they perform.

Another important question is whether the autoantibodies might be playing a pathogenetic role. In light of recently accumulated information on the structure and function of autoantigens, it is difficult to conceive that autoantibodies are pathogenic by inhibiting the function of their corresponding antigens, which are intracellular. However, they might be pathogenic under special circumstances (e.g., when immune complexes of DNA and antibody form in extracellular fluids and are deposited in small blood vessels of various organs). These immune complexes might initiate several inflammatory cascades, including activation of the complement sequence with the production of anaphylatoxins such as C3a and C5a.

However, these effects of autoantibodies are not related to inhibition of the function of their corresponding antigens. Moreover, the example of DNA and antibody might be a special case and not widely prevalent. The evidence that immunogens are subcellular particles and that the functioning sites of these particles stimulate the immune responses raises the possibility that activated states of these particles might actually be the immunogens. The intracellular particles could become immunogenic under the action of precipitating factors, including chemicals, infectious agents, or drugs. In fact, evidence now shows that procainamide, which is used for the treatment of cardiac arrhythmias, is capable of inducing autoantibodies to histones and that mercuric chloride given to mice and rats is capable of inducing autoantibodies to the nucleolar antigen fibrillarin.

## ACKNOWLEDGMENTS

These studies were supported by National Institutes of Health Grants AI10386, AR32063, and AR38695 and by a Senior Distinguished U.S. Scientist Award from the Alexander von Humboldt Foundation, Federal Republic of Germany.

## BIBLIOGRAPHY

Ben-Chetrit, E., Chan, E. K. L., Sullivan, K. F., and Tan, E. M. (1988). A 52 kD protein is a novel component of the SS-A/Ro antigenic particle. *J. Exp. Med.* **167**, 1560–1571.

Brinkley, B. R., Zimkowski, R. P., Mallon, W. L., Davis, F. M., Pisegnama, M. A., Pershouse, M., and Rao, P. N. (1988). Movement and segregation of kinetochores experimentally detached from mammalian chromosomes. *Nature (London)* **336**, 251–254.

Chan, E. K. L., Sullivan, K. F., and Tan, E. M. (1989). Ribonucleoprotein SS-B/La belongs to a protein family with consensus sequences for RNA-binding. *Nucleic Acids Res.* **17**, 2233–2244.

Habets, W., Sillekens, P. T. G., Hoet, M. H., McAllister, G., Lerner, M. R., and van Venrooij, W. J. (1989). Small nuclear RNA-associated proteins are immunologically related as revealed by mapping of autoimmune reactive B-cell epitopes. *Proc. Natl. Acad. Sci. USA* **86**, 4674–4678.

Krapf, A. R., von Mühlen, C. A., Krapf, F. E., Nakamura, R. M., and Tan, E. M. (1996). "Atlas of Immunofluorescent Autoantibodies." Urban and Schwarzenberg Publishers, Munich, Germany.

Lerner, M. R., and Steitz, J. A. (1979). Antibodies to small nuclear RNAs complexed with proteins are produced by patients with systemic lupus erythematosus. *Proc. Natl. Acad. Sci. USA* **76**, 5495–5497.

Morris, G. F., and Mathews, M. B. (1989). Regulation of proliferating cell nuclear antigen during the cell cycle. *J. Biol. Chem.* **264**, 13856–13864.

Query, C. C., Bentley, R. C., and Keene, J. D. (1989). A common RNA recognition motif identified within a defined U1 RNA binding domain of the 70 K U1 snRNP protein. *Cell* **57**, 89–101.

Rubin, R. L. (1989). Autoimmune reactions induced by procainamide and hydralazine. *In* "Autoimmunity and Toxicology" (M. E. Kammuller, N. Bloksma, and W. Seinen, eds.), pp. 117–144. Elsevier, Amsterdam.

Stollar, D. D. (1986). Antibodies to DNA. *CRC Crit. Rev. Biochem.* **20**, 1–36.

Tan, E. M. (1989). Antinuclear antibodies: Diagnostic markers for autoimmune diseases and probes for cell biology. *Adv. Immunol.* **44**, 93–151.

van Venrooij, W. J., and Maini, R. N. (1996). "Manual of Biological Markers of Disease," sections A–C. Kluwer Academic Publishers, Dordrecht, Netherlands.

# Antisense Inhibitors

RUSSELL T. BOGGS
*Cardiff, California*

NICHOLAS M. DEAN
*Isis Pharmaceuticals, Inc.*

## GLOSSARY

**Occupancy** Method of action whereby an antisense oligonucleotide disrupts the function of a targeted RNA by passively occupying a location

**RNase H** Family of enzymes that recognize RNA/DNA duplexes and cleave the RNA strand

ANTISENSE OLIGONUCLEOTIDE DRUGS CAN BE readily designed to specifically reduce the expression of a targeted gene. Since, in theory, the only requirement for the basic design of an antisense drug is the sequence of the gene, a researcher or pharmacologist can rapidly design and synthesize potential drugs with a high probability of reducing the expression of the gene in question. These compounds have proven useful as research tools and several are currently in clinical trials as promising therapeutic agents. Chemical modifications are being developed to make them even more effective.

## I. INTRODUCTION

The development of antisense nucleotide therapeutics may prove to be a watershed event in the history of rational drug design. This approach holds the possibility of allowing the design of drugs capable of reducing the expression of a specific gene product while minimizing unwanted side effects. Most traditional drugs act by inhibiting the activity of a targeted protein; antisense drugs, however, work by directly reducing the expression of a targeted gene. In doing so, it is possible to take advantage of the uniqueness of each gene's genetic code, and thus selectively inhibit the expression of even very closely related proteins, something that is very difficult to achieve with traditional protein targeting drugs.

Using antisense oligonucleotide drugs also allows for an extremely fast transition between the determination of a gene's sequence, which frequently is known before the protein is isolated and fully characterized, and the design and testing of a drug capable of reducing the expressed activity of the encoded protein. One valuable benefit of this approach is that it should facilitate mining the vast amount of sequence data expected from the Human Genome Project for potential drugs and thereby quicken the accrual of practical benefits.

## II. DEFINITION

Antisense technology is based on the ability of a single-stranded nucleic acid to hybridize to its complementary strand in an antiparallel fashion via the same base-pair interactions that hold DNA in a duplex and thus allow the exact copying of genes into RNA transcripts (Fig. 1). In general, the target for antisense intervention is the messenger RNA (mRNA) transcript of a protein-encoding gene (the "sense" strand). The challenge is to select an appropriate and effective complementary strand ("antisense") and deliver it to

ENCYCLOPEDIA OF HUMAN BIOLOGY, Second Edition, VOLUME I.   Copyright © 1997 by Academic Press.   All rights of reproduction in any form reserved.

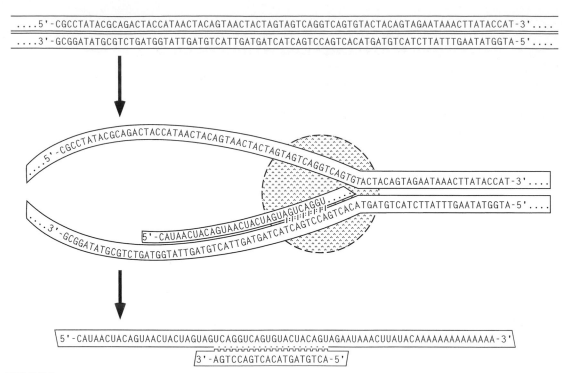

FIGURE I    Messenger RNA production and attack by antisense oligonucleotide. A hypothetical protein-encoding gene is represented at the top. The start site for RNA polymerase (TATA) is included as well as the polyadenylation signal (AAUAAA). In the middle of the figure, the gene is being transcribed by RNA polymerase II. At the bottom, the polyadenylated messenger RNA is being bound by an antisense DNA molecule.

the target mRNA in order to disrupt the function of the selected target.

This approach bypasses the mode of action of traditional drugs, which usually act by inhibiting the activity of a targeted protein, and instead aims to reduce the level of the protein through suppressing its expression rather than inhibiting its activity. In most examples, this is accomplished by preventing the synthesis of the targeted protein via interfering with the function of its messenger RNA(s). [See Protein Targeting, Specific Mechanisms.]

The antisense field can be subdivided into two areas, depending on where and how the antisense molecule is produced: trans- and cis-genetic. In the trans-genetic approach, an antisense oligonucleotide (ODN), composed of a short piece of DNA (generally 15–30 bases in length), is synthesized externally from the body and administered through normal drug delivery methods. In the cis-genetic approach, the antisense molecule is indirectly provided as an intact gene complete with the transcription and processing signals: typically a double-stranded DNA sequence (although this would depend on the vector used) to be

transcribed by the cell's own machinery into an active antisense molecule consisting of the transcribed RNA strand. Usually, the inserted gene is fairly long, and therefore this approach has a great deal in common with the field of gene therapy. Cis-genetic antisense and gene therapy differ, however, since most antisense strategies attempt to block the action of an inappropriately expressed gene rather than replacing a defective gene. This article addresses the former topic, trans-genetic antisense. For an overview of the area of cis-genetic antisense therapy, refer to the article by T. Mukhopadhyay listed in the bibliography.

## III. BACKGROUND

The life of a mammalian cell is directed by molecules of nucleic acids: DNA, which encodes the genes required by the cell to function, and RNA, responsible for converting the information contained in the genes into functional products. The DNA of a cell is primarily found in a double-stranded state within the nucleus, with one strand tightly bound to its complementary strand. In this state, the ability to affect the gene by using an exogenously supplied complementary strand is problematical. (Triplex strategies are available to intervene at this level, but they are outside the scope of this article. See the article by J. Cohen and M. Hogan listed in the bibliography.) On the other hand, most of the RNA used by a cell is found in a single-stranded form and hence should be readily accessible to short exogenously synthesized antisense molecules.

In most cases, the RNA molecule to be targeted by an antisense agent will be one transcribed by the enzyme complex known as RNA polymerase II, for example, a messenger RNA molecule destined for translation into proteins in the cytosol. Transcripts produced by RNA polymerases I and III generally code for ubiquitous expressed and utilized structural RNA molecules, and their reduction or elimination would probably prove toxic to all cells. On the other hand, by targeting the RNA polymerase II-derived transcript, one could hope to reduce the levels of a protein proven to be critical for the progress of a disease.

After synthesis by RNA polymerase II, mRNA transcripts are rapidly (usually) processed by capping, splicing, and polyadenylation in the nucleus, followed by export to the cytosol for translation into protein. Potentially, if an artificially produced complementary ("antisense") strand was available for binding to an RNA molecule using the same base-to-base hydrogen bonding that holds the DNA duplex together, then this binding should be theoretically able to interfere with the function of the mRNA molecule. At the same time, if the antisense molecule is sufficiently long and avoids using certain repetitive sequences, it should have a single unique binding site within the entire human genome. The theoretical length for a unique sequence within the human genome is 17 bases, however, if you restrict consideration to only the translated mRNA molecules, then a series of 13 bases is probably sufficient to attain a unique sequence. This is the basis of antisense therapy, binding an artificial complementary nucleic acid strand ("antisense" drug) to a target molecule of RNA (the "sense" strand) in order to specifically block the function of the targeted RNA molecule. By using the base-pairing interactions between the complementary strands, we take advantage of the extensively characterized thermodynamics of this interaction to achieve a predictable specificity, and allow a direct conversion of sequence information into drug design.

The idea of antisense drugs is not new, at least in terms of the 33-year history of molecular biology. Experiments in the early 1980s showed that cells themselves occasionally use a cis-antisense approach to regulate the expression of genes. Certain organisms were demonstrated to control the expression of some of their genes via an antisense transcript. This finding provided a theoretical basis to apply similar approaches to inhibiting gene expression in model systems. This approach, initially the artificial insertion of reversed genes into organisms such as the slime mold *Dictyostelium*, was shown to be effective in reducing the expression of the gene of interest, although the technique did not prove to be infallible. Still, the promising results were obtained that served to validate the concept, at least for cis-genetic antisense.

A key step in the development of trans-genetic antisense was the discovery that the stability of synthetic DNA antisense ODNs toward intracellular nucleases could be increased by the replacement of one of the nonbridging oxygen atoms of the phosphate backbone with a sulfur atom. (See Fig. 2 for a comparison of modified backbones to a phosphodiester backbone.) Fortunately, this modification does not prevent the ODN from hybridizing specifically to its complementary sequence (although it does slightly decrease the affinity). A second key development in the technology was the automation of the process of oligonucleotide synthesis. Automation led the way to the ability to synthesize and screen large numbers of antisense ODNs, and also brought with it the potential of reducing the price of the final drug to an affordable level. Phosphorothioate ODNs are the most readily available and inexpensive compounds to screen for an active compound, and these can certainly be considered the workhorses of the field.

## IV. MECHANISMS OF ACTION

Currently, there are two primary mechanisms whereby antisense ODNs are thought to act. Either they can act by occupying a site required for the function of an RNA transcript, or they act by inducing the degradation of the RNA transcript, usually through the intervention of RNase H.

Possibly the easiest example of the first type ("occupancy") to imagine would be an ODN binding to an mRNA in the cytosol with sufficient affinity to block the movement of ribosomes along the mRNA strand. However, the presence of strong helicase activities

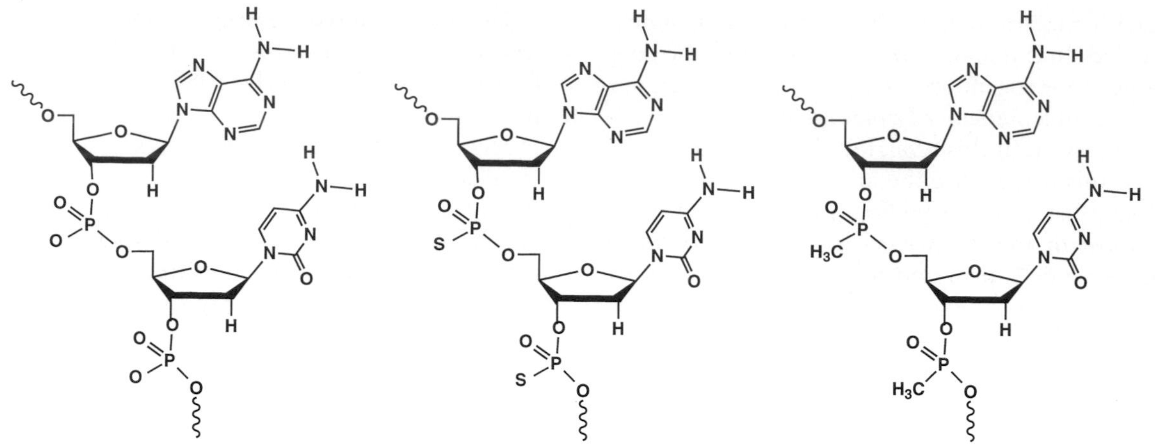

**FIGURE 2** Three oligonucleotide backbone structures. The backbone on the left is a phosphodiester backbone, that in the middle is a phophorothioate backbone, and that on the right is a methylphosphonate backbone.

associated with the assembled ribosome probably rules out any potential of sterically blocking the movement of the ribosome along the mRNA once the ribosome is assembled. So we need to look elsewhere for viable targets, rather than just a random spot in the open reading frame of an mRNA.

It is surprisingly easy to find localized sites on an RNA transcript that might prove critical for its function and that are of the appropriate length to be masked by an antisense ODN. Before being translated into protein, a messenger RNA molecule needs to go through numerous intricate processing steps. Many of these events in the lifetime of a messenger RNA molecule appear to be controlled by "motifs" or sequences that are similar (or in some cases identical) from message to message, and signal the cell's machinery to operate on the message at that particular site in a regulated manner. A great deal of effort in molecular biology has been directed toward the identification of these motifs, that is, short lengths of conserved sequences lying within the sequence of RNA and DNA molecules that are thought to direct synthesis and processing of the mRNA from its transcription from the gene to its translation in the cytosol. RNA motifs are thought to figure in all of these processes; indeed, the existence of these conserved motifs strongly implies that these regions are important points of recognition by the cell's machinery in the processing and utilization of RNA. Broadly defined, such motifs include that start codon (AUG), splice signals, and turnover signals (for examples and their relative location on an mRNA molecule, see Fig. 3). Therefore, since all of these motifs involve specific nucleic acid base

sequences, these are possible targets for antisense intervention; one could use a complementary ODN to bind to the recognition site of the RNA and thereby hide it from the cell's machinery. Operating against this is the secondary structure of the RNA (the tendency of a single-stranded RNA molecule to fold back on itself and form transient double-stranded structures), the presence of helicases (enzymes designed to unwind double helices), and the presence of generalized RNA-binding proteins that normally appear to be associated with molecules of RNA present within a cell. Any of these three things could operate to reduce the effectiveness of an antisense drug.

Three areas of RNA processing appear to be particularly promising for antisense intervention, and the first to be discussed is splicing. After transcription the message must undergo splicing to eliminate noncoding, intronic sequences. These introns are defined by short motifs at the 5' and 3' ends of the RNA sequence to be excised. These motifs, although they display a consensus, are still sufficiently unique when viewed in the context of their surrounding bases to allow for a specific ODN to possibly block splicing without affecting unrelated genes. As an example, an exon/intron boundary of the human betaglobin gene is shown in Fig. 4. If an ODN sequence was selected as indicated in the following, it comprises a sequence that would not be expected to occur in any other human gene (although we cannot be certain until the human genome is fully sequenced), so that we could be reasonably certain that it would not interfere with the splicing of any other genes in spite of containing the complement of a motif sequence. [A search of

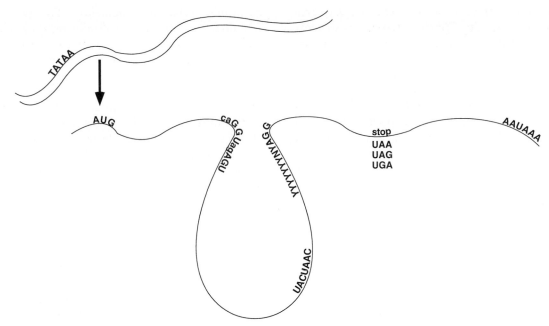

**FIGURE 3** Schematic of mRNA motifs. After transcription by RNA polymerase II, the mRNA molecule contains signals for the start of translation (AUG), the end of translation (UAA or UAG or UGA), and polyadenylation (AAUAAA). An intron before splicing is represented by the large loop, with the 5′ splice site shown as caGGUagAGU. The lowercase letters indicate that both of the indicated bases are commonly found at that position; cleavage of the RNA strand occurs between the first and second G's. The branch point of the intron is indicated by UACUAAC, and the 3′ splice site is shown as YYYYYYYNYAGG, where Y represents either a C or a U, N represents any base, and the cleavage is between the G's.

GenBank using the indicated (theoretical) ODN turns up no similar sequences in unrelated genes.]

Zbigniew Dominski and Ryszard Kole at the University of North Carolina, working with an *in vitro* splicing system, have shown that it should be possible to correct genetic defects in splicing using antisense ODNs. Some cases of β-thalassemia arise for genetic mutations that cause incorrect splicing of the beta-globin gene in humans. Using ODNs that do not support RNase H cleavage of RNA (see the next section), they were able to redirect the splicing of certain thalassemic pre-mRNAs to the correct forms by antisense

```
                cAGGUaAGU
                |.|.||..|.|
5'...UGGUGGUGAGGCCCUGGGCAGguuggguaucaagguuacaagacagguuua...3'
                ~~~~~~~~~~~~~~~~~~~
              |3'-ACCCGTCCAACCATAGTTCC-5'|
```

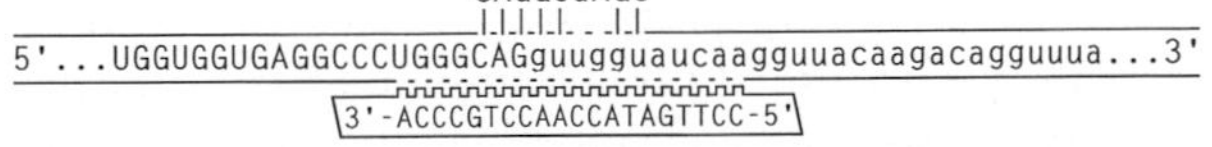

**FIGURE 4** Exon/intron boundary from human betaglobin gene. The mRNA molecule is represented by the sequence in the middle, with uppercase letters being exon sequences and lowercase letters being intron sequences. Above the messenger RNA molecule is the consensus motif for the splice site, and below is a hypothetical antisense DNA molecule.

ODNs designed appropriately to interfere with the splicing process. In one case, where a mutation results in the appearance of an inappropriate 3′ splice site, they were able to restore correct splicing by masking the normally used branch point site (UACUAAC in Fig. 3) and forcing the cell's splicing machinery to find another branch point; when this new branch point was utilized, correct splicing occurred.

Another popular site for antisense intervention has been the start codon (AUG) of an mRNA's open reading frame. The theoretical basis behind this selection is that by binding to the AUG, the ODN would interfere with the recognition of the AUG by ribosomes and hence their assembly, thereby blocking translation. However, many of these early studies may have measured only a nonspecific blockage of translation brought about by the addition of a sulfur-containing polyanion. This provides an excellent example of the need for careful controls when developing antisense compounds. Phosphorothioate-backboned ODNs have been documented to bind to non-nucleic acid targets, resulting in unanticipated effects. Therefore, measuring secondary effects, such as an ODN-mediated reduction in the rate of proliferation, may produce misleading results. For this reason, it is strongly recommended that anyone wishing to design antisense ODNs screen a battery of sequences and test for their direct effects on message RNA levels, if planning to pursue a RNase H mechanism of action. The specificity of an antisense ODN should be confirmed by a specific reduction of the targeted species, whether it be protein or RNA, and this reduction should not occur with either the scrambled or "sense" control.

Another site that could be used to affect the stability of mRNAs is the cap assembly on the message, a structure involved in mRNA stabilization and its transport out of the nucleus. Work at Isis Pharmaceuticals led by Brenda Baker has shown that this is potentially a fruitful site for antisense intervention, either by passively occupying the 5' end of the RNA molecule and preventing its capping, or by designing an ODN that carries with it an accessory group capable of cleaving the cap from the messenger RNA. Both antisense ODN strategies should result in a destabilized mRNA molecule.

As mentioned earlier, structural RNA genes are generally transcribed by RNA polymerase I and RNA polymerase III. It is unlikely that these would prove to be useful targets because of their ubiquitous expression and utilization in cellular housekeeping functions; hence down-regulation of these genes would probably prove to be toxic. There are notable exceptions to this generalization, however, and structural RNA molecules in some cases may prove to be fruitful targets. An example of this would be telomerase, an intriguing target for antisense therapy since it comprises a set of proteins built around an RNA template. So an antisense drug could directly intervene against this enzyme complex by disabling the RNA component. Telomerase itself may not prove to be a successful target for technical reasons, but it is known to be frequently required by cancerous cells for growth. [*See* Telomeres and Telomerase.]

## V. RNase H

In spite of the conceptual clarity of directly interfering with RNA processing and/or translation with an antisense ODN, perhaps the most successful antisense oligonucleotides to date have relied on the fortuitous presence of RNase H. RNase H is a broad term for a family of enzymes thought to function primarily in the DNA replication process where their role appears to be elimination of the short pieces of RNA used

as primers (Okazaki fragments). To carry out this function, the enzyme specifically recognizes double-stranded hybrids in which one strand is RNA and the other DNA, then follows this recognition/binding by cleaving the RNA strand. How many distinct proteins display this activity is unknown, as is the possibility of other roles of the enzymes, but their effect is easy to demonstrate by using DNA ODNs to direct the cleavage of RNA strands during *in vitro* assays using nuclear extracts. Fortunately, ODNs with phosphorothioate backbones function as DNA for the RNase H enzymes, and therefore an extremely successful antisense strategy has been to design and screen ODNs for their ability to cause the degradation of the messenger RNA molecule in tissue culture cells. At Isis Pharmaceuticals, we have had numerous successes at designing ODNs functioning through RNase H, some examples of which are given in the next section on targets. We normally evaluate from between 12 to 40 ODNs of 20 bases in length to search for the compound that displays the greatest ability to bring about the reduction of the targeted mRNA after 24 hr (routinely assayed via the Northern blot technique). Normally, we expect this procedure to result in one or two active compounds capable of reducing mRNA levels by at least 90%. It is thought that the high number of inactive ODNs revealed by these screens arises from the secondary structure of mRNA molecules or perhaps RNA-binding proteins interfering with the accessibility of the target RNA by the antisense molecule.

The screen also serves the purpose of providing an automatic, internal control, as the ODNs that prove to be inactive demonstrate that the reduction in mRNA levels is not due to nonspecific effects. Once an active ODN is selected from the screen, its activity is confirmed in comparison to specifically designed controls, in most cases, ODNs designed as having the same base composition as the active ODN but in a random order. The activity is then confirmed, when possible, by measuring the resultant reduction in protein levels, usually via the Western blot technique. And of course, other unrelated genes are checked to ensure that their mRNA levels are not being reduced concomitantly.

During a routine assay, we use cationic lipids to increase the uptake of ODNs and alter their subcellular distribution in our favor. Many cell lines commonly used in the laboratory require cationic lipids to allow the functional delivery of ODNs inside the cell, and some cell lines appear to be totally refractory to ODN uptake even in the presence of cationic lipids (usually these same cell lines are difficult to transfect with DNA plasmids). Examples of successful anti-sense ODNs are found in the next section.

## VI. TARGETS

A major arena for the endeavor of antisense drug design and discovery is the field of cancer therapeutics. Transformed cells (i.e., cells that are capable of inducing tumors in a nude mouse model system) have been intensively studied to determine if their transformed phenotype can be traced back to a single genetic defect. In many cases it is possible to ascribe the transformed defect to a mutation in a single gene such as *ras, erb,* or *src.* [*See* Oncogenes and Proto-oncogenes.]

Because of the apparent one-for-one correlation between a mutated gene and the transformed phenotype, these genes should be attractive targets for antisense intervention. In the clinic, however, the picture is not so simple. Nevertheless, mutations are found in certain oncogenes, such as *ras* or *erb,* in a high percentage of certain types of cancers. Thus, in theory, selective elimination of the appropriate gene should result in a reversal of the transformed phenotype in the lab and it is hoped, the elimination of the cancer *in vivo*.

Additionally, because some of these oncogenic mutations occur with a great degree of regularity between clinical isolates, it is possible to design an ODN specifically for the mutated messenger RNA. The ability to specifically reduce the mutant message with respect to the wild-type message has been demonstrated in tissue culture by Brett Monia at Isis Pharmaceuticals for the *ras* oncogene and its common codon-12 mutation. Other genes have been implicated in the occurrence and progression of cancer, and thus may also prove to be valuable targets. One such target is protein kinase C (PKC), a family of closely related kinases involved in intracellular signaling. PKC is known to bind the tumor-promoting agents, phorbol esters, and therefore is strongly implicated in the initiation and promotion of cancer. For this reason it has been chosen as an attractive candidate for antisense therapeutics. The PKC family, however, consists of 12 closely related genes, and traditional attempts at designing a therapeutic inhibitor of PKC protein function affect all or at least several of the isozymes. However, with antisense therapeutics, we can easily achieve isozyme-specific inhibition of the PKC family members because even the most closely related isozymes can be differentiated on the basis of the DNA sequences

of their respective genes. In this case, however, the practice has outdistanced the theory, as we cannot predict the differential effects of selective elimination of members of the PKC family. Antisense drugs may allow us to discern the differences in the family members that have been heretofore difficult to distinguish by allowing researchers to selectively inhibit each in turn during controlled experiments.

Furthermore, since cancer represents the inappropriate growth of cells, interfering with the mitogenic signaling pathway may down-regulate the growth of cells and hence hold cancer in check. Therefore, any member of this pathway could be considered a possible target for antisense intervention. In this line, several groups have implicated the kinase c-*raf*-1 to be central in the transmission of the growth signal from the membrane to the nucleus. By eliminating this protein, we should be able to block or at least slow the growth of cells within a tumor. This approach may also allow us to control inappropriate growth in other disease conditions involving hyperproliferation, such as psoriasis.

The three anticancer therapeutic ODNs described here, targeting *ras,* c-*raf*-1, and PKC-$\alpha$, have now been tested in a nude mouse xenograft model, where they were all shown to be efficacious in reducing tumor growth. This model uses transformed cells derived from human cancers to start subcutaneous tumors in mice that have been bred for a compromised immune system and thus cannot reject the implanted tumor. The ODNs used are in all cases specific for the human gene sequence, and so should be selectively eliminating the target gene in the tumor mass while not affecting the copies of the host gene.

Surprisingly, in the animal models tested so far, there is not a requirement for cationic lipids to aid in the uptake of the ODNs. For example, an antisense molecule designed against the murine PKC-$\alpha$ sequence shows a marked reduction in PKC-$\alpha$ message levels in the liver, and to a lesser extent in the kidney, following treatment of mice with this ODN.

## VII. MEDICINAL CHEMISTRY

Up until now, we have been considering ODNs synthesized using slightly modified phosphate backbones. Alternate chemistries are available that could improve features such as ODN stability or uptake without disrupting the base pairing required for activity. Two such backbones are shown in Fig. 5: methylene-methylimino (MMI) and peptide nucleic acid (PNA), both

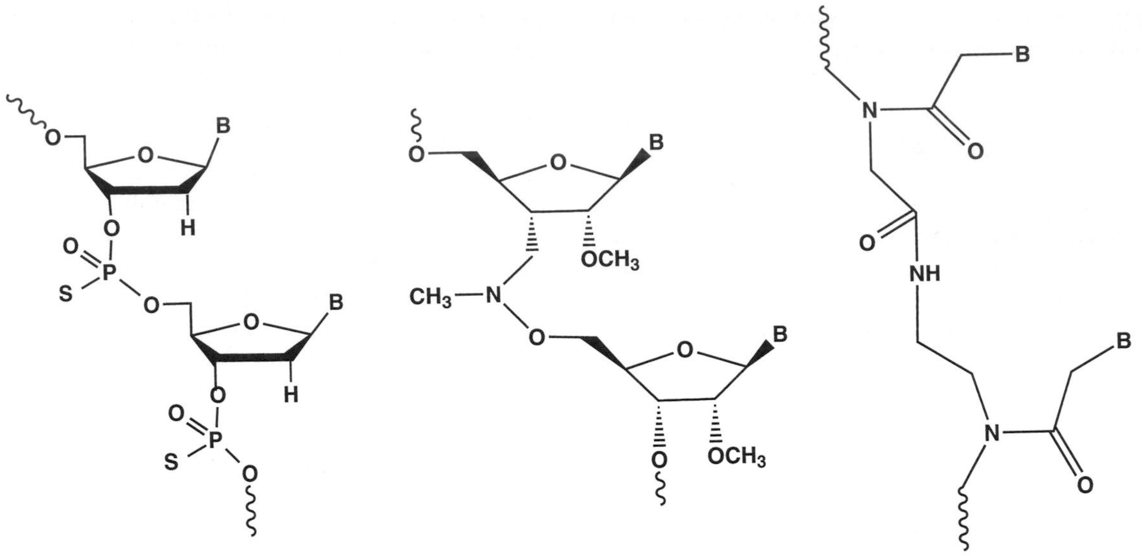

**FIGURE 5**  Alternate backbone chemistries. On the left is the standard phosphorothioate backbone, in the middle is the methylene-methylimino (MMI) backbone, and on the right is the peptide nucleic acid (PNA) backbone.

of which are currently under development as antisense nucleotide drugs. By using an alternate backbone, the drug avoids the sugar–phosphate linkage in the backbone, and this dramatically increases the stability of the ODN.

Unfortunately, none of these backbones supports RNase H activity, so their activity, if synthesized as fully modified backbone, will have to rely on the occupancy mechanisms described earlier. However, an alternate design strategy is available. Chimeric ODNs can be synthesized in which the ends of the ODN use an alternate backbone, such as MMI, to increase stability, whereas the center of the ODN retains its phosphorothioate backbone and so remains a substrate for RNase H. This takes advantage of the increase in stability afforded by these alternate backbones, but still allows utilization of the robust RNase H-driven enzymatic reduction of mRNA levels. Additionally, these modifications may improve affinity and/or availability of the ODNs.

The antisense ODNs can also be readily modified at the 2' position of the ribose (and elsewhere, but the 2' position is frequently selected) in a search for improved binding affinity and stability, and to reduce the nonspecific effects of the ODNs. Many of these modifications to produce second-generation ODNs are currently under development and several appear to be extremely promising.

## VIII. PHARMACOKINETIC AND TOXICOLOGY OVERVIEW

Several phosphorothioates have been extensively studied in pharmacokinetic and toxicology studies. These compounds are rapidly bound to proteins present in the serum, and the binding has a low enough affinity (roughly the same as other common drugs such as aspirin or penicillin) so as to provide a reservoir of the drug without having it immediately excreted by the kidneys. Phosphorothioates are rapidly absorbed by numerous tissues, especially the liver and kidneys, but there is no evidence of it crossing the intact blood–brain barrier. A single dose is cleared by the body with a half-life of approximately 2 days. All of this is well within acceptable parameters for a functional drug.

Toxicology studies have turned up few significant problems in primate and rodent models. At high doses in primate models, ODNs can cause complement activation, but this can be overcome by slow infusion. Some ODNs can cause a broad-based stimulation of the immune system, reflected as splenomegaly in rodent models, where the immune system appears to be responding to the presence of foreign DNA. None of these side effects should prove to be a significant problem in using these drugs in the clinic. Finally, it should be noted that, in recent years, several phosphorothioate ODNs have been tested in clinical trials and have been well-tolerated in numerous patients.

## IX. SUMMARY

Conceptually, the antisense paradigm has the potential of allowing the pharmacologist to rapidly move from the sequence of a gene to a drug capable of modulating the expression of that gene. When coupled to the vast amount of sequence information being generated by the Human Genome Project, it puts the potential of modifying the expression of any gene in the human body on the horizon.

Given this, within 20 years, it is possible that any clinical isolate of cancer could be genotyped to determine the specific mutations responsible for the cancerous phenotype, and the patient will then be treated with a specific cocktail of antisense ODNs tailored specifically, and with minimal side effects, to combat his or her cancer.

## BIBLIOGRAPHY

Cohen, J., and Hogan, M. (1994). The new genetic medicines. *Sci. Am.* **271**, 76–82.

Crooke, S. (1996). Progress in antisense therapeutics. *Med. Res. Rev.* **16**, 319–344.

Crooke, S., and Bennett, C. (1996). Progress in antisense oligonucleotide therapeutics. *Annu. Rev. Pharmacol. Toxicol.* **36**, 107–129.

Dean, N., McKay, R., Miraglia, L., Geiger, T., Muller, M., Fabbro, D., and Bennett, C. (1996). Antisense oligonucleotides as inhibitors of signal transduction: Development from research tools to therapeutic agents. *Biochem. Soc. Trans.* **24**, 623–629.

Mukhopadhyay, T., and Roth, J. (1995). Antisense therapy for cancer. *Cancer J.* **1**, 233–242.

# Appetite

NORI GEARY
*Cornell University Medical College*

## GLOSSARY

**Anorexia** Reductions in appetite and food intake due to abnormal interference with physiological hunger or satiety processes

**Appetite** The urge to eat; hunger; sometimes limited to the urge to eat aroused by anticipation of sensory pleasure (incentive motivation); it is referred to as an intervening variable because it is not directly observable and must be measured indirectly by verbal reports or behavioral techniques

**Classical conditioning (Pavlovian conditioning)** Form of associative learning in which presentation of a neutral conditional stimulus before an unconditional stimulus causes the conditional stimulus to elicit a conditioned response, usually a response similar to the unconditioned response elicited by the unconditional stimulus

**Flavor** Olfactory, gustatory, and tactile response to food during ingestion, which includes both a discriminative dimension (i.e., sensory intensity and quality) and a hedonic dimension; the sensory dimension appears wholly innate and varies little, whereas the hedonic dimension has both innate and learned aspects and is affected by both the physiological and environmental variables.

**Food** Object of appetite, whether macronutrient, micronutrient, or nonnutritive material, in pure or combined forms

**Hedonics** Psychology of pleasant and unpleasant subjective states, including associated behaviors and underlying physiological mechanisms; sensory perception typically involves an hedonic dimension partially independent of discriminative intensity and quality

**Hunger** Synonym for appetite that is sometimes limited to the urges to eat elicited by the physiological consequences of nutrient depletion or of body weight loss (homeostatic or regulatory motivation); according to drive reduction theory, hunger is an aversive state that we seek to escape

**Palatability** Hedonic aspect of flavor that affects liking, choice, or ingestion of food

**Satiety** Satisfaction of appetite brought about by food ingestion; in common, rather than scientific, usage, satiety often includes the feeling of surfeit caused by gluttony

APPETITE, A TERM WIDELY USED IN PHYSIOLOGY and psychology, has neither a rigorous theoretical definition nor an operational empirical definition. The pioneering physiologist W. B. Cannon proposed that appetite and hunger are fundamentally different, appetite being the learned anticipation of the pleasures of eating, and hunger being the unpleasant visceral sensation elicited by fasting. This dichotomy has not stood the test of time. The variety and complexity of both the perceptual–ideational and the gastrointestinal–metabolic antecedents and consequences of eating defy simple categorization. Furthermore, many of the most interesting aspects of appetite result from interactions of experience and physiology. Finally, appetite is inferred from behavioral as well as subjective phenomena. Therefore, appetite's plain English definition remains most appropriate. Appetite is the urge to eat, whether expressed subjectively or behaviorally. This article outlines current understanding of the physiology of normal appetite in humans and animals, and its relation to human eating disorders.

ENCYCLOPEDIA OF HUMAN BIOLOGY, Second Edition, VOLUME I.   Copyright © 1997 by Academic Press.   All rights of reproduction in any form reserved.

## I. VARIETIES AND MEASURES

Appetite is a multidimensional phenomenon. The pleasure of eating is certainly an important control of appetite. For example, if unexpectedly offered dessert after having finished a meal to satisfaction, most of us can begin again with relish. Sensory pleasure also has long-term influences. More food is consumed and body weight may increase when a larger variety of foods or more palatable foods are available. Other observations, however, reveal that regulation of energy balance, as well as hedonics, influence appetite. If the caloric density of a staple diet is changed, the amount consumed changes proportionally. Similarly, caloric intake compensates for changes in energy expenditure brought about by work or thermoregulatory requirements.

Testing of choices of foods that vary in nutrient composition reveal appetites for particular nutrients. Such appetites may operate continuously to help balance the intake of different types of nutrients (e.g., of protein and carbohydrate) or may be expressed only under special circumstances. Salt appetite is a dramatic example of the latter type of specific appetite. Humans and animals generally do not ingest large amounts of very salty food but do so avidly if in a state of sodium depletion. Although most specific appetites of this type involve learning, salt appetite appears to be an innate capacity. Specific appetites wax and wane relatively independent of each other and independent of the sensory and regulatory appetites described earlier. [*See* Salt Preference in Humans.]

A number of methods are used to measure appetite. In humans, classic psychophysical methods produce objective and reliable data. The visual analogue scale is perhaps the most common rating technique. Subjects are asked to judge their appetite by placing a mark on a line of standard length, about 10 cm, which is usually appropriately anchored (e.g., by labeling one end "most possible" and the other end "not at all"). The division lengths are then analyzed quantitatively. This kind of technique can be used to rate overall hunger, the appetite for particular foods, which may be simply named or actually tasted, and a number of other subjective aspects of appetite.

As some of the preceding examples demonstrate, appetite is often inferred from ingestive behavior. This, of course, is the only alternative in animal research. The definition of appetite as an urge, or motive, gives behavioral measures face validity. The simplest consummatory measure of the intensity of

appetite is amount ingested. Amount ingested, however, is influenced by postingestional consequences of the food as well as by the initial appetite. An important measure that can be adapted for either psychophysical or behavioral tests is the preference for different foods. Postingestional consequences can be minimized in preference tests by allowing subjects only brief tastes of the food or, in animals, by implanting esophageal or gastric cannulas to remove ingested food from the digestive system.

Considering appetite as both a subjective experience and a motivational urge and using so many methods to measure it raise practical and theoretical questions: Which is the most appropriate measure, and how are different measures to be compared? The relation of the motivational properties of appetite to its subjective dimensions does not appear simple. For example, psychophysical measurements of premeal appetite do not always accurately predict meal size. Similarly, many indirect indices of the motivation to eat that have been developed in animal research, such as running speed in straight alleys, rate of performance of operant tasks, general activity level, or the degree of acceptance of unpalatable foods, are differentially affected by changes in appetite (i.e., changes in pretest food deprivation). Thus, there is no unambiguous single measure of appetite in either humans or animals.

Finally, enthusiasts of the mind–body problem may ponder the relationship among the conscious experience of appetite, a mental phenomenon, and the physics of behavior. Does the conscious experience of hunger cause the act of eating? Consider the behavior of rats with midcollicular decerebrations. In these animals the mouth and gut are disconnected from the cerebral hemispheres, the presumed seat of consciousness. They do not actively forage for food; however, if food is presented to them, not only do they eat, but their ingestion is in many respects normal. For example, they increase intake after food deprivation, and decrease intake after manipulations that are thought to elicit satiety.

## II. FOOD PREFERENCES

### A. The Nature of Flavor

Flavor stimuli, especially their olfactory components, elicit different kinds of experiences than other stimuli. The sensory response to gustatory and olfactory stimuli includes a hedonic dimension, i.e., a perception of pleasantness, goodness, or liking. Inherent in this experience is an urge to approach or to avoid the stimulus. With the exception of nociception, this hedonic–motivational dimension is lacking in other sensory modalities. The unique neurology of the olfactory system may account for this special aspect of flavor. Olfaction is the only sensory system in which perception does not originate in a thalamic sensory projection to a primary neocortical sense area. The olfactory system projects to several phylogenetically older forebrain structures, including many of the structures of the limbic system, that appear to be involved in arousal, emotional, and motivational processes. [See Olfactory Information Processing.]

The consequences of anosmia (loss of the sense of smell) illustrate the special role of olfaction in appetite. Flavor perception is impoverished, food becomes monotonous, and appetite is reduced in ansomics. The transient partial anosmia accompanying head colds is familiar. Permanent ansomia is also, unfortunately, a frequent result of head injuries. The short, thin fibers of the olfactory nerve enter the cranium via small perforations in the cribiform plate at the base of the skull and are easily sheared off if the brain moves after a blow to the head. Anosmia is not considered an important clinical problem. Nevertheless, the resulting disturbance of appetite often produces lasting weight loss, and victims sometimes must consciously force themselves to eat enough to maintain good health.

Although some flavor stimuli have innately determined hedonic values, most likes and dislikes are learned. Some of the learning processes involved appear to reflect the unique access of flavor stimuli, especially olfactory stimuli, to limbic system structures that are involved in memory functions. Presumably as a result, odor memories are more resistant to interference and evoke much more immediate and powerful emotional responses than other types of memory. It was probably no coincidence that a flavor stimulus energized Proust's "Recherche du Temps Perdu."

## B. Innate Mechanisms of Preference and Selection

### 1. Innate Flavor Preferences

Relatively little of the hedonic value of flavor seems to be due to innate biases. The liking for sweet tastes is the most well-known innate taste bias. Neonates respond to sweet foods with positive facial expressions and consume them enthusiastically, as do ani-

mals with no prior experience with sweetness. Conversely, bitter and sour tastes and trigeminal irritation elicit rejection and negative facial expressions. The association of sweet tastes with nutritive value in natural foods (i.e., fruits) and of bitter tastes with dangerous foods (i.e., alkaloids) may have resulted in the evolution of these preferences. There is also an innate liking of the taste of salt, although it is potently expressed only in states of physiological sodium depletion (see Section II,C,4). In contrast to these innate preferences, preferences for olfactory stimuli, even the odors of feces or putrefaction, appear learned.

More is known about the responses to purified simple food stimuli than to the complex flavors of foods that humans typically eat. This is unfortunate because, for example, the hedonics of pure sugar solutions do not accurately predict the hedonics of the same sugar concentrations in real foods.

### 2. Neophobia

The avoidance of novelty, or neophobia, is especially pronounced in the case of food. Sampling new foods in small amounts would seem to have the selective advantage of minimizing the risk of poisoning.

## C. Learned Preferences

### 1. Culture

Cultural influences account for more of the variation in human food preferences and food selection than any other single factor. Because repeated exposure increases liking, we tend to like and eat the foods we are raised on. Culture therefore determines which potential foods or flavors are considered acceptable at all (e.g., insects, dog meat, chili pepper), how frequently foods are eaten (e.g., how often meat is eaten versus fish or other protein sources), how foods are processed (e.g., whether or not fish or meat is eaten raw; whether cultured or fresh milk products are used), and the particular characteristic flavor combinations of various cuisines (e.g., Hungarian paprika versus Italian garlic versus French herbs). Culture also engenders nonappetitive reactions to food, such as the important religious, symbolic, and social meanings of food in Indian and Islamic cultures. How best to determine when appetite ends and other cognitive influences begin is not entirely clear. The potency of cultural determinants of eating is evidenced by how frequently they override strong biological biases (e.g., the innate dislikes of bitter or spicy food).

The interface between biological and cultural determinants of appetite is also poorly understood. Some mechanisms, however, can be intuited. The structures of dining rituals may classically condition food preferences. For example, if some nutrient-rich foods are flavored in characteristic ways, the pleasantness of satiety may be associated with those flavors and, thus, increase liking and preference for any similarly flavored food. Conditioned flavor preferences of this kind have been demonstrated in animals. If initially neutral or nonpreferred flavors are paired with intragastric or systemic administration of protein, carbohydrate, or fat, the flavors are subsequently preferred.

Either physiological stimuli resulting from the metabolic utilization of the nutrients or hedonic stimuli may serve as unconditional stimuli for flavor preference learning. An apparent metabolic association has been demonstrated in diabetic animals. Diabetic rats do not learn to prefer flavors associated with administration of carbohydrates, which they cannot utilize, but do still develop fat preferences. Similar feedbacks from the nutritive consequences of foods may influence the cultural evolution of cuisines. For example, the consistent combination of beans and rice in Hispanic cuisine may be related to the fact that this combination has an amino acid pattern that is much more biologically useful than that of either food alone. On the other hand, pairing of initially unpreferred flavors with very pleasant flavors, such as a sweet taste, also appear to suffice for conditioned preference learning.

## 2. Conditioned Taste Aversions

Aversive consequences of foods, especially novel foods, produce strong dislikes of the food. This associative process appears to be a unique form of classical conditioning in which the aversive consequence, usually gastrointestinal sickness, is the unconditional stimulus and the flavor is the conditioned stimulus. The special preparedness for conditioned taste aversion learning is evidenced by the limited range of appropriate conditional stimuli (taste is most effective in mammals, whereas food color may be more effective in some birds), by the rapidity of acquisition of the conditioned response (one pairing usually suffices, in comparison to scores of trials in typical Pavlovian paradigms), by the long interval between the conditional and unconditional stimuli (hours rather than seconds), and by the extremely slow decay of the conditioned association. Conditioned aversions may also include gastrointestinal motor responses and vomiting as well as decreased preference. The rapidity of acquisition and the extreme degree of conditioned avoidance differentiate conditioned taste aversions

from the learned preferences described earlier. [*See* Conditioning.]

The combination of neophobia and conditioned taste aversion learning accounts for the legendary difficulty of eliminating rat populations by poisoning. Taste aversions presumably also help humans avoid toxins, although they probably more frequently decrease the preference for perfectly nutritious foods whose consumption was linked only by chance to nausea.

An interesting distinction has been drawn between the food dislikes based on conditioned taste aversions and the rejection of foods known to be dangerous. Only taste aversions render flavors unpleasant and disliked. People may still find the flavors of foods they know to be dangerous pleasant and, if they only could, would eat dangerous foods that previously elicited headaches, allergic reactions, or even some types of lower intestinal cramps. Lactose-intolerant people, for example, still tend to be tempted by ice cream and other milk products. [*See* Lactose Malabsorption and Intolerance.]

## 3. Specific Appetites

The physiological need for particular nutrients for the maintenance of homeostasis can elicit nutrient-specific appetites or hungers. Thus, when the utilization of a nutrient increases (e.g., when calcium secretion increases during lactation; when the body's reserves of an essential nutrient are depleted by, for example, feeding a vitamin-deficient diet), the selection and the consumption of any available source of the nutrient increase.

Salt appetite is the best understood specific appetite. Animals and humans normally find dilute concentrations of salt palatable but dislike and avoid extremely salty foods. Sodium-deficient animals or people, however, avidly ingest salty foods or even pure salt. Although salt appetite is potentiated by experience, it is innate. It appears as soon as a salt source is made available during the first episode of sodium deficiency.

Salt appetite results from a change in the palatability of salt. The normal preference for dilute rather than strong salt solutions is reversed, although the sodium need could be met by sufficient ingestion of either. The palatability shift appears to result from the resetting of a central neural mechanism rather than from a change in the sensitivity of salt taste receptors. Recent work suggests that the brain is signaled to activate salt appetite by the increases in blood levels of the hormones aldosterone and angiotensin

II, which are caused by sodium depletion. Increases in one hormone in the absence of increases in the other fail to elicit salt appetite, but their simultaneous action elicits the complete response. In endocrinological terms, this response is both activational, in that the behavioral effect depends on the hormones' presence, and organizational, in that the hormones elicit a lasting change in the brain, as evidenced by the potentiation of future responses.

Specific appetites exist for several, but not all, micronutrients. Most specific appetites, however, differ from salt appetite in that the taste of the needed nutrient is not innately recognized. Rather, animals learn to identify and select dietary sources of the nutrient based on their physiologically beneficial effects. For example, vitamin B-deficient animals learn within a few days to increase their intake of vitamin B-containing foods only if those foods are distinctively flavored. If the flavor marker of the vitamin B-containing food is then shifted to a vitamin B-poor food, the animals prefer that food to the still available, but unmarked, vitamin B-rich food. Thus, the appetite is for the flavor marking the nutrient source rather than for the flavor of the nutrient, and the appetite is learned only if the nutrient consistently occurs in a recognizable food source. This kind of specific appetite, like the conditioned taste aversions discussed earlier, is an example of biological preparedness for learning special types of ingestive responses. Opportunities for this learning are also facilitated by the apparent reduction of neophobia and increase in neophilia during nutrient depletion.

There may be similar specific appetites for macronutrients. When animals are offered foods that differ in protein quantity or quality (i.e., amino acid pattern), they learn, within limits, to select a diet that contains adequate amounts and qualities of protein. Appetites for protein and carbohydrate may be activated on a meal-to-meal basis. Carbohydrate ingestion decreases and protein ingestion increases brain levels of the neurotransmitter serotonin (5-hydroxytryptophan). Serotonin administration, in turn, seems to inhibit carbohydrate appetite. The observation that rats tend to alternate meals of carbohydrate-rich and protein-rich foods suggests that this mechanism may be an important control of nutrient-specific appetite. Excessive carbohydrate craving may also be related to the development of obesity in some humans.

Specific appetites are compensatory responses to perturbations in nutrient homeostasis. Such regulatory systems evidence what Cannon called the wisdom of the body. This wisdom is formidable, but not absolute. First, essential micronutrients exist for which specific hungers do not appear. Second, whether or not the wiser course is always followed is unclear. We may, for example, be too attracted by the variety and palatability of foods available in our culture to eat what we should.

## III. HUNGER AND MEAL INITIATION

### A. Underlying Theory

The categorization of particular controls of appetite as hunger signals or satiety signals is, to a great extent, arbitrary. One common distinction is that hunger signals cause meal initiation or an increase in meal frequency, whereas satiety signals decrease meal size. In fact, however, determining whether or not eating results from increases in hunger or decreases in satiety is almost always impossible. Therefore, a parsimonious alternative conception is that no fundamental difference between types of signals exists. Hunger may arise from the decay of a satiety signal as easily as from the onset of a hunger signal. Therefore, the terms hunger and satiety are retained here only for convenience and, unless otherwise noted, should not suggest that the signal cannot also influence appetite in the opposite way.

### B. Physiological Controls

The classic hypotheses of hunger are that it is stimulated either by gastric "hunger contractions" or by decreased glucose availability. These theories are no longer tenable.

Although hunger sensations are often referred to the epigastric area, modern techniques (i.e., psychophysical measures that are free of bias and motility measures that do not themselves affect motility) demonstrate that no relationship exists between gastric motility and the intensity of hunger. In fact, humans appear incapable of detecting their own gastric motor activity even with training. Nevertheless, as described below, some satiety signals may arise in the stomach. The epigastric focus of hunger sensations may originate in these signals.

The original form of the glucostatic theory (i.e., central nervous system receptors sensitive to their own glucose utilization normally signal hunger) has been rejected on the basis of decades of research. Hunger may be stimulated by extreme decreases in glucose utilization, such as occur during the biochemical hypoglycemia that can be elicited by insulin injection, but this is an emergency mechanism rather than a normal control of appetite.

Metabolism may nevertheless contribute to the control of appetite. Pharmacological antagonism of cellular oxidation of metabolic fuels, including fat metabolites as well as glucose, stimulates feeding in rats, suggesting that these metabolic pathways are normally involved in the control of feeding. The effect appears to originate in the liver because most of the metabolic fuels affected are primarily oxidized in the hepatocytes and because hepatic vagotomy blocks the effect. Furthermore, hepatic portal infusions of metabolic fuels inhibit feeding more effectively than do infusions via other parenteral routes of administration.

How might liver cells integrate changes in different metabolic pathways into a single control signal? A possible link among different pathways is their common function of supplying adenosine triphosphate (ATP) to drive the membrane sodium potassium pump $Na^+/K^+$ ATPase. Any change in $NA^+/K^+$ ATPase activity will change the membrane transport of $Na^+$ and $K^+$ and, consequently, will change the hepatocyte's membrane potential. This, in turn, would affect the neural activity of any vagal afferents connected to the hepatocyte membrane. Several observations support this possible mechanism. Inhibition of $Na^+/K^+$ ATPase with ouabain reduces the hepatic membrane potential, increases the discharge rate of hepatic vagal afferents, and stimulates feeding. The effect of ouabain is also blocked by selective lesion of the hepatic branch of the vagus, further indicating that this signal originates in the liver.

Finally, peripheral blood glucose levels per se may also contribute to meal initiation. A transient (about 15 min), small (about 10%) decline in systemic blood glucose precedes and apparently signals meals in *ad libitum* fed rats. If the decline is blocked by glucose infusion, rats do not begin meals, whereas if a similar decline is pharmacologically induced, they do begin meals. This signal may also be sensed in the liver. Transient blood glucose declines still occur after hepatic vagotomy, but they no longer predict meals. This signal appears distinct from the metabolic signal described earlier because the premeal blood glucose decline is too small to affect glucose metabolism. Pharmacologically induced blood glucose declines that are too large apparently fail to stimulate feeding, suggesting that the effective signal is a particulate pattern of blood glucose change. Interestingly, the transient blood glucose decline appears to signal meal initiation but not to predict meal size or the duration of the subsequent intermeal interval.

## C. Endogenous Rhythms

Internal physiological clocks produce rhythmic changes in appetite. The daily, or circadian, rhythm is the most prominent of these. Appetite shows dramatic circadian rhythms. Rats caged with constant access to food do not space their meals regularly, as might be expected if the capacity of the gut to store food or metabolic utilization of food were the only determinants of hunger. Rather, rats eat about four times as many meals during the dark as during the light. Humans consume about two to four meals and snacks each day, and fast much longer overnight than during the day. Although cultural and social factors certainly influence this pattern, so do internal clocks. For example, humans living in laboratory environments in the complete absence of temporal cues increase the duration of their "free-running" sleep–wake cycles by several hours. These subjects often also increase their intermeal intervals so that they eat about the same number of meals per sleep–wake cycle. This suggests that the physiological mechanisms regulating sleeping and waking rhythms also contribute to the timing of meals.

Another endogenous rhythm of appetite is related to the ovarian cycle. In many animal species, females markedly reduce food intake prior to ovulation. A preovulatory decrease in appetite has also been observed in women. This change in appetite may be caused by the cyclic variation in secretion of the ovarian hormone estradiol. In rats, this is because ovariectomy increases food intake and eliminates this feeding rhythm, whereas estradiol administration decreases food intake.

## D. Learned Controls

### 1. Conditioned Hunger

Common sense suggests that hearing the clock strike seven, seeing the set table, and similar cues regularly associated with eating should classically condition hunger. Recent research demonstrates that this is indeed a strong control of feeding in animals. If a tone or a light signals each presentation of food to otherwise food-deprived rats, subsequent presentation of the exteroceptive stimulus when the rats are fed *ad libitum* reliably causes the initiation of large meals. The unconditional stimulus in this paradigm is presumably the normal hunger that stimulates feeding after food deprivation.

Stimuli conditioned in this way also affect the size of ongoing meals, i.e., if the stimulus is presented after the beginning of spontaneous meals, meal size increases. Finally, this associative process also affects food choice. If more than one food is available when the conditioned hunger stimulus is presented, the food paired with the stimulus during the conditioning will be chosen. Conditioned hunger can undoubtedly play similar roles in humans.

## 2. Conditioned Taste Aversions

Conditioned taste aversions may influence the amount eaten as well as the selection of food. In animals, if a diet to which a taste aversion has been formed is the only available food, intake decreases and weight is lost. Taste aversions have been demonstrated to accompany the anorexia that is caused by implantation of tumors in rats. Thus, conditioned taste aversions may cause or potentiate the anorexia that is frequently associated with cancer, gastrointestinal surgery, and anorexia nervosa, each of which is associated with some physical signs, such as nausea or gastrointestinal cramps, that might serve as such unconditional stimuli for taste aversion learning.

## E. Appetite and the Regulation of Body Weight

Adults' maintenance of body weight at nearly constant levels, the dramatic compensatory hyperphagia following forced weight loss, the compensatory hypophagia after weight gain, and the adjustments of amount eaten when the caloric density of the diet is altered are incontrovertible evidence that the regulation of energy balance contributes to the control of appetite. Because feeding occurs as meals, the feedback signals related to energy balance and body weight must have access to the mechanisms controlling individual meals. The nature of this control is unknown. Because a strong relationship exists between adiposity and basal plasma insulin levels, insulin has been suggested as a feedback signal linking energy balance to appetite. Several aspects of the physiology of brain insulin are consistent with this. For example, an active transport mechanism delivers insulin from the blood into the brain, insulin receptors are localized in parts of the brain implicated in the

control of appetite, and chronic slow infusion of insulin into the brain decreases food intake.

## IV. SATIETY AND MEAL SIZE

### A. Physiological Controls

#### 1. Gastrointestinal Factors

Satiety signals may be related to both the volume and the nutrient content of the gastric contents. Experiments in rats indicate that gastric distension appears to be effective only when relatively large volumes are ingested. In humans, comfortable gastric capacity increases when large meals are ingested frequently. Thus, a reduction in the contribution of gastric distension to appetite may facilitate the capacity for binge eating in bulimia. A satiety signal related to the starch or oil concentration of the stomach contents independent of volume has also been demonstrated in rats. The gastric distension signal appears to be relayed to the brain via the vagus nerve, whereas the gastric content signal is apparently mediated by the splanchnic nerve.

Gastric emptying may also play an important role in appetite. Gastric emptying obviously affects gastric distension. Also, it usually is the rate-limiting step in the delivery of ingested nutrients from the stomach to the postabsorptive compartment. Thus, effects on gastric emptying will secondarily affect any intestinal or postabsorptive control of appetite. Unfortunately, attempts to manage food intake by controlling gastric function surgically or pharmacologically have to date met with little success.

The presence of nutrients in the small intestine, especially fats, also elicits potent satiety signals. The satiating mechanism of fat is clearly preabsorptive as it occurs before absorbed fats appear in the circulation. Both fat-sensitive vagal sensory fibers and cholecystokinin appear to contribute to this mechanism.

#### 2. Peptide Satiety Signals

A great deal of research indicates that satiety may be signaled by peptides (i.e., chains of up to about 100 amino acids) that are synthesized by specialized cells in various gut organs and are secreted during and after meals. Many peptides may function as peripheral satiety signals; present evidence for a normal physiological role is most compelling, however, for three: cholecystokinin (CCK), secreted by the upper small intestine; glucagon, secreted by pancreatic $\alpha$ cells; and bombesin-family peptides (BFP), localized in the stomach wall. Administration of these peptides at meal onset reduces meal size in a variety of species and conditions. This inhibition is behaviorally specific in that the amount and sequence of postprandial behaviors (exploring, grooming, resting, etc.) is not disrupted. Furthermore, they do not inhibit other appetitive behaviors. For example, although feeding is inhibited in hungry animals, drinking is not affected in thirsty animals. Remotely controlled, meal-contingent infusions of CCK, BFP, or glucagon also reduce the size of spontaneous meals in free-feeding rats, suggesting that their satiety effects are not limited to the usual test situation involving food deprivation and the stress of handling and injection. Finally, in humans, administration of these peptides inhibits eating without any change in the perception of satiety and without any physical or subjective side effects.

Endogenous CCK and glucagon may also be necessary participants in normal satiety, at least in rats. This is because meal size is increased by antagonism either of the action of endogenous CCK by administration of CCK receptor-blocking drugs or of the action of endogenous glucagon by administration of glucagon antibodies. In the case of glucagon, this again includes increases in the size of spontaneous meals in free-feeding rats.

### 3. Variety

Variety influences eating independent of differences in nutrient composition or palatability. Humans and animals eat more when a choice of several foods of differing flavors is available than when only a single food is offered. The effect is largest when more varied foods are offered, as in the more elaborate meals of all cuisines. It also occurs, however, when differently flavored foods of the same palatability and nutrient composition are offered. This phenomenon is labeled sensory-specific satiety. It appears unrelated to gustatory or olfactory adaption and may reflect a fundamental urge to maximize sensory pleasure.

Sensory-specific satiety produces large intake differences during single meals; however, how much variety determines amount eaten in the long term is not yet clear. When meals consisting of the same small number of foods are offered for several weeks, palatability ratings of the foods (especially the initially less-preferred foods) decrease, but it has not yet been shown that food intake or body weight decreases. In rats, on the other hand, caloric intake and body weight increase markedly when a variety of palatable foods are offered instead of just the stock diet. This suggests that minimizing the variety and maximizing the monotony of the diet might be an effective weight-loss regimen.

## V. BRAIN MECHANISMS OF APPETITE
### A. Hunger and Satiety

It has been known for a half century that small lesions in particular parts of the hypothalamus, an area at the base of the brain, can dramatically disrupt normal food intake. After lateral hypothalamic damage, rats stop eating voluntarily and may starve themselves to death if not carefully nursed. Subsequently, it was observed that healthy rats eat immediately if the lateral hypothalamus is electrically stimulated. Lesion and stimulation of the ventromedial hypothalamus produced the opposite results. These findings gave rise to the idea that these areas are the brain's "hunger center" and "satiety center," respectively. This idea now seems inappropriate. These areas do not appear directly involved in the processing of normal physiological controls of appetite. Rather, their feeding effects may be indirect consequences of disturbances in digestion and metabolism caused by ventromedial or lateral hypothalamic manipulations. Further, the concept of motivational centers also appears grossly oversimplified in light of current knowledge. Rather, appetite and other motivational functions appear to be mediated by diffuse, complex neural networks.

Neuropharmacological techniques have provided many advances in knowledge of brain mechanisms of appetite. Subpopulations of neurons in discrete areas can be manipulated by local application of neurotransmitter substances or specific agonists and antagonists in order to identify and analyze neural pathways underlying particular behaviors. The picture that emerges from this work is of a neural network involving scores of areas and particular neurotransmitters distributed widely through both the hindbrain and the forebrain. Progress in understanding these systems suggests that it may be possible to develop specific pharmacological therapies for the control of appetite. A great deal of effort in the drug development industry is now focused on brain serotonin, opiate, and benzodiazepine modulators. In contrast, amphetamines are of decreasing interest because they appear to be nonspecific anorectics with dangerous side effects.

Numerous hypothalamic areas participate in brain appetite mechanisms. For example, NPY neurons (i.e., neurons that secrete neuropeptide Y as their transmitter substance) that project from the arcuate nuclei to the paraventricular nuclei of the hypothalamus appear to be involved in meal initiation. Microinfusion of small amounts of NPY into the paraventricular nuclei in sated rats immediately elicits voracious eating. Some evidence suggests that the insulin adiposity feedback signal described earlier may be linked to the NPY feeding mechanism.

Much less is known about the contribution of more rostral brain areas to appetite. Studies of forebrain lesions suggest that the hippocampus and amygdala, two structures deep in the temporal lobe of the cerebral hemispheres, contribute to the perception of internal signals related to appetite. Patients (or rats) with damage to these areas do not seem to know whether they are hungry or satiated, in that their verbal or behavioral measures of hunger state are not influenced by eating. Although patients with this type of brain damage also suffer from amnesia, this deficit does not appear to result from their memory loss. The result of their insensitivity to satiety cues combined with their failure to remember previous meals is that if offered a second meal shortly after clearing away the remains of an identical meal just finished, they eat almost as much again.

## B. Food Reward

Neuropharmacological studies implicate several brain neurotransmitters in food reward. Of these, the role of dopamine (DA) is most firmly established. DA antagonists reduce both appetitive and consummatory responses maintained by food reward. Although central DA also contributes to brain control of movement, the effect of DA antagonism on food reward can be obtained without any signs of a motor deficit. Nor does DA antagonism produce an aversive effect. DA antagonists also have been demonstrated to specifically reduce the hedonic intensity of sucrose solutions without affecting their discriminative or sensory intensity. Neurochemical measurements also indicate that DA is released in several brain areas thought to be part of the neural reward networks, including various hypothalamic areas and the amygdala. Finally, DA may contribute to the mediation of aspects of reward specific to food as well as to more general reward functions produced by rewarding stimuli other than food.

# VI. APPETITE AND EATING DISORDERS

## A. Restrained Eating

Dieting is not a uniquely human phenomenon. Animals whose body weight has been experimentally increased (e.g., by force-feeding; by repeated insulin injections) voluntarily restrain their food intake until body weight is renormalized, but only humans appear to deny hunger in the absence of physiological reasons. Social and cultural pressures have made obsession with body image and dieting a prevalent feature of Western cultures. Does appetite function differently in dieters?

Various questionnaire scales distinguish the attitudes and habits of most dieters. "Cognitive restraint" is indicated by positive responses to statements such as "I count calories," or "I consciously hold back at meals." Most overweight people and a substantial number of normal-weight people are restrained eaters. The characteristic attitudes of these people correlate with unusual eating patterns. Paradoxically, in some situations, restrained eaters actually eat *more* than unrestrained eaters. For example, eating itself may temporarily lift restraint. If unrestrained eaters consume food preloads (the experimental version of a first course), their subsequent intake is reduced in proportion to the amount of the preload. Restrained eaters do not display this compensation. To the contrary, they counterregulate, eating more after large preloads than after small preloads. Perhaps the restrained eater, feeling his or her self-imposed limit is exceeded by the preload, gives up any further attempt at dieting for the time being. Other observations further support this cognitive restraint interpretation. First, the counterregulation occurs even with low calorie preloads if the restrained eaters believe that they are high calorie loads. Second, restraint is maintained despite preloading when the experimenter remains in the room, presumably reminding the dieter of the social norms that caused restraint to develop originally. [*See* Food Choice and Eating-Habit Strategies of Dieters.]

The restrained eating phenomenon suggests that chronic dieting is a psychological stressor that may reduce the capacity for normal behavioral self-regulation and facilitate the development of serious eating disorders or influence functioning in noneating contexts.

## B. Obesity

Human obesity is clearly not a uniform syndrome with a single etiology and course. Nevertheless, it is commonly assumed that overeating is a frequent cause of overweight, including the extreme overweight of obesity. Neither psychological nor behavioral measures, however, provide strong evidence for abnormal operation of hunger or satiety in overweight or obese people. In animal obesity syndromes, such as the obesity produced by lesions of the ventromedial hypothalamic area, physiological hunger and satiety signals appear to operate normally. Fewer such tests have been done in human obesity, although CCK has been shown to reduce meal size similarly in obese and normal weight males. This suggests that clinically useful treatments for obesity may arise from better understanding of the normal physiological controls of appetite.

The hedonic response to flavor may differ in obesity. Obese as well as formerly obese normal-weight subjects tend to judge sweet–fat mixtures as more pleasant and maximally prefer a much fattier mixture than do normal-weight subjects without a history of overweight. Such mixtures, of course, occur in ice cream, chocolate, and other desserts that the obese are advised to avoid. The extent to which such hedonic responses may stimulate increases in food intake and contribute to the development of obesity remains to be established.

The heritability of body size, adiposity, and basal metabolic rate have focused attention on genetic contributions to obesity. Several single gene mutations have been identified in experimental rodents that result in an obese phenotype. For example, the obese (ob) mutation in rats and mice results in a complex syndrome that includes extreme adiposity and type II diabetes. The mouse ob gene had been cloned, and a very similar DNA sequence appears in humans. The actual contribution of this gene to the pathophysiology of ob/ob rodents, as well as the roles of this and other genes in human obesity, however, remains to be determined.

More than mild degrees of overweight are associated with increased mortality and morbidity. Body fat distribution is a crucial factor in these risks, with abdominal adiposity more dangerous than comparable degrees of gluteal-femoral adiposity. The health risks of a mild overweight are frequently exaggerated. The deleterious effects of restrained eating discussed earlier as well as the psychological effects of prejudice against the overweight in our society may be more significant than mild overweight itself. [See Obesity.]

## C. Anorexia Nervosa

Anorexia nervosa is characterized by a disturbed body image, intense fear of obesity, and refusal to eat enough to maintain minimal weight. Anorexia nervosa is extremely serious: Often its course is unremitting and progresses to death. Pathophysiology of ap-

petite is not thought to cause anorexia nervosa, although apparently identical symptoms have been associated with hypothalamic lesions.

Hunger ratings are generally lower and satiety ratings higher in anorectics, although the internal validity of such ratings is especially poor in anorexia patients. Hunger and satiety changes may not be reciprocal, and hunger may increase late in meals or just after meals. It is possible that anorectics are insensitive to the normal experience of appetite or that they deny normal appetite. Interestingly, however, disturbed appetite ratings appear to persist even after body weight and eating patterns are normalized. The crucial question, of course, is whether or not appetite differences predate the development of anorexia and, thus, possibly cause or predispose people to it.

Flavor hedonics also appear disturbed in anorexia. Anorectics differ from normal-weight people by preferring more dilute fats to more concentrated fats. More surprisingly, they prefer concentrated sweets more than normal weight people do. This is paradoxical given their extreme avoidance of sweet foods. These patterns also persist through weight restoration.

## D. Bulimia Nervosa

Bulimia is a disorder of both normal-weight and anorectic people characterized by frequent, prodigious binges in which 2000–10,00 calories may be eaten at a sitting. Binge size often appears limited only by gastric capacity. Although much of the food ingested during binges is eliminated by vomiting or purging, even normal-weight bulimics eat few meals between binges. Binges are most often triggered by depression or anxiety.

Analyses of bulimics' appetite have produced apparently inconsistent results. For example, bulimics report no satiety during binges, yet during structured experimental meals, their perception of satiety appears relatively normal (although sensory-specific satiety may be absent). The contribution of palatability to appetite may also be decreased during binges in that unpalatable, nearly tasteless foods, such as partially defrosted frozen food, are often consumed in large quantities during binges. On the other hand, tasty high-fat and high-carbohydrate foods are more typically binged, and in psychophysical tests bulimics have shown exaggerated preferences for intensely sweet stimuli.

As in the other eating disorders, the extent to which

these differences reflect changes in appetite is not clear. It is also not clear whether such differences are causes, correlates, or consequences of the disorders. Interestingly, the increased preference for carbohydrate in bulimics, like the hedonic changes in anorectics, apparently persists through restoration of normal eating. Thus, some appetite differences may reflect a trait marker of eating disorders. [*See* Eating Disorders.]

# BIBLIOGRAPHY

Anderson, G. H., and Kennedy, S. H. (eds.) (1992). "The Biology of Feast and Famine: Relevance to Eating Disorders." Academic Press, San Diego.

Barker, L. M. (ed.) (1982). "The Psychobiology of Human Food Selection." AVI PUblishing Co., Westport, CT.

Björntorp, P., and Brodoff, B. N. (eds.) (1992). "Obesity." J. B. Lippincott, Philadelphia.

Blundell, J. (1991). Pharmacological approaches to appetite suppression. *Trends Pharmacol. Sci.* **12**, 147.

Geary, N. (1990). Pancreatic glucagon signals postprandial satiety. *Neurosci. Biobehav. Rev.* **13**, 323.

Hoebel, B. (1988). Neuroscience and motivation: Pathways and peptides that define motivational systems. *In* "Stevens' Handbook of Experimental Psychology" (R. C. Atkinson, R. J. Herrnstein, G. Lindzey and R. D. Luce, eds.), 2nd Ed., p. 547. Wiley, New York.

Langhans, W., and Scharrer, E. (1992). Metabolic control of eating. *In* "World Review of Nutrition and Dietetics" (A. P. Simopoulos, ed.), Vol. 70, p. 1. Karger, Basel.

Rolls, B. J. (1986). Sensory-specific satiety. *Nutr. Rev.* **44**, 93.

Rolls, B. J., Federoff, I. C., and Guthrie, J. F. (1991). Gender differences in eating behavior and body weight regulation. *Health Psychol.* **10**, 133.

Rozin, P., and Vollmecke, T. A. (1986). Food likes and dislikes. *Ann. Rev. Nutr.* **6**, 433.

Scharrer, E., and Langhans, W. (1990). Mechanisms for the effect of body fat on food intake. *In* "The Control of Body Fat Content" (J. M. Forbes and G. R. Hervey, eds.), p. 63. Smith-Gordon Publishing, London.

Schwartz, M. W., Figlewicz, D. P., Baskin, D. G., Woods, S. C., and Porte, D. J., Jr. (1994). Insulin and the central regulation of energy balance: Update 1994. *Endocr. Rev.* **2**, 109.

Smith, G. P. (1994). Dopamine and food reward. *In* "Progress in Psychobiology and Physiological Psychology" (A. R. Morrison and S. J. Fluharty, eds.), Vol. 15, p. 83. Academic Press, New York.

Smith, G. P., and Gibbs, J. (1992). The development and proof of the cholecystokinin hypothesis of satiety. *In* "Multiple Cholecystokinin Receptors in the CNS" (C. T. Dourish, S. J. Cooper, S. D. Iversen and L. L. Iversen, eds.), p. 166. Oxford Univ. Press, Oxford.

Stellar, J. R., and Stellar, E. (1985). "The Neurobiology of Motivation and Reward." Springer-Verlag, New York.

Stricker, E. M. (ed.) (1990). "Handbook of Behavioral Neurobiology," Vol. 10. Plenum, New York.

# Articular Cartilage and the Intervertebral Disc

ALICE MAROUDAS
*Technion—Israel Institute of Technology*

JILL URBAN
*Oxford University*

Revised by
KLAUS E. KUETTNER
*Rush University*
ALAN GRODZINSKY
*Massachusetts Institute of Technology*

## GLOSSARY

**Compressive stress** Compressive force per unit area

**Creep** Progressive deformation of a material when the applied stress is constant

**Diffusion coefficient** Mobility of a solute under the effect of a concentration gradient

**Extrafibrillar space** Space in the matrix outside the collagen fibrils

**Fibrillation** State in which the articular surface no longer appears smooth and intact

**Fixed-charge density** Net concentration of fixed-charged groups in tissues

**Hydration** Water content

**Intrafibrillar compartment** Space inside the collagen fibrils

**Matrix** Noncellular components of tissue

**Partition coefficient** Ratio of concentration of a solute in cartilage to that in the surrounding solution, at equilibrium

**Swelling** Process of gaining fluid

**Swelling pressure** Pressure needed to be applied to a tissue at a given hydration to prevent is from swelling

ARTICULAR CARTILAGE AND THE INTERVERTEBRAL disc are the two structures that support the highest compressive stresses in the body, while retaining their flexibility. The pressure on the lumbar intervertebral discs can reach 1 mega-Pascal (MPa), and the pressure on the hip can vary from average values of 3 MPa to peak values up to 10 MPa. The remarkable ability of articular cartilage and the disc to perform for several decades under such demanding conditions is due to a special combination of properties of the three major constituents, namely, collagen fibrils, proteoglycans (PGs), and water. The aim of this article is to link the functional properties of these tissues to their structure and chemistry.

## I. MATRIX: STRUCTURE AND PHYSICOCHEMICAL PROPERTIES

The matrix of cartilaginous tissues consists of a relatively coarse network of collagen fibrils, capable

ENCYCLOPEDIA OF HUMAN BIOLOGY, Second Edition, VOLUME I. Copyright © 1997 by Academic Press. All rights of reproduction in any form reserved.

of resisting tensile, but not compressive, stresses. The collagen network serves to immobilize the PG macromolecules, which form a concentrated solution of high osmotic pressure and high resistance to flow. This PG–water gel confers on the tissues their ability to resist the high compressive stresses encountered in life. Briefly, the basic unit of the PG molecule consists of a protein core to which side chains of the glycosaminoglycans (GAG) keratan sulfate (KS) and chondroitin sulfate (CS) are attached. The most important feature of PGs is that they are negatively charged: CS carries two negatively charged groups per disaccharide (i.e., one ester sulfate and one carboxyl), whereas KS carries one negatively charged group (i.e., an ester sulfate). These fixed-charged groups convey a polyelectrolyte character on cartilaginous tissues and are responsible for a number of important properties, which are briefly reviewed in this article. [*See* Cartilage; Collagen, Structure and Function.]

A schematic view of the matrix of a cartilaginous tissue is shown in Fig. 1. This figure shows that PGs cannot penetrate into the collagen fibril because of their size. Thus, the matrix consists of two compartments: the spaces between the collagen molecules within the fibril, from which PGs are excluded, and the extrafibrillar space, whose properties are determined chiefly by the presence of PGs. At physiological pH, the intrafibrillar compartment has no effective charge, while the extrafibrillar compartment has a high concentration of negatively charged groups. [*See* Proteoglycans.]

The proportion of water in the intrafibrillar compartment is controlled partly by steric factors and intermolecular repulsion forces, whose precise nature is not clear at present, and partly by the osmotic pressure gradients between the outside and the inside of the fibril. The proportion of extra- to intrafibrillar water varies, depending on the particular tissue as well as on the concentration of PGs and, hence, the

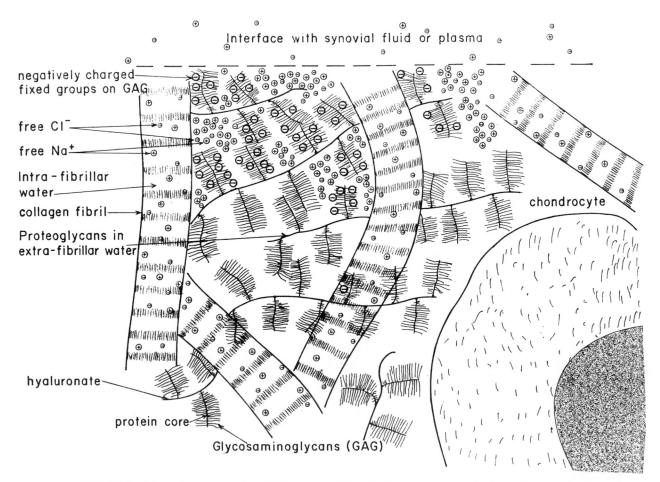

**FIGURE I**  Schematic representation of the extra- and intrafibrillar compartments in the cartilage matrix.

osmotic pressure in the extrafibrillar space. The PG–water gel filling the extrafibrillar compartment can be visualized as consisting of a meshwork of very thin, "solid" branches with liquid-filled pores.

The cells in human adult cartilage and disc occupy a very small fraction of the tissue's volume—<1%—so they do not contribute to the functional properties of the tissue; however, their role in maintaining the matrix at its optimum composition throughout life is clearly essential. Cell nutrition and cell metabolism are therefore of fundamental importance. Because articular cartilage and the intervertebral disc are avascular, the transport of nutrients to the cells takes place through the matrix and therefore must depend on the properties of the latter. In many respects, cell activity also is controlled by the microenvironment, which again depends on the concentration of various species in the matrix.

The factors that govern the transport of solutes through the matrix are the partition coefficient of the solute, the rate of movement of solutes through the matrix, and the permeability of the interface.

## II. PARTITION OF SOLUTES

Although the major constituent of the matrix is water, the concentration of solutes in the tissue water is not necessarily the same as that in the surrounding synovial fluid or plasma. The solute partition coefficient defines the ratio of the concentration of a solute in the tissue to that in the surrounding medium when the two phases are in equilibrium.

### A. Uncharged Solutes

With a mean pore radius in normal cartilage around 2 nm (20 Å), the penetration of small uncharged solutes (radius 0.3–0.5 nm) such as oxygen, glucose, glycine, or proline into the cartilage matrix is uninhibited. Practically all the water in cartilage, whether extra- or intrafibrillar, is available to small solutes.

The penetration of the larger solutes into the extrafibrillar compartment is limited by the volume available within the "pores" of the PG–water network: the partition coefficient is sensitively dependent on both the size of the particular solute and the concentration of PGs in the tissue. It should be noted that for solutes the size of serum albumin, the partition coefficient for normal cartilage can be as low as 0.002–0.01.

As for the access of the larger solutes into the intrafibrillar compartment, with intermolecular distances within the collagen fibril <1 nm, molecules whose radius exceeds that value cannot penetrate into it at all.

### B. Ionic Solutes and Fixed-Charge Density

The fixed negative charge within the matrix determines the concentration of ionic solutes in the tissue. Negatively charged solutes are excluded to a certain extent and have a lower concentration in the tissue than in the surrounding plasma. Thus, for anions such as chloride, the partition coefficient is always less than unity. The opposite is true for cations such as sodium and calcium; their concentration in the tissue is always greater than in the surrounding medium, with their partition coefficient greater than unity.

## III. MOVEMENT OF SOLUTES THROUGH THE MATRIX AND THEIR DIFFUSION COEFFICIENTS

In principle, solutes can move through the matrix by two mechanisms: by molecular diffusion and by being carried along with the fluid, which is pumped in and out of the tissue as the load changes. For large solutes, both modes of transport are of the same order of magnitude. For small solutes, on the other hand, diffusion is by far the faster process.

Solutes diffuse more slowly in the tissue than in free solution. This is due to the presence of solids within the tissue, which act as obstacles and lead to an increased tortuosity: the solute must move through a longer path to cover a given distance than it would in free solution. The diffusion coefficients of small solutes in cartilage and in the disc have values equal to about half of their value in water. Tortuosity increases with the concentration of solids in the matrix.

For large solutes, the diffusion coefficients are even further reduced because, in addition to tortuosity, frictional effects also tend to retard the motion of these molecules within the matrix.

## IV. SUPPLY ROUTES

In principle, articular cartilage has two possible routes through which it can exchange solutes (i.e., receive nutrients and dispose of the waste products): (1) the articular surface, which when unloaded is completely in contact with synovial fluid, and (2) the bone–cartilage interface. In the adult human, the bone–cartilage interface, apart from a few blood vessels that penetrate it, acts as a solid barrier to transport. Because the effective area of contact between these vessels and cartilage is small, the contribution of the subchondral route to cartilage nutrition is very small compared with the synovial route.

In the intervertebral disc, the situation is rather different. As shown on the basis of dye penetration in the human disc *in vitro* and on the basis of radioactively labeled solutes in the canine intervertebral disc *in vivo,* the disc receives its nutrients through two routes: via the end plate, which is most permeable in the region of the nucleus (about 85% permeable in the case of the dog), and via the periphery of the annulus fibrosus.

It should be noted that because of its large size, the disc has regions where the concentrations of some nutrients are very low, whereas those of some products of metabolism such as lactic acid are very high. Thus, the concentration oxygen in the center of the canine nucleus is only about 5 mm Hg, as compared with about 50 mm near the end plate, whereas that of lactic acid is around 8 $\mu$moles/g, as compared with 2 $\mu$moles/g near the end plate. The resulting lowering of pH of one unit or more might in turn adversely affect cell metabolism in the central nucleus and might also enhance the activity of pH-dependent degradative enzymes, such as cathepsin D or B.

Apart from the specific effects of low pH, it should be noted that, in a general sense, the cells in the central part of a large disc are in an especially precarious situation for the following reason. Since in that region a low oxygen tension prevails, an increased rate of glycolysis appears to be needed to satisfy the energy requirements of the cells. This, in turn, implies a greater demand for glucose. However, since this is the very region that is farthest removed from blood vessels, particularly in a large disc, the extra glucose may not be forthcoming and the cells may therefore be unable to satisfy their energy requirements fully.

## V. OSMOTIC PRESSURE OF PROTEOGLYCANS

Within the physiological range of concentrations, the major contribution to the osmotic pressure of PGs comes from the ionic (Gibbs–Donnan) effect, that is, it is simply due to the excess ions resulting from the presence of the negatively charged groups of the component glycosaminoglycans. The size and degree of aggregation of PGs thus have no measurable bearing on the value of the osmotic pressure. No semipermeable membrane is required for the manifestation of this osmotic pressure, because the PG solution forms a separate "phase" as a result of its entanglement in the collagen network.

The hydration of cartilaginous tissues results from a balance between the tendency of PGs to imbibe fluid and the resistance to swelling offered, internally, by the tensile stiffness of the collagen fiber network and, externally, by compressive forces due to body weight and/or muscle and ligament tension.

The swelling tendency, due to the osmotic pressure of PGs, is the same in the disc as in articular cartilage, provided the PG content in the extrafibrillar space is the same. However, normal articular cartilage can imbibe very little fluid over and above its normal water content present under unloaded conditions. This is because in cartilage the collagen network is very "tight," practically inextensible. The disc, on the other hand, has a far looser collagen fiber organization and will swell considerably when excised from the body and immersed in physiological saline.

This difference between the behavior of cartilage and that of the disc can also be understood in relation to the respective functional requirements of the two tissues. To confer flexibility on the spine, the disc itself needs to be pliable and flexible, and a rigid collagen network would interfere with this requirement. At the same time, the disc does not depend on a stiff collagen network for the control of its swelling tendency because the latter is counteracted by compressive stresses due to muscles and ligaments; these stresses are always present, even when a person is lying down.

In contrast, joint cartilage must rely on a stiff collagen network to prevent swelling because most joints are unloaded in the position of rest and no external forces are then acting. On the other hand, cartilage itself does not have to be as flexible and "giving" as the disc because in synovial joints flexibility and mobility are ensured by the relative motion of the two articulating surfaces rather than by the inherent pliability of the tissues. [*See* Articulations, Joints.]

## VI. MECHANICAL DEFORMATION OF CARTILAGE AND FLUID FLOW

When cartilage is subjected to an applied compressive load, the deformation that one observes is due to a combination of the following two effects: (1) the rearrangement of the polymer network, which leads to a change in the shape of the specimen at constant volume, and (2) the loss of fluid from the matrix, which leads to a decrease in tissue volume. The latter effect is much slower than the former, since it involves the movement of fluid relative to the very fine meshwork formed by the PG molecules.

Because both articular cartilage and the disc respond to changes in applied load by adjusting their hydration level, fluid flow plays a major role in the behavior of these tissues. As long as the applied pressure is superior to the tissue swelling pressure of the matrix, fluid will be squeezed out of the tissue. Fluid loss leads to a higher concentration of the PG in the extrafibrillar space and this, in turn, results in a higher osmotic (swelling) pressure and a lower hydraulic permeability. Outward flow continues until the osmotic pressure of the PG in the extrafibrillar compartment has increased sufficiently to balance the externally applied stress. At this stage no further fluid movement will take place between the extrafibrillar tissue compartment and the outside solution. At the same time, equilibrium will also be established between the osmotic pressure in the extrafibrillar compartment and compression-resisting "hydration forces" in the intrafibrillar compartment; thus no further fluid flow will take place between those two tissue compartments. It is possible today to form a rough estimate of the equilibrium hydration of both the extra- and intrafibrillar compartments, corresponding to a given external pressure. For instance, for a pressure of 50 atm (5 MPa)—a value within the upper range of physiological pressures—we have calculated that at equilibrium the amount of water retained within both matrix compartments in human femoral head cartilage will be approximately 0.9 g per gram dry weight. This means that the total volume of the tissue will be reduced by some 50% of its initial value.

It should be borne in mind, however, that *in vivo*, except if a person is standing still for a long time, equilibrium is never achieved. During normal activities, tissues are subjected to dynamic conditions. If an increased load is placed, for instance, on the spine, the discs lose fluid and become flatter. When the load is removed, the discs reimbibe fluid and return to their original height. Thus, in addition to defining equilibrium conditions, it is essential to be able to provide a description of what happens during the non-steady-state processes such as the "creep" phase of the deformation or the recovery phase.

The actual rate of fluid flow during these phases is governed by two factors. The first is a "driving force"—the difference between the applied pressure and the tension in the collagen fibrils on the one hand and the osmotic pressure of the PGs on the other. The second factor is the hydraulic permeability—the smaller the resistance to fluid flow in the tissue, the larger the hydraulic permeability coefficient and the faster fluid flows for any given driving force. The resistance of the matrix to fluid flow depends on the viscous interactions between the fluid and the solid constituents of the matrix. Since the collagen fibrils are very coarse compared to the PG molecules, the hydraulic resistance is chiefly due to the friction between the PGs and the permeating fluid.

It should be mentioned that the cumulative creep occurring in all the discs of the spine throughout the day is probably responsible for the fact that a person of average height is about 1–2% shorter at the end of the day than in the morning.

## VII. AGING AND DEGENERATION

From what has been discussed earlier, it is clear that the special functional characteristics of articular cartilage and the disc are largely determined by the high concentration of PGs.

One of the obvious questions that arises, therefore, is how do the PG and the water content in these tissues change with aging and degenerative processes?

In the case of the intervertebral disc, it is not possible to distinguish between the changes that characterize degeneration from those that accompany aging. In both processes the concentration of PGs decreases, and the swelling pressure decreases as a consequence; hence, the discs become less hydrated, thinner, and less capable of supporting external loads.

In the case of articular cartilage, distinction between the differences that are found purely as a result of aging and those that are observed in cartilage showing degenerative changes are clear. Thus, in *intact* cartilage, PG content increases and hydration decreases with age; both of these changes result in an increased osmotic pressure and an improved resistance to compression. Thus, aging itself does not appear to affect cartilage adversely.

On the other hand, cartilage that shows signs of fibrillation, however minimal, exhibits an increased water content and swells in solution—changes that are thought to be due to incipient damage of the collagen network and that lead to a deterioration in a whole range of functional properties.

## VIII. PHYSICAL AND BIOLOGICAL REGULATORS OF METABOLISM

Given that the primary function of cartilage is to support mechanical loads during joint motion, it is important to mention recent studies showing that physiological mechanical forces can regulate chondrocyte behavior and matrix metabolism. For example, cartilage *in vivo* has been found to be thickest and the PG concentration highest in regions of cartilage that are habitually subjected to high loads. In contrast, if loads are removed from a joint, cartilage thins and PGs are lost. Exercise studies have shown that *in vivo* alterations in loading pattern lead to changes in cartilage composition and cell morphology. The few studies performed on intervertebral disk show similar responses to changes in mechanical loads.

To understand the mechanisms by which loads produce a cellular response, investigators have subjected cartilage specimens to mechanical forces *in vitro*. It is known that compression of cartilage results in deformation of cells and matrix, hydrostatic pressure gradients, interstitial fluid flow, and electric fields called streaming potential fields (associated with the motion of ions relative to GAG charges during compression). These mechanical, chemical, and electrical phenomena can regulate cell behavior. Thus, static loads *in vitro* decrease PG and collagen synthesis, whereas dynamic loads at walking frequency can stimulate synthesis (similar to trends found *in vivo*). Higher-amplitude compressions can increase the rate at which matrix molecules undergo catabolic degradation and loss from the tissue. Current research focuses on the possibility that the metabolic effects of growth factors, cytokines, and matrix-degrading enzymes might be tightly coupled to the physical environment of the chondrocytes by mechanical forces *in vivo*.

## BIBLIOGRAPHY

Grodzinsky, A. J., and Urban, J. P. G. (1995). Physical regulation of metabolism in cartilaginous tissues: Relation to extracellular forces and flows. *In* "Interstitium Connective Tissue and Lymphatics" (R. K. Reed, N. G. McHale, J. L. Bert, C. P. Winlove, and G. A. Lain, eds.), pp. 67–84. Portland Press, London.

Maroudas, A. (1988). Nutrition and metabolism of the intervertebral disc. *In* "The Biology of the Intervertebral Disc" (P. Ghosh, ed.), Vol. II, pp. 1–37. CRC Press, Boca Raton, Florida.

Maroudas, A., and Urban, J. (1980). Swelling pressures of cartilaginous tissues. *In* "Studies in Joint Disease" (A. Maroudas and J. Holborow, eds.), Vol. 1, pp 87–116. Pitman Medical, London.

Maroudas, A., Katz, E. P., Wachtel, E. J., Mizrahi, J., and Soudry, M. (1986). Physicochemical properties and functional behaviour of normal and osteoarthritic human cartilage. *In* "Articular Cartilage Biochemmistry" (K. Kuettner, R. Schleyebach, and V. C. Hascall, eds.), pp. 311–327. Raven, New York.

Maroudas, A., Mizrahi, J., Ben Haim, E., and Ziv, I. (1987). The role of swelling pressure in fluid transport in cartilage. *In* "Interstitial-Lymphatic Liquid and Solute Movement" (N. C. Staub, J. C. Hogg, and A. R. Hargens, eds.), *Advances in Microcirculation,* Vol. 13. Karger, Basel.

Maroudas, A., Schneiderman, R., and Popper, O. (1992). The role of water, proteoglycan, and collagen in solute transport in cartilage. *In* "Articular Cartilage and Osteoarthritis" (K. E. Kuettner, R. Schleyerbach, J. G. Peyron, and V. C. Hascall, eds.), pp. 355–371. Raven, New York.

Mow, V. C., Kuei, S. C., Lai, W. M., and Armstrong, C. G. (1980). Biphasic creep and stress relaxation of articular cartilage in compression: Theory and experiments. *J. Biomech. Eng.,* **102,** 73–84.

Sah, R. L.-Y., Grodzinsky, A. J., Plasas, A. H. K., and Sandy, J. D. (1992). Effects of static and dynamic compression on matrix metabolism in cartilage explains. *In* "Articular Cartilage and Osteoarthritis" (K. E. Kuettner, R. Schleyerbach, J. G. Peyron, and V. C. Hascall, eds.), pp. 373–392. Raven, New York.

# Articulations, Joints

ANTHONY J. GAUDIN
*California State University, Northridge*

## GLOSSARY

**Amphiarthrosis** Joint allowing slight, limited movement between two bones

**Bursa** Fluid-filled cavity (often continuous with a joint cavity) that reduces friction between muscles, muscles and ligaments, muscles and bone, ligaments and bone, and skin and bone

**Diarthrosis** Freely moveable joint

**Ligament** Tough, fibrous tissue that connects bones together at a moveable joint

**Meniscus** Curved, cartilaginous pad found in the knee and other joints

**Mesenchyme** Network or cells in the embryonic mesoderm that produce connective tissue, the lymphatic system, and the blood vessels of the circulatory system

**Synovial** Pertaining to the synovial membrane, a layer of tissue that secretes synovial fluid, a lubricant, into the space of moveable joints

THE BONES OF THE BODY ARE USUALLY THOUGHT of as structures that fit into a framework providing body support. In addition to providing structural support, most of the bones of the skeleton are connected to one another by *articulations* or movable joints. Contraction of the muscles connected to the bones causes movement at these joints. This action provides for movement of the body from one place to another and manipulation of objects in the environment. *Arthrology,* the study of skeletal joints, takes two general approaches. One approach is based on function whereas the other is based on structure.

## I. FUNCTIONAL CLASSIFICATION

A functional classification of articulations is based on the amount of movement between the bones involved. In some articulations, the skeletal elements are held so tightly that virtually no movement is possible. These joints are called *synarthroses*. The *sutures* between the skull bones are synarthroses. *Amphiarthroses,* another type of joint, permit a small amount of movement between the bones. An example of an amphiarthrosis in the spine is the joint between the bodies of adjacent vertebrae and the intervening intervertebral cartilage. While only a slight amount of movement is possible between two adjacent vertebrae, considerable bending and twisting of the vertebral column is possible when these movements are added up along the entire length of the spine.

*Diarthroses* are articulations that allow a considerable amount of movement. Freely movable joints such as the knee, elbow, wrist, and shoulder are good examples of diarthroses.

## II. STRUCTURAL CLASSIFICATION

While a functional classification is useful in understanding the amount of movement possible between two bones, it does not explain the reason for the movement associated with each type of joint. To understand the structural differences that make this movement possible, it is necessary to study the histology (nature of the tissues) of the joints.

During the development of the skeleton, a space is present between adjacent embryonic bones. This

ENCYCLOPEDIA OF HUMAN BIOLOGY, Second Edition, VOLUME I. Copyright © 1997 by Academic Press. All rights of reproduction in any form reserved.

space is the *joint cavity*. In some cases, this cavity becomes filled with fibrous connective tissue, and the joint becomes a *fibrous joint*. In other cases the joint cavity fills with cartilage, becoming a *cartilaginous joint*. In the majority of joints, however, the cavities persist as fluid-filled spaces called *synovial cavities*, and these joints develop into *synovial joints*. [*See* Cartilage; Connective Tissue.]

## A. Fibrous Joints

In fibrous joints, embryonic mesenchyme tissue between developing bones produces a dense network of short, sturdy, collagenous fibers. These fibers stretch across the joint cavity and hold the bones securely together, thus restricting their movement. In some instances, the two bones grow close to one another in an interlocking pattern, and the cavity is filled with very short fibers. The *sutures* found between the skull bones are examples of such fibrous joints (Fig. 1). Sutures normally assume their adult form by the time an individual reaches the age of 25 years. At this time, they are considered to be true *synarthroses*, or immovable joints. After about age 50, the fibrous connective tissue in the sutures is slowly replaced by bone and the skull bones slowly fuse in a condition

known as *synostosis*, resulting in an obliteration of the suture in elderly people. This ossification process also occurs in certain cartilaginous joints. The growth (epiphyseal) plates of long bones and the three bones that make up each os coxa (hip bone) are examples of synostoses.

The joint between the tibia and fibula at their distal ends also contains dense, fibrous connective tissue, but the fibers here are longer and denser than those found in a suture (Fig. 2). This type of joint is an example of a *syndesmosis*. The syndesmosis of the tibia and fibula is essentially immovable and is functionally a synarthrosis. Other syndesmoses are found along the borders of the tibia and fibula and along the radius and ulna. Fibers here are long, and the amount of movement is sufficient for these two joints to be classified as amphiarthroses.

The articulation of the teeth within sockets in the maxilla and mandible constitutes a special type of fibrous joint called a *gomphosis*. In this joint, one skeletal element fits into another like a peg in a pegboard. The fibrous material in the joint cavity forms the *periodontal ligament*.

## B. Cartilaginous Joints

During development of cartilaginous joints, the cavity becomes filled with hyaline cartilage. Such a joint is known as a *synchrondrosis*. One example in the human skeleton is the epiphyseal plate. During growth in the length of long bones, the shaft of the bone is held to the ends (epiphyses) by the epiphyseal plate, a flattened area of hyaline cartilage. This connection constitutes an immovable synchrondrosis. This joint persists for many years while the bone increases in

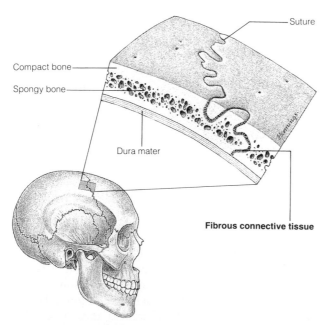

Suture

Compact bone

Spongy bone

Dura mater

Fibrous connective tissue

**FIGURE 1**   Sutures are narrow joints between the bones of the cranium. Source: Gaudin, A. J., and Jones, K. C. (1989). "Human Anatomy and Physiology," p. 166. Harcourt Brace Jovanovich, San Diego.

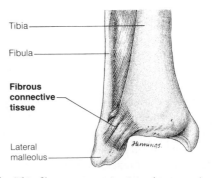

Tibia

Fibula

**Fibrous connective tissue**

Lateral malleolus

**FIGURE 2**   This fibrous connective joint between the tibia and fibula is called a syndesmosis. Source: Gaudin, A. J., and Jones, K. C. (1989). "Human Anatomy and Physiology," p. 166. Harcourt Brace Jovanovich, San Diego.

length, but eventually it is eliminated when the epiphyseal plate ossifies and the bone ceases to elongate. Another example of synchondroses are the costal cartilages that connect the ribs to the sternum (breastbone); these persist throughout an individual's life.

A *symphysis* (Fig. 3) is a cartilaginous joint in which the joint cavity has become filled with fibrocartilage. The intervertebral discs found in the vertebral column maintain this kind of joint between the bodies of adjacent vertebrae. Another example is the pubic bones in the pelvis, which are joined by the pubic symphysis. Both of these symphysis joints have slight movement and thus are functionally classified as amphiarthroses. Generally speaking, a symphysis is more flexible than a synchondrosis.

## C. Synovial Joints

In a *synovial joint* the fluid-filled joint cavity persists. Figure 4 shows a diagrammatic view of an idealized synovial joint. It is more complex than either fibrous or cartilaginous joints. The ends of two bones connected by a synovial joint are covered by a thin layer of hyaline cartilage (called the *articular cartilage*), whose free surface lacks the fibrous connective tissue layer (perichondrium) that covers most other cartilaginous structures. This articular cartilage protects the ends of the two participating bones from the trauma of continual contact. The joint cavity is enclosed in a capsule composed of dense, fibrous connective tissue. The capsular fibers penetrate the fibrous layer surrounding the bone (called the periosteum of the bones)

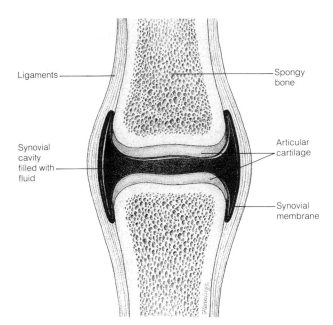

FIGURE 4 An idealized synovial joint. Source: Gaudin, A. J., and Jones, K. C. (1989). "Human Anatomy and Physiology," p. 167. Harcourt Brace Jovanovich, San Diego.

creating a strong bridge between them. The inner surface of the capsule is modified into a *synovial membrane,* which is provided with an abundant blood supply. The membrane secretes *synovial fluid* composed of lymph (also called tissue fluid) and a viscous component called hyaluronic acid. Synovial fluid serves as a lubricant reducing friction and general wear and tear on the articular cartilages of bones.

Synovial joints usually allow free and sometimes extensive movement between the bones involved and are thus classified as diarthroses. *Ligaments* are elongate straps of fibrous connective tissue that reinforce the joint capsule. They help to prevent hyperextension (extension beyond the anatomical position, as in bending the head backward) of a joint that could result in injury to the synovial membrane. The ends of the ligaments are also continuous with the periosteum of the bones in the joint. In some joints, such as the mandibular and knee joints, pads of fibrocartilage present between the bones serve as additional shock absorbers and may also increase the mobility of the joint.

In addition to the synovial cavity, several sacs, or *bursae,* are present near synovial joints. They are usually continuous with the synovial cavity and are filled with synovial fluid. Bursae are located between muscles, muscles and ligaments, muscles and bones, liga-

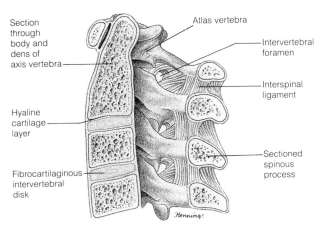

FIGURE 3 The fibrocartilage intervertebral disks of the spine form symphysis joints. Source: Gaudin, A. J., and Jones, K. C. (1989). "Human Anatomy and Physiology," p. 167. Harcourt Brace Jovanovich, San Diego.

ments and bones, and skin and bone. They increase the freedom of movement of body parts by reducing friction. *Tendon sheaths* are special bursae that surround elongate tendons, fibrous bundles that attach a muscle to a bone.

## III. ANATOMY OF SYNOVIAL JOINTS

### A. Gliding Joints

*Gliding joints,* or *arthrodial joints,* allow a simple type of back-and-forth and side-to-side sliding movement between two bones (Fig. 5). The articular surfaces

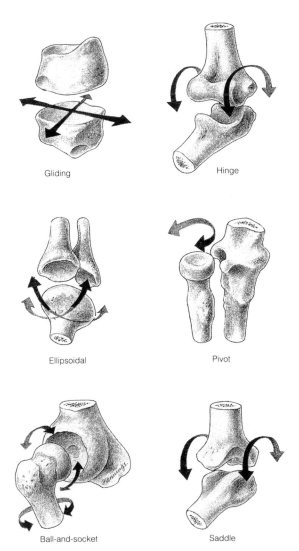

Gliding

Hinge

Ellipsoidal

Pivot

Ball-and-socket

Saddle

**FIGURE 5**  Types of synovial joints. Source: Gaudin, A. J., and Jones, K. C. (1989). "Human Anatomy and Physiology," p. 169. Harcourt Brace Jovanovich, San Diego.

involved are usually small and flat. Gliding joints are found between the articulating processes of vertebrae, and between carpals and tarsals, the small bones of the wrist and ankle, respectively.

### B. Hinge Joints

*Hinge joints,* or *ginglymus joints,* allow the motion typical of a hinge on a door, specifically, movement through an arc in one plane. The articular surface of one bone in a hinge joint is usually concave while that of the other is convex. This produces a complementary surface that facilitates the hinge-like movement of the joint. Typical hinge joints include the knee, elbow, ankle, and the interphalangeal joints of the fingers and toes.

### C. Ellipsoidal Joints

*Ellipsoidal joints,* or *condyloid joints,* involve contact between an oval-shaped condyle on one bone and an elliptical depression on another. Condyloid joints allow movement sometimes referred to as *biaxial movement,* which means movement through two arcs in two planes. Examples of condyloid joints include the major wrist joint between the radius/ulna and the carpals, and between the metacarpals and the proximal phalanges.

### D. Pivot Joints

*Pivot joints,* or *trochoid joints,* allow rotation of one bone relative to another. The articulating surfaces vary in shape, but they usually consist of a roughly cylindrical bone segment that rotates within a concave depression in the other bone. Examples of pivot joints are the articulation between the first two vertebrae in the neck (the atlas and axis) and the articulation between the head of the radius and the radial notch of the ulna at the elbow.

### E. Ball-and-Socket Joints

*Ball-and-socket joints* consist of a spherical knob or ball on one bone inserted into a concave spherical socket on another bone. Ball-and-socket joints allow the maximum movement in a joint. The shoulder and hip joints are ball-and-socket types.

### F. Saddle Joints

In *saddle joints,* also called *sellaris* or *biaxial joints,* the articulating ends of the bones are saddle-shaped,

TABLE I
A Survey of Important Joints in the Human Body[a]

| Joint | Structural classification | Functional classification | Movements |
|---|---|---|---|
| **Skull** | | | |
| Suture | Fibrous | Synarthrosis | None |
| Temporomandibular | Synovial (two synovial cavities separated by a fibrocartilage disk) | Hinge diarthrosis | Flexion, extension, abduction and adduction |
| **Vertebral column** | | | |
| Atlantooccipital | Synovial | Gliding diarthrosis | Slight lateral, and forward nodding |
| Atlantoodontoid process | Synovial | Pivot diarthrosis | Rotation |
| Articular process | Synovial | Gliding diarthrosis | Gliding |
| Vertebrae-intervertebral disk | Cartilage | Amphiarthrosis | Very slight |
| **Pectoral girdle** | | | |
| Both clavicular | Synovial | Gliding diarthrosis | Gliding |
| Shoulder | Synovial | Ball-and-socket diarthrosis | Flexion, extension, abduction, adduction, rotation, and circumduction |
| **Thoracic** | | | |
| Rib-vertebra | Synovial | Gliding diarthrosis | Gliding |
| Sternocostal | | | |
| First rib | Synchondrosis | Synarthrosis | None |
| Remaining ribs | Synovial (cavity becomes filled with cartilage in older people) | Gliding diarthrosis | Gliding |
| Manubrium-sternum | Cartilage (synostosis fuses this joint in older people) | Amphiarthrosis | Slight |
| Sternum-xiphoid process | Cartilage (synostosis usually fuses this joint in older people) | Amphiarthrosis | Slight |
| **Upper extremity** | | | |
| Elbow | Synovial | Hinge diarthrosis | Flexion and extension |
| Proximal radioulnar | Synovial | Pivot diarthrosis | Rotation |
| Distal radioulnar | Synovial | Pivot diarthrosis | Rotation |
| Wrist | Synovial | Saddle diarthrosis | Biaxial |
| Carpal | Synovial | Gliding diarthrosis | Gliding |
| Metacarpal 1-trapezium | Synovial | Saddle diarthrosis | Biaxial |
| Carpometacarpal | Synovial | Gliding diarthrosis | Gliding (slight) |
| Metacarpophalangeal | Synovial | Ellipsoidal diarthrosis | Biaxial |
| Interphalangeal | Synovial | Hinge diarthrosis | Flexion and extension |
| **Pelvic girdle** | | | |
| Sacroiliac | Part fibrous and part synovial | Part amphiarthrosis and part diarthrosis | Very slight |
| Pubic symphysis | Cartilage | Amphiarthrosis | Very slight |
| Femur-acetabulum | Synovial | Ball-and-socket diarthrosis | Flexion, extension, abduction, adduction, rotation, and circumduction |
| **Lower extremity** | | | |
| Knee | Synovial | Hinge diarthrosis | Flexion and extension |
| Proximal tibiofibular | Synovial | Gliding diarthrosis | Slight gliding |
| Distal tibiofibular | Fibrous | Amphiarthrosis | Very slight |
| Ankle | Synovial | Hinge diarthrosis | Dorsiflexion and plantar flexion |
| Tarsal | Synovial | Gliding diarthrosis | Slight |
| Tarsometatarsal | Synovial | Gliding diarthrosis | Slight |
| Metatarsophalangeal | Synovial | Ellipsoidal diarthrosis | Flexion, extension, abduction, and adduction |
| Interphalangeal | Synovial | Hinge diathrosis | Flexion and extension |

[a]From Gaudin, A. J., and Jones, K. C. (1989). "Human Anatomy and Physiology," Harcourt Brace Jovanovich, San Diego.

and this type of joint allows movement in two planes. The articulation between two wrist bones, the metacarpal of the thumb and the trapezium carpal is a saddle joint, and it permits opposing the thumb against other fingers on the hand, in a motion such as picking up a coin from a desk top.

Table I is a summary of the functional and structural classifications of the major joints in the human body.

## IV. THE KNEE: A SPECIAL JOINT

The knee is the largest joint in the body, and is composed of three joints that work together: an anterior *patellofemoral joint* between the patella (kneecap) and the patellar surface of the femur, an inner, *tibiofemoral joint* between the medial condyles of the femur and tibia, and a lateral *tibiofemoral joint* between their corresponding lateral condyles (*See* Color Plate 2).

Two fibrocartilage disks, the *medial* (inner) and *lateral* (outer) *menisci*, lie between the medial and lateral condyles as shown in the color plate. They prevent the bones from rubbing together when moving and they also act as shock absorbers. The menisci are connected to each other by a *transverse ligament* and to the head of the tibia by *coronary ligaments*.

The patella is surrounded by the *patellar ligament*, which is a portion of the lower end of the tendon of the quadriceps femoris muscles of the thigh. Three bursae separate the patella from surrounding tissues (*See* Color Plate 3). These fluid-filled sacs contribute to the relatively free gliding movements of the patella

by reducing friction between the bone and adjacent ligaments, tendons, and skin.

Several other ligaments also connect the femur, tibia, and fibula. The *tibial collateral ligament* lies on the medial side of the joint and is attached to the medial meniscus. The *fibular collateral ligament* lies on the lateral side and, unlike its tibial counterpart, is external to the capsule. These two ligaments prevent side-to-side movements of the joint. The *oblique popliteal ligament* and the *arcuate popliteal ligament* are located on the posterior side of the knee. They strengthen the lower posterior portion of the joint. The *anterior cruciate* and *posterior cruciate ligaments* are *intraarticular ligaments* because they are located within the joint. They connect the tibia and fibula and prevent twisting movements of the knee.

In addition to ligaments, the tendons of the quadriceps muscles hold the bones together at their anterior surfaces. These tendons are the *medial* and *lateral patellar retinacula*.

## BIBLIOGRAPHY

Allman, W. F. (1983). The knee. *Science* **83**.
Calabro, J. J. (1986). Rheumatoid arthritis: Diagnosis and management. *Clin. Symp.* **38**, No. 2.
McMinn, R. M., and Hutchings, R. T. (1988). "Color Atlas of Human Anatomy," 2nd Ed. New York Medical Publishers, Chicago.
Melloni, J. L. (1988). "Melloni's Illustrated Review of Human Anatomy." J. P. Lippincott Co., Philadelphia.
Simon, W. H. (ed.) (1978). "The Human Joint in Health and Disease." University of Pennsylvania Press, Philadelphia.
Sonstegard, D. A., Matthews, L. S., and Kalufer, H. (1978). The surgical replacement of the human knee joint. *Sci. Am.* **238**, 44.

# Artificial Cells

THOMAS MING SWI CHANG
*McGill University*

I. Introduction
II. Artificial Cells Containing Adsorbents and Immunoadsorbents
III. Artificial Cells as Drug Carriers
IV. Other Applications

## GLOSSARY

**Hemoperfusion** Process in which blood (hemo) perfuses through a column of biologically active particles

**Immobilized bioreactants** Biotechnological technology for retaining enzymes, cells, microorganisms, and other biologically active materials; artificial cells based on microencapsulation is one of four major approaches; the other three include adsorption, covalent linkage, and matrix entrapment

**Microencapsulation** Procedure in which materials are encapsulated within microscopic containers; biological and artificial cells are typical examples in which enzymes and intracellular organelles are microencapsulated within the cell membrane

ARTIFICIAL CELLS ARE PREPARED IN THE LABORAtory. They are usually about the same size as biological cells. Each artificial cell contains biologically active materials. Different types of artificial cells are now available for use in medicine and biotechnology. In medicine, artificial cells containing adsorbents are already a routine form of treatment; this includes treatment for acute poisoning, high blood aluminum and iron, kidney failure, and some types of acute liver failure. Artificial cells formed from hemoglobin are being tested in humans for use as red blood cell substitutes for transfusion. Still other types of artificial cells

contain living cells. These are being tested in animals and in humans for the treatment of diabetes, liver failure, and others. Artificial cells containing enzymes are being tested for treatment in hereditary enzyme deficiency diseases and other diseases. Artificial cells containing complex enzyme systems can convert wastes such as urea and ammonia into useful amino acids. Lipid-membrane artificial cells are extensively explored for use as drug carriers and for targeting to specific sites. In biotechnology, artificial cells are used for the production of monoclonal antibodies, interferons, and other biotechnological products. They are also being investigated for use in other applications in biotechnology, chemical engineering, and medicine.

## I. INTRODUCTION

Cells are the basic units of all humans, animals, and other living organisms. They are complex in their structures and functions; however, this should not prevent us from using some of their simpler properties to prepare artificial cells for use in medicine and biotechnology. The first artificial cells were prepared by the author in 1957. They were not exact replicates of biological cells; however, they did have some of the simpler properties of biological cells. It should not be too surprising that many possible medical and biotechnological uses exist for artificial cells (Table I); after all, biological cells are important functional units of all living matter. [*See* Cell.]

The basic idea is to prepare in the laboratory artificial cells of the same size as or smaller than biological cells (Fig. 1). Each artificial cell contains biologically active materials (Fig. 2). The artificial cell membrane separates these materials from the outside. Biologi-

ENCYCLOPEDIA OF HUMAN BIOLOGY, Second Edition, VOLUME 1.
Copyright © 1997 by Academic Press. All rights of reproduction in any form reserved.

## TABLE I

Medical and Biotechnological Applications of Artificial Cells

1. Treatment of acute poisoning (routine clinical application)
2. Treatment of aluminum and iron overload (routine clinical application)
3. Treatment of end-stage kidney failure (routine clinical application as a supplement to hemodialysis)
4. Treatment of liver failure (routine clinical application for certain types of acute liver failure)
5. Red blood cell substitutes for transfusion (clinical trial ongoing)
6. Blood group antibody removal (clinical application and clinical trial)
7. Treatment of hereditary enzyme deficiency (clinical trial started)
8. Clinical laboratory analysis (clinical application)
9. Production of monoclonal antibodies (routine use in industry)
10. Treatment of diabetic mellitus and other endocrine diseases (clinical trial ongoing)
11. Drug delivery systems (clinical application in some cases)
12. Conversion of cholesterol into carbon dioxide (experimental for dairy products and blood cholesterol)
13. Bilirubin removal (experimental)
14. Production of fine biochemicals (industrial application)
15. Food and aquatic culture (industrial application)
16. Conversion of wastes into useful products (experimental)
17. Other biotechnological and medical applications (in progress)

cally active materials retained inside the artificial cells can act on outside molecules, which can cross the membranes. Permeant substances produced inside the artificial cells can leave if they can pass through the artificial cell membranes.

Since 1957, artificial cell preparation has undergone many extensions and modifications, including the use of many types of synthetic polymer membranes, protein membranes, lipid membranes, and even biological cell membranes. Artificial cells can also contain many types of biologically active materials in different combinations (Fig. 3). With these large variations in membrane materials and contents, many types of artificial cells are now possible. Many of these are still in experimental and laboratory stages. Generally speaking, the simpler the system, the less time required to develop the system for clinical application. Two very simple systems in actual clinical use are used as examples in this article. These examples are followed by a summary of the more sophisticated and futuristic systems (Table I) based on biotechnology.

## II. ARTIFICIAL CELLS CONTAINING ADSORBENTS AND IMMUNOADSORBENTS

### A. Artificial Cells Containing Activated Charcoal

The simplest type of artificial cells contains adsorbents. This simplicity is such that this system has been used routinely for treating patients since the early 1980s. Activated charcoal can remove chemicals, toxins, drugs, and waste products; however, by itself, charcoal granules release harmful particles into the blood and also adversely affect blood cells. The principle of artificial cells was applied to solve these problems (Fig. 4). Coating each small activated charcoal granule with an ultrathin (200 Å) polymer membrane prevents the release of charcoal powder. The artificial cell membrane also separates the activated charcoal from blood cells. As a result, activated charcoal in artificial cells no longer has any adverse effects. On the other hand, the ultrathin membrane allows for rapid exchanges. Thus, toxins, waste products, drugs, chemicals, poisons, and other materials can cross the membrane rapidly to reach the enclosed activated charcoal. In this way, artificial cells containing activated charcoal can effectively remove unwanted materials from blood.

The modern hemoperfusion device contains 70 g of artificial cells (Fig. 5). Each artificial cell consists of a 100-$\mu$m-diameter charcoal granule coated with an ultrathin polymer membrane. Screens placed on both sides of the device retain millions of these artificial cells. The patient's blood circulating through the column comes in direct contact with the artificial cells. It is effective in removing unwanted molecules from the blood of patients perfusing through the column. This procedure is hemoperfusion and is used in the following conditions.

### 1. Acute Poisoning

This approach is especially effective in the treatment of patients with suicidal or accidental poisoning. Hemoperfusion can remove the drugs much more rapidly than standard hemodialysis. As a result, this is a routine treatment for patients with acute poisoning. It can remove drugs that charcoal can adsorb. It is effective where the drugs are available in the circulating blood for removal.

### 2. Aluminum or Iron Overload

High aluminum blood levels can cause severe central nervous system and bone problems. Activated char-

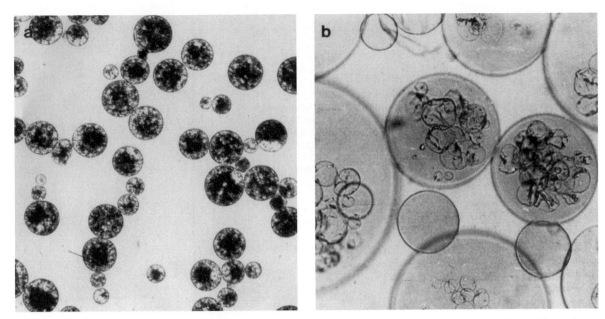

FIGURE I  (a) Artificial cells, 20 μm mean diameter, containing proteins and multienzyme systems. Artificial cell membrane formed from ultrathin cellulose nitrate membrane of 200 Å thickness. (b) Artificial cells, 100 μm mean diameter, containing smaller ones to form "intracellular compartments." Artificial cell membrane formed from ultrathin polyamide nylon membrane of 200 Å thickness. Contains protein and multienzyme systems. [Reproduced, with permission, from T. M. S. Chang (1972). "Artificial Cells." Thomas, Springfield, IL.]

coal cannot remove aluminum; however, desferoxamine, a chelating agent, can bind aluminum from the body. Hemoperfusion can then remove the desferoxamine together with aluminum. This is effective for removing elevated aluminum in dialysis patients. Hemoperfusion can also remove excess iron in the same way.

### 3. Kidney Failure

In severe kidney failure, waste products, salts, and water accumulate in the body. The standard treatment is by hemodialysis. The artificial cell approach can remove most of the organic waste metabolites faster than hemodialysis, resulting in improved symptoms in the patients; however, hemoperfusion does not re-

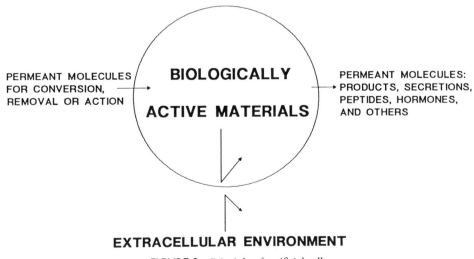

FIGURE 2  Principle of artificial cells.

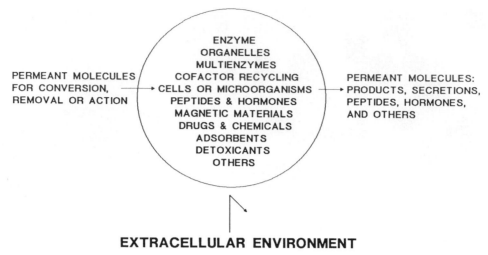

**EXTRACELLULAR ENVIRONMENT**

FIGURE 3 Examples of biologically active materials enclosed within artificial cells, individually or in combinations.

move urea, salts, and water. Combining hemoperfusion with standard dialysis can cut down the time of treatment because dialysis is effective in removing salt, water, and urea. Another method is to combine hemoperfusion with a small ultrafiltrator. This results in a very small and portable device, if a suitable urea removal system is available.

### 4. Liver Failure

Liver is an organ with very complex functions. A complete artificial liver is not yet available to replace all the functions of the natural liver. Hemoperfusion can remove toxic products in deeply comatose pa-

tients with liver failure, resulting in temporary recovery of consciousness. In acute liver failure, the liver has the potential to recover to full function; thus, hemoperfusion can increase the survival rates of patients with certain types of acute liver failure. However, it is not conclusively established whether or not this is more effective than good intensive care in specialized liver units. Livers in end-stage chronic liver failure cannot recover as this requires continuing support of all liver functions. Detoxification by hemoperfusion only supports one of the many functions of the liver; therefore, for end-stage chronic liver failure, additional systems are required.

## B. Artificial Cells Containing Other Adsorbents and Immunoadsorbents

Artificial cells containing activated charcoal are one example of the simplest form of artificial cells. Here the membrane is a synthetic polymer, and the content is an activated charcoal. Artificial cells containing activated charcoal result in a routine treatment method for different conditions in patients.

This simple approach has been applied to other adsorbents. Ion-exchange resin-like Amberlites are potentially useful in hemoperfusion for the removal of specific drugs. However, Amberlites also have adverse effects on blood cells. One approach is to use artificial cells to form a coating of albumin around each individual resin. This improves blood compatibility, allowing this system for use in hemoperfusion. Immunoadsorbents can remove antibodies or antigens from the circulating blood. However, most immunoadsor-

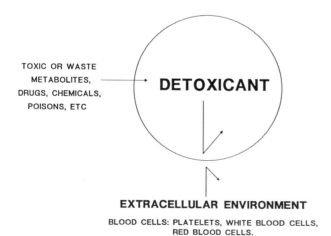

**EXTRACELLULAR ENVIRONMENT**

BLOOD CELLS: PLATELETS, WHITE BLOOD CELLS, RED BLOOD CELLS.
OTHER COMPONENTS

FIGURE 4 Artificial cells containing adsorbents as detoxicants prevent adsorbents from releasing particles or damaging blood cells. Ultrathin membranes allow rapid removal of toxins and wastes.

FIGURE 5 Modern hemoperfusion device containing 70 g of artificial cells. Each artificial cell consists of a 100-$\mu$m-diameter-activated charcoal sphere with an ultrathin polymer coating 200 Å thick. Screens on both sides of the device retain the artificial cells. Blood from the patient perfuses through the lower entry. It enters the device to come in direct contact with the artificial cells. After removal of waste or toxins, blood returns to patients by upper exit port.

bents also have problems similar to activated charcoal of release of particles and adverse effects on blood. A polymer membrane-coated immunoadsorbent no longer has these problems but can continue to remove antibodies. For example, a membrane-coated synthetic immunoadsorbent can remove blood group antibodies from the blood of patients. This basic principle allows the potential use of direct blood perfusion through immunoadsorbent.

## III. ARTIFICIAL CELLS AS DRUG CARRIERS

The second example is also a very simple form of artificial cells. Artificial cells are used to retain medica-tions, hormones, or peptides. After injection, the materials are slowly released into the body. Its simplicity allows this approach to be used already in patients.

## A. Artificial Cells Prepared from Biodegradable Polymers

Biodegradable polymers (e.g., polylactic acid) can be used to microencapsulate drugs or hormones or other peptides. After implantation, the contents of the artificial cells are released slowly into the body. After the release of the contents, the polymer membrane is degraded in the body to carbon dioxide. This eliminates the undesirable accumulation of the polymer in the body. One application is in the injection of encapsulated slow-release contraceptive hormones. Artificial cells can be prepared from biodegradable polymers in submicron dimension nanocapsules. This has further increased its potential applications as drug carriers.

## B. Lipid-Membrane Artificial Cells

Lipid-membrane artificial cells containing enzymes and protein were first prepared with a bilayer lipid complexed to cross-linked protein or polymer membranes. Permeability is comparable to the biological cell membrane. The permeability can be modified by adding carrier systems such as valinomycin and adenosine triphosphatase and it has been used to retain cofactors in multienzyme systems. Liposomes are an onion-like, multilayer lipid-membrane structure. These were originally used for basic membrane research. In this form, they are not the type of artificial cells discussed in this article. Liposomes consist of a large proportion of lipid and a relatively small amount of encapsulated material. Recent attempts to improve this has now resulted in liposomes that more closely resemble artificial cells. It now consists of a single bilayer lipid membrane enclosing biologically active materials. The stability is improved by preparing very small ones, usually in the range of 0.2–0.7 $\mu$m. Thus, liposomes are now being modified into lipid-membrane artificial cells. Others are starting to incorporate protein and other materials into the membrane. These are now in the form of submicron lipid membrane artificial cells; however, many researchers as a matter of habit still use the term liposomes.

Lipid-membrane artificial cells fulfill important functions in a way that is different from the other types of artificial cells. Its most important function is in drug delivery. The lipid membrane is only permeable to lipid-soluble substances. As such, when enzymes and most drugs are microencapsulated into lipid-membrane artificial cells, they do not act on external permeant molecules. Instead, the contents must be released for action. Therefore, the most important function of this approach is for applications as carriers for drugs, vaccines, and other materials for targeting to specific sites. On reaching the specific site, the contents are then released to carry out its action. This way, patients are protected from toxic or undesirable effects of the content. Most of its effects are then able to act locally at the desired sites of actions. Areas of possible application include tumor imaging, chemotherapy, antimicrobial therapy, antiparasitic therapy, treatment of fungal diseases, immunomodulation, and carrier for vaccines. Many major research groups and industries are actively developing this important area for medical applications.

## IV. OTHER APPLICATIONS

The above two examles are the simplest types of artificial cells that are already being used in patients. Many other examples exist, some of which are based on biotechnological approaches. These are more complicated and futuristic. Because of lack of space, only a brief survey of these areas is made here. Further details are available in the review articles cited in the bibliography.

### A. Red Blood Cell Substitutes

The potential problem of acquired immunodeficiency syndrome in donor blood has stimulated mounting activities in the research and development of artificial red blood cells. There are two major approaches. The first one is modification of hemoglobin by microencapsulation or by cross-linking. These modifications prevent the hemoglobin, a tetramer, from breaking down into a dimer in the circulation. Also, the modifications improve the ability of the hemoglobin to release the oxygen it carries (oxygen affinity). In this form, there is no need for cross-matching since there is no blood group antigen. This allows for ease of use, especially in emergencies, major disasters, or wars. They can be stored for a long time in the lyophilized form. Before modifying hemoglobin, the hemoglobin can be sterilized to remove infective microorganisms like HIV, and hepatitis viruses. Other sources of hemoglobin for use in modified hemoglobin have

been studied, including bovine hemoglobin, recombinant human hemoglobin from microorganisms, human hemoglobin from transgenic animals, and synthetic heme. These preparations are effective for treating hemorrhagic shock and other conditions in experimental animals. The earlier problems of adverse reactions in humans have now been solved. Several groups have carried out Phase I clinical trials in humans with no major adverse effects. By 1996 several Phase II clinical trials in humans for efficacy have been ongoing. Another approach is the use of the synthetic oxygen-carrying material fluorocarbon. This has been tested for specific applications. Further research is being carried out on fluorocarbons to increase its oxygen-carrying capacity and its removal after leaving the circulation.

### B. Artificial Cells Containing Enzymes and Multienzyme Systems

Enzymes in biological cells are responsible for many of the major functions of biological cells. Artificial cells are prepared to contain a single enzyme system or complex multienzyme systems (Figs. 1–3). Substrate can diffuse into the artificial cells to be acted on by the enzyme. The product can either remain in the artificial cell or diffuse to the outside. Artificial cells containing urease and ammonia adsorbent have been used to remove urea; this has been tested in patients. In animal studies, artificial cells have been used in hereditary enzyme defects (e.g., catalase in acatalasemia and phenylalanine ammonia lyase in phenylketouria). Artificial cells containing xanthine oxidase have been tested clinically in a patient with hypoxanthinuria. Most enzymes in biological cells function as complex enzyme systems. Artificial cells have been prepared to contain multienzyme systems with cofactor recycling. This has been studied for converting metabolic wastes (e.g., urea and ammonia) into essential amino acids required by the body. Artificial cells containing two enzyme systems can also be prepared to remove bilirubin.

### C. Artificial Cells Containing Living Biological Cells

Islets enclosed inside artificial cells are prevented from immunorejection after implantation into animals. They remained viable and can control the glucose levels of diabetic rats and in recent clinical trials in humans. Liver cells enclosed inside artificial cells are also prevented from immunorejection. They can lower the high bilirubin level in the Gunn rats, whose liver cannot remove bilirubin. Artificial cells containing hybridoma have also been used in biotechnology for monoclonal antibody production. Artificial cells containing a microorganism that can remove cholesterol have been studied in a bioreactor system. Artificial cells containing a genetically engineered microorganism can remove metabolic wastes, urea, and ammonia.

## D. Other Applications

A magnetic field applied outside the body can direct artificial cells containing magnetic materials and medications to some specific sites. These magnetic artificial cells are also used in bioreactors. Another type of artificial cell is used in aquatic culture for filter feeders such as shrimp. Others are used in laboratory diagnosis. Space is not available to describe these and other applications. With increasing interests in biotechnology, one looks forward with anticipation to further extensions and modifications of artificial cells. This will result in other applications in medicine, biotechnology, and other areas.

## BIBLIOGRAPHY

Chang, T. M. S. (1964). Semipermeable microcapsules. *Science* **146**, 524–525.

Chang, T. M. S. (1972). "Artificial Cells." Thomas, Springfield, IL.

Chang, T. M. S. (1988). Methods in the medical applications of immobilized proteins, enzymes and cells. *In* "Methods in Enzymology" (K. Mosbach, ed.), Vol. 137, pp. 444–457. Academic Press, San Diego.

Chang, T. M. S. (1992). "Blood Substitutes and Oxygen Carriers." Dekker, New York.

Chang, T. M. S. (1995). Artificial cells in biotechnology. *Biotechnol. Ann. Rev.* **1**, 267–296.

Chang, T. M. S., Reiss, J., and Winslow, R. (eds.) (1994). Blood Substitutes. *Artificial Cells Blood Substitutes Immobilization Biotechnol.* **22**, 123–271.

Chang, T. M. S., Bourget, L., and Lister, C. (1995). Amino acids enterorecirculation in depletion of unwanted amino acids using oral enzyme-artificial cells. *Artificial Cells Blood Substitutes Immobilization Biotechnol.* **25**, 1–23.

Goosen, M. (ed.) (1993). Fundamentals of Animal Cell Encapsulation and Immobilization. CRC Press, p. 326.

Gregoriadis, F. (1989). "Liposomes as Drug Carriers: Recent Trends and Progress." Wiley, New York.

Praskan, S. and Chang, T. M. S. (1995). Microencapsulated genetically engineered *E. coli. J. Biotechnol. Bioeng.* **46**, 621–626.

# Artificial Heart

WILLEM J. KOLFF
*University of Utah*

## GLOSSARY

**Artificial heart** Pump to substitute for the function of the natural heart

**Cardiac output** Amount of blood pumped out by the left ventricle, usually expressed in liters per minute

**Compliance sac** Elastomeric sac that takes care of the displaced volume of air, gas, or liquid, usually inside the chest

**Left ventricular assist device** Pump to substitute for the left ventricle, used primarily outside the body

**Percutaneous** Going through the skin with a tube or wire (e.g., to provide energy for artificial heart)

**Pulmonary artery–pulmonary artery shunt** Allows testing of a ventricle without damaging the subject

**Right ventricular assist device** Pump to substitute for the right ventricle

**Roller pump** Flexible tube compressed by rollers to pump blood

**Starling's Law** Level of venous return determines cardiac output

**Systole/diastole** Contraction and relaxation (filling) of the heart

**Thrombosis** Clot formation during life, which becomes an embolus when let loose

**Transcutaneous** Going through the intact skin for either energy or signals to or from the artificial heart

IN THE UNITED STATES ALONE, 35,000 PEOPLE WILL die from irreparable heart disease each year. Attempts are being made to help these people with either left and right ventricular assist devices (LVADs, RVADs), total artificial hearts (TAH), or heart transplantations.

For the next few years, clinically used artificial hearts will be driven by compressed air, which enters the body through tubes and special skin buttons that reduce the risk of infection. The artificial heart is presently used mainly as a bridge to donor heart transplantation.

The restoration of the patient in critical heart failure once the artificial heart is implanted can be dramatic. The number of donors for heart transplantation is limited (at most 2000 per year in the United States), so that gradually more and more patients will be discharged from hospitals with artificial hearts.

ENCYCLOPEDIA OF HUMAN BIOLOGY, Second Edition, VOLUME I. Copyright © 1997 by Academic Press. All rights of reproduction in any form reserved.

A totally implantable artificial heart would be ideal, but the public is not prepared for an atomically driven artificial heart, which could be developed. Fortunately, enough energy can be transmitted through the intact skin to drive an artificial heart. The artificial heart is then driven electromechanically (usually a pusher-plate mechanism) by either electromagnetic or electrohydraulic means. Artificial hearts can be controlled by Starling's law (i.e., venous return determines the cardiac output).

Durability is improving, and as long as totally implantable artificial hearts are not yet available, patients can be provided with air-driven artificial hearts. Elastomeric valves patterned after the natural aortic valves are being introduced. The occurrence of thromboembolism is reduced by better blood-flow design and more compatible elastomers.

The cost of implanting an artificial heart (approximately $125,000, surgery and short hospital stay included) is not excessive if it is compared with the cost of slowly dying in a constant care unit.

The author is convinced that a happy, useful life will be possible for a person with an artificial heart.

## I. MECHANICAL SUPPORT FOR THE FAILING CIRCULATION

### A. Balloon Pumps

There are several ways to support the failing circulation system when drugs alone are no longer effective. The most frequently used mechanical support for a failing circulation is the intraaortic balloon pump. This consists of a long, slender balloon in the aorta (the main artery of the body), which is rhythmically inflated between two heartbeats (Fig. 1).

One estimate is that 400,000 intraaortic balloons are used every year in the United States alone. The intraaortic balloon can support about one-third of the circulation, but it is not enough if the heart failure is severe. An interesting variation of the intraaortic balloon pump, the intraventricular balloon pump, has been introduced by Dr. Spyros Moulopoulos in Greece, but as yet it has not been clinically applied.

### B. Left and Right Ventricular Assist Devices

For severe heart failure, LVAD and RVAD pumps can be used (Fig. 2).

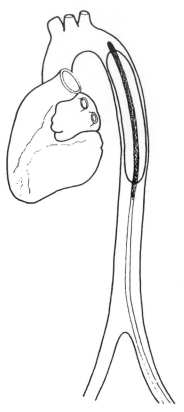

FIGURE 1 Diagram of the intraaortic balloon pump. It is rhythmically inflated between two heartbeats and then expands, pressing the blood out of the aorta. For the next heartbeat, the aorta is found almost empty at a lower pressure, so that the failing left ventricle can empty itself easily into the aorta.

In 45% of the patients, bilateral support is necessary. These LVADs and RVADs are usually aimed at short periods of support. Indeed, they are able to postpone death, and in a few cases the natural heart can recover. Of 85 cases, 13 patients could be discharged from the hospital.

A very small screw-type blood pump has been introduced on a long cable that sits between the left ventricle and the aorta and propels the blood, even if the left ventricle cannot pump at all. It has saved some lives (Fig. 3a).

## II. LVAD OR TOTAL ARTIFICIAL HEART

Over the years, the National Institutes of Health (NIH) has favored the development of intracorporeal LVADs over the development of total artificial hearts,

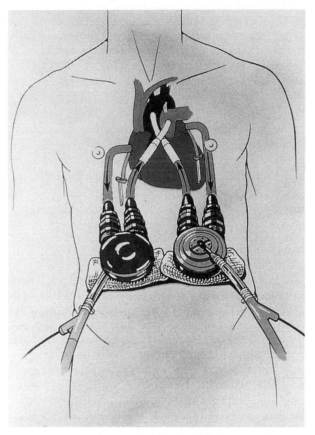

**FIGURE 2** Biventricular assist device of Pierce–Donachy design. Forty-five percent of patients on LVADs require biventricular assist devices.

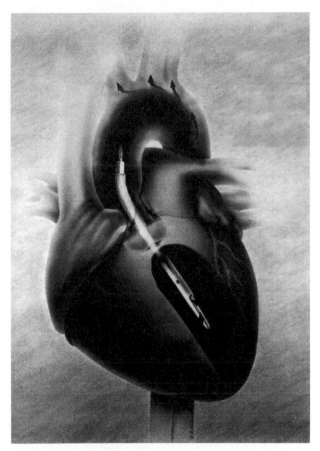

**FIGURE 3a** Richard Wampler, inspired by the Archimedes screw pumps he saw in Egypt, developed this pump. It is inserted on a long rotating shaft from the femoral artery. The Archimedes screw pump goes through the aortic valve and sits between the ventricle and the aorta. The rotor turns at 5000 RPMs. [Reproduced with permission from Johnson & Johnson Interventional Systems.]

although no scientific evidence proves that the LVAD is safer as a support for longer than a few days. An LVAD plus an RVAD connected on either side of the natural heart could leave the natural heart in place so that it might act as a temporary standby in case of failure of the VADs. In case the natural heart recovers, as is possible with acute viral myocarditis, the LVADs and RVADs can later be removed (Fig. 4). LVADs or TAH are used as a bridge to transplantation. In 1994, the survival rate of patients who had received LVADs, RVADs, or TAH were virtually the same.

## A. Extracorporeal LVAD

The most used LVAD is the Pierce–Donachy pump. It is an air-driven reciprocating pump (see Fig. 2) and uses Medtronic valves. Dr. William Pierce has indicated from the onset that one should be prepared to do a biventricular assist with this LVAD. It is sold

by Thoratec (about \$12,000). Of 151 patients who were supported, 98 received a transplant and 82 survived.

A variation of the LVAD by Abiomed is attached to a I.V. pole placed next to the bed; by pumping it higher or lower, the amount of suction can be varied.

W. J. Kolff has made a variation of this with a collapsible atrium which he first used back in 1949. It prevents the aspiration of air when the blood pump "sucks in" too hard. These pumps are less expensive because they use elastomer valves.

Abiomed has introduced the bedside LVAD which is attached to an I.V. pole placed next to the patient's bed. A variation is shown in Fig. 3b. A very simple extracorporeal, magnetically activated drive system for air-driven LVADS is made by Dr. Robert L. Whalen at Whalen Biomedical, Inc.

**BABY EXTACORPOREAL L.V.A.D.**

**WITH ATRIUM**

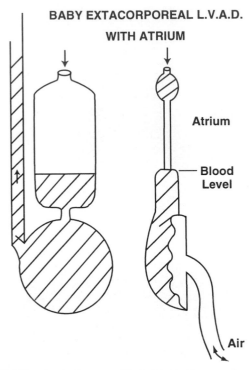

Atrium

Blood
Level

Air

FIGURE 3b   An extracorporeal baby LVAD (the same design has been made for the 40-ml LVAD). A collapsible atrium indicates the level of the venous inflow from the patient. The entire aggregate can be moved up and down on a pole placed next to the bed. If, for some reason, the blood flow is reduced or stopped, no blood can be aspirated through the collapsed atrium, which offers protection against the aspiration of air or "sucking in" of the atrial wall against the cannula.

## B. Simple Roller Pumps

Simple roller pumps, commonly used for blood oxygenators during open-heart surgery, are now being used, apparently without approval from the Food and Drug Administration (FDA). Combined with an oxygenator, they can be used between a femoral vein, for the outlet, and a femoral artery, for the return of the blood. These systems have limited use in support of a failing heart but are spectacularly lifesaving for newborn babies who, for some reason, fail to breathe on their own. Indeed, some clinics do not allow newborn babies to die from respiratory failure without having tried the ECMO (Fig. 5).

A heart/lung machine for babies has been designed and prototyped by Stephen R. Topaz for DLP, Inc., in Grand Rapids, Michigan. It has a rigid blood volume of 41 ml, a built-in heat exchanger, pressure-controlled diaphragm pumps, remote-controlled tubing clamps, and a flow capacity of 1.51 liter/min (Fig. 5). This device is meant to be placed next to the baby/small child, not on the floor.

## C. Centrifugal Extracorporeal Pumps

Centrifugal extracorporeal pumps are becoming more popular as short-term cardiac assist pumps because they are relatively safe and inexpensive (disposable blood pumps ±$300). They cannot rupture a tube if the outlet is occluded, and they spin ineffectively when air comes into the system. There are several types: (1) Medtronic-Sarns, (2) Biomedicus (Fig. 6), and (3) Centrimed. Because these pumps do not have valves, if they stop, reverse flow may occur from the artery to the oxygenator. The siphon created by the column of blood in the arterial line causes aspiration of air

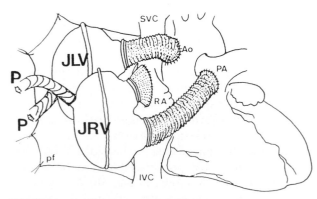

FIGURE 4   Arrangement (suggested by Dr. Jacques G. Losman) with both an LVAD and an RVAD located in the right side of the chest. When the heart recovers, they can be removed.

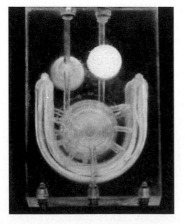

FIGURE 5   The prototype of the pump oxygenator using baby ventricles. One pump is seen with its driveline; the oxygenator is "U"-shaped and consists of capillary fibers through which O₂ is "sucked." The essential advantage of this baby pump oxygenator is its very small unchanging blood volume.

FIGURE 6  Example of an centrifugal extracorporeal pump.

into the aorta, which has killed patients. This can be prevented by a simple check valve placed in the arterial line.

## III. IMPLANTABLE LVADs

Because the NIH funded five implantable LVADs, they are highly developed. The one developed by Dr. Peer Portner of Novacor has reached clinical application (Fig. 7).

All totally implantable LVADs and RVADs are still dependent on energy from the outside (see later). Because of their complexity, the cost of the blood pump inside the chest is between $12,000 and $15,000 and the driving console is around $50,000.

A much simpler approach to an implantable LVAD is one developed in Florida. Its driving mechanism has no moving parts other than a permanent magnet. The magnetic flux lines are controlled so that a large excursion of the magnet is possible (Fig. 8).

A flexible wall or compliance sac must be provided when a pusher plate moves in systole of the artificial ventricle because it creates a vacuum behind it if no compliance is available.

## IV. MUSCLE-POWERED LVADs

In dynamic cardiomyoplasty, the musculus latissimus dorsi is wrapped around the ailing human heart with great care so that neither the blood supply nor the nerve is damaged. The nerve is stimulated with a kind of pacemaker that gives a burst of stimulation (made by Medtronic, Inc.). In about 8 weeks, the skeletal muscle adopts the capacity of a heart muscle so that it can contract 60 times per minute without becoming tired. Cardiomyoplasty is being done in several medi-

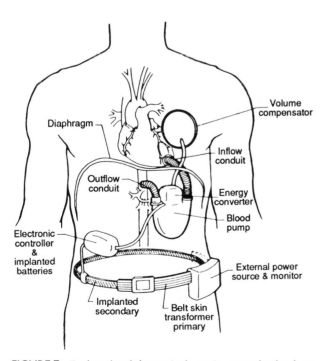

FIGURE 7  Pusher plate left ventricular assist pump that has been clinically applied.

FIGURE 8  A magnetically driven left assist pump developed by Dr. Stephen G. Kovacs. It is the most simple design available.

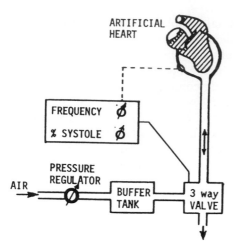

**FIGURE 9a** Schematic for the most used drive system for artificial hearts (only one ventricle is shown). The three-way valve should provide easy outflow of the air to the atmosphere so that the filling of the ventricle with blood during diastole is unencumbered.

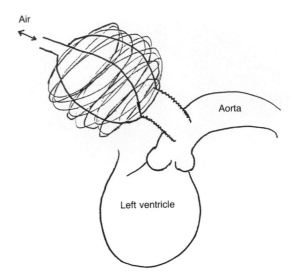

**FIGURE 9b** A pouch is connected with the root of the aorta. It has an air (or rather $CO_2$) chamber inside that can be inflated in counterpulsation like an intraaortic balloon pump, but it is much larger (uses the same drive system, e.g., Datascope Driver, which is available in all major hospitals). Pumping can begin on the operating table and thus the patient should be out of heart failure immediately. The musculus latissimus dorsi is then wrapped around the pouch and stimulation of its nerve can begin sometime later. After 8 weeks, the muscle is trained and the pneumatic power can be discontinued. The $CO_2$ chamber can be partially filled to reduce dead space in the blood pump and the $CO_2$ line can be buried under the skin. The patient is then tether free.

cal centers around the world. Many heart patients have benefited from this procedure but it is a serious operation with considerable mortality and it takes 8 weeks before any benefit is determined. It cannot be undone. Dr. Larry Stephenson and Dr. W. J. Kolff are applying muscle power in combination with pneumatic power. One way in which this can be done is described in Fig. 9b.

## V. THE TOTAL ARTIFICIAL HEART

### A. Progress

In 1987, 46 patients were treated with air-driven artificial hearts while they were waiting for a transplant. Of those 46 patients, 36 left the hospital with functioning transplants—an excellent record when one considers that all these patients were on the verge of death when treatment began.

### B. Drive Systems Outside the Chest (Fig. 9a)

At the time of this writing, practically all drive systems of artificial hearts are driven with compressed air either by a large, cumbersome, but safe, console or by a small, portable, battery-powered drive system. The most sophisticated extracorporeal drive system is the Heimes' Driver (Fig. 10). This system allows recipients to maintain reasonably normal lives.

Having the drive system on the outside of the chest is advantageous in that it can be replaced or repaired so that the demands on durability are far less. Moreover, the weight of an air-driven pump inside the chest is much less than that of a mechanically driven artificial heart. Drive systems need not be very complicated or expensive. A very simple reciprocating pump is shown in Fig. 11.

A four-phase gasoline engine driven by an electromotor can be used as a pump by changing the camshaft that lifts the valves. Outlet valves open one-fourth for systole; inlet valves open three-fourths for diastole. The driveline is connected to the hole for the spark plug (Fig. 12).

## VI. THE BLOOD PUMP INSIDE THE CHEST

Most air-driven artificial hearts closely resemble each other. Presently, they are used mainly as a bridge toward transplantation and they are expensive. An

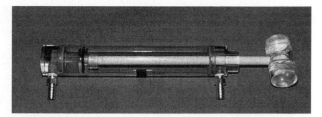

FIGURE 11 A very simple reciprocating drive pump. The positions of the holes in the cylinder determine the ratio between systole and diastole.

is used for only a few days or even a few weeks while the patient is waiting for a donor heart transplant.

## VII. SURGICAL IMPLANTATION

A patient who needs an artificial heart is usually connected to a heart–lung machine or extracorporeal pump oxygenator. The blood is taken with two cannulas from the vena cavae (the largest veins in the body) and is put into an oxygenator where the $CO_2$ is removed and $O_2$ is added. The bright red oxygenated blood is pumped back into the aorta, and the tissues never know where the blood came from. The patient's own lungs, with nothing to do, are from time to time slightly inflated by pumping air into the trachea. [*See* Heart–Lung Machine.]

With the circulation provided for, the patient's heart can be cut out and discarded. With most currently used artificial hearts, short artificial aorta, pulmonary artery, and wide atrial cuffs are sutured to their natural counterparts. This is not difficult because the chest is already wide open, usually via a sternal split.

The suture lines are then tested for leaks. The cuffs have the female part of the quick connects, and the artificial ventricles have the male parts. The ventricles are snapped in place. The famous medical illustrator (Netter[1]) shows an artificial heart after implantation in the human chest in Fig. 16. All connections are completed and the heart has been snapped in. The artificial heart is now functioning and the bypass tubes have been removed.

Unfortunately, the quick connects have proven to be a source of thromboemboli. Heart transplant sur-

FIGURE 10 Leif Stenberg of Sweden, wearing the portable drive system that drives his artificial heart, is returning from a restaurant where he has just served himself from a smorgasbord. He has an air-driven artificial heart inside his chest, and the portable Heimes' Driver is suspended from his shoulder. Neither the waiters nor the customers realized that he had an artificial heart. He sent a telegram to the United States: "I am the happiest man in Europe!"

inexpensive, polyurethane artificial heart is now being made at the University of Utah (Figs. 13–15). The university has also developed inexpensive polyurethane tricusp semilunar valves (see later also in Fig. 20). It is hard to justify the expense of using four commercially available valves in an artificial heart costing between $2000 and $4000 each if that heart

---

[1]Netter came out of retirement to do what he described as his "most important work."

FIGURE 12   A two-cylinder gasoline engine energized by an electromotor can be used to drive an artificial heart. The driveline to the artificial heart comes from the hole for the spark plug.

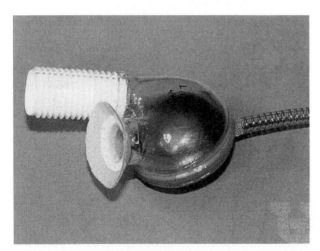

FIGURE 13   Philadelphia-type I artificial heart (left ventricle). The quick connects were eliminated to help avoid the occurrence of thromboembolism. In the Philadelphia-type I model, the aorta and the atrium are directly connected to the ventricle, and the surgeon can, at the operating table, sew in the valve of his choice, even if he wants to use a tissue valve. However, the Food and Drug Administration did not approve this and required the manufacturer to insert the valves before the artificial hearts are delivered.

FIGURE 14   Another view of the Philadelphia-type I artificial heart (right ventricle). These ventricles are called "Philadelphia-type" because Dr. Jack Kolff in Philadelphia designed the soft compressible ventricles and implanted them successfully in calves. The FDA approved two human implantations.

**FIGURE 15** The Philadelphia-type II artificial hearts do not have quick connects either. They contain tricusp semilunar valves, which are fixed into the inflow and outflow ports at the factory. The ventricules are soft and the figure shows 20-, 50-, 65-, and 85-cc ventricles. Note that the ports are provided with sinus valsalvae. These half-circular cavities on the outflow side of each cusp make a revolving blood flow possible to wash the backside of the valve.

geons feel that the quick connects would not be needed if the artificial ventricles could be made soft and pliable. The Philadelphia-type artificial hearts (Figs. 13–15) are soft and pliable and can be easily inserted.

## VIII. TESTING THE ARTIFICIAL HEART SYSTEM

The various compartments of the artificial heart can be tested separately; however, one needs to know the function of the entire system. This is done in our so-called mock circulation, which is totally self-regulating and also demonstrates the shifts in blood volume that occur when there is an imbalance between the left and the right ventricles. When this occurs in a patient, particularly when the blood is shifted from the large systemic space to the much smaller pulmonary space, the results can be overloading of the pulmonary circulation with pulmonary edema as a result. This can be deadly in a short period of time.

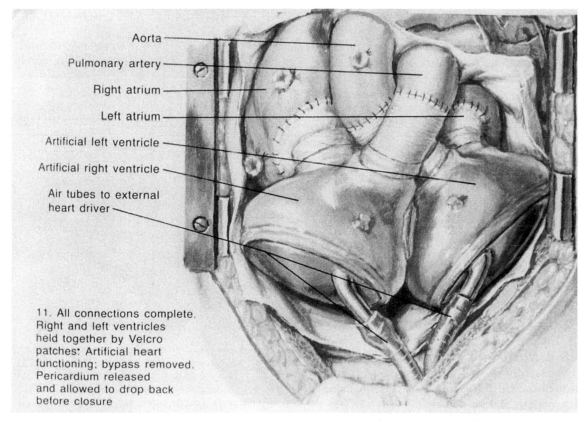

**FIGURE 16** Illustration of the chest wide open after implantation of an artificial heart. The connectors have been sewn to the aorta, to the pulmonary artery, and to the atria. The left ventricle was snapped in first and then the right ventricle. The clamps have been removed, and the artificial heart is pumping. [Drawing by D. Netter. Reproduced from W. C. DeVries and L. D. Joyce, CIBA.]

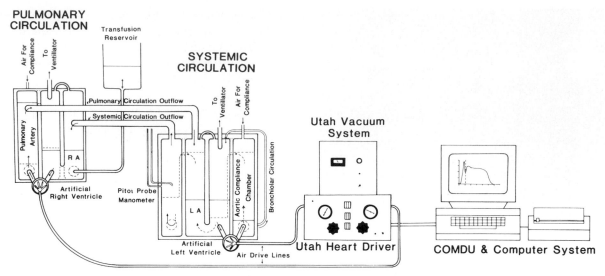

**FIGURE 17** The mock circulatory system has two series of water cascades to represent the systemic and pulmonary circulations. They provide the typical physiologic afterload pressures to the ejecting artificial ventricles. Air chambers in the cascade provide compliance. With its transparent walls, the mock circulatory system vividly displays the flow movement and balance of fluid volume between the lesser (pulmonary) and greater (systemic) circulations. Many of the circumstances that arose with the first human recipient can be demonstrated with this system. Other pathologic states can also be created for the education of physicians.

A cascade automatic mock circulation is described in Fig. 17. The most common test for an artificial heart is to draw a Starling's curve (Fig. 18). The cardiac output (i.e., the amount of blood pumped per minute) is accurately registered with the cardiac output monitor and diagnostic unit (COMDU) (Fig. 19); this information is advantageous for cardiologists who, with seriously ill patients, can only guess at cardiac output unless they catheterize the patient.

## IX. THROMBOSIS

The primary fear of the use of artificial hearts is that they will give rise to thrombosis, and the thrombus may get loose and form an embolus. If the embolus goes to the brain, the patient has a stroke. Indeed, the lives of four of the first five patients treated with air-driven artificial hearts were severely debilitated by strokes caused by emboli to the brain.

The Philadelphia-type artificial heart avoids the use of connectors because the connectors and mechanical valves are the main source of thrombosis and then emboli. Tissue valves are virtually free of thrombosis (Fig. 20). We are trying to match that with elastomer valves (Fig. 21). To avoid thrombosis, the flow pattern inside the artificial heart should minimize turbulence, high sheer stress, and dead or stagnating areas.

## X. INTIMA OF ARTIFICIAL HEARTS

The best surface for the inside of an artificial heart is the subject of various and abundant research.

### A. Biolyzed Surfaces

Using carefully prepared pericardium or gluteraldehyde-fixed gelatin for the inner lining of the artificial

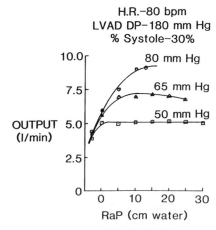

**FIGURE 18** Function curves of the artificial heart. A classic experiment is the development of a Starling's curve of the artificial heart by increasing the filling pressure, which results in increased cardiac output: DP, driving pressure; H.R., heart rate; RaP, Right atrial pressure. This artificial heart had a maximum stroke volume of 100 ml. The figures 80, 65, and 50 mm Hg indicate the drive pressure for the right ventricle: 50 mm Hg is obviously not enough.

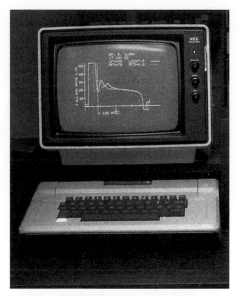

FIGURE 19 The COMDU uses the air expelled during diastole to calculate cardiac stroke volume and cardiac output. It is accurate to 10% and is noninvasive because it is totally outside the body.

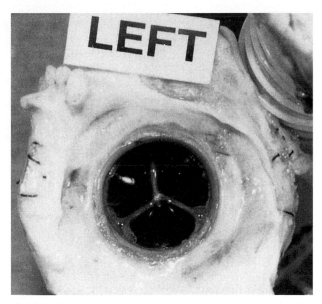

FIGURE 21 Tricusp semilunar valve (an elastomer valve) after having pumped for 62 days in an artificial heart in a calf, shown here from the atrial side. It is virtually free of thrombosis.

heart and for the valve prevents the formation of blood clots. With this method, anticoagulants are not needed. The drawback is that this surface is hard to make and perhaps some calcification will occur later.

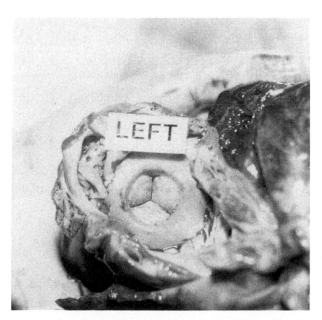

FIGURE 20 A tissue valve, which was in the inflow port of an artificial heart, is shown from the left atrium 3 days after insertion. It is free of thrombosis. These valves were made by Sorin Biomedica in Italy.

## B. Smooth Polymer Surfaces

Polyurethane, polyurethane-siloxane, silicone rubber without silica at the surface, or highly polished methylmethacrylate can be made so smooth that very few thrombocytes (blood platelets) will adhere to it. When they do, the thrombocytes are swept away before the clusters become too large. Dr. Jack Kolff and his associates in Philadelphia have examined the inside of a polyurethane artificial ventricle with electron microscopy and have found no difference in appearance between 4 hr and 4 weeks postoperation. We presently make polyurethane ultrasmooth by painting the surface with dimethylacetamide.

## C. Rough Surfaces

Dacron fibrils, Dacron velours, or a composite of Lycra spandex and nylon are covered with fibrin that becomes a smooth, glistening surface, to which thrombocytes do not adhere. Sometimes the surface is covered with endothelium but not usually. Although Dacron artificial aortas show a raw surface to the bloodstream, they apparently do not cause trouble. On nonmoving surfaces, a surface consisting of tiny titanium beads is used.

## D. Anticoagulants

The coating of surfaces with heparin or prostaglandin PgE has been attempted by many researchers and has produced no consistent advantages. Dr. Chisato Nojiri, at Terumo Corp. in Japan, has coated our artificial ventricles with heparin after pretreatment with ozone. The occurrence of thrombosis in the blood bag test, which was already low, was further reduced. The administration of anticoagulants to the recipients, either heparin or Coumadin (Warfarin) and aspirin (which inhibits platelet aggregation), is widely used and is mandatory when mechanical valves are used.

## XI. *EX VIVO* AND *IN VIVO* TESTING

To test for thrombogenicity, a real implantation is the ultimate test, but it is very expensive. The pulmonary artery–pulmonary artery shunt (PAPAS) offers the possibility of testing an artificial ventricle with valves outside the body of a calf without doing damage to that animal. Thus, a series of artificial ventricles can be tested, removed, and studied (Fig. 22).

To provoke thrombosis, we use our blood bag test system. Heparinized blood is pumped by an artificial ventricle in a blood bag while the heparin is neutralized with protamine sulfate. Consequently, thrombosis occurs at places where it would occur if the artificial heart were in an animal or a patient (Fig. 23).

There is no way around *in vivo* testing. Animals of adult human size must be used; thus, we use adult sheep and calves. The surgical approach is *not* via a sternal split because the animals normally lie on their sternum, but through the right chest wall. The first

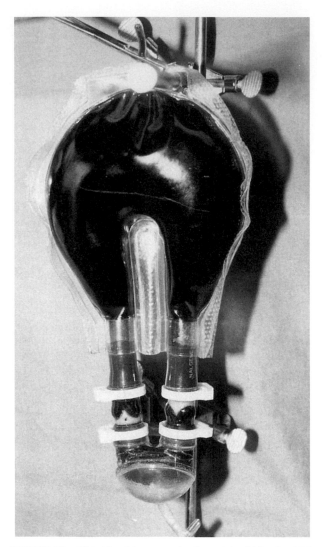

FIGURE 23 This blood bag contains heparinized blood, which is pumped around in a blood sac while the heparin is slowly neutralized by protamine sulfate. This procedure provokes thromosis which will occur in certain places. It indicates what changes in design or surface must be made. The artificial heart with the valves is at the bottom of the photograph. Courtesy of the Laboratory of Dr. Fazal Mohammed, University of Utah.

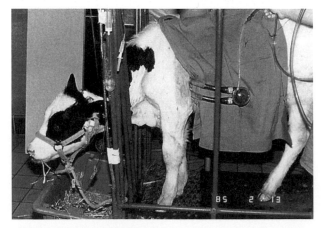

FIGURE 22 The PAPAS system shows an artificial heart connected between two cannulas in the pulmonary artery of a calf, which allows a series of artificial hearts to be tested in the same calf without discomfort or damage to the animal.

rule of such testing is kindness and excellent care for the animal (see Color Plate 4 and Fig. 24).

Blood trauma is damage to the blood caused by an artificial heart. It is most easily expressed in the percentage of free plasma hemoglobin (HB), which derives from broken red blood cells. It should never be >5 mg/100 ml of plasma. Rough surfaces on the intima give higher free HB in the beginning. Blood damage is tested using the methods described for thrombosis; thrombocytes and leukocytes are also counted. Complement, a blood component, may be

FIGURE 24   The barn at the University of Utah's Artificial Heart Research Laboratory where the calves with artificial hearts are taken care of. The drive systems are placed on top of the cages and can be lifted with a crane when the animal does its regular exercises on the treadmill.

somewhat activated but without the grave consequences as seen in organ transplantation. There is no immune rejection since there is no foreign protein in artificial hearts.

## XII. INFECTION

Infection is another reason for concern: air drivelines offer an opportunity for the entry of microorganisms into the body. However, more sophisticated skin buttons coated with velour have been developed in the hope that the skin will attach firmly to the velour, forming a barrier to infection (Fig. 25).

In many cases, infection of the drivelines does not happen.

## XIII. POWER SOURCE INSIDE THE BODY

Ideally, the power plant for the artificial heart would be inside the body, but this can really only be accom-

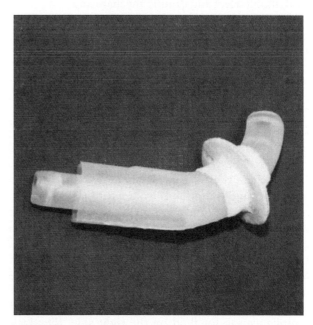

FIGURE 25   A skin button produced by Stephen and Peter Topaz, which allows the passage of the drivelines of the artificial heart through the skin. It is provided with a flange and Dacron velour to prevent the propagation of infection along the driveline.

plished with an atomically powered artificial heart. Indeed, the Atomic Energy Commission (which later became ERDA and then evolved into the Department of Energy) financed the development of artificial hearts driven by atomic power. In 1977, a calf survived for 37 days with a prototype of such an artificial heart. The artificial heart was not driven by an atomic hot finger, but by an electric motor (see Color Plate 5). Ten years later, that record for a mechanically driven heart was broken in the laboratory of Dr. William Pierce at Hershey, Pennsylvania.

The NIH has continued the development of the Stirling engine, a hot-air engine to power artificial hearts that ultimately should be powered by an atomic heat source. These thermal engines have become very small and reliable and can be linked with hydraulic or mechanical connects to the blood sacs that pump the blood. Ideally, it should power axial flow pumps suspended in the bloodstream by magnetic forces, but none are being seriously pursued at this time.

Purely electromechanical drive systems are usually heavy and cumbersome. Perhaps the simplest (but still heavy) systems are the magnetically driven artificial ventricles, such as that shown for the LVAD earlier in Fig. 8.

The blood handling elastomeric pumps are basically the same as those for the pneumatically driven heart; the external pneumatic drive is replaced by a totally implantable electrohydraulic mechanism.

**FIGURE 26a** Instead of an air-drive system, the ventricles can have an electrohydraulic pump placed in between then (S. Topaz); this pumps saline alternately from left to right and then right to left. (The ventricles in the picture can also be driven with compressed air.)

## XIII. THE ELECTROHYDRAULIC ARTIFICIAL HEART

Stephen R. Topaz, in Kolff's Laboratory at the University of Utah, has placed a small flat motor with a hole in the center between the right and left ventricles (Fig. 26a). An impeller sits inside the hole which pumps hydraulic fluid (not blood) from left to right and vice versa. The motor (and the impeller) is guided by back electromotive force. The hydraulic fluid is saline (no danger if leaking should occur). The impeller blades are designed by NASA engineers who concentrate on small pumps (Fig. 26b). The efficiency is high and the size is small. The difference between right and left cardiac output (left is larger) is provided by a small compliance sac for fluid on the right side. (Dr. Jack Kolff proved years ago that further protection against imbalance is provided by a small hole in the atrial septum.)

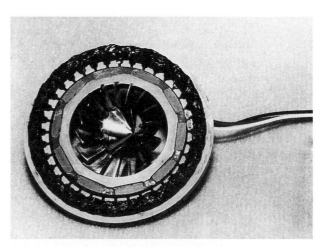

**FIGURE 26b** Electrohydraulic motor with hole in its center (for further details see text).

## XV. PERCUTANEOUS (WIRES) OR TRANSCUTANEOUS SUPPLY OF ENERGY

The energy to drive the heart must still enter the body. This can be done easily with electrical wires, or the energy can be transmitted through the intact skin (Fig. 27).

## XVI. CONTROL OF ARTIFICIAL HEARTS

Because 87% of our normal heart action is guided by Starling's law, the control system for the artificial hearts developed in the author's laboratory is based on this law. When the body exercises, more blood returns to the heart and then is automatically pumped out. When the body, which still has its neuromuscular blood vessel control, sends less blood to the heart, it automatically pumps less (see Fig. 18).

## XVII. DURABILITY

The question of durability of the artificial heart is difficult. A required durability of 10 years may be too optimistic for the near future. To quote from Dr. Dwight Harken, a Boston surgeon, "A device is safe when it is safer than the disease it corrects and is the best available." An artificial heart is applied only in people whose life expectancy is otherwise counted in hours or days.

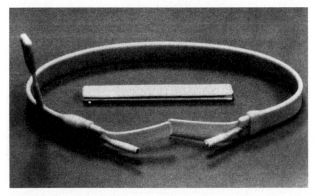

**FIGURE 27** Belt skin transformer that transmits energy for the artificial heart through the intact skin. The primary coil is like a belt around the middle. The secondary coil is encapsulated in silicone rubber and is under the skin.

Artificial hearts are first tested in mock circulations under realistic flow and pressure conditions and in real time (Fig. 28). There are also accelerated test systems. One cannot extrapolate their results to real time, but we have found that the mode of failure in diaphragms and valves is often the same in accelerated testing and in real time testing. Therefore, both systems are used.

## XVIII. MATERIALS

At present, all artificial hearts are made of polyurethanes, although the use of silicone rubber is being reintroduced. All polyurethanes change with age, whereas silicone rubber does not. Silastic HP-100 (Dow Corning) is most promising.

## XIX. ARTIFICIAL HEARTS AS A BRIDGE TO TRANSPLANTATION AND THEIR EXPANSION

Currently, the artificial heart is used almost exclusively as an interim or as a bridge while awaiting a donor for a heart transplant. Many surgeons have witnessed patient's dying from heart failure (who had secondary failure of their lungs, kidneys, and liver; who were too sick to brush their teeth; and who were filled with edematous fluid) recover within 2 or 3 days. But the waiting time for donor hearts is becoming longer and longer. In Philadelphia, for example, the waiting period averages 4 months and one-third of the patients die before a donor heart becomes available. Thus, it is likely that more and more patients, sustained with artificial hearts, waiting for a donor heart, after 3 or 4 weeks will ask the doctor, "Can I go home?" Indeed, they feel perfectly normal.

## XX. THE NEED FOR ARTIFICIAL HEARTS

In the United States, 35,000 people per year are estimated (probably conservatively) to be in need of replacement of their irreparably sick hearts. At most, 2000 human donor hearts will be available for transplantation, leaving 33,000 people with a choice between death or the use of an artificial heart. However, before an artificial heart, or any artificial organ, is applied, we should ask ourselves, "Will this device

FIGURE 28    A room full of mock circulations for the left side (which is the high pressure side). Each of them has an artificial ventricle with inlet and outlet valves. These hearts pump day and night to test durability.

restore the patient to an acceptable, if at all possible, happy existence?" If it only prolongs misery, it should not be applied. If there is less than a reasonable chance that the patient will ever return to a happy existence, it should not be applied.

## XXI. THE COST TO SOCIETY

The author of this article will not accept the notion that the United States is unwilling to spend money to save its fellow citizens in need. If the cost of an artificial heart would be the same as for a heart transplant (approximately $120,000), then that cost should be compared with the cost of dying slowly and miserably in an intensive care unit. Many patients return to these care units several times before the end finally comes, and the families will tell you that the cost may well exceed $120,000.

I believe that the artificial heart can restore the recipient to a worthwhile and happy existence.

## XXII. DR. BARNEY CLARK

In December 1982, Dr. Barney Clark volunteered to be the first patient to receive a "permanent" artificial

heart (see Fig. 29). Dr. Clark lived for 112 days and died from a fulminating infection of his colon. There was no infection of his drivelines and none inside or outside his artificial heart. We already knew that the

FIGURE 29    Photograph of Dr. Barney Clark who was the first recipient of a permanent artificial heart in December 1982. We remain indebted to him forever and to his wife, Una Loy Clark, for their contribution. We cannot say when, but it seems certain that hundreds of thousands of recipients of future artificial hearts will benefit from their courage.

artificial heart could sustain the circulation from our work in animals, but we learned from Dr. Clark that (1) the artificial heart causes no pain or discomfort in the chest, (2) the noise of the drive system did not bother him, (3) his sense of humor remained intact, (4) he still loved his family and, (5) his desire to help and to serve humanity remained strong: all essential functions of the human mind were preserved.

## BIBLIOGRAPHY

DeVries, W. C., and Joyce, L. D. (1983). The artificial heart. *Ciba Clin. Symp.* **35**(6), 1–21.

Farrar, D. J., and Hill, J. D. (1993). Univentricular and biventricular thoratec VAD support as a bridge to transplantation. *Ann. Thorac. Surg.* **55**, 276–282.

Kolff, W. J. (1983). Artificial organs—Forty years and beyond. *Trans. Ann. Soc. Artif. Intern. Organs* **XXIX**, 6.

Kolff, W. J. (1987). The future of artificial organs and of us all. *In* "Artificial Organs" (J. D. Andrade, J. J. Brophy, D. E. Detmer, S. W. Kim, R. A. Normann, D. B. Olsen, and R. L. Stephen, eds.). VCH Publishers, Inc., New York.

Kolff, W. J. (1988). The Tenth Hastings Lecture: Experiences and practical considerations for the future of artificial hearts and of mankind. *Artif. Organs* **12**(1), 89.

Kolff, W. J., DeVries, W. C., Joyce, L. D., Olsen, D. B., Jarvik, R. K., Nielson, S., Hastings, L., Anderson, J., Anderson, F., and Menlove, R. (1983). Lessons learned from Dr. Barney Clark, the first patient with an artificial heart. *In* "Progress in Artificial Organs—1983" (Y. Nose, C. Kjellstrand, and P. Ivanovich, eds.). ISAO Press, Cleveland.

Kolff, W. J., and Lawson, J. (1975). Status of the artificial heart and cardiac assist devices in the United States (special address). *Trans. Am. Soc. Artif. Intern. Organs* **XXI**, 620.

Kolff, W. J., and Stephenson, L. W. (1993). Total artificial hearts, LVADs or nothing? And muscle and air-powered LVADs. *In* "Heart Replacement, Artificial Heart 4" (T. Akutsu and H. Koyanagi, eds.), pp. 3–11. Springer-Verlag, Tokyo.

Moulopoulos, S. D., Topaz, S. R., and Kolff, W. J. (1962). Extracorporeal assistance to the circulation intraaortic balloon pumping. *Trans. Am. Soc. Artif. Intern. Organs* **VIII**, 85.

Pennington, D. G., Kanter, K. R., McBride, L. R., Kaiser, G. C. Barner, H. B., Miller, L. W., Naunheim, K. S., Fiore, A. C., and Willman, V. (1988). Seven years' experience with the Pierce-Donachy ventricular assist device. *J. Thorac. Cardiovasc. Surg.* **96, 901.**

Portner, P. M., Jassawalla, J. S., Oyer, P. E., Chen, H., Miller, P. J., LaForge, D. H., Ramasamy, N., Lee, J., Billich, J., Beering, F. K., Conley, M. G., Sohrab, B., Ryan, M., Daniel, M. A., Strauss, L. R., Brugler, J. S., Ream, A. K., and Shumway, N. E. (1988). Ventricular assist devices. *In* "ASAIO Primers in Artificial Organs" (A. Kantrowitz, ed.), Vol. 3, p. 57. J. B. Lippincott Co., Philadelphia.

Rosenberg, R., Snyder, A. J., Landis, D. L., Geselowitz, D. B., Donachy, J. H., and Pierce, W. S. (1984). An electric motor-driven total artificial heart: Seven months survival in the calf. *Trans. Am. Soc. Artif. Intern. Organs* **XXX,** 69.

Smith, L. M., Olsen, D. B., Sandquist, G., Crandall, E., Gentry, S., and Kolff, W. J. (1975). A totally implantable mechanical heart. *In* "Proceedings of ESAO" (E. S. Bucherl, ed.), Vol. II, p. 150. Westkreuz-Druckerel und Verlag, Berlin.

Smith, L., Sandquist, G., Olsen, D. B., Arnet, G., Gentry, S., and Kolff, W. J. (1975). Power requirements for the A.E.C. artificial heart. *Trans. Am. Soc. Artif. Intern. Organs* **XXI,** 540.

Verhoef, C., Topaz, P., Topaz, S., Golub, D., Bishop, D., Shelton, A., and Kolff, W. J. (1993). Corrugated diaphragms for adult and baby-size artificial ventricles. *In* "Heart Replacement, Artificial Heart 4" (T. Akutsu and H. Koyanagi, eds.), pp. 93–100. Springer-Verlag, Tokyo.

Yarnoz, M. D., and Kovacs, S. G. (1983). Magnetically actuated LVAD. *Trans. Am. Soc. Artif. Intern. Organs* **XXXIX**, 574.

# Artificial Kidney

WILLEM J. KOLFF
*University of Utah*

## GLOSSARY

**Cascade membrane plasmapheresis** First separation of some plasma from the blood, and then further treatment of this plasma in a cascade of different filters so that only the undesirable components are removed

**Complement activation** Immunological response of the body to contact with foreign substances. Acute symptoms include chest pain, shortness of breath, headaches, itchiness, nausea, fever, and sometimes shock

**Hemodialysis** Purification of blood by dialysis through a membrane

**Hemoperfusion** Perfusion of blood over a cartridge with sorbent material, usually charcoal or resins, to adsorb substances to be removed from the blood, such as poisons and potassium

**Membrane plasmapheresis** Separation of some plasma from the other components of the blood, using a membrane for the separation, not centrifugation

**Peritoneal dialysis** Purification of blood using the peritoneal membrane by instilling dialyzing fluid into the peritoneal cavity

**Plasmapheresis** Separation of some plasma from the other components of the blood, usually by centrifugation

**Single-needle dialysis** Technique where inflow and outflow from the patient are alternated in a pulsating manner; the only place where to and fro movement of the blood flow occurs is in the needle

WHEN KIDNEYS DO NOT FUNCTION, PATIENTS DIE. Hemodialysis and peritoneal dialysis are designed to compensate for the excretory function of the kidneys and also to correct electrolyte disturbances. Because electrolytes pass through the membranes in both ways, an automatic adjustment takes place when the dialyzing fluid is selected correctly. With the addition of glucose or other molecules, additional water can be extracted from the blood through the osmotic gradient. [*See* Kidney.]

Plasmapheresis is an extension of the dialyzing techniques by which some plasma is removed from the blood. This plasma can be treated in a variety of ways without damaging the patient. The plasma can be led over sorbents or it can be subjected to a cascade of membranes, thereby removing specific molecules and returning everything else to the patient.

Dialysis has become more expensive than it needs to be. There has been and still is a negative incentive to home dialysis because it is less profitable for physicians and dialysis units than in center dialysis. Life and death committees or their modern equivalent—cost containment—should be condemned if the result is that we cannot take care of our fellow citizens in need.

## I. HISTORICAL BACKGROUND

In 1913, Abel, Rowntree, and Turner described the first artificial kidney. It was a machine that exposed

ENCYCLOPEDIA OF HUMAN BIOLOGY, Second Edition, VOLUME 1.   Copyright © 1997 by Academic Press.   All rights of reproduction in any form reserved.

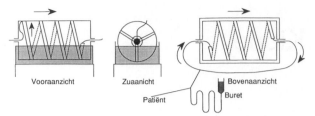

**FIGURE 1**    Diagram of a rotating drum artificial kidney. A cellophane tube is wound like a spiral around a large cylinder. The blood in the cellophane tube always sinks to the lowest point. If the drum rotates, the blood moves from left to right.

blood to dialysis. Dialysis means purification through a membrane; it was first described around 1880 by Thomas Graham in Glasgow, Scotland. Abel, Rowntree, and Turner used collodion tubes as dialyzing membranes. They had to extract the heads of thousands of leeches to obtain the hirudin to prevent the blood from clotting. They described exactly what hemodialysis (the dialysis of blood) would do and used the term artificial kidney. They even treated one patient, but, thereafter, hemodialysis became obsolete until 1943. That year, during the German occupation of The Netherlands, this author treated the first patient, who was in chronic renal failure, with the so-called rotating drum artificial kidney (Figs. 1–3). This was the first device that was clinically useful because it exposed a small volume of blood to a large surface area of dialyzing membrane—in this case, cellophane tubing, which is regenerated cellulose (actually artificial sausage skin). The author used heparin to prevent the blood from clotting.

In this machine, 0.5 liter of the patient's blood was dialyzed. After 1 or 2 days, 1 liter of blood was dialyzed then 1.5, 3, and, finally, 20 liters of blood during one session (Fig. 3). After 12 dialyses, no more blood access ports were available and the patient died.

During the following years, the artificial kidney proved to be useful mainly for the treatment of acute renal failure when the renal damage was reversible. This is the case for patients exposed to nephrotoxic poisons such as antifreeze or bichloride of mercury, for those affected by acute glomerulonephritis with renal shutdown, or with the often fatal kidney failure after criminal abortion.

Around 1960, Dr. Belding Scribner and Mr. Wayne Quinton in Seattle developed the arteriovenous shunt (Fig. 4) which made chronic, repeated dialysis possible and opened the era of effective treatment of patients with chronic renal failure. One cannula was inserted

into the radial artery in the arm and another one in a vein in the underarm. The two cannulas were connected with a Teflon shunt. The flow through this shunt was high enough to prevent the blood from clotting, and, whenever the patient needed dialysis, the shunt was disconnected and the arterial and venous cannulas were connected with the artificial kidney. Presently, it is common to create a connection (fistula) between an artery and a vein in the arm. The vein subsequently distends, and, for each dialysis, one or two catheters are inserted in the distended vein.

## II. PRINCIPLE OF HEMODIALYSIS

In artificial kidneys, three things take place:

1. Removal of small molecules by dialysis (urea, creatinine, uric acid, etc.)
2. Removal of water by ultrafiltration through the membrane
3. Equilibration of electrolytes in the blood plasma with dialyzing fluid

### A. The Dialysate

The electrolytes—ions of sodium, chloride, potassium, etc.—will go through the membrane back and forth, in and out of the blood, whereas larger molecules will stay in. If the concentration of the dialyzing fluid is selected appropriately, the electrolytes of the patients are normalized. For example, if the sodium is too high, it is lowered; if it is too low, it is raised (Fig. 5).

The resistance to dialysis is caused by the membrane that separates blood and dialysate, but also by the boundary layers in contact with the membrane, both in the blood and in the dialyzing fluid. If, for example, the blood-containing tubes have a large diameter, the distance to be traveled by urea molecules to the membrane becomes so large that dialysis becomes prohibitively slow.

To make hemodialysis practical, three things are needed: (1) a small volume of blood outside the body, (2) a large surface area, and (3) a constant movement of blood and dialyzing fluid.

### B. The Membranes

Regenerated cellulose is still the membrane used in most artificial kidneys. One excellent type of cellulose

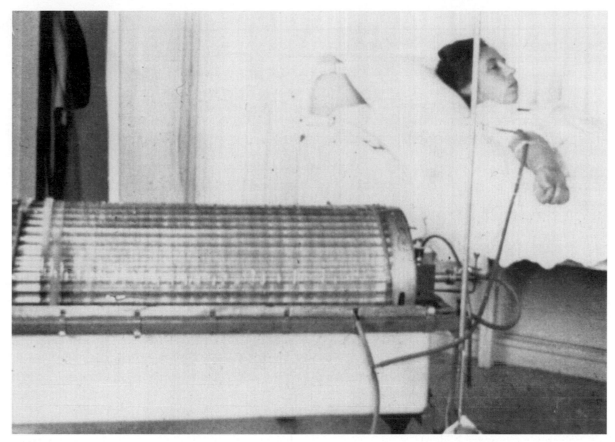

**FIGURE 2** A rotating drum artificial kidney. Although this picture was a simulation with head nurse Sister ter Welle as the would-be patient, it demonstrates how the rotating drum artificial kidney was run in the beginning. It used a buret, not a blood pump, which could be raised or lowered to let the blood in or out of the patient and in or out of the artificial kidney. Twenty meters of cellophane tubing (artificial sausage skin) is wrapped as a spiral around the rotating drum. When the drum rotates, the blood, always seeking the lowest point, runs from left to right through the cellophane tubing. It enters and leaves the drum through the hollow axle. The lower part of the drum rotates through a bath of dialyzing fluid. This was the first artificial kidney with a large enough surface area and a small enough blood volume to make it clinically useful.

is Cuprophane, made by Enka-Glanzstoff in Wuppertal, West Germany.

A number of complications, sometimes severe, have been attributed to "complement activation."

The (−OH) group of regenerated cellulose (linear polymer of D-glucose) is blamed for activating the complement and attempts are being made worldwide to replace it or to come up with other membranes. Some examples include

1. Cuprophan (e.g., Baxter, CF): 8 $\mu$m
2. Cellulose acetate (e.g., Baxter, CA): 70% of (−OH) groups replaced with acetate groups
3. Cellulose triacetate (e.g., Baxter, CT): 90% of (−OH) groups replaced with acetate groups (larger pores: 15 $\mu$m)

Patients who went into shock or had great difficulty breathing when dialyzed with Cuprophane had no symptoms when dialyzed with polyacrylonitrile membranes. Because polyacrylonitrile membranes are much more expensive than a Cuproplane dialyzer, dialysis centers are not earger to use it. A list of some synthetic membranes follows.

1. Polysulfone (e.g., Fresenius F60) Vinyl monomer + sulfur dioxide (vinyl chloride makes PVC) Polysulfone is hydrophobic: add polyvinylpyrrolidone (PVP) Few (−OH) groups
2. Polyacrylonitrile (e.g., Hospal AN-69) petroleum product compare: propylene and "Orlon" Said to activate complement but to bind the toxin

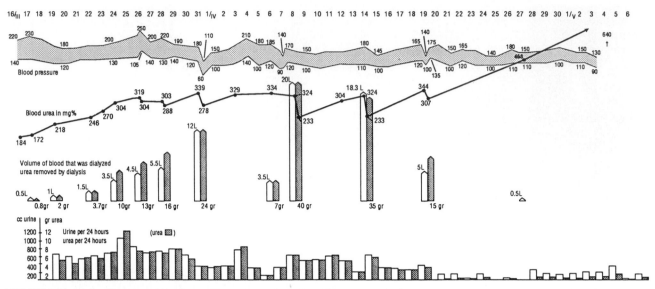

**FIGURE 3** Graph of patient, a 29-year-old woman, suffering from a malignant hypertension with contracted kidneys. She was the first patient to be treated with an artificial kidney. At the top, the systolic and diastolic blood pressures are shown, the difference between them is shaded. The content of urea of the blood is indicated by a line. It decreases after each large dialysis (e.g., from 339 to 278 mg/100 cc). The dialyses are reproduced by columns. The white column indicates the quantity of blood dialyzed; the shaded one shows the quantity of urea removed by dialysis. At the bottom, the quantities of urine passed per 24 hr are represented by white blocks; the quantities of urea excreted with this urine are indicated by shaded blocks.

3. Polymethylmethacrylate (compare plexiglass)
4. PVP-Polyamide (compare Nylon)

The (−OH) groups are blamed but there is no close correlation between the numbers of the (−OH) groups and the symptoms.

The artificial kidneys in use at present ordinarily consist of hollow fibers or artificial capillaries,

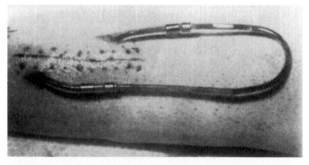

**FIGURE 4** A slight modification of the Scribner–Quinton shunt. A Teflon cannula, with a bend near the tip, is inserted into the radial artery and is connected with a Teflon shunt to another Teflon cannula, which is inserted into a vein. The blood flow through the shunt is fast enough to prevent clotting. During dialysis, the arterial cannula is connected with the inflow line to the artificial kidney, and the blood is returned through the venous cannula.

through which the blood flows, surrounded by the dialyzing fluid (Figs. 6 and 7). There is a trend to replace regenerated cellulose with other membranes that are more permeable, therefore allowing the passing of larger molecules.

## C. Dialysis versus Convection

Whereas dialysis mainly uses diffusion through the membrane, certain artificial kidneys remove substances by convection. They use membranes with larger pore sizes, and both water and substances move out of the bloodstream with filtration pressure. The advantage of this procedure is that large and small molecules move at practically the same rate; however, one does not want to remove molecules of the size of albumin (molecular weight 60,000 $M_r$). Convection, therefore, removes the so-called middle molecules (larger than urea and smaller than albumin) well (see later). A disadvantage of membranes with large pore size is the loss of large volumes of water, which must be replaced by reinfusion or compensated for; this requires a more complex apparatus than simple dialysis machines (Fig. 8).

Clinical blood purification through convection was introduced in the United States by Dr. Lee Henderson.

Although it was very cumbersome in the beginning, it now can be used to make truly wearable artificial kidneys.

## D. The Uremic Toxin

Many efforts have been made to define "the uremic toxin," which is responsible for the clinical picture of uremia, characterized by drowsiness, failing appetite,

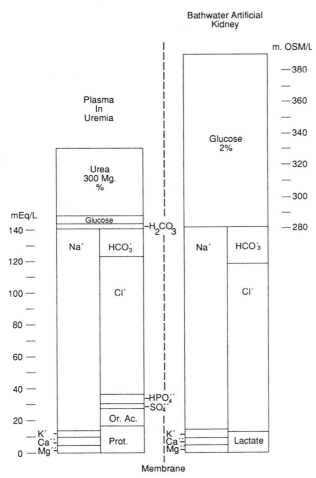

FIGURE 5 Gamble's diagram adapted to the condition of a uremic patient with the urea on top and with 2% glucose on top of the dialysate. The scale for the milliequivalents of the composition of the electrolytes is indicated on the left. The scale for milliosmoles of the nonelectrolytes is indicated on the right. Because the sum total of the dialyzing fluid with the glucose is higher than that of the patient's blood plasma, water is moved from the patient's blood into the dialyzing fluid by osmosis. Similar considerations are used for peritoneal dialysis to remove water from the patient by osmosis. This diagram does not take into account the filtration pressure. If the blood is under pressure, filtration forces extra water from the blood compartment into the dialyzing fluid.

FIGURE 6 Most hollow fibers are packaged in a round cartridge. This package of dialyzing fibers is packaged in a flat square to make it even smaller for the wearable artificial kidney.

vomiting, and general malaise. One possible substance might be urea, the end product of protein metabolism, but it is often argued that urea has very low toxicity and does not need to be removed. Under a normal diet, humans form about 20 g of urea per day. If urea produced day after day were not removed, in about 4 years we would be nothing but a column of urea. A Scottish investigator took so much urea by mouth that his blood level became around 100 mg/100 ml (upper limit of normal is 40 mg/100 ml). He then tried to write a scientific article and found that he was unable to do so, apparently due to toxicity of the urea.

Even if urea itself was not toxic, chemical reactions preceding its formation might all be shifted back if it is allowed to accumulate. The theory of the middle molecules found ardent defenders. It was argued that because urea is not very toxic, the clinical picture of uremia must be caused by another toxin. Molecules larger than urea or creatinine and uric acid and smaller than albumin, the so-called middle molecules, would have to be blamed. Dr. Carl Kjellstrand was the first to seriously question the theory. The clinical picture of uremia must be explained by the total effects of all the products that are accumulated and by the chemical reactions that are retarded or changed by this accumu-

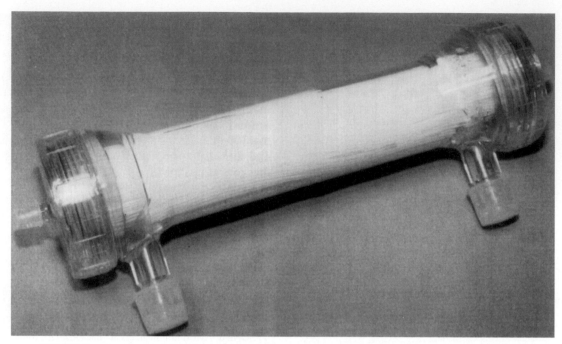

FIGURE 7   A commercial hollow fiber dialyzer. They come in different sizes with different membranes.

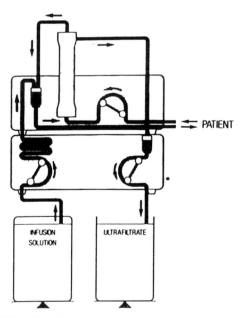

FIGURE 8   Hemofiltration. One pump pushes the patient's blood through the filter. The filtrate is removed by a second pump to be discarded. An infusion pump replaces the same volume of fluid and adds it to the return cannula, which returns the blood to the patient. Both the infusion solution tank and the ultrafiltrate tank are on a scale to measure their contents.

lation. In addition, there are the effects of electrolyte imbalance, overhydration, and hypertension.

Fortunately, we did not have to wait until everyone agreed on the theory of what was toxic in uremia. In fact, hundreds of thousands of patients are being maintained in a reasonable state of health by dialysis, although we do not know exactly what essential factors are removed.

## E. Shorter Dialysis Time

Using these more highly permeable membranes, there is a tendency to reduce the dialysis time. With the conventional artificial kidneys, dialysis was needed three times per week for approximately 5 hr. Presently, with the high-flux artificial kidneys, there is a tendency to dialyze three times per week for 3 hr or less.

Presently, the survival of dialysis patients in Japan is much better (94%) than in the United States (84%) per year. The reimbursement policy in the United States leads to shorter dialysis time, dialysis with a smaller surface area, and membranes that are cheaper and of less quality. The Japanese also have steam-sterilized dialyzers with sophisticated on-line monitoring capabilities; Americans do not. Dr. Yuki Nose points out that he and the author had better results

25 years ago at the Cleveland Clinic than the average results found in the United States today!

## III. DISPOSABLE ARTIFICIAL KIDNEYS

In 1955, the first disposable artificial kidney was made. It consisted of coils of plastic window screening and, in between, cellophane or regenerated cellulose tubing. These so-called twin-coil kidneys made dialysis possible worldwide (Fig. 9).

The first model of the wearable artificial kidney was developed at the University of Utah. Whereas the previous models used two needles, one to withdraw the blood and the other to return it to circulation, this system employs a single needle (Fig. 10). Single-needle dialysis was described earlier. In this approach, the flow alternates from the vein and into the vein; using a short needle with a volume of no more than 0.5 ml, only this amount goes back and forth without reaching the artificial kidney, reducing the overall blood flow very little. This is true as long as the total net blood flow through the single needle is at least 200 ml per minute.

This wearable artificial kidney machine was used widely in our Dialysis in Wonderland trips (see later). However, the tubing set for this type of single-needle

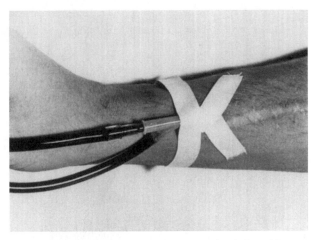

FIGURE 10  The single-needle arrangement for dialysis. The outflow cannula is shown at the top of the figure whereas the return cannula is at the bottom. Inflow and outflow alternate in a pulsating manner. The only place where the blood goes to and fro is in the needle (covered by the adhesive tape). Single-needle dialysis is widely used in Europe but very little in the United States.

dialysis was more expensive than the standard sets, and the additional cost caused this type of wearable artificial kidney to be discontinued. Then, a new machine with less extensive tubing sets was introduced by Dr. Steve Jacobsen. But when a Japanese firm

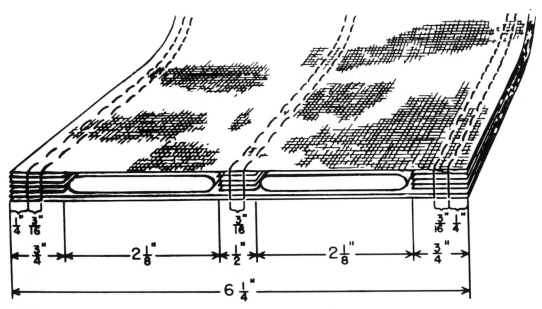

FIGURE 9  Cross section of the first disposable so-called twin-coil artificial kidney. Two cellophane tubes separated by screens are distended in the center. Additional stacks of screens form the spacers that allowed enough space for the blood to run through the tubing. Large rolls were made on a special sewing machine and then wrapped around tin cans to make the first disposable artificial kidneys.

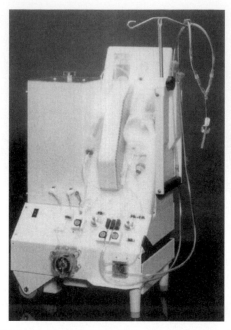

FIGURE II    Dr. Stephen Ash's BioLogic-HD machine. This machine is designed to provide greater simplicity, safety, portability, and middle molecule removal in dialysis. It is single-access dialysis without a blood pump and with a simplified blood tubing set.

agreed to manufacture the machine, the changes they made again increased the cost.

A novel portable dialysis machine (developed by Dr. Steve Ash) (Fig. 11) needs little dialyzing fluid because it uses a sorption column to absorb urea and other metabolites. This column is the same as the one used for years by the REDY system. The REDY system is manufactured by Organon Teknika. It regenerates a small volume of dialyzing fluid by first converting the urea with urease to $NH_3$ and subsequently adsorbing the $NH_4OH$ on zirconium compounds.

## IV. NUMBER OF PATIENTS ON DIALYSES AND ECONOMIC IMPACT

"To be dialyzed three times per week—I would rather be dead." This was the common answer of physicians who made decisions for their patients and did not authorize dialysis even though the patients would die without it. Almost one-half of a million patients on dialysis worldwide prove that this opinion was unjustified

In 1988, 90,000 patients in the United States were maintained with some form of dialysis. In Japan, with less than half the population of the United States, 60,000 patients were on dialysis. This may indicate several things:

1. There is more kidney disease in Japan than in the United States. This does not appear to be true.
2. In Japan, too many patients are dialyzed. Not true.
3. In the United States, not all people who need dialysis are dialyzed.

In 1973, the United States passed a law (PL 92-603) that recognized end stage renal disease (ESRD) as a financially catastrophic disease. Everyone covered by social security can receive reimbursement of 80% of the cost of treatment. Unfortunately, no stringent control system was set up to contain the cost. Excessive concern by the Food and Drug Administration (FDA), and fear of the FDA, made equipment more expensive. The cost of treatment per patient per year in a dialysis center is $40,000, if there are no complications. This amounts to 3 or 5 billion dollars per year in the United States alone.

## V. LIFE AND DEATH COMMITTEES

In the early 1960s, there were not enough artificial kidneys. Dr. Scribner in Seattle originated the life and death committees. A medical committee would ask, "Is this patient an emotionally mature adult, etc.?" A lay committee would ask, "Is he employed, married, or divorced? Does he have children? Does he go to church, etc.?" Only when the answers were positive would the patient qualify. Not much has been heard about these committees lately, but a new term was coined with the same purpose called cost containment.

## VI. HOME DIALYSES

Home dialysis gives the patient a much better opportunity to fit dialysis into his or her own daily schedule. Either the patient or a member of the family can be instructed to work with the dialysis machine; it is not that difficult. However, home dialysis is being used less and less in the United State because it is more expensive for the patient.

## VII. RESTORATION OF THE PATIENT'S HAPPINESS

Restoration of an acceptable way of life and, if at all possible, restoration of the patient's happiness should be the goal of any artificial organ. For a patient to learn that for the remainder of his life he is dependent on either a kidney transplant or a dialysis machine three times per week is a severe blow. If kidney transplantation works, it is more desirable for the patient. According to Terasaki and Cecka, in 1993 there were 315,737 kidney transplants worldwide. However, quite a number of patients still lose their transplants by immune rejection. Some have tried as often as four times. Therefore, a large number of patients depend on dialysis for life. We have to try to make that life as good as possible for them.

To prove to people dependent on dialysis for life that such a life still has much to offer, we (Jacobson and associates) built a wearable artificial kidney which made traveling possible. Groups of various patients went to Utah's Canyonlands National Park, to Hawaii, or to St. Johns in the Virgin Islands. Many patients were sent down the Colorado or the Salmon River on rafts with these artificial kidneys. They dialyzed themselves in the evening on the shores of the river (Fig. 12). On one of these trips, two patients fell into the Salmon River; they were fished out and can now tell their friends that being dependent on dialysis still allows for a full life.

FIGURE 12  Teresa Peterson dialyzing herself with a wearable artificial kidney on the shores of the Colorado River. After 12 years of hemodialysis, Teresa has now received a kidney transplant and is doing well.

## VIII. COMPLICATIONS

As was expected, hemodialysis has brought some complications to the patients who use it. Sometimes the complications are a consequence of the primary disease, sometimes a consequence of dialysis.

### A. Copper Intoxication

Copper heating coils of machines that mix the dialysis salts in the proper proportion can cause complications from copper intoxication. Such complications can stem from the copper in the water pipes of one's house as well. It causes severe hemolysis (i.e., the breakdown of red blood cells) and, consequently, severe anemia in some patients. It is very difficult to detect because if the blood is centrifuged, no copper is found in the plasma. It is only found in the ghosts of the red blood cells, which are present in a small area between the plasma and the blood cells after centrifugation.

### B. Calcium Problems

Gradually, we are learning to take care of the disturbances in calcium metabolism. Broken bones and calcium deposits in the tissues can now be prevented. Phosphate retention leads to disturbance of the phosphate–calcium equilibrium in the blood plasma. To prevent this complication, an aluminum preparation was given by mouth to bind the phosphates, but it led to aluminum intoxication, with consequent irreversible dementia dialytica. Parathyroid glands are often resected for the control of the calcium level.

### C. Hypertension

High arterial blood pressure, or hypertension, can be severe and difficult to control when it is caused by the hardly functioning, contracted kidneys of the patient. Their removal makes control of blood pressure easier, with restriction of NaCl (kitchen salt) and water. [See Hypertension.]

## D. Diet Restrictions

With adequate dialysis, severe restriction of protein (meat) is not necessary. The use of excessive amounts of potassium ($K^+$)-containing foods, such as strawberries or orange juice, must be limited because they can cause acute cardiac arrest.

## E. Amyloid

Deposits of amyloid have been recognized as a late complication as the cause of arthritis and carpal tunnel syndrome. This syndrome is caused by narrowing of the tunnel at the base of the wrist through which the nerves to the hand run. Amyloid is apparently caused by deposits of $\beta_2$-microglobulin, which can now be removed by dialysis ultrafiltration, using highly porous membranes produced by Asahi in Japan and a new adsorbent used for direct hemoperfusion.

## F. Anemia

Anemia literally means not enough blood but actually implies deficiencies of red blood cells, although the blood volume may be adequate. It is a common complication of renal failure due to lack of a hormone called erythropoietin, which is made by the kidney. Concerning blood volume, the earlier artificial kidneys, such as the rotating drum artificial kidney, required priming with 2 pints of blood, thus increasing blood volume. The patients were often a little overloaded with blood volume and felt quite comfortable when some blood was removed. The newer artificial kidneys, whether coil or hollow fiber, do not require priming with blood. A possible complication is that water from the blood during the treatment is removed from the patient by ultrafiltration. If too much is removed, the patient will go into shock (i.e., not have enough blood to fill the vascular system).

Concerning blood hemoglobin, it was an old rule that one should not try to elevate it to more than 7 g per deciliter, because if one did, the patient in kidney failure would not regenerate any blood cells himself. An exception is found in some patients with a disease called polycystic kidneys, who often produce erythropoietin, although they otherwise are in renal failure. Erythropoietin is available as a drug and can be used to normalize the hemoglobin level of patients treated with artificial kidneys.

## G. Hemochromatosis

Those of us with long experience also recall the patients who, after years of multiple dialyses, developed a bronze-like color to their skin called hemochromatosis. This was caused by the breakdown of red blood cells, with an accumulation of the blood pigment in the skin and also in some internal organs. To reduce this phenomenon, only fresh blood was accepted from the blood bank to prime the artificial kidney, and not outdated blood, which would have a high free hemoglobin to begin with. Of course, it is even better if one does not use any blood at all, as is presently the case.

## H. Lack of Taste and Smell

Actually, there is more a change of taste, called dysgeusia, than a lack of taste and smell. Quite often, uremic patients have a distaste for meat, thereby avoiding it. This is not discouraged by their physicians because meat, containing protein, is a source of retention products such as urea, creatinine, and uric acid. Meat, however, seems to be the main source of zinc; lack of zinc in these patients causes the dysgeusia. A test kit has been developed that makes it easy to put a drop of substance—sweet, sour, salt, bitter (urea)—on the patient's tongue. Its use clearly shows a deficiency of taste ability, which could be reversed by the administration of zinc. This is a neglected area of research. Even more neglected is the testing of smell, which possibly suffers in the same way as the ability to taste. [See Copper, Iron, and Zinc in Human Metabolism.]

## IX. RESEARCH OPENED BY DIALYSIS

### A. Removal of Electrolytes

It has always been easy in patients or experimental animals to increase an electrolyte in the blood by giving it either by mouth or intravenously. In contrast, the removal of electrolytes is extremely cumbersome but can be done very rapidly by hemodialysis. In this way, low potassium has proven to affect the heart by altering the cardiogram; the effect is aggravated by digitalis (which is used to treat cardiac insufficiency). The combination of low blood potassium and digitalis might cause cardiac arrest.

### B. Hypertension

Ever since Goldblatt put his silver clip on the renal artery, kidneys have been believed to cause hypertension. These organs produce a hormone called renin, which in turn generates another hormone, angiotensin; the latter hormone causes hypertension. Indeed, removal of contracted kidneys in malignant hypertension makes its treatment easier. Hypertension, however, can occur without kidneys, as shown in animals without kidneys maintained with either hemodialysis or peritoneal dialysis. This so-called renoprival hypertension is greatly dependent on water and salt content. Perfusion of the blood through a pair of normal kidneys could reduce the renoprival hypertension in a matter of hours.

## C. Dialysis of Schizophrenics

After very optimistic reports, it was hoped that dialysis, or rather plasmapheresis, might be a means to treat schizophrenic patients—certainly not all schizophrenics—but perhaps some. These hopes were augmented when it was reported that a slightly abnormal endorphin was found in these dialysates. The endorphin story has not been confirmed and was found to be erroneous. Since, most likely, schizophrenia is a disease dependent on inherited factors and is probably of metabolic origin, it is possible that some cases of beginning schizophrenia may be aborted in the future by removal of an as yet unknown substance from the bloodstream.

## X. REMOVAL OF POISONS

For the treatment of an overdose of most sleeping pills, one has a choice between dialysis and hemoperfusion. Hemoperfusion leads the blood over an adsorbent (e.g., activated charcoal). The adsorbent removes barbiturates and many other poisons quickly if they are freely circulating in the bloodstream. This treatment may wake up a patient in hours, whereas it may otherwise be necessary to keep him or her on a respirator for several days.

The removal of bromide is one example where the artificial kidney is more effective than the natural kidney. The natural kidney removes bromide as a fraction of the total halide present in the blood because the natural kidney cannot distinguish between the bromide and the chloride ion. In contrast, the artificial kidney removes each halide according to its own gradient. In the dialysis of a patient with a high bromide content in the blood, the bromide is removed very rapidly because there is no bromide in the dialyzing

fluid. At the same time, the chloride content, which is low in these patients, is replenished from the chloride present at a normal level in the dialyzing fluid.

## XI. ALTERNATE METHODS OF BLOOD PURIFICATION

### A. Hemosorption or Hemoperfusion

Hemosorption or hemoperfusion perfuses blood directly over charcoal or resins, and, in many instances, can remove toxic substances, such as barbiturates (sleeping pills). Most spectacular perhaps is the removal of low-density lipoprotein (LDL) cholesterol in patients with familial hypercholesterolemia, who have a very high incidence of heart attacks and strokes. The LDL can be reduced by an intensive treatment with either cascade membrane plasmapheresis (using membranes from Asahi) or by hemoperfusion. After initial intensive treatment, treatment once every 2 months may be sufficient. It is hoped that this treatment will stand the test of time.

### B. Regular Peritoneal Dialysis

Peritoneal dialysis, or peritoneal lavage, uses the peritoneal membrane and the tissue right under it as the dialyzing membrane. The peritoneal cavity has a large

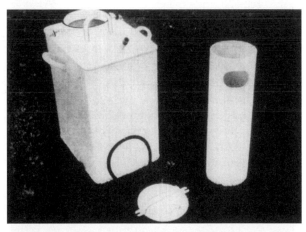

FIGURE 13 A tank developed in the Netherlands in 1948 to prepare 28 liters of sterile peritoneal dialysis fluid on the kitchen stove. The tank has two compartments so that the glucose and calcium chloride are separated from the other minerals and the sodium bicarbonate. After cooling overnight, the tank is tilted, and the smaller compartment empties into the larger compartment, readying the dialyzing fluid. It is the only dialysis method that can be afforded in developing countries.

surface area and is richly vascularized. By instilling a dialyzing fluid into the peritoneal cavity, an exchange takes place between the blood in the capillaries and the dialyzing fluid.

In the Netherlands, a 28-liter tank was developed that could be sterilized on a kitchen stove (Fig. 13). By tilting the tank, the two components of the dialyzing fluid—one containing glucose and calcium chloride, the other containing the remaining electrolytes, including sodium bicarbonate—could be mixed. Glucose did not caramelize, and calcium carbonate did not form during sterilization. When it was sufficiently cooled, the tank could be used to run 28 liters of fluid through the patient's peritoneal cavity (Figs. 14 and

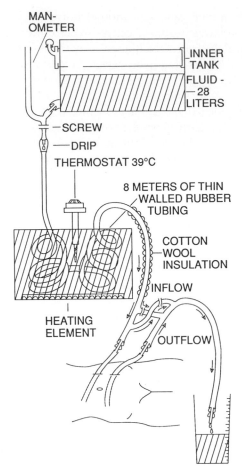

**FIGURE 15**   Diagram of peritoneal lavage. The inflow and outflow can be interchanged at a moment's notice. At any time, a loop of gut or a piece of omentum may block the outflow. Of course, the same tank can be used for continuous ambulatory peritoneal dialysis in developing countries.

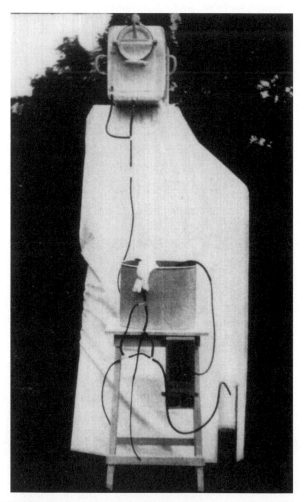

**FIGURE 14**   Peritoneal dialysis tank ready for use. After cooling, the reservoir is tilted 90° and the contents of the inner tank flow out into the outer tank and are thoroughly mixed by shaking. A tube from the spout is led through a warm water bath to bring it to body temperature.

15). To remove water from the peritoneal cavity, a water attractant, usually glucose, must be added to the dialyzing fluid because the blood plasma contains proteins that provide a colloid osmotic pressure, attracting water (cf. Fig. 5). Peritoneal dialysis is a lifesaver for patients who run out of blood access ports for hemodialysis.

## C. Recirculating Peritoneal Dialysis

A vastly improved method is recirculating peritoneal dialysis, where a small volume of sterile dialyzing fluid is rapidly circulated through the peritoneal cavity (Fig. 16). This small volume of dialyzing fluid is, in turn, dialyzed with the conventional, disposable artificial kidney against a much larger volume of fluid that

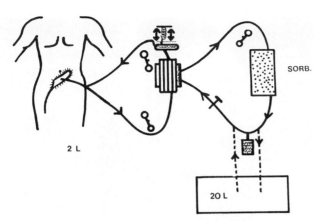

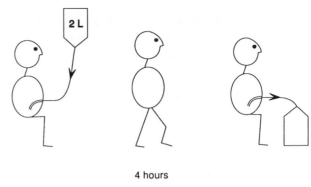

4 hours

FIGURE 16 Recirculating peritoneal dialysis has the advantage of a closed system, and because the small volume of sterile fluid is recirculated, no albumin is lost. Reciprocating or recirculating peritoneal dialysis can be used with sorbents or with dialyzing fluid. In the abdominal wall, the so-called "mouse" is implanted under the skin. The mouse has a heavy Dacron velour body (any good mouse has a furry body) and a smooth tail with many side holes, which goes into the peritoneal cavity. To gain access to the peritoneal cavity, a needle is inserted through the skin into the body of the mouse.

FIGURE 17 CAPD is presently the most used method. The patient instills a volume (e.g., 1.2 liters) of sterile fluid into his abdominal cavity and puts the empty container under his belt and lets the fluid out after 4 hr, replacing it with a new batch of fresh, sterile fluid.

does not need to be sterile. The advantage of the recirculating peritoneal dialysis is that it forces a closed system, thereby reducing the risk of infection. The pain often experienced in the regular peritoneal dialysis when the last bit of fluid was removed from a patient's abdomen and the first of a new batch instilled was completely eliminated. This method causes no loss of albumin.

## D. Continuous Ambulatory Peritoneal Dialysis (CAPD)

A careful mathematical analysis resulted in a very effective method of dialysis (Fig. 17). Nolph and Popovitch determined that if the peritoneal cavity is filled with about 1.2 liters of a dialyzing fluid and is left to equilibrate for 4 hr, a near optimal exchange is obtained. The fluid could be removed and a new batch could be instilled. The patient was free to do what he wanted; he could even perform his regular work duties while dialyzing. There is a famous conversation between two patients—one on hemodialysis who said, "I am being dialyzed three times per week. How often are you dialyzed?" The CAPD patient answered, "All the time."

The very high expectation that CAPD would remove the middle molecules more effectively than he-

modialysis has not been confirmed by an obvious clinical improvement of the patients. The main drawback of CAPD is a considerable loss of protein from the patient's abdominal cavity. This may be a distinct disadvantage for patients who prefer not to eat a lot of meat. (For the manufacturers of intravenous and other parenteral fluids, however, it was a bonus too good to be true since they could sell large amounts of salt water to thousands of patients for a price the manufacturer could determine.)

Peritonitis is one of the known complications of peritoneal dialysis but should not be feared as much as it generally is. When peritonitis is caused by perforation of the intestine, it is dangerous or deadly, but as a complication of peritoneal dialysis, it can be easily treated by instillation of an antibiotic or even a little iodine, plus postponement of dialysis for a few days. Another complication is that a reduction of the dialyzing capacity of the peritoneal cavity decreases with time, perhaps as a result of chronic, repeated stimulation of the peritoneal membrane. This should be an argument for reactivation of the recirculating peritoneal dialysis technique, which undoubtedly causes fewer symptoms.

## XII. RELATED METHODS OF BLOOD TREATMENT OUTSIDE THE BODY

### A. Plasmapheresis

Plasmapheresis is a method by which some plasma is separated from the blood cells. The plasma can be discarded or treated in various ways. This separation

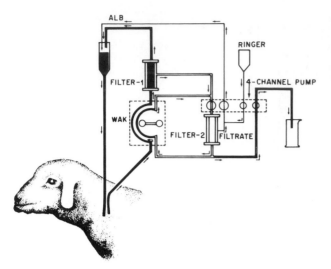

**FIGURE 18** Cascade membrane plasmapheresis. The sheep's blood passes through filter-1, which separates some of the plasma. Diluted or not, this plasma is transferred through filter-2, which, according to its pore size, can separate albumin from globulin, etc. The albumin is returned to the sheep; the globulins are discarded. The plasma can be treated in various ways: it can be let over sorbents, it can be irradiated, or it can be refrigerated. After treatment, the desirable components are returned to the patient.

takes place with centrifugation (as is now commonly done); a newer way is to use suitable membranes.

## B. Cascade Membrane Plasmapheresis

This process is discussed here because it uses the same kind of filters and pumps that are used for artificial kidneys. Cascade membrane plasmapheresis is far su-

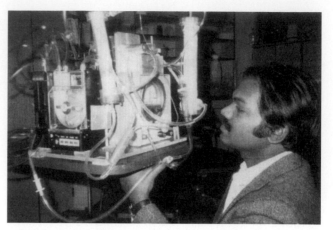

**FIGURE 20** Dr. Udipi Shettigar with a portable plasmapheresis machine. The left side of the machine shows the blood pump and automatic controls of the newest wearable artificial kidney. Next to it is the plasma filter, which separates plasma from blood. The long cylindrical filter on the right separates plasma components according to their molecular size.

perior over the older methods (Fig. 18). It is used regularly in Switzerland.

Cascade membrane plasmapheresis has opened an entirely new field in the treatment of blood that has been barely explored. Whole blood can easily be damaged, but plasma separated from cells can be treated in many ways. Fractions can be separated and returned to the patient. It does not require expensive blood plasma or albumin to substitute what is thrown away, as in exchange plasmapheresis. Specific elements can be removed from plasma.

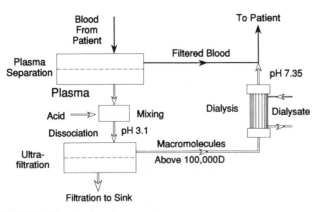

**FIGURE 19** Another diagram of cascade membrane plasmapheresis. When hydrochloric acid is added to the plasma, antigen dissociates from immune complexes. The antigen is removed by ultrafiltration and dialysis, which also removes the excess acid.

**TABLE I**

Diseases for which Plasma Exchange Is Reimbursable under Medicare

Myasthenia gravis

Waldenstrom's macroglobulinemia

Hyperglobulinemia associated with multiple myeloma, cryoglobulinemia, hyperviscosity syndrome, and other causes

Thrombotic thrombocytopenic purpura, last resort

Rheumatoid vasculitis, last resort

Goodpasture's syndrome, life-threatening

Renal failure or pulmonary hemorrhage secondary to antiglomerular basement membrane antibodies, life-threatening

Pruritis of cholestatic liver disease (plasma perfusion over charcoal)

Chronic relapsing polyneuropathy, severe or life-threatening

Scleroderma and polymyositis, life-threatening

**FIGURE 21** Patients dependent on dialyses (three times per week) go down the Colorado River on a raft. In the evening, they dialyze with the wearable artificial kidney on the shore. This proves to them that although they are dependent on dialysis, they can still have a wonderful life.

Controlled trials of plasmapheresis in the United States have been reported in immunological disorders, such as multiple sclerosis, rheumatoid arthritis, systemic lupus erythematosus, and acute vascular rejection in kidney transplantations. All the controlled studies, except one in multiple sclerosis, performed only a few plasmapheresis treatments; they showed that both true and placebo groups improved in the beginning of the treatment but were too short to eliminate the effects of placebo and hospitalization. In general, the results were poor. Better results were obtained in uncontrolled trials with longer and more frequent treatments. Studies of longer duration are necessary to establish the value of plasmapheresis in immunological disorders.

No one questions the efficacy of plasmapheresis in hyperviscosity and related hyperglobulinemic syndromes. Although no controlled studies have yet been reported, results in thrombotic thrombocytopenic purpura and Goodpasture's syndrome (except in anuric cases) are so superior to historic controls that it is reasonable to continue the treatment of patients with these disorders with plasmapheresis. A diagram of plasmapheresis with acidification to cause dissociation of immune complexes and the removal of the antigens is shown in Fig. 19; the apparatus is shown in Fig. 20. Table I lists the disorders that are eligible for reimbursement under Medicare, an indication of some acceptance of plasmapheresis in the United States.

The aim of an artificial kidney is to rehabilitate people. Although not everyone is restored to a happy existence, many people are (Fig. 21).

## BIBLIOGRAPHY

Abel, J. J., Rowntree, L. G., and Turner, B. B. (1913). On the removal of diffusible substances from the circulating blood of living animals by dialysis. *J. Pharmacol. Exp. Ther.* **5**, 275.

Akizawa, T., Kinugasa, E., Kitaoka, T., Koshikawa, S., Nakabayashi, N., Watanabe, H., Yamawaki, N., and Kuroda, Y. (1987). Removal of beta-2-microglobulin by direct hemoperfusion with a newly developed adsorbent. *Trans. Am. Soc. Artif. Intern. Organs* **XXXIII**, 532.

Ash, S. R., Baker, K., Blake, D. E., Carr, D. J., Echard, T. G., Sweeney, K. D., Handt, A. E., and Wimberly, A. L. (1987). Clinical trials of the BioLogic-HD. *Trans. Am. Soc. Artif. Intern. Organs ASAIO* **10**(3), 524.

Evans, D. H., Sorkin, M. I., Nolph, K. D., and Whittier, F. C. (1981). Continuous ambulatory peritoneal dialysis and transplantation. *Trans. Am. Soc. Artif. Intern. Organs* **XXVII**, 320

Hampl, H., Lobeck, H., Bartel-Schwarze, S., Stein, H., Eulitz, M., and Reinhold, P. L. (1987). Clinical, morphologic, biochemical and immunohistochemical aspects of dialysis-associated amyloidosis. *Trans. Am. Soc. Artif. Intern. Organs* **XXXIII**, 250.

Henderson, L. W., Quelhorst, E. A., Baldamus, C. A., and Lysaght, M. J. (eds.) (1986). "Hemofiltration." Springer-Verlag Berlin, Heidelberg, Germany.

Jacobsen, S. C., Stephen, R. L., Bulloch, E. C., Luntz, R. D., and Kolff, W. J. (1975). A wearable artificial kidney: Functional description of hardware and clinical results. *In* "Proceedings of the Fifth Annual Meeting of Clinical Dialysis and Transplant Forum," pp. 65–71. New York.

Kolff, W. J. (1987). Renoprival hypertension: Obscure projects. *In* "Artificial Organs, Proceedings of International Symposium on Artificial Organs, Biomedical Engineering and Transplantation" (J. D. Andrade *et al.*, eds.), p. 5. VCH Publishers, Inc., New York.

Kolff, W. J. (1987). Economics, diabetics, schizophrenics: The future of artificial organs and of us all. *In* "Artificial Organs, Proceedings of International Symposium on Artificial Organs, Biomedical Engineering and Transplantation" (J. D. Andrade *et al.*, eds.), p. 723. VCH Publishers, Inc., New York.

Kolff, W. J., and Watschinger, B. (1956). Further development of a coil kidney. *J. Lab. Clin. Med.* **47**(6), 969.

Kopp, K. F., Gutch, C. F., and Kolff, W. J. (1972). Single needle dialysis. *Trans. Am. Soc. Artif. Intern. Organs* **XVIII**, 75.

Nosé, Y. (1994). Why do we kill so many patients on hemodialysis in the U.S.? *Artif. Organs* **17**(11), 893–894.

Shettigar, U. R., Kolff, W. J., and Anstall, H. B. (1984). Membrane cascade plasmapheresis: Theoretical analysis and in vitro studies. *Uremia Invest.* **8**(1), 25.

Thies, K., Prigent, S. A., and Heuck, C.-C. (1988). Selective removal of low-density lipoproteins from plasma by polyacrylate-coated fractogel in vitro and in experimental extracorporeal perfusion. *Artif. Organs* **12**(4), 320.

Woffindin, C., Hoenich, N. A., and Matthews, J. N. S. (1992). Cellulose-based haemodialysis membranes: Biocompatibility and functional performance compared. *Nephrol. Dial. Transplant.* **7**, 340–345.

# Artificial Skin

JONATHAN MANSBRIDGE
*Advanced Tissue Sciences, Inc.*

---

## GLOSSARY

**Allogeneic** Tissue or cells for grafting derived from an individual of the same species other than the graft recipient

**Autologous** Tissue or cells for grafting derived from the graft recipient

**Basement membrane** Protein structure that marks the boundary between the dermis and the epidermis and maintains structural integrity of the skin at this point. It contains many specialized proteins, including laminins, collagen IV, and collagen VII. Epidermal keratinocytes have specialized structures called hemidesmosomes by means of which they adhere to the basement membrane, which is, in turn, bound to the dermis through anchoring fibers made of collagen VII

**Collagen** The major protein of the dermis which forms fibrils that provide the mechanical strength of the dermis

**Collagenase** An enzyme of the metalloprotease class that degrades collagen

**Dermis** The inner layer of the skin that provides structural strength to the surface of the body

**Epidermis** Outermost cellular layer of the skin, including the stratum corneum that provides the primary permeability barrier

**Fibroblast** The major cell of the dermis that is capable of laying down and maintaining the collagen structure of the dermis

**Keratinocyte** The major cell type in the epidermis that continuously divides and differentiates to form and replace the stratum corneum

**Tissue Engineering** Synthetic application of the culture of living cells, usually using three-dimensional culture techniques, in order to produce organ analogs

ARTIFICIAL SKIN COMPRISES A RANGE OF DEVICES with properties of human skin that have been applied in experimental and therapeutic settings and used for the testing of commercial products to predict the human response and possible adverse reactions. Many of the earliest applications were intended to provide dressings that would cover large denuded areas of the body, e.g., burn victims, and promote healing. These have been developed from occlusive nonbiological dressings to structures that include both biological molecules and live cells. Several of the tissue-engineered systems grew out of experimental models of skin that were developed to explore the interaction between the dermis and the epidermis. Artificial skin produced by three-dimensional tissue culture techniques is now used both to cover large areas of skin loss and to treat chronic wounds such as ulcers. The experimental skin models that have been developed have the ability to replace much of the routine testing of commercial products in animals. At this time, the field is still developing and future directions include the possibility of including genetically modified cells in therapeutic and testing applications.

## I. BACKGROUND

Artificial skin of one kind or another for covering wounds has been in the medical compendium since the earliest times. Included was everything from spiders' webs and pitch to bandages, adhesive patches, permeable and occlusive dressings, sophisticated, controlled

ENCYCLOPEDIA OF HUMAN BIOLOGY, Second Edition, VOLUME I.   Copyright © 1997 by Academic Press.   All rights of reproduction in any form reserved.

permeability membranes, and, most recently, tissue-engineered products. Most of the earlier products were intended to protect the wound from exposure to infection and mechanical damage and were not intended to replace the skin or interact with the wound healing system. In terms of therapeutic devices, this discussion will be restricted to the more recent preparations derived from biological molecules or cultured tissues that are designed to integrate into the wound healing process or form a component of the final skin. In addition to the therapeutic applications, various artificial skin devices, which have found use in experimental and testing applications, have been developed in parallel.

The organ that is replaced by artificial skin forms the outermost barrier between the internal structures of the animal and its environment. As such, it has to be able to withstand mechanical, chemical and microbial encounters and to be able to recruit the inflammatory and immune systems in the event of injury or infection. One of its most important characteristics is its impermeability to many substances. In addition, the skin carries a variety of structures concerned with sensation, sweating, hair, lipid secretion, and other specialized functions, such as lactation.

Structurally, the skin is formed from two major layers: the outer epidermis and the inner dermis. Most of the mechanical strength is supplied by the dermis, largely through collagen and elastin fibers. Chemical resistance and impermeability is conferred by the stratum corneum that is continually renewed by cell proliferation and differentiation in the epidermis. [*See* Skin.]

Various artificial skin systems have been directed toward primarily replacing the epidermis, the dermis, or both. In the case of a severe injury, because both dermis and epidermis must ultimately be replaced, restoration of the permeability barrier with an artificial epidermis may provide the immediate solution to water loss and protection against infection. As healing proceeds, however, the role of the dermis becomes more important. Regeneration of the epidermis by the host is greatly enhanced by dermal components, and large third-degree burns frequently develop severe scarring that may be disabling as well as disfiguring. Dermal replacements have been developed with the intention of supplying a seed dermis to minimize the load on the host dermal regeneration system and hence to reduce scar formation. Currently, despite its commercial potential, no major attempt has been made to develop artificial skin including structures such as hair, and no artificial system to date has har-

nessed the immune or inflammatory systems in a useful way. [*See* Thermal Injuries.]

## II. THERAPEUTIC APPLICATIONS OF ARTIFICIAL SKIN

Therapeutic applications include the treatment of both acute and chronic wounds. In acute wounds, the goal is to cover a severely denuded area to prevent fluid loss and infection and ultimately to allow normal wound healing to proceed to complete closure. Applications of this type include third-degree burns, major plastic surgery excisions, removal of giant hairy or pigmented nevi, some blistering diseases, autograft skin donor sites, and, in some cases, second-degree burns. In contrast, chronic wounds have frequently persisted for a long period without healing and have reached a condition where progression to a normal acute wound healing process is prevented. The wound is unable to close and shows many abnormalities when compared to an acute wound. An ulcer bed is depleted of growth factors and cellular growth factor receptors and contains abnormally high concentrations of proteases, including collagenases. The vascular supply is frequently compromised and the structure of local capillary blood vessels is abnormal. Ulcers are frequently infected and contain inflammatory cells such as macrophages and leukocytes in addition to granulation tissue. The role of the artificial skin device may be considered to supply depleted wound healing components and to reset the healing process so that it can proceed along the normal acute pathway.

### A. Nonbiological Dressings

In 1972, Winters obtained experimental evidence that wounds maintained in a moist condition healed more rapidly than dry, provided that infection could be controlled. This view gained acceptance over the ensuing decade and gave rise to a series of related wound dressings that were essentially temporary replacements for the epidermis. These wound dressings were constructed of polyurethane and had water permeabilities in the range of $500-2000 \text{ g m}^{-2} \text{ day}^{-1}$, which is comparable to human skin. Such dressings have been used widely and successfully and include such products as Opsite, Xeroform, and Duoderm. Calcium alginate has been used similarly for some applications. In a wound, fluid exudation may be very much higher, leading to blister formation that may compromise healing. To overcome this problem, hydrophilic polyurethane films have been developed that are capable of much higher water transpiration, which enables excessive liquid to be released, but still acts as a permeability barrier to maintain a moist environment suitable for wound healing.

In a parallel development, artificial skin devices have been developed that place an emphasis on dermal development. One, Biobrane, developed in the early 1980s by Woodroof, used a silastic permeability barrier on a knitted nylon fabric that was coated with porcine collagen peptides to encourage the ingrowth of granulation tissue and the development of a dermis-like structure. This was used to cover large wounds. In cases that required grafting, the silastic and nylon had to be excised from the wound bed before the autograft could be applied. A second device, Integra, that was developed at about the same time by Yannas, used a biological scaffold consisting of bovine collagen and shark cartilage chondroitin sulfate attached to the silastic permeability barrier. This was placed on a wound site such as a third-degree burn and was allowed to develop into a dermal structure. The silastic was then removed and the site was covered with autograft. While having some success in themselves, these products have formed the basis for tissue-engineered artificial skin systems that include living cells or their products. This process has been developed into a commercial product and has been the basis for further modifications. Among these, Hansbrough, Boyce, Matsuda, and others have added living cells by seeding fibroblasts into the collagen, chondroitin sulfate sponge and subsequently seeding autologous keratinocytes to the surface, providing a living skin equivalent containing only autologous cells that could be applied to the patient. One of the advantages of this technique was that it avoided the extreme fragility of cultured epidermal autografts, discussed later.

An alternative approach used de-epidermized dermis, a concept pioneered by Prunieras. When seeded on de-epidermized, acellular dermis, keratinocytes proliferate and form a histologically well-formed epidermis. The acellular dermis resembles cadaver skin in many ways, but avoids the immunogenic problems of allogeneic cellular components. In this case, the dermal structure becomes incorporated into the wound and forms part of the developing dermis.

## B. Biological Dressings

Many of the applications of artificial skin in a therapeutic setting are related to covering parts of the body from which the natural skin has been lost, as in burns and trauma. Winters pointed out in 1962 that the best wound covering was skin itself, if not available from the patient, then from cadaver sources. Although not strictly an artificial skin, fresh allograft skin has been the treatment of choice as a temporary covering of severe burns, despite immunological rejection of the epidermis, which occurs after about 2 weeks, and the risk of disease transmission. Allograft skin has also been developed as a permanent integument replacement by placing small, autologous biopsies into the allograft that provide keratinocytes that replace those lost by rejection. Patients treated in this way show minimal overt rejection.

## C. Keratinocyte Sheets

A biological approach to providing an epidermis was the use of cultured keratinocyte sheets. Human keratinocytes can be grown under special conditions to provide extensive sheets of cells that can be grafted to a patient. In general, because allogeneic keratinocytes cause an immune response that leads to rejection, epidermal sheets used therapeutically are autologous and are called cultured epithelial autografts. When cultured keratinocyte sheets are placed over a third-degree burn site, following careful excision of all damaged tissue including the dermis, a neodermis can be formed under epidermis, ultimately generating a full thickness replacement skin. The use of these devices has achieved some spectacular successes, but is technically demanding, expensive, and the overall experience has been very mixed. Moreover, the culture period required to obtain adequate cell expansion may be as long as 3 weeks. The technique remains important because of its potential in conjunction with other artificial skin devices. [See Keratinocyte Transformation.]

## C. Fibroblasts in Collagen Gels

A major source both of model systems for studying the cell biology of skin cells and therapeutic products has been the use of fibroblasts grown in collagen gels. First studied by Ehrmann and Gey in 1956, the system was extensively developed by Bell during the late 1970s and 1980s. Fibroblasts cast in a collagen gel show properties quite different than their monolayer counterparts. The cells cease to proliferate and organize the structure of the matrix. The size of the gel is reduced to one-twentieth to one-fiftieth of its initial value, increasing the concentration of collagen and forming a dermis-like structure. Collagen synthesis is markedly inhibited by comparison with monolayer cultures; however, a much larger proportion of the collagen made in the collagen gel is actually deposited into the matrix. In monolayers, most of the collagen remains soluble. This is related to a collagen-processing reaction in which a C-terminal peptide is removed, reducing the solubility of the collagen by four orders of magnitude. Collagenases are markedly induced in this system, apparently as part of the collagen reorganization process. Because some characteristics, such as metalloprotease synthesis, are sensitive to the shear state of the gel, the system has found application to investigation of the responses of fibroblasts to mechanical stimuli. The contracted collagen gels are capable of supporting the growth and differentiation of keratinocytes to a state that is histologically very similar to normal skin, although examination of gene expression shows a persistent wound healing phenotype. Such a preparation is capable of providing both dermal and epidermal components of the acute wound healing process. It has been grafted to wound sites in animals and has been applied successfully in the treatment of chronic venous stasis ulcers.

A problem with many of the biological artificial skin systems is that they include components derived from animals. Reservations have been expressed about the possible immunological consequences of exposure of human wounds to nonhuman proteins. Although attributable incidents are rare and are questionably associated with animal products, some effort has been made to reduce their presence in artificial skin systems. Thus Auger has developed a living skin equivalent system that uses human collagen instead of bovine. [*See* Collagen, Structure and Function.]

## D. Artificial Skin Grown on Nonbiological Scaffolds

Animal components can be eliminated by growing fibroblasts on a nonbiological scaffold as pioneered by Naughton. In this system, human dermal fibroblasts, grown in a three-dimensional system, lay down an extracellular matrix that consists of collagen, glycosaminoglycans, and a variety of matrix-associated proteins that closely resemble papillary dermis. The properties of this system are in sharp contrast to those of the living dermal equivalent system developed by Bell (see earlier discussion). During the first week of culture, the cells adhere to the scaffold and grow exponentially in contrast to the dermal equivalent system where growth is arrested. Following the first week,

growth slows and cells begin to enter a stationary phase. During the second week, cells start to deposit collagen at a rate that increases until the time of harvest when collagen comprises about 10% the tissue wet weight. At harvest time, the rate of collagen synthesis is still high and will reach a value comparable to skin (30%) if grown longer. This collagen is processed by removal of the C-terminal peptide so that it is deposited into the developing cell-generated extracellular matrix. Collagenase activity in this system is very low. Fibroblasts grown in the Naughton three-dimensional culture system differ markedly from those in the Bell dermal equivalent, both in their high rate of collagen synthesis and in their collagenase activity.

This three-dimensional fibroblast culture system has been applied in a variety of forms to the covering of severe burns, as a replacement for cadaver skin, for the treatment of diabetic foot ulcers, and as a test system that replaces, in many cases, the use of animals. The temporary skin replacement Dermagraft-TC[1] uses a knitted nylon fabric with a silastic backing (Biobrane II) as the scaffold and contains, at harvest, about 1–1.5 million cells and 0.8–1.5 mg collagen/cm$^2$ of surface. The material adheres to a third-degree burn wound bed and appears to discourage a wound healing granulation response until such a time as autograft skin can be obtained. Viability of the cells is unnecessary for this effect, and the device appears to provide signals to the underlying wound bed that healing is proceeding and that further wound healing activity by the host, such as the generation of granulation tissue, is unnecessary and may be placed on hold. On autografting, normal epidermal regeneration and subsequent wound healing events proceed normally. In addition to its value in areas where the entire dermis and epidermis are lost, Dermagraft-TC markedly accelerates the healing of partial thickness burns. Dermagraft-TC provides an environment free of toxic antimicrobial agents that is conducive to epidermal regeneration, and healing proceeds without interference (see Color Plate 6).

Although nonviable Dermagraft-TC may be adequate for the majority of severe wounds, it may be noted that certain types of acute injury may compromise the remaining cells beyond obvious margins of the wound site, thus hindering healing. Examples include chemical and radiation burns. Such burns tend

---

[1]Dermagraft and Dermagraft-TC are registered trademarks of Advanced Tissue Sciences, Inc.

to heal poorly and ulcerate; in these instances, the use of a tissue-engineered product containing viable fibroblasts may well have special value.

The form of Dermagraft used for diabetic foot ulcers employs a biodegradable scaffold knitted from vicryl. The tissue formed by the fibroblasts is very similar to Dermagraft-TC, but the cells are viable as the tissue is applied to the patient. The mechanism of action of Dermagraft is not known, but the viable fibroblasts in the tissue are growing, as in wound healing. They are also able to respond to the wound environment in an appropriate manner, unaffected by the possible abnormalities of the cells of their diabetic host. In addition, they possess the mechanisms to present growth factors and other molecules of wound healing in their correct context. The tissue contains newly synthesized collagen together with other matrix-associated molecules characteristic of the wound healing process, including fibronectin, decorin, tenascin, and SPARC.

Although the fibroblasts used in tissue-engineered products are generally allogeneic, rejection has not been observed. In the case of Dermagraft, cells derived from foreskin dermis are male and can, therefore, be detected in female patients using male-specific DNA markers. By this means it has been shown that the allograft fibroblasts persist for at least 6 months in an ulcer site. These cells are very extensively tested for possible infectious contaminants and are shown to be free of them. [*See* Immunobiology of Transplantation.]

## III. *IN VITRO* SKIN MODELS

Although it is easily accessed for experimental purposes, the physiological mechanisms that maintain the integument normally in a stable condition, but also allow it to respond dramatically to injury, are not fully understood. Artificial skin models have been very valuable as experimental tools for exploring cellular interactions, particularly the dermal–epidermal interaction and basement membrane formation. A variety of such systems, based largely on the Bell system using fibroblasts cast in collagen gels, the Prunieras system using keratinocytes grown on acellular dermis, and the Naughton system using a cultured dermis (as discussed earlier) seeded with keratinocytes, have been used extensively for these purposes. Areas in which these systems have provided substantial insight include the interaction between dermis and epidermis,

the formation and role of the basement membrane, and the control of collagenase activity.

## IV. COMMERCIAL TESTING APPLICATIONS

In addition to its therapeutic use, artificial skin has also found application for testing products designed to be applied to the skin, including pharmaceuticals, sunscreens, cosmetics, cleansers, and other household and industrial products. The major advantages of this approach are that the tissue investigated is of human origin, the methods are highly reproducible, and animals are not involved in testing. The last point results in cost savings and avoidance of restrictions on animal experimentation. The disadvantages are that the artificial skin preparations only approximate normal skin, particularly in permeability (barrier function), and do not, at the date of this article, incorporate elements of the inflammatory, immune, or pigmentary systems or adnexal structures such as hair follicles or sweat glands. However, the fibroblasts and keratinocytes of the skin are capable of detecting many noxious agents and responding appropriately, and the skin model systems provide a reliable indicator of the effects seen *in vivo*.

The most advanced system of this kind to date is the Skin² range, which includes a dermal equivalent, an ocular model, and a system including a fully formed epidermis and stratum corneum. Each system is based on human dermal fibroblasts grown on a woven nylon scaffold. The cells lay down a collagen-based extracellular matrix, on which human foreskin keratinocytes are seeded to form the ocular and full thickness models. This system has been applied extensively in testing the irritance and corrosivity of many products, potentially replacing corresponding animal testing. Applications for this system include testing the efficacy of sunscreens. Extension of this type of skin model, including other components of the skin, such as Langerhans cells, has the potential of replacing the majority of the skin and ocular testing that is currently performed in animal systems.

Several related artificial skin testing systems have been developed. A similar system, Testskin, was based on Bell's skin equivalent system and provides a comparable test capacity. Simpler models using only epidermal or dermal components may provide an adequate model in some applications because much of the response to noxious stimuli lies in the epidermis.

## V. GENE THERAPY APPLICATIONS

Early experiments on the expression of the growth hormone gene in keratinocyte sheets grafted to nude mice pointed to the possible use of genetically modified artificial skin as a delivery system for gene therapy applications. A major advantage is that the tissue would be easy to apply and remove in the event of an adverse reaction. Both keratinocytes and fibroblasts have been used as targets for transfection, and in both cases it has been shown that cellular proteins can be formed and that secreted proteins reach the bloodstream. Keratinocytes have advantages in ease of application but are liable to rejection unless autologous cells are used or are modified to evade the immune system. Studies using modified versions of the Dermagraft system have shown that cells genetically modified with the marker gene $\beta$-galactosidase can be detected in a graft site by expression of the marker for at least 6 months. Fibroblasts have been more difficult to apply to a patient, but the availability of dermal and full thickness three-dimensional therapeutic systems make this an increasingly attractive possibility. In some cases, the use of genetically modified cells may be unnecessary. Because fibroblasts appear to survive for considerable periods in wound sites, it may be possible in some cases to supply allogeneic normal cells to a patient with a genetic disease to overcome the genetic defect.

## ACKNOWLEDGMENTS

The author thanks Jennifer Vickerman for supplying the illustration and Dr. Gail Naughton for helpful comments and criticism.

## BIBLIOGRAPHY

Arons, J. A., Wainwright, D. J., and Jordan (1992). The surgical applications and implications of cultured human epidermis: A comprehensive review. *Surgery* **111**, 4–11.

Cairns, B. A., deSerres, S., Peterson, H. D., and Meyer, A. A. (1993). Skin replacements: The biotechnical quest for optimal wound closure. *Arch. Surg.* **128**, 1246–1252.

Hansbrough, J. F. (1995). Status of cultured skin replacements. *Wounds* **7**, 130–136.

Naughton, G. K., and Mansbridge, J. N. (1997). Tissue engineering: Synthetic biodegradable polymer scaffolds. *In* "Synthetic Biodegradable Polymer Scaffolds" (A. Atala and D. J. Mooney, eds.). Birkauser, Boston.

Naughton, G. K., and Tolbert, W. R. (1996). Tissue Engineering: Skin. *In* "Yearbook of Cell and Tissue Transplantation" (R. P. Lanza and W. L. Chick, eds.). Kluwer Academic Publishers, The Netherlands.

# Asbestos

JACQUES DUNNIGAN
*University of Sherbrooke*

## GLOSSARY

**Asbestosis** Disease of the lungs characterized by the presence of scar tissue (i.e., fibrosis) resulting specifically from the inhalation of asbestos fibers

**Mesothelioma** Rare forms of malignancy arising from the mesothelium, which is the epithelial tissue enveloping the lungs and lining the thoracic cavity; also lining the abdominal cavity and covering most of the viscera contained therein

ASBESTOS IS A COMMERCIAL TERM APPLIED TO A group of naturally occurring silicate minerals which separate into fibers that are used in a variety of applications. There are two main groups: serpentine (chrysotile asbestos), from the Latin *serpentinous* or "resembling a serpent," in reference to its mottled shades of green, and amphiboles (crocidolite, amosite, tremolite, actinolite, and anthophyllite asbestos), from the Greek *amphibolos*, "ambiguous, or doubtful," in reference to its many varieties.

Because of its fibrous shape and chemical composition, asbestos displays a unique combination of technical properties (e.g., flexibility, heat resistance, and chemical inertness) and therefore is suitable for a great variety of uses, such as yarns, cloth, paper, friction materials, tiles, insulation, cement, and plastics, in which nonconducting, incombustible, chemically resistant, strong, and light composite products are re-

quired. Chrysotile, amosite, and crocidolite are the only types of asbestos of significant industrial use.

There is no universal consensus on a definition for asbestos. A workshop convened by the American Society for Testing and Materials in 1984 formulated the following definition: "A term applied to six naturally occurring minerals exploited commercially for their desirable physical properties, which are in part derived from their asbestiform habit. The six minerals are the serpentine mineral chrysotile and the amphibole minerals grunerite asbestos (also referred to as amosite), riebeckite asbestos (also referred to as crocidolite), anthophyllite asbestos, tremolite asbestos, and actinolite asbestos. Individual mineral particles, however processed and regardless of their mineral name, are not demonstrated to be asbestos if the length-to-width ratio is less than 20:1."

A length-to-width ratio of 20:1 would certainly exclude the so-called nonfibrous tremolite (apparently unrelated to health effects). However, hygienists now seem to prefer a 5:1 ratio for monitoring purposes due to uncertainties in determining the innocuity of nonfibrous tremolite. In other words, a tremolite particle 10 $\mu$m long with a diameter of 0.1 $\mu$m (aspect ratio 10:1) could well be related to health effects. Indeed, tremolite particles with an 11:1 aspect ratio have been found in the lungs of mesothelioma patients.

Adverse health effects (e.g., asbestosis, lung cancer, and mesothelioma) have been associated with exposure to "respirable" asbestos fibers. Fiber dimensions, which relate to the respirability (i.e., length and diameter) of the fibrous particles, and physicochemical parameters, which relate to the relative persistence of fibers in tissues, are considered the key parameters of biological activity. Differences in these parameters could explain the markedly different potencies among different fiber types.

ENCYCLOPEDIA OF HUMAN BIOLOGY, Second Edition, VOLUME I.   Copyright © 1997 by Academic Press.   All rights of reproduction in any form reserved.

# I. DEFINITION AND USES

## A. Chemical Composition

The word "asbestos" is the term applied to all of the serpentine and amphibole varieties to describe fibrous silicate minerals which readily separate into thin strong fibers used industrially. Significant mineralogical and chemical differences among the asbestos fibers might account for their different uses as well as their different pathological potentials (see Table I and Fig. 1).

Chrysotile asbestos is a magnesium silicate whose theoretical formula closely approximates $Mg_3Si_2O_5(OH)_4$, where magnesium oxide can form 38–42% of the total weight. Traces of aluminum, iron, calcium, and sodium oxides are also found. Exposure to acid causes the liberation of magnesium ions and the formation of a siliceous residue; chrysotile is more resistant to alkali than any of the amphiboles. High temperatures (600–650°C) cause gradual dehydroxylation, and this can be followed by recrystallization to nonfibrous forsterite and silica at about 820°C.

Of the amphiboles, amosite asbestos, also referred to in the mineralogical literature as grunerite, has a theoretical chemical formula approximating $(Mg, Fe^{2+})_7 \cdot Si_8O_{22}(OH)_2$, where the iron oxides can contribute up to 45% of the total weight, with only

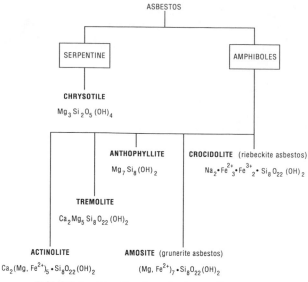

**FIGURE 1** Classification of asbestos fiber types.

5–7% for magnesium oxide. The decomposition temperature is higher (600–800°C) than for chrysotile. Crocidolite asbestos, also known to mineralogists as riebeckite, has the following approximate chemical formula: $Na_2 \cdot Fe^{2+}_3 \cdot Fe^{3+}_2 \cdot Si_8O_{22}(OH)_2$, where iron oxides can contribute up to 39% of the total weight and sodium oxide, up to 8%. This is the most acid resistant of commercial asbestos fiber types.

Chrysotile asbestos separates into silky, flexible, tough fibers, whereas amphibole asbestos fibers are more brittle. Chrysotile is also sometimes referred to as white asbestos; amosite, as brown asbestos; and crocidolite, as blue asbestos.

## B. Uses

Asbestos fibers are used in a wide range of applications. Current world production (Table II) is about 4 million metric tons, of which chrysotile represents 98% and amphiboles (mainly amosite and crocidolite), 2%.

Asbestos cement construction materials represent 70–75% of end uses around the world, automotive products represent about 15%, and the remainder is found in textiles, flooring and roofing materials, coatings, and sealants, papers, as filler in plastic composites, and in other specialty applications (e.g., diaphragms in electrolytic cells and spacecraft components). These proportions are different for the United States (Table III).

TABLE I

Average Chemical Composition of Commercial Asbestos Fiber Types

| Component | Chrysotile (wt. %) | Crocidolite (wt. %) | Amosite (wt. %) |
|---|---|---|---|
| $SiO_2$ | 38.75 | 50.90 | 49.70 |
| $Al_2O_3$ | 3.09 | None | 0.40 |
| **$Fe_2O_3$** | **1.59**[a] | **16.85** | **0.03** |
| **FeO** | **2.03** | **20.50** | **39.70** |
| MnO | 0.08 | 0.05 | 0.22 |
| **MgO** | **39.78** | **1.06** | **6.44** |
| CaO | 0.89 | 1.45 | 1.04 |
| $K_2O$ | 0.18 | 0.20 | 0.63 |
| $Na_2O$ | 0.10 | 6.20 | 0.09 |
| $H_2O^+$ | 12.22 | 2.37 | 1.83 |
| $H_2O^-$ | 0.60 | 0.22 | 0.09 |
| $CO_2$ | 0.48 | 0.20 | 0.09 |

[a]Boldfaced numbers point to the principal differences in composition between commercial asbestos fiber types.

**TABLE II**

World Production[a] of Asbestos

| | Metric tons produced | | |
|---|---|---|---|
| | 1991 | 1993 | 1995 |
| USSR ('91); CIS ('93, '95) | 2,500,000 | 1,700,000 | 1,000,000 |
| Canada | 670,000 | 510,000 | 511,000 |
| Brazil | 200,000 | 250,000 | 80,000 |
| South Africa | 160,000 | 130,000 | 100,000 |
| Zimbabwe | 160,000 | 150,000 | 145,000 |
| China | 200,000 | 250,000 | 250,000 |
| Others | 152,000 | 124,000 | 118,000 |
| Total | 4,042,000 | 3,114,000 | 2,304,000 |

[a]Estimates (Source: Natural Resources Canada).

The need for fibrous materials dates far back in history, when humans began to use them for the reinforcement of clays with straw, grasses, hair, flax, etc. When the need for structural composites exceeded these easily available organic materials, new man-made composites were prepared, using naturally occurring fibers of mineral types such as asbestos. Wholly man-made mineral fibers eventually followed, including glass and mineral wools and, more recently, metallic wiskers, ceramic fibers, and newly developed organic fibers such as aramid. Despite substantial research development, many applications remain for which asbestos is the only satisfactory choice. Whether cost-effectiveness or societal risk–benefit analysis is used, industry researchers believe that chrysotile asbestos is still unmatched in such applications as automotive friction materials and reinforced cement.

## II. ASBESTOS-RELATED HEALTH EFFECTS

### A. Sources of Environmental and Occupational Exposure

Asbestos fibers released into the air and water come from two distinct sources. Much comes from natural sources because asbestos minerals are widely distributed thoughout the world. Chrysotile is the most abundant and is found in most serpentine rock formations in the earth's crust. The distribution of amphibole asbestos is far more limited, mainly in South Africa and Australia. Analyses of the mineral content of the Greenland ice cap have shown that chrysotile was present in the atmosphere long before it was used industrially. Other analyses of asbestos in water supplies indicate that its presence is due to the natural erosion of rock formations.

It is generally agreed that the total amount of asbestos released into the environment from natural sources far exceeds that emitted from the relatively more recent industrial sources. Thus, asbestos is present practically everywhere in the air and water and in the living organisms that feed off them. Almost everyone inhales several thousand fibers every day and, among these, ingests hundreds of thousands, possibly millions, of asbestos fibers. Drinking water typically contains 200,000 to 2 million asbestos fibers per liter.

**TABLE III**

Major Uses of Asbestos[a]

| | United States (%) | Western Europe (%) |
|---|---|---|
| Asbestos cement | 21 | 80 |
| Friction materials | 25 | 5 |
| Textiles | 5 | 3 |
| Others (e.g., roofing, products, gaskets and packings, coating compounds, and additives in plastics) | 49 | 12 |

[a]Estimates from the Asbestos Institute, Montreal, Quebec, Canada, for 1988.

Artificial sources of asbestos fibers are contributed by human activities. Road construction, excavation, and tilling of the soil are one source. Mining and milling of asbestos is another. A third source is the manufacture of asbestos-containing products. A fourth source is the handling of these products in construction activities, which include removal, demolition, and repair. Finally, transportation and disposal of wastes are sources of asbestos fibers. In the past, all of these activities contributed to high exposure levels in the workplace. More recently, strict enforcement of regulations in the safe use of asbestos has contributed to dramatically reduce emissions in the workplace and the environment. For example, reports by the U.S. National Institute for Occupational Safety and Health show that in 1988 practically all industrial activities effectively complied with the 1986 permissible exposure level of 0.2 fiber per milliliter of air, a level which could be two or three orders of magnitude lower than occupational exposure levels often measured in the past.

## B. Asbestos-Related Diseases

Asbestos has been associated with diffuse pulmonary fibrosis (i.e., asbestosis), bronchial carcinoma (i.e., lung cancer), and primary malignant tumors of the pleura and the peritoneum (i.e., mesothelioma). Cancers at other sites have been reported (e.g., gastrointestinal and laryngeal cancers), but the causal relationship with asbestos exposure is less well established. No substantial evidence exists for cancers at other sites. Potencies vary among fiber types, with amphibole types (e.g., crocidolite and amosite) being more potent than chrysotile, especially for mesothelioma. Responses vary with the dose (i.e., intensity, duration, and time since the first exposure) and also with the size of the inhaled particles; long thin fibers (i.e., a length of more than 5 $\mu$m, a diameter of less than 3 $\mu$m) are more potent than shorter fibers. These biologically relevant parameters are discussed in Section III. [*See* Pulmonary Pathophysiology; Toxicology, Pulmonary.]

### 1. Asbestosis

Asbestosis is a slowly developing chronic restrictive lung disease resulting from the inhalation of asbestos. The disease is characterized by scarring (i.e., diffuse interstitial fibrosis), and the most prominent symptom is breathlessness associated with "crackling" noises (i.e., crepitations) in the lung. Asbestosis is neither malignant nor necessarily fatal, but it can lead to respiratory impairment and to a higher risk for the development of other diseases (e.g., cardiorespiratory complications and lung cancer).

Restrictive lung changes cause reduced lung functions in the volumes of gas exchanged. Typical radiographic changes of asbestosis are linear, irregular small opacities in the lower and middle lung zones. The fibrotic changes (i.e., scarring) are irreversible.

With the gradual introduction of dust control measures in the workplace since the early 1970s, the incidence of asbestosis has shown a substantial decrease. According to the 1984 Royal Commission on Matters of Health and Safety Arising from the Use of Asbestos in Ontario, under present regulations asbestosis will become a disease of the past. There is no evidence of asbestosis as a result of environmental exposure.

### 2. Lung Cancer

Unlike asbestosis, lung cancer is not specifically related to asbestos. The most important cause of lung cancer is undoubtedly smoking, responsible for at least 85% of the lung cancer cases.

Although an excess incidence of lung cancer has been reported in asbestos workers who do not smoke, most people exposed to asbestos who develop cancer are smokers. The association of cigarette smoking and exposure to asbestos in the induction of lung cancer has received much attention, and studies have shown a synergism of action, resulting in a multiplicative effect. Thus, workers who were exposed to asbestos (i.e., of mixed fiber types) and who do not smoke have shown a fivefold greater risk of dying from lung cancer than people not exposed to asbestos who do not smoke, whereas asbestos workers who do smoke have shown a fivefold greater risk than the smoking control population. Taking into account that smokers are at 11 times greater risk for lung cancer than nonsmokers, the risk for smoking workers who are also exposed to asbestos is approximately 55 times that of the nonsmoking control population. Actual numbers from studies show death rates of 11.3, 58.4, 122.6, and 601.6 per 100,000 men for nonexposed nonsmokers, asbestos-exposed nonsmokers, nonexposed smokers, and asbestos-exposed smokers, respectively (Table IV).

### 3. Mesothelioma

Mesothelioma is a rare form of cancer of the cells lining the pleura and the peritoneum. Generally, cases

TABLE IV

## TABLE IV

Effects of Smoking and Occupational Exposure to Asbestos on Lung Cancer Death Rate[a]

| Groups | | Lung cancer death rate (per 100,000) | Ratio |
|---|---|---|---|
| Exposure[b] | Smoking | | |
| Control | No | 11.3 | 1.0 |
| Asbestos workers | No | 58.4 | 5.1 |
| Control | Yes | 122.6 | 10.8 |
| Asbestos workers | Yes | 601.6 | 53.2 |

[a]From E. C. Hammond, I. J. Selikoff, and H. Seidman (1979). *Ann. N.Y. Acad. Sci.* **330**, 473–490.

[b]Insulators were exposed to mixed fiber types.

of mesothelioma are rapidly fatal. For a time mesothelioma was thought to be exclusively associated with exposure to asbestos, but more recent reviews indicate that 30–40% of the cases have occurred in the absence of any known asbestos exposure. The reported outbreak of mesothelioma in rural Turkey has been associated with fibrous zeolites found in such regions. Other possible causes have been reported, such as exposure to ionizing radiation and to biogenic silica in sugar cane. Several agents, both organic and inorganic, have been shown to induce malignant mesothelioma in experimental animals. With regard to asbestos-related mesothelioma, it is generally agreed that it is strongly associated with exposure to amphiboles (e.g., crocidolite and amosite). The tumor is comparatively rare in chrysotile-exposed populations; in these cases the occasional contamination of chrysotile with tremolite asbestos has been mentioned as the likely cause.

## C. Animal Experimentation

The majority of studies published in the field of animal experimentation, following administration of different asbestos fiber types by inhalation or injections, have not distinguished significant differences in the potencies among asbestos fiber types. The reported effects were not consistent with epidemiological observations, indicating that amphibole types are markedly more potent than chrysotile in inducing asbestosis, lung cancer, and mesothelioma. However, in the great majority of experimental protocols, the comparison of effects (e.g., fibrogenicity and tumor yield) has

traditionally been based on a gravimetric basis, i.e., the effects produced by an equal mass of tested minerals were compared (e.g., a 20-mg dose by intraperitoneal injection or 20 mg/m$^3$ by inhalation).

In recent years the development of sophisticated techniques for tissue mineral analysis has led to posssible explanations of the inconsistencies between animal experimental data and epidemiological evidence. These differences might be attributable to the different durabilities of mineral fibers in lung tissues and the different relative persistence of these fibers in laboratory animals and humans. Another explanation could also reside in the basis for reporting the observed effects (i.e., whether per equal mass or per equal number of fibers).

In fact, limited retrospective attempts to transform gravimetric doses into fiber number doses have indicated that, if based on fiber number, the pathogenicity would show fiber for fiber that chrysotile is less pathogenic than the other asbestos fiber types, and possibly less than some man-made mineral fibers as well. One *in vitro* study on the comparison of mass versus number of fibers showed that chrysotile requires a significantly higher number (i.e., 50-fold) of fibers than crocidolite to produce an equivalent degree of toxicity in cultures of lung cells. Thus, it would appear that results reported on the basis of fiber number, as opposed to fiber mass, are consistent with the results of epidemiological studies. Indeed, more research is needed to resolve this important issue. The role of durability is discussed in Section III,B.

## D. Exposure to Asbestos and Disease Risk

### I. Workplace

It is generally agreed that asbestos is a workplace hazard and that the risk for the general population due to the presence of asbestos in the environment and inside buildings is much smaller. It has also been shown that the risk in occupational settings is dose related and increases with duration of exposure. While this is certainly true for the relatively high occupational exposure levels of the past (in some cases over 100 fibers per milliliter of air), the question as to whether present low exposure levels (i.e., fewer than 2 fibers per milliliter) are still associated with some residual risk is still debated. The problem is that the carcinogenic risk at low exposure levels has not been based on actual observations, but, rather, on extrapolation from observed risk at much higher exposure levels. There is no direct epidemiological evidence for the existence or absence of a threshold below which there is no risk. For regulatory purposes it is assumed that there is no such threshold and that any exposure will always carry some risk, however small.

Studies have shown that no excess risk of cancer (at any site) could be detected in cohorts of workers at asbestos cement and friction materials manufacturing plants using exclusively chrysotile asbestos. In these studies, reported in the early 1980s, the exposure levels were approximately one or two fibers per milliliter of air. The most recent update of chrysotile-only exposed workers (miners and millers) was published in 1993. The study involved the fifth follow-up of a cohort of 11,000 men born between 1891 and 1920, employed for at least 1 month in the chrysotile mines and mills of Québec. The authors established that there was no significant excess lung cancer deaths of these men at exposure levels of 300 million particles per cubic foot × years (roughly equivalent to 50 fibers per milliliter for 20 years) nor evidence of a trend. For workers exposed to amphibole asbestos, or to a mixture containing it, the situation is very different. In practically all cohorts, mesothelioma cases have been reported, even in occupational settings where crocidolite or amosite asbestos had been used for relatively short periods. Studies on asbestos cement plant workers, carried out on two cohorts totaling 3102 people in England and on one cohort of 1216 people in Sweden, were reviewed in 1986. A study on friction material workers was carried out on 13,460 workers at a factory founded in 1898. In all of the studies mentioned, the results were statistically significant at a 95% confidence limit.

Developments in the techniques of mineral analysis of lung tissue have brought some explanation for the different pathogenic potentials among asbestos fiber types. It has been found that amphiboles are highly concentrated (i.e., 100- to 300-fold) in the lungs of patients with mesothelioma compared to control populations, whereas chrysotile concentrations are not increased. It has also been reported that amphibole concentrations in lung tissue vary with the severity of asbestosis, with no change seen in chrysotile concentrations. Animal experimentation has also shed some light, showing that while inhaled chrysotile asbestos fibers tend not to remain in the lungs for long periods of time, amphibole fibers are insoluble and remain in the lungs for practically a whole life span. Thus, even at low but constant exposures, the amphibole asbestos fibers tend to accumulate in lung tissue. The few re-

ported mesothelioma cases associated with household exposures in relatives of plant workers using amosite, and also those found in populations living in the neighborhood of crocidolite asbestos mines in South Africa, could be explained by this phenomenon.

## 2. General Population

### a. Ingestion

The presence of asbestos fibers in drinking water, and perhaps also in food, has received much attention with regard to the potentially increased incidence of gastrointestinal tract cancers in populations exposed over many years. Typically, drinking water in North America cities contains asbestos concentrations ranging from below detection to 2 million fibers per liter, with the fibers being predominantly chrysotile. Concentrations of 10 million fibers are not uncommon.

Epidemiological studies have been conducted in areas with relatively high concentrations, such as Asbestos and Thetford Mines, chrysotile mining communities in the eastern townships of Quebec, where drinking water can be as much as 100 times more polluted with chrysotile asbestos than in most North American and European cities. No statistically significant excess mortality rate was found. In general, all studies provide little convincing evidence of an association between asbestos in public water supplies and cancer induction. Even experimental studies have failed to provide evidence that ingested asbestos is carcinogenic to animals.

### b. Inhalation

Asbestos fibers are found almost everywhere in the environment, from both natural and industrial sources. Fiber measurements in cities in the United States, Canada, England, the Federal Republic of Germany, and Switzerland show a range of values between less than 0.001 (the average) and 0.01 fibers per cubic millimeter in some rare localized areas. In the United States an average concentration of 0.0006 fibers per milliliter has been reported. Inside buildings the nonoccupational indoor levels are generally within the range found in ambient air. At these low levels, according to the World Health Organization, the risk is undetectably low. However, workers engaged in repair or removal activities in buildings could become exposed to much higher levels. In fact, some studies have shown that removal activities can result in high levels which persist for several weeks after completion of the work inside buildings where preremoval levels were below the detection limit. The U.S. Environmental Protection Agency has determined a postabatement (i.e., clearance) level of 0.01 fiber per milliliter before allowing occupancy of a building after removal activities.

## III. RELEVANT PARAMETERS OF BIOLOGICAL POTENCY

### A. Size

The potencies of asbestos fibers vary according to their dimensions. Whereas long (i.e., more than 5 $\mu$m) fibers are certainly associated with biological activity, there is doubt as to whether the short (i.e., less than 5 $\mu$m) fibers, which form the majority of fibers found in the general environment, contribute to risk. In animal experiments short fibers have been cleared rapidly from the lungs, most likely by the phagocytic (i.e., ingestion) action of scavenger cells called macrophages, and possibly also by the rapid dissolution by biological fluids in the case of chrysotile asbestos.

For long fibers the phagocytosis by macrophages is incomplete, which is thought to induce the release of fibrogenic factors from macrophages. Scarring (i.e., fibrosis) results if enough macrophages are constantly challenged for long periods. Carcinogenicity is also dependent on fiber dimensions; long thin fibers are more potent than short thick fibers. Fibers with a maximum carcinogenic potency are longer than 8 $\mu$m and less than 1.5 $\mu$m in diameter. Thus, the dimensions of fibers causing fibrogenicity or carcinogenicity are similar. Some authors have suggested that fibrosis might be a prerequisite for carcinogenicity. This is a debated issue for which further extensive analysis is needed.

### B. Durability

A number of reasons for the different potencies in causing pulmonary diseases between chrysotile and the amphiboles have been proposed and investigated. One reason relates to the striking differences among asbestos fiber types with regard to their relative persistence in tissues. This observation has been possible by studying fibers extracted from known amounts of lung tissue using recent analytical methods, such as energy-dispersive analysis of X-rays (EDAX) and selected-area electron diffraction (SAED) on a transmission electron microscope with direct computer readout. These modern techniques allow determination of the chemical composition and the crystalline structure of individual particles as small as 1 $\mu$m. Data from this type of investigation are known as "lung burden" analysis.

These methods have shown a predominance of amphibole fibers, mainly crocidolite and amosite and occasionally tremolite. For example, studies published in the early 1980s were conducted on lung tissue samples from workers whose deaths were considered to be asbestos related compared to controls exposed to various levels of urban pollution. The results showed that amphiboles were present in patients with levels up to 100 times greater than in the controls, but the amounts of chrysotile were similar. Furthermore, in asbestotic cases the amounts of amphiboles, but not of chrysotile, related well quantitatively to the severity

TABLE V

Asbestos Content by Fiber Type in Lung Tissue: Correlation with Severity of Asbestosis[a]

| Grade of asbestosis | Total number of cases | Mean fiber count ($10^6$) | Fiber number ($10^6$) per gram of dry lung tissue | | |
|---|---|---|---|---|---|
| | | | Amosite | Crocidolite | Chrysotile |
| None | 19 | 22.94 | 1.34 | 2.44 | 9.62 |
| Minimal | 36 | 42.13 | 2.44 | 14.47 | 12.89 |
| Slight | 69 | 59.47 | 14.94 | 21.30 | 12.25 |
| Moderate | 49 | 343.15 | 94.98 | 73.88 | 13.75 |
| Severe/marked | 16 | 2550.29 | 266.21 | 2224.04 | 17.34 |
| All grades | 189 | 336.9 | 53.21 | 218.21 | 12.93 |

[a]From J. C. Wagner, C. B. Moncrieff, R. Coles, D. M. Griffiths, and D. E. Munday (1986). *Br. J. Ind. Med.* **43**, 391–395.

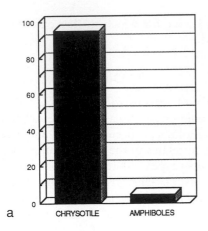

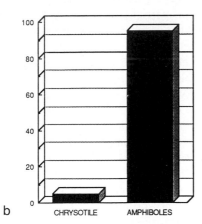

**FIGURE 2** (a) Proportion (%) of asbestos fiber types used commercially. (b) Fiber type composition (%) in asbestos body cores.

of the disease. The results of one such study are shown in Table V.

Another line of evidence comes from analysis of "asbestos bodies." Once inhaled, asbestos fibers can be deposited peripherally, even if their length exceeds 200 $\mu$m, because fibers line up with their long axes parallel to the direction of air flow. For this reason their deposition pattern (i.e., location) is determined mainly by their diameter rather than by fiber length. When deposited in the lung, the fibers become engulfed by the macrophage cells, which deposit a coating of iron–protein–mucopolysaccharide material. This process leads to the formation of the so-called asbestos bodies, dumbbell-shaped golden brown structures characteristic in appearance. This process is not specific for asbestos fibers, as other inhaled

structures can also become coated (e.g., talc, glass fibers, ceramic fibers, and other nonasbestos fibers). For this reason a more generic term, "ferruginous bodies," is also used.

Analysis of the composition of the fibrous core of ferruginous bodies in autopsy lung tissue from the general population, using EDAX and SAED techniques with the electron microscope, has revealed that most ferruginous bodies with a thin transparent core have amphibole asbestos cores, although the asbestos fiber type used commercially is predominantly (i.e., 95%) chrysotile (Fig. 2). Thus, while chrysotile tends to be gradually cleared from the lungs, amphiboles persist and accumulate over time.

Many published epidemiological studies have shown that this difference in potencies is particularly striking for inducing mesothelioma. These studies have shown that the mesothelioma cases are correlated with vastly increased lung burdens of amphiboles, but not chrysotile. For example, two reports were published in 1980 by the International Agency for Research on Cancer. One report showed that mesothelioma patients in England had more amphibole, but not chrysotile, fibers in their lungs than did control cases. A similar result was obtained with 274 fatal cases of mesothelioma in North America during 1972 and, more recently, with many mesothelioma cases in Canada.

A 1982 study indicated that patients in the general population who have pleural plaques (i.e., thickening of the tissue lining the pleural surfaces) have about the same total number of fibers as do controls, but their lungs contain about a 50-fold increase in long amphibole fibers of commercial origin. Other data published in 1984 on the mineral analysis content of lung tissue from insulation and shipyard workers with mesothelioma show a 250-fold increase in amphibole fibers over those found in the general population.

These results of tissue burden analysis are generally accepted as a key element in pointing to those fiber types that have long durability (i.e., residence time) in lung tissue, with the possible consequence of disease induction.

## IV. CONCLUSION

Asbestos has left a sad legacy of diseases as a result of past misuse (e.g., the spraying of loose asbestos) and high uncontrolled exposures in the workplace. Fortunately, the risk for the general population appears negligible. In 1986 the International Labour

Organization adopted a Convention Concerning Safety in the Use of Asbestos (Convention 162), which, among its articles, states that "spraying of all forms of asbestos shall be prohibited" (article 12.1), and "the competent authority shall prescribe limits for the exposure of workers to asbestos or other exposures criteria for the evaluation of the working environment" (article 15.1).

With regard to the use of alternatives (i.e., the so-called asbestos substitutes), the Convention (article 10 a) indicates that they must be evaluated by the competent authority as harmless or less harmful. In view of the relevant parameters of the biological potency of fibrous materials discussed here, all fibers, natural or synthetic, should be considered potentially carcinogenic if they are long, thin, and durable, i.e., when they have an aspect ratio (i.e., length over diameter) of 5:1, are longer than 5 $\mu$m, are thinner than 1 $\mu$m, and can persist in the tissue for more than 2 years.

## BIBLIOGRAPHY

Churg, A. (1988). Chrysotile, tremolite, and malignant mesothelioma in man. *Chest* 93, 621–628.
Davis, J. M. G., and Jones, A. D. (1988). Comparisons of the pathogenicity of long and short fibers of chrysotile asbestos in rats. *Br. J. Exp. Pathol.* 69, 717–737.
Davis, J. M. G., and McDonald, J. C. (1988). Low level exposure to asbestos: Is there a cancer risk? *Br. J. Ind. Med.* 45, 505–508.
Gardner, M. J., and Powell, C. A. (1986). Mortality of asbestos cement workers using almost exclusively chrysotile fibre. *J. Soc. Occup. Med.* 36, 124–126.
Hammond, E. C., Selikoff, I. J., and Seidman, H. (1979). Asbestos exposure, cigarette smoking and death rates. *Ann. N.Y. Acad. Sci.* 330, 473–490.
McDonald, J. C., Liddell, F. D. K., Dufresne, A., and McDonald, A. L. (1993). The 1891–1920 birth cohort of Québec chrysotile miners and millers: Mortality 1976–88. *Br. J. Ind. Med.* 50, 1073–1081.
Nonoccupational Exposure to Mineral Fibers (1989). IARC Sci. Pub. No. 90 (J. Bejnore, J. Peto, and R. Saracci, eds.), Lyon, France.
Newhouse, M. L., Berry, G., and Skidmore, J. W. (1982). A mortality study of workers manufacturing friction materials with chrysotile asbestos. *Ann. Occup. Hyg.* 26, 899–909.
Pelnar, P. V. (1988). Further evidence of non-asbestos-related mesothelioma. *Scand. J. Work Environ. Health* 14, 141–144.
"Report of the Royal Commission on Matters of Health and Safety Arising from the Use of Asbestos in Ontario" (1984). Ontario Ministry of the Attorney General, Toronto.
Ross, M., Kuntze, R. A., and Clifton, R. A. (1984). A definition for asbestos. *ASTM Spec. Tech. Publ.* 834.
Wagner, J. C. (ed). (1986). The biological effects of chrysotile. *Accomplish Oncol.* 1.
Wagner, J. C., Moncrieff, C. B., Coles, R., Griffiths, D. M., and Munday, D. E. (1986). Correlation between fibre content of the lungs and disease in naval dockyard workers. *Br. J. Ind. Med.* 43, 391–395.
World Health Organization (1986). Asbestos and other natural mineral fibers. *Environ. Health Criter.* 53.

# Ascorbic Acid

ROBERT B. RUCKER
FRANCENE M. STEINBERG
*University of California, Davis*

## GLOSSARY

**Carnitine** Metabolite derived from L-lysine, which is responsible for transporting fatty acids from the cytoplasm into the mitochondria

**Collagen** One of the main constituents of connective and vascular tissues, e.g., bone, skin, blood vessels, representing about one-third of the body's protein

**Complement** Group of about 20 serum proteins that interact in response to innate and adaptive immunological signals; some complement proteins contain hydroxyproline

**Enediol** Compound with two adjacent double-bonded carbon atoms, each with an associated hydroxyl group. A strong redox potential is a characteristic of enediols. The hydroxyl groups can also disassociate their hydrogen ions in a manner characteristic of acids

**Hydroxyproline** L-Proline is hydroxylated to the nonessential amino acid L-hydroxyproline. Hydroxyproline is found in extension (plant) or collagen (animal) structural proteins and in proteins that contain collagen-like sequences, e.g., acetylcholine esterase, Clq-complement

**Mixed-function oxidation** Reactions involving the direct utilization of $O_2$ wherein a part of the molecule is utilized in one product (R—OH) and the other part combines with $H_2O$ to form hydrogen peroxide ($H_2O_2$)

**Nitrosamine** Nitrites and secondary amines are precursors of nitrosamine. The nitrosamines ($R_2N$—$N$=$O$) are of concern since they can act as carcinogens

**Osmotic pressure** Measure of the number of dissolved particles per unit volume. The pressure-related properties of liquid compartments, when separated by semipermeable membranes, are influenced by the concentration of solute (dissolved substances) in a given liquid compartment

**Oxidation** Chemical reaction in which oxygen is gained, or hydrogen or one or more electrons is lost (see Reducing Agent)

**Oxo function** Carbon–oxygen double-bonded structures, e.g., a ketone

**P450-dependent hydroxylases** P450-dependent hydroxylases transfer electrons to cytochrome-like redox proteins that are characterized by their absorbance at 450 nm

**Reducing agent** Agent that facilitates the removal of oxygen from a compound or provides hydrogen or electrons for chemical processes

**Transition element** Element found in the third or fourth periods of the periodic table, e.g., zinc, copper, iron, and manganese; many transition elements can catalyze redox reactions

ASCORBIC ACID IS A VITAMIN FOR HUMANS AND other primates, guinea pigs, and certain fruit-eating birds. It serves chemically as a strong reducing agent and as cofactor for a number of microsomal enzymes. One such enzyme is prolyl hydroxylase, which is important in connective tissue protein assembly and maturation. For example, ascorbic acid deprivation, or scurvy, results in impaired wound healing because of defective collagen maturation. Ascorbic acid is also important to adrenal gland function and various microsomal oxidations.

ENCYCLOPEDIA OF HUMAN BIOLOGY, Second Edition, VOLUME I.   Copyright © 1997 by Academic Press.   All rights of reproduction in any form reserved.

# I. INTRODUCTION

Ascorbic acid (i.e., vitamin C) plays important roles as a chemical reducing agent, an antioxidant, and an enzymatic cofactor. Ascorbic acid is an essential dietary factor, i.e., a vitamin, for many insects, invertebrates, and some avian and mammalian species, including humans. Moreover, when ascorbic acid is required, the nutritional need is relatively high compared with other vitamins. Consequently, it is possible to become ascorbic acid deficient, or scorbutic, by consuming a monotonous diet or a diet composed of highly processed foods if they are low in ascorbic acid.

Human history has been shaped periodically by ascorbic acid deficiency, or scurvy. Throughout the period of intense exploration of the Americas and the search for new passages to India, more explorers are thought to have died of scurvy than of any other cause. For example, Magellan lost four or five ships to scurvy in his 1519–1522 expeditions. The failure of many of the arctic expeditions has also been attributed to scurvy. It has been speculated that the development of the Pacific Rim and the western coast of the Americas would be considerably different if scurvy had not been such a predominant factor to early expeditions.

By the end of the 19th century, the incidence of scurvy had declined as a result of increased acceptance and availability of foods containing ascorbic acid, such as potatoes, citrus fruits, and sauerkraut. It was also during this period that advances in chemistry and biology accelerated our conceptual understanding of nutrient function. As an example, studies by A. Holst and F. Fröhlich in 1907 showed that guinea pigs developed scurvy-like symptoms when fed a diet devoid of ascorbic acid. These studies were among the first to use an animal model to examine nutrient deficiency and, in this case, to test foods for antiscorbic properties.

# II. CHEMISTRY OF ASCORBIC ACID

In 1933, A. Szent-György and W. M. Haworth received Nobel prizes in medicine and chemistry, respectively, for work that focused in part on ascorbic acid chemistry. By this time, the structures given in Fig. 1 were known, as well as some of the chemical properties of ascorbic acid. It was also during this period that chemical synthesis of ascorbic acid was achieved.

Ascorbic acid is a 2,3-enediol-L-gulonic acid. Both of the hydrogens of the enediol groups disassociate, which results in the acidity of ascorbic acid. Enediols are often excellent reducing agents; facilitating chemi-

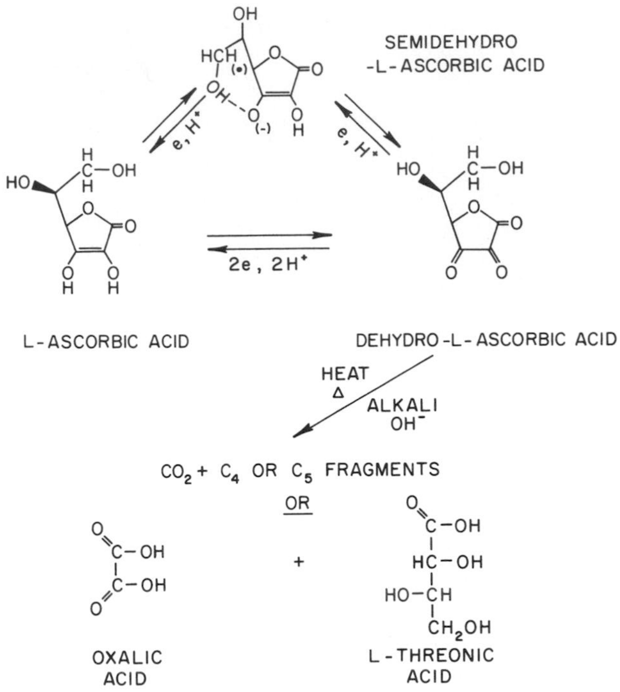

**FIGURE 1**   Oxidation states of L-ascorbic acid. The dehydro or 2,3-dioxo form of L-ascorbic acid is cleaved in a strong base (alkali) or by heat to $CO_2$ and various smaller carbon fragments, most often "onic" acids, such as L-threonic acid. Oxalic acid may also be produced.

cal reductions is a major function of ascorbic acid. Moreover, ascorbic acid is an effective quencher of singlet oxygen, a particularly reactive form of oxygen. Transition metals, such as copper and iron, are also reduced in the presence of ascorbic acid.

Reactions involving ascorbic acid often occur in a stepwise fashion with monodehydroascorbic acid as a semiquinoid intermediate (Fig. 1). This form of ascorbic acid is a strong acid ($pK = -0.45$). It is also an oxidizing agent, which disproportionates to ascorbic acid and dehydroascorbic acid. Dehydroascorbic acid is not as hydrophilic as ascorbic acid since it exists in a completely deproated form. This form of ascorbic acid is more easily transported across cell membranes than is L-ascorbic acid. The dehydro form is susceptible to cleavage in alkali, e.g., to oxalic acid and L-threonic acid (Fig. 1). Some other chemical and physical properties of ascorbic acids are given in Table I.

# III. OCCURRENCE IN FOODS AND TISSUES

Many of the chemical assays for L-ascorbic acid are based on its reducing capacity. The dehydro form may

## TABLE I
### Selected Physical and Chemical Properties of L-Ascorbic Acid

| | |
|---|---|
| Molecular weight | 176.13 |
| Melting point | 190–192°C |
| Redox potential | $E_0^1$ + 0.166 V at pH 4.0 |
| Absorption spectra | |
| pH 2 (1%, 1 cm) | $E_{max}$ 695 at 245 nm |
| pH 4 (1%, 1 cm) | $E_{max}$ 940 at 265 nm |
| Dissociation constants | |
| $pK_1$ | 4.2 |
| $pK_2$ | 11.6 |
| Solubility (g/dl) | |
| Water | 33 |
| Ethanol | 2 |
| Ether | 0 |
| Alkanes | 0 |
| Chloroform | 0 |
| Benzene | 0 |

be estimated by taking advantge of the reactivity of the 2,3-dioxo functions. Other methods are based on separation of L-ascorbic acid and related metabolites by high-performance liquid chromatography. Ascorbic acid is the only water-soluble vitamin for which no microbiological assay exists.

Many fresh fruits contain high concentrations of ascorbic acid (Table II)—some as much as 0.1–0.2% on a fresh weight basis (100–200 mg/100 g). Ascorbic acid may be lost easily from foods because of its solubility and its ease of oxidation during cooking or contact with transition metals, such as iron or copper. The minimal ascorbic acid requirement, however, is usually met if a variety of foods are consumed. For most countries, the recommended intakes of ascorbic acid range from 35 mg/day (for infants) to 75–100 mg/day for nursing women. These amounts are achieved by consuming a typical Western diet, although consumption of monotonous diets, e.g., an overly cooked soup or gruel stock, for weeks at a time can lead to compromised status and even depletion of the body pool of ascorbic acid.

## IV. METABOLISM

### A. Synthesis

The pathways for ascorbic acid biosynthesis and degradation are shown in Fig. 2. To iterate, ascorbic acid synthesis is compromised in insects, invertebrates, and most, but not all, fish (e.g., carp) because

of a reduction or complete absence of the enzyme L-gulonolactone oxidase. In lower animals (amphibians and reptiles) and birds, ascorbic acid is made in the kidney. The egg-laying mammals also synthesize ascorbic acid in their kidneys. Mammals and higher-order birds usually synthesize ascorbic acid in the liver. Marsupials appear to make L-ascorbic acid in both the kidney and liver. In humans, however, evidence suggests that a dietary source is essential. This

## TABLE II
### L-Ascorbic Acid in Fresh Foods

| | mg/100 g[a] |
|---|---|
| Fruits | |
| Apple | 3–30 |
| Banana | 5–15 |
| Blackberry | 8–12 |
| Currant, red | 20–50 |
| Currant, black | 150–200 |
| Grape | 2–5 |
| Lemon | 40–60 |
| Melon | 10–30 |
| Orange | 30–60 |
| Peppers | 50–300 |
| Rose hips | 300–1100 |
| Strawberry | 40–80 |
| Tomato | 20–30 |
| Other plants and vegetables | |
| Brussels sprouts | 50–150 |
| Cabbage | 30–70 |
| Carrot | 5–10 |
| Lettuce | 15–20 |
| Onion | 5–15 |
| Parsley | 100–300 |
| Potatoes | 4–30 |
| Radishes | 10–30 |
| Spinach | 40–80 |
| Animal products | |
| Beef | 1–3 |
| Pork | 1–2 |
| Veal | 1–2 |
| Ham | 1–2 |
| Liver (bovine) | 5–15 |
| Kidney (bovine) | 5–15 |
| Milk | |
| Bovine | 0.5–2 |
| Human | 3–6 |
| Crustaceans | |
| Crab muscle | 1–3 |
| Lobster muscle | 3 |
| Shrimp muscle | 1–2 |
| Fish | 0.5–2 |

[a] Note that on a dry weight basis, the relative L-ascorbic acid content may be as high as 1–5% of the dry weight.

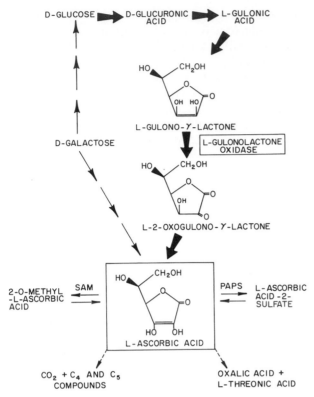

**FIGURE 2** Pathways for L-ascorbic acid synthesis. In many mammals, synthesis is from the direct oxidation of D-glucose (heavy arrows). In many lower animals and plants, the pathway can involve the direct oxidation of D-galactase. L-Ascorbic acid-2-sulfate and 2-O-methyl-L-ascorbic acid are important cellular metabolites. 3'-Phosphoadenosine 5'-phosphosulfate (PAPS) is an important cellular sulfating agent. S-Adenosyl-methionine (SAM) is an important cellular methylation agent.

versy exists regarding the relationship between intake and uptake of ascorbic acid, most careful studies indicate that 80–90% of the vitamin is absorbed within the range of normal intakes (up to 200 mg/day). The absorption is characterized by saturation kinetics and appears to be carrier dependent.

In humans, when amounts are consumed in the range of 1 g or more, intestinal absorption is significantly decreased. A case may also be made that when consumed above the physiological requirements, the ascorbic acid that is not utilized is either excreted by the kidneys or is chemically modified or degraded (Fig. 2). Amounts of 10 g or more per day can cause an osmotic diarrhea and gastric irritation.

With respect to tissue distribution, the highest concentrations of ascorbic acid are found in the adrenal and pituitary glands (30–50 mg/100 g), followed by liver, thymus, brain, and pancreas (10–20 mg/100 g). Cell uptake of ascorbic acid may occur by simple diffusion if ascorbic acid is in the dehydro form. The uptake of ascorbic acid by red cells probably occurs by this mechanism. Other cells, e.g., lung, liver, and intestinal cells, take up ascorbic acid primarily by active processes. Ascorbate transport is dependent on Na(+)-dependent, energy-driven processes. In some, but not all, cellular systems, ascorbic acid transport competes with glucose. Specificity of L-ascorbic acid uptake is also suggested by the observations that ascorbic acid analogs, such as isoascorbic acid (erythorbic acid), are not enriched in tissues to the same extent as L-ascorbic acid when fed in diets or added to cell cultures. An interesting note is that the ascorbate content of tissues in diabetics appears depressed, which suggests that factors that respond to hyperglycemic states can compromise ascorbic acid tissue status.

is also true for all higher-order primates. Again, the controlling factor is the presence or absence of L-gulonolactone oxidase. For example, when a stabilized form of L-gulonolactone oxidase is injected into guinea pigs, which normally require dietary ascorbic acid, there is protection from scurvy. The cloning and chromosomal mapping of the functional gene for L-gulono-γ-lactone oxidase has been achieved. A missense mutation of L-gulono-γ-lactone oxidase has been shown to be the cause for the inability of the scurvy-prone animal to synthesize L-ascorbic acid.

## B. Absorption and Tissue Distribution

Dietary ascorbic acid is absorbed from the duodenum and proximal jejunum, and measurable amounts also can cross the mucous membranes of the mouth and the gastric mucosa. Although considerable contro-

## C. Cellular Regulation

Ascorbic acid is maintained in cells by several mechanisms. Active ascorbic acid reductases can maintain L-ascorbic acid in its reduced form, which prevents leakage from the cell as dehydroascorbic acid. Significant amounts of ascorbic acid may also exist as 2-O-methyl or 2-sulfate derivatives. In rats, about 5% of a labeled dose of ascorbic acid is recovered as the 2-O-methyl ascorbic acid in urine. Obviously, the ability to modify ascorbic acid to the 2-sulfate or 2-O-methyl derivative could have considerable impact on the ability of cells to compartmentalize or modulate the reducing potential of ascorbic acid. Moreover, cells also have the ability to degrade ascorbate acid

(Fig. 2) to either four-carbon or five-carbon metabolites with the concomitant formation of oxalic acid or carbon dioxide, respectively. Since the human body has no large reserve of ascorbic acid, the ability to control functional ascorbic acid levels is important. Without tissue storage there are few ways to buffer the body against toxic intakes of ascorbic acid by storing or sequestering an excess. Rather, protection from an excess of ascorbic acid depends on renal excretion or tissue catabolism.

The body pool of ascorbic acid is normally 1–2 g. Of this amount, about 2–4% of the body pool is degraded or excreted per day, i.e., 20–80 mg/day. This percentage remains relatively constant even as the body is depleted. Consequently, when the body pool is decreased 0.5–1.0 g, excretion may be as little as 10 mg per day. It is at this point, i.e., a body pool below 500 mg, that signs of scurvy are observed in experimental subjects. These observations are also consistent with a half-life of L-ascorbic acid in humans of about 10–20 days.

## V. CHEMICAL, COENZYMATIC, AND REGULATORY FUNCTIONS OF ASCORBIC ACID

Ascorbic acid plays several important roles, such as in mixed-function oxidations that involve the incorporation of molecular oxygen into substrates. In humans, mixed function oxidations that utilize L-ascorbic acid include (1) hydroxylation of proline and lysine and collagen; (2) hydroxylation of proline in other proteins, such as elastin, $C1_q$ complement, and acetylcholine esterase; (3) hydroxylation of steroids, drugs, xenobiotics, and various lipids via P450-dependent hydroxylases; (4) hydroxylation steps in the biosynthesis of carnitine; and (5) hydroxylation of tyrosine in the formation of catecholamines. Ascorbic acid can also act as a chemical reductant and chelating agent for metals such as iron. Moreover, studies have shown that the administration of ascorbic acid to glutathione (GSH)-deficient newborn rats and guinea pigs prevents toxicological responses to agents that are normally inactivated by GSH-sensitive systems. Mortality is decreased and there is an increase in tissue and mitochondrial GSH levels. The sparing effect of GSH in scurvy may be mediated through an increase in the reduction of dehydroascorbate, which would otherwise be degraded. GSH is essential for the physiological function of ascorbate because it is required in vivo for the reduction of dehydroascorbate by ascorbic acid reductases (see Section IV,C) and there is metabolic redundancy and overlap of the functions of these reducing agents.

Ascorbic acid can also increase dietary iron availability by maintaining iron in the intestinal lumen in its reduced state. Reduced iron ($Fe^{2+}$) complexes are more soluble and tend to move more easily through cell membranes and compartments. The chemical reducing potential of ascorbic acid also protects against the formation of nitrosamines. This protection takes place primarily in the stomach and intestine. As such, the reactions may be viewed as an "external" protective chemical modification since it occurs in the lumen of the stomach rather than in internal tissues.

Ascorbic acid may play a role as a regulatory agent. L-Ascorbic acid influences histamine metabolism in humans. There is an inverse correlation in man between ascorbic acid levels in serum and histamine levels. Second, steps in the transcriptional regulation of certain proteins, e.g., Type I collagen, appear to be influenced by the presence or the absence of ascorbic acid in cell culture systems in vitro and in tissue biosynthesis in vivo.

The primary function of ascorbate, however, is as a coenzyme. L-Ascorbic acid is a cofactor for a number of oxygenases and hydroxylases. Prolyl hydroxylases (prolyl-4- and prolyl-3-hydroxylases) and lysyl hydroxylase that catalyze important posttranslational chemical modifications of the collagens are perhaps the best studied. The two prolyl hydroxylases recognize prolyl residues in recurring triplet sequences found in most collagens. Ascorbic acid functions in the reaction as a reducing agent and also acts to retard inappropriate oxidations of prolyl hydroxylase itself, which occur during catalytic cycles. Prolyl hydroxylation is essential for the intrachain hydrogen bonding of collagen. When underhydroxylated, collagen intrachain helical formation does not occur and the collagen may be susceptible to degradation. When lysyl residues in collagen are underhydroxylated, the number and stability of some of the lysine-derived crosslinks in collagen are decreased. [See Collagen, Structure, and Function; Enzymes, Coenzymes, and the Control of Cellular Chemical Reactions.]

## VI. SCURVY

Scurvy is the consequence of perturbing biochemical functions that are dependent on ascorbic acid. Abnormal collagen formation, changes in fatty acid metabolism, and altered brain and adrenal functions are measurable events in scorbutic subjects. Increased susceptibility to infection and profound fatigue are also characteristics of scurvy. Fatigue and infections occur in part because of the inability to control tissue iron mobilization, complement activity, and histamine homeostasis. Extracellular matrix structures in bone, cartilage, teeth, and other connective tissues are weakened because of decreased prolyl hydroxylase functional activity. Eventually there may be hemorrhage due to capillary fragility, loss of teeth, rheumatic pain, degeneration of muscles, and impaired wound healing as the result of scurvy.

## VII. PHARMACOLOGICAL USES OF ASCORBIC ACID AND SUPPLEMENTATION

Pharmacological doses of ascorbic acid in humans (500 mg or more per day) are sometimes indicated in genetic disorders, wherein ascorbic acid-requiring enzymes need higher than normal tissue concentrations of L-ascorbic acid for enzyme saturation. Pharmacological doses may also be useful to enhance wound repair after trauma, surgery, or optimal recovery from peridontal diseases. However, the success of ascorbic acid supplementation in these situations is variably effective since ascorbic acid homeostasis is under such tight metabolic control. Fortunately, ascorbic acid is reasonably well tolerated in high dosages because of the well-elaborated systems for controlling tissue levels. The exceptions are related to the induction or activation of so-called microsomal P450 activities. Activation of the microsomal P450 hydroxylase can cause accelerated drug metabolism and thereby alter given therapeutic protocols involving certain drugs.

Whether or not excessive intakes are efficacious in protection or treatment of viral infections, common colds, cancer, or heart disease is also speculative. For example, although a case can be made that ascorbic acid is an effective antihistamine, its role in the actual prevention of colds and viral infections is not proven, only hypothesized. None of the recent clinical trials involving ascorbic acid supplementation and colds has proved efficacy as it relates to actual prevention. With respect to specific cancers, epidemiological studies do link reduced risk for esophageal, gastric, and colorectal cancers with consumption of foods that are high in vitamin C and antioxidants, e.g., $\beta$-carotene

and vitamin E. In studies with experimental animals, ascorbic acid protects against tumor induction by given carcinogens and does retard the formation of nitrosamines and other N-nitro compounds. These studies affirm the importance of ascorbic acid as an antioxidant and in the detoxification of specific compounds that may enhance the promotion or induction of some cancers. Populations with a long-term consumption of ascorbic acid at levels that are higher than the RDA are also reported to have reduced risks of cardiovascular disease. The lowered cardiovascular disease risk is based on increases in HDL and decreases in LDL oxidation, blood pressure, and cardiovascular mortality. To iterate, however, the results are often variable. The best case is made for those whose ascorbic acid intake is consistently below normal intakes. For example, ample evidence suggests that cigarette smokers and alcoholics may have increased requirements for ascorbic acid. A current recommendation is that adult smokers consume 90–100 mg per day compared to the current recommended allowance of 60 mg per day for healthy individuals.

## VIII. SUMMARY

Ascorbic acid is one of the most important reducing agents of the cell. In many animals, L-ascorbic acid is derived from the direct oxidation of glucose. In humans the absence of the enzyme L-glulonolactone oxidase requires that ascorbic acid be supplemented as an essential nutrient. Ascorbic acid serves as a cofactor for important hydrolylases that are involved in protein, amino acid, and lipid modifications. Scurvy can result from a severe and prolonged deficiency of ascorbic acid. Further advances will undoubtedly involve a better understanding of the role of ascorbic acid as a metabolic regulator, as well as its role in modulating important posttranslational protein modifications, such as prolyl and lysyl hydroxylations in collagens.

## BIBLIOGRAPHY

Bendich, A., and Langseth, L. (1995). The health effects of vitamin C supplementation: A review. *J. Am. College Nutr.* **14**, 124–136.

Chatterjee, I. (1973). Evolution and the biosynthesis of ascorbic acid. *Science* **182**, 1271–1273.

England, S., and Seifer, S. (1986). The biochemical function of ascorbic acid. *Ann. Rev. Nutr.* **6**, 365–406.

Franceschi, R. T., Wilson, J. X., and Dixon, S. J. (1995). Requirement for Na(+)-dependent ascorbic acid transport in osteoblast function. *Am. J. Physiol.* **268**, C1430–C1439.

Kawai, T., Nishikimi, M., Ozawa, T., and Yagi, K. (1992). A missense mutation of L-gulono-gamma-lactone oxidase causes the inability of scurvy-prone osteogenic disorder rats to synthesize L-ascorbic acid. *J. Biol. Chem.* **267**, 21973–21976.

Meister, A. (1994). Glutathione-ascorbic acid antioxidant system in animals. *J. Biol. Chem.* **269**, 9397–9400.

Nishikimi, M., Fukuyama, R., Minoshima, S., Shimizu, N., and Yagi, K. (1994). Cloning and chromosomal mapping of the human nonfunctional gene for L-gulono-gamma-lactone oxidase, the enzyme for L-ascorbic acid biosynthesis missing in man. *J. Biol. Chem.* **269**, 13685–13688.

# Asthma

IAN M. ADCOCK

*Imperial College School of Medicine at the National Heart and Lung Institute*

## GLOSSARY

**Airway hyperresponsiveness** Twitchiness of the airway to stimulation by asthma triggers such as allergens

**Allergen** Agent that causes the release of inflammatory mediators from cells within the airway such as mast cells

**Bronchi** Airway tubes containing various amounts of muscle and cartilage which enable changes in the lumen size to occur

**Cytokines** Class of small inflammatory mediators released from most cells within the airway of which the number and relative concentrations are thought to control the inflammatory response

ASTHMA IS NOT A NEW DISEASE AND WAS RECOGnized by the ancient Egyptians in the second millennium B.C. The term derives from the Greek for panting or short-drawn breath, the characteristic symptom of asthma, and was recognized as a distinct respiratory disease by Hippocrates (460–375 BC). Asthma is now recognized as a chronic inflammatory respiratory disease characterized by breathlessness due to the reversible narrowing of airways within the lung and which affects between 5 and 10% of the world population. Asthma symptoms are worsened by exposure to viruses, air pollution, and allergens. This reversibility of airway narrowing or obstruction may occur spontaneously or as a result of treatment. Treatments currently used either treat the symptoms of the disease, airway narrowing, or the underlying causes of the inflammation. The two major classes are β-adrenergic agonists which relax airway smooth muscle increasing the airway lumen size and corticosteroids which reduce the underlying inflammation. Increasing evidence shows a rise in asthma severity and mortality in many countries despite the increasing awareness of asthma and the use of better treatments, underlying the importance for a greater understanding of this complex disease.

## I. SYMPTOMS AND PHYSIOLOGY

### A. Structure of the Airway

The trachea is approximately 12 cm long and 2.5 cm across and lies in front of the esophagus. It descends from the larynx until it divides into the left and right main bronchi. Incomplete (horseshoe-shaped) cartilage rings rigidify the trachea with muscle connecting the two ends of the cartilage. This muscle can vary the cross-sectional area of the trachea by contracting and relaxing. As the bronchi enter the lung, they divide into secondary bronchi to serve the lobes (three on the right, two on the left). This is repeated, forming smaller airways until they reach the areas where gaseous exchange takes place. In the transition from trachea to terminal bronchi the smooth muscle decreases in thickness and becomes a complete circular layer. [*See* Respiratory System, Anatomy.]

Asthma is a disease of the bronchi that is marked by wheezing, breathlessness, and sometimes cough and expectoration of mucus. Asthma may be caused by an allergic reaction, infection in persons with susceptible bronchi, or malfunction of the autonomic nervous system. In allergic asthma the bronchial tubes have increased reactivity to various stimuli, such as allergens, cigarette smoke, or bronchial infection. The

ENCYCLOPEDIA OF HUMAN BIOLOGY, Second Edition, VOLUME I. Copyright © 1997 by Academic Press. All rights of reproduction in any form reserved.

bronchial tubes react to the stimuli by becoming narrowed or obstructed, blocking the flow of air and causing difficulty in breathing. Blockage commonly involves contraction of the bronchial muscles and the swelling of bronchial walls, which narrow the passageways, and the secretion of a clogging mucus from bronchial glands. More severe and chronic asthma may be associated with the accumulation of inflammatory cells, basement membrane thickening, mucous gland enlargement and proliferation, loss of epithelial lining cells, and filling of the airway with thick mucus (Fig. 1). The narrowing is usually temporary and reversible but it can change in severity either over time or as a result of therapy. Typically, asthma is an intermittent disease, but a small proportion of patients suffer continuously if not treated. If severe enough, asthma may become resistant to conventional treatment and hospitalization is necessary.

An asthmatic attack may start at any time but it most frequently occurs during the early hours of the morning. In some cases this may be related to specific allergens such as feathers in pillows, but usually it is just part of the pattern of daily variation in asthma which has characteristic dips in symptoms at this time. The patient may notice a choking sensation, tightness in the chest, shortness of breath, and audible wheezing. In a severe attack the breathing rate and the pulse

rate increase and the patient may find it easier to breathe sitting up with the shoulders raised. The face is pale and anxious, the voice is gasping, speech is difficult, the eyes become prominent, the lips may take on a bluish hue due to reduced oxygen in the blood, and beads of cold, clammy perspiration are seen on the forehead and face; breathing is seen to be labored and extremely difficult. The attach may reach its apex in a few minutes and persist for a brief period of time or it may persist for hours or days. During asthma attacks, air can enter the lungs more readily than it can leave, so the lungs become overinflated. Once it is over, the person feels perfectly well, except for fatigue, until the onset of the next attack.

Asthma that is provoked by allergens and certain other stimuli may show a biphasic response. Bronchoconstriction occurs in a few minutes and is often followed by a late-phase reaction in 4–6 hr. The late-phase reaction appears to be the consequence of mediators that have chemotactic and other inflammatory roles; these mediators include leukotrienes, cytokines, and chemokines.

The amount of airway narrowing can be measured as the peak expiratory–flow rate (PEFR) by a simple meter. The hallmark of asthma is the variability of airway narrowing; this may be intermittent mild breathlessness often precipitated by an obvious cause

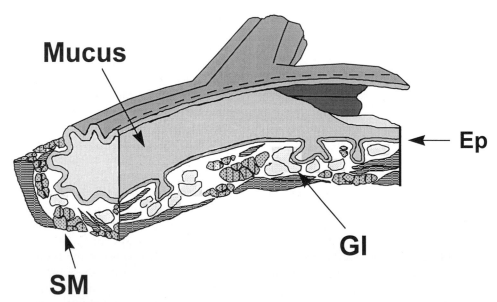

**FIGURE 1**    Characteristics of the bronchus during an asthma attack. Bronchoconstriction of the airway occurs following smooth muscle (SM) contraction, leading to folding of the epithelium (Ep). There is marked enlargement of the mucus-secreting glands (GI) and increased fluid within the airway wall (edema). The bronchial lumen is completely obstructed as a result of smooth muscle contraction, airway edema, and thick mucus secretion, resulting in the characteristic symptoms of wheeze and chest tightness.

such as exercise or severe life-threatening attacks. Some asthma attacks have a sudden onset, but routine recordings of PEFR show that there is often a period of gradual decline for some days before the acute attack. A change in treatment during this decline will often prevent a severe attack and hospital admission. When asthma is poorly controlled for a long time in childhood, the chest may be left overexpanded. Chronic severe asthma may eventually result in airway narrowing that is no longer reversible in response to treatment but has become fixed.

Numerous theories have been advanced as to the mechanism of precipitating an asthma attack. The best evidence indicates that initially there is a release of certain chemical substances (histamine, acetylcholine) from mast cells within the lung tissue which in turn cause the muscles of the bronchi to contract, producing bronchial obstruction. The extent of the asthmatic attack and its perpetuation are controlled by the release of other mediators from a number of resident and infiltrating cells stimulated by this first

wave of mediators. This may cause the further recruitment of inflammatory cells and/or activation of cells already present in the airways (Fig. 2). Although there seems to be a hereditary predisposition to the disease in the vast majority of cases, the actual attacks themselves seem to be triggered by a person's exposure to inhaled allergens. [*See* Pulmonary Pathophysiology.]

## II. ASTHMA AS AN INFLAMMATORY DISEASE

Inflammation describes the processes occurring in tissue containing blood vessels following local injury which results in the accumulation of fluids and blood cells at the focus of the tissue damage. Examination of autopsy specimens resulted in the acceptance that asthma was no longer primarily a disorder of smooth muscle contraction, but an ongoing inflammatory disease involving intricate interactions between inflammatory mediators and local and infiltrating cells

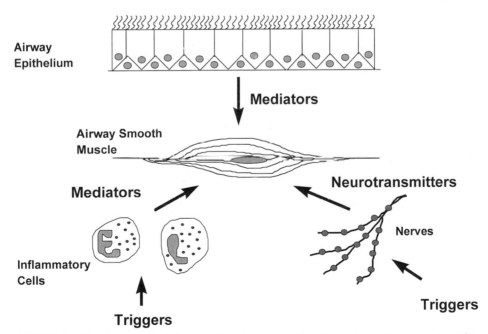

FIGURE 2   The airway constriction seen in asthma is a result of smooth muscle contraction. This contraction is a result of numerous interactions between local and infiltrating cells which release inflammatory mediators that act on the airway smooth muscle. The release of inflammatory mediators by the airway epithelium or local mast cells following an airborne insult results in smooth muscle contraction and recruitment of circulating inflammatory cells to the airway. These infiltrating cells, including neutrophils, lymphocytes, and eosinophils, in turn release inflammatory mediators which affect smooth muscle contraction. The release of neurotransmitter substances such as acetylcholine, nitric oxide, and substance P from airway nerves after stimulation by asthmatic triggers may modulate smooth muscle contraction.

causing the characteristic pattern of asthmatic inflammation. These infiltrating cells include neutrophils, lymphocytes, and monocytes, but mainly eosinophils. These inflammatory events are now known to occur even in recently diagnosed asthmatics, and the damage caused by inflammation may be quite marked even before the onset of symptoms.

Chronic inflammation of the bronchial tubes is brought about by chemical substances (cytokines, chemokines, leukotrienes, histamines, etc.) collectively termed mediators, whose release from certain local and infiltrating cells (mast cells, epithelial cells, macrophages, and lymphocytes) is triggered by a wide array of stimuli.

## A. The Inflammatory Reaction

The early phase reaction, induced by allergen inhalation, is predominantly the consequence of the direct action of mast cell-derived mediators and metabolites on the hyperresponsive smooth muscle. A reaction during which peak airway obstruction is achieved occurs in 20 to 30 min.

The late response reaction when a second wave of airway obstruction occurs is thought to be related to the release of specific cytokines simultaneously with or soon after that of the short-lived mediators responsible for the early phase. These cytokines have a systemic effect that is able to diffuse into the general circulation from the airway blood vessels where they recruit effector cells (chiefly eosinophils, plus lymphocytes, neutrophils, and monocytes) of the inflammatory response to the lungs, activating them in the process. Once in the lungs they release both preformed and newly synthesized factors, including cytokines, chemokines, enzymes, and proteins, that enhance and perpetuate the inflammatory processes. [*See* Inflammation.]

## B. Inflammatory Cells

Many different inflammatory cells are involved in asthma, although the precise role of each cell type is unknown. Some cells are predominant in the asthmatic inflammation, but no single cell can account for the complex pathophysiology seen in asthma.

Mast cells are important in initiating the acute responses to allergens but their role in chronic inflammation is disputed. Cells involved in the chronic inflammatory processes underlying asthma include macrophages, eosinophils, T lymphocytes,

and airway epithelial cells. Macrophages, which are derived from blood cells, are attracted into the airways and may be activated by the allergen to release a host of mediators, including a range of cytokines. The type of cytokine released and their relative concentrations may determine the pattern of inflammation. Macrophages may also present the allergen to T lymphocytes, thus activating these cells. Eosinophil infiltration is characteristic of asthmatic inflammation and differentiates asthma from other inflammatory conditions of the airway. Allergen inhalation results in the production of eosinophil attractants which recruit eosinophils into the airway. Once present, eosinophils release a number of cytokines, proteins, and oxygen-derived free radicals that are involved in the development of airway hyperresponsiveness.

## C. Inflammatory Mediators

A great diversity of mediators are released during the initial inflammatory response. These include presynthesized mediators, such as histamine, from granules within mast cells, all of which induce an immediate exaggerated bronchoconstriction due to enhanced baseline airway hyperresponsiveness. The further enhancement of the inflammatory response depends on the release of a large number of other mediators, released from various cell types, which act to increase inflammation leading to the symptoms previously described (Fig. 3).

During the degranulation process of the mast cell, the membrane is broken down by enzymes to produce specific lipids that can be further converted into mediators such as prostaglandins and leukotrienes (eicosanoids). As well as having a bronchoconstrictor effect, these newly formed mediators also induce a transient airway hyperresponsiveness.

Cytokines evoke many biological processes, overlapping and interacting with the roles of lipid mediators within the cell. They regulate eicosanoid actions by increasing the synthesis of eicosanoids, by increasing enzyme synthesis, and by acting in synergy with other mediators to exert an effect. [*See* Cytokines and the Immune Response.]

The different cytokines and mediators present in a particular network can create both pro- and anti-inflammatory effects at the level of the tissue by the induction of either a stimulatory or an inhibitory pathway. *In vivo*, a cell is exposed to a soup of these various cytokines which may modulate the synthesis

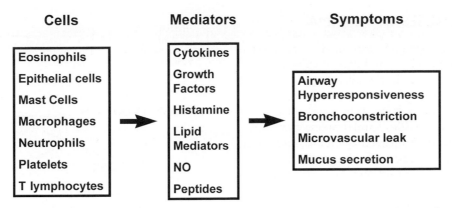

| Cells | Mediators | Symptoms |
|-------|-----------|----------|
| Eosinophils<br>Epithelial cells<br>Mast Cells<br>Macrophages<br>Neutrophils<br>Platelets<br>T lymphocytes | Cytokines<br>Growth Factors<br>Histamine<br>Lipid Mediators<br>NO<br>Peptides | Airway Hyperresponsiveness<br>Bronchoconstriction<br>Microvascular leak<br>Mucus secretion |

**FIGURE 3** The symptoms of asthma are due to a combination of effects on various cells within the airway, including bronchoconstriction, mucus hypersecretion, enhanced airway hyperresponsiveness, and increased leakage of plasma into the airway (microvascular leak). These symptoms are due to the release of a large number of pro-inflammatory mediators from both infiltrating and resident airway cells and their subsequent action on these or other inflammatory cells.

and release of the various eicosanoid mediators. This in turn probably determines the distinct type of inflammation seen in asthma as compared with other inflammatory diseases.

The biologically active prostanoids prostaglandin (PG)$E_2$, $PGF_{1\alpha}$, prostacyclin, and thromboxane $A_2$ are produced by the enzymatic conversion of membrane lipids in a cell-specific manner, with each cell type producing usually only one of the prostanoids in abundance. There is a wide spectrum of prostanoid inflammatory activity. Conversely, prostaglandins may also show important *anti*-inflammatory effects.

Histamine and leukotrienes are probably involved in both intrinsic and extrinsic asthma. Leukotrienes are generated by several types of inflammatory cells in addition to mast cells. Many other mediators may be involved, including several enzymes, various chemotactic factors, and some prostaglandins. These chemotactic factors attract eosinophils, which may aggravate asthma by generating more inflammatory mediators and by releasing proteins that may cause loss of the protective bronchiolar epithelial lining characteristic of asthmatic inflammation.

Nitric oxide (NO) is an exceedingly unstable gas which is now thought to be one of the most important mediators of cellular responses. Nitric oxide is produced by the enzyme nitric oxide synthase (NOS) which exists in inducible (iNOS) or constitutive forms (cNOS). Inflammatory cells in the resting state do not express iNOS; however, once stimulated, nearly every

tissue in the body has the capacity to synthesize this enzyme. The iNOS stimulants are cell specific but are almost always a type of inflammatory cytokine, microbe, or their products, which often combine to have a synergistic effect.

The role of nitric oxide in inflammation has been described as a "double-edged sword." Nitric oxide produced in small amounts by cNOS activation acts as a messenger molecule in a variety of physiological actions; large local nitric oxide output through the activation of the inducible NOS can result in either a protective *or* a damaging effect on the host. The final result depends on the levels of nitric oxide and its degree of conversion to other products. Inhibition of endogenous nitric oxide production causes a marked decrease in airway inflammation and protects against lung injury. Asthmatic patients exhale three to four times more NO than normal subjects. However, in asthmatic patients inhaling corticosteroids the NO levels are similar to those in normal subjects, suggesting that the increased levels in asthma are a direct reflection of iNOS expression and may reflect the underlying level of inflammation.

Nitric oxide is also thought to have a contributory effect on inflammation-induced tissue damage, as it can be both cytotoxic and cytotactic not only for the invading microorganisms but for the synthesizing and surrounding cells as well. These toxic effects can be a consequence of direct NO action or, sometimes, the effects of the product of NO reaction with oxygen-derived free radicals.

## III. CAUSES OF ASTHMA

The most common triggers of asthma are viral illness and allergy. There is a large number of substances to which the asthmatic subject may be hypersensitive and contact with these may be responsible for an attack. These include pollens, skin scales of certain animals such as cats, dogs, and horses, the house dust mite, and certain industrial agents. Exercise is a common cause of asthma attacks, but swimming very rarely causes asthma. Cigarette smoke is also a powerful trigger in some patients. Other causal factors include sinusitis (inflammation or infection of the sinuses), weather changes (e.g., cold air), dietary factors, and the use of aspirin. Although often thought of as a cause of asthma, air pollutants are better described as asthma triggers in that they may exacerbate an attack in a susceptible subject but do not actually cause asthma in a subject without the disease.

People with allergic asthma have acquired a specific sensitivity to common inhaled substances. The allergy is mediated by antibodies of the immunoglobulin E (IgE) class. These antibodies are present in the blood plasma in very low concentrations, but are strongly bound to the surface of tissue mast cells and blood basophils, where they exert their strong biological effects. On contact with an allergen, mast cells are activated, resulting in the release or synthesis of highly potent chemical mediators such as histamine and membrane lipid-derived leukotrienes. Allergic asthma is likely to vary in severity according to the degree of exposure to allergens, although this is not always the case, and often runs in families. The genetic basis for the determination of susceptibility and severity of asthma is currently under investigation (Fig. 4). [*See* Allergy.]

Intrinsic asthma that is not IgE antibody mediated is more likely to be chronic and is often associated with or initiated by lung or sinus infections. Most adult-onset asthma is of this type. Some asthmatic individuals, for unknown reasons, become extremely intolerant of aspirin, which has produced fatal reactions. The problem usually develops in adults, especially those with a history of sinus disease and nasal polyps, but occasionally appears in children with allergic asthma as well. An intolerant person may also have trouble with other drugs that resemble aspirin in their action.

## IV. OCCURRENCE

Asthma commonly has its onset during the first few years of life, appearing in up to 8% of children before

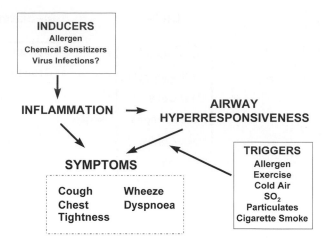

FIGURE 4 A combination of airway inflammation and hyperresponsiveness results in the production of common symptoms of asthma. The underlying inflammation may lead to symptoms on its own or enhance symptoms following exposure to airborne triggers by increasing the susceptibility of the airways to such triggers of asthma. The inducers that cause inflammation and the triggers of hyperresponsiveness may in some cases be the same (allergens) or distinct entities, e.g., viruses are thought to cause airway inflammation and have no effect on hyperresponsiveness. In contrast, cold air, traffic pollutants, and cigarette smoke do not cause asthma *per se* but may cause an increase in asthma symptoms by stimulating the asthmatic airways.

the age of 5. Until puberty, it is more common in boys than in girls, but with increasing age this trend reverses itself. Currently, somewhat more than 5% of the American population, or about 8 million people, have or have had asthma at some time in their lives. More than a third of these are children under age 15. Figures for asthma incidence worldwide vary from 5 to 10% depending on the country and degree of urbanization, although the highest incidence of asthma occurs in New Zealand and Australia, which are relatively rural countries.

Significant numbers of children with asthma undergo remission, or a lessening of the disease, during puberty, but few actually outgrow their asthma entirely. Spontaneous remission is rare for adult-onset asthma. About 50 to 75% of children with relatively mild asthma become free of wheezing in early puberty and remain so until early adult life. The majority of children with persistent symptoms in childhood will continue to have asthma into adult life.

Mortality from asthma appeared to be increasing in the 1980s throughout the world despite the increased awareness of asthma and the better treatment available.

## V. TREATMENT

Asthma is presently incurable, but successful management allows the asthmatic to live a normal life. Current therapies involve allergen avoidance, immunotherapy, and drug treatment. The use of personal peak flow meters and asthma symptom score cards has enabled the asthmatic to monitor their disease on a day-to-day basis. Self-management is helpful in increasing the compliance with treatment and improving the control of asthma.

### A. Avoidance

All allergens, irritants, and precipitating factors such as house dust mites and animal dander that serve as triggers should be identified and exposure to them reduced or avoided. This is sometimes possible where a specific animal, occupational exposure, or foodstuff is involved, but more often the problem is a widespread allergen, such as pollen or house dust mites, which are difficult or impossible to avoid. Exposure to specific allergens may produce a prolonged reaction in the airways which remain more susceptible to other precipitants for days afterwards. There is a small group of patients who are aspirin-sensitive asthmatics and who should avoid all contact with this drug. Although exercise may provoke asthma, it does not increase problems encountered with other substances and it is beneficial for asthmatics to stay fit. Exercise should continue with appropriate medication such as a β-agonist taken beforehand.

### B. Immunotherapy

Allergen injections are sometimes given to asthmatics in order to induce desensitization to the allergen. This is performed by trying to produce tolerance by giving small injections of the substance and gradually increasing the dose. Antibodies which block the asthmatic response may develop. Severe reactions can occur occasionally and it is becoming clear that its effectiveness is limited, especially in the long term.

### C. Drug Management

There are two main forms of drug treatment. Bronchodilator drugs are used to dilate the airways and relieve breathlessness. They may also prevent problems when taken before exposure. The most useful group of bronchodilators are the β-adrenergic agonists. There are a range of these and the most common drugs are salbutamol and terbutaline. These drugs act by relaxing bronchial muscle, which in turn results in a widening of the airways. Also included in this group of drugs are theophylline and its derivatives, which also act to relax airway muscles, and anticholinergics (such as ipratropium bromide), which alter the control of airways. Theophylline may also have some anti-inflammatory properties. These drugs should be used to control symptoms and not as regular therapy on their own.

When bronchodilators are used more than once a day regularly, then the second group of drugs, the anti-inflammatory agents, which help reduce swelling and inflammation, should be considered. These are inhaled glucocorticoids, sodium cromoglycate, and nedocromil. Inhaled glucocorticoids are the most effective anti-inflammatory drugs and can be given to both children and adults. These prophylactic drugs must be taken regularly to be effective in reducing airway inflammation and swelling. Because there is often no immediate benefit discerned by the patient, careful patient preparation and explanation by the physician are essential. Early treatment of asthmatics can result in a marked suppression of the inflammation seen in asthmatic airways. Current guidelines suggest treatment with a high dose of glucocorticoids initially before stepping down to a minimal level of effective drug.

All of the drugs mentioned, apart from theophyllines, are best taken by inhalation into the respiratory tract. This means that only a small dose is necessary, little of the drug gets into the rest of the body, and few side effects occur. Oral corticosteroids are reserved for severe asthma.

Newer drugs are being developed to deal with various specific aspects of the inflammatory process in asthma. These include the more specific theophylline-like drugs known as PDE inhibitors which relax airway smooth muscle and inhibit certain inflammatory cells. Another class of drugs in current development are specific mediator antagonists or synthesis inhibitors. These include antagonists against leukotrienes and thromboxanes (Fig. 5). Leukotriene antagonists protect against some bronchoconstrictor challenges and improve clinical symptoms. These drugs also have the benefit of oral administration. These drugs may be useful in mild asthma but their place in therapy is not established. It seems unlikely that antagonizing a single mediator will prove as effective as therapies with a broad spectrum of activity in most asthmatics, but certain types of asthmatic subjects may respond to these specific therapies, e.g., aspirin-sensitive asth-

| Anti-inflammatories | Bronchodilators | Mediator Inhibitors |
|---------------------|-----------------|---------------------|
| Corticosteroids | β-agonists | Leukotrienes |
| Cromoglycate | Long-acting Beta β-agonists | PAF |
| PDE Inhibitors | Prostaglandin E$_2$ | Thromboxane |
| Theophylline | PDE inhibitors | Bradykinin |
| Nedocromil | Theophylline? | Tachykinins |
| Methotrexate | Calcium antagonists | Cytokines |
| Cyclosporin A | | |
| Lipocortin | | |

FIGURE 5  Drugs used in the treatment of asthma. These drugs can be separated into anti-inflammatory agents, bronchodilators, and mediator inhibitors.

matics may respond particularly well to leukotriene antagonists.

At present, inhaled corticosteroids are the most effective anti-inflammatory drug used to treat asthma and are relatively free of adverse effects. They should be considered when a β-agonist such as salbutamol (Ventolin) is being used more than once every day. The newer, long-acting β-agonists may also provide better control of asthma symptoms, at least in mild/moderate asthmatics. [*See* Anti-Inflammatory Steroid Action.]

## BIBLIOGRAPHY

Barnes, P. J. (1993). Anti-inflammatory therapy in asthma. *Annu. Rev. Med.* **44,** 229–249.

Barnes, P. J., Rodger, I. W., and Thomson, N. C. (eds.) (1992). "Asthma: Basic Mechanisms and Clinical Management," 2nd Ed. Academic Press, London.

Clark, T. J. H., Godfrey, S., and Lee, T. H. (eds.) (1992). "Asthma," 3rd Ed. Chapman Hall, London.

International Asthma Management Project. (1992). "International Consensus Report on Diagnosis and Treatment of Asthma," U.S. Department of Health and Human Services, Public Health Service, National Institutes of Health, Bethesda, MD.

# Astrocytes

ANTONIA VERNADAKIS

*University of Colorado School of Medicine*

## GLOSSARY

**Cell models** Cells (neurons or glial cells) that under *in vitro* (in culture) conditions duplicate (or assimilate) some of the characteristics of properties of cells *in vivo*. Examples are neuroblastoma cells as a neuronal model and C-6 glioma as a glial cell model

**Cell passage** Transfer or transplantation of cells from one culture vessel to another. Usually cells are diluted and a known, predetermined number of cells are transferred. Cell passages are or should be recorded and thus a relative cultural age is ascertained

**Ependyma** Structure that forms the lining of the ventricular system and consists of special columnar-to-squamous epithelium, whose constituent cells are polarized and show a luminal side, a lateral side, and a basal side

***In vitro* vs. in culture** This terminology has been interchangeable throughout the literature. Both terms imply live cells maintained or growing under artificial conditions outside of the organism from which they were derived

**Microenvironment** Immediate local extracellular milieu of both neurons and glial cells. In early literature, it is referred to as the "milieu interieur"

**Phenotypic expression** Indicates morphological, biochemical, and functional cell characteristics or properties associated with or assigned to differentiated cells

**Pinocytotic** Ability of cells to engulf (drink) foreign substances. Microglia in the CNS show pinocytotic property

**Primary culture** Culture that has started from cells, tissue, or organs directly derived from the organism. When a primary culture is subcultured, it becomes a "cell line"

**Satellite cells** Neuroglia cells including Schwann cells are frequently referred to as satellite cells to neurons. The term implies a functional relationship

**Trophic factors** Substances necessary for growth of cells (neurons or glia). They are produced by various nonneuronal (glial) cells and are secreted in the microenvironment, where they then exert their effects on other cells. They include a large range of substances (e.g., ions, glucose, amino acids, neurotransmitters, peptides, hormones)

**Vimentin** Intermediate filament protein that has been detected in both neuronal and glial precursors. It remains in glial precursors and astrocyte precursors until GFAP is present

ALTHOUGH GLIAL CELLS (NEUROGLIA) WERE noted as early as 1856, by Virchow ("Nervenkitt"), it is only in the last decade that intense focus has been directed to the role of neuroglia cells in the function of the nervous system and, in particular, during development.

It is now established that glial cells, and particularly astrocytes, are intimately involved in neuronal function. As Hyden proposed in 1961, neurons and glia function as a unit. Early in neurodevelopment, neuroglia and, specifically, astrocytes provide guidance for neuronal migration and also secrete factors important for neuronal differentiation and maturation. Astroglia regulate the ionic microenvironment and act as spatial buffers for neuronal function and homeostasis. They accumulate and release neurotransmitter substances and thus regulate the amount of a neurotransmitter at the synaptic cleft. Their role in aging remains to be further defined, but they appear to be involved in "reactive synaptogenesis," "activated gliosis," and regeneration.

ENCYCLOPEDIA OF HUMAN BIOLOGY, Second Edition, VOLUME I. Copyright © 1997 by Academic Press. All rights of reproduction in any form reserved.

## I. GLIOGENESIS AND CELL LINEAGE

W. His, in 1889–1890, was the first to note the existence of two separate cell types within the germinal layer during the early stages of neurogenesis: "Keimzellen," which give rise to neurons, and "Spongioblasten," which give rise to glial cells. It is now established that stem cells in the ventricular zone (and possibly in the subventricular zone) give rise to progenitor cells, the glioblasts, which can contribute cells to more than one macroglial cell lineage.

The pathway(s) from the ventricular glioblast to the mature astrocyte continue to be debated. The most accepted precursors of astrocytes are the radial glial cells (cells with radially oriented long processes extending from the ventricular lining to the pial basal lamina). Radial glia, upon losing their ventricular contacts, transform into radial astrocytes that can either persist into adulthood (radial astrocyte) or transform into nonradial astrocyte (astrocyte).

Glial fibrillary acidic protein (GFAP), the astrocyte marker, has been identified within the radial glial fibers of the human fetal cerebrum at 10–18 weeks of ovulatory age. Vimentin, an intermediate filament protein found primarily in mesenchymal cells, is also expressed in these early embryonic glial cells. There has been some debate in the literature as to the presence or absence of GFAP in radial glia of prenatal rodent forebrain. A correlative light microscopic, ultrastructural, and immunocytochemical study in mouse brain shows radially oriented glial cells within the neopallium by embryonic Day 12. These early radially organized glial cells may be the ultimate source of all macroglial cell types, including astrocytes, ependymal cells, and oligodendrocytes. This proposal is based on the presence of "transitional" cells with cytological, ultrastructural, and immunohistochemical features intermediate between those of astroglia and oligodendroglial cells, and the close relationship that develops just before the onset of myelination has lent support to the view that oligodendroglia may also be derived from radial glial cells either directly or from intermediate astroglia forms.

*In vitro* (culture) studies have provided evidence that there may be subpopulations of glioblasts that persist throughout embryonic development that are capable of giving rise to oligodendroglia and astroglia. Two immunologically distinct types of astrocytes have been identified in the rat optic nerve and more recently in the rat brain. Anti-GFAP antibody and lectins are useful for identifying astrocytes and for subdividing astrocytes in primary cultures into two subpopulations: A2B5, monoclonal antibody specific to certain gangliosides located on the surface and also in cytoplasm; and monoclonal antibody to neural antigen 2 (Ran-2), also situated on the cell surface. The cells that are GFAP$^+$ but do not bind A2B5 monoclonal antibody are named type 1 astrocytes, and GFAP$^+$ cells that do are named type 2 astrocytes. Type 1 astrocytes proliferate in culture, have a fibroblast-like appearance, and do not bind the monoclonal antibody A2B5. Type 2 astrocytes divide only infrequently in culture, have numerous processes, and bind A2B5 antibody. Type 1 and type 2 astrocytes are biochemically and developmentally different. When cultured in the presence of fetal calf serum, precursor cells develop into type 2 astrocytes, but when cultured in the absence of serum, they have the ultrastructural characteristics of oligodendrocytes and express galactocerebroside (GC), an oligodendrocyte marker. It is advocated that there are two distinct glial cell lineages that diverge prenatally in the rat optic nerve, with one lineage forming type 1 astrocytes and the other, consisting of progenitor cells, forming cells giving rise to both oligodendrocytes and type 2 astrocytes.

A common lineage for astrocytes and oligodendrocytes from radial glial cells has been suggested during spinal cord gliogenesis from studies using a variety of specific antibodies: A2B5 described in glial precursor cells in the optic nerve; AbR24 directed against GD3 ganglioside and binding to immature neuroectodermal cells and to developing oligodendrocytes in forebrain and cerebellum; and the intermediate filament antibody vimentin. With time, two different populations emerge, both of which seem to be derivatives of radial glial cells. One cell type expresses GFAP, in addition to vimentin, and eventually takes the form of astrocytes. The other type expresses carbonic anhydrase, an enzyme characteristic of oligodendrocytes and enriched in myelin. Carbonic anhydrase-labeled cells eventually develop into small cells with oligodendrocyte morphology. The GD3$^+$/vimentin$^+$/A$_2$B$_5$$^+$ radial glia observed in spinal cord may be analogous to the O-2A precursors, giving rise to GFAP$^+$ cells.

## II. MORPHOLOGICAL AND TOPOGRAPHICAL CHARACTERISTICS: ASTROCYTE-RELATED CELLS

### A. General Introduction

Differences in the temporal and regional appearance of astrocytes have led to the classification of a "family of astrocytes." Whether all cells of the astrocyte family belong to a branch of the main astrocyte stem of whether they represent a number of independent pathways of development that have certain aspects in common has not been determined.

## B. Fibrous and Protoplasmic Astrocytes

Glial cells in the white and gray matter are distinct. The main features of the astrocytes of the white matter of fibrous astrocytes are the smooth appearance of the processes that run between myelinated fibers and the presence of vascular end feet. The fibrous astrocytes appear to correspond to the type 2 astrocytes. The astrocytes of the gray matter or protoplasmic astrocytes have many short, crimped processes that are most often ramified. The protoplasmic astrocytes appear to correspond to the type 1 astrocyte. The question has been raised whether this classification is meaningful since these cells show many similarities in ultrastructure characteristics and share many antigenic determinants (except for A2B5). The prevailing view is that astrocytes can acquire specialized characteristics and functions according to their location in the central nervous system.

## C. Tanycytes

In adult submammals, process-bearing ependyma cells are found intermingled with common cuboidal-to-columnar ependyma that line the walls of the lateral, third, and fourth ventricles and the cerebral aqueduct. Because of the elongated form of these process-bearing ependyma, they were first named "tanycytes" or "stretch cells" and were considered to form an important link between the cerebrospinal fluid and the underlying structures in the brain wall. The original idea that tanycytes develop from radial glia and, therefore, with little capacity for postnatal differentiation has been argued with more recent evidence. The consensus is that tanycytes develop morphologically and probably functionally in relation to the postnatal development of those areas along the ventricular walls where they are found. The functional role of tanycytes has been debated. Foremost is the notion of a transport capacity, particularly in tanycytes located in the floor ventricle at the level of the median eminence of the hypothalamus. This morphological localization of the tanycytes has implicated them in the neuroendocrine regulation of the adenohypophysis. Synaptoid contacts between tanycytes and nerve fibers in the median eminence have been described.

## D. The Muller Cell

H. Muller (1851) first noted the principal glial elements of the vertebrate retina named the "radial fibers of Muller." Muller cells are distinct in their location within the retina; the nucleus of Muller cells is situated in the inner nuclear layer and their cellular processes are radially oriented, extending between the inner and outer limiting membrane. The first signs of Muller cell differentiation occur at an early stage of retinal development when the inner retinal and outer pigment epithelium layers of the eyecup become closely opposed. Vimentin is present in Muller cells, but GFAP is not normally present in Muller cells of most species, although it has been reported in the goldfish and in the rat after trauma to the eye. Also, Muller cells express GFAP-immunoreactive intermediate filaments after genetically or experimentally induced degeneration of retinal neurons, suggesting a functional state of Muller cells. Classically, Muller cells were thought to function principally as a structural framework for the neural elements of the retina. Recent evidence suggests that these cells have several physiological functions, including regulating extracellular $K^+$ levels in the retina, inactivation of several putative neurotransmitters [glutamate, $\gamma$-aminobutyric acid (GABA), acetylcholine], conversion of $CO_2$ to bicarbonate, and possible involvement in the visual cycle of synthesis and renewal of visual pigments. In cultures from rat retinas, 95–100% of the Muller cells have shown tetanus toxin (TT) binding, a marker frequently used for identifying neuronal cells in cultures. *In vivo* TT binding is expressed 4 days postnatally; this phenotypic alteration coincides with a period of rapid functional maturation of the retina. [*See* Retina.]

## E. Pituicytes

Pituicytes are the predominant nonneuronal elements of the neural lobe constituting 25–30% of the tissue. Besides these cells, perivascular connective tissue cells, endothelial cells of blood vessels, and a varying number of microglial cells, astrocytes, and oligodendrocytes are present. Attempts have been made by light microscopy to classify pituicytes into fibrous and protoplasmic.

The spatial relationship between neurosecretory fibers, pituicytes, and perivascular space implicates these cells in playing a role in neurosecretion. Like tanycytes in the median eminence, synaptoid contacts between nerve fibers and pituicytes have been reported to be present in several mammals. These synaptoid contacts differ from interneural synapses and can be compared to synapses on smooth muscle cells or glandular cells. The functional role of pituicytes in neurosecretion is based on: (a) enclosure and release of neurosecretory axons; (b) metabolic transport and storage of macromolecules released from axon terminals; and (c) regulation of extracellular fluid composition. It has been demonstrated immunocytochemically that nerve fibers containing leucine enkephalin and making synaptoid contacts with somata and processes of pituicytes and opiate receptors in the neural lobe are mainly correlated with pituicytes. The speculation is that opioid peptides, and also vasopressin and oxytocin, might influence the release of vasopressin and oxytocin via pituicytes by modulating the degree of enclosure of vasopressin or oxytocin terminals. A similar cellular phenomenon is also proposed for catecholaminergic nerve fibers contacting pituicytes.

## F. Pineal Astrocytes

Like the rest of the nervous system, the human pineal gland is made up of parenchymal cells, the pinealocytes, and supportive cells. Whether or not the supportive cells are astrocytes is controversial. In the human fetal pineal gland, astrocytes are seen in the older (32 weeks of gestation) fetal brain. In the human, differentiation of the pineal primitive neuroepithelial cells starts early, but it is only around the 32nd week of gestation that clusters of differentiated cells containing both pinealocytes and astrocytes can be easily seen with routine histological stains. Of interest is the astrocytic gliosis observed in the pineal gland between 30 and 60 years of age. It is believed that astrocytic gliosis follows degeneration of pinealocytes. In addition, astrocytic plaques have been described in the human pineal gland at autopsy. [*See* Pineal Body.]

The presence or absence of astrocytes in the pineal gland in lower species, especially in rodents, remains controversial. The functional significance of pineal astrocytes other than their supportive role is not understood.

## G. Cerebellar Astrocytes: Golgi Bergmann Glia and Velate Astrocytes

Bergmann fibers can be observed at embryonic Day 15 in the mouse and at embryonic Day 17 in the rat. GFAP immunofluorescence can be detected at postnatal Day 4 in Bergmann glial processes of the rat cerebellum. From embryonic Day 15 to postnatal Day 2, vimentin is expressed throughout the cerebellar anlage in a number of cell types, including radial glia and Bergmann fibers. In the human, radial fibers are present at 9 weeks of ovulation age with features of astroglial differentiation. With advancing age (20 weeks), glia traversing the molecular layer demonstrate an increasing resemblance of Bergmann fibers with the cell bodies located below the Purkinje cells. It has been advocated that Bergmann glia are modified radial glial cells. Considerable evidence favors the view that Bergmann glia originate directly from the primary germinal zone over the fourth ventricle and not from the external granule cell layer.

The consensus of opinion is that although divisions of both Bergmann glia and velate astrocytes occur in the Purkinje cell layer and internal granule cell layer, precursors of these cells migrate from the primary germinal zone along with the immature Purkinje and Golgi neurons early in cerebellar development. On the other hand, in cerebellar mutants, movement of neurons has been shown to take place even in the presence of disturbed cytoarchitecture.

## H. Astrocytes in the Optic Nerve: "Optic Nerve Glia"

Recent circumstantial coincidence suggests that progenitor cells in the optic nerve do not develop from neuroepithelial cells of the optic stalk but instead migrate into the developing optic nerve from the brain. Clusters of stem cells appear in the optic stalk by fetal Day 16 in the rat, and their cytoplasmic extensions reach the basal portion of the optic stalk where axons are growing in. These elongated cells will differentiate into optic nerve glia. GFAP can be demonstrated in mouse optic nerve as early as embryonic Day 17. A small number of astroglia undergo their final cell division as early as 15.5 days of gestation in the rat optic nerve, but the majority of cells do so during the first postnatal week. Fully differentiated, fibrous astrocytes are not present in the rat optic nerve before the 14th postnatal day.

As discussed earlier, different subtypes of astrocytes are formed in the optic nerve, based primarily on staining differences of optic nerve astrocytes with the monoclonal antibody A2B5. By Day 28, more than 50% of astrocytes in the optic nerve are GFAP[+] + A2B5[+], resembling fibrous astrocytes. These astrocytes are transversely oriented, and their processes often end on the blood vessels. Astrocytes at the periphery of the nerve forming the glia limitans stain weakly with GFAP and not with A2B5, resembling protoplasmic astrocytes.

## I. Astrocytes in the Olfactory Bulb

The olfactory bulb is a bilateral, oblong structure attached to the rest of the brain by means of the olfactory peduncle. The olfactory nerve fiber layer is composed of densely packed unmyelinated olfactory axons (of the olfactory receptor cells located in the olfactory epithelium of the nasal cavity) and glial cells; these axons terminate in the glomerular layer of the olfactory bulb. In the olfactory bulb of the rat and rabbit and in the accessory olfactory bulb of the rat, there are two morphologically distinct types of glial cells: typical astrocytes that form end feet both on blood vessels and at the surface of the bulb, but do not ensheathe the axons, and glial cells solely responsible for ensheathement of olfactory axons in the olfactory nerve fiber layer, sometimes referred to as "ensheathing cells." Astrocytes and ensheathing cells differ only in terms of their cytoplasmic density and numbers of filaments, suggesting that the ensheathing cell is a morphological variant of the typical astrocyte, with the function of supporting or encouraging axonal growth.

## J. The Perinodal Astrocyte

Nodes of Ranvier represent an important specialization of myelinated fibers in both the peripheral nervous system (PNS) and the central nervous system (CNS). The morphology and physiology of nodes of Ranvier have been classically considered to involve only the axon and the myelin-forming oligodendrocyte. However, a series of studies, since 1971, have demonstrated the presence of perinodal astrocyte processes as an integral component of the central nodes. A unique relationship between astrocytic processes and the nodes of Ranvier has been demonstrated by freeze-fracture and electron microscopic studies of the optic nerve. Perinodal astrocytes form numerous interastrocytic gap junctions and also gap junctional complexes with neighboring oligodendrocytes. A multiplicity of functions have been attributed to the perinodal astrocyte and include formation of a nodal gap substance, regulation of the extracellular ionic environment, participation in action potential electrogenesis, and a role in the development and/or maintenance of nodal membrane, providing extraneuronal sites for the synthesis of ionic channels, which are subsequently transferred to nodal regions of the axon membrane. Moreover, in view of the presence of gap functions between the perinodal astrocyte and paranodal oligodendrocyte, the perinodal astrocyte may also play a role in the maintenance of the integrity of the oligodendrocyte and its associated myelin and axon–glial paranodal junction by preventing activity-related swelling. Thus it has been suggested that, within the CNS, the oligodendrocyte and the perinodal astrocyte constitute a counterpart of the Schwann cell, which subserves both functions in the peripheral nervous system. This possibility correlates well with the finding that type 2 astrocytes and oligodendrocytes arise, at least in the optic nerve, from a common precursor (the O-2A cell). The perinodal astrocyte appears to derive from type 2 astrocytes.

## K. Ependyma, Ependymoglia Astroglia, and Oligodendroglia: A Functional Syncytium

The ependymal cells constitute the ependyma, which forms the lining of the ventricular system. When the basal processes of these cells establish direct contact with the pia matter, such cells resemble in form radial glia and are thus referred to as ependymoglial cells to distinguish them from the tanycytes bound by tight junctions. Ependymoglial cells are absent or rare in most regions of the CNS with a large ventricular–pial distance. In the thin-plated cerebellum and in the spinal cord of certain amphibia and reptiles, practically all ependymal cells occur in the form of ependymoglial cells, and astrocytes are absent, or scarce. In the turtle cerebellum, there are no Golgi–Bergmann cells; all Bergmann fibers are provided by nontight junction-forming ependymoglial cells, whereas myelin is provided by oligodendrocytes. Ependymoglial cells may subserve the functions of both the ependymal cells and the astrocytes but not those of the oligodendrocytes.

In mammals, individual ependymal, ependymoglial, and astroglia cells (of all subtypes) form numerous gap junctions and adherent junctions, in contrast to oligodendrocytes which also form tight junctions. Adherent junctions, that is, septate and desmosomoid types, provide some mechanical attachment between cells. The intercellular gap of adherent junctions is partially porous to extracellular tracers. Gap junctions provide minuscule intercellular channels (hollow proteins on one cell membrane dock across the 2-nm-wide extracellular gap with an analogous protein on the adjoining cell membrane to create a charged cytoplasmic channel) through which small molecules can pass from one cell to the next. By the use of fluorescent probes impermeable to the plasma membrane, it has been demonstrated that intercellular channels can provide passage not only to ions, but also to small solutes of molecular weight of the order of 1000 (oligosaccharides, amino acids, nucleotides, small peptides, catabolic products, transmitter molecules, low-molecular-weight trophic factors). Tight junctions are not as well defined as gap junctions. Some authors propose that the subunit particles in the membranes are glycoproteins and others argue that there is also a lipid component.

In the ependymal lining, each cell seems to be connected by homologous gap junctions with all of its neighbors, thus the lining can be considered a "functional syncytium." It is assumed that a similar relationship is present among ependymoglial cells. The available evidence also suggests that astrocytes form homologous gap junctions with all neighboring cells coming in; thus it can be assumed that astrocytes can represent a functional syncytium. Evidence deriving from thin sections (anterior medullary system, dorsal cochlear nucleus) shows the occurrence of gap junctions between ependymal cells and astrocytes (ependyma–fibrous astrocyte relation; ependyma–protoplasmic astrocyte relation). Evidence from freeze-fracture has confirmed a high frequency of astrocyte-to-oligodendrocyte gap junctions in both gray and white matter. Thus, the hypothesis has been proposed that ependyma, ependymoglia, astroglia, and oligodendroglia, by virtue of their gap junctions, form a potentially uninterrupted and comprehensive network that can be termed supporting functional syncytium. In this context, gap junctions among neural sustaining cells form an intracellular pathway for ions (and small solutes), especially potassium, that enter the glia during sustained neuronal activity, and thus ionic equilibrium throughout the nervous tissues is maintained and regulated.

## III. BIOCHEMICAL, PHYSIOLOGICAL, AND PHARMACOLOGICAL CHARACTERISTICS

### A. Voltage-Gated and Neurotransmitter-Gated Ion Channels

It has become an accepted dogma that the satellite cells of the nervous system express a vast array of voltage-gated ion channels. In culture, astrocytes express the greatest diversity of ion channels of glial cells, including four types of $K^+$ channels, two forms of $Na^+$ channels, L-, and T-types of $Ca^{2+}$ channels, anion channels, and stretch-activated channels. The expression of voltage-sensitive $K^+$ channels could be of functional importance for the differentiation of precursor cells to astrocytes or oligodendrocytes. It has been demonstrated that the presence of rectifying $K^+$ channels is linked to a proliferative state of cells. Culture studies have indicated the existence of two distinct $Na^+$ channel types in astrocytes that are expressed in morphologically differing astrocyte types. Several hypotheses have been advanced for the functional role of astrocyte $Na^+$ channels. Astrocytes have been suggested to provide extraneuronal sites of $Na^+$ channel synthesis, which are subsequently transferred into axons. Astrocyte $Na^+$ channels may also be involved in ionic homeostasis of extracellular space by regulation of $Na^+/K^+$-ATPase activity.

Voltage-activated $Ca^{2+}$ channels were among the first channels described of rat cerebral astrocytes. The recent discovery of intracellular $Ca^{2+}$ waves in astrocytes following neurotransmitter stimulation or mechanical stimulation have opened the way for new and possibly important roles for intracellular $Ca^{2+}$ as a signal in glia networks. Anion channels are probably the least studied in astrocytes. High-conductance anion channels have also been found in Schwann cells, in mouse and rat astrocytes. Voltage-activated $Cl^-$ channels have been proposed to be involved in the regulation of $[K^+]_0$ by glia cells. It has been proposed that depolarization may simultaneously activate $Cl^-$ and $K^+$ channels, resulting in a net uptake of KCl by glial cells. Such $K^+$ buffering may occur in pathological conditions where $[K^+]_0$ can rise to 80 m$M$, resulting in sufficient depolarization for $Cl^-$ channel activation.

As discussed in Sections III,B and III,C, astrocytes possess not only high-affinity uptake sites for GABA but also a GABA receptor-coupled $Cl^-$ channel. The glial GABA receptor shares many similarities with the neuronal $GABA_A$ receptor. Type 2 astrocytes possess quisqualate- and kainate-receptor channels but lack receptors for $N$-Methyl-D-aspartic acid (NMDA). These glutamate channels exhibit multiple conductance levels that are similar in amplitude to neuronal glutamate channels. The importance of glutamate-activated ion channels in type 2 astrocytes may play a significant role at the nodes of Ranvier, where type 2 astrocytes are in close approximation.

The conventional accepted way that neurons communicate with glia are via the events at the synaptic cleft through the release of neurotransmitters. However, evidence is beginning to attribute a possible direct influence of astrocytes on neurons and neuronal activity. Interstitial regulation of ions, such as $K^+$, $Ca^{2+}$, and $Na^+$, and neurotransmitter release and uptake by astrocytes at the synaptic cleft appear to control synaptic events. An additional signaling astrocyte to the neuron pathway is the release of glutamate from astrocytes, which causes an increase in neuronal $Ca^{2+}$. Thus perturbation of neuronal intracellular $Ca^{2+}$ by local astrocytes can contribute to information processing in the CNS and regulation of neuronal physiology and pathology.

## B. Neurotransmitter and Other Related Receptors

Numerous studies have shown that both glial cell lines and primary cultures of astrocytes exhibit a variety of neurotransmitter-associated receptors and other receptors. Astrocytes have $\beta$- and $\alpha$-adrenergic receptors, muscarinic cholinergic receptors, glutamate receptors $\gamma$-amino-3-hydroxy-5-methyl isoxazole-4-proprionic acid (AMPA/kainate), GABA$_A$ receptors, and benzodiazepine receptors; also, receptors for hormones and peptides are found and include vasoactive intestinal peptides, somatostatin, adrenocorticotropic hormone (ACTH) and melanocyte-stimulating hormone (MSH), and glucocorticoid and thyroid hormone receptors. Astrocytes appear to be depolarized by glutamate, which gates a $Na^+$-permeable ion channel through the activation of a receptor that shares many properties with neuronal kainate/AMPA receptors. Of significance is the report that precursor cells express AMPA-type glutamate receptors. The presence of these receptors in precursor cells would implicate them in perhaps playing a role in glial precursor differentiation into astrocytes or oligodendrocytes.

Extensive evidence has demonstrated that various neurohormones, including various transmitters, substance P, somatostatin, secretin, glucagon, adrenocorticotropin, and others, regulate the intracellular concentration of cAMP in astroglia-enriched cultures derived from murine brain. Using adenylate cyclase as one effector system, it has been clearly shown that astroglia cells are highly susceptible to several peptides and thus have to be considered as delicately regulated by neurohormones. In the context of neuron–glia interactions, the responsiveness of glial cells to neurohormones and neurotransmitter substances reflects the role of neuronal signaling in providing a link be-

tween neurons and glial cells. [See Neurotransmitter and Neuropeptide Receptors in the Brain.]

## C. Monoamine and Amino Acid Uptake

### 1. Monoamines

Extensive evidence derived primarily from astrocytes in culture indicates that astrocytes can take up both catecholamines and serotonin by high-affinity systems and thus have the potential of competing with neuronal terminals for uptake of the released transmitters. The uptake system in astrocytes behaves pharmacologically like the uptake in various brain preparations. The likely fate of the transmitter monoamines taken up into astrocytes would be metabolism and removal. A low-affinity norepinephrine uptake, uptake$_2$, has also been reported in glial cells in culture, and its significance can only be speculated: it provides a safety ratio at the synaptic cleft of possible buildup of norepinephrines (NE). Thus, specific inhibitors for uptake$_2$, including glucocorticoids, may cause a shift in extracellular NE and ultimately in neuronal firing. Since the uptake system in astrocytes is also sensitive to clinically effective antidepressants, the therapeutic effects of such agents may be mediated to some extent by their actions on astrocytes.

### 2. Choline Uptake

The transport of choline has been extensively studied using clonal cell lines of neuronal and glial origin. Both neuronal and glial cells possess high-affinity transport systems for exogenous choline. However, in glial cells, choline transport is not $Na^+$ dependent. It is speculated that *in vivo* choline uptake in glia cells may occur during depolarization of presynaptic terminals. In contrast to the high-affinity choline uptake reported in glial cell lines, glial cells in primary cultures exhibit low-affinity choline uptake. The conclusion that has been drawn from these reports is that the maturation state of the glial cells plays a vital role in their functional state as reflected in transport mechanisms.

Since both acetylcholinesterase and butyrylcholinesterase are found in glial cells, it is very likely that glial cells are intimately involved in acetylcholine metabolism. The hypothesis has been proposed that choline enters directly into astrocytes from the capillary lumen through end-feet processes. About 50% of the free choline derives from transport and 50% from phospholipid recycling through inversion of the Kennedy pathway reactions and the enzyme glycerophosphocholine diesterase, preferentially localized in glial cells.

## 3. GABA, Glutamate, Aspartate, and Glutamine

High-affinity GABA uptake has been demonstrated in various nonneuronal constituents of the nervous system: astrocytes, satellite (glia) cells in peripheral ganglia, Muller cells, primary cultures of astrocytes derived from cerebral hemispheres, and glial cell lines. Regional differences in astrocyte GABA uptake exist, since the uptake capacity for GABA into cortical astrocytes is considerably higher than into cerebellar astrocytes. Astrocytes possess not only high-affinity uptake sites for GABA but also a GABA receptor-coupled Cl$^-$ channel.

Glutamate is taken up much more intensely into astrocytes than into neurons. It is suggested that glutamate is primarily inactivated by the uptake into astrocytic elements. This view is substantiated by the fact that glutamate synthetase (GS), the synthesizing enzyme of glutamine, is localized predominantly in astrocytes (see Section III,E). Thus, glutamine synthesis may be exclusively an astrocytic phenomenon. Glutamine released from astrocytes would then be taken up into neurons. It appears that no major difference exists between GABAergic neurons, glutamatergic neurons, and astrocytes in their ability to accumulate glutamine, since in both cases there is only a relatively low-affinity uptake. However, the consensus is that unless there is a considerable net uptake of glutamine from blood to the CNS, glutamine cannot be the only precursor for transmitter glutamate and GABA.

The uptake and release processes of D- and L-aspartate have been compared to that of L-glutamate in primary astrocyte cultures and cerebellar granular cells. A high-affinity uptake system for L- and D-aspartate has been found in both cell types.

Differences between neuronal and astrocytic glutamate carriers have been implicated by the differences in Na$^+$ dependency of the two uptake systems. The astrocytic glutamate uptake is characterized by requiring Na$^+$ and glutamate in a one-to-one ratio, whereas the neuronal uptake system requires two sodium ions per glutamate molecule transported. The uptake of glutamate, an excitatory amino acid, into astrocyte is of paramount significance in health but also in such conditions as anoxia and ischemia, where a dramatic increase of extracellular glutamate concentration is observed. Glutamic acid excess has been implicated

in neuronal cell death in aging and neurodegenerative diseases.

## D. Lipids of Astrocytes

Lipids are ubiquitously present in all tissues as integral membrane components. The major portion of the lipids is represented by the phospholipids, the sphingoglycolipids are the least abundant, and cholesterol is quantitatively intermediate. Many investigations have compared the composition of astrocytes to oligodendrocytes or neurons. It appears that astrocytes have a greater quantity of total lipids than do neurons but a similar quantity compared to oligodendrocytes. [*See* Lipids.]

## E. Enzymes

### 1. Glutamine Synthetase

Immunocytochemical studies of tissue sections have established that GS in the brain is predominantly, if not exclusively, located in astrocytes and their retinal counterparts, the Muller cells. The importance of GS in the retina is substantiated by its dramatic rise during development, its hormonal inducibility, and the fact that higher levels are found in the retina than in any other organs. GS is not distributed uniformly among astrocytes within a brain area as well as from region to region. This heterogeneity has also been observed in primary brain cell cultures, where only 50% of the GFAP-positive astrocytes immunostain with GS antibody. Different levels of GS activity have also been observed among strains of mice genetically different in alcohol sensitivity.

The developmental accumulation of GS in various brain regions is associated with the differentiation of astrocytes in close association to glucocorticoid regulation. *In vivo* and *in vitro* regulation of GS resembles that of another brain glucocorticoid-inducible enzyme, glycerol phosphate dehydrogenase (GPDH), which is an enzyme marker for oligodendrocytes. GS activity is also influenced by other hormones and factors: $\beta$-adrenergic agonists and dibutyryl cyclic AMP induce GS in C6 glioma cells and in chick neural retina; thyroid hormone increases GS of rat cerebellar glial cells; insulin and thyroid hormone increase GS levels in primary cerebral astrocyte cultures; and both fibroblast-conditioned medium and fibroblast cell substratum increase GS in astrocytes of primary mouse cultures. However, neuron-conditioned medium from differentiated neuronal-enriched cultures lowers GS activity in glial cultures, and this is in contrast to an increase in cyclic nucleotide phosphohydrolase, a marker for oligodendrocytes. Finally, GS activity markedly increases in both primary glial cultures and C6 glioma with increasing cell passage, and this phenomenon has been interpreted to assimilate aging *in vitro*. High levels of GS activity are also expressed in cultures derived from aged mouse brain.

## 2. Carbonic Anhydrase

In the CNS, carbonic anhydrase (CA) is restricted to glial cells in the choroid plexus. Immunohistochemical studies at the light and electron microscopic level show that CA II (isozyme) is primarily localized in oligodendrocytes. However, GFAP-positive astrocytes will also stain if the cultures are treated with dibutyryl cyclic AMP, an agent that increases CA II activity. It has been suggested that astrocytes have low but inducible levels of CA II and that this enzyme is physiologically involved in the regulation of astroglial cell volume. Treatment of glial cell cultures with acetazolamide, a carbonic anhydrase inhibitor, produces a marked effect on glial cell $pH_2$, reducing from 7.1 in control to 6.88, which represents a significant increase in intracellular $H^+$ concentration. This finding further implicates glial cells in regulating the neuronal microenvironment.

## 3. Butyrylcholinesterase

Butyrylcholinesterase (BuChE), a pseudocholinesterase enzyme, has been one of the earlier glial markers, since first detected in the CNS. The rapid increase in BuChE activity in the chicken brain between 18 days of embryogenesis and 3 months after hatching is associated with marked glial cell proliferation during this age. Moreover, a marked rise in BuChE in the aged chicken brain again coincides with astrogliosis associated with aging. Thus, this enzyme marker appears to reflect a functional responsiveness of glial cells and adaptability to the microenvironment.

## IV. FUNCTIONAL ROLES

### A. Contribution to Neuronal Survival, Neurite Growth, and Phenotypic Expression

In 1928, Ramon y Cajal put forward the hypothesis that nonneuronal cells play an important physiological role in the trophic support of neurons. In the past 15 years, the search for factors secreted from glial cells has been extensive. Several glial cell populations of peripheral and central nervous system origin have been investigated and appear to produce neuroactive substance. Some of these factors have been shown to be nerve growth factor (NGF), whereas others do not interact with antibody against NGF or are capable of influencing NGF-insensitive cell populations. A 43-kDa glia-derived protein with both promoting activity and serine protease inhibitory activity has been isolated. This protein is able to inhibit the plasminogen activator activity, which has been associated with neurite extension in neuroblastoma cells and cell migration. Other factors have been isolated from serum-free conditioned medium over astroglia cells and include a survival-supporting low-molecular-weight neurotrophic activity that can be mimicked by pyruvate, a high-molecular-weight protein that has neurite-promoting activity and also shares antigenic determinants with laminin, an as yet unidentified soluble component of conditioned medium that promotes neurite elongation and survival. It has been reported that cultured rat astrocytes produce a factor with biological and immunological properties similar to those of insulin. This factor may in part mediate the observed neurotrophic effects of astrocyte-conditioned medium. The addition of insulin in chemically defined medium has been reported as essential for neuronal survival and differentiation. Astrocytes in culture synthesize and presumably secrete unprocessed angiotensinogen, which is thought to function in the brain, as in the periphery, as an extracellular reservoir of angiotensin peptides. [*See* Laminin in Neuronal Development; Nerve Growth Factor.]

Studies in coculturing neurons and glial cells have provided additional evidence that cell contact or membrane-bound factors may also be involved in neuronal survival and growth. For example, cerebellar astrocytes ("velate" protoplasmic astrocytes) have been implicated in 30 to 50% survival of granule cells when cocultured. This is in contrast to "fibrous" astrocytes, which only 5% of neurons survive when cocultured.

Using cocultures of dispersed embryonic transmitter-identified neurons (serotonin, dopamine) and monolayers of astrocytes or fibroblasts demonstrates clear differences in the behavior of cells growing on these monolayers. More specifically, the expression of tyrosine hydroxylase in immunoreactive dopaminergic neurons from substantia nigra is higher on astrocyte monolayers than on fibroblasts, indicating a significant enhancement of biochemical differentiation of dopaminergic neurons. Another related study

on the possible role of glia factors on neuronal phenotypes found that in neuroblast-enriched cultures from
3-day-old chick embryo, both cholinergic and
GABAergic neuronal expression, assessed biochemically and immunocytochemically, is enhanced by the
culture substratum, that is, collagen versus polylysine,
favoring the growth of flat cells (presumptive glia).
Whether glial-secreted factors or cell membrane–
bound factors are involved in this cellular phenomenon cannot as yet be distinguished.

## B. Neuronal Migration and Guidance

Glial cells have been suggested to provide a framework for organization of the nervous system and a
"blueprint" for axonal guidance along stereotyped
pathways. Thus the migrating granule cells are related
to the Bergmann glial fibers in the rhesus monkey cerebellum.

The optic system has also been used as an anatomical model to examine neuron–target relationships and
the role of glial cells. Muller cells (the counterparts
of astroglia in the retina) offer favorable growth surfaces to the mass of retina ganglion cell axons as they
emerge from the undifferentiated mouse optic cup
around embryonic Days 11.5 to 12 and are segregated
into discrete portions of the optic stalk.

*In vitro* studies have provided some cellular clues
to the issue of neuronal migration and the role of glial
cells. For example, in cultures derived from 3-day-
old chick embryo, neuroblast-enriched cultures, the
characteristic growth pattern is neuronal aggregation
and neurite fasciculation; long neurite processes are
aligned with a continuous chain of interconnected
cells. These cells are either Schwann cells beginning
to ensheathe an axonal-type neurite or radial glial
guiding the neurite. Also, in cultures derived from
cerebella of mice 5–7 days after birth, cell migration
along the arms of Bergmann-like astroglia has been
described. In this system, most neurons associated
with astroglia are granule neurons. In contrast to neurons observed in other studies, neurons do not migrate
along the arms of the stellate astroglial forms but
rather remain in place and extend growth cones. Two
populations of stained astroglia are described in these
cultures. The first, constituting ~10–20% of stained
cells present in the cultures, had a smaller cell body
and thinner, longer processes; several unstained neuronal cells bound to the arms of these astroglial distal
to the cell body. On the basis of their shape and
pattern of association with neurons, these cells have
been identified as Bergmann-like glia. A second type

of AbGF-positive cells (AbGf, a glial filament protein)
has a slightly larger cell body and stellate, shorter
process. Among the arms of these astroglia nestle several dozen unstained neurons with very fine short
processes. These cells resembled astrocytes of the cerebellar granular layer as well as those of the white
matter. When cells are cultured from the neurological
mutant weaver, a mouse that suffers failure of granule
neuron migration in concert with abnormalities in
the shape and alignment of the Bergmann glia, the
following observations have been described: very few
granule cells survived and little neurite outgrowth was
seen, the morphology of the astroglia was abnormal,
and neuron–glia associations were mostly absent.
"Mixing and matching" neurons and astroglia purified from weaver and normal mice demonstrated that
weaver neurons fail to migrate on wild-type astroglial
processes *in vitro* and that they impair astroglial differentiation; in contrast, normal neurons associate
with weaver in astroglia, inducing their differentiation
and forming tight appositions like those seen in migrating neurons *in vivo*. These studies suggest that
the granule neuron may be a primary site of action
of the weaver gene and further demonstrate that neuron–glia interactions may regulate astroglia differentiation.

## C. Synaptic Function and Neurotransmission

There is a rich background literature on the intimate
relationship between neuroglia and nerve cells.

Synaptic contacts between neuronal processes and
glial cells have been demonstrated during development *in vivo* as well as in neural cell culture. Axoglial
synapses have been observed in the mouse spinal cord
11–14 days postnatally, and synaptic contacts between neuronal and glial cell processes have been described in chick spinal cord and cerebellar cortex cultures. In the adult nervous system, synapse-like
contacts between neurites and glial cells seem to be
limited to those of the ependymal cells lining the ventricle and to tanycytes or astrocytes of certain periven-
tricular regions such as the pituitary, subcommissural
organ, medial vascular prechiasmatic gland, and median evidence. The ultrastructural relationships between axons, glial cells, and connective tissue have
been examined in the walking leg nerves of crabs
and support the view that all these cell units together
participate in the maintenance of the microenvironment around axons.

Using the parasympathetic avian ciliary ganglion as

a model, it has been proposed that there is competition between glial cells and afferent fibers for apposition to neuronal surfaces. The postsynaptic neuron, the presynaptic elements, and the satellite glia continually adjust themselves to some sort of balance during both development and aging.

Astrocytes have been implicated in the regulation of synaptic density. Although the evidence is not conclusive, it appears that astroglia ensheathement of axons after neuronal insult may be neuronally directed and could be the physical element provoking a reduced number of synapses. This concept has serious implications in regeneration, to be further discussed in the following section.

The close neuron–astrocyte morphological interrelation, together with evidence discussed earlier of the ability of astrocytes to accumulate neurotransmitter substances and ions and also that astrocytes express several receptors, establishes an intimate relationship between neurons and glial cells in neurotransmission processes. According to a model on the metabolism and turnover of putative amino acid transmitters, there would be two metabolic compartments, "large" and "small," but three cell types, that is, neurons, consisting of synaptic (nerve endings) and nonsynaptic (e.g., perikarya) parts, astrocytes and oligodendrocytes. The main products of the synaptic parts, that is, glutamate, GABA, and aspartate, are released as transmitters; glutamate and aspartate are accumulated efficiently and almost quantitatively into astrocytes. GABA is taken up the astrocytes with less efficiency so that about one-half of the released GABA is reaccumulated into neuronal perikarya, from where it may travel back to nerve endings by axonal transport. All three amino acids are metabolized in the astrocytes either to carbon dioxide and water or to glutamine, which may be released from the astrocytes. The model assumes that transmitters flow from a "large" to a "small" compartment. Other models have proposed glutamine flow from the "small" to the "large" compartment, and the difference is primarily in the metabolic compartmentation of glutamine, astrocytic versus synaptosomal. The implications of such a metabolic model are significant. For example, disruption of these close metabolic interactions under pathological conditions (e.g., when the astrocytic glutamate uptake is reduced under ischemia or when the oxidative metabolism of glutamate is impaired during exposure to elevated ammonia concentrations) may contribute to the aberrations in brain function under such conditions.

The role of astrocytes in synaptic plasticity and function has become more compelling with the evidence of signals from astrocytes to neurons. Studies in culture have shown that glutamate released from astrocytes causes an increase in neuronal $Ca^{2+}$. Thus astrocytes regulate neuronal $Ca^{2+}$ levels through the $Ca^{2+}$-dependent release of glutamate. Therefore, perturbation of neuronal intracellular $Ca^{2+}$ by local astrocytes contributes to information processing in the CNS and regulation of neuronal physiology and pathology.

As also discussed in Section III,C, the uptake of monoamines by astrocytes may play a fundamental role in synaptic function in that it may regulate the amount of neurotransmitter present in the synaptic cleft. This function would be of particular significance under conditions of excessive transmitter release. Another role of the monoamine uptake and metabolism system in astrocytes would be to protect the CNS against systemically derived monoamines entering the CNS via the capillaries in the case of failure of the blood–brain barrier.

The uptake of choline into astrocytes described earlier has several implications in neuron–glia interaction in general and cholinergic neurons in particular. During nerve activity, free choline is furnished into the intercellular spaces by the following suggested pathways: (a) through hydrolysis of released acetylcholine and (b) through the degradation of polyphosphoinositol, the induction of phospholipase A, and the production of oleate followed by the activation of phospholipase D, which hydrolyzes phosphatidylcholine into diacylglyceride and choline. Choline thus produced is consequently retaken up into neurons after reestablishment of ionic gradients (also involving the role of glia) or into glial cells for stocking or for "efflux" from the brain. Finally, the trophic role of glial cells in choline metabolism of neurons *in vivo* under steady and stimulating conditions has been speculated upon and remains to be confirmed.

## D. Gliosis, Reactive Gliosis, and Regeneration

Gliosis represents a major effect of injury to the CNS and has been frequently viewed as being an important determinant of the extent and quality of neural repair in mammals. Because of the apparent limitation it imposes on regeneration, glial reactivity has been traditionally viewed in a negative context. However, a number of studies provide evidence against the view of glial barriers to regeneration, and a beneficial role of glial cells in regeneration has been given favorable consideration.

Studies in the peripheral nervous system using peripheral nerve grafts have concluded that severed axons can regrow when they can interact with nonneuronal components of the PNS. Several factors may be involved in these regenerative processes, including changes in the neuronal perikaryon, perhaps triggerred by signals received through growth cones in the damaged tip of axons from molecular cues secreted by the ensheathing cells, or present on cell surfaces and/or the extracellular matrix.

The role of astrocytes changes during development, regenerative failure, and induced regeneration upon transplantation. Thus, when the cerebral midline is lesioned in the mouse embryo or neonate, the would-be callosal axons form neuromas. An implanted Millipore bridge inserted between the neuromas in young acallosal animals can support the migration of immature astrocytes that, in turn, support the *de novo* growth of commissural axons between the hemispheres. Moreover, there is a critical period for the glia to promote reconstruction of malformed axon pathways. In acallosal postnatal mice given Millipore implants on or before postnatal Day 8, GFAP-positive, stellate-shaped astrocytes migrate and attach to the implant by inserting foot processes into the pores of the filter; this form of gliotic response is established on axon growth-promoting substratum within 24–48 hours of implantation. Of interest is the observation that, during this age period, there is no evidence of scar formation or necrosis at or around the implant surface. However, when acallosal mice are given implants on or later than postnatal Day 14, extensive tissue degeneration occurs, and a mixed population of astrocytes and fibroblasts invades the surface of the filter, producing a dense scar. Reactive cells within the scar do not promote axonal outgrowth. Furthermore, when filters coated with glia from 8-day-old mouse forebrains are transplanted into the brains of 14-day-old acallosal animals, glial scarring in the host is reduced and axonal regeneration is enhanced. Thus, axonal regeneration in the CNS at postcritical periods may be stimulated by reintroducing an immature glial environment at the lesion site. It appears that the immature glial environment within such embryonic grafts supplies essential growth and guidance cues to the regrowing axons. In contrast to adult reactive gliosis, the gliotic response in neonatal animals is an active rather than reactive phenomenon; activated gliosis can be considered a beneficial and constructive

process. Phenomenologically similar activated astrocytes have been described in the developing or regenerating optic nerve of *Xenopus laevis* and along the regenerating olfactory nerve in adult rats. The failure to regenerate after a critical period has been attributed to at least two factors: the presence of ectopic (out of place) basal lamina and connective tissue that form a mechanical impediment to axons and the physical state of glial cells that migrate onto the implant.

*In vitro* studies using rat sciatic nerve stumps that regenerate through an empty silicone chamber and astroglia cell populations grown in cell culture and inoculated within the silicone chamber have provided further supportive evidence of the changing role of astroglia with cell differentiation. Adult astrocytes seem to down-regulate axonal growth; presumably, their function is to confine the neurites within designated structural and functional boundaries. It has been suggested that the inhibitory role that the mature astrocytes exert on nerve regrowth may be mediated in part by the lack of plasminogen activator activity, namely, by the lack of migratory capacity of the mature astrocytes.

On the other hand, the use of the cryogenic model of spinal injury as a means for studying the relationship of astrocytes and regrowing axons has provided evidence that astrocytes are not obstructive and in fact may be supportive of axonal regrowth through the damaged area. For the time period of 60 days following injury (a temperature of −8°C was applied to the tissue for 15 minutes), it was observed that from an undifferentiated group of cells that appears early, there develops a population of astrocytes that is associated with the restoration of neural structure, including the return of axons to the damaged area. In this model, the absence of obstacles such as glia limitans, tissue gaps, and cavities provides an unimpeded pathway for the development of this astrocyte–axon relationship. Thus, astrocytes have the opportunity to provide growth signals similar to those in developing immature systems, repair of inframammalian systems, growth up to peripheral nerve bridges, fetal transplantation, and tissue culture. However, the intrinsic capacity of the axon for regrowth also appears to be implicated in this axon–astrocyte relationship.

The role of neuroglia in regeneration has also been implicated in fetal CNS grafts into adult host nervous system. Behaviorally, fetal CNS grafts reduce deficits from a specific lesion. However, this return to normal function remains an enigma. It is not clear if the functional return is due to the neuronal synaptic contact of the host with the graft or vice versa. Perhaps the return to normal function is due to trophic effects of transplanted neuroglia.

## E. Astrocytes and the Blood–Brain Barrier

The anatomical relationship between the end feet and cerebral blood vessel has led to numerous speculations on the role of astrocytes in the blood–brain barrier, including the possibility that they are the site of the barrier and that they transport nutrients and metabolites between the blood and the neurons. However, the role of astrocytes, if any, in the blood–barrier continues to be speculative. It has been suggested that astrocytes might act to inhibit the pinocytotic mechanism in contiguous endothelial cells and thus seal the barrier; in another way, the glial sheath may act to induce and maintain the development of barrier characteristics in the contiguous endothelial cells. Strong circumstantial evidence supports this role of astrocytes. For example, the formation of the glial sheath coincides with the functional "tightening" of the barrier during development. Evidence from pathological studies shows that when cerebral vessels grow into tumors in the CNS, they lose their glial sheath and their barrier characteristics. However, whether this is due to the loss of the glial sheath or to the induction of a different phenotype in the endothelium by the tumor cells is a debatable point. [See Blood–Brain Barrier.]

From all available evidence, it appears that astrocytes cannot be the only cell type with barrier characteristics. Blood–brain barriers with similar permeability properties are found in the iris, peripheral nerve, retina of the immature rabbit, and frog brain surface capillaries, where glial sheaths are absent. It is possible that the blood–brain barrier has different levels of barrier function that are accounted for by different inducers. For example, the one feature common to blood–brain barriers is the continuous bands of tight junctions between contiguous endothelial cells. Without such a seal, functional activities of the barrier would be ineffective, since concentration gradients across the wall could not be maintained. This "basic" feature may be induced by any neuroepithelial cell derivative, whereas the more specialized functional endothelial characteristics may be determined by astrocytes.

## F. Changes in Astrocytes with Aging

Only in the last decade has consideration been given to glial cells in the aging brain, and the role of glial cells in the neuronal aging process is far from being understood.

Studies using Cajal's gold stain for astrocytes and light microscopic analysis have revealed that major astrocytic reactivity (hypertrophy and increased numbers of thickened processes) occurs in the hippocampus, caudate nucleus, a number of major myelinated fiber tracts, and various other brain regions during aging. Whereas the density of astrocytes exhibiting hypertrophy in the hippocampus increases dramatically and progressively in rats of three ages (4 months, 15 months, and 25 months) and is significantly elevated even in middle-aged (15-month-old) animals, the total population of astrocytes is not significantly elevated. Also the relative distribution of hypertrophied glial cells and total number of glial cells remain constant at the three ages studied. The following conclusions have been put forward.

1. Hypertrophy of astrocytes appears to occur "in place," and migration of hippocampal astrocytes does not appear to be a major factor.
2. Glial hypertrophy does not occur uniformly throughout the hippocampus with age, and it appears more pronounced in synaptic terminal fields.
3. Glial hypertrophy appears to have its onset in middle age.

Several studies using cell cultures as models have explored cell changes with aging. Immunocytochemical studies using cultures derived from aged mouse cerebral hemispheres and maintained for several cell passages (up to 45) have shown the presence of astrocytes at various stages of maturation, some oligodendrocytes, and more importantly the presence of glia precursors. The presence of immature glial cells may be "residual radial glial" cells and may have a special function, such as providing a reservoir for astrocyte proliferation occurring under various conditions including aging and regeneration.

In the context of neuron–glia interactions, the marked activity in GS-containing astrocytes with aging is of importance. As discussed in Section III,C,3, astrocytes are intimately involved in the compartmentation of glutamate–glutamine–GABA. Increases in GS activity would be expected to lead to increases in both cellular and extracellular glutamine released from astrocytes. Glutamine is a precursor for GABA release. Thus, changes in GS activity in GS-containing astrocytes will be reflected in both GABAergic and glutamatergic neuronal activity. The consequences of such shifts in cellular activity during aging of the CNS have only recently been considered. Considerable evidence exists that synaptic loss and decline in synaptic function occur in the senescent brain and may play a role in the age-related decline in brain function. The possibility that compensation for neuronal loss by synapse growth in the aged brain is being vigorously explored, and reactive synaptogenesis or axon sprouting has been reported to occur in a complex series of events. Glial cells are proposed to play a role in the initiation of growth and axon and/or dendritic growth. The intimate astrocytic–synaptic relationship (discussed earlier) appears to be a key factor in synaptic degeneration and remodeling. An imbalance in this relationship would impede synaptic turnover and consequently affect brain plasticity.

# BIBLIOGRAPHY

Bignami, A., and Dahl, D. (1973). Differentiation of astrocytes in the cerebellar cortex and the pyramidal tracts of the newborn rat. An immunofluorescence study with antibodies to a protein specific to astrocytes. *Brain Res.* **49**, 393–402.

Black, J. A., and Waxman, S. G. (1988). The perinodal astrocyte. *Glia* **1**, 169–183.

Cajal, Ramon y (1928). "Degeneration and Regeneration of the Nervous System" (translated by R. M. May). Oxford Univ. Press, London.

Choi, B. H. (1988). Prenatal gliogenesis in the developing cerebrum of the mouse. *Glia* **1**, 308–316.

Fedoroff, S., and Vernadakis, A. (eds.) (1986). "Astrocytes. Vol. 1. Development, Morphology, and Regional Specialization of Astrocytes." Academic Press, Orlando, FL.

Fedoroff, S., and Vernadakis, A. (eds.) (1986). "Astrocytes. Vol. 2. Biochemistry, Physiology and Pharmacology of Astrocytes." Academic Press, Orlando, FL.

Fedoroff, S., and Vernadakis, A. (eds.) (1986). "Astrocytes. Vol. 3. Cell Biology and Pathology of Astrocytes." Academic Press, Orlando, FL.

Hertz, L., and Schousboe, A. (1986). Role of astrocytes in compartmentation of amino acid and energy metabolism. *In* "Astrocytes" (S. Fedoroff and A. Vernadakis, eds.), Vol. 2, pp. 179–208. Academic Press, Orlando, FL.

Hirano, M., and Goldman, J. E. (1988). Gliogenesis in rat spinal cord: Evidence for origin of astrocytes and oligodendrocytes from radial precursors. *J. Neurosci. Res.* **21**, 155–167.

His, W. (1890). Histogenese und Zusammenhang der Nervehelemente. *Arch. Anat. Physiol.* (Suppl.), 95–117.

Hosli, E., and Hosli, L. (1993). Receptors for neuro-transmitters on astrocytes in the mammalian central nervous system. *Prog. Neurobiol.* **40**, 477–506.

Raff, M. C., Miller, R. H., and Noble, M. (1983). A glial progenitor cell that develops *in vitro* into an astrocyte or an oligodendrocyte depending on culture medium. *Nature (London)* **303**, 390–396.

Rakic, P. (1971). Neuron–glia relationship during granule cell migration in developing cerebellar cortex. A Golgi and electron microscopic study in *Macacus rhesus. J. Comp. Neurol.* **141**, 283–312.

Schnitzer, J. (1988). Immunocytochemical studies on the development of astrocytes, Muller (glial) cells and oligodendrocytes in the rabbit retina. *Dev. Brain Res.* **44**, 59–72.

Somjen, G. G. (1988). Nervenkitt: Notes on the history of the concept of neuroglia. *Glia* **1**, 2–9.

Vernadakis, A. (1988). Neuron–glia interrelations. *Int. Neurobiol. Rev.* **30**, 149–223.

Vernadakis, A., and Roots, B. I. (eds.) (1995). "Neuron–Glia Interrelations during Phylogeny. I. Phylogeny and Ontogeny of Glial Cells. II. Plasticity and Regeneration." Humana Press, Totowa, NJ.

# Ataxia-Telangiectasia and the ATM Gene

YOSEF SHILOH

*Sackler School of Medicine, Tel Aviv University*

## GLOSSARY

**cDNA** Complementary DNA, an artificial copy of a mature messenger RNA (mRNA) molecule obtained in the laboratory using a viral enzyme, reverse transcriptase

**Complementation cloning** Strategy for identification and molecular cloning of genes with unknown protein products, which determine a distinct cellular phenotype. An attempt is made to modify the phenotype by introducing foreign DNA into the cells and identifying the piece of DNA responsible for the resultant phenotypic change. That piece of DNA is expected to represent the relevant gene

**Cosmid** Vector for cloning and propagating DNA sequences up to tens of thousands of basepairs in bacterial cells

**Linkage analysis** Method of determining the location and relative distance between genes or DNA sequences that vary between individuals. It is based on the coinheritance of two genetic elements that reside on the same chromosome within a certain distance from each other

**Molecular cloning** Technique for obtaining an unlimited number of copies of a specific DNA fragment. It is a prerequisite for molecular analysis of the structure and function of a DNA sequence

**Positional cloning** Strategy to identify disease genes with unknown protein products. Genetic analysis locates the responsible gene to a specific chromosomal region, and all the genes residing in that region are isolated. The disease gene is identified by screening patients for mutations in each of those genes

**Vector** Artificial DNA molecule, constructed from sequences of self-replicating molecules (such as plasmids or viruses), that retains the ability to self-replicate in bacterial or eukaryotic cells. Vectors are the main vehicles in molecular cloning experiments, where they are used to propagate the cloned sequences

**Yeast artificial chromosome** Vector based on yeast genetic elements used to clone and propagate DNA segments up to hundreds of thousands of basepairs in yeast cells

ATAXIA-TELANGIECTASIA (A-T) IS A HUMAN genetic disorder inherited in an autosomal recessive manner and characterized by progressive neuromotor dysfunction, immunodeficiency, premature aging, chromosomal instability, cancer predisposition, and radiosensitivity. A basic cellular feature of A-T is a defect in cell-cycle checkpoints that are supposed to halt the life cycle of the cell following the induction of certain types of DNA damage. It has been suggested that A-T heterozygotes are moderately cancer prone and radiosensitive. The responsible gene, ATM, was located on chromosome 11 and identified using a positional cloning approach. This gene encodes a protein whose molecular mass is 350 kDa, and contains a region similar to the catalytic domain of phosphatidylinositol 3-kinase, a mediator of mitogenic signals in mammalian cells. The ATM protein is similar to several proteins in other species that share this domain and are responsible for conveying regulatory signals to various cellular systems following DNA damage induction. This protein provides a missing link between DNA metabolism, regulation of the cell's life cycle, and cancer development, and its discovery may have wide implications in many areas of biomedical research. [*See* Cell Cycle; Cancer Genetics.]

ENCYCLOPEDIA OF HUMAN BIOLOGY, Second Edition, VOLUME I.   Copyright © 1997 by Academic Press.   All rights of reproduction in any form reserved.

## I. THE GENETIC DISORDER ATAXIA-TELANGIECTASIA

### A. Clinical and Laboratory Features

Since its establishment as a clinical entity in 1957, ataxia-telangiectasia has presented a biological, medical, and human challenge to clinicians and researchers. The mutations responsible for A-T seemed to inactivate an unidentified physiological junction linking the development of various tissues, genome stability, DNA metabolism, cell-cycle control, cellular aging, and neoplastic transformation. Hence, identifying the site of these mutations had long been expected to have far-ranging effects in several areas of biomedical research.

A-T is inherited in an autosomal recessive manner and is found worldwide, with patient frequencies of about 1 : 100,000 in Caucasian populations. The disease makes its appearance initially as a neurological disorder. Cerebellar ataxia begins in infancy and progresses steadily, confining the patient to a wheelchair by the beginning of the second decade of life. Other major neurological signs are involuntary movements, diminished or absent deep reflexes, slow and unfocused eye movements, and slurred speech. The neuropathological hallmark of A-T is cerebellar degeneration involving primarily the Purkinje and granular cells; degenerative changes have also been noted in the spinal cord and ganglia, brain stem, and peripheral nerves. The second clinical hallmark of A-T, which typically appears between ages 3 and 6, is telangiectases (dilation of blood vessels, making them more prominent) in the eyeballs and conjunctivae, sometimes spreading over sun-exposed areas of the skin. Some 50–80% of patients show the third clinical hallmark of A-T, recurrent sinopulmonary infections. Serum levels of IgA, IgG2, and/or IgE are reduced, the number of circulating lymphocytes is diminished, and mitogen response is poor. The thymus appears malformed and sometimes absent. Serum levels of two proteins usually associated with development and malignancies—$\alpha$-fetoprotein and carcinoembryonic antigen—are consistently higher in A-T patients. Somatic growth and sexual maturation are usually retarded, with female hypogonadism being almost uniform. Progeric changes typically appear in the hair and skin, marking premature senescence. Intelligence is usually normal.

Another cardinal feature of A-T is profound cancer predisposition, which becomes evident in about 10% of patients during childhood. Lymphomas and acute lymphocytic leukemias constitute over 85% of all cancers in A-T and appear primarily in younger patients. The incidence of other cancers, mainly epithelial, rises steadily with age. Early attempts to treat these malignancies by radiotherapy resulted in acute radiation reactions, revealing another feature of A-T—a profound sensitivity to the cytotoxic effects of ionizing radiation. The course of A-T is progressive and relentless, and patients usually die with respiratory failure or malignancy during the second or early in the third decade of life. There is no effective way to retard the progression of the disease.

The primary diagnostic laboratory finding in A-T is chromosomal instability, evidenced by high rates of chromosomal breaks, usually in peripheral lymphocytes or fibroblasts. Lymphocyte cultures show cell clones containing specific chromosomal translocations, involving particularly the chromosomal regions 7p14, 7q35, 14q12, and 14q32, which harbor the T-cell receptor and immunoglobulin heavy chain genes. Such clones often precede the onset of lymphoreticular malignancies and undergo clonal expansion as malignancy progresses. Molecular analysis of several translocation breakpoints showed that the immune system genes residing in these regions were indeed involved in these aberrations. [*See* Chromosome Anomalies.]

### B. Cellular Characteristics

The cellular phenotype of A-T further reflects the complexity of this disorder. Besides chromosomal instability, A-T cells show a reduced life span in culture, higher requirements for unspecified serum growth factors, abnormalities in the shape and arrangement of cytoskeletal actin fibers, accelerated shortening of the telomeres (chromosome ends), and abnormal content of a variety of extracellular surface proteins. A major cellular characteristic of A-T, which has become diagnostic, is the profound sensitivity of the cells to the cytotoxic and clastogenic effects of ionizing radiation and chemicals that mimic the action of X rays on the DNA (often called "radiomimetic chemicals"). Among such chemicals, A-T cells are particularly sensitive to those that produce hydroxyl radicals capable of inducing strand scissions. The overall kinetics of single- and double-strand break repair in A-T cells has been found to be normal in most studies, although some studies noted an elevation in the residual amount of unrepaired strand breaks in irradiated A-T cells and suggested a subtle defect in handling DNA strand breaks in A-T. An abnormally high rate of intrachromosomal recombination was also seen in A-T cells, whereas interchromosomal recombination remained normal.

Contrary to expectations, semiconservative DNA synthesis in A-T cells was found to be more resistant to the inhibitory effect of the DNA damaging agents to which A-T cells are sensitive, most notably ionizing radiation. This phenomenon, called "radioresistant DNA synthesis" (RDS), reflects a defect in a control mechanism (checkpoint) responsible for halting the progression of the cell cycle at the S phase following the formation of DNA damage. The two other cell-cycle checkpoints, at the G1 and G2 phases, also are defective in A-T cells, resulting in a decrease in the initial postirradiation inhibition of progression of the cell cycle. Radiation-induced G1 arrest is mediated by a signal transduction system that involves a rise in the cellular level of the protein produced by the tumor suppressor gene p53. The elevation in p53 level is indeed slower in A-T cells. These observations led to the suggestion that A-T cells may harbor a defect in a protein or a protein complex involved in a signal transduction system responsible for cell-cycle arrest and enhancement of DNA repair following radiation damage. [*See* DNA Repair; Tumor Suppressor Genes, p53.]

## C. Variants, Associated Disorders, and Genetic Heterogeneity

The clinical and cellular characteristics of A-T have been used to delineate phenotypic variants among patients. Patients with somewhat milder clinical signs, later age of onset, and slower progression of the disease were found in several countries, and in some patients this phenotype was correlated with milder radiosensitivity, and sometimes reduced or absent RDS.

Other disorders share features with A-T, such as immunodeficiency coupled with chromosomal instability. The combination of microcephaly, growth retardation, immunodeficiency, chromosomal instability, radiosensitivity, and RDS but no telangiectases has been particularly related to A-T. Patients with this syndrome, sometimes associated with mental retardation, were reported in several ethnic groups, and some were classified as "Nijmegen breakage syndrome" (NBS). NBS patients tend to develop lymphoreticular malignancies at a higher rate than classic A-T patients.

Attempts have been made to delineate the possible genetic heterogeneity of A-T and its relationship with related syndromes by fusing cells from different patients and measuring RDS, or radiation-induced chromosomal aberrations in the fusion products (heterokaryons). These studies revealed four complementation groups in classic A-T, designated A, C, D, and E, and two complementation groups, V1 and V2, among patients with NBS. No correlation was found between complementation group assignment and clinical variation in A-T, and it was unclear whether these complementation groups represented different genes, or different mutations within one gene.

## D. A-T Heterozygotes

A-T heterozygotes have always received special attention. In a sense A-T is not entirely recessive, since carriers may mildly manifest two of the disease characteristics, cancer predisposition and radiosensitivity. Epidemiological studies have suggested that A-T heterozygotes exhibit a higher rate of certain cancers, especially breast cancer in women. It was estimated that the cancer tendency among male A-T heterozygotes was 3.8-fold higher than among the general population, whereas that of female carriers was estimated to be 3.5-fold higher. However, the relative risk for breast cancer in women alone was estimated to be 5.1-fold higher than in a control population. It was further suggested that up to 8.8% of American white female patients with breast cancer may be A-T carriers.

These findings were accompanied by repeated reports of moderate sensitivity of cells from A-T carriers to ionizing radiation, as measured by survival and cytogenetic assays. These results, which imply that A-T heterozygotes might face special hazards from medical procedures involving radiation, stimulated attempts to develop laboratory assays for the detection of A-T carriers in the general population. Though obligatory A-T heterozygotes could be clearly distinguished from controls in some studies, controls and A-T heterozygotes marginally overlapped in other samples. The wide range of radiation responses among control groups has thus reduced the reliability of this parameter as an assay for carrier detection.

## II. IDENTIFICATION OF THE A-T GENE

### A. Complementation Cloning

In the absence of any recognizable protein unequivocally defective in A-T cells, attempts to elucidate the

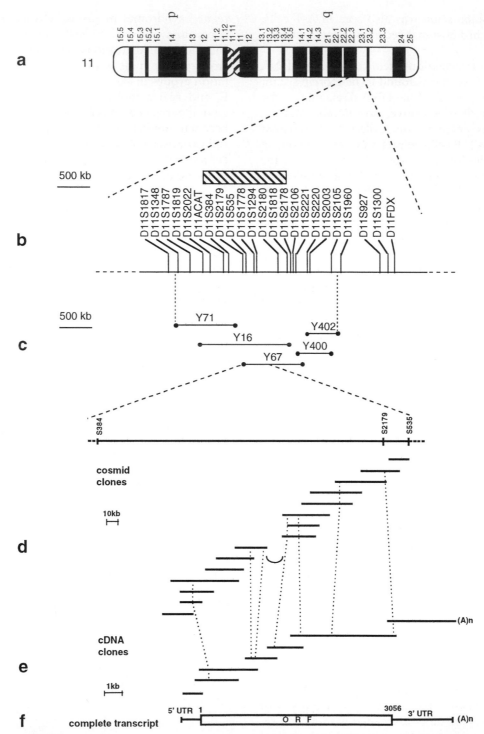

FIGURE 1   Positional cloning of the ATM gene. (a) Assignment of the A-T genetic locus to chromosome 11, region q22–23, was based on linkage analysis. p, the short chromosomal arm; q, the long chromosomal arm. The numbers denote chromosomal bands. (b) A high-density map of genetic markers was subsequently constructed within this region. These markers are denoted by the prefix "D11" followed by a number. FDX, the adrenal ferredoxin gene; ACAT, the acetoacetyl-coenzyme A acetyltransferase gene. Additional genetic analysis of A-T families based on these markers led to confinement of the disease gene into an interval of 1.5 megabases of genomic DNA

molecular basis of this disorder focused primarily on identifying the gene harboring A-T mutations. Since the common handle to gene identification, namely, the sequence of the gene's protein product, could not be used to identify the A-T gene, alternative methods had to be sought. The sensitivity of A-T cells to radiation and radiomimetic chemicals seemed to enable the application of a unique strategy for gene identification—functional cloning by complementation of the cellular phenotype. In this approach, exogenous DNA is introduced into the cells, and selection is applied to identify cell clones in which this sensitivity has been "corrected." An attempt is then made to identify the piece of DNA supposedly responsible for this effect, which is expected to represent a normal allele of the disease gene. Functional cloning by this strategy is appealing since it circumvents the more labor-intensive positional cloning (see Section II,B). This strategy proved successful in isolating genes involved in other human disorders showing sensitivity to DNA damaging agents, such as xeroderma pigmentosum and Fanconi's anemia. Several laboratories have invested considerable effort for more than a decade in applying this approach to A-T.

Using this technique, three laboratories identified complementary DNA (cDNA) clones that corrected the radiomimetic sensitivity of A-T cells of various complementation groups. But many of these clones did not represent complete transcripts of the corresponding genes and were derived from a large variety of genes, none of which mapped to the genetic locus of the A-T gene on chromosome 11 (see the following). These studies led to the conclusion that the biological end points that define the two A-T phenotypic hallmarks—radiosensitivity and RDS—can be modulated by high expression of a number of sequences, not necessarily full-length transcripts. In such a situation, complementation cloning may suffer from a too low signal-to-noise ratio. Attempts to identify the elusive A-T gene thus shifted to another strategy, positional cloning.

## B. Positional Cloning

The basic steps in the positional cloning strategy include the localization of a disease locus to a specific chromosomal region by linkage analysis; extensive generation of highly polymorphic markers in the region and narrowing the locus by repeated genetic analysis; long-range cloning and physical mapping of the disease locus; identification of transcribed sequences ("gene hunting"); and, finally, a search of the candidate genes for mutations in patients.

The genetic heterogeneity of A-T presented a potential obstacle to linkage analysis, should several A-T genes reside in different locations. This problem was conveniently skirted by conducting initial linkage analysis on a 61-member Amish A-T kinship assigned to complementation group A. Significant linkage was identified in this family between the disease and several markers on the long arm of human chromosome 11, at a chromosomal region designated 11q22–23 (Fig. 1a). Further analysis with additional A-T families substantiated this finding and suggested that the mutations responsible for all other complementation groups might be located in the same chromosomal region. A consortium-based analysis of an increasing number of families, concomitant with saturation of the A-T region with highly informative genetic markers, finally narrowed the A-T interval to 1.5 megabases of genomic DNA (Fig. 1b). As the boundaries of the A-T locus were moving closer together, systematic cloning of the region was accomplished in several laboratories by constructing a series of overlapping clones (contigs) spanning the region, in yeast artificial chromosome (YAC) and cosmid vectors (Figs. 1c and 1d). These contigs were used in the next step in the positional cloning scheme, gene hunting.

---

(stippled box). (c) Part of a contig (series of overlapping DNA fragments) constructed across this region in yeast artificial chromosomes. (d) Part of a contig of DNA fragments cloned in a cosmid vector spanning a portion of the A-T interval in which the A-T gene was most likely to be found according to genetic analysis. The arch between the ends of two cosmids in the middle of this contig represents a gap that was subsequently bridged. (e) A series of partly overlapping complementary DNA (cDNA) clones that together encompass the entire transcript of the ATM gene. Dotted lines drawn between the cDNA and cosmid contigs indicate colinearity of sequences. (A)n, a poly(A) tail typical of eukaryotic mRNAs. (f) The diagram of the complete ATM transcript, obtained by alignment of the above cDNA clones. This transcript is composed of an open reading frame (ORF) encoding a protein of 3056 amino acids and two untranslated regions (UTRs).

Gene hunting, the search for genes in a given genomic domain, typically relies on two commonly used methods to identify transcribed sequences in genomic DNA: the direct selection method is based on hybridization of genomic DNA with cDNA collections of various tissues; the exon trapping method identifies and clones gene exons (the portions of genes that are finally included in mature mRNAs), by virtue of their ability to get spliced to each other. In direct selection experiments, cosmid and YAC clones served to capture cross-hybridizing sequences in cDNA collections from placenta, thymus, and fetal brain. The cosmids were used in parallel in exon trapping experiments. The captured cDNA fragments and trapped exons were mapped back to the A-T region. An extensive transcriptional map of the A-T region was thus constructed, and pools of adjacent cDNA fragments and exons, expected to converge into the same transcriptional units, were used to search for full-length cDNA clones derived from the corresponding genes.

Figure 1e shows a contig of several partly overlapping cDNA fragments that together span a transcript of 13 kilobases (Fig. 1f) encoded by a gene in the A-T interval. A search for gross rearrangements in the corresponding gene in a number of A-T families revealed a homozygous deletion spanning a large portion of this gene in affected members of a Palestinian Arab A-T family. This finding spurred the search for point mutations in other A-T families. Such mutations were indeed found in A-T patients representing all four complementation groups, suggesting that the corresponding gene alone was responsible for all A-T cases. This gene was subsequently designated ATM (A-T, mutated).

Interestingly, two patients of different complementation groups were found to share the same mutation, ruling out the possibility that the different groups represented mutations in different domains of the ATM protein. How then could the phenomenon of complementation groups be explained? A-T patients were assigned to complementation groups based on measurements of RDS, or radiation-induced chromosome breakage in heterokaryons. However, it has been shown subsequently that RDS and radiosensitivity in A-T cells can be dissociated from each other and modulated separately by *in vitro* manipulations of the cells, in particular by gene transfer. Thus, the biological end points of cellular sensitivity and RDS appear to be modulated by a variety of unknown genes and physiological factors unrelated to A-T. In retrospect, they may have been unsuitable experimental clues to the A-T genetic defect.

## III. THE ATM GENE AND PROTEIN

### A. Genomic and cDNA Organization

The ATM gene extends over 150 kilobases (kb) of genomic DNA and contains 66 exons. Complete cloning and sequencing of the ATM gene transcript that totals about 13 kb (Fig. 1f; GenBank accession number U33841) revealed an open reading frame (ORF) of 9168 nucleotides. Several untranslated regions (UTRs) of varying lengths precede the ORF, and a long 31 UTR of about 3.6 kb follows the termination codon.

### B. Protein Sequence, Motifs, and Similarity to Other Proteins

The predicted product of the ATM ORF is a large protein of 3056 amino acids, with an expected molecular mass of 350.6 kDa. A search for functional domains in its sequence revealed a region spanning 350 amino acids at the carboxy terminus of this protein, which shows strong similarity to the catalytic domain of the 110-kDa subunit of the signal transduction mediator phosphatidylinositol 3-kinase (PI 3-kinase) of mammalian cells, and the corresponding yeast protein VPS34. PI 3-kinase is a key enzyme in mediating cellular responses to several mitogenic growth factors, to factors triggering cellular differentiation, and to insulin. It is also involved in processes such as the oxidative burst in neutrophils, membrane ruffling, and glucose uptake. Another motif in the ATM protein is a potential leucine zipper, a sequence of amino acids that is usually a hallmark of protein dimerization or interaction with other proteins. No nuclear localization signal or transmembrane domains were observed in this protein.

The PI 3-kinase-like domain is common to all the members of a rapidly growing family of large proteins in various species (Fig. 2). The homology of ATM to these proteins is highest at the PI 3-kinase domain, and to a lesser extent beyond this region.

The product of the TEL1 gene in the baker's yeast, *Saccharomyces cerevisiae,* shows the highest homology to ATM (Fig. 2). The TEL1 protein is involved in controlling the length of chromosome ends (telomeres) in this organism. TEL1 deletion mutants are viable and not X-ray sensitive, but have shortened telomeres and elevated levels of mitotic recombination and chromosome loss. MEC1 is an essential gene of *S. cerevisiae* required for arrest in G2 after DNA damage, and for arrest in S phase when replication is

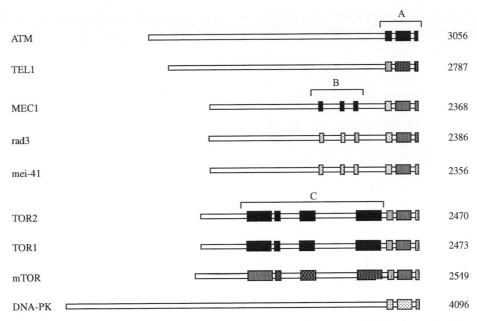

**FIGURE 2** Schematic diagram of the similarity domains in the ATM and related proteins. The length of each protein, in amino acids, is shown on the right. The different patterned boxes represent varying degrees of amino acid identity measured by pairwise comparisons between the specific protein and a prototype protein for each region. Boxes in region A are located within the PI 3-kinase domain, and pairwise comparisons in this region were performed against the ATM protein throughout the entire protein family. Boxes in region B were compared to the MEC1 protein. Boxes in region C belong to the TOR subgroup and were compared with TOR2. ■, 80–100% identity; ■, 50–60%; ■, 40–50%; ▨, 30–40%; ☐, 20–30%. [Reprinted from K. Savitsky *et al.* (1995b), with permission.]

incomplete. Mutations in the MEC1 gene cause hypersensitivity to a variety of DNA damaging agents, such as UV, ionizing radiation, and hydroxyurea. These mutants also have defects in DNA repair and meiotic recombination and show elevated mitotic instability. TEL1 seems to be functionally related to MEC1, since overexpression of this gene rescues the viability and reverses the radiosensitivity of mec1 mutants to DNA damaging agents. In addition, tel1 mec1 double mutants are synergistically sensitive to ionizing radiation, UV, hydroxyurea, and radiomimetic drugs.

rad3 mutants in the fission yeast, *Schizosaccharomyces pombe*, resemble mec1 mutants in *S. cerevisiae*. They are hypersensitive to a broad range of DNA damaging agents, show elevated mitotic chromosome instability, and seem to be defective in a DNA repair process. The rad3 gene product is required for cell-cycle checkpoint controls at G2 and S phases, which respond to damaged or unreplicated DNA.

The product of the mei-41 gene of *Drosophila* is required for repair of DNA damage and for mitotic and meiotic stability. It is involved in late cell-cycle arrest, most likely at G2, and may also play a role in

a G1/S or S checkpoint. mei-41 mutants are highly sensitive to a wide range of DNA damaging agents. mei-41 is also required in embryonic cell divisions and, like MEC1, is involved in meiotic recombination.

Another important member of this protein family is the catalytic subunit of the DNA-dependent protein kinase (DNA-PK$_{cs}$). DNA-PK, which has been studied primarily in humans, is a serine/threonine protein kinase that is activated by DNA double-strand breaks. DNA-PK may be a modulator of transcription, since it phosphorylates several transcription factors *in vitro*, including p53, fos, jun, and Sp1, and is a potent inhibitor of transcription by RNA polymerase I. It is a heterotrimer consisting of a very large polypeptide of approximately 460 kDa and the Ku antigen (a dimer of 70- and 80-kDa subunits). The former contains the kinase domain and serves as the catalytic subunit, and the latter interacts with double-strand DNA breaks and provides the DNA targeting component. Mutations in the mouse gene encoding the DNA-PK$_{cs}$ in this organism cause the *scid* phenotype (severe combined immunodeficiency) characterized by immunodeficiency, hypersensitivity to ionizing radiation, cancer predisposition and deficiencies in both DNA double-strand break repair and somatic recombination of the immune system genes.

The TOR proteins of *S. cerevisiae,* TOR1 and TOR2, and their mammalian homolog, mTOR, form a separate subgroup within this protein family, based on functional and sequence similarities. These proteins, which show very high similarity to each other, were identified as the targets of the immunosuppressant drug rapamycin, and were shown to bind the complex formed between this drug and the cellular protein FKBP12. Rapamycin inhibits a late step in the pathway leading to activation of T cells by interleukin-2 by preventing T-cell progression from G1 to S phase of the cell cycle. Additional studies have shown that rapamycin blocks cellular events associated with G1/S transition. Rapamycin is also a potent antifungal compound and, as in T cells, it also blocks progression of yeast cells from G1 to S phase.

## C. The Possible Function of the ATM Protein

What can we learn about the function of the ATM protein from its sequence similarities? Functionally, the ATM protein seems to be close to the products of the MEC1, rad3, mei-41, and DNA-PK$_{cs}$ genes, since mutations in these genes confer phenotypes that share characteristics with A-T cells, such as radiosensitivity, mitotic instability, and defects in the processing of DNA damage. ATM, MEC1, rad3, and mei-41 are involved in cell-cycle checkpoints. No checkpoint abnormalities have been detected in *scid* cells, which are defective in DNA-PK$_{cs}$. However, this could be due to the existence of backup mechanisms for signaling the presence of double-strand breaks in mammalian cells. A-T and *scid* cells differ substantially from mec1, rad3, and mei-41 mutants in that they are sensitive only to ionizing radiation, whereas mutations in the latter three confer hypersensitivity to a large variety of DNA damaging agents. Another similarity between *scid* mice and A-T patients is the appearance of combined immune deficiency in both cases, although it is not clear whether somatic recombination of the immune system genes is defective in A-T patients.

It is of interest that the TEL1 protein is more similar to ATM than are the other proteins in this group. One of the functions of TEL1 is the control of telomere length in budding yeast. Short telomeres are typical of senescing human cells. It has been suggested

that loss of telomeric sequences may be associated with senescence, and that chromosome ends with shortened telomeres may activate a checkpoint pathway that inhibits cellular proliferation. Short telomeres might also cause the end-to-end chromosome associations frequently observed in normal senescing cells and in A-T cells, which are also known to senesce prematurely and exhibit telomere shortening. These observations suggest that A-T cells, like tell mutants, might have a defect in the control of telomere length, which could be associated with the premature aging observed in A-T. On the other hand, MEC1, TOR1, and TOR2 are not involved in telomere length in the budding yeast.

There may not be a perfect ATM homolog in non-mammalian species. DNA damage surveillance in mammalians might be divided among a large number of proteins, each one sensitive to a specific type of damage. The different members of this family of proteins are most probably involved in the detection of certain types of DNA damage, incompletely replicated DNA, or abnormal telomere length, and may subsequently transmit that information to regulators of repair machineries and of cell-cycle progression. This mechanism should allow the necessary repair to take place before DNA replication or cellular division. In unicellular organisms, such as yeast, defects in some of these processes may result in nonviability.

In mammalian cells, a key cell-cycle regulator recruited by the ATM protein is likely to be p53. This protein is encoded by a tumor suppressor gene and has been defined as "guardian of the genome." It may play a central role in activating DNA repair mechanisms while temporarily halting cell-cycle division, in response to a variety of DNA lesions. This protein may also promote the damaged cell to self-destruct (a process known as programmed cell death, or apoptosis) rather than pass on its damaged DNA. In complex multicellular organisms, failure to repair DNA lesions prior to DNA replication or cell division might eventually result in mutations and genomic instability, leading to uncontrolled cell proliferation or apoptotic death. The latter should be manifested particularly in postmitotic cells and may be responsible for the neurodegeneration observed in A-T patients. [*See* Apoptosis].

It is not clear whether the ATM protein is a lipid kinase or a protein kinase or both. DNA-PK is an example of a protein with a PI 3-kinase-like domain that acts as a protein kinase, and it has been suggested that the TOR proteins also have protein kinase activity. By analogy, other proteins of this family may not actually act as PI 3-kinases, but rather display different enzymatic activities, such as protein kinase.

## IV. MUTATIONS IN THE A-T GENE

The ATM transcript was scanned for mutations in cell lines from an extended series of A-T patients. A large variety of mutations were found, and most of them were unique to single families. These mutations are dominated by deletions and insertions expected to inactivate the ATM protein by truncating it or by causing large deletions in its amino acid sequence. A missense mutation was identified, which leads to substitution of an extremely conserved glutamic acid residue for glycine within the PI 3-kinase-like domain. This mutation profile suggests that the classic A-T phenotype is caused by homozygosity or compound heterozygosity for "null alleles," and hence is probably the most severe expression of defects in the ATM gene. By inference, the missense mutation that substitutes a conserved amino acid at the PI 3-kinase-like domain points to the importance of this domain for the biological activity of the ATM protein. This domain most likely contains the catalytic site of this protein. Mutations with milder effects on the protein's activity might therefore be found in phenotypes different from classic A-T, such as "A-T variants" or related disorders showing only part of the A-T phenotype. Most notable disorders of this type are cerebellar ataxias coupled with various degrees of immunodeficiency. Screening for mutations in this gene in such cases may reveal wider boundaries for the molecular pathology associated with the ATM gene.

## V. CLINICAL AND SCIENTIFIC IMPLICATIONS OF THE A-T GENE

The identification of a single gene responsible for all A-T cases has immediate practical ramifications: it will enable clinical geneticists to offer reliable diagnostic tests, like prenatal diagnosis and carrier detection, to all A-T families based on the chromosome 11 markers (Fig. 1b), with less concern about A-T mutations elsewhere in the genome. As the mutations that cause the disease in individual families are identified, such diagnostic tests become simpler and easier.

However, the ATM gene is responsible not only for the devastating, rare disease A-T, but possibly also for a subpopulation of heterozygous individuals who may be cancer prone and radiosensitive. In fact, the carrier phenotype brings the ATM gene into the realm of public health. While the consequences of heterozygosity for ATM mutations are being reassessed using molecular assays, it is highly likely that screening of specific populations for A-T carriers may be conducted in the future. Their identification may lead to improved surveillance for early signs of cancer and alert physicians to expected sensitivity to conventional radiotherapy. At the same time, it may allow the use of higher radiation doses in treatment of other cancer patients. The implications for the well-being and lifestyle of individuals found to carry these mutations are far-ranging, making screening a matter with serious ethical, psychosocial, and legal considerations.

Development of new and better means to treat A-T patients will obviously depend on further understanding of the function of the ATM protein and the signal transduction pathway in which it is involved. This pathway supplies a missing link between DNA damage, cell-cycle progression, genome stability, and neoplastic transformation, whose existence has been long assumed in view of the A-T phenotype. Complete understanding of this pathway should have wide implications in cell biology and particularly cancer research. The normal life cycle of the cell depends on a delicate balance between numerous interconnected cellular systems. Disturbance of one of the delicate relays allowing continuous flow between these systems may drive the cell out of balanced growth and lead to cell death or the development of cancer. A-T is a striking example of both of these consequences, underscoring the importance of the signal transduction system that is defective in this disease. Investigation of the ATM gene and its protein product is thus expected to have a major impact on many areas of biomedical research.

## BIBLIOGRAPHY

Harnden, D. G. (1994). The nature of ataxia-telangiectasia: Problems and perspectives. *Int. J. Radiat. Biol.* 66, S13–S19.
Savitsky, K., Bar-Shira, A., Gilad, S., Rotman, G., Ziv, Y., Vanagaite, L., Tagle, D. A., Smith, S., Uziel, T., Sfez, S., Ashkenazi, M., Pecker, I., Frydman, M., Harnik, R., Patanjali, S. R., Simmons, A., Clines, G. A., Sartiel, A., Gatti, R. A., Chessa, L., Sanal, O., Lavin, M. F., Jaspers, N. G. J., Taylor, A. M. R., Arlett, C. F., Miki, T., Weissman, S., Lovett, M., Collins, F. S.,

and Shiloh, Y. (1995a). A single ataxia telangiectasia gene with a product similar to PI-3 kinase. *Science* **268,** 1749–1753.

Savitsky, K., Sfez, S., Tagle, D., Ziv, Y., Sartiel, A., Collins, F. S., Shiloh, Y., and Rotman, G. (1995b). The complete sequence of the coding region of the ATM gene reveals similarity to cell cycle regulators in different species. *Hum. Mol. Genet.* **4,** 2025–2032.

Sedgwick, R. P., and Boder, E. (1991). Ataxia-telangiectasia. *In* "Handbook of Clinical Neurology" (P. J. Vinken, G. W. Bruyn,

and H. L. Klawans, eds.), Vol. 16, pp. 347–423. Elsevier, New York/Amsterdam.

Shiloh, Y. (1995). Ataxia-telangiectasia: Closer to unraveling the mystery. *Eur. J. Hum. Genet.* **3,** 116–138.

Taylor, A. M. R., Byrd, P. J., McConville, C. M., and Thacker, S. (1994). Genetic and cellular features of ataxia telangiectasia. *Int. J. Radiat. Biol.* **65,** 65–70.

Zakian, V. A. (1995). ATM-related genes: What do they tell us about functions of the human gene? *Cell* **82,** 685–687.

# Atherosclerosis

KEITH E. SUCKLING
*SmithKline Beecham Pharmaceuticals*

I. Disease of Atherosclerosis
II. Natural History of the Developing Atherosclerotic Plaque
III. Risk Factors in Humans
IV. Therapy of Atherosclerosis
V. Future Prospects

## GLOSSARY

**Bile acids** Biological detergents synthesized from cholesterol in the liver and secreted into the intestine in bile to help in solubilizing and digesting fat; the only way in which cholesterol can be transformed and eliminated from the body in quantity

**Cholesterol** A lipid obtained in the diet or synthesized in the body, especially in the liver; an essential component of cell membranes and the precursor of steroid hormones and bile acids

**Hyperlipoproteinemia** A pathological state in which excess lipoprotein is present in the blood; several types are distinguished, a number of which are correlated with the occurrence of atherosclerosis

**Macrophage** A scavenger white blood cell that forms upon activation of a monocyte

**Plasma lipoprotein** Lipid transport particle in the blood, several types of which can be distinguished by their buoyant density. The lipoproteins are spherical, with a core of triacylglycerol or cholesteryl ester and an outer monolayer of phospholipid and cholesterol. The monolayer is penetrated by specific proteins, apolipoproteins, that allow the particle to be recognized by receptors on cell surfaces and by enzymes in the blood

**Platelets** Small blood cells that make the primary response to injury in a blood vessel, aggregating to cover the injured area in response to stimuli secreted by the first platelets to adhere

**Receptor** A protein that binds a ligand; in the present context, a cell surface protein that binds specifically to an apolipoprotein

**Smooth muscle cell** Contractile cell found in the middle layer of the artery wall (or media)

**Triacylglycerol** Ester of long-chain fatty acids and glycerol; the main component of fat

ATHEROSCLEROSIS, TOGETHER WITH CANCER, IS the largest cause of death in the developed world. Atherosclerosis accounts for about one-third of human mortality, with much of it due to coronary artery disease. In 1992, 42% of all deaths in the United States were due to coronary heart disease. Eleven million Americans had symptomatic coronary heart disease, and many more were undiagnosed. The cost of the disease to the United States has been estimated to be over $60 billion a year. Atherosclerosis is characterized by a thickening of the walls of medium-sized and large arteries, and is particularly dangerous and life-threatening when it occurs in the coronary arteries. Raised areas of inflammation, deposition of debris, and cellular proliferation, known as atherosclerotic plaques, are formed slowly over the early decades of life. The mature plaque (Fig. 1) may consist of a fibrous cap covering a core of dead cells and other debris, including cholesterol crystals and calcium salts. Surrounding the core are cells derived from the artery wall, smooth muscle cells in great number, endothelial cells, and macrophages and T lymphocytes derived from the blood. The core has a gruel- or porridge-like appearance, from which the disease got its name. Other terms used, often interchangeably, for this disease are "arteriosclerosis" and, for the lesions, "atheroma."

The consequences of human atherosclerosis can be sudden and dramatic. Blood flow may be blocked or severely restricted by the plaque, by fragments broken off from it, or by a thrombus, or blood clot, forming at the site of a plaque. Atherosclerosis in coronary arteries can cause angina, myocardial ischemia, myo-

ENCYCLOPEDIA OF HUMAN BIOLOGY, Second Edition, VOLUME I. Copyright © 1997 by Academic Press. All rights of reproduction in any form reserved.

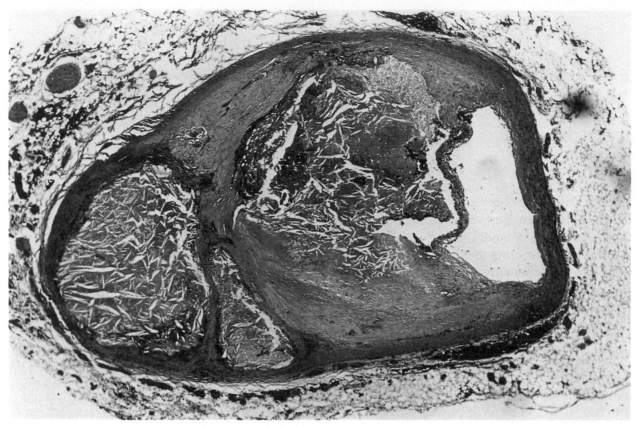

**FIGURE I**    A coronary artery severely occluded by atherosclerosis. The much reduced residual lumen is eccentrically placed on the right. The intima on the right side shows little thickening, whereas on the opposite side there is massive thickening with a very large pool of atheromatous debris, including crystals of cholesterol. [Reprinted from N. Woolf (1987). *Eur. Heart J.* 8 (Suppl. E), 3–14 by permission of the author.]

cardial infarction, and death. The blood supply to the brain may be blocked by fragments derived from the atherosclerotic plaques, resulting in a stroke. Atherosclerosis is associated with several other chronic diseases (see Section III), such as hypercholesterolemia, hypertension, obesity, and diabetes mellitus.

## I. DISEASE OF ATHEROSCLEROSIS

Atherosclerosis is a characteristic disease of the developed world. Many life-style factors are thought to contribute to this, including diet (high saturated fat and cholesterol contents), smoking, and a low level of physical activity (see Section III). The significance of these cultural factors can be judged by the fact that Japanese people, who, in their native country, do not have a high incidence of atherosclerosis, suffer from the disease to the same extent as Americans when

they migrate and adopt an American life-style. An interesting contrast is the Masai of East Africa, who consume a diet rich in saturated fat from their cattle. These people do develop atherosclerosis, but, perhaps because of their active life-style, they do not develop blockage of their arteries. The arterial lumen tends to remain open to allow good circulation of the blood, despite the presence of a thickened artery wall. Genetic differences also have a significant effect on the development of atherosclerosis (see Section III).

With a disease of such significance it would be helpful if early diagnosis were possible to allow treatment to begin before a fatal scenario is reached, which may not be until middle age or later. Simple routine methods for this are not available, so medical attention focuses on assessing the risk of developing atherosclerosis according to several more or less well-understood risk factors (see Section III). Until recently, the only way of measuring atherosclerosis directly in a living patient was by coronary angiography. This surgical procedure involves injecting a radio-opaque dye into the circulation and photographing the X-ray image of the coronary circulation on cine film. From selected frames of the film, a pathologist can assess the extent to which an artery is blocked. More recent computer-based techniques allow a less subjective evaluation to be made. This technique has provided evidence for the effects of drug treatment on the progression of disease (see Section IV,A). A further development of angiography, intracoronary or intravascular ultrasound, produces valuable images which show the kind of tissue that is present in plaque, revealing, for example, fatty or calcified regions of the artery wall. This technique is very promising, but has to be developed further in clinical studies of progression of atherosclerotic disease and of effects of drugs.

A noninvasive technique, B-mode ultrasound, can be used to image atherosclerosis in superficial arteries (e.g., carotid or femoral). This technique also allows a more detailed assessment of the nature of the plaque, together with some estimate of the blood flow, a factor that may be important in thrombus formation. Coronary atherosclerosis cannot be measured by B-mode ultrasound. A number of large-scale clinical studies

using ultrasonography have been reported. It has been estimated from ultrasound studies of large groups of people that 20% of the population of the United States between the ages of 45 and 65 has 60% stenosis, or blockage, of the cross-sectional area of some arteries. This is enough to present a clinical problem. Other techniques for imaging atherosclerosis include nuclear magnetic resonance (or magnetic resonance imaging), which has been able to show the development of atherosclerosis in cholesterol-fed rabbits over time, and which holds great promise for human studies and scintigraphy, which reveals the nature of the lesion using radioactively labeled lipoproteins.

## II. NATURAL HISTORY OF THE DEVELOPING ATHEROSCLEROTIC PLAQUE

Since atherosclerosis develops over decades of life in humans, it is not possible to observe its development directly in an individual patient. Many studies by pathologists have led to the following probable sequence of events (Fig. 2).

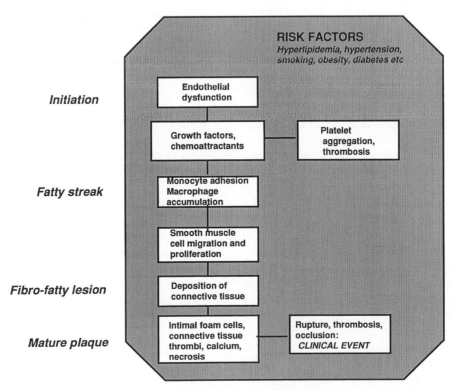

**FIGURE 2**   Sequence of events in the pathogenesis of atherosclerosis.

Atherosclerosis is thought to be initiated by some dysfunction in the endothelial cells, the thin layer of cells that lines the inner surface of the artery. For example, a slight deformation due to shear stress in response to blood flow or a slight injury may cause a temporary association of platelets at the site. The platelets may secrete peptide growth factors, such as platelet-derived growth factor (PDGF). Such growth factors are secreted by other cells, including endothelial cells, macrophages, and smooth muscle cells, as the process continues. These and other stimuli cause monocytes in the bloodstream to migrate to the endothelium where they adhere through interaction with specific cell surface binding proteins known as adhesion molecules. The monocytes migrate through the endothelial cell layer and mature into macrophages. In this environment, an excess uptake of cholesteryl ester from plasma lipoproteins, particularly low-density lipoprotein (LDL) and its derivatives, occurs, leading to the appearance of a characteristic type of cell, the foam cell, so called because of the massive accumulation of intracellular lipid droplets. Their appearance gives rise to the earliest detectable form of the lesion, the fatty streak. [*See* Macrophages.]

Alternative views of the initiation of the atherosclerotic lesion point to the formation, at an early age, of a diffuse thickening of the arterial intima, from which the lesions may then develop. The earliest lesions form at the flow divider in bifurcations in arteries on the lateral angles of the junction. This localization emphasizes the importance of hemodynamic factors in determining where lesions develop. The severity of the lesions appears to be highest where the pressure of the blood flow is greatest. These factors are not completely understood and could affect other initiating factors.

Following the initiation of the lesion, smooth muscle cells begin to migrate to the inner layer of the artery, the intima, as they change their phenotype from that of a contractile muscle cell to that of a proliferating cell. Interestingly, smooth muscle cell proliferation, not migration, is characteristic of changes that occur in small blood vessels in patients with hypertension, which itself is a major risk factor for atherosclerosis (see Section III). Smooth muscle cells also secrete growth factors that stimulate their division and migration to the site of the developing plaque. Other secretory products from smooth muscle cells add to the bulk of the plaque. These include collagen and glycosaminoglycans, the latter being able to associate with LDL and further increase uptake, leading to a more lipid-rich lesion. As these processes continue, the initial raised lesion, the fatty streak, develops into a larger growth. Cells at the center of the plaque die, and the surface of the plaque becomes coated with connective tissue (a fibrous cap). Such a fibrous plaque may remain stable for many years or it may rupture, an event with serious consequences. The surface of the plaque also provides a substrate on which thrombosis can occur and with a well-occluded artery only a small thrombus is sufficient to stop the supply of blood. Plaques may remain for many years without causing trouble, especially if they are diffusely spread around the artery wall. Those that bulge out in an eccentric manner are especially dangerous. The fibrous cap tends to fracture and ulcerate, and the core of dead cells and other debris forms a large mass of material which is highly thrombogenic and whose release causes further complications. [*See* Smooth Muscle.]

To study the disease process, a number of animal models have been widely used, many of which require diet-induced hypercholesterolemia. Thus, rabbits and monkeys in particular, but also hamsters and certain strains of mice and rats, in about 6 months develop lesions with characteristics similar to the lipid-rich lesions in humans. Eight genes, of which four relate to high density lipoprotein metabolism, have been identified in strains of mouse that are susceptible to atherosclerosis. The response of the artery wall to injury has been tested in rabbits using a range of methods that remove, or slightly damage, a small area of endothelial cells. Pigs and pigeons have also been popular species for study.

Some strains of animal develop atherosclerosis spontaneously. Two strains of rabbit have characteristics similar to certain types of hyperlipidemia found in humans (see Section III,A). Of particular interest to the genetics of atherosclerosis has been the development of strains of transgenic mouse that are susceptible to the development of atherosclerosis. For example, mice can be genetically manipulated so that they do not express apolipoprotein E (apoE), a protein that is essential for the clearance of lipids from the blood (see Section III). These mice (apoE knockout mice) develop atherosclerosis more readily than mice of the same strain that do have apoE. If a gene for apoE is inserted into the apoE knockout mice, they once again become more resistant to atherosclerosis.

## III. RISK FACTORS IN HUMANS

Because of the difficulty of early diagnosis in humans, approaches to treating atherosclerosis have focused on determining who is at greatest risk. Guidelines prepared for physicians in the United States and Europe now emphasize the importance of assessing the complete risk profile of a patient. Studies have been performed to understand the biological basis for the risk factors (see Table I). From epidemiological data it is possible to calculate the probability of a combination of several factors in one individual that could lead to the development of coronary heart disease. Clinical risk factors include hyperlipidemia (see Section III,A), perhaps the most publicized risk factor in recent years. High blood pressure carries with it a twofold increased risk for all forms of atherosclerotic disease. Diabetes mellitus is also associated with a twofold increase in risk. Certain components of the blood clotting system (e.g., factors VII, VIII, plasminogen activator inhibitor-1, and fibrinogen) have also been closely associated with increased risk of coronary heart disease. Studies have drawn attention to the possible role of viruses in inducing atherosclerosis. Many of these conditions derive from a genetic predisposition to the disease, factors of which are discussed in Section III,D. In addition, cultural characteristics or life-style can play a major part in increasing the risk of disease.

Smoking is perhaps the most avoidable of all of the life-style risk factors for atherosclerosis and coronary heart disease. Another factor is obesity, which may be due in part to a genetic predisposition, but perhaps also to the diet and a lack of physical exercise. Cholesterol- and fat-rich diets that contain certain saturated fatty acids and low levels of physical exercise also clearly contribute to the incidence of atherosclerosis. In contrast, a low level of alcohol consumption may be beneficial. [*See* Cholesterol; Obesity.]

**TABLE I**

Risk Factors for Atherosclerosis

Clinical factors
  Hyperlipidemia
  Hypertension
  Diabetes mellitus
  Blood-clotting factors VII and VIII and fibrinogen
  Lipoprotein(a)
  Obesity
Life-style factors
  High-cholesterol high-saturated fat diet
  Smoking
  Lack of physical exercise
Genetic predisposition

## A. Hypercholesterolemia

The risk factor that has attracted the most attention in recent years is hypercholesterolemia, although many subjects with atherosclerosis do not have a clear hypercholesterolemia. Hypercholesterolemia has been defined as over 200 mg of cholesterol per deciliter, a criterion that includes over 50% of the adult population of the United States. Hypercholesterolemia is one of a group of diseases known as hyperlipidemias, characterized by the presence of an excess of certain particles that transport lipid in the blood, known as plasma lipoproteins. The components of these particles have two main sources: They can derive exogenously from the diet or endogenously from the liver. Two interlocking cycles of lipid transport are present (Fig. 3). In the "exogenous" cycle, cholesterol from the diet is absorbed in the intestine, where it is packaged as cholesteryl ester into large triacylglycerol-rich particles, known as chylomicrons. These are secreted by the gut and travel through the lymph system before entering the bloodstream. As these particles pass through the capillaries of adipose tissue and muscle, the triacylglycerol is hydrolyzed by an enzyme known as lipoprotein lipase, which is bound to the endothelial cells of the capillary. The resulting chylomicron remnant is taken up by the liver. [*See* Plasma Lipoproteins.]

In the internal, or endogenous, pathway, the liver secretes smaller triacylglycerol-rich particles, known as very low-density lipoproteins (VLDL) into the blood. In a way similar to that of the chylomicrons, the triacylglycerol is hydrolyzed, producing a remnant particle known as intermediate density lipoprotein (IDL). Most of the IDL particles are taken up by the liver, but in humans a large number are further modified, gaining cholesteryl ester from high-density lipoproteins (HDL) to become LDL. It is the latter particle which carries most of the cholesterol in human blood and which is associated with the increased risk of cardiovascular disease.

## B. Clearance of LDL from the Blood

The key to our understanding of the removal of LDL from the blood has come from the work of the Nobel Prize winners Michael Brown and Joseph Goldstein of the University of Texas Southwestern Medical Center at Dallas. They found that cells take up lipoproteins through specific receptor proteins on the cell surface. They characterized a receptor for LDL which recognizes the particle by its specific protein compo-

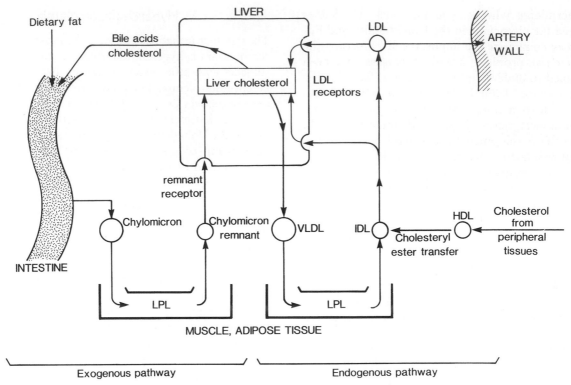

**FIGURE 3** Endogenous and exogenous cycles of lipoprotein and cholesterol metabolism. The central roles of the liver in taking up cholesterol from each of the cycles and in the elimination of cholesterol as cholesterol and as bile acids into the bile are apparent.

nent, known as apolipoprotein B (apoB). ApoB is an enormous protein of molecular weight 514,000. The LDL receptor also recognizes another apoprotein, apolipoprotein E, which is found on chylomicrons, chylomicron remnants, VLDL, and IDL. After the particle is bound to the receptor, it is taken up by the cell through a process known as receptor-mediated endocytosis. In the lysosomes, enzyme-rich vesicles within the cells, the particle is hydrolyzed to its component lipids, cholesterol, fatty acids, and amino acids. The increase in the cholesterol content of the cell results in an inhibition of the synthesis of cholesterol in the cell, principally through inhibition of the rate-limiting enzyme of the pathway of cholesterol biosynthesis, 3-hydroxy-3-methylglutaryl coenzyme A (HMG-CoA) reductase. At the same time the influx of LDL cholesterol into the cell is reduced by a reduction of the synthesis of new LDL receptors (down-regulation). In humans and other species, most of the removal of cholesterol from the blood occurs through hepatic uptake mediated by the LDL receptor. Increased concentrations of LDL in the blood can arise from increased synthesis of the precursor particle,

VLDL, or impaired removal through deficient or absent LDL receptor activity. The latter case is found in an extreme form in the congenital disease known as familial hypercholesterolemia, in which the absence of LDL receptor activity leads to massive hypercholesterolemia, premature atherosclerosis, and death. More recent work has demonstrated the existence of other receptors that may play a role in the removal of remnant particles from the circulation as well as a family of receptors known as scavenger receptors (see below).

LDL has been implicated as the major atherogenic lipoprotein, but, in contrast, HDL is thought to exert a protective role. There is a strong negative correlation of the plasma concentration of HDL with the incidence of coronary heart disease. This may be due to the role of HDL in returning cholesterol to the liver from peripheral tissues (reverse cholesterol transport) or may be associated with the more efficient metabolism of triacylglycerol-rich lipoproteins, chylomicrons, and VLDL. Evidence now supports the concept that a subclass of LDL particles, those of smaller size and higher density, particularly atherogenic because they are cleared from the blood more slowly. The presence of small dense LDL with low levels of HDL in the plasma is recognized as an atherogenic phenotype, one which is often associated with a mild hypertriglyceridemia (e.g., in noninsulin-dependent diabetic patients).

## C. Uptake of Cholesterol in the Artery

The uptake of LDL by cells through the LDL receptor is a highly regulated process which would not in itself be expected to lead to the deposition of cholesterol in the artery wall. Evidence is increasing that in the region of a developing atherosclerotic plaque LDL becomes modified in such a way that it is taken up by scavenger cells (macrophages) through an unregulated pathway by binding to receptors known as scavenger receptors. A number of scavenger receptors have been cloned and expressed. Several types of modified LDL have been described, including oxidized LDL, LDL complexed with glycosaminoglycans (secreted by smooth muscle cells in the plaque), and LDL complexed with cellular debris found around the developing plaque. Studies of these phenomena are strengthening the link between hypercholesterolemia and the development of atherosclerosis.

## D. Genetic Factors

The genetic defect that causes familial hypercholesterolemia—deficiencies in the LDL receptor—is the most well defined at the molecular level and is also extremely common. One person in 500 has a defect in one of the two copies of this gene in his or her DNA. However, it is only when both copies are defective, the homozygous situation, that the catastrophic symptoms of familial hypercholesterolemia are apparent. [*See* Coronary Heart Disease, Molecular Genetics.]

Other variations in lipoproteins are linked with an increased incidence of atherosclerosis, although the biological mechanism is less clear. ApoE, a basic protein associated with certain subfractions of HDL and with IDL, VLDL, chylomicrons, and chylomicron remnants, exists in several isoforms, caused by single substitutions of amino acids. The apoE2 form is particularly associated with atherosclerosis, probably due to differences in its ability to associate with the LDL (apoB/E) receptor on cells. Interestingly, the apoE2/2 phenotype is found in a large number of French Canadians, a relatively closed and stable population for over 150 years. In this population, there is also a higher incidence of defects in the LDL receptor.

The incidence of defects in one of the two copies of the LDL receptor gene is as low as 1 in 154 in northeastern Quebec compared with 1 in 500 for most of the United States and Europe. Interestingly, another isoform of apoE, apoE4, has recently been correlated with a early development of Alzheimer's disease.

Another lipoprotein closely correlated with atherosclerosis is lipoprotein(a). This plasma lipoprotein particle is similar to LDL but has a further large protein structure known as apolipoprotein(a) linked to apoB through a disulfide bridge. Part of the structure of apolipoprotein(a) has a striking similarity to plasminogen, which, when activated to plasmin, catalyzes the conversion of fibrinogen to fibrin, a late stage in the process of blood clotting. Some of the most promising hypotheses about how lipoprotein(a) is connected with increased coronary heart disease center on the potential of apolipoprotein(a) to interact with proteins that would normally interact with plasminogen. Thus apolipoprotein(a) has been shown to interfere with the process of thrombolysis in transgenic mice expressing human apolipoprotein(a). [Mice have no endogenous apolipoprotein(a).] The presence of this lipoprotein has been known for many years, but recently its correlation with cardiovascular disease and its possible connection between atherosclerosis and thrombosis have become more apparent. Lipoprotein(a) is a major target for drug treatment, but at the time of this writing, no drugs are able to reduce its plasma concentration.

Techniques of molecular biology have helped define the association of genes that may be involved in atherosclerosis and other disorders known to be risk factors. Studies of restriction fragment-length polymorphisms (RFLPs) show that variations in the apoB gene are found in subjects with obesity, hypertension, hypercholesterolemia, and coronary heart disease. Such variations can also be correlated with the rate of clearance of LDL from the blood, which reflects the efficiency of the LDL receptor population. The locations of some of the genes that may be related to atherosclerosis on the chromosome have been determined. The genes for apolipoproteins E, C-I, and C-II are closely linked on chromosome 19. The gene coding for the LDL receptor, which is mutated in familial hypercholesterolemia, is also on chromosome 19, but at a greater distance from the apolipoprotein cluster. Chromosome 11 has the genes for apolipoproteins A-I, A-IV, and C-III closely linked. In human populations and in groups of patients and their families, the complex patterns of genetic variation in these gene clusters and in other genes associated with lipoprotein metabolism are being studied. The goal is to identify the defective genes that contribute to atherosclerosis and to determine the molecular basis for such effects. Overall, data suggest that the connection between variations in apolipoprotein genes, determined using RFLP techniques, and coronary heart disease is weak and may not be the same in different populations. Detailed genetic studies are also being performed in mice in which genes predisposing the animal to atherosclerosis when on a high-cholesterol high-fat diet have been located. Genes on these two chromosomes stand out as possible contributors to atherosclerosis. The availability of transgenic mice now allows a much more incisive study of the genetics and pathogenesis of atherosclerosis.

It is important to realize that only in some extreme case, as in familial hypercholesterolemia, do defects in one gene provide the major cause of atherosclerosis. In most patients, a combination of genetic factors, together with elements of their life-style, provide the background to the disease.

## IV. THERAPY OF ATHEROSCLEROSIS

The main strategies used to combat atherosclerosis have been to target those risk factors that can be modified with reasonable ease. Special education programs have been devised to encourage people to eat a more healthy diet, restricting the intake of cholesterol and saturated fat, both of which tend to increase the plasma cholesterol concentration. Exercise tends to increase the relative amounts of HDL to LDL and so promotes the flux of cholesterol to the liver from peripheral tissues, a process known as reverse cholesterol transport. Smoking has so many adverse effects on health that it is a prime target for elimination.

### A. Hypercholesterolemic Drugs

Several major clinical trials have shown that reduction in the plasma cholesterol concentration leads to a reduction in mortality due to coronary heart disease. A reduction of 10% in plasma cholesterol is correlated with a one-sixth reduction in the probability of disease. Therefore, if dietary and other changes of lifestyle are not successful in reducing plasma cholesterol, treatment with drugs is indicated. The mode of action of current hypercholesterolemic drugs can be understood from the previous discussion.

Bile acid sequestrants, such as cholestyramine and cholestipol, are anion-exchange resins that bind bile acids in the intestine. Bile acids are synthesized in the liver from cholesterol. They are the only way in which cholesterol can be transformed and eliminated from the body. Therefore, it is advantageous to increase the conversion of cholesterol to bile acids. The binding of bile acids to a sequestrant interrupts their return to the liver, a process that is usually 95% efficient. The result is stimulation of the synthesis of bile acids. A further consequence is an increase in the expression of LDL receptors on the liver, which causes increased uptake of LDL from the plasma. Current sequestrants are not very palatable, but are known to be safe and effective.

HMG-CoA reductase inhibitors (or statins) are a recent class of drugs and have become the leading drug in the class of hypolipidemics. They inhibit cholesterol synthesis, primarily in the liver, which then responds by synthesizing more LDL receptors. The resulting reduction in plasma cholesterol is striking. Use of a bile acid sequestrant in combination with an HMG-CoA reductase inhibitor has proved to be a powerful way of reducing the plasma cholesterol concentration. In clinical studies of groups of patients with preexisting heart disease, HMG-CoA reductase inhibitors have been shown to cause a small but measurable reduction in the size of the atherosclerotic plaques determined by coronary angiography. Even more strikingly, these drugs cause a reduction in mortality to coronary heart disease that is much greater than would be expected from the angiographically observed changes in atherosclerosis. Recent clinical trials with statins in patients with either documented coronary heart disease (secondary prevention) or with no previous history of disease (primary prevention) have shown that a 30% reduction in mortality can be achieved. Results from studies such as these provide great encouragement that atherosclerosis can be halted and even reversed.

Probucol is an interesting drug that lowers plasma LDL as well as HDL. Its main interest is that it appears to associate with LDL and to prevent the oxidized forms of LDL thought to be highly connected with the formation of atherosclerotic plaques. It may thus have a direct action on the artery where the plaque is forming.

Other drugs in common use include niacin and fibrates. Niacin probably works by inhibiting VLDL synthesis. Its use is limited because of its common unpleasant side effects. Gemfibrozil, the most widely used of the class of fibrates, is effective in lowering the plasma triacylglycerol concentration and has a proven effect on coronary heart disease. The primary action of fibrates is to stimulate the activity of lipoprotein lipase. It should be kept in mind that most patients presenting with coronary heart disease are not hypercholesterolemic. A substantial majority of them are mildly hypercholesterolemic and hypertriglyceridemic. At present there is no drug that can treat both aspects of the hyperlipidemia to the desired extent.

## B. Other Forms of Treatment

Other treatments for atherosclerosis require hospitalization. LDL can be removed from the blood of a patient with familial hypercholesterolemia by a process known as LDL apheresis, which resembles dialysis for patients with kidney disease. The blood is passed through a column which contains an antibody to LDL attached to a solid support or an ion-exchange resin. The column binds to the apoB in the LDL and removes it from the blood. In other severe cases, a coronary artery blocked by an atherosclerotic plaque may be bypassed by surgically diverting the blood flow through a piece of vein taken from the patient's leg. Alternatively, the lumen of the artery may be physically widened by expanding a small device at the site of narrowing, a technique known as angioplasty.

## IV. FUTURE PROSPECTS

While many of the factors that contribute to the pathogenesis of atherosclerosis have been established, there are many aspects of the disease that require deeper understanding. Of particular importance is understanding the microenvironment in the artery that results in the adhesion of monocytes or macrophages to the endothelium, followed by their migration into the subendothelial space. In this new environment, the conditions that determine whether smooth muscle cells will migrate and proliferate into the intima must be defined more closely. The changes that take place in the participating cells in response to signals from their environments must be understood. It is necessary, for example, to understand the way in which growth factors such as PDGF are synthesized and are secreted, the timing of the secretion of these factors, and the intracellular mechanisms by which the target

cells respond to them. Such information may indicate additional ways in which the development of the plaque can be inhibited. Some of the newer classes of drugs that may emerge will be targeted directly at the plaque rather than, as at present, at a risk factor that may be closely connected with some aspect of the plaque, such as hypercholesterolemia.

Interestingly, certain existing classes of drugs may well have a beneficial effect on atherosclerosis. Results obtained in animals and in the laboratory suggest that the calcium channel blockers, a class of drugs used for many years to treat hypertension, may be effective in treating atherosclerosis as well; they are being tested clinically. Results have been only partially successful. The problem of an unhealthy life-style is generally being taken more seriously. However, much more can be done to eliminate the substantial contribution of the most obvious risk factors, such as smoking. Many workers in the field agree that atherosclerosis is not an inevitable consequence of the aging process; more can be done to prevent it and, in time, treat it.

## BIBLIOGRAPHY

Breslow, J. L. (1993). Genetics of lipoprotein disorders. *Circulation* **87**, 16–21.

Fears, R., Suckling, K. E., and Poste, G. (1994). Novel agents in the treatment of atherosclerosis. *Expert Opin. Invest. Drugs* **3**, 181–491.

Grundy, S. M., Bilheimer, D., Chait, A., Clark, L. T., Denke, M., Havel, R. J., Hazzard, W. R., Hulley, S. B., Hunninghake, D. B., Kreisberg, R. A., Drisetherton, P., Mckenney, J. M., Newman, M. A., Schaefer, E. J., Sobel, B. E., Somelofski, C., Weinstein, M. C., Brewer, H. B., Cleeman, J. I., Donato, K. A., Ernst, N., Hoeg, J. M., Rifkind, B. M., Rossouw, J., Sempos, C. T., Gallivan, J. M., Harris, M. N., and Quintadler, L. T. I. (1993). Summary of the second report of the National Cholesterol Education Program (NCEP) expert panel on detection, evaluation, and treatment of high blood cholesterol in adults (adult treatment panel II). *J. Am. Med. Assoc.* **269**, 3015–3023.

Ross, R. (1993). The pathogenesis of atherosclerosis: A perspective for the 1990s. *Nature* **362**, 801–809.

Stary, H. C., Chandler, A. B., Glagov, S., Guyton, J. R., Insull, W., Rosenfeld, M. E., Schaffer, S. A., Schwartz, C. J., Wagner, W. D., and Wissler, R. W. (1994). A definition of initial, fatty streak, and intermediate lesions of atherosclerosis: A report from the Committee on Vascular Lesions of the Council on Arteriosclerosis, *Am. Heart Assoc. Arterioslcer. Thrombosis* **14**, 840–856.

# Atherosclerosis: From Risk Factors to Regulatory Molecules

TRIPATHI B. RAJAVASHISTH
ARTHUR H. LOUSSARARIAN
*UCLA School of Medicine and Harbor–UCLA Medical Center*

## GLOSSARY

**Atherosclerosis** A chronic arterial disease characterized by infiltration of the tunica intima with leukocytes, smooth muscle cells, lipids, connective tissue, and calcium. Spontaneous atherosclerosis develops over a span of decades, predominantly in individuals with risk factors. An accelerated form of atherosclerosis, which closely resembles but is not pathologically identical to the spontaneous form, arises over a span of months to years following angioplasty, coronary artery bypass grafting, or cardiac transplantation

**Fatty streak** Accumulation of lipid-laden macrophages and smooth muscle cells in the arterial intima; the earliest pathologic lesion observable in atherosclerosis

**Homeostasis** Tendency of biologic systems to maintain a state of physiologic equilibrium through a series of active processes

**Regulatory molecules** Stimulatory or inhibitory molecules that modulate a given process. In the case of atherogenesis, these include cytokines, chemokines, enzymes, growth factors, integrins, and transcription factors

**Risk factor** Any clinical characteristic observed in an apparently healthy individual that can be used to predict the probability for developing a particular disease

**Transcription factor** Nuclear protein that binds to specific recognition sites in the regulatory regions of a gene and thereby alters its transcription rate

ATHEROSCLEROTIC CARDIOVASCULAR DISEASE, A leading cause of morbidity and mortality in the United States, is responsible for 1 million deaths each year from myocardial infarction, stroke, and complications from peripheral vascular disease. A similar and related process, that of accelerated atherosclerosis following angioplasty, coronary artery bypass grafting, or cardiac transplantation, inflicts significant morbidity and mortality in patients who undergo these procedures. Fortunately, important advances have been made in recent years in understanding the molecular basis of atherosclerosis that have yielded insight into possible strategies to treat this disease. Human population studies have established clinical risk factors for coronary artery disease (CAD) and, in doing so, have provided some clues about the triggering events in atherogenesis. Studies of diseased arteries from humans and animal models have defined cellular and molecular components that contribute to atherosclerosis. Advances in molecular genetics have identified nuclear factors that regulate the expression of genes encoding atherogenic molecules.

These studies have revealed that atherosclerosis is an active process of inflammation and cell proliferation that occurs when normal vascual functions go awry. Under healthy circumstances, a dynamic equilibrium exists between pro- and anti-atherogenic processes in the blood and the vessel wall. Vascular cells,

ENCYCLOPEDIA OF HUMAN BIOLOGY, Second Edition, VOLUME 1.   Copyright © 1997 by Academic Press.   All rights of reproduction in any form reserved.

by elaborating specific molecules, perform diverse functions that enable them to buffer abrupt changes in their local environment and maintain homeostasis. Atherogenic stimuli in the blood can disturb this homeostasis, causing an alteration in the normal interaction of blood and vessel wall. The result is an inflammation of the endothelium, an "endotheliopathy," that marks the initiation and progression of atherosclerosis.

## I. VASCULAR HOMEOSTASIS

The wall of a normal blood vessel is composed of three layers: the intima, media, and adventitia. The intima is a single layer of endothelial cells (ECs) that lines the lumen of the blood vessel and serves as an interface between blood and the rest of the vessel wall. The vascular intima is bounded by an internal elastic lamina (IEL) that separates it from the tunica media, a multicellular layer consisting of smooth muscle cells (SMCs), fibroblasts, and extracellular matrix (ECM). An external elastic lamina (EEL) separates the media from the outermost layer, the adventitia, which is composed mainly of connective tissue. By virtue of their anatomic position, ECs sense chemical, humoral, and mechanical changes in their local milieu and effect stimulatory and inhibitory responses to these changes. ECs, attached to one another by tight junctions, form an impermeable barrier to the passive penetration of blood cells and macromolecules. At the same time, the endothelium selectively transports macromolecules through active microcytosis or receptor-mediated endocytosis. The endothelium helps to maintain the integrity of the ECM by secreting collagens, laminin, and other structural proteins. Conversely, by secreting collagenases, elastases, and matrix-metalloproteinases (MMPs), the endothelium regulates local breakdown and turnover of the ECM. This process allows leukocyte infiltration and penetration in areas of infection and SMC migration at sites of wound healing. The endothelium regulates vascular tone by secreting the vasodilators prostacyclin and nitric oxide (NO), also known as endothelium-derived relaxing factor (EDRF), as well as the vasoconstrictors endothelin and angiotensin I. Angiotensin converting enzyme (ACE), expressed on the surface of ECs, inactivates the vasodilator bradykinin and is also responsible for the conversion of angiotensin I to angiotensin II. [See Cardiovascular System, Anatomy.]

The endothelium serves as a mechanosensor and can detect vascular distention and changes in local shear forces and can alter the expression of cellular proteins in response to these changes. The endothelium provides a nonthrombogenic surface for circulating blood by acting as a barrier between tissue factor (TF) and other procoagulants in the subendothelium and platelets and clotting factors in blood, and by expressing heparan sulfate (an activator of antithrombin III). In addition, the endothelium secretes the anticoagulants protein C, protein S, and thrombomodulin and the thrombolytics tissue- and urokinase-like-plasminogen activator, as well as inhibitors of platelet aggregation and adhesion, prostacyclin and nitric oxide, respectively. In contrast, the endothelium can promote thrombosis at sites of vascular injury by secreting the coagulants high-molecular-weight kininogen, factors V and VIII, and TF, and by binding several coagulation factors on its cell surface. The endothelium also secretes the antithrombolytic plasminogen activator inhibitor, platelet aggregator thromboxane A2, and platelet adhesive von Willebrand factor. The endothelium regulates inflammatory and immunologic processes by producing the leukocyte adhesion molecules [vascular cell adhesion molecule-1 (VCAM-1), intercellular adhesion molecule-1 and -2 (ICAM-1 and -2), and the endothelial leukocyte adhesion molecule-1 (ELAM-1, also called E-selectin)], the interleukins IL-1 and IL-6, and the cytokines monocyte chemotactic protein-1 (MCP-1) and monocyte-colony-stimulating factor (M-CSF). These molecules influence the attachment, demargination, differentiation, proliferation, survival, and activation of blood leukocytes at sites of infection or injury. Furthermore, the endothelium synthesizes growth regulatory molecules that stimulate or inhibit activity of SMCs in the neighboring tunica media. Platelet-derived growth factor (PDGF) stimulates migration, and fibroblast growth factor (FGF) stimulates proliferation, of SMCs. Transforming growth factor-$\beta$ (TGF-$\beta$) is an inhibitor of SMC mitosis but a potent stimulator of collagen synthesis.

Thus, the endothelium actively and continuously effects a variety of "pro-" and "anti-" processes in order to achieve and maintain vascular homeostasis (Fig. 1). In performing these dual and opposing functions, the vascular endothelium maintains a delicate balance between permeability and impermeability, growth promotion and growth inhibition, vasoconstriction and vasodilation, leukocyte adherence and nonadherence, and anticoagulation and procoagulation. Atherogenic activation of the endothelium leads to an impairment in its ability to maintain vascular homeostasis and may allow one process to dominate

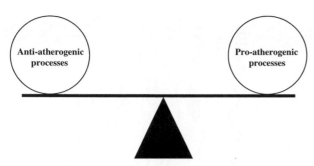

FIGURE 1  Under normal conditions, the endothelium maintains vascular homeostasis by balancing the pro- and anti-atherogenic processes always present in its environment.

over its counterpart, thus disturbing the normal balance (Fig. 2). The result is altered vascular tone, adhesiveness, thrombogenicity, and cell proliferative activity. These processes are collectively termed "endothelial dysfunction." [*See* Cardiovascular System, Physiology and Biochemistry.]

## II. RISK FACTORS

The major risk factors for atherosclerotic coronary artery disease have become known from prospective epidemiological studies in the United States and Europe. As shown in Fig. 3, these are elevated serum low-density lipoprotein (LDL), reduced serum high-density lipoprotein (HDL), cigarette smoking, elevated blood pressure, diabetes mellitus, age, male sex, and family history. Recently, additional rarer and more clinically subtle factors that associate with CAD have been described and added to this list. They in-

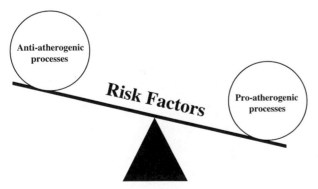

FIGURE 2  Atherogenic risk factors tip the delicate balance between pro- and anti-atherogenic processes and impair the ability of the endothelium to maintain vascular homeostasis.

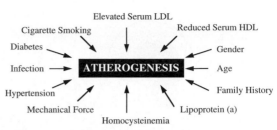

FIGURE 3  The risk factors for atherosclerosis.

clude elevated serum homocysteine, elevated serum lipoprotein (a), and infection with cytomegalovirus or chlamydia. The risk factors exhibit additive and dose-dependent properties. That is, an individual having multiple risk factors has a greater likelihood of developing atherosclerosis than someone with one risk factor or a short exposure to a given risk factor. Some risk factors can be modified by life-style change or medical intervention to reduce an individual's likelihood for developing CAD. Cessation of smoking and reduction of serum cholesterol have both been shown to reduce the risk of coronary events. Treatment of hypertension also likely reduces coronary risk, but evidence for this is not unequivocal.

Although the clinical risk factors for atherosclerosis are diverse and seemingly unrelated, experimental data suggest that they may share a common pathophysiologic mechanism. Each risk factor is believed to inflict a form of injury to the vascular endothelium and bring about endothelial dysfunction. Indeed, early changes in endothelial function have been demonstrated in asymptomatic human subjects having risk factors for atherosclerosis. Experimentally, injection of acetylcholine into the normal coronary artery leads to vasodilation. This phenomenon requires an intact endothelium and is mediated by release of NO. Injury to the endothelium leads to an impairment in its function and results in a vasoconstrictive response to acetylcholine. Patients with normal coronary angiograms but with one or more risk factors for atherosclerosis (hypertension, diabetes, hypercholesterolemia, or history of smoking) have an attenuated vasodilator response or an actual vasoconstrictor response to acetylcholine. On a more detailed analysis, these patients are found to have reduced resting and inducible bioavailability of NO in the coronary circulation.

The nature of injury to the endothelium may be chemical (oxidants or other toxins), immunologic (histocompatibility interactions between donor heart and host immune system), infectious (viral or bacterial), or mechanical (local blood shear stresses or

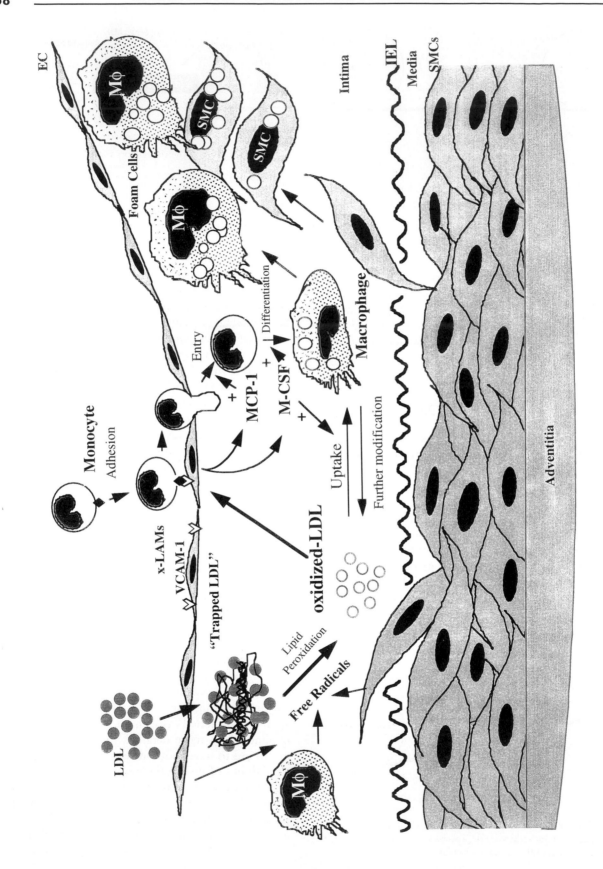

trauma induced by angioplasty). The most convincing evidence for the risk factor-induced injury hypothesis is in the case of hyperlipidemia. Circulating LDL in the bloodstream crosses the intact endothelium of the vessel wall and becomes trapped in the mesh of extracellular matrix fibers in the subendothelial space. In this microenvironment, LDL is exposed to cellular free radical products and is oxidatively modified to varying degrees. Highly oxidized-LDL (ox-LDL) is lethal to the vascular endothelium, but mildly oxidized LDL activates the endothelium and induces production of inflammatory cytokines and adhesion molecules that attract and bind monocytes to the area of injury and induce their transmigration into the subendothelium. The ox-LDL also activates monocytes and stimulates their phagocytic activity for extracellular lipid. As discussed in Section III (also see Fig. 4), these are early events in the process of atherosclerosis. Elevated serum LDL is thus a risk factor for atherosclerosis because it results in the production of ox-LDL, a molecule that interacts with the vascular endothelium and induces endothelial dysfunction. Immunostaining has documented the presence of ox-LDL in the atherosclerotic lesions of humans and animals.

The precise effects of other risk factors on the vascular endothelium are not as well understood. Some evidence indicates that HDL exerts a protective effect against atherogenesis by preventing the oxidative modification of LDL. In addition, HDL removes lipids from the artery wall and delivers them to the liver in a "reverse cholesterol transport" mechanism. Cigarette smoke contains many toxins that are capable of interacting with ECs or oxidatively modifying LDL. Smokers also have higher levels of LDL and lower levels of HDL than do nonsmokers. The chronic hyperglycemic state in diabetes mellitus results in the nonenzymatic amino glycosylation of proteins, leading to advanced glycosylation end products (AGEs). AGEs may have atherogenic properties by generating oxygen free radicals and quenching NO activity and have been shown to impair endothelium-dependent relaxing activity. Monocytes have receptors for AGEs (RAGE) and accumulate at sites of AGE deposition. Some evidence indicates that AGEs can stimulate monocyte chemotaxis and activation as well as promote secretion of growth factors such as IL-1, tumor necrosis factor (TNF), and PDGF. Coronary flow disturbances and local shear stresses cause subtle deformations of endothelial cells. Recently, the transcription of several growth factors and adhesion molecules has been shown to be shear inducible. Atherosclerosis occurs in a nonrandom distribution, having a predilection for vessel branch points and regions of curvature, areas characterized by disturbed blood flow patterns. It is possible that the hypertensive state increases such alterations in local shear forces and in this way leads to endothelial dysfunction. Lipoprotein (a) [Lp(a)], which resembles plasminogen in structure but lacks its proteolytic activity, has been proposed to act as a competitive inhibitor for clot lysis and could thus favor the formation of microthrombi at the arterial endothelium. Fibrin degradation products and platelet-derived factors from these thrombi may act as mitogens and stimulate cell proliferation. Lp(a) also may interfere with plasminogen-mediated activation of TGF-$\beta$, an inhibitor of cell proliferation. Homocysteine has been shown to induce oxidation of LDL and to impair endothelial production of NO. It also has proliferative effects on vascular SMCs. [See Plasma Lipoproteins.]

## III. CELLULAR EVENTS

Vascular wall activation as a result of exposure to atherogenic risk factors initiates a process of inflammation and cell proliferation that involves the interaction of monocytes, ECs, and SMCs and leads to the formation of the fatty streak (Fig. 4; Color Plate 8). One of the earliest events observed in atherogenesis is the adherence of leukocytes to the vascular endothe-

FIGURE 4   A model for the role of lipids in atherogenesis. Circulating LDL enters the subendothelial space and becomes trapped in a meshwork of structural proteins. In this microenvironment, the LDL is exposed to free radical products of ECs, SMCs, and macrophages (M$\phi$'s) and becomes oxidized to varying degrees. Mildly oxidized forms of LDL stimulate ECs to produce adhesion molecules (X-LAMs and VCAM-1), cytokines (M-CSF), and chemokines (MCP-1). These molecules mediate adherence, transmigration, differentiation, and activation of monocytes. M-CSF promotes scavenger receptor-mediated endocytosis of ox-LDL by M$\phi$'s, which become foam cells as they accumulate lipid. SMCs migrate from the vascular media to the intima and synthesize extracellular matrix structural proteins. SMCs also phagocytose ox-LDL and become foam cells. These events lead to the formation of the fatty streak, the earliest pathologic lesion in atherogenesis. [See also Color Plate 8.]

lium. Groups of monocytes adhere to dysfunctional regions of endothelium and, influenced by local chemokines, transmigrate between ECs to enter the subendothelial space. In the subintima, cytokines and growth factors produced by vascular ECs and SMCs stimulate monocytes to proliferate and mature into macrophages. These macrophages phagocytose LDL particles that have become trapped and oxidized in the subendothelium and are transformed into lipid-laden foam cells. This endocytotic process is mediated by the macrophage scavenger receptor that recognizes ox-LDL but not native LDL. Activated ECs secrete cytokines that stimulate SMCs to migrate from the neighboring tunica media into the intimal layer. As they migrate, the SMCs undergo a phenotypic change from a contractile to a synthetic phenotype and begin secreting ECM proteins, thereby increasing the bulk of the atheromatous lesion. In the intima, the SMCs also accumulate lipid and become foam cells. As the lesion is progressively engorged with lipid, its core may become necrotic. Active collagen synthesis at the surface of the lesion leads to the formation of a tough fibrous cap that separates the highly thrombogenic "gruel" content of the lesion from platelets and clotting factors in the blood. Extensive monocyte-macrophage infiltration often occurs at the edge, or "shoulder region," of the fibrous cap. Mesenchymal cells, either already present in the vascular wall or imported through the blood, proliferate and synthesize bone matrix in response to local bone morphogenic factors, leading to calcification of the lesion.

The formation of fatty streaks is a ubiquitous process in the human vascular system and begins to occur as early as in the first decade of life. However, only a very small percentage of fatty streaks progress to mature atherosclerotic lesions. The precise mechanisms responsible for such progression are not well understood but likely relate to repeated exposure to the various risk factors. In addition, the progression of fatty streaks may not be a continuous process but, rather, may occur in sequential waves of high activity separated by periods of quiescence.

## IV. REGULATORY MOLECULES

Endothelial dysfunction in early atherogenesis is marked by changes in endothelial cell adhesiveness to monocytes. These changes are thought to be mediated by the augmented expression of endothelial cell adhesion molecules and selectins. VCAM-1 and ICAM-1 and -2 are membrane-bound glycoproteins that be-

long to the immunoglobulin gene superfamily. VCAM-1 binds to monocytes or lymphocytes that express the $\alpha4\beta1$ (also called very late antigen-4, VLA-4) and $\alpha4\beta7$ integrins. VCAM-1 is not constitutively expressed by ECs but can be up-regulated in response to certain stimuli, such as tumor necrosis factor-$\alpha$ (TNF-$\alpha$) or modified lipids, to modulate monocyte adhesion. ICAM-1 binds to $\alpha M\beta2$ and is inducible by inflammatory cytokines, whereas ICAM-2 binds $\alpha L\beta2$ and is expressed constitutively.

E-selectin is a lectin-like molecule that binds to sialylated carbohydrate epitopes related to the sialyl Lewis[x] antigen on the surface of leukocytes. E-selectin is not constitutively expressed by ECs but its expression is induced by cytokines such as IL-1$\beta$ and TNF-$\alpha$. The expression of these mediators of cell adherence in response to blood stimuli causes monocytes and lymphocytes to attach to the vascular endothelium. The subsequent activation, chemotaxis, transendothelial migration, and proliferation of monocytes in the vascular wall appear to involve the action of several chemokines and growth factors. Monocyte chemotactic protein, produced by ECs and SMCs, is a potent chemoattractant for monocytes and has been shown *in vitro* to be a key mediator of monocyte transmigration. Monocyte-colony stimulating factor is a cytokine that regulates the differentiation, proliferation, chemotaxis, and survival of mononuclear phagocytes. M-CSF also stimulates scavenger receptor-mediated uptake of lipids by activated macrophages. The expression of MCP-1 and M-CSF is induced by cytokines and blood toxins, including ox-LDL. Atherosclerotic lesions derived from humans and rabbits contain increased levels of MCP-1 and M-CSF mRNA and proteins. VCAM-1 expression has been found in arterial ECs covering early foam cell lesions in both diet-induced and Watanabe heritable hyperlipidemic rabbits. VCAM-1 expression in these lesions is not uniform but is concentrated at the lesion edge and overlies areas of monocyte penetration. Induction of VCAM-1 expression is an early event in atherogenesis, occurring approximately 1 week following the initiation of a hypercholesterolemic diet and preceding detectable intimal monocyte accumulation. In human autopsy specimens, VCAM-1 and ICAM-1 expression has been found in advanced atherosclerotic plaques.

The migration of SMCs from the vascular media and their proliferation in the intima are key components of the spontaneous atherosclerotic process as well as in the accelerated form of atherosclerosis following angioplasty. Several cytokine mediators of this process have been identified. PDGF is a mitogen and chemoattractant produced by macrophages, ECs, and SMCs. In animal models of atherosclerosis induced by vascular injury, PDGF has been found to stimulate the migration of SMCs to the intima. Basic fibroblast growth factor (bFGF) is a potent mitogen for SMCs and may stimulate their proliferation in the vascular intima. M-CSF has also been implicated in the regulation of SMCs migration and proliferation. The receptor for M-CSF, c-fms, is not expressed by SMCs in the normal vessel media, but intimal SMCs isolated from an experimental rabbit model of atherosclerosis do express c-fms. The activity of angiotensin converting enzyme is thought to be important in cell proliferation. The action of ACE leads to the production of angiotensin II, a molecule that activates FGF and PDGF. ACE also degrades bradykinin, which has vasodilator effects on the endothelium by stimulating prostacyclin biosynthesis and NO release. In human atherosclerotic lesions, increased ACE activity has been noted.

Regulation of ECM production and remodeling are central processes in the formation and evolution of the atheromatous lesion. PDGF and TGF-$\beta$ both stimulate synthesis of interstitial collagens and thus may play important roles in the elaboration of ECM components and formation of the fibrous cap in atherosclerotic lesions. TGF-$\beta$ also has inhibitory effects on SMC replication. The lymphocyte product interferon-$\gamma$ (IFN-$\gamma$) inhibits collagen synthesis as well as SMC proliferation. T lymphocytes frequently predominate at sites of plaque rupture, and some evidence shows that these lymphocytes produce IFN-$\gamma$. Activity of these lymphocytes could thus weaken the fibrous cap. Matrix-metalloproteinases are a family of proteolytic enzymes that degrade collagen, proteoglycans, and other structural proteins in the ECM and thus likely function in normal morphogenesis and wound healing, processes that require breakdown of matrix and migration of cells. Many types of MMPs are expressed under basal circumstances by SMCs and ECs, but their activity is tightly regulated by the concomitant expression of tissue inhibitors of metalloproteinases (TIMPs). Exposure of SMCs to cytokines such as IL-1 or TNF induces them to express some forms of MMPs not expressed in the basal state but does not change the expression of TIMPs. In addition, foam cells isolated from atheromatous lesions also express these MMPs, whereas alveolar macrophages do not. MMPs may mediate matrix breakdown necessary for monocyte infiltration and SMC migration in atherogenesis. It has been proposed that

MMPs from macrophages concentrated at the edges of atheromatous lesions may cause weakening of the fibrous cap and play a role in plaque rupture, leading to the acute coronary syndromes.

NO, as discussed earlier, is an important modulator of vascular tone. In addition, NO has many actions that are antiatherogenic. NO inhibits platelet aggregation and adhesion, smooth muscle cell proliferation, and leukocyte adhesion to the endothelium. NO down-regulates expression of VCAM-1 as well as suppresses expression of MCP-1 and attenuates ox-LDL-induced M-CSF expression. Endothelial dysfunction is marked by dramatic reduction in the release of NO.

## V. GENETIC CONTROL

Identification of the molecules associated with dysfunctional endothelium has revealed that the expression of inflammatory genes is a key event in atherogenesis. The transcriptional factors that positively or negatively modulate these inflammatory genes are therefore important regulators of atherogenesis. Several transcription factors belonging to the nuclear factor-$\kappa$B (NF-$\kappa$B) family have been isolated that bind as heterodimers to DNA and stimulate gene expression (Fig. 5; Color Plate 9). In the inactive state, NF-$\kappa$B is present in the cell cytoplasm bound to an inhibitor molecule, I$\kappa$B. Activators of NF-$\kappa$B cause it to dissociate from I$\kappa$B and move to the nucleus, where it binds to specific *cis*-DNA elements present in the regulatory region of a set of endothelial cell genes important in inflammation. Activation of NF-$\kappa$B leads to increased transcription of genes encoding adhesion molecules such as VCAM-1, E-selectin, and ICAM, the chemokine MCP-1, the cytokines M-CSF, IL-1, IL-6, and IFN-$\beta$, as well as tissue factor.

A variety of stimuli lead to the activation of NF-$\kappa$B. An atherogenic diet or intravenous injection of ox-LDL induces NF-$\kappa$B in mice. *In vitro*, NF-$\kappa$B is activated by ox-LDL, the cytokines TNF-$\alpha$, IL-1$\alpha$, and -$\beta$, and bacterial lipopolysaccharide (LPS). In addition, secondary messengers such as protein kinase C stimulate NF-$\kappa$B. Mechanical trauma to the arterial endothelium, such as balloon injury, has also been shown to activate NF-$\kappa$B. Recently, infection with cytomegalovirus (CMV) was shown to involve NF-$\kappa$B-mediated transcription. The CMV gene *ie1* activates binding of host NF-$\kappa$B to the CMV enhancer, thus stimulating viral gene expression. It is possible that viral-activated NF-$\kappa$B in infected cells could also increase expression of host NF-$\kappa$B-dependent genes

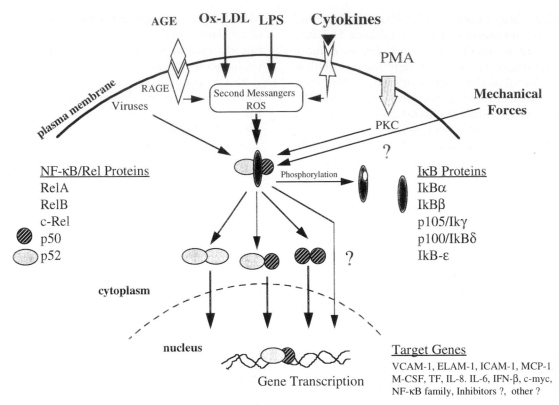

**FIGURE 5** The activation of genes by NF-κB. The NF-κB is a family of DNA-binding proteins consisting of the members RelA, RelB, c-Rel, p50, and p52. NF-κB is normally bound in the cytoplasm by its inhibitor, IκB. Phosphorylation of IκB allows NF-κB to dissociate from the cytosolic complex and move to the nucleus, where it binds to its recognition sequence in the regulatory regions of specific genes and thereby activates gene transcription. Various stimuli act in a receptor-dependent or -independent pathway and lead to the production of reactive oxygen species (ROS) in the cytoplasm. ROS serve as secondary messengers and activate NF-κB. Some stimuli may activate NF-κB directly or by mechanisms other than through ROS. NF-κB is involved in the activation of many genes that may play a critical role in inflammation. [*See* also Color Plate 9.]

important in atherogenesis. A virus in the herpes simplex family that has been shown to cause an atherosclerosis-like disease in chickens also activates NF-κB. AGEs, which have been described in atherosclerotic lesions of diabetics but not in lesions of nondiabetics, can also induce NF-κB.

The intracellular signals involved in the activation of NF-κB are not well known, but one event that seems to play a critical role is oxidative stress. Activation of NF-κB by an atherogenic diet *in vivo* or by TNF-α *in vitro* involves the generation of intracellular reactive oxygen species (ROS) such as superoxide anion. AGEs are also known to generate oxygen free radicals. Agents that inhibit formation of reactive oxygen molecules, such as the antioxidant N-acetylcysteine, attenuate activation of NF-κB. NO, which acts as a scavenger for superoxide anion, also inhibits activation of NF-κB. In animal experiments, activation

of NF-κB has been demonstrated in strains of mice that exhibit a propensity for extensive accumulation of hepatic lipid peroxidation products in response to a high-fat diet. Interestingly, these mouse strains are also prone to the development of fatty streaks.

Although NF-κB is the best-studied transcription factor in atherogenesis, evidence is emerging for the involvement of other nuclear factors. The transcription factor *egr-1* has recently been shown to be activated by balloon injury. Putative nucleotide recognition elements for *egr-1* appear in the promoters of several growth factor genes and genes important in the clotting cascade, including M-CSF, TGF-β1, PDGF-A, PDGF-B, TF, and u-TPA. Arterial balloon injury has been shown to induce the expression of these growth regulatory molecules in vascular cells. In addition, a cis-acting shear-stress-response element (SSRE) has been described in the promoters of the genes that code for PDGF-B and ICAM-1. The transcription of these and other growth factors and adhesion molecules has been shown to be shear inducible. Early evidence indicates that at least one of the factors that bind SSRE is NF-κB. Thus, activation of specific endothelial transcription factors may represent a molecular link between the diverse clinical risk factors and agents of endothelial injury to endothelial dysfunction and initiation of atherogenesis. [*See* DNA and Gene Transcription.]

## VI. FUTURE PROSPECTS

A first step in combating atherosclerosis was the identification of clinical risk factors associated with the disease. Even prior to understanding the molecular mechanisms of atherogenesis, attempts to modify these risk factors yielded early successes. Reduction in serum cholesterol and cessation of smoking were shown to halt the progression of atherosclerosis and reduce mortality from coronary heart disease. As additional risk factors are identified, those amenable to modification may be associated with similar successes. A promising example is in the case of hyperhomocysteinemia. Approximately 20% of Americans have homocysteine levels above 16 $\mu$mol/liter, the level associated with a threefold increased risk of myocardial infarction. Although dramatically elevated homocysteine levels are due to rare defects of enzymes in the homocysteine metabolic pathway, more moderate serum elevations of homocysteine may be due to less severe alterations in these enzymes or to dietary deficiencies of folate or vitamin B12. Treatment of patients with 1 to 2 mg/day of oral folate supplementation is sufficient to reduce homocysteine levels in most individuals. Whether reduction of serum homocysteine will lead to a concomitant reduction in cardiovascular risk remains to be seen.

Insight into the molecular mechanisms of atherogenesis and identification of key regulatory molecules have yielded strategies for intervening in the atherosclerotic process. Realization of the important role of oxidative stress in atherogenesis has led to the rationale of using antioxidant therapy to slow the progression of atherosclerosis. Experiments in hyperlipidemic rabbits have shown that the antioxidant probucol markedly reduces the development of fatty streak lesions. Human trials with antioxidant agents have not been as definitive, with some trials showing reduced risk of coronary events in patients consuming antioxidant vitamins and other trials showing no benefit of such agents. Discovery of the antiatherogenic actions of NO and the fact that endothelial dysfunction is characterized by a dramatic reduction in endothelial NO release has inspired experiments using L-arginine, a metabolic precursor of NO, to restore endothelial function and block the progression of atherosclerosis. A recent experiment has shown that dietary supplementation of L-arginine in hyperlipidemic rabbits led to a striking reduction in aortic and coronary atheromatous lesions. The possible antiatherogenic effects of ACE inhibitors, which increase NO release and block production of angiotensin II, are being tested. Experiments have shown that the ACE inhibitor captopril is effective in reducing aortic intimal lesions as well as cholesterol content and cellularity of the lesions in hyperlipidemic animals. A currently ongoing trial is assessing the effect of ACE inhibition on ischemic events in patients with documented coronary disease but with normal blood pressure, serum LDL, and left ventricular function.

Additional advancements in understanding the mechanism of atherogenesis will come in the wake of novel methods of molecular genetics. These include the technique of targeted mutagenesis of genes encoding pro- and anti-atherogenic molecules. Gene knockout animals, principally mice, will provide new insights into how these genes are expressed and the function they serve in the whole animal. Transgenic techniques that allow production of animals overexpressing specific genes should provide a better understanding of the role of genes in the intact animal. Genetically altered animals are, in effect, living test tubes that allow investigators to molecularly dissect pathophysiologic processes. For example, a knockout mutation in the apolipoprotein E (apo E) gene in C57BL/6J mice results in markedly elevated serum cholesterol levels, revealing a central role of apo E in lipid metabolism. Moreover, the mutant mice readily develop atheromatous lesions and have provided a new model for the study of atherosclerosis. Further study design can be employed by crossbreeding mice lacking one factor with mutant mice overexpressing or with null mutations of various pro- and anti-atherogenic factors of interest to test predictions in the etiologic and protective roles of genes in the disease process. [*See* Gene Targeting Techniques.]

In addition to providing a powerful tool for delineating disease mechanisms, sophisticated molecular genetic techniques are now entering the realm of disease treatment in the form of *ex vivo* and *in vivo* gene insertion and antisense oligonucleotide-mediated disruption of gene expression. In *ex vivo* gene therapy, cells from a diseased individual are harvested and grown in tissue culture. A recombinant gene is introduced into these cells and its incorporation and expression are assayed. The cells are then infused back into the host individual. With this "gene replacement" technique, a normal copy of a gene can be introduced into cells of an individual having the defective allele(s) of that gene, thereby replacing its function. This technique is being used to insert the normal LDL receptor gene into hepatocytes of patients with familial hypercholesterolemia, a disease characterized by markedly elevated serum cholesterol and premature atherosclerosis secondary to homozygosity for a defective LDL receptor gene. The technique of *in vivo* gene therapy involves direct introduction of a gene into cells of the host. This method has been proposed to deliver recombinant angiogenic factor genes coated onto balloon catheters to specific segments of arteries in individuals with peripheral vascular disease. Antisense gene therapy uses oligonucleotides to suppress the function of specific genes. In this treatment, synthetic oligonucleotides complementary to the mRNA sequence of a gene are introduced into a tissue. The oligonucleotides then bind to their complementary sequences on DNA or RNA and interfere with the expression of the gene by inhibiting gene transcription, mRNA splicing, mRNA transport to the nucleus, or gene translation. [*See* Gene Therapy.]

The future promises many exciting developments. The steady, chronic progression of the atherosclerotic lesion is an intimidating problem, but should be measured against the remarkably rapid progression of scientific knowledge and innovation of new molecular techniques. A deeper understanding of the molecular mechanism of atherogenesis should yield many new strategies for intervening in the process of atherosclerosis.

## BIBLIOGRAPHY

Berliner, J. A., Navab, M., Fogelman, A. M., Frank, J. S., Demer, L. L., Edwards, P. A., Watson, A. D., and Lusis, A. J. (1995). Atherosclerosis: Basic mechanisms, oxidation, inflammation, and genetics. *Circulation* **91**, 2488–2496.

Collins, T. (1993). Biology of disease: Endothelial nuclear factor-$\kappa$B and the initiation of the atherosclerotic lesion. *Lab Invest.* **68**(5), 499–508.

Davies, M. G., and Hagen, P-O. (1993). The vascular endothelium, a new horizon. *Ann. Surgery* **218**(5), 593–609.

DiCorleto, P. E., and Gimbrone, M. A., Jr. (1996). Vascular endothelium. *In* "Atherosclerosis and Coronary Artery Disease" (V. Fuster, R. Ross, and E. J. Topol, eds.), Vol. 1, p. 387. Lippincott–Raven, Philadelphia.

Libby, P. (1995). Molecular bases of the acute coronary syndromes. *Circulation* **91**(11), 2844–2850.

Ross, R. (1993). The pathogenesis of atherosclerosis: A perspective for the 1990's. *Nature* **362**, 801–809.

Springer, T. A., and Cybulsky, M. I. (1996). Traffic signals on endothelium for leukocytes in health, inflammation, and atherosclerosis. *In* "Atherosclerosis and Coronary Artery Disease" (V. Fuster, R. Ross, and E. J. Topol, eds.), Vol. 1, p. 511. Lippincott–Raven, Philadelphia.

# ATP Synthesis in Mitochondria

PETER L. PEDERSEN
*Johns Hopkins University*

---

## GLOSSARY

**Acetylcoenzyme A** Common breakdown product of carbohydrate, fat, and protein metabolism; represented biochemically as $CH_3C-SCoA$

**Metabolism** The sum of the chemical steps by which a particular substance is handled within a living cell or within a living organism

**Mitochondria** Organelles found in eukaryotic cells that are involved in ATP synthesis from ADP and inorganic phosphate via the process known as oxidative phosphorylation

**Organelle** Cellular structure composed of one or more membranes surrounding a soluble compartment. Mitochondria, nuclei, and lysosomes are among those organelles found in eukaryotic cells

**Proton** An elementary particle that is identical with the nucleus of the hydrogen atom

**Reducing equivalent** A proton plus an electron; represented here as (H)

**Respiration** Oxygen consumption by mitochondria resulting from the transport of electrons from respiratory substrate to molecular oxygen

**Respiratory substrate** Substrate that gives up electrons directly or indirectly (i.e., via NADH or FADH₂, respectively) to the electron transport chain

**Tricarboxylic acid cycle** Cycle catalyzed by a set of enzymes located in the mitochondrial matrix which convert acetyl-coenzyme A to $CO_2$ while making reducing equivalents available for the oxidative phosphorylation system

OXIDATIVE PHOSPHORYLATION IS THE TERM USED to describe the process by which mitochondria make ATP in eukaryotic cells from ADP and inorganic phosphate ($P_i$). All steps involved in oxidative phosphorylation are localized in the inner membrane of mitochondria. The overall process involves a sequence of oxidative reactions dependent on the presence of oxygen, followed by a subsequent phosphorylation event in which ADP is phosphorylated, with $P_i$ converting the ADP to ATP. The oxidative part of the process is essential for providing the energy essential for driving the reaction $ADP + P_i \rightleftharpoons ATP + HOH$ in the direction of net ATP synthesis. Newly synthesized ATP is used to meet almost all energy needs of the cell. In meeting these needs, ATP is converted to ADP and $P_i$, which are immediately recycled back to ATP by the mitochondrial oxidative phosphorylation system.

## I. OVERVIEW OF MITOCHONDRIAL OXIDATIVE PHOSPHORYLATION

### A. Purpose

The purpose of oxidative phosphorylation in human biology is to complete the transfer of chemical energy stored within the food we consume (Fig. 1A) to ADP and $P_i$ in order to give ATP (Fig. 1C). ATP is the cell's useful form of energy which is available upon demand. It is used to drive directly or indirectly almost every energy-requiring reaction in the body. Among these processes are muscle contraction, nerve conduction, vision, and the biosynthesis of macromolecules such

ENCYCLOPEDIA OF HUMAN BIOLOGY, Second Edition, VOLUME I. Copyright © 1997 by Academic Press. All rights of reproduction in any form reserved.

A

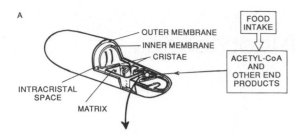

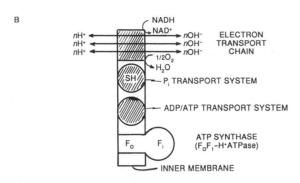

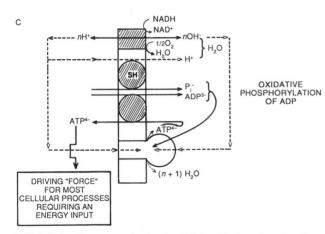

FIGURE 1 Overview of mitochondrial oxidative phosphorylation. Oxidative phosphorylation takes place exclusively in the mitochondrial inner membrane (A). NADH derived from the tricarboxylic acid cycle is used to support a resting state of respiration in the absence of ADP or $P_i$ (B). Under these conditions, an electrochemical gradient of protons is generated across the mitochondrial inner membrane. Three different regions of the electron transport chain (complexes I, III, and IV) participate in generating this ion gradient. Upon addition of ADP and $P_i$, the electrochemical proton gradient is used to drive the uptake of $P_i$ and ADP and to synthesize ATP from these substrates (C). (Note that only one of the three sites involved in generating the proton gradient is shown.)

as DNA, RNA, protein, fat, and carbohydrate. [*See* Adenosine Triphosphate (ATP).]

ATP used to energize these processes is converted to ADP and $P_i$ and is then "recycled" to ATP via the

oxidative phosphorylation process (Fig. 1C). A 70-kg adult at rest might turn over as much as 50% of his or her body weight in ATP per day, a value that might increase under working conditions to almost 800 kg, or about 1 ton.

## B. Location

In animal cells, oxidative phosphorylation takes place exclusively in the mitochondria (Fig. 1A). These organelles consist of four major components: an outer membrane; an inner membrane; a space between the outer and inner membranes, called the intracristal space; and the space enclosed by the inner membrane, called the matrix. The inner membrane serves as the site of oxidative phosphorylation. The matrix and outer compartment serve important ancillary roles by processing metabolic end products essential for sustaining the oxidative phosphorylation event.

During cellular metabolism, food supplied in the diet is processed to fuel oxidative phosphorylation by first undergoing conversion to smaller molecules, particularly acetylcoenzyme A. The latter are converted to $CO_2$ and reducing equivalents (H) at the level of the tricarboxylic acid cycle localized in the mitochondrial matrix (Fig. 1A). The major carrier of reducing equivalents from the tricarboxylic acid cycle to the oxidative phosphorylation system is NADH, although another carrier, $FADH_2$, is also involved.

## C. Essential Parts

Oxidative phosphorylation in intact coupled mitochondria involves the efficient function of four different inner membrane components (Fig. 1B): the electron transport chain, the $H^+/P_i$ transporter, the ADP/ATP transporter, and the ATP synthase complex.

The process of oxidative phosphorylation can take place in purified inner membrane vesicles with the involvement of only the electron transport chain and the ATP synthase complex. However, this is not a normal physiological situation, because such vesicles are "inside-out" and therefore do not require transport of the oxidative phosphorylation substrates ADP and $P_i$.

## D. Function: Critical Role of an Electrochemical Gradient of Protons

The electron transport chain, supplied with electrons from reducing equivalents, carries out a series of electron transfer events, ultimately resulting in the reduction of oxygen (Fig. 1B). At three different locations along the electron transport chain, there is an effective splitting of water across the inner membrane, resulting in the net accumulation of protons on the outside and a net accumulation of hydroxyl ions on the matrix side. Because the mitochondrial inner membrane has a low permeability to protons and hydroxyl ions, potential energy in the form of an *electrochemical gradient* of protons is set up across the membrane.

Part of the energy is stored in the form of membrane potential, which results from the fact that the electron transport process separates charged ions (i.e., positively charged protons on the outside and negatively charged hydroxyl ions within the matrix). The remaining part of the potential energy is stored as a chemical gradient, as the proton concentration on the outside is higher than that in the matrix.

In the absence of ADP or $P_i$, the potential energy cannot be used to drive ATP synthesis. Consequently, an equilibrium, or resting, state of respiration occurs in which both the respiratory substrate and oxygen are used at a low rate. In the presence of both ADP and $P_i$, the mitochondrial oxidative phosphorylation process is activated (Fig. 1C). The catalytic moiety ($F_1$) of the ATP synthase complex now gains access to the electrochemical proton gradient. The largest portion (i.e., about 75% of this gradient) is used to drive ATP synthesis, while the remaining 25% is used to drive the transport of $P_i$ and ADP. Provided ADP, $P_i$, and a source of reducing equivalents (i.e., NADH, $FADH_2$, or some other reductant) are in ample supply, mitochondrial electron transport and ATP synthesis proceed unabated at a high rate. However, if either ADP or Pi becomes limiting, the rate of electron transport returns to its normally low resting state.

### E. Definition and Importance of Acceptor Control

In viable animal cells, mitochondria must be capable of using the oxidative phosphorylation system in the manner indicated in the previous section. Thus, when a given cell type recovers from a sudden burst of work (e.g., physical, electrical, or biosynthetic), the mitochondria must be capable of switching rapidly from the resting to the active state in order to resynthesize ATP from ADP and $P_i$. Mitochondria capable of undergoing this transition are said to have acceptor control of respiration and can be assigned an acceptor control ratio (ACR), where the ACR is equal to the rate of respiration with ADP divided by the rate of

respiration without ADP. This ratio is determined in the presence of excess respiratory substrate and $P_i$. [*See* Mitochondrial Respiratory Chain.]

Mitochondria that exhibit an ACR value much greater than 1 (usually $\geq 5$ in the rat liver) are said to be well coupled, whereas those that approach an ACR value of 1 might be either uncoupled (i.e., have no net capacity to make ATP) or loosely coupled (i.e., have a very low capacity to make ATP). In effect, the ACR value is an index of the degree to which the inner membrane is "leaky" to protons, and therefore the degree to which oxidation and phosphorylation events are coupled. Thus, when the mitochondrial inner membrane maintains the electrochemical gradient of protons with minimal leakage back into the matrix, the ACR ratio is much greater than 1, and oxidation events are well coupled to ATP synthesis. However, in cases in which the mitochondrial inner membrane is rendered leaky to protons, oxidation events become either loosely coupled to ATP synthesis or uncoupled entirely, depending on the magnitude of the leak.

## II. ELECTRON TRANSPORT CHAIN

### A. Electron Transport

The "mainstream" electron transport chain in mitochondria consists of three transmembrane complexes (I, III, and IV)[1] and two mobile molecules: ubiquinone (i.e., coenzyme Q) and cytochrome *c* (Fig. 2). When the electron transport chain is operative, these electron carriers function in a sequence that is based on their capacity to give up electrons in the reduced state.

Under physiological conditions, electrons enter the electron transport chain at two major locations; at the level of complex I and at the level of ubiquinone (Fig. 2). NADH (derived from $NAD^+$-requiring dehydrogenases) gives up electrons to complex I. $FADH_2$ and $FMNH_2$ (derived from dehydrogenases requiring either FAD or FMN) give up electrons to ubiquinone. Some of the FMN- and FAD-linked dehydrogenases are located on the inner surface of the inner membrane, while others are located on the outer surface. Electrons from bound $FADH_2$ or $FMNH_2$ at these different locations are believed to be able to enter the mainstream electron transport chain because of the

---

[1]There is also a complex II, which is not a member of the mainstream electron transport chain, but is involved in supplying this chain with electrons.

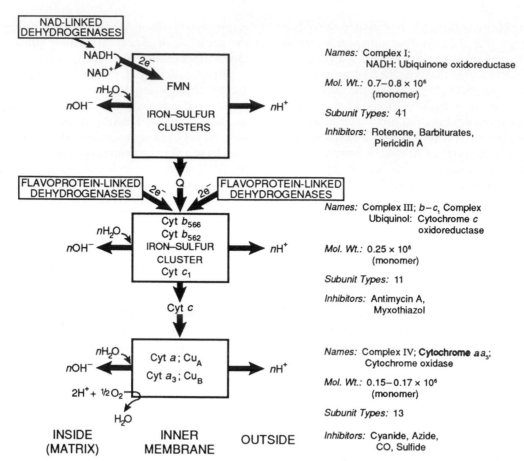

**FIGURE 2** The mitochondrial electron transport chain. The mitochondrial electron transport chain consists of three types of transmembrane complexes (I, III, and IV) and two mobile electron carriers: coenzyme Q and cytochrome $c$. Electrons are supplied to the electron transport chain by dehydrogenases which yield either NADH or $FADH_2$ (or $FMNH_2$). Dehydrogenases which yield $FADH_2$ or $FMNH_2$ are called flavoprotein-linked dehydrogenases and are located both on the inner surface of the inner membrane and on the outer surface of the inner membrane. Each of the transmembrane complexes (I, III, and IV) represents a site where an electrochemical gradient of protons is generated during electron transport. (See text for a detailed discussion.) $n$ = number of $H^+$ or $OH^-$, which remains controversial.

mobility, or "shuttle," capacity of the lipid-soluble ubiquinone molecule. Finally, it should be noted that some compounds, such as ascorbic acid (i.e., vitamin C), can give up electrons to cytochrome $c$, but it is unclear to what extent this is a physiologically important event.

Complexes I, III, and IV are three of the most complex proteins known, consisting, respectively, of 41, 11, and 13 polypeptides (Fig. 2). Within these complexes, the electron flow from NADH to molecular oxygen proceeds through organic and inorganic molecules. These are FMN, iron–sulfur clusters, ubiquinone, and heme groups (present in all cytochromes) as well as copper. The electron-carrying atom of heme groups is iron, which in all cases is

sandwiched within a tetrapyrrole ring system. Iron is also the electron-carrying species of the iron–sulfur clusters:

$$\left[\begin{array}{c} \text{Fe} \overset{\text{S}}{\underset{\text{S}}{<}} \text{Fe} \overset{\text{S}}{\underset{\text{S}}{<}} \text{Fe} \overset{\text{S}}{\underset{\text{S}}{<}} \text{Fe} \overset{\text{S}}{\underset{\text{S}}{<}} \text{Fe} \end{array}\right]$$

which operate as mini-electron transport chains within complexes I and III. Here the sulfur is supplied by cysteine residues of specific protein subunits.

Electron transport via the electron transport chain is frequently written as shown in Fig. 2, with two electrons being given up to reduce complex I and two electrons ultimately being used to reduce an oxygen atom (i.e., $\frac{1}{2}O_2$) at the level of complex IV (cytochrome oxidase). However, complex IV interacts under physiological conditions with an oxygen molecule ($O_2$) which is believed to form a bridge across the iron of heme $a_3$ and a nearby copper called $Cu_B$.

To reduce the bound oxygen molecule completely requires four electrons. Therefore, for every oxygen molecule that is ultimately reduced, four electrons must get to complex IV. As complex I appears to be restricted to two electron transfers, an obvious dilemma exists as to how four electrons are ultimately made available to completely reduce molecular oxygen. One possibility is that one atom of oxygen bound to complex IV is reduced by one pass of two electrons through the electron transport chain, and the second atom is reduced by another pass. In any event, two NADH molecules are required to ultimately reduce a single molecule of oxygen.

## B. Proton Translocation

As emphasized in Section I,D, the major role of the electron transport chain is to generate an electrochemical gradient of protons across the mitochondrial inner membrane that can be used to drive ATP synthesis and the associated transport of $P_i$ and ADP (Fig. 1). A central unanswered question in mitochondrial bioenergetics concerns the mechanism by which electron flow is coupled to proton translocation. What is known is that each of the three transmembrane complexes (I, III, and IV), when isolated and incorporated into phospholipid vesicles (liposomes), is capable of catalyzing proton translocation coupled to electron flow. At the level of complex III, coenzyme Q is believed to be the major vehicle involved in chemically translocating protons from the matrix surface to the outer surface of the mitochondrial inner membrane. Two recent x-ray structures of complex IV provide new insights into potential proton paths in this terminal complex. One such path may involve both a critical "swinging" histidine associated with the heme $a_3/CU_B$ center and several other amino acid functional groups. Much more work is necessary to clearly establish the mechanism(s) by which complexes I, III, and IV catalyze electron flow coupled to proton translocation across the mitochondrial inner membrane.

## C. P/O and $H^+$/O Ratios

The P/O ratio refers to the number of molecules of ATP formed during oxidative phosphorylation di-

vided by the number of atoms of oxygen consumed. The P/O ratio is near 3 when NADH is the electron donor, near 2 when $FADH_2$ (or $FMNH_2$) is the electron donor, and near 1 when ascorbic acid is the electron donor. These values refer to measurements made in intact coupled mitochondria in the presence of excess respiratory substrate, ADP, $P_i$, and oxygen. These numbers indicate that there is a sufficient free energy change as electrons flow through each of the three major complexes (I, III, and IV) to drive the phosphorylation reaction

$$ADP + P_i \rightleftharpoons ATP + HOH.$$

Under standard conditions this reaction has a free energy change near $+7.3$ kcal/mol, indicating that the equilibrium lies far to the left of the equation, in favor of ADP and $P_i$. To shift the equilibrium in favor of ATP synthesis, the free energy change derived from electron flow through each of the major complexes must be at least $-7.3$ kcal/mol.

The $H^+$/O ratio refers to the number of protons translocated to the mitochondrial medium per atom of oxygen consumed. Values are measured by adding a pulse of oxygen to mitochondria maintained anaerobically with nitrogen and then monitoring both oxygen consumption and the pH decrease in the external medium. $H^+$/O ratios as high as 12, 8, and 4 have been reported for substrates which give up electrons to complexes I, III, and IV, respectively.

Measurements of both P/O and $H^+$/O ratios strongly imply that there are three regions of the electron transport chain: the transmembrane-spanning regions of complexes I, III, and IV, which are involved in proton translocation coupled to ATP synthesis. Such measurements further imply that as many as four $H^+$ might be associated with the synthesis of one ATP. Since one $H^+$ is essential for the transport of one $P_i$ into the mitochondria, the remaining three would be available to the ATP synthease complex for driving ATP synthesis.

## III. ADP/ATP and $H^+$/$P_i$ TRANSPORT SYSTEMS

In molecular terms the ADP/ATP and $H^+$/$P_i$ transport systems are among the simplest functional proteins contained within the mitochondrial inner membrane. Both proteins have been studied kinetically, purified to homogeneity, functionally reconstituted in phospholipid vesicles, and completely sequenced.

Functionally, the $P_i/H^+$ and ADP/ATP transport systems are coupled to one another and to the electrochemical proton gradient (Fig. 1C). During oxidative phosphorylation, $P_i^-$ is transported into mitochondria on the $H^+/P_i$ transporter together with an $H^+$ (symport), while the ADP/ATP transport system transports $ADP^{3-}$ into the mitochondria while exporting $ATP^{4-}$ out (i.e., antiport). Four net negative charges enter the mitochondrial matrix, while four negative charges are transported out. The net effect is that the substrates of oxidative phosphorylation (i.e., ADP and $P_i$) are made available to the ATP synthase complex while maintaining charge balance across the mitochondrial inner membrane. The $H^+$ entering with the $P_i$ drives the net uptake of ADP and $P_i$ entry and ATP exit when it reacts with an $OH^-$ within the matrix to form $H_2O$. This $H^+$ does not enter into the calculation of net charge balance.

The $H^+/P_i$ transporter is a medium-sized protein containing 313 amino acid residues (in the bovine heart), with an apparent size of 34 kDa. It has a single highly reactive cysteine residue located at position 42 from the N-terminal end. The covalent reaction of $N$-ethylmaleimide and many other sulfhydryl reactive reagents with this cysteine completely inhibit the phosphate transport capacity of the $H^+/P_i$ transporter and, therefore, oxidative phosphorylation.

The ADP/ATP transport system is a protein similar in size to the $H^+/P_i$ transporter. It contains 297 amino acid residues (in the bovine heart) and exhibits an apparent size of 30 kDa. The two characteristic inhibitors of this transporter are atractyloside and bongkrekic acid, both of which act to inhibit ADP entry (or ATP exit) and therefore oxidative phosphorylation.

Both the $H^+/P_i$ and ADP/ATP transport proteins are predicted to span the mitochondrial inner membrane six or seven times. The two transporters have a number of amino acid homologies. These proteins are believed to have arisen during evolution from a common gene which split into three segments and then diversified, respectively, into the ADP/ATP transporter, the $H^+/P_i$ transporter, and an $H^+$ transporter found only in brown adipose tissues.

The mechanism by which the $H^+/P_i$ and ADP/ATP transporters catalyze the uptake of ADP and $P_i$ into mitochondria is currently unknown. Studies of the ADP/ATP transporter indicate that it functions as a dimer within the membrane. A model has been proposed in which six transmembrane helices (three from each monomer) are depicted as forming an anion

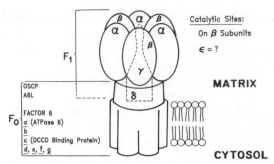

**FIGURE 3** Structure of the mitochondrial ATP synthase complex, illustrating the possible locations of the many different subunit types. DCCD, dicyclohexylcarbodiimide. Two recent x-ray structures of $F_1$ reveal the alternating arrangement of the $\alpha$ and $\beta$ subunits and the central location of the $\gamma$ subunit.

channel for negatively charged ions through the mitochondrial inner membrane.

## IV. ATP Synthase Complex

The mitochondrial ATP synthase complex ($F_0F_1$) has a size of >450 kDa and consists in higher eukaryotes of 16 different subunit types (Fig. 3). Five of these are associated with the $F_1$ moiety, which is water soluble, has a molecular mass of 360–380 kDa, and projects into the mitochondrial matrix. The remainder are associated with $F_0$, which spans the mitochondrial inner membrane and helps comprise the "stalk" at the membrane interface. $F_1$ subunits are referred to as $\alpha$, $\beta$, $\gamma$, $\delta$, and $\varepsilon$ and in rat liver exhibit apparent sizes of 62, 57, 36, 12.5, and 7.5 kDa, respectively. They are present in the ratio $\alpha_3\beta_3\gamma\delta\varepsilon$. $F_0$ subunits are called a, b, c, d, e, f, g OSCP, A6L, and factor 6. Subunit c is the predominant $F_0$ subunit, consisting of several copies per ATP synthase complex.

The five types of $F_1$ subunits are required for ATP synthesis. The catalytic sites lie on $\beta$ subunits, near $\alpha$–$\beta$ interfaces. $\beta$ subunits alone exhibit little or no catalytic capacity and are believed to require interaction with both $\alpha$ and $\gamma$ subunits in order to catalyze maximal rates of ATP synthesis. The $\gamma$ subunit might specify which of the three $\beta$ subunits will be "tagged" for catalysis. It might also function either as a "proton gate," regulating the flow of protons to the catalytic sites, or as a transducer of a conformational change from $F_0$ to $F_1$. Finally, one role of the $\delta$ subunit appears to be to bind $F_1$ to $F_0$, although other roles have not been ruled out (see below). Roles of the $\varepsilon$ subunit are unknown.

## V. MECHANISM OF ATP SYNTHESIS

The reaction cycle summarized in Fig. 4A depicts the way ATP might be synthesized on the surface of the mitochondrial ATP synthase complex and then released. In this scheme, ADP and $P_i$ bind in sequence to a single $\beta$ subunit. Water is spilled out as ATP is synthesized on the enzyme surface. At this point the electrochemical gradient of protons interacts with the enzyme in such a way as to release newly synthesized ATP. The reaction cycle emphasizes that energy from the electrochemical proton gradient is not necessary to effect synthesis of bound ATP at a single catalytic site. Rather, it is necessary to effect the release of newly synthesized ATP.

Several types of experiments tend to support this reaction cycle. First, ATP has been shown to be synthesized from ADP and $P_i$ on the $F_1$ moiety of the ATP synthase molecule. Second, the reaction ADP + $P_i \rightleftharpoons$ ATP + HOH on the enzyme surface can go equally well in either direction. Third, ATP, once formed, has been shown to be bound with a very high affinity. Finally, generation of an electrochemical proton gradient by adding a respiratory substrate to inner mitochondrial membrane vesicles has been shown to alter the binding of ATP bound to $F_1$, presumably at a catalytic site.

The scheme depicted in Fig. 4A accounts only for ATP synthesis at a single catalytic site (i.e., on a single $\beta$ subunit). For all three $\beta$ subunits to be involved in a sequential manner, it has been suggested that some type of subunit movement (or rotation) must be involved (Fig. 4B). This might entail in the simplest context the repositioning of one or more of the smaller subunits $\gamma$, $\delta$, and $\varepsilon$ from one $\beta$ subunit to the other. Conformational changes induced in $\beta$ subunits by smaller subunit repositioning could serve both to facilitate binding of ADP and $P_i$ to a new catalytic site and to redirect the flow of protons from one $\beta$ subunit to the other.

A final aspect of the mechanism of ATP synthesis that deserves discussion is the manner by which protons are translocated through $F_0$ to $F_1$. In this regard two $F_0$ subunits—namely, subunits c and a—are thought to be major players. Subunit c contains a single conserved aspartic acid residue which, when blocked by the agent dicyclohexylcarbodiimide, prevents both proton flow and ATP synthesis. Because there are several copies of subunit c and therefore multiple copies of the conserved aspartic acid residue, it is possible that a charge relay network (or "proton wire") allows protons to transverse $F_0$ to $F_1$, i.e., protons may "hop" from one aspartic acid residue to

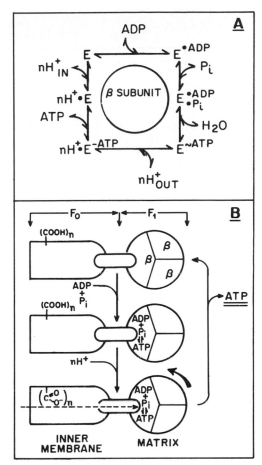

**FIGURE 4** (A) Reaction cycle catalyzed by a single $\beta$ subunit within the mitochondrial ATP synthase complex. The synthesis of ATP from ADP and $P_i$ is believed by some investigators to proceed on the enzyme surface with an equilibrium constant near 1 (i.e., equally well in either direction). Thus, energy is not believed to be necessary to make enzyme-bound ATP. Rather, the energy from the electrochemical proton gradient is believed to be required to effect a conformational change in the enzyme, which in turn lowers the binding affinity of bound ATP and causes its release. (B) Model of the ATP synthase complex, emphasizing the possible involvement of all three $\beta$ subunits or $\alpha$–$\beta$ pairs in catalysis and the role of the electrochemical gradient of protons. The model is an extension of that shown in A. When ATP synthesis has been completed on a single $\beta$ subunit and then released, the other two $\beta$ subunits are thought to become sequentially involved. This "alternating site" view for ATP synthesis was first proposed by Repke and Schön in 1974 and was developed in greater detail by Boyer and colleagues in 1982. Whether the larger subunits $\alpha$ and $\beta$ actually rotate so that they have access in turn to the proton gradient or whether the smaller subunits actually undergo some repositioning (i.e., detaching from one $\beta$ subunit and reattaching to another) is unknown. The model also shows how the conserved aspartic acid residues on subunit c might initiate proton translocation by switching from the lipid phase of the inner membrane to the interior of $F_0$.

TABLE I

Commonly Used Inhibitors of Mitochondrial Oxidative Phosphorylation

| Site of action | Inhibitor | Inhibition constant or $K_D$[a] |
|---|---|---|
| Complex I | Rotenone | 0.037–0.07 nmol/mg of protein ($I_{50}$) |
| | Amytal[b] | $10^{-4}$ to $10^{-3}$ $M(I_{50})$ |
| | Piericidin A | 0.036 nmol/mg of protein ($I_{>95}$) |
| Complex III | Antimycin A | 1 antimycin A/cytochrome $c_1$ ($I_{>95}$) |
| Complex IV | Hydrogen cyanide | $<1 \times 10^{-7}$ $M(K_D)$ |
| | Hydrogen azide | $4–8 \times 10^{-7}$ $M(K_D)$ |
| $P_i/H^+$ transporter | N-Ethylmaleimide | 175 nmol/mg of protein ($I_{100}$) |
| | Mersalyl | 10–12 nmol/mg of protein ($I_{100}$) |
| ADP/ATP transporter | Atractyloside | 0.22–0.30 nmol/mg of protein ($I_{50}$) |
| | Carboxyatractyloside | 0.21–0.26 nmol/mg of protein ($I_{50}$) |
| | Bongkrekic acid | 0.02–0.8 $\mu M(K_i)$ |
| ATP synthase $F_0$ moiety | Oligomycin | 1 $\mu$g/mg of protein ($I_{>80}$) |
| | Venturicidin | 1 $\mu$g/mg of protein ($I_{>80}$) |
| | Dicyclohexylcarbodiimide | 1 $\mu M(I_{>80})$ |
| ATP synthase $F_1$ moiety | Aurovertin | 1 $\mu$g/mg of protein ($I_{>80}$) |

[a] $I_n$, Concentration giving $n$% inhibition; $K_D$, dissociation constant; $K_i$, inhibition constant, kinetically determined; mg of protein, mitochondrial protein (rat liver or bovine heart).

[b] Trademark preparation of amobarbital.

another or from one amino acid functional group to another. As changes in subunit a also prevent proton translocation, this proton relay network might involve not only aspartic acid residues on subunit c, but also critical amino acid residues on subunit a.

Finally, it should be noted that there is much to be learned about the exact steps involved in ATP synthesis. First, it remains to be established with certainty whether the site shown to make ATP on the enzyme surface is a true catalytic site. Second, there has been much debate as to whether two or three $\beta$ subunits actually participate in ATP synthesis. Third, there is the question as to whether subunit rotation actually occurs in the physical sense or whether subunit interactions suffice as signals to permit their sequential involvement in ATP synthesis. Finally, there is the question as to whether protons are delivered directly to a catalytic site on $F_1$ or to the $F_1$ moiety at all. Some investigators believe a conformational change induced in $F_0$ by the electrochemical proton gradient induces a conformational change in $F_1$, which in turn lowers the binding affinity of ATP.

## VI. INHIBITORS AND UNCOUPLERS

Table I lists the agents most commonly used to inhibit ATP synthesis catalyzed by intact mitochondria. Inhi-

bition can occur at three different levels in the overall process of oxidative phosphorylation: at the level of the electron transport chain, at the level of the transport systems for ADP and $P_i$, and at the level of the ATP synthase complex. Consequently, Table I includes inhibitors which act at these levels.

Table II lists some of the more commonly used

TABLE II

Commonly Used Uncouplers of Mitochondrial Oxidative Phosphorylation

| Abbreviation | Chemical name | Uncoupling concentration ($\mu M$)[a] |
|---|---|---|
| DNP | 2,4-Dinitrophenol | 7.4 |
| CCP | Carbonyl cyanide phenylhydrazone | 0.4 |
| CCCP | Carbonyl cyanide 3-chlorophenylhydrazone | 0.06 |
| FCCP | Carbonyl cyanide 4-trifluoromethoxy phenylhydrazone | 0.04 |
| S-13 | 2,5-Dichloro-3-tert-butyl-4'-nitrosalicylanilide | 0.005 |

[a] Concentration giving 50% uncoupling in rat liver or bovine heart mitochondria.

mitochondrial uncoupling agents. These agents also inhibit mitochondrial ATP synthesis, but are referred to as uncoupling agents (or uncouplers) because they prevent oxidation events (i.e., electron transport) from being coupled to phosphorylation of ADP. Unlike the inhibitors of oxidative phosphorylation listed in Table I, uncouplers do not exert their primary site of action on protein complexes. Rather, because of their lipid solubility and weak acid character, uncouplers catalyze the translocation of protons through the mitochondrial inner membrane, preventing the formation of an electrochemical gradient of protons. If such a gradient has been previously established, uncouplers dissipate the gradient. Without an electrochemical proton gradient, there can be no driving force for ATP synthesis.

## BIBLIOGRAPHY

Abrahams, J. P., Leslie, A. G. W., Lutter, R., and Walker, J. E. (1994). Structure at 2.8 Å resolution of $F_1$-ATPase from bovine heart mitochondria. *Nature* **370**, 621–628.

Aquila, H., Link, T. A., and Klingenberg, M. (1987). Solute carriers involved in energy transfer of mitochondria from a homologous protein family. *FEBS Lett.* **212**, 1–9.

Belogrudov, G., Tomich, J. M., and Hatefi, Y. (1996). Membrane topography and near neighbor relationships of the mitochondrial ATP synthase subunits e, f, and g. *J. Biol. Chem.* **371**, 20340–20345.

Bianchet, M., Ysern, X., Hullihen, J., Pedersen, P. L., and Amzel, L. M. (1991). Mitochondrial ATP synthase: Quarternary structure of the $F_1$ moiety at 3.6 Å determined by x-ray diffraction analysis. *J. Biol. Chem.* **266**, 21197–21201.

Capaldi, R. A., Aggeler, R., Turina, P., and Wilkins, S. (1994). Coupling between catalytic sites and the proton channel in $F_1F_0$-type ATPases. *Trends in Biochem. Sci.* **19**, 284–289.

Fillingame, R. H., Girvin, M. E., Fraga, D., and Zhang, Y. (1993). Correlations of structure and function in $H^+$ translocating subunit C of $F_0F_1$ ATP synthase. *Ann. N.Y. Acad. Sci.* **671**, 323–334.

Gogol, E. P., Johnston, E., Aggeler, R., and Capaldi, R. A. (1990). Ligand-dependent structural variations in *E. coli* $F_1$ ATPase revealed by cryoelectron microscopy. *Proc. Natl. Acad. Sci. USA* **87**, 9585–9589.

Gresser, M. S., Meyers, J. A., and Boyer, P. D. (1982). Catalytic site cooperativity of beef heart mitochondrial $F_1$ adenosine triphosphatase: Correlation of initial velocity, bound intermediate, and oxygen exchange. Measurements with an alternating three-site model. *J. Biol. Chem.* **257**, 12030–12038.

Grubmeyer, C., Cross, R. L., and Penefsky, H. S. (1982). Mechanism of ATP hydrolysis by beef heart mitochondrial ATPase: Rate constants for elementary steps in catalysis at a single site. *J. Biol. Chem.* **257**, 12092–12100.

Hatefi, Y. (1985). The mitochondrial electron transport chain and oxidative phosphorylation system. *Annu. Rev. Biochem.* **54**, 1015–1069.

Iwata, S., Ostermeier, C., Ludwig, B., and Michel, H. (1995). Structure at 2.8 Å resolution of cytochrome C oxidase from *Paracoccus denitrificans*. *Nature* **376**, 660–669.

Mitchell, P. (1967). Proton-translocation phosphorylation in mitochondria, chloroplasts and bacteria: Natural fuel cells and solar cells. *Fed. Proc. Fed. Am. Soc. Exp. Biol.* **26**, 1370–1379.

Pedersen, P. L. (1996). Frontiers in ATP synthase research: Understanding the relationship between subunit movements and ATP synthesis. *J. Bioenerg. Biomemb.* **28**, 389–395.

Pedersen, P. L., and Amzel, L. M. (1993). ATP synthases: Structure, reaction center, mechanism, and regulation of one of nature's most unique machines. *J. Biol. Chem.* **268**, 9937–9940.

Pedersen, P. L., and Carafoli, E. (1987). Ion motive ATPases. II. Energy coupling and work output. *Trends Biochem. Sci.* **12**, 186–189.

Repke, K. R. H., and Schön, R. (1974). Flip-flop model of energy interconversion by ATP synthetase. *Acta Biol. Med. Ger.* **33**, K27–K38.

Senior, A. E. (1988). ATP synthesis by oxidative phosphorylation. *Physiol. Rev.* **68**, 177–231.

Tsukihara, T., Aoyama, H., Yamashita, E., Tomizaki, T., Yamaguchi, H., Shinzawa-Itoh, K., Nakashima, R., Yaono, R., and Yoshikawa, S. (1996). The whole structure of the 13-subunit oxidized cytochrome C oxidase at 2.8 Å. *Science* **272**, 1136–1144.

Wilson, D. (1976). IV. Mitochondria. *In* "Biological Handbooks. I. Cell Biology" (P. Altman and D. D. Katz, eds.), pp. 143–230. *Fed. Am. Soc. Exp. Biol.*, Bethesda, MD.

# Atrial Natriuretic Factor

ADOLFO J. de BOLD

*University of Ottawa and the University of Ottawa Heart Institute at the Ottawa Civic Hospital*

ATRIAL NATRIURETIC FACTOR IS ONE OF THREE polypeptide hormones referred to as natriuretic peptides. These are: atrial natriuretic factor (ANF, ANP), brain natriuretic peptide (BNP), and "C" type natriuretic peptide (CNP) and some of their processing variants. These hormones participate in the regulation of blood pressure and blood volume. In the adult mammal, the bulk of ANF and BNP are normally produced and stored by cardiac muscle cells of the atria of the heart (atrial cardiocytes), from where they are secreted into the bloodstream to interact with specific receptors, thus establishing the existence of an endocrine function for the heart. In the mammalian fetus and in nonmammalian vertebrates, in addition to atrial cardiocytes, ventricular cardiocytes normally express such secretory function, as does the ventricular myocardium in adult mammals during hypertrophy. The natriuretic peptide CNP is produced mainly by the vascular endothelium and central nervous tissue.

## I. INTRODUCTION

Natriuretic peptides participate in the maintenance of cardiovascular homeostasis through modulator actions on several systems that are involved in the regulation of blood pressure and blood volume, such as the renin–angiotensin–aldosterone system and the sympathetic nervous system, and on various functions such as renal salt excretion and the thirst mechanism, among many others. [*See* Cardiovascular System, Physiology and Biochemistry.]

The discovery of ANF introduced the concept that the mammalian heart can intrinsically adapt to changing workload using two types of mechanisms: (1) mechanisms based on the mechanical properties of the heart muscle, including, for example, the Frank–Starling mechanism, and (2) an intrinsic endocrine mechanism that regulates preload and afterload based on the endocrine function of the atria.

Natriuretic peptide genes are expressed in many noncardiac tissues, albeit at levels much lower than those found in the mammalian cardiac atria. The sites and mode of expression of natriuretic peptide in noncardiac tissues hint at paracrine and autocrine interactions of these hormones with a variety of other systems, as different from each other as the vascular endothelium, some gonad-specific cells, and neurons in the central nervous system.

## II. FUNCTIONAL MORPHOLOGY OF THE ENDOCRINE HEART

A morphological comparison of cardiocytes in the ventricle of the mammalian heart with those of the atria reveals common elements in these cells. Namely, atrial and ventricular cardiocytes share many cellular components generally expected of striated muscle although some relative parameters, such as mitochondrial volume, cell diameter, or abundance of the T system, tend to be, on the average, lower in atrial cardiocytes. A most striking difference between atrial and ventricular cardiocytes in mammals is the fact

ENCYCLOPEDIA OF HUMAN BIOLOGY, Second Edition, VOLUME I.    Copyright © 1997 by Academic Press.    All rights of reproduction in any form reserved.

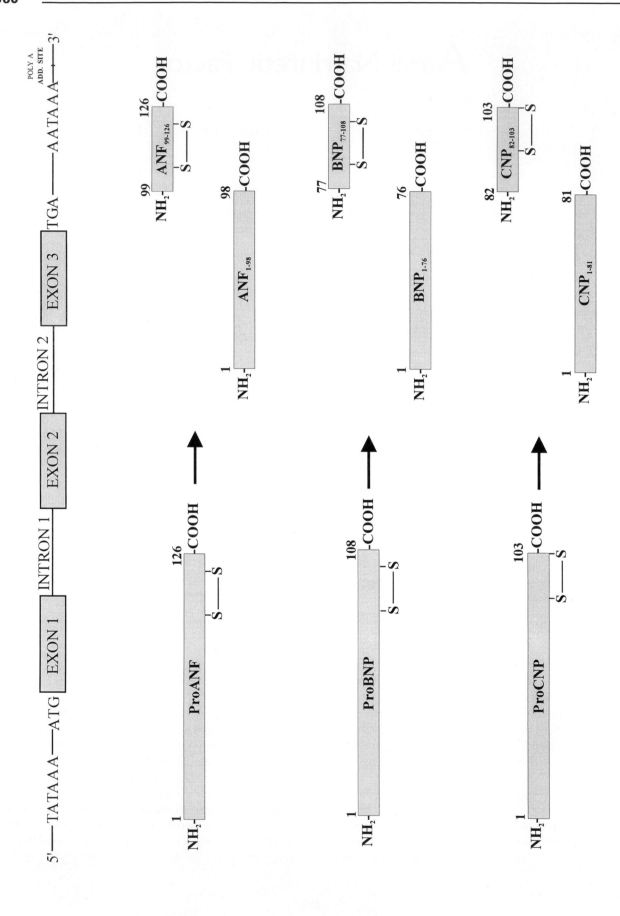

that the former cell type displays, in addition to a muscle phenotype, morphological features found in polypeptide hormone-producing cells, including the presence of storage granules known as *specific atrial granules*. This finding, published in the mid-1950s, led to intensive morphological, histochemical, and functional studies culminating in 1981 with the description of a potent activity promoting renal sodium excretion (natriuresis) associated with the injection of atrial extracts into bioassay rats. Thus the name atrial natriuretic factor or ANF was proposed for the chemical species responsible for the observed natriuretic activity. In addition, this seminal work demonstrated that atrial extracts also have hypotensive and hemoconcentrating effects. The discovery of these effects was followed in 1983 by the isolation and sequencing of a 28-amino-acid peptide, $ANF_{99-126}$, with the aforementioned properties of the crude atrial extracts and with the immunocytochemical localization of its precursor within the specific atrial granules. On this basis, BNP and CNP were subsequently identified. BNP was first isolated from porcine brain, thus the designation brain natriuretic peptide, but it was soon realized that BNP is produced and costored in the specific granules of mammalian atrial cardiocytes.

The atrial cardiocyte population constitutes a phenotypically heterogeneous group of cells that contain obviously varying numbers of specific granules and, in the extreme, cells of the sinoatrial node do not display a secretory-like phenotype at all. To best conceptualize these observations, it is useful to group cardiocytes in functional–morphological terms. In doing so, cardiocytes may be divided into those mainly differentiated for mechanical work ("working" cardiocytes: the bulk of ventricular cardiocytes), secretion and mechanical work (the bulk of atrial cardiocytes), excitation (cardiocytes of the sinoatrial node), and conduction (cardiocytes of the atrial and ventricular conducting system of the heart).

The separation of atrial and ventricular cardiocyte lines is one of the earliest events of mesodermal differentiation and the expression of the ANF gene is an early marker of commitment to the cardiac phenotype of cardiac myoblasts. ANF gene expression is seen in the fetal and early postnatal development throughout the heart and, though in mammals ventricular expression quickly falls after birth, in nonmammalian vertebrates ventricular expression of the ANF gene and the secretory phenotype remains in adult life. An exception to this general view is the fact that parts of the conducting system in the adult ventricle of mammals normally display the secretory phenotype, including specific granules that stain positively for ANF by immunocytochemistry.

## III. BIOCHEMISTRY OF NATRIURETIC PEPTIDES

Each natriuretic peptide is encoded by a specific gene and shares in common a 17-amino loop created by a disulfide bridge (Fig. 1). ANF and BNP are produced by atrial cardiocytes as preprohormones, as is the case for other polypeptide hormones. These are processed during transit from the endoplasmic reticulum to mature secretory granules to yield prohormones and, eventually, hormones that are released into the circulation. Human proANF is a 126-amino-acid polypeptide ($ANF_{1-126}$) that is processed to yield $ANF_{1-98}$ and $ANF_{99-126}$. The latter is the biologically active portion. Human proBNP is 108 amino acids long and the bioactive form is the 32-amino-acid $BNP_{77-108}$. [*See* Cardiovascular Hormones.]

Some controversy exists regarding whether the amino-terminal portion resulting from ANF processing (i.e., $ANF_{1-98}$) possesses biological activity or whether it represents another example of a cryptic

---

**FIGURE I**    Schematic representation of the structure of natriuretic peptide genes, human natriuretic peptide prohormones, and their processing products. The structure of natriuretic peptide genes corresponds to the consensus structure of polypeptide hormone genes. The 5′ flanking region contains the promoter and several consensus sequences (see text). The TATAAA box specifies the start of transcription at a site referred to as a cap site (not shown) proximal to the ATG codon (methionine) in which translation of the mRNA starts. Natriuretic peptide genes contain three exons that are translated up to the stop codons, which generally may be TGA (shown), TAA, or TAG. A few bases in the 3′ direction from the consensus sequence AATAAA, a polyadenylate tract is added during formation of the mRNA precursor. The human CNP gene does not contain this consensus polyadenylation signal. This has been interpreted as possibly indicating the presence of a long second intron.

The product of translation of the natriuretic peptide genes are preprohormones that are processed to yield prohormones and, eventually, the circulating hormones. Atrial natriuretic factor and brain natriuretic peptide are mostly produced by the atria of the heart in humans and in other mammals. Natriuretic peptide type "C" is mostly produced by the vascular endothelium and the central nervous system. The biologically active portion of natriuretic peptides is the carboxyl-terminal fragment that results from proteolytic processing of the prohormone. All the hormones thus contain a loop formed by a disulfide bridge (indicated), which is essential for biological activity.

peptide often seen as a nonbiologically active portion of the precursor of a polypeptide hormone.

Only small amounts of ANF are processed during the maturation of the granules, so that the main ANF species found in the atrial specific granules is proANF. There is substantial evidence indicating that proANF processing to yield $ANF_{99-126}$ takes place at the time of secretion and not in the circulation. The enzyme responsible for this processing at the monobasic site $Arg_{98}$–$Ser_{99}$ has not been identified. Unlike ANF, BNP exists in the specific granules processed to a significant extent, but details of the proteolytic cleavage leading to the mature, circulating BNP hormone also remain to be elucidated.

The circulating form of ANF for all mammalian species so far studied, $ANF_{99-126}$, has a high degree of homology between species. In these species, all amino acids are conserved except residue 110, which is isoleucine in rodents but methionine in humans, dogs, and cows. This is a change brought about by a single nucleotide change in the relevant codon. Circulating BNP peptides, on the other hand, have a much lower degree of homology, even in the biologically key sequence within the disulfide loop, and differ in the number of amino acids of the prohormone and of the circulating peptide. The human, canine, and porcine circulating BNP is 32 amino acids long, whereas the rat and mouse sequences are 45 amino acids long. Given that the number of amino acids in the prohormone is different in each of these cases, sequences with the same number of amino acids do not share the same residue allocation between species as is the case for ANF. Although not a product of the endocrine heart, it is worth mentioning that CNP has the highest degree of homology between species.

The general structure of natriuretic peptide genes is similar in all mammals: three exons and two introns, with the first two exons containing most of the protein-coding sequences (see Fig. 1). The 3' noncoding sequence of the human CNP gene does not contain the consensus polyadenylation signal found at the 3' end of most eukaryotic genes, possibly indicating the presence of a long second intron.

The ANF gene has been the most studied and, as expected from the general physiological role of ANF, experimental procedures known to affect blood volume or blood pressure modulate the level of its expression. For example, sodium-free diets and water deprivation induce a decrease in ANF mRNA levels, whereas deoxycorticosterone acetate (DOCA)-salt treatment and the potent vasoconstrictor endothelin can enhance the levels of ANF gene expression. The

mechanism whereby these stimuli induce changes in gene expression is an active area of research, involving studies on cis-regulatory elements and their interactions with transacting factors. This research has been carried out not only to elucidate the mechanism of control of gene expression of ANF and BNP as cardiac hormones, but also as a model of control of gene expression for cardiac-specific proteins, including tissue-specific expression and developmental and metabolic regulation. As already stated, one of the earliest signs of commitment of myoblasts to the cardiac line is the expression of the ANF gene. Its down-regulation in the mammalian ventricle following birth is a dramatic example of developmentally associated change in expression. Moreover, a hallmark of the hypertrophic process in the cardiac ventricles, which involves the reexpression of some of the fetal isoforms of cardiac-specific proteins, is the reexpression of the ANF gene.

Important advances have been made in the field of tissue-specific expression and developmental regulation of the ANF gene through studies based on the production of transgenic mice and, *in vitro,* using transfection of cells with constructs containing reporter genes attached to different-sized fragments of the 5' flanking region of the ANF gene.

Evidence obtained through this approach suggests species differences in the location of the sequences necessary to direct transcription as well as differences in cis-acting sequences between species, a fact exemplified by the presence of a putative corticoid-binding sequence in the rat ANF gene that is absent in the human gene. However, in spite of these differences in the location of cis-acting sequences, ANF gene tissue expression is similar between species. This is not the case for the BNP gene, as reflected by the fact that BNP is essentially absent in the rat brain and is present in the pig brain.

The upstream regulatory sequences driving the rat ANF gene include the characteristic TATAA box, essential for transcription of most eukaryotic genes, located approximately 30 base pairs (bp) upstream of the transcription initiation site. The other sequences that are important for ANF gene expression are found in the 3-kilobase (kb) sequence preceding the protein-coding sequences. Sequences have been found that are responsible for the tissue-specific expression and for $\alpha_1$-adrenergic inducibility. Other elements include sequences with homology to the serum-responsive element (SRE)/CArG, Egr-1, AP-1, AP-2, and the cAMP-responsive element (CRE), as well as other sequences that modulate positively or negatively gene expression.

Promoter analyses of the BNP gene, although not as extensive as those for ANF, have shown that, in the rat, the elements required for cardiac-specific expression are limited to a 114-bp region upstream of the transcription start site. An AP-1-like element has been found in this region that confers much of the promoter activity. In addition, a GATA motif is also present that is important for basal expression of BNP.

## IV. BIOLOGICAL PROPERTIES OF NATRIURETIC PEPTIDES

The natriuretic peptides possess a remarkably wide spectrum of biological actions mainly concerned with, although not restricted to, the regulation of blood pressure and blood volume. In many ways, the hormonal products of the endocrine heart, ANF and BNP, share practically all biological properties and appear as modulators of several systems, such as the sympathetic nervous system, the renin–angiotensin–aldosterone system, and other determinants of vascular tone and renal function. All of these systems can have opposite actions to those of ANF and BNP by increasing blood pressure and blood volume.

It is not clear why the heart normally produces two hormones that share biological properties and receptors, but in the pathophysiological condition the mode and rate of activation or induction of their genes differ (see the following).

Through various, often overlapping mechanisms, ANF and BNP are potent hypotensive and natriuretic agents. CNP, which lacks the C-terminal sequence of ANF and BNP, is not an effective natriuretic peptide. Critical for the biological action of all natriuretic peptides is the sequence forming the 17-amino-acid disulfide loop. Disruption of this structure leads to loss of biological activity

ANF and BNP are potent vasorelaxants whereby they can function as hypotensive agents and can induce a decrease in cardiac output by decreasing preload. At the same time they can suppress the sympathetic baroreceptor response to decreased blood volume. In addition, ANF infusions lead to increased hematocrit, thus reducing intravascular volume, by inducing increased capillary permeability or capillary pressure gradients.

ANF can induce an increase in glomerular filtration rate and filtration fraction and inhibit renin synthesis, and $Na^+$ reabsorption in the renal tubules, mainly at the level of the collecting ducts. ANF-induced increases in vasa recta hydraulic pressures may contrib-

ute to an unfavorable gradient for $Na^+$ and fluid reabsorption. ANF increases the glomerular filtration rate and filtration fraction probably as the result of dilation of afferent arterioles and constriction of efferent arterioles, and by counteracting the effects of angiotensin II. ANF may also relax glomerular mesangial cells, thus increasing filtration surface area.

In keeping with its perceived role of antagonizing the renin–angiotensin–aldosterone system, the endocrine heart, through ANF and BNP, acts to directly inhibit aldosterone release and synthesis from adrenal cortical glomerulosa cells. By reducing renin secretion, ANF decreases circulating levels of angiotensin II and aldosterone.

ANF and BNP have important actions on the synthesis, release, and biological effects of the potent vasoconstrictor endothelin 1 (ET-1). ANF can decrease ET-1 synthesis and release from endothelial cells and from mesangial cells. The synthesis and secretion of CNP is stimulated by ANF in endothelial cells, suggesting that CNP may act as a mediator on the relaxant properties of ANF on vascular smooth muscle.

The peripheral actions of circulating ANF and BNP just described are only part of the functional targets for these widely distributed peptides. Both centrally and peripherally, natriuretic peptide genes are expressed by many cell types, thus contributing in paracrine or autocrine fashions to regulation processes that, for the most part, are not yet elucidated. Pituitary function is significantly affected by natriuretic peptides as suggested by numerous actions described on oxytocin, gonadotropins, adrenocorticotrophic hormone, and prolactin. For example, ANF inhibits antidiuretic hormone secretion in response to hemorrhage and chronic dehydration and inhibits vasopressin-neuron activity. Intracerebroventricular infusion of ANF leads to significant decreases in thirst and salt appetite. Finally, important effects of natriuretic peptides on central nervous system catecholamine metabolism have been described in areas known to be involved in cardiovascular control.

All natriuretic peptides have antigrowth properties in the vasculature. Hypertrophy of cultured vascular smooth muscle cells is inhibited by ANF, as is growth of cultured endothelial cells and of cardiac fibroblasts.

Two "biologically active" receptors, A and B, have been described for natriuretic peptides. The intracellular portions of these receptors are homologous to protein kinase and guanylyl cyclase domains. Interaction of natriuretic peptides with the A (which binds approximately equally well ANF and BNP) and B (which binds CNP) receptors results in the generation of cGMP. A "C" type receptor contains a short intracellular domain and is often considered to be "biologically silent" in terms of coupling to an intracellular signal transduction system, although it has been associated with a signal pathway through G proteins. The C receptor binds all natriuretic peptides and its main function is believed to be that of clearance. ANF can also interact directly with ion channels as observed in the modulation of aldosterone production by adrenal glomerulosa cells. The natriuretic peptide receptors may be ranked by order in terms of potency of binding as follows: A type = ANF $\geq$ BNP $\gg$ CNP; B type = CNP $\gg$ ANF $\geq$ BNP; C type = ANF > CNP $\geq$ BNP.

Intracellularly, cGMP can interact with cGMP-dependent protein kinases, cGMP-gated ion channels, and cGMP-regulated cyclic nucleotide phosphodiesterases. The activation of guanylyl cyclase in target tissues, as observed, for example, after injection of ANF *in vivo,* is followed by an elevation of cGMP in plasma and urine.

## V. CONTROL OF ANF AND BNP SECRETION FROM THE HEART

ANF and BNP are released from the heart at a basal rate that may be significantly changed by mechanical or neuroendocrine stimuli with or without a concomitant change in their synthetic rate. Sudden increases in central venous pressure as achieved by head-out water immersion leading to stretch of the atrial wall or simply stretch of atrial muscle *in vitro* produce a significant increase in the rate of release of both ANF and BNP. This "stretch-secretion coupling" phenomenon has an unknown molecular basis. In the short term, the burst of natriuretic peptide following acute atrial wall stretch is dependent on stored hormone and does not involve a detectable increase in gene transcription as judged by steady-state mRNA levels. Longer periods of stretch are accompanied by an increase in ANF mRNA levels. *in vivo* this situation is accompanied by changes in other parameters such as overall hemodynamics and neurohumoral balance, so that it is unlikely that pure muscle stretch may be responsible for long-term increases in natriuretic peptide synthesis and stimulated release.

Under chronic overload conditions of the heart, as observed, for example, in long-standing human hypertension, ANF and BNP synthesis is increased in both atrial and ventricular cardiocytes. The increase in natriuretic peptide transcripts in the ventricle usu-

ally correlates with the degree of hypertrophy. The response of the ventricles to overload is characterized by the fact that BNP is recruited more strongly from the ventricles and would thus appear to be more of a ventricular hormone, at least in conditions of disease. In human subjects with congestive heart failure, ventricular BNP levels are at almost double the atrial levels. The activation of natriuretic peptide genes both *in vivo* and *in vitro* is seen in association with the activation of the early response protooncogenes *c-myc, c-fos, c-jun, junB,* and *nur77* and the growth response gene *Egr-1*. These genes code for DNA-binding proteins that could be involved in activating other genes, such as those for contractile protein or for natriuretic peptide. [*See* Hypertension.]

Chronic overload conditions such as chronic congestive heart failure are characterized by profound changes in the neurohumoral balance. Changes in this balance, along with hemodynamic simuli, likely contribute to the observed changes in natriuretic peptide gene expression and release. Indeed, experimental evidence has shown that a number of hormones and neurotransmitters can significantly affect natriuretic peptide synthesis or release or both. These include ET-1, norepinephrine, epinephrine, acetylcholine, gluco- and mineralocorticoids, arginine vasopressin, prostaglandins, angiotensin II, transforming growth factor $\beta$1, acidic or basic fibroblast growth factor, $\alpha$-thrombin, hydroxyvitamin D3 and retinoids, and thyroid hormone, among others.

It is of interest to point out that *in vitro*, acute atrial muscle stretch results in increased expression of the early response genes *c-fos, Egr-1,* and *c-myc,* whereas ET-1 strongly induces *Egr-1* expression. These specific changes in early response gene expression suggests the existence of different pathways leading to the stimulation of atrial gene expression by mechanical and neuroendocrine stimuli.

## VI. CONTRIBUTION OF THE ENDOCRINE HEART TO CARDIOVASCULAR HOMEOSTASIS

The importance of the function of the endocrine heart may be gleaned from experiments showing that blockade of guanylyl cyclase-coupled natriuretic peptide receptors results in impairment of cardiorenal homeostasis. This is particularly evident in departures from the normal water and electrolyte balance as seen in mineralocorticoid escape or congestive heart failure.

Also, transgenic mice homozygous for a disruption of the gene encoding for the "A" natriuretic peptide receptor can develop salt-insensitive high blood pressure.

Overall, the endocrine response of the heart to changes in pressure or volume load can result in three types of responses. First, the acute response to stretch that may occur, for example, after acute volume expansion, is based on stretch-secretion coupling resulting in enhanced secretion of ANF stored in the atria sufficient to increase plasma levels of the hormone for a relatively short period of time. BNP, which undoubtedly is released in increased amounts from the granules together with ANF, does not do so in sufficient amounts to significantly increase its plasma levels. The release after an acute challenge is made at the expense of a depletable pool, and synthesis of the natriuretic peptide is not affected. Second, in a subacute type of stimulation of natriuretic peptide production as seen during the mineralocorticoid escape phenomenon, ANF and BNP secretion and gene transcription are secondary to volume overload. In this situation, plasma ANF but not plasma BNP is significantly elevated. Neither ANF nor BNP gene expression is stimulated in the ventricles. Third, with chronic stimulation, as seen in the DOCA-salt model of hypertension, the cardiac fetal gene program is activated and stimulation of ANF and BNP takes place in both atria and ventricles.

In summary, the available evidence indicates that the endocrine heart responds to different hemodynamic challenges in either acute or chronic conditions, with specific changes in translation, posttranslational processing, storage, and release of ANF and BNP. Human clinical applications of these findings have been developed. The measurement in plasma of the C- or N-terminal portions of ANF or BNP appears to be a useful indicator for therapy, and has been used to help in the diagnosis of symptomless left ventricular dysfunction, in the prognosis of mortality in cardiac failure, and after acute myocardial infarction.

The human therapeutic use of ANF or BNP is being pursued vigorously, including applications in chronic congestive heart failure and intrinsic acute renal insufficiency.

## BIBLIOGRAPHY

de Bold, A. J., Bruneau, B. G., and Kuroski-de Bold, M. L. (1996). Mechanical and neuroendocrine regulation of the endocrine heart. *Cardiovasc. Res.* **31**, 7–18.

Kishimoto, I., Dubois, S. K., and Garbers, D. L. (1996). The heart communicates with the kidney exclusively through the guanylyl cyclase-A receptor: Acute handling of sodium and water in response to volume expansion. *Proc. Natl. Acad. Sci. USA* **93**, 6215–6219.

Ogawa, T., Linz, W., Stevenson, M., Bruneau, B. G., Kuroski-de Bold, M. L., Chen, J. H., Eid, H., Schölkens, B. A., and de Bold, A. J. (1996). Evidence for load-dependant and load-independant determinants of cardiac natriuretic peptide production. *Circulation* **93**, 2059–2067.

Yokota, N., Bruneau, B. G., Kuroski-de Bold, M. L., and de Bold, A. J. (1994). Atrial natriuretic factor significantly contributes to the mineralocorticoid escape phenomenon. Evidence for a guanylate cyclase-mediated pathway. *J. Clin. Invest.* **94**, 1938–1946.

# Attention

MICHAEL I. POSNER
KEVIN A. BRIAND
*University of Oregon*

## GLOSSARY

**Computerized tomography** Imaging method in which X-rays are used by a computer to construct a map of measured brain densities

**Magnetic resonance imaging** Imaging method based on changes in the magnetic resonance of atoms, which is then used by a computer to construct a map of the brain structure

**Orienting reflex** Constellation of autonomic and central nervous system changes first proposed by Pavlov as an unconditioned response to a novel sensory event

**Parallel organization** Information flow that occurs simultaneously at several nodes of a distributed anatomical or functional network

**Positron emission tomography** Imaging method in which the decay products of radioactively labeled compounds are counted by an array of external detectors and used to contruct an image of blood flow, metabolism, or other brain process

**Serial organization** Information flow that must access one node of a distributed anatomical or functional network prior to activating another node

THE PROBLEM OF SELECTIVE ATTENTION IS ONE OF the oldest in psychology. William James wrote at the turn of the century, "Everyone knows what attention is. It is the taking possession by the mind in clear and vivid form of one out of what seem several simultaneous objects or trains of thought." This article deals with two aspects of attention. The first is the selection of information for conscious processing and action. The second is the maintenance of the alert state required for attentive processing.

The dominance of behavioral psychology postponed research into the internal mechanisms of selective attention in the first half of this century. The finding that the integrity of the brain stem reticular formation was a necessity to maintain the alert state provided some anatomical reality to the study of an aspect of attention. The quest for information-processing mechanisms to support the more selective aspect of attention began following World War II with studies of selective listening. A filter was proposed which was limited for information (in the formal sense of information theory) and was located between highly parallel sensory systems and a limited-capacity perceptual system.

Selective listening experiments supported a view of attention that suggested early selection of a relevant message, with nonselective information being lost to conscious processing. However, on some occasions it was clear that unattended information was processed to a high level since evidence suggested that an important message on the unattended channel might interfere with the selected channel.

In the 1970s psychologists began to distinguish between automatic and controlled processes. It was found that words could activate other words similar in meaning (their semantic associates), even when the person had no awareness of the words' presence. These studies indicated that the parallel organization found for sensory information extended to semantic processing. Thus, selecting a word meaning for active attention appeared to suppress the availability of other word meanings. Attention was viewed less as

ENCYCLOPEDIA OF HUMAN BIOLOGY, Second Edition, VOLUME I.   Copyright © 1997 by Academic Press.   All rights of reproduction in any form reserved.

an early sensory bottleneck and more as a system for providing priority for motor acts, consciousness, and memory.

Another approach to problems of selectivity arose in work on the orienting reflex. The use of slow autonomic systems (e.g., skin conductance as measures of orienting) made it difficult to analyze the cognitive components and neural systems underlying orienting. Since the mid-1970s there has been a steady advancement in our understanding of the neural systems related to visual orienting from studies using single-cell recordings in alert monkeys. This work showed a relatively restricted number of areas in which the firing rates of neurons were enhanced selectively when monkeys were trained to attend to a location. At the level of the superior colliculus (i.e., the midbrain), selective enhancement could only be obtained when eye movement was involved, but in the posterior parietal lobe of the cerebral cortex, selective enhancement occurred even when the animal maintained fixation. An area of the thalamus, the lateral pulvinar, was similar to the parietal lobe in containing cells with the property of selective enhancement.

Until recently, there has been a separation between human information processing and neuroscience approaches to attention using nonhuman animals. The former tended to describe attention either in terms of a bottleneck which prevented limited-capacity central systems from overload or as a resource that could be allocated to various processing systems in a way analogous to the use of the term in economics. On the other hand, neuroscience views emphasized several separate neural mechanisms that might be involved in orienting and maintaining alertness. Currently, there is an attempt to integrate these two within a cognitive neuroscience of attention. For example, studies of visual search have incorporated a modern neuroscience view of a multichannel visual system with separate mechanisms for dealing with color, form, and motion, with the cognitive idea of a separate visual attention system needed for integrating information from these channels when the target requires it. The following section places emphasis on this integrated viewpoint.

## I. CURRENT STATE

### A. Methods

An impressive aspect of current developments in this field is the convergence of evidence from various methods of study. These include performance studies using reaction time, recordings from scalp electrodes, and lesions in humans and animals, as well as various methods for imaging and recording from restricted brain areas, including individual cells.

Current progress in the anatomy of the attention system rests most heavily on two important methodological developments. First, the use of microelectrodes with alert animals allowed evidence for the increased activity of cell populations with attention. Second, anatomical (e.g., computerized tomography or magnetic resonance imaging) and physiological (e.g., positron emission tomography) methods of studying parts of the brain allowed more meaningful investigations of localization of cognitive function in normal people. The future should see a combined use of localizing methods (e.g., positron emission tomography) with noninvasive methods of tracing the time course of brain activity in the human subject based on scalp electrical and magnetic recordings. This combination should provide a convenient way to trace the rapid time-dynamic changes that occur in the course of human information processing. [*See* Magnetic Resonance Imaging.]

### B. Principles

Three fundamental working hypotheses characterize the current state of efforts to develop a combined cognitive neuroscience of attention. First, an attentional system of the brain is anatomically separate from various data-processing systems that can be activated passively by visual and auditory input. Second, attention is carried out by a network of anatomical areas. It is neither the property of a single brain area nor is it a collective function of the brain working as a whole. Third, the brain areas involved in attention do not carry out the same function, but specific computations are assigned to different areas.

It is not possible to specify the complete attentional system of the brain, but something is known about the areas that carry on three major attentional functions: orienting to sensory stimuli, particularly locations in visual space; detecting target events, whether sensory or from memory; and maintaining the alert state.

### C. Orienting

Usually visual orienting is defined in terms of the foveation[1] of a stimulus. Foveation improves the effi-

---

[1]Foveation is a movement of the eyes that brings information onto the fovea which is the highest acuity part of the retina.

ciency of processing targets in terms of acuity, but it is also possible to change the priority, given a stimulus, by attending to its location covertly, without any change in eye or head position. When a person or a monkey is cued to attend to a location, events that occur at that location are responded to more rapidly, give rise to enhanced scalp electrical activity, and can be reported at a lower threshold. This improvement in efficiency is found within the first 150 msec after an event occurs at the location to which the subject is to attend. Similarly, if people are asked to move their eyes to a target, an improvement in efficiency at the target location begins well before the eyes move. This covert shift of attention appears to function as a way of guiding the eyes to appropriate areas of the visual field. Brain injury to any of the three areas that have been found to show selective enhancement of neuronal firing rates causes a reduction in this ability to shift attention covertly. However, each area seems to produce a somewhat different deficit. Damage to the posterior parietal lobe has its greatest effect on the ability to disengage from attentional focus to a target located in a direction opposite to the side of the lesion.

The effects of the parietal lobes of the two cerebral hemispheres are not identical. Damage to the right parietal lobe has a greater overall effect than does damage to the left parietal lobe. Disputes arise about the reasons for the asymmetries. One account supposes that the right parietal lobe is dominant for spatial attention and controls attention to both sides of space whereas the left parietal lobe plays a subsidiary role. According to another account, the right parietal lobe is influenced more by the global aspects of figure whereas the left parietal lobe is more influenced by local aspect. A third view argues that the ability to disengage is handled symmetrically by each hemisphere, but the maintenance of the alert state is asymmetrical. Testing these theories requires comparisons between separate populations of patients with left and right lesions, and thus it is difficult to arrive at a clear choice. Of course, more than one theory could be correct.

Lesions of the superior colliculus and the surrounding midbrain areas also affect the ability to shift attention. However, in this case the shift is slowed whether or not attention is first engaged elsewhere. This finding suggests that a computation involved in moving to the target is impaired. In addition, patients with damage in this midbrain area also return to former target locations as readily as to fresh locations that have never been attended to. Normal subjects

and patients with parietal and other cortical lesions show a reduced probability of returning attention to an already examined location.

Patients with lesions of the thalamus and monkeys with chemical lesions of one thalamic nucleus (the pulvinar) also show difficulty in covert orienting. This difficulty appears to be in selective attention to a target on the side opposite the lesion, so as to avoid responding in error to distracting events that occur at other locations. A study of patients with unilateral lesions of this thalamic area showed a slowing of responses to a cued target on the side opposite the lesion, even when the subject had plenty of time to orient there. This contrasted with the results found with parietal and midbrain lesions, in which responses are nearly normal on both sides once attention has been cued to the location. Alert monkeys with chemical lesions of this area made faster than normal responses when cued to the side opposite the lesion and given a target on the side of the lesion, as though contralateral cues were ineffective in engaging their attention. They were also slower than normal when given a target on the side opposite the lesion, irrespective of the side of the cue. Data from normal human subjects, required to filter out irrelevant visual stimuli, showed selective metabolic increases in the pulvinar. [*See* Thalamus.]

These findings make two important points. First, they confirm the idea of anatomical areas carrying out individual cognitive operations. Second, they suggest a particular hypothesis of the circuitry involved in covert attention shifts. The parietal lobe first disengages attention from its present focus, then the midbrain is active to move the index of attention to the area of the target, and the pulvinar is involved in restricting input to the indexed area.

While the circuitry just described remains speculative, it is clear that patients with parietal lesions have difficulties in pattern recognition, implying that somehow the parietal lobe damage comes to affect the processing of patterns. The dorsal pathway extending from the primary visual cortex to the parietal lobe appears to mediate selective visual attention. Considerable anatomical data suggest that a second ventral cortical pathway, leading from the striate cortex to the infratemporal cortex, is involved in processing color and form during pattern recognition. There is evidence from single-cell recordings in alert monkeys that visual spatial attention affects this pattern recognition system. Attention to a visual location affects the processing of stimuli within the receptive fields of neurons of the V4 area. This area lies along the ventral pattern recognition pathway known to be active when monkeys are processing color and form information. While it is not known how attention gains access to V4, one likely candidate is via the pulvinar, which has close connections to both the parietal system and V4. [*See* Cortex.]

Cognitive studies of normal humans have been important in exploring how attention influences pattern recognition processes. A major distinction is between the processing of simple features (e.g., line orientation and color) and that of items defined by a combination of features (e.g., a red vertical line). Simple features appear to be processed in parallel, i.e., the search time is not affected by the number of nontarget items in the display. When targets are defined by a combination of attributes (e.g., the red vertical line) that are located within displays of highly similar nontargets (e.g., red horizontal lines and green vertical lines), the search appears to be a serial process and takes longer as the number of distractors increases. There is evidence that the visual orienting system just described is also involved in visual search.

One theory of how attention affects pattern recognition it that it works to combine separate features into unitary percepts. According to this view, simple features are not combined until one orients attention to them. It is for this reason that attention is necessary to search for a conjunction of features. When a target is made of features that are also present in distractors, illusory conjunctions become apparent due to an improper conjunction of elements from different locations. Such illusory combinations should be avoided so that one attends selectively to each item present in the array.

There is a second aspect to the visual orienting attention system. Just as we can attend to a spatial location, we can also attend to a small or large object. If one views a large letter composed of small ones, it is possible to attend either to the overall form or to its constituents. The size of feature selected is a general property of visual system cells related to the type of sine wave to which they will be most sensitive (spatial frequency). When attending to local objects, people are relatively good at detecting high-frequency probes, but when attending to global objects, they do relatively better for low-frequency stimuli.

There is evidence from both normal people and patients that the right hemisphere is biased toward global processing whereas the left is biased toward local processing. When given a large letter made of small letters, patients with right parietal lesions copy the local letter, but miss the global organization, whereas patients with left hemisphere lesions copy the global orientation while missing the local constituents.

We have concentrated on visual orienting since that has been the area for which integration between cognitive and neuroscience studies has been most advanced. However, the earliest studies of selective attention used the ears or both the eyes and the ears as channels for the presentation of sensory information. There is good evidence that one can bias processing toward one ear or one particular frequency. When this is done, the electrical signal from the selected channel is amplified with respect to information on unselected channels. When required to do so, subjects do quite well in attending to several channels at once. However, an exception to this generally good parallel processing arises when targets occur on more than one channel. The interference between targets can happen between as well as within sensory channels. The reasons for this form of sensory interference are discussed in the next section.

## D. Cognitive Control

A persistent issue in cognitive psychology is whether one should think of an executive exercising voluntary control. In one sense this raises the issue of a homunculus and the possibility of an infinite regress. Despite this problem, there appears to be little doubt that there is some central control over our behavior and thought patterns. In particular, the study of human expertise in problem solving and other behavior has always considered a central executive system that can describe at least a significant portion of the mental operations involved in problem solving. [*See* Problem Solving.]

However, there is evidence of a limited capacity at some level of the system relatively close, but not identical, to motor actions. If a person must make two rapid responses, there is interference between them, even when the stimuli are in different modalities, and the responses differ so that one is vocal and the other is manual. In fact, this kind of central interference does not depend on any immediate response requirement. As described in the previous section, evidence suggests that a person can monitor several channels for rare targets with relatively little interference between the channels. However, if two targets occur at once, even if the subject's only task is to note whether there was one or two targets and he need not make any immediate response, there is a great deal of interference between the targets. This finding underlies the idea that a limited-capacity system is involved whenever a signal (sensory or memorial) is to be consciously noted. There is also a good deal of evidence that the storage of recently presented information, the generation of ideas from long-term memory, and other such acts interfere with the detection of new signals. In this sense a unified executive system appears to be involved in many forms of cognitive and motor actions.

Similar issues about attention arise in the study of the neural mechanisms underlying attention. Should attention be thought of as a unified system of many anatomical areas (e.g., as found for the visual system) or are there many independent attentional systems? Close anatomical and behavioral ties clearly exist between the posterior areas mediating visual spatial attention and areas involved in the control of eye movements and manual movements into the surrounding space. Although both cellular and performance studies argue that attention enhances the visual input in several posterior areas, it seems likely that much of its effect lies in differential access to more motor regions of the brain. Indeed, studies of patients with parietal lesions and of normal people have examined whether a shift of covert attention can be performed when the person is engaged in an attention-demanding language task. Apparently there is specific interference between the two tasks, as though they involved a single system.

Some evidence indicates that areas of the frontal lobe are active during both language and spatial tasks. Studies of blood flow and metabolism have shown frontal activation during tasks involving language and spatial imagery. Studies of normal subjects processing individuals words show changes in blood flow for frontal midline areas, including the cingulate gyrus and the supplementary motor, when subjects were required to process the input actively. Moreover, experimental studies show that the degree of blood flow in the anterior cingulate increases regularly as the number of targets to be detected increases. Thus, this area appears to be sensitive to the mental operations of target detection.

The anterior cingulate has an internal organization that shows alternating bands of cells with close connections to the dorsolateral frontal cortex and the posterior parietal lobe. This organization suggests an integrative role because studies have implicated the lateral frontal cortex in semantic processing while, as we have seen, the posterior parietal lobe is important for spatial attention. The anterior cingulate might provide an important connection between widely different aspects of attention (e.g., attention to semantic content and visual location). Unfortunately, both cognitive and anatomical theories of this type of cognitive control remain highly speculative.

## E. Alerting

The earliest anatomy of attention involved maintenance of the alert state. Cognitive psychologists have studied changes in alerting, both by using long boring tasks with low target probability, such as is required by the military when monitoring radar screens for possible enemy planes or missiles, and by the use of warning signals, such as used in foot races to get the runners to prepare to move quickly from the start position. In both of these situations, there is evidence that an increase in alertness improves the speed of target processing, and in some conditions alertness also serves to reduce the accuracy of responding to the targets. The trade-off between improved speed and reduced accuracy with warning signals has led to a view that alerting does not act to improve the buildup of information concerning the nature of a target, but instead acts on the attentional system to enhance the speed of actions taken toward the target.

There has been some improvement in our understanding of the neural systems related to alerting. Patients with lesions of the right frontal area have difficulty in maintaining the alert state. In addition, experimental studies of blood flow in normal people during tasks that demand sustained vigilance show right frontal activation.

The neurotransmitter norepinephrine appears to be involved in maintaining the alert state. This norepinephrine pathway arises in the midbrain, but the right frontal area appears to have a special role in its cortical distribution. Among posterior visual areas in the monkey, norepinephrine pathways are selective for areas involved in visual spatial attention. At least one study shows that, during the maintenance of vigilance, the metabolic activity of the anterior cingulate is reduced over a resting baseline value. These anatomical findings would support the subjective observation that, while waiting for infrequent visual signals, one has to be prepared to orient, but also has to empty one's head of any ideas that might interfere with detection.

## II. APPLICATIONS AND FUTURE DIRECTIONS

Much remains unknown concerning the macroanatomy of attention, particularly the anterior portions of the system. Studies of blood flow and metabolism in normal people should be adequate to provide candidate areas involved in aspects of attention. It will then be possible to test further the general proposal that these constitute a unified system and that constituent computations are localized.

We have a start on understanding the circuitry that underlies the posterior attention system. However, more detailed cellular studies in monkeys are necessary to test these hypotheses and to understand more completely the time course and the control structures involved in covert shifts of attention. Even more fascinating is the possibility that the microstructure of areas involved in attention will be different somehow in organization from those areas carrying out passive data processing. Such differences could give us a clue as to the way in which brain tissue might relate to subjective experience.

Even in our current state of knowledge, ideas about attention have proved useful in integrating aspects of social developmental psychology with psychopathology.

The idea of attention as a network of anatomical areas makes relevant study of both the comparative anatomy of these areas and their development in infancy. In the first few months of life, infants develop nearly adult abilities to orient to external events, but the cognitive control produced by the anterior attention system requires many months or years of development. Studies of orienting and motor control are beginning to lead to an understanding of this developmental process. As more about the maturational processes of brain and transmitter systems is understood, it should be possible to match developing attentional abilities with changing biological mechanisms. The neural mechanisms of attention must support not only common development among infants in their regulatory abilities, but also the obvious differences among infants in their rates and success of attentional control.

There are many disorders that are often supposed to involve attention, including neglect, closed-head injury, schizophrenia, and attention deficit disorder. The specification of attention in terms of anatomy and function might be useful in clarifying the underlying bases for these disorders. The development of theories of deficits might also foster the integration of psychiatric and higher-level neurological disorders, both of which might affect the attentional system of the brain.

## ACKNOWLEDGMENTS

Research for this review was supported in part by Office of Naval Research Contract N00014-89-J-3013 and by National Institute of Mental Health Grant 43361. We appreciate the help of Mary K. Rothbart in writing this review.

## BIBLIOGRAPHY

Broadbent, D. E. (1958). "Perception and Communication." Pergamon, London.

DeRenzi, E. (1982). "Disorders of Space Exploration and Cognition." Wiley, New York.

Goldman-Rakic, P. S. (1988). Topography of cognition: Parallel distributed networks in primate association cortex. *Annu. Rev. Neurosci.* **11**, 137–156.

Hillyard, S. A., and Picton, T. W. (1987). Electrophysiology of cognition. *Handb. Physiol. Sect. 1: Nerv. Syst. [Rev. Ed.]* **5**, 519–554.

Kahneman, D. (1973). "Attention and Effort." Prentice Hall, Englewood Cliffs, NJ.

Mesulam, M.-M. (1981). A cortical network for directed attention and unilateral neglect. *Ann. Neurol.* **10**, 309–325.

Moruzzi, G., and Magoun, H. V. (1949). Brainstem reticular activation of the EEG. *Electroencephalogr. Clin. Neurophysiol.* **1**, 445–473.

Näätänen, R. (1992). "Attention and Brain Function." LEA, Hillsdale, NJ.

Posner, M. I. (1978). "Chronometric Explorations of Mind." Erlbaum, Hillsdale, NJ.

Posner, M., and Dehaene, S. (1994). Attentional networks. *Trends Neurosci.* **17**, 75–79.

Posner, M. I., and Petersen, S. E. (1990). The attention system of the human brain. *Annu. Rev. Neurosci.* **13**, 25–42.

Treisman, A., and Schmidt, H. (1982). Illusory conjunctions in the perception of objects. *Cognit. Psychol.* **14**, 107–141.

Wurtz, R. H., Goldberg, M. E., and Robinson, D. L. (1980). Behavioral modulation of visual responses in the monkey: Stimulus selection for attention and movement. *Prog. Psychobiol. Physiol. Psychol.* **9**, 43–83.

# Attitudes as Determinants of Food Consumption

HELY TUORILA

*University of Helsinki*

## GLOSSARY

**Affection** Emotional response expressed nonverbally or verbally (e.g., reports of liking)

**Attitude** Disposition to respond in a consistently favorable or unfavorable manner with respect to a given object

**Cognition** Mental process involved in achieving awareness or knowledge of an object (e.g., beliefs about an attitude object)

**Conation** Behavioral tendency or actual behavior

ATTITUDES CAN BE DEFINED AS CONSTRUCTS INcluding affective and cognitive components. Attitudes as determinants of food consumption have been the focal point of interest in two traditions of research: Research on food acceptance has concentrated on affective responses to food, whereas research on nutrition education has focused on attitude change and has been more cognitively oriented. Both approaches are discussed here. Examples of empirical models describing the role of attitudes in food consumption are presented. Since the definition and measurement techniques are in a central position when the significance of attitudes is evaluated, particular attention is paid to methodology.

## I. FACTORS AFFECTING FOOD CONSUMPTION

In most Western societies the scarcity of food has changed into an excessive supply within a relatively short period. Marketing imposes a pressure to consume, while the scientific evidence of the contribution of diet to health often causes opposite pressures regarding the amount and quality of diet. In a contradictory situation of this kind, individual decisions become important. Attitudes are assumed to be predictors of actual behavior and thus may play an important role in an individual's own decision-making. Because of their relative stability, attitudes have been regarded as potential indicators of future trends in consumption.

Several models for the description of physiological, psychological, and social determinants of food intake have been developed. A common feature among these models is that these determinants have been divided into three subgroups: related to food, to the person, and to the social, cultural, and economic context. Possible pathways among the variables are shown in Fig. 1. Attitudes have been categorized partially as responses to perceptual properties of food and partially as results of psychological and environmental inputs. Research on attitudes belongs mainly to the domain of social psychology.

## II. ATTITUDES AND THEIR MEASUREMENT

### A. Concept

Attitudes have been defined in a vast number of ways. Some common characteristics recur in the majority of

ENCYCLOPEDIA OF HUMAN BIOLOGY, Second Edition, VOLUME 1.   Copyright © 1997 by Academic Press.   All rights of reproduction in any form reserved.

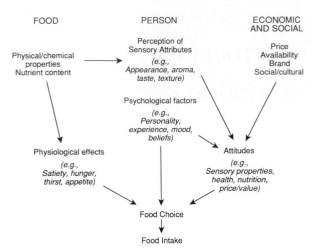

FOOD     PERSON     ECONOMIC AND SOCIAL

**FIGURE I** Factors affecting food consumption. [From R. Shepherd, Factors influencing food preferences and choice. *In* "Handbook of the Psychophysiology of Human Eating" (R. Shepherd, ed.). Wiley, New York. Copyright © 1989 John Wiley & Sons, Ltd. Reprinted by permission of John Wiley & Sons, Ltd.]

definitions: (1) attitudes are considered to be affective dispositions toward an object, (2) they are often considered products of a learning process, and (3) they are believed to be relatively consistent within an individual.

The affective dimension is the core dimension. However, attitudes are closely associated with cognitions. Theoretically, this association has been described using two different approaches. The first is to consider attitudes as tripartite constructs with affective, cognitive, and conative dimensions. These three dimensions should be intercorrelated to act as indicators of a single attitude.

In the second approach the interpretation of a general attitude (expressed in affective terms) is derived from units of salient beliefs about an object, weighted by the importance of each belief. Thus, cognitive inputs (i.e., beliefs) would contribute to a general attitude only if beliefs contain relevant information. Furthermore, in this approach a general attitude is considered an antecedent of a behavioral tendency, which, in turn, is considered the best predictor of actual behavior. This approach is most explicit in the Fishbein–Ajzen model of reasoned action (see Section V).

Thus, the fundamental difference between the two approaches is that the first treats affection, cognition, and conation as components of a single structure, whereas the second presents a hierarchical order among these variables.

For example, a person might like cream in his or her coffee (affective component), might consider it fattening (cognitive component), and might or might not intend to add it to the coffee (conative component). According to the first approach, a person might or might not actually add cream to his or her coffee with all of these factors in mind. According to the second approach, the attribute of being "fattening" and other relevant beliefs act as moderators of a general attitude toward adding cream to coffee. If the attribute "fattening" is of crucial importance for this person, it controls the general attitude toward adding cream to coffee; if a person does not categorize foods in terms of their fattening effect, the belief remains insignificant for the overall attitude. The general attitude, then, guides the final act of adding or not adding cream.

Attitudes can be broad or narrow in scope. We cannot assume that there is always a certain attitude corresponding to a certain overt act. Instead, it is possible that several attitudes lead to the same behavior, or one attitude can underlie several behaviors.

In empirical research on food and nutrition, it is often appropriate and generally agreed to exclude behavioral tendencies from the concept of attitude. The behavioral component is then used as the dependent variable in studies in which the food consumption is explained or predicted by affectively and/or cognitively loaded attitudes. Only for infants and small children are consumption rates, facial responses, and other behavioral tendencies used as indicators of affection.

## B. Measurement

### 1. General

Behavioral constructs (e.g., attitudes) are latent, not directly observable. Therefore, various verbal and nonverbal responses meant to capture the essential features of the theoretical construct are used as indicators of an attitude (Fig. 2). These operational definitions might reflect different aspects of an attitude. Thus, any interpretation of results in attitude research requires a full description of how the instrument has been constructed. For example, responses to verbal statements can be evaluated only when the precise verbal expression is reported.

The affective response could be measured by the registration of reactions of the autonomic nervous system (nonverbal) or by ratings of liking or disgust (verbal). Verbal responses are used in the majority of measurements. The measurement of cognitions, be-

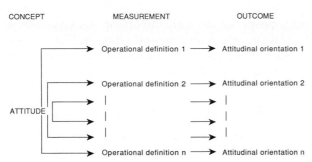

CONCEPT      MEASUREMENT      OUTCOME

**FIGURE 2** The measurement of attitudes. An attitude can be measured (i.e., operationalized) in ways which emphasize various aspects of the concept.

liefs, or information concerning the object is almost always verbal.

## 2. Hedonic Responses

In psychophysical and food acceptance research, the affective reaction is called a hedonic response. This response is determined using various rating scales based on appropriate verbal labels (e.g., "dislike"–"like" or "unpleasant"–"pleasant") or as pairwise comparisons for the relative liking (Table I). The stimulus is either a physical sample of the product in question or the stimulus name. It is important to realize that actual products and their verbal labels can elicit completely different uncorrelated affective responses. Responses to stimulus names generally correlate better with other attitudinal responses than those to samples of food. [*See* Food Acceptance: Sensory, Physiological, and Social Influences.]

## 3. Summated Attitude Scales

The verbal measurement is usually conducted by presenting subjects with a set of pretested statements concerning the object to be rated on a structured cate-gory rating scale. The scales can range from "disagree" to "agree" (the Likert scale) or from "unlikely" to "likely" (the Fishbein–Ajzen model), with five or seven verbally based categories. Unstructured verbal approaches, in which subjects are interviewed for their opinions and beliefs and the interviews are then analyzed for their contents, have been applied only minimally as the final step of attitude measurement. However, the unstructured phase is a necessary beginning for the elicitation of attitude statements and beliefs that are used in the structured form described earlier.

A large set of items (e.g., 10–100 statements on the "disagree"–"agree" scale) is reduced by means of correlational and/or multivariate statistical techniques to fewer or only one summated score reflecting one or more attitudinal orientations. These summated indices are typically mixtures of affective and cognitive dimensions. It is essential to prove that individual items of a summated scale reflect the same attitudinal orientation. Various indices (e.g., Cronbach's $\alpha$, which reflects interitem correlations) are used as a measure of internal consistency, or reliability. Typical and acceptable $\alpha$ value range from 0.6 to 0.9 (or higher).

## III. AFFECTIVE RESPONSES TO FOOD

### A. Liking and Aversions

Sensory stimuli, particularly those related to smell and taste, readily arouse affective responses. The intensity of an affective response can vary from absolute dislike to extreme liking. The degrees of liking vary among foods, individuals, and situations.

Although varying degrees of liking characterize most interactions between a person and food, it is usual for a person to also have one or more complete

## TABLE I
Techniques for Measuring Attitudes to Food and/or Nutrition

| Object of measurement | Type of scale | Examples of statements used in measurement |
|---|---|---|
| Liking for food | Hedonic scale | Dislike extremely–like extremely, unpleasant–pleasant |
| Attitude to sugar | Likert scale | "Use of sugar is not as harmful as health educators say" (disagree–agree)<br>"Every effort should be made toward restriction of the use of sugar" (disagree–agree) |
| Beliefs concerning the addition of table salt to food | Probability rating | "Food tastes better if I add salt" (unlikely–likely);<br>"My adding salt to food increases the risk of high blood pressure" (unlikely–likely) |

dislikes, called aversions. Surveys and clinical trials have shown that aversions are commonly caused by gastrointestinal upset associated with the particular food. Once manifested, such aversions can be extremely persistent. The occurrence of food aversions in individuals often increases in connection with pregnancy, chemotherapeutic treatment, and other conditions in which gastrointestinal upset is likely to occur.

## B. Genetic Dispositions

Although genetically determined factors are not learned and thus, according to some definitions, are not considered attitudes, such factors should be taken into account in the discussion of liking for food. However, solid scientific evidence on genetically determined affective responses is only available for the modality of taste. Based on facial reflexes, sucking patterns, and consumption rates, newborn babies exhibit an innate preference for sweetness. The facial reflexes also indicate their dislike for sour- and bitter-tasting foods.

On the other hand, infants do not initially show any affective response to salty taste, but the preference for saltiness appears to develop gradually during the first months of life. By the age of 2–3 years, children display a like for salty taste in foods that are typically salty in their own culture.

## C. Exposure

Human subjects are assumed to have an inherent predisposition to react ambivalently to novel foods. A negative reaction, called neophobia, is believed to act as a safeguard against overly "daring" experiments with new and potentially dangerous foods. On the contrary, the positive reaction, neophilia, is associated with a will to try potential new foods. This is biologically rationalized by the requirement for flexibility in food selection since chances to survive in conditions under which the food supply is scarce or unpredictable are thus enhanced.

Mere exposure to any object, including food, induces positive affections. Thus, exposure also serves to counteract neophobia. Members of a given food culture exhibit, at least to some extent, similar preferences for foods. In a child-rearing situation, other mechanisms (e.g., as the persuasion used with children to make them accept the available supply of food) contribute to the exposure effects. Social learning (i.e., model learning from parents, peers, and other relevant groups) and social pressure exerted by parents and

other people involved are important. Also, associative learning in which an unknown stimulus (i.e., a food or a flavor) is paired with one having a positive connotation (e.g., sweetness) enhances liking.

## D. Shifts in Affective Responses

Hunger versus satiety explains some of the shifts observed in affective responses to food. A hungry subject responds more favorably to sugar solutions than a satiated one. It is unclear to what extent cognitive factors (i.e., the acquired knowledge of the link between sweetness and calories) or learned satiety effects play a role and to what extent the association is purely physiological.

Sensory-specific satiety provides another explanation for transient shifts in affective responses. This concept refers to decreased pleasantness ratings or to decreased consumption rates of foods consumed by a person prior to rating the pleasantness.

Few systematic experiments have aimed to determine the conditions under which affective responses to individual ingredients, particularly sodium chloride, can be modified and how much time is needed for a change. Preference for lower saltiness in foods seems to develop after sodium restriction of approximately 3 months.

## IV. COGNITIVE INFLUENCES

### A. Background

Moving from affective to cognitive influences on food consumption means switching to a very different field of research. Cognitive influences have mainly been of interest to nutrition educators, whose primary concern has been to change habits of food consumption. The focus has been not so much on the consumption of individual foods, but on the general quality of the diet. In this tradition cognitive changes have been emphasized, whereas the affective component of attitude has been neglected. This view is gradually changing, and nutrition educators are beginning to recognize the complex nature of dietary change.

### B. Information Processing

Cognitive inputs concerning food and nutrition derive from any information available concerning the origin, composition, perceived sensory qualities, nutritional consequences, health impact, and economic considerations. The information is spread by the cultural tradition, mass media, the school system, health professionals, product labels, and other more or less formal sources. The information is intensified in specific campaigns or when directed toward specific population groups. The content, style, and emphasis of information, particularly that distributed by the mass media, are entirely culture dependent. Therefore, cognitions related to food also vary according to culture.

The way a communication is received depends on the source (e.g., attractiveness and credibility), the message (e.g., content and style), the channel (e.g., pictures and written text), the receiver (e.g., motivation and abilities), and the nature of the target behavior (e.g., immediate or long term). The way the information is processed can be divided into several stages. First, a person needs to be exposed to information, to attend to it, and to become aroused (i.e., interested) in order to comprehend it. Only a small proportion of the information available ever reaches the stage of creating arousal. Once comprehended and agreed with, parts of the information might be forgotten and others modified. To affect behavior, the information must be recalled at the point of decision-making and used as a basis for actual behavior. Finally, consolidation, accomplished by fitting behavior and beliefs together, helps the individual continue acting in the new way.

## C. Knowledge, Attitudes, and Behavior

### 1. Traditional Approach

The triangle of knowledge, attitudes, and behavior has long been one of the cornerstones of nutrition education. The notion that nutritionally relevant information offered to consumers will change their perceptions and consequently their attitudes, finally resulting in nutritionally meaningful changes in food consumption, has been the impetus for many education campaigns.

The results of studies following this paradigm have often been disappointing, as correlation coefficients among knowledge, attitudes, and behavior have been low. However, an analysis integrating several hundreds of published findings showed that nutrition education positively affects knowledge, attitudes, and nutrition behavior. The analysis also showed that behavior correlates positively, although not highly, with attitudes and knowledge, but that knowledge and attitudes are not correlated. Thus, the paradigm is viable to some extent, but suffers from low predictive power.

### 2. Reasons for Limited Success

There are several possible explanations for this limited success. First, food selection, like any other human behavior, is complex. Many factors control the feeding behavior of an individual. The motivation within an individual varies with time, alternatives, and social situations. Many components of the complex motivational matrix (e.g., pleasure-seeking or feelings of hunger) can counteract a predisposition toward "good nutrition." Therefore, it might not be realistic to assume that a single factor could act as the only or dominating determinant.

Second, the process whereby nutritional information is cognitively registered provides us with opportunities for interpreting the low correlations. The fact that people become exposed to information does not guarantee any progress beyond this stage. Furthermore, the nutritional information might arouse interest and might be comprehended, but later forgotten or modified to fit better into the conceptual framework already existing in a person's mind. Even if the information becomes integrated into the belief structures, its retrieval in a real decision-making situation might fail. In this light the simple input–output schema often assumed by educators is inadequate.

Third, the poor correlations among the three variables can be due to poor or inadequate methods and measurement techniques. The diverse techniques used to measure attitudes might not always quantify the phenomena they are meant to measure. The contents of instruments used to measure "a favorable attitude to nutrition" vary widely in their composition, number of items, internal consistency, and validity. Even if the different instruments for attitude are identically labeled, they might actually measure different and possibly uncorrelated orientations. A specific problem brought up by some researchers is that attitudes might be measured at a general level, and therefore attempts to relate them to a specific behavior inevitably lead to low correlations.

### 3. Causal Relations

Even if correlations among knowledge, attitudes, and behavior do exist, a comment should be made concerning the causal connections among them. Behavioral change need not result from attitudes that have been modified by information. An alternative would be to change an attitude first, leading to the absorption of knowledge; this would subsequently or simultaneously lead to changed behavior. Another pathway could be from changed behavior to the adoption of new information and appropriate attitudes.

According to the attribution theory, people infer why they behave the way they do, developing attitudes and cognitive structures which justify their behavior. The mechanism is related to the step of consolidation in information processing: Supportive structures promoting the formation of balance between feelings, thoughts, and acts are developed. Consistency theories, which emphasize the coherent intercorrelated systems of values, attitudes, and beliefs, are also in accordance with this view.

Data related to the consumption habits of certain beverages (e.g., milk or soft drinks) show that once a person has a reasonably long history of using a certain option (e.g., a nonfat milk or diet soda), the attributes of the chosen alternative are rated positively, and negative aspects are perceived more leniently than they are by those who have selected another option. This type of rationalization is feasible in the case of food consumption which is repeated several times a day. If every decision concerning food were treated as unique, the selection of food would require a great deal of mental resources. From the point of view of the individual, it is economic to rationalize such repeated behaviors.

## V. EMPIRICAL MODELS

### A. Background

In recent research the role of attitudes as related to other variables controlling human food consumption has received increasing attention. In particular, the relative importance of attitudes and subjective norms (i.e., the pressure a person receives from important reference groups and the motivation to comply with them) has been studied using the Fishbein–Ajzen model of reasoned action as a frame of reference. This model states that a person's behavioral intention can be predicted from attitudes (which derive from salient beliefs concerning the consequences of behavior and their value for a person) on the one hand and from subjective norms on the other. In a regression model for intention, standardized regression coefficients of each component should reflect their relative importance, provided that the two predictors are not intercorrelated. The model represents the hierarchical organization of the components of attitudes, discussed in Section II.

Examples of studies using the Fishbein–Ajzen model and certain other approaches in the prediction

of consumption are presented in Table II. Milk was chosen as an example because several results concerning milk have been published in recent years. Essentially the same features have been observed in predictive equations for other foods (e.g., cheese, meat, ice cream, chocolate, and regular or diet soda). The major observations are discussed in the following sections.

### B. Predictive Power

The models predict up to 52% of the total variation of the dependent variable. Considering the highly variable and generally somewhat unpredictable nature of human behavior, including feeding behavior, the predictions obtained are relatively high. However, the dependent variable is either the reported frequency of consumption or the intention to consume, not the actual consumption. The reported behavior and behavioral intention generally correlate better with attitudes than does the actual behavior; diverse situational factors affect the actual behavior. Therefore, the predictive powers shown in Table II should be taken as comparative rather than absolute values.

### C. Role of Attitudes

In Equations 1–5 in Table II, attitudes are major predictors, whereas in Equation 6 enjoyment of taste and a commitment to using low-fat foods (which was a positive predictor for low-fat milk and a negative predictor for whole milk) are dominant predictors. The most likely reason for the discrepancy lies in the differences between the operational definitions of attitudes. The definitions used in Equations 1–5 are evaluative, whereas the last definition is less direct and more cognitive in nature. The different definitions and the predictive powers offer significant evidence for what was discussed in Section II,B about the critical role of the operational definition.

The powerful predictors of Equation 6, in fact, confirm the results of Equations 1–5. "Enjoyment of taste" is the variable that most likely reflects liking. Also, since there are many attitudinal orientations, commitment could also be regarded as a type of attitude, as it is a favorable disposition toward a general class of products (e.g., low-fat milk).

### D. Subjective Norms

Perceived pressure from other people, the subjective norm, is unimportant for food consumption. This is

TABLE II

Predictive Models for the Intention to Consume and the Reported Consumption of Milk

| No. | Equation[a] | Predictive power (%) | | Operational definition of attitude |
|---|---|---|---|---|
| | | Entire model | Attitude alone | |
| 1 | $I = 0.58$ attitude $+ 0.11$ subjective norm | 24 | 23 | Drinking milk is pleasant and beneficial |
| 2 | $I = 0.55$ attitude $+ 0.09$ subjective norm | 35 | 34 | Drinking milk is pleasant and good for you |
| 3 | $B = 0.59$ attitude $+ 0.15$ subjective norm | 45 | 44 | Drinking milk is pleasant and good for you |
| 4 | $I = 0.62$ attitude $+ 0.09$ subjective norm | 42 | 41 | Liking for milk |
| 5 | $I = 0.51$ attitude $+ 0.28$ beliefs about nutrients $- 0.20$ beliefs about weight gain $+ 0.18$ beliefs about suitability for various purposes | 52 | 42 | Liking for milk |
| 6 | $B = 0.20$ attitude $+ 0.45$ taste enjoyment $+ 0.27$ commitment to using low-fat foods | 47 | — | Milk is important for good nutrition |

[a]In Equation 1, 481 subjects were British; in Equations 2 and 3, 236 subjects were Finns; in Equations 4 and 5, 100 subjects were Americans; and in Equation 6, 693 subjects were Americans. $I$, intention to consume milk; $B$, reported consumption of milk.

a consistent finding in all models explaining food consumption. However, it does not mean that social factors were altogether unimportant. Rather, it is likely that social factors play an important role at certain stages of the formation and changes of attitude. Later, their influence is likely to become integrated into the general concept of "liking" or "attitude," the origin of which becomes irrelevant and is forgotten. Repeated frequent consumption is perhaps not possible before social pressure has been integrated into liking.

## E. Beliefs

In contrast to the Fishbein–Ajzen model, which considers beliefs to be antecedents of general attitudes, Equation 5 demonstrates a 10% improvement in predictive power when beliefs are included in the model. Although this is only one example, it might reflect the growing importance of cognitive influences. It seems possible that, with an abundant supply of information, human food selection will be increasingly guided by a diverse and often conflicting set of cognitive inputs.

## F. Further Predictors

During the past few years, the potential of other predictors of food choice and consumption has also been investigated. These predictors include perceived con-

trol (the degree to which a person feels that she or he is in control of a behavior), habit, moral obligation (e.g., responsibility for the nutrition of other family members), and self-identity (e.g., whether a person considers himself as a "green" consumer). If perceived control is used as a predictor in addition to attitudes and subjective norms, the approach is called the theory of planned behavior. According to some recent data, perceived control is a relevant predictor of food choice. The use of habit as a predictor can be justified by the fact that food consumption is a frequently repeated behavior. In the course of time, such behavior becomes automatic. Although relevant to food-related behavior, the satisfactory measurement of habit remains a challenge.

Overall, it is important to note that all behaviors are not similar in nature: what predicts behavior in nonfood area may be irrelevant for the prediction of food choice.

## BIBLIOGRAPHY

Dawes, R. M., and Smith, T. L. (1985). Attitude and opinion measurement. *In* "Handbook of Social Psychology" (G. Lindzey and E. Aronson, eds.), 3rd Ed., Vol. I, Random House, New York.

Johnson, D. W., and Johnson, R. T. (1985). Nutrition education: A model for effectiveness, a synthesis of research. *J. Nutr. Educ.* **17** (Suppl.).

Lewis, C. J., Sims, L. S., and Shannon, B. (1989). Examination of specific nutrition/health behaviors using a social cognitive model. *J. Am. Diet. Assoc.* **89,** 194.

MacFie, H. J. H., and Thomson, D. M. H. (eds.) (1994). "Measurement of Food Preferences." Blackie, London.

McGuire, W. J. (1985). Attitudes and attitude change. *In* "Handbook of Social Psychology" (G. Lindzey and E. Aronson, eds.), 3rd Ed., Vol. II, Random House, New York.

Rozin, P., and Vollmecke, T. A. (1986). Food likes and dislikes. *Annu. Rev. Nutr.* **6,** 433.

Shepherd, R. (ed.) (1989). "Handbook of the Psychophysiology of Human Eating." Wiley, New York.

Stafleu, A., de Graaf, C., van Staveren, W. A., and Schroots, J. J. F. (1991/92). A review of selected studies assessing social-psychological determinants of fat and cholesterol intake. *Food Qual. Pref.* **3,** 183.

Thomson, D. M. H. (ed.) (1988). "Food Acceptability." Elsevier, London.

Tuorila, H., and Pangborn, R. M. (1988). Prediction of reported consumption of selected fat-containing foods. *Appetite* **11,** 81.

# Attraction and Interpersonal Relationships

ELLEN BERSCHEID
*University of Minnesota*

## GLOSSARY

**Affiliation** Fact of interaction and association with another or the expressed desire to interact with another; may or may not be associated with attraction

**Attitude** Predisposition to respond in favorable or unfavorable ways toward the object of the attitude

**Close relationship** Association in which the partners are highly interdependent on each other, where interdependence is evidenced by the fact that the activities of each partner are strongly influenced by the activities of the other; usually but not always associated with attraction

**Closed-field interaction settings** Interaction settings in which interaction with a specific other is virtually mandated by the characteristics of the setting, especially the social norms governing it

**Hypothetical construct** Internal state of the individual assumed to exist for explanatory and predictive purposes; inferred on the basis of various kinds of observable evidence

**Interpersonal attraction** A hypothetical construct that refers to the positivity of an individual's cognitions about, feelings and emotions precipitated by, and overt behaviors directed toward another person; closely identified with the construct of attitude

**Liking** Mild form of interpersonal attraction, usually evidenced by a positive attitude toward the other, often associated with the belief that the other possesses positive qualities

**Relationship schema** Hypothetical construct that refers to a relatively stable cognitive structure that represents a person's knowledge of their relationship partner, of themselves as they characteristically interact with the partner, and of the interaction scripts (the sequences of the partners' interaction behaviors) typical of the relationship

INTERPERSONAL ATTRACTION THEORY AND research, traditionally conducted largely within the discipline of social psychology and directed toward understanding the antecedents and consequences of an individual's predisposition to respond to another in favorable or unfavorable ways, has helped form the nucleus of a rapidly expanding multidisciplinary field addressed to understanding the dynamics of *close relationships*. The phenomenal growth in the past decade of relationship theory and research is a result of increased recognition that many of the questions about human behavior traditionally addressed by the social, behavioral, and health sciences directly engage questions about the individual's close relationships, especially the consequences of these for the mental and physical health of the individual and for society. Prominent current contributors to the field of interpersonal relationships include the discipline of psychology, especially social, clinical, and developmental psychology, and the disciplines of sociology, marital and family therapy, and communication. Although many relationship phenomena are the subject of current theory and research, those associated with the partners' attraction to each other remain central.

## I. BIOLOGICAL ROOTS OF INTERPERSONAL ATTRACTION

Positive and negative sentiments toward others have their source in the human's most basic and fundamental biological needs because the satisfaction of these needs from birth to death critically depends on the

ENCYCLOPEDIA OF HUMAN BIOLOGY, Second Edition, VOLUME I.   Copyright © 1997 by Academic Press.   All rights of reproduction in any form reserved.

actions of other people. The importance of others for the individual's survival and well-being is reflected in the facts that humans are among the most social creatures in the animal kingdom and that they possess a finely honed propensity for making evaluative (good or bad) judgments of other people. These evaluative judgments are primarily a consequence of the individual's belief that the other will act to further the individual's well-being or, conversely, might harm it. Matters of interpersonal attraction are thus of the highest importance to the welfare and survival of the species because in order to survive, the members of a species need to find food, avoid injury, reproduce, and rear the young. For humans, all of these adaptive behaviors engage issues of interpersonal attraction for the individual and present concerns for human society.

Most human behavior, then, takes place in a social context permeated with the causes and consequences of interpersonal attraction. Although attempts to identify the laws governing attraction began at least with Aristotle and continued through the early philosopher psychologists, the use of systematic observation to uncover these laws is relatively recent. J. L. Moreno's development of sociometry, a self-report questionnaire technique designed to assess an individual's preferences for interacting with others within a specific group, was one of the first empirical investigations of attraction. With the 1934 publication of Moreno's book, "Who Shall Survive?," sociometric measures became an integral part of the emerging discipline of social psychology. Interest in the antecedents, consequences, and correlates of interpersonal attraction surged in the 1960s as social psychologists recognized that attraction is central to virtually all social psychological phenomena. Today, it is increasingly recognized that because almost all human behavior takes place in a social context, no science of behavior will be complete without an understanding of the dynamics of interpersonal relationships.

Until recently, most work on attraction was (1) theoretically derived from general theories of social behavior rather than from specific theories of attraction; (2) directed toward understanding the antecedents and consequences of attraction toward strangers in closed-field settings where the individual has little alternative but to interact with another, as opposed to open-field settings which permit personal choice of interaction partners; and (3) thus conducted primarily within laboratory settings with college students and outside the context of ongoing relationships in naturalistic settings. These efforts resulted in a robust body of theory and investigation, which is de-

scribed in many reviews and usually occupies a standard chapter in contemporary social psychology texts. In the late 1970s, however, attraction theorists and researchers began to shift their attention toward examining the role attraction plays in the formation and the maintenance or dissolution of ongoing relationships. This shift came about partially because it was the obvious next step after examining initial encounters between strangers. However, it also was a result of increased public and governmental demand for information about close relationships, a demand spurred by a dramatic rise in divorce, in serial marriage, and in the growing incidence of nontraditional family forms, including cohabitation, single parent families, and binuclear families made up of the remnants of two or more former nuclear families. As it became increasingly difficult to identify a family through the use of traditional criteria (e.g., blood relationship, the existence of a marital contract), it became evident that generic knowledge about close interpersonal relationships, not directly tied to any one form of relationship, was needed.

## II. CONCEPTUALIZATION AND MEASUREMENT OF ATTRACTION

Because most research on interpersonal attraction was initially conducted within social psychology where the construct of attitude has been central to an understanding of social behavior, it perhaps was inevitable that attraction would be defined initially as an attitude. The conceptualization of attraction as an attitude toward another resulted in attraction often being measured just as attitudes are measured, typically through verbal self-report. Responses to such questions as "How much do you like X?" are usually facilitated by the presentation of simple bipolar scales, whose anchors may range from "like very much" to "dislike very much." Another common means of measuring attraction asks the respondent to evaluate the person's properties (e.g., kind, honest, warm) and then the known positivity of the affective loadings of each property attributed to the other are summed or combined in some way to arrive at an attraction assessment.

The conceptualization and measurement of attraction as an attitude carried with it a number of assumptions that retarded the investigation of several important attraction issues. First, the attitude construct carries the assumption of stability; thus, how attraction toward another waxes or wanes over substantial time frames, especially within enduring close relationships, is only now receiving attention. Second, it has been assumed that the constellation of cognitions and behaviors that comprise an attitude toward another are affectively homogeneous or that the properties the other is believed to possess, the emotions and feelings precipitated by the other, and the favorability of actions taken toward the other are all of the same level of positivity. In fact, ambivalence toward another is common. Ambivalence is manifested in affectively mixed views of the other's qualities, in experiencing a range of positive and negative feelings in association with him or her, and in behaving in both favorable and unfavorable ways toward the other. The need to take a more complex view of affective responses to the other has become evident as researchers have turned to questions of attraction in long-term relationships where ambivalence is often apparent.

Yet another consequence of the identification of attraction with the attitude construct, as well as with the investigation of attraction in brief encounters between strangers, was a focus on the mild forms of attraction, principally liking or disliking. Recent interest in attraction phenomena as they occur within the context of many different types of ongoing close relationships has fueled theory and research on the many forms attraction may take, especially such strong forms as romantic love. [*See* Attitude and Attitude Change.]

## A. Attraction as Emotion and Feeling

The strong forms of attraction are often associated with the experience of such strong emotions as passionate love, hatred, jealousy, contempt, and joy. The renaissance of theory and research in emotion within psychology in the 1970s, along with an interest in attraction as it is manifested in close relationships, has led many attraction theorists and researchers to focus on the antecedents and consequences of emotions as they are experienced in ongoing relationships. Although emotion is also treated as a hypothetical construct that refers to an internal state inferred on the basis of various kinds of observable evidence, unlike the construct of attitude the construct of emotion does not carry the assumption of stability; in fact, the experience of an emotion is recognized to be short-lived. Moreover, emotion is not regarded as a state always reportable by the individual, and thus self-reports of emotion are not taken at face value. Consequently, other kinds of evidence, including discharge of the sympathetic component of the autonomic ner-

vous system, changes in facial musculature and temperature, and other nonvoluntary and nonverbal behaviors as well as more easily observable and volitional actions, are often deemed necessary to the measurement of emotion. In addition, because virtually all contemporary theories of emotion emphasize the adaptive function of emotion in enhancing the individual's welfare and survival, special attention is given to the individual's motivations (e.g., his or her needs, plans, and goals and their facilitation or frustration by others) as they underlie affective phenomena, as well as to the social context in which the emotion is experienced. Furthermore, emotion researchers have documented that affectional space is better defined as two relatively independent dimensions, one positive and one negative, rather than as one bipolar dimension ranging from positive to negative; recent research in psychological neuroscience supports the nonsymmetry of positive and negative emotion at the neurophysiological level as well. Thus, ambivalence, or the near-simultaneous experience of both positive and negative feelings toward another, is more easily recognized by those who approach attraction phenomena from the perspective of emotion theory and research. Finally, although emotion researchers search for the common denominators underlying all the emotions and a single theory that will account for them all, they typically are careful to preserve the distinctions among them.

Some attraction theorists have attempted to extend emotion theory and research to account for the experience of positive and negative emotions in close relationships. Others have sought to determine how such long-lasting feeling states as moods, precipitated by events extraneous to the relationship with another, influence attraction for that other. Currently, many relationship researchers are attempting to identify the typical emotional interaction patterns of couples and the correlates of these patterns. For example, evidence indicates that marital dissolution is associated with an affect reciprocity pattern, primarily negative affect reciprocity where the partner's negative affective response is immediately reciprocated by the other partner, but positive affect reciprocity as well is often observed in distressed couples.

## B. Other Views of Attraction: Affiliation

Attraction has been assessed in a number of other ways over the years. Galton, for example, believed attraction to one's dinner partner could be measured by the lean toward the partner, or the degree of deviation from a 90° plumb line. Pupil dilation, duration of eye gaze, and relaxation of skeletal musculature are among the many means by which investigators have hoped to measure attraction. Each is influenced by factors other than attraction, however, thus proving to be an unreliable indicator when taken by itself.

Attraction was initially assumed to be both necessary and sufficient for affiliation, or for an individual to attempt to interact with another. Thus, Moreno's sociometric technique, where each member of a group indicates with which other members he or she would like to engage in a specific activity, was used for many years as the primary measure of attraction. As a result, the sociometric literature and the interpersonal attraction literature were simply two different labels for the same body of research. Two separate issues were gradually distinguished, however: (1) the role attraction plays in attempts to interact with another, and (2) the identification of determinants of affiliation other than attraction.

Attraction is not a necessary condition for affiliation: many factors other than liking another prompt attempts to affiliate and interact with another. Environmental conditions, for example, often facilitate and sometimes virtually mandate interaction with others. Such closed-field social contexts as work settings and classrooms are examples. In an open-field context, such as a cocktail party, more personal choice may be exercised and, thus, there is closer correspondence between attraction and affiliation.

Theory and research on ingratiation, or attempts to interact with another to augment or maintain power in a relationship by inducing the other to like oneself, describe another set of conditions under which affiliation may occur without attraction. A number of other social motivations may lead to affiliation without attraction. One important factor has been identified as the need for social comparison, or the desire to evaluate the correctness of one's opinions and beliefs through comparison with those held by others. It has been demonstrated, for example, that when people are uncertain about the validity of their opinions and when there is no physical or nonsocial means of validation, others will be actively sought out for social comparison. With whom one will attempt to affiliate under these conditions of uncertainty encompasses a very large body of theory and research.

Much theory and research on affiliation currently focus on the effects of loneliness, a state that is associated with poor physical and psychological health. The development of scales to measure loneliness and of therapies designed to ameliorate it are relatively recent by-products of loneliness research. An interest in social networks, or the number and nature of persons with whom an individual actually interacts and the frequency of their interactions, has been spurred by research investigating the efficacy of social support in reducing the harmful effects of physical and psychological stress. Indeed, much of the current interest in the dynamics of close relationships has been stimulated by increasing recognition that relationships play a central role in human happiness and physical and mental health. Although findings from large-scale epidemiological studies of the determinants and distribution of disease continue to strongly implicate social relationships in mortality and morbidity, the precise causal mechanisms involved are still poorly understood, including the role close relationships appear to play in the functioning of the immunological system. It has become clear that more observational and experimental evidence is needed to explicate the mass of studies that focus on the correlates of perceived social support. Moreover, the fact that not all close relationships are supportive, that some are dysfunctional and induce rather than reduce stress, has only recently begun to be incorporated into theoretical models.

Attraction also is not a sufficient condition for affiliation. Even in open-field settings people do not always attempt to interact with those to whom they are attracted. One of the most important factors governing the attraction–affiliation relationship is the anticipation of acceptance or rejection by the other should an interaction attempt be made. For example, anticipation of social rejection by persons possessing higher social desirability than one's own often leads to a choice to interact with persons of approximately equal social desirability. This and other mechanisms, including the greater availability of partners similar rather than dissimilar to oneself along myriad dimensions, produces de facto "matching" of social partners along social desirability as well as other dimensions.

## III. THEORETICAL APPROACHES TO UNDERSTANDING ATTRACTION

### A. Cognitive Consistency Theories

The cognitive consistency theories, which dominated research in social psychology through the 1960s and early 1970s when experimental research on interpersonal attraction burgeoned, assume that people try to keep their thoughts about themselves and other people and objects in a psychologically consistent relationship and, thus, will strive to achieve consistency and to reduce inconsistency if it occurs. Heider's balance theory, for example, proposes that sentiment (e.g., liking or disliking) for another and feelings of "belongingness" with that other tend toward a harmonious state such that people tend to like those with whom they perceive they are joined in some way and, if they do not, either liking will be induced or an attempt will be made to break the associative bond. Supporting this hypothesis, it has been demonstrated that the mere prospect of interacting with another over a substantial time period induces liking for that other prior to the interaction and prior to information about the other's attributes. On the basis of experimental findings, balance theory was subsequently modified to account for the fact that some situations the theory assumed to be pleasant and balanced are, in fact, regarded by most people as unpleasant. Examination of the nature of these situations indicates that people prefer to like others rather than to dislike them, even when disliking the other formally may present a more cognitively balanced situation.

Newcomb's strain toward symmetry theory adopted Heider's general position but focused on attraction as it occurred within a group. His classic longitudinal study of patterns of attraction in male students residing in a college dormitory examined the hypothesis that cognitive balance in a group will increase with members' length of acquaintance. As reported in his classic book "The Acquaintance Process" (1961), Newcomb initially found little relationship between the men's actual attitude similarity and their liking for one another; over time, however, the pattern of attraction among the men became cognitively balanced in that they came to like those who shared their own attitudes and beliefs and to dislike or become indifferent toward those men who did not. This was the first study to demonstrate the importance of attitude similarity in generating attraction, now one of the most well-documented findings in the attraction area.

Festinger's theory of cognitive dissonance also produced insights into the antecedents of attraction. One of the most important experimental findings derived from dissonance theory was that although it is popularly believed that an individual's liking for another is a response to the other's positive characteristics and behaviors, the direction of causation may run in the opposite direction, i.e., the individual's liking or enmity for another may be a result of the individual's own behavior toward the other, behavior that may be under the control of forces external to the individual's relationship with the other and independent of the other's attributes. It has been demonstrated, for example, that if an individual accidentally harms another, the dissonance generated by such cognitions as "I am a kind person" and "I have unjustly hurt another" may be reduced by derogating the victim of the harmful act, thus coming to believe that the other deserved the injury. In such cases, the insult of derogating the victim's personality, character, or other attributes follows, not precedes, the harmful act and is a consequence of dissonance reduction.

## B. Social Exchange Theories

Reinforcement theories, which have dominated all of psychology, have been most influential in attraction research. The fundamental assumption of these theories is that people like those who reward them, and the more an individual is rewarded by another, the more attraction that will be generated. The reinforcement theories were primarily developed and refined with research on infrahuman animals, mostly rats and pigeons. When they were translated for application to human social behavior, they became known as social exchange theories because they assume that to understand social phenomena in general and attraction in particular, one must understand the principles that underlie the rewards and punishments people exchange when they interact.

One of the first social exchange theories was developed by Homans who adapted Skinner's theory of operant behavior to humans and incorporated some concepts from economics as well. This theory views people as reward-seeking and punishment-avoiding creatures who try to maximize their rewards and minimize their costs in social interaction to obtain the most "profit" they can. An expression of esteem for another is regarded as a "generalized reinforcer," one that people give to those who reward them to help ensure that the other will continue to provide rewards. Of particular interest was Homan's principle of "distributive justice," which stated that people expect that the rewards received by each person in a social interaction should be proportional to his or her costs and that each person's profits should be proportional to his or her investments (e.g., time, money, and other personal assets, including social status and talent) such that the greater the investments, the greater the profit. When this principle is violated, Homans predicted that anger and an attempt to obtain social justice would result.

The equity theories subsequently elaborated the circumstances under which individuals would regard their exchanges with another as inequitable and detailed how attempts to regain equity would be made. Inequity is theorized to exist when an individual's ratio of outcomes to inputs is unequal to the relationship partner's outcome–input ratio. It has been shown that such situations often lead an individual to increase or decrease inputs, depending on whether the ratio is personally advantageous or disadvantageous. Research derived from the equity theories has demonstrated that although people are motivated to obtain high profits in their social interactions, they are also sensitive to social justice, often modifying their behavior when injustice exists. Early equity research focused on social exchange in work settings or other situations in which money was the principal reward exchanged. Equity predictions have now been extended to a wide array of settings including close relationships, where the nature of the rewards exchanged are diverse. Many of these studies have pitted equity against the sheer magnitude of rewards received from another to predict attraction and relationship growth. Although the matter is not settled, and although it is apparent that equity is a consideration, it appears that the amount of reward received from another often predicts relationship status better than equity does. Moreover, equity is not the only exchange rule people use in close relationships; people may distribute their rewards according to need rather than desert, using a "communal" exchange rule, or distribute rewards according to an "equality" exchange rule. The conditions under which different exchange rules will be adopted have been extensively investigated and it is clear that different exchange rules are adhered to in different types of relationships and sometimes at different stages of the same relationship (e.g., equity early and communal later in a close relationship).

The most influential of the exchange theories has been Thibaut and Kelley's theory of social interdependence, which focuses on the profitability, or "outcomes," associated with each behavior an individual may perform at any given time in interaction with another. It predicts that the behavior an individual will actually choose to perform from an array of behavioral options is a function of not only of the individual's own outcome matrix but also of the outcome matrix of the partner, or of the configuration presented by *both* partners' matrices and the extent to which each partner has the power to influence the other's outcomes (and thus the other's choice of behavior) by varying his or her own behavioral choices. Social interdependence theory also formally separated the issue of attraction to another from the issue of the stability of the relationship. It hypothesized that attraction is a function of the degree to which the individual's profits from the relationship are above his or her comparison level, defined as the level of profits the individual believes he or she deserves, but the stability of a relationship is a function of the comparison level for alternatives, defined as the level of profits available in the individual's best alternative relationship. Thus the theory predicts that people may dissolve a satisfying relationship with a person they like to enter an even more profitable relationship, and they may also stay in relationships with people they do not like and in which they receive poor outcomes because they have no better alternative. Social interdependence theory has been extended and elaborated several times to better account for interaction in close relationships. For example, the theory now recognizes that people will often "transform" the raw outcome matrix of rewards and costs by the adoption of exchange rules, such as the communal rule or the equality rule, both of which benefit the less powerful person in the relationship who could not induce the partner to perform such rewarding behaviors simply through varying their own behavior.

Although the social exchange theories have provided many insights into attraction and other relationship phenomena, there has been some dissatisfaction with these theories because, although useful for a general understanding of social behavior, their theoretical precision proves illusory in practice. The central problem lies in determining what is a reward for whom under what circumstances and then in quantifying the degree of reward represented in each exchange. People exchange many different kinds of rewards and punishments in social interaction (e.g., love, information, status, services) and reducing these to a common metric for quantification and prediction purposes is difficult if not impossible. In response to dissatisfaction with the exchange theories of attraction, numerous other theoretical approaches have been developed. These are typically addressed to a particular kind of attraction (e.g., love) to a particular phenomenon (e.g., relationship stability; the development of closeness in a relationship), or to attraction as it occurs in a particular kind of relationship (e.g., courtship; marital).

## IV. VARIETIES OF ATTRACTION

At the present time, there is no generally accepted taxonomy of the specific forms attraction to another may take; however, theory and research within social psychology have focused largely on four types of attraction. Although conceptually separable and differing in their causal antecedents and consequences, several of these varieties of attraction are no doubt blended in many attraction experiences in naturalistic situations.

## A. Liking

Despite the fact that the general theories guiding most attraction research were assumed to be applicable to all varieties of attraction across all social relationships and contexts, they primarily have been used to predict a mild form of attraction—liking—and, until relatively recently, often liking between two strangers in their initial encounter with one another. In early attraction theory and research, liking was virtually synonymous with the word attraction, especially given the definition of attraction as a positive attitude toward another. Several factors associated with liking have been extensively investigated.

### 1. Physical Proximity

Many studies have found an association between the sheer physical distance between two people and the extent to which they like each other. In classrooms and apartment houses, for example, physical proximity predicts friendship development. Distance is not only a causal factor in attraction but it is also, of course, a consequence, as studies of affiliation have documented. The physical distance between two people usually approximates the degree to which they are accessible to each other for interaction and it is this accessibility, which is usually but not always associated with physical distance, that is facilitative of attraction, primarily because it facilitates the exchange of rewards. Moreover, physical proximity lowers the costs of the exchange; physical distance in a relationship usually adds costs, thereby reducing the profitability of the relationship. Furthermore, and as previously discussed, cognitive consistency processes are likely to produce attraction to those in close physical proximity, especially if they are compelled by extraneous forces to interact with each other. And, finally, evidence indicates that familiarity itself, which increases with proximity, leads to attraction, as will be discussed later. Just as physical proximity promotes attraction, physical distance between partners in an established relationship should be associated with relationship instability. This hypothesis has received increasing examination in recent years as the number of "long-distance" romantic relationships has dramatically increased due to the demands of partners'

dual careers. In accord with the general physical proximity–attraction principle, the available evidence suggests that the relationships of many couples cannot surmount the additional costs physical separation imposes and that these relationships are more likely to dissolve.

Although accessibility for interaction may be a necessary condition for the exchange of rewards, interaction sometimes results in the exchange of punishments. Thus, the "contact hypothesis" (i.e., that bringing hostile groups together to interact with each other may reduce their hostility) has not always been confirmed; greater attention is now given to providing an interaction context conducive to the exchange of rewards rather than punishments when this means of reducing prejudice and enmity between groups is pursued.

### 2. Similarity

Few variables have been so thoroughly investigated in the attraction literature as similarity, especially attitude similarity. Correlational studies of friends and spouses have shown that they tend to be more similar to each other than chance would dictate on virtually every dimension investigated, including background, religion, education, height, and political affiliation. At least part of this association can be accounted for by the fact that people are more frequently thrown together in time and space with those who are similar rather than dissimilar to themselves and that similar people are thus more accessible for interaction. Similarity also has been shown, however, to be a causal antecedent of attraction. For example, many experiments have demonstrated that attraction, at least to a stranger, is associated with the proportion (rather than the number) of similar, as opposed to dissimilar, attitudes shared with another. A number of factors help account for the similarity–attraction relationship: (1) similarity often signals that rewards will be forthcoming in the relationship, especially similarities that signal shared preferences for activities and people; (2) similarity of attitude provides social validation for the individual's own views, which itself has been shown to be rewarding; and (3) people assume that similar people will like them and that dissimilar people will not, and it is known that the anticipation of being liked by another generates attraction in return, in accord with the "reciprocity-of-liking" principle that states that attraction breeds attraction.

### 3. Physical Attractiveness

Another factor shown to generate attraction is physical attractiveness. Researchers have shown a renewed interest in identifying the principles underlying judgments of another's physical attractiveness, and some evidence now suggests that with respect to facial physical attractiveness, those faces whose characteristics are congruent with the modal value of facial characteristics in their gender and age group (i.e., show little deviance from the average face as determined by a composite of many faces) are more likely to be judged attractive. Whatever it is people are responding to, there is substantial agreement about who is physically attractive and who is not, agreement that ranges across age, sex, educational, and social groups.

A large body of evidence documents that the physically attractive, from infancy to late adulthood, receive preferential treatment from others and, other things being equal, will be liked more than the physically unattractive. One important mediating factor of preferential treatment and attraction is a physical attractiveness stereotype such that a number of desirable, but less visible, qualities are typically inferred from a physically attractive appearance. For example, a meta-analysis of many of these studies indicates that physical attractiveness often leads to inferences of positive attributes associated with social competence (e.g., poised, warm, and responsive); in fact, evidence suggests that physically attractive people are indeed more competent in social interaction, perhaps as a result of their history of preferential social treatment. Finally, association with physically attractive persons has been shown to increase an individual's status in the eyes of others under some circumstances.

Physical attractiveness is an especially potent factor in date and mate selection, with the most physically attractive persons initially preferred by all but with assortative pairing eventually taking place on this dimension (e.g., spouses' physical attractiveness is positively correlated). Findings in this culture that men are more likely than women to report that physical attractiveness is important to them in date and mate selection and that women are more likely than men to report that they value economic status factors have been replicated across many cultures and have been interpreted as supporting evolutionary models of mate selection. Evolutionary models assume that humans have been selected to maximize gene replication and thus reproductive success, but that differential biology of reproduction results in gender differences along dimensions associated with mate preference. For example, women are assumed to invest more resources in their offspring and so are theorized to be especially sensitive to resource availability and thus to prefer mates who can provide resources; men's gene replication, on the other hand, is assumed to be limited mostly by access to females who are likely to reproduce, resulting in a sensitivity to female characteristics that reflect reproductive capacity, such as youth which is associated with physical attractiveness. The differential preferences of men and women report can be accounted for in other ways, however, especially when the universal lower status and economic power of women is considered along with the fact that marriage may present the primary means by which they may improve these. The matter is further complicated by the fact that when the actual preference behavior of men and women is observed experimentally rather than through self-report, women often, no less than men, show a preference for the physically attractive over the financially solvent; thus, compared to men, there is likely to be a greater discrepancy between what women say they prefer and what they behaviorally reveal they prefer in the date and mate selection arena. Date and mate preference, particularly gender differences in these, continues to receive attention.

The impact of physical attractiveness beyond the early stages of a relationship has not been investigated as much as is now warranted by evidence that the receipt of individuating information about another attenuates the influence of social stereotypes. Influences of other morphological factors on attraction also have not been extensively investigated, the exceptions being male height (often associated with financial and political success) and obesity (with even very young children imputing a negative value to overweight).

## 4. Familiarity

Two different lines of investigation suggest that, other things being equal, familiarity leads to attraction. Much research has supported the hypothesis that simple repeated exposure to a person (or any other stimulus) is a sufficient condition for increased attraction to him or her. The explanation for the familiarity–attraction association is assumed to be that familiar people are safe, unlikely to harm the individual. Although exposure to any initially neutral stimulus generally leads to a long-term increase in attraction under a wide range of conditions, there may be a short-term decline immediately following many sustained exposures. Moreover, although repeated exposure to another may be associated with increases in liking, it may be negatively associated with another form of attraction in humans, sexual attraction, just as familiarity has been shown to be negatively associated with sexual attraction in infrahuman animals (i.e., the Coolidge effect).

Another line of research suggesting that familiarity should lead to attraction stems from attachment theory. Attachment, in fact, is now viewed by some theorists as a type of attraction, different from liking in a number of respects, including the fact that one may exhibit attachment behavior toward persons one does not like.

## B. Attachment

Bowlby's theory of attachment maintains that the primary purpose of affectional bonds is to promote the individual's physical proximity to other members of the species who will act to protect the individual from survival threats. That most animals are born with an instinctive proximity-promoting mechanism that leads them to stay physically close to those who can provide food, shelter, and protection was first demonstrated by Lorenz in his classic imprinting experiments. The importance of this mechanism was later demonstrated by Harlow with monkeys, where failure of the affectional system to develop normally in the infant–mother relationship was shown to impair later socioemotional adjustment.

Human infants also quickly develop a strong preference for a person who meets certain minimum visual, tactile, and auditory requirements and who is also the first person with whom they have become familiar, and they exhibit attachment behavior toward this person. Attachment behavior has two distinctive features: (1) it results in the restoration or maintenance of proximity to a specific other, and (2) it is usually exhibited toward an older, stronger, or more dominant individual by a younger or weaker one. Attachment behavior, which has security as its goal and is believed to be experienced subjectively as an emotional bond to the other, is especially evident between a child's first and third years but it may characterize people of all ages and is likely to be displayed when the individual is threatened, ill, or otherwise distressed. Attachment is conceptually different from liking, at least liking as defined as a favorable attitude toward another. Individuals may become "attached" to people whom they do not like and who are, in fact, punishing. Moreover, attachment and liking are different in their causal antecedents, with familiarity presumed to be the major determinant of attachment but the receipt of rewards the most important determinant of liking.

Investigators are currently attempting to trace how later relationships are influenced by the child's quality of security in the early child–caretaker relationship (i.e., whether the quality of attachment was "secure" or "insecure," which is presumed to depend on whether the caretaker is consistently responsive to the child's needs). These longitudinal studies are revealing that the initial attachment pattern is often associated with differential behaviors in later relationships, although many intervening factors may modify the early relationship's influence.

Working solely with adults, many investigators are investigating the correlates of "adult attachment style," which is assessed by asking people to endorse one of three statements as characteristic of their relationships with others. These statements are regarded as adult analogues of the three classic attachment classifications derived from observation of a child's reactions to episodes of separation and reunion with the caregiver (i.e., secure, insecure/avoidant, and insecure/anxious-ambivalent). Which of the three paragraphs the adult endorses as best describing his or her typical relationships with others seems to be associated with his or her retrospective memories of childhood, memories of experiences in romantic relationships, and many belief, attitude, and personality assessments. Trust of a particular partner and security in a specific relationship and the roles these play in relationship dynamics and outcomes has been the focus of a good deal of other recent research derived from a variety of theoretical positions. Taken together, this evidence strongly suggests that trust and security may be central properties of a relationship—as central, for example, as attraction, a property with which it may (but need not) be associated.

## C. Altruistic Love

The phrase "altruistic love" commonly refers to behavior that is primarily intended to promote the welfare of another as contrasted to satisfying one's own needs or to facilitate the future receipt of rewards from the other. The circumstances under which people will act to further another's welfare have been extensively investigated and are usually reported in most introductory social psychology texts in a chapter entitled "Prosocial Behavior." Altruistic behavior has been shown to have many more determinants than attraction to the other; social norms and sanctions, for example, are particularly important determinants. Thus, attraction is not a necessary condition for a person to exhibit altruistic love toward another; people often act to benefit people whom they do not know, do not like, and would not care to interact with. Attraction also is not a sufficient condition for altruistic acts; although providing the other with benefits is among the factors that discriminates behavior exhibited toward acquaintances and loved persons, people who love each other (as determined through self-report or observation of behavior on other dimensions) do not always unselfishly promote each other's welfare and, not infrequently, behave in ways severely destructive of the other's well-being. Altruistic love, then, is conceptually distinguishable from other forms of attraction although it is often associated with them.

## D. Romantic Love

As previously noted, theorists and investigators initially assumed that the causal antecedents of the strong forms of attraction were the same in kind and simply differed in magnitude from the mild forms. It became clear, however, that, at least in heterosexual dating relationships, greater and greater liking typically leads to a lot of liking, not love, and especially not romantic love. As a consequence, theorists gradually recognized that liking and romantic love are two different phenomena with distinctly different causal antecedents.

Interest in the romantic variety of attraction has continued to increase for a number of reasons, including evidence that romantic love not only remains a prerequisite for marriage in Western culture but that it has increased in importance over the past few decades, especially for women. The view that romantic love is a necessary condition for marriage has extended to Japan, China, and other Eastern cultures, and cultural differences in the experience of romantic love are now of special interest. Although some researchers hypothesize that Western individualistic culture is more conducive to romantic love than collectivist cultures, some studies have found remarkable similarities among, for example, American, Japanese, and Russian experiences of love in heterosexual relationships. Some theorists currently argue that romantic love is virtually universal, occurring at all ages and at all times across all cultures, and anthropologists are beginning to document this view.

Many theories of romantic love have been offered. They often differ from general theories of attraction in their special emphasis on the role that deprivation plays in making a social stimulus event a reward, as well as in their emphasis on the role fantasy, idealization of the other, and anticipation of reward (as opposed to the actual receipt of reward) play in generating this variety of attraction. Unlike liking, which is often referred to as "companionate love" in heterosexual relationships, romantic love is also associated with the experience of strong emotions, both positive and negative, and thus general theories of emotion have been applied to understand the phenomenon. Moreover, unlike liking, which tends to increase with familiarity, romantic love appears to decrease in intensity over time, with its symptoms virtually disappearing after 6 years of frequent interaction, according to one study that examined couples in Japan and the United States. Several sociobiological theories of romantic love have been offered as well, with their general thesis being that the behaviors generated by romantic love cause people to attract mates, reproduce, and invest in the survival of offspring; thus, reproductive success is believed to be associated with the experience of romantic love.

In recent years researchers have relied heavily on an inductive psychometric approach to the investigation of love. Self-reports of experiences of persons in putative love relationships are obtained and subjected to factor analytical techniques in order to identify the dimensions underlying these experiences. For example, based largely on factor analyses of self-reports of dating relationship experiences, Sternberg has developed a "triangular theory" of love which proposes that many popularly recognized subtypes of love (e.g., "infatuation") can be accounted for by different weightings of three components: intimacy, passion, and decision/commitment. However, other investigators who have factor analyzed self-report responses to many love scales, including Sternberg's scales and the popular Hendrick and Hendrick Love Attitudes Scale, have concluded that love apparently means different things to different people in different relationships at different times. Many of those who have examined the concept of love as it is used in everyday language have made a similar observation. Given the sustained degree of interest in the topic, however, it seems unlikely that the matter will be left here and that researchers will abandon their attempts to better understand love.

## V. ATTRACTION AND OTHER RELATIONSHIP PHENOMENA

As previously noted, theory and research on interpersonal attraction shifted in the early 1980s from a focus on identifying the causal determinants of attraction between people in their first encounter with each other to examining the role attraction plays in a number of phenomena characteristic of ongoing relationships in naturalistic situations.

## A. Theories of Adult Relationship Development

Relationship researchers are especially concerned with understanding relationship development, or identifying the factors associated with the relationship's growth and its later maintenance or dissolution. Although people develop relationships with only a small portion of people to whom they feel initial attraction, attraction is universally regarded as an important condition facilitating the growth of a relationship, especially under open-field conditions where people can choose with whom they interact. Moreover, the same conditions that promote an individual's attraction to another, especially the individual's evaluation of the rewards and costs currently received in the relationship as well as forecasts of future profits, are the same conditions theorized to promote relationship growth and development. All theorists recognize, however, that attraction need not be characteristic of a developing relationship nor even a characteristic of an established relationship, for reasons previously discussed.

Most theorists of relationship development view relationships as proceeding from a superficial stage, such as that typical of the initial encounter, to, in some cases, a very close relationship. The specific stages through which a relationship may pass, and the kinds of processes and events presumed to be critical to each stage, differ among theorists. Appearing in several theories, however, are (1) an explorative stage where people try to identify the rewards that may be available to them in the relationship, often through processes of mutual self-disclosure; (2) a conflict, bargaining, and negotiation stage where the partners try to mesh their goals and objectives and set the terms of the relationship; and (3) a commitment stage where each partner expresses, directly or indirectly, an intention to continue the relationship and to cease exploration of alternative relationships. Many of the processes believed to be associated with these stages have been the subject of a good deal of theory and research in themselves, with self-disclosure, conflict resolution, and commitment processes each representing robust literatures.

Relationship researchers' special interest in identifying the determinants of marital stability continues, partly because of the high rate of marital dissolution in this country (with some demographers forecasting that about two-thirds of all first marriages will now end in divorce). Marital quality (e.g., satisfaction, adjustment) remains an ubiquitous variable in investigations of the marital relationship, deriving much of its import from its presumed association with marital stability even though recent reviews of the marital satisfaction literature reveal that there is sparse empirical evidence to support the connection. The immense and largely atheoretical literature on marital satisfaction is characterized by a host of problems, including the correlational nature of most of the studies and severe measurement problems associated with the variable of interest.

Many researchers are now undertaking longitudinal, or prospective, studies of relationship stability. Sometimes, as in the case of dating relationships, the elapsed time between the initial and follow-up assessments is only a matter of months, but much is being learned about the factors associated with stability in this notoriously unstable type of relationship. These studies are highlighting, for example, the importance of assessing not just factors internal to the relationship (e.g., satisfaction), but its social context (e.g., the alternative relationships available to each partner and the support important others in the partners' social network provide the relationship).

Longitudinal studies of marital relationships are more rare, but several are in progress. At least one such ongoing study has documented that, as suggested by a host of correlational studies, there are significant declines in love and marital satisfaction over the first year of marriage. This decline traditionally has been attributed to the advent of parenthood, but it now appears that declining satisfaction over the first and ensuing years of marriage is simply associated with the relationship's advancing age. A host of cross-sectional correlational studies show that the age of this type of relationship is inversely correlated with satisfaction, with the largest decreases taking place in the first few years of the relationship. Relationship age is an important factor to consider in interpreting any relationship finding. For example, relationship age previously was not considered in interpretions of findings showing a negative association between marital stability and premarital cohabitation; it now appears that the association is not necessarily due to the act of cohabitation or to the characteristics of people who cohabit but, rather, to the fact that spouses who cohabited prior to marriage generally are in an older relationship than are those who have been married for an equal length of time but did not cohabit.

## B. Other Recent Developments in Relationship Theory and Research

As a result of decline in the traditional family form and public demand for information about close relationships of all types, scholars in a variety of disciplines that traditionally have been concerned primarily with one type of relationship are combining their efforts to develop a superordinate body of knowledge about the dynamics of interpersonal relationships that transcends relationship type. Reflecting the youth of this multidisciplinary effort, however, the matrix of interpersonal relationship knowledge is still severely fractured along the lines of relationship type.

Sociologists, for example, typically examine the associations between macrosocietal forces and the changing forms and stability of family relationships, attending particularly to such outcomes as marital satisfaction. Researchers in marital and family therapy are concerned primarily with identifying the sources of distress within marital relationships and in devising effective intervention strategies. Communication researchers often attempt to identify dysfunctional communication patterns within close relationships, especially marital relationships. Developmental psychologists continue to address the parent–child relationship but in recent years have extended their focus to include child–peer relationships; an improved understanding of the developmental significance of a child's social relationships on later social and emotional functioning has been, in fact, characterized as one of the most significant advances within psychology since the mid-1970s. Finally, social psychologists, who have typically examined young adult relationships, often premarital courtship relationships as the bulk of theory and research on interpersonal attraction reflects, continue to pursue an understanding of relationships within this population, although recent years have seen a growing interest in friendship and in life span developmental issues that engage relationships, as well as in extending theory and research to the marital relationship.

As a consequence of their effort to develop a generic relationship knowledge base that transcends both relationship type and disciplinary boundaries, concepts that are applicable to all relationships have become

of particular interest to theorists and researchers. One such concept is relationship "closeness," a relationship property that is believed to underly many important relationship phenomena. The definition of closeness proposed by Kelley *et al.*, who presented a conceptual and methodological framework to facilitate the study of close relationships in their 1983 book "Close Relationships," has been generally accepted; a close relationship is viewed as one in which the partners are highly interdependent on each other as revealed in the behavioral interaction pattern characteristic of their relationship. Behavioral interdependence has been shown to be both conceptually and empirically independent of the partners' subjective view of the closeness of their relationship. For example, partners whose relationship is not close in terms of mutual behavioral influence may believe that they are close, especially if they have positive feelings and attitudes toward the other because these appear to be closely associated with subjective feelings of closeness. Conversely, partners may not believe their relationship to be close even though independent observations reveal high interdependence between the partners' behaviors (as negative affect reciprocity in distressed couples illustrates). Thus, contrary to popular assumption, close relationships are not necessarily happy or healthy relationships, although most probably are.

Other concepts that are applicable to all relationships and that are associated with important relationship phenomena include trust and security, as previously discussed, and intimacy. Intimacy is now viewed as a process in which the individual expresses, either verbally or nonverbally, important self-relevant feelings and information to another and, as a consequence of the other's responses, comes to feel known, validated (in terms of world view and self-worth), and cared for; intimacy, of course, is strongly associated with attraction to the partner.

Many relationship researchers currently are taking a new path to the development of generic relationship knowledge, one that promises to illuminate the dynamics of all relationships. Building on the advances made in cognitive psychology, especially those in the area of social cognition, these researchers are attempting to gain an understanding of the cognitive processes that underlie partners' interaction patterns. Thus, attention is turning to relationship schemas, relatively stable cognitive structures that represent an individual's knowledge of the partner, of themselves as they characteristically interact with that partner, and the interaction "scripts" that detail the sequence of behaviors typical of their relationship. These cognitive schemas represent expectations about the partner and his or her behavior in the relationship. In interaction, these expectations tend to act as self-fulfilling prophecies. They direct an individual's attention toward the partner's behaviors that confirm the expectation and away from those that do not, and they also influence interpretation of the causes, meaning, and significance of the partner's behavior.

Dysfunctional but firmly entrenched relationship schemas have been shown to be implicated in many distressed relationships. For example, much research has shown that during conflictful interactions, partners in distressed relationships have difficulty in accurately identifying their partner's affect, goals, and intentions; moreover, the partner's behavior, even well-intentioned and objectively positive behavior, tends to be attributed to negative and harmful motives. Therapeutic intervention in long-term relationships is made difficult by the fact that the individual's relationship schema, which acts as a filter through which the individual views the interaction with the partner and which guides the individual's attentional processes and interpretations of the meaning of the partner's behavior, functions "automatically" (i.e., quickly and involuntarily, without conscious awareness and control). Moreover, the schema itself is often not accessible to consciousness. This is one of the many reasons why intervention into distressed relationships is likely to be more successful early, rather than later, in the relationship.

The processes guiding memories of relationship events are also of current interest. "Account narrative" methodology, whereby the individual tells the "story" of the relationship in unstructured form, is increasingly used to assess these memories. Evidence suggests that in heterosexual relationships, women tend to be the relationship historians, recalling relationship events with greater accuracy, detail, and vividness than men do. It also has been demonstrated that an individual's current satisfaction with the relationship influences his or her memory of relationship events; thus, in distressed relationships, even events that were originally experienced as positive cannot be counted on to attenuate current dissatisfaction with the relationship because these events often have been subject to the mind's tendency to write a "revisionist" history of the relationship, a reconstruction that represents the past as affectively congruent with present feelings toward the partner and the relationship.

Although an integration of the several literatures devoted to different types of relationships has not yet been achieved, the tremendous amount of current activity in the relationship area assures that the 21st century will witness such an integration as well as continued acceleration of the development of a science of interpersonal relationships.

## BIBLIOGRAPHY

Berscheid, E. (1994). Interpersonal relationships. *Annu. Rev. Psychol.* **45,** 79–129.

Berscheid, E. (1985). Interpersonal attraction. *In* "Handbook of Social Psychology" (G. Lindzey and E. Aronson, eds.), 3rd Ed., pp. 413–484. Random House, New York.

Cate, R. M. (1992). "Courtship." Sage, Newbury Park, CA.

Clark, M. S., and Reis, H. T. (1988). Interpersonal processes. *Annu. Rev. Psychol* **39,** 609–672.

Collins, W. A., and Gunnar, M. R. (1990). Social and personality development. *Annu. Rev. Psychol.* **41,** 387–416.

Coyne, J. C., and Downey, G. (1991). Social factors and psychopathology: Stress, social support and coping process. *Annu. Rev. Psychol.* **42,** 401–425.

Fincham, F. D., and Bradbury, T. N. (eds.) (1990). "The Psychology of Marriage: Basic Issues and Applications." Guilford, New York.

Fletcher, G. J. O., and Fincham, F. D. (eds.) (1991). "Cognition and Close Relationships." Erlbaum, Hillsdale, NJ.

Hartup, W. W. (1989). Social relationships and their developmental significance. *Am. Psychol.* **44**(2), 120–126.

Kelley, H. H., Berscheid, E., Christensen, A., Harvey, J. H., Huston, T. L., Levinger, G., McClintock, E., Peplau, L. A., and Peterson, D. (1983). "Close Relationships." W. H. Freeman, New York.

Sternberg, R. J., and Barnes, M. L. (eds.) (1988). "The Psychology of Love." Yale University Press, New Haven.

# Autism

*University of North Carolina at Chapel Hill*

## GLOSSARY

**Developmental disability** A chronic, severe disability that is caused by a physical impairment, manifested during childhood, likely to continue indefinitely, and that requires lifelong, individualized interdisciplinary care

**Cognitive** Relating to functions of the brain such as thinking, organizing, perceiving, sequencing, and remembering

**Differential diagnosis** Process of determining which of two or more related, though distinct, diagnostic categories best fits a particular mental or physical disability

**Etiology** Causes of a disease or abnormal condition

**Epidemiology** Incidence, distribution, and control of a disease or abnormal condition in a population

AUTISM IS THE MOST SEVERE OF THE DEVELOP-mental disabilities with an incidence of approximately 1 per 1000 live births. Originally conceptualized as an emotional disorder caused by inadequate mothering, its organic basis is now recognized. The primary dysfunction in autism is in the way the brain processes and integrates information, resulting in problems of social interaction, communication, learning, and behavior. Several causes and a variety of neurological mechanisms have been identified. Although autism is a severely handicapping condition, outcomes for autistic people are improving as more effective interventions and more appropriate community resources are developed. Most autistic people have a normal lifespan; if they are to achieve their potential, there is a compelling need for more community-based, cost-effective resources.

## I. PRIMARY CHARACTERISTICS

The primary characteristics of autism include impaired reciprocal social interactions, impaired communication, and restricted behaviors. The impaired social interactions, among the most conspicuous of the autistic deficits, caused Leo Kanner to use the term "autism" when he first described the syndrome in 1943. Social difficulties of autism are gross and sustained including impaired use of multiple nonverbal social behaviors; a failure to develop peer relationships; lack of spontaneous seeking to share enjoyment, interests, or achievements with others; or lack of social or emotional reciprocity.

Communication and language problems are also marked and sustained, affecting both verbal and nonverbal skills. Approximately 40% of autistic people do not develop meaningful communicative language and most have difficulties with other forms of communication as well. Verbal autistic youngsters are frequently ecolalic, stereotypic, or repetitive in their use of language and sustaining conversation is difficult. Nonverbal youngsters have difficulty understanding or being understood and, consequently, often retreat from interactions with others. Play is impaired in most autistic children, who lack the spontaneous, make-believe or social and imitative aspects generally seen in nonhandicapped children.

The third primary characteristic of autism is their restricted range of behaviors, activities, and interests. Lower-functioning autistic people frequently engage in repetitive bodily movements, self-stimulatory behaviors, and sometimes even self-abuse. Higher-functioning autistic people may be preoccupied and show

ENCYCLOPEDIA OF HUMAN BIOLOGY, Second Edition, VOLUME I.   Copyright © 1997 by Academic Press.   All rights of reproduction in any form reserved.

restricted patterns of interests that are abnormal in intensity and focus. Rigid adherence to specific, nonfunctional routines and rituals, and persistent preoccupation with parts of objects are other examples of this third characteristic.

## A. Associated Features

Autism manifests itself through delays or abnormal functioning in at least one of these three areas prior to age 3 years. In 10 to 15% of cases, parents report regression of language development after a child has acquired 10–15 words. In addition to the primary characteristics that define the autism syndrome, there are associated features that are frequently present as well: abnormalities in the development of cognitive skills; abnormalities of posture and motor behavior; unusual sensory responses; abnormalities in eating, drinking, or sleeping; abnormalities of mood; and self-injurious behaviors. Although these features are not essential for a diagnosis of autism, they are often observed in this group and can have important implications for the treatment of autistic children.

Several cognitive abnormalities are frequently observed in autistic youngsters: distractibility, poor organizational ability, difficulties with abstractions, and a strong focus on details. Mental retardation is an additional cognitive disability in about 70% of autistic people. Often they exhibit an uneven cognitive profile with some skills being strong while other aspects of cognitive functioning are quite limited.

Abnormalities of posture and motor behavior include stereotypes like arm-flapping and grimacing, abnormal gaits, and odd posturing with the hands. Under- and overresponsivity to sensory input are common; some autistic people resist being touched while others ignore sensations, like pain. Many autistic people are fascinated by specific sounds or tastes.

Abnormalities of drinking, eating, and sleeping behavior and fluctuations of mood are also common. Eating, drinking, and sleeping problems often resolve themselves by adolescence, but can be troublesome prior to then. Eating a limited variety of foods and staying up all night are among the most difficult of the ongoing problems parents face with autistic youngsters. Lability of mood is observed in several variations: giggling or weeping for no apparent reason, absence of emotional responses or reactions to danger, excessive fearfulness, or generalized anxiety.

Self-injurious behaviors, such as head banging and finger or hand biting, are the most extreme and frightening of the behaviors accompanying autism. These occur in less than 10% of the population, but can be the most difficult to control or remediate. In their most extreme form, these behaviors require hospitalization.

Although age of onset is no longer a diagnostic criteria, autism begins early in life; almost always before age 3 years and rarely after age 5 years. Most autistic children show signs of the disability from birth, although there are some cases where early normal development is followed by a deterioration of social, cognitive, behavioral, or communicative skills. In these instances of apparent normal development, deterioration following normal language development is usually the first indication of a problem.

## B. Differential Diagnosis

One of the greatest frustrations among parents and professionals is the confusion about diagnostic issues. Diagnosing autism can be difficult because it resembles other disabilities of behavior, communication, and learning. Because autism is also a low incidence disorder, most professionals do not see enough cases for them to consistently identify subtle distinctions between this syndrome and related disorders. The historical confusion between autism and emotional difficulties has further clouded the diagnostic picture. Over time, autism has been misdiagnosed as many different disabilities: mental retardation, schizophrenia, schizoid personality, developmental language problem, hearing impairment, or pervasive developmental disorder, not otherwise specified.

The most recent "Diagnostic and Statistical Manual of the American Psychiatric Association" (DSM-IV) has established several categories of pervasive developmental disorders that are similar to autism, but unique and different in important ways. Rett's disorder is diagnosed only in females and shows a characteristic pattern of head growth deceleration, loss of previously acquired purposeful hand skills, and poorly coordinated gait and trunk skills. Childhood disintegrative disorder shows similar characteristics to autism except the difficulties only appear after at least 2 years of normal development. Asperger disorder can be distinguished from autism by the lack of any delay in language development. Pervasive developmental disorder not otherwise specified (PDDNOS) is perhaps the most puzzling category related to autism. A diagnostic category used solely when others are not more appropriate, PDDNOS does not have any specific diagnostic criteria. In DSM-IV the definition of PDDNOS reads:

This category should be used when there is a severe and pervasive impairment in the development of reciprocal social interaction or verbal and nonverbal communication skills, or when stereotyped behavior, interests, and activities are present, but the criteria are not met for a specific pervasive developmental disorder, schizophrenia, schizotypal personality disorder, or avoidant personality disorder. For example, this category includes "atypical autism" presentations that do not meet the criteria for autistic disorder because of late age at onset, atypical symptomatology, or subthreshold symptomatology, or all of these.

The relationship between autism and mental retardation has been a source of confusion for several decades. Many have noted that intellectual impairments in people with autism resemble the limitations of mentally retarded people. Compared with mentally retarded people, however, individuals with autism have more intellectual strengths, which can even be above average in some areas, and a wider spread between their skills and deficits. Gross motor skills of autistic children tend to be stronger as well. Mentally retarded children, on the other hand, generally have better social and communication skills in relation to their overall developmental levels. A major source of the confusion between these two disabilities has been the historical notion that autism is a relatively "pure" disability that cannot coexist with other syndromes, like mental retardation.

The relationship between autism and mental retardation has been clarified more recently with the acknowledgment that autism, as a behavioral syndrome, can and does coexist with other disabilities. The most common of these concurring disabilities is mental retardation. Current estimates are that approximately 70% of individuals with autism have an additional diagnosis of mental retardation.

In identifying autism, Leo Kanner described it as the earliest form of childhood schizophrenia because of the similarities he observed between the conditions. Today, autism and schizophrenia are seen as distinct and different; autism is viewed as a developmental disorder and schizophrenia is classified as a mental illness. The major differences are the hallucinations and delusions in schizophrenia are absent in autism and the earlier onset of autism (almost always before age 5 years) compared to the abrupt onset of schizophrenia, typically during adolescence. [*See* Schizophrenic Disorders.]

Investigators have recently identified other important distinctions between autism and schizophrenia. The family histories of children in these diagnostic groups are generally different; children with autism more frequently have family histories of developmental disabilities whereas families of schizophrenics more frequently have histories of personality, affective, and other emotional disorders. Autistic children are physically healthier and have better motor skills, on the average. While autistic children never form appropriate interpersonal relationships, schizophrenia is viewed as a withdrawal from presumably unsatisfactory relationships, often because of a particular traumatic event. Schizophrenics generally have higher IQs than children with autism and, finally, they also have periods of remission, not seen in autism, when their behavior returns to near normal.

Language and hearing impairments can also be confused with autism. Language impairments in children with autism include delayed development of vocal expression and language comprehension, echolalia, pronoun reversals, and problems with sequencing. These communication difficulties can occur in, and sometimes limit the social relationships of, children with language impairments though not nearly as much as in children with autism. Compared with autistic children, those with communication handicaps use alternative forms of communication more effectively (e.g., gestures), have higher IQs, engage in more imaginative play, and have a better prognosis. Language impairments are not typically associated with restricted, repetitive, or stereotyped patterns of behavior.

Nonresponsive and indifferent to others, children with autism can also be misdiagnosed as hearing impaired. Recent advances in testing have reduced the frequency of this problem. There is also growing awareness that nonresponsiveness does not necessarily mean that a child cannot hear. Other differences between hearing impaired and autistic children include higher IQs, better social relationships, better nonverbal communication, and a better prognosis for children with hearing impairments.

## C. Demographics and Epidemiology

DSM-IV estimates the prevalence of autism as 2 to 5 per 10,000; other estimates range closer to 1 per 1000. About 70% of those diagnosed as autistic function intellectually within the mentally retarded range, with IQ scores as stable and accurate in predicting later academic performance as those of nonhandicapped children.

Autism occurs more frequently in males than females; the sex ratio is approximately three and one-half to one, which is similar to other developmental disabilities. A lower percentage of females than this ratio predicts, however, appear to be higher functioning. Although the distribution in social class was once thought to differ, with autism more frequent at the higher levels, recent studies have invalidated this assumption: autism is equally distributed among all social classes, ethnic and racial groups, and nationalities.

## D. Etiology

It is generally accepted that autism is not a single entity, but a series of behaviors with multiple causes and neurological mechanisms. One of the most important known causes is genetic with several possible transmissions. Twin studies have shown a concordance rate for autism of greater than 50%. Other studies have demonstrated an increased risk of related language, speech, and developmental problems in families with an autistic child. Autism is one of a number of possible outcomes for children with this genetic predisposition for communication or learning problems. Fragile X is another genetically transmitted form of autism. Although all children with this chromosomal abnormality do not have autism, 5–10% probably do. [See Fragile X Syndromes.]

Other identified causes of autism are infectious diseases, metabolic disorders, and structural abnormalities. Rubella is one prenatal infection that is a proven cause of autism and others are thought to exist as well. Metabolic disorders causing autism are phenylketonuria (PKU) and celiacs disease and it is suspected that high uric acid levels and difficulties in metabolizing purines could be implicated. In addition, structural abnormalities such as hydrocephalus can cause autism; developing technology in brain scanning equipment makes it likely that other specific structural deficits will be identified in the near future. One such deficit might relate to underdevelopment of the cerebellum. Although preliminary, this hypothesis is the first to propose a specific neurological structure underlying autism and support it with convincing empirical evidence. Data on the underdeveloped cerebellum are thus far limited to higher-functioning individuals with autism.

## E. Neurological Correlates

Research evidence suggests that autism relates to specific forms of neurological dysfunction, although the precise nature of the neurological impairments remains elusive. Several investigators have identified neurological correlates of autism whose specific relationships to the disability are unclear.

It is well documented that autistic children have more soft neurological signs than nonhandicapped control groups; studies report 40 to 100% of autistic children show at least one of these signs. Although there is considerable disagreement as to the relevance of neurological soft signs, some believe that they are indications of subtle brain damage, immaturity, or poor organization.

Autistic people also have a higher incidence of abnormal electroencephalograms (EEGs). Studies have reported abnormal EEGs in 20 to 65% of autistic children with abnormalities characterized by focal slowing, spiking, or paroxysmal spike-wave discharges. Abnormal EEGs are not the only "hard" evidence of neurological problems, autistic children also have seizure disorders more frequently than the general population. Current estimates are that 25 to 33% of autistic people develop seizure disorders, most frequently during adolescence. The typical onset of seizure disorders during adolescence is unique to this group. There is a negative correlation between IQ and seizures; seizures are more common in autistic people with lower IQs. [See Seizure Generation, Subcortical Mechanisms.]

Finally, autism is often found in association with the following nervous system difficulties: retrolental fibroplasia, tuberous sclerosis, congenital syphilis, PKU, and neurolipidosis. The incidence of autism is much higher in children with these neurological conditions than in the normal population.

## II. THEORIES OF AUTISM

Most early theories of autism were psychogenic, emphasizing the role of parents in causing this severe disability of behavior and development. Psychogenic theorists argued that parents of autistic children were intelligent, obsessive, and lacking in warmth. These theorists identified the cause of autism in the family environment and described possible mechanisms including lack of maternal communication, pathological parent–child interaction, inadequate stimulation, or reactions to parental rejection. The psychodynamic theorists, however, have never generated supporting evidence. The only empirical studies of the emotional status of parents find that the extreme stress of parenting an autistic child can cause emotional difficulties. In the infrequent situations when these reactions occur, however, they are in reaction to having an autistic child and have in no way been shown as a cause.

Following the decline of the psychodynamic theories, several organic theories have emerged to explain dysfunctions in autism. Theorists have identified four possible neurological explanations: overarousal of the reticular system, perceptual inconstancy associated with brain stem dysfunction, dysfunction of the limbic system, and left hemisphere dysfunction.

Possible problems with the reticular system were hypothesized when autism was first recognized as an organic disorder in the 1960s. Investigators speculated that a chronically high level of nonspecific activity in the reticular system might be responsible for the bizarre behaviors observed in autistic children. These behaviors were viewed as attempts to maintain continuity in the presence of this overarousal. Recent studies, however, have not shown any relationships between arousal and overt behaviors.

Another early organic theory of autism was perceptual inconstancy. This theory sees autism as an inability to regulate sensory input, making it impossible for autistic children to develop coherent or meaningful concepts of external reality. Perceptual inconstancy has some advocates today, although they have never been able to identify the precise nature of the "perceptual instabilities."

A more recent theory of autism suggests that it might be similar to amnesias arising from lesions in the limbic system. Animal models have demonstrated similarities between behaviors resulting from hippocampal lesions and those observed in autism. Although these similarities suggest consistent underlying mechanisms, there are substantial behavioral differences as well, making this theory incompatible with the current state of our knowledge.

Other investigators have noted similarities between the specific cognitive and language impairments in autism and functions that are associated with the left hemisphere of the brain. Many autistic children also show superior abilities in right hemisphere functions. Because of the early onset of autism, however, left hemispheric dysfunctions should be compensated for by the right hemisphere of the brain. The apparent lack of right hemispheric compensation for left-sided dysfunction suggests a bilateral dysfunction in autism.

Our present theories of autism are inadequate and incomplete. One complicating factor is that autism is not a single entity; different causes and neurological dysfunctions are responsible for this disability in different people. As our sophistication and understanding of brain functioning increase, we can expect that the neurological theories will be more fully developed.

## III. TREATMENT

There have been three major approaches to treatment for children with autism over the years: psychodynamic, medical, and behavioral. Psychodynamically oriented therapies dominated the early work when autism was viewed as an emotional disorder and some of these interventions are still used today, although on a very limited basis. Biological interventions have included drug and vitamin therapies and also restricted diets. Behavioral approaches have followed the prinicple of teaching people with autism appropriate behaviors and helping them to eliminate inappropriate ones. Behavioral approaches emphasize special education, focusing on the development of academic and school-related skills.

### A. Psychodynamically Oriented Therapy

Bruno Bettelheim, the main proponent of psychodynamically oriented therapy for people with autism, indicted cold and rejecting parents as the main cause of the disability. He advocated removing children from their parents' homes to place them in residential settings. His interventions combine removal from parental control with therapeutic, residential milieus. He recommends individual, psychodynamically oriented therapy for the children as well as for the parents. A few psychodynamically oriented therapists, following Bettelheim, are currently practicing these interventions; however, psychodynamically oriented therapies are not commonly used with autistic children today because of the accumulating evidence refuting the assumptions upon which these approaches are based. In fact, studies on the effectiveness of psychodynamically oriented therapies have shown no advantages for treated children compared with untreated controls. Today, autism is no longer thought to result from inadequate parenting, but rather from undefined brain dysfunctions.

### B. Biological Interventions

Although autistic children are idiosyncratic responders to medication and most are not helped by drug treatments, a small percentage do seem to benefit and are treated pharmacologically. This is in addition to the 25 to 30% who are helped with anticonvulsant medications, which affect autistic children in the same ways as the nonhandicapped population.

Amphetamines sometimes reduce the hyperactivity accompanying autism; they can improve attention spans and reduce excessive activity in these youngsters. Although reports of improvement with amphetamines are encouraging, several published studies show deterioration of behavior in certain children with these medications.

Phenothiazines have proven useful, though unpredictable, in reducing anxiety, severe aggression, and self-injurious behaviors. Haldol is the most thoroughly researched of these drugs, although Mellaril is also commonly used. Unfortunately, phenothiazines have been shown to increase learning deficits and must be carefully monitored for several possible side effects including tardive dyskinesia, reduced seizure thresholds, and excessive weight gain.

Lithium, generally prescribed for manic-depressive patients, has been used with autistic children, especially those exhibiting episodic aggressive behaviors and who have not been responsive to other forms of drug treatment. Lithium has been effective with some of these youngsters, especially those showing family histories of cyclical affective illnesses. Lithium is especially difficult to monitor because there is a narrow range between therapeutic and toxic levels.

A 1982 study on fenfluramine sparked controversy by claiming remarkable improvements in two autistic children. Although designed to facilitate weight loss in nonhandicapped adults, fenfluramine also reduces blood serotonin levels in the brain, leading many investigators to believe it might be helpful for people with autism. The 1982 study led to a large, multicenter trial of fenfluramine which was unable to replicate earlier positive results. Subsequent studies have also shown serious side effects associated with this medication. Although fenfluramine might produce positive changes in isolated cases, its lack of general effectiveness and serious side effects make it less desirable than most other alternatives.

Naltrexone, an opiate receptor blocker, is generating considerable interest among investigators. Based on the theory that a major problem in autism is elevated brain opioid activity, this intervention has only been used on an experimental basis. Although reports on its effectiveness have been mixed, the most enthusiastic accounts are from those using opiate receptor blockers with children who are severely self-injurious. More research is definitely needed with this exciting new biological intervention.

Prozac is a new medication prescribed for depression in the nonhandicapped population and there has been recent success with people with autism. Prozac has ameliorated some of the effects of depression, especially among adolescents and young adults with autism for whom this is a serious problem. Further, Prozac has been helpful with extreme obsessive and compulsive behaviors. Anafranil is another new medication that has proven helpful in modifying the extreme obsessive and compulsive behaviors that can accompany autism.

Weaker and less potent than other medications, megavitamins have been administered and evaluated in several studies. Although the evidence on the effectiveness of megavitamins is mixed, several studies show modest improvements. It appears that some autistic children, although clearly not all and probably not even the majority, benefit from these interventions. The improvement rate with megavitamins is similar to the other biological interventions which makes this approach preferable, according to some professionals, because there are fewer side effects. Dietary modifications, assuming that allergic reactions cause or at least aggravate symptoms of autism, are also being advocated by some professionals.

## C. Behavioral Interventions

Behavioral interventions are effective in improving behaviors of people with autism. Generated from learning theory, these techniques are strongly influencing programs for people with autism and related developmental handicaps. Although originally limited to the systematic administration of rewards and punishments, behavioral interventions have increased and diversified. Today, there are several different behavioral systems for working with handicapped people: operant, cognitive, social learning, and educational.

### 1. Operant Approaches

Operant training techinques are the straightforward application of the principles of learning theory. The major principles of reward and punishment are clear and direct: behaviors paired with positive events or consequences become more positive whereas those paired with negative events or consequences become more negative. The principles of reward and punishment are central to operant approaches with the goal of developing and increasing positive behaviors while eliminating or decreasing less productive behaviors.

Operant approaches have been effective in developing cognitive and social behaviors in children with autism and related developmental handicaps. Finding appropriate rewards is often a challenge with nonresponsive autistic youngsters, but investigators able to build them up have been effective in improving skills and behaviors. Operant behavioral techniques have been effective in decreasing some of the most troublesome and severe behavior problems accompanying autism such as aggression and self-injurious behaviors. Successful techniques for reducing behaviors have been withdrawal of reinforcements, such as attention; time-out procedures requiring isolation; and overcorrection (following an undesirable behavior with activities designed to correct the damage). Although many of these procedures for reducing inappropriate behaviors have been effective and are common practice, many professionals are now discouraging their use in favor of more positive approaches.

## 2. Cognitive Approaches

Cognitive behavioral approaches are also effective with autistic children. Like the operant techniques, they follow learning theory and emphasize observable behaviors. Unlike operant learning theory, however, cognitive approaches do not dismiss all unobservable variables as unsuitable for meaningful study. Although unobservable cognitions are difficult to measure, thoughts and ideas are central for cognitive theorists who believe these cognitive processes follow basic rules of learning and behavior.

Structured teaching techniques, based on cognitive theory, have been widely used with autistic people. These techniques are similar to operant approaches in emphasizing behaviors rather than underlying psychodynamic processes. Structured teaching differs from the operant techniques, however, in stressing the autistic person's understanding of what is expected rather than the principle of positive reward. The focus is on how well an autistic person can understand the environment and its expectations for him. To the extent that rewards and punishments clarify what is expected—and in many cases they do—these are useful and important for structured teaching. Nevertheless, several other techniques are considered to be equally important: organizing the physical environment to help visually clarify tasks and boundaries; establishing developmentally appropriate schedules; performing careful, individualized assessments; and establishing meaningful routines.

Relaxation training is another cognitive approach that is helpful for autistic clients. Because anxiety is so frequently associated with autism, helping autistic people stay calmly in control is an important priority. Relaxation training focuses on an autistic person's cognitions, using deep breathing, muscle relaxation, and visual imagery to neutralize anxiety. Biofeedback is sometimes used with those who cannot understand the basic aspects of relaxation training.

## 3. Social Learning Approaches

Social learning theory examines behaviors in their social contexts and the implications for personal functioning. Because social interaction is a central deficit in autism, this approach has much to offer those working with autistic people and their families.

Social learning approaches have emphasized the importance of social skills training. Targeting specific skills for remediation and practicing those skills in natural settings are important aspects of these efforts. Techniques like modeling, role playing, and rehearsal are frequently used to highlight and teach more appropriate social behaviors.

## 4. Educational Approaches

Special education programs stress behavioral interventions with autistic children. The most effective of these identify specific, individualized behavioral goals and develop behavioral interventions to achieve them. Educational interventions emphasizing individualized assessment strategies and the development of meaningful environments are especially effective. The need for community-based training and close parent–professional collaboration are also recognized as important.

Several new trends are emerging in special education programs for students with autism and related disabilities. First, there is a movement toward community-based instruction outside the classroom and in community settings to teach skills necessary for effective adult functioning. Examples are teaching shopping skills in an actual grocery store or teaching mobility skills by learning to ride the neighborhood bus. Because the goal for autistic children is to function as adults in their own communities, community-based instruction has become an important way to prepare them.

Another new trend is to provide opportunities for autistic children to be with nonhandicapped peers for portions of the day. Recent investigations have demonstrated the effectiveness of nonhandicapped peers in teaching social and play skills to autistic children. Exposure to nonhandicapped peers also provides autistic students with more appropriate models of acceptable behavior. Progressive, potentially helpful programs providing contact with nonhandicapped peers are only effective to the extent that they are carefully planned and well organized.

Although there is general agreement about the value of interactions with nonhandicapped peers, there is some disagreement as to the best way to implement programs. Some argue for special classes in regular public schools where autistic children can get the specialized instruction they need, but still be exposed to nonhandicapped students. Others advocate for including students with autism in regular education.

Another current trend is the emphasis on vocational training with less attention to traditional academic subjects. This change from the former practice is a direct result of experiences with autistic adults. Many successful graduates of special education programs, given adequate support and training, are now working at competitive jobs in their communities. Their success has been a major source of pride. These graduates have challenged prevailing educational practices because their successes have resulted from strong vocational training and not from traditional academic school programs.

## IV. SUMMARY

Autism, the most severe of the developmental disabilities, has been carefully studied by researchers and clinicians since Leo Kanner first identified the syndrome in 1943. Defined primarily by difficulties in communication, social relationships, and by a narrow range of interests, several causes and possible neurological mechanisms have been identified. Although there is no cure on the horizon, behavioral, biological, and educational interventions have been instrumental in diminishing its devastating effects. Current practices are emphasizing increased community involvement throughout the lives of children and adults with this disability.

## BIBLIOGRAPHY

Mesibov, G. B., and Dawson, G. (1986). Pervasive developmental disorders and schizophrenia. *In* "Behavior Disorders in Infants, Children, and Adolescents" (J. M. Reisman, ed.), pp. 117–152. Random House, New York.

Mesibov, G. B., Schopler, E., Schaffer, B., and Michal, N. (1989). Use of the Childhood Autism Rating Scale (CARS) with autistic adolescents and adults. *J. Child Adolescent Psychiat.* **28**, 538–541.

Rutter, M., and Schopler, E. (eds.) (1978). "Autism: A Reappraisal of Concepts and Treatment." Plenum, New York.

Schopler, E., and Mesibov, G. B. (eds.) (1983). "Autism in Adolescents and Adults." Plenum, New York.

Schopler, E., and Mesibov, G. B. (eds.) (1984). "The Effects of Autism on the Family." Plenum, New York.

Schopler, E., and Mesibov, G. B. (eds.) (1985). "Communication Problems in Autism." Plenum, New York.

Schopler, E., and Mesibov, G. B. (eds.) (1986). "Social Behavior in Autism." Plenum, New York.

Schopler, E., and Mesibov, G. B. (eds.) (1987). "Neurobiological Issues in Autism." Plenum, New York.

Schopler, E., and Mesibov, G. B. (eds.) (1988). "Diagnosis and Assessment in Autism." Plenum, New York.

Schopler, E., and Mesibov, G. B. (1994). "Behavioral Issues in Autism." Plenum, New York.

# Autoantibodies

SENGA WHITTINGHAM
MERRILL ROWLEY
*Monash University, Australia*

## GLOSSARY

**Adjuvant** Substance that can increase a specific immunologic response to an antigen

**Complement** Heat-labile substance in serum that binds to antibody in the antigen–antibody complex and initiates a series of reactions resulting in lysis of a cell

**Epitope** Part of the antigen molecule with which antibody binds

**Lysis** Rupture of a cell when its membrane is breached and its medium is disturbed as occurs when an antibody binds to the cell in the presence of complement

**Receptor** Structure on the membrane of a cell that receives information from molecules such as hormones which bind to the receptor and signal the cell to perform a special function

THE HUMAN IMMUNE SYSTEM IS A HIGHLY EFFICIENT defense system designed to protect an individual from the damaging effects of infectious agents such as viruses, bacteria, and parasites. One defense tactic of this immune system is to produce specific antibodies to potentially damaging infectious organisms. These antibodies have the facility to inactivate or destroy microorganisms that are recognized as nonself.

Immunological tolerance or nonaggression against self is essential to well-being. When the immune system of an individual produces antibodies against components of its own body (self), these are termed autoantibodies. The development of any immunological attack against self would seem undesirable, although apparently harmless, weakly reactive autoantibodies exist in the circulation without jeopardizing health, and may play a role in removing dead or damaged cells and tissue. However, the autoantibodies characteristic of autoimmune diseases have the same aggressive capacity as the antibodies developed in response to infection. Such autoantibodies can inflict life-threatening damage, e.g., autoantibody destruction of red blood cells in autoimmune hemolytic anemia.

## I. NATURE OF AUTOANTIBODIES

Autoantibodies, like other antibodies, are immunoglobulins (Ig). The basic unit of all immunoglobulins is made up of two distinct types of polypeptide chains linked together by disulfide bonds to give a four-chain structure made up of two small (light) polypeptides and two larger (heavy) polypeptides. According to the composition of the heavy chains, autoantibodies are divided into five *classes:* IgG, IgM, IgA, IgD, and IgE. IgG, IgD, and IgE occur only as monomers of the four-chain unit, IgA occurs in both monomeric and polymeric forms, and IgM occurs as a pentamer with five linked four-chain subunits.

The precise nature of autoantibodies depends on the composition of the chains which are divided into constant regions that show little variability between antibody molecules and highly variable regions produced by gene mutations within the antibody-producing cells which, through their differences, determine

ENCYCLOPEDIA OF HUMAN BIOLOGY, Second Edition, VOLUME I. Copyright © 1997 by Academic Press. All rights of reproduction in any form reserved.

the specificity of the autoantibody. Thus there is the capacity to produce the same infinite number of variants of autoantibodies as there is in the normal response designed to defend the body against an unknown multitude of infectious agents in a hostile environment.

Autoantibodies are usually polyvalent, i.e., more than one class is present. Most autoantibodies are predominantly IgG and as such are potentially damaging because antibodies of class IgG are the most abundant in blood, have high affinity for antigen, bind complement and hence cause cell lysis, bind to scavenger monocytes via Fc receptors, and pass through the placenta from mother to child. Autoantibodies that are predominantly class IgM or IgA are less common, exceptions being, respectively, the "cold" autoantibodies of autoimmune hemolytic anemia and rheumatoid factor of rheumatoid arthritis, and the endomysial autoantibodies of celiac disease and dermatitis herpetiformis which react with primate reticulin.

Class IgG is further divided into four *subclasses:* IgG1 through IgG4. Like the classes, these are isotypic variants that do not vary between subjects and, individually, have properties of special functional importance. For example, whereas antibodies of subclass IgG1 or IgG3 bind complement, antibodies of subclass IgG4 do not, and indeed may protect from complement-mediated lysis. Examples of autoantibodies that are IgG subclass restricted are the antinuclear antibody, anti-La, of primary Sjögren's syndrome, which is mainly IgG1; the antimitochondrial antibody of primary biliary cirrhosis, which is predominantly IgG3; and the basement membrane antibody of bullous pemphigoid, which is chiefly IgG4.

In contrast to these isotypic variants, variants also exist among immunoglobulins that confer differences between individuals and are known as *allotypes.* These variants include regions on the constant region of the heavy chains of IgG (Gm) which are genetically determined. Light chain constant regions also have two isotypic forms, $\kappa$ and $\lambda$, and a genetically determined allotype on the $\kappa$ chain, Km. Predisposition to some autoimmune diseases is genetically determined, and particular allotypes have been shown to be associated with these diseases and their corresponding autoantibodies.

The autoantigen-binding sites which contribute to autoantibody specificity on the Ig molecules are located at the hypervariable regions at the tips of the variable regions. Located at these hypervariable regions and independent of the isotypic and the allotypic structures are *idiotypes* that are unique to individual

antibody molecules and behave as antigens in that they can induce antibodies to them. [*See* Idiotypes and Immune Networks.]

## II. AUTOANTIBODIES IN DISEASE

Autoantibodies specific for characterized autoantigens are associated with autoimmune diseases that can be arbitrarily classified into those that are multisystem (Table I) and those that are organ specific (Table II). Autoantibodies associated with *multisystem autoimmune diseases* usually react with autoantigens that are widely distributed within the body, such as autoantibodies to the various nuclear antigens, e.g., DNA, Ro(SS-A), La(SS-B), histones, ribonucleoproteins, centromere proteins, or topoisomerase 1. As may be expected in these multisystem autoimmune diseases, many organs are affected by the autoimmune reaction.

Autoantibodies associated with *organ-specific autoimmune diseases* usually react with autoantigens that are specific to a particular tissue, e.g., insulin-producing cells of the pancreas in insulin-dependent diabetes mellitus or thyroid peroxidase in thyroiditis.

Because the grouping is broad, some diseases do not "fit" the group description readily (Tables I and II). In the multisystem group there is a gradation from those that are truly multisystem affecting many organs and associated with a range of autoantibodies, such as systemic lupus erythematosus, to those like primary biliary cirrhosis which mainly affects the liver yet has a highly specific autoantibody reactive with an enzyme of the pyruvate dehydrogenase complex that is found in all cells of the body. [*See* Autoimmune Disease.]

## III. INDUCTION OF AUTOANTIBODIES

With few exceptions the development of autoantibodies cannot be attributed to any particular precipitating event. Infectious agents, drugs, cancer, and trauma may induce an autoantibody response (Table III) which in some is due to molecular mimicry or the inability of the immune system to distinguish the inducing agent from self. However, the majority of cases of autoimmunity have no obvious cause and autoantibodies appear to develop spontaneously. How self-tolerance is maintained in normal healthy subjects is still unclear and it is evident from studies in animals that it is not easily broken. Autoantibodies are only induced by an autoantigen when it is injected with

**TABLE I**
Autoantibodies in Multisystem Autoimmune Disease

| Autoantibodies to | Disease | Frequency in Caucasians (%) |
|---|---|---|
| dsDNA | Systemic lupus erythematosus | >90 |
| Histones | | 50 |
| Sm | | 5 (30% Blacks and Asians) |
| PCNA (proliferating cell nuclear antigen) | | <2 |
| $P_0$, $P_1$, $P_2$ (ribosomes) | | 10 |
| Ro(SS-A) | Subacute cutaneous lupus | 70 |
| | Congenital heart block | 70 |
| Histones | Drug-induced lupus | >90 |
| Erythrocyte antigens | Autoimmune hemolytic anemia | 100 |
| La(SS-B) | Primary Sjögren's syndrome | 90 |
| Ro(SS-A) | | 90 |
| Jo-1 (histidyl tRNA synthetase) | Polymyositis | 30 |
| Mi | Dermatomyositis | 35 |
| Cardiolipin | Anticardiolipin syndrome ±SLE | >90 |
| Lupus anticoagulant | | >90 |
| Ribonucleoprotein | Mixed connective tissue disease | >90 |
| Immunoglobulin (rheumatoid factor) | Rheumatoid arthritis | 70 |
| Collagen type II | | 70 (early disease) |
| Granulocyte nuclei | | 70 (low titer) |
| Histones | | 20 |
| Topoisomerase I (Scl-70) | Scleroderma—diffuse variant (systemic sclerosis) | 80 |
| PM/Scl | | 8 |
| Ku | | 5 |
| Centromere proteins | Scleroderma—CREST variant | 70 |
| Granulocyte nuclei | Autoimmune hepatitis Type I | 70 (high titer) |
| Histones | | 70 |
| Smooth muscle (actin) | | 70 (high titer) |
| Asialoglycoprotein receptor | | 75 |
| LKM 1 (liver kidney microsomes) | Autoimmune hepatitis Type II | 70 |
| Mitochondria (pyruvate dehydrogenase E2 complex) | Primary biliary cirrhosis | >90 |
| Basement membrane of skin (hemidesmosomes) | Bullous pemphigoid | >90 |
| Intercellular substance of squamous epithelium (desmosomes) | Pemphigus vulgaris | >90 |
| Glomerular basement membrane (collagen type IV) | Goodpasture's syndrome | >90 |
| ANCA (neutrophil cytoplasm) | Wegener's granulomatosis Vasculitis | >80 |
| Reticulin | Celiac disease Dermatitis hepetiformis | 100% (active disease) |
| Heart | Dressler's syndrome Cardiomyopathy | 70 |
| Purkinje cells (PC Ab or Yo) | Paraneoplastic cerebellar degeneration | High frequency |
| Neuronal nuclei (ANNA or Hu) | Paraneoplastic peripheral neuropathy or encephalomyeloradiculopathy | High frequency |

## TABLE II
### Autoantibodies in Organ-Specific Autoimmune Diseases

| Autoantibodies to | Target organ | Disease | Frequency in Caucasoids (%) |
|---|---|---|---|
| TSH receptor | Thyroid | Graves' hyperthyroidism | 90 |
| Thyroid peroxidase (TPO) | | | 50 (low titer) |
| Thyroglobulin | | | 30 (low titer) |
| Thyroid peroxidase | Thyroid | Hashimoto's thyroiditis | 100 (high titer-acute phase) |
| Insulin | Pancreas | Insulin-dependent diabetes mellitus | 70 (in children) |
| Islet cell | | | 70 (at diagnosis) |
| GAD (glutamic acid decarboxylase) | | | 70 |
| IA-2 (ICA512 or 40 K) | | | 65 |
| Neuronal voltage-gated calcium channels | Neuromuscular junction | Lambert–Eaton syndrome | High frequency |
| Acetylcholine receptor | Neuromuscular junction | Myasthenia gravis | 90 |
| Striated muscle | | | 30 (associated with thymoma) |
| $H^+K^+$ATPase (gastric parietal cell) | Stomach | Pernicious anemia | 90 |
| Gastric intrinsic factor | | | 70 |
| Adrenal | Adrenal | Primary Addison's disease | 80 |
| ACTH receptor | Adrenal | Nodular adrenal hyperplasia (Cushing's syndrome) | Not known |
| Ovary | Ovary | Premature ovarian failure + Addison's disease | >90 |
| Pituitary | Pituitary | "Hypophysitis"/hypopituitarism | Not known |

strong immunological adjuvants known to enhance the antibody response. From animal studies, usually performed in mice, it is clear that autoimmunity is strongly influenced by genetic effects; some strains of mice spontaneously develop autoimmune diseases that are equivalent to human diseases. In humans, susceptibility to autoimmunization is influenced by a

## TABLE III
### Inducers of an Autoantibody Response

| Inducer | Examples | Autoantigen |
|---|---|---|
| Infectious agents | β hemolytic *Streptococci* | Heart specific |
| | *Trypanosoma cruzi* | Heart specific |
| | *Mycoplasma pneumoniae* | Erythrocyte |
| Drugs | Hydralazine | Histones |
| | Procainamide | Histones |
| Tumors | Ovarian | Erythrocyte |
| | Small cell carcinoma of lung | Neuromuscular junction |
| | Ovarian/breast | Brain related |
| | Lymphoid malignancies | Skin keratinocytes |

range of genetic factors, including gender, race, and age. There is a striking preponderance of females among postpubertal patients with autoimmune disease; the ratio of females to males reaches 10 to 1 in some diseases. Most autoimmune diseases occur rarely before puberty and when they do, there is no female preponderance, suggesting that female hormones enhance the autoimmune response. Insulin-dependent diabetes mellitus is an example of an autoimmune disease that occurs in childhood and, in childhood, has no female preponderance.

Particular autoimmune diseases occur more frequently in some races than in others and within some families than others. While this can be clearly attributed to genetic susceptibility to some autoimmune diseases, for others, such as autoimmune gastritis where the association with genetic markers is weak, the primary event and the associated potentiating factors are not well understood. Also, the concordance of autoimmune diseases in identical twins is lower than can be explained in genetic terms alone, implying that the influential factors at play may be environmental or due to somatic mutations. Noteworthy, but still unexplained, is the steady increase in the frequency of autoantibodies throughout life, with the peak occurring in the advanced age of apparently healthy people. [*See* Genetic Diseases.]

## IV. MECHANISMS FOR AUTOANTIBODY-MEDIATED INJURY

Autoantibodies employ the same mechanisms to inflict damage on self as antibodies do in their attack on foreign invaders. The early stages of induction of human autoantibodies have not been studied. After infection, the first antibodies produced are class IgM, and these IgM antibodies, which appear in blood within 3 or 4 days of infection, peak and fall with time. By 10 days, class IgG antibodies are detected and these also peak and fall with time. For autoantibodies, the Ig class, which is usually IgG, has been well established when autoantibody is first detected and persists until the tissue containing the autoantigen is destroyed or the causal agents (Table III) are withdrawn. [*See* Antibody-Antigen Complexes: Biological Consequences.]

Autoantibodies initiate damage to tissue by binding to epitopes on the autoantigen. If the autoantigen is present on the cell surface, the result of the interaction between autoantibody and multiple autoantigenic sites on the cell surface may cause damage directly. As early as the turn of the century, it was recognized that antibodies reacting with antigens on the surface of erythrocytes could initiate a cascade of reactions, which culminated in the rupture of the erythrocyte, spillage of its contents, and collapse of the cell (Fig. 1). Destruction of large numbers of erythrocytes in the body by this means resulted in life-threatening hemolytic anemia.

The first described autoantibody was an autoantibody of this type, the Donath–Landsteiner antibody, detected in patients with paroxysmal cold hemoglobinuria (PCH), the descriptive term given to the disease because people who had the autoantibody passed red urine (blood-stained) following exposure to cold. It was shown that the symptoms were due to the fact that erythrocytes, chilled in the presence of serum from patients with PCH, lysed when they were warmed to 37°C. This accounted for the paroxysm in which the passing of red urine followed exposure to cold. Previous inactivation of the patient's serum at 56°C prevented lysis. Although it was many years before the process was understood, it is now known that autoantibody in the serum of patients with PCH binds to the surface of the erythrocyte in the cold. On warming, the erythrocytes lyse in the presence of

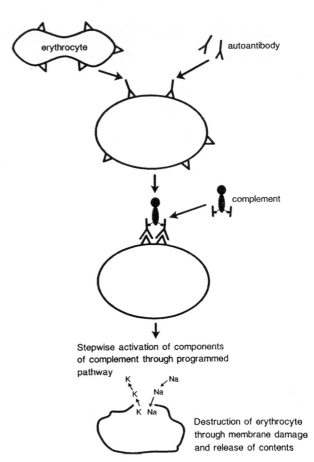

**FIGURE 1** When autoantibody reacts with the normally biconcave erythrocyte, the erythrocyte becomes spherical in shape. Complement present in serum reacts with autoantibody now firmly bound to the surface of the erythrocyte and initiates a series of reactions which result in the rupture of the membrane of the erythrocyte, expulsion of its contents, and collapse of the erythrocyte.

the serum component, complement, and the contents of the erythrocytes are excreted in the urine.

Other autoantibodies may initiate cell damage by binding to receptors on cell membranes. If the autoantigen is associated with a membrane receptor on a cell, the function of the receptor may be impaired because normal signal transmission is blocked by autoantibody and the receptor is inactivated or there may be cross-linking of receptors with or without lysis by complement, resulting in destruction of the cell (Fig. 2). The first demonstration of this was the detection of an autoantibody that reacted with a receptor for thyroid-stimulating hormone (TSH). Autoantibody blockade of the TSH receptor interrupted signals from the pituitary gland to the thyroid, resulting in failure of the thyroid cell to secrete the thyroid hor-

Functional impairment of cell receptors

(A) Antibody binds to receptor      (B) Antibody blocks ionic channel

cell with receptors

(C) Antibody ⋏ binds to receptor binding substance ⚬̇
then a second antibody forms which fortuitously
"looks like" the receptor binding substance ⤳
so binds to the receptor

Destruction of cell receptors

(D) Antibodies cross-link receptors      (E) Complement ⋏ binds to antibodies on receptors and damaged cell is ingested by the scavenger cells

**FIGURE 2** Functional impairment of a cell may occur because (A) autoantibody bound to the receptor blocks the reception of the signal necessary to initiate function, (B) autoantibody blocks the ionic channel necessary for the transmission of the signal, or (C) autoantibody induces the formation of another antibody (idiotyic antibody) which binds to the receptor because it resembles the natural binding substance. Destruction of the cell may occur if (D) autoantibodies cross-link receptors, thereby damaging the membrane, or (E) complement binds to autoantibody on the receptors and scavenger cells scavenge the cell as they would a microorganism that has bound antibody and complement.

mone thyroxine. [*See* Thyroid Gland and Its Hormones.]

Another chronic, debilitating illness in which autoantibodies interfere with the normal transmission of signals between cells is myasthenia gravis, a disease characterized by weakness, rapid onset of fatigue, and paralysis manifesting itself as a drooping eyelid or generalized weakness. Symptoms are explained by autoantibody blockade of the muscle cell receptors, which normally bind to the signaling substance acetylcholine, discharged from the nerve endings to initiate muscular activation (Fig. 2). The receptor on muscle is known as the acetylcholine receptor, and the sig-

nificance of autoantibodies in myasthenia gravis was recognized when rabbits injected with acetylcholine receptors produced antibodies associated with muscular weakness similar to that of patients with myasthenia gravis. [*See* Myasthenia Gravis.]

Although the means by which these autoantibodies may cause disease is clear, the effect of other autoantibodies is less clear. An enigma is the presence of autoantibodies to intracellular autoantigens. Many autoantibodies react with molecules of great functional importance that lie in organelles deep within the cell. Experimentally these autoantibodies can be shown to inhibit cell function *in vitro* but these antibodies have no access to the contents of living cells so it is unclear as to whether these autoantibodies can impair the function *in vivo*.

Finally, there is a role of circulating autoantibody–autoantigen complexes which are due to binding of autoantibodies with circulating autoantigens. Such autoantibody–autoantigen complexes are particularly associated with some of the multisystem autoimmune diseases, e.g., systemic lupus erythematosus. These immune complexes may be deposited in skin, kidneys, joints, and small peripheral blood vessels, leading to inflammation and damage of the organs involved.

## V. ASSAYS FOR AUTOANTIBODIES

The serologic assays for detecting autoantibodies are based on classical assays developed early in the century for the detection of antibodies (Ab) produced in response to infections or by immunization of animals with selected antigens (Ag). All assays are based on attempts to visualize the simple equation

$$Ag + Ab \Leftrightarrow AgAb \text{ complex.}$$

This may be detected directly by precipitation or agglutination or by indirect methods in which antihuman globulin with an appropriate tag binds to the complex so that the reaction can be read. Such tags include fluorescent dyes (immunofluorescence), enzymes (ELISA), or radioactive isotopes (radioimmunoassay).

Historically, the first serologic tests were applied to the detection of autoantibodies to erythrocytes. These employed either lysis (Fig. 1) in the presence of complement or agglutination of the autoantibody-coated erythrocytes by antihuman globulin made in a rabbit.

The most frequently requested test referred to the modern clinical immunologic diagnostic service laboratory is that for antinuclear antibodies (ANA), and the test most widely applied to the detection of this and other tissue autoantibodies such as gastric parietal cell antibodies and smooth muscle antibodies is immunofluorescence. Immunofluorescence involves the application of serum containing autoantibody to either a monolayer of fixed cells or a frozen section of tissue on a microscope slide followed by antihuman globulin tagged with a fluorescent dye, usually fluorescein isothiocyanate, before examination microscopically using a source of ultraviolet light. By this means, autoantibodies can be localized to particular tissues or to particular locations within cells (see Color Plate 7). As a procedure for detecting autoantibodies, this versatile, sensitive assay is, in the hands of an experienced reader, the most informative screening test in the service laboratory. Given a clue to the specificity of the autoantibody, the assayist can then confirm specificity by further tests employing particular antigens.

Gel diffusion or enzyme-linked immunosorbent assays are favored for the definition of the saline-soluble ANAs, latex particle agglutination for antithyroid peroxidase and antithyroglobulin, and immunoprecipitation employing isotopes for the detection of antibodies to receptors as in myasthenia gravis or glutamic acid decarboxylase as in insulin-dependent diabetes mellitus. Which assay is selected depends on the availability of the autoantigen, the ease with which the reaction can be detected, and the cost. Many assays of a more sophisticated nature require further adaptation and simplification for use in the diagnostic service laboratories and are still performed only in special laboratories. One assay popular in the research laboratory is immuno- or Western blotting. For this assay a cell extract or pure autoantigen is subjected to electrophoresis in a gel to separate the component proteins according to their molecular size. After separation the proteins are transferred from the gel to a solid support of nitrocellulose or nylon membrane which is then exposed to serum containing autoantibodies. The reaction with antigen is detected by antihuman globulin tagged with an enzyme, a fluorescent dye, or a radioactive isotope.

## VI. AUTOANTIBODIES IN HEALTH

Autoantibodies present in apparently healthy people may be arbitrarily divided into natural autoantibodies with no known disease association and autoantibodies which have a well-recognized association with autoimmune diseases.

Natural autoantibodies are usually present in low titers and are directed at autoantigens on the cytoskeleton or nucleus of the cell. Using the sensitive technic of immunoblotting for the detection of natural autoantibodies, it has been shown that some autoantibodies in the serum of healthy people are, like fingerprints, individual specific. These natural autoantibodies are class IgG and pass the placenta from mother to child so that the maternal specificities are present at birth but are lost and replaced by the individual's own specificities which persist for a long period of time, possibly life.

Autoantibodies associated with well-recognized autoimmune diseases may arise transiently, persist for a period of weeks or months, and then disappear. Examples are rheumatoid factor, ANA, and smooth muscle autoantibodies that are frequently detected during the acute and convalescent stages of infection by Epstein–Barr virus. These autoantibodies are usually low in titer and contrast with the chance finding in health of a high-titer autoantibody such as gastric parietal cell autoantibody which has a known association with a particular autoimmune disease. A gastric parietal cell autoantibody detected in an apparently healthy subject may point to a family history of the autoimmune gastritis of pernicious anemia or to one of a cluster of autoimmune diseases with which it is associated. In the absence of such family history, it may signal the impending onset of symptoms of pernicious anemia. In over 90% of subjects, the autoantibody is associated with some impairment of gastric parietal cell function and chronic inflammation of the body of the stomach, but the period taken for the disease to progress to pernicious anemia may be months to many years. Thus the presence of persistent, high-titer autoantibodies in a healthy person alert the clinician to look for warning signs of the autoimmune disease with which the autoantibody is associated. The rapidity of evolution of disease varies greatly between diseases and between individual people, suggesting that the control of maintenance of self-tolerance is dependent on multiple and complex biological mechanisms and/or there is regeneration of specialized tissues concomitant with the autoimmune attack.

## VII. CONCLUDING REMARKS

The association of a specific autoantibody or group of autoantibodies with disease provides a valuable clue to diagnosis. Autoantibodies can be used as predictors of disease or as confirmatory evidence of established disease. For example, the detection of islet cell antibodies in an apparently healthy child of parents with a family history of insulin-dependent diabetes mellitus points to the need to monitor the child for signs of diabetes, and the detection of gastric parietal cell and intrinsic factor antibodies in a woman of late middle age with anemia will alert her clinician to a diagnosis of latent pernicious anemia.

The contribution of autoantibodies to organ damage is not always clear. It is obvious when cell injury or impairment of cell function can be shown to be due to the direct effect of autoantibody on the living cell. More controversial is the role of autoantibodies when the autoantigen is intracellular. Then it is assumed that T cells which are prominent in the cellular infiltrate in the inflamed organ, and not antibody, are responsible for damage. However, presently the performance of assays for the T-cell response is not cost-effective in the diagnostic laboratory because these assays are imprecise, time-consuming, and the results are difficult to interpret because of the high background in normal subjects. Until the complex issue of how immunological tolerance to self is broken and the role of T cells in the autoimmune response is amenable to simple analysis, autoantibodies will remain the diagnostic tools of epidemiologists and clinical immunologists.

## BIBLIOGRAPHY

Roitt, I. (1994). "Essential Immunoloy," 8th Ed. Blackwell Scientific Publications, Oxford.

Rose, N. R., and Mackay, I. R. (eds.) (1992). "The Autoimmune Diseases," Vol. II. Academic Press, San Diego.

# Autoimmune Disease

NOEL R. ROSE
*The Johns Hopkins University*

I. Discrimination between Self and Nonself
II. Experimental Autoimmune Disease
III. The Spectrum of Human Autoimmune Disease
IV. Common Features of Autoimmune Disease

## GLOSSARY

**Activated lymphocytes** Lymphocytes that have been stimulated, specifically (by a corresponding antigen) or nonspecifically (by mitogens), to produce soluble products (cytokines)

**Activated macrophages** Macrophages that have been stimulated by cytokines and are, thereby, metabolically active

**Adhesion molecule** Molecules on the cell surface that promote the binding of one cell to another

**Adjuvant** Substance capable of potentiating an immune response

**Allele** One of a number of alternative genes at a particular locus

**Alloantigen (or isoantigen)** Antigen that exists in alternative forms in the same species and, thus, may induce an immune response in the same species

**Anergy** A state of unresponsiveness to an antigen

**Antibody** Protein produced as a result of the introduction of an antigen and capable of combining with that particular antigen

**Antigen** Substance that can induce an immune response

**Antigen-presenting cell** Cell that prepares antigen so that it is capable of initiating an immune response

**Apoptosis** Programmed cell death

**Autoantibody** Antibody directed to self-antigen

**B cell** Lymphocyte that acquires antigen-specific recognition without need of a functioning thymus

**CD** Clusters of differentiation; identifying molecules located at the lymphocyte cell surface

**Cell-mediated immunity** Immune response based on lymphocytes and macrophages with no participation of antibody

**Class I major histocompatibility antigen** Cell-surface antigen present on all nucleated cells and responsible for tissue graft rejections; encoded in humans at HLA-A, -B, or -C loci

**Class II major histocompatibility antigen** Surface antigen on immunologically active cells responsible for regulating immune responses; encoded in humans at HLA-DR, -DP, or -DQ loci

**Clone** Uniform population of cells descended from a single progenitor

**Complement** System of plasma proteins that, following activation by antigen–antibody complexes, is able to lyse susceptible cells

**Cytokines** Soluble products of cells that affect the behavior of other cells

**Cytotoxic T cells** T cells that can kill other cells; usually MHC class I-restricted CD8 T cells

**Fc** Crystallizable portion of the antibody molecule responsible for many of its biological properties

**Helper T cells** T cells that can help B cells to make antibody; MHC class II-restricted CD4 T cells

**Idiotype** Unique sites of an antigen-specific recognition structure (receptor) of a B cell or a T cell

**Major histocompatibility complex (MHC)** Primary genes that encode alloantigens preventing tissue transplantation; termed HLA in humans. Class I MHC products are found on the surface of virtually all nucleated cells in the body. Class II MHC products are found on immunologically active cells, such as antigen-presenting cells

**Plasmaphoresis** Removal from body of blood, separation of plasma, and readministration of blood cells to the same donor to reduce levels of circulating antibody

**T cell** Lymphocyte that acquires antigen-specific recognition during maturation in the thymus

ENCYCLOPEDIA OF HUMAN BIOLOGY, Second Edition, VOLUME 1. Copyright © 1997 by Academic Press. All rights of reproduction in any form reserved.

ONE OF THE BASIC RULES OF IMMUNOLOGY IS that animals do not react to their own bodies; however, exceptions to this rule are apparent in the form of autoimmune disease, where immunity to an antigen of one's self is responsible for disease. This article considers the ways by which the body normally distinguishes self from nonself and then examines the exceptions to that rule, which result in autoimmune disease.

## I. DISCRIMINATION BETWEEN SELF AND NONSELF

The most striking characteristic of the immunological system is its limitless capacity to respond to any foreign substance (referred to as antigen). Upon entering the body, an antigen will encounter a lymphocyte with a complementary recognition structure (receptor) to which the antigen binds. Specialized genetic mechanisms permit the generation of a sufficient diversity of receptors to accommodate any conceivable antigen. Binding to its corresponding antigen activates the lymphocyte so that it proliferates, eventually producing a clone of lymphocytes bearing the same receptor as its ancestor (Fig. 1). One major limitation in immunological responses is that the host seems incapable of responding to antigens naturally present in its own body.

The practical importance of the body's ability to discriminate accurately between self and nonself is illustrated in the practice of blood transfusion. Individuals of blood group A are incapable of producing antibodies to the A substance, even though they are perfectly capable of responding to the closely related B substance. Conversely, blood group B individuals produce antibody to the A substance (anti-A) but never anti-B. The difference between A substance and B substance rests entirely in the terminal monosaccharide units of a large polysaccharide–lipid complex. This small difference, barely demonstrable by chemical means, is unerringly distinguished by the immunological system.

A similar situation prevails with respect to the major histocompatibility complex (MHC). THe human MHC is referred to as HLA. These cell-surface structures are the primary barrier impeding transplantation of tissues and organs from one person to another. Patients rapidly reject kidneys that are greatly different in their HLA type but will accept tissues that are HLA identical. Thus, the common practice of tissue matching for organ transplantation is based on recognition of self. [See Immunobiology of Transplantation.]

The rule of unresponsiveness to self-antigens is not absolute. In fact, all normal human beings have autoantibodies to many constituents of their own bodies. These "natural" autoantibodies usually belong to the IgM class and have low affinity for their corresponding antigen. They are not associated with any disease. It has been suggested that they have a normal "housekeeping" function in ridding the body of damaged or effete constituents of organs and tissues that enter the bloodstream. [See Autoantibodies.]

## A. The Basis of Self-Recognition

Several explanations have been offered to account for the ability of the immunological system to distinguish foreign antigens from self-antigens. F. M. Burnet first proposed that exposure of immature lymphocytes to their corresponding antigens during embryonic development results in the elimination of these lymphocytes due to programmed cell death or apoptosis. Because most antigens of the self category are present during embryonic development, the precursors of self-reactive lymphocytes would be deleted (Fig. 2). This view was supported by experiments on acquired immunological tolerance which showed that mice could be made unresponsive to tissues of genetically different mice by introducing spleen cells from the potential donors during embryonic life. The presence of the foreign cells at an early stage of immunological development led to a form of false self-recognition by eliminating the precursors of lymphocytes potentially reactive with histocompatibility antigens of the donor.

When it was later realized that the production of an immunological response depends on cooperation of at least three cell types, the suggestion was offered that to achieve tolerance to self-antigens only one of the necessary cell types needs to be silenced. The first cell in the sequence, the antigen-presenting cell, takes up antigen, cleaves it into fragments, and introduces one of the fragments to the next cell in line, the helper (CD4) T cell. For the T cell to recognize it, the antigen fragment must be joined to the HLA. The T cell, one of the two major T lymphocyte populations involved in the immune response, acquires its ability to recognize a particular antigen fragment during its development in the thymus. It recognizes antigen as a sequence of amino acids only in the context of compatible Class II MHC gene products (Fig. 3). The cytotoxic (CD8) T cell is responsible for cell-mediated immunity. It recognizes antigen fragments in conjunction with Class I MHC. Another lymphocyte population, B cells, responds to particular antigenic determinants

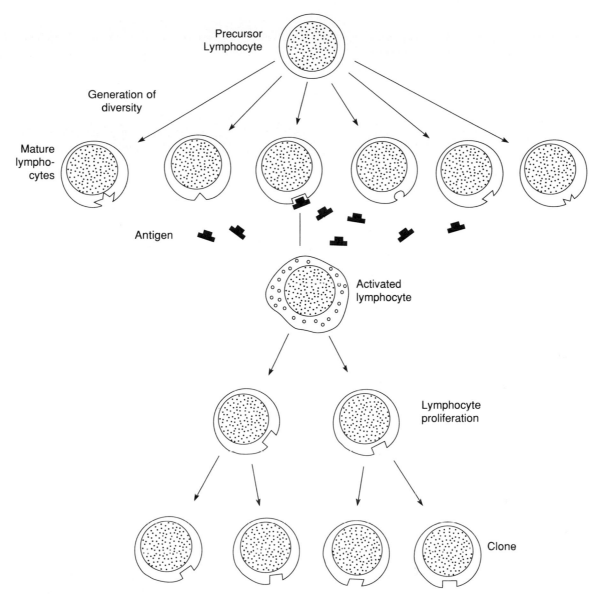

FIGURE I  Development of lymphocyte clone reactive with corresponding antigen.

on the basis of their three-dimensional configuration. B cells usually require help from T cells to differentiate into antibody-secreting plasma cells. [*See* Lymphocytes.]

Substantial evidence now shows that antigen-presenting cells are fully capable of processing and presenting self-antigens as proficiently as foreign antigens. B cells reactive with many self-antigens are also demonstrable in normal animals and are the source of the many natural autoantibodies referred to earlier. Faced with large amounts of their corresponding anti-

gen, however, B cells may be unresponsive. The situation with respect to T cells is still controversial. Young T cells destined to react with prominent cell-surface constituents, including important histocompatibility antigens, are deleted in the thymus (clonal deletion). On the other hand, T cells reactive with late-developing antigens, with organ-specific antigens, and with inaccessible antigens can be found in the blood of normal individuals.

It has been discovered that mature T cells require two types of signals for activation. A specific signal

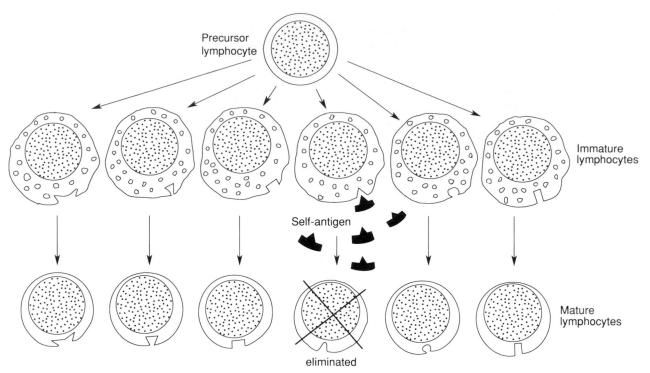

**FIGURE 2**  Deletion of immune lymphocytes reactive with self-antigens.

results when the T cell receptor is occupied by its corresponding antigen. Nonspecific stimuli are also necessary to instigate clonal proliferation; these costimuli are produced by antigen-presenting cells as well as by other lymphocytes. In the absence of the costimulatory signals, occupancy to the antigen-specific receptor induces a state of prolonged unresponsiveness referred to as clonal anergy in the T cell. This mechanism may produce tolerance to self-antigens in T cells that have escaped elimination in the thymus. [*See* T-Cell Receptors.]

Other evidence suggests that in many cases self-tolerance may depend on active suppression rather than on passive deletion or anergy. This view is based on a better understanding of normal immunological homeostasis due to the regulatory factors that govern the immunological response. Some immunologists think that one major regulatory cell inhibiting immunological responses is the suppressor T cell, which is able to control both antibody synthesis and cell-mediated immunity. T-suppressor cells are still a very controversial subject. Whether or not an immunological response occurs would depend on the balance of various T cells and their soluble products, especially between suppressor function versus the T cells that promote an immunological function (Fig. 4). Research has redefined the regulatory function of T-cell popula-

tions. Two functional subpopulations of CD4 helper T cells have been distinguished. One subpopulation, $T_H1$, produces cytokines IL-2, IFN-$\gamma$, and lymphotoxin and favors the production of cell-mediated immunity and cytotoxic T cells. The other subpopulation, $T_H2$, helps B cells produce antibodies. This subpopulation produces cytokines IL-4, IL-5, and IL-10. The two subpopulations are mutually inhibitory since the $T_H1$ cytokine IFN-$\gamma$ inhibits $T_H2$, whereas the $T_H2$ cytokine IL-10 inhibits $T_H1$ cells. $T_H1$ helpers promote the inflammatory response that marks many of the autoimmune diseases affecting solid tissues. $T_H2$ are involved in the production of autoantibodies which may mediate damage to blood cells.

In the early 1970s, N. K. Jerne pointed out that the receptor sites for antigen on antibody molecules and on T cells may themselves be antigenic. The selective portions of these receptors are called idiotypes, and anti-idiotypic antibodies and anti-idiotypic lymphocytes have been shown to abrogate particular immune response. Anti-idiotypes, then, may serve as natural regulators of immunity (Fig. 5). [*See* Idiotypes and Immune Networks.]

These various mechanisms for establishing and maintaining self-tolerance are not mutually exclusive. In fact, in any function as important as main-

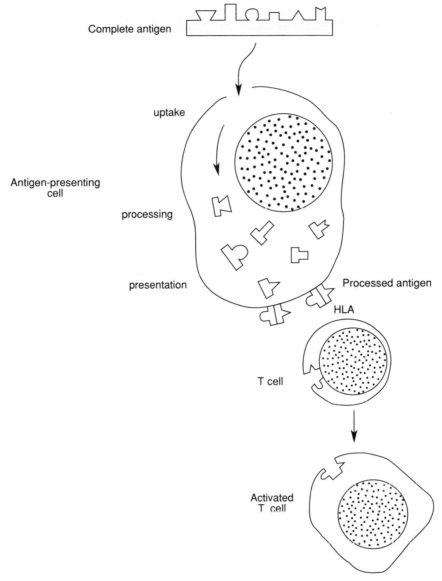

**FIGURE 3**   Processing and presentation of antigen in conjunction with major histocompatibility antigen.

taining self-recognition, several lines of defense are likely. Perhaps active suppression, for instance, represents a major mechanism for preventing self-directed immunological reactions when clonal deletion mechanisms fail to eliminate completely all self-reactive cells.

## B. Induction of Autoimmunity

Despite the existence of several mechanisms to obviate self-directed immunological responses, autoimmunity can develop and, on occasion, is associated with dis-

ease. Several different methods for inducing autoimmunity have been obtained from experimental studies of animals. Some of these mechanisms are listed in Table I.

## I. Altered Antigen

A variety of events may disturb autologous antigens and cause the body to respond to them as if they were alien. In a few instances, the change in antigen is mainly locational (e.g., antigens that are normally sequestered and do not come in contact with the immunological system). An example is the production

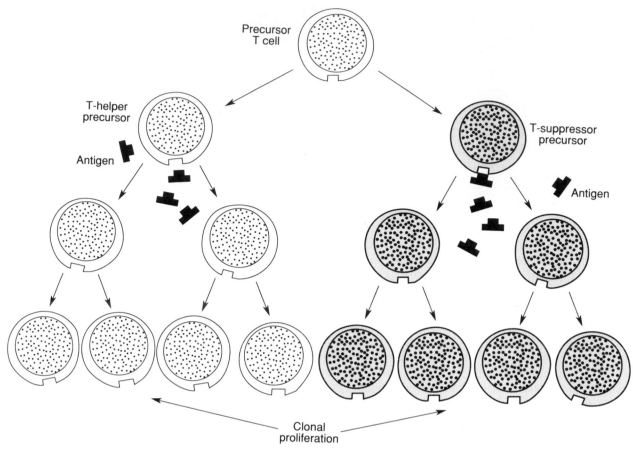

**FIGURE 4** Clonal balance. The immune response is governed by the rate of proliferation of helper versus suppressor T lymphocytes.

of autoantibodies to sperm following vasectomy. Normally, men have natural autoantibodies to sperm in low titers only. However, following vasectomy, sperm production continues and the spermatozoa are absorbed into the body, eliciting an increased sperm antibody production. There is no evidence of autoimmune disease due to vasectomy.

Another way in which autoantibodies can be induced is to administer autologous antigens with a suitable adjuvant. Freund's adjuvant, a mixture of mineral oil and acid-fast bacilli, is commonly used in the experimental laboratory. Many naturally occurring substances, including bacterial lipopolysaccharides, also function effectively as immunological stimulators and, therefore, may promote autoantibody production. It is quite possible that microbial products in nature can play the role of adjuvant and confer autoimmune potency on self-antigens that come into contact with the immunological system. The orchitis (testicular inflammation) that occasion-

ally follows mumps infection is believed to be caused by the adjuvant effect of the virus infection on testicular antigens. Other investigations have suggested that the encephalitis occasionally following measles is caused by adjuvant activity of the measles virus. However, the issue of how virus infection precipitates an autoimmune response is still unsettled.

Another possibility is that foreign antigens mimic antigens of the body and, thereby, initiate an autoimmune response. Thus, the cell surface of $\beta$-hemolytic streptococcus contains antigenic determinants similar to human cardiac muscle. Antibodies to the streptococcus, in principle, could affect the heart and perhaps cause disease such as rheumatic fever. Evidence for this possibility is still circumstantial.

In experimental animals, self-antigens can be modified chemically and thereby can induce an autoimmune disease. Such events may occur in clinical situations (e.g., the hypotensive drug $\alpha$-methyldopa is known to alter the surface of red blood cells). A cer-

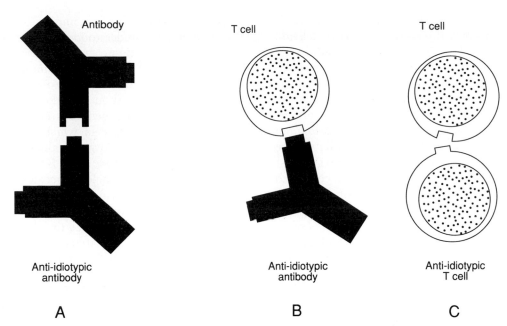

**FIGURE 5**  Anti-idiotypes. (A) Anti-idiotypic antibody directed to antigen-specific receptor to antibody complex. (B) Anti-idiotypic antibody directed to antigen-specific receptor of T cell. (C) Anti-idiotypic T cell-directed to antigen-specific receptor of T cell.

tain number of patients receiving this medication develop a hemolytic anemia in which autoantibodies react with the red cell surface. It may well be that the drug induces a sufficient change in a red cell surface determinant to initiate an autoimmune response.

Many macromolecules in the body are constantly being broken down and reformed. In the course of proteolytic breakdown, unfamiliar fragments of autologous constituents possibly are exposed and engen-

der an immunological response. Native collagen can induce an autoimmune reaction in experimental animals. These findings have raised the speculation that such a reaction may underlie the development of diseases such as rheumatoid arthritis or other collagen disorders.

## 2. Altered Immune Response

Another possibility is that an autoimmune reaction is induced by a change in the immunological system. One of the first mechanisms suggested to account for this change is the appearance of somatic mutations in the cells responsible for immunological recognition. It is as if foreign lymphocytes suddenly appear and survive in the body. An experimental analogy is found in graft vs host disease. In this disorder, immunologically competent cells from a genetically different donor are introduced into a recipient that is incapable of rejecting the transferred lymphoid cells. The transferred cells then react to the antigen of their new host. The symptoms that arise from such an encounter bear an uncanny resemblance to certain human systemic diseases of autoimmune origin, especially systemic lupus erythematosus, hemolytic anemia, and even some dermatological and gastrointestinal disorders.

A second method of altering the immunological system is to remove the thymus at a predetermined

**TABLE I**

Possible Methods for Inducing Autoimmunity

Altered antigen
- release of sequestered antigen
- combination with adjuvant
- microbial infection
- molecular mimicry
- chemical modification of self-antigen
- abnormal proteolysis

Altered immunological response
- somatic mutation
- thymectomy
- heightened immune response
- decreased suppression
- target organ defect
- chemical or physical modification of lymphocyte populations

point in its development. The thymus is the site of T-cell apoptosis, so that thymectomy impairs deletion and clonal balance. The loss of thymic regulation may then lead to disproportionately greater decrease in suppressor cells compared with helper cells or interfere with deletion of anti-self clones and give rise to the spontaneous appearance of autoimmune disorders. Experimental evidence indicates that irradiation and certain drugs that mimic the effect of irradiation are capable of altering the balance of helper and suppressor lymphocyte populations. Thymectomy combined with irradiation is an effective means of producing autoimmune disease in certain genetically predisposed animals. [*See* Thymus.]

Genetically selected animal strains in which disease occurs spontaneously have been one of the most effective tools for the study of autoimmune disease. The New Zealand black (NZB) strain of mice is remarkable for the many autoimmune phenomena the animals develop during their short life span. At 3 months of age, they start to show hemolytic anemia due to the autoimmune destruction of their red blood cells. Later, they develop enlarged spleen and lymph nodes and have multiple autoantibodies in their serum, including some that react with cell nuclei. If the mice live long enough, many of them develop cancer of the lymphatic system. New Zealand white (NZW) mice develop no autoimmune disease and have a normal life span. When NZB and NZW strains are crossed, however, the hybrid strain, NZB/W, shows severe autoimmune disease. The hemolytic anemia is less prominent than it is in the NZB parent, but the kidney lesions are more severe. Antinuclear antibodies eventually appear in almost all of the animals. The disease is a close facsimile of human systemic lupus erythematosus. The antibodies are analogous to the antinuclear antibodies found in human lupus, and premature death of these mice is usually due to kidney failure caused by antigen–antibody deposits in the glomeruli of the kidney.

The origin of the autoimmune disease in New Zealand mice has long been a subject of controversy. Some investigators claim they have isolated viruses that cause lupus-like symptoms. The abnormal development of the immune system of New Zealand mice is shown by experiments in which disease can be transferred to normal young mice by primitive cells taken from the bone marrow of older NZB/W mice. Attention has shifted to the B cells of the NZB/W mice, which have proved to be hyperreactive before the onset of any autoimmune disorder. Such B-cell abnormalities are present in NZB/W mice that lack a thymus, so they do not result from overstimulation by

T cells. Although the intrinsic impulse that drives these B cells is not yet understood, they are certainly responding to a genetic program. This inborn error in the developmental program of immunologic control results in multiple autoimmune responses.

Additional lessons about the inheritance of autoimmune disease can be learned from the study of a genetically determined disease in which the autoimmune response assails a single organ. Valuable insights have come from studies of obese strain (OS) chickens which suffer from a thyroid deficiency because of severe thyroiditis. By demonstrating the presence of autoantibodies to thyroglobulin in the serum of OS chickens, it was possible to prove that the disease is due to autoimmunity. Several distinct genetic defects have been identified in the OS flock that combine to predispose the chickens to spontaneous development of autoimmune thyroiditis. First, the birds respond vigorously to chicken thyroglobulin. This trait is inherited in conjunction with their major histocompatibility complex (i.e., the chicken's analogue of HLA). This trait is probably an immunoregulatory gene coding for the response to some particular determinant on thyroglobulin. Second, the chickens have a generalized defect in the maturation of their thymus because T cells with the ability to induce cell-mediated immunologic responses leave the thymus at an earlier age than they do in comparable normal chickens and take up their residence in peripheral lymphoid tissues. Finally, the OS chickens have an intrinsic aberration of thyroid function so that their thyroidal uptake of iodide is greater than normal. Iodine is a substituent of the thyroglobulin molecule and enhances its antigenic potency.

Thus, the spontaneous appearance of autoimmune disease in the OS chicken depends on the conjunction in this selected flock of at least three independent genetic traits: a strong immunologic response to thyroglobulin, a thymic abnormality, and a defect in the thyroid gland. Lessons learned from the study of spontaneous autoimmune disease in animals seem directly applicable to understanding autoimmune disease in human beings.

## II. EXPERIMENTAL AUTOIMMUNE DISEASE

A number of autoimmune diseases can be induced by deliberate immunization of animals with tissue antigens, especially organ-specific antigens that are distinctive for a particular tissue, together with Freund's adjuvant. The adjuvant not only magnifies but redirects the immunologic response, favoring cell-mediated immunity, so that pathological damage occurs to the target organ. In the absence of an appropriate adjuvant, it is often possible to induce autoantibodies to organ-specific antigens without the simultaneous appearance of lesions in the appropriate organ.

Examples of some experimentally induced autoimmune diseases are given in Table II. Allergic encephalomyelitis is a demyelinating disease of the central nervous system, which can be produced in animals by injection of brain antigen from the same or another species. The antigen responsible for the encephalomyelitis has been identified as a basic protein of myelin, and the particular amino acid sequence within the protein that is responsible for the demyelinating changes has been pinpointed. This encephalogenetic determinant differs somewhat among species. Although antibodies to myelin basic protein are produced, the disease itself seems to be attributable to T-cell-mediated immunity.

A very similar disease used to be seen in humans receiving the old Pasteur vaccine for the prevention of rabies. This vaccine, now no longer in use, was prepared from rabbit brain and presumably induced the same sort of autoimmune response that one sees in experimentally immunized animals. The encephalitis that occasionally follows infection by certain viruses may also exemplify an autoimmune response. Some investigators draw an analogy between experimental encephalomyelitis and human multiple sclerosis. Although both represent demyelinating disorders, clinical and pathological features are different.

Peripheral neuropathy or Guillain–Barré syndrome results from demyelinization of the peripheral nervous system. An antigen has been isolated from peripheral nerve myelin that is similar to, but distinct from, the myelin basic protein of the central nervous system. It has been shown to produce peripheral nerve lesions in some species of animals. Based on this evidence, therefore, some investigators have claimed that Guillain–Barré syndrome is an example of an autoimmune response to peripheral nerve antigen.

In many ways, thyroiditis represents the most complete example of human autoimmune disease. Antibodies to the major protein of the thyroid gland, thyroglobulin, are found in most of the patients with classical chronic lymphocytic thyroiditis. An experimental analogue of thyroiditis can be induced by injecting thyroglobulin into experimental animals. The animals then develop autoantibodies to thyroglobulin and produce histological lesions, which are basically similar to those seen in human thyroiditis patients.

Immunization of genetically susceptible strains of mice with murine myosin produces cardiac lesions resembling human myocarditis (inflammation of the heart muscle) and cardiomyopathy (enlargement of the heart). The antigen in this experimental disease is strikingly organ specific because cardiac myosin induces the response but myosin isolated from skeletal muscle does not. Following infection by coxsackievirus, Group $B_3$, the same genetically susceptible mice develop a similar progressive myocarditis accompanied by the production of cardiac myosin-specific autoantibodies. In other resistant strains of mice, coxsackievirus $B_3$ causes acute, self-limited myocarditis without evidence of autoimmunity.

## III. THE SPECTRUM OF HUMAN AUTOIMMUNE DISEASES

### A. Organ-Specific Autoimmune Diseases

Many human disorders are associated with an autoimmune response. In some cases, the autoimmunity may be incidental; in other instances, it actually contributes to the pathogenesis of the disease. Even in those instances where autoimmunity is incidental, the appearance of autoantibodies is an important diagnostic feature.

It is convenient to group human autoimmunities into three major categories: organ specific, systemic, and receptor specific. However, the delineation among the three groups is somewhat arbitrary so that it may be more reasonable to think of these diseases as a spectrum ranging from the most localized type to the most generalized diseases (Table III).

**TABLE II**

Some Examples of Experimentally Induced Autoimmune Diseases and Possible Human Equivalents

| Disease | Antigen | Human equivalent |
|---|---|---|
| Allergic encephalomyelitis | Myelin basic protein | Postviral encephalitis Multiple sclerosis (?) |
| Peripheral neuropathy | Peripheral nerve basic protein | Guillain-Barré syndrome |
| Thyroiditis | Thyroglobulin | Chronic lymphocytic thyroiditis |
| Myocarditis | Cardiac myosin | Postviral myocarditis Cardiomyopathy |

## TABLE III
### Some Human Autoimmune Diseases

| Disease | Antigen(s) |
| --- | --- |
| **Organ specific** | |
| Chronic lymphocytic thyroiditis | Thyroglobulin; thyroid peroxidase |
| Insulin-dependent diabetes mellitus | Unknown antigen of the islet B cell |
| Pernicious anemia | Intrinsic factor; gastric mucosal antigen |
| Chronic active hepatitis | Actin |
| Primary biliary cirrhosis | Pyruvate dehydrogenase |
| Sympathetic ophthalmia | S-antigen of uvea |
| Postpericardiotomy and postinfarction syndrome | Heart antigens |
| Postinfection myocarditis | Heart antigens |
| Pemphigus vulgaris | Intercellular substance of epidermis |
| Bullous pemphigoid | Cutaneous basement membrane |
| Autoimmune glomerulonephritis | Glomerular basement membrane |
| Immune complex glomerulonephritis | Various |
| **Systemic** | |
| Systemic lupus erythematosus | Various, especially native DNA |
| Rheumatoid arthritis | Fc of immunoglobulin G |
| Sjögren's disease | Ribonucleoproteins La(SS-B) and Ro(SS-A) |
| Autoimmune hemolytic anemia | Red blood cell surface (e.g., Rh antigens) |
| Paroxysmal cold hemoglobinuria | P or p blood group antigens |
| Idiopathic thrombocytopenia | Platelet surface |
| Leukopenia | White blood cell surface |
| **Receptor specific** | |
| Myasthenia gravis | Acetylcholine receptor |
| Graves' disease | Thyrotropin receptor |

Prototypes of the organ-specific diseases are the autoimmune endocrinopathies, such as chronic lymphocytic thyroiditis described previously, and the insulin-dependent form of diabetes mellitus (also referred to as Type-1 diabetes).

In the case of the autoimmune endocrinopathies, the antigen targeted by the autoimmune response is specific for a particular organ, and the pathological changes are largely confined to a single organ or part of it. Not uncommonly, two or more organ-specific autoimmune responses may be found in a particular individual. For instance, patients with chronic thyroiditis generally produce autoantibodies to both thy-

roglobulin and thyroid peroxidase. The latter antigen, unique for the thyroid, is a cellular enzyme responsible for the incorporation of iodine into thyroglobulin. Patients with thyroiditis frequently have additional antibodies to other endocrine organs, such as the pancreatic islet or the adrenal gland, as well as to the gastric mucosae, the source of intrinsic factor.

An important endocrine disease associated with autoimmunity is insulin-dependent diabetes mellitus. This form of the disease was formerly referred to as juvenile-onset diabetes, but we now realize that the disease can begin at any age. A major feature of this disease is the production of autoantibodies to pancreatic islet cells, sometimes accompanied by lymphocytic infiltration of the islets. The antigen responsible for diabetes has not yet been identified but is presumed to be specific for the insulin-producing $\beta$ cells of the islet.

Among the gastrointestinal diseases in which autoimmunity is a prominent feature, pernicious anemia is outstanding because it serves as a bridge between the autoimmune endocrinopathies and the autoimmune gastrointestinal diseases. The autoimmune response is directed to the gastric mucosae, but two organ-specific antigens are involved. One is a component of the gastric mucosal cell itself and the other is to the soluble product of the mucosal cells—the intrinsic factor. The intrinsic factor is essential for absorbing vitamin $B_{12}$, so that patients with this antibody develop $B_{12}$ deficiency. Considerable overlap exists between pernicious anemia and thyroiditis because about half of the patients with chronic thyroiditis develop antibodies to gastric mucosae and many of them also show some clinical signs associated with gastric autoimmunity. Conversely, about 80% of patients with pernicious anemia have one or another of the antibodies to thyroid antigens.

The immunological picture in two important liver diseases, chronic active hepatitis and primary biliary cirrhosis, is quite different from that seen in the typical organ-specific diseases. Only the liver is affected because the pathology is confined to the hepatic parenchymal cell on the one hand or to the biliary tree on the other; however, the autoantibodies are not organ specific. In the case of chronic active hepatitis, antibody to smooth muscle is a common finding. This antibody is directed to actin, a contractile protein of muscle and other tissues. In primary biliary cirrhosis, one finds antibody to several mitochondrial proteins, which are present in virtually all tissues. One of the mitochondrial antigens has been identified as the widely distributed enzyme pyruvate dehydroxygenase. Thus, we are left with the paradoxical situation of organ-specific disease associated with nonorgan-specific autoantibodies. The specificity of T cells from patients with these diseases has not yet been studied.

Immunologically, the eye occupies a special position because it seems to be sequestered from the main channels of the immunologic response. In fact, the lens of the eye was the first tissue shown to possess a unique organ-specific antigen and the first to which autoimmunity could be produced by experimental autoimmunization. Autoantibodies to the uveal tract are found in sympathetic ophthalmia, suggesting that autoimmunity contributes to the pathology. This disease typically follows injury to the uveal tract of one eye. Inflammation, and eventually blindness, occurs in both the injured and the uninjured eye, supporting the premise that the disease is due to an organ-specific autoimmune response. In experimental animals, uveitis has been produced by immunization by a unique protein of the retina, S-antigen.

Following cardiac surgery or myocardial infarction, a certain proportion of patients develop late symptoms consisting of fever and leukocytosis. Often, these patients have antibodies to heart muscle. The explanation generally offered is that the necrotic heart tissue releases partially degraded constituents, which are able to stimulate an autoimmune response.

As described previously, several types of viruses, especially Group B coxsackieviruses, may produce myocarditis. Chagas' disease, which is caused by the protozoan parasite *Trypanosoma cruzi*, is the principal cause of myocarditis in South America. Some myocarditis patients develop autoantibodies to heart muscle antigens.

Several important skin diseases are associated with autoimmune processes. Pemphigus vulgaris is a potentially lethal disease characterized by massive blister formation within the epidermis. Bullous pemphigoid is a less serious disease in which blisters form at the junction of the epidermis with the underlying dermis. In pemphigus, one finds antibodies to the intercellular substance of the stratified squamous epithelium, recognizable by immunofluorescence, whereas in bullous pemphigoid antibodies are directed to the basement membrane between the epidermis and dermis. In both cases, experimental evidence has shown that the antibodies can produce pathological changes *in vivo*.

The kidney is another organ that is often affected by autoimmune disease. One form of immunological kidney disease is autoimmune glomerulonephritis, in which autoantibodies are produced to the glomerular basement membrane of the kidney. This antibody is recognized by its smooth, linear deposition along the glomerular basement membrane. Often the antibody reacts with basement membranes of other tissues such as the lung so that some patients show pathological changes in both kidney and lung, referred to as Goodpasture's syndrome.

In the more common form of immunologically mediated renal disease, immune complex glomerulonephritis, the long-standing persistence of antigen, perhaps one derived from an infectious organism, results in the production of antibody and the development of antigen–antibody complexes. These complexes localize in various parts of the body, including the lung and the skin, but they are most harmful in the renal glomerulus. Progressive immune complex disease of the kidney is associated with systemic lupus erythematosus, where antinuclear antibody combines with nuclear antigen liberated from dying cells to produce circulating immune complexes. The antibody can be seen in the kidney, where it takes on a granular, irregular appearance.

## B. Systemic Autoimmune Diseases

Just as chronic thyroiditis can be thought of as the prototype of organ-localized autoimmune diseases, systemic lupus erythematosus (SLE) serves as the paradigm of generalized autoimmune disease. Some basic dysfunction in the immune regulation causes the patient to produce antibodies to a great variety of self-antigens that are widely distributed throughout the body and easily accessible to the immunological system. Why these particular antigens should be the target of the autoimmune response in SLE is still uncertain. Lupus patients show antibodies to red cell-surface antigens, leukocytes, platelets, clotting factors, and cytoplasmic components of many tissues, including mitochondria and microsomes. The most prominent antibody found in SLE is the antinuclear antibody, which really represents a family of antibodies to different constituents of the cell nucleus. It has become possible to define many of these nuclear antigens more precisely and to carry out tests that determine the pattern of antibody response in individual patients with SLE or a related disease. Virtually all patients with SLE have an antibody to denatured or single-stranded DNA, but this antibody is also found in many other patients with related or even unrelated diseases. When present, antibody to native or double-stranded DNA is indicative of SLE.

Most patients with rheumatoid arthritis also produce several kinds of autoantibodies. The most characteristic one is an antibody to Fc (i.e., the constant portion of the immunoglobulin molecule) and is referred to as rheumatoid factor. It is not peculiar to rheumatoid arthritis because patients with other chronic diseases, including infectious diseases like subacute bacterial endocarditis, may develop such antibodies. They presumably result from the long-term production of primary antibody to the infectious agent, followed by production of secondary antibody to the primary antibody. Interaction of antigen with the primary antibody produces a conformational change in the antibody molecule so that its Fc portion is altered. Such a change may create a novel antigen that will, in turn, provoke the secondary immune response resulting in rheumatoid factor. Circulating antigen–antibody complexes involving the rheumatoid factor may localize in the joints of patients with rheumatoid arthritis and contribute to the pathology. The rheumatoid synovium also shows evidence of immunoglobulin synthesis, suggesting local formation of antigen–antibody complexes in the fluids of the joints.

Another autoimmune response described in rheumatoid arthritis is directed to a type of collagen, and cell-mediated immunity to the type II collagen may play a role in the inflammatory lesions in the joints.

T cells isolated from the joints of patients with rheumatoid arthritis have been found to react with a heat-shock protein. These proteins are produced by cells in response to a stress such as heating. They are remarkably conserved so that similar proteins are found in bacterial and animal cells, providing a striking example of molecular mimicry. [*See* Heat Shock.]

Sjögren's disease is a striking example of a disorder that is intermediate between the organ-specific autoimmunities and the systemic autoimmunities. This syndrome of dry eyes and dry mouth is often found in patients with rheumatic or connective tissue diseases, and many patients have rheumatoid factor or antinuclear antibody. The autoantibodies most characteristic of Sjögren's disease are directed to nucleoprotein antigen, referred to as Ro(SS-B) and La(SS-B). In addition, antibody to thyroglobulin, often in high titers, is found in a few patients with Sjögren's syndrome. Finally, antibodies to the ductular epithelium or salivary gland have been described in cases of Sjögren's syndrome and perhaps are more relevant to the pathological focus of the disease.

Autoimmunity can be a major cause or a contributor to hemolytic anemia. Whether induced initially by an invading virus or by a drug–erythrocyte complex, antibodies to red cell-surface antigens are found in patients with acquired hemolytic anemias. These anti-

bodies are often directed to particular blood group antigens on the surface of the red cell. In the common forms of hemolytic anemia, the antibody is often specific for one of the Rh antigens. The antibody in paroxysmal cold hemoglobinuria combines with the red cell at low temperature but produces complement-mediated lysis when the sensitized red cells are warmed. These antibodies are directed to the P system of blood group antigens. Remarkably, alloantigens of the red blood cell frequently serve as autoantigens under pathological conditions.

Other blood cells may also be targets of immune attack. Idiopathic thrombocytopenic purpura is accompanied by antibodies to the platelet, whereas some forms of leukopenia are associated with leukocyte agglutinins.

## C. Receptor-Specific Autoimmune Diseases

Myasthenia gravis is the best example of an autoimmune disease involving specific cell receptors because autoantibodies to the acetylcholine (Ach) receptors at neuro-muscular junctions are characteristic of the disease. These antibodies interfere with the transmission of nerve–muscle impulses, perhaps by blocking the Ach receptor or by causing more rapid breakdown of it. The levels of antibody correlate well with the clinical symptomatology, and reduction of circulating antibody by plasmaphoresis provides temporary remission of the disease. [*See* Myasthenia Gravis.]

Another disease caused by antireceptor antibodies is Graves' disease, in which hyperactivity of the thyroid is the major abnormality. Graves' disease patients produce antibodies that bind to the thyrotropin receptor on the surface of the thyroid epithelial cell. The antireceptor antibodies actually simulate the actions of thyrotropin by stimulating the thyroid gland, as thyrotropin would do. This effect gave rise to an earlier term for this autoantibody—long-acting thyroid stimulator.

## IV. COMMON FEATURES OF AUTOIMMUNE DISEASE

### A. Effector Mechanisms

The spectrum of human autoimmune diseases extends across all organ systems of the body. Their clinical manifestations depend on the location of the antigenic target, the vulnerability of that target, and the effector mechanisms involved. Some autoantigens are directly accessible to antibody. In the hemolytic anemias, for example, autoantibodies to the surface coat the red blood cells. Some antibodies can activate complement, which directly lyses the cells, whereas other antibodies promote ingestion of the coated red blood cells by phagocytic (scavenging) monocytes of the spleen. Antibodies to physiologically important receptors are able to block receptor function or, sometimes, to stimulate the receptors in an unregulated fashion. Antibodies to soluble constituents of cells may bind to the corresponding antigen when the latter are released following cell injury or death, producing circulating immune complexes that can localize in and damage the renal glomeruli. In all of these instances, autoantibodies themselves produce autoimmune disease, but the ancillary pathogenetic mechanisms vary greatly.

If the antigenic target of an autoimmune response is located in a solid tissue, it is rarely accessible to antibody, but cellular immunity may mediate disease. Cytotoxic T cells attack their targets directly by reacting with cell-surface antigens. Other subpopulations of T cells respond to antigen stimulation by producing lymphokines, which are soluble products of activated lymphocytes. Some lymphokines injure surrounding cells directly; other lymphokines activate macrophages, prompting them to injure tissue. Long-term lymphokine production may lead to granuloma formation with a consequent destruction of surrounding tissues. Some lymphocyte and monocyte populations express receptors on their surface for the Fc portion of immunoglobulin on their surface. Such mononuclear cells have the potential to act cooperatively with an antibody specific for a particular cell-surface antigen resulting in antibody-dependent, cell-mediated cytotoxic damage. [*See* Lymphocyte-Mediated Cytotoxicity.]

All of these various immunopathological mechanisms resulting from the encounter of a self-reactive T cell with its corresponding antigen have been implicated in autoimmune disease. In many diseases, both humoral and cellular immunity probably contribute to the final outcome. The effector mechanisms in autoimmune pathology are considered similar to those brought into play to defend the body against infection by invading pathogenic microorganisms.

## B. Genetics

In studies of autoimmune disease, as in so many other investigations of biological phenomena, both inherent and environmental agents play a role. The genetic predisposition to autoimmunity becomes obvious when monozygotic twins are compared with dizygotic twins. In the case of insulin-dependent diabetes, if one monozygotic twin is found to be diabetic, the other identical twin is about 50% more likely to be positive for the disease. The comparable percentage for nonidentical twins is 10%. The difference between the two percentages is a minimal estimate of the considerable role of inheritance in leading to this autoimmune disease. At the same time, the finding that the concordance between identical twins is less than 100% emphasizes the importance of environmental factors in inducing autoimmune disease.

Insulin-dependent diabetes is closely associated with the genes that control HLA. Population studies have shown that the prevalence of diabetes in individuals with HLA types HLA-DR3 and HLA-DR4 (Class II MHC gene products; see below) is much greater than in the population at large. Yet, the incidence of autoimmune diabetes is low even in DR3/DR4-positive individuals. On the other hand, diabetes does occur in HLA-DR3/DR4-negative subjects. Several possible explanations for this incomplete association have been proposed.

First, environmental as well as genetic factors interact in initiating autoimmune disease. Infection, diet, toxic chemicals, irradiation, and stress are all known to trigger autoimmune responses in genetically predisposed individuals. Second, penetration of genetic traits depends on internal regulatory factors. Autoimmune disease, in general, is more frequent in females than in males, and autoantibody production increases with age. Very likely, these differences reflect the interaction between the regulatory mechanisms of the body and immunological systems.

On the genetic level, evidence from studies on animals (described previously) has shown that a number of genes acting through entirely different pathways determine the onset of autoimmunity. Some genes produce abnormalities in the target organ, whereas others affect immunological recognition. The association of a particular MHC specificity with a particular disease, such as diabetes, likely reflects some directive function of the MHC molecule in the autoimmune response.

With respect to their participation in the immune response, the HLA antigens are divided into two classes. Class I MHC antigens are expressed on virtually all nucleated cells in the body; they are the main barriers to successful tissue transplantation. The Class I HLA gene products, referred to as HLA-A, -B, or -C, combine with antigen fragments to provide the actual target for cytotoxic T cells. Class II MHC gene products appear primarily on immunologically active cells, including B lymphocytes and antigen-presenting cells. They also bind particular antigen fragments to produce the molecular complex that activates antigen-specific helper T cells and initiates the immune response. It is, then, noteworthy that autoimmune diseases are usually more closely associated with the Class II loci, HLA-DR and HLA-DQ, than they are with the Class I loci.

The HLA-DR4 specificity is present in approximately 70% of patients with insulin-dependent diabetes, compared with about 30% of controls. It has been shown that the DR4 locus possesses different alleles (see glossary) that can be distinguished based on their base sequence. The DR4 association with diabetes corresponds with specific alleles of the DR locus. Rheumatoid arthritis and other autoimmune diseases with DR4 association associate with a different allele at the DR locus.

The basis of Class II MHC association is not known. Possibly, the particular autoantigenic fragment responsible for the response is presented better to T cells by some Class II products than by others. Experimentally, this situation has been described using fragments of myelin basic protein to produce allergic encephalomyelitis. An alternative explanation for the MHC association is based on the finding that the repertoire of T-cell specificities is shaped by MHC expression in the thymus. In brief, the T-cell population may be biased on its responsiveness to particular antigenic determinants depending on the HLA-DR type of the thymic epithelium of the host. Regardless of which explanation is correct, it should be possible to predict, with reasonable accuracy, those individuals at greatest risk of developing a particular autoimmune disease.

## C. Treatment

Understanding that autoimmunity contributes significantly to the pathogenesis of human diseases has fostered different philosophies of treatment.

### 1. Nonspecific

In the case of a life-threatening illness, such as SLE, it is sometimes necessary to depress the entire immunological system by the use of powerful immunosuppressive drugs, such as antimetabolites, corticosteroid, and cyclosporine—the same drugs used to prevent rejection of transplanted organs. Obviously, such drastic immunosuppression is hazardous because it leaves the patient vulnerable to microbial infection. A high degree of clinical skill is needed to apply such therapy. Similar in concept is the use of an antiserum, referred to as antithymocyte serum, directed specifically to the T-cell population. A somewhat more selective effect is produced by a monoclonal antibody that is specific for the helper subpopulation of T cells.

### 2. Replacement

In the case of organ-specific autoimmune disease, a particular physiological function is abnormal. Treatment can be aimed at correcting the particular altered function. For example, insulin replaces the product of the damaged islet cell in diabetes, whereas thyroid hormones are used to replace the deficient thyroid cell products in thyroiditis. On the other hand, thyrotoxic drugs, radioactive iodine, or surgical removal of the thyroid reduce the hyperthyroidism characteristic of Graves' disease.

### 3. Specific Immunosuppression

The more desirable method of treating autoimmune disorders is precise elimination of the clones of damaging autoimmune lymphocytes only. This goal requires a greater understanding of the normal mechanisms of immunological regulation. One possible mechanism is the production of antigen-specific suppressor cells. It has been found that antigen-specific suppressor T cells are induced by oral administration of antigens. Such cell populations may act directly or through soluble factors to reduce an autoimmune response. Another possible strategy is to develop anti-idiotypic sera or lymphocytes that react specifically with the T-cell receptor involved in autoimmune recognition. Immunization with the pathogenetic T-cell clone in an inactivated form may evoke an active anti-idiotypic response to the particular T-cell receptors. This actively elicited anti-idiotypic response, which involves antibody or cellular effectors or both, would be especially promising if individuals at great risk of developing autoimmune disease can be identified in advance through MHC or other genetic traits.

Finally, in experimental animals, antibodies directed to particular MHC Class II specificities can arrest those immune responses dependent on the particular MHC specificity. Such antibodies have been used to prevent autoimmune disease in animals. The ideal target of such antibodies would be not the HLA Class II determinant but rather the complex of HLA with the self-antigen fragment responsible for initiating the autoimmune response.

## BIBLIOGRAPHY

Bigazzi, P. E., and Reichlin, M. (eds.) (1991). "Systemic Autoimmunity." Dekker, New York.

Bigazzi, P. E., Wick, G., and Wicher, K. (eds.) (1990). "Organ-Specific Autoimmunity." Dekker, New York.

Bona, C. A., and Kaushik, A. K. (eds.) (1992). "Molecular Immunobiology of Self-reactivity." Dekker, New York.

Cohen, I. R., and Miller, A. (eds.) (1994). "Autoimmune Disease Models: A Guidebook." Academic Press, San Diego.

Rose, N. R., and Bona, C. (1993). Defining criteria for autoimmune diseases (Witebsky's postulates revisited). *Immunol. Today* **14**, 426–430.

Rose, N. R., and Mackay, I. R. (eds.) (1992). "The Autoimmune Diseases," Vol. II. Academic Press, San Diego.

Talal, N. (ed.) (1991). "Molecular Autoimmunity." Academic Press, San Diego.

# Autonomic Nervous System

IAN GIBBINS
*Flinders University of South Australia*

---

I. Organization of the Autonomic Nervous System
II. Autonomic Neurotransmission
III. Autonomic Control of Effector Tissues

## GLOSSARY

**Autonomic ganglion (plural: ganglia)** Aggregation of autonomic nerve cell bodies lying completely outside the central nervous system

**Enteric neurons** Autonomic neurons with cell bodies located within the walls of the gastrointestinal tract

**Myogenic activity** Intrinsic contractile activity of smooth or cardiac muscle which is not dependent on nervous or hormonal stimulation

**Parasympathetic pathways** Autonomic pathways with preganglionic neurons in the brain stem or sacral spinal cord

**Peripheral autonomic neurons** Neurons with their cell bodies located within autonomic ganglia. They include the final motor neurons in sympathetic and parasympathetic pathways, where they often are called ganglion cells or postganglionic neurons.

**Preganglionic neurons** Autonomic neurons with cell bodies in the central nervous system and axons making direct synaptic input to peripheral autonomic neurons or to adrenal chromaffin cells

**Sympathetic pathways** Autonomic pathways with preganglionic neurons in the thoracic and lumbar levels of the spinal cord

**Tonic activity** Ongoing or continuous activity of neurons or muscle cells

AUTONOMIC NERVOUS PATHWAYS PLAY A FUNdamental role in maintaining the stability of the internal environment of the body (homeostasis). For example, neurons in autonomic pathways innervate effector tissues involved in digestion, thermoregulation, fluid and electrolyte balance, control of the circulation, excretion, reproduction, and several other functions. Autonomic pathways can excite or inhibit contraction of smooth and cardiac muscle and regulate the secretory activities of many epithelia and glands. They also may influence the metabolic activity and growth of certain cell types. Some autonomic neurons in the gastrointestinal tract have a sensory function. Most target tissues are innervated by specific populations of autonomic neurons lying in precisely connected circuits. These circuits are regulated by sensory inputs, either directly to autonomic neurons themselves or into areas of the central nervous system controlling autonomic activities. Several different classes of peripheral autonomic neurons exist. They can be identified by their locations and connections, their neurotransmitter content, and their functions.

Autonomic pathways generally are activated with a minimum of conscious control. Nevertheless, there are extensive pathways in the brain and spinal cord that regulate autonomic activity in response to a wide variety of sensory inputs. This article deals only with the peripheral autonomic pathways lying outside the central nervous system. Unless stated otherwise, all the information presented here pertains to normal humans. Readers interested in clinical and pathological aspects of autonomic function should consult the books by Bannister and Appenzeller cited in the bibliography.

## I. ORGANIZATION OF THE PERIPHERAL AUTONOMIC PATHWAYS

Peripheral autonomic pathways usually are classified into three divisions, which are characterized primarily by anatomical criteria. They were named the *sympa-*

ENCYCLOPEDIA OF HUMAN BIOLOGY, Second Edition, VOLUME I.   Copyright © 1997 by Academic Press.   All rights of reproduction in any form reserved.

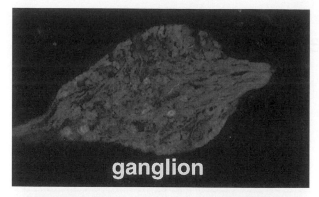

**ganglion**

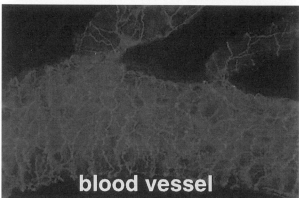

**blood vessel**

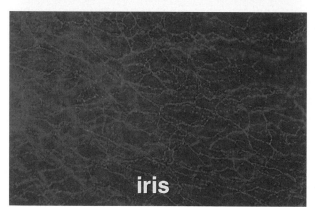

**iris**

FIGURE 1 Immunohistochemical preparations of sympathetic neurons of guinea pigs labeled to show the distribution of a neuropeptide cotransmitter, neuropeptide Y (NPY). (Top) A small ganglion from the lumbar sympathetic chain. Many of the ganglionic neurons contain NPY. Most of these cells also contain noradrenaline and are vasoconstrictor neurons whereas the rest of them innervate the pelvic organs. Most of the neurons without NPY (dark areas of the ganglion) do contain noradrenaline and mostly are pilomotor neurons. The neurons range from 15 to 45 $\mu$m in diameter. (Middle) A small artery and its branches supplying the brain. The artery is surrounded by a dense meshwork of fine varicose fibers. They are the terminal axons of sympathetic vasoconstrictor neurons. The varicosities are the sites from which transmitters (in this case, noradrenaline, NPY, and perhaps ATP) are released to cause constraction of the underlying vascular smooth muscle. (Bottom) The dilator muscle of the iris and its innervation by sympathetic pupillodilator neurons. Once again, they contain both noradrenaline and NPY. Compare with Fig. 9.

*thetic, parasympathetic,* and *enteric* divisions by J. N. Langley in the early part of this century. Within each division, there are discrete neuronal pathways, each with a characteristic function. Nevertheless, most autonomic neurons have many structural features in common.

## A. Peripheral Autonomic Neurons

Peripheral autonomic neurons lie completely outside the central nervous system. Such neurons constitute the final motor neurons of most sympathetic and parasympathetic pathways. Their nerve cell bodies are usually grouped into ganglia containing from just a few to over a million neurons (Fig. 1). The cell bodies of

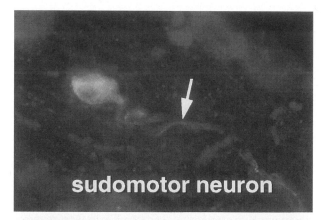

**sudomotor neuron**

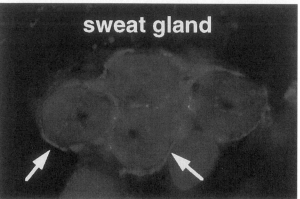

**sweat gland**

FIGURE 2 (Top) A sudomotor neurons in a human lumbar sympathetic chain ganglion that has been labeled immunohistochemically to show one of its transmitters, VIP. This neuron also contains ACh, CGRP, and somatostatin as cotransmitters. The arrow indicates the axon of the neuron that ultimately will project to sweat glands in the skin. (Bottom) A section through a sweat gland in a sample of human skin. The main secretory regions (large circles) are surrounded by varicose fibers labeled for VIP (arrows). These fibers are the terminal axons of sudomotor neurons like that above. When stimulated, these neurons cause sweating and cutaneous vasodilation by the combined effects of several transmitters on the secretory cells and nearby blood vessels. Compare with Fig. 8.

autonomic neurons typically range from about 20 to 60 μm in diameter. Most of them bear up to 10 or 12 dendrites, which may be many hundreds of micrometers long. With the possible exception of some enteric neurons, each neuron has a single 0.3- to 2.0-μm-diameter axon (Fig. 2), which is usually unmyelinated. The axons travel to their target tissue in nerve trunks which almost always contain the axons of other types of neurons such as preganglionic autonomic neurons, sensory neurons, or somatic motor neurons (for an example, see Fig. 4).

Once in or near their target tissues, autonomic axons branch extensively and become varicose in appearance (Figs. 1 and 2). The terminal branches of a single axon may have a total length of more than 20 cm and bear more than 20,000 varicosities. Within the varicosities are aggregations of membrane-bound vesicles which contain neurotransmitters. Although

morphologically well-defined synapses between peripheral autonomic neurons and their target cells are rare, varicose terminal axons often lie only 20–80 nm away from their target cells. These varicosities probably have specialized transmitter release sites similar to those seen in presynaptic neurons elsewhere in the nervous system. Neurotransmitters are released from the varicosities and diffuse across this narrow space to reach their receptors on the effector cells. In many cases, the receptors on the effector cells immediately adjacent to the varicosities are functionally different from those at more distant sites (see Fig. 3).

## B. Preganglionic Inputs

The final motor neurons in the sympathetic and parasympathetic ganglia receive direct excitatory synaptic input from preganglionic neurons with cell bodies in

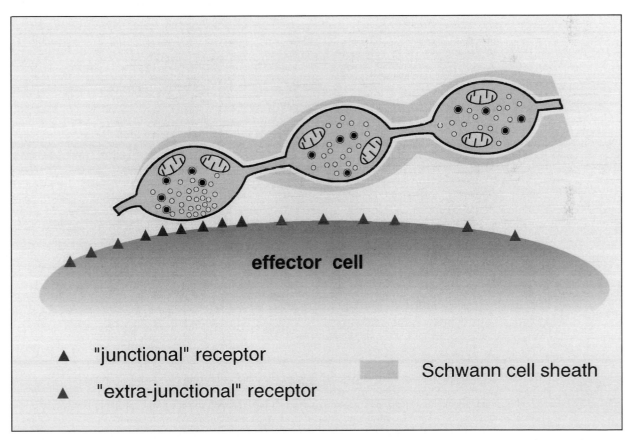

▲ "junctional" receptor

▲ "extra-junctional" receptor

　　　Schwann cell sheath

FIGURE 3　A typical autonomic neuroeffector junction. The terminal axons of autonomic motor neurons consist of a series of swellings ("varicosities") that contain small and large vesicles. The vesicles release their transmitters from sites on the varicosities close to the surface of the effector cells. In these regions, the varicosities are at least partially free of their covering of supporting Schwann cells. At low rates of stimulation of the nerve fiber, the transmitter interacts with specialized "junctional" receptors on the effector cells near the release sites. With greater rates of stimulation, the transmitter can diffuse to interact with more widely spread "extrajunctional" receptors. Each type of transmitter interacts with its own set of specific receptors. Once activated by the transmitters, the receptors trigger the appropriate responses of the effector cell.

the brain stem or spinal cord. The exact location of the preganglionic neurons is different for sympathetic and parasympathetic pathways. Indeed, each of these divisions is defined largely by the location of its preganglionic neurons as well as by its final motor neurons. The arrangement for the sympathetic division is shown in Fig. 4.

The axons of most preganglionic neurons are myelinated. As they enter a ganglion, they lose their my-

elin and branch extensively. The terminal axons of preganglionic neurons tend to enmesh the cell bodies of ganglionic neurons and form synapses onto the cell bodies and the dendrites. Each ganglion cell may receive excitatory synaptic inputs from up to 20 or more individual preganglionic neurons. This is known as *convergence* of synaptic input (Fig. 4). Some individual inputs may be sufficiently strong to trigger action potentials in the ganglion cell. In other cases, the

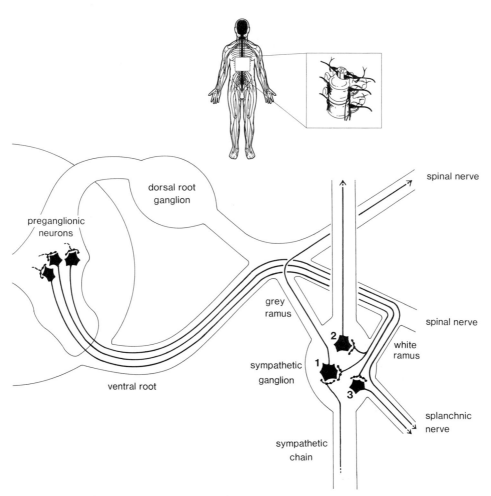

FIGURE 4   Generalized arrangement of preganglionic and final motor neurons (ganglionic neurons) in sympathetic pathways. (Inset) The thoracic spinal cord and the adjacent sympathetic ganglia. The cell bodies of the preganglionic neurons are located in the intermedio-lateral column of the spinal cord. Their axons leave the cord via the ventral roots and enter the sympathetic chain via a white communicating ramus. Some synapse with ganglionic neurons immediately. Each axon may synapse with more than one ganglionic neuron (divergence; e.g., neurons 1 and 2 receive inputs from the same preganglionic axon). Neuron 1 receives inputs from two different preganglionic neurons (convergence), one of which originates from a different spinal segment. Other preganglionic axons continue along the chain or out the splanchnic nerve. The axons of ganglionic neurons may run out the spinal nerves via a gray communicating ramus (neuron 1), they may travel along the sympathetic chain before exiting (neuron 2), or they may run out the splanchnic nerve (neuron 3).

postsynaptic excitatory potentials arising from several different preganglionic inputs need to summate before an action potential will be initiated in the postganglionic neuron. Each preganglionic neuron may form synapses with several hundred ganglionic neurons. This is called *divergence* (Fig. 4). Both convergence and divergence are important in the integration of preganglionic inputs to the final motor neurons.

Many enteric neurons do not receive direct preganglionic inputs from the central nervous system. However, most of them have synaptic inputs from other enteric neurons. Nevertheless, some enteric neurons can be activated by preganglionic neurons in the brain stem or the sacral spinal cord. Consequently, these neurons can be activated by the appropriate parasympathetic pathways.

## C. The Sympathetic Division

The anatomical arrangement of the sympathetic division is shown in Fig. 5. The cell bodies of sympathetic final motor neurons are grouped mainly into two sets of ganglia: the paravertebral ganglia and the prevertebral ganglia. The paravertebral sympathetic ganglia are connected to each other by nerve trunks to form a sympathetic chain on each side of the vertebral column. Each chain runs from cervical to sacral levels and contains about 24 ganglia arranged in an approximately segmental fashion for much of its length. The prevertebral sympathetic ganglia lie within a plexus found ventral to the abdominal aorta where they occur at or near the origins of the major arteries supplying the abdominal organs. Inferiorly, the prevertebral sympathetic plexuses extend into the pelvic autonomic plexuses associated with the pelvic viscera. Altogether, an estimated 7 to 10 million ganglionic neurons are found within the sympathetic pathways of adult humans.

By definition, all preganglionic neurons with their cell bodies located in the thoracic and lumbar segments of the spinal cord are considered to lie in sympathetic pathways. Preganglionic axons leave the spinal cord in the ventral roots (see Fig. 4). Once in the sympathetic chain, the preganglionic fibers may form synapses with neurons in the ganglion corresponding to the same spinal segment. Alternatively, they may travel several segments cranially or caudally before they form synapses onto neurons in other ganglia. Consequently, the ganglionic neurons in any particular paravertebral ganglion receive inputs from preganglionic neurons located in several contiguous segments of the spinal cord. Preganglionic fibers reach the

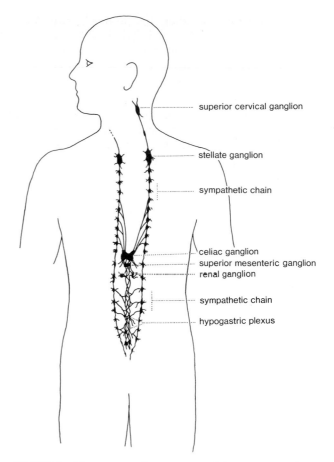

**FIGURE 5** The anatomical arrangement of the main sympathetic ganglia and their interconnections. The sympathetic chain communicates with the celiac, superior mesenteric, and renal ganglia and the hypogastric plexus via splanchnic nerves. These ganglia constitute the prevertebral ganglia and are found near the origins of the arteries with the corresponding names.

prevertebral ganglia via the splanchnic nerves which originate from the mid-thoracic to lumbar sympathetic ganglia (Figs. 4 and 5). The splanchnic nerves also contain axons of sympathetic motor neurons innervating the abdominal viscera and blood vessels (Fig. 4) as well as the axons of different classes of sensory neurons innervating the abdominal organs. [*See* Spinal Cord.]

The paravertebral sympathetic neurons innervate a wide variety of target tissues, including much of the cardiovascular system (Fig. 1), sweat glands (Fig. 2) and pilo-erector muscles in the skin, adipose tissue, and many internal organs. The cervical sympathetic ganglia supply many structures in the head, including the iris, blood vessels (Fig. 1), salivary glands, and the skin. The prevertebral ganglia project to the abdominal organs, including the stomach, intestines (Fig. 12), liver, spleen, and kidneys. In addition to innervating smooth muscle, cardiac muscle, and secretory glands, sympathetic motor neurons exert effects within parasympathetic and enteric ganglia, where they suppress synaptic transmission. Individual targets are innervated by discrete populations of sympathetic neurons lying in pathways that can be activated independently or in combination as part of coordinated reflex responses to specific sensory stimuli.

Also considered along with the sympathetic neuronal pathways are the adrenal medullary chromaffin cells. These cells are aggregated in the core of the adrenal gland and are structurally and functionally more like endocrine cells than neurons (see below). Nevertheless, adrenal chromaffin cells receive preganglionic inputs from the lumbar spinal cord and they share developmental and biochemical characteristics with many sympathetic neurons. Similar cells, known as extra-adrenal chromaffin cells, may be present in sympathetic, thoracic parasympathetic, and pelvic ganglia.

## D. The Parasympathetic Division

There are two distinct sets of parasympathetic pathways: the cranial and the sacral. The anatomical arrangement of these pathways is shown in Fig. 6. The cranial pathways innervate tissues in the head, thorax, and upper abdomen whereas the sacral pathways innervate the pelvic viscera, such as the urinary bladder and internal genital organs. In many cases, the parasympathetic ganglia are located relatively close to the organs they innervate and may even occur within the walls of the organ itself.

There are four main pairs of parasympathetic ganglia in the head. A ciliary ganglion lies behind each eye and innervates the iris (Fig. 9) and the ciliary muscles. Its preganglionic axons travel out with the oculomotor nerve. Below each orbit is a sphenopalatine (pterygopalatine) ganglion, which provides vasodilator and secretomotor innervation to the lacrimal gland, nasal mucosa, and mucosa of the palate. The preganglionic axons to this ganglion arise from neurons in the brain stem and travel with the facial nerve. The facial nerve also provides preganglionic axons to the submandibular ganglion located near the submandibular salivary gland. The submandibular ganglion cells provide vasodilator and secretomotor innervation to this gland and the adjacent sublingual salivary gland. The fourth main ganglion is the otic ganglion, located deep behind the articulation of the jaw on each

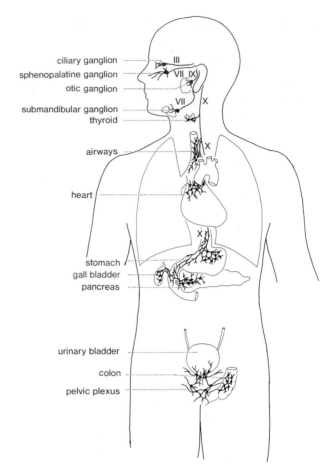

**FIGURE 6**  The anatomical arrangement of the main parasympathetic ganglia and their interconnections. Only the cranial ganglia have specific names. The locations of the small ganglia associated with the various organs are variable and are illustrated schematically here. III, oculomotor nerve; VII, facial nerve; IX, glossopharyngeal nerve; X, vagus nerve. Note that the pathways running to the gut intermingle with the intrinsic enteric plexuses.

side. It receives preganglionic inputs from neurons in the brain stem predominantly via the glossopharyngeal nerve. It also is a source of secretomotor and vasodilator innervation, this time to the parotid gland and blood vessels in the lower jaw. Scattered along the facial and glossopharyngeal nerves are many small parasympathetic ganglia that probably innervate local glands and blood vessels.

The vagus nerve provides a pathway for the parasympathetic control of diverse structures in the thoracic and upper abdominal regions. Long preganglionic axons arise from neurons in the brain stem and travel in the vagus to motor neurons located in small ganglia lying near or within the heart, airways (Fig.

10), pancreas, and thyroid gland. Some enteric neurons in the esophagus, stomach, gall bladder, and the proximal part of the small intestine also receive synaptic inputs from vagal preganglionic neurons. Within each of the organs innervated by the vagal autonomic pathways, there are mixed populations of final motor neurons which have a variety of different, and sometimes opposite, functions, e.g., contraction or relaxation of the stomach.

The sacral parasympathetic pathways contain collections of small ganglia throughout the pelvic plexus, often in close association with the pelvic viscera. For example, they occur in or near the walls of the bladder, ureters, and the cervix in females and around the vas deferens and the base of the penile erectile tissues of males.

The cell bodies of the sacral parasympathetic preganglionic neurons are located in the sacral spinal cord. Their axons run out in the pelvic nerves to the pelvic plexus. Some pelvic ganglia contain distinct populations of neurons with preganglionic inputs from cells in the lumbar spinal cord: these neurons lie in sympathetic pathways. Thus, pelvic autonomic ganglia may have neurons lying in both sympathetic and parasympathetic pathways. Indeed, some individual pelvic ganglion cells may receive convergent synaptic inputs from both sympathetic and sacral parasympathetic preganglionic neurons.

## E. The Enteric Division

The third division of the peripheral autonomic nervous system consists of the intrinsic neurons of the gastrointestinal tract. There are approximately 100 million enteric neurons, far more than are found in all the rest of the autonomic nervous system. The cell bodies of these neurons are grouped into an interconnected network of small ganglia that extend nearly the full length of the gut. The ganglia are located in two main plexuses, shown in Fig. 7. The myenteric plexus (or Auerbach's plexus) lies between the longitudinal and circular layers of the external smooth muscle coat and extends from the esophagus to the rectum. The enteric plexuses also continue out to the gall bladder. The submucous plexus (or Meissner's plexus) lies in the layer of connective tissue between the circular smooth muscle and the muscularis mucosae, which underlies the mucosal epithelium of the gut. Extensive interconnections exist between the two plexuses. The submucous ganglia are rare or absent in the esophagus and stomach.

The enteric neurons are extremely diverse in their functions. Within the myenteric plexus, different populations of neurons act as sensory neurons, interneurons (associative neurons), or as motorneurons projecting to the smooth muscle layers (Figs. 7 and 12). In the stomach, the myenteric plexus also contains secretomotor neurons (Fig. 11). Each of these populations is further subdivided on functional and microanatomical grounds. For example, both inhibitory and excitatory motor neurons innervate the smooth muscle of the stomach and intestines. Similarly, the submucous plexus contains distinct sets of neurons that control secretory activity of the mucosal epithelium, local blood flow, and the level of contraction of smooth muscle in the muscularis mucosae. The submucous plexus is also likely to contain sensory neurons and interneurons.

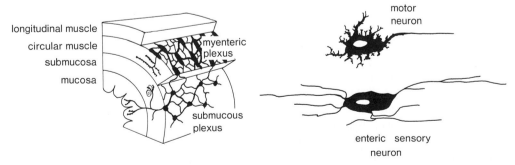

**FIGURE 7** The location of the myenteric and submucous plexuses within the wall of the intestine. The myenteric plexus lies between the longitudinal and circular layers of smooth muscle, whereas the submucous plexus lies within the submucosa. Two different types of enteric neurons are also illustrated. A motor neuron projecting to smooth muscle has many short dendrites and a single long axon. The sensory neuron has many long fine processes. The enteric division is probably the only part of the autonomic nervous system where sensory neurons occur.

The enteric neurons are interconnected so that they are capable of generating a variety of reflexes involving both extrinsic and intrinsic neuronal circuits. The intrinsic sensory neurons respond to various stimuli, including distension of the gut wall or changes in the chemical composition of the luminal contents. These reflexes can control motility, secretion, and local blood blow. Central modulation of the activity of the enteric neurons is achieved via preganglionic vagal inputs to some neurons, mainly in the esophagus, stomach, and gall bladder, and via sacral parasympathetic preganglionic inputs to other neurons, mainly in the distal colon and rectum. These extrinsic inputs mostly affect motor neurons to the smooth muscle or to secretory epithelia of the gut mucosa. It is likely that the final neurons in these pathways also participate in intrinsic enteric reflexes. The central nervous system can also influence the overall activity of the enteric nervous system via the inhibitory effects of sympathetic pathways on intrinsic enteric motor neurons (Fig. 12).

## II. AUTONOMIC NEUROTRANSMISSION

### A. Neurotransmitters in Autonomic Neurons

Autonomic neurotransmission to effector tissues is chemical. Many different neurotransmitters are employed by various autonomic neurons. Furthermore, a single autonomic neuron may utilize several neurotransmitters, which then act as cotransmitters. The precise combinations of transmitters found in different populations of autonomic neurons often are highly correlated with the targets that they innervate.

Noradrenaline (NA; norepinephrine) is long established as a transmitter of the majority of sympathetic final motor neurons. However, in many situations it may be a cotransmitter with other compounds (Fig. 1). Neurally released NA acts through two main classes of receptors: $\alpha$-adrenoceptors and $\beta$-adrenoceptors. Each receptor type has been classified into subtypes on the basis of their pharmacological properties. Other types of adrenoceptors with distinct pharmacological characteristics may exist. The effect of noradrenaline activating a particular type of receptor varies with the target tissue. Activation of $\alpha$-receptors contracts vascular smooth muscle but inhibits the activity of enteric neurons. Similarly, activation of $\beta$-receptors increases cardiac muscle activity but inhibits activity of the smooth muscle in the airways.

Noradrenaline also occurs in about 12% of adrenal medullary chromaffin cells. The rest of the adrenal medullary cells contain the closely related compound, adrenaline (epinephrine). The adrenaline and noradrenaline released from the adrenal chromaffin cells enter the bloodstream to act as hormones, affecting the activity of many different tissues throughout the body.

Acetylcholine (ACh) is a transmitter in several different populations of autonomic neurons, many of which also contain other cotransmitters. These populations include many parasympathetic motor neurons, sympathetic neurons innervating sweat glands (Fig. 2), and a variety of functionally distinct enteric neurons. ACh released from final motor neurons acts via muscarinic receptors to cause a wide variety of effects, depending on the tissue, e.g., inhibition of the activity of cardiac muscle, relaxation of vascular smooth muscle, contraction of smooth muscle of the gut and airways, and stimulation of secretion from epithelial cells in sweat and salivary glands. There are also muscarinic receptors on the terminals of some peripheral autonomic neurons where they act to reduce transmitter release.

ACh is the principal transmitter mediating excitatory synaptic transmission between preganglionic and ganglionic neurons in sympathetic and parasympathetic pathways. It also mediates synaptic transmission between some enteric neurons. In each case, nicotinic receptors are involved.

Increasing evidence has accumulated for the use of two other small molecules as transmitters by autonomic neurons. Adenosine triphosphate (ATP) or a pharmacologically similar compound is probably a cotransmitter with NA in some sympathetic neurons including those causing contraction of smooth muscle in many blood vessels and in the male reproductive tract. ATP also may contribute to parasympathetic excitatory transmission to the urinary bladder and to inhibitory transmission from enteric neurons to gastrointestinal smooth muscle.

Nitric oxide (NO) or a similar substance now appears to be a major inhibitory cotransmitter in many vasodilator neurons and other autonomic neurons causing relaxation of smooth muscle in the gut, airways, and the urinary and genital tracts. In vasodilator neurons, NO coexists with ACh (Fig. 8), but in the gut it occurs in noncholinergic neurons (Fig. 12), some of which may use ATP as a cotransmitter. NO also coexists with ACh in some preganglionic neurons.

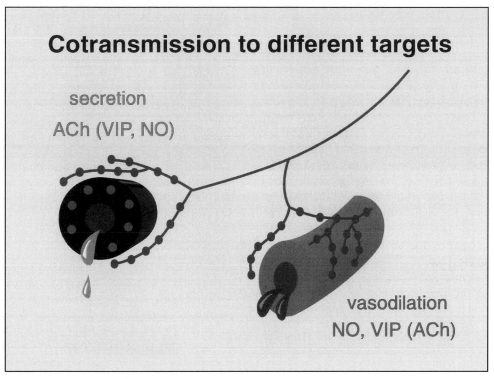

**FIGURE 8** Cotransmission from autonomic neurons innervating secretory glands and their associated vascular supply. A single population of neurons contains many transmitters, including ACh, NO, and VIP. ACh has a major role in stimulating fluid secretion from the glands, whereas NO and VIP causes vasodilation of nearby blood vessels, thereby increasing the delivery of fluid to the secretory tissue itself. VIP, and perhaps other peptides, can modify the chemical makeup of the secretions, whereas ACh can have presynaptic effects on the vascular neurotransmission. This arrangement, with various transmitter combinations, occurs in parasympathetic pathways to salivary glands and the exocrine pancreas and in sympathetic pathways to the sweat glands (see Table I for specific examples).

A variety of other low molecular weight compounds may be transmitters in autonomic neurons, although the evidence is incomplete. They include serotonin (5-hydroxytryptamine), dopamine, and γ-aminobutyric acid.

In addition to low molecular weight transmitters, a high proportion of autonomic motor neurons and many preganglionic neurons contain one or more neuropeptides (Figs. 1 and 2), many of which are neurotransmitters. Neuropeptides are small polypeptides ranging from 3 to more than 35 amino acids in length, with a wide range of biological actions. The neuropeptides found so far within human autonomic neurons include vasoactive intestinal peptide (VIP), neuropeptide Y (NPY), somatostatin (SOM), enkephalin (ENK), substance P, gastrin-releasing peptide (GRP), galanin, and calcitonin gene-related peptide (CGRP). All these neuropeptides also occur in populations of neurons outside the autonomic pathways.

Many neuropeptides coexist in individual neurons with nonpeptide transmitters. For example, NPY coexists with NA and probably ATP in sympathetic neurons innervating much of the cardiovascular system and the smooth muscle of many organs (Fig. 1). Similarly, VIP coexists with ACh and NO in many cranial parasympathetic neurons. Some neurons contain several different neuropeptides. For example, sympathetic neurons innervating the sweat glands contain VIP, CGRP, and probably SOM, in addition to ACh (Fig. 2). Although specific pharmacological agonists and antagonists for the various subclasses of NA and ACh receptors have been of enormous value both experimentally and therapeutically, comparable substances are not yet widely available for most neuropeptide receptors. Nevertheless, there is increasing evidence that neuropeptides do take part in autonomic neurotransmission (see below). [*See* Neurotransmitter and Neuropeptide Receptors in the Brain.]

## C. Transmitter Release

Autonomic neurotransmitters are released from terminal varicosities of the final motor neurons. The amount of transmitter that is released is dependent on the frequency of firing of the neuron. Indeed, the release of some neuropeptides may be effective only at higher stimulation frequencies. Autonomic motor neurons have been recorded to fire irregularly in short bursts. Within each burst, the instantaneous frequency of action potentials may be as high as 60 Hz. However, the average firing rate tends to be much less, typically in the order of 1 to 4 Hz. Neurons in many autonomic pathways are tonically active for long periods, e.g., the sympathetic neurons maintaining arterial constriction or the vagal neurons controlling resting heart rate. Ganglionic neurons normally are not spontaneously active in the absence of their preganglionic inputs. Thus, most of the tonic activity seen in sympathetic and parasympathetic motor neurons is generated within the central nervous system and represents the output from diverse autonomic reflexes.

Transmitter release from varicosities along autonomic axon terminals can be controlled by receptors located on the terminals themselves. Presynaptic inhibition of release has been demonstrated for many transmitters, including NA, ACh, and neuropeptides such as NPY, SOM, and ENK. Not only can these transmitters modulate their own release, but they can also influence the release of other transmitters from nearby axons. Transmitter release from autonomic neurons can be influenced further by circulating hormones such as adrenaline and angiotensin or by locally formed agents such as prostaglandins.

Once released, transmitters such as NA and ACh may be degraded by specific extracellular enzymes. Neuropeptides are also broken down to varying degrees by extracellular peptidases. In addition, some transmitters, or their metabolites, can be actively transported back into the nerve terminals. All these processes combine to restrict the time course of the effector response.

## III. AUTONOMIC CONTROL OF EFFECTOR TISSUES

The details of autonomic regulation vary considerably from tissue to tissue. Most of the target cells of the autonomic nervous system show some degree of spontaneous or nonneurogenic activity that can be influenced by alterations in the patterns of autonomic stimulation. Many effector tissues receive both sympathetic and parasympathetic innervation. Depending on the tissue, the effects of sympathetic and parasympathetic stimulation may be opposing or complementary. A single effector tissue may be regulated by more than one sympathetic or parasympathetic pathway and each pathway may produce opposite effects. Some tissues are innervated predominantly by only sympathetic or parasympathetic neurons and autonomic control is achieved by alterations in their firing rate.

The following examples, together with the information in Table I, illustrate some of the different ways in which the autonomic control of effector tissues operates. They are not intended to be comprehensive functional descriptions of these tissues. All of the circuits described here can be activated independently of the others. In each case, the appropriate patterns of autonomic activity are regulated by sensory inputs leading to specific reflex pathways. Nevertheless, none of these pathways operates in isolation from each other or the rest of the body and its control systems. Ultimately, the coordinated output of the autonomic pathways alters the behavior of its effectors to maintain the homeostatic integrity of the body.

## A. The Iris: Dual Controls

The smooth muscle of the iris is innervated by sympathetic neurons originating in the superior cervical ganglion and by parasympathetic neurons originating in the ciliary ganglion (Fig. 9). Parasympathetic stimulation constricts the pupil via the release of ACh. ACh contracts the circularly arranged smooth muscle cells of the iris sphincter and relaxes the radially arranged cells of the iris dilator muscle. Both actions are mediated by muscarinic receptors. Sympathetic stimulation dilates the pupil by relaxing the sphincter muscle, via NA acting on $\beta$-adrenoceptors, and by contracting the dilator muscle, via the action of NA on $\alpha$-adrenoceptors. The sympathetic neurons also contain NPY, whose function here is unknown.

Parasympathetic activity to the iris can be stimulated by two distinct reflexes, each of which requires the pupil to constrict. Reflex constriction of the pupil occurs in response to bright light, when it reduces the amount of light reaching the retina. It also occurs in response to focusing on nearby objects, when the reduced pupillary diameter creates an increased depth of field. In this case, the parasympathetic regulation of the iris is coordinated with the parasympathetic stimulation of the ciliary smooth muscles which adjust

TABLE I
Autonomic Innervation of the Effector Tissues (Humans)

| Organ and tissue | Autonomic division[a] | Souce ganglion | Probable transmitters[b] | Usual effects |
|---|---|---|---|---|
| **Eye and associated structures** | | | | |
| Iris | Symp | Superior certical | NA, NPY | Dilate pupil |
| | Parasymp | Ciliary | ACh | Constrict pupil |
| Ciliary muscles | Parasymp | Ciliary | ACh | Contraction, focus lens closer |
| Choroid vessels | Parasymp | Sphenopalatine | ACh, NO, VIP | Vasodilation |
| Levator palpebrae smooth muscle | Symp | Superior cervical | NA | Contraction, hold upper eyelid open |
| Lacrimal gland | Parasymp | Sphenopalatine | ACh, NO, VIP | Fluid secretion, vasodilation |
| | Symp | Superior cervical | NA | Protein secretion |
| **Salivary glands** | | | | |
| Parotid | Parasymp | Otic | ACh, NO, some VIP | Fluid secretion, vasodilation |
| | Symp | Superior cervical | NA | Protein secretion |
| Submandibular | Parasymp | Submandibular | ACh, NO, VIP | Fluid secretion, vasodilation |
| | Symp | Superior cervical | NA | Protein secretion |
| Sublingual | Parasymp | Submandibular, local | ACh, NO, VIP | Secretion, vasodilation |
| **Airways** | | | | |
| Nasal mucosa | Parasymp | Sphenopalatine | ACh, VIP | Secretion, vasodilation |
| | Symp | Superior cervical | NA | Vasoconstriction |
| Smooth muscle | Parasymp | Local vagal | ACh | Contraction |
| | Parasymp | Local vagal | NO, VIP | Relaxation |
| | Symp | Stellate, cervical chain | NA, NPY | Inhibit local ganglia, relaxation |
| Tracheal mucosa | Parasymp | Local vagal | ACh, VIP | Secretion |
| **Heart** | | | | |
| Cardiac muscle | Symp | Stellate | NA, NPY | Increase force and rate of beat, inhibit parasympathetic activity |
| | Parasymp | Local vagal | ACh, some SOM | Decrease force and rate of beat, reduce atrioventricular conduction, inhibit arrhythmias |
| | Parasymp | Local vagal | ACh?[c] VIP | Increase rate from very low levels, vasodilation? |
| Coronary vessels | Parasymp | Local vagal | ACh | Vasoconstriction |
| | Symp | Stellate | NA, NPY | Vasodilation, vasoconstriction? |
| **Thyroid gland** | | | | |
| Follicular epithelium | Symp | Middle cervical | NA, NPY | Thyroid hormone secretion |
| | Parasymp | Local vagal | NO, VIP | Secretion, vasodilation |
| **Pancreas** | | | | |
| Acinar cells | Parasymp | Local vagal | ACh | Fluid, bicarbonate, and enzyme secretion |
| Islet cells | Parasymp | Local vagal | ACh, VIP, GAL | SOM and pancreatic polypeptide secretion |
| Blood vessels | Parasymp | Local vagal | VIP?[c] | Vasodilation |
| | Symp | Celiac | ATP, NA, NPY | Vasoconstriction |
| **Liver** | | | | |
| Parenchymal cells | Symp | Celiac | NA, NPY | Glucose mobilization, gluconeogenesis |
| **Spleen** | | | | |
| Capsule, sinusoids | Symp | Celiac | NA, NPY | Contraction |
| **Kidney** | | | | |
| Blood vessels | Symp | Renal | NA, NPY | Constriction |
| Juxtaglomerular apparatus | Symp | Renal? | NA, NPY | Renin release |
| **Urinary Bladder** | | | | |
| Detrusor muscle (body) | Parasymp | Pelvic, intrinsic | ATP, ACh | Contraction |
| Trigone muscle (base) | Parasymp | Local | NO, VIP | Relaxation? |
| | Symp | Lumbosacral | NA, NPY | Contraction |
| Ganglia | Symp | Lumbosacral | NA, NPY | Relaxation, by inhibiting parasympathetics |
| **Urethra** | | | | |
| Smooth muscle | Symp | Pelvic | NA, NPY | Contraction |
| | Parasymp | Pelvic | NO, VIP, (?)[d] | Relaxation |
| **Vas deferens, seminal vesicles, prostate gland** | | | | |
| Smooth muscle | Symp | Lumbosacral, hypogastric | NA, ATP, NPY, ENK?[c] | Contraction |
| Epithelium | Parasymp | Local? | NO, VIP | Secretion |
| **Uterus, oviducts, vagina** | | | | |
| Smooth muscle | Symp | Lumbosacral, paracervical | NA, NPY | Nonpregnant: contraction; pregnant: relaxation |
| | Parasymp | Paracervical | (?),[d] VIP | Nonpregnant: relaxation |
| Epithelium, mucosal glands | Parasymp | Paracervical | (?),[d] VIP | Secretion? |

*(continued)*

TABLE I (*Continued*)

| Organ and tissue | Autonomic division[a] | Souce ganglion | Probable transmitters[b] | Usual effects |
|---|---|---|---|---|
| **Genital erectile tissue** | | | | |
| Smooth muscle | Parasymp | Pelvic, local | NO, ACh, VIP | Relaxation, vasodilation causing erection |
| | Symp | Pelvic, lumbosacral | NA, NPY | Contraction, preventing erection |
| **Skin** | | | | |
| Sweat glands | Symp | Paravertebral | ACh, VIP, CGRP, SOM | Secretion, vasodilation |
| Piloerector muscles | Symp | Paravertebral | NA | Contraction |
| Blood vessels | Symp | Paravertebral | NA, NPY | Constriction (arteries and veins) |
| | Symp | Paravertebral | ACh, VIP | Vasodilation (arteriovenous anastomoses) |
| **Vasculature** | | | | |
| Systemic | Symp | Paravertebral Prevertebral | NA, NPY | Constriction, increase wall stiffness |
| Skeletal muscle beds | Symp | Paravertebral | NA, NPY | Constriction, possible vasodilation in smallest vessels |
| | Symp | Paravertebral | ACh | Vasodilation |
| Renal | Symp | Renal | NA, NPY | Vasoconstriction |
| Splanchnic | Symp | Paravertebral Prevertebral | NA, NPY, ATP | Constriction |
| | Parasymp | Pelvic, local | NO, VIP | Local vasodilation |
| | Enteric | Submucous | VIP | Local vasodilation |
| Cranial | Symp | Superior cervical | NA, NPY | Vasoconstriction, vasodilation in facial veins |
| | Parasymp | Cranial | NO, ACh, VIP | Vasodilation |
| **Adipose tissue** | | | | |
| Adipocytes | Symp | Paravertebral | NA | Lipolysis |
| **Periosteum** | | | | |
| ?[c] | Symp | Paravertebral | VIP | ? |
| **Gastrointestinal tract** | | | | |
| Smooth muscle (nonsphincter) | Parasymp[e] | Myenteric | ACh | Contraction (esophagus, stomach, gall bladder, colon) |
| | Parasymp[e] | Myenteric | NO, VIP | Relaxation (esophagus, stomach, gall bladder, colon) |
| | Enteric | Myenteric | ACh | Contraction |
| | Enteric | Myenteric | NO, VIP, ATP | Relaxation, inhibition of spontaneous activity |
| Enteric ganglia | Symp | Prevertebral | NA | Inhibit excitatory enteric neurons |
| Gastric mucosa | Parasymp[e] | Myenteric | ACh | Acid secretion |
| | Parasymp[e] | Myenteric | GRP | Gastrin secretion |
| | Parasymp[e] | Myenteric | ACh | SOM secretion (inhibits gastrin secretion) |
| | Parasymp[e] | Myenteric | ACh? | Bicarbonate secretion |
| | Parasymp[e] | Myenteric | ACh? | Pepsinogen secretion |
| Intestinal mucosa | Enteric | Submucous | ACh | Secretion |
| | Enteric | Submucous | (?),[d] VIP, NPY?[c] | Secretion |
| Submucous ganglia | Symp | Prevertebral | NA | Inhibit enteric secretomotor neurons |
| Blood vessels | Enteric | Submucous | ACh, VIP | Vasodilation |
| | Symp | Prevertebral | NA, NPY, ATP?[c] | Vasoconstriction |

[a]Parasymp, parasympathetic; Symp, sympathetic.

[b]ACh, acetylcholine; ATP, adenosine triphosphate; CGRP, calcitonin gene-related peptide; ENK, enkephalin; GRP, gastrin-releasing peptide; NA, noradrenaline; NO, nitric oxide; NPY, neuropeptide Y; SOM, somatostatin; VIP, vasoactive intestinal peptide.

[c]Evidence for a transmitter role in this tissue is incomplete.

[d]The nature of the transmitter is unknown.

[e]The final motor neurons in these pathways are also likely to be the final neurons in intrinsic enteric pathways.

the shape of the lens during close focusing. Sympathetic activity is stimulated by suddenly moving from light into darkness.

In the iris, therefore, the opposing effects of sympathetic and parasympathetic stimulation actually work together to exert fine control of the diameter of the pupil in a variety of different circumstances.

## B. The Heart: Regulating Myogenic Activity

Intrinsic cardiac pacemakers keep the heart beating regularly in the absence of any outside stimulation. However, the rate of beat, force of beat, and the conduction velocity of the pacemaker signal all can

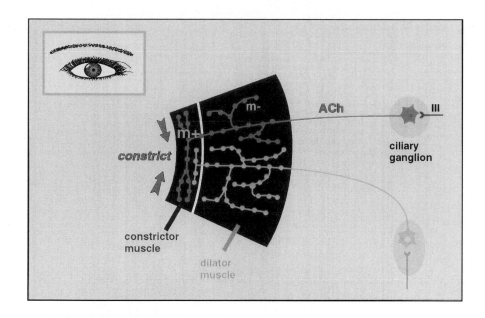

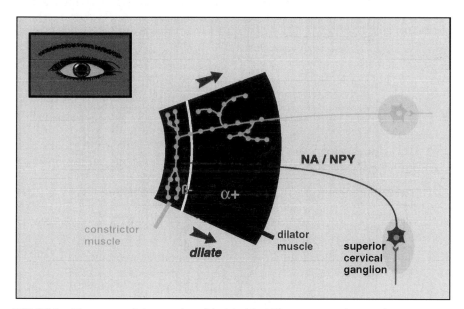

**FIGURE 9**    The autonomic innervation of the iris. (Top) The parasympathetic pathway responsible for constricting the pupil in bright light. The pupilloconstrictor neurons have their cell bodies in the ciliary ganglion and receive preganglionic inputs via the oculomotor nerve (III). They release acetylcholine (ACh) which contracts the constrictor smooth muscle and relaxes the dilator smooth muscle, both by its action on muscarinic receptors (m). (Bottom) The sympathetic pathways responsible for dilating the pupil when suddenly entering the dark. The pupillodilator neurons have their cell bodies in the superior cervical ganglion. They release both noradrenaline (NA) and neuropeptide Y (NPY). The noradrenaline contracts the dilator muscle via $\alpha$-adrenoceptors and relaxes the constrictor muscle via $\beta$-adrenoceptors.

be affected by the autonomic nervous system. As in the iris, the net effects of sympathetic and parasympathetic stimulation tend to be in opposite directions. At rest, vagal parasympathetic neurons innervating the heart fire tonically to reduce the rate and force of

the heart beat. They act primarily via the effects of ACh on muscarinic receptors. Some of these neurons probably also release somatostatin, which may reduce the tendency for cardiac arrhythmias to develop at low rates of beating. Sympathetic stimulation of the heart is usually recruited when an increase in cardiac output is required, e.g., during exercise. These neurons release NA and probably NPY. The NA acts via $\beta$-adrenoceptors to increase both the rate and the force of beat from resting levels. Sympathetic activation is usually accompanied by a decrease or cessation of basal vagal cardioinhibition. At least some of this decreased vagal effect may be due to NPY, released along with NA, reducing neurotransmission from the vagal nerve terminals. Conversely, at the completion of exercise, the heart rate returns to normal resting levels as a result of decreased sympathetic activity and increased vagal parasympathetic activity. [*See* Heart, Anatomy.]

The autonomic neurons controlling the heart participate in several different reflexes which are activated by changes in a variety of cardiovascular parameters such as systemic blood pressure, right atrial distension, and the partial pressures of oxygen and carbon dioxide in the blood. In addition to direct nervous control, cardiac output can be increased in stressful situations by the release of adrenaline from the adrenal glands into the bloodstream. Electrophysiological evidence suggests that the cardiac $\beta$-adrenoceptors activated by circulating adrenaline form a population distinct from that activated by neuronally released noradrenaline.

## C. Blood Vessels: Integration of Independent Pathways

The tone of vascular smooth muscle is controlled by many different factors. Circulating hormones such as adrenaline, angiotensin, and arginine–vasopressin play important roles in maintaining blood pressure under different circumstances. Local blood flow can be modified by metabolites formed within tissues as well as by vasoactive substances originating in the vascular endothelium such as NO (endothelial-derived relaxing factor). Furthermore, myogenic mechanisms can alter vascular resistance in at least some locations. It is against this background that the vasculature is regulated by autonomic pathways.

Most of the vasculature is innervated by sympathetic pathways, many of which are tonically active most of the time we are alive. In general, stimulation of these pathways results in vasoconstriction mediated largely by NA acting on $\alpha$-adrenoceptors. NPY and ATP also may contribute to the generation of vasoconstrictor responses, especially in smaller vessels. The level of vasoconstriction in many of these vessels is related directly to the level of sympathetic activity.

However, in some vascular beds, neurally released NA activates $\beta$-adrenoceptors to cause vasodilation, e.g., in facial veins and some coronary arteries.

Sympathetic vasoconstrictor activity can be regulated separately for each of the major vascular beds, including those supplying the skin, the muscles of the limbs, the kidneys, the gastrointestinal tract, and the brain. Consequently, blood flow can be redistributed to different organ systems, while maintaining overall blood pressure at appropriate levels. For example, during exercise, sympathetic stimulation selectively constricts the mesenteric circulation so a higher proportion of cardiac output can be redirected from the stomach and intestines to working muscles. Crucial to the efficient operation of this differential control are sensory inputs to the central nervous system which provide measures of many different variables such as systemic blood pressure, blood volume, the concentration of carbon dioxide in the blood, body temperature, and the degree of muscle activity. The sensory information is integrated in the brain stem, which then regulates the activity of the specific sets of preganglionic neurons associated with each sympathetic vasoconstrictor pathway.

Sympathetic pathways also provide vasodilator innervation to skin and to the muscles of the limbs. The final motor neurons in these pathways do not contain NA. In the skin, the sympathetic vasodilator neurons contain ACh, VIP, and probably CGRP. They are activated in heat stress and are primarily associated with the sweat glands: the ACh mediates sweat secretion whereas the VIP and CGRP are probably responsible for the accompanying vasodilation (Figs. 2, and 8).

Parasympathetic vasodilator neurons containing VIP, and probably also ACh and NO, innervate much of the cranial, thoracic, and pelvic circulation, particularly the smaller vessels associated with secretory tissues (Fig. 8). They are prominent in the erectile tissues. The relative contributions of ACh, NO, and VIP to the vasodilator responses vary from one vascular bed to another. Vasodilator neurons also occur within the submucous plexus of the intestines, where their activity is coordinated with that of the secretomotor neurons. Not all parasympathetic vascular neurons are vasodilators: some vagal vasomotor neurons innervating the coronary circulation release ACh to cause vasoconstriction. Probably all vascular beds innervated by parasympathetic or enteric neurons also are innervated by sympathetic neurons with opposing effects on vascular resistance. The interactions between these systems in the control of local circulation are not well understood.

## D. Airway Smooth Muscle: Opposing Parasympathetic Pathways

The smooth muscle of the airways contributes to the regulation of airway resistance. The muscle has little if any intrinsic myogenic activity, but it can be contracted by many different agents, most notably histamine. Two functionally distinct vagal parasympathetic pathways innervate the airway smooth muscle (Fig. 10). One population causes contraction of the muscle via the action of ACh on muscarinic receptors whereas the other population causes relaxation of the muscle, probably mediated by NO or a similar compound. The excitatory pathway is normally tonically active and its activity may be increased reflexly by irritation of the airways. It is not clear when the inhibitory pathway is used, but it is capable of reversing the effects of airway constrictors like histamine.

There is little direct sympathetic innervation of airway smooth muscle in humans. However, sympathetic motor neurons terminate around parasympa-thetic vagal neurons located in local microganglia. Sympathetic stimulation inhibits excitatory vagal neurons, mainly via the effects of NA on $\alpha$-adreno-ceptors on the vagal neurons themselves. Circulating adrenaline from the adrenal glands can relax airway smooth muscle directly via an action on $\beta$-adreno-ceptors.

## E. Secretion: Complementary Controls

Autonomic neurons control secretion from many different glands as well as from secretory epithelia lining the gut and airways. In most places, a subtle interplay between the different autonomic pathways results in the fine control of the quantity and constituents of the secretions.

Secretion from the salivary and lacrimal glands is stimulated both by parasympathetic and sympathetic pathways. Parasympathetic stimulation produces a watery secretion, mediated mainly by ACh acting on muscarinic receptors. However, most parasympathetic secretomotor neurons also contain and release

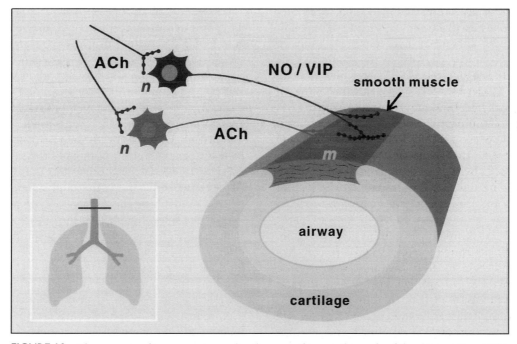

**FIGURE 10**  The presence of two separate vagal pathways to the smooth muscle of the airways. (Inset) The level of the trachea illustrated by the main diagram. One population of local ganglion neurons releases acetylcholine (ACh) which contracts the muscle via its action on muscarinic receptors (*m*). The other population of local ganglion neurons relaxes the muscle using nitric oxide (NO) and perhaps vasoactive intestinal peptide (VIP) as transmitters. Both sets of neurons receive excitatory preganglionic input from neurons releasing ACh which acts on nicotinic receptors (*n*). The vagal preganglionic neurons have their cell bodies in the brain stem. The sympathetic pathways are not shown here.

NO and VIP, both of which contribute to a concomitant vasodilation of the local vasculature (Fig. 8). VIP also may increase the sensitivity of the epithelial cells to ACh and modify the exact composition of the secretions. Sympathetic stimulation of the glands produces a secretion that is rich in proteins, including enzymes (e.g., amylase in saliva). This effect is largely mediated by NA acting on $\alpha$-adrenoceptors. In general, sympathetic secretomotor neurons do not contain NPY. Autonomic secretomotor activity to salivary and lacrimal glands can be stimulated by a variety of reflexes, including irritation of associated structures such as the oral mucosa or cornea, respectively, resulting in salivation or tear production. Psychogenic and emotional stimuli also can activate these pathways.

Within the gastrointestinal tract, secretion is regulated directly by intrinsic enteric neurons. The activity of these neurons is in turn influenced by the reflex activation of parasympathetic, sympathetic, and intrinsic enteric pathways. The integration of these inputs can be illustrated by the autonomic control of acid secretion from the stomach and fluid secretion from the small intestine.

Gastric acid secretion is controlled by the coordinated actions of autonomic neurons and locally released hormones as illustrated in Fig. 11. Preganglionic vagal fibers activate at least two populations of intrinsic secretomotor neurons in the stomach. One population of neurons releases ACh, which directly stimulates acid secretion via muscarinic receptors on parietal cells. The other population of neurons releases GRP to stimulate the secretion of a hormone, gastrin, from specialized epithelial cells. The gastrin is carried through the blood and then stimulates the secretion of acid from the acid-secreting cells. These pathways can be activated by several reflexes, including chemical and mechanical stimulation by food in the stomach.

Vagal stimulation may also inhibit acid secretion. Some intrinsic neurons receiving vagal inputs stimulate the release of a locally hormone, somatostatin, from a distinct class of epithelial cells. Somatostatin inhibits the release of gastrin, thereby reducing gastric acid output. Thus, the vagus nerve controls acid secretion by a combination of direct and indirect inputs, which not only stimulate secretion, but also provide a mechanism for inhibiting an excessive release of gastric acid. Acid secretion also is stimulated by locally formed histamine. Moreover, the acid itself contributes to the feedback regulation of its production by directly inhibiting gastrin secretion

and exciting somatostatin release. The net amount of acid secretion is determined, therefore, by a constant interplay between neural and hormonal influences.

Fluid and electrolyte secretion from the small intestine can be elicited by several different stimuli, including mechanical irritation, distension, or chemicals as diverse as glucose or bacterial toxins within the intestine. Secretory reflexes are initiated by intrinsic sensory neurons which activate secretomotor neurons located in the submucous plexus. Although some secretomotor neurons release ACh, most enteric secretomotor neurons probably use VIP as their main transmitter. As part of the overall control of blood volume, submucosal secretomotor neurons tend to be inhibited tonically by sympathetic pathways with their final neurons in the prevertebral ganglia. These sympathetic neurons also are activated by excitatory synaptic inputs from intrinsic enteric neurons forming an inhibitory feedback loop to directly regulate intestinal secretion. This circuit is shown in Fig. 12.

## F. The Urinary Bladder: Sympathetic Modulation of Parasympathetic Activity

The smooth muscle of the body of the urinary bladder is innervated almost entirely by parasympathetic pelvic neurons. Most of these neurons cause contraction of the muscle mediated by ACh and, in some circumstances, ATP or a similar compound. The excitatory pathway is activated by a spinal reflex or a voluntary central command which results in the initiation of micturition (urination). As in the airways, there is little direct innervation of most of the smooth muscle by sympathetic neurons. Instead, noradrenergic sympathetic neurons innervate the postganglionic pelvic neurons and inhibit their activity. They also innervate specialized smooth muscle at the outlet of the bladder which they cause to contract like a sphincter. The net effect of activating the sympathetic pathways to the bladder is to enhance the retention of urine by allowing the body of the bladder to expand, while preventing the passage of urine from the bladder to the urethra. The sympathetic pathways probably are inactive during micturition. Urination itself requires the activity of autonomic motor pathways to the bladder and urethra to be coordinated with somatic motor pathways innervating the striated muscles of the lower abdominal and pelvic regions, including the external sphincters of the urinary tract. A similar degree of coordination between autonomic and somatic path-

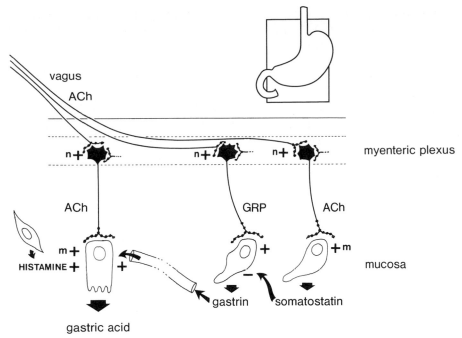

**FIGURE 11** The vagal control of gastric acid secretion by cells in the lining of the stomach. Preganglionic axons synapse with three different populations of secretomotor neurons. In each case, the synaptic transmission is mediated by acetlycholine (ACh) acting on nicotinic receptors (*n*). One population of secretomotor neurons causes release of acid from parietal cells via ACh acting on muscarinic receptors (*m*). Another population causes the release of gastrin; the principal transmitter in these neurons is gastrin-releasing peptide (GRP). The gastrin travels in the local blood supply to further stimulate acid release. A third population of secretomotor neurons stimulates the release of somatostatin via ACh acting on muscarinic receptors. The somatostatin inhibits gastrin release. Note that this is only the minimum circuit involved in controlling acid secretion. Histamine released from local cells (histaminocytes) is also a powerful stimulant of acid secretion, and there are many additional feedback loops.

ways occurs during defecation. [*See* Urinary System, Anatomy.]

## G. The Male Genital Tract: Cooperation between Pathways

The autonomic control of the male genital tract is controlled by the cooperative action of sympathetic and pelvic parasympathetic pathways. Erection is mostly maintained by pelvic parasympathetic pathways which mediate vasodilation of the penile vasculature and relaxation of the smooth muscle of the erectile tissue, allowing engorgment of the penis with blood. The transmitters involved here include NO, ACh, and VIP.

Ejaculation requires the activation of sympathetic pathways to the vas deferens, seminal vesicles, and the prostate gland. Most of the sympathetic motor neurons involved here are located in ganglia within the pelvic plexuses. They cause contraction of the smooth muscle of these organs, largely via the combined effects of NA, ATP, and NPY. Thus, the whole sequence of erection and ejaculation requires both parasympathetic and sympathetic inputs. These pathways can be activated both by spinal reflexes and by psychogenic stimuli. It is likely that other sympathetic and pelvic parasympathetic circuits are involved in the cessation of erection, the secretory activities of epithelia lining the genital tract, local blood flow changes, and the coordination of genital activities with those of the bowel and the urinary bladder. A similar coordinated sequence of activation of parasympathetic and sympathetic pathways probably takes place in females during sexual activity, although the details are much less well understood.

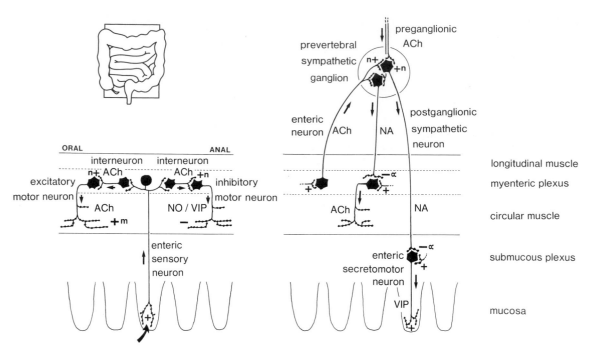

**FIGURE 12** Examples of neuronal circuits in the intestine. (Left) The minimum connections required for the generation of the peristaltic reflex. Stimulation of an enteric sensory neuron (e.g., by distending the gut wall) activates interneurons projecting in each direction along the gut. They release ACh which acts on nicotinic receptors ($n$) to stimulate excitatory motor neurons in the oral direction and inhibitory motor neuron in the anal direction. The excitatory neurons utilize ACh acting on muscarinic receptors ($m$), but may also release other cotransmitters. The transmitters for the inhibitory neurons include ATP, NO, and VIP. The contraction behind the distension and the relaxation ahead of it allow the bolus to be propelled along the intestine. (Right) The interaction between sympathetic pathways and enteric neurons. Sympathetic neurons with cell bodies in the prevertebral ganglia release noradrenaline (NA) which acts via $\alpha$-adrenoceptors to inhibit enteric secretomotor neurons and excitatory motor neurons to the smooth muscle. The sympathetic neurons receive excitatory inputs from preganglionic neurons and from enteric neurons. In each case, the principal transmitter is ACh acting on nicotinic receptors ($n$), although various neuropeptides also may be released. Inhibition of synaptic transmission by sympathetic postganglionic neurons also occurs in many parasympathetic ganglia. The net result of activating these pathways is to reduce the activity of enteric neurons, perhaps under conditions when body fluid needs to be retained.

## H. Gut Smooth Muscle: Intrinsic and Extrinsic Reflexes

The control of the activity of smooth muscle in the gut involves a large amount of intrinsic and extrinsic neuronal circuitry, some of which is shown in Fig. 12. Ultimately, however, most of the circuits converge upon one of two main functional classes of intrinsic enteric neurons. One class of neurons causes contraction of the gut musculature by releasing ACh, which activates muscarinic receptors. These neurons may use substance P as an excitatory cotransmitter. The other class of neurons causes relaxation of gut smooth muscle by using a variety of transmitters, including ATP, NO, VIP, and related peptides. An example of an extrinsic reflex is the receptive relaxation of the

stomach. Distension of the esophagus is detected by vagal sensory neurons. They in turn activate a specific population of vagal preganglionic neurons which synapse onto intrinsic inhibitory neurons in the myenteric plexus of the stomach. These neurons relax the gastric smooth muscle, thereby increasing the capacity of the stomach and enabling it to receive food from the esophagus. The same vagal inhibitory pathway also can be stimulated via another population of vagal sensory neurons responding to distension of the stomach wall. This is an accommodation reflex that adjusts the capacity of the stomach to the volume of its contents.

The best studied intrinsic reflex is the peristaltic reflex of the small intestine illustrated in Fig. 12. Distension of the small intestine stimulates a population of intrinsic enteric sensory neurons that form synapses onto interneurons in the myenteric plexus. Some of the interneurons project orally and synapse onto excitatory motor neurons whereas other interneurons project anally and synapse onto inhibitory motor neurons. The net effect of the reflex is to contract the smooth muscle oral to a bolus of intestinal contents while relaxing the muscle anal to the bolus. Consequently, the bolus is moved along the gut in an oral to anal direction. The peristaltic reflex can take place in a piece of intestine completely isolated from the central nervous system. Nevertheless, the central nervous system can influence the activity of the enteric neurons involved in the reflex. For example, stimulation of the prevertebral sympathetic neurons projecting to the myenteric plexus inhibits the intrinsic excitatory neurons and thus slows or prevents peristalsis.

## I. Concluding Remarks

The foregoing examples illustrate the variety of ways in which the autonomic neuronal pathways can influence acutely the activity of their effector tissues. Autonomic neurons also may affect their targets over a much longer time course. For example, they may alter rates of cell division or the expression of particular genes. Autonomic neurons in turn are subjected to trophic influences from the target tissues themselves and from other neurons. The mechanisms mediating most of these interactions are complex and only now are beginning to be unraveled. Nevertheless, the presence of such interactions reinforces the physiological and anatomical evidence that autonomic pathways are both highly organized and well coordinated for their vital role in regulating a wide range of bodily functions.

## BIBLIOGRAPHY

Appenzeller, O. (1990). "The Autonomic Nervous System: An Introduction to Basic and Clinical Concepts," 4th Ed. Elsevier, Amsterdam.

Bannister, R., and Mathias, C. J. (ed.) (1993). "Autonomic Failure: A Textbook of Clinical Disorders of the Autonomic Nervous System," 3rd Ed. Oxford University Press, London/New York.

Bowman, W. C., and Rand, M. J. (1980). "Textbook of Pharmacology," 2nd Ed., Chaps. 9–11 and 22–29, Blackwell Scientific Publications, Oxford.

Burnstock, G., and Hoyle, C. H. V. (eds.) (1992). "Autonomic Neuroeffector Mechanisms," Harwood Academic, Reading, United Kingdom.

Furness, J. B., and Costa, M. (1987). "The Enteric Nervous System." Churchill Livingstone, Edinburgh.

Furness, J. B., Morris, J. L., Gibbins, I. L., and Costa, M. (1989). Chemical coding of neurons and plurichemical transmission. *Annu. Rev. Pharmacol. Toxicol.* **29**, 289–306.

Gabella, G. (1976). "Structure of the Autonomic Nervous System." Chapman and Hall, London.

Gibbins, I. L. (1989). Peripheral autonomic nervous system. *In* "The Human Nervous System" (G. Paxinos, ed.), pp. 93–123. Academic Press, New York.

Hendry, I. A., and Hill, C. E. (eds.) (1992). "Development, Regeneration and Plasticity of the Autonomic Nervous System," Harwood Academic, Reading, United Kingdom.

Kandel, E. R., Schwartz, J. H., and Jessell, T. M. (1991). "Principles of Neuroscience," 3rd Ed. Appleton & Lange, East Norwalk, Connecticut.

Maggi, C. A. (ed.) (1993). "Nervous Control of the Urinogenital System," Harwood Academic, Reading, United Kingdom.

McLachlan, E. (ed.) (1995). "Autonomic Ganglia," Harwood Academic, Reading, United Kingdom.

Nilsson, S. (1983). "Autonomic Nerve Function in the Vertebrates." Springer-Verlag, Berlin/New York.

Nilsson, S., and Holmgren, S. (eds.) (1994). "Comparative Physiology and Evolution of the Autonomic Nervous System," Harwood Academic, Reading, United Kingdom.

Pick, J. (1970). "The Autonomic Nervous System: Morphological, Comparative, Clinical and Surgical Aspects." Lippincott, Philadelphia.

Robertson, D., and Biaggioni, I. (eds.) (1995). "Disorders of the Autonomic Nervous System," Harwood Academic, Reading, United Kingdom.

# Bacterial Infections, Detection

FRED C. TENOVER

*Centers for Disease Control and Prevention, Atlanta*

## GLOSSARY

**Acid-fast bacteria** Group of organisms with a high lipid content in their cell wall that stain only with the acid-fast stain and not with the Gram stain

**Antibody** Substance produced by the immune system in response to a specific stimulus (an antigen) that is capable of binding specifically to the stimulus

**Gram stain** Biologic stain that differentiates types of bacteria based on the composition of the cell wall and their overall size and shape

**Nucleic acid hybridization** Pairing and hydrogen bonding of two complementary strands of nucleic acid

**Pathogen** Organism that causes an infectious disease in the host

**Selective medium** Agar-based growth medium that contains substances that inhibit the growth of certain groups of bacteria

THE DIAGNOSIS OF BACTERIAL INFECTIONS IN-volves a cooperative interaction between the physician and the diagnostic microbiology laboratory. The primary functions of the microbiology laboratory are to detect and identify the presence of pathogenic microorganisms in specimens collected from patients and to determine the antimicrobial susceptibility profile of those organisms. This article explores the methods and techniques used by the laboratory to diagnose bacterial infections.

## I. PATHOGENS VERSUS NORMAL FLORA

Although often considered to be free of microbial contamination, the human body serves as host to billions of microorganisms, the majority of which are bacteria. These microorganisms are normally confined to specific areas of the body. The skin, the mouth and upper airway, the intestines, the mucous membranes of the urethra, and the external genitalia all have a characteristic resident population of organisms. This resident flora, often referred to as the normal flora, is beneficial to the host in a variety of ways, producing vitamins, maintaining skin tone, and, more important, helping to protect the host from invasions of pathogenic microorganisms. Pathogenic bacteria are those that invade the host, cause disease, and require eradication to bring the host back to health. At times, when the host is debilitated or when the immune system is not fully functional, some of the normal flora may take on the role of pathogens and cause disease. These organisms are deemed opportunistic pathogens. The predominant sites of infection are the skin, the gastrointestinal and respiratory tracts, and those areas of the body that are normally sterile or free of bacterial contamination. These sites include the bloodstream, the cerebrospinal fluid (CSF), the pleural cavity, the lungs, and the bladder and upper urinary tract. When these sites are invaded by bacteria, the host becomes ill. While some bacterial infections are easily eradicated with antimicrobial

ENCYCLOPEDIA OF HUMAN BIOLOGY, Second Edition, VOLUME I.  Copyright © 1997 by Academic Press.  All rights of reproduction in any form reserved.

therapy, other infections, such as those of the CSF and blood, can be life-threatening and frequently necessitate hospitalization of the patient.

To diagnose an infectious process, the physician assesses the physical signs and symptoms of the patient to determine whether or not an infection is present and, if so, its site and type. For instance, if the patient complains of a burning sensation when urinating and the need to urinate frequently, the physician may suspect an infection of the urinary bladder. To confirm the diagnosis and help decide the appropriate antibiotic therapy, the physician will ask the laboratory (1) to examine the urine for the presence of bacteria and inflammatory cells (to confirm the diagnosis), (2) to determine the identity of the organism if bacteria are present, and (3) to determine the antimicrobial susceptibility pattern of the bacteria.

## II. IMPORTANCE OF THE MICROBIOLOGY LABORATORY

### A. Accurate Diagnoses

The key reason for involving the laboratory in the diagnostic process is to ensure that an accurate diagnosis is made. A wide variety of bacteria can cause similar types of illnesses that cannot always be differentiated by clinical signs and symptoms. While many bacteria have a similar appearance, their susceptibilities to antibiotics are quite different. Also, certain bacteria are more likely to spread from the initial site of infection to other organs or tissues, thereby causing more serious disease. Thus, the identification of bacterial pathogens often is critical to the successful management of infections. Sometimes the laboratory confirms the clinical suspicions of the physician and isolates an expected organism, such as *Streptococcus pyogenes* from a patient with strep throat; at other times, the laboratory is instrumental in establishing an unsuspected diagnosis, such as isolating an unusual organism from a blood culture.

### B. Guide to Antimicrobial Chemotherapy

The second reason to seek laboratory assistance is to get an accurate picture of the antimicrobial susceptibility of the pathogen. Not all antimicrobial agents are effective against all bacteria. Using a standardized methodology, the laboratory can assess the ability of various antimicrobial agents to inhibit the growth of the pathogen. These results are given to the physician who, after assessing the patient's clinical status and possible allergies, chooses the mode of therapy that is most appropriate. [*See* Antimicrobial Drugs.]

### C. Guide Public Health Activities

Some bacterial pathogens, such as *Mycobacterium tuberculosis*, the cause of tuberculosis, are highly contagious and are easily spread throughout a community. Other organisms, such as *Salmonella* species, are acquired from the consumption of contaminated food or from water sources that may serve thousands of people. Still others, such as *Neisseria gonorrhoeae*, the cause of gonorrhea, require tracing of sexual contacts so that they can be treated with antibiotics to help stop the spread of the disease. These types of infections are the concern of public health officials who have the responsibility of controlling infectious diseases. The diagnostic microbiology laboratory plays a pivotal role in public health, not only by identifying the presence of bacterial pathogens in food and water, but also by determining whether or not a single strain is responsible for an outbreak in a community. The ability of the laboratory to recover bacterial agents from water and food as well as from human specimens is of considerable importance. [*See* Infectious Diseases, Pediatrics.]

### D. Expand Understanding of Organisms and the Diseases They Cause

The last reason for seeking laboratory assistance in recovering the pathogenic bacteria that cause human illness is to learn more about the organisms themselves and the types of patients in which they cause disease. Some of the organisms recovered by the laboratory may not have been associated with disease before and, therefore, may be "new" pathogens, such as the cause of Legionnaire's disease, *Legionella pneumophila*. If we are to control disease-causing bacteria, we must understand where they lie dormant when they are not causing infection, i.e., their reservoirs; we must understand how they move from these reservoirs to the host; and we must learn how they invade and damage the host. The more frequently these organisms are isolated from infected individuals, the more the complex story of pathogenesis for each of these human afflictions can be unraveled.

## III. MODERN DIAGNOSTIC METHODS

### A. Rapid Detection of Bacterial Pathogens

#### I. Microscopy

Direct microscopy remains the best method of rapidly screening patients' specimens for diseases of bacterial origin. The Gram stain, which differentiates types of bacteria on the basis of their size, shape, and staining characteristics, often provides critical information regarding the type of organisms present in the specimen.

#### 2. Antigen Detection

A diagnostic test that uses antibodies directed against targets (antigens) on the surface of pathogenic bacteria can be used to identify the presence of the organism in clinical specimens in situations in which the Gram stain is not helpful. The antibody-based test fills a void left by culture and Gram stain.

#### 3. Nucleic Acid Probes

The search for more sensitive and specific tools with which to identify bacterial pathogens has led to the use of various sequences of nucleic acid contained within bacterial cells as fingerprints for their identification. Because DNA is a double-stranded molecule, the strands can be pulled apart, or denatured. Once separated, the strands of DNA seek to join together again with their complementary strand. Small segments of DNA that match the sequence of the complementary strands will bind to the target strand just as well as the original complementary strand (hybridization). By using short pieces of DNA as probes for the presence of the unique sequences contained within pathogenic microorganisms, it is possible to detect that DNA, and hence the organism, directly in a patient's specimen. Probes are labeled with chemical substrates or enzymes that makes them detectable after binding to the target sequence. [*See* DNA Markers as Diagnostic Tools.]

### B. Antibody Detection

Sometimes the laboratory is unable to isolate the bacterial pathogen that is causing disease in a given patient, and probes and antigen tests are either negative or unavailable. In each of these situations, the laboratory can turn to the patient's production of a specific antibody to indicate the cause of the disease. Antibody to an organism normally will not be present in a patient unless the patient has been infected with that organism. The drawback is that antibodies to an organism often do not arise until late in the course of the infection. This often can be too late to help in the management of the patient's illness.

### C. Culture Methods

The major focus of the diagnostic microbiology laboratory is on growing the bacteria from patients' samples on agar media. This allows the laboratory to perform biochemical tests to identify the organism and to determine its antimicrobial susceptibility profile.

To separate one organism from another, substances are added to agar plates to enhance or inhibit the growth of certain groups of bacteria. These media are classified as enriched, differential, or selective, and each has its special role in the recovery of pathogenic microorganisms.

#### I. Enriched Media

First, it is important to obtain an overview of all the types of bacteria that are present in the specimen. To accomplish this, agar media containing high amounts of vitamins, nutrients, and growth factors are employed. Such media often contain either sheep or horse blood, as well as digests of casein or infusions of animal organs, such as beef hearts. These supplements provide a rich source of nutrients and vitamins to support the growth of almost all bacteria.

#### 2. Selective and Differential Media

The search for specific bacterial pathogens follows different strategies depending on the nature of the specimen. If this specimen is taken from an anatomic site that does not contain a resident bacterial flora, such as the blood, then media containing substances inhibitory to most normal flora are not necessary; however, for samples from the bowel, respiratory tract, or skin, the normal flora needs to be suppressed if the pathogens are to be discerned. To accomplish this goal, selective media are required.

Selective media contain inhibitors that suppress the growth of certain groups of bacteria. If several groups of bacteria are present in a sample, they can be easily separated from one another using a series of agar plates, each containing a different inhibitory substance. For instance, MacConkey agar, which contains the dye crystal violet as well as bile salts, serves to suppress the growth of organisms that normally grow on the skin (gram-positive organisms) and allow bowel organisms (gram-negative) to flourish. In contrast, colistin–nalidixic acid agar suppresses the growth of gram-negative organisms common to the bowel, permitting gram-positive organisms to grow.

To differentiate two types of gram-negative bacilli present in a sample, dyes and chemical indicators are incorporated into the media. These compounds change color under different pH conditions so that as different organisms grow the colonies take on different colors. Such media are called differential because they allow one group of organisms, such as those that ferment the sugar lactose, to be differentiated from other bacteria, such as those that do not ferment lactose.

To identify pathogenic microorganisms, individual colonies are isolated in pure culture (i.e., making sure only a single type of organism is present) and biochemical tests are performed to identify the genus and species of the organism. The results, based on the site from which the specimen was collected, will determine the significance of the organism's presence in that specimen.

### 3. Sensitivity and Specificity

The ability to detect a pathogen in specimens from patients who truly have an infectious disease is referred to as the sensitivity of the test, whereas the ability to correctly classify patients without disease as negative is referred to as its specificity. Tests with high sensitivity are usually used to screen patients for a particular disease, whereas tests with high specificity are used to confirm a suspected diagnosis.

Culture techniques frequently are the most sensitive techniques for the detection of bacterial pathogens and also have high specificity.

## D. Site-Specific Bacteriology

### 1. Blood

One of the most critical samples examined by the clinical bacteriology laboratory is blood. In patients who are septic (i.e., have bacteria multiplying in their blood), the number of potential pathogens is large and the disease can be fatal if not treated rapidly with the appropriate antimicrobial agents. Thus, blood-borne pathogens must be identified as rapidly and accurately as possible. [See Hemoglobin.]

Because the number of organisms present in the bloodstream of a septic patient can vary from 1 to >1000 organisms/ml, blood samples (5–10 ml per sample) are placed in broth media to enhance the

growth of the organisms and to increase the possibility of detecting low numbers. Samples plated directly on agar media may not produce colonies when so few organisms are present. Both aerobic (oxygen-containing) and anaerobic (without oxygen) broth cultures are inoculated, and the aerobic bottle or vial is shaken during the first 48 hr to enhance oxygenation and to encourage the growth of aerobic bacteria. Bacterial growth can be detected in the bottles by turbidity, by inoculating a sample of the broth onto solid media and incubating the plates overnight to enhance growth, or through the use of automated equipment that detects the presence of $CO_2$ produced by organisms as they metabolize the glucose present within the vial. Once organisms have been detected and subcultured to solid agar, they are identified to genus and species by biochemical or serological methods.

### 2. Urine

Urinary tract infections (UTI) are among the most common bacterial infectious illnesses in the United States (approximately 1 million cases per year). Young women are the predominant population infected, although children and men also can develop such infections. The urinary tract is clinically divided into the upper (kidney and ureters) and lower (bladder and urethra) regions; different groups of bacteria may cause infections in each of these distinct areas. It is important to differentiate between the two infections because upper urinary tract infections are more severe, may result in permanent kidney damage, and have a greater tendency to lead to infection of the bloodstream.

The detection of bacterial pathogens in the urinary tract is primarily accomplished by culture on solid media. Several additional techniques, both automated and nonautomated, can facilitate the rapid detection of UTI in patients. These include the staining of urine samples with Gram's method, use of dip-stick tests that detect enzymes present in bacteria and white blood cells (an indication of the body's response to infection), and automated procedures that detect bacteria in the urine. The problem with these methods is that the former methods lack sensitivity, whereas many of the automated methods lack in specificity.

The presence of ≥10,000 bacteria of a single type per ml of urine usually indicates that the patient has a UTI, but lesser quantities can cause infection in certain instances.

### 3. Cerebrospinal Fluid

Bacterial meningitis is another serious infectious illness that requires prompt recognition and appropriate therapy if the patient is to survive without sequelae. The three most common causes of bacterial meningitis are (1) *Streptococcus pneumoniae,* particularly in neonates and elderly patients; (2) *Haemophilus influenzae* in children ages 3 months to 6 years; and (3) *Neisseriae meningitidis,* which is more common in older children and young adults. A Gram stain performed on CSF obtained by lumbar puncture is the key test used to make a rapid diagnosis and guide therapy while the culture results are pending. Culture of the organism on solid, enriched media remains the gold standard for diagnosis. Antigen detection tests for the organisms just mentioned can be used to assist in the rapid diagnosis of bacterial meningitis in cases where the Gram stain of CSF fails to reveal any organisms. This situation can occur if patients have been treated with antimicrobial agents before their lumbar puncture was performed. Antimicrobial susceptibility testing of bacteria isolated from CSF is important for ensuring that the patient is receiving optimal therapy.

## 4. Infections of the Respiratory Tract

### a. Upper Respiratory Tract

The respiratory tract also can be divided into two clinically distinct regions, the upper and lower tracts, each of which is host to different types of bacterial infections. The upper tract, composed of the nose, mouth, and throat, is primarily the site of viral infections, although several well-known bacterial pathogens also can infect this area. These include *S. pyogenes,* also known as Group A β-hemolytic streptococcus; *Bordetella pertussis,* the cause of whooping cough; and *Corynebacterium diphtheriae,* the cause of diphtheria. Although other bacteria can cause disease in this region of the body, we will focus on these three because the techniques to recover other pathogens are similar. [*See* Respiratory System, Anatomy.]

*Streptococcus pyogenes* (the genus of a bacterial name is often abbreviated in medical literature, spelling out only the species name for clarity) is the cause of strep throat, a common childhood malady that is also seen, on occasion, among adults. The back of the throat is swabbed vigorously to collect infected material, especially from whitish areas suggestive of infection. The throat swabs are rolled onto the surface of agar plates, usually containing 5% sheep blood, and the sample is spread around the plate using a sterile inoculating loop. After 24 hr of incubation, the plates are examined for colonies typical of streptococci. Group A streptococci, which produce distinctive gumdrop-shaped colonies on agar, also are capable of lysing (hemolyzing) the sheep red blood cells contained within the medium, producing zones of clearing that are readily recognizable (Fig. 1). A limited number of biochemical or serologic tests are needed to confirm the identification of the organism. Because *S. pyogenes* is not part of the normal flora, its isolation from the throat of a patient with an inflamed pharynx confirms the diagnosis of strep throat.

A rapid testing procedure is available that uses latex beads coated with antibodies to *S. pyogenes.* If the organism is present in the material from the swab, the beads clump together (Fig. 2). Although not 100% sensitive, the latex bead test is widely used because it can be done in a physician's office, is rapid, and is highly specific.

The isolation of *S. pyogenes* in culture is an example of the use of a nonselective medium to isolate a pathogen. Although other bacteria may grow on the blood agar plate, *S. pyogenes* can be identified by virtue of its distinctive colony morphology and by its ability to lyse red blood cells. Because other hemolytic organisms, some of which may resemble *S. pyogenes,* can be present, biochemical identification is required.

With *B. pertussis,* the cause of whooping cough, a different situation exists. This organism is quite fastidious, i.e., it requires a large number of cofactors, vitamins, and nutrients; grows slowly; and is difficult

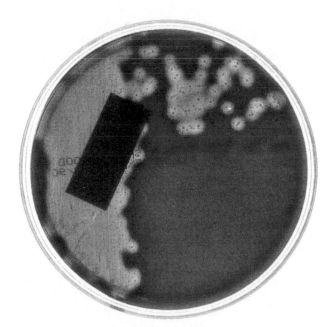

**FIGURE I** Blood agar plate inoculated with a swab taken from a patient with strep throat. The plate demonstrates a clearing of the red blood cells (hemolysis) around colonies of *Streptococcus pyogenes* but not around other organisms.

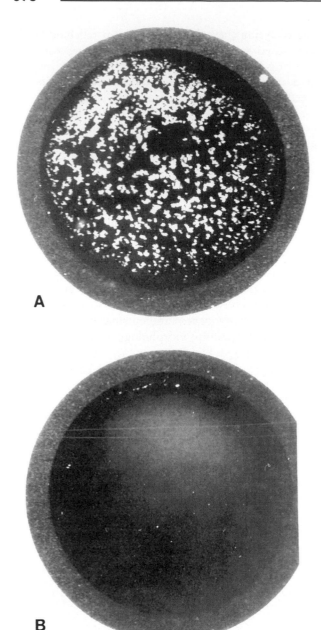

**A**

**B**

FIGURE 2 Results of a rapid latex agglutination test for the presence of *S. pyogenes* from a throat swab. (A) The test shows clumping of the latex particles indicating that *S. pyogenes* was present in this patient's throat sample. Thus, the patient probably has "strep throat." (B) The test shows a faint haze at the left of the circle, but no clumping. This is a negative test for the presence of *S. pyogenes*. Thus, the patient probably does not have "strep throat."

to recognize among the normal flora of the nasopharynx (passageway that leads from the nose to the throat). To recover *B. pertussis* from a child suspected of having whooping cough, a thin cotton-tipped wire

is passed through the nose to sample the nasopharynx. This material is then inoculated onto a highly nutritious selective medium containing antibiotics that inhibit the growth of normal flora but do not inhibit *B. pertussis*. Because some normal flora may grow in spite of the antibiotics, bacteria growing on the plate are identified as *B. pertussis* using a fluorescein-labeled antibody that will bind specifically to *B. pertussis*. After the antibody binds to the organism, it makes the organism glow when examined with a special fluorescent light. This is an example of the importance of selective media in suppressing the normal flora to allow detection of pathogens.

Selective media also are used to isolate *C. diphtheriae* from throat samples. Although diphtheria is no longer common in the United States, it is still a major health problem in many parts of the globe. The detection of this gram-positive organism in throat samples depends on the use of one of several special agar media that will support the growth of the organism. After isolation of this bacillus on agar, it is identified by biochemical tests. However, this organism is unusual in that its pathogenicity is due to the production of a toxin. Thus, an antibody test must be performed on extracts of the organism to determine whether or not the isolate produces diphtheria toxin.

### b. Lower Respiratory Tract

The major bacterial infection of the lower respiratory tract is pneumonia, which is inflammation of the lung tissue. A variety of bacterial pathogens can cause pneumonia, including *S. pneumoniae, L. pneumophila, Mycoplasma pneumoniae,* and several species of gram-negative bacilli. Because of the wide spectrum of agents that could account for the disease, a combination of selective and enriched media must be used to isolate the pathogen responsible.

If the patient expectorates infected material from the lower respiratory tract (sputum), it is immediately cultured. If the patient is not producing sputum, secretions are collected after saline is introduced into the patients lower airway to help dissolve and loosen infected material. The material is plated on a variety of enriched, selective, and differential media in an effort to isolate the responsible pathogen. Any pathogens growing on the agar plates must be differentiated from contaminating normal flora. Because *S. pneumoniae* and a few other bacterial pathogens may be present in low numbers in the normal flora, the quantity of the organisms present and a Gram stain of the present spectum are often the only clues for determining whether or not it is the cause of the disease.

To circumvent the problem of upper airway contamination, a sample may be collected directly from the lung; however, this technique poses some risk to the patient and is not undertaken unless expectorated sputa are uninterpretable.

Another organism that can infect the lower respiratory tract is *Mycobacterium tuberculosis,* the cause of tuberculosis. This organism poses special challenges because it grows very slowly in culture, sometimes taking as long as 4–6 weeks to produce visible colonies on solid medium. Agar media containing eggs and fatty acids often are employed in its isolation. This organism also is resistant to the usual stains used in microbiology but can be recognized using an acid-fast method.

Tuberculosis can be diagnosed more rapidly by incorporating $^{14}C$-labeled fatty acids into the broth culture medium and detecting $^{14}C$-labeled $CO_2$ produced as the organism grows. This approach, coupled with the use of nucleic acid probes to confirm the identity of organisms grown either in broth or on solid media, reduces the time of detecting *M. tuberculosis* in clinical samples to 1–2 weeks. Still, this pathogen, which remains a major public health problem throughout the world, is a difficult organism to isolate in the laboratory.

## 5. Fecal Cultures

Diarrhea, while often nothing more than an inconvenience in the United States, is a serious human illness in many parts of the globe. Diarrheal diseases are a major cause of death in many developing nations, especially during times of natural disasters, when normal means of sanitation are disrupted.

The most frequent bacterial causes of diarrheal disease in the developed world are *Salmonella* species, *Campylobacter* species, and *Shigella* species. The latter organisms also can cause dysentery, a disease characterized by the presence of mucus and blood in the feces, signifying invasion and inflammation of the intestinal wall rather than just the outpouring of fluid. While infrequent in the United States, *Vibrio cholerae,* the cause of cholera, still occurs in many parts of the world. In the past, global outbreaks, called pandemics, have infected whole populations. [*See* Salmonella.]

The key to the detection and identification of diarrheal agents is the use of selective media. Because the number of bacteria required to cause disease is often low for organisms such as shigella (about 200 organisms can cause infection), the recovery of these organisms from stool can be a challenge. Therefore, the laboratory uses a combination of enrichment broths and selective and differential agars. The purpose of an enrichment broth is to enhance the growth of pathogens, thought to be in low concentration in a specimen, while simultaneously suppressing the organisms of normal bowel flora. After 18 hr of enrichment, samples of the broth are plated on selective agar to enhance the detection of the pathogens. Other diarrhea-causing pathogens, such as *Campylobacter* species or *Yersinia enterocolitica,* require different types of selective media and different temperatures of incubation for recovery.

## 6. Skin and Wound Cultures

The most common causes of skin and wound infections are *Staphylococcus aureus* and *S. pyogenes.* Both bacteria grow readily on a variety of growth media and can be easily identified.

Among other bacteria that can be involved in skin and wound infections are anaerobic bacteria, especially *Clostridium perfringens,* which is a cause of gas gangrene, and certain species of mycobacteria. Anaerobic organisms require culture conditions devoid of oxygen, but rich in $CO_2$ and $H_2$. Chemical packs that produce hydrogen and carbon dioxide are used to create anaerobic atmospheres within jars with tight-fitting lids to enhance the growth of anaerobes. The genus and species of the anaerobes are then identified by biochemical testing. [*See* Anaerobic Infections in Humans.]

Mycobacteria are organisms related to *M. tuberculosis,* the respiratory tract pathogen. Both *Mycobacterium ulcerans* and *Mycobacterium marinum* are important causes of nonhealing skin lesions, particularly in people who encounter seawater (*M. marinum*) frequently. These organisms grow slowly and require special media, much like *M. tuberculosis.* An acid-fast stain may be helpful in early diagnosis of such wound infections. *Mycobacterium leprae,* the cause of leprosy, also is often found in skin and tissue lesions and can be detected with acid-fast stains, but cannot be cultured on solid or broth media. Inoculation into animals such as armadillos is required to propagate the organism.

## 7. Genital Tract Cultures

The most important and widespread of the agents that are transmitted primarily by sexual contact between humans include *N. gonorrhoeae,* the cause of gonorrhea; *Treponema pallidum,* the cause of syphilis; and *Chlamydia trachomatis,* the cause of nongonococcal urethritis. Each pathogen requires unique assays for detection due to their fastidious nature. [*See* Sexually Transmitted Diseases (Public Health).]

*Neisseria gonorrhoeae* is a gram-negative, coffee bean-shaped organism. It is usually seen in pairs (diplococci) in a Gram stain when material is taken from the urethra of a male or from the cervix of a female with gonorrhea (Fig. 3). In males, the presence of gram-negative diplococci within inflammatory cells is diagnostic of gonorrhea. This is not the case in the female, however, where other nonpathogenic *Neisseria* species may be present in the sample. Thus, the "gold standard" for diagnosis of gonorrhea is culture. To isolate gonococci (a pseudonym of *N. gonorrhoeae*), enriched culture medium containing hematin and serum is required. Often, antibiotics to which gonococci are resistant are added to the medium to inhibit the growth of other organisms that may be present in the urethra or cervix. Colonies that grow on the selective medium are identified biochemically or by using specific antibodies to *N. gonorrhoeae*.

Chlamydia infections are more difficult to diagnose because the causative agent *C. trachomatis* must live inside of a cell to survive and, thus, cannot be grown on culture media like other bacteria. Therefore, samples taken from the genital tracts of males or females are inoculated into a special cell culture. After incubation for 48–72 hr, the cultures are stained with labeled antibodies to *C. trachomatis* to confirm its growth. An alternate method of detection is direct staining of urethral and cervical material with a fluorescein-labeled antibody.

The third organism, *T. pallidum*, poses a unique diagnostic problem because it cannot be readily isolated in cell culture or on plates. Thus, the diagnosis of syphilis is made by visualizing the organism directly in aspirates of syphilitic lesions. *Treponema pallidum* is a spiral-shaped organism known as a spirochete. Its very slender width does not permit its detection by light microscopy, but requires dark-field microscopy.

Although dark-field microscopy is rapid, it is technically difficult and only successful in the earliest stages of disease. Thereafter, the laboratory diagnosis of syphilis is made by detecting the presence of antibodies to the organism. The test takes advantage of the

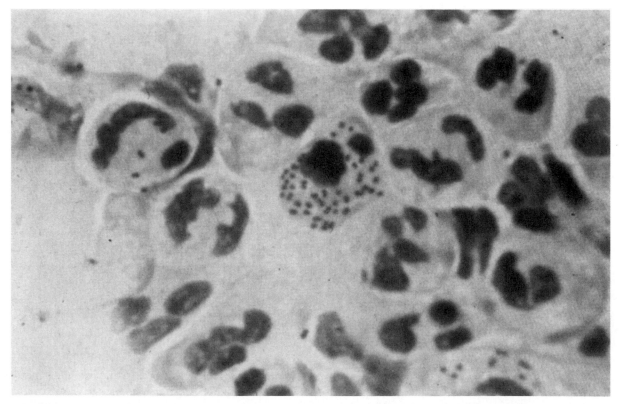

**FIGURE 3** Gram stain of a urethral discharge from a male with gonorrhea. Many white blood cells (leukocytes) can be noted, which signifies inflammation of the urethra. They contain kidney bean-shaped organisms that appear red in the stain. They are gram-negative and characteristic of *N. gonorrhoeae*.

fact that the antigens of *T. pallidum* are similar to an antigen of heart muscle known as cardiolipin. An inexpensive test that detects antibodies to cardiolipin, known as Venereal Disease Research Laboratory (VDRL) test, is readily available in a commercially prepared kit known as the Rapid Plasma Reagin (RPR) test. This screening test, if positive, must be confirmed by additional tests that assess the presence of antibodies specific for *T. pallidum* in the patient's blood. These latter tests are referred to as antitreponemal tests.

## IV. ANTIMICROBIAL SUSCEPTIBILITY TESTING

Once a pathogen is isolated and identified, the final step is to determine the antimicrobial agents to which it is susceptible. Susceptibility to an antimicrobial agent can be determined in four ways: agar dilution, broth dilution, disk diffusion, or an automated method.

### A. Agar and Broth Dilution Susceptibility Testing

In the agar dilution method of susceptibility testing, antimicrobials are incorporated into an agar medium at various concentrations. The plates are inoculated with the bacterium to be tested and incubated for 24 hr. The lowest concentration of an antimicrobial agent that prevents colonies of the pathogen from forming is referred to as the minimal inhibitory concentration (MIC). Broth dilution susceptibility testing is similar except that the absence of turbidity in the broth is used to indicate the end point. If the organism is inhibited at a concentration of antibiotic easily achieved in the bloodstream, the organism is deemed to be "susceptible" to that drug, otherwise it is said to be "resistant." Organisms are often tested against several different types of antimicrobials.

### B. Disk Diffusion Testing

A simpler method of susceptibility testing is to spread a suspension of the organism on a large agar plate and place disks containing known amounts of antibiotics on its surface. As the organism grows, it will encounter the antibiotics diffusing out of the disks and into the agar medium and, if susceptible, will be inhibited from growing. This will produce a "zone of inhibition" around the disk (Fig. 4), the size of which correlates with the degree of inhibition. The size of the zone is compared with those produced by a series of standard laboratory strains with known antimicrobial susceptibility. In this procedure, called the Kirby-Bauer disk diffusion assay, up to 12 antimicrobial agents can be tested against a single organism on one plate. The test requires 18 hr of incubation before the zone diameters, signifying the degree of growth inhibition, can be read and interpreted.

### C. Automated Method of Susceptibility Testing

Because both the agar dilution and Kirby-Bauer methods are labor intensive, automated machines have been developed to replace agar and broth dilution methods. The major drawback to automated methods is their expense, which can be twice that of the Kirby-Bauer method. However, the speed with which the answers are determined (2–6 hr) may make a considerable difference in the management of a seriously ill patient.

### D. Antimicrobial Resistance

Some microorganisms tend to be very resistant to antibiotics and require additional tests with an expanded array of antimicrobial agents. Usually, some antimicrobial can be found that will inhibit the growth of the organism. Combinations of drugs are often used against bacteria to prevent the development of resistance that occurs while the patient is being treated. The laboratory, however, is unable to reproduce exactly the growth conditions that the bacteria encounters in the human body, and, consequently, susceptibility testing is limited to a certain degree in its utility.

## V. REPORTING THE RESULTS AND QUALITY CONTROL

Reporting is always the last step of testing and is greatly facilitated by computers. However, with the use of more automated methods in the laboratory, there is growing concern about the accuracy of the results that are generated. How accurate are machines versus their human counterparts? At present, the bulk of the critical work in microbiology laboratories is still done by technologists. This includes the plating

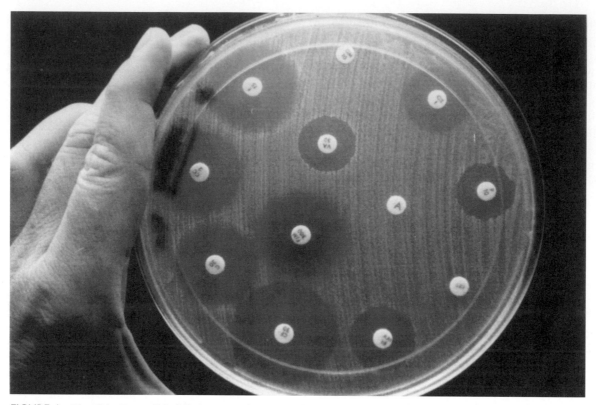

**FIGURE 4** The Kirby-Bauer disk diffusion assay for testing the susceptibility of a bacterium to antimicrobial agents. Note the zones of inhibition of growth around the various disks containing the antibiotics. The size of the zone correlates with the degree of susceptibility of the organisms to that antibiotic.

of specimens on agar media, reading and interpretation of Gram stains, and picking of isolates from the different agar plates for further work. However, with the increased use of direct detection systems, such as labeled monoclonal antibodies and DNA probes, it may be a challenge to maintain the same levels of sensitivity and specificity afforded by culture methods. The need for additional quality control methods to ensure the accuracy of the rapid tests is critical.

## VI. FUTURE DIRECTIONS

The future of clinical microbiology lies in rapid testing methods, such as DNA probes and antibody-based tests, which rely less on the traditional skills of the microbiologist to recognize colony types, staining characteristics, and unusual and exotic smells, and more on direct detection of pathogenic microorganisms in clinical samples. The question of the next decade regarding the diagnostic microbiology laboratory is how much sensitivity and specificity the physi-

cian will be willing to sacrifice to obtain test results more rapidly.

## BIBLIOGRAPHY

Finegold, S. M., and Baron, E. J. (1986). "Bailey and Scott's Diagnostic Microbiology," 7th Ed. Mosby, St. Louis.
Isenberg, H. (ed.) (1992). "Clinical Microbiology Procedures Handbook." American Society for Microbiology, Washington, D.C.
Jorgenson, J. (ed.) (1987). "Automation in Clinical Microbiology." CRC Press, Boca Raton, FL.
Mandell, G. L., Bennett, J. E., and Dolin, R. (eds.) (1994). "Principles and Practice of Infectious Diseases." Churchill Livingston, New York.
Murray, P. R., Baron, E. J., Pfaller, M. A., Tenover, F. C., and Yolken, R. H. (eds.) (1995). "Manual of Clinical Microbiology," 6th Ed. American Society for Microbiology, Washington, DC.
Persing, D. H., Smith, T., Tenover, F. C., and White, T. (eds.) (1993). "Diagnostic Molecular Microbiology: Principles and Applications." American Society for Microbiology, Washington, D.C.
Tenover, F. C. (ed.) (1988). "DNA Probes for Infectious Diseases." CRC Press, Boca Raton, FL.

# B-Cell Activation

NORMAN R. KLINMAN

*The Scripps Research Institute*

I. B-Cell Repertoire of Antibody Specificities
II. B-Cell Development
III. B-Cell Subpopulations
IV. B-Cell Triggering
V. Origins of B-Cell Memory Responses
VI. Basis of Self–Nonself Discrimination by B Cells

## GLOSSARY

**Antibody** Globular protein produced in response to administration of a suitable antigen that combines specifically with that antigen

**Antigen** Protein or carbohydrate that, when introduced into the body, can induce the production of antibody or reactive T cells

**Antigenic determinant** Component of an antigen molecule that constitutes the recognition site for an antibody

**Clonotype** Unique heavy- and light-chain variable region amino acid sequence of a particular B cell or B-cell clone

**Cross-reactivity** Binding of antibody to an antigen other than that used for immunization

**Idiotype** Antigenic determinant within the variable region of an antibody molecule

**Immunoglobulin** Family of globular proteins that includes antibodies of all isotypes as well as other antibody-like serum proteins without known antigen recognition

**Tolerance** Specific inactivation, by an antigenic determinant, of B cells specific for that determinant

THE TERM "B CELL" REFERS TO THOSE LYMPHOID cells that, upon antigenic stimulation, give rise to a clone of antibody-forming cells. The function of B cells is to aid in the defense against viruses, bacteria, and parasites by the production of antibody molecules that inactivate or facilitate the removal of these patho-

gens. The characteristics, development, and function of B cells are similar in humans and most other mammalian species; however, most of our present knowledge has come from studies of B cells from mice because of the availability of inbred murine strains which have permitted studies on a controlled genetic background. Adult mice have $2-4 \times 10^8$ B cells, and humans have approximately $10^{11}-10^{12}$ B cells. These cells are distributed throughout the body in interstitial tissues, the peritoneal cavity, and the lymphatic and circulatory systems and are localized in organs of the reticuloendothelial system, representing 10–30% of cells in lymph nodes and 40–50% of nucleated cells in the spleen. In adults, B lineage cells are generated primarily in the marrow of large bones and represent 20–30% of all nucleated bone marrow cells. B cells range in size from 7 to 12 mm in diameter and express numerous cell surface antigens, some of which have been defined as receptors for immunoglobulins (lgs), interleukins (ILs), or hormones. Most characteristic is the expression on the surface of B cells of high levels of major histocompatibility complex (MHC) class II antigens, which are essential for the stimulatory interaction with T cells, and Ig, which serves as the primary receptor for antigen.

## I. B-CELL REPERTOIRE OF ANTIBODY SPECIFICITIES

### A. Antibody Molecule

The function of B cells is mediated through the production of antibody molecules. The structure and molecular events responsible for the biosynthesis of antibodies are presented elsewhere in this encyclopedia. In brief, antibodies are globular proteins made up of pairs of heavy (H) and light (L) polypeptide chains.

ENCYCLOPEDIA OF HUMAN BIOLOGY, Second Edition, VOLUME I. Copyright © 1997 by Academic Press. All rights of reproduction in any form reserved.

Both H and L chains are composed of domains that are approximately 12 kDa and contain an internal disulfide loop. The amino-terminal domains of both the H and L chains are termed variable (V) domains because their sequence varies from antibody to antibody, particularly within three areas termed hypervariable regions. The V domains constitute the antigen-binding site, which presents as a surface area of approximately 800 $Å^2$, with most of the amino acid residues that contact antigen residing within the hypervariable regions of both the H and L chains. [*See* Antibody–Antigen Complexes: Biological Consequences.]

L chains are composed of two domains: one V and one constant (C). There are two types of L chains, $κ$ and $λ$, which arise from separate chromosomes and differ in both their V and C domains. H chains are composed of one V domain and two to four C domains. There are 8 to 10 isoforms (isotypes) of the H chain, each sharing the same set of V domains, but differing in their C domains. The particular set of C domains that comprise a given isotype also determines the size of the antibody molecule, the number of H- and L-chain pairs, and much of the biological activity of the molecule. For example, IgM can exist as a cell surface receptor molecule with two H and L pairs, its carboxy-terminal H-chain C domain ($C_H$) having a membrane insertion sequence. Alternatively, IgM can be expressed as a secreted antibody, having five H- and L-chain pairs with no hydrophobic tail, but with an additional 15-kDa polypeptide chain [joining (J) chain]. IgG exists as a two H- and L-chain pair structure with three $C_H$ domains, which are modified to maximize complement binding and placental transfer. IgA generally exists as a four H- and L-pair structure with a J chain and $C_H$ domains that facilitate secretion by mammary gland cells and intestinal epithelium. *In toto* the entire set of antibodies comprises the immunoglobulin family.

## B. Molecular Basis of Antibody Diversity

In mice the random association of H and L V regions enables the generation of over $10^7$ antibody specificities. The expression of this vast array of antibody-combining sites and its clonal distribution among B cells are the products of a complex set of molecular events. Briefly, $V_H$ is the product of the rearrangement of members of three gene segment families arranged in the order $V_H$-$D_H$-$J_H$ ("D" representing the diversity region) on chromosome 12 of the mouse. The carboxy terminus is encoded by 1 of 4 or 5 JH regions and is rearranged to 1 of 10–20 $D_H$ regions. The $D_H$ region,

in turn, is rearranged to one of several hundred $V_H$ regions. The enzymatic process responsible for the rearrangement of genetic segments depends on various enzymes, including the recombinase enzymes RAG-1 and RAG-2, and is accomplished by the juxtaposition of homologous sequences flanking the various gene segments that permit looping, recombination, and excision of the looped regions. Diversity in the $V_H$ sequence is created by the random association of different $V_H$, $D_H$, and $J_H$ segments as well as imprecisions at the joining regions (junctional diversity). Junctional diversity is created by nucleotide deletions, and the addition of nucleotides either by inclusion of nucleotides encoded by the opposite DNA strand during excision (P nucleotides) or by the addition of nucleotides (N addition) by the enzyme terminal deozynucleotide transferase (TdT). In most B cells both chromosomes display rearranged H chain V region segments but there is usually only one productively rearranged $V_H$-$D_H$-$J_H$ per cell.

$V_L$ diversity is generated in much the same way as $V_H$ diversity, although there are only two relevant gene segment families, $V_L$ and $J_L$, and junctional diversity rarely includes N additions. Similar to the H chain, only one successfully rearranged $V_L$-$J_L$ is expressed per cell. Diversity in $κ$ L chains is the product of the random association of one of several hundred V$κ$ gene segments with one of four J$k$ segments and diversity at the junction. $λ$ L chains are less diverse, since in mice only one or two V$λ$s or J$λ$s are available for each of the two to three $λ$ isoforms

Since antibodies are composed of H- and L-chain dimers, antibody diversity is achieved as the random product of highly diverse $V_H$ and $V_L$ partners. In all, this association enables the generation of a vast repertoire of specificities, which has been estimated to exceed $10^7$ V regions. Among these individual H + $V_L$ combinations (i.e., clonotypes) are a few that are expressed at an unusually high frequency in all mice of an inbred strain. Since the generation of these "predominant" clonotypes is reproducible and can be identified in newly generating B cells, this phenomenon implies that repertoire generation and V region gene segment selection and rearrangement are not totally random processes.

## C. Cellular Basis of Antibody Repertoire Expression

The broad repertoire of available antigen-combining sites provides the potential for the production of serum antibodies that discriminately recognize any of an enormous array of antigenic determinants. The mechanism by which this repertoire is accessed by antigen is dependent on the clonal distribution of each antibody within the B-cell population so that any given B cell has the capacity to produce antibodies expressing only a single V region (clonotype). Since each B cell expresses a different clonotype, the repertoire of available antibodies is equivalent to the repertoire of available B cells or B-cell clones.

Microorganisms generally present a multiplicity of antigenic determinants, each capable of being recognized by numerous different antibody molecules expressed as B-cell surface receptors. Those B cells whose surface Ig binds a recognized antigenic determinant with sufficient affinity are stimulated. Usually, this includes numerous B cells having different combining sites, each having in common only the capacity to bind well to the same determinant. Since most microorganisms express several protein or carbohydrate antigens, each of which might exhibit multiple recognized determinants, responses generally encompass numerous sets of responding B cells. Any selected antibody binds well to one determinant of the stimulatory antigen and should, in general, bind poorly, if at all, to the universe of other determinants, thus displaying discriminatory recognition. However, most antibodies will bind, at lower affinity, to some determinants that bear a three-dimensional resemblance to the stimulatory antigen, and some antibodies might display binding to certain unrelated antigenic determinants.

This lack of uniqueness in antibody binding is evidenced by "cross-reactivity" to antigens other than the one used for immunization. Since microorganisms display multiple determinants generating numerous sets of antibodies, cross-reactivity can be extensive. However, since (1) only a small minority of antibodies would cross-react with any determinant that is not homologous to a determinant of the stimulating antigen, (2) most cross-reactions would be of low affinity, and (3) the totality of all of the antibodies of all of the sets specific for determinants of the stimulating antigen would bind well to the stimulating antigen; in composite, antibody responses are highly specific. Nonetheless, it is the specificity of responding B cells at the population level, rather than the specificity of any given B cell, that accounts for the specificity of an antibody response.

## II. B-CELL DEVELOPMENT

The first functional B cells of the murine fetus are found in the liver at the end of the second trimester of fetal development. Although B cells continue to develop in the liver until birth, the major source of B cells in late fetal and early neonatal development is the spleen. Soon after birth B-cell development increases rapidly in the marrow of large bones. After the first week of neonatal life and throughout adulthood, the bone marrow is the major site of B-cell development. The continuous generation of B cells is necessary because although many mature B cells appear to survive for several weeks, most newly generated B cells appear to have a half-life of only several days. This is consistent with the generation in mice of approximately $1–5 \times 10^7$ B cells per day from the bone marrow throughout adulthood, which is sufficient to replace 5–25% of the total B-cell population on a daily basis.

Reproduction experiments have demonstrated that B-cell development can proceed through a series of proliferative and differentiative steps from the cells within the population of multipotential bone marrow stem cells. The microenvironment of a differentiating stem cell appears to influence whether its progeny will become lymphoid cells and subsequently T cells or B cells. It remains unclear whether most B cells are derived directly from self-renewing multipotential stem cells or, alternatively, from a self-renewing population of more differentiated cells further along in the lymphoid or B-cell lineage.

The earliest discernible events in the commitment of precursors to the B-cell lineage involve the expression of a set of transcription factors which is soon followed by the expression of TdT, RAG-1, RAG-2, and a set of cell surface antigens including B220. At this stage, genes encoding the antibody H chain, particularly $D_H$ and $J_H$ gene segments, initiate recombination. This is followed by the rearrangement of $V_H$ to a $D_H$–$J_H$ complex, the expression of H-chain mRNA, cytoplasmic expression of the H-chain mRNA, and the cytoplasmic expression of the H chain itself. Cells at this stage are considered "pre-B cells."

The initial H-chain RNA transcript includes both coding and noncoding sequences (i.e., exons and introns). It contains a leader sequence, which does not code for proteins, the $V_H$-$D_H$-$J_H$ regions, and C-region exons for the C domains of the $\mu$ and $\delta$ chains. This RNA is then processed to produce the mature H-chain mRNA, which includes the leader, $V_H$-$D_H$-$J_H$, and preferentially $C\mu$ rather than $C\delta$. Then there follows, in the immature B cells, the successful recombination of genes encoding L, $V_L$ and $J_L$, and the production of L-chain polypeptide. Finally, an intact Ig molecule of the IgM isotype is formed and is expressed at the surface of the cells. After L-chain expression most developing bone marrow B cells cease dividing. The repertoire of antibody specificities appears to be largely established at this stage of B-cell development. In general, the diversity of the antibody repertoire is unchanged as these cells mature and populate peripheral lymphoid tissue, with the exception that some clonotypes are eliminated by tolerance or anti-idiotypic recognition (see Section VI).

Newly developed bone marrow B cells express surface IgM but no surface IgD and are termed "immature B cells." These cells express MHC class II antigen and can be stimulated by T-cell-dependent antigens under appropriate conditions, but are unique in that they are highly tolerance susceptible (see Section VI). With further maturation, B cells become more resistant to tolerance induction, begin to express surface IgD, and egress from the bone marrow to populate the peripheral lymphoid system.

The molecular events that accompany B-cell development in the fetal liver and the spleen during early neonatal development are similar to those described earlier for the adult bone marrow. However, the repertoire of antibody specificities present during late fetal and early neonatal development is much more limited (i.e., $10^4$–$10^5$ clonotypes rather than over $10^7$ in adult mice) and is similar in all individuals of an inbred murine strain. Additionally, repertoire diversity is acquired in a reproducible fashion during neonatal development. This pattern of B-cell repertoire expression during fetal and neonatal development might, in part, be accounted for by restrictions in the utilization of $V_H$- and $V_L$-region genes during development. In particular, $V_H$ gene segments in mice that are located on chromosome 12 most proximal to the $D_H$ and $J_H$ regions are preferentially rearranged in fetal and neonatal B cells. Furthermore, TdT is not expressed in developing fetal and early neonatal B cells so that the H chain V regions of these cells do not have N additions which greatly reduces their diversity. An additional characteristic of B-cell clones developing in the neonatal spleen is that they are tolerance susceptible for several days rather than several hours, as is the case for B cells developing in the adult bone marrow (see Section VI).

## III. B-CELL SUBPOPULATIONS

B cells exist as at least three distinct subpopulations. These subpopulations can be distinguished by certain cell surface markers, but most importantly they differ functionally.

### A. "Conventional" Primary B Cells

This subpopulation includes the vast majority of B cells in peripheral tissues. In adults these cells are generated in the bone marrow as described in Section II and, at some time in their life cycle, express high levels of both surface IgM and surface IgD. They also express intermediate to high levels of heat-stable antigen (HSA). These cells are responsible for the bulk of antibody responses following primary antigenic stimulation.

### B. CD.5 B Cells

In mice these cells represent a small minority of B cells in the spleen and are rare in lymph nodes, but are enriched among cells in the peritoneal cavity. They are prevalent in the neonatal spleen, but are not normally generated in the bone marrow of adults. Rather, mature cells of this subset might self-renew, especially within the peritoneal cavity. Phenotypically, these cells are HSA$^{hi}$ and express high levels of surface IgM, but only low levels of surface IgD. Unique among B cells, they often express low levels of CD.5 (also called Ly-1 in mice, Leu 1 in humans), an antigen that is present at high levels on T cells, but is absent from most B cells. Some of these cells also express Mac-1, normally a macrophage cell surface antigen.

Functionally, CD.5 B cells respond to primary antigenic stimulation and appear to be particularly prone to respond to certain bacterial antigens such as phosphorylcholine. Because they are prevalent in certain autoimmune murine strains and often give rise to antibodies that cross-react with numerous autoantigens, they have been implicated in autoimmune states. [*See* Autoantibodies.]

### C. Memory B Cells

The response to antigenic stimulation is characterized by both the production of antibody and the generation of a population of B cells specific for the immunizing antigen that appears to be largely responsible for responses to subsequent contacts with the same antigen (see Section V). These so-called "memory" B cells differ substantially from the two aforementioned primary B-cell subsets. Like conventional primary B cells, these cells are found in the spleen and lymph nodes; however, these cells can also be found as recirculating cells in the lymphatic system. Memory B cells generally appear to be longer lived than most primary B cells, express low levels of HSA, and often express surface Ig receptors of isotypes other than IgM or IgD.

Functionally, memory B cells differ from primary B cells in parameters of their stimulation (see Section V) and, generally, memory B cells produce antibodies with higher affinity for the immunizing antigen than do primary B cells. Most striking is the accumulation of numerous somatic mutations in the genes encoding $V_H$ and $V_L$ regions of antibodies of memory B cells. These mutations are particularly prevalent in the hypervariable regions and appear to reflect antigen selection since their presence is often accompanied by an increase in the binding affinity for antigen. Antigen selection and somatic mutation, which are relatively unique to the generation of memory B cells, appear to be at least partially responsible for differences in the antibody repertoire of memory versus primary B cells.

Memory B cells appear to be generated primarily from a small subset of primary precursor cells within the population of cells generated from the adult bone marrow. These cells bear IgM and high levels of IgD, but are $HSA^{lo}$, like their secondary B-cell progeny. The majority of precursor cells that are isolated in nonimmune mice, as $HSA^{lo}$, do not give rise to antibody-forming cell (AFC) clones upon primary antigenic stimulation, but instead give rise to secondary B cells. Since conventional ($HSA^{hi}$) primary B cells give rise to primary AFC clones and not secondary B cells, it appears likely that the small subset of precursor cells that are $HSA^{lo}$ might be uniquely responsible for memory generation and might be unique in their ability to somatically hypermutate after antigenic stimulation.

## IV. B-CELL TRIGGERING

### A. Antibody as a Clonally Distributed Antigen Receptor

The specificity of antibody responses is dependent on (1) the availability of a vast repertoire of antibody specificities distributed clonally, one specificity per B cell, and (2) the capacity of an antigenic determinant to selectively stimulate those B cells whose surface antibody can bind with high affinity. These antibodies act as receptors for antigen, revealing the potential antibody product of the B cell. These receptors are generally monomer IgM and IgD and are produced by alternate splicing of the RNA transcript of the rearranged Ig gene. Other isotypes in memory B cells can also be expressed as cell surface receptors; they are also produced by splicing the RNA transcript to incorporate a membrane insertion region. Regardless of the surface isotype, each cell expresses approximately $10^5$ surface Ig receptor molecules.

The Ig receptor functions in two ways in antigen-mediated B-cell triggering. First, under certain circumstances, surface Ig acts as a triggering receptor itself. Second, the surface Ig receptor serves to capture antigen, thus facilitating both T-cell-dependent and -independent B-cell stimulation.

When the surface Ig receptor per se, without T-cell cooperation, acts to trigger B cells, there is a requirement for receptor cross-linking on the cell surface. In general, only antigens or antibodies that bind multivalently and with high affinity are capable of triggering B cells through their surface Ig receptors. In primary B cells the principle triggering receptor appears to be surface IgM, although surface IgD can also participate. Surface Igs form part of a receptor complex that also includes two molecules, Ig$\alpha$ and Ig$\beta$, that are integral to the signaling pathway. Interlinking of surface Ig receptor complexes initiates their tyrosine phosphorylation and activates the phosphoinositide pathway (e.g., the activation of phospholipase C which cleaves phosphatidylinositol 4,5-bisphosphate, yielding both inositol 1,4,5-triphosphate, which can cause $Ca^{2+}$ release from internal stores, and diacylglycerol, which can activate protein kinase C). Although evidence suggests a contribution of surface Ig receptor-mediated triggering in the stimulation of B cells to proliferation and differentiation to antibody-secreting cells, this process seems primarily driven by engagement of other receptors, such as mitogen receptors or MHC class II which mediates interactions with T cells. However, the surface Ig receptor is the primary triggering receptor for antigen- or antibody-driven B-cell inactivation or tolerance induction (see Section V).

### B. Antigen Presentation and T-Cell–B-Cell Collaboration

The principal mechanism for B-cell stimulation is the collaborative interaction of antigen-specific B cells and T helper ($T_H$) cells. $T_H$ cells recognize protein antigenic determinants presented as peptides in association with the MHC class II molecule on the surface of an antigen-presenting cell. B cells, like other cells that can present antigen to T cells, express MHC class II molecules on their surface. In addition, antigen-presenting cells have the capacity to take up antigens and to process them, enabling the association of peptides derived from the antigen with MHC class II antigen molecules as these are expressed on the cell surface. B cells represent efficient antigen-presenting cells since their surface Ig antigen receptors allow for the specific concentration of appropriate antigens on the B-cell surface for subsequent internalization, processing, and presentation. Thus, B-cell stimulation by T-cell-dependent antigens is the result of a complex sequence of events wherein the surface Ig receptor of the B cell binds to a determinant of an antigen which is then processed. Another determinant on the same antigen in the form of a peptide associated with the MHC class II molecule of the B cell is then presented to T cells. The presentation of antigen by the B cell not only serves to capture T cells specific for that peptide–MHC complex, but also initiates the interaction of a series of coreceptors and their ligands, some of which are up-regulated, on both the interacting B cells and T cells. Among the coreceptors on B cells is the CD40 molecule whose interactions with a ligand on T cells is crucial for B-cell stimulation. B cells express B7.1 and B7.2 whose recognition by the CD28 coreceptor on T cells is important for T-cell activation. These activated T cells in turn provide various ILs that facilitate the activation of the antigen-presenting B cell. Thus, antigen mediates a truly collaborative interaction between T cells and B cells that generally recognizes different determinants on the antigen molecule.

## C. Nonspecific B-Cell Activation

Interactions of B cells with certain bacterial and viral molecules, such as lipopolysaccharide, can lead to B-cell stimulation. This form of stimulation is considered mitogenic since it usually leads to proliferation and occasionally to antibody synthesis. At a high concentration of the mitogen, a large percentage of B cells can be stimulated in this fashion irrespective of the kind of surface Ig they carry, and a broad distribution of antibody specificities results (i.e., polyclonal activation). However, when the surface Ig receptor of a B cell recognizes a determinant on, or attached to, a molecule with mitogenic activity, the mitogen can be concentrated to its surface. In this case low concentrations of mitogens can serve as specific stimulators of relevant B cells. Such stimulation might play an important role in B-cell responses to certain bacterial antigens,

particularly when there is a paucity of determinants that could be recognized by T cells.

## D. B-Cell Differentiation to Antibody-Forming Cell Clones

When antigen interacts at sufficient affinity with the surface Ig receptor of a B cell, several events can ensue. If the antigen itself can act as a mitogen, B-cell stimulation might proceed merely through the concentration of the mitogen to mitogen receptors on the cell surface. For antigens that require B-cell–T-cell interaction, antigen binding to the surface Ig receptor initiates two processes. First, the B cell might be "activated" through the triggering function of its surface Ig receptors, particularly if the antigen cross-links these receptors. A consequence is an increased expression of MHC class II antigen on the B-cell surface.

The second process is the internalization and processing of the bound antigen. As peptides of the antigen are presented on the B-cell surface, together with MHC class II molecules, recognition by $T_H$ cells and activation of $T_H$ cells become possible. The activated T cell releases a variety of protein factors (i.e., ILs) that participate in both the proliferative and the differentiative B-cell events. The activated B cell initiates mitosis, with cell division occurring every 8–16 hr thereafter for several days. The newly generated progeny begin to splice their RNA transcripts of the H-chain gene so as to favor the production of secreted IgM and disfavor the expression of surface IgM and IgD. Additionally, there is a marked increase in H- and L-chain mRNA and protein production. Within 3–4 days of stimulation, IgM secretion is initiated from the clone of dividing cells. [*See* Mitosis.]

During the course of proliferation and differentiation, further rearrangement of the DNA-encoding H chain takes place such that a switch region present in the rearranged gene downstream from the $C\mu$ domain can become juxtaposed to a switch region upstream from the first domain of another isotype. Then, primarily by switch recombination and excision of intervening DNA, $V_H$ is expressed with the C region of the other isotype. Each daughter cell has the potential to switch to the production of any isotype, with the exception that once the DNA encoding an upstream isotype is excised, the cell can no longer switch to the production of that isotype because the relevant genetic information is lost. Although the clonal progeny of any B cell might be capable of switching to any isotype, the actual isotypes expressed by the cells of any given clone are greatly influenced by the type of $T_H$ cells encountered and the ILs it produces.

High concentrations of certain ILs have been shown to favor the production of various isotypes, apparently by triggering an increase in the accessibility of particular switch regions. Thus, for example, IL-4 favors production of IGE and $IgG_1$, and IL-4 plus IL-5 favor $IgG_4$ production in humans. As antibody-forming cell clone development proceeds, the cells express decreasing amounts of surface Ig, increased polyribosomes, and smaller nuclei. These cells gradually assume the appearance of plasma cells, which are ultimately nondividing cells that might survive several days *in vivo* and several weeks *in vitro*.

## E. Regulation of B-Cell Responses

As discussed earlier, the quality of B-cell responses can be greatly affected by the collaborating $T_H$ cells. As discussed elsewhere in this encyclopedia, there are at least two types of $T_H$ cells—$T_{H1}$ and $T_{H2}$—that appear to preferentially express various ILs. Both make IL-2, which is required mainly for the multiplication of T cells, but $T_{H2}$ cells release IL-4 and IL-5, which in humans favor $IgG_4$ and IgE expression, whereas $T_{H1}$ cells release interferon-$\gamma$, which can suppress these isotypes and favor $IgG_2$ expression. [*See* Interferons; Interleukin-2 and the IL-2 Receptor.]

Responses of B cells can also be regulated by other mediators and cell–cell interactions. For example, transforming growth factor $\beta$, which is present in various tissues, suppresses the production of most isotypes, but favors IgA production. The interaction of complement components and their B-cell receptors has been found to augment B-cell proliferative responses, and IL-6, which is released by $T_H$ cells as well as other tissues, aids in the terminal differentiation of antibody-forming cells. $T_H$ cell function per se can be enhanced or suppressed by interactions with various factors and other T cells. Finally, antibody responses can be stimulated or suppressed by the recognition of clonotype-specific antigenic determinants of the V region of the B-cell surface Ig (idiotypic recognition) by antibodies or T cells. The suppression of B-cell responses through anti-idiotypic recognition can be induced as a consequence of a primary immune response and is considered a regulatory mechanism dampening primary responses. Anti-idiotypic recognition might also play a role in regulating the development and expression of the B-cell repertoire. [*See* Transforming Growth Factor $\beta$.]

## V. ORIGINS OF B-CELL MEMORY RESPONSES

In addition to the generation of primary AFC clones, primary T-cell-dependent antigenic stimulation induces numerous changes in the responding B-cell population. Most notably, as the response progresses, the generated antibodies have increased affinity for the immunizing antigenic determinant, a process termed "immunological learning." One of the mechanisms responsible for this increasing affinity is a progressively selective stimulation of B cells whose surface Ig affinity is highest for the antigen. This is presumably a result of competition for antigen and T-cell help as antigen concentrations decrease and serum antibody increases. The major mechanism of immunological learning is the rapid accumulation of somatic mutations within the V regions of B cells responding to T-cell-dependent stimulation. These mutations appear to be the result of a hypermutation mechanism initiated within several days of the initial stimulation. Mutations occur throughout the $V_H$ and $V_L$ regions at a rate approaching one mutation per $10^3$ bp per generation.

Since most of the mutations cause amino acid replacements, there is the potential for strong selective forces acting on new sequences. Many mutations are likely to be deleterious to the production or selection of antibody and some might lead to antiself reactivities, which are likely to be inactivated (see below). However, some mutations lead to sequence changes that increase the affinity of the antibody for the stimulating antigen. Such mutations provide B cells with a selective advantage since, during the course of a response, mutations that favor amino acid replacements in the complementarity-determining regions of $V_H$ and $V_L$ accumulate preferentially. Thus, during the final stages of an initial response to antigenic stimulation, the antibodies that are produced are generally of isotypes other than IgM, display a relatively high affinity for antigen, and include numerous V-region somatic mutations.

Subsequent to a primary response, a second contact with the same antigen yields a more rapid and vigorous response which reflects both the isotype and high affinity of the antibodies produced during the final stages of the primary response. This accelerated response is called a secondary, or memory, response. Secondary responses can be induced fairly early in the course of a primary response or months to years after the primary contact. Secondary responses are also characterized by having a greater longevity than primary responses; however, subsequent antigen contact can lead to tertiary or quaternary responses, which are phenotypically similar to secondary responses, but can display even higher affinity antibodies.

The ability to obtain memory responses is dependent on the availability of a $T_H$ cell population already preselected for determinants of the immunizing antigen presented as peptides in the context of the MHC class II of the responding B cells. Most important are numerous alterations in the repertoire of B cells responsive to the immunizing antigen consequent to the primary immunization. For most antigenic determinants the frequency of responsive B cells is increased after primary immunization. Additionally, the characteristics of most B cells responsive to the original immunizing antigen in immune individuals (i.e., secondary B cells) are quite different from those of nonimmune individuals (i.e., primary B cells).

Primary B cells have IgM and IgD as their surface receptors, whereas many secondary B cells bear receptors of other isotypes. The V regions of primary B cells and their receptors are devoid of somatic mutations, whereas those of secondary B cells generally have accumulated multiple V-region somatic mutations. In part, because of this, the repertoire of secondary B cells responsive to a given antigen often differs from that of primary B cells. Secondary B cells bear many of the same cell surface markers as primary B cells, but express relatively low levels of HSA.

Functionally, secondary B cells can be more easily stimulated than primary B cells by cross-reacting antigens and are more resistant to inhibition by anti-idiotypic recognition. Additionally, once responses are initiated, secondary B cells give rise to larger AFC clones. In general, primary B cells are stationary cells within the lymphoid tissues, whereas secondary B cells appear earliest in germinal centers, where they are generated after stimulation, but eventually appear to be relatively mobile and recirculate throughout the lymphatic system.

To a large extent, the existence of more secondary than primary B cells responsive to a given antigen and the differences in the biological characteristics of these cells account for the disparities between primary and secondary humoral responses. The generation, via antigenic stimulation, of a B-cell subpopulation that differs so substantially from B cells of nonimmunized individuals is largely the result of the recruitment of cells from a separate precursor cell subpopulation whose function appears to be the generation of memory B cells. These cells, as isolated from the spleens of nonimmune individuals, have many characteristics in common with secondary, as opposed to primary, B cells. For example, these precursors to secondary B cells bear low levels of HSA and their V-region repertoire parallels that of secondary B cells. Importantly, upon stimulation, most of these cells do not give rise to primary AFCs, but give rise to secondary B cells. Additionally, the progeny of these precursor cells rapidly accumulate V-region somatic mutations, whereas the AFC progeny of primary B cells do not. Because of this, it is likely that the subpopulation of precursors to memory B cells not only accounts for the establishment of secondary B cells, but is likely to be responsible for the events characteristic of ongoing primary responses such as immunological learning. Thus, via this scenario, primary antigenic stimulation yields an initial IgM response from the abundant population of antigen-specific primary B cells. These responses mature to give rise to multiple isotypes, but the responses are self-limiting and do not include restimulation, selection, and immunological learning. Primary immunization also enlists precursors to secondary B cells, which proliferate and yield both secondary B cells and more precursor cells, both of which begin to hypermutate. A persistent antigen can encounter these newly generating cells and selectively stimulate those of highest affinity. This would ultimately lead to antibodies of increasingly higher affinity and the continued generation of memory B cells that can respond to further challenges.

## VI. BASIS OF SELF–NONSELF DISCRIMINATION BY B CELLS

The vast repertoire of antibody specificities enables the generation of populations of antibodies specific for the universe of foreign antigens. However, it is of equal importance that the potential for reactivity of any of these antibodies against the large variety of autoantigenic determinants be minimized. Although it is possible that evolutionary forces select against the occurrence of certain specificities, the basis for nonrecognition of self appears to be the specific elimination of antiself specificities by a process called tolerance induction.

B-cell responsiveness *per se* can be decreased by anti-idiotype recognition or by blocking of the B-cell receptors with nonstimulatory forms of antigen (i.e., those that cannot be effectively processed and are not mitogenic). Furthermore, since B-cell responses generally depend on T-cell help, nonresponsiveness among $T_H$ cells for a given antigen can often preclude

B-cell responses as well. Additionally, certain antigens (e.g., soluble monomer forms of Ig or nondegradable polymers) appear to be capable of inactivating B cells whose receptors are bound by them.

However, for most antigens, mature B cells appear to be quite resistant to inactivation. Thus, even when antigen–B-cell receptor interactions are not stimulatory, mature B cells retain their capacity to subsequently respond. This is not true for immature B cells either in neonates or in the bone marrow of adults. If the surface Ig receptors of immature B cells are cross-linked by high-affinity binding to any antigen that presents determinants multivalently, the immature B cell is inactivated and, in most cases, undergoes apoptosis. Although it is possible that some cells inactivated in this way survive, they are unable to respond to antigenic stimulation. By these events the B-cell repertoire can be purged of any potential high-affinity antiself specificities by the inactivation of B cells as they develop.

The period during which developing B cells are susceptible to tolerance induction might be several days for neonatal cells, but is only a few hours for developing bone marrow B cells. Additionally, immature B cells can be stimulated rather than inactivated if T-cell help is available during the B-cell–antigen interaction. Therefore, the tolerance of developing B cells would not apply to all antigens. For example, in adults, tolerance would apply only to antigenic determinants that are present in a multivalent form in the bone marrow milieu of developing B cells. This would apply to certain autoantigens (e.g., cell surface antigens), but not others. Additionally, only high-affinity specificities would be eliminated. Thus, it could be anticipated that low-affinity antiself B-cell responses would be possible, particularly if T-cell help were available.

Although newly generating "conventional" B cells are continuously purged to remove self-reactive cells as they mature, Ly-1 B cells present a distinct challenge. Many of these cells are the product of self-renewal from cells already expressing surface Ig. In this case it is likely that much of the repertoire is established and antiself specificities are eliminated during neonatal development. This neonatally selected specificity repertoire would then persist through adulthood.

As with Ly-1 B cells, memory B cells present a unique challenge insofar as the elimination of antiself reactivities. In this case, even if the initial precursor cell population were devoid of high-affinity antiself specificities, new, potentially antiself, specificities would likely be generated as the result of somatic hypermutation. However, it is likely that secondary B cells bearing newly generated antiself specificities are inactivated, since, during the course of secondary B-cell generation from antigen-stimulated precursors to secondary B cells, newly generating secondary B cells become highly susceptible to tolerance induction. Thus, although neither mature primary or secondary B cells nor nonimmunized precursors to secondary B cells can be readily tolerized, newly generating secondary B cells are tolerance susceptible. In this way antiself specificities generated by somatic mutation can be eliminated.

For all tolerance-susceptible B-cell populations the relevant trigger appears to be the surface Ig receptor per se. Ancillary signals, mitogenic signals, or $T_H$ cell collaboration and ILs do not contribute to the tolerance signal, but might act to circumvent it. Because the trigger is the surface Ig receptor, tolerance induction can be highly selective and affinity dependent. This selectivity is critical in that it permits elimination of the highest affinity and potentially most harmful antiself specificities, while permitting the expression of a vast repertoire of antibodies to encounter the universe of foreign antigens.

## BIBLIOGRAPHY

Coffman, R. L., Seymour, B. W. P., Lebman, D. A., Hiraki, D. D., Christiansen, H. A., Sharader, B., Cherwinski, H. M., Savelkoul, H. F. J., Finkelman, F. D., Bond, M. W., and Mosmann, T. R. (1988). The role of helper T cell products in mouse B cell differentiation and isotype regulation. *Immunol. Rev.* **102,** 5–28.

Kantor, A. B., and Herzenberg, L. A. (1993). Origin of murine B cell lineages. *Annu. Rev. Immunol.* **11,** 501.

Klinman, N. R., and Linton, P.-J. (1988). The clonotype repertoire of B cell subpopulations. *Adv. Immunol.* **42,** 1–93.

Linton, P.-L., and Klinman, N. R. (1992). The generation of memory B cells. *In* "Seminars in Immunology" (D. Gray, ed.), Vol. 4, pp. 3–9. Saunders, London.

Scott, D. W. (1993). Analysis of B cell tolerance *in vitro. Adv. Immunol.* **54,** 393.

Vitetta, E. S., Fernandez-Botran, R., Myers, C. D., and Sanders, V. M. (1989). Cellular interactions in the humoral immune response. *Adv. Immunol.* **45,** 1–105.

# Behavioral Development, Birth to Adolescence

LAWRENCE V. HARPER

*University of California, Davis*

I. Introduction
II. Age-Related Developmental Changes
III. Issues in Describing and Explaining Developmental Change

## GLOSSARY

**Categorical perception** Responding more similarly to discriminably different stimuli within a class (e.g., color category "blue") than to otherwise physically similar stimuli belonging to another class (e.g., color category "green")

**Habituation** Gradually reduced responsiveness to a repeated stimulus which can be reversed by exposure to a different stimulus

**Phoneme** Elementary sound patterns that are used to convey differences in meaning within a language, e.g., "p" *vs* "b"

BEHAVIORAL DEVELOPMENT REFERS TO AGE-related changes occurring in the motor, intellectual, and social activities of children. These changes tend to increase youngsters' abilities to alter their surroundings and influence the activities of other people in ways that would be advantageous to them.

## I. INTRODUCTION

It is generally recognized that all development ultimately derives from biological processes that are based on the activity of genes and that gene action is controlled by the environment of the cell in which the genes are contained. However, the pathways from gene action to behavior are not yet clearly identified and there is no consensus as to how best to conceptualize these relationships. In principle, where variation exists in the genetic substrates for the development of sensory, neural, and endocrine function, one would expect to see corresponding variations in behavior. In practice, with the exception of pervasive conditions such as cretinism, it has been difficult to specify precisely how genetic variation relates to behavioral differences. Behavior is the product of the coordinated activity of many organ systems; there must be flexibility in that coordination to permit functionally equivalent outcomes in different circumstances (e.g., reaching for an object whether standing or seated, and adjusting for the size and weight of the object). That is, there are many "degrees of freedom" in how even a simple act can be accomplished. Therefore, simple relations between allelic variation at a single genetic locus and individual differences in behavior are unlikely.

In analyzing development, the issues become more complex. Partly because of degrees of freedom in task–performance, there often are age-related differences in the "strategies" children employ in order to deal with events. Since gene expression changes throughout the life cycle, when the substrates for age-related strategies develop over time, different patterns of gene expression (and nervous system function) might be expected to influence performance of the same task at different ages. With developmental changes in children's sensory, motor, and mental functioning, one also can expect corresponding shifts in what they can experience and how events affect them. Children respond to their surroundings selectively and, to the extent that their circumstances permit,

693

ENCYCLOPEDIA OF HUMAN BIOLOGY, Second Edition, VOLUME 1. Copyright © 1997 by Academic Press. All rights of reproduction in any form reserved.

choose the physical and social contexts in which they grow up. Moreover, even in the first few postnatal months an infant's behavior can influence what it is exposed to and how it is treated by others.

There is a great deal of ongoing research devoted to the identification and analysis of environmental inputs—from nutrients to patterns of receptor stimulation—and the ways in which these inputs influence behavioral development. In general, the average child can develop normally under a fairly wide range of circumstances; severe maladjustments typically are found only when several "risk factors" such as premature birth, poverty, and parental maladjustment combine. Our knowledge of both the range of individual differences and young children's behavioral capacities and limitations and how they change over time continues to expand. It is now clear that the failure to display a capacity does not mean that the underlying mechanism is not available; developmental change often involves changes in the use of one's abilities. Advances are being made by adherents of very different, often contradictory, theoretical positions; no generally accepted theory of development exists for any aspect (such as language), yet alone behavior in general.

## II. AGE-RELATED DEVELOPMENTAL CHANGES

The behavioral development of children may be conveniently, if arbitrarily, described in terms of patterns of abilities and activities that are manifested at different age periods.

### A. Birth to 3 Months of Age

By the end of the typical 40-week gestation period, a normal baby is ready to adapt to the extrauterine physical environment and to its human caregivers. Neonates are capable of coordinating breathing, sucking, and swallowing. They are most sensitive to sounds in the frequency range of human voices and are especially attentive to visual configurations resembling human faces; they will "follow" moving objects with their eyes. They seem to focus most clearly at ranges approximating the face-to-face distance during feeding. Babies who are delivered without pain-killing medication appear to be alert and attentive for several hours postpartum and will turn and "look toward" the source of a sound. Many parents report pleasure

in seeing their newborn offspring "look at them." Neonates' cries in response to pain and hunger can be informative signals for caregivers. "Pain" cries begin with a prolonged wail followed by a pause; hunger cries begin with rhythmic crying.

From birth, infants differ in terms of the frequency and intensity of their body movements, and also in the degree to which they protest to such events as interruptions of feeding, diapering, and changes in routine or the physical setting. Some babies are highly sensitive to even mild stimulation, others are relatively unresponsive to low levels of sound, small changes in illumination, etc. Infants also differ in terms of the degree to which they relax when being held by a caregiver and, when upset, how readily they can be soothed by parental ministrations. Some babies show fairly predictable cycles of sleep and waking or activity and shift from one state to another as one might expect from the situation (e.g., quiet alertness followed by drowsiness after a feed). Other infants, especially those who have had difficult deliveries, may behave in less predictable ways. Extremes of such "temperamental" variation, particularly unpredictability of response, can lead to disturbances in the parent–child relationship.

In the first weeks infants' sleep–wake cycles vary from 3 to 4 hr in length; they do not follow a diurnal pattern and seem to be less differentiated than those of older children and adults. Within bouts of sleep, newborns' basic rest–activity cycles are much more brief than the adult 90-min rhythm. Only three states can be identified clearly: awake, rapid eye movement (REM) sleep, and non-REM sleep. Sleep accounts for 16 to 17 hr of the infant's day. By the end of the third month, non-REM sleep becomes more differentiated, body temperature begins to cycle, and most infants' sleep–wake patterns are entrained to the day–night cycle. Interestingly, situations or conditions that might be considered stressful (e.g., circumcision) lead to decreases in crying wakefulness and increases in quiet sleep in the days immediately after their occurrence. Some evidence suggests that as more time is spent awake, REM sleep declines. [*See* Sleep.]

Newborns' repertoires of "voluntary" motor activities appear to be limited. Neonates display a number of reflexive responses, including manual grasping to pressure in the palm, and the "Moro reflex," a sweeping movement of the arms and extension of the legs to sudden stimulation and loss of head support. The functional significance of infantile reflexes is not apparent in modern Western child care contexts, but some of them may still be adaptive in other cultures.

Although many early actions of infants appear to be "random," or unpatterned, detailed analyses of leg movements suggest that most, if not all, early behavior will ultimately prove to be organized and constrained, although in very different ways than the activity of older infants and children. Despite some evidence of the early existence of reflex-like reaching toward objects and facial "imitation" of tongue protrusion and mouth opening, neonates' most obviously voluntary motor actions involve head turning, visual fixation, and feeding (sucking).

Under controlled conditions, using head turning, visual fixation, or measurements of suck rate or amplitude as indicators, very young infants have been shown to "work" to gain access to auditory and visual stimulation when they are awake and alert. Even in the first few postnatal weeks infants are sensitive to events whose occurrence is contingent upon their looking or sucking behaviors such as the onset of speech or the presentation of visual displays; they appear to come to "expect" that previously experienced contingencies will obtain in the same context(s). Using simple learning situations or by examining the course of habituation (and conditions required to reinstate full responsiveness), infants' sensitivities to their environments are being documented in some detail. Very young infants have been shown to be able to distinguish sweet from bitter and sour tastes and to make rudimentary visual brightness and hue discriminations. By the end of the first postnatal week, infants can distinguish familiar from unfamiliar human odors. Although they are less sensitive than adults to sounds, especially those with frequencies below 1000 Hz, very young babies are responsive to differences in the pitch and intensity of sounds and distinguish familiar from unfamiliar voices. At birth, they are particularly responsive to their mothers' voices and speech patterns, suggesting some effect of prenatal experience. Within weeks of birth, infants can discriminate among many, if not all, the phonemes used by natural, human languages and perceive them categorically, as do adults.

Between 5 and 10 weeks after birth, smiling, which initially appeared primarily during REM sleep, becomes regularly elicited by human voices and, a week or so later, by the sight of human faces and even rough schematic drawings of faces. Eye-like images seem to be critical; even the mother's face in profile view or with the eyes occluded by opaque skin-colored inserts in glasses will be ineffective. However, if the opaque inserts are replaced by *pictures* of eyes, babies will respond with smiling. Even though they begin to

show some abilities to abstract or categorize the salient features of particularly attractive stimuli by as early as 3 months of age, they continue to smile to familiar, unfamiliar, and even caricatured human faces. During this phase, noncrying, vowel-like "cooing" vocalizations are uttered in response to people and by 3 months of age, infants begin to laugh in playful exchanges with adults.

## B. Three to 10 Months of Age

Responsiveness to other people increases in intensity and selectivity after the third postnatal month. Although 4-month-old infants clearly distinguish their mothers' faces from strangers', obvious differences in response to familiar and unfamiliar persons do not become manifest until after the sixth month. By 8 months some babies may become visibly upset by an unfamiliar adult's attention. In the second half of the first year, most infants seem to be more readily soothed by a particular caregiver and show enduring signs of distress if they are put in the care of a different person. By 8 or 9 months of age, most infants show clear emotional "attachments" to specific people. Babies cry or manifest other indications of distress at brief separations from such people, especially in unfamiliar surroundings, and many display obvious signs of pleasure upon reunion. Whereas most infants appear to be behaviorally inhibited in novel settings, the presence of an attachment figure can facilitate exploration and play. Infants tend to become attached to a familiar adult who is responsive to the baby's actions and who displays signs of emotional involvement with the infant; ministering to the infant's needs for food and warmth is not necessary for attachment formation. Many infants form attachments to several people and even to older playmates. Should adult attachment figures become unresponsive to their infants, babies tend to become more demanding, less readily soothed when attention is forthcoming, and less likely to use the adult as a "secure base" from which to engage their surroundings.

Around the fourth postnatal month, infants' noncrying vocalizations become more complex, including consonant-like utterances, e.g., "da." Subsequently, these vocalizations become more varied and complex so that, by 6 months, these elementary consonant–vowel "syllables" become combined into repeated strings "da–da." Such utterances can provide the focus for caregiver–infant mutual imitation (although the adult typically imitates the infant at this stage). As infants approach 10 months of age, their spontaneous "babbling" may have the prosodic features of adult speech.

By 6 months of age, most infants show clear evidence of being attentive to the exaggerated pitch contrasts and other features of caregiver "baby talk." Given increasingly precise capacities to localize sounds in space, they also appear to "expect" that voices emanate from people who appear to be speaking. Moreover, by 6 months of age, despite themselves having only a limited repertoire of syllable-like utterances, they selectively watch video presentations of speakers whose mouth movements correspond to the particular sounds that they hear. This sensitivity to correspondences across sensory modalities is not limited to speech perception. By 5 months, sound intensity is linked with visual distance, and by 6 months, infants also respond to similar patterns of stimulation involving combinations of visual, tactual, and auditory simuli, e.g., three-flash as opposed to two-flash, visual displays that correspond to three-tap tactual input.

The latter capacities perhaps reflect a sensitivity to numerosity, or the elaboration of perceptual "classes." Newborns can detect the difference between displays of two or three items; by 5 months of age, infants seem to "expect" a corresponding change in the number of objects when they see one element either added to, or taken from, a set. By 4 months of age, infants' color perception seems to be fairly well developed and babies respond to the "categorical" boundaries between hues (e.g., blue and green) much as do adults. By 4 months of age, they appear to "expect" three-dimensional objects to maintain the same shape and act as if common motion of contiguous shapes defined an object. By 7 months of age, many infants show clear evidence of visually recognizing a three-dimensional object despite changes in its spatial orientation, and by 8 to 10 months they seem to "categorize" elements in a series of visual presentations (e.g., items of furniture) in that they show apparent surprise should an incongruent element (e.g., a banana) be presented.

Between the third and seventh postnatal months, infants' visual depth perception becomes increasingly adult-like. By 3 months, many infants seem to be responsive to increases in the proportion of the visual field subtended by an image that specifies the approach of objects. Within another month or so, stereoscopic vision and the use of motion, parallax, the changes in a visual image resulting from head movement, can be demonstrated as cues to depth perception and, by 5 months, babies appear to use "pictorial" cues (e.g., relative size of familiar objects).

Between the third and fourth months after birth, infants show increasing interest in events occurring around them and better control over their own actions. Before 4 months of age, infants "play" with their own hands, and by $5\frac{1}{2}$ months of age display visually guided attempts to grasp objects. Suspended objects are batted with the obvious intent of setting them in motion. Partially hidden objects may be retrieved, and dropped objects are followed with the eye, suggesting that some idea of "things" is being formed. After 8 months of age, infants begin to search for hidden objects, but seem to be easily confused with respect to where objects may be located. While they clearly can distinguish their own movements from movements of their surroundings by 6 months and have some awareness of the location of their limbs, they do not seem to fully appreciate the fact that the object seen and the object felt must be the same until around 9 months of age.

Insofar as they are gaining increased voluntary control over their limbs, and spending more time awake and alert, infants seem to become more sensitive to the outcomes of their own actions than they were in their first 3 months and are thereby capable of a wider range of learning. The lessons learned, however, are selective. Babies focus on the consequences of those motor activities over which they are gaining voluntary control. Thus, whereas 1 to 3 month olds readily learn to turn their heads or to suck in order to receive a "reward," older infants seem to be more "interested" in what they can gain from the use of their limbs and hands. Such gains are facilitated by a variety of objects upon which infants can act to "make things happen" in surroundings that provide sufficient order to permit them to clearly distinguish "cause and effect."

## C. Ten to 20 Months

Toward the end of the first year, before they begin talking, many infants begin to develop some understanding of speech. In this same period, between 8 and 12 months of age, their responsiveness to phonemic contrasts becomes restricted to those which are characteristic of their own language community. When they start approximating words, between 10 and 14 months, babies' efforts are largely limited to combinations of sounds that were already utilized in their spontaneous babbling. (It may be no coincidence that this is the period at which they begin to imitate new combinations of body movements.) In general, infants understand more language than they can produce. Early efforts at vocal communication include both brief, phrase-like elements which seem to have been learned by rote and words (initially largely nouns) whose specific meanings have been deduced by more analytic processes. Individual babies differ with respect to which approach dominates in their early language learning. In the early phases of language acquisition, caregivers' appreciation of many of the "meanings" conveyed by infants depend on the context of physical gestures. Indeed, studies of the apparently spontaneously developed sign language of some deaf children indicate that verbal language input is not required for the development of elementary aspects of human language, including the process of combining elements to convey meaning. Moreover, early language acquisition is not a simple function of the quantity of input received; despite wide cultural variations in parental speech to infants, normal babies in all cultures learn to speak at about the same rates.

Although different aspects of linguistic and cognitive competence develop somewhat independently, under normal circumstances, communicative advances partly reflect an increase in infants' understanding of the natures of objects, situations, and states, and a sense of causality beyond the outcomes of their own actions. In the middle of their second year, there tends to be a significant increase in infants' vocabularies that just precedes their first attempts to combine words to produce primitive sentences. These utterances do not simply express desires or needs (although "no" appears at this time), rather they involve commentaries upon the existence, location, possession, or the states or attributes of objects. When they lack an appropriate lexical element in their vocabularies, babies (and older children) often will try to "force" another, somewhat related word into the appropriate context.

As they gain the capacity to walk, infants begin to utilize their knowledge of their surroundings in different ways. Although much younger babies *can* use fixed landmarks as points of reference for locating objects or activity sites, they tend just as often to orient relative to their own position (left, right, etc.). Similarly, despite the fact that almost all infants perceive depth and distinguish familiar objects and people from unfamiliar ones, many do not seem to act on this information. With the onset of the capacity to locomote on their own, whether independently or with the aid of a walker, babies begin to use these cues to govern a wider range of activities. Precipices are avoided, landmarks become prefered cues for governing movements or locating objects, and strangers and unfamiliar objects evoke more wariness. It seems that being able to move about on their own provides infants with a powerful stimulus for adopting new "strategies" for governing their behavior. The flexibility of such strategies is limited, however. In an enclosure such as a small room, when "turned around" and getting their bearings, infants this age essentially ignore obvious landmarks and focus on the shape of the enclosure to locate hidden objects.

In this period, babies develop a sense of the "permanence" of physical objects and will follow displacements as one moves an object from one to another hiding place. By 20 months of age, infants will not abandon a search for a hidden object even if it is not found where they expected to find it. Rather, they continue to look in places where the lure *might* be found. Paralleling these increases in their understanding of their surroundings and their developing language abilities, babies' play patterns suggest the development of the capacity to think "symbolically." By 15 months of age, babies begin to "pretend" to carry out real or imaginary activities themselves; by 20 months, they move from using representational toys (e.g., cars) appropriately to using an object as if it were something else (e.g., a leaf as a boat). At this age, mental advance is facilitated opportunities to explore and by variety vs sheer number of stimuli.

By about 18 months of age, babies evidence self-recognition in a mirror or a photograph. At about the same time, young children clearly recognize that they are "little" and can distinguish children from adolescents and older people. Their social overtures to others also reflect this pragmatic knowledge. Infants turn to age mates or siblings primarily for playful exchanges and seek adults when scared or hurt. If adult caregivers are unresponsive, or are inconsistently responsive, an "insecure" relationship may develop; insecurely attached infants may have difficulties in later social relationships. On the other hand, a good, early relationship with an adult caregiver can provide a "buffer" against subsequent adversities.

Although infants and toddlers do show interest in other youngsters and, especially in the case of twins, appear to become attached to other infants, most babies' interactions with their age mates are infrequent and brief, seldom exceeding 10 to 20 sec in duration. Even youngsters who have become "friends" as a result of spending substantial amounts of time together in day care simply play near one another and engage in occasional idiosyncratic routines involving mutual imitation or alternating turn-taking activities.

Children who are confronted with age mate strangers seldom display much more than curiosity, although they may show or offer a toy to another child. This failure to engage others in ways that depend upon an appreciation of the interactant's feelings or point of view does not mean that toddlers are *incapable* of doing so. Observations of interactions of 10 to 20 month olds with their older siblings suggest that even some 14 month olds can sympathize with or deliberately behave so as to upset their elder siblings.

## D. Twenty to 36 Months

In the latter half of the second year, toddlers begin to combine the words at their command to convey shades of meaning. This step in language development usually is a function of an increase in their use of verbs. The expansion of children's verbal competence has been attributed to a combination of several factors, including their understanding of adult utterances, the rote learning of brief functional phrases, and more analytic learning of specific words (typically nouns). Grammatical advances usually begin with the addition of utterances that seem to function as predicates. As they pass their second birthdays, most toddlers start to employ words and phonological markers systematically. However, these early grammatical productions often violate the rules that govern their native languages and, although intelligible, do not seem to mirror the speech that toddlers commonly hear. Moreover, when coaxed to repeat a statement "correctly" after an adult, young children translate the adults' utterance into their own grammatical terms. The early grammatical constructions of children who grow up in very different language communities often are quite similar despite marked differences in the grammatical conventions of their parents' languages. Such cross-cultural observations suggest that some pan-human strategies may exist that govern the ways in which young children segment and interpret the linguistic stream to which they are exposed. Indeed, comparative studies of the apparent ease with which children attain competence in expressing certain kinds of meanings indicate that some grammatical conventions are more readily mastered than others.

As youngsters progress through their second and third years their syntax becomes more like that of their language community. The errors that they commit suggest that their utterances are governed by their understanding of grammatical rules. For example,

after a period of correct use, many children occasionally overextend the rules for expressing the past tense of regular verbs to irregular forms, e.g., uttering "goed" instead of "went."

In this period, children's practical understanding of their own capacities and the nature of familiar objects continues to expand. By the end of the second year, toddlers can see how executing one action may permit them to subsequently perform an entirely different response in order to attain a goal. Not only can toddlers imitate novel patterns of activities, but they now do so in the absence of the model, suggesting that they are capable of storing information about other people's actions and the outcomes of such actions. Indeed, it is during this period that young children not only display symbolic play with objects, but begin to involve other people in make-believe activities such as playing "house."

This phase of development seems to span the decline of developmental plasticity in at least two areas of experience. Several lines of evidence suggest that "critical periods" for the development of binocular depth perception and the detection of line orientation come to a close during the third year. Clinical studies also suggest that young children during this same phase, accept their identities as boys or girls with a tenacity that far exceeds their intellectual appreciation of gender differences. That is, although young infants appear to be more attentive to the activities of a member of their own, as opposed to a member of the opposite sex, and prefer to play with "sex appropriate" toys before they reach 18 months of age, their knowledge of sex differences is minimal. The same child who adamantly insists that she or he is a girl or a boy is not at all certain whether she or he will grow up to be a "daddy" or whether a male playmate could grow up to be a mother. The rapidity of this aspect of self-identification, the importance it holds for children, and the difficulty encountered in attempts to reassign children after 18 to 20 months of age (in cases of ambiguous external genitalia) have convinced some investigators that there may be a critical period for the establishment of gender identity.

With their expanding locomotor skills, toddlers become subject to increasing parental demands and restrictions. Compliance with parental requests for restrictions, although unpredictable, is facilitated by prior parental responsiveness and willingness to accommodate to the child's legitimate interests. Young children's dawning recognition that they are expected to abide by conventions or regulations is also manifested by their adoption of culturally prescribed forms of address to adults.

## E. Three to 5 Years of Age

During this "preschool" phase of development, young children's language competence continues to expand. English-speaking children begin to master the use of articles and auxiliary verbs. They develop facility with more complex sentence structures such as negation, the use of inverted word order in questions, and the incoporation of subordinate clauses in sentences. Toward the end of this period they also begin to use conjunctions to join simple sentences. However, they do not usually coordinate sentences with one another when attempting to tell a story. Although they do seem to be aware that others, particularly younger children, may not have all the necessary information to understand a situation, their narratives are not only disconnected, but typically fail to include the contextual information necessary to fully inform the listener.

Some of their communicative difficulties may reflect the fact that children this age are still refining the conceptual categories that facilitate thinking and communicating about events. At this phase, preschoolers seem to be developing their understanding of such relational concepts as "bigger than," "more," and acquiring the vocabulary to describe attributes of objects such as shape or color. Although they have an appreciation for sequences of events in time, their sense of time per se is poorly developed. Interestingly enough, even before they have learned "to count" correctly, to the extent that they know some number words, even 3 year olds understand certain rules of counting: They count each item in a set only once, assign every item a unique label, and use the last label assigned to define the number of items in the set.

Similarly, even before they understand the measurement of linear distance, preschoolers will employ a rod or pencil to assess the (relative) depth of small holes. Sometimes it seems that failures in formal reasoning tasks involving understanding of the invariance of such attributes as quantity (despite changes in the configuration of a test array) involve confusion in the distinction between the number of items in a row and the length of the row. When shown two rows of five candies, many preschoolers may say that the longer one contains "more." However, when told they may eat them and given a choice between a longer row containing only five candies and a shorter one

containing six, even 3 year olds typically choose the greater *number* of candies. Thus, it seems that, at least by the age of 4 or 5 years, young children are capable of conceptualizing and utilizing the information required for some kinds of adult-like logical reasoning. However, in order to elicit these abilities, problems must be presented in ways that highlight the relevant distinctions.

Preschool children's capacities for perceiving information are approximately adult-like by about 4 years of age; however, they seem to have difficulty in encoding their experiences economically. Thus it has been suggested that, although capable of forming and using "mature" concepts, young children need time and repeated exposures to a variety of objects, materials, and situations in order to fully develop and become proficient in the application of their abstract knowledge. It is as if they must first focus on all the sensory ramifications of an idea such as volume before the concept itself can become a usable abstraction that can "free-up" the mind to ponder its implications.

Around their third birthday, whether or not they have had much experience with other youngsters, preschoolers begin to display increasing interest in interacting with other children. Although prior experience and familiarity with age mates are both associated with greater amounts of social interaction among 3 to 5 year olds, age alone also accounts for some of the differences among preschoolers' sociability with peers. Among preschoolers, the decision to play alone seems to be an alternative to *any* degree of proximity to or engagement with age mates. Children enrolled in half-day group care or "nursery school" programs spend about one-third of their time playing alone; another 40 to 50% is spent in proximity to other children, in similar activities, but in the absence of mutual coordination or other kinds of direct exchanges. Preschoolers who are enrolled in group care differ in the degree to which they seek social exchanges with adults as opposed to age mates; however, from one day to the next, most children of both sexes treat interactions with peers as alternatives to interactions with adults.

There is a tendency for preschool boys to spend more time in physical play interacting with peers and less time interacting with adults than girls. Boys also spend more time outdoors than girls during free play. By this age youngsters of both sexes, especially boys, are acutely aware of what is deemed appropriate behavior for males and females. They are also gaining an expanding comprehension of societal and group norms. Preschoolers often explain regulations as if they were absolutes and apparently confuse conventional routines with ethical mandates. However, when they are discussing recently witnessed events, even 3 year olds can distinguish between conventions peculiar to a specific setting, such as a day-care center, and more general, ethical rules of conduct. Nevertheless, their moral judgments tend to focus on actions or outcomes; adult authority is explained in terms of relative power and, when they are personally involved, preschoolers' solutions to ethical questions are typically self-serving. At this age, helpful behavior is most often evidenced when youngsters appear to feel secure that their own needs will be met.

When they are asked to describe themselves, preschoolers tend to talk about their age, sex, possessions, and favorite activities. They seldom use evaluative terms or utilize even simple descriptions of character traits.

## F. Five to 8 Years of Age

In many cultures, children this age are expected to begin to contribute to their family's welfare. Some cross-cultural research suggests that the kinds of tasks that children are assigned may have widespread influences on their behavior. For example, boys who are assigned to care for younger siblings are more nurturant and less aggressive and boastful than boys who are assigned more traditionally "masculine" tasks. In Western societies, children typically enter public school at this age. For many youngsters this may be the first time that they have been confronted with the necessity to meet culturally defined standards of accomplishment. Moreover, it may also represent their first experiences in assessing their competence relative to a group of other children who are about the same age. At first, these new experiences may not fully impact on youngsters. Many do not distinguish between effort and (personal) ability. Their immediate focus is on the acts of performing assigned tasks rather than the degree to which their performances have met some criterion of success.

In this period, most youngsters are just becoming aware that their thoughts are personal and private. Many are still uncertain as to whether their dreams have tangible referents or are simply mental phenomena. Despite these limitations, they are becoming aware of language forms and can identify when something is (grammatically) wrong or nonsensical in a sentence. They are better able to reduce redundancies in complex utterances and, in narratives, begin to apply the appropriate transitional words or phrases to produce a more cohesive train of thought. They begin to use appropriately such conjunctions as "if," "because," "when," and "then." After about the age of 6 years, it becomes progressively more difficult to fully master both the phonology ("accent") and grammar of a second language.

Between 5 and 7 years of age, children's understanding of relational concepts (e.g., "bigger than") increases and they become able to employ such notions to solve simple problems. For example, when presented with a series of learning tasks such as one in which choosing the larger of two items is the "correct" response and the criterion is suddenly reversed (i.e., the smaller item must be chosen), children come to recognize that relative size is the key to the solution and adjust within one or two trials. Similarly, between the ages of 6 and 8 years, most children become more adept at using such concepts as numerosity, length, or volume in solving problems involving logical reasoning. For example, when the liquid contents of a short, wide container are poured into a tall, narrow one, children recognize that the amount of fluid has not changed despite the fact that its configuration was altered.

In commonly encountered practical problems (or as a result of exposure to schooling), children become sensitive to the possibilities of discovering "principles" that may apply to the solution of a specific problem. For example, even unschooled street vendors, 7 or 8 years of age, may recognize that numbers (money) may be broken down into their components and use this knowledge to simplify calculations. (Linguistic conventions that emphasize regularities, such as counting systems using the base 10, facilitate the early acquisition of basic mathematical principles.) The search for and application of principles for problem-solving is not widely deployed, however. Practical importance and/or instructional emphases determine which principles are acquired first. Frequently, children under 8 years of age simply fail to realize how they can apply the knowledge that they possess. For example, although naming or classifying objects can facilitate later recall, children this age often fail to employ this strategy in memory tasks. Even after experiencing success after having been instructed to use naming, many fail to continue to use this tactic in similar situations.

Between 5 and 8 years of age, children become increasingly interested in mutually regulated activities or games with age mates of the same sex; play groups are almost completely segregated. Although children's games are often governed by conventions or rules, these regulations are simple (e.g., tag, hide and seek). Just as often, as in the case of preschoolers' make-believe play, only a very general consensual framework obtains and players essentially pursue their own themes with only loose coordination among participants.

Although they have fully mastered most of the basics of their native tongues and are capable of recognizing that others may not share their understanding of a situation, children this age still often try to explain a situation or tell a story without providing the necessary contextual information. Moreover, when they receive such an incomplete message, they fail to appreciate what additional information is required to properly understand the situation. (When questioned appropriately, they *can* supply the necessary details.)

During this phase of growth, most children are reasonably compliant with adult authority, although they will readily emulate deviant behavior when no sanctions are forthcoming. In individual, game-like situations, they will follow rules explained or modeled by adults so long as the rules are clearly presented and attainable. When asked to evaluate the conduct of people in stories, they tend to evaluate actions in terms of the benefit or harm actually done rather than the actor's intent. Authority figures such as political leaders or school principals are seen as wise and powerful, not to be questioned or challenged. Adherence to parental authority is justified in terms of one's duty to repay debts, or deference to parental expertise. On average, children this age are most likely to be obedient and follow established norms when they are neither personally threatened nor unduly tempted. Although the causal linkages are uncertain, well-behaved children typically come from families in which parents are accepting and affectionate, set clear and consistent standards, explain why certain actions should or should not be performed, and are willing to discuss and negotiate at least some expectations.

Activities that are subject to ethical evaluations are diverse, and knowledge of a child's behavior in one area does not necessarily predict how that youngster will act in another. For example, a child who shares food with a friend may not donate "to charity" or work to help others in need. Similarly, lying about accomplishments, stealing small trinkets, and cheating on school examinations all seem to be essentially independent domains of behavior. Although the abilities to sympathize with other people's feelings and appreciate their predicaments may be essential pre-

conditions for the display of altruistic behavior, they probably are not sufficient.

## G. Eight Years to Preadolescence

In the middle elementary school years children become more aware of their own thought processes and are better able to focus their attention. They develop an appreciation of their abilities and the strategies that they must employ to perform different intellectual tasks. For example, if instructed to remember an array of objects, children in this phase might spontaneously name and attempt to categorize the items or relate them to some already well-known configurations. They are increasingly interested in and adept at discovering "principles" that can be applied to the solution of both abstract and practical problems. Occasionally, they appear to become so enamored with a newly acquired skill or discovery that they will apply it in inappropriate contexts leading to what, at first glance, appears to be a loss of skill.

Up to the onset of this phase, most physically healthy youngsters seem to develop roughly comparable intellectual capacities regardless of culture or educational opportunities so long as they encounter a variety of objects, materials, and situations. However, the intellectual advances made in Western society at this point and subsequently seem to owe much to formal education. In the middle elementary years, youngsters develop more complete understanding of the logical implications of such concepts as quantity, volume, and, eventually, mass. The explanations that they give to support their conclusions become more detailed and elaborate. For example, in explaining why a given volume of fluid remains constant regardless of the dimensions of the vessel containing it, they are able to relate changes in fluid level to a trade-off between the height of a container and the area of its base. Similarly, they now can reason—verbally—that if $A > B$ and $B > C$, then $A > C$. By the end of this phase, when presented with a tangible problem, youngsters can select problem-solving strategies in advance and anticipate the kinds of input that are needed to confirm or disconfirm hypotheses related to the problem.

During this period, youngsters become more aware of the rules governing grammar and the full range of meaning of the words in their vocabularies. Although they have difficulty in recognizing logical inconsistencies and cannot identify the most and least important ideas in a written text until preadolescence, their own utterances become more coherent. Their narratives become more cohesive and elaborate. Moreover, youngsters in the middle grade school years become able to appreciate the distinction between a spoken message and the speaker's attitude toward the topic of conversation; they become capable of understanding sarcasm and the possibility that an utterance may have a double meaning. When youngsters attempt to communicate something to an age mate there is a greater exchange of information; the receiver will demand clarification if the message is not clear and, with practice, the sender will become skilled in providing the necessary input.

In this phase youngsters become involved in competitive group games governed by mutually accepted rules. Presumably as a result of such experiences and as a consequence of negotiating changes in rules to meet special conditions, children gain a deeper understanding of the degree to which custom and convention are arbitrary and depend on consensus to maintain their force. The rationales that youngsters give to questions about the necessity for obeying societal norms tend to emphasize the importance of recognizing other people's feelings and, toward preadolescence, the idea that society requires common guidelines in order to function. Their judgments in hypothetical moral dilemmas increasingly involve considerations of intent, mitigating circumstances, and other situational considerations. However, their moral behavior remains a mosaic of relatively independent behavioral predispositions. Juvenile delinquents often come to the attention of the authorities in the middle grade school years. These children tend to present their parents with problems from an early age which contribute to reduced parental affection, erratic and harsh discipline, and lax supervision. However, when parents can maintain positive relations, consistent, reasoned discipline, and close supervision, the prognosis even for "difficult" children is positive.

From about 8 years of age, youngsters begin to distinguish effort from ability and gain a sense of their own capacities and limitations relative to those of others. Around the age of 9 years children become self-critical and assess their own productions according to societal standards. During this same period, they also become increasingly aware that adults and other youngsters are forming opinions about them. By the time they reach preadolescence, they are fully aware that they are earning a "reputation" with those around them. During this period, youngsters will discuss their hopes, fears, and feelings with trusted friends. Friends are expected not only to re-

spect the confidence of these revelations, but to remain supportive and loyal even though one's darker side is revealed.

## III. ISSUES IN DESCRIBING AND EXPLAINING DEVELOPMENTAL CHANGE

### A. Unresolved Issues

Although this synopsis has portrayed behavioral development essentially as a series of phases of growth, development is more accurately described in terms of age-related changes in patterns of organization of the processes underlying *specific aspects* of behavior. However, debate continues whether it will be more useful to attempt to analyze broad, general domains of behavior (e.g., cognition) or to seek to isolate and study the "modular" component processes presumed to contribute to complex activities.

There is growing agreement that changes in central nervous system function must account for the development of behavioral ontogeny. However, there is no generally accepted model that leads to specific predictions as to what must change or how function must be altered in order to promote behavioral growth. Similarly, there is no concensus on how to conceptualize the relationship between "maturation" and "experience."

### B. New Directions

Many behavioral scientists are coming to the conclusion that the development of behavior ultimately will prove to be the outcome of a number of developmental processes that parallel one another in time without necessarily being closely synchronized or regulated by common factors. Increasing evidence points to the early existence of latent capacities which, although demonstrable under controlled conditions, are not manifest in more typical surroundings until much later in development. Another conclusion that appears to be demanded by the facts is that nervous systems, including the neonate's, are so organized that certain patterns of receptor excitation are more or less automatically interpreted in particular ways. Without such assumptions, it seems that both "simple" perceptual illusions and such apparently prodigious tasks as language acquisition would be inexplicable. In a similar vein, there is increasing support for the view that children's learning capacities are developmentally constrained; that certain lessons are most readily learned at (or after) particular points in the life cycle. Moreover, recent findings indicate that maturation of the brain involves not only increases in the growth of myelin sheaths around nerve processes, but also a *reduction* in the absolute numbers of nerve processes and/or the number and kinds (excitatory or inhibitory) of synaptic connections they make with other cells. These data suggest that some developmental advances cannot be understood simply in terms of increases in the number or extensity of neural connections; rather, they may reflect reductions of "noise" in the system.

Finally, it is becoming clear that many investigations of the putative antecedents of the development of individual differences in behavior must be interpreted cautiously. The vast majority of these studies have identified patterns of upbringing techniques and child characteristics were assessed at the same point in time, thereby leaving open the possibility that preexisting child characteristics contributed to the observed relationships. In addition, the majority of these studies evaluated relationships occurring in biological families. Where heredity affects adults' choices of activities and surroundings, the environments in which children grow up are not independent of their genetic makeup, except in cases of out of family adoptions. Even in the latter cases, when such factors as temperament and appearance have heritable components and when such phenotypic attributes evoke common responses from members of a child's community, children's experiences will be influenced by their heredities. Thus, antecedents and outcomes cannot be unambiguously identified. Unfortunately, conclusions drawn from even the few long-term studies that are available also must be qualified by the fact that essentially all of them followed the growth and behavioral development of children who were being reared in their biological families, leaving the questions of the role of shared heredity unresolved.

## BIBLIOGRAPHY

Bates, E., Bretherton, I., and Snyder, L. (1988). "From First Words to Grammar." Cambridge University Press, New York.

Bloom, P. (ed.) (1993). "Language Acquisition: Core Readings." MIT, Cambridge, MA.

Harper, L. V. (1989). "The Nurture of Human Behavior." Ablex, Norwood, NJ.

Karmiloff-Smith, A. (1992). "Beyond Modularity: A Developmental Perspective on Cognitive Science." MIT, Cambridge, MA.

Mussen, P. H. (ed.) (1983). "Handbook of Child Psychology," 4th Ed. Wiley, New York.

Osofsky, J. D. (ed.) (1987). "Handbook of Infant Development," 2nd Ed. Wiley, New York.

Perlmutter, M. (ed.) (1986). Perspectives on intellectual development. "Minnesota Symposia on Child Psychology," Vol. 19.

Thelen, E. (1985). Developmental origins of motor coordination: Leg movements in human infants. *Dev. Psychobiol.* 18, 1–22.

Various authors. (1994). Special issue on genes and behavior. *Science* 264, 1686–1739.

Various authors. (1994). Special issue: Children and poverty. *Child Dev.* 65, No. 2.

Werner, E. E., and Smith, R. S. (1992). "Overcoming the Odds: High Risk Children from Birth to Adulthood." Cornell, Ithaca, NY.

# Behavioral Effects of Observing Violence

RUSSELL G. GEEN
*University of Missouri*

BRAD J. BUSHMAN
*Iowa State University*

## GLOSSARY

**Arousal** Activation of physiological processes associated with the autonomic nervous system (e.g., heart rate, skin conductance, respiration), usually accompanied by the subjective experience of increased excitation

**Aversive condition** Any condition of the person that is experienced as an unpleasant state of feeling

**Cross-lagged panel correlation** Correlation of data on one variable from one observation period with data on another variable from a later observation period; most likely cause–effect relationships are inferred from relative magnitudes of several such correlations

**External validity** The extent to which one can generalize the findings from a study to other people, settings, and times

**Habituation** Loss of reactivity with successive presentations of a stimulus

**Meta-analysis** An approach to reviewing the literature in which the results from independent studies that investigate the same phenomenon are statistically combined

**Moderator variable** A variable that influences the strength and/or direction of the effect of the independent variable on the dependent variable

**Social comparison** Drawing of conclusions regarding one's abilities, opinions, motives, and/or emotional states from observations of other people in similar conditions

A LARGE BODY OF EVIDENCE INDICATES THAT OB-servation of violence can elicit or facilitate the expression of aggressive behavior in the viewer. The original theory linking observation of violence to aggressive behavior was built on the concepts of modeling and imitation. Although these explanatory concepts persisted until recent times, they have been superseded by other constructs, such as arousal, cognitive labeling, and cognitive association.

## I. HISTORICAL BACKGROUND

Social scientists have long been interested in the question of whether the mass media of communication play a role in the acquisition or maintenance of aggressive behavior. With the rapid spread of informational technology in the 19th century, people interested in the social bases of mental and behavioral problems began to regard such media as newspapers, magazines, and popular books as potential sources of antisocial influence. The French physician Paul Aubrey (1858–1899) was instrumental in propagating a theory of mental contagion that explained some psychoneurotic behaviors in terms of these influences. A similar viewpoint was articulated by the sociologist

ENCYCLOPEDIA OF HUMAN BIOLOGY, Second Edition, VOLUME 1.   Copyright © 1997 by Academic Press.   All rights of reproduction in any form reserved.

Jean-Gabriel de Tarde (1843–1904), who observed that "epidemics of crime follow the line of the telegraph," thereby indicating a belief that the reporting of acts of violence in the news may disseminate violent images that can elicit aggression.

In a manner consistent with prevailing social theories, observers such as Tarde attributed this apparent cause–effect relationship between symbolic presentation of violence and subsequent aggressive behavior to processes of suggestion and imitation. It was commonly thought by social and medical theorists in 19th-century France, for example, that certain types of persons who inherit "weak" or degenerate dispositions are especially susceptible to nervous trauma induced by vivid images of criminal or antisocial behavior and are also likely to act out the ideas elicited by imagery without interference from other ideas. Eventually the theory of nervous trauma was abandoned. However, the concepts of imitation, inhibition, and individual differences implied in that theory were assimilated into more modern approaches to the investigation of mass media effects.

With the advent of radio, motion pictures, and television in the 20th century, even more powerful images of violence became widely disseminated in both news broadcasting and the entertainment media. During the 1930s, numerous investigations were carried out to explore a possible effect of movies on violent and criminal behavior. These early explorations were addressed to a number of issues that would be studied again in the 1960s and 1970s, such as the possible identification of the viewer with the actors in films, the role of maturation and development in responses to movies, the factors involved in individual preferences for aggressive film content, and the possible role of individual differences as moderators of film-induced effects. Unfortunately, the results of these early investigations yielded few reliable results, mainly because of their relatively unsophisticated methodology.

Interest in the study of media violence increased with the advent of widespread home television in the 1950s. Although some interest in movie research remains, over the years the problem of "media" violence has become defined mainly as one involving the effects of violent television programs on aggressive behavior. There is little question that television is a major purveyor of violent stimuli. Ongoing studies by the National Coalition on Television Violence and by social scientists at the Annenberg School of Communication at the University of Pennsylvania show a high level of violence in prime-time programming that has remained fairly constant for many years.

## II. IMITATION–MODELING HYPOTHESIS

Research efforts on effects of media violence during the late 1950s turned with increasing frequency to the use of experimental designs as investigators sought not only to discover whether media violence had effects on the viewers, but also what the intervening variables in any such effects might be. Initial attempts at theorizing used the familiar concepts of suggestion and imitation, recast in the conceptual language of social learning theory. This approach led to a number of interesting and ingenious experiments in the early 1960s by Bandura in which it was shown that children responded to an aggressive adult female model by imitatively aggressing against an inanimate clown doll. The research showed the operation of two processes in imitative aggression: (1) the learning by observation of new aggressive responses, and (2) the emission, ostensibly due to the model's influence, of aggressive behaviors already in the subject's behavioral repertoire. The latter is usually attributed to a reduction of social restraints against aggressing.

In several subsequent elaborations of his original modeling experiments, Bandura explored such matters as the effects of quasi-cartoon figures on imitative aggression (the model wore a cat costume), the influence of models presented in films rather than live, the effects of observing punished as well as rewarded aggression, and the role of inducements in acting out aggressive behaviors learned through imitation. In general, cartoon-like models elicited less imitative aggression than either live models or realistic ones shown on the screen. Furthermore, whereas punishment was found to inhibit the expression of modeled aggression, it did not hinder the learning of the punished acts. Given strong enough incentives to perform, children imitated a model's punished aggression as much as rewarded aggression. More recent studies of modeling have shown that observed aggression is more likely to be acted out by children when adults present at the time of viewing express approval of the acts or when other children act out the observed behaviors in the viewer's presence.

## III. MEDIA AS A SOURCE OF INFORMATION

### A. Justification of Observed Violence

Despite the adequacy of the concept of observational learning to account for the acquisition of novel or unusual aggressive behaviors, such learning alone does not explain the more general phenomenon of media-induced aggression. Most experiments on the effects of observing violence show that such observation increases the level of many aggressive responses that bear little resemblance to the ones observed. Obviously, processes other than observational learning are involved in this sort of aggression. Observing violence in media presentations may influence aggression by providing information about aggressive behavior. It may, for example, inform the observer that aggression is a permissible or even desirable means of solving interpersonal conflicts and, by so doing, helps reduce any inhibitions about aggressing that the person may have. Several studies have shown that when violence is presented as morally justified behavior because the victim deserves the attack, it elicits aggression, whereas violence that is described as being morally unjustified either has no effect or may produce an inhibition of aggression.

## B. Social Comparison with the Observed Aggressor

Judgments concerning the motives of the observed aggressor may also influence the ways in which media violence elicits aggression. Of the many motives that may animate aggressive behavior, vengeance is one that most people would probably agree is at least somewhat morally justified. A series of experiments by Geen and his associates has shown that when violence is described as motivated by a desire for revenge, it elicits more aggression from an observer than does the same violence attributed to other motives, provided that the observer has first been insulted, attacked, or otherwise provoked to act aggressively. Provoked persons who observe a scene of revenge also report feeling less restrained in aggressing than similarly provoked who observe violence due to other causes. However, when provoked persons observe a media portrayal of an unsuccessful attempt at revenge, they are later less aggressive than those who see successful vengeance.

It is important to note that in these studies, as in others showing the aggression-facilitating effects of observing revenge, only participants who had first been angered behaved aggressively after observing violence. These findings suggest that one function of observing portrayals of revenge in the media is the facilitation of a social comparison process. The prospects of attacking another person may ordinarily raise inhibitions and aggression anxiety in angry participants, thereby prohibiting retaliation. If, however,

participants are able to observe in the media an angry character who successfully exacts revenge, they may consider their own desire to retaliate to be more appropriate. In the same way, observation of an unsuccessful attempt at revenge may remind participants that retaliation can have punishing consequences and may thereby reinforce inhibitions.

It should also be noted that the studies showing that aggression is influenced by the perceived motives of an observed aggressor were all carried out with young adult viewers. The cognitive processes involved in this effect are moderated by the age and developmental level of the viewer. Whereas very young viewers such as kindergarteners tend to comprehend the violence they see only in terms of its aggression level and ultimate consequences and to be relatively unaffected by the aggressor's motives, older children and adults are more likely to judge violence in terms of its motivation than in terms of its intensity or consequences.

The social comparison hypothesis is further supported by studies that have shown that when participants are instructed to identify with the winner of an act of observed violence, their aggression against a victim is enhanced. These studies suggest that "identification with the aggressor," or covert role-taking, facilitates the expression of media-engendered aggression. As proposed earlier, such covert role-taking may facilitate a social comparison process wherein participants interpret the correctness of their motives to aggress on the basis of what is seen on television or in a motion picture.

## C. Reality of Observed Violence

When violence in the media is thought to be real it elicits more aggression than when it is thought to be fictitious. Realistic violence may be processed as a more intensive informational input than is fictitious violence. As a consequence, realistic violence may be more likely to occupy the observer's attention than fictitious violence.

## D. Sex Differences

Differences in responses of male and female participants to observed violence have not been studied systematically in laboratory experiments. A recent meta-analysis of over 200 studies by Paik and Comstock, however, found that male viewers are slightly more affected by violent media than are female viewers.

## E. Age Differences

Social scientists have studied media-related aggression in viewers ranging in age from nursery school children less than 5 years old to adults over 21 years old. In their meta-analysis, Paik and Comstock found a strong negative relation between the age of viewers and the magnitude of the effect of media violence on aggression. Younger viewers are more affected by media violence than are older viewers, perhaps because they are more susceptible to the influence of role models than are older viewers.

## IV. EFFECTS OF OBSERVED VIOLENCE ON AROUSAL

### A. Arousal and Elicitation of Aggression

Observation of violence may facilitate the expression of aggression by causing an increase in autonomic arousal. Three processes may be suggested as causes for this facilitation. First, autonomic activation produced by watching violence may simply raise the person's overall activity level and strengthen any responses. If the person has been provoked or otherwise instigated to aggress at the time this increased activation occurs, aggression will be a likely outcome. This line of reasoning is strengthened by the fact that most studies reporting a positive relationship between observation of violence and subsequent aggression find that relationship only in persons who have previously been provoked. A second, and as yet untested, possibility is that high levels of arousal may be aversive to the observer and may therefore stimulate aggression in the same way as other aversive or painful stimuli.

A third possibility is that arousal elicited by media portrayals of aggression may be mislabeled as anger in situations involving provocation, thus producing anger-motivated aggressive behavior. This mislabeling process has been demonstrated in several studies by Zilimann, who has named it "excitation transfer." Zilimann's theory is especially interesting because it addresses a problem not covered by some of the viewpoints reviewed earlier: the length of time during which a person is likely to be affected by observation of violence. Whereas most experiments show an immediate effect that persists for a very short period after observation, the notion of excitation transfer suggests that the effect may be extended over a longer period. Even after the media-induced arousal has dissipated, the observer may remain potentially aggressive for as long as the self-generated label of anger persists.

### B. Habituation with Repeated Exposure

As is the case with arousal elicited by any stimulus, that elicited by media violence habituates with repeated exposure. Some investigators have therefore suggested that a "desensitization" to observed aggression may be the result of long-range viewing. At the level of psychophysiology, reduced activation as a function of exposure to violent videotapes has been shown in laboratory settings. Evidence also indicates that people who watch large amounts of televised violence are generally less reactive to new depictions of violence than are those who watch less violence. The effects of such desensitization on behavior are, however, not obvious and data bearing on the question are scarce. At the present we cannot conclude whether the habituation to violence that occurs with extended television viewing has any effect on aggressive behavior in response to such viewing.

## V. COGNITIVE PROCESSES IN MEDIA-RELATED AGGRESSION

Terms like "observational learning," "imitation," and "modeling" are little more than labels for the emission of aggressive responses in specific settings and they hardly explain the wide range of complex effects found in studies of the effects of observing violence. Likewise, the several other explanations that have been offered—those involving disinhibition, arousal, and the acquisition of aggressive attitudes—are not sufficient to explain the general processes that underlie the results of the research studies. What is needed is a level of theorizing that rises above the invention of labels and links the processes involved in observation of media violence to more general psychological constructs. Two such theoretical explanations for the effects of media violence have been offered. Both provide more comprehensive explanations than do earlier viewpoints. Both, moreover, are derived from the premises of contemporary cognitive psychology.

### A. Retrieval of Violent Scripts

Huesmann has proposed that when children observe violence in the mass media, they thereby learn complicated scripts for social behavior. The fundamental element in a script is the vignette, defined as "an encoding of an event of short duration," consisting of both a perceptual image and a "conceptual representation" of the event. A simple vignette might consist, for example, of an image of one person hitting another (image) in anger over something the other person has done (a conceptual representation). A script consists of a sequence of vignettes. Such scripts define situations and guide behavior: the person first selects a script to represent the situation and then assumes a role in the script. Once a script has been learned, it may be retrieved at some later time as a guide for behavior.

How does one know which of the many scripts in a person's memory will be retrieved on a given occasion? Granted that some aggressive scripts may be learned from observation of violence, why should one of them, and not some other, be recalled when the person has been provoked and is in a state of interpersonal conflict? One answer that has been suggested to these questions involves the principle of encoding specificity. According to this theory, the recall of information depends in large part on the similarity of the recall situation to the situation in which encoding occurred. As a child develops, she or he may observe cases in which violence has been used as means of resolving interpersonal conflicts. Such events are common in television programming. The information is then stored, possibly to be retrieved later when the child is involved in a conflict situation. Retrievability will depend partly on the similarity between cues present at the time of encoding and those present at the time of retrieval.

Certain stimulus conditions may determine what happens during the encoding process. Any characteristic of observed violence that makes a scene stand out and attract attention should enhance the degree to which that scene is encoded and stored in memory. One such characteristic may be the perceived reality of violence: acts of aggression seen as real may be regarded as more instrumental to the solving of future conflicts than less realistic ones. As noted earlier, several studies have shown that when portrayals of violence are said to be of real events they elicit more aggression than when they are described in less realistic terms.

## B. Priming of Aggressive Associations

A line of reasoning similar to that of Huesmann has been followed by Berkowitz, according to whom the aggressive ideas elicited by a violent stimulus can prime other semantically related thoughts. This in turn increases the probability of the viewer's having other aggressive ideas. Berkowitz bases this "priming" hypothesis on the notion of spreading activation from cognitive psychology: thoughts send out radiating activation along associative pathways, thereby activating other related thoughts. In this way, ideas about aggression that are not identical to those observed in the media may be elicited by the latter. In addition, thoughts are linked, along the same sort of associative lines, not only to other thoughts but also to emotional reactions and behavioral tendencies. Thus, observation of violence can engender a complex of associations consisting of aggressive ideas, emotions related to violence, and the impetus for aggressive actions. The hypothesis of cognitive priming may also help explain the finding that the presence of weapons is sometimes associated with elevated levels of aggression. This finding, first reported by Berkowitz and since replicated and extended by others, may indicate that weapons, because of their associations with violence, prime aggressive thoughts, emotions, and behavioral dispositions that facilitate the expression of aggressive behavior in frustrating or provocative settings.

## VI. TELEVISED VIOLENCE AND FEAR

Viewing violence in the media may have long-range effects on behavior that do not have any obvious connection with immediate effects. In an extensive series of studies, Gerbner and his associates have described one such long-range consequence of viewing violence. Briefly stated, the hypothesis of Gerbner and his colleagues is that extended watching of television brings a person into contact with a high level of violence and that this violence fosters attitudes of fear, suspicion, and distrust, as well as a distorted view of the world in which violence is given an importance disproportionate to its prevalence. Some studies conducted in both the United States and Europe have shown that fear may be an immediate consequence of watching presentations of violence.

Whether fear engendered by televised violence translates into influences on aggression is not known. However, it is possible that fear and associated beliefs that the world is a threatening place could lead people to feel a need to protect themselves against a clear and present danger. Among the results could be public demands for punitive justice, authoritarian control, and vigilantism. For example, one finding reported by Gerbner and his colleagues is that heavy users of television are more likely than lighter users to believe that too little money is being spent on fighting crime.

## VII. PORNOGRAPHY AND AGGRESSION

Certain kinds of pornography, in which persons are sexually assaulted, tortured, or placed in bondage, represent a special case of violent stimuli. Although the content of such pornographic displays is sexual, sex is usually not an end in itself but rather a means whereby the real purpose of the drama is served: aggression against a human victim, who is almost always a woman. This matter is especially important in that it offers a line of approach toward answering the question of whether observation of pornography engenders or facilitates the expression of aggression against women.

The role of pornography in violence against women has been studied extensively and there is some evidence that observation of pornographic stimuli is associated with violence against female victims. Most of this evidence comes from controlled laboratory experiments in which videotapes are carefully selected to depict sexual activity carried out in either an aggressive or a nonaggressive context. Allen and his colleagues conducted a meta-analysis of 33 studies on the effects of nonviolent and violent pornography on aggression. The results showed that both types of pornographic material increase aggression, but the effects are larger for violent pornography than for nonviolent pornography. Nonviolent pornography may increase aggression by increasing arousal levels in viewers. Pornographic scenes also elicit more aggression against female victims than against male victims.

Research has shown that when rape is depicted as something the female victim seems to enjoy, aggression by men against women is further enhanced. In addition, belief in this "rape myth"—that women secretly like to be sexually abused and assaulted by men—is fostered by exposure to violent pornography.

Men show considerable individual differences in their attitudes toward women and in the likelihood of aggressing against them. Likelihood of raping (LR) is a motivational variable that has been shown to moderate some of the situational effects associated with the presentation of pornographic stimuli. It is measured by means of a simple five-point rating scale by which men indicate how likely they would be to commit rape if they knew that they would not be punished for it. The LR variable is associated with both aggression toward women and attitudes about women. Men who score high on this variable are more likely than low scorers to report that they have used force against females for sexual purposes and that they will probably do so again. Under controlled laboratory conditions, men high in LR have also been shown to be more aggressive toward women than men low in LR, but not toward other men. LR also moderates the amount of arousal that men experience when exposed to certain types of pornography. Men high in LR have been found to become more excited while listening to an audio tape in which a woman apparently becomes sexually aroused while being raped than when she is aroused during consenting sex.

The effects of pornography on male aggression toward women appear therefore to involve several of the processes noted earlier in this article. Presentations of pornography are likely to elicit such aggression to the extent that they provide information about the acceptability of such behavior (i.e., the consent of women to be so attacked), that they reduce inhibitions against aggressing in men normally disposed to commit such acts, and that they arouse male viewers.

## VIII. NATURALISTIC STUDY OF MEDIA EFFECTS

### A. Types of Naturalistic Study

Experimental research such as that reviewed in the preceding sections allows investigators to test subtle hypotheses and effects. By controlling sources of extraneous variation in their studies, experimental researchers are able to isolate certain other variables (such as arousal, disinhibition, and social comparison processes) and to test whether such variables are in fact antecedents of aggression. For this reason, experimental studies constitute an important part of the literature on the effects of media violence. However, experiments have limitations as well as advantages. They examine behavior over short periods of time only. They involve behaviors that usually bear little similarity to the sort of aggressive behavior in which people engage on a daily basis. The samples of television programs that they use represent only a small portion of the wide array of programs seen on television. A more complete analysis of media effects on aggression must therefore include results from studies carried out under natural conditions. Three such types of study have been carried out: field experiments, longitudinal correlational investigations, and archival analyses. In many cases, these studies are designed to test hypotheses derived from the findings of laboratory research and therefore provide a test of the external validity of the latter type of study.

## B. Field Experiments

In a field experiment, independent variables are manipulated and controlled, and dependent variables measured, much as they are in a laboratory experiment. The entire procedure takes place in a natural setting, however. Random assignment of individual participants to conditions is usually difficult or impossible in field experiments. Usually, the field experimenter must resort to randomly assigning pre-existing groups of participants to conditions. The degree of control is not as high as it is in the laboratory, but, because the event takes place in a more realistic setting, external validity is perhaps greater. For example, the context of a study conducted by Josephson was a game of field hockey played by young boys. The young boys saw either a violent program or an exciting but nonviolent program prior to the game. Some of the boys were then interviewed by people who carried walkie-talkies, an instrument that had been carried by the aggressors in the violent program just seen. Josephson predicted that boys who saw the violent program and who later saw a walkie-talkie would be more aggressive during the field hockey game than those who saw the violent program but not the walkie-talkie. Josephson reasoned that the walkie talkie, which had been associated with violence in the program, would serve as a cue to prime other aggressive thoughts and emotions, in the way described by Berkowitz. This prediction was supported, but only among highly aggressive boys. This finding is not surprising if we consider that highly aggressive boys possess a relatively larger network of aggressive associations that can be activated by a cue.

In their meta-analytic study, Paik and Comstock reported that media violence significantly enhanced viewers' aggressive behavior. However, the effects from field experiments were smaller than the effects from laboratory experiments.

## C. Longitudinal Studies

Longitudinal studies involve the repeated measurement of television viewing and aggressive behavior under real-life conditions over a lengthy period of time. For example, one major longitudinal study has been reported by Eron and his associates. This work began with a study of third-grade children in a rural county in the state of New York in 1960, in which each child's aggressiveness was assessed through ratings made by the child's peers and parents and by the children themselves. Each child's preference for violent television programs was also measured. Ten years later, measures of the same variables were obtained for a large number of the children used in the original sample. Data from the two periods were analyzed by means of cross-lagged panel correlations. This analysis revealed that preference for television violence among boys in the third grade was positively and significantly correlated with aggressiveness 10 years later, whereas aggressiveness in the third grade was not correlated with preference for televised violence a decade later. This pattern of correlations supports the hypothesis that, for boys, observation of television violence in childhood contributes to aggressiveness in young adulthood. Additional analyses showed that the pattern of results was not due to differences in the level of aggressiveness among children who did or did not like violent television in third grade. Across all levels of aggressiveness—high, moderate, and low—in the third grade, an early preference for violent television was correlated significantly with aggressiveness 10 years later. This relationship was not weakened by the controlling of several possible contaminating variables, such as socioeconomic status of the boys' parents, the boys' intelligence, parental aggressiveness, and the total number of hours of television watched. All of these results pertained only to boys. Among girls, preference for violent television in the third grade was not significantly related to aggressiveness in young adulthood.

In a more recent report, Eron and his associates noted the results of a second follow-up study, involving 295 people from their original pool, 22 years after the original study. Data were gathered from interviews with the participants, both face-to-face and mailed, interviews with spouses and children of the participants, and archival records including criminal justice files. Childhood aggression was shown to be a predictor of both aggression and criminal behavior even 22 years later. Furthermore, the seriousness of the crimes for which males were convicted by age 30 was shown to be significantly related to the amount of television they watched as 8-year-old boys. The findings of this investigation are therefore consistent with those of the 10-year follow-up cited earlier.

Additional evidence of a connection between viewing of televised violence and aggression comes from a series of cross-cultural longitudinal studies carried out by Huesmann and Eron in five countries: the United States, Australia, Finland, Poland, and Israel. Two samples were studied in Israel: one from an urban setting and the other from a rural kibbutz. The time period of the study was 3 years, and more than 1000 boys and girls were tested. Aggressiveness was studied as a concomitant of such variables as the violence of preferred programs, overall viewing of violence, identification with televised aggressors, and judgments of realism of televised violence. In general, the evidence from the cross-cultural studies was consistent with that obtained from American samples in earlier research. Early television viewing was associated with aggressiveness among boys in the United States, Finland, Poland, and urban Israel. Among girls, early television viewing was related to aggression in the United States and urban Israel. Moreover, early aggression was associated with an increased viewing of televised violence for boys and girls in the United States and Finland, and for girls in urban Israel.

In their meta-analytic study, Paik and Comstock reported that viewing violence was positively related to later acts of aggression in longitudinal studies. As might be expected, the effects for longitudinal studies were smaller than the effects for laboratory and field experiments.

## D. Archival Analysis

Experimental and longitudinal studies are different in many respects but in one way they are similar: at least some of the persons involved (e.g., the experimental participants, the parents who monitor a child's television viewing, the teacher who rates a child's aggression) realize that they are taking part in an investigation. To some extent, therefore, the measurements are intrusive in that they break into the ongoing flow of behavior and possibly remind the participants of the purpose of the study. Another type of naturalistic study avoids this by involving the measurement of media effects and aggression after the fact and unknown to the participants. Usually this type of study consists of the analysis of public records and archival data. Inferences of cause and effect may then be drawn from evidence of a spatial or temporal relationship between media violence and aggressive reactions. For example, several writers have proposed that certain aggressive acts may occur in clusters following some spectacular violent incident, such as an airplane hi-

jacking or mass murder, that is widely reported in the communications media. The proximity of the aggressive acts to the preceding violent events may suggest the possibility that the latter somehow elicit, or facilitate the occurrence of, the former.

An ambitious and systematic attempt to account for real-world aggression with a hypothesis derived from experimental research has been made in a series of investigations by Phillips. Dealing entirely with archival data, Phillips has sought to show a relation between violence-related events shown on television or reported in other news media and increments in aggressive acts among the public in the immediate aftermath of the reports. Phillips based his studies on a hypothesis of observational learning and imitation derived from Bandura's experimental work. For example, Phillips has shown that the incidence of suicides increases immediately after a suicide has been reported in the newspapers, reaching a peak during the month immediately following the report. In addition, the increase in suicides appears to be directly related to the amount of publicity given the publicly reported suicides. Phillips has described this phenomenon as the "Werther effect," after the character in Goethe's novel whose self-inflicted death was said to have elicited many real suicides among readers.

In related studies, Phillips has shown a correlation between publicized suicides and the incidence of automobile fatalities. Interpretation of this finding rests on the assumption that some motor vehicle accidents may in fact be due to suicidal intentions on the part of drivers involved. For example, in one study the number of motor vehicle deaths occurring over an 11-day period following each of 23 front page suicide stories was compared with the average number of such deaths during four control periods in which no suicides were reported. In the case of all but 5 suicides, the number of automobile fatalities was greater following the story than during the control periods. Overall, the average number of fatalities increased significantly following suicide stories. The peak in incidence of traffic deaths occurred, on average, 3 days after the suicide stories were reported. In still another study Phillips reported a positive relationship between suicides shown in televised fiction and acts assumed to indicate suicidal motives. Thirteen suicides shown in televised soap operas during 1977 constituted the eliciting stimuli. The number of motor vehicle deaths, nonfatal traffic accidents, and suicides all increased, compared to rates during a control period consisting of the latter part of the weeks in which the fictitious suicides were shown.

Finally, in a study that pertains more directly to evidence from experimental research, Phillips has reported a positive relationship between televised heavyweight prize fights and the incidence of homicides in the United States over a 10-day period following each fight. It should be noted that some of Phillips's findings involve the operation of variables not yet accounted for theoretically and that the findings and conclusions of this research have been criticized on methodological grounds. Nevertheless, it represents an interesting extension of a hypothesis developed in the experimental laboratory to aggressive behavior in real-life settings.

## IX. THE MODERATING ROLE OF TRAIT AGGRESSIVENESS IN THE EFFECTS OF VIOLENT MEDIA ON AGGRESSION

People in the mass media claim that media violence only has harmful effects on certain individuals who are highly aggressive by nature. Bushman conducted three studies to test this hypothesis. In each study, aggressiveness was measured using a standardized personality test. Sample items from the test include "Once in a while I cannot control my urge to strike another person" and "I have threatened people I know." In the first study, participants read violent and nonviolent film descriptions and then chose a film to watch. High aggressive people were more likely to choose a violent film to watch than were low aggressive people. Although causal inferences cannot be made from these correlational results, the results suggest that the relationship between media violence and aggression might be reciprocal. Not only does media violence increase aggression, but aggressive people also like to watch media violence. The results also are consistent with the idea that habitual exposure to media violence can lead to the development of extensive aggressive cognitive-associative networks in highly aggressive people.

In the second experiment, participants reported their mood before and after the showing of a violent or nonviolent videotape. High aggressive people reported feeling more angry after watching the violent video than did low aggressive people, even though they were not provoked. These results suggest that viewing violence may be especially likely to prime aggressive feelings in highly aggressive people.

In the third experiment, participants first viewed either a violent or nonviolent videotape and then competed with an ostensible "partner" on a reaction time task. Participants were told that they and their partner would have to press a button as fast as possible on each trial, and whoever was slower would receive a blast of noise. Each participant was permitted to set in advance the intensity of the noise that the other person would receive (between 60 and 105 decibels) if the other lost. A nonaggressive no-noise setting (0 decibels) was also offered. The results showed that the violent videotape was more likely to increase aggression in high aggressive people than in low aggressive people.

The results from Bushman's studies suggest that the negative effects of media violence are more pronounced for individuals who are highly aggressive by nature than for their nonaggressive counterparts. These results should not, however, be interpreted to mean that media violence does not affect most members of society. But even if viewing violence increases aggression in only a small subset of the population, the negative consequences to society would not be trivial. For example, if 1 million people watched a violent movie, and only 1% are likely to behave aggressively as a result of this exposure, then the movie could increase aggression in 10,000 people. The victims may be actors, the blood may be ketchup, the bullets may be blanks, but for some viewers, the image on the screen is all too real.

## BIBLIOGRAPHY

Allen, M., D'Alessio, D., and Brezgel, K. (1995). A meta-analysis summarizing the effects of pornography. II. Aggression after exposure. *Hum. Commun. Res.* **22,** 258–283.

Berkowitz, L. (1984). Some effects of thoughts on anti- and prosocial influences of media events: A cognitive-neoassociationist analysis. *Psychol. Bull.* **95,** 410–427.

Bushman, B. J. (1995). Moderating role of trait aggressiveness in the effects of violent media on aggression. *J. Personal. Soc. Psychol.* **69,** 950–960.

Eron, L. D., Huesmann, L. R., Dubow, E., Romanoff, R., and Yarmel, P. W. (1987). Aggression and its correlates over 22 years. *In* "Childhood Aggression and Violence" (D. Crowell, I. M. Evans, and C. R. O'Donnell, eds.), pp. 249–262. Plenum, New York.

Eron, L. D., Huesmann, L. R., Lefkowitz, M. M., and Walder, L. O. (1972). Does television violence cause aggression? *Am. Psychol.* **27,** 253–263.

Geen, R. G. (1983). Aggression and television violence. *In* "Aggression: Theoretical and Empirical Reviews" (R. G. Geen and E. I. Donnerstein, eds.), Vol. 2. Academic Press, New York.

Geen, R. G. (1990). "Human Aggression." Open University Press, Milton Keynes, UK.

Huesmann, L. R. (1986). Psychological processes promoting the relation between exposure to media violence and aggressive behavior by the viewer. *J. Soc. Issues* **42,** 125–139.

Huesmann, L. R., and Eron, L. D. (eds.) (1986). "Television and the Aggressive Child: A Cross-National Comparison." Erlbaum, Hillsdale, NJ.

Josephson, W. L. (1987). Television violence and children's aggression: Testing the priming, social script, and disinhibition predictions. *J. Personal. Soc. Psychol.* **53,** 882–890.

Paik, H., and Comstock, G. (1994). The effects of television violence on antisocial behavior: A meta-analysis. *Commun. Res.* **21,** 516–546.

Phillips, D. P. (1986). Natural experiments on the effects of mass media violence on fatal aggression: Strengths and weaknesses of a new approach. *In* "Advances in Experimental Social Psychology" (L. Berkowitz, ed.), Vol. 19. Academic Press, New York.

Zillmann, D. (1988). Cognitive-excitation interdependencies in aggressive behavior. *Aggress. Behav.* **14,** 51–64.

# Behavior: Cooperative, Competitive, and Individualistic

GEORGE P. KNIGHT
*Arizona State University*

## GLOSSARY

**Resource distribution preference** Desire to obtain a specific division of some valuable commodity between oneself and one or more others

**Socialization** Process of the transmission of cultural and family values, prescriptions, and prohibitions

**Sociobiology** Branch of ethology concerned with the study of the biological basis of social behavior

COOPERATIVE, COMPETITIVE, AND INDIVIDUALIS-tic behaviors frequently occur in a wide variety of contexts in our daily social interactions. Because of the frequency of these types of behavior patterns, there has been a substantial theoretical and empirical effort to investigate the causal mechanisms that lead to these behaviors.

Three classes of theoretical perspectives have guided the research: sociobiological, social psychological, and developmental. Although these three classes of theoretical perspectives have focused on quite different causal mechanisms, in many cases these different mechanisms represent different levels of proximal (i.e., more immediate) versus distal (i.e., more ultimate) causation.

## I. DEFINITIONAL ISSUES

Summarizing the current state of knowledge regarding the causes of cooperative, competitive, and individualistic behaviors is complicated by a variety of definitional issues that are not completely resolved. Much of the literature available contains either relatively vague or inconsistent definitions of these behaviors. However, most of this literature has relied on definitions as either a preferred interaction style or as a resource distribution preference.

## A. Preferred Interaction Styles

In the preferred interaction style view, cooperation is a preference for working together with another person in a coordinated fashion, competition is a preference for working against another person, and individualism is a preference for working alone. It is important to note that across different situations these different interaction styles can have quite different consequences. For example, in a situation in which the availability of resources is not fixed and finite and the task requires some coordinated effort, cooperation will likely allow the person to obtain the available resources. In a situation in which the resources are finite, such as when there is only one resource available to the person in the situation, competition will likely allow the person to obtain the available resource. In a situation in which acquiring resources is dependent on consistent individual performance and another person represents only a social distraction, individualism will likely allow the person to obtain the available resources.

ENCYCLOPEDIA OF HUMAN BIOLOGY, Second Edition, VOLUME I.   Copyright © 1997 by Academic Press.   All rights of reproduction in any form reserved.

## B. Resource Distribution Preferences

In contrast, the resource distribution preference view is concerned with how individuals prefer resources to be allocated to themselves and to one or more other persons, and with the social values reflected in those preferences. In this view cooperative, competitive, and individualistic behaviors are defined as preferences for specific forms of outcomes based on the individual's value system.

At least six resource distribution preferences are likely to occur in any given real-world context. Three of them are relatively prosocial or cooperative: equality, group enhancement, and altruism. Equality is a preference for resource distributions that minimize the difference between one's own and another's resources. Group enhancement is a preference for resource distributions that maximize the resources of two or more individuals as a group, irrespective of the specific distribution of resources to oneself or another. Altruism is a preference for resource distributions that maximize the resources of another. [See Altruism.]

Two relatively competitive preferred resource distributions also occur: superiority and rivalry. Superiority is a preference for resource distributions that maximize one's relative resources compared to those of another. Rivalry is a preference for resource distributions that minimize the resources of another.

Finally, there is one individualistic preferred resource distribution: individualism. Individualism is a preference for resource distributions that maximize one's own resources without regard for the impact on others.

Although a number of theoretical frameworks have suggested these as the most likely resource distribution preferences, the measurement of these preferences has proved somewhat difficult. This difficulty is a result of the forced interrelationships among the various resource distributions within any given circumstance. There are contexts in which behaving in a manner which maximizes one's own resources simultaneously maximizes a peer's resources, creates parity in the resources for oneself and a peer, and maximizes the resources of the group. In contrast, there are also real-world contexts in which maximizing one's own resources simultaneously leads to minimizing the resources of a peer and maximizing relative resources compared to those of a peer. Given these kinds of motivational confounds, it has proved somewhat difficult to precisely identify the resource distribution preferences of individuals. In some real-world contexts it is difficult to determine whether a person is

distributing resources in a way that is individualistic or cooperative, whereas in other contexts it is difficult to determine whether a person is distributing resources in an individualistic or competitive manner.

One solution to this problem of inferring the individual's preferred resource distribution pattern has been to look at the individual's resource distributions across a variety of contexts or situations and to infer the preferred resource distribution from this pattern of distributions. For example, if a person chooses a resource distribution that provides individualism, altruism, equality, and group enhancement in one situation, but chooses a resource distribution that provides individualism, rivalry, and superiority in another situation, it is somewhat reasonable to infer that the person has been attempting to express an individualistic preference (i.e., attempting to maximize his or her own resources across situations). Indeed, there has been considerable research of this nature. On the basis of this research, there are more recent views regarding the actual occurrence of different preferred resource distributions. In particular, it appears that some individuals do prefer individualistic, equality, group enhancement, or superiority resource distributions; but there is less evidence, and mixed evidence, regarding the preferences for altruism or rivalry.

Although the research on human cooperative, competitive, and individualistic behaviors has defined these behaviors in the two ways noted earlier, there has been relatively little consideration of the links between these behaviors. That is, there has been little attention to how different social interaction styles might serve to create different resource distributions in different situations. Understanding of the relationships between these two types of conceptualizations might be useful in understanding human cooperative, competitive, and individualistic behaviors and the different theoretical approaches described later.

## II. SOCIOBIOLOGICAL PERSPECTIVES

Cooperative, competitive, and individualistic behaviors are considered important social behaviors within the sociobiological framework because behaviors such as these are directly related to the acquisition of resources necessary for reproductive success. The sociobiological perspective suggests that human social behavior is basically self-interested. A behavior has adaptive significance if it somehow contributes to the actor's reproductive success. Natural selection will ensure the continuation of a given behavior if that behavior in some way enhances the actor's contribution to the gene pool. Thus, the sociobiological perspective considers that the individual is motivated to obtain resources if those resources are necessary for or enhance the individual's contribution to the gene pool. In effect, the sociobiological perspective represents an attempt to describe the ultimate, or most distal, cause of cooperative, competitive, and individualistic behaviors.

Clearly, an individualistic resource distribution preference is quite compatible with the sociobiological perspective. Maximizing one's own resources is an adaptive behavior if those resources are ultimately necessary for reproductive success. However, there are also sociobiological mechanisms that would make either cooperative or competitive resource distributions adaptive. Kin selection or genetic similarity might be one sociobiological mechanisms leading to cooperative or altruistic resource distribution preferences. An individual might distribute resources in a manner favorable to another person to the extent that such a resource distribution increases the fitness and potential reproductive success of individuals who are genetically related. That is, one might distribute resources to others who are genetically related if the resources allow the benefactor to contribute to the gene pool and if the benefactor has some significant subset of one's own genes.

Some indirect evidence of the kin selection explanation of cooperative behavior derives from investigations of monozygotic (i.e., identical) and dizygotic (i.e., fraternal) twins (as well as individuals related in other ways). Monozygotic twins share the same genes, whereas dizygotic twins (and nontwin siblings) share one-half of their genes in common. A sociobiological perspective leads to the expectation that the greater genetic similarity among monozygotic twins than dizygotic twins should lead to more frequent cooperative behavior among monozygotic twins than among dizygotic twins. Indeed, this is precisely what is found. In addition, a more frequent preference for cooperative resource distributions has been found among siblings, who share one-half of these genes, than among acquaintances, who share a much smaller proportion of genes.

Reciprocal altruism might also be a sociobiological mechanism that leads to cooperative or altruistic resource distribution preferences. An individual might distribute resources in a manner favorable to another person if the two individuals associate long enough to exchange roles frequently and thereby enhance each

individual's potential for reproductive success. That is, an individual might distribute resources between himself (or herself) and another cooperatively if he expects that the other person will later distribute resources to him in a favorable manner and if the total resources he can obtain are greater in this case than if he simply distributed resources individualistically at every opportunity. Indirect evidence compatible with this reciprocal altruism comes from a variety of studies demonstrating more frequent prosocial behavior among friends and persons who anticipate future interactions. For example, a more frequent preference for cooperative resource distributions has been found among close friends than among acquaintances. Reciprocal altruism could explain this finding if the individual expects that future interactions, particularly those of a reciprocating nature, are more likely to occur with a friend than with an acquaintance.

Competitive resource distribution preferences can occur frequently in daily interactions, when the resources, available in limited supply, are necessary for reproductive success.

Finally, the sociobiological perspective suggests links between the interaction styles and resource distribution preference notion of cooperative, competitive, and individualistic behaviors. Individuals' resource distribution preferences and the characteristics of the situation probably jointly determine the interaction styles displayed. For example, much of the cooperation observed in daily interactions occurs in situations in which working together is necessary to produce a product from which each individual profits. That is, two or more individuals might work together to accomplish some task, but they might be doing so to satisfy individualistic motives or values. In contrast, in a situation in which only one resource is available, individuals might engage in competition and work against one another, perhaps even actively inhibiting or sabotaging the other's work. However, these individuals might be engaging in competition to satisfy individualistic motives and values.

## III. SOCIAL PSYCHOLOGICAL PERSPECTIVES

### A. Measurement

By far a majority of the research on cooperative, competitive, and individualistic behaviors has been social psychological in nature. In much of this research individuals are placed in a social dilemma requiring some

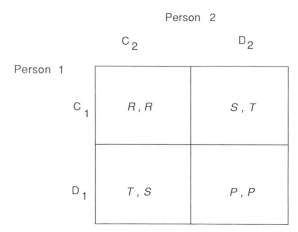

Person 2

FIGURE 1 The payoff matrix for the prisoner's dilemma is defined by the inequalities $T > R > P > S$ and $R > (S + T)/2$. In each cell the left number represents the payoff for person 1 and the right number represents the payoff for person 2. For an example prisoner's dilemma matrix substitute the following payoffs: $T = 18$, $R = 12$, $P = 16$, and $S = 0$.

decision or behavior that provides some resources or payoffs for themselves and another person. Thus, the situation is one of mutual interdependence.

A commonly used dilemma is the prisoner's dilemma, in which each of the two participants receives some payoff, depending on the decision to cooperate or defect (see Fig. 1). In the prisoner's dilemma the payoffs are defined such that $T > R > P > S$ and $R > (S + P)/2$. In effect, $R$ is the reward both individuals receive for mutual cooperation, $P$ is the punishment both individuals receive for mutual competition, $T$ is the temptation to defect or compete if one believes the other person will cooperate, and $S$ is the "sucker's payoff" the cooperator receives if the other person competes. This social dilemma is named after the following hypothetical dilemma that exemplifies the contingencies just described:

Two suspects are questioned separately by the district attorney. They are guilty of the crime of which they are suspected, but the D.A. does not have sufficient evidence to convict either. The state has, however, sufficient evidence to convict both of a lesser offense. The alternatives open to the suspects, "A" and "B," are to confess or not to confess to the serious crime. They are separated and cannot communicate. The outcomes are as follows. If both confess [defect], both get severe sentences [P], which are, however, somewhat reduced because of the confession. If one confesses (turns state's evidence), the other gets the book thrown at him [S], and the informer goes scot free [T]. If neither confess, they cannot be convicted of the serious crime, but will surely be tried and convicted for the lesser offense [R]. (Anatol Rapaport)

The dilemma is whether or not each individual should attempt to obtain his best outcome. Each indi-

vidual's best outcome results from defecting (i.e., confessing), but only if the other does not defect. However, if both reason that defecting is the only way to obtain the best possible outcome (i.e., no jail time), then both will defect and each will receive more jail time than if they had trusted each other. Prisoner A's situation is as follows: If prisoner B fails to implicate him, then he should implicate prisoner B in order to get $T$ instead of $R$; and if prisoner B implicates him, then he should implicate prisoner B in order to get $P$ instead of $S$. Of course, prisoner B's situation is precisely the same as that of prisoner A.

Although much of the social psychological research has assessed cooperative and competitive behavior with individual's choices in the prisoner's dilemma, a wide variety of dilemmas have been used. These dilemmas differ in the contingencies associated with cooperative and competitive decisions. An example is the maximizing difference dilemma. Here, the defining inequalities are $R > T > P = S$ and $R > (S + T)/2$ (see Fig. 1). In effect, the contingency structure of the maximizing difference dilemma makes the cooperative alternative the mathematically rational choice for both individuals. Various forms of interdependence can be created through subtle manipulations of the contingencies (i.e., the defining inequalities) in these types of social dilemmas.

Much of the more recent social psychological research has relied on decomposed dilemmas to measure cooperative and competitive preferences. In a decomposed dilemma the individual is presented with a choice of two or more alternatives which completely define some resource distribution for himself and one or more other individuals (see Fig. 2). This is in contrast to the standard form of this type of social dilemma, in which the distribution of resources is dependent on the choices of two or more individuals. Thus,

FIGURE 2 The payoff matrix for a decomposed dilemma would represent a decomposed prisoner's dilemma if $Y > W$, $X > Z$, and $W + X > Y + Z$. This decomposed dilemma would be analogous to the prisoner's dilemma matrix in Fig. 1 if $W = 6$, $X = 6$, $Y = 12$, and $Z = 6$. (Note that $T = X + Y$, $R = W + X$, $P = Y + Z$, and $S = W + Z$.) C, cooperate; D, defect.

in the decomposed dilemmas there is a more direct connection between the individual's choice and the distribution of resources, thereby reducing the need for strategic behaviors aimed at influencing other participants. These decomposed dilemmas can be constructed such that they conceptually present the same dilemmas as the more standard forms of this type of dilemma. For example, the numerical resource distributions are such that two individuals making choices in the decomposed prisoner's dilemma jointly receive a better outcome if both select the cooperative alternative, jointly receive a poorer outcome if both select the competitive alternative, individually receive the best possible outcome if one selects the competitive alternative while the other selects the cooperative alternative, and individually receive the worst possible outcome if one selects the cooperative alternative while the other selects the competitive alternative.

Further, some researchers have been assessing cooperative, competitive, and individualistic social values by having individuals rate the desirability of a number of resource distributions. These researchers present a resource distribution, such as one row of a decomposed dilemma, and then asked the individual to rate that resource distribution on, for example, a seven-point scale ranging from very desirable to very undesirable. The systematic analysis of these desirability ratings allows more accurate inferences of the social values underlying the individual's resource distribution preferences.

These types of measures have been used to investigate the social factors that might influence the frequency of cooperative and competitive behaviors. These factors can be classified into one of three types of influences: those due to characteristics of the target individual whose cooperative, competitive, and individualistic behaviors are being assessed; those due to the characteristics of the individual(s) who will receive some resources based on the target individual's behavior; and those due to the characteristics of the situation in which the individuals find themselves.

## B. Individual Differences in Social Values

The research on the influence of characteristics of the target individual on their cooperative, competitive, and individualistic behaviors has investigated characteristics such as individual and group differences in social values and expectations of others' behavior. This research has led to the conclusion that there are somewhat stable social values that lead to somewhat

consistent resource distribution preferences across situations and differential expectations of the behavior of others.

Several lines of evidence support this notion of individual differences in social values. Thus, individuals with different social values (i.e., altruists, cooperators, individualists, and competitors) make different choices in a variety of social dilemmas, such as those described earlier. Individuals with these different social values respond to cooperative and competitive overtures in quite different ways. Cooperators cooperate unless the partner is extremely competitive; competitors are competitive regardless of the behavior of the partner; and individualists are competitive unless the partner mimics their behavior and mutual cooperation occurs. It can also be shown that cooperators expect that others will be cooperative and that competitors expect that others will be competitive. These findings, in conjunction with evidence that children's peers can reasonably accurately predict which classmates will make cooperative, competitive, or individualistic resource distributions, clearly demonstrate the individual differences in social values. Further, cross-cultural, cross-ethnic, cross-race, and cross-class research has demonstrated group differences in social values.

## C. Recipient Influences

The research on the influence of the characteristics of the partner on the cooperative, competitive, and individualistic behaviors of the subject has primarily considered the behavior of the partner. This research has led to the conclusion that conditionally cooperative behavior on behalf of the partner leads to a mutually cooperative interaction. That is, if the partner responds to the target individual's cooperative resource distribution preferences with cooperative resource distribution preferences, the target individual will display more cooperative preferences. For example, a tit-for-tat response strategy (i.e., copying the previous trial behavior of the target individual) on behalf of the partner leads to the highest rates of mutual cooperation in the prisoner's dilemma.

## D. Situational Influences

The research on the influence of contextual or situational factors has investigated factors such as the motivational structure of the situation, individuals' involvement in the situation, the communication opportunities in the situation, the public versus private nature of behavior in the situation, and group size. This research has led to the conclusion that certain motivational structures, high involvement in the situation, communication, public behavior, and small group size lead to more cooperative and less competitive preferences.

The motivational structures created by the motivational confounds described earlier might have a substantial influence on the apparent frequencies of cooperative and competitive preferences. Specifically, in some social dilemmas, and indeed in some real-world settings, individualism is confounded with cooperative behavior, whereas in other dilemmas individualism is confounded with competitive behavior. That is, in some situations individuals can maximize their own resources through apparently cooperative responses, whereas in other situations they can maximize their own resources through apparently competitive responses. Thus, individualistic values might lead to quite different behaviors, depending on the motivational structure of the situation. A number of studies have demonstrated more frequent cooperative behavior when such behavior also results in an individualistic resource distribution. In this case, however, some individuals might be engaging in cooperative behavior to satisfy equality or group enhancement values, whereas other individuals might be engaging in cooperative behavior to satisfy an individualism value.

Evidence of the influence of involvement in the situation comes primarily from studies of the effect of the value of the resources available. Although there are some mixed findings, when the assessment procedure involves resources of substantial value (usually substantial amounts of money), there is often an increase in the frequency of cooperative behavior observed. Possibly, the mixed findings are a function of the motivational structure of the situation. As the value of the resources to be distributed increases, then those individuals who value individualism might more often moderate their behavior in accord with the situation: behaving cooperatively or competitively if that is what maximizes their own resources.

The research on the opportunity for, and the occurrence of, communication has consistently indicated that communication among participants resulted in an increased frequency of cooperative behavior. Group discussion enhances cooperation because it arouses group-regarding motives. That is, the communication among individuals in the situation creates motivation to want other group members to do well in the situation.

Finally, as noted earlier, there is consistent evidence

that there is likely to be more cooperation and less competition when there is public disclosure of one's behavior and when a small group is involved in the situation. This greater frequency of cooperative behavior under these circumstances might be a function of norms and social sanctions. Clearly, in these cases cooperation is the socially desirable behavior.

## IV. DEVELOPMENTAL PERSPECTIVES

A substantial body of recent research has addressed the issue of the development of cooperative, competitive, and individualistic behavioral styles. Most of this research has assessed the age differences in cooperative, competitive, and individualistic preferences across the 3- to 12-year-old age range. Although there are some inconsistencies in the research findings, probably due to measurement difficulties, recent research has clarified the development patterns. It appears that the most accurate description of the age differences is that there is a developmental shift from individualistic to cooperative or competitive preferences across the 3- to 8-year-old age range and a shift toward increasing cooperative preferences after 8 years of age. From 3 to 8 years of age, there is a shift from individualistic to equality or superiority preferences, and after 8 years of age there are increasing equality and group enhancement preferences. Furthermore, some research attempts to explain these developmental patterns as a function of either cognitive development or learning.

### A. Measurement

The research addressing the development of cooperative, competitive, and individualistic social values and behaviors has primarily relied on experimental game measures, choice card measures, or decomposed dilemmas. These experimental game measures usually put two or more children in a situation in which they must complete some task in order to receive some desirable resource, such as a toy.

The cooperation board is a representative example of this assessment method. It consists of a flat posterboard with a circle drawn on each side and an eyelet at each corner, a ring which holds a marker in an upright position, with four strings attached to the ring and each passing through one eyelet. A child stands at each side of the board holding onto one string. By coordinating the tension on each of their strings, the children can move the marker around the board. The children are told (at least in some conditions) that they will receive a toy every time the marker passes through the circle in front of them within a limited time frame.

Note that this is clearly an interaction-style measure of cooperation and competition, as are nearly all of the experimental game measures. Cooperation is defined as the coordinating of actions during the task. Unfortunately, inference of social values and resource distribution preferences is difficult from behavior in this measure, as it is for most of the experimental game measures, because of the inherent motivational confounds and the group nature of this measure. That is, individuals can only maximize their own resources (i.e., express an individualism value) by cooperating. In addition, the presence of only one uncooperative child results in all children in the task scoring low on cooperation (i.e., it takes all four children to coordinate the movement of the marker).

The choice card measures are similar to the decomposed dilemma measures described earlier. They are different, however, in that they are usually designed to assess specific resource distribution preferences and might provide the child with more than two alternatives to choose from. The social orientation choice card (see Fig. 3) is a representative example. In this case a child is given a choice among three resource distributions that represent the theoretically prescribed social values. These type of measures, and the decomposed games described earlier, are more useful for identifying resource distribution preferences and inferring social values. However, some inherent difficulties exist even with the relatively good measures. For example, the middle alternative in the social orientation choice card clearly offers individualism, but it also offers some superiority. Because of these motivational confounds within a constrained set of alternatives, developmental researchers have also moved

toward inferring social values from resource distribution preferences across a variety of choice cards or decomposed dilemmas that differentially confound the theoretically proposed social values.

## B. Influence of Cognitive Development

It has been suggested that the age differences in these social behaviors are a function of the developmental decreases in egocentrism and centration. Egocentrism refers to the tendency of young children to see the world from their own viewpoint with no awareness of the existence of other viewpoints. Centration of thought refers to the tendency of the young child to be able to attend to only one dimension of an object or situation at a time. As thought becomes less egocentric and decentered and the child becomes capable of role-taking, this role-taking ability should lead to more frequent other-oriented cooperative behavior. Unfortunately, there is little direct empirical evidence addressing this causal mechanism. While several studies have examined the relationship between perspective-taking and cooperative/competitive behaviors, this research has generally produced inconsistent findings. In addition, the research assessing the relationship between a variety of forms of prosocial behavior and role-taking has also led to somewhat inconsistent findings.

It has been suggested that the age differences in cooperative, competitive, and individualistic social behaviors are a function of cognitive–numerical development. The numerical operations necessary for the expression of fairness rules (e.g., equality and equity), as well as other comparisons (e.g., superiority), are the abilities to make quantitative judgments of more than, less than, and equal. Thus, as the child's understanding of these mathematical concepts develop, the child is more likely to engage in social behaviors such as equality and superiority because these decisions require such comparative information. There might also be links between mathematical abilities and social behaviors.

Indeed, there is some evidence quite consistent with this cognitive–numerical development explanation. From 6 to 14 years of age, resource allocations go from equality to ordinal equity to proportional equity. In a study in which children were allowed to distribute resources to two hypothetical children who had contributed in varying degrees to the product for which the children were to be rewarded, it was found that the youngest children generally divided the resource equally between the two hypothetical children. In con-

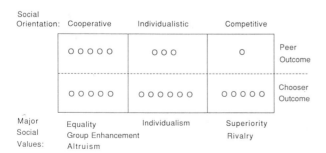

**FIGURE 3** The social orientation choice card. The circles within each box represent valuable resources, usually money or tokens, that can be used to trade for desirable toys.

trast, the middle-aged children generally divided the resources such that the child who contributed the most to the product received the most resource (ordinal equity). Finally, the older children divided the resources such that each of the two children received a number of resources proportional to their contribution to the product (proportional equity). Further, these age differences occurred at the same age as the development of the logical mathematical concept of proportionality.

It has been suggested that the age differences in cooperative, competitive, and individualistic preferences might be a function of the development of the executive control functions that guide the use of mathematical abilities. The expression of individualistic preferences requires the same mathematical operations as the expression of equality or superiority preferences. That is, the identification of the individualistic resource distribution requires making the same kind of more than, less than, and equal judgments that are required in identifying the equality or superiority resource distributions. However, identification of the equality and superiority resource distributions requires using these mathematical abilities repetitively in a systematic and functional order. Thus, young children might make individualistic decisions because the need to sequence mathematical operations appropriately to make an equality or superiority decision taxes their abilities.

The evidence supporting this executive control explanation is based on a series of studies which indicate (1) that it takes longer and is more difficult to accurately identify equality and superiority resource distributions than individualism distributions; (2) that those children who spontaneously make individualism decisions have greater difficulty identifying equality and superiority distributions than those children who spontaneously make equality or superiority decisions; and (3) that reducing or eliminating the relatively greater executive control demands of the equality and superiority decisions reduces or eliminates the age differences in these preferences. Further support comes from two studies in which children completed a cooperation/competition task in either a cooperative or a competitive incentive condition and with either a cooperative or a competitive partner. In the first study the frequency of cooperative behavior among young children was influenced by the partner's behavior but not by the incentive manipulation. In contrast, the behavior of the older children was influenced by both the partner's behavior and the incentive manipulation. In the second study the frequency of coopera-

tive behavior among young children was influenced by both the partner's behavior and the incentive manipulation when the children were instructed to use specific information-processing strategies. Thus, it appears that the young children were capable of understanding and using the contextual features of the situation unless the cognitive demands of the situation exceeded their cognitive process capabilities.

## C. Influence of Socialization Experiences

There have also been attempts to link the age differences in cooperative, competitive, and individualistic social values and behaviors to socialization experiences. One of the most significant functions of socialization is the transmission of cultural, societal, and familial values. Socialization is a broad range of experiences that control the process through which prescriptions and prohibitions are transmitted to members of a social group. Thus, the family and society create opportunities for the child to learn values and to utilize those values in day-to-day behaviors. Cooperative, competitive, and individualistic values among the many societally based values children internalize through socialization. Although there has been relatively little research directly investigating the socialization of cooperative, competitive, and individualistic values and behaviors, there is evidence consistent with a socialization model. The differences in the behaviors across ethnic groups, racial groups, socioeconomic status groups, genders, and birth orders could well be the result of differential socialization experiences associated with group membership. Further, there is substantial literature describing investigations of the socialization of other prosocial and antisocial values and behaviors. However, our current view of the socialization of cooperative, competitive, and individualistic social values and behaviors must be considered speculative because of the limited empirical base.

The socialization mechanisms that have been linked to social behaviors similar to cooperative, competitive, and individualistic behaviors can be grouped into those involving direct instruction, modeling, regulation practices, and/or affective relationships. Instruction involves the use of verbal prompts or commands to induce specific behaviors. Exhortations are similar verbal attempts to influence future behaviors by communicating what the child ought to do and why the child ought to behave in this manner. Modeling involves exposure to the behaviors of others and provides an opportunity for the child to acquire information about specific behaviors. Specifically, the observation of others might increase the salience of specific behavioral alternatives and social norms and provide information about the appropriateness of behavioral alternatives.

Regulation practices involve a variety of processes involving the reward or reinforcement of desired behaviors, the assignment of responsibilities that foster desired behaviors, and the discipline of undesired behaviors. The reward or reinforcement of specific behavioral alternatives provides the child with guidance regarding the expected mode of behavior. The assignment of responsibilities that foster specific behaviors provides the child with specific opportunities to engage in the desired behaviors. The discipline of specific behavioral alternatives provides the child with guidance regarding which behavioral alternatives are prohibited. Relationships with a socialization agent include affective qualities such as acceptance, affection, and nurturance. A positive affective relationship with the key socialization agents is essential for emotional security, which enhances the occurrence of prosocial behaviors by lessening the child's preoccupation with his own needs.

Thus, the child will likely come to value cooperative behaviors if his parents and other socialization agents instruct the child in how, when, and why the child should behave cooperatively; model cooperative behaviors before the child; reward cooperative behaviors; assign roles and responsibilities that require cooperative behaviors; discipline behaviors that are incompatible with cooperative behaviors; and have a positive affective relationship with the child. Further, the specific combination of practices might be particularly important because these mechanisms function interactively or jointly in the socialization of the child. For example, in addition to the intended effects of discipline, the form that the discipline takes might affect behavior. Thus, victim-centered or psychological discipline might help put the feelings and thoughts of the other into the child's consciousness and thus help guide the child's behavior in a prosocial direction. When this discipline style suggests concrete acts of reparation, the socialization agent is not only focusing attention on the other, but is also providing the child with a prosocial model and instruction on how to be prosocial. In contrast, when the agent uses physical punishment, the agent is focusing the child's attention on himself and at the same time is providing an antisocial model. Thus, disciplinary practices might also influence the child through one or more of the other socialization mechanisms. Similarly, accep-

tance, affection, and nurturance might foster identification with and imitation of the socialization agent, while also providing a prosocial model.

## D. Integration of Cognitive Development and Socialization

The cognitive development and socialization explanations of the age differences and individual differences in cooperative, competitive, and individualistic social values and behaviors are not necessarily incompatible. The age differences are consistent with the possibility that the complexity of the child's resource distribution preferences is largely determined by the child's cognitive abilities and the cognitive demands of the context, whereas the prosocial/antisocial quality of those preferences is largely determined by the child's socialization history. That is, perhaps the young child's behavior in many contexts is individualistic because the cognitive demands of equality or superiority tax their cognitive abilities. In contrast, the older child's behavior might be more flexible because his cognitive abilities allow him to engage in equality, superiority, or individualistic behaviors. Once the child has sufficient cognitive abilities, his specific behavioral preferences might be based on the social values he has acquired through socialization. The child might encounter socialization experiences which direct him toward valuing cooperative behaviors, and this socialization pressure might lead to internalized cooperative values once the child has the necessary cognitive abilities. Perhaps perspective-taking, decentered thought, and prosocial/moral reasoning are among the requisite skills for acquiring these values. Further, once these cooperative values are acquired, they might be selectively applied to various contexts, depending on the cognitive demands of the context. Perhaps cognitive–numerical or executive control skills are the requisite skills for enacting these internalized values.

While this model emphasizes largely different roles for cognitive development and socialization, this is not to imply that these are independent processes. Indeed, as described earlier, cognitive development likely impacts on socialization. Further, it is likely that socialization impacts on cognitive development. For example, parents who consistently use combinations of socialization practices designed to enhance the child's understanding will most likely have children who acquire social values at an early age. Parents who consistently reward equal sharing of resources and at the same time verbally describe the appropriateness of equally sharing will have children who understand and adopt a sharing value at an earlier age than parents who use only reward contingencies. In the former case the verbal information might help the child to correctly abstract the appropriate rule rather than requiring the child to make inferences about the reasons for rewards and punishments.

While socialization probably impacts on cognitive development, it is not likely that the socialization of specific cooperative or competitive values is the mechanism through which the requisite cognitive abilities are acquired. That is, while we assume that socialization does influence cognitive development, and vice versa, we do not believe that exposure to pressure to value cooperative or competitive behaviors in and of itself causes the child to acquire the specific cognitive skills discussed earlier. The more likely case is that the development of those types of cognitive skills is the result of interaction with the environment in the broadest sense. There is a vast array of socialization experiences that create momentum for cognitive development to proceed. The socialization pressure to value either cooperative or competitive behaviors is but a small part of this vast array of pressures, and the presence or absence of any small set of specific socialization pressures is likely to be inconsequential for cognitive development.

## V. FUTURE DIRECTIONS

Although there is a substantial data base for understanding human cooperative, competitive, and individualistic behaviors, there are potential research directions that could be fruitful. The following research directions are a few possibilities.

First, although there is considerable research investigating individual differences in cooperative, competitive, and individualistic values and behaviors, there is limited evidence regarding the ecological validity of the measures used in this research. Evidence suggests that the resource distribution preferences assessed in these measures are related to verbal reports of preferred resource distributions and to other measures of social motivation in a theoretically consistent manner. Perhaps the most impressive evidence of the validity of these measures is the significantly better-than-chance agreement rates between the sociometric evaluations of peers and the peers' resource distribution preferences. More future research efforts assessing the ecological validity of these types of measures are needed.

Second, the role of affect, or feelings, in cooperative, competitive, and individualistic preferences has been relatively ignored. There is a rapidly developing literature relating empathy and other affective responses to social cognitions (e.g., perspective-taking and prosocial reasoning) and prosocial behaviors. Empathy might induce prosocial and moral judgments, which might, in turn, impact on behavior. Empathy, sympathy, and personal distress might also mediate the link between cognition and social behavior; conversely, cognition might mediate the link between affect and social behavior. Ultimately, understanding of the development of cooperative, competitive, and individualistic values and behaviors might require future research effort addressing the role of affect in these types of social behaviors.

Third, although the nature of the developmental differences in cooperative, competitive, and individualistic behaviors appears to be reasonably well demonstrated, the exclusive reliance on cross-sectional methods could be problematic in a number of ways. For example, the age differences across the 3- to 12-year age range could reflect developmental changes or artifactually produced differences. That is, in studies sampling children across the 3- to 12-year age range, the 3 to 5 year olds generally are sampled from preschools and day-care centers, whereas the 5 to 12 year olds are generally sampled from elementary schools. Differences beween these age groups could result from development or from any nonequivalence in the representativeness of the two samples. Thus, it is necessary to verify the nature of the developmental pattern with longitudinal methods in which the developmental changes in cooperative, competitive, and individualistic social values and behaviors are observed within individuals as they develop.

Fourth, there have been few direct attempts to investigate the effects of the development of cognitive abilities and the effects of socialization on the development of cooperative, competitive, and individualistic preferences. While there is a small but growing empirical literature on the relationship between cognitive development and cooperative, competitive, and individualistic behaviors, there is little research assessing the empirical relationships between socialization and these behaviors. Thus, it is necessary for future research to investigate these types of causal mechanisms more thoroughly.

## BIBLIOGRAPHY

Dawes, R. M. (1980). Social dilemmas. *Annu. Rev. Psychol.* **31**, 169–193.
Derlega, V. J., and Grzelak, J. (eds.) (1982). "Cooperation and

Helping Behavior: Theories and Research." Academic Press, New York.

Knight, G. P., Bernal, M. E., and Carlo, G. (1995). Socialization and the development of cooperative, competitive, and individualistic behaviors among Mexican American children. *In* "Meeting the Challenge of Linguistic and Cultural Diversity in Early Childhood Education (E. Garcia and B. M. McLaughlin, eds.). Teachers College Press, New York.

Knight, G. P., and Chao, C.-C. (1991). Cooperative, competitive, and individualistic social values among 8–12 year old siblings, friends, and acquaintances. *Personal. Soc. Psychol. Bull.* **17**, 201–211.

MacDonald, K. B. (ed.) (1988). "Sociobiological Perspectives on Human Development." Springer-Verlag, New York.

Trivers, R. (1985). "Social Evolution." Benjamin/Cummings, Menlo Park, CA.

Wilke, H. A., Messick, D. M., and Rutte, C. (eds.) (1986). "Experimental Social Dilemmas." Verlag Peter Lang, New York.

# Behavior Measurement in Psychobiological Research[1]

MARTIN M. KATZ
SCOTT WETZLER
*The University Hospital for the Albert Einstein College of Medicine*

I. Role of Behavior in Biological Research
II. Methods for Measuring Emotion, Behavior, and Cognition
III. Examples of the Multivantaged Approach in Psychopathology

## GLOSSARY

**Behavior** Manner of acting, reacting, functioning, or performing. Usually refers to the action of the individual as a unit. In a more general sense, it refers to the process of perception, thinking, emotion, and action. The science of behavior is the science which investigates the nature of such psychological phenomena across species

**Componential approach** An orientation based on the assumption that all psychological states can be conceived of as patterns of perceptual, thought, emotional, and behavioral components. The components that make up the structural pattern are conceived of as independent, or partially so, and should be measured as separate factors in themselves

**Emotion** A complex and usually strong subjective response such as love or fear. The response involves physiological changes as a preparation for action. The part of the consciousness that involves feeling and sensibility

**Medical model** A model or an orientation based on the assumption that disorders or illnesses are whole entities with distinct configurations of characteristics, qualitatively different from each other. In psychiatry, the unit of measurement is usually the disorder itself and not a single type of behavior or emotion

**Mental disorder** Conceptualized as a clinically significant behavioral or psychological syndrome that occurs in an individual and that typically is associated with either a painful symptom (distress) or impairment in important areas of functioning (disability). In addition, there is an inference that a behavioral, psychological, or biological dysfunction exists and that the disturbance is not confined to a relationship between the individual and society

**Multivantaged approach** An approach to psychological assessment which recognizes that measurement of many of the components of psychological functioning is still at an early stage of development and is subjective in quality and therefore requires more than one perspective. The recommended approach to measurement of such cognitive, behavioral, and emotional components uses self-reports, performance tests, and multiple observers making behavioral ratings in different settings

THE DISCIPLINE OF PSYCHOBIOLOGY IS CONcerned with the interaction of neural processes and psychological functioning and includes studies of the actions of psychotropic drugs on human behavior. Great advances have been made since the mid-1960s in identifying discrete neurotransmitter systems within the central nervous system, and attempts have been made to link their functioning to that of specific emotions such as anxiety. At the same time, basic investigations in psychopharmacology determine how drugs bring about rapid recovery from such complex mental disorders as severe depression. It is necessary in all such research to apply measures of cognition, emotion, and behavior which are as precise and valid

[1]Reprinted from *Encyclopedia of Human Behavior*, Volume 1 Copyright © 1994 by Academic Press, Inc. All rights of reproduction in any form reserved.

ENCYCLOPEDIA OF HUMAN BIOLOGY, Second Edition, VOLUME I.

as those which measure biological factors. This article deals with how psychologists define behavior, how they approach problems of measurement, and various instruments for measuring the relevant components of psychological functioning. Examples are then provided of research in which such methods are applied to the investigation of the interaction of biology, drugs, and behavior.

## I. ROLE OF BEHAVIOR IN BIOLOGICAL RESEARCH

### A. Introduction

Developments in biology since the mid-1960s have radically changed our views of the causes of the major mental disorders and of the bases of normal human behavior. We now have psychotropic drugs which can control and, in some cases, completely resolve such serious, formerly intractable, disorders as mania, severe depression, and schizophrenia. These discoveries have contributed to advances in the understanding of central nervous system functioning, particularly the identification of distinct neurotransmitter systems and their roles in these disorders and normal emotions.

Despite these advances in knowledge about the new drugs, little has been learned about the underlying links between their behavioral actions on the one hand and specific drug–neurochemical interactions on the other. The future of research on the roots of the mental disorders lies with our capacity to identify the links between the functioning of neurobiological systems and human behavior.

### B. "Medical" and "Componential" Models

Given the advances in biological technology, it may appear unusual that more progress has not been made in establishing these links. The fact is, however, that in the field of mental disorder, the major mode of thinking tends to be in terms of *classes* of illness, referred to as the "medical model." This model is based on the assumption that disorders or illnesses are whole entities with distinct configurations of characteristics, qualitatively different from each other, which appear and disappear as whole entities. The unit of measurement, therefore, is the disorder itself, and researchers try to establish links between the entire disorder and neurochemical systems. The unit is

not a single type of behavior or type of emotion. We believe that although useful in the clinical framework, this categorical or holistic model of thinking tends to impede rather than advance research.

Establishing the presence of a specific disorder in a patient is a complex judgmental process based on clinical history and patterns of behavior, emotions, and cognition and therefore makes linking with precisely measured biological factors quite difficult. When an association between a disorder and a biological factor is demonstrated, there is still great difficulty in identifying which of the many aspects that make up the complex disorder are responsible for the relationship.

This model, as noted, is a traditional one in medical research; it has on occasion been successfully applied in psychiatric research and has succeeded even more frequently in other specialities of medicine. If the primary goal is clinical, i.e., discovering new treatments, then it is understandable that medical research takes illness as its point of departure. If, however, the goal is to uncover the specific behavioral mechanisms underlying the disorder or to determine how drugs which prove to be efficacious bring about significant changes in these disorders, then more basic questions are involved. These questions would focus, for example, on uncovering which behaviors, emotions, and cognitions are linked specifically with the functioning of those neurotransmitter systems that are directly affected by these drugs.

We therefore conclude that to advance the science for investigating biological–behavioral interactions in normal and abnormal states, we must apply an alternative or complementary model which we identify as the "componential approach." In this latter strategy, we assume with the medical model that all human psychological states can be conceived of as patterns of perception, thinking, emotion, and behavior. For the purposes of scientific investigation, however, the components that make up this complex structure are conceived of as independent, or partially so, and thus should be measured as separate factors in themselves. The units in this approach are, therefore, components which describe the various aspects of the disordered state and permit their independent measurement. This permits a quantitative approach, resulting in precise and articulated measurement, more suited to the main problems of uncovering the relationships between biological and psychological factors. The next section presents the background of this approach in psychology and then describes a system for accomplishing its aims.

## C. Applying the Componential Approach

The measurement of psychological attributes, many of which are subjective in nature, is difficult, and the field is in a continuing state of development. Psychologists find it useful when considering issues of measurement to catgorize psychological functions such as perception, emotion, cognition, and social behavior.

The area of emotion exemplifies the multiform problems of measurement. Emotions are, by definition, experiential or subjective states—states that only the person experiencing them can "know," even if he or she does not always have the words with which to describe them. Through the use of inventories of descriptors, psychologists try to present the person with the right words, but these terms can only be approximations of what any patient is likely to feel. When the person's own report is used as a vehicle, there are other barriers to attaining accurate measurement of a "feeling" state. It turns out that subjects may have constraints of various types that prevent them from providing a frank picture of their internal state. For example, additional complications arise when dealing with disturbed patients, who may in many cases be unable to indicate how they feel.

Investigators, therefore, seek other ways of assessing emotional states. In most instances of clinical research, they will ask experts to examine the patient and then, on the basis of these observations, to judge the extent to which a given state (e.g., anxiety or depression) exists. The experts in this case are psychiatrists, psychologists, and nurses who see the patients in different interview and ward settings. Psychologists will also use other instruments, a self-report or performance measures, in the course of their studies to assist them in gauging the intensity of a particular affect, such as anxiety.

## D. The Multivantaged Method

The use of several perspectives acknowledges that from not one of these vantages is it possible to obtain a comprehensive and reliable estimate of the depth and extent of an affect or emotional state. Validity and reliability of measurement of the state are therefore enhanced by combining these perspectives. In addition, combining such measurements assumes that the more intensive or severe a state, such as anxiety, the more likely it is that it will be manifested in more situations. It also assumes the converse—that is, to the extent that the state is only mildly present, it will be less likely to appear in all situations and may be detectable only in one or two of those to be described.

In measuring psychological qualities, we also use such terms as "constructs." A construct is an abstraction; it implies that from a physical or objective standpoint the quality itself may not be directly measurable but that there are aspects of it in the subject's overt behavior or in his or her description of that state that appear to occur together and that help to define it. Thus, "anxiety" is something that we cannot actually see but that we believe exists because there are a number of aspects occurring together (e.g., subjective fear, uncomfortable physical feelings, restlessness) that help to define it. From the multiperspective approach, a patient can usually describe the fear, but it requires outsiders to observe the agitation and the other aspects of its physical expression.

Many of the constructs we use in psychopathology are, in addition, overinclusive and thus quite complex. Clinicians, for example, use such terms as "cognitive impairment" to describe serious thinking disorders. In examining aspects of this construct, we note the inclusion of such varying disturbances as difficulties in concentration and in memory, loss of insight, confusion, and poor judgment. Some of these impairments can only be discovered by examining the performance of the patient, others only by expert observation, and still others only by the patient's own report.

Thus, in all of these areas of psychological functioning, we find it necessary to rely on a multiperspective approach to attain both validity and comprehensiveness, since in regard to the latter, certain characteristics of the construct can only be measured by applying a certain type of instrument. Before we indicate how the multivantaged approach is applied, we provide an overview of the types of methods that are used to measure the various aspects of these and other psychological constructs.

# II. METHODS FOR MEASURING EMOTION, BEHAVIOR, AND COGNITION

## A. Self-Report Tests

The subject's or patient's own description of his or her psychological state or attitudinal set is a crucial vantage for the assessment of human behavior. He or she has unique access to his or her internal feelings—states which outside observers can only infer. While this information may be subject to various biases, it is also the most truly "phenomenological" of all vantages. It conveys how the subject views his or her psychological state; the subject is presumed to know his or her feelings and attitudes better than any one else. Thus, self-report tests must be included in any kind of comprehensive assessment approach.

Self-report tests consist of lists of simply worded items or questions to which the subject responds in a true/false (e.g., "I like mechanics magazines"), yes/no (e.g., "Do you enjoy meeting new people?"), or numerical rating (e.g., "How much were you distressed by feeling sad during the past week, rated on a 0–4, not at all to extremely, scale?") format. They may be relatively brief, covering only 15 items (e.g., the Beck Depression Inventory) or as many as 566 items (e.g., the Minnesota Multiphasic Personality Inventory-2). In general, items are then summed together to create dimensional scores. These total raw scores are then converted to *standardized scores* based on normative data. For example, a standardized *T* score of 70 on a depression scale would mean that the subject was 2 standard deviations above the mean (or in the top 3% of the population) and that he or she may be described as being significantly clinically depressed.

There are hundreds of different self-report tests which may be used to measure feelings, behavior, and attitudes in psychobiological research. It is beyond the scope of this article to outline them all or to describe any in great detail. Brief summaries of the most popular tests will give an impression of the wide range of item content. The first decision in test selection is determining whether the aim is to study *normal* or *abnormal behavior*. A single psychological test is rarely able to focus on both kinds of human behavior. One of the most widely used tests of normal behavior is the California Psychological Inventory, which measures socially desirable qualities on 20 scales (e.g., sociability, tolerance, flexibility). Another is the Personality Research Form, which measures 15 personality traits relevant to an individual's functioning (e.g., impulsivity, dominance, nuturance).

Within the domain of abnormal behavior or psychopathology, there are three kinds of tests: (1) psychiatric symptom and mood checklists which measure *transient clinical states,* (2) questionnaires which measure *long-standing personality dysfunction,* and (3) inventories which measure *maladaptive behaviors.*

The most widely used self-rating scale in psychopathological and psychopharmacological research is the Symptom Checklist 90, which measures the distress caused by 90 common psychiatric symptoms scoring on nine dimensions of psychopathology (e.g., depression, anxiety, hostility, psychoticism). Related to the symptom checklists are mood scales which measure disturbed affects. For example, one of the most sensitive indicators of change during a psychotropic drug treatment study is the Profile of Mood States, which measures six affective dimensions (e.g., tension/anxiety, vigor, fatigue). The Symptom Checklist 90 and the Profile of Mood States both provide multidimensional profiles of a patient's clinical status. Other symptom and mood checklists (e.g., the Beck Depression Inventory, the Taylor Manifest Anxiety Scale) may focus on only one critical emotion and thus provide a single unidimensional score.

The Minnesota Multiphasic Personality Inventory is one of the oldest self-report personality questionnaires, measuring 10 dimensions of psychopathology (e.g., hysteria, schizophrenia, hypochondriasis). A revision with new norms has recently been established (MMPI-2). Another personality questionnaire that has been recently developed and is widely popular in both research and clinical settings is the Millon Clinical Multiaxial Inventory-II. This test measures 11 personality disorder (e.g., narcissistic, schizoid) and 9 clinical syndrome (e.g., anxiety, thought disorder) scales. Finally, advances in behavior theory and behavior therapy based on models of conditioning have generated a number of behavioral inventories such as the Attributional Style Questionnaire, which measures a subject's tendency to have a depressive world view (e.g., pessimistic, helpless).

## B. Observational Rating Scales

When the focus in an investigation is on a construct (e.g., anxiety) rather than merely on an individual's own perception of his or her state, then it would be a mistake to rely solely on the self-report vantage. The self-report vantage has intrinsic weaknesses when behavior is at issue, e.g., the patient's potential lack of insight concerning the nature of his or her own behavior. Other vantages (e.g., doctor, nurse, family) may offer more valid information on particular facets of functioning or, at least, offer additional information. Behavior may be observed and rated in a variety of settings. A number of different observers with different qualifications may be enlisted to observe and rate behavior in those settings. The wider the range of settings and the greater the number and kind of observers, the more valid the measurement of that particular construct.

We have identified three kinds of observers with different qualifications who make their ratings in specific settings. Of particular interest are rating scales completed by a doctor based on an interview with the patient (either live or viewed on videotape). These ratings are considered to be from the *"expert's" vantage,* based on the doctor's accumulated clinical knowledge of psychopathology and prior experience with psychiatric patients. However, these interviews are relatively brief (1–2 hr) and therefore are based on little contact with the patient. For psychiatric inpatients, *nurses' ratings* of ward behavior are based on more extended direct observation of the patient's functioning and are made by professionally trained staff. Finally, for psychiatric outpatients, *family ratings* of the patient's behavior in the community are also based on direct observation (although by untrained observers) and are the most ecologically valid of the three kinds of observation. That is to say, observations of behavior during a brief interview or in a psychiatric hospital are less useful predictors of behavior in the community than is the direct observation of behavior in the community. Needless to say, it is much more difficult for professional raters to observe behavior in the community, a task that is delegated to family members.

The most commonly used structured interviews and observational rating scales are the Schedule for Affective Disorders and Schizophrenea (SADS) and the Structured Clinical Interview DSM-III-R (SCID), which generate diagnoses for the major psychiatric syndromes. With proper training, these diagnoses are made in a highly reliable fashion, and ratings define homogeneous patient populations for inclusion in a research study. Another popular observational rating scale is the Hamilton Depression Rating Scale, which provides an example of this type of methodology when applied to a single syndrome. It covers the full range of depressive symptoms and measures the severity of a depressive disorder. One example of a nurses' rating scale is the Global Ward Behavior Scale, which measures general features of a patient's functioning (e.g., agitation, retardation). Finally, the Katz Adjustment Scale is a family rating instrument that is completed by a well-informed source (e.g., parent, sibling, child, significant other) and measures the patient's social behavior and functioning in the community (e.g., withdrawal, belligerence, bizarreness).

## C. Objective Performance Instruments

Both self-report tests and observational rating scales require a subjective evaluation of feelings, attitudes, and behavior. Ideally, we would like to have objective measures of these constructs as well. For example, the objective measurement of galvanic skin response represents one component in the assessment of a construct such as anxiety, or the measurement of finger tapping speed as a component of psychomotor retardation. In general, it is possible to measure an individual's objective performance on a number of cognitive and motor tasks and to draw conclusions about certain components of human behavior. These measures of objective performance may then be integrated with the subjective rating described earlier to provide a fuller picture of each construct.

Significant advances in cognitive and experimental psychology now permit the accurate and objective measurement of many different functions. In addition, a new field has developed, called "neuropsychology," which emphasizes brain–behavior relationships. While the reader is best directed to standard texts in these fields for a more in-depth description of assessment strategies and methods, suffice it to say that neuropsychological tests are commonly used in conjunction with sophisticated radiological techniques for the determination of how brain structure and function are related to specific neuropsychological abilities and functions.

Three multipurpose test batteries have been developed which cover many different intellectual, cognitive, and neuropsychological functions: Wechsler Adult Intelligence Scale—Revised, Halstead–Reitan Battery, and Luria–Nebraska Neuropsychological Battery. In addition to these comprehensive test batteries, there are more than a hundred tests which measure specific cognitive functions. In this brief overview, six areas of cognitive functioning and examples of neuropsychological tests which measure those functions are mentioned: (1) sensation and perception, (2) motor speed and coordination, (3) language, (4) memory, (5) higher integrative functions, and (6) general intellectual functioning. Representative tests include the Speech Sounds Perception Test and the Line Bisection Test for sensation and perception; the Finger Tapping Test and Purdue Pegboard Test for motor speed and coordination; the Western Aphasia Battery for language; the Wechsler Memory Scale—Revised for memory; the Trailmaking Test and Wisconsin Card Sorting Test for higher integrative functions; and the Wechsler Adult Intelligence

Test—Revised for IQ scores and for general intellectual functioning.

## III. EXAMPLES OF THE MULTIVANTAGED APPROACH IN PSYCHOPATHOLOGY

To illustrate how the componential or multivantaged approach is applied in biobehavioral research, we draw illustrative examples from the Collaborative Study of the Psychobiology of Depression sponsored by the National Institute of Mental Health (NIMH). That study was designed to test hypotheses which implicated disturbances in central nervous system chemistry as the basis for depressive disorders. It involved study of a diverse sample of patients and normal controls. In the course of that research, two ancillary issues were addressed:

1. Investigating relationships between neurochemical variables and emotions associated with the disorder of depression
2. Examining the psychological changes brought about by effective tricyclic drugs in the treatment of depression

These issues are of basic importance in understanding the mechanisms underlying serious psychopathology and in determining the ways in which drugs bring about improvement in biological and behavioral pathology.

### A. Associating the Functioning of Neurotransmitter Systems with Specific Emotions or Behaviors

One of the most influential biological theories of the nature of a specific mental disorder in recent years has been the "catecholamine" theory of depression. It postulated a deficiency in the nervous system of certain monoamine neurotransmitters at the central synapses as responsible for depression and an excess as responsible for mania. A great deal of inferential evidence and results from small studies provided early support for this intriguing theory. Later, however, conflicting evidence appeared, some studies reporting higher, others normal concentrations of monoamine metabolites in depressed patients. [*See* Catecholamines and Behavior; Depression.]

It is now accepted that a substantial proportion of severely depressed patients have higher than normal concentrations. Such findings leave open the question of whether the differences in monoamine metabolite levels between depressions and normals are due to a specific emotional state such as depressed mood or anxiety, both closely associated with the disorder, or result directly from the biochemical pathology of the illness itself. Determining the possible relationships among the emotional states of depressed mood, anxiety, and hostility as a function of central monoamine neurotransmitters was an important focus of the NIMH Collaborative Study of the Psychobiology of Depression. The study examined the relationships (before treatment) between behavioral and emotional measures, hypothesized to be associated with the functioning of three brain monoamine neurotransmitters, norepinephrine, dopamine, and serotonin.

Measuring the associations between these emotions and the principal metabolites of these neurotransmitter systems might indicate whether the behavioral differences can be accounted for by the biochemical differences between depressed patients and normals. Such studies also suggest ways in which specific emotions can be traced directly to the pattern of functioning of specific neurotransmitters in the central nervous system.

Adopting the componential approach, it was possible to test several hypotheses in this study concerning the relationships of anxiety and depressed mood to specific neurotransmitters and their metabolites. The study example included 233 subjects (134 depressed, 19 manic patients, and 80 normal controls) from six hospital settings. During an initial 2-week period before the initiation of treatment, 174 had had complete analyses of cerebrospinal fluid metabolites and the set of components measuring behavioral and emotional states. Although the results showed modest relationships between the metabolite (MHPG) from the norepinephrine system with both depressed mood and anxiety, more extensive analyses led to the conclusion that concentrations of this neurotransmitter metabolite were significantly related to an anxiety–agitation factor which included such somatic aspects of that factor as sleep disorder. This relationship, which was highly significant in depressed patients, was also found in urinary MHPG–behavioral analyses conducted within the smaller sample of manic patients whose behavior is, in most respects, quite opposite to that of depression. There is strong reason, therefore, to suggest that the norepinephrine neurotransmitter system may be a neural structure for "anxiety" or alarm states in severely mentally ill patients rather than being specific to "depressive" illness.

The findings also support, as do other results in this study, the suggestion that differences in the functioning of neurotransmitter systems are more likely to be traced to variations in specific emotional states rather than to "whole" disorders. Given these promising results in linking emotions and monoamine neurotransmitter concentrations, it would appear that applying the componential approach to these problems would be a highly effective strategy in developing a sound scientific base for further work.

## B. Evaluating the Effects of Drugs on Behavior and Emotions

One of the most remarkable findings since the mid-1960s in the field of psychopharmacology is the discovery that tricyclic drugs, specifically imipramine and amitriptyline, are highly effective in the treatment of a large proportion of the severe depressive disorders. These disorders, which can extend over months, sometimes years, and were thought to be due mainly to psychological factors, were found in upward of 50% of cases to be resolved or nearly resolved after 4–6 weeks of drug treatment.

Despite many years of study of the clinical and biological mechanisms underlying the process of recovery brought about by these drugs, the basic mechanisms remain obscure. Much has been learned about the process of neurochemical change in the central nervous system effected by the drugs, which precedes recovery; very little is known about the psychological changes that accompany these neurochemical effects. The imbalance of research effort appears to have come about as a result of the assumption that the drugs are specific for the state of depression, making it appear unnecessary to investigate their specific actions on certain of the other affective components known to be present in most depressive disorders, notably anxiety and hostility.

Further, more recent theory about the bases of depression suggests that the core of the illness may lie in certain cognitive attitudes about life rather than in the pattern of emotional functioning. For these reasons, more recent studies have begun to examine the nature of drug effects very early in the treatment of patients in an attempt to identify their specific actions on the emotional, cognitive, and behavioral components of the state of depression.

Recent studies, one of which followed from the already described NIMH Collaborative Program on the Psychobiology of Depression, were able to differentiate between actions of the drug which appear to affect all patients, whether or not they will eventually recover, and those effects which appear to occur in the drug responders only. For example, it was clear from several studies that the sleep problems associated with the depressive disorder are markedly reduced by the drugs in all patients, regardless of whether they eventually get better. Early reductions in anxiety, hostility, and cognitive impairment, however, only occurred in those patients who would later recover within the 4-week treatment period. These changes in the drug responders were effected quite early, before there was much evidence of overall improvement in these patients. Thus, such early effects appear to be the "triggering" actions of the drugs, those that initiate the recovery process.

It is difficult in such studies, however, to distinguish the effects actually due to the drugs from those in the responders which may be due to placebo factors or simply to being in the hospital. Current research is now utilizing new methods and additional control groups to identify those actions that are due specifically to the tricyclic drugs. The findings currently make it appear, for example, that the drugs are more specific to anxiety than to depression, in accord with other clinical research which shows the drugs to be equally effective for generalized anxiety disorders. Such results have many implications for further work in psychopharmacology and for the treatment of mental disorder.

The components of psychopathology, such as anxiety, depression, and hostility, pervade almost all the serious mental disorders. The componential approach permits the linking of behavioral elements with specific changes in the functioning of the critical neurotransmitter systems. The results from drug–behavioral studies than, in conjunction with what is already known about the sequence of drug actions on the functioning of neurotransmitter systems in the central nervous system, will contribute to uncovering the mechanisms through which these drugs bring about their remarkable effects in the depressive disorders. Such studies will then make possible more effective planning in psychopharmacology as investigators in that field seek new drugs for the wide range of serious mental disorders for which we still do not have effective treatments.

## BIBLIOGRAPHY

Eysenck, H. J. (1973). "Handbook of Abnormal Psychology," 2nd Ed. Pitman, London.

Hersen, M., Kazdin, A. E., and Bellak, A. S. (1991). "The Clinical Psychology Handbook," 2nd Ed. Pergamon Press, New York.

Katz, M. M. (1987). The multivantaged approach to the measurement of affect and behavior in depression. *In* "The Measurement of Depression" (A. J. Marsella, R. M. A. HIrshfeld, and M. M. Katz, eds.), pp. 297–316. Guilford, New York.

Lezak, M. (1983). "Neuropsychological Assessment," 2nd Ed. Oxford University Press, New York.

Maas, J. W., Koslow, S., Davis, J., Katz, M. M., Mendels, J.,

Robins, E., Stokes, P., and Bowden, C. (1980). Biological component of the NIMH Clinical Research Branch Collaborative Program in the Psychobiology of Depression. I. Background theoretical considerations. *Psychol. Med.* **10**, 759–776.

Mitchell, J. V. (ed.) (1985). "Ninth Mental Measurements Yearbook." Buros Institute Mental Measurements, Highland Park, NJ.

Wetzler, S., and Katz, M. M. (eds.) (1989). "Contemporary Approaches to Psychological Assessment." Brunner/Mazel, New York.

# Bile Acids

NORMAN B. JAVITT
*New York University Medical Center*

## GLOSSARY

**Amphipathic molecule** A molecular structure with the dual property of having a region that is hydrophilic and a region that is hydrophobic. For bile acids, the methyl groups are a hydrophobic region and the hydroxyl groups impart hydrophilicity

**Cytochrome P450** A very large group of proteins that subserve the function of enzyme catalyzed biologic oxidations that have in common a strong absorption band at 450 nm after exposure to carbon monoxide

**Enterohepatic** Recycling of compounds between the intestines (enteron) and the liver governed by the existence of specific transport processes in both organs

**Lipolysis** Splitting of the ester bonds of triglycerides to yield fatty acids by enzymes referred to as lipases

**Micelle** Property of molecules in free solution to begin to form soluble aggregates as their concentration increases (critical micellar concentration). Bile acid aggregates or micelles can solubilize cholesterol to form a mixed micelle

**Stereospecific** A fixed spatial arrangement of the consituent atoms of a molecule in relation to each other

**Steroid** A wide variety of chemical species that have in common a core structure containing four carbon rings that are always joined together in a structural unit referred to as cyclopentanoperhydrophenanthrene

**Triglyceride** Chemical name for a molecule with three fatty acid molecules attached to one glycerol molecule via an ester bond; commonly referred to as a fat and classified as a type of lipid

BILE ACIDS ARE THE MAJOR END PRODUCTS OF cholesterol metabolism in humans. They should more properly be referred to as steroid acids to distinguish them from the steroid hormones, which are also derived from cholesterol. However, because they were first discovered in gallbladder bile, they became known as bile acids before their origin from cholesterol was known. [*See* Cholesterol.]

For many years it was thought that bile acid synthesis began and ended in the liver, culminating with their secretion into bile. However, it is now known that two different metabolic pathways exist: one originates in the liver with the $7\alpha$-hydroxylation of cholesterol and the other commences with the 27-hydroxylation of cholesterol in various tissues throughout the body. The steroid intermediates produced in these tissues enter the bloodstream and are transported to the liver, where they are removed from the circulation and further metabolized to bile acids. These bile acids mix with those formed directly from cholesterol in the liver and, after being further processed to glycine or taurine conjugates, are excreted into the bile and stored in the gallbladder between meals. [*See* Liver.]

Ingestion of foods, particularly those containing fat, causes the release of cholecystokinin, a hormone synthesized and stored in the duodenum; it stimulates contraction of the smooth muscle in the wall of the gallbladder and entry into the intestines of the bile, which contains several grams of conjugated bile acids. The amphipathic structure of the bile acids gives them detergent properties which facilitate both the hydrolysis of triglyceride to monoglycerides and fatty acids and their absorption in the jejunum or midportion of the small intestines. After the fat in the meal has been

ENCYCLOPEDIA OF HUMAN BIOLOGY, Second Edition, VOLUME 1.   Copyright © 1997 by Academic Press.   All rights of reproduction in any form reserved.

hydrolyzed and absorbed, the bile acids are reabsorbed in the terminal portion of the small intestines referred to as the ileum. Active transport processes in the liver and in the ileum maintain a highly efficient episodic enterohepatic cycling of bile acids initiated by the ingestion of food. Less than 10% of this recirculating bile acid pool either bypasses the liver to enter the general circulation or escapes ileal reabsorption to undergo metabolism by bacteria in the large intestines and excretion as fecal bile acids.

## I. HISTORICAL BACKGROUND

Because of its bitter taste and deep yellow color, bile has been recognized as a unique fluid since antiquity and was classified as one of the four humors by the ancient Greeks. The detergent properties of bile were recognized and utilized for washing clothes long before any scientific understanding of its physiologic function was developed. An even longer time interval elapsed before the chemical composition of bile acids became known.

In the course of the studies that he did with his patient Alexis St. Martin in the 1830s, William Beaumont noted that bile was specifically useful for assisting the digestion of oily food, thus focusing on its specific role in fat digestion. By the end of the 19th century, the term bile acid had been introduced and some knowledge of its acidic properties had been obtained. However, it remained for Rosenheim and King in the 1930s to work out the steroid ring structure of cholesterol and bile acids and for Konrad Bloch, with his mentor Rudolph Schoenheimer, to develop tracer methodology and to establish the precursor–product relationship between cholesterol and bile acids.

## II. BIOSYNTHESIS OF BILE ACIDS

### A. Initial Hydroxylations

Once the precursor–product relationship between cholesterol and bile acids had been established, the next major advance was to determine the sequence of events by which the stepwise metabolic transformations occurred. Initially it was thought that only a single enzyme, referred to as cholesterol $7\alpha$-hydroxylase, initiated bile acid synthesis by the stereospecific oxidation of the $C_7$ carbon atom in the sterol ring

of cholesterol (Fig. 1). However, exceptions to this paradigm eventually led to the recognition of a second initial site of oxidation, the stereospecific $C_{27}$ atom at the end of the cholesterol side chain. Of the total 27 carbon atoms that comprise a cholesterol molecule, it is only at these two sites, -$C_7$ and -$C_{27}$, that initial modifications lead to bile acid synthesis.

The two enzymes that catalyze these initial oxidation steps, cholesterol $7\alpha$-hydroxylase and cholesterol 27-hydroxylase, can be classified together as belonging to the large multigene family of cytochrome P450 enzymes but differ in many other respects. In addition to their different substrate specificities, the genes coding for their synthesis in humans are located on different chromosomes. Other differences are the subcellu-

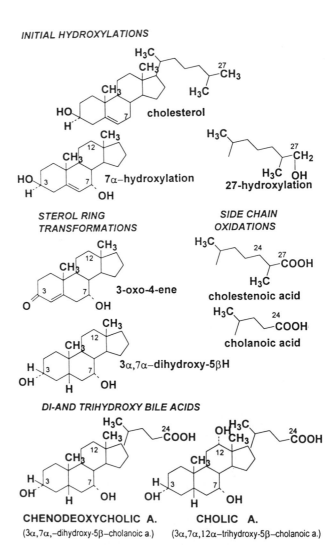

**FIGURE I** Biosynthesis of bile acids from cholesterol.

lar locations of the enzyme, the tissue distribution, and the substrate specificities.

The C-7α-hydroxylating enzyme is located subcellularly in the endoplasmic reticulum and requires different cofactors than the C-27-hydroxylating enzyme, which is in the mitochondria. Perhaps the greatest differences in the two enzymes are (1) the very narrow substrate specificity for the 7α-hydroxylase which is limited only to cholesterol and a few other closely related sterols and (2) its expression only in hepatocytes, the major cell type of the liver. In contrast, the C-27-hydroxylase is found in many tissues such as the vascular endothelium lining of blood vessels and the macrophages that travel in the bloodstream as monocytes, oxidizes a wide variety of C-27 sterols, and is also essential for the 25-hydroxylation of vitamin D.

The full biological significance of why these two very different enzymes have in common a pathway for bile acid production is not yet fully understood. The 27-hydroxylation pathway is active in fetal life, remains active after birth, and reduces the accumulation of cholesterol in tissues. In contrast, the activity of the 7α-hydroxylase pathway in the liver becomes detectable after birth and rises rapidly when the newborn begins to nurse. Thus, although acting at different sites, both enzymes initiate a sequence of events that decrease the accumulation of cholesterol in the body by metabolism to bile acids.

## B. Further Metabolic Transformations

In the liver, the 7α-hydroxylation entry step is followed by a series of very rapid, enzymatically catalyzed reactions that yield a final product within seconds (Fig. 1). Key to these transformations are the stereospecific conversion of the steroid ring from a cholest-5-ene to the 5β-cholane configuration and the further oxidation of the side chain to a C-24 carboxylic acid. Normally the two final products are chenodeoxycholic acid and cholic acid; the only difference between these two major bile acids is the presence of an additional hydroxyl group at C-12 in cholic acid, making it a trihydroxy bile acid, whereas chenodeoxycholic acid is a dihydroxy bile acid.

In nonhepatic tissues, the initial 27-hydroxylation step is followed by either further oxidation to the C-27 acid or 7α-hydroxylation by a novel enzyme different from cholesterol 7α-hydroxylase. Because of these multiple possibilities, a variety of sterol and acidic intermediates normally circulate in plasma and are extracted by the liver. These intermediates, by a series of analogous enzymatic reactions, also yield chenodeoxycholic and cholic acids. In addition, intermediates that are not 7α-hydroxylated are further metabolized to a unique group of monohydroxy bile acids.

Because of the rapidity at which all the enzymatic reactions normally occur, the concentration of all the intermediates in the hepatocytes is very low and the final products are rapidly transferred into bile. When any of these enzymatic steps are blocked, the intermediates accumulate to much higher levels in the cells and follow alternate pathways for their excretion.

## C. Conjugation of Bile Acids

Prior to their excretion into bile, an additional reaction, in which the carboxylic acid group of the bile acids condenses with the amino group of either glycine or taurine, takes place. This enzymatically catalyzed reaction splits out a molecule of water to form a peptide-like bond, and the glyco- or tauro- adduct is termed a conjugated bile acid. More than 95% of the bile acids secreted by the hepatocyte are conjugated bile acids.

## D. Esterification and Glucuronidation of Bile Acids

Other derivatives of bile acids also occur but normally do not account for more than a few percent of the total amount. Either a sulfate group or glucuronic acid may be joined to one of the hydroxyl groups on the steroid ring to form an ester-type linkage. These reactions are also enzymatically catalyzed and most typically occur with the hydroxyl group attached to C-3 that is common to all bile acids. Virtually all monohydroxy bile acids form a C-3-ester sulfate or, to a lesser extent, a glucuronide. Esterification of dihydroxy or trihydroxy bile acids represents only a very minor fraction in bile; however, similar esterifications also occur in the kidney and provide a much higher proportion of these types of esters in urine.

## III. BILE ACID SECRETION AND BILE FLOW

Hepatic bile is the secretory product of the liver and the bile acids are the major organic constituents. The secretion of bile acids by hepatocytes into the canalicular system of the liver initiates and maintains bile flow. The small canals merge into larger and larger channels to form ductules and ducts, until finally the two major hepatic ducts that drain the two major lobes of the liver join to form the common bile duct. The secretory process is classified as active because energy is required to develop the high concentration of conjugated bile acids that occur in bile, several hundredfold greater than in the hepatocyte.

Specialized carriers, located in the surface of the hepatocyte that is modified to form a canaliculus, are essential for developing the concentration gradient between canalicular fluid and the surrounding cells. The transport system is often characterized as an organic anion transport system. The carriers for bile acids are different than the carriers for other organic anions that are secreted by the hepatocyte, and no other organic anions attain the high concentration of the bile acids.

## IV. GALLBLADDER BILE AND CHOLESTEROL GALLSTONES

Between meals most of the continually flowing hepatic bile enters the gallbladder rather than flowing directly into the intestines. The gallbladder, an organ lined with epithelial cells that continually extract inorganic salts together with water from the entering hepatic bile, further increases the concentration of the conjugated bile acids. In this manner, most of the hepatic bile formed between meals can be stored as a highly concentrated fluid (Table I).

The major function of this highly concentrated solution of conjugated bile acids is to maintain the solubility of cholesterol in bile. Phospholipids, such as lecithin which are also present in bile, participate in maintaining cholesterol solubility. The complex that is formed is sometimes termed a mixed micellar solution, although other complex associations occur. The continual entry of hepatic bile and continual reabsorption of inorganic salts and water maintain these complex associations in a state of flux. Periodic emptying of most of the gallbladder contents with meals is important in preventing the formation of insoluble aggregates containing cholesterol crystals. Only in humans and a few nonhuman primates does the concentration of cholesterol attain levels producing a supersaturated solution that is unstable, leading to crystal formation, aggregation, and gallstone formation.

**TABLE I**
Major Components of Human Bile

| | Hepatic (mmol/liter) | Gallbladder mmol/liter | Percentage |
|---|---|---|---|
| Conjugated bile acids total | 13–25 | 75–95 | |
| Glycocholate | | | 31 |
| Glycochenodeoxy-cholate | | | 28 |
| Glycodeoxycholate | | | 17 |
| Taurocholate | | | 10 |
| Taurochenodeoxy-cholate | | | 9 |
| Taurodeoxycholate | | | 5 |
| Phospholipids | 3.2–5.6 | 26.0–33.1 | |
| Cholesterol | 3.6–4.6 | 10.1–10.4 | |
| Bilirubin | 0.6–1.1 | 3.6–5.1 | |
| Inorganic salts | | | |
| Sodium | 146–149 | 179–209 | |
| Chloride | 100–117 | 59–66 | |
| Bicarbonate | 25–30 | 17–19 | |
| Potassium | 4.7–4.8 | 12.8–24.5 | |
| Calcium | 2.6–5.6 | 3.7–10.8 | |

## V. FAT DIGESTION AND ABSORPTION

Upon release of the hormone cholecystokinin from the walls of the small intestines in response to the entry of food, the smooth muscle in the wall of the gallbladder contracts and squeezes its contents into the intestines. The system is designed to add several grams of conjugated bile acids to the meal that is entering the first section of the small intestines, the duodenum.

### A. Lipolysis

The bile acids have detergent properties which means that they will interact with fat and other water-insoluble molecules to form a complex that interfaces with water. To describe this detergent property sometimes the terms amphipathic or emulsifier are used. The former term refers to any molecule that has both a surface that interacts with water and another surface that interacts with lipid, thus creating a close connection between the aqueous and the lipid phases. The latter term refers to the actual process that forms a dispersion between bile acids and lipids. It accelerates the rate at which lipase, an enzyme secreted by the pancreas, hydrolyzes triglyceride to monoglycerides and fatty acids. [*See* Fatty Acids.]

## B. Absorption

The conjugated bile acids coat the intestinal epithelial lining and thus raise the solubility of the monoglycerides and fatty acids, greatly accelerating their absorption across the intestinal mucosa. To maximize their effectiveness, most of the conjugated bile acids are not absorbed until they reach the last part of the small intestines, termed the ileum, where special transporters in the surface of the cells lining the ileum exist. In this segment the bile acids are rapidly transferred into the blood flowing through the intestines and are returned to the liver where another special group of transport molecules remove them from the blood and transfer them to the hepatocyte.

## VI. ENTEROHEPATIC CIRCULATION

The presence of specialized active transport systems both in the surface of the liver cells that face the plasma and in the lumenal surface of the cells lining the ileum limits the bile acids to the liver and the intestines rather than their being generally distributed in all tissues. Because of the high efficiency of the system, the bile acids are very rapidly recirculated and several grams of bile acids effectively function as a much larger amount.

Characteristically, diseases of the liver or of the ileum lead to a loss of efficiency of the enterohepatic circulation with very characteristic consequences. Diseases of the ileum are associated with an increased amount of bile acids entering the large intestines, characteristically causing watery diarrhea. Diseases of the liver cause an increase in the amount of bile acids in plasma and other tissues and are associated with itching of the skin.

## VII. INTESTINAL BILE ACID METABOLISM

Bacteria are normally present in both the small and the large intestines. The small intestine has relatively few bacterial populations, and the enzymes that they produce catalyze mostly the hydrolysis of the conjugated bile acids. These nonconjugated bile acids, sometimes referred to as "free" bile acids, can be absorbed to a greater extent than the conjugated bile acids before they reach the ileum.

The large intestines contain bacterial populations that modify the steroid ring. A large variety of bacterial transformations occur, giving rise to more than 50 species of bile acids. However, the major type of transformation, accounting for the predominant species of bile acids found in feces, focuses on the $7\alpha$-hydroxyl group. Removal of this group results in the formation of deoxycholic acid from cholic acid and of lithocholic acid from chenodeoxycholic acid. Deoxycholic acid is reabsorbed from the intestines and joins the general pool of intestinal bile acids that are returned to the liver for processing to conjugated bile acids. Lithocholic acid is very insoluble in water and binds to fecal residues so that very little enters the recirculating bile acid pool.

Oxidation of the $7\alpha$-hydroxyl to a keto group yields an intermediate that is also recirculated to the liver, where it is reduced to a 7-hydroxylated bile acid. This reduction step is not stereospecific and therefore gives rise to a new species of $7\beta$-hydroxylated bile acid, referred to as ursodeoxycholic acid, which comprises a portion of the conjugated bile acids normally found in bile.

The terms primary, secondary, and tertiary bile acids are sometimes used respectively to distinguish those formed initially from cholesterol, from those formed from the primary bile acids by intestinal bacterial flora, and those formed from secondary bile acids by further processing in the liver.

## VII. INBORN ERRORS OF BILE ACID SYNTHESIS

It has been recognized for many years that certain abnormalities in bile acid synthesis, which often present with jaundice in the newborn period, have a familial basis, implying a genetic basis. The structure of the bile acids peculiar to each disease has been identified, which has led to the identification of the deficient enzymatic step.

### A. 27-Hydroxycholesterol 7α-Hydroxylase Deficiency

Lack of this enzymatic activity results in the production of excessive amounts of monohydroxy bile acids via the cholesterol 27-hydroxylase pathway. Monohydroxy bile acids in excess diminish bile flow in contrast to di- and trihydroxy bile acids which augment bile flow.

## B. 3$\beta$-Hydroxy-C$_{27}$-Steroid Dehydrogenase/Isomerase Deficiency

Lack of this enzymatic activity results in the accumulation of bile acids that have the same chol-5-en ring structure as cholesterol.

## C. $\Delta^4$-3-Oxosteroid 5$\beta$-Reductase Deficiency

Lack of this enzyme results in the accumulation of 3-oxo-bile acids with a sterol ring structure intermediate between the chol-5-ene structure of cholesterol and the normal 5$\beta$-cholane configuration of bile acids.

## D. Cholesterol 27-Hydroxylase Deficiency

In contrast to the enzyme deficiencies just listed, this deficiency is not associated with liver disease. However, because of the wide normal tissue distribution of the enzyme, a profound metabolic disorder referred to phenotypically as cerebrotendinous xanthomatosis occurs. Cholesterol metabolism is profoundly disturbed both in the central nervous system and in the vascular system, leading to both neurologic disturbances and premature atherosclerosis.

## E. Peroxisomal Disorders

The peroxisomes are subcellular organelles that accumulate many different enzymes, particularly those that catalyze the oxidation of long chain fatty acids and the oxidation of C$_{27}$ to C$_{24}$ bile acids. An increase in the proportion of C$_{27}$ bile acids is often the hallmark of a peroxisomal disorder and this finding can be helpful diagnostically. However, the severe metabolic disturbances that occur are unrelated to the changes in bile acid metabolism.

## BIBLIOGRAPHY

Anderson, K. A., and Javitt, N. B. (1974). Bile formation. *In* "The Liver" (F. F. Becker, ed.). Dekker, New York.

Bergstrom, S., and Danielsson, H. (1968). Formation and metabolism of bile acids. *In* "Handbook of Physiology" (C. F. Code and W. Heidel, eds.). Waverly Press, Baltimore, MD.

Coon, M. J., and Koop, D. R. (1983). P-450 oxygenases in lipid transformation. *In* "The Enzymes" (P. D. Boyer, ed.). Academic Press, New York.

Javitt, N. B. (1994). Bile acid synthesis from cholesterol: Regulatory and auxiliary pathways. *FASEB J.* **8,** 1308.

Jones, M. N., and Chapman, D. (1995). "Micelles, Monolayers, and Biomembranes." Wiley-Liss, New York.

Reichen, J., and Paumgartner, G. (1980). Excretory function of the liver. *In* "Liver and Biliary Tract Physiology I (N. B. Javitt, ed.). University Park Press, Baltimore, MD.

Setchell, K. D. R., and O'Connell, N. C. (1994). Inborn errors of bile acid metabolism. *In* "Liver Disease in Children" (F. J. Suchy, ed.). Mosby, St. Louis.

# Binocular Vision and Space Perception

CLIFTON SCHOR

*University of California, Berkeley*

---

## GLOSSARY

**Accommodation** Change in focal length or optical power of the eye produced by change in power of the crystalline lens as a result of contraction of the ciliary muscle

**Baseline** Line intersecting the entrance pupils of the two eyes

**Binocular fusion** Act or process of integrating percepts of retinal images formed in the two eyes into a single combined percept

**Binocular parallax** Angle subtended by an object point at the nodal points of the two eyes

**Binocular rivalry** Temporal alternation of perception of portions of each eye's visual field when the eyes are simultaneously presented with targets composed of dissimilar colors or different contour orientations

**Convergence** Inward rotation of the two eyes

**Divergence** Outward rotation of the two eyes

**Egocentric direction** Body-referenced direction

**Exocentric direction** Direction with reference to an external object

**Fronto-parallel plane** Plane that is parallel to the baseline and orthogonal to the primary position of gaze

**Haplopia** Perception of a single target by the two eyes in an identical visual direction (single vision)

**Horizon line** Line connecting two or more vanishing points projected within the retinal image from a single object plane or surface

**Horopter** Locus of points in space whose images are formed on corresponding points on the two retinas

**Midsagittal plane** Plane that is perpendicular to and bisects the baseline

**Motion parallax** Apparent relative displacement or motion of one object or texture with respect to another that is usuallly produced by two successive views by a moving observer of stationary objects at different distances

**Motor inflow** Information about eye position obtained from stretch receptors of the extraocular muscles

**Motor outflow** Information about eye position obtained from innervational control of the extraocular muscles

**Object motion** Motion of an external object or feature

**Oculocentric direction** Direction relative to the visual axis of the eye

**Optic flow** Pattern of retinal image movement

**Perceptual invariance** Constant perception of rigid objects in three-dimensional (3-D) space viewed from various vantage points which yield variable two-dimensional (2-D) retinal image transforms

**Perspective** Variation of perceived size, separation, and orientation of objects in 3-D space from a particular viewing distance and vantage point

**Primary position of gaze** Position of the eye when the visual axis is directed straight ahead and is perpendicular to the fronto-parallel plane

**Principal visual direction** Line of sight which serves as a reference for oculocentric (retinal based) direction

**Retinal disparity** Difference in the angles formed by two targets with the entrance pupils of the eyes

**Shear** Differential displacement during motion parallax of texture elements along an axis perpendicular to the meridian of motion

**Station point** Location in space or position from which an observer views a 2-D representation or view plane of a 3-D spatial transformation

**Tilt** Amount of angular rotation in depth about an axis in the fronto-parallel plane—synonymous with orientation

**Unconscious inference** Assumption about the organization of physical space that is utilized without conscious awareness in perception but that is not necessarily represented in the 2-D retinal image

ENCYCLOPEDIA OF HUMAN BIOLOGY, Second Edition, VOLUME 1.   Copyright © 1997 by Academic Press.   All rights of reproduction in any form reserved.

**Vanishing point** The point in the 2-D projection onto the retinal image plane where two parallel lines in 3-D space appear to intersect one another

**Vantage point** Position in space of the entrance pupil through which 3-D space is transformed by an optical system to a 2-D image or view plane

**Vernier acuity** Form of hyperacuity in which the alignment of two targets is judged

**Visual capture** Utilization of a rigid-invariant stimulus in 3-D space as a frame of reference for space perception

**Visual plane** Any plane containing the fixation point and entrance pupils of the eyes

SPACE PERCEPTION INCLUDES THE PERCEPTION OF direction, distance, orientation, and shape of objects. If temporal variations are considered, the topic also encompasses motion perception. These perceptions arise from several sensory modalities, predominantly visual, auditory, vestibular, and tactile. Each of these sensory modalities yields perceptions of a three-dimensional (3-D) space. The topic of binocular vision usually accompanies discussions of space perception because it provides a highly reliable and extremely sensitive percept of the third dimension, depth. However, it does not provide the sole source of depth information for vision. A wealth of monocular "artist's" cues for depth account for our ability to continue perceiving a 3-D world when one eye is occluded. In some cases, these cues can override the binocular information. This article is restricted to visual aspects of space perception with the omission of object motion perception. [*See* Perception.]

## I. PARAMETERS OF 3-D PERCEPTION (DIRECTION ORIENTATION, DISTANCE)

### A. Optical Transformation of 3-D to 2-D Space

As mobile creatures, we are constantly observing our environment from different or changing vantage points. Each of these vantage points is associated with a unique projection of a three-dimensional world upon a two-dimensional surface, the retina. Yet the world appears nearly the same from all vantage points, despite the many different transformations of the retinal image. This perceptual invariance or stability illustrates that our perceptions are not unique to the vantage point of a single observer, but are largely independent of our specific position in space at any given moment in time.

### B. Unconscious Inference

Because the retinal image is a 2-D transformation from a 3-D space, there is some loss of information about the 3-D object. In order to perceive a 3-D space we must regain that information either by comparing successive views from different vantage pionts or by making assumptions about the object that are not based on information contained in the retinal image (unconscious inference). For example, we usually assume that shading and shadows result from a light source located above and not below an object. This inferential theory considers perception to be an intelligent process that reasons from experience. A variation of this theme considers interpretations of the 2-D image to arise from an evolutionary process whereby an organism develops automatic inferences that are based on the physical laws that govern our world. For space, these laws are described using Euclidean geometry. Some general examples include observations that the size of the retinal image increases with proximity, that an object that appears to be partially obscured by another object must be farther away than overlapping unoccluded objects, and that orientation variations occur with shape variations. Borders of objects occur at discontinuities of surface texture, size, motion, disparity, and color. Some of these assumptions are more reliable than others; for example, overlap is a more consistent cue than variations in surface texture density and, in our perceptual judgments of distance, these cues are weighted accordingly.

### C. Stimulus Covariance

Another major problem in space perception is the discrimination between changes in the 2-D transformed retinal image that are produced by changes in our position and changes in shape and motion of a nonrigid object. For example, the same retinal image motion occurs when the eyes remain stationary and look directly at the side of a moving train as when walking beside a stationary train. This ambiguity is resolved by knowledge of our own eye and body movements. Sensed motion of the head by the vestibular organs in the middle ear as well as innervation controlling eye movements (motor outflow) is correlated with ongoing transformations of the retinal image. The observer anticipates transformations of the retinal image due to eye and body motion. When these predictions are not realized, the object appears to move and change shape and orientation. When the predictions are realized, all of the transformation is attributed to self (body) motion rather than object motion.

## D. Frame of Reference

Some objects in our environment are assumed to remain invariant, such as the sky and horizon, or the vertical orientation of walls in a building. These highly reliable objects form a frame of reference that can always be considered as immobile and invariant in shape and orientation. Transforms of the retinal image of these objects are always interpreted as due to body or self motion (visual capture). Under certain unusual circumstances, such as being in a boat or watching a wide-screen movie, motion of the surrounding field imparts a false sense of body motion known as vection. This illusion can produce some of the discomfort reported in motion sickness when there is a conflict in visual and vestibular sense of body motion.

## II. PERCEIVED DIRECTION

### A. Coordinate Systems

Resolving the ambiguous translation of the retinal image into components of object motion and self motion requires additional knowledge about eye motion. Image motion is described with reference to the retina as oculocentric. Images are mapped onto a polar coordinate system with the fovea as its origin. This oculocentric map is represented throughout the visual system in the brain centers through which the retinal image is processed. These centers include the lateral geniculate nucleus, visual cortex, frontal eye fields, and superior colliculus. The map is used primarily to describe the correspondence between areas of these sites and those of the retina. Whenever the eye or an object in space moves, the retinal image of the object also moves. When both the eye and object move, the component of the retinal image motion belonging to the object can be derived either by subtraction of eye movements from retinal image motion or by a stimulus covariance function that anticipates the retinal image motion produced by eye motion and residual image motion corresponding to the object.

## B. Role of Eye Movements

There are two potential sources of eye position information that could be used to perceive object motion. One is the sensed movement of the eye (motor inflow, reafference) and the other is the sensed innervation to move the eye that Hering and Helmholtz referred to as the "effort of will" (motor outflow, efference copy). Additional information, either from successive egocentric views or from inference or assumption, is required to perceive objects with reference to objects other than the observer's body; an example requiring such information is map reading (exocentric direction). [*See* Eye Movements.]

## C. Resolution

Resolution limits of perceived direction are task dependent and range from several degrees of arc when judging the absolute straight-ahead position of a single target to several seconds of arc when judging differences in direction or relative position of several targets. This latter category comes under the heading of hyperacuity because the visual system is capable of detecting a spatial offset smaller than the diameter of a single light-detecting cell on the retina (photoreceptor). Two such hyperacuity tasks are judging the vertical alignment of two lines (vernier acuity) and judging the midpoint in the gap between two lines (bisection acuity). Both of these are far superior to the standard gap resolution acuity in which a space between two lines is detected, as in standard clinical tests of visual acuity, for example.

## III. ORIENTATION SENSE

### A. Tilt and Slant

Orientation and direction are two closely related properties of space perception. As just described, we normally think of direction in terms of horizontal and vertical angular deviations from the line of sight. Surface tilt is described as some pivot or rotation of the surface about an axis contained in the frontoparallel plane (FPP). The complete description of tilt requires information about the axis of rotation and the magnitude of rotation. If FPP is used as a reference, all tilted planes will intersect the FPP along a line that defines the axis of surface rotation (see Fig. 1). The amount the surface plane is rotated away from the FPP about that axis is referred to as tilt. Both the

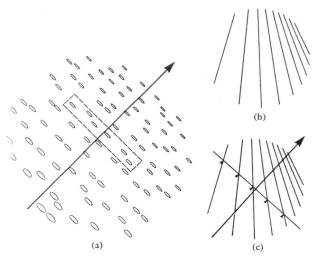

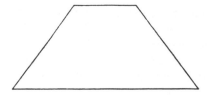

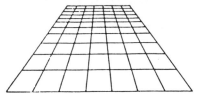

**FIGURE 1** The tilt of a surface is the direction in which it is slanted away from the viewer. (a) If the surface bears a uniform texture, the projection of the axis of tilt in the image indicates the direction in which the local density of the textures varies most, or, equivalently, it is perpendicular to the direction in which the texture elements are most uniformly distributed (dotted rectangle). Either technique can be used to recover the tilt axis, as illustrated. Interestingly, however, the tilt axis in situations like (b) can probably be recovered most accurately by using the second method, i.e., searching for the line that is intersected by the perspective lines at equal intervals. This method is illustrated in (c). (Reprinted by permission from K. Stevens (1979). "Surface Perception from Local Analysis of Texture and Contour." Ph.D. thesis, Department of Electrical Engineering and Computer Science, Massachusetts Institute of Technology.)

**FIGURE 2** Outline and textured backgrounds used by Epstein to test the optical adjacency explanation of the relation between relative height and perceived distance.

amount of tilt and its axis can be sensed globally from the perspective of the complete (global) outline of an object, whereas the axis of tilt but not its amount can be sensed from local surface texture. Specific portions of the retinal image can be utilized for perception in isolation (local analysis) or in context with additional information at other retinal locations (global analysis).

## B. Global Cues

Perspective is referred to here as global because the entire edge of a target is extrapolated until it intersects another extrapolated edge of the same target (Fig. 2). When these edges are parallel, they meet at a point in the retinal image plane called the vanishing point (VP). Normally we do not perceive the vanishing point directly because it needs to be derived by extrapolation. An intrinsic assumption is that all extrapolated lines that intersect a common VP are parallel to one

another. A powerful assumption is that the line of sight directed at a given VP is also parallel to the edges of a rectangular surface, such as a roadway, that extrapolate to the same VP. By sensing the direction of our eye while fixating the VP, it is possible to acquire egocentric information about the direction of a road. Three separate vanishing points can be derived from the three sets of parallel lines forming the edges of a cube or even a rhomboid. The axis of tilt of the surface containing the rectangle or rhomboid is the line connecting two of the vanishing points (horizon line).

The ground plane, and all planes parallel to it, forms a horizontal horizon line at eye level. Planes tilted about a horizontal axis will form a horizon line that is either above or below the horizon by an amount proportional to the amount of slant. Vertical planes tilted about vertical axes have vertical horizon lines that are displaced laterally from the midsagittal plane by the amount of slant. The amount of slant equals the angle subtended at the eye by the shortest line segment between the horizon line and primary position of gaze. Thus, information derived from vanishing points and horizon lines can be used by the brain to compute both shape and slant of surfaces that have rhomboid outlines.

## C. Local Analysis

Local analysis of direction and orientation relies heavily on two assumptions about surface texture: (1) shape and size of texture elements remain uniform across the scene, and (2) uniform spacing between texture elements. Normally neither of these assumptions can be met for an entire scene. However, they may be met over restricted ranges within a scene. For example, the foliage on trees is statistically uniform in size and evenly spaced. Surface texture can be used to analyze surface shape, direction, and orientation within the bounds of the foliage.

The local analysis requires that two independent factors be isolated. These factors are overall magnification or scaling of texture in relation to distance of the observer from the object and the slant of the tilted surface with respect to the FPP. The perceived size of texture elements, as well as the spacing between them, varies inversely with the viewing distance. Tilt will produce a compression of the texture elements that is greatest along the axis perpendicular to the horizon line (axis of tilt).

The most effective texture for characterizing slant, which causes compression within the retinal image, is a set of evenly spaced lines parallel to the axis of tilt. The lines in this set are referred to as the transversals, as opposed to perspective lines that are perpendicular to the horizon line (orthogonals) (Fig. 2). The implied assumption is that the image spacing between transversals decreases proportionally with the square of target distance. For example, consider the texture compression distribution of a checkerboard floor pattern with tile edges on the orthogonal and transversal axes (Fig. 2). As viewing distance increases, the spacing between orthogonal lines will be scaled down or compressed proportionally with viewing distance, but spacing between transversal lines will be compressed with the square of viewing distance. This will produce a shape distortion whereby tiles under the observers feet will appear square, whereas tiles that are farther away will become progressively shrunken trapezoids. There are two local ways of finding the axis of tilt using texture. One way is to find the line that is perpendicular to the axis of maximum texture compression (Fig. 1a). The second way is to identify the transversal axis directly as the axis in which the images of texture elements remain evenly spaced (Figs. 1b and 1c).

Tilt angle can be quantified from measures of the relative rate of change or gradient of compression along the orthogonal axis (i.e., axis in the meridian of slant) as depicted by the arrow in Fig. 1a. However the eye is much better at computing the amount of slant with global information, such as outline perspective, than with local information, such as variations in texture density. In contrast, local variations in texture provide a robust cue for identifying the axis of tilt.

## D. Perceived Distortion of 3-D Space in 2-D Projections

An interesting problem related to this topic is the interpretation of distance and orientation of objects seen, for instance, in photographs or video monitors that portray 3-D perspective cues on 2-D displays. The problem arises from discrepancies between the camera's vantage point with respect to the 3-D scene, which involves both object distance and lens orientation, and the station point of the observer who is viewing the 2-D projected image after it has been transformed by the optics of the camera lens. Errors of both observer distance and additional tile of the picture plane from the FPP contribute to errors of the observer station point. The same discrepancies occur when people observe artists paintings from a vantage point not used by the artist, only here the situation is complicated further by the artist who paints different figures within the composition at different vantage points. For example, a landscape may be painted from a remote distance and a figure from a short distance, although the figure is represented in the painting at a remote distance. Additional distortion occurs when figures are painted from 2-D photographs that have their own station point distortions. The printing process in photography also confounds the problems of finding the correct station point if the enlarger selects the center of the print from an eccentric point in the film plane.

The station point represents the position of the observer with respect to the view plane or projected image. To see the picture in the correct perspective, the station point should be the same as the location of the artist or the camera lens with respect to the film plane. Two violations of these requirements are errors in viewing distance and translation errors (i.e., horizontal or vertical displacements). Tilt of the view screen can be described as a combination of distance and translation error. In the case of the photograph, the correct distance for the station point is the lens focal length multiplied by the enlargement scaling factor. Thus, when a 35-mm negative taken through a 55-mm lens is enlarged seven times to a 10-in. print, it should be viewed at a distance of 38.5 cm or 15 in. ($7 \times 5.5$ cm). Assuming the painter holds up a thumb to get constant proportional scale for all images on the canvas, then the viewer should be at the same observation distance from the canvas as the painter to find the correct station point.

Station point errors consisting of incorrect distance from the view plane produce uniform magnification or reduction of the entire image, resulting in perceptual distortions of shape and tilt. For example, the retinal image of a circle lying on a horizontal floor plane has an elliptical shape when viewed at a remote distance and a more circular shape when viewed from a shorter distance. When a remote circle is viewed through a telescope, the uniformly magnified image appears to be closer and appears squashed in the vertical direction to form an ellipse (shape distortion). This uniform magnification also distorts texture-spacing gradients from one texture element to another such that the floor plane containing the texture elements is perceived as inclined or tilted upward toward the FPP (tilt distortion). Normally, the change in texture gradient is greatest for near objects because when a distant object is magnified, it appears nearer but its low texture density gradient makes the plane appear tilted upward toward the FPP by an angle equal to the angle of slant multiplied by the amount of magnification. Magnification affects slant derived from global outline perspective in the same way. For these reasons, a tiled rooftop appears to have a steeper pitch when viewed through binoculars or a telescope, the pitcher-to-batter distance appears reduced in telephoto view from the outfield, and the spectators may appear larger than the players in front of them in telephoto lens shots.

## IV. KINETIC CUES FOR DIRECTION

Tasks involving locomotion of an observer with respect to stationary obstacles or a stationary observer and moving obstacles that require avoidance of collision with other moving objects, as well as tasks that attempt capture of a moving object such as a baseball, all require an acute sense of egocentric direction in 3-D space. Several kinetic cues are available for these judgments of direction. Two prominent cues considered here are binocular motion in depth and monocular size looming.

### A. Binocular Motion in Depth

Binocular assessment of motion in depth utilizes the amplitude ratio and relative direction of horizontal motion displacement produced by the optical transform of moving objects onto the retinal image. A target moving in the FPP produces retinal displacement that is equal in amplitude and direction in the two eyes (Fig. 3, top left). Targets that move in depth on a trajectory that does not intersect with the observer produce displacements of the retinal image of unequal amplitude but equal direction (Fig. 3, bottom right). Targets that move directly toward one eye produce no displacement of the image in that eye and a retinal displacement directed toward the other eye where trajectories intersect. Targets with trajectories that intersect the head between the eyes produce retinal image displacements toward the outside in both eyes, i.e., in opposite directions (Fig. 3, bottom left). It is possible to demonstrate four channels for the brain's analysis that are tuned selectively to specific ranges of motion in depth. Two channels encode motion in depth of objects with trajectories that will collide with the observer and two channels encode motion in depth of objects with trajectories that will not collide with the head. Half of these detect motion approaching to the left of the midsagittal plane, a vertical plane along the axis of the skull between the eyes, and the other half detect motion approaching to the right of the midsagittal plane. In addition to these four channels, four others sense motion away from the observer. Each of these channels is tuned to a large range of directions ($>15°$), yet we are able to discriminate differences in direction as small as $1°$.

### B. Looming and Optic Flow

Size expansion and contraction (looming) is a robust monocular cue for both motion in depth and changing size. Receptive fields for looming detectors used in the analysis of motion in depth are approximately $1.5°$ in diameter in the fovea. Larger ranges of motion are utilized by detectors which sense changing size. Direction of motion in depth is derived from the combination of uniform displacement and loom of the retinal image which are each processed by a separate encoding mechanism. Motion in depth, stimulated by size change and binocular disparity, responds equally well to rapid stimuli, but motion in depth stimulated by slow depth changes is more sensitive to binocular disparity than to changing size. Separate processes underlie motion in depth stimulated by size expansion and size contraction.

In textured fields, motion of the body produces an optic flow pattern of the retinal image. If the position of the eye in the head and body does not change, the optic flow expands from a point in space (expansion point) that is in the direction of body motion (Fig.

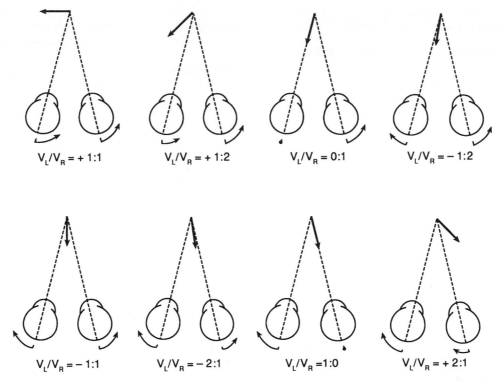

**FIGURE 3** Relative velocities of left and right retinal images for different target trajectories. When the target moves along a line passing between the eyes, its retinal images move in opposite directions in the two eyes; when the target moves along a line passing wide of the head, the retinal images move in the same direction, but with different speeds. The ratio ($V_L/V_R$) of left- and right-eye image velocities provides an unequivocal indication of the direction of motion in depth.

4b). If the eye rotates or pursues a portion of the optic flow pattern, the expansion point is shifted away from the direction of body motion to the direction of the line of sight. Rotation of the eye produces a constant displacement or translation of all texture elements, independent of their distance from the observer. The direction of body motion as well as the point of impact, for example, of an aircraft approaching a landing strip (Fig. 4b), can then be computed by subtracting a constant displacement equal to the velocity of eye rotation, caused by body motion, from all motion displacements within the optic flow pattern. Indeed we are able to judge our direction of motion accurately using the optic flow pattern while allowing our eyes to move.

The velocity of optic flow, produced by movement of the observer in depth toward the expansion point, increases as the target is approached. The velocity of retinal image motion of a given target from the expansion point is proportional to the time to reach contact. The time of impact with the expansion point can be estimated from the distance to the impact point divided by the velocity of the moving observer. Estimates of the time of impact to within 0.8 sec are useful for adjustments of body rotation such as swerving to avoid impact with stationary obstacles. Such reactions have a delay of 0.05 sec.

## C. Motion Parallax

Nonuniform motion within the retinal image of texture in the visual field provides information about both depth and form. These cues from movement can be isolated from other cues by presenting a pattern whose texture elements move differentially (Fig. 4a). Differential lateral shifts of retinal images (shearing motion) at different viewing distances normally result from translational movements such as driving a car and viewing scenery from the side window. When a single point in the field is fixated (pivot point) by the moving observer, this point and all other points at the same distance remain stationary on the retina. Objects farther away appear to move in the same direction as the observer's translation while nearer

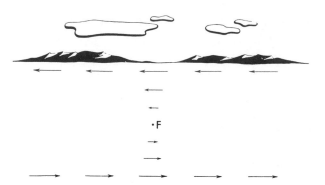

**FIGURE 4a**  The optical flow of motion parallax. Assume that an observer moving toward the left fixates a point at F. Objects nearer than F will appear to move in a direction opposite to that of the movement of the observer; objects farther away than F will appear to move in the same direction as the observer. The length of the arrows signifies that the apparent velocity of the optical flow is directly related to the distance of objects from the fixation point.

**FIGURE 4b**  Motion perspective. The optical flow in the visual field as an observer moves forward. The view is that as seen from an airplane in level flight. The direction of apparent movement in the terrain below is signified by the direction of the arrows; speed of apparent movement is indicated by the length of the arrows.

objects appear to move in the opposite direction. In this example, distance can only be interpreted correctly with knowledge of the direction of body translation. Unlike binocular disparity, which only yields depth percepts from disparity in the horizontal direction, shearing motion parallax yields depth from parallax in any direction (i.e., vertical and oblique) in which body translation occurs.

Rotation of the eye does not change the optical shear of the retinal image; it only subtracts a constant amount of displacement from all points within it. During following eye movements, eye rotation reduces retinal image displacement, and sensitivity to shear at the distance of the pivot point is optimized.

Perceived depth variations also result from dynamic compression of motion in one meridian of the retinal image. Shear is produced when depth variations are in a meridian perpendicular to the meridian of translation. For example, when looking out of the side window of a moving train or car, the horizon and road are seen to shear laterally about their vertical displacement in the visual field (Fig. 4a). In contrast, when looking out of the front window of a moving car, depth variations occur along the axis of translation of the car. This produces gradients of velocity or compression that vary with target distance in the meridian of translation (Fig. 4b).

Sensitivity to spatial variations in velocity gradients is different in response to compression and shear. Low spatial variations in depth, produced, for example, by gradual depth variations in hills by a highway, are more easily detected using shear than compression. Interestingly the same loss of sensitivity for low spatial

variations in depth is found for monocular motion parallax and binocular stereopsis produced by disparity compression. In contrast, depth sensitivity from motion parallax and stereopsis are more sensitive to abrupt depth variations such as steps when conveyed by compression than by shearing motion of the retinal image.

## V. BINOCULAR VISION

The principal benefit of binocular vision is stereopsis, the perception of relative depth that is stimulated by horizontal retinal image disparity. Stereopsis assists us in several types of spatial tasks, including judgment of relative depth or separation between objects and perception of forms hidden by texture camouflage (such as branches in tree foliage), and, as presented earlier, it is related to perceived direction of motion in depth. Stereopsis is an extremely acute mechanism, boasting one of the lowest visual thresholds—several seconds of arc—which places it high in the ranks of the hyperacuities. This highly sensitive process requires the support of several elaborate binocular systems. These include an organization of binocular correspondence that signals precise alignment of the eyes to a target distance, a sensory fusion process that integrates visual directions, a stereoscopic process that senses differences in depth, and an inhibitory process which suppresses diplopic and overlapping images of objects that lie in front or behind the fixation plane.

## A. Binocular Disparity

The slightly different perspective views that the two eyes have of the world yield two retinal images whose features have differences in relative position (Fig. 5). Classically, the position of each feature within the image is described on the retina by directions relative to the eye's visual axis. These directions are compared in the brain by a system of binocular correspondence which encodes as disparity any differences in direction relative to the visual axes of similar image features in the two eyes. This disparity depends on viewing distance to the target and the angle of convergence of the eyes. The angle subtended by a target point at the nodal points of the two eyes is called binocular parallax. The difference between binocular parallax and the angle at the intersection of the two visual axes (convergence) is called absolute disparity. When several targets are in the field of view, the difference in their absolute disparities forms relative disparity which is the stimulus for stereoscopic depth discrimination.

Our depth sensitivity to these two forms of disparity is vastly different. An isolated object which presents a single absolute disparity is usually seen at a depth of 1–1.15 m, regardless of its true distance from the observer (specific distance tendency). Even if the isolated target is moving in depth, for example, the headlamp of a moving vehicle seen at night, it appears to remain motionless. However, when a second stationary reference target is placed in the field along with the moving test stimulus, depth perception becomes highly sensitive to differences in target distance.

## B. Binocular Correspondence

The issue of binocular correspondence deals primarily with the encoding of absolute disparity from which relative disparity is subsequently derived. The topic can be divided into three sections: (1) the transducer which encodes disparity, (2) the raw data or stimulus features to be transformed, and (3) the algorithms or inferences used by the visual system to determine what features in the two eyes are to be compared (i.e., the matching problem).

### 1. Transducer

Absolute disparity, which is the difference in oculocentric directions of a target, is computed by the visual system in much the same way as separation or difference in direction of multiple targets are seen by one eye. The retinal topographical maps of the two eyes are merged in the visual cortex on an approximate point–point basis. Targets seen by each eye in identical visual direction (IVD) are imaged on corresponding retinal points, whereas targets seen by each eye in different visual directions are imaged on noncorresponding points. The physiological organization of binocular correspondence forms the basis of a comparator that allows the computation by the brain of absolute disparity of objects seen in the visual field. The comparator has a limited range of disparity sensitivity, and the corresponding distances or depths in 3-D space of stimuli, to which it can respond.

The spatial topography of corresponding retinal loci is described by the horopter or horizon of vision. The horopter is defined operationally as the locus of all points in space whose images are formed on corresponding retinal loci (i.e., points with zero disparity). A plane parallel to the baseline that passes through the fixation point (FPP) perpendicular to the primary direction of gaze approximates a zero-disparity plane. If the topography were such that corresponding retinal points were evenly spaced from their respective corresponding foveas (regions of maximal visual acuity), then the horopter would not be the FPP but would be a circle passing through the fixation and nodal points of the two eyes (Vieth-Müller circle, geometrical, or theoretical horopter). Usually the empirically measured horopter falls somewhere between these two boundaries (Fig. 6).

The vertical dimension of the horopter is described as a vertical straight line passing through the midline.

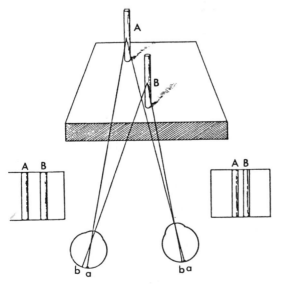

**FIGURE 5**   Disparateness of retinal images, producing stereopsis.

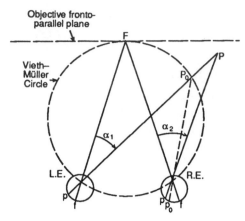

**FIGURE 6** The spatial asymmetry between the organization of retinal points corresponding to points on a horopter that lies outside the Vieth-Müller circle.

All other points in the visual field that lie to the left or right of the midline are physically nearer to one eye. The unequal ocular magnification produced by this difference in proximity causes points above or below the visual plane to be imaged with a vertical disparity and, by definition, they do not lie on the horopter. These veritcal disparities increase with both convergence of the eyes to near targets and lateral displacement of the test target from the primary position of gaze. Normally the potential diplopic appearance of targets above and below the visual plane is minimized by binocular sensory fusion as well as a slight declination of the vertical horopter.

## 2. Stimulus for Binocular Sensory Fusion

Even for a given correspondence task, such as sensory fusion and stereopsis, the range of disparities that meets the judgment criterion will change with stimulus parameters such as contrast, luminance, spatial frequency, exposure duration, and the presence of other stimuli in the visual field. For example, the sensory fusion range known as Panum's fusional area (PFA) varies both in shape and in size with the coarseness or size of the test target. Size is described here in terms of spatial frequency which is the reciprocal of size or width in degrees. For targets whose spatial frequency composition is above 2.5 cycles per degree, PFA has an elliptical shape with a horizontal axis (10 arc min) that is twice the vertical. However, PFA becomes circular in shape when coarser lower spatial frequencies are used and the size of PFA increases proportionally with the coarseness of the fusion stimulus. Accordingly, the thickness of the horopter can

vary from one-fourth to 6° at the fovea when the fusion range is tested with fine or coarse features, respectively. These variations illustrate that disparity is encoded by neurons that also process limited ranges of size, and their range of disparity sensitivity is scaled proportionally with their range of sensitivity to size. Target spacing and depth also influence the sensory fusion range. As lateral spacing of targets is reduced to less than 0.25°, the range of sensory fusion (PFA) becomes reduced. Fusion becomes impossible when target spacing is less than the differences in depth between two adjacent targets. This crowding effect has been described in terms of a disparity gradient limit for fusion (change in disparity per change in separation). As will be seen later, this gradient limit serves as a way of simplifying the problem of correctly matching like features in the two eyes when presented with complex textured patterns such as tree foliage.

In addition to spatial disparity gradients, there are temporal variations of disparity that influence the range of fusion. The more rapidly that disparity varies in time, such as with motion in depth, the smaller the range of sensory fusion will be. The fusion range remains near its peak until temporal variations in depth exceed 0.25 Hz; then the fusion range reaches its minimum of 2–3 arc min at temporal frequencies about 2.5 Hz. Interestingly, the temporal frequency variation does not apply to fusion of vertical disparity and the vertical dimension of PFA remains constant over the frequency range from 0 to 5 Hz. Accordingly, the size and shape of PFA changes from large and elliptical at low temporal frequencies to smaller and circular at high temporal frequencies (Fig. 7). The variation in fusion range with stimulus parameters is attributed to spatial resolution limits or separation acuity with coarse detail and also to some plasticity in the point–point relationships in binocular correspondence. A small degree of plasticity has been demonstrated for binocular correspondence under conditions where convergence movements of the two eyes are unable to completely align similar targets on the two foveas. Under these conditions, a small shift (<0.25°) in retinal correspondence can complete the sensory fusion response.

## 3. Algorithms for Solving the Matching Problem

On the surface, the process of fusion and computation of absolute disparity appears to be a straightforward problem of determining the position difference or disparity between like features in the two ocular images. The problem becomes ambiguous, however, when

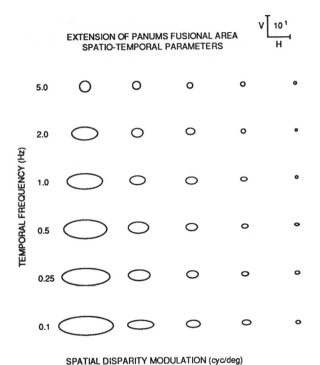

EXTENSION OF PANUMS FUSIONAL AREA
SPATIO-TEMPORAL PARAMETERS

SPATIAL DISPARITY MODULATION (cyc/deg)

**FIGURE 7** The size and shape of Panum's area varies with both temporal and spatial characteristics of disparity modulation. The horizontal but not the vertical diameter of the area vaires from a maximum of 20 to a minimum of 1.5 arc minutes. At low temporal and spatial modulation frequencies, the area is elliptical with a ratio of 2.5 : 1 between the horizontal and vertical extents. At high temporal frequencies, the horizontal extent is reduced to equal the fixed vertical extent and the shape becomes circular. Both the horizontal and vertical extents are reduced to 1.5 arc minutes by increasing spatial modulation frequency to an upper limit of 2 cycles per degree.

several features are present in the visual field and potential matches are possible between a given feature in one eye and several features in the other eye. This problem is exacerbated by camouflage surface texture such as tree foliage where there is little spatial uniqueness of any particular texture element in the visual field. The number of possible matches is $N!$ where $N$ equals the number of features. For example, with only 10 points there are 3.6 million possible combinations, or matches, that could be made between the two eyes. This false matching error occurs frequently when viewing vertically oriented repetitive patterns and is termed the wallpaper illusion, in which the vertical bars of the wallpaper appear to float out of the depth plane of the wall. The visual system utilizes several inferential shortcuts to limit the possible matches to a manageable number.

A local solution for reducing ambiguous binocular matches is the coarse-to-fine strategy that relies on a size-disparity correlation. Earlier it was shown that Panum's fusional limit was greater for coarse than for fine features. Similarly, the range of stereoscopic depth increases to larger depth intervals when judging depth between coarse than between fine features. In addition, stereoacuity (i.e., the smallest disparity that can stimulate depth) is much finer with high than with low spatial frequency targets. The visual system utilizes this property by initially attending to low spatial frequency components of the images which guide convergence movements of the eyes in order to null or reduce large disparities in the visual field. The motor convergence of the eyes aligns smaller features also contained in the target with the horopter so that small variations in disparity can be resolved. Once overall disparity is reduced, the visual system can attend to finer features in order to process the smaller remaining variations in disparity near the horopter where stereosensitivity is greatest. Additional global processes assist in the reduction of the number of possible matches made between features in the two retinal images. One of these mentioned earlier is the disparity gradient limit. This process will automatically prevent matches between closely spaced features that would subtend abrupt changes in disparity with their neighboring features. A second global process, referred to as the principle of depth continuity, is one in which the perceived depth in an unambiguous region of the visual field biases the depth seen in an ambiguous region to minimize the difference or change in depth across the visual field. In cases where exposure time is too brief to form a clear solution to the matching problem, unambiguous solutions of depth from edges of a pattern are simply filled into the intervening surface where matches have not yet been made.

## C. Stereopsis

As described earlier for binocular sensory fusion, stereoscopic depth discrimination thresholds vary markedly with target features including size, spatial frequency, contrast, spacing, motion, and field location. The most influential of these variables are the absolute and relative locations of targets in the visual field. Stereosensitivity to depth between two points varies with their sagittal distance from the fixation plane. Sensitivity to depth increments is highest at the horopter or fixation plane, where the disparity of one of the comparison stimuli is zero. The Weber fraction, describing the ratio of increment stereo threshold (arc sec) over the background disparity (arc min) (3 sec/min), is fairly constant (5%) with background depth amplitudes up to 2°. Stereoscopic depth is not perceived beyond this distance from the horopter or fixation plane. Monocular depth cues must be relied on beyond this distance to judge relative depth.

Unlike the reduction of stereosensitivity away from the horopter, stereoacuity remains uniformly at its peak in the central 5° of the tangent plane about the fixation point. Curiously, monocular vernier thresholds, measured with the same targets and at the same retinal eccentricities as used for testing stereopsis, are markedly elevated (reduced sensitivity), while stereoacuity is unaffected by small retinal eccentricities. This difference illustrates that different factors limit these two forms of hyperacuity. It has been proposed that vernier sensitivity is a precursor of sensing retinal image disparity, i.e., that relative retinal image disparities are computed from differences between spatial separations seen by each eye. However, this is clearly not the case. A more likely computation of relative disparity performed by the brain comes from a comparison of absolute disparities as described earlier. Vernier acuity is apparently reduced by a central (nonretinal) process that does not limit stereopsis in the fronto-parallel plane.

Stereo thresholds are influenced by the size or spatial frequency of the depth stimulus. Stereo thresholds are lowest and remain relatively constant for fine features composed of higher spatial frequencies above 2.5 cycles per degree. Thresholds decrease proportionally with coarser targets composed principally of lower spatial frequencies. Even though stereo threshold varies markedly with target coarseness, suprathreshold disparities needed to match the perceived depth of a standard disparity (i.e., 40 arc min) are independent of spatial frequency, constituting a form of stereo-depth constancy. The spatial dependence of stereopsis and fusion on spatial frequency can result in the simultaneous perception of diplopia, fusion, and stereopsis in complex targets composed of a broad range of spatial frequencies. For example, leaves within the foliage of an extended limb of a tree can be seen diplopically, while clusters of leaves appear single. Clearly, a single shape and size for Panum's area for binocular sensory fusion and a single threshold for stereopsis are incompatible with this observation.

## D. Stereo-Depth Scaling

Stereoscopic depth is stimulated by angular disparities which are optically transformed from linear depth intervals in object space. As with foreshortening in linear perspective, the linear depth interval associated with a fixed angular disparity increases proportionally with the square of the viewing distance. Thus, a 1-cm depth interval at a distance of 1 m subtends the same angular disparity as 1 m at a distance of 10 m. Despite this growth of the linear depth interval for a fixed disparity, we perceive stereo-depth veridically. That is, 1 cm in the sagittal plane appears to be the same size depth interval whether viewed as 0.1 m or 10 m even though its retinal image disparity is reduced proportionally with the square of viewing distance. To accomplish this, the visual system needs information about absolute distance in order to perceive relative depth veridically. The same problems occur when scaling depth responses to monocular cues such as motion parallax. Convergence and accommodation are possible kinesthetic sources which provide absolute distance information used to scale binocular depth perception. In addition, binocular vertical disparity, caused by perspective distortion and monocular pictorial cues, such as retinal image size of familiar objects, also provide information for scaling disparity in order to obtain the veridical sense of depth. There are upper limits to the range of distances that can be scaled from familiar size or size–distance constancy. For example, when taking off in an airplane, ground objects (e.g., cars, houses, and trees) appear normal in size and then suddenly appear to shrink to a dollhouse scale at an altitude of 100 m. Another example of limited size distance scaling is the perceived distance of extremely distant objects like the sun and moon, which are perceived at a distance of the sky dome or heavenly vault. Kinesthetic cues from sensed motor responses by accommodation and vergence become ineffective at the same remote viewing distances.

## E. daVinci Stereopsis

Normally, a nearby opaque object in our field of view will occlude parts of the distance background scene. Under binocular viewing conditions, Leonardo da Vinci noted that the near object occluded different parts of the background seen by the two eyes, leading to a compelling sense of depth. This form of depth perception, also referred to as occlusion stereopsis, is of interest because it illustrates that binocular depth can be seen even though there are no binocular corresponding luminance edges to match for the occluded portions of the background. Physiological studies have only identified binocular neurons in the primary visual cortex which respond to disparity of matched luminance edges but none have reported neurons that respond to unmatched edges resulting from occlusion. This may be because the occlusion disparity is processed elsewhere in the brain or that visual cells have not been tested with appropriate occlusion stimuli.

## F. Autostereograms

Autostereograms, first illustrated in drawings by Alfons Schilling in 1974 and in computer-generated displays by Christopher Tyler in 1983, have become a popular form of decorative art in which a hidden 3-D image emerges from a random dot or patterned picture when it is viewed binocularly. The autostereogram is a curiosity because it contains separate views for the two eyes in a single picture. This is possible because the visual axes of the eyes actually meet or converge in front or behind the plane of the picture so that each eye is directed at horizontally separated regions of the picture.

The process is a refinement of the classic wallpaper effect discovered in 1844 by Sir David Brewster. If a repetitive pattern such as used in Victorian wallpaper is viewed with the eyes converged inaccurately, the plane of the paper appears at a false depth in the plane of convergence. Binocular disparity is produced in the autostereogram with slight horizontal shifts or disparities in the pattern contained in adjacent repetitive columns. Depth variations across the picture are produced by changing the repetition width of individual patterns as they appear in adjacent columns. Depth is changed in a specific region simply by varying the repetition width in that part of the stereogram from the width in adjacent regions. The misaligned eyes each view and match separate columns and the resulting disparity is perceived in stereoscopic relief. Figure 8 is a replication of the first successful autostereogram developed by Professor Tyler.

## G. Interocular Suppression and Binocular Rivalry

Simultaneous cues for fusion, stereopsis, and diplopia, as well as clear and blurred vision, occur in normal complex scenes, where several objects are positioned at different depths along a single direction, or when viewing a near object displaced laterally from the primary position of gaze, where one eye's image is slightly out of focus. The visual system is capable of selectively suppressing the diplopic and blurred images while retaining the percept of a single in-focus image in stereoscopic depth. There are several suppression mechanisms which respond to different stim-

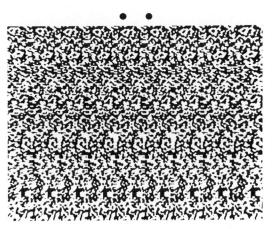

**FIGURE 8** A checkerboard pattern can be seen in the random dot stereogram by either crossing or uncrossing your eyes until the two dots above the figure appear as three dots. When the eyes are not properly aligned, you will see either two or four dots at the top of the figure. When three dots are seen, shift your attention to the random dot pattern and concentrate on the checkerboard pattern that emerges in stereoscopic depth.

uli. Binocular orientation rivalry is stimulated by differently shaped images that are formed on corresponding retinal points. Only the dissimilar form is suppressed, allowing simultaneous stereoscopic depth perception stimulated by similar images. This simultaneous sensory integration and suppression at the same retinal locus indicates that binocular rivalry and stereopsis involve independent and parallel pathways. The suppression of unequal blur in the two eyes (anisometropia) involves a different binocular inhibitory process that eliminates the blur of one retinal image while the eyes continue to fuse and perceive depth in the anisometropic image. The retention of fusion and stereopsis indicates that binocular suppression of anisometropic blur occurs more centrally in the brain than the site at which retinal image disparity is encoded.

## H. Neurophysiology of Binocular Vision

The physiological basis for mechanisms that encode binocular disparity has been identified in the visual cortex of monkey. Four classes of binocular cortical neurons have been described (Fig. 9). Tuned excitatory cells are sensitive to small disparities from 0 to 0.25°. These cells are likely to be involved in the process of binocular sensory fusion. Another cell class, the tuned inhibitory neurons, are inhibited by the presence of small disparities and are likely to be related to binocular rivalry. A third and fourth class of

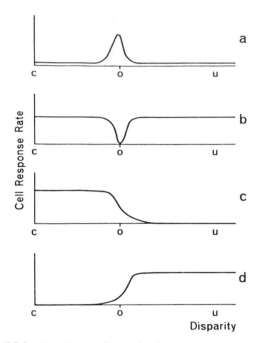

**FIGURE 9** Four classes of binocular disparity sensitivity in monkey cortex: (a) binocular facilitation; (b) binocular occlusion; (c) "monocular" crossed; and (d) "monocular" uncrossed sensitivities. [Reprinted, by permission of the publisher, from C. W. Tyler and A. B. Scott (1979). Binocular vision. *In* "Physiology of the Human Eye and Visual System" (R. Records, ed.). Harper and Row, Hagerstown.]

cells are near and far neurons that are stimulated by proximal and distal disparities, respectively. They are also inhibited by distal and proximal disparities, respectively. These cell types are likely to be involved in the perception of diplopia as well as the initiation of large changes in convergence eye movements from one viewing distance to another.

## I. Postnatal Development of Binocular Vision

Binocular neurons in the visual cortex require stimulation during the first few years of development, otherwise they become sensitive to only one or the other eye's input. In their absence, stereopsis, fusion, and binocular eye alignment may not function properly. However, after this critical period of development, the organization of binocular vision is immune to the effects of visual deprivation. Under conditions of normal development, stereoscopic depth perception

usually emerges postnatally between 6 and 12 weeks of life. Binocular vergence eye movements begin to respond to small disparities or depth intervals during the same period of time. This rapid development is more abrupt than the maturation of visual acuity, which does not approach adult levels until 18–24 months. Prior to the development of binocular vision, the infant is still able to utilize monocular cues for distance and direction, such as looming, which appears to be present at birth. These early processes utilize coarse stimulus features whereas processes that develop later require postnatal growth of the eye and development of contrast sensitivity and visual acuity. The early cues to space perception must play a vital role in the calibration and scaling of processes underlying space perceptions that emerge later in development.

## BIBLIOGRAPHY

Aslin, R. M. (1987). From sensation to perception. *In* "Handbook of Infant Perception" (P. Salapatek and L. Cohen, eds.), Vol. I. Academic Press, New York.

Bishop, P. O. (1987). Binocular vision. *In* "Adler's Physiology of the Eye," 8th Ed. Mosby, St. Louis.

Boff, K. R., Kaufman, L., and Thomas, J. P. (eds.) (1986). "Handbook of Perception and Human Performance," Vol. I. Wiley, New York.

Brady, J. M. (1981). "Computer Vision." North-Holland, Amsterdam.

Cutting, J. E. (1986). "Perception with an Eye for Motion." MIT Press, Cambridge, MA.

Dember, W., and Warm, J. (1979). "Psychology of Perception." Holt, Rinehart, and Winston, New York.

Epstein, W. (ed.) (1977). "Stability and Constancy in Visual Perception: Mechanisms and Processes." Wiley, New York.

Haber, R. N. (ed.) (1968). "Contemporary Theory and Research in Visual Perception." Holt, Rinehart, and Winston, New York.

Hein, A., and Jeannerod, M. (eds.) (1983). "Spatially Oriented Behavior." Springer-Verlag, New York.

Held, R., Leibowitz, H. W., and Teuber, H. C. (eds.) (1978). "Handbook of Sensory Physiology," Vol. VIII. Springer-Verlag, Berlin.

Howard, I. P. (1982). "Human Visual Orientation." Wiley, New York.

Kaufman, L. (1974). "Sight and Mind: An Introduction to Visual Perception." Oxford University Press, New York.

Liu, L., Stevenson, S. B., and Schor, C. M. (1994). Quantitative stereoscopic depth without binocular correspondence. *Nature* **367,** 66–69.

Marr, D. (1982). "Vision." Freeman, New York.

Ogle, K. N. (1950). "Researchers in Binocular Vision." Saunders, Philadelphia.

Regan, D., Frisby, J., Poggio, G., Schor, C., and Tyler, C. The perception of stereo-depth and stereo-motion: Cortical mechanisms. *In* "Visual Perceptions: The Neurophysiological Foundations" (L. Spillman and J. Werner, eds.). Academic Press, San Diego.

Rogers, B., and Bradshaw, M. F. (1993). Vertical disparities, differential perspective and binocular stereopsis. *Nature* **361,** 253–255.

Schor, C. M. (1987). Spatial factors limiting stereopsis and fusion. *Optics News* **13,** 14–17.

Schor, C. M., and Cuiffreda, K. (eds.) (1983). "Vergence Eye Movements: Basic and Clinical Aspects." Butterworth, Boston.

Spillman, L., and Wooten, B. (eds.) (1984). "Sensory Experience, Adaptation, and Perception." Erlbaum, London.

# Bioenergetics and the Adrenal Medulla

DAVID NJUS
JOEL P. BURGESS
PATRICK M. KELLEY
*Wayne State University*

I. Structure of Secretory Vesicles
II. Synthesis of Secreted Material
III. Energetics of Secretory Vesicles
IV. Evolution and Development of Secretory Systems
V. Conclusion

## GLOSSARY

**Catecholamine** One of the biogenic amines containing the catechol ring structure (i.e., dopamine, norepinephrine, or epinephrine)

**Chromaffin cell** Catecholamine-secreting cells of the adrenal medulla and other peripheral tissues

**Chromaffin vesicle** Catecholamine-storing organelles of chromaffin cells

**Cytochrome $b_{561}$** Integral membrane protein that mediates the regeneration of ascorbic acid in chromaffin and other secretory vesicles

**Vacuolar ATPase** One of a class of enzymes that hydrolyzes cytosolic adenosine 5′-triphosphate (ATP) and transports $H^+$ across membranes of the vacuolar system

**Vacuolar system** Organelles and vesicles involved in trafficking material from the endoplasmic reticulum to the plasma membrane and back again (e.g., endoplasmic reticulum, Golgi, secretory vesicles, lysosomes, endosomes)

**Vesicular monoamine transporter** Protein found in secretory–vesicle membranes and catalyzing the $H^+$-linked transport of catecholamines or another biogenic amine into the vesicle

THE ADRENAL MEDULLA REGULATES THE ENERGY metabolism of the whole body, and its own metabolic requirements may seem insignificant by comparison.

However, because the adrenal medulla is a highly specialized tissue, some aspects of its energy metabolism are greatly magnified and provide unique insights into cell biology. Of special significance are the chromaffin vesicles, the organelles that store and secrete catecholamines. Their abundance makes the adrenal medulla a Rosetta Stone for deciphering not only mechanisms of hormone storage and secretion, but also principles of energy metabolism in organelles. This article focuses on the bioenergetics of chromaffin vesicles. We will discuss the energy metabolism of these vesicles particularly as it applies to catecholamine storage and secretion and, more generally, as it relates to the energetics of secretory and synaptic vesicles in endocrine, neuronal, and other secreting cells. Because adrenal tissue is most readily available from cattle, much is known about bovine adrenal medulla. Consequently, the bovine model will be presented here with confirmatory or differing observations of human tissue cited where available.

The adrenal medulla, which stores and secretes the catecholamines epinephrine and norepinephrine, is a relatively homogeneous tissue. It consists of two types of chromaffin cells: one type synthesizes, stores, and secretes epinephrine, whereas the other stores and secretes norepinephrine because it lacks the enzyme phenylethanolamine-N-methyltransferase, which converts norepinephrine to epinephrine. In humans, approximately 90% of the cells produce epinephrine and about 10% secrete norepinephrine. In both cell types, the catecholamines are stored in membrane-bound organelles, the chromaffin vesicles. The secretion of the catecholamines occurs by exocytosis. Upon stimulation of a chromaffin cell, the chromaffin vesicles move to the cell periphery. The vesicular and plasma membranes fuse, and the soluble contents of

ENCYCLOPEDIA OF HUMAN BIOLOGY, Second Edition, VOLUME I. Copyright © 1997 by Academic Press. All rights of reproduction in any form reserved.

the vesicle matrix are released into the extracellular space.

Efficient release of the catecholamines is achieved by packaging them at awesome concentrations: the catecholamine concentration in chromaffin vesicles is ~0.55 $M$, or roughly $10^4$ times higher than the concentration in the surrounding cytosol. This implies that the vesicle membrane has a very effective catecholamine uptake system. Catecholamine transport into chromaffin vesicles is driven indirectly by a proton-translocating adenosine triphosphatase (ATPase) (see Section III). This dependence on a proton-translocating ATPase to power membrane functions is shared by many other organelles, including the Golgi, lysosomes, endosomes, and secretory vesicles of other types. In contrast, the plasma membrane of the chromaffin cell, like those of other animal cells, relies on the $Na^+/K^+$ ATPase to drive its transport activities. Thus, the adrenal medulla clearly illustrates the difference in the energy-transducing strategies of the organelle and plasma membranes of the cell.

## I. STRUCTURE OF SECRETORY VESICLES

A chromaffin cell contains about 30,000 chromaffin vesicles having an average diameter of 0.3 $\mu m$. The vesicles are so numerous that they contain about 13% of the cell's volume and their total membrane area exceeds that of the cell's plasma membrane by one order of magnitude. Consequently, the chromaffin vesicles are a dominant factor in the metabolism of the cell.

## A. Matrix Composition

Before discussing the energetics of the membrane, the composition of the chromaffin vesicles should be reviewed. The function of the vesicle requires that it store catecholamines at a high concentration. Because these amines are positively charged, a high concentration of negative charges must also be present to maintain overall electroneutrality. However, the high solute concentration cannot be reflected in the osmolality of the matrix, otherwise the vesicles would be osmotically unstable. The vesicle satisfies these conflicting requirements by storing components that interact to reduce their osmotic activities. The principal constituents are the catecholamines (0.55 $M$), ATP (0.12 $M$), and protein (169 mg/ml). The protein (mostly

acidic chromogranins) and ATP contribute the negative charge needed to counter the catecholamines. Moreover, interaction between these components reduces their chemical activities and keeps the vesicle matrix in osmotic equilibrium with the cytosol. It should be emphasized that the interactions in the matrix are strong enough to reduce chemical activities by a factor of about 3, but they are not strong enough to create a crystalline complex. Nuclear magnetic resonance studies have shown that the chromaffin–vesicle matrix may be a very nonideal solution in the thermodynamic sense; nevertheless it is a solution.

The chromaffin–vesicle matrix contains about 77% of the vesicle protein; the rest is membrane bound. The soluble proteins are mostly chromogranins and proteolytic fragments derived from them. In human adrenal medulla, chromogranins A and B appear to be present in equal amounts. Human chromogranin A has 439 amino acid residues ($M_r$ 48,918) and chromogranin B has 657 ($M_r$ 76,295). Secretogranin II (formerly known as chromogranin C) has 587 amino acids. The chromogranins are extremely acidic (~25% glutamic and aspartic acids) and their negative charge causes them to appear anomalously large on sodium dodecyl sulfate–polyacrylamide gel electrophoresis (SDS-PAGE). In chromaffin vesicles, these proteins act as low osmolarity counterions to the catecholamines. This is probably not their only function, however, since they have been found in a variety of other secretory vesicles, including many in which cation neutralization does not seem to be an important consideration. The possibility that the chromogranins contain active peptides is gaining credence as pancreastatin, a 49 amino acid peptide from chromogranin A, has been isolated from porcine pancreas and found to inhibit the stimulated secretion of insulin. Moreover, secretoneurin, a 33 residue peptide derived from secretogranin II, seems to function as a neurotransmitter in the brain. Chromaffin vesicles also contain precursors for known peptide hormones including enkephalins and neuropeptide Y.

The major enzyme in chromaffin vesicles is dopamine $\beta$-monooxygenase, which is responsible for the synthesis of norepinephrine. Dopamine $\beta$-monooxygenase exists both in a soluble form in the matrix and in an insoluble state attached to the inner surface of the vesicle membrane. Also present in the matrix are processing enzymes for the peptide hormones: peptidyl-glycine $\alpha$-amidating monooxygenase, carboxypeptidase H, and a trypsin-like proteinase. Carboxypeptidase H, like dopamine $\beta$-monooxygenase, is a glycoprotein found in the vesicles both in a soluble and in a membrane-bound form. Ascorbic acid (vitamin C), which serves as a cofactor for both dopamine $\beta$-monooxygenase and peptidyl-glycine $\alpha$-amidating monooxygenase, is also found in the matrix. Its intravesicular concentration (20 mM) is about four times its concentration in the cytosol.

Finally, the vesicle matrix is notable for its acidic pH (5.5). The low pH stabilizes the catecholamines, which would decompose by oxidation much more rapidly at a higher pH. The matrix pH may also regulate the activity of intravesicular enzymes, some of which have pH optima near 5.5. Perhaps the most important consequence of the intravesicular acidity, however, is the $H^+$ concentration gradient created across the vesicle membrane. As described in Section III, this pH gradient plays a central role in the transport activities of the vesicle membrane. The intravesicular pH is stabilized by the high buffering capacity of the matrix ($\sim$100 mM $H^+$/pH unit). The major contributors to this buffering capacity are intravesicular protein and ATP. The p$K$ of the $\gamma$-phosphate of ATP, although normally higher, is shifted down to 5.6 in the environment of the chromaffin–vesicle matrix. This coincidence between the intravesicular pH and the p$K$ of its $\gamma$-phosphate makes ATP an especially significant intravesicular buffer. [*See* Adenosine Triphosphate (ATP).]

## B. Membrane Composition

The chromaffin–vesicle membrane is 33% protein, 51% phospholipid, and 16% cholesterol. Most of the phospholipid is contributed by phosphatidylcholine (27.5%) and phosphatidylethanolamine (31.8%). Two unusual features of the phospholipid composition are the high amounts of lysophosphatidylcholine (16.7%), most of which is in the inner monolayer of the membrane, and sphingomyelin (12.8%). The anionic phospholipids phosphatidylinositol (8.2%) and phosphatidylserine (2.5%) contribute to the overall negative surface charge density ($-1.38 \times 10^{-6}$ $C/cm^2$) of the chromaffin vesicle.

The membrane contains at least 40 proteins, as revealed by relatively intense and reproducible bands on SDS-PAGE. The two major proteins are the membrane-bound form of dopamine $\beta$-monooxygenase, which represents about 20% of the total membrane protein, and cytochrome $b_{561}$, which comprises about 10%. Two well-characterized transport activities, the $H^+$-translocating ATPase and the vesicular monoamine transporter, probably account for only about 1% of the membrane protein. The major glycopro-

teins of the chromaffin–vesicle membrane are dopamine $\beta$-monooxygenase and glycoproteins II and III. Glycoprotein III also occurs in a soluble form, which comprises 0.25% of the protein in the matrix. A minor membrane-bound glycoprotein is carboxypeptidase H. [*See* Proteins.]

Like most biological membranes, the chromaffin–vesicle membrane is more permeable to anions than to cations. The membrane is essentially impermeable to $Na^+$, $K^+$, and $Ca^{2+}$. The $H^+$ permeability is $\sim$$10^{-5}$ cm/sec. The chromaffin–vesicle membrane is somewhat permeable to anions in the order $SCN^-$, $I^- >$ $Br^- > Cl^- > F^-$. The $Cl^-$ permeability ($10^{-10}$ cm/sec) is similar to that of a pure lipid bilayer membrane.

## II. SYNTHESIS OF SECRETED MATERIAL

### A. Catecholamines

An interesting aspect of catecholamine biosynthesis is the cellular location of the synthetic enzymes (Fig. 1). The first two are cytosolic. The third, dopamine $\beta$-monooxygenase, is found within the chromaffin vesicles. The final step occurs back in the cytosol. Consequently, the catecholamines must be repeatedly transported across the vesicle membrane. As discussed in Section III,B, this is accomplished by a transporter that exchanges these amines for protons. Thus, the proton gradient provides energy for dopamine uptake. This energy is recovered when norepinephrine effluxes but is reinvested in the storage of epinephrine.

The biosynthetic pathway for catecholamines begins with the amino acid tyrosine. The first and rate-limiting step is the hydroxylation of tyrosine to 3,4-dihydroxyphenylalanine (L-dopa). In this reaction, the enzyme tyrosine hydroxylase incorporates one atom of $O_2$ into tyrosine and reduces the other to $H_2O$ using reducing equivalents donated by the cofactor tetrahydrobiopterin. The rate of catecholamine synthesis may be increased by activating existing tyrosine hydroxylase, by synthesizing new enzyme, or by increasing the concentration of the cofactor. Because secretion creates a demand for catecholamine synthesis, it is not surprising that depolarization of the cell membrane, which triggers secretion, is also the primary stimulus for tyrosine hydroxylase activation. The enzyme is activated by phosphorylation catalyzed by one of three protein kinases: cAMP dependent, $Ca^{2+}$-calmodulin dependent, and protein kinase C. Four N-terminal serine residues seem to be phosphor-

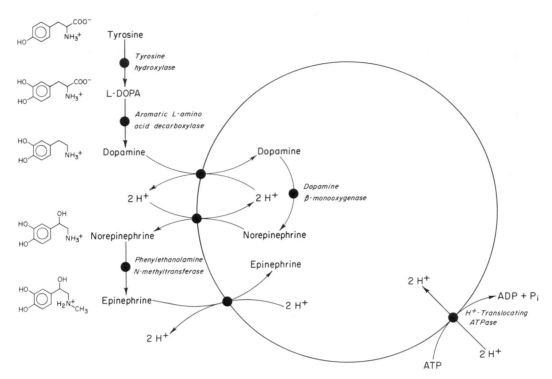

**FIGURE 1** Pathway of catecholamine synthesis and storage. L-Dopa, 3,4-dihydroxyphenylalanine; $P_i$, inorganic phosphate.

ylated. Different phosphorylation patterns can cause short-term (acute), long-term, or no activation of tyrosine hydroxylase activity. Activation increases the maximum reaction velocity of the enzyme and reduces product inhibition. In addition, activation makes tyrosine hydroxylase more sensitive to regulation by the concentration of the cofactor tetrahydrobiopterin (Section II,C,2).

L-Dopa is rapidly decarboxylated to dopamine by the action of aromatic L-amino acid decarboxylase. As its name implies, this enzyme has a broad specificity for aromatic amino acids. This reaction is not rate limiting, so there is no accumulation of L-dopa unless the enzyme is inhibited.

The conversion of dopamine to norepinephrine is catalyzed by dopamine β-monooxygenase. Like tyrosine hydroxylase, this enzyme incorporates one atom of molecular oxygen into the substrate and reduces the other to $H_2O$. Unlike tyrosine hydroxylase, it uses ascorbic acid instead of tetrahydrobiopterin as the physiological reductant. Moreover, ascorbate functions as a one-equivalent donor so two molecules of ascorbate are needed for each molecule of substrate, and two molecules of the ascorbate radical anion, semidehydroascorbate, are formed. Dopamine β-

monooxygenase has a low substrate specificity and attacks side chains of a variety of aromatic compounds. The mature enzyme is found in the vesicles either as a soluble tetramer with four identical 70- to 73-kDa subunits or as a membrane-bound tetramer with two 70-to-73-kDa subunits and two 75- to 77-kDa subunits. The subunits of human dopamine β-monooxygenase are apparently synthesized as identical polypeptides ($M_r$ 64,862), which are then glycosylated and modified to yield distinct soluble and membrane-bound subunits. The two larger subunits are presumed to contain an uncleaved signal sequence that anchors the enzyme to the inner surface of the membrane. About half of the enzyme is membrane associated and half is soluble.

The final step in the synthesis of epinephrine is the N-methylation of norepinephrine. The enzyme, phenylethanolamine N-methyltransferase transfers methyl groups from S-adenosylmethionine to the N terminus of norepinephrine. The presence of phenylethanolamine N-methyltransferase distinguishes the adrenergic cells of the adrenal medulla from the noradrenergic cells of the sympathetic nervous system. Most chromaffin cells continue to express this enzyme because of their exposure to glucocorticoids from the adrenal cortex early in development. This is one functional reason for the juxtaposition of the adrenal cortex and medulla.

Chromaffin cells can take up extracellular catecholamines as an alternative to *de novo* synthesis. This reuptake occurs by $Na^+$-catecholamine cotransport and is inhibited by cocaine and tricyclic antidepressants such as desipramine. Because the function of the adrenal medulla is to put catecholamines into the circulation, reuptake is probably less significant in chromaffin cells than in sympathetic neurons. The $Na^+$-linked mechanism for catecholamine transport across the chromaffin–cell membrane contrasts, however, with the $H^+$-linked mechanism for transport across the chromaffin–vesicle membrane (Section III,B).

## B. Purine Nucleotides

A considerable amount of ATP is stored in the adrenal medulla and is secreted along with the catecholamines. While the chromaffin cell has the capacity to synthesize adenine nucleotides *de novo*, this is an ener-

getically expensive process. A more efficient strategy is to salvage and reuse circulating precursors. ATP secreted by the chromaffin cell is hydrolyzed to adenosine by ectonucleotidases. In addition, as discussed in Section II,C,3, adenosine is produced in the chromaffin cell as a byproduct in the synthesis of epinephrine. S-Adenosylmethionine, which serves as the methyl donor for phenylethanolamine N-methyltransferase, is formed from ATP and methionine and converted to homocysteine and adenosine (Fig. 2). Thus, the salvage of adenosine is an essential metabolic function of the chromaffin cell.

Chromaffin cells take up adenosine via a high-affinity adenosine transporter. To convert adenosine to AMP, the cells have a high level of adenosine kinase activity. Adenylate kinase then phosphorylates AMP to ADP. Finally, ADP may be phosphorylated to ATP in the usual way (oxidative phosphorylation or glycolysis).

Chromaffin cells also have the capacity to produce ATP from adenine. An adenine phosphoribosyltransferase activity forms AMP and pyrophosphate from adenine and 5-phosphoribose-1-pyrophosphate. Ade-

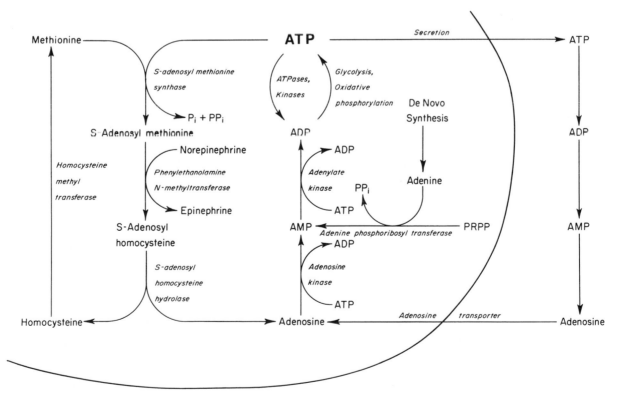

**FIGURE 2** Major pathways of ATP metabolism in the adrenal medulla. $PP_i$, pyrophosphate; PRPP, 5-phosphoribose-1-pyrophosphate.

nine may be taken up by the cells, although an adenine transport activity has not yet been reported. Adenine may also be synthesized *de novo*. This would be unusual, however, because the normal pathway for purine nucleotide biosynthesis proceeds through inosine to inosinic acid, which is then converted directly into AMP or guanosine monophosphate.

## C. Cofactor Metabolism

### 1. Ascorbic Acid

Ascorbic acid (vitamin C) is used by dopamine $\beta$-monooxygenase in the production of norepinephrine. Because the chromaffin cell cannot synthesize the vitamin, it must be able to take up circulating ascorbate. This uptake is energy and temperature dependent and requires $Na^+$, implying the existence of a $Na^+$-linked ascorbate transporter in the plasma membrane. This transporter maintains an ascorbate concentration of about 5 m$M$ in the cytosol of chromaffin cells. A higher ascorbate concentration (20 m$M$) is attained inside the chromaffin vesicles. The mechanism by which ascorbate is initially imported into the vesicles is not known, but intravesicular ascorbate is regenerated after its oxidation by dopamine $\beta$-monooxygenase. The product, semidehydroascorbate, is reduced back to ascorbate allowing the cofactor to recycle (Section III,C). [*See* Ascorbic Acid.]

### 2. Tetrahydrobiopterin

Tetrahydrobiopterin ($BH_4$) serves as the reductant for tyrosine hydroxylase and is oxidized to the dihydro compound $BH_2$. $BH_2$ is reduced back to $BH_4$ by a nicotinamide adenine dinucleotide phosphate-dependent enzyme, dihydrobiopterin reductase, so the cofactor cycles between the fully reduced $BH_4$ and the oxidized $BH_2$. In the cell, biopterin is kept mostly in the reduced form so the degree of reduction is probably not important in regulating cofactor availability. Instead, the cofactor concentration is controlled by regulating biopterin synthesis. Indeed, catecholamine depletion induced by reserpine treatment leads not only to an increase in tyrosine hydroxylase activity, but also to an increase in biopterin content. Biopterin is synthesized from GTP, and its synthesis is probably regulated at the level of GTP cyclooxygenase.

### 3. S-Adenosylmethionine

The sustained synthesis of epinephrine requires a constant supply of the methyl group donor *S*-adeno-

sylmethionine. Transfer of its methyl group to norepinephrine converts *S*-adenosylmethionine to *S*-adenosylhomocysteine (Fig. 2). *S*-Adenosylhomocysteine cannot simply be remethylated. Instead, *S*-adenosylhomocysteine is cleaved by *S*-adenosylhomocysteine hydrolase into homocysteine and adenosine. The homocysteine is then remethylated by homocysteine methyltransferase, which transfers a methyl group from $N^5$-methyltetrahydrofolate in a reaction requiring a derivative of vitamin $B_{12}$. This newly formed methionine can then be reincorporated into *S*-adenosylmethionine at the expense of another molecule of ATP. The energy for coupling methionine and adenosine is derived from the hydrolysis of ATP. The initial products of this reaction, triphosphate and *S*-adenosylmethionine, are tightly bound. Their release is effected by the further hydrolysis of triphosphate to pyrophosphate and phosphate.

## III. ENERGETICS OF SECRETORY VESICLES

### A. H$^+$-Translocating ATPase

Energy transduction by the chromaffin–vesicle membrane is driven by an H$^+$-translocating ATPase, a proton pump. This enzyme hydrolyzes cytosolic ATP (not the ATP within the vesicle matrix) and directs a flow of protons (H$^+$) to the vesicle interior. It establishes a pH gradient (inside acidic) and a membrane potential (inside positive) across the chromaffin–vesicle membrane.

The H$^+$-translocating ATPase in chromaffin vesicles is very similar (perhaps identical) to those found in synaptic vesicles, lysosomes, Golgi, and other membranes of the vacuolar system. In recognition of this, these proton pumps have been collectively classified as vacuolar, or V type, ATPases. This distinguishes these enzymes from the mitochondrial ATPase (F type) and those ATPases having a phosphorylated intermediate (P type).

### 1. Mechanism

Because the energetics of the chromaffin–vesicle membrane are governed by the transmembrane difference in the electrochemical potential of H$^+$, a brief review of the thermodynamics involved is appropriate. The electrochemical potential of a proton inside the vesicle relative to outside is

$$\Delta\mu_{H^+} = F\,\Delta\psi - 2.3RT\,\Delta pH.$$

This so-called proton-motive force ($\Delta\mu_{H^+}$) depends on the pH gradient ($\Delta$pH) and the membrane potential ($\Delta\psi$) both measured inside relative to outside. F, R, and T are the Faraday constant, gas constant, and absolute temperature, respectively. For a membrane potential of +60 mV and a pH gradient of −1.5 units (pH 5.5 inside; pH 7.0 outside), $\Delta\mu_{H^+}$ amounts to 14.7 kJ/mol (3.5 kcal/mol). This is the energy required to transport a hydrogen ion into the vesicles. It is released when $H^+$ leaves the vesicle and may be captured to drive catecholamine transport, ascorbic acid regeneration, and other functions of the membrane. Movement of $H^+$ alone has a substantial impact on the membrane potential but a negligible effect on the pH gradient, with several important ramifications. First, the intravesicular pH changes very little, even when the external $H^+$ concentration is varied over a wide range. In the absence of active $H^+$ translocation, $H^+$ leakage brings the membrane potential into equilibrium with the pH gradient:

$$\Delta\psi = (2.3RT/F)\,\Delta pH.$$

This establishes a Donnan equilibrium in which the concentration-driven diffusion of protons out of the vesicle is exactly balanced by the potential-driven movement of protons into the vesicle. If, for example, the external pH is 7.0 and the internal pH is 5.5, the membrane potential will be −90 mV. If the external pH is raised to 8.0, the internal pH will remain about 5.5, but the membrane potential will change to −150 mV.

When ATP is added, the active transport of protons into the vesicles changes the membrane potential from negative to positive. Again, in the absence of permeant ions, the buffering capacity of the matrix nullifies any significant change in the pH gradient. If a permeant anion (e.g., $Cl^-$) is present in the external medium, however, then the anion will move into the vesicles driven by the ATP-dependent membrane potential. This anion influx will dissipate the membrane potential and permit a sustained $H^+$-anion influx, which will lead to acidification of the matrix. Thus, in the presence of a permeant anion, the energy derived from ATP hydrolysis is invested in $\Delta$pH. In the absence of a permeant anion, the energy is invested in the membrane potential.

## 2. Structure

The $H^+$-translocating ATPase from chromaffin vesicles, when reconstituted into liposomes, catalyzes both ATP hydrolysis and proton transport, mimicking the activity seen in the native membrane. The different polypeptides comprising this enzyme are divided between a catalytic section and a membrane section. Structural and functional similarities exist between the chromaffin–vesicle V-type ATPase and the F-type mitochondrial ATP synthase, and this may reflect a common ancestry (Section IV). The catalytic section of the V-type ATPase contains five subunits designated A, B, C, D, and E (72, 57, 44, 34, and 26 kDa, respectively). These correspond to the $\beta$, $\alpha$, $\gamma$, $\delta$, and $\varepsilon$ subunits of the $F_1$ complex of the mitochondrial enzyme. The V-type ATPase is thought to have the same subunit stoichiometry as the $F_1$ complex with three A (catalytic) subunits, three B subunits, and one each of the C, D, and E subunits. The membrane section consists of six c subunits (16 kDa) and possibly one 20-kDa polypeptide, designated a. Some V-type ATPases seem to be associated with additional peptides. In particular, several secretory vesicle V-type ATPases have a 116-kDa protein. A peptide of this size is known to be required for ATPase assembly in yeast.

The active site of the chromaffin–vesicle ATPase has been assigned to the A subunit which corresponds to the $\beta$ subunit of the F-type ATPases. Both ATP and the inhibitor N-ethylmaleimide (NEM) label the A subunit with high affinity. ATP protects against labeling by [³H]NEM and against inhibition by NEM in the native membrane. A peculiar feature of inhibition by NEM is that proton pumping is abolished at a very low concentration (around 20 $\mu$M NEM), whereas complete inhibition of ATPase activity occurs only at concentrations that are 10–50 times higher. This suggests that the inhibitor may react at two different sites: one associated with proton movement and the other with ATPase activity.

The six c subunits of the chromaffin–vesicle ATPase are thought to form a channel conducting protons across the membrane. N,N'-dicyclohexylcarbodiimide (DCCD), which inhibits the $H^+$-translocating ATPase, reacts with these subunits. This parallels the situation in mitochondria where DCCD inhibits the F-type ATPase by reacting with the c subunit. The mitochondrial c subunit constitutes the proton channel of the $F_0$ portion of that enzyme. The 8-kDa c subunit in mitochondria is half the size of the chromaffin–vesicle c subunit. Sequence comparison suggests that the two subunits have a common ancestry with the vesicular c subunit arising through gene duplication.

## B. H⁺-Linked Transport

The chromaffin vesicles act as storage sites for the large concentrations of catecholamines released by the adrenal medulla. The accumulation of catecholamines into the vesicles occurs via a vesicular monoamine transporter coupled to the $H^+$-translocating ATPase (Fig. 1). The transporter exchanges internal protons for external catecholamine, utilizing the proton gradient established by the ATPase to drive catecholamines against a substantial concentration gradient. The coupling of transport to the proton gradient distinguishes vesicular transport from the $Na^+$-linked or facilitated transport typical of plasma membrane transporters.

## 1. Structure of the Monoamine Transporter

The vesicular monoamine transporter is specifically inhibited by reserpine, and reserpine-inhibitable transport is diagnostic of transporter activity. Proton-linked, reserpine-inhibitable transport is exhibited by platelet-dense granules, secretory granules in mast cells, and noradrenergic synaptic vesicles. In fact, the vesicular amine transporter appears to be responsible for the storage of biogenic amines in the secretory or synaptic vesicles of all cells that release either serotonin (5-hydroxytryptamine) or one of the catecholamines (dopamine, norepinephrine, epinephrine).

Vesicular monoamine transporters have been cloned from rat, bovine, and human sources. The transporters are 55- to 60-kDa peptides with 12 apparent transmembrane domains. Additional mass is contributed by glycosylation probably attached to a large lumenal loop between transmembrane domains I and II. The vesicular monoamine transporters have been divided into two classes (VMAT1 and VMAT2) primarily on the basis of sequence homology. VMAT1 was cloned from rat PC12 cells and is found in rat adrenal medulla. VMAT2 was cloned from RBL 2H3 cells and is found in rat brain. Curiously, the bovine adrenal transporter appears to be of the VMAT2 type, like the rat brain transporter, but unlike its counterpart in rat adrenal. The only human transporter cloned to date is also of the VMAT2 type. A functional difference between the two transporter types may be sensitivity to tetrabenazine. Transport mediated by rat VMAT1 is an order of magnitude less sensitive to tetrabenazine than is transport mediated by either rat or bovine VMAT2.

Unlike most transport enzymes, the vesicular monoamine transporter has a surprisingly broad substrate specificity. Although the catecholamines are the normal substrates in the adrenal medulla, hydroxy-phenylethylamines and hydroxyindoleamines (serotonin) are also transported. In addition, the dopaminergic neurotoxin 1-methyl-4-phenylpyridinium ($MPP^+$) and the adrenal imaging agent *m*-iodobenzylguanidine are also good substrates. It appears that the vesicular monoamine transporter will carry a wide spectrum of aromatic amines and that cellular specificity for a particular biogenic amine depends instead on the cell's biosynthetic machinery and plasma membrane amine transporter.

## 2. Mechanism of the Monoamine Transport

The vesicular monoamine transporter mediates an exchange of two $H^+$ for each protonated substrate molecule. The 2 : 1 stoichiometry is important because it determines the energy available for transport and thus the concentration gradient that can be achieved. Given the pH gradients (1.5 units) and membrane potentials (30–50 mV) typically found across vesicle membranes, a 2 : 1 stoichiometry means that intravesicular amine activities can be as much as 10,000 times greater than cytosolic concentrations. *In vivo*, the concentration gradient must be close to this; the catecholamine concentration is 550 m$M$ in the matrix and <100 $\mu M$ in the cytosol. To transport catecholamines against a concentration gradient of this magnitude, the vesicular monoamine transporter must cycle between two states: one having a high-affinity, externally exposed amine binding site and the other having a low-affinity internally accessible site (Fig. 3). This permits the transporter to bind external substrate, despite its low concentration, and to release the substrate internally in the face of a high concentration.

Potent specific inhibitors of the vesicular monoamine transporter include reserpine, tetrabenazine, and ketanserin. Reserpine appears to be a competitive inhibitor of catecholamine transport; its binding to the chromaffin–vesicle membrane is competitively inhibited by transporter substrates such as dopamine, norepinephrine, and serotonin. Reserpine binding is enhanced by ATP and is inhibited by the uncoupler carbonyl cyanide *p*-trifluoromethoxyphenylhydrazone, indicating that binding is promoted by the proton gradient across the vesicle membrane. Because the reserpine-binding site (designated R1) appears to be on the outside of the membrane and has a high affinity for catecholamines, we believe that it is the externally exposed substrate-binding site. Thus, the transporter may rest in the inward-facing conformation but be brought to the outward-facing conformation by energization and be stabilized in this conformation by reserpine binding. The dependence on the proton gra-

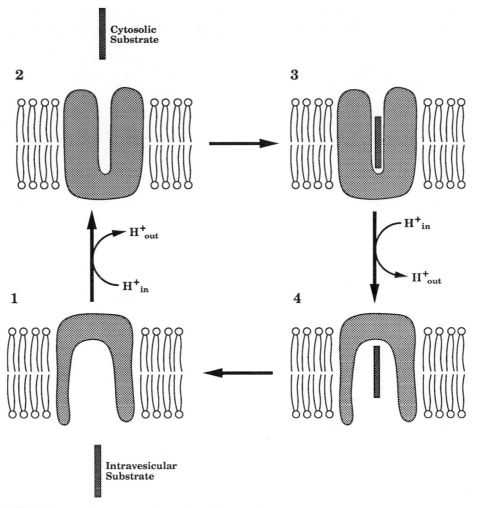

**FIGURE 3** Hypothesized catalytic cycle of the vesicular monoamine transporter. (1) Resting state with internally exposed binding site. (2) State with externally exposed R1-binding site. (3) Externally bound amine. (4) Internally bound amine.

dient indicates that this step is coupled to the efflux of one of the two protons exchanged for the amine (Fig. 3). The other proton may be coupled to the substrate translocation step.

Tetrabenazine and ketanserin are also inhibitors of catecholamine transport, but they bind to a different site (designated R2). Binding to the R2 site is not dependent on either ATP or the proton gradient. In addition, substrates such as norepinephrine are poor inhibitors of ligand binding to the R2 site.

The vesicular monoamine transporter's ability to carry a broad spectrum of aromatic amines is reminiscent of the proton-linked multidrug transporters in prokaryotes and the ATP-driven multidrug transporters in animals. In fact, Schuldiner has noted that vesic-

ular monoamine transporters and prokaryotic multidrug transporters exhibit considerable sequence homology and a common sensitivity to reserpine. These transporters all must have a binding site that binds aromatic compounds tightly when accessible from the cytosol but weakly when exposed to the other side of the membrane. This can be rationalized by a hydrophobic-binding site that fits closely around the aromatic compound in the high-affinity, cytosolically accessible conformation. In the other conformation, when the binding-site exposure shifts to the other side of the membrane, the hydrophobic pocket may simply expand so that substrate binding no longer protects the hydrophobic surfaces from water. There is still much to be learned about the structure and mechanism of the vesicular monoamine transporter, but it clearly has significant implications for transport mechanisms in general.

### 3. Other H$^+$-Linked Transport Systems

Catecholamines are not the only compounds stored by chromaffin vesicles, and it would not be surprising to find that other components are also accumulated by H$^+$-linked transport. In fact, chromaffin vesicles may contain a nucleotide transporter that mediates ATP-dependent uptake not only of ATP but also of GTP, uridine triphosphate, sulfate, phosphate, and phosphoenolpyruvate. ATP uptake has been reported to be coupled to the membrane potential generated by the H$^+$-translocating ATPase but not to the pH gradient. However, because ATP uptake was observed using intact chromaffin vesicles, the possibility that it represents exchange for preexisting intravesicular ATP rather than net accumulation cannot be excluded.

## C. Ascorbate Regeneration

Ascorbic acid, present in the chromaffin–vesicle matrix at a concentration of ~20 mM, functions as a cofactor for the enzymes dopamine $\beta$-monooxygenase (Section II,A) and peptidyl-glycine $\alpha$-amidating monooxygenase. The amount of product formed by dopamine $\beta$-monooxygenase alone exceeds the amount of intravesicular ascorbate by more than one order of magnitude, yet chromaffin vesicles do not take up ascorbic acid at a detectable rate. Consequently, the internal ascorbate must somehow be recycled. This is accomplished via a simple redox system involving the membrane protein cytochrome b$_{561}$; however, before discussing how this system works, it is instructive to review some properties of ascorbic acid.

### 1. Properties of Ascorbic Acid

Ascorbic acid is a good reducing agent and generally functions as a one-equivalent donor. Dopamine $\beta$-monooxygenase uses ascorbate in this way as do peroxidase and ascorbate oxidase. The radical thus formed, semidehydroascorbate, is also anionic so ascorbic acid under physiological conditions is a donor of single hydrogen atoms rather than electrons.

Semidehydroascorbate is a relatively stable free radical. Its usual mode of decay is disproportionation: two molecules react, one being reduced to ascorbate and the other oxidized to dehydroascorbate. This disproportionation is a crucial property because it keeps the semidehydroascorbate concentration low and makes ascorbate a better reducing agent.

### 2. Mechanism of Regeneration

Regeneration of intravesicular ascorbic acid occurs according to the following cycle. Dopamine $\beta$-monooxygenase uses internal ascorbate as a one-equivalent donor. The resulting radical anion, semidehydroascorbate, must then be reduced back to ascorbate. This is done by taking an electron from reduced cytochrome b$_{561}$ (Fig. 4). The now-oxidized cytochrome b$_{561}$ accepts an electron from external ascorbate and the intravesicular cycle begins anew. The external ascorbate is regenerated by reducing equivalents transferred from nicotinamide adenine dinucleotide, reduced, in a reaction catalyzed by the mitochondrial enzyme semidehydroascorbate reductase. According to this mechanism, cytochrome b$_{561}$ mediates the transmembrane transfer of an electron from external ascorbate to internal semidehydroascorbate.

The regeneration of intravesicular ascorbate at the expense of external ascorbate results in H$^+$ consumption internally and H$^+$ release externally (Fig. 4). Thus, the cytochrome mediates the equivalent of ascorbate regeneration coupled to H$^+$ efflux. The H$^+$-translocating ATPase provides the driving force for this ascorbate regeneration by maintaining both a pH gradient and a membrane potential favoring H$^+$ efflux.

### 3. Cytochrome b$_{561}$

The function of cytochrome b$_{561}$ is to carry electrons across the membrane, and it contains a single noncovalently bound heme. Cytochrome b$_{561}$ has little sequence homology to other cytochromes, indicating that it may represent a novel class of cytochromes independently evolved to perform its unique function. The human protein consists of 251 amino acids with a molecular weight of 27,600. The primary structure indicates that the cytochrome is a very hydrophobic protein with five or six transmembrane regions and very little extramembranous protein. A relatively large fraction of aromatic amino acids (16%) may allow the single heme to transfer electrons efficiently across a relatively long distance (across the membrane). Finally, clusters of cationic amino acids on either side of the membrane may facilitate interaction of the cytochrome with ascorbate and semidehydroascorbate.

The ascorbate-binding sites are particularly important in the function of cytochrome b$_{561}$. Because ascorbic acid is a hydrogen atom donor, it must lose a proton as it gives an electron to the cytochrome's

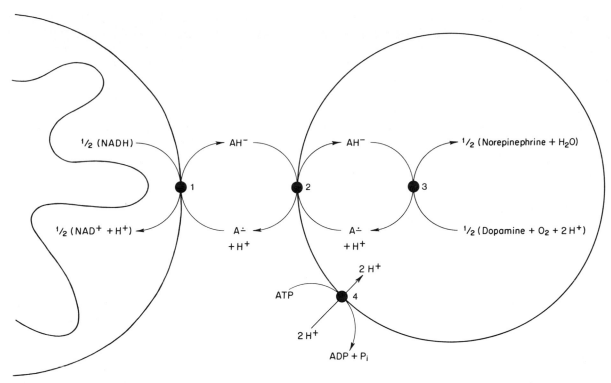

**FIGURE 4** Mechanism for regeneration of intravesicular ascorbic acid. (1) Semidehydroascorbate reductase on the outer mitochondrial membrane. (2) Cytochrome $b_{561}$ in the chromaffin–vesicle membrane. (3) Dopamine $\beta$-monooxygenase in the chromaffin vesicle. (4) $H^+$-translocating ATPase in the chromaffin–vesicle membrane. $AH^-$, ascorbic acid; $A^{\pm}$, semidehydroascorbate.

heme. The ascorbate-binding site facilitates the removal of this proton, thus catalyzing concerted $H^+/e^-$ transfer between ascorbate and the heme. This mechanism is a paradigm for the many biologically significant reactions between organic redox compounds (which gain or lose hydrogen atoms) and metalloproteins (which donate or accept electrons).

## IV. EVOLUTION AND DEVELOPMENT OF SECRETORY SYSTEMS

The chromaffin cells of the adrenal medulla have become a popular model for studies of secretion and secretory vesicles. The cells are embryonically derived from the neural crest and have much in common with sympathetic neurons. Thus, they are good models for neurotransmitter release from nerve terminals as well as for endocrine secretion. Both processes occur by exocytosis. It is not surprising, therefore, that the bioenergetic mechanisms of chromaffin vesicles apply to secretory and synaptic vesicles in general. The recent

cloning and sequencing of the bioenergetic machinery from different types of secretory vesicles and different organisms have made it possible to speculate about the evolution and the developmental regulation of the various secretory systems. Where did the genes for these activities originate? How is their expression controlled to produce cells storing and secreting particular products?

It appears that all secretory vesicles have a vacuolar ATPase (Table I). The vacuolar ATPase probably does not differ from tissue to tissue, at least in its major subunits. By virtue of its universality, the vacuolar ATPase offers a record of the evolutionary origin of secretory and synaptic vesicles. Eukaryotic cells are believed to have arisen through the assimilation of eubacterial symbionts (progenitors of mitochondria and chloroplasts) by archaebacterial hosts. Consequently, the vacuolar organelles might be expected to derive from the archaebacteria. Supporting this concept is the fact that the vacuolar ATPase shows sequence homology to the archaebacterial ATP synthase. V-type ATPases are found in the vacuolar or-

TABLE I

Bioenergetic Machinery of Secretory and Synaptic Vesicles

| Vesicle (secreted product) | Vacuolar ATPase | Vesicular neurotransmitter transporter | Cytochrome $b_{561}$ |
|---|---|---|---|
| Adrenal chromaffin vesicle (epinephrine) | + | Monoamine | + |
| Adrenergic synaptic vesicle (norepinephrine) | + | Monoamine | + |
| Platelet-dense granule (serotonin) | + | Monoamine | − |
| Synaptic vesicle (acetylcholine) | + | Acetylcholine | − |
| Synaptic vesicle (glutamate) | + | Glutamate | − |
| Synaptic vesicle (glycine) | + | Glycine | − |
| Sytnaptic vesicle [γ-aminobutyric acid (GABA)] | + | GABA | − |
| Neurohypophyseal secretory vesicle (amidated peptides) | + | — | + |

ganelles of eukaryotic cells of all kinds, including those of plants and fungi. Thus, it appears that the V-type ATPase has been the primary power source for the vacuolar organelles of eukaryotic cells from the beginning and that the secretory vesicles have inherited this machinery.

In contrast to the ATPase, the transporters in secretory vesicles differ from cell type to cell type. The vesicular monoamine transporter carries catecholamines, serotonin, and possibly histamine, and is found in secretory and synaptic vesicles storing those compounds. As mentioned, at least two varieties of this enzyme exist in rat and this may prove to be true in humans as well. Different enzymes are responsible for the transport of acetylcholine, glutamate, γ-aminobutyric acid, and glycine in synaptic vesicles. Of course, these neurotransmitter transporters have different substrate and inhibitor specificities, and they may mediate substrate : proton exchanges with different stoichiometries. They do, however, all catalyze the transport of a substrate into the secretory vesicle in exchange for protons. Moreover, the proteins are structurally similar in that they all seem to have 12 transmembrane segments with a large lumenal loop between transmembrane domains I and II. In terms of sequence, the proteins appear to constitute a family. Interestingly, this family includes toxin-extruding antiporters found in bacteria. Functionally, this makes some sense because the enzymes perform similar functions: transport of a broad range of compounds out of the cytoplasm in exchange for protons. It is not known, however, whether the neurotransmitter transporters evolved from toxin-extruding proteins originating in ancestral prokaryotes and con-

served throughout evolution or whether they were introduced in some other way. The search for homologous proteins in simple eukaryotes will help to resolve this question.

It does appear, however, that the various neurotransmitter transporters originated by duplication of a common ancestral gene. They then evolved separately to transport particular substrates into their respective synaptic and secretory vesicles. Different transporters are, of course, expressed in different cell types (Table I). Thus, the individual neurotransmitter transporters differ not just in substrate specificity but in cell-specific control of expression. The developmental regulation required for expression of the appropriate transporter in the appropriate cell is an important question that may be clarified as the sequences surrounding the various genes are sequenced and the regulatory elements are identified.

Finally, cytochrome $b_{561}$ appears to be unique to secretory and synaptic vesicles. It is found in vesicles containing ascorbic acid and monooxygenases. Thus, cytochrome $b_{561}$ is found in catecholamine-storing vesicles where it supports dopamine β-monooxygenase and in peptide hormone-storing vesicles such as those in the pituitary and pancreas where it supports peptidyl glycine α-amidating monooxygenase. Here, too, regulation of expression is an important issue.

## V. CONCLUSION

The adrenal medulla illustrates the fundamental role of ion gradients in storing energy for membrane functions and the contrasting energy-coupling strategies employed by the cell's plasma membrane and the membrane of the secretory vesicles. Each membrane has its own ion pumps responsible for powering its unique functions. The plasma membrane has a $Na^+/K^+$ ATPase. The $Na^+$ and $K^+$ concentration gradients it establishes provide energy for depolarization and excitability of the cell membrane. Moreover, the $Na^+$ fgradient supplies energy for $Ca^{2+}$ removal as well as for the cellular uptake of ascorbate, catecholamines, and other metabolites. The chromaffin–vesicle membrane has an $H^+$-translocating ATPase. The $H^+$ gradient drives vesicular catecholamine accumulation and ascorbate regeneration.

The adrenal medulla has also established a general model for the energetics of secretory and synaptic vesicles. The V-type $H^+$-translocating ATPase is fundamental, and the same ATPase is found in vesicles in all types of secretory cells. The proton gradient generated by this ATPase drives transport into the vesicles via $H^+$/substrate exchange. The vesicular neurotransmitter transporters are related proteins, but a given cell expresses only the one transporting the appropriate substrate. Cytochrome $b_{561}$ uses the proton gradient to regenerate intravesicular ascorbic acid. It too is expressed only where it is needed: in cells with intravesicular ascorbate-requiring enzymes. How is the expression of these different components coordinated to match the phenotypes of the different secretory cells? The adrenal medulla has yielded many insights but has not yet conceded all of its secrets.

## BIBLIOGRAPHY

Henry, J. P., Botton, D., Sagne, C., Isambert, M. F., Desnos, C., Blanchard, V., Raisman-Vozari, R., Krejci, E., Massoulie, J., and Gasnier, B. (1994). Biochemistry and molecular biology of the vesicular monoamine transporter from chromaffin granules. *J. Exp. Biol.* **196**, 251–262.

Johnson, R. G., Jr. (1988). Accumulation of biological amines in chromaffin granules: A model for hormone and neurotransmitter transport. *Physiol. Rev.* **68**, 232–307.

Nelson, N. (1992). Evolution of organellar proton-ATPases. *Biochim. Biophys. Acta* **1100**, 109–124.

Njus, D., and Kelley, P. M. (1993). The secretory-vesicle ascorbate-regenerating system: A chain of concerted $H^+/e^-$ transfer reactions. *Biochim. Biophys. Acta* **1144**, 235–248.

Njus, D., Kelley, P. M., and Harnadek, G. J. (1986). Bioenergetics of secretory vesicles. *Biochim. Biophys. Acta* **853**, 2347–2265.

Schuldiner, S., Shirvan, A., and Linial, M. (1995). Vesicular neurotransmitter transporters: From bacteria to humans. *Physiol. Rev.* **75**, 369–392.

Srivastava, M., Gibson, K. R., Pollard, H. B., and Fleming, P. J. (1994). Human cytochrome $b_{561}$: A revised hypothesis for conformation in membranes which reconciles sequence and functional information. *Biochem. J.* **303**, 915–921.

Winkler, H. (1993). The adrenal chromaffin granule: A model for large dense core vesicles of endocrine and nervous tissue. *J. Anat.* **183**, 237–252.

# Bioethics

LAWRENCE J. SCHNEIDERMAN

*University of California, San Diego*

## GLOSSARY

**Autonomy** Freedom to make choices and act according to a person's values, goals, and plans

**Beneficence** Act of conferring benefits on others. This act is ordinarily nonobligatory; that is, members of society generally do not have a duty to confer benefits. Exceptions occur, however, within certain roles and relationships. For example, the physician–patient relationship imposes a duty on the part of the physician to act benefi cently in the patient's best interests

**Distributive justice** Morally valid distribution of benefits (e.g., health care) and burdens (e.g., costs) in society

**Duty** Morally valid obligation to others

**Ethics and morals** Terms coming from Greek and Latin roots, respectively, meaning "character," "behavior," and "custom." Today they refer to the systematic examination of actions by some explicit standard of value which attempts to judge the actions as right or wrong, good or evil

**Nonmaleficence** Obligation shared by all members of society to avoid causing risk or harm to others

**Paternalism** Controlling of another person's actions—by such means as coercion or deception—for what is believed to be that person's own good

**Right** Morally valid claim on others (including privacy, i.e., the right to be left alone)

BIOETHICS DEALS WITH THE ETHICAL IMPLICA-tions of developments in the life sciences and health care.

## I. MATTERS OF LIFE AND DEATH

When does a human life begin and when does it end? To most people these might seem like simple questions. For the pregnant woman life often begins with the powerful emotional bonding that occurs with "quickening" (i.e., the first detectable fetal movement, which occurs at about the 15th week of pregnancy). And traditionally, for the family gathered around the death bed of the elderly matriarch, life ended when the pulse could no longer be felt and breath no longer stirred a feather. Thus, spontaneous physical movement can constitute the borders of human existence.

Philosophers, theologians, and judges have not found these matters to be so simple, however. Their questions would include those in the following section.

### A. Human Life's Beginning

When does the entity within a woman's womb become a moral agent, imposing on others its right to life? When is it entitled to protection from harm, claims to property, and the honors and rituals (e.g., baptism, naming, and burial ceremonies) expected by all full members of society? Do these rights begin at conception, or even prior to conception (which therefore imposes limits on contraception)? Or do these rights begin sometime between conception and birth, and, if so, when should the line be drawn— when the first primitive brain waves appear (at about the 7th week of development), or when the waves begin to take on a "human" pattern (after the 20th week)? When the heart starts its first primitive beats (in the 3rd or 4th week), or when these beats take on a more "human" pattern (after the 12th week)? When fetal movements occur in reaction to stimuli (in the 7th or 8th week) or when they are spontaneous (after the 10th week)?

ENCYCLOPEDIA OF HUMAN BIOLOGY, Second Edition, VOLUME 1.   Copyright © 1997 by Academic Press.   All rights of reproduction in any form reserved.

Strong human impulses of protectiveness are aroused when the developing fetus begins to "look human," an observation used to great effect by opponents of abortion. However, efforts to establish a universally accepted beginning of human life have failed. At one point members of the U.S. Congress tried to enlist the scientific community to agree that it was a "scientific fact" that human life began at conception. This effort failed, however, because scientists hold beliefs that are as varied as the community at large.

In fact, defining the beginning of human life is a matter to which scientists can only contribute empirical observations. Human life ultimately is a philosophical construct with strong religious, social, and psychological overtones.

Proposed criteria have tended to fall into one or more categories:

1. Religious: In the Catholic tradition, for example, human life begins at ensoulment. The leading 12th century theologian St. Thomas Aquinas, influenced by the Aristotelian theory of embryonic development as a transformation of vegetable soul to rational soul, proposed that ensoulment occurred when the fetus took on a human shape. Centuries of debate followed regarding the severity of sin attached to early (i.e., prior to "vivification") and late abortion before the Church settled on its present position with respect to ensoulment and its absolute opposition to contraception and all abortion.

2. Psychological: Some argue that to be a human being involves internal capacities of sentience (i.e., the capacity for feeling, or affect, and consciousness). More stringent criteria of personhood have also been advocated; that is, human life should date its onset from the time that self-awareness and self-determining thoughts and actions—in other words, autonomy—are possible.

3. Social: Those who advocate social criteria look to external capacities of the individual to communicate and interact with the environment and the surrounding society, and would require these manifestations to qualify for human life designation.

4. Biological: A number of notions have been proposed, including (a) species specificity, that anything with the normal human complement of chromosomes should be regarded as human; (b) viability, that human life should be dated from the moment it is able to live free of its mother's womb; (c) potentiality, that human life should be acknowledged when, in the normal course of events, a product of conception would be expected to develop into a fully mature

human being; and (d) individuality, that a fully recognized and protected member of society should contain not one but many attributes, including the following: the human genome, developed mental singleness, functional unity, behavioral performance, sentience, and social interaction.

As is evident, depending on which construct one places on the beginning of human life, many divergent consequences follow. For example, is abortion destruction of a human being at one point in the pregnancy but not at another? If human life begins at conception, should women's menses (which could represent an undetected spontaneous miscarriage) be examined so that the early embryo can be named, baptized, and buried appropriately? Are infants born with a condition known as anencephaly (i.e., absence of most of the brain) deserving of the same life-prolonging treatment as a normally developed newborn?

It is possible to begin life by a variety of nontraditional techniques, including artificial insemination by a donor, *in vitro* fertilization, and surrogate embryo transfer, sometimes involving frozen material. These methods raise additional questions. Who should be granted custody of frozen eggs fertilized in the laboratory in the event of divorce or death of one or both parents? Should fertilized eggs or other early embryonic forms be made available for research? Is their destruction equivalent to abortion? Is it consistent with American values to allow a "free market," with renting of wombs and buying and selling of eggs and sperm?

## B. Human Life's Ending

Agreement has been easier to reach at the other end of life. Although in the late 1960s and early 1970s confusion and controversy were caused by newly developed technologies, including mechanically supported respiration and circulation—which could keep organs pumping as long as electricity was provided—in 1981 the President's Commission for the Study of Ethical Problems in Medicine and in Behavioral and Biomedical Research established what is now regarded as the Uniform Determination of Death: "An individual who has sustained either irreversible cessation of circulatory and respiratory functions, or irreversible cessation of the entire brain, including the brain stem is dead."

This definition reflects the current consensus that a person's existence is centered in the brain. The total maintenance of vital functions (e.g., heart beat and respiration) by some external means does not constitute life. Conversely, any evidence of brain function disqualifies a conclusion of death.

Nevertheless, questions remain: What about the patient whose brain damage does not cause death, but is severe enough to cause permanent unconsciousness, called a permanent vegetative state? Such individuals, if they are provided with nutrients and fluids through the veins and by means of tubes to the stomach, can survive for decades without giving evidence of many of the qualities and behaviors we associate with being human. Should such patients be treated as equal members of the human community and be given the same life-prolonging treatment as a fully conscious person? Does a person's right to make autonomous life choices include the right to commit suicide? Is a patient with severe painful illness entitled to request and obtain treatment which deliberately shortens his or her life, sometimes called euthanasia? Is withholding life-saving treatment ethically the same as or different from treatment deliberately given to shorten life?

## II. HUMAN EXPERIMENTATION

Long before scientists codified ethical standards of research, physicians were reciting the Hippocratic oath, a text which emerged from scientific and ethical writings of a "school" of physicians on the island of Cos, possibly in the late fourth and fifth century BC. Since then the oath has become the accepted creed throughout the world of medicine, recognized by Jewish, Christian, and Moslem physicians, holding sway through diverse times, including the Medieval period, the Renaissance, and the Enlightenment, up to the present.

Thus, for historical reasons alone, the Hippocratic oath is a remarkable document. Also remarkable is that it was the first instance in which a powerful figure, the physician, swore a duty to someone in his (all were men) power: the patient. Unlike other traditions of fealty (e.g., between royalty and subject), the oath imposed no reciprocal obligation.

In the Hippocratic oath the physician swears, "I will use treatment to help the sick according to my ability and judgement, but never with a view to injury and wrong doing. . . . Into whatsoever houses I enter, I will enter to help the sick and I will abstain from all intentional wrong-doing and harm, especially from abusing the bodies of man or woman, bond or free. And whatsoever I shall see or hear in the course of my profession, as well as outside my profession in my intercourse with men, if it be what should not be published abroad, I will never divulge, holding such things to be holy secrets."

This strong abjuration to serve the patient's best interests and abstain from wrongdoing and harm proved to be insufficient to prevent physicians from conducting forced experiments on human beings which in the physicians' view served to advance medicine and mankind. At the Nuremberg trials of the late 1940s following World War II, the world was shocked to discover the scale and brutality of experiments carried out by Nazi physicians on concentration camp inhabitants. Later, even in the United States, abuses were discovered, such as depriving poor unconsenting blacks of medication to observe the natural history of untreated syphilis, injecting unwitting patients with cancer cells, and misleading Spanish-speaking women into thinking they were obtaining contraceptives, thus causing unwanted pregnancies. These events drew attention to the long and until then commonly accepted practice of scientists' performing research on people without informing them or gaining their permission.

Since then strict standards for scientific research on human subjects have been established throughout the world based on the principle of autonomy and informed consent. As stated in the Nuremburg Code:

> The voluntary consent of the human subject is absolutely essential.
>
> This means that the person involved should have legal capacity to give consent; should be so situated as to be able to exercise free power of choice, without the intervention of any element of force, fraud, deceit, duress, over-reaching or other ulterior form of constraint or coercion; and should have sufficient knowledge and comprehension of the elements of the subject matter involved as to enable him to make an understanding and enlightened decision. This latter element requires that before the acceptance of an affirmative decision by the experimental subject there should be made known to him the nature, duration, and purpose of the experiment; the method and means by which it is to be conducted; all inconveniences and hazards reasonably to be expected; and the effects upon his health or person which may possibly come from his participation in the experiment.
>
> The experiment should be such as to yield fruitful results for the good of society, unprocurable by other methods or means of study and not random and unnecessary in nature. . . .
>
> The experiment should be conducted so as to avoid all unnecessary physical and mental suffering and injury. . . .
>
> The degree of risk to be taken should never exceed that determined by the humanitarian importance of the problem to be solved by the experiment. . . .

These principles have been reinforced by the Declaration of Helsinki and regulations issued by the U.S. Department of Health and Human Services. Although the Nuremburg Code asserts the right of the human subject "to bring the experiment to an end if he has reached the physical or mental state where continuation of the experiment seems to him to be impossible," regulations of the Department of Health and Human Services assert an even stronger claim: "The subject may discontinue participation at any time without penalty or loss of benefits to which the subject is otherwise entitled."

## III. GENETIC AND HUMAN ENGINEERING

Research in human biology has advanced to the stage that scientists have now embarked on the ambitious project of characterizing the structure and function of all of the estimated 100,000 or so genes and other components in the human genome. Techniques include the ability to cut and splice genetic material—a process known as genetic engineering—which provide opportunities for scientists to not only study, but even correct, some genetic defects, a process which might be called human engineering. Although this prospect is greeted with amazement and enthusiasm by many, others are alarmed by the variety of ethical problems that will inevitably occur. As society becomes more proficient in detecting and treating genetic disorders, the very nature of what constitutes a normal human being will come under scrutiny. Will society, tempted by the possibility of human perfection, begin to view not merely serious or life-threatening diseases as intolerable, but also minor variations (e.g., below average height or less than average physical beauty), further distorting health and illness into matters of aesthetics?

Through genetic manipulation in the laboratory, large amounts of biosynthetically manufactured human growth hormone have already become available. In the past this hormone was available only in scanty supply from animal pituitary gland extraction, and its use was confined to the treatment of children with severely impaired growth due to hormone deficiency. Now, however, with vastly increased supplies, physicians could use the hormone to increase the height of any child; already it has proved to be tempting for parents who want their sons to be tall to gain the perceived advantage such increased height gives to men in this society. If tall men have such advantages, can parents be condemned for trying to help their sons? As society's expectations change will men whose heights are acceptable now be considered defective or abnormal in the future? Along with developing

technology will come the temptation to insert into a child's somatic chromosomes the gene itself that promotes growth (or some other desirable trait). And already there are those who advocate inserting selected genes into germ line chromosomes to allow future generations to inherit currently favored traits. What are the ethical as well as biological implications of making choices for unconsenting descendents, meanwhile tampering with processes that have resulted in our species' survival after eons of natural selection?

One can also foresee that employers and health insurers, under economic pressures and in the pursuit of perfection, will be tempted to require preemployment and preinsurance genetic screening analogous to drug testing to exclude individuals predisposed to undesirable and costly conditions, say heart disease and cancer. Will individuals with such genes thus face the double jeopardy of job and insurance discrimination? Will their screening status be protected from invasion of privacy? Will society put pressure on "abnormal" individuals either for treatment or to avoid reproducing in order to reduce health care costs? What will be the effect of all of these considerations on society as its tolerance for human diversity narrows? Will it continue to support freedom of choice and maintain compassion for the less fortunate?

History has shown that these are not idle questions. Not long ago, screening programs to detect sickle cell anemia in American blacks proved harmful to those individuals who were identified as heterozygous carriers of the innocuous sickle cell trait. Many were unfairly denied life insurance and employment opportunities. And, of course, Nazi Germans used a eugenic rationale of "racial purification" to justify the extermination of physically and mentally handicapped persons as well as the slaughter of millions of Jews. Because scientific research and technology are moving so swiftly, these potential conflicts among science, society, and the individual which pose problems of unprecedented ethical complexity must be addressed as soon as possible by an enlightened, thoughtful, knowledgeable—and vigilant—society.

## IV. RATIONING AND RESOURCE ALLOCATION

Ideally, research in human biology—like all scientific research—seeks knowledge free from any political agenda. Unavoidably, however, limited resources lead to the establishment of priorities. Because of the close relationship between research in human biology and health care, society will want to make the most rational use of such limited resources to achieve human benefits toward curing disease. Ironically, as technology has succeeded in achieving remarkable advances on a variety of frontiers, the costs of this technology have come under increasing scrutiny. Scientists and the public are in constant debate as to whether dollars should be channeled into basic research, on the chance that future breakthroughs will occur, or devoted to applications and refinements of treatments currently available.

Decisions such as these will affect conditions as diverse as the acquired immunodeficiency syndrome (AIDS), cancer, Alzheimer's disease, heart disease, diabetes, birth defects, and kidney failure. For example, should more clinical effort be devoted to clinical studies of drugs now available for managing AIDS, which could provide temporary benefits, but no long-term cure? Or should more resources be devoted to the development of vaccines, which might or might not be feasible in the near future, if at all? Or should research efforts be concentrated at an even more fundamental level, emphasizing basic biological questions about viruses and immunology? What principles should guide such decisions: Maximizing present benefits? Maximizing future benefits? Should priorities be based on disease prevalence? As the proportion of the aged in the population increases, should more research be devoted to diseases affecting the elderly? Or should children always be favored over the elderly in order to give them a chance at a full life? Should principles of social utility be adopted or principles of social compassion? In other words, should resources be aimed at helping the more productive members of society or the more handicapped? Or should "free market" principles continue to operate, which tend to favor benefits for the wealthy over the poor?

All of these questions are subsumed under the term "distributive justice." When a society allocates a large sum of money to one technology or service, it is reasonable to ask what might be accomplished if the same investment were spent on an alternative technology or service. That is, what are the opportunity costs of these technologies, i.e., what better opportunities are being missed by choosing one over the other? Similarly, one might ask what constitutes a benefit? That is, what does society owe to its citizens? In terms of health care, it has been argued that a society owes a normal opportunity range for individuals to exercise their skills and talents to the fullest, and that if a person's limitations are the results of disease and disability—not merely the result of normal variation in individual talents and abilities—society should make an effort to correct for the inequalities since health care institutions are among the basic institutions of a civilized society, and fair equality of opportunity should be the goal of a just society.

## BIBLIOGRAPHY

Arras, J., and Rhoden, N. (eds.) (1989). "Ethical Issues in Modern Medicine," 3rd Ed., Mayfield, Mountain View, CA.
Bulger, R. E., Heitman, E., and Reiser, F. J. (1993). "The Ethical Dimensions of the Biological Sciences." Cambridge Univ. Press, New York.
Daniels, N. (1985). "Just Health Care." Cambridge Univ. Press, New York.
Gert, B. (1988). "Morality: A New Justification of the Moral Rules." Oxford Univ. Press, Oxford, England.
Grobstein, C. (1988). "Science and the Unborn." Basic Books, New York.
Weil, W. B., Jr., and Benjamin, M. (eds.) (1987). "Ethical Issues at the Outset of Life." Blackwell, Boston.

# Biofeedback

JOHN P. HATCH
*The University of Texas Health Science Center at San Antonio*

---

## GLOSSARY

**Classical conditioning** Formation of a conditioned reflex through the temporal pairing of an unconditioned stimulus, which reflexively elicits a response, and a neutral conditioned stimulus. After the pairing the formally neutral conditioned stimulus acquires the ability to elicit a response similar to the one elicited by the unconditioned stimulus

**Operant conditioning** Presentation of a stimulus dependent on the occurrence of a response for the purpose of modifying the strength or frequency of the response

**Reinforcer** In operant conditioning any stimulus which when presented or removed dependent on the occurrence of a response results in the modification of the response. In classical conditioning the presentation of a conditioned stimulus and an unconditioned stimulus in close temporal proximity

BIOFEEDBACK IS THE TECHNIQUE OF USING ELEC-tronic equipment to display to people some aspect of their biological functioning, usually in the form of auditory or visual signals, for the purpose of teaching them to modify some physiological event. The physiological event to be modified can be either the normal or the abnormal activity of an organ or system. The first step in the biofeedback process is the detection of the physiological signal with an appropriate trans-ducer. The detected signal is then amplified and con-verted into a form that is accessible to the external senses. Finally, the signal is fed back to the person who then uses the information in his or her attempts at gaining voluntary control over the targeted physio-logical event. Usually, the targeted physiological event is one that is normally not available to conscious awareness such as blood pressure level or brain wave activity. However, biofeedback also may be used in modifying a normally voluntary function over which the person has lost voluntary control because of dis-ease or injury. Although the psychological mecha-nisms involved are not fully understood, it generally is assumed that the development of voluntary physio-logical control with biofeedback represents a learn-ing process.

## I. UNDERLYING LEARNING THEORY

Learning theorists traditionally have distinguished be-tween two types of learning. One type, called *classical conditioning*, involves the pairing in close temporal proximity of two stimuli. One of the two stimuli, known as the unconditioned stimulus, normally elicits some reflex behavior such as salivation, eye blink, or change in skin conductance. The other stimulus, known as the conditioned stimulus, is neutral prior to the pairing but acquires the ability to elicit a response similar to the unconditioned response after the pairing process. The second type of learning recognized by learning theorists is known as *operant conditioning*. In operant conditioning the response is generally thought of as emitted on a more or less voluntary basis rather than being reflexively elicited by any particular stimulus. Also, the operant conditioning process in-volves the pairing of a stimulus dependent on the

ENCYCLOPEDIA OF HUMAN BIOLOGY, Second Edition, VOLUME I.   Copyright © 1997 by Academic Press.   All rights of reproduction in any form reserved.

occurrence of a response. The paired stimulus is known as a positive reinforcer if its presentation leads to an increase in the frequency or magnitude of the response, and it is known as a negative reinforcer if its removal leads to an increase in the frequency or magnitude of the response. [*See* Conditioning.]

Between 1928, when the two types of learning were formally distinguished, and the early 1960s it was generally assumed that operant and classical conditioning were mutually exclusive processes. All voluntary skeletal motor behavior was assumed to be subject to operant conditioning, while visceral and glandular responses mediated by the autonomic nervous system were assumed to be involuntary and modifiable only through classical conditioning. This view was retained until the early 1960s when groups of researchers in the Soviet Union, Canada, and the United States began a series of experiments designed to show that responses mediated by the autonomic nervous system can in fact be brought under operant control.

## II. EARLY DEVELOPMENT

### A. Experiments on Humans

It was first shown that human subjects could learn to voluntarily dilate blood vessels of the finger in order to avoid or escape an electric shock if they were provided with visual information about their vasomotor activity. Subjects provided with an amplified auditory representation of their heartbeat also were trained to accelerate their heart rate in order to avoid an electric shock to the ankle. Other responses generally considered involuntary were brought under voluntary control by providing subjects with feedback information about the targeted response and some form of operant reinforcement. Studies conducted in the 1960s also demonstrated operant conditioning of the skin conductance response and brain wave activity. Voluntary control over the firing of single motor units was demonstrated by providing subjects with auditory and visual displays of individual myoelectric potentials recorded from fine wire intramuscular electrodes.

### B. Experiments on Animals

At about the same time that voluntary control of autonomically mediated behavior was being demonstrated in humans, parallel lines of research were in progress using various animal species. For example, thirsty dogs were trained to increase and decrease the flow of saliva using water as a reinforcer. Bidirectional changes in heart rate and blood pressure were also shown to be subject to operant conditioning in monkeys and rats.

## III. SOMATIC MEDIATION

The results of the human and animal studies just described were replicated many times, and the conclusion that autonomically mediated responses can be brought under voluntary control through operant conditioning is now firmly established. However, needed information about the biological mechanisms involved in such learning was not available, and the question of whether autonomic responses were being directly controlled or whether they were being indirectly affected by some change in somatic responding was hotly disputed. For instance, in the early studies involving heart rate and vasomotor conditioning in humans, the subjects were frequently observed to alter their respiration or muscle tension level. It could not be determined from these studies whether subjects were learning to control their heart rate directly or whether they were learning to alter their respiration or their level of muscular exertion, which in turn caused their heart rate to change in the appropriate direction.

### A. Experiments on Curarized Animals

In order to address the issue of somatic mediation, a series of studies were carried out on animals that had their skeletal muscles paralyzed with the drug *d*-tubocurarine. This experimental preparation was believed to control for changes in skeletal motor or respiratory activity that might be mediating the physiological response. Using direct electrical stimulation of the brain as a positive reinforcer and escape from electric shock as a negative reinforcer, operant conditioning of many autonomic responses was demonstrated in over 20 experiments. Soon, however, the magnitude of the results gradually declined to the point where the original experiments could no longer be replicated. Although a total of 2500 additional rats were studied under curare, the original experiments could not be replicated, and the reasons for the failure to replicate were never discovered. Therefore, the existence of operantly conditioned autonomic responding by rats under curare must be considered unproved at this time.

## B. Specificity of Biofeedback Effects

Despite the difficulties in proving that operant conditioning of autonomic responses is not somehow mediated by somatic responding, research on the voluntary control of autonomic responses has continued. Under the rubric biofeedback much has been learned about the ability of human beings to voluntarily alter many aspects of their biological function. Some of these voluntary visceral acts achieve a high degree of response specificity, which suggests that they are not simply a part of a more general somatic activation response. For example, when patients paralyzed from the neck down were trained with biofeedback to raise their blood pressure, most did so without significant alteration in heart rate. Normal subjects also were trained to alter heart rate without affecting blood pressure and to alter blood pressure without affecting heart rate. Subjects also were trained to raise and lower heart rate and blood pressure together or in opposite directions simultaneously. In biofeedback studies designed to train people to modify hand temperature, it was found that people can reliably produce a difference in the temperature of the two hands, and that as training progresses they can control the temperature of increasingly specific areas of skin. If these responses were secondary to somatic activity, then such a high degree of response specificity would be unexpected.

Studies in which somatic activity in the form of gross movement, muscular electrical activity, or respiratory activity has been recorded and correlated with change in the targeted physiological response have produced mixed results. Some studies showed parallel changes in somatic and operantly conditioned visceral responses while others did not. It also was shown that subjects can alter their heart rate with biofeedback even while pacing their respiration rate to a frequency set by the experimenter. On the whole, the available evidence suggests that somatic responding and visceral responding often show parallel change during biofeedback training. Thus, some change in somatic behavior may be necessary for voluntary visceral responding to occur. However, changes in somatic responding do not appear to be a sufficient explanation for changes in visceral responding that occur during biofeedback training. The extent to which somatic responses are involved in the voluntary regulation of visceral functions remains an important theoretical question. However, to view the somatic and autonomic nervous systems as independent would be to take a narrow view of the physiology involved. Neither somatic nor autonomic responses occur in a vacuum, and it probably is the case that neither type of response can be well understood without considering the other type as well.

## IV. COGNITIVE MEDIATION

Questions also were raised about possible cognitive mediation in biofeedback studies with human subjects. Because people have voluntary control over their cognitive processes and because certain types of thoughts, images, and emotions are known to affect physiological function, it was considered possible, for example, that a subject who was operantly conditioned to increase and decrease heart rate could do so by creating exciting or relaxing mental images.

In contrast to the issue of somatic mediation, the question of cognitive mediation of visceral responding has received relatively little research attention. In a number of investigations subjects were simply asked what they were thinking about at the time that they demonstrated the desired response. In general, subjects did tend to report more active and arousing images during training to speed heart rate and more relaxing and passive images during training to slow heart rate. However, when subjects were asked to engage in arousing or relaxing imagery, the corresponding changes in physiological function were small compared to those observed during biofeedback. These limited findings would suggest that, like somatic mediation, cognitive mediation is probably not a sufficient explanation for the voluntary control that is achieved with biofeedback training.

Although biofeedback research has been strongly influenced by operant conditioning, other theoretical models have influenced the field. One such model is the *ideomotor model*. According to the ideomotor model, voluntary acts are mediated by mental images stored in the brain of the movements to be carried out. An issue that continues to be debated is the question of whether the biofeedback signal should be conceptualized as a reinforcing stimulus or as a source of information that the subject uses in establishing voluntary control. The two points of view are not mutually exclusive since a reinforcer always conveys some information and the presentation of information under the right circumstances can reinforce behavior.

The most well-developed version of the ideomotor theory in the field of biofeedback is the conceptualization of learned heart rate control. According to this model, the repeated occurrence of a response results in the establishment of a motor image of the response. When a person later performs the act he does so by constantly comparing the immediate sensations consequent upon the response with the stored image of that act. This theory would predict that as a response is learned, the learner would develop a greater ability to sense the occurrence of the response. The ideomotor theory has spawned a considerable amount of research on the discriminability of heart rate with biofeedback training, but the evidence does not strongly support the hypothesis that biofeedback enhances the discriminability of heart beats. Although myocardial contraction is presumed to be the stimulus for the heartbeat sensation, the accuracy of heart beat detection is not a simple function of stroke volume.

## V. BIOFEEDBACK PARAMETERS AFFECTING LEARNING

Studies have also focused on the various parameters that affect the acquisition and performance of voluntary visceral responding with biofeedback. Many of the factors that affect learning with biofeedback also affect the learning of other voluntary motor acts. Either visual or auditory signals may be used as the feedback stimulus, and no clear advantage has been found for either type. A distinction also can be made between binary and analogue biofeedback. With *binary biofeedback*, the information presented indicates only whether the magnitude of the target response is above or below some threshold level. For example, a tone may be turned on whenever hand temperature equals or exceeds 90°F and turned off whenever it falls below that temperature. The subject is informed only whether or not the response meets the criterion at any moment. With *analogue biofeedback*, the feedback signal varies proportionately with the magnitude of the target response. For example, hand temperature might be continuously displayed on a meter so that the subject receives information not only about the direction of the response but also about the magnitude and the form of the response. Many studies have demonstrated superior performance of subjects with analogue feedback as compared to binary feedback, but there are exceptions. For instance, binary feedback can produce as large a decrease in heart rate as analogue feedback. Almost all current studies use some form of analogue feedback signal, some of which can become quite elaborate with the use of computerized graphics displays.

Another feedback parameter that affects learning is the immediacy of the feedback signal with respect to a change in the target response. Similar to the reinforcer in operant conditioning, any change in the biofeedback stimulus should occur as soon as possible following a change in the target response for optimal learning. Feedback that is delayed by only a few seconds is not as effective as immediate feedback.

Sometimes the feedback signal is the only form of reinforcement used in biofeedback training, but tangible reinforcers such as money or prizes for appropriate performance may also be used. For most subjects the knowledge of results provided by the biofeedback plus their own desire to master the task is sufficient to motivate learning. However, an additional motivational effect of tangible reinforcers has been shown.

## VI. BIOLOGICAL MECHANISMS

Relatively little is currently understood about what changes occur in the biological mechanisms that underlie biofeedback-assisted voluntary control of visceral responding. In monkeys, operant heart rate control depends on activity in the sympathetic and parasympathetic cardiac innervations. In humans, different physiological response profiles develop during voluntary heart rate speeding and slowing, which suggest differences in the relative degree of sympathetic and parasympathetic activation. The available evidence suggests that during biofeedback-assisted heart rate speeding there is both a decrease in cardiac parasympathetic neural tone and an increase in sympathetic neural tone. Voluntary heart rate slowing, however, seems to be primarily dependent on an increase in cardiac parasympathetic activity. Operant heart rate control is possible in human subjects with an intact innervation at the level of the arterioventricular (AV) node, but operant conditioning of heart rate has not been demonstrated in patients suffering from complete heart block. It also has recently been discovered that the vasodilation produced during biofeedback-assisted hand temperature warming is mediated through a nonneural, $\beta$-adrenergic mechanism. The vasoconstriction that accompanies voluntary hand temperature cooling, however, is mediated by an efferent sympathetic, nervous pathway.

## VII. CLINICAL APPLICATIONS

As soon as it was discovered that autonomically mediated responses could be brought under voluntary control, great interest was generated over possible clinical applications of biofeedback. It was speculated that patients who demonstrated dysregulation of various response systems might be trained to regulate their physiological functioning in a direction of improved health. Although large-scale, controlled clinical trials are lacking, there have been a large number of small studies that support the clinical efficacy of biofeedback therapy for a variety of disorders.

In general, there are two rather broad philosophies that are followed in the clinical application of biofeedback therapy. One attempts to take advantage of the specific effects that can be produced in a response system. Following this approach, the therapist attempts to teach the patient to reverse specific pathophysiological events. For instance, sinus tachycardia patients were trained to slow their heart rate with biofeedback, and patients suffering from muscle contraction headache were trained to relax hypertonic scalp and neck muscles. The second philosophy contends that biofeedback can be used to promote voluntary control over a general psychophysiological state, which is thought to result from physical or psychological stress. Because many common disorders are believed to be caused or aggravated by stress, biofeedback is used to teach patients to relax and to cope more effectively with stress. For example, biofeedback often is used to teach migraine headache sufferers to raise their hand temperature. Even though warming of the hands is not known to directly or specifically alter the pathophysiology of migraine headache, several studies demonstrated a reduction in headaches using this technique. It is assumed that the hand-warming response assists the patient in achieving a state of deep relaxation, which in turn somehow interferes with the relationship between stress and headache.

Biofeedback has been applied to a wide variety of medical disorders. Within the cardiovascular system heart rate biofeedback has been used in the treatment of sinus tachycardia, cardiac arrhythmias, and anxiety. The vasodilation response that accompanies biofeedback-assisted hand warming is beneficial in the treatment of Raynaud's disorder, which involves the painful constriction of blood vessels in the hands and feet. Hand-warming biofeedback also is used in the treatment of migraine headache and hypertension. Direct blood pressure biofeedback has been used in the treatment of hypertension with limited success, but it has met with greater success as a treatment for orthostatic hypotension in patients who have suffered spinal cord injury.

Feedback of skeletal muscle activity in the form of electromyographic biofeedback is used in treating many stress-related disorders. Electromyographic biofeedback is widely used in the treatment of patients with muscle contraction headache, temporomandibular joint disorders, and low back pain. Electromyographic biofeedback also is extensively applied as part of a physical rehabilitation program in the treatment of patients suffering from cerebral palsy, spinal cord injury, stroke, and spasmodic torticollis. As a component of a generalized relaxation training program, electromyographic biofeedback is used in treating chronic pain syndromes, hypertension, asthma, irritable bowel syndrome, and many other stress-related or psychosomatic disorders.

One of the most successful clinical applications of biofeedback is the use of manometric feedback from the anal sphincter to restore voluntary control in patients with fecal incontinence. Manometric biofeedback is assumed to operate by increasing the strength of the external anal sphincter muscle, by teaching the patient to coordinate external and internal anal sphincter activity, and by assisting the patient in learning to recognize the sensations associated with rectal distention. Biofeedback of brain wave activity is used in treating epilepsy. Epileptic patients provided with biofeedback information about their production of 12–14 Hz (sensory motor rhythm) activity can selectively augment their production of this brain wave. Patients treated in this way often show a reduction in the frequency of their epileptic seizures.

## BIBLIOGRAPHY

Basmajian, J. V. (1989). "Biofeedback: Principles and Practice for Clinicians," 3rd Ed. Williams & Wilkins, Baltimore.

Engel, B. T., and Schneiderman, N. (1984). Operant conditioning and the modulation of cardiovascular function. *Annu. Rev. Physiol.* **46,** 199.

Finley, W. W., and Jones, L. C. (1992). Biofeedback with children. *In* "Handbook of Clinical Child Psychology" (C. E. Walker and M. C. Roberts, eds.), 2nd Ed. Wiley, New York.

Gatchel, R. J. (1988). Clinical effectiveness of biofeedback in reducing anxiety. *In* "Social Psychophysiology and Emotion: Theory and Clinical Applications" (H. L. Wagner, ed.), Wiley, Chichester, England.

Hatch, J. P., Fisher, J. G., and Rugh, J. D. (eds.) (1987). "Biofeedback: Studies in Clinical Efficacy." Plenum, New York.

Miller, N. E. (1978). Biofeedback and visceral learning. *Annu. Rev. Psychol.* **29,** 373.

Schwartz, M. S. (1987). "Biofeedback: A Practitioner's Guide." Guilford Press, New York.

# Bioimpedance

HENRY C. LUKASKI

*United States Department of Agriculture*[1,2]

## GLOSSARY

**Body cell mass** Living cells of the body; contains the oxygen-utilizing, energy-producing, potassium-rich, work-performing component of the body

**Densitometry** Method for determination of whole-body density by using measurements of mass in air and body volume determined by submersion in water and corrected for residual lung volume; synonymous with underwater weighing

**Electrolyte** Any ion or compound that in a water solution conducts an applied electrical current

**Extracellular fluid** Water and electrolytes bathing the outside of cells; chemical composition is principally sodium and chloride ions and water

**Extracellular mass** Tissues and fluids of the body that are wholly outside of cells; includes extracellular fluid, bone and connective tissue; functions to provide mechanical support via the skeleton and connective tissue and to transport substrates and metabolites in the plasma and fluid outside of cells

**Fat** Triglycerides stored in fat cells or adipocytes

**Fat-free mass** Non-fat-containing tissues in the body; contains only water, potassium, protein, bone, and connective tissue

**Membrane capacitance** Amount of electrical charge stored by a cell membrane; may be related to the electrochemical gradient or separation of electrolytes by the cell membrane

**Percent body fat** Fat mass of the body expressed as a percentage of total body mass

THE USE OF METHODS TO ASSESS HUMAN BODY composition extends across many scientific disciplines and physiological states. Human nutritionists estimate fat-free mass (FFM), fat mass (FM), and percent body fat (%BF) in evaluations of whole-body energy status in malnutrition, obesity, pregnancy, and lactation. Animal scientists seek to evaluate in a nondestructive manner the carcass composition (fat and lean) in meat-producing livestock used for human food consumption. Physiologists rely on noninvasive methods of body composition assessment in physical fitness appraisal, athletic counseling, and the establishment of standards for metabolic variables such as energy expenditure. Clinical scientists require techniques that facilitate the assessment of compositional changes (body cell mass and body fluid distribution) associated with pathology and the serial evaluation of the effects of therapeutic intervention on body composition. Each of these applications utilizes costly, laboratory-based instrumentation requiring skilled technical support.

In contrast, the physical anthropologist and human biologist require methods that are not limited to the research laboratory but are amenable to epidemiological and field use to identify population differences in body composition and to describe structural adaptations associated with growth, development, aging, and acculturation. Such a method must be safe, relatively noninvasive, portable, rapid to use, and cost-

---

[1]U.S. Department of Agriculture, Agricultural Research Service, Northern Plains Area, is an equal opportunity/affirmative action employer, and all agency services are available without discrimination.

[2]Mention of a trademark or proprietary product does not constitute a guarantee or warranty by the U.S. Department of Agriculture and does not imply its approval to the exclusion of other products that may also be suitable.

ENCYCLOPEDIA OF HUMAN BIOLOGY, Second Edition, VOLUME 1.   Copyright © 1997 by Academic Press.   All rights of reproduction in any form reserved.

effective. The method also must be sensitive and specific for the measurement of the important compositional variable of interest. Whereas anthropometric measurements of the body have a general appeal because of the ease of performing the measurements, the accuracy of estimation of FFM and %BF with anthropometry is subject to errors, particularly in the obese and the elderly, because of differences in adipose tissue distribution and skin compression. Although many methods and techniques are currently available, the majority of them do not possess all of these desirable characteristics. One method, tetrapolar bioelectrical impedance analysis (TBIA), apparently meets these criteria.

This review describes the physical basis of the impedance method and its application in the assessment of human body composition. A discussion of the development of new models of TBIA use for assessment of complex body compositional variables in health and disease states also is presented.

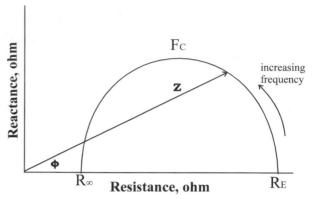

**FIGURE 1** Impedance plot showing the geometric relationships between resistance, reactance, impedance ($Z$), and phase angle ($\phi$). The critical frequency ($F_C$) is defined as the frequency at which reactance is maximal. Estimates of the resistance of the extracellular fluid ($R_E$) and the total body water ($R_\infty$) are derived by using curve-fitting techniques. The resistance of the intracellular fluid ($R_I$) is calculated as $R_I = [(R_E \cdot R_\infty)/(R_E - R_\infty)]$.

## I. TETRAPOLAR BIOELECTRICAL IMPEDANCE ANALYSIS

### A. General Principles

The TBIA method relies on measurements of basic electrical variables (impedance, resistance, and reactance) to index specific body compositional parameters. Impedance ($Z$) is the opposition of a conductor, whether it be a living organism or a metallic wire, to the flow of administered alternating electrical current. Impedance has two geometric components, resistance and reactance (Fig. 1). Resistance ($R$) is the real opposition of the conductor to the flow of current. In practical terms, $R$ is the inverse of conductance or the ability of a conductor to transmit an electrical current; it is equal to the voltage divided by the current according to Ohm's law. Reactance ($Xc$), which is the reciprocal of capacitance, represents the brief and transient storage of electrical charge associated with several types of polarization (separation of electrical charges and electrochemical gradient) that may be produced by cell membranes. Importantly, capacitance causes the current to lag after the applied voltage and creates a phase shift that is characterized as the arctangent of the ratio of $Xc$ to $R$, or the phase angle ($\phi$) in Fig. 1.

The geometric relationships among $Z$, $R$, $Xc$, and $\phi$ depend on the frequency of the applied current (Fig. 1). At low frequencies, the $Z$ of the cell membranes and tissue interfaces is too large for conduction of the

current into the cells to occur. As a result, the current is conducted only through the fluid and electrolytes bathing the cells, the extracellular fluid ($R_E$). Thus, the measured $Z$ is considered resistive with no $Xc$ component. As the frequency is increased, the current transverses the cell membranes and tissue interfaces, thereby causing an increase in $Xc$, a decrease in $R$, and a concomitant increase in $\phi$. For cells suspended in physiological saline, the magnitude of $Z$ is equal to a vector defined as $Z^2 = R^2 + Xc^2$.

A specific attribute of any conductor is the characteristic or critical frequency ($F_C$), which is defined as the frequency at which $Xc$ is maximal (Fig. 1). At frequencies exceeding the $F_C$, the $Xc$ decreases as cell membranes and tissue interfaces lose their capacitive ability and the applied current penetrates all cells. At very high current frequencies, $Z$ is equivalent to only $R$ ($R_\infty$), which is considered an index of total body water.

### B. Electrical Models for Body Composition Assessment

The theoretical basis for the use of TBIA variables for the *in vivo* assessment of human body composition is the simple electrical circuit model (Fig. 2). This model includes an $R$ attributable to an extracellular conduction path ($R_E$ or $R_I$) that depends on the amount and composition of the extracellular fluid and a complex intracellular component that includes cell

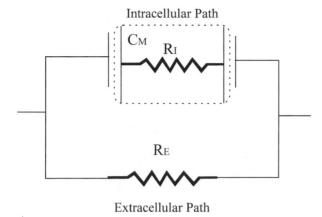

Intracellular Path

FIGURE 2 Circuit-equivalent model for cells suspended in physiological saline.

membrane capacitance ($C_M$) and intracellular fluid ($R_I$ or $R_2$).

In reality, the simple electrical circuit shown in Fig. 2 is an oversimplification of the complex electrical configurations that exist in the living organisms. The human body consists of many diverse arrangements of multicellular systems characterized by resistors and capacitors in series and parallel configurations. Awareness of these diverse electrical configurations has prompted an evolution of electrical models for the estimation of various body compositional variables.

## I. Electrophysical Model

The first applications of TBIA to monitor body composition used single-frequency impedance devices and assumed that the body was a simple geometric conductor characterized by variable length (e.g., standing height), uniform cross-sectional area, and homogeneous composition. It was hypothesized that the estimated electrical conductor volume was equivalent to the biological conductor volume. This hypothesis relied on basic conductor theory that $Z$ is proportional to conductor length ($L$) and inversely related to conductor cross-sectional area ($A$): $Z = \rho L/A$, where $\rho$ is the resistivity of the conductor. If both sides of the equation are multiplied by $L/L$, then $Z = \rho L^2/V$. Rearrangement yields $V = \rho L^2/Z$. Application of this model assumes that fluids and cells exist in a series electrical configuration of resistors and capacitors, thus $Z^2 = R^2 + Xc^2$.

## 2. Multiple-Frequency Models

Awareness that the electrical configuration of conductors (fluids and cells) and capacitors (cell membranes)

and their orientation in the body were more complex than the simple model of cells suspended in saline led to the recognition that a parallel electrical model was needed to estimate conductors of electrical current in the body. In this model (Fig. 2), the reciprocal of $Z$ squared is equal to the sum of the reciprocals of the squares of $R$ and $Xc$: $Z^{-2} = R^{-2} + Xc^{-2}$.

The availability of impedance devices with the capability of administering alternating current at frequencies ranging from a few kHz to 1 MHz and measuring $Z$ and $\phi$ led to the assessment of fluid distribution in animals and humans.

Two distinct approaches have been developed and implemented. In one approach, fixed or *a priori* determined frequencies were assumed to index different fluid distributions. It was hypothesized that $Z$ at a low frequency (e.g., 5 kHz) predicted extracellular water (ECW) and $Z$ at a high frequency (e.g., 1 MHz) estimated total body water (TBW). When standing height was squared and then divided by these individual $Z$ determinations, these values were used with linear regression to develop mathematical models for prediction of ECW or TBW.

A second experimental strategy was to model $Z$ data as a semicircle (Fig. 1) derived by scanning a wide range of frequencies from a few kHz to MHz during the TBIA measurement. This approach acknowledges that the human body has variable amounts and distributions of heterogeneous components (e.g., muscle, adipose tissue, organs) that cause electrical interfaces that retard and release electrical charges when an alternating current is applied. It has been hypothesized that this approach will permit determination of the relative contribution of each component in the body. Use of specific $R_0$ and $R_\infty$ values derived for an individual from the Cole–Cole plot permits calculation of the resistance of the intracellular fluid volume, a quantity that cannot be measured directly with standard body composition methods.

## II. INSTRUMENTATION AND MEASUREMENT

Measurement of TBIA is performed by using four surface electrodes and an impedance device (Fig. 3). The use of four, as compared to two, electrodes is selected to minimize contact impedance and skin-to-electrode interactions. For TBIA measurement, individuals wear clothing, but no shoes or socks, and lie supine in a horizontal position on a nonconducting surface with limbs not touching the torso or each

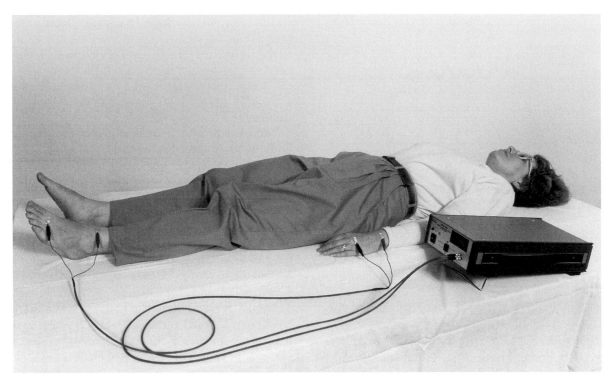

**FIGURE 3** Measurement of tetrapolar bioelectrical impedance using an impedance instrument (model 103, RJL Systems, Mt. Clemens, MI).

other. Measurements generally are performed on patients 2 to 4 hours after eating a light meal. Patients are asked to avoid alcohol and exercise for 24 hours before scheduled testing. Measurements are not performed if an individual has sweat accumulated on the skin.

Spot, adhesive electrodes are placed in the middle of the dorsal surfaces of the hands and feet proximal to the metacarpal–phalangeal and metatarsal–phalangeal joints, respectively, and also medially between the distal prominences of the radius and ulna and between the medial and lateral malleoli at the ankle. The current-introducing electrodes are placed a minimum distance of the diameter of the wrist or ankle beyond the paired detector electrode. Depending on the TBIA instrument used, an excitation current of about 800 (at 50 kHz) or 100 $\mu$A (multiple-frequency instruments) of alternating current is introduced into the individual at the distal electrodes of the hand and foot, and the voltage drop is detected at the proximal electrodes.

TBIA instruments yield different impedance measurements. Single-frequency instruments give measurements of $R$ and $Xc$, or $Z$ and $\phi$. Multiple-frequency devices furnish measures of $Z$ and $\phi$.

## III. BODY COMPOSITION ASSESSMENT

Body composition may be described in various terms (Fig. 4). In the simplest model, the body consists of two components, fat and fat-free. The fat-free component consists of water, protein, and bone. Because the FFM is a heterogeneous quantity, it may be further divided into the cellular and extracellular components. The body cell mass is unique because it contains protein, potassium, water, and some electrolytes. In contrast, the extracellular mass contains bone, connective tissue, water, and other electrolytes. Components of the body that behave as biological conductors have been estimated with TBIA.

### A. Fat-Free Mass

Knowledge that water and electrolytes are found only in the biological conductors of the body stimulated

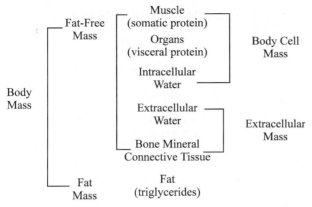

**FIGURE 4** Schematic representation of the relationship of body composition variables.

researchers to develop and use mathematical relationships, including standing height or stature and impedance variables ($Z$, $R$, and $Xc$), to estimate some components of body composition. It is important to recognize that TBIA has the potential to estimate many body compositional variables.

Because of the interest in assessing body energy status, concerted efforts have been made to estimate FFM with TBIA among healthy adults. With densitometry as the reference method and 50-kHz TBIA measurements, mathematical prediction models were derived and validated in independent samples of adults. The principal independent variable was height squared divided by $R$ ($Ht^2/R$), which accounted for more than 90% of the variability in the prediction of FFM. Secondary independent variables included body weight ($Wt$) and sex, which contributed less than 10% of the variability in predicting FFM. Interestingly, $Xc$ also was a significant, albeit small predictor of FFM (about 5% of the variability in predicting FFM). Cross-validation trials indicated comparability among measured and predicted FFM values with an average error of 3 to 5%, which is within the theoretical error of the densitometric method.

The TBIA method also has been used to develop models of estimating FFM in children. A multicomponent body composition model (body density, TBW, and mineral density) was used to determine reference body composition. Comparisons of TBIA-estimated FFM with reference FFM values in another group of children indicated no difference. Thus, in healthy populations of adults and children, TBIA is a reliable predictor of FFM.

## B. Body Fatness

The TBIA method can be used to indirectly estimate %BF. Using predicted values of FFM from measurements of $R$ at 50 kHz, body weight, and sex, %BF can be calculated. This approach has been used to demonstrate that impedance estimates of %BF are similar to those determined with appropriate densitometric procedures.

Comparisons among TBIA and anthropometric estimates of %BF also have been made. Although some studies have reported statistically significant differences between the methods, one must consider that the estimate of %BF from TBIA is an indirect estimate because TBIA is used to estimate conductor volume and fat is a nonconductor. In addition, because %BF is calculated from TBIA-predicted FFM, biological and technical errors associated with each technique (densitometry and TBIA) are compounded by propagation of errors.

## C. Obesity

Several investigators have suggested that TBIA overestimates %BF in lean individuals and underestimates it in obese people. A significant correlation coefficient ($r = -0.80$) was reported between %BF and FFM residual scores, calculated as the difference between measured and predicted FFM values derived using a manufacturer's prediction equation. Later, the investigators confirmed the previous finding of a fatness-dependent bias or error in predicting FFM determined densitometrically in a large cross-validation study. They also proposed and validated sex- and fatness-specific FFM prediction equations based on $Ht^2$, $R$ determined at 50 kHz, $Wt$, and age.

In contrast to these findings, only a minor influence of body fatness on the TBIA prediction of FFM was reported elsewhere. A weak relationship ($r = -0.33$) was observed between %BF and residual FFM scores in a sample of 161 adults whose FFM and %BF ranged from 30 to 97 kg and 5 to 50%, respectively. This finding indicates that body fatness accounts for only 10% of the variation in the bias in predicting FFM. Thus, other factors appear to be more important in explaining the discrepancy between measured and predicted FFM data.

To further investigate the hypothesis that TBIA, at least TBIA determined at 50 kHz, is influenced by degree of %BF, it will be necessary to use multicomponent body composition models or radiologic reference body composition methods. These approaches will minimize errors associated with assumptions regarding the chemical composition of the fat-free body. [*See* Obesity.]

## D. Weight Loss

There appears to be controversy regarding the validity of TBIA estimates of body composition during weight loss. Early studies with series-equivalent determinations of R at 50 kHz reported either a lack of concurrence or an acceptable concordance between measured and TBIA-predicted FFM values in obese patients after short-term weight loss. An explanation for the discrepancy is the disproportionate loss of water relative to the structural components of the fat-free body during early weight loss. Whether this limitation is a function of the use of a series-equivalent model and will be overcome with alternative TBIA approaches remains to be investigated.

## E. Total Body Water

Because of the concern about body fatness as a health indicator, availability of equipment, and ease of performing densitometric measurements, early applications of TBIA principally at 50 kHz focused directly on FFM and indirectly on %BF. However, the more appropriate use of TBIA is to monitor TBW and its distribution in health and disease. In this regard, there is consistency in the use of TBIA.

The use of TBIA at 50 kHz has produced relatively consistent results. Although many investigators have proposed a variety of regression models to estimate TBW in healthy people, there has emerged a consistent relationship between the independent variable $Ht^2/R_{50}$ and TBW determined by deuterium or $^{18}O$ dilution in subjects ranging from infants to adults. This important observation indicates the validity of TBIA determined at 50 kHz to predict TBW.

This point has been supported by studies of the use of TBIA at 50 kHz to estimate change in TBW during a variety of physiological conditions. During longitudinal studies of controlled weight loss in obese patients and serial studies of women before and during pregnancy and lactation, prediction models using $R_{50}$ found no difference between TBW predicted by TBIA at 50 kHz and that estimated with deuterium dilution. These results indicate the validity of TBIA at 50 kHz to monitor changes in TBW in adults.

## F. Body Fluid Distribution

Although dilution methods are available to assess ECW and TBW separately, the homeostatic control of fluid distribution or the ratio of ECW/TBW in healthy people results in an interdependence in these fluid volumes. That is, ECW is very highly correlated with TBW ($r = 0.90$), and vice versa. Therefore, studies in healthy volunteers may be expected to yield similar TBIA predictors for TBW and ECW, particularly when single-frequency, series-equivalent TBIA data are used. This observation has been reported by independent research groups. Therefore, the need for more sophisticated approaches is indicated. Some novel experimental TBIA models have been used to address this problem.

Multiple-frequency TBIA has been proposed to distinguish ECW and TBW. Using impedance spectroscopy to identify the Z values where $Xc$ is 0 ($Z_0$) and $Xc$ is maximal ($Z_C$), it has been shown that reference dilution volumes can be predicted with acceptable accuracy in animal models by the independent variables of the square of the length between electrodes ($L^2$) divided by $Z_0$ and $Z_C$ for ECW and TBW, respectively. Similar approaches have been used in human subjects with the exception that $Ht^2$ was divided by $Z_0$ and $Z_\infty$ to estimate ECW and TBW, respectively.

This approach has been expanded to eliminate the use of regression analysis to estimate fluid distribution in humans. Multiple-frequency TBIA measurements are made over a range of frequencies from 5 to 600 kHz and the R and $Xc$ data are fitted using nonlinear curve fitting to the Cole–Cole model (refer to Fig. 1). Estimates of ECW and intracellular water (ICW) are calculated from TBIA data using general mixture theory that describes the effect of a conducting medium with a nonconductive material in it. This approach permits estimation of TBW and its distribution as ECW and ICW when changes in ECW occur independently of ICW.

This modeling technique of TBIA data has been used to monitor the accuracy of serial determinations of TBW and ECW in women before, during, and after pregnancy. No differences were found between dilutional and TBIA values (Fig. 5). However, the variability of the TBIA predicted values was almost twice as large as that for the reference values.

## G. Body Cell Mass

One innovative application of TBIA is the assessment of body cell mass (BCM) and hence nutritional status

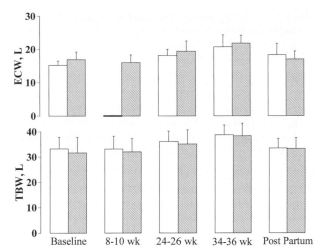

**FIGURE 5** Relationships between repeated estimates of total body water (TBW) and extracellular water (ECW) determined by dilution methods (open bars) and with bioelectrical impedance spectroscopy (hatched bars) of women before and during pregnancy and postpartum. [From M. D. Van Loan *et al.* (1995). *J. Appl. Physiol.* **78**, 1037–1042.]

in patients with catabolic disease. Because BCM is a critical predictor of survival in many clinical illnesses, there has been a recent attempt to derive and validate a new TBIA model to assess this clinically important variable.

The hypothesis that parallel $Xc$ can be used as a discriminating predictor of BCM is based on studies of vegetables measured before and after cooking (Fig. 6). Recalling that impedance variables are inversely related to conductor volume, the 100% decrease in $Xc$ with a 76% decrease in $R$ and an 85% reduction

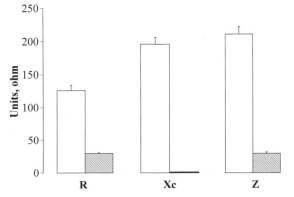

**FIGURE 6** Measurements of resistance ($R$), reactance ($Xc$), and impedance ($Z$) with a 50-kHz impedance device in Russet potatoes before (open bars) and after (hatched bars) cooking.

in $Z$ suggests that capacitance of the cells of the potato was eliminated. The lack of a measurable $Xc$ implies that the physicochemical barrier separating intra- from extracellular fluid was removed with cooking. Thus, $Xc$ may be hypothesized to be an index of ICW and, hence, BCM, whereas $R$ might be expected to reflect ECW.

Reports of the utility of TBIA to monitor BCM have utilized a transformation of $Xc$ determined with 50-kHz impedance instruments from a series- to a parallel-equivalent value. If one measures series-equivalent $Xc$ and $R$, then one can transform the measurements into parallel-equivalent values by using the basic formulae

$$R_p = R_s + [(Xc_s)^2/(R_s)],$$
$$Xc_p = Xc_s + [(R_s)^2/(Xc_s)],$$

where subscripts s and p refer to seris and parallel values, respectively.

The use of TBIA to assess serial changes in body composition has been initiated in patients with acquired immunodeficiency syndrome (AIDS). Using a model based on parallel $Xc$, TBIA was shown to discriminate long-term survival of HIV-infected patients on the basis of changes in BCM. This finding suggests that routine use of TBIA may be useful to monitor disease progression.

A similar TBIA approach has been implemented to assess the nutritional status of patients on maintenance hemodialysis. The reproducibility of TBIA was very high (coefficient of variability = 0.6%), suggesting that TBIA is a reliable measurement in patients with end-stage renal disease. Estimates of BCM by TBIA were highly correlated with reference determinations of BCM ($r = 0.92$). These initial findings indicate that TBIA is a useful tool to monitor nutritional status of patients on maintenance renal dialysis and that TBIA estimates of BCM represent a clinically important parameter that should be monitored to avoid physiological impairments associated with modest decreases in BCM in these patients.

## H. Regional Fluid Accumulation

Although TBIA has been shown to be a valid method for determination of whole-body fluid volumes in most individuals, there has been some concern about its ability to discriminate regional fluid accumulation. Early studies in animals and patients with ascites or localized edema reported that TBIA underestimated

TBW as compared to values determined by traditional dilution methods or paracentesis. The solution to this problem was use of a different electrode placement.

A modified TBIA in which the distal electrode placements for the source electrodes were unchanged and the detector electrodes were moved proximally toward the regional fluid accumulation proved to be useful. Studies in humans and rats indicated increased accuracy and precision in detection of acute changes in TBW. Applications in humans undergoing liver transplants used detector electrodes placed on the abdomen and the parallel model to accurately estimate abdominal fluid accumulation with an error of less than 1%.

## IV. SUMMARY AND CONCLUSION

The TBIA method offers a variety of potential applications for noninvasive assessment of human body composition because the method is safe, convenient, and easy to use. As the evolution of the TBIA method progresses, comparisons of the accuracy and precision of the various experimental approaches (impedance spectroscopy and parallel model) will facilitate decision making regarding the most appropriate model for use of TBIA for routine assessment of human body composition and clinical needs for management of patient care. The prospect that TBIA could be used to monitor the nutritional status of patients with catabolic disease will be a major focus of future research activities.

## BIBLIOGRAPHY

Anonymous. (1992). Bioelectrical impedance and body composition. *Lancet* **340**, 1511.

Cha, K., Hill, A. G., Rounds, J. D., and Wilmore, D. W. (1995). Multifrequency bioelectrical impedance fails to quantify sequestration of abdominal fluid. *J. Appl. Physiol.* **78**, 736–739.

Chertow, G. M., Lowrie, E. G., Wilmore, D. W., Gonzales, J., Lew, N. L., Ling, J., Leboff, M. S., Gottlieb, M. N., Huang, W., Zebowski, B., College, J., and Lazarus, J. M. (1995). Nutritional assessment with bioelectrical impedance analysis in maintenance hemodialysis patients. *J. Am. Soc. Nephrol.* **6**, 75–81.

Cornish, B. H., Ward, L. C., and Thomas, B. J. (1992). Measurement of extracellular and total body water of rats using multiple frequency bioelectrical impedance analysis. *Nutr. Res.* **12**, 657–666.

Kushner, R. F., Schoeller, D. A., Fjeld, C. R., and Danford, L. (1992). Is the impedance index ($Ht^2/R$) significant in predicting total body water? *Am. J. Clin. Nutr.* **56**, 835–839.

Lukaski, H. C. (1991). Assessment of body composition using tetrapolar bioelectrical impedance analysis. *In* "New Techniques in Nutritional Research" (R. G. Whitehead and A. Prentice, eds.). Academic Press, New York.

Lukaski, H. C., and Scheltinga, M. R. M. (1994). Improved sensitivity of the tetrapolar bioelectrical impedance method to assess fluid status and body composition: Use of proximal electrode placement. *Age Natur.* **5**, 123–129.

Lukaski, H. C., Siders, W. A., Nielsen, E. J., and Hall, C. B. (1994). Total body water in pregnancy: Assessment by using bioelectrical impedance. *Am. J. Clin. Nutr.* **59**, 578–585.

Ott, M., Fischer, H., Polat, H., Helm, E. B., Frentz, M., Caspary, W., and Lembke, B. (1995). Bioelectrical impedance analysis as a predictor of survival in patients with human immunodeficiency virus infection. *J. Acquir. Immune Defic. Syndr.* **9**, 20–25.

Sluys, T. E. M. S., van der Ende, M. E., Swart, G. R., van den Berg, J. W. O., and Wilson, J. H. P. (1993). Body composition in patients with acquired immunodeficiency syndrome: A validation study of bioelectrical impedance analysis. *J. Parenter. Enter. Nutr.* **17**, 404–406.

Van Loan, M. D., Kopp, L. E., King, J. C., Wong, W. W., and Mayclin, P. L. (1995). Fluid changes during pregnancy: Use of bioimpedance spectroscopy. *J. Appl. Physiol.* **78**, 1037–1042.

Zillikens, M. C., van den Berg, J. W. O., Wilson, J. H. P., and Swart, G. R. (1992). Whole-body and segmental bioelectrical impedance analysis in patients with cirrhosis of the liver: Changes after treatment of ascites. *Am. J. Clin. Nutr.* **55**, 621–625.

# Biomechanics

*Henry Ford Hospital*

---

I. Human Cancellous Bone Mechanics
II. Human Motion Analysis
III. Occupant Dynamics
IV. Conclusion

## GLOSSARY

**Dynamics** Study of the effect of variable loads on a material body

**Kinematics** Study of motion without consideration of loading

**Statics** Study of the effect of unchanging loads on a material body

BIOMECHANICS IS OFTEN DEFINED AS THE APPLICA-tion of *mechanics* to living creatures. Mechanics is a branch of physics that deals with the motion of material bodies and the action of forces upon them. Classical mechanics includes *dynamics,* the study of variable forces on a moving body, *statics,* the study of forces on a motionless body, and *kinematics,* the study of motion without consideration of forces.

A less classical and more functional definition is that biomechanics is the use of *applied mechanics* for the understanding of living systems. The value of this definition is that the course catalog from any applied mechanics division of a large mechanical engineering department provides a list of specialty areas of mechanics that can be applied to living systems. In no particular order, a partial list of these specialties includes: experimental stress analysis, strength of materials, mechanics of composite materials, vibration analysis, kinematics, dynamics, statics, elasticity, plasticity, viscoelasticity, theory of elastic stability, fracture mechanics, contact mechanics, fluid mechanics,

turbulent flow, thermodynamics, and optimal structural design. It would be a mistake, however, to assume that persons interested in biomechanics are ordinarily (or even predominately) engineers. There is broad application of biomechanics by workers in the fields of medicine, exercise science (physical education), ergonomics, zoology, botany, agriculture, automotive design, and other fields.

For economic reasons, biomechanics is becoming more widely used. In agriculture, there is an ongoing emphasis on efficiency of production for both animal and plant products. This increases the need to better understand the living portions of agricultural systems to increase yield without injury to the crop or to the farmer. A similar economic emphasis on efficient factory production results in a need to understand the mechanics of how workers interact with their workplace. [*See* Ergonomics.] A third economically driven application is the use of biomechanical analysis in an attempt to obtain maximum possible performance from highly paid professional athletes and amateur Olympic athletes. Similar economic issues affect the application of biomechanics to medicine. For example, increasing the mechanical efficiency and reliability of orthopedic implants is an area of intense governmental and private development efforts. A strong additional factor causing an increase in the application of mechanics to biological systems is the increasingly powerful ability to use methods of molecular biology to modify the mechanical properties of living tissues. For example, the possibility of growing artificial skin, bone, and cartilage that have the same (or superior) mechanical properties as natural tissues may soon revolutionize the treatment of burns, bone fractures, arthritis, and other musculoskeletal diseases.

In the academic arena, interest in biomechanics is increasing in part because of development of the digital computer. The computer makes it possible to apply mechanics to realistic biological systems. For exam-

ENCYCLOPEDIA OF HUMAN BIOLOGY, Second Edition, VOLUME 1.   Copyright © 1997 by Academic Press.   All rights of reproduction in any form reserved.

ple, computer-based methods now make it possible to calculate the forces underlying gastrulation of an embryo, predict the formation of reactive wood in an overloaded tree branch, and predict shear stresses on blood cells during turbulent flow in the heart. The application of biomechanics in academic research has also been growing owing to increased appreciation of the importance of mechanical factors to the function and evolution of living systems. For example, information resulting from biomechanical analysis of the gait of bipedal dinosaurs has been used in the debate over whether dinosaurs were warm- or cold-blooded.

The immense breadth of the applications of biomechanics makes it impossible to provide a comprehensive overview of the subject. However, to help the reader appreciate the methods used in the field, three applications will be presented. These three, human cancellous bone mechanics, human motion analysis, and occupant dynamics during an automobile crash, represent basic science, clinical/sports application, and industrial application of biomechanics, respectively. There are a multitude of other topics that could have been included, such as measurement of blood flow dynamics, fracture mechanics of human ligaments, and energy absorption characteristics of the motorcycle helmet–head complex. The author hopes that no one will feel slighted if their particular area was ignored.

## I. HUMAN CANCELLOUS BONE MECHANICS

The human skeleton contains two primary hard tissue types. Cortical bone has a small volume of pores (approximately 5%) and composes the shafts and the thin cortices of the ends of long bones. Cancellous bone has a greater pore volume (65–95%) and fills the large ends of bones, the interior of the pelvis, and the interior of the vertebral bodies and is the "filling" of the sandwich construction forming the dome of the skull.

The amount of bone of both types declines with age. In some cases, bony material is lost to the extent that bone fractures occur, a condition known as *osteoporosis*. A plurality of osteoporotic bone fractures occur in the vertebrae of the spine. Loss of the cancellous bone of the interior of the vertebrae contributes to the prevalence of spine fracture (Fig. 1). Consequently, the biomechanics of human vertebral cancellous bone are of medical and basic science interest. [*See* Bone Density and Fragility, Age-Related Changes.]

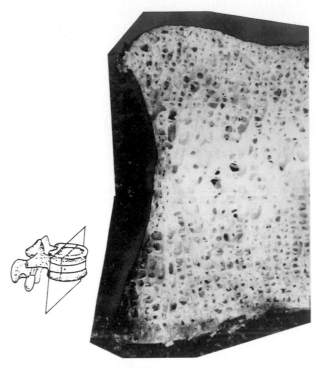

**FIGURE I**  Cross section of a human lumbar vertebral body showing the cancellous bone structure. Note that it is composed primarily of pores.

The weight of the upper body and contraction of spinal muscles loads the spine during normal daily activities. Some of the applied load is carried by the thin outer shell of the vertebra (cortical shell) and the remainder is borne by the cancellous bone of the interior. Under normal circumstances, the vertebrae are sufficiently strong to support the loads without damage to the hard tissues. However, as bone is lost with aging, the normal activities of daily living produce forces sufficiently large to damage the material and eventually to cause the vertebra to collapse. A biomechanical understanding of the process of vertebral cancellous bone damage would be useful to the understanding of this disease.

Apparent weight density is a typical measure of the material density of vertebral cancellous bone. The apparent density is the weight of a bone specimen divided by the volume of the geometrical shape that the specimen fills (the apparent volume of the specimen). A separate, but related, measure of the amount of material is the bone volume fraction (BV/TV). This parameter is the ratio of the actual volume of hard tissue (BV) divided by the apparent volume (TV) of the tissue. The normal range of

BV/TV for human vertebral cancellous bone is approximately 0.08 to 0.39, with an average of 0.16 and standard deviation of 0.05 (BV/TV is normally distributed).

The compressive strength of cancellous bone can be estimated by crushing a regularly shaped specimen between metal platens and measuring the force during displacement. The maximum force attained divided by the cross-sectional area of the specimen is the *strength* of the tissue. Not surprisingly, the strength of cancellous bone is related to the amount of tissue in the specimen (BV/TV). For human vertebral cancellous bone with $0.08 < BV/TV < 0.39$, the relationship is linear:

$$Strength = 30.5 \times (BV/TV - 0.047),$$

$$r^2 = 0.76, \quad (1)$$

where strength is measured in Newtons per square millimeter (60 specimens were used in the study: 13 black male, 15 white male, 17 black female, 15 white female). Studies of cancellous bone tissue that combine data primarily from the knee and hip have found a nonlinear relationship between bone strength and BV/TV. This difference from the relationship found for the spine may be a result of the different microarchitectures of the tissue between the sites.

An additional issue of cancellous bone tissue mechanics is determining the mechanisms of damage in the tissue. Examination of the broken tissue under low-power stereomicroscopy ($30\times$) shows that the small rods and plates that form the tissue break at their attachments to one another, as would be expected from studies of man-made low-density foam materials. Examination of the bone using high-power ($250\times$) microscopy and stain enhancement of the tissue cracks shows that the damage occurs in a manner consistent with engineering experience of fiber-reinforced composites (small microcracks caused by shear and tensile strains).

This concise description of human vertebral cancellous bone mechanics could be taken as a paradigm for the examination of any tissue strength relevant to human injury. For example, the rupture of knee ligaments, particularly the anterior cruciate ligament (ACL), is common during sporting activities. A project to biomechanically determine the strength of the ACL, develop an ability to predict ACL strength, and develop an understanding of the mechanisms of ACL rupture would have virtually the same steps as the study of cancellous bone. In the literature

of human biomechanics there are studies of the strength of cortical bone, many different ligaments, skin, hair, cartilage, tooth enamel, and so on that all follow the paradigm: measure breaking strength, develop a predictive equation for strength based on engineering principles, and determine the microscopic features of the failure process. This is one of the fundamental investigative approaches to the understanding of tissue mechanics.

## II. HUMAN MOTION ANALYSIS

People walk, build equipment, drive cars, do almost everything by moving their bodies. [*See* Movement.] Measuring how the body moves and measuring the forces and muscle activities used to move it are the subjects of human motion analysis. The purposes of motion analysis are many: to understand the unusual walking styles of patients with cerebral palsy, to predict the likelihood that an elderly person may be at risk of falling and breaking a hip, to determine whether a high jumper is using the most efficient strategy to clear the bar, to define the difference in arm motions between throwing a curve ball and a slider. A complete list of such activities would fill the remainder of this volume of the encyclopedia.

The primary method of motion analysis used now is video motion analysis with retro-reflective markers (Fig. 2). In this technology, rings of light-emitting diodes around the lenses of one or more video cameras illuminate the subject. Spherical markers coated with retro-reflective glass beads are taped to the subject to define important points on her body. The retro-reflective glass beads cause the light from the camera to reflect directly back to the camera lens. As a consequence of this return reflection, the camera "sees" the marker as a very bright circular spot on the video image. It is possible to track the motion of the marker automatically using a computer program since it is so much brighter than the rest of the image. The motions of the markers in the video images are combined to determine the position, velocity, and acceleration for the limbs, torso, and head.

The greatest power of motion analysis comes from using two (or more) cameras. If only one camera is used, it is not possible to determine the exact position of the marker in three-dimensional space. If the marker can be seen by two cameras, however, then the position of the marker can be determined using the lines that pass from the two cameras through the center of the marker. In theory only two cameras are

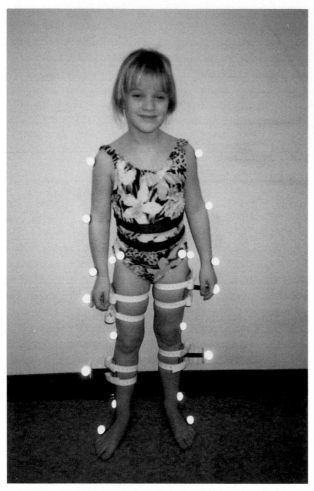

**FIGURE 2** Motion analysis subject with retro-reflective markers attached. Note the much greater brightness of the markers compared to the background due to back-reflection of the photographic flash.

needed. In practice it is often necessary to use five or six cameras to ensure that the markers are always in direct view of at least two cameras during a complex motion. The primary technical problem is that the markers on one limb may pass repeatedly behind the torso or another limb. Nevertheless, when the three-dimensional path of the markers is known, then the three-dimensional motion of the limbs and body is also known.

One application for motion analysis is to measure the walking motions of a patient with cerebral palsy, to help guide an orthopedic surgeon in the surgical modification of the muscles and bones of the patient. It is quite common for a child with cerebral palsy to have some difficulty walking. This is due, in part, to muscle spasticity (uncontrolled and permanent muscle

contraction), which limits the motions of the ankle, knee, and hip in ways unique to the particular patient. If the surgeon can move the muscle attachments and change the shapes of the bones in the right way, it is possible to help the patient walk in a more normal manner.

Motion analysis can be useful to the surgeon because human walking is a quite complex activity. If one joint or muscle is not working normally, the patient will change his use of *all* of his other joints and muscles to compensate for the original deficit. To understand this in a personal way, try walking normally with your right calf fully contracted. To walk in the most efficient (perhaps least tiring) way, it will be necessary to change the motions of your other leg, pelvis, spine, and arms. Human motion analysis records the motions of body parts in an objective and repeatable way to allow comparison against standards. Motion analysis also makes possible an objective comparison of a patient's pre- and postoperative gait to determine whether surgery was effective.

At the same time that the patient's gait is measured, it is also common to measure the forces between the foot and the floor and the intensity of the electrical (nerve) signals causing muscle contraction. These measurements can also be compared to normal standards and be used to compare pre- and postoperative gait. The techniques used to measure these values will not be discussed here.

The use of human motion analysis for other purposes is widespread. The key features of the method are (1) the ability to record the motions of significant points on the human body and (2) calculation of important body motions from the motions of the marked points. The details of application of the method to particular problems are peculiar to the application and not amenable to a detailed discussion in this article.

## III. OCCUPANT DYNAMICS

When one automobile strikes another, the occupants of both vehicles can be subjected to injuriously large accelerations and impact forces within 100 milliseconds. The extraordinary cost in human suffering and death due to automobile crashes has caused the federal government and the automobile manufacturers to develop safety guidelines for design that reduce the risks of crashing. The highest standard for an automobile design used by the government is experimental reproduction of an actual crash. Every automobile design sold in the United States must pass several federally mandated safety tests. One of these is a barrier crash in which a prototype of the new automobile containing crash dummy occupants is rammed at 30 mph into a solid wall. If the accelerations and impact forces on the dummies are too large (as mandated by the regulations), then the design cannot be sold. A design that fails certification can be devastatingly expensive since development of a new automobile can cost as much as $1 billion.

In the past, the approach to automobile safety testing used by all manufacturers was based on the use of actual experimental crashes. These tests, in which a physical prototype of the new design is destroyed, provide the best data on crashworthiness. Unfortunately, physical model tests are extremely expensive. The automobile that is crashed is handmade and can easily cost $400,000 to $750,000 to produce. This cost does not even begin to cover the cost of crash facilities and data collection associated with a physical crash test. Clearly, manufacturers would like to avoid physical testing. However, because of the extreme cost of failing the government certification tests, the manufacturers are forced to perform preliminary testing of designs as they are developed.

Fortunately, using modern digital computer technology and sophisticated programs, it is now possible to simulate prototype car crashes. This can result in significant savings to the automobile company and increase the probability that the design will pass the certification test. Computer simulations cost many times less than a physical test and allow closer monitoring of the effects on safety of even minor design changes.

The process of simulation begins with the building of a computer model of the automobile and occupants (Fig. 3). The development of the model is facilitated by using computer-aided design tools that permit direct visualization of the model. Inputs to the model include the geometric shapes and locations of the automobile interior and occupants, descriptions of how the joints of the occupants can move, a numerical description of the air bag and safety belt, and definition of the stiffness and strength of the various materials in the model. After the model is developed, it can be "crashed" into a wall or "crashed" into another car to obtain the resulting forces and accelerations on the occupant as he smashes into the instrument panel or is restrained by the belt and bag.

After the computer model runs without generating errors, it is time to determine whether it produces results similar to those of actual tests. Validation is a crucial part of developing a computer model, and one way is to compare its predictions to results obtained in the past. This type of validation, called *retrospective* testing, is useful for developing a candidate model, but cannot actually tell you whether the final model is correct. The reason for this is that the analyst

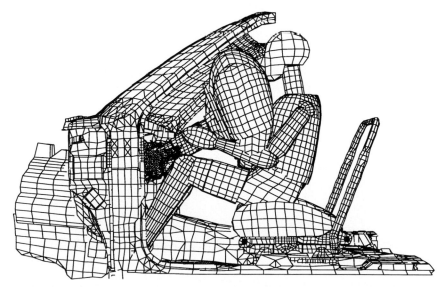

**FIGURE 3**   Dynamic computer model of a vehicle occupant at the moment of impact with the air bag. A great deal of computational power is needed for occupant dynamic simulation because of the contact, sliding, geometric, and material nonlinearities.

can (and will!) adjust the model until it matches previously known results. The acid test of a biomechanical model is a *prospective* test in which the results of a physical test are predicted without any knowledge of the actual results. If the simulation matches reality, then the model works. If they do not, the model does not work. Finally, after the computer model of the new automobile design is validated in general it can be used to predict the safety consequences of variations in the design. For example, if the angle of the steering column with respect to the ground is changed, does the acceleration of the driver's head increase when in contact with the air bag? Similarly, simulating different charges for the air bag inflator affects the rate at which the bag expands to its final configuration. If the bag is too slow, the occupant will hit the steering column and be thrown violently backward by bag inflation. If bag inflation is too rapid, the bag will begin to deflate before the occupant hits the bag, decreasing its effectiveness. Computer modeling of the crash process allows one to pick the optimal configuration for the automobile interior, speed of inflation of the air bag, stiffness of the seat belt (restraint system), and so on. This application of biomechanics has a large positive effect on the health and safety of humanity.

## IV. CONCLUSION

The breadth of application of biomechanics is immense. Its coverage includes everything from the design of human total joint arthroplasties to analysis of paleontological data to help determine if dinosaurs were warm-blooded. The ability to manipulate biological systems is increasing and along with it the need for a biomechanical understanding of humans, animals, and plants. It is expected that biomechanics will be one of the more rapidly growing areas of applied mechanics over the next 50 years.

## BIBLIOGRAPHY

Alexander, R. M. (1994). "Bones." Macmillan, New York.

Fung, Y. C. (1990). "Biomechanics: Motion, Flow, Stress and Growth." Springer-Verlag, New York.

Gage, J. R. (1991). "Gait Analysis in Cerebral Palsy." MacKeith Press/Blackwell Scientific Publications, New York.

Mow, V. C., and Hayes, W. C. (1991). "Basic Orthopaedic Biomechanics." Raven, New York.

Muybridge, E. (1989). "The Human Figure in Motion." Bonanza Books, New York.

Pauwels, F. (1980). "Biomechanics of the Locomotor Apparatus" (translated by P. Macquet and R. Furlong). Springer-Verlag, New York.

# Biomineralization and Calcification

KENNETH P. H. PRITZKER
*Mount Sinai Hospital and University of Toronto*

## GLOSSARY

**Biologically controlled mineralization** Biomineralization in which cell-mediated processes control the site, volume, and physical properties of the mineralization

**Biologically induced mineralization** Biomineralization in which biologic processes facilitate mineralization. The volume and physical properties of the mineral are partially controlled by cell-mediated processes and partially by physical characteristics of the environment

**Biomineralization** Processes by which minerals form within tissues

**Calcification** Process by which insoluble calcium salts form within tissues. Calcium salts may be inorganic (e.g., calcium phosphate) or organic (e.g., calcium oxalate). Physiologic calcification is usually biologically controlled, whereas pathologic calcifications is usually biologically induced. Calcifications may form as amorphous solids lacking long-range atomic structure or as crystal aggregates

**Crystal** Solid composed of atoms arranged in an orderly repetitive array

**Heterogeneous nucleation** Nucleation in the presence of surfaces or foreign particles

**Homogeneous nucleation or "spontaneous nucleation"** Nucleation in the absence of surfaces

**Matrix vesicle** Cell-derived, membrane-bound structure found in the extracellular matrix of many calcifying tissues

**Mineral** Solid inorganic compound having an orderly internal structure and characteristic chemical composition, crystal form, and physical properties

**Nucleation** Process of forming a crystal phase from a solution across a free energy barrier

**Pathologic calcification** Calcification that occurs in disease processes. Usually the properties of both the solid phase and the tissue site differ from those of normal biomineralization

**Primary nucleation** Process encompassing homogeneous nucleation and heterogeneous nucleation with foreign surface interfaces

**Secondary nucleation** Nucleation in the presence of crystals from the solute

BIOMINERALIZATION AND CALCIFICATION ARE overlapping but not synonymous terms. Biomineralization includes the processes by which insoluble inorganic crystal salts deposit in tissues. Biomineralization processes can be classified into biologically controlled and biologically induced mineralization. Biologically controlled mineralization involves close regulation of the mineralization process and implies mineralization under physiologic conditions. Biologically induced mineralization involves precipitation of mineral salts in tissues under less stringent conditions and, in humans, implies pathologic mineralization processes. In human biology, the distinction between biologically controlled and biologically induced mineralization relates to the closeness of regulation rather than to lack of regulation. If mineralization processes were totally unregulated, we would become transformed into pillars of salt, like Lot's wife of the biblical story. Most but not all biomineralization processes in humans involve calcium salt formation. An important exception is hemosiderin, a biomineral containing iron. Calcifi-

ENCYCLOPEDIA OF HUMAN BIOLOGY, Second Edition, VOLUME 1. Copyright © 1997 by Academic Press. All rights of reproduction in any form reserved.

cation, on the other hand, is the process of calcium salt deposition within tissues and encompasses the deposition of both mineral (inorganic crystals containing calcium) crystallized salts of calcium and organic compounds, as well as amorphous, inorganic, and organic compounds containing calcium.

It is well known that tissues such as bone and teeth contain calcium phosphate mineral salts. That these salts form under very restricted physical, chemical conditions and that these substances have very specific physical and chemical properties are seldom appreciated. Although other forms of biominerals can exist in humans, this article focuses on the general biology of the major types of calcific biominerals and amorphous calcifications that are found in humans. The specificity and variation in structures and physical properties of biominerals, together with their tissue distribution, will be described. Subsequently, the functions of biominerals and the principal processes that govern their formation and resorption will be addressed. Finally, we will consider the processes that lead to pathologic calcification and the contributions of such calcification and mineralization to disease processes.

## I. CALCIFIC BIOMINERALS: STRUCTURE AND PROPERTIES

Table I lists the principal tissue sites for mineral deposition, the mineral phases involved, and a few physical properties of the biominerals. In most tissues, biomineralization consists of the calcium phosphate crystal phase, or calcium apatite. The name apatite is derived from the Greek word ἀπάτη (*apate*, "disguise or deceit") because this mineral was often mistaken for other minerals. Even casual perusal of Table I indi-

cates that, depending on the tissues, calcium apatite crystals vary in size, association, and alignment to organic matrices. In biomineralization, the "deceit" is that apatite is a group of poorly crystalline minerals that, in properties and mode of deposition, vary from tissue to tissue. Furthermore, within tissues, calcium apatite can demonstrate variable substitution of the hydroxyl ions with other ions such as fluoride ($F^-$) or bicarbonate ($HCO_3^-$). Within each tissue, the crystal phase, habit (relative growth of crystal faces), particle size, and packing of crystals within the matrix are relatively uniform. Each property of the mineral is under biologic control and has specific implications for biologic function. Some properties of biominerals and amorphous calcifications and their implications for biological function are shown in Table II.

## II. FUNCTIONS OF BIOMINERALS

Biominerals serve as reservoirs for storage of ions and inorganic compounds, as a detoxification mechanism for storage of potentially toxic substances (including calcium!), as surfaces for reactions, as constituents of composite structural materials, and as components for biosensors. Examples of these functions are listed in Table III. It is important to recognize that, within tissues, biominerals may have multiple functions at the same site. The presence and the composition of the biomineral are controlled by the complex interaction of each function with its physiologic regulation.

The storage function of biominerals for the principal ions (e.g., $Ca^{2+}$ and $PO_4^{3-}$) may be obvious but it is not trivial. The functions of cells and extracellular processes are highly dependent on maintenance of the extracellular calcium concentration $[Ca^{2+}]$ in a very narrow range around 1 m$M$. Calcium apatite biomin-

### TABLE I
Tissues and Biominerals

| Tissue | Mineral phase | Chemical composition | Approximate crystal size | Matrix association |
|---|---|---|---|---|
| Bone | Calcium apatite | $Ca_{10}(PO_4)_6OH$ | $40 \times 20 \times 4$ nm | Oriented, on Type I collagen |
| Cartilage | Calcium apatite | $Ca_{10}(PO_4)_6OH$ | $40 \times 20 \times 4$ nm | Isotropic, Type II collagen |
| Enamel, tooth | Calcium apatite | $Ca_{10}(PO_4)_6OH$ | $100,000 \times 50 \times 25$ nm | Parallel to surface, layers of crystals |
| Dentin, tooth | Calcium apatite | $Ca_{10}(PO_4)_6OH$ | $35 \times 25 \times 4$ nm | Oriented, on Type I collagen |
| Otoconia, inner ear | Calcite | $CaCo_3$ | $5000 \times 1000 \times 1000$ nm | Amorphous acid proteins in a mucin derived from neuroepithelial cells |

### TABLE II
Biominerals and Amorphous Calcifications: Some Physical Properties and Biologic Implications

| Physical property | Biologic implication |
| --- | --- |
| Phase | |
|   Amorphous component | Increased solubility |
|   Crystal component | Specific formation/resorption conditions |
|   Crystal surface component | Adsorption of substances |
|   Lattice substitution (e.g., $F^-$) | Hardness, resistance to dissolution |
| Crystallinity (i.e., long-range order) | Hardness, resistance to dissolution |
| Particle size | Surface available for adsorption |
| Habit | Face-specific chemical reactivity |
| Crystal packing | Compressive strength |
| Association with organic molecules | Nucleation, promotion, or inhibition of mineralization, alignment with macro-molecules, alignment with other crystals |

eralization, principally in bone, provides an essential homeostatic mechanism that acts as a source during calcium depletion and as a sink for calcium excess. Similarly, the storage of phosphate serves to conserve a compound essential for energy metabolism. Furthermore, on the surface hydration shell of bone mineral, 80% of $HCO_3^-$, 80% of citrate, and 35% of sodium

### TABLE III
Functions of Biominerals

| Function | Examples |
| --- | --- |
| 1. Homeostasis (storage) | |
|   Ion reservoir | Primary ions $Ca^{2+}$, $PO_4^{-3}$ |
| | Secondary ions, $Mg^{2+}$, $HCO_3^-$ $Na^+$, citrate |
|   Essential trace elements | Zinc, copper |
|   Detoxification | Lead, cadmium |
|   Organic compounds | Growth factors, hormones, acidic proteins complex lipids |
| 2. Specific reactions | |
|   Physical | $HCO_3^-$ exchange |
|   Chemical | Surfaces for phosphatase and ? other enzyme-mediated reactions |
| 3. Structure | Compressive strength, fracture resistance, cell and tissue protection |
| 4. Sensor | |
|   Gravity | Otoconia |
|   Magnetic fields | Magnetite |
|   Sound and vibration | Bone mineral |

($Na^+$) found in the body are stored. An extension of the concept that biominerals can store excess ions is the property of biominerals to store by adsorption, lattice substitution, or coprecipitation other ions that are potentially toxic to the body. These include substances such as fluoride and strontium, which in small concentrations can increase the structural strength of biominerals, as well as other substances such as lead or cadmium, which, though having some adverse effects on normal mineralization, are quite toxic for other cellular processes. The small size of calcium apatite crystal particles implies an enormous surface area with both positive and negative ionic charges. Calcium apatite surfaces adsorb many different kinds of molecules with varying affinity. By adsorption of small ions such as $HCO_3^-$, the biomineral provides a pH buffer to extracellular fluid. Peptide hormones, paracrine substances, and complex lipids also become adsorbed. With incremental mineralization, these substances can be trapped and stored intact beneath mineral surfaces, awaiting release and activation when the mineral is resorbed. [See Bone Regulatory Factors.]

Biomineral surfaces provide a restricted space with defined hydration for enzymatic and other catalytic reactions. Although this general concept has been recognized, scientific attention to date has been focused mainly on the effects of organic substances on the face specificity of mineral formation rather than on the effects of the mineral surface on organic molecule composition and functions. Nonetheless, the face-specific molecular arrangement of the mineral surface can affect properties such as enzyme activity.

As a structural material in bone and teeth, calcium apatite crystals usually align in a pattern to accommodate the normal mechanical forces, providing compressive strength for the tissue. For example, in cartilage, both the mineral and the compressive forces are isotropic. [See Cartilage.] In contrast, mineral in bone is aligned along collagen fibers; the compressive forces are similarly anisotropic. The arrangement of biominerals also reflects other characteristics of structural adaptation. For example, the small particle size in bone limits crack propagation, thus reducing fracture risk. Furthermore, in bone, with increasing age after mineral formation, the mineral crystals become more closely packed. This increase in density correlates with increasing compressive strength. In tooth enamel, calcium apatite crystals are arranged as parallel layers of prisms that can reach more than 0.1 mm in length. These large crystals have more inherent strength than smaller crystals. The surface coverage with large crystals greatly limits the available space between the crystals, rendering the tissue beneath the enamel relatively impermeable to the external environment.

At cell, tissue, and organ levels, the encasement by biominerals has protective functions. Osteocytes within bone maintain not only bone structure but also calcium homeostasis protected from the external environment by their mineral surroundings. At the tissue level, the protection of bone marrow tissue by cortical bone and fragile parenchymal organs by complex mineralized bony structures, such as the cranium, thorax, vertebral column, and pelvis, is both obvious and fundamental to human existence.

Biomaterials also contribute to biosensors for many ambient physical forces. In many species, magnetite (iron oxide) particles form part of the mechanism that detects external magnetic fields. This biologic system assists animals with navigation. Such particles are postulated to exist in humans, but both their presence and current function are questioned. In humans and other vertebrate species, change in position is detected and modulated by gravity position biosensors located in the vestibular labyrinth (semicircular canals) in the inner ear. The biosensors consist of neuroepithelium that secretes a mucinous organic matrix that forms a membrane between the cells and the vestibular fluid. Within the matrix, otoconia or "ear dust" forms. Otoconia are composed of crystals of calcite (calcium carbonate). The movement of the otoconia relative to the cilia tufts of neuroepithelial cells signals the direction of the applied force. Low-frequency sound and intrasonic vibration are transmitted through mineral within bone. The bone mineral surrounding the middle ear is particularly dense and therefore well adapted to detect low-frequency sound and infrasonic vibrations.

## III. FORMATION OF BIOMINERALS

Under normal circumstances, biologic mineral forms in tissues at specific sites with very specific mineral composition and with a very specific spatial organization over a time interval that is also specific and well defined. Simultaneously adjacent to the mineralization sites, many other biological processes take place, yet only some of these processes interact with biomineralization. These features imply that biologically controlled mineralization occurs under very tightly controlled conditions indeed. These conditions, summarized in Table IV, include: (1) the provision of an environment that restricts diffusion of all substances involved in the mineralization process, that is, solvent, solutes, nucleation inhibitors, and promoters; (2) the initiation of mineralization by solute concentration, by tissue dehydration, and by provision of nucleating surfaces; (3) the orderly propagation (crystal growth, crystal aggregation/agglomeration) and limitation of mineralization controlled by the availability of solutes, nucleators, inhibitors, and promoters; and (4) the maturation of the mineral by rearrangement of the mineral particles, by tissue dehydration, and by adsorption of solutes from the extracellular fluid.

### A. Tissue Environment for Biomineralization

The site specificity for biomineral formation is determined by cellular activity that elaborates an extracel-

**TABLE IV**
Biomineral Formation

| Process | Events |
|---|---|
| Biomineralization environment | Confined space, matrix that restricts diffusion, solute concentration |
| Initiation of biomineralization | Tissue dehydration, nucleation—primary (homogeneous, heterogeneous) and secondary |
| Propagation and limitation of biomineralization | Supply of solutes, availability of promoters and inhibitors, confinement and alignment of matrix molecules |
| Maturation of biomineralization | Mineral/matrix interaction, further dehydration, adsorption of soluble molecules |

lular matrix capable of restricting the diffusion of solvent and solutes. Typically, the initial matrix consists of soluble, highly negatively charged macromolecules, proteoglycans, which are focally adherent to the cell membranes. These molecules have properties that restrict diffusion of substances with molecular weights above a few hundred daltons and negatively charged groups bind positively charged ions such as $Ca^{2+}$. The charged groups and bound ions restrict water diffusion. Subsequently, protein macromolecules, such as collagens in bone and cartilage or phosphoproteins and amelogenins in enamel, are elaborated. The mineralizing environment changes dynamically with modifications in the extracellular matrix structure. In turn, extracellular matrix structure is closely modulated by cell biosynthetic and proteolytic activity. Tissue hydration is controlled by association of water molecules to charged groups, principally carboxylates, sulfates, and phosphates, on macromolecules and by ionic solute concentration. Mineral formation itself is a dehydrating event as the mineral salts are less hydrated than their participating ions. The biomineralizing environment facilitates concentration of solutes, principally calcium and phosphate. Although the tissue ionic calcium, $[Ca^{2+}]$, is maintained constant at 1 m$M$, the total tissue calcium can increase to more than 40 m$M$ without mineralization by binding of calcium ions to the negatively charged sulfate and carboxylate groups on the proteoglycans. Phosphate concentration increases by enzymatic phosphorylation of macromolecules as well as by matrix binding of phosphoproteins and phospholipids that diffuse into the mineralizing environment from adjacent cells or the circulation. Many noncollagenous proteins, such as osteonectin, osteopontin, osteocalcin, and bone sialoprotein, originally thought to be associated exclusively with bone mineralization, are recognized to occur at mineralizing sites in other tissue locations such as blood vessel walls. Nonetheless, at certain tissue sites, specific proteins or protein fragments may be found. These include dental phosphorylin in dentin and chondrocalcin, the C-terminal propeptide of Type II collagen, in calcifying epiphyseal cartilage. Within many sites of tissue mineralization, matrix vesicles derived from cell membranes are present. These vesicles contain phospholipids and enzymes that are thought to participate in mineralization processes.

## B. Initiation of Biomineralization

In simple chemical solution systems, solute supersaturation concentrations are constant and mineralization is initiated by nucleation alone. In biologically controlled mineralization, the ambient supersaturation by solute ions is relatively small. Biomineralization is facilitated by processes that transiently increase solute concentrations, namely, local release of participating solutes from large molecules by catabolic activity and tissue dehydration. The precise mechanisms are not yet clear but may involve proteoglycan degradation, an event that supplies calcium ions and facilitates outward diffusion of proteoglycan degradation fragments. Similarly, phosphatase activity, particularly alkaline phosphatase, supplies phosphate from proteins, lipids, and macromolecules. The removal of both proteoglycans that contain hydrated ions and hydrated phosphate ions from organic molecules serves to dehydrate the extracellular matrix.

Biomineralization is initiated when molecules in solution combine to form insoluble solid precipitates. Theoretically, at a molecular level, the process could occur at elevated supersaturations by primary homogeneous nucleation in the absence of a solid interface. Conceptually, the energy required for homogeneous nucleation could be supplied by sudden mechanical force. However, for a given supersaturation the free energy required to facilitate homogeneous nucleation is higher than that for heterogeneous nucleation processes. The abundance of complex molecules in the mineralizing environment with affinity for calcium suggests that primary nucleation occurs as heterogeneous nucleation with some of the ambient organic molecules serving as the "solid" interface. Certain noncollagenous proteins such as bone sialoprotein may be heterogeneous nucleators. Once mineralization is initiated, the mineral that is formed can act as a secondary nucleator.

## C. Propagation and Limitation of Biomineralization

Once biomineralization is initiated, the chemical kinetics are such that mineralization would proceed to a point where the solutes are no longer saturated in the surrounding solution. However, the extracellular fluid is usually supersaturated and is constantly resupplied with solutes by physiologic homeostatic mechanisms. If no other mechanisms prevailed, mineralization would progress indefinitely. Mechanisms that control the rate and amount of mineralization include the restriction of solute diffusion through the extracellular space and the presence of mineralization inhibitor molecules. Mineralization inhibitors range from secondary ions such as $Mg^{2+}$ to proteins such as osteopontin. Inhibitors act in a variety of ways, including sequestration of the principal solutes by ionic binding, increasing the hydration of the mineralizing environment, or inhibition of secondary nucleation by adherence to the biomineral. Pyrophosphate is a most interesting example of the latter type of inhibitor. It is derived from nucleotides such as adenosine triphosphate by enzymatic pyrophosphohydrolase activity and binds to calcium apatite mineral, thus limiting further crystallization. Alkaline phosphatase, an enzyme commonly present at biomineralization sites, has high pyrophosphatase activity at pH 7.4. The inhibition effect of pyrophosphate on calcium apatite mineralization can be removed by the pyrophosphatase activity of alkaline phosphatase.

For most tissues, biomineralization is limited not by the properties of one substance but by the structure of the organic matrix as a material. It is the structure of the organic matrix that both limits mineral formation and organizes mineral deposition in tissue-specific isotropic or anisotropic patterns.

Active biomineralization can be detected clinically by the use of probes that are adsorbed to mineral in a similar fashion to calcium. These probes include the fluorescent antibiotic tetracycline and technetium ($Tc^{99}$) methylene bisphosphonate. The latter is a radioactive substance that is used to image body sites where there is active mineral accretion. Tetracycline can be detected in histologic sections by fluorescence microscopy.

## D. Maturation of Biomineralization

Once formed, biominerals continue to undergo changes in their arrangement and properties in response to their environment. Because the chemical milieu is relatively constant, the composition and size of the mineral crystals remain stable. However, environmental conditions do drive the aggregation of crystals and the adsorption of ambient substances to the biomineral. The most studied example is calcium apatite in bone. With time, bone calcium apatite crystals become packed more closely together. This reflects progressive dehydration of bone tissue. In bone, dentin, and enamel, subsequent waves of biomineralization result in a mineralization architecture with layers of crystals arranged in a constant relationship to layers above and below. Bicarbonate is an important constituent adsorbed onto bone mineral. The amount and concentration of bicarbonate reflect the metabolic state of the acid–base balance in the body. Bicarbonate adsorption acts not only to provide buffering for mechanisms against circulating bicarbonates, but also to modify the physical properties of the bone mineral itself.

## E. Cell Biology of Biomineral Formation

In dissecting the chemical molecular details of biomineral formation, it should be noted that all biomineralization processes are controlled by cells. Many processes, including those that regulate extracellular fluid ion concentrations and those that involve differentiation of cells that directly participate in biomineralization, are located in organs and tissues remote from biomineralization sites. An overview of the cell functions in biomineral formation is given in Table V.

The calcium ion concentration, $[Ca^{2+}]$, is among the most exquisitely regulated of all body systems. Extracellular calcium ion concentration is maintained between 1 and 1.2 m$M$ by an endocrine cell-mediated system that involves calcium absorption from the gut, excretion from the kidney, and tissue-buffering mechanisms, including ion exchange with the mineral phase of bone. Phosphate and magnesium ions also

TABLE V

Cell Biology of Biomineral Formation

| Cell | Functions |
| --- | --- |
| Cells remote from biomineralization sites (e.g., gastrointestinal tract, liver, kidney, endocrine glands) | Extracellular fluid ion homeostasis, formation/secretion of circulating ion binding proteins and biomineralization inhibitors, regulation of cell differentiation and cell function at biomineralization sites |
| Cells adjacent to biomineralization sites [e.g., osteoblasts, odontoblasts (dentin), ameloblasts (enamel), chondrocytes] | Formation and regulation of extracellular matrix structure, matrix vesicle elaboration, apoptosis, secretion of nucleators, biomineral inhibitors/promoters, and enzymes regulating biomineralization |

participate in calcium ion regulation, but their extracellular concentrations, although of the same magnitude as $[Ca^{2+}]$, vary over a wider concentration range. Extracellular homeostasis is maintained by the complex interaction of hormones, principally parathyroid hormone and $1,25(OH_2)$ vitamin $D_3$. Cells remote from biomineralization sites also secrete circulating proteins such as albumin and $\alpha_2$ HS glycoprotein that bind calcium. Furthermore, hormones and growth factors secreted by remote cells regulate differentiation and function of cells at biomineralization sites. This regulation is complex and involves interaction between hormones and locally derived paracrine and autocrine factors. Many of these factors adhere to newly formed mineral surfaces, effectively increasing their concentration at biomineralization sites.

The functions of cells adjacent to biomineralization sites have been described in earlier sections. It is sufficient here to summarize the role of local cells in biomineral formation. First, local cells are involved in the formation and regulation of the specialized extracellular matrix that provides the spatial environment for biomineralization. This environment has two stages of development. The first stage precedes biomineralization, whereas the second involves deposition of incremental biomineral on a previously mineralized surface. Considerable controversy exists about the mechanisms of initial biomineralization. The controversy revolves around the relative roles of specialized substances, particularly acidic proteins or collagen, compared to cell-derived structures, namely, matrix vesicles, in the initiation of mineralization. Matrix vesicles are structures that are derived from the budding of cell membranes from the cell surface. They are bound by phospholipid membrane and contain enzymes and phospholipids that in isolated form, are implicated in biomineralization. Matrix vesicles are seen in many tissues involved in biomineralization and are present before the tissue mineralizes. In some instances, initial mineralization may take place inside matrix vesicles. Much controversy is also centered on whether the tissue mineralization is propagated from vesicle mineralization or whether matrix vesicles provide components (including ions from unstable amorphous calcium phosphate precipitates) for subsequent mineralization at sites associated with matrix proteins such as collagen. [*See* Bone Cell Genes, Molecular Analysis.]

Apoptosis, or controlled cell death, may play a role similar to that of matrix vesicles in the initiation of calcific biomineralization. In most mineralizing tissues, cells involved in matrix synthesis prior to mineralization and cells involved in regulating biomineral formation can undergo apoptosis. Cellular degradation products of apoptosis include membrane phospholipid, intracellular organic phosphates, and possibly residual ectoenzymes, all of which can contribute to biomineralization.

In the second stage, where mineral becomes accreted to preexisting mineral surfaces, local cells elaborate appropriate extracellular matrix, secrete heterogeneous nucleators and mineralization inhibitors, and provide enzymes that promote or inhibit mineralization. The specific nucleating substances are not yet well defined but include bone sialoprotein and certain phospholipids such as calcium phospholipid–phosphate complex. Mineralization inhibitors include proteins such as osteopontin and osteocalcin, which bind calcium avidly.

Alkaline phosphatase is an ectoenzyme present in the cell membrane of cells involved in biomineralization. Though it is recognized that alkaline phosphatase plays an important role in biomineralization, the enzyme is also found in cells of nonmineralizing tissues, for example, kidney and liver. Furthermore, the mechanism by which alkaline phosphates acts in biomineralization remains unknown. Suggested roles for alkaline phosphatase include increasing the availability of inorganic phosphate by its phosphatase activity on organic molecules, destruction of the calcification inhibitor, pyrophosphate, by its pyrophosphatase activity, or transport of calcium to the biomineralization sites by its calcium binding activity. Alkaline phosphatase itself is subject to complex regulation, including maintenance of transcription activity by hormones such as glucocorticoids, $1,25(OH)_2$ vitamin $D_3$, and estradiol, release from cell membranes by phospholipases, and activity inhibition by excess $[Ca^{2+}]$.

## IV. RESORPTION OF BIOMINERALS

Under physiologic conditions, biomineral resorption can occur rapidly and selectively as part of ion homeostasis or more slowly in association with other matrix components as part of skeletal tissue remodeling. Regarding ion homeostasis, cells adjacent to mineralized surfaces can respond to systemic stimuli to facilitate mineral resorption, without tissue matrix resorption. These stimuli include decreased $[Ca^{2+}]$ or increased circulating acid load, such as inorganic phosphates from dietary animal proteins or inorganic phosphates from cola drinks.

In the special circumstance of primary dentition, the mineralized tissue is shed by resorption of the adjacent bone tissue. This process is in fact part of bone modeling. Biomineral resorption associated with bone and calcified tissue modeling and remodeling has several features in common. First, a confined space is formed at the site of mineral resorption, usually within the extracellular space adjacent to a chondroclast or osteoclast. These cells, derived from macrophages, have the capacity to seal off the environment between the cell membrane and the mineralized tissue. First, the cells secrete proteolytic enzymes that digest organic material on the mineral surfaces. Subsequently, the cells activate a proton pump mechanism in the cell membranes that creates an acidic environment favorable to mineral dissolution. The cells also elaborate acid phosphatase ectoenzymes capable of destroying resorption inhibitors such as pyrophosphate, which may be adherent to the mineral surface. There are now two classes of pharmacologic compounds that can directly inhibit mineral resorption. Bisphosphonates are pyrophosphate analogs that bind to mineralize surfaces. These substances, which cannot be destroyed by phosphatases, act to inhibit mineral resorption perhaps by interfering with cell adhesion to the biomineral. A second group consists of proton pump inhibitors that prevent the acidification of the extracellular fluid adjacent to the biomineral resorption site. Both types of compounds have applications in which it is desirable to inhibit biomineral resorption, for example, osteoporosis.

# V. PATHOLOGIC BIOMINERALIZATION AND PATHOLOGIC CALCIFICATION

Traditionally, pathologic calcification has been classified by pathogenesis as *dystrophic calcification*, that is, associated with tissue injury; *metastatic calcification*, associated with systemic abnormalities of calcium ion metabolism; and *calcinosis*, local calcification at soft tissue sites without prior tissue injury. This type of classification ignores the specificity of biomineral and amorphous calcific deposits. Therefore, traditional classifications did not identify pathologic biomineralization within mineralizing tissues and could not discriminate among the many different pathologic processes leading to pathologic calcification. In the light of current knowledge, the classification offered in Table VI is more useful. In tissues exhumed postmortem, especially under archaeological conditions, diagenic changes, that is, calcifications

**TABLE VI**
Pathologic Biomineralization and Calcification

| Type | Examples |
|------|----------|
| Pathologic biomineralization | |
|   Inadequate mineralization | Vitamin D deficiency, aluminum toxicity |
|   Hypermineralization | Fluorosis, necrotic bone |
|   Demineralization | Hyperparathyroidism, inflammation |
|   Normal mineralization, inappropriate site | Heterotopic bone formation, cement line reduplication, calcified cartilage |
| Pathologic calcification | |
|   Systemic pathologic calcification | |
|     Abnormal ion metabolism | Hypercalcemia, hyperphosphatemia |
|     Excess circulating promoters | Trypsin, acute pancreatitis |
|   Focal pathologic calcification | |
|     Cell injury | Mitochondrial calcification, myocyte calcification, psammoma bodies, thyroid tumor cells |
|     Tissue necrosis | Inflammation, infarction |
|     Heterogeneous nucleation | Cholesterol crystals, atherosclerosis |
|     Microbial metabolism | Calcium oxalates, *aspergillus* infection, dental plaque, calcium phosphates |
|     Metabolic disorder leading to solute excess | Pyrophosphate, oxalate |

forming or changing their characteristics after death, must be distinguished from pathologic biomineralization and pathologic calcification.

Pathologic biomineralization includes processes of inadequate mineralization, hypermineralization, and demineralization in mineralizing tissues, as well as normal mineralization at physiologically inappropriate sites. Inadequate mineralization can result from failure to supply minerals, such as occurs in vitamin D deficiency and abnormalities in vitamin D or phosphate metabolism, as well as from failure of the matrix to calcify in conditions such as aluminum toxicity. Hypermineralization occurs where the mineral phase is altered by incorporation of ions such as fluoride or strontium and where there is loss of cell control such as in necrotic bone. Demineralization can follow from hormonal (e.g., hyperparathyroidism) or paracrine (e.g., inflammation) stimulation of mineral-resorbing cells or from microbial environments that produce acid and appropriate enzymatic conditions (e.g., dental caries). Pathologic biomineralization can also involve normal mineralization where an extracellular matrix capable of biomineralization becomes established at a site that does not normally mineralize. One example is heterotopic new bone, where tissue metaplasia facilitates osseous tissue formation at a site where bone does not usually form. A more common and more limited example is reduplication of the hypermineralized cement line or "tide mark" between calcified and uncalcified cartilage. In this situation, the pathologic mineralization of preexisting matrix appears to be adaptive to a shift of mechanical forces upon the cartilage. With the exception of necrotic bone in these examples, biomineralization remains fully controlled by the adjacent cells.

Most pathologic calcifications can be classified as biologically induced calcifications. Pathologic calcifications demonstrate greater variation in the solid phases formed, including amorphous as well as crystal phases. There is greater variability in particle size and orientation patterns than in normal biomineralization. The calcification may enlarge to destroy the structure in which it is formed. Furthermore, because the solid phases and the tissue pattern of calcific salt deposition are abnormal, pathologic calcifications are less easily resorbed by biologic mechanisms that are adapted to normal biominerals.

Systemic pathologic calcification can result either from metabolic arrangements that increase effective $[Ca^{2+}]$ hypercalcemia or from pathologic conditions associated with increased availability of calcification promoters. With hypercalcemia, the sites of mineral-

ization include the connective tissue of blood vessel walls and cytoplasmic sites of acid excretion, namely, lung parenchymal cells, gastric parietal cells, and kidney tubular cells. Typically, the calcification formed is an amorphous calcium phosphate. When cell injury occurs in association with hypercalcemia, focal cytoplasmic calcification can occur at susceptible sites, particularly striated and myocardial muscle. Pathologic calcification with hyperphosphatemia usually occurs in soft connective tissues, including the dermis, blood vessel wall, cornea, band keratopathy, and periarticular bursa. In fulminant acute pancreatitis, increased circulating trypsin and other proteases produce systemic fat necrosis. At these sites, calcifications form by calcium precipitating with fatty acids derived from degraded lipid (saponification).

In contrast to extracellular fluid, which is saturated with respect to $[Ca^{2+}]$ and $[PO_4^{3-}]$, the intracytoplasmic milieu is intrinsically nonmineralizable. This is related to the very high $[Mg^{2+}]$ and the very low ($\approx 10^{-3}$ mM) $[Ca^{2+}]$ present in the cytoplasm. Cell membrane injury of any type allows ingress of $Ca^{2+}$ ions, which are then transported externally by active metabolic pumps that utilize adenosine triphosphate. Focal failure of these pumps permits sequestration of $Ca^{2+}$ as Ca–Mg–phosphate complexes in cytoplasmic compartments containing abundant phospholipid (e.g., mitochondrial membrane). [*See* Cell Membrane Transport.] In other circumstances, cell injury results in focal reduplication of cell membranes rich in phospholipid or concentric formation of cell membrane-like structures. When these structures undergo calcification, intracytoplasmic inclusions that resemble grains of sand form. In cells such as those in thyroid epithelial tumors, these structures have been termed "psammoma bodies."

Pathologic calcification, usually amorphous calcium phosphate or very poorly crystalline calcium apatite, is found at sites of tissue necrosis secondary to inflammation or infarction. Typically, the necrotic tissue is rich in phospholipid derived from membranes and cytoplasmic structures of necrotic cells that have been incompletely resorbed. Precipitation of insoluble calcium salts further limits tissue repair. In some cases where calcification persists for many years, for example, subcutaneous tissues in scleroderma, the calcifications undergo an annealing process and transform into very large crystals of calcium apatite, further limiting resorption.

Pathologic calcification at sites of soft tissue injury, particularly if the injury is associated with metal deposition, is termed calcergy. Soft tissue calcification that occurs following soft tissue injury in situations where there is hypercalcemia related to increased vitamin D is termed calciphylaxis.

Pathologic calcification in atheromatous blood vessels has been commonly ascribed to tissue necrosis. However, there is now considerable evidence that the initial calcification can occur epitaxially on the surfaces of cholesterol crystals prior to tissue necrosis or fibrosis. In this situation, the cholesterol crystal surfaces act as heterogeneous nucleators. Pathologic calcification can occur in microbial infections related to microbial metabolism. Calcium oxalate crystal formation in aspergillosis is an example of this condition.

An important set of pathologic calcifications occurs when there is an accumulation of a metabolite that, in the presence of calcium [$Ca^{2+}$], forms insoluble salts. The metabolic disorder may be restricted to particular tissues. For example, in articular tissue such as hyaline cartilage, under certain pathologic conditions, pyrophosphate can accumulate and subsequently precipitate as calcium pyrophosphate dihydrate crystals. This pathologic process illustrates the specificity of pathologic mineralization. Of more than 30 similar crystal phases known, only 2 crystal phases form, namely, calcium pyrophosphate dehydrate

monoclinic and calcium pyrophosphate dehydrate triclinic. Furthermore the crystal formation appears to be restricted to the extracellular matrix adjacent to chondrocytes or to the matrix of cells undergoing chondroid metaplasia. Although the conditions are not as well defined, calcium oxalate crystals can form extracellularly in thyroid colloid and, in granulomatous inflammation, intracellularly within macrophages. Massive calcium oxalate deposits also occur in the urinary tract as urinary calculi. The pathogenesis of urinary calcium stone formation includes nucleation, often on an organic molecule or cell membrane, followed by extensive waves of crystal aggregation and agglomeration.

Pathologic calcifications are seldom mere markers of disease. Rather, pathologic biomineralization and pathologic calcifications contribute actively as etiologic agents with crystal specificity as a major determinant of the disease process. Examples of the disease processes involved are provided in Table VII.

When pathologic biomineralization occurs in mineralized tissues, hypoplasia (inadequate mineralization), hypertrophy (hypermineralization), or atrophy (demineralization) can result. In certain pathologic conditions, the extracellular matrix forms but mineral

## TABLE VII
Disease Processes Involving Pathologic Biomineralization and Calcification

| Disease process | Tissue | Disease | Mineralization/calcification |
|---|---|---|---|
| Hypoplasia (mineralization failure) | Cartilage | Rickets | Calcium apatite |
| | Bone | Osteomalacia | Calcium apatite |
| Atrophy (demineralization) | Bone | Osteoporosis | Calcium apatite |
| Hypertrophy (hypermineralization) | Bone | Osteosclerosis, osteopetrosis | Calcium apatite |
| Duct obstruction | Gallbladder and biliary tract | Calculi | Calcium carbonate, Calcium bilirubinate |
| | Kidney and urinary tract | Calculi | Calcium oxalate, Calcium phosphate |
| | Salivary duct | Calculi | Calcium apatite |
| Acute inflammation | Joints | Crystal associated arthritis | Calcium pyrophosphate, dihydrate, calcium apatite |
| Chronic inflammation | Connective tissues | Calcified scars (previous acute inflammation or infarction) | Amorphous calcium, phosphate |
| Fibrosis | Soft connective tissue | Hemorrhage | Hemosiderin (iron molecular complex) |
| Degeneration | Joints | Degenerative arthritis | Calcium pyrophosphate, dihydrate, calcium apatite |
| | Blood vessels | Arteriosclerosis | Amorphous calcium phosphate |
| Storage disorders | Thyroid | Crystal deposition in colloid | Calcium oxalate |

fails to be deposited. In cartilage, this hypoplastic condition is known as rickets; in bone, the disorder is termed osteomalacia. These diseases are usually secondary to disorders in which there is insufficient supply of calcium or phosphate or alternatively to conditions where mineralization inhibitors are present in the extracellular matrix. Hypertrophy or hypermineralization disorders include conditions where the accreted mineral is particularly dense (e.g., fluorosis) and where the cell regulation of mineralization is abnormal (e.g., osteopetrosis). Caries in teeth and osteoporosis in bone are examples of atrophic demineralization disorders. In both conditions, both the mineral and the associated matrix are resorbed. Although the pathogenesis of these conditions is very different, the mineral can be made more resistant to resorption if, at the time of formation, the mineral, calcium apatite, contains incremental fluoride.

From Table VII, it can be observed that pathologic biomineralization and pathologic calcification can participate in a variety of disease processes. Although there are differences in intensity, progression, and residual effects, different kinds of calcifications can take part in the same disease process. Conversely, although the process may occur at different sites or sequentially, calcifications of the same crystal phase can be involved in multiple disease processes. Often, the key to treatment and/or prevention of the disease is the precise diagnosis of the pathologic calcification involved.

## BIBLIOGRAPHY

Bobryshev, Y. V., Lord, R. S. A., and Warren, B. A. (1995). Calcified deposit formation in intimal thickenings of the human aorta. *Atherosclerosis* **118**, 9–21.

Bonucci, E. (ed.) (1992). "Calcification in Biological Systems." CRC Press, Boca Raton, Florida.

Boskey, A. L., and Boyan, B. D. (1996). "Connective Tissue Research," Vol. 34, No. 4 and Vol. 35, Nos. 1–4. Gordon and Breach Science Publishers, Amsterdam.

Driessens, F. C. M., and Verbeeck, R. M. H. (eds.) (1990). "Biominerals." CRC Press, Boca Raton, Florida.

Grynpas, M. D. (1993). Age and disease-related changes in the mineral of bone. *Calcif. Tissue Int.* **53**(Suppl. 1), 57–64.

Hukins, D. W. L. (ed.) (1989). "Calcified Tissue." CRC Press, Boca Raton, Florida.

Hunter, G. K. (1996). Interfacial aspects of biomineralization. *Curr. Opin. Solid State Materials Sci.* **1**, 430–435.

Kim, K. M. (1995). Apoptosis and calcification. *Scanning Microsc.* **9**(4), 1137–1178.

Pritzker, K. P. H. (1994). Calcium pyrophosphate dihydrate crystal deposition and other crystal deposition diseases. *Curr. Opin. Rheumatol.* **6**, 442–447.

Ross, M. D., and Potc, K. G. (1981). Some properties of otoconia. *Philos. Trans. Roy. Soc. London Ser. B* **304**, 445–452.

# Birth Control

ANA-ZULLY TERÁN
*Medical College of Georgia*

## GLOSSARY

**Amenorrhea** Lack of menses. Primary: Absence of the onset of menstruation at the time of puberty. Secondary: Cessation of menses once they have started

**Dysmenorrhea** Pain associated with menstrual bleeding

**Menarche** First period ever; the initiation of a woman's sexual and reproductive development

**Menorrhagia** Loss of blood occurring every 28 days for 3–5 days; normally seen in women between 12 and 50 years of age and is due to necrosis and periodic reparation of the endometrium under the influence of ovarian hormones

**Metrorrhagia** Bleeding from the uterus at any time other than during the menstrual period; may be caused by lesions of the cervix uteri; its occurrence should lead one to suspect and search for a malignancy in the genital tract, specifically cancer of the cervix

**Oral contraceptives** Called birth control pill, a combination of estrogen and progestogens given in small amounts to prevent ovulation

FROM THE BEGINNING OF TIME, WHEN HUMANS became aware of the relation between coitus and pregnancy, they searched for a method that would prevent conception that was magical, mechanical, or medicinal. Egyptian women inserted pessaries fabricated from crocodile dung into their vagina. Chinese women drank liquid of fried quicksilver as a contraceptive barrier or swallowed 14 live tadpoles 3 days after menstruation. The Greek women of the second century made vaginal plugs of wool soaked in sour oil and honey, pomegranate, fig pulp, and cedar gum. In the Middle Ages, women used potions prepared from willow leaves, iron rust, clay, and the kidney of a mule. The Europeans used to tell their brides to sit on their fingers while riding in their coaches.

Although picturesque, these methods, which were used for family planning prior to our time, show us that preventing pregnancy has been of great concern throughout the history of humankind. Later, a number of reasonably effective means of contraception were described and are still used. The condom has been known since the 16th century. In the beginning of the 19th century, the cervical cap was first used and, later, the diaphragm and vaginal spermicides were introduced. Intrauterine devices (IUDs) were introduced in the 20th century. However, the most accurate birth control has been the use of the female sex steroids: estrogen and progesterone, better known as the PIU. A review of nonhormonal methods will not be assessed at this time.

## I. BACKGROUND

Birth control by hormonal methods was sought as early as 1897, when progesterone was postulated as inhibiting ovulation during pregnancy. In 1940, inhibition of ovulation was introduced into clinical practice as a therapeutic tool. However, ovulation still occurred every other month, making this regimen only partially effective. In 1942, stilbesterol suppositories were used throughout the cycle as therapy to alleviate dysmenorrhea, and in 1949 continuously conjugated estrogens were used to inhibit ovulation for 1 year's time, interposing short cycles of a progestin to induce menses.

ENCYCLOPEDIA OF HUMAN BIOLOGY, Second Edition, VOLUME I.   Copyright © 1997 by Academic Press.   All rights of reproduction in any form reserved.

During the late 1950s, successful inhibition of ovulation was obtained using a potent progestogen (Norethynodrel); however, breakthrough bleeding was observed. As a result, the addition of an estrogen was proposed to maintain the integrity of the endometrium. These studies and a preliminary report presented and initiated by the Family Planning Association of Puerto Rico on the efficacy and safety of Norethynodrel and mestranol for conception control led to further investigations to prove the utility and safety of the estrogen–progestogen combination pill. In the early 1960s, the first-generation oral contraceptive was approved by the Food and Drug Administration (FDA) in the United States. Still today, the ideal hormonal agents for conception control deprived of side effects have not been found. Hopefully, because continuous efforts in improving technology are achieved, the untoward reactions of oral contraceptives will be lessened and severe adverse reactions eliminated, while efficacy is maintained. [*See* Steroids.*]

Usually, combination drugs are seen as nonpharmacologic, and the FDA looks with disfavor upon their use. There is, however, one field of endeavor where combination drugs are of prime importance: conception control. The combination estrogen–progestogen is based on physiological grounds. It mimics the menstrual cycle, but, preventing ovulation, properly taken oral contraceptives are almost 100% effective and are preferred by a great majority of patients, although they have many transient metabolic and undesirable effects. In many instances, these effects have caused discontinuation and failure as a contraceptive method. The FDA and the International Planned Parenthood Federation recommended that women over the age of 40 years should utilize forms of contraception other than the pill. Appropriately, contraceptive choice varied with age. According to a 1976 survey of several thousand American couples, oral contraceptives were still the first choice among women younger than 29 years of age, condoms were second choice, while diaphragms and IUDs were about equal in popularity. In 1983, for couples 30 years old and over, surgical sterilization was the leading method of fertility control. For many, the availability of legal abortions since 1973 has meant that couples would not worry about pregnancy and use less precaution.

Newer forms of birth control have emerged in the first part of the nineties. The Norplant system is a long-acting subdermal progesterone implant system that uses only levonorgestrel, a second-generation progestogen. Its mechanism of action as a contraceptive is suppression of ovulation. It thins the endometrial lining and thickens the cervical mucus. Although this method appears to be invasive compared to other methods, like intrauterine devices, many women prefer it because of its 5-year effect.

## II. EFFECTIVE AND SAFE BIRTH CONTROL

The requirements of a good birth control method are effectiveness and safety. It should also be reversible and not linked to coitus. To be effective, four factors should be determined: fecundability of the user, biological effectiveness of the method, care with which the method is used, and coitus frequency. The more frequently a couple has intercourse, the greater the chance of conception, therefore increasing a woman's sense that she needs to use some sort of contraception. Often, she is more inclined to consider a method like the pill, which needs to be taken continually, or the IUD, which acts continually. Proper use of the contraceptive method is necessary to prevent conception, and it is necessary to evaluate different influencing factors such as education, age, socioeconomic and marital status, religion, race, and experiences with that particular contraceptive method. It is also important that the method offer some protection against infectious diseases, such as HIV.

The time when women are sexually active but not interested in having children and are not in a stable relationship puts the most demands on a method. It must be effective not only in preventing pregnancy but also in protecting against sexually transmitted diseases (STD) and pelvic inflammatory diseases (PID), which can reduce fertility. Condoms and barriers do offer some protection against PID and STD, but they have higher failure rates as a contraceptive than the pill and are coitus-dependent. In contrast, other methods now under development, vaccines and male methods, for example, offer effective pregnancy prevention, coitus independence, and reversibility, but do not offer protection against STDs or PID. These methods would probably be less appropriate than the pill for most of the younger population. Consequently, many of them are at risk of contracting an STD or HIV infection. Many also do not want to use a continual method like the pill or the IUD. In fact, the latter is not a good method since it does not prevent STDs or HIV and may even increase the likelihood of infection in those who contract an STD while using it, thereby increasing the risk of infertility.

Fecundability measures the probability of pregnancy when no contraception is used. This factor strongly depends on age. It begins in girls at menarche, but it is less probable due to anovulatory cycles. It is highest when women are in their late teens and early 20s; thereafter, it slowly declines. Therefore, legal abortion is more common in women between 18 and 25 years of age. Nonuse of any contraception by sexually active, nonsterile women who are not trying to become pregnant is highest among never-married and formerly married women. It is moderate, however, among women who are married and planning to enlarge their family in the near future and among women who, independent of marital status, want no more children or no children at all. Because they are more mature and assume that they are less fertile or because of their life status or professional life, it is probable that these women have sex less frequently.

If a method is safe as a birth control, it should be evaluated in view of health risk either from complications due to its use or from pregnancy occurrence, which should be considered a failure of the method. Age is an important factor because failures of contraception are more common in the younger population, but women in the older population are at a higher risk of complications and are more likely to die as a result.

In recent years, smoking has been identified as an important factor associated with increased incidence of complications of oral contraceptive use, such as cardiovascular and thromboembolic disorders. Women are now smoking more than they did in the past two decades. Studies performed to evaluate the relationship between OC complications and smoking have indicated that OC users who smoke are at higher risk for thromboembolic and cardiovascular disease. The increased risk is related to the dose of the estrogen (ethinyl estradiol) and to the type and dose of the progestogen. Another well-documented factor that must be taken into consideration is the body mass index. There appears to be a correlation between a body mass index of more than 30 kg/m² and venous thrombosis.

## III. COMPLICATIONS OF BIRTH CONTROL

### A. Cardiovascular

Data obtained in 1983 on the risks of death associated with different fertility control practices show that for women under 35 years of age, any contraception method is safer than no method at all. However, because there are many noncontraceptive benefits to taking birth control pills, women who do not smoke are encouraged to use oral contraceptives into their forties and smokers are encouraged to stop smoking so they can reap the benefits of OC use.

In the seventies, several reports showed that acute myocardial infarction is more than three times greater in oral contraceptive users than in controls. These reports also showed an alarming incidence of cerebrovascular accidents (cerebral and subarachnoid hemorrhage and thrombosis). The increased incidence is especially evident in older women and in women who smoked. Therefore, the suggested age limit was decreased to 35 years and in some cases 30 years. However, this entire topic is still controversial. The increase risk is related to the dose and the type of estrogen (ethinylestradiol) and the type and dose of the progestogens. Naturally occurring estrogens such as 17-β-estradiol in the form of subcutaneous pellets can be used as a form of birth control. They have been used in Europe since the forties with much success, but in the United States they are more frequently used as homonal replacement therapy for menopausal women. They appear to lower low-density lipoprotein levels and to increase high-density lipoproteins, the protective lipoproteins, therefore reducing the risk of myocardial infarction (see Fig. 1).

In general, many reports throughout the years have implicated oral contraceptives as responsible for the markedly increased incidences of myocardial infarction, thromboembolic disease, and liver and gallbladder disorders. The newer oral contraceptives with

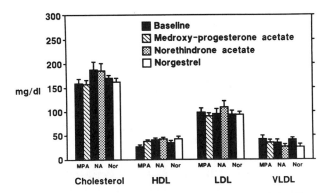

**FIGURE 1** Comparison of different progestogens and the effect on lipoproteins in patients who received hormone replacement in the form of pellets for contraception. HDL, high-density lipoproteins; LDL, low-density lipoproteins; VLDL, very low-density lipoproteins.

lower estrogen content and different progestogen combination aim to minimize cardiovascular and thrombotic side effects while maintaining good cycle control. Recent studies have shown that use of oral contraceptives containing low estrogen content (30 $\mu$m of ethinyl estradiol) is associated with a lower risk of thrombotic disease compared with use of higher dose pills (50 $\mu$g). In October 1995, a warning was issued by the Committee on Safety of Medicines in the United Kingdom to all British doctors in reference to the newer third-generation progestogen in oral contraceptives. These pills contained the third-generation progestagens desogestrel, gestodene, and norgestimate. The warning was issued in reference to the higher risk of developing nonfatal venous thromboembolism (VTE) for users of oral contraceptives in this group compared to the risk for users of pills with the second-generation progestagens levonorgestrel and norgestrel. The increased incidence is thought to be due to the fact that these newer pills were prescribed to women at an anticipated increased thrombotic risk. When prescribing these new forms, doctors must take into consideration any clinical history of varicose veins and any possible previous hypertension during pregnancy or during use of the pill, as well as smoking habits and family history of VTE. Consequently, these patients should probably be prescribed the newer third-generation pills with lower estrogen content.

## B. Lipids and Lipoproteins Changes of Birth Control Pill

It is well known that estrogens, natural or synthetic, alter lipid levels. The lipids and lipoproteins have been implicated in the development of atherosclerosis and coronary heart disease. Studies by the Lipid Research Clinics Program have shown that plasma triglycerides are approximately 50% higher in oral contraceptive users than in nonusers; cholesterol concentrations are also higher, primarily in the low-density lipoprotein (LDL) fraction, whereas the high-density lipoprotein cholesterol (HDLc) shows minor differences. It has been suggested that the estrogenic component of the oral contraceptives, especially ethinylestradiol or mestranol, decreases the values of HDLc and increases the risk of atherosclerosis. However, studies have drawn attention to decreased levels of HDL when progestins are used alone or when their proportions are higher in the combination pill; in contrast, pure estrogens seem to increase the levels of HDLc, thus acting indi-

rectly as a protective factor. [*See* Atherosclerosis; Cholesterol; Lipids.]

This latter form of contraception is with natural estrogen in the form of crystalline pellets. Implantation of 17$\beta$-estradiol has been used for hormone replacement therapy since 1938. When hard pellets of pure crystalline steroids are implanted in the subcutaneous tissue, the hormone is slowly absorbed, resulting in continuous release in small quantities. The amount of the absorption will depend on the surface area of the pellet exposed to the action of the body fluids. The rate of absorption will depend on the number of pellets implanted, and the duration of its effectiveness will depend on the weight of the pellets. This method of contraception may be important for women >35 years because the oral forms have been questioned for their atherogenicity. The use of four pellets of 17$\beta$-estradiol (25 mg each) every 6 months reportedly provides excellent ovulation suppression. On the basis of previous experience with ovulation suppression with conjugated estrogen in a monthly step-down method, the pellets were reduced one each time, establishing the step-down course 4-3-2-1. A step-down course 4-2-1-1 gave successful inhibition of ovulation and, therefore, effective contraception.

In 1981, it was suggested that the androgenic progesterone of the 19-testosterone series reverses the beneficial effect of estrogen on the lipoproteins, whereas the dehydroxyprogesterone derivative medroxyacetate has no such effect. At that time, a group of 312 patients on hormone replacement therapy in the form of pellets for birth control was evaluated clinically, and lipids and lipoproteins were measured at the beginning of the trial and then every 6 months.

The patients were divided into three groups according to the progestogen used: medroxyprogesterone acetate, norethindrone acetate, and levonorgestrel. All patients showed an increase in cholesterol, but higher increases were found with the 19-nortestosterone derivative than with the medroxyacetate derivative. Compared with the other two progestogens, LDLc and the very-low-density lipoprotein cholesterol, the HDLc increase in patients with medroxyprogesterone acetate showed decreased levels with the medroxy-progesterone derivative. A statistically significant ($P < 0.05$) decrease was observed in the risk ratio (Chol/HDL-c and LDL/HDL) for the medroxyprogesterone derivative compared with the other two forms (Figs. 1 and 2). In conclusion, the medroxyprogesterone derivative does not wholly offset the positive effect of estrogen on prevention of heart disease.

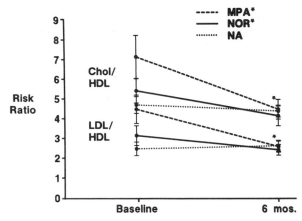

**FIGURE 2** Patients who had received estradiol pellets for conception control and medroxy-progesterone acetate. The risk ratios were decreased and shown to be statistically significant (*$P < 0.05$). MPA, medroxy-progesterone acetate; NA, norethindrone acetate; NOR, norgestrel.

## C. Thromboembolism

Changes implicated in the increased risk of thromboembolic disorders in oral contraceptive users are alterations in certain coagulation and fibrinolytic enzymes, resulting in increased generation and dissolution of the fibrin clot. Studies of markers of coagulation endpoints such as fibrinopeptide A suggested that an increased activation of the coagulation system did indeed occur in pill users. Norris and Bonnar in 1996 showed that the reduction of estrogen in the pill in combination with newer progestogens can decrease the activation of the coagulation cascade found in patients taking birth control pills.

Factors that prevent clot formation are antithrombine III and Protein C. The synthetic estrogen in oral contraceptives has a lowering effect on these factors as well as on circulating levels of fibrinogen and platelet activator inhibitor, increasing platelet aggregation. Other factors implicated in adverse effects of oral contraceptives are decreased levels of factors VII and X. These results agree with those from previous studies suggesting that mestranol in the oral contraceptive pill may provide a thrombogenic hazard. Similarly, greater hemostatic changes have been observed in women taking pills with higher estrogen contents than in women taking pills with lower estrogen. It has also been reported that the progestogen component of the pill may modify the estrogenic effect. This contrasts with the effect of natural estrogens. As concluded at the 1978 International Congress on the Menopause, "despite the increase in clotting factors associated with aging, natural estrogens influence relatively little change in most of the coagulation factors studied." [*See* Hemostasis and Blood Coagulation.]

## D. Depression

One other adverse reaction of oral contraceptives is depression. The control of mood is thought to be produced by biogenic amines, such as serotonin, noradrenaline, and dopamine, which are synthesized from the amino acids precursors tryptophan and tyrosine (Fig. 3). Any metabolic effect that lowers the plasma ratio of these amino acids will decrease the rate of synthesis of these amines in the brain neurons. Oral contraceptives have been found to lower the concentrations of these amino acids, especially tryptophan. Another effect is on the pyridoxal phosphate enzyme, which converts 5-hydroxytryptophan to 5-hydroxytryptamine (serotonin). This enzyme has been found to be in competition with the estrogen conjugates, possibly resulting in a decreased formation of serotonin. In a recent publication, it was mentioned that the newer progesterone implants, Norplants, are associated with the development of major depression within 1 to 3 months after the insertion of the implants and with resolution of the depressive episodes after its removal. This contrasts with previous studies, in which depressive episodes were found to diminish or even disappear after estrogen implants were used (Fig. 4). In reference to the mechanism of induction of depression linked to oral contraceptives, possible explanations include a progesterone-induced decrease in norepinephrine concentration in the brain; an increase in monoamine oxidase activity in the brain,

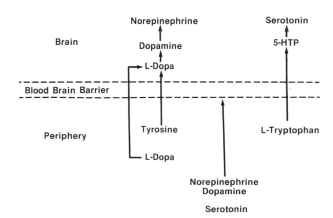

**FIGURE 3** Representation of different brain neurotransmitters (NT) and selective effect of the blood–brain barrier through which the NT concentrations are modified in the brain.

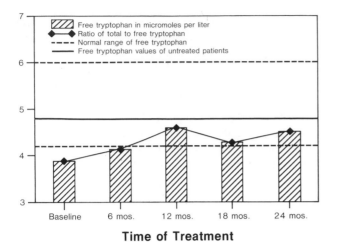

**Time of Treatment**

**FIGURE 4**   Tryptophan fractions before and after estradiol pellet implantation for conception control.

which would lower levels of 5-hydroxytryptamine; and a decrease in tyrosine availability. Therefore, catecholamine deficiency likely plays a major role in the development of depressive symptoms in some oral contraceptive users. These patients may benefit from the use of vitamin $B_6$ and tryptophan to compensate for the deficiency and to maintain their metabolic pathway (Fig. 5). [*See* Depression; Depression, Neurotransmitters and Receptors.]

Another method of contraception, once well accepted but slowly declining since the introduction of the birth control pill, is the condom, known since the 16th century. "Armour Against Enjoyment and a Spider Web Against Danger" was the description of a condom by Madame de Sevigne in a letter to her daughter in 1671. In the 18th century, condoms were

used extensively in brothels. They were made of animal intestine, and today this type is still sold and preferred over rubber ones because some say they are better at transmitting sensation. Their effectiveness and safety depend on the ability to hold the seminal fluid. The possibility that the condom may rupture has been one objection to the use of this method. Another objection is that it interferes with the spontaneity of the couple and reduces both partners' sensation. The condom has advantages over other contraceptive methods in the reduction of the risk for sexually transmitted diseases (e.g., acquired immunodeficiency syndrome) and of cervical cancer. Other methods such as coitus interruptus and the rhythm method have a very small success rate because of the disadvantage of their use. The latter method is used more commonly as part of the infertility workup in which the basal temperature is determined daily to detect ovulation manifested by the LH surge elevation and persistency of a high temperature above 98° (e.g., 98.2, 98.4, etc.). The production of progesterone by the corpus luteum maintains the temperature at this level for about 14 days in a normal ovulatory cycle.

Nursing an infant has been postulated as another form of contraception. Women with high prolactin, due to nursing, will have secondary amenorrhea and anovulatory cycles. The effect of antiovulatory action will vary according to the frequency and duration of nursing. Usually, women who do not nurse will have a menstrual period within the first 3 months after delivery, whereas women who nurse for 18 months may have amenorrhea for about 8–13 months after delivery. Ovulation will return eventually, despite continuation of nursing. As soon as women start to reduce the nursing time because they feed the infant artificially, ovulation will again occur; therefore, precautionary contraceptive methods should be advised.

## IV. BIRTH CONTROL BENEFITS

In addition to contraception, the birth control pill has several other important functions and benefits: it alleviates dysmenorrhea, it controls menometrorrhagias, it regulates menstrual cycles, it controls hirsutism, and it treats endometriosis. Also, in cases such as the Stein–Leventhal syndrome, in which continuous formation of ovarian cysts produces a variety of symptomatology (hirsutism, obesity, amenorrhea), the birth control pill will prevent some ovarian cysts that

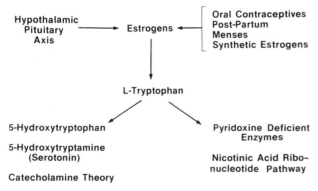

**FIGURE 5**   Different estrogen components can affect tryptophan metabolism. The addition of vitamin $B_6$ and tryptophan can compensate for this deficiency.

otherwise would require surgery. Among the most important benefits are those associated with the presence of less thyroid disease, less rheumatoid arthritis, greater bone density, and less benign breast disease. In reference to fertility, the use of the pill is associated with the occurrence of less pelvic inflammatory disease (PID), therefore preserving fertility.

## A. In Ovarian and Endometrial Cancer

Prolonged use of oral contraceptives is now known to reduce a woman's risk of developing ovarian cancer or a cancer of the endometrium. In 1983, the Center for Disease Control's Steroid Hormone Study estimated that as many as 1700 deaths from ovarian cancer and 2000 deaths from endometrial cancer are prevented each year in American women because of oral contraceptives use. [*See* Endometriosis.]

## B. In Cervical Cancer

The risk for cervical cancer has been reduced in women who use the diaphragm and other barriers as well, presumably due to reduction in the transmission of causative agents such as papilloma viruses. Pill users in comparison have a higher risk: An Oxford study has found that these women have a tendency to develop cervical dysplasias and cancer at a more rapid rate than a control group of IUD users. Pill users reduce the chance of developing pelvic inflammatory disease by 50%. Diaphragm use appears to somewhat increase the risk of bladder infections.

## C. In Breast Cancer

The specter of breast cancer in hormonally treated women is constantly being raised. Carcinoma of the breast is the most frequent malignancy in females, constituting about 26% of all cancers, although it is now second to lung cancer as a cause of death in women. It is also the leading cause of death from cancer, responsible for 18% of all female cancer deaths in the United States. It accounts for about 10 times the number of deaths from endometrial cancer. Unlike the incidence of endometrial cancer, that of breast cancer continues to rise progressively throughout life. This differs from cervical, endometrial, and ovarian cancers, which peak in the menopausal years. In other words, the older a woman lives, the greater her risk for breast cancer. Family history, early menarche, age at first delivery after age 30, number of children, lack of breast feeding, obesity, anovulation, and

late menopause are known risk factors of developing breast cancer. In 1970, Fechner published an extensive analysis of this problem regarding breast carcinoma. He stated that of all breast cancers, 4% occur in patients younger than 35. This is the same figure reported prior to the availability of the antiovulants.

Over the years there has been a division of opinion among scientists about the influence of estrogen on the development of breast cancer. The general conclusion is that most studies do not find an association between breast cancer and oral contraceptive use, except possibly in women younger than 45 years. Use of oral contraceptives has been found not to alter the risk of parity, family history, or the age of first birth in breast cancer. It appears that it does produce a protective effect compared to other measures of birth control. Oral contraceptives should be encouraged in younger women desiring family planning, particularly those with fibrocystic disease of the breast and, paradoxically, those patients with anovulatory cycles or infertility problems, becuase opposing estrogen therapy with cyclic progesterone therapy reduces the breast cancer risk to a level lower than that of the untreated population. Birth control pills reduce benign breast disease, which may be precancerous lesions in some women, and probably afford protection from subsequent carcinoma of the breast.

## V. CONCLUSIONS

Thus far, the inhibition of ovulation by hormonal contraception seems to be the most effective method of contraception. The birth control pill appears to be the best agent, although the ideal pill, devoid of side effects, has not yet been developed. Mechanical barriers such as cervical caps, diaphragms, and condoms offer good alternatives for those women who cannot tolerate the hormonal form of contraception. In many instances, because of the side effects of the pill (nausea, headaches, metrorrhagia, loss of libido, and depression), the inconvenience of mechanical barriers (cystitis, PID, and irritation), or the varied feasibility of surgical sterilization, an alternative hormonal way to produce anovulatory cycles, and therefore contraception, would be to use estradiol pellets or the newer long-term progesterone—the Norplant system—for subcutaneous implantation. This method may be particularly suitable for the premenopausal woman who may wish to continue the low-dosage estrogen (one pellet every 6 months of 17β-estradiol) in her advancing years. In this way, preventing the onset of hot flushes and sweats, minimizing the tendency to osteo-porosis, and, in some cases, decreasing annoying mood changes and depressive episodes or the frequency and severity of migranoid headaches seen commonly at this stage of a woman's life are possible. In any situation, any form of birth control must be carefully evaluated by the patient and her physician, and the advantages must be weighed against the risk in any particular or individual case. [*See* Estrogen and Osteoporosis.]

# BIBLIOGRAPHY

Abbate, R., Pinto, S., Rostagno, C., *et al.* (1990). Effects of long term gestodene-containing oral contraceptive administration on haemostasis. *Am. J. Obstet. Gynecol.* **163**, 424.

Daly, L., and Bonnar, J. (1990). Comparative studies of 30 μg ethinylestradiol combined with gestodene and desogestrel on blood coagulation, fibrinoysis and platelet function. *Am. J. Obstet. Gynecol.* **163**, 430.

Fechner, R. E. (1970). Breast cancer during oral contraceptive therapy. *Cancer* **26**, 1208.

Gambrell, R. D. (1984). Oral contraceptives, postmenopausal oestrogen–progestogen use and breast cancer. *J. Obstet. Gynaecol.* (Suppl 2) **4**, S121–S127.

Gram, J., Munkvad, S., and Jespersenm, J. (1990). Enhanced generation and resolution of fibrin in women above the age of 30 years using oral contraceptives low in estrogen. *Am. J. Obstet. Gynecol.* **163**, 438.

Hannaford, P. C., Croft, P. T., and Kay, C. R. (1994). Oral contraception and stroke: Evidence from the Royal College of General Practitioners: Oral Contraception Study. *Stroke* **25**, 935.

Lidegaard, O. (1993). Oral contraception and risk of a cerebral thromboembolic attack results of a case control study. *Br. Med. J.* **306**, 956.

Lidergaard, O., and Mill, Rom I. (1996). The pill: The controversy continues. *Acta. Obstet. Gynecol. Scand.* **75**, 93.

Meade, T. W. (1988). Risk and mechanism of cardiovascular events in users of oral contraceptives. *Am. J. Obstet. Gynecol.* **158**, 1646.

Norris, L. A., and Bonnar, J. (1996). The effect of oestrogen dose and progestogen type on haemostatic changes in women taking low dose oral contraceptives. *Br. Med. J.* **103**, 261.

Shaarawy, M., Gayad, M., Nagyi, A. R., *et al.* (1982). Serotonin metabolism and depression in oral contraceptives users. *Contraception* **26**, 193.

Stadel, B. V. (1981). Oral contraceptives and cardiovascular disease. *N. Engl. J. Med.* **305**, 612 and 672.

Terán, A. Z., and Greenblatt, R. B. (1986). A study of estradiol pellets and contraception update. Lecture presented at the 18th annual meeting of the International Society of Reproductive Medicine. Rancho Mirage, California.

Tomasson, H., and Toasson, K. (1996). Oral contraceptives and risk of breast cancer. A historical prospective case-control study. *Acta. Obstet. Gynecol. Scand.* **75**, 157.

Wagner, K. D. (1996). Major depression and anxiety disorders associated with Norplant. *J. Clin. Psychiatry* **57**, 152.

Wagner, K. D., and Berenson, A. B. (1994). Norplant associated major depression and panic disorder. *J. Clin. Psychiatry* **55**, 478.

World Health Organization (1995). Collaborative study on cardiovascular disease and steroid hormone contraception. Venous thromboembolic disease and combined oral contraceptives: Results of international multicentre case-control study. *Lancet* **346**, 1575.

World Health Organization (1995). Collaborative study on cardiovascular disease and steroid hormone contraception. Effect of different porgestagens in low estrogen oral contraceptives on venous thromboembolic disease. *Lancet* **346**, 1582.

# Birth Defects

FRANK GREENBERG
*Baylor College of Medicine*

---

## GLOSSARY

**Association** Nonrandom occurrence in one or more individuals of defects that are not thought to be pathogenetically related and do not fit the category of a sequence or field defect

**Deformations** Abnormalities in form or position of body parts due to extrinsic mechanical forces

**Disruption** Defect of an organ or body part resulting from a breakdown of or interference with what was originally a normal developmental process

**Dysmorphology** Study of birth defects, as well as the extreme variability of normal physical features

**Dysplasia** Defects in organs or body parts due to cellular disorganization resulting in benign or malignant tumors or hamartomas

**Field defect** Pattern of multiple anomalies derived from the disturbance of a developmental field, usually affecting one specific area of the body

**Major malformation** Malformation that requires surgical correction or interferes with normal function

**Malformations** Defects of an organ or body part due to an intrinsically abnormal developmental process

**Minor malformation** Malformation that does not interfere with normal function and may be unnoticed by a casual observer

**Sequence** Pattern of multiple anomalies that derives from the single known or presumed defect in a cascade-like manner

**Syndrome** Pattern of multiple anomalies that are known to be or thought to be pathogenetically related

**Teratology** Study of abnormal development resulting from the exposure to chemicals or environmental agents

BIRTH DEFECTS OCCUR IN 2–3% OF ALL NEWBORNS and are a major cause of neonatal deaths. Defects may occur as an isolated abnormality or may occur as groupings of multiple defects. They can occur as part of genetic disorders, they can be due to exposure to environmental agents encountered during the pregnancy, or they can be sporadic. The care and treatment of birth defects are costly not only from a financial but also from a psychosocial viewpoint. Although some types of birth defects can be prevented, at the present most cannot. Some defects can be detected during pregnancy by prenatal diagnosis. Current research on birth defects centers on their causes on a molecular and embryologic level, on their occurrences at the epidemiologic level, and their causes and treatment on a clinical level.

## I. TYPES OF BIRTH DEFECTS

### A. Pathogenesis

Birth defects can occur as a result of abnormalities that occur during the developmental process. If development begins abnormally and progresses in an abnormal fashion, the resulting defects are referred to as malformations. Examples of malformations are cleft lip and spina bifida (open spine). If the developmental process begins normally but is subsequently interfered with, the defect is designated as a *disruption*. The most common types of disruption are congenital amputations caused by amniotic tissue bands.

ENCYCLOPEDIA OF HUMAN BIOLOGY, Second Edition, VOLUME I.   Copyright © 1997 by Academic Press.   All rights of reproduction in any form reserved.

If development is initially normal but mechanical forces change the form or position of body parts, the defects are referred to as *deformations*. Deformations usually occur because of cramping inside the uterus or because of decreased fetal movement. Some deformations resolve spontaneously after delivery or may be corrected with physical therapy. Common examples of deformations include most types of club feet, congenital hip dislocations, and congenital torticollis (wry neck). *Dysplasias* represent disturbances in cellular or tissue structure that usually lead to the development of benign or malignant tumors, to the disorganization of cells, or to the presence of cells in inappropriate places.

Malformations and dysplasias, as well as some disruptions, occur through several different mechanisms during development. These include abnormalities of cell growth and proliferation, abnormalities in cell migration, or abnormalities in programmed cell death. Some defects have animal models that can be studied embryologically. However, most have no animal analogue and are difficult to study from a pathogenetic standpoint. Thus, the pathogenesis of many human malformations is hypothetical.

## B. Classification of Patterns of Multiple Anomalies

Although some birth defects occur as isolated defects, many occur in association with or combined with other abnormalities. In order to understand the possible causes, the prognosis, and recurrence risks, it is necessary to classify multiple defects into several categories. These are syndromes, sequences, field defects, and associations.

A *syndrome* is defined as a pattern of multiple defects, usually affecting several parts of the body or organ systems, which are thought to be pathogenetically related. In some cases the cause of a particular syndrome may be known, such as trisomy 21 as the cause of Down's syndrome or prenatal isotretinoin exposure as the cause of isotretinoin teratogenicity syndrome. In other cases the cause may be unknown. Some syndromes are inherited in a specific Mendelian fashion, such as Marfan's syndrome, achondroplasia, or X-linked aqueductal stenosis. Other syndromes are nongenetic in origin and usually occur in a sporadic fashion. [*See* Down's Syndrome, Molecular Genetics.]

A *sequence* is a pattern of anomalies that derives from a single known or presumed defect. For example, a child with spina bifida (or open spine defect) is likely to develop other problems such as club feet,

hydrocephalus, and hydronephrosis due to neurologic abnormalities affecting the rest of the nervous system, the lower extremities, and the urinary tract.

A *field defect* is a pattern of defects derived from the disturbance of a developmental field. A developmental field is defined as a section of the embryo that is anatomically and sequentially related. Field defects may have many different causes but each operates through a common pathway to produce various defects. An example of a field defect is the DiGeorge anomaly, which consists of complete or partial absence of the thymus gland, complete or partial absence of the parathyroid gland, and conotruncal cardiac defects. DiGeorge anomaly occurs because of a defect in the development of the third and fourth branchial arches, which gives rise to the anomalous organs. Field defects can be produced by chromosome abnormalities, single-gene defects, exposure to drugs or chemicals during pregnancy, or as sporadic events of unknown causation.

An *association* is described as the nonrandom occurrence in two or more individuals of defects that are not thought to be pathogenetically related. In general, groups of defects that form associations do not fit into any of the previously mentioned categories. An example is the VACTERL association. VACTERL is an acronym for *v*ertebral anomalies, *a*nal atresia, *c*ardiac defects, *t*racheo*e*sophageal fistula, *r*enal defects, and *l*imb defects. Two or more of these defects are more likely to occur together in an individual than expected by chance.

## II. CAUSES OF BIRTH DEFECTS

### A. Single-Gene Defects

Numerous single-gene defects have been shown to cause specific defects, specific malformations, or malformation syndromes. In some cases the family history helps to delineate the inheritance pattern; in others, a single-gene defect may occur as an isolated case in a family, usually as a result of a new mutation. Single-gene defects, sometimes referred to as inborn errors of morphogenesis, may be inherited in a classic Mendelian fashion as autosomal dominant, autosomal recessive, or X-linked recessive.

An example of an autosomal dominant malformation syndrome is achondroplasia, one of the most common forms of dwarfism; it occurs in about 1 in 20,000 births. The basic defect is in the development of cartilaginous bone, but the precise biochemical defect is unknown. About 90% of infants born with achondroplasia represent new mutations with a mutation rate of $1.9 \times 10^{-5}$ per generation. The remaining 10% of cases are inherited from either parent.

An example of an autosomal recessive malformation syndrome is the Meckel-Gruber syndrome, which consists of posterior encephalocele, polydactyly, and polycystic kidneys as well as cleft palate. This disorder is lethal; most infants seldom survive more than a few weeks. Its cause is unknown. The parents of an affected patient are carriers of the genes for this disorder but show no recognizable expression. There is a 25% recurrence risk for subsequent pregnancies.

X-linked hydrocephalus is due to aqueductal stenosis, a narrowing of the canal through which cerebrospinal fluid exits the ventricles of the brain. This is one of the more common X-linked recessive malformation syndromes. Like hemophilia or Duchenne muscular dystrophy, this disorder primarily occurs in males born to female carriers. Hydrocephalus is usually apparent at birth. In addition, many of the affected males also have short, flexed thumbs. The disorder is usually associated with significant mental retardation. Several families are now known with multiple males in several generations affected with this disorder.

## B. Chromosome Abnormalities

The presence of an extra chromosome or an absence of a chromosome is likely to lead to birth defects, usually associated with mental retardation. In addition, chromosome rearrangements, consisting of duplication or deletion of parts of chromosomes, may also cause birth defects and mental retardation. For that reason, chromosome studies are one of the main diagnostic studies done on children with birth defects. Approximately 0.5% of all newborns are found to have some type of chromosome abnormality. Some result in clinically recognizable syndromes, whereas others may be difficult to recognize clinically because they do not have recognizable findings or are so rare that few cases have been described. Many chromosome abnormalities are likely to result in a miscarriage; up to 50% of all miscarriages may be due to nonviable chromosome abnormalities. [See Chromosome Anomalies.]

Down's syndrome or trisomy 21 is the most common chromosome abnormality among liveborn infants, with three chromosomes 21 instead of the usual two. The typical physical findings observed in the newborn are a relatively flat facial profile, redundant neck skin, upslanted palpebral fissures, epicanthal folds, relatively small ears, protruding tongue, close-spaced nipples, a single palmar crease, and an increase space between the first and second toes. Most infants with Down's syndrome have decreased muscle tone. Approximately 40% of these infants have some type of cardiac defect, and 2% have some type of significant gastrointestinal defect. Individuals with Down's syndrome are mentally retarded to varying degrees. The mean IQ is between 50 and 60.

Overall, Down's syndrome occurs in approximately 1:1000 births. The risk of having a baby with Down's syndrome increases with the age of the mother, being approximately 1:2000 at 20 years of age, 1:360 at age 35, and 1:100 at age 40. The reasons for this maternal age association are still unknown. However, this finding is the basis for offering prenatal diagnostic studies routinely to women who are 35 years of age or older at the time of pregnancy. Most cases of Down's syndrome occur sporadically within the families. However, once a couple has had a child with Down's syndrome there is a 1–2% risk of having another child with Down's syndrome in a subsequent pregnancy. In addition, there are also inherited forms of Down's syndrome associated with translocation of chromosome 21 and another chromosome. In families with such a translocation, the recurrence risk for subsequent pregnancies will be higher.

Another relatively common chromosome abnormality is trisomy 18. Trisomy 18 infants are very small for gestational age, have relatively small facial features, and show clenching of the fists. Most of them have congenital heart defects as well as other significant birth defects. They usually die before 6 months of age and survivors are profoundly mentally retarded. The condition is due to the presence of an extra chromosome 18. As for trisomy 21 there is an increasing risk of this abnormality with maternal age.

A third common chromosome abnormality is trisomy 13, due to an extra chromosome 13. Infants with trisomy 13 have multiple anomalies including brain malformations, eye abnormalities, cleft lip and/or cleft palate, cardiac defects, and extra digits. Survival is short, and survivors are mentally retarded. Like trisomy 21 and trisomy 18, there is also an association of increased risk with maternal age.

## C. Polygenic/Multifactorial Causes

The majority of isolated birth defects and many deformations fall into this category. The term polygenic refers to defects that are caused by several genes, usually with one major gene and one or more modifying genes. Multifactorial defects occur as a result of interactions between a single gene or multiple genes and environmental factors. In most cases neither the genetic nor the environmental factors are easy to detect or delineate. Specific information about multifactorial defects is obtained from epidemiologic studies. The most common multifactorial defects observed in humans are neural tube defects and cleft lip and/or cleft palate.

Neural tube defects are defects in the closure of the neural tube, which gives rise to the brain and spinal cord during the fifth week of development. Defects in the closure of the anterior portion of the neural tube result in anencephaly in which the brain and skull do not form. Anencephaly occurs in approximately 1 in 1000 pregnancies and is a lethal condition. Few infants born with anencephaly survive for more than a week. In contrast, spina bifida or open spine is due to a defect in the closure of the posterior neural tube forming the spinal cord, leading to a defect in the spinal cord somewhere on the back. The spine may be either open to the outside or covered by membrane. Spina bifida causes paralysis below the level of the defect and is accompanied by hydrocephalus in approximately 80% of children. Because of the associated paralysis, children may have club feet and bowel and bladder problems. Although spina bifida may be compatible with normal life expectancy with a mild to moderate physical handicap, some children with spina bifida are severely handicapped and mentally retarded. It is difficult to predict how any child will do after birth. However, several risk factors may play a role in prognosis. The closer to the head the defect is, the greater the complications. Open defects have greater complications than those that are covered by membrane or skin. The presence of hydrocephalus or other malformations at birth is more likely to indicate a poor prognosis. The spine defect requires surgical closure shortly after birth and the development of hydrocephalus requires a shunt procedure to remove excess fluid from the brain.

Anencephaly and spina bifida are caused by both genetic and environmental factors. Both defects have a similar geographic distribution, being more frequent in the British Isles than in the United States and in the eastern United States than in the western United States. Their frequency is higher during certain time periods. Recent studies suggest that neural tube defects are more likely to occur in women with insufficient intake of vitamins, particularly folic acid, so nutrition may play a role. It is now recommended that all women of child-bearing age should take 0.4 mg of folic acid a day to reduce the occurrence of neural tube defects. Women who have had a previous affected child should take 4.0 mg of folic acid a day prior to conception. Genetic factors are also involved. The incidence is higher in whites and there are differences between various ethnic groups. Couples who have previously had a child with anencephaly or spina bifida have a 3–5% recurrence risk in a subsequent pregnancy. In addition, the greater the number of affected individuals within a family, the higher the recurrence risk.

Cleft lip and/or cleft palate, when they occur as isolated defects, may be of multifactorial origin. These clefts, which occur either together or individually, result from a failure of lip or palate closure around the sixth week of development. Cleft lip causes cosmetic disfigurement but usually causes little disability and is surgically repaired at 2 to 3 months of age. In contrast, cleft palate may cause problems such as feeding difficulties, speech difficulties, and high frequency of ear infections and hearing problems. Affected infants usually require special feeding techniques. The defect is usually repaired surgically at 1 to 1½ years of age. Like neural tube defects, clefts also have varying epidemiologic patterns of occurrence with differences over time and by geography, and in different racial and ethnic groups. Nutrition may also play a role. The recurrence risk for couples who have had one child with cleft lip or cleft palate is approximately 2–4% in a subsequent pregnancy. If multiple family members have clefts, the recurrence risk may be higher.

## D. Environmental/Teratogenic Causes

This category includes exposure to various drugs, chemicals, or physical agents during pregnancy. Drug exposures may include prescribed drugs, over-the-counter medications, or drugs of abuse including alcohol. Chemical exposures may be either occupational or environmental. Other environmental exposures may include biological agents such as viruses or parasitic infections as well as physical agents, such as radiation or heat. Maternal disorders such as diabetes, mellitus, hypothyroidism, and maternal phenylketonuria are also classified in this category.

Exposures to these various agents, referred to as teratogens, may cause various types of defects, including obvious structural defects, histological defects, fetal death, fetal growth retardation, biochemical abnormalities, behavioral defects (including functional abnormalities, neurological abnormalities, mental retardation, or learning disabilities), or subsequent malignancy. Exposure to a single agent may lead to a combination of these defects. [See Teratogenic Defects.]

Several conditions determine whether an exposure will be teratogenic. The first is timing. For example, an exposure that can lead to cleft palate must occur close to the time of palate closure around the sixth week of development. After the palate is closed, this defect cannot occur. The dosage of the exposure is important. Theoretically, the higher the dosage, the greater the risk of defects and/or severity of the defects. Another factor is how well the agent crosses the placenta and, therefore, how much of the agent reaches the fetus. Genetic factors controlling the metabolism of various drugs or chemicals are also involved, as are other factors such as nutrition.

Although drug or environmental exposures actually cause only a small percentage of birth defects, they tend to have a very high emotional impact on individuals or populations. Contrary to popular opinion, specific drug or environmental exposures do not uniformly increase the risk of birth defects in the exposed population. Exposure to a specific agent usually produces a specific defect or a syndrome.

The best known teratogen is undoubtedly thalidomide, a sedative that was available in Western Europe in the early 1960s. The drug was never approved for marketing in the United States and was quickly withdrawn from use in all countries. Thalidomide was documented to cause not only limb defects such as phocomelia (in which the hands and feet are attached directly to the trunk) but other limb abnormalities, cardiac defects, and ear defects as well. Exposure to lithium tends to cause congenital heart defects. Exposure to isotretinoin during the pregnancy (used to treat acne) results in ear, heart, and brain malformations. Rubella exposure during pregnancy causes deafness, blindness, growth problems, and mental retardation. Occupational or environmental exposure to methyl mercury causes neurological handicaps with a cerebral palsy-type picture. Exposure to alcohol during the pregnancy causes fetal alcohol syndrome, seen most commonly in infants born to alcoholic mothers. The greater the alcohol intake during the pregnancy, the greater the risk of defects, which include prenatal and postnatal growth retardation, craniofacial abnormalities, and heart and kidney defects as well as mental retardation. As mentioned previously, certain ma-

ternal illnesses can also cause specific defects. For instance, insulin-dependent diabetic mothers have higher rates of birth defects, especially for the neural tube and the heart.

### E. Other Causes

In the case of many birth defects there is no evidence of chromosomal or genetic alteration or known teratogenic exposures during the pregnancy; their cause is unknown. Therefore, unless the defect is previously known to be multifactorial, it is classified as sporadic and of unknown etiology. These defects may occur as stochastic (random) events interfering with the normal developmental process and can be viewed as a chance occurrence.

## III. BIRTH DEFECT EPIDEMIOLOGY

### A. Method of Studying Birth Defect Epidemiology

The frequency of birth defects in the population can be studied in several ways. Many, but not all, are reported on birth certificates so that their use leads to underascertainment of the incidence of birth defects. Records of discharge diagnoses on infants from the nursery provide information regarding most obvious birth defects that are diagnosed at the time of birth. This approach, however, will miss defects that are diagnosed only later in life. Probably the most complete way of ascertaining birth defects is by combining discharge diagnoses with diagnoses made through various clinics and laboratories (such as birth defect and genetics clinics or cytogenetic laboratories). This approach has now been adopted in many countries, including the United States. In the United States most of these activities are done through the Centers for Disease Control in Atlanta, Georgia. Table I shows the frequency of some of the more common birth defects studied by these methods.

### B. Risk Factors for Birth Defects

Various epidemiologic studies have been helpful in showing an association of birth defects with various risk factors. The risk factors found to increase a couple's risk of having a child with birth defects are listed in Table II. Such risk factors are helpful to couples and to obstetricians in management of the pregnancy

**TABLE I**

Birth Prevalence of Some Common
Birth Defects[a]

| Defect | Rate[b] |
| --- | --- |
| Anencephaly | 6.5 |
| Spina bifida | 8.7 |
| Cleft palate without cleft lip | 5.7 |
| Cleft lip with or without cleft palate | 10.6 |
| Renal agenesis | 1.8 |
| Anal atresia | 4.1 |
| Down's syndrome | 9.8 |
| Club foot | 38.3 |
| Heart defects (all) | 77.6 |
| Ventricular septal defect | 19.4 |

[a]Data from the Metropolitan Atlanta Congenital Defects Program, "Congenital Malformations Surveillance Report 1981–83," issued September 1985.
[b]Per 10,000 total births.

and in helping to make decisions regarding approaches for prenatal diagnosis of birth defects in pregnancies at risk.

## IV. DIAGNOSIS OF BIRTH DEFECTS

When an infant is born with birth defects the physician must determine their cause in order to obtain information regarding prognosis, treatment, and recurrence risk. As for any other medical problem, the diagnosis of birth defects requires accurate personal

**TABLE II**

Risk Factors for Birth Defects[a]

1. Previous child with birth defect or parent with birth defect
2. Other family history of birth defect
3. Consanguinity
4. Maternal age >35 years at delivery
5. Chronic illness in mother[b]
6. Acute illness during pregnancy[b]
7. Medication during pregnancy[b]
8. Alcohol or drug abuse in mother
9. Occupational exposure to chemicals[b]

[a]These risk factors require additional evaluation to determine actual risk to a specific couple of pregnancy.
[b]Certain types may be of greater risk than others.

and family history, physical examination, and appropriate laboratory tests. The physician will concentrate on any complications, illnesses, or exposures that the mother may have had during the pregnancy, when the problems occurred, and how severe they were. Obtaining a family history is important because it provides information about the occurrence of birth defects or other genetic conditions within the family, especially birth defects similar to those of the affected child. Physical examination should include a thorough description of the obvious defect present as well as measurement of various body parts (anthropometry). During this physical examination the physician will look for the presence of other birth defects as well as any minor dysmorphic features that may be helpful in reaching a clinical diagnosis. Laboratory studies generally will involve chromosome analysis when a chromosome abnormality is suspected and various imaging studies (including routine X-rays, computerized tomography, ultrasound, and magnetic resonance imaging). Other studies may be required depending on the particular defects present. On the basis of this information, the physician can determine the most likely diagnosis and possible cause for the defects.

## V. TREATMENT OF BIRTH DEFECTS

### A. Medical

The care of most children with birth defects is usually carried out by a pediatrician or family practitioner, although frequently various specialists will be involved. For many birth defects no specific treatment may be available and the physician must work with the family to treat specific problems on a symptomatic basis or to help deal with chronic medical problems associated with these conditions.

### B. Surgical

Some defects are amenable to surgical repair and can be completely corrected. These include cleft lip and palate. In other cases surgery is done as a palliative procedure to prevent further damage or to reduce disability. An example is the closure of spina bifida. Some defects require reconstructive surgery, occasionally in stages, to improve appearance or function. This would include craniofacial surgery or reconstructive surgery of the hands. The timing of any of these surgeries depends on various factors. Only defects that are considered to be life threatening are likely to be repaired shortly after birth; all other surgical procedures will be done later on an elective basis.

## C. Habilitative

Many children with birth defects require occupational and physical therapy; some may require prosthetic or orthotic devices to develop better function. Children with mental retardation or learning disabilities may require special educational resources.

## D. Psychosocial

The birth of a child with a birth defect is likely to be emotionally devastating to a couple. Couples must deal with a high level of guilt regarding causation of the defect and go through a grieving process because they mourn the loss of the normal child they expected. This process goes through the stages of denial, anger, bargaining, and acceptance. Couples may also suffer from significant marital strife as a result of this stress, as shown by the higher divorce rate among couples of children with birth defects than in the general population. These are all issues that must be dealt with by the physician and ancillary health-care workers including nurses, social workers, and psychologists, not to mention the need for support from family and friends. In addition, genetic counseling is also available to couples.

## E. Financial

Birth defects have a significant financial impact on families. Some support may be available to families from federal, state, and social service agencies to assist them with the care of an affected child.

## VI. PREVENTION OF BIRTH DEFECTS

## A. Education

Information about birth defects and the risk factors that cause them is probably the most important aspect of preventing birth defects. Such education started by organizations such as the March of Dimes is helpful in teaching women how to avoid dangerous exposures during pregnancy as well as helping couples to determine whether they may be at increased risk of birth defects.

## B. Genetic Counseling

For couples who are known to be at risk, genetic counseling may be helpful in providing information about the defects, their recurrence risks, and methods that are available to families to prevent the birth of subsequent affected children. Information is provided in a nondirective fashion so that the couple may make their own decision regarding reproduction and the utilization of various methods of prenatal diagnosis. It is hoped that such information would be available to the couple before a subsequent pregnancy. [See Genetic Counseling.]

## C. Identification of Pregnancies at Risk

Some couples may already be aware of their increased risks of birth defects through education, genetic counseling, or from their family history. In other circumstances it is the obstetrician or family practitioner who first identifies the risk factors. The physician would then recommend specific testing to determine whether there is a risk or whether a defect is present. One of the most common indications for pregnancies at risk is maternal age; therefore, most women over age 35 will be referred for prenatal diagnostic studies. Tests that are available for diagnosing these defects include amniocentesis, chorionic villus sampling (CVS), maternal serum $\alpha$ fetoprotein and other maternal serum analytes, and fetal ultrasound.

## D. Prenatal Diagnosis

Amniocentesis is usually done between 16 and 18 week of pregnancy. A small amount of amniotic fluid from the sac surrounding the baby is obtained by a needle inserted through the maternal abdomen under ultrasound guidance. Cells from the fluids are grown in the laboratory for chromosome analysis. In addition, biochemical studies can be done on the fluid to determine levels of $\alpha$ fetoprotein and other biochemicals. Amniocentesis is useful for detecting chromosome abnormalities, many biochemical defects, neural tube defects, other defects associated with elevated $\alpha$ fetoprotein levels, and more recently for DNA analysis of some genetic disorders. The risk of amniocentesis is approximately 0.5%, primarily through miscarriage.

CVS is a more recent procedure that is done at about 9 to 11 weeks of pregnancy. A small amount of chorionic villi is obtained either transcervically or transabdominally under ultrasound guidance. Cells from chorionic villi are cultured in the laboratory for chromosome analysis, biochemical analysis, or DNA studies. The risk of CVS (primarily through miscarriage) appears to be approximately 0.5–1%. CVS is not useful for detecting many birth defects including neural tube defects (which are revealed by biochemical changes in the amniotic fluid or serum).

Maternal serum $\alpha$ fetoprotein (MSAFP) determination has been useful for identifying women at higher risk for neural tube defects or for Down's syndrome. Pregnancies of neural tube defects have higher $\alpha$ fetoprotein levels in maternal serum and amniotic fluid. Pregnancies with Down's syndrome fetuses have lower levels of $\alpha$ fetoprotein in maternal serum. Women who have elevated MSAFP levels or low MSAFP levels can then be offered amniocentesis for more definitive diagnosis. MSAFP studies are recommended to women at 16 to 18 weeks into the pregnancy. MSAFP levels are not diagnostic of birth defects in general and cannot detect all cases of Down's syndrome or neural tube defects.

Recently, elevated levels of human chorionic gonadotropin (HCG) and lower levels of unconjugated estriol (UE3) in maternal serum in the second trimester have also been associated with Down syndrome pregnancies. The combined use of MSAFP, HCG, and UE3 along with maternal age detects Down syndrome in 65% of affected pregnancies. This compares favorably to MSAFP alone which can detect only about 30% of Down syndrome pregnancies. Women who are calculated to have an increased risk of Down syndrome using those four parameters similarly can be offered amniocentesis for more definitive diagnosis.

Ultrasound is another useful diagnostic tool. Its ability to detect defects depends on the size of the defect, the degree of resolution of the ultrasound equipment, and the experience of the ultrasonographer.

The use of the described methods of prenatal diagnosis depends on several factors. First is the type of defect for which a couple is at risk. Chromosome abnormalities are best detected by CVS or amniocentesis. Neural tube defects are best detected through $\alpha$ fetoprotein analysis of serum or amniotic fluid and through ultrasound. Some types of defects are detectable by ultrasound alone, whereas others are not amenable to any type of prenatal diagnosis and become apparent only after delivery.

When a fetal defect is found by prenatal diagnosis, few options are usually available to a family. In most cases they must decide whether to terminate or continue the pregnancy. If the pregnancy is continued, knowledge of the defect may influence subsequent obstetrical management and prenatal and postnatal care.

## E. Prenatal Treatment

At present, few types of treatment can prevent the occurrence of birth defects. Recent information suggests that some neural tube defects and other defects may be related to nutritional or vitamin deficiencies. Theoretically, some of these defects may be prevented or at least limited by use of multivitamins or nutritional supplementation prior to conception. A medical treatment in use for congenital adrenal hyperplasia (21-hydroxylase deficiency) involves administration of corticosteroids during the pregnancy. Surgical repair of certain defects such as hydrocephalus or urinary tract defects by shunting have been attempted.

## VII. CURRENT RESEARCH ON BIRTH DEFECTS

### A. Clinical Research

Most research is based on case reports or case series of children born with specific defects or syndromes. Difficulties are due to the rarity of many defects and the ethical problems of monitoring human pregnancies while doing prospective studies. The information obtained is helpful in generating hypotheses on the specific causes of birth defects, which then need to be studied further by other approaches. A very useful direction of clinical research is to determine the natural history of various disorders and the best modalities of treatment.

### B. Epidemiological Research

These studies are helpful in tracking the occurrence of birth defects in the population because they may identify specific environmental factors. They are also useful for testing hypotheses generated by clinical observations.

### C. Embryological Research

The study of birth defects in animal species or the effects of various exposures during animal pregnancy have provided much information about the likely sequence of events leading to birth defects in humans. The use of the scanning electron microscope has greatly aided the study of both normal and abnormal embryologic development and has provided a great deal of information of the pathogenesis of some defects.

## D. Molecular Genetics Research

Investigation of the effects of certain single-gene defects in animals will be helpful in studying these defects on a molecular level. The study of genes involved in development in drosophila, mice, and other organisms may eventually provide clues to the genetic control of normal development in humans and the abnormalities that lead to birth defects. [*See* Chromosomes, Molecular Studies.]

## BIBLIOGRAPHY

Borgaonker, D. S. (1984). "Chromosomal Variation in Man: A Catalog of Chromosomal Variants and Anomalies," 4th Ed. A. R. Liss, New York.

Buyse, M. L. (1990). "Birth Defects Encyclopedia." Blackwell Scientific Publications, Cambridge, MA.

Cohen, M. M. (1982). "The Child with Multiple Birth Defects." Raven Press, New York.

de Grouchy, J., and Turleau, C. (1984). "Clinical Atlas of Human Chromosomes," 2nd Ed. Wiley, New York.

Gorlin, R. J., Cohen, M. M., and Levin, L. S. (1990). "Syndromes of the Head and Neck," 3rd Ed. Oxford University Press, New York.

Jones, K. L. (1988). "Smith's Recognizable Patterns of Human Malformation," 4th Ed. Saunders, Philadelphia.

McKusick, V. A. (1992). "Mendelian Inheritance in Man," 10th Ed. Johns Hopkins University Press, Baltimore.

Nyhan, W., and Sakati, N. (1976). "Genetic and Malformation Syndromes in Clinical Medicine," Year Book Medical Publishers, Chicago.

Shepard, T. H. (1989). "Catalog of Teratogenic Agents," 6th Ed. Johns Hopkins University Press, Baltimore.

Spranger, J. S., Langer, L. D., and Wiedemann, H. R. (1974). "Bone Dysplasias: An Atlas of Constitutional Disorders of Skeletal Development," Saunders, Philadelphia.

Stevenson, R. E., Hall, J. G., and Goodman, R. M. (1993). "Human Malformations and Related Anomalies." Oxford University Press, New York.

Warkany, J. (1971). "Congenital Malformations: Notes and Comments." Year Book Medical Publishers, Chicago.

ISBN 0-12-226971-3

90018

9 780122 269714